Birkhäuser

Monografie Matematyczne

Volume 67

Starting in the 1930s with volumes written by such distinguished mathematicians as Banach, Saks, Kuratowski, and Sierpinski, the original series grew to comprise 62 excellent monographs up to the 1980s. In cooperation with the Institute of Mathematics of the Polish Academy of Sciences (IMPAN), Birkhäuser now resumes this tradition to publish high quality research monographs in all areas of pure and applied mathematics.

Henryk Żołądek

The Monodromy Group

Second Edition

Birkhäuser

Henryk Żołądek
Institute of Mathematics
University of Warsaw
Warsaw, Poland

ISSN 0077-0507 ISSN 2297-0274 (electronic)
Monografie Matematyczne
ISBN 978-3-031-91272-6 ISBN 978-3-031-91270-2 (eBook)
https://doi.org/10.1007/978-3-031-91270-2

Mathematics Subject Classification (2020): 30B40, 32B10, 33C05, 34C08, 22F05, 37C27, 53C12, 14B05, 14B07, 51F15, 54H15, 55N10

This book is published under the imprint Birkhäuser, www.birkhauser-science.com by the registered company Springer Nature Switzerland AG
The registered company address is: Gewerbestrasse 11, 6330 Cham, Switzerland

Contents

Preface

The origins of the monodromy theory lie in the works of B. Riemann on functions of complex variable and on complex linear differential equations. Riemann formulated one of fundamental problems in the monodromy theory (now called the Riemann–Hilbert problem): having given singularities and corresponding monodromy transformations find the differential equation which realizes these data. The monodromy groups of linear differential equations and systems were intensively studied in the nineteen century by F. Klein, G. Stokes, H. Schwarz, L. Schlesinger, L. Pochhammer, E. Picard, R. Garnier, P. Painlevé, H. Poincaré, R. Fuchs and others. Schwarz associated the monodromy group of the hypergeometric equation with the spherical triangle groups, generated by inversions with respect to the sides of a spherical triangle. Schlesinger investigated deformations of differential systems with fixed monodromy. Fuchs associated the equation Painlevé 6 with the isomonodromic deformation equation. Stokes discovered the strange phenomenon (the Stokes phenomenon) of non-uniqueness of constants in the asymptotic expansions of systems near irregular singularities.

In his talk at the 1900 International Congress of Mathematicians, D. Hilbert included the Riemann problem into his famous list of problems for the twentieth century mathematics (as the $\mathrm{XXI}^{\mathrm{th}}$). This problem was solved independently by J. Plemelj and by H. Röhrl in the class of systems with regular singularities. Only in the 80's A. Bolibruch discovered that the Riemann–Hilbert problem may have no solutions in the Fuchs class of systems with first order poles. Recently some relations between linear differential systems and quantum field theory were revealed (M. Sato, T. Miwa, M. Jimbo, B. Dubrovin).

Near the end of the nineteen century E. Picard and E. Vessiot created an analogue of the algebraic Galois theory in the case of linear differential systems. The corresponding differential Galois group consists of symmetries of the system and is identified with an algebraic subgroup of the linear group of automorphisms of a complex vector space (E. Kolchin). In the regular case it is the Zariski closure of the monodromy group (Schlesinger). The fundamental result in this theory states that the differential system is solvable in quadratures and algebraic functions iff the identity component of the differential Galois group is solvable. An analogous result holds in the topological Galois theory, which is represented by the monodromy group of an algebraic function. A. Khovanski generalized the latter result to a wider class of multivalued holomorphic functions.

The Stokes phenomenon, which occurs in the case of irregular singularity, found complete explanation. Firstly, it was proved that there exists a formal normal form, which is diagonal and contains only finite number of terms with rational powers of the 'time' (M. Hukuhara, A. Levelt, H. Turrittin). Next, the change, reducing the system to its formal normal form, turns out to be analytic in sectors.

This is done either by solving some integral equation (H. Poincaré, W. Wasow), or by showing that the normalizing series belongs to some Gevrey class and is summable in sectors. The moduli of analytic classification are cocycles in some cohomology group with values in certain Stokes sheaf and are represented by 'differences' between the normalizing maps in adjacent sectors.
After publication of the "Analysis situs" by H. Poincaré, investigation of the topology of algebraic varieties began. That time the first variant of the Picard–Lefschetz formula, describing the change of topology of a family of algebraic varieties as the parameter varies around a critical value, appeared. The further rapid progress in this field occurred in the 1960s–70s. The research proceeded in two parallel streams.

J. Milnor proved that the local level of a holomorphic function near an isolated critical point has the homotopy type of a bucket of spheres. R. Thom, J.-C. Tougeron, J. Martinet and V. Arnold developed theory of normal forms of holomorphic functions and began the classification of singularities. The (topological) monodromy groups of some singularities turned out to be isomorphic with certain Coxeter groups generated by reflections. Relations with the classification of semi-simple Lie algebras were revealed.

The other approach was more algebraic and based on the cohomology theory of coherent sheaves developed by J. Leray, A. Grothendieck, P. Deligne and others. Another tool was the theorem about resolution of singularity proved by H. Hironaka. People studied families of algebraic varieties which degenerate, as the (complex) parameter approaches a critical value. Here the critical points can be isolated and non-isolated as well. After resolution of singularity, the singular variety becomes an union of smooth divisor with normal crossings in the ambient complex space. Information about multiplicities of these divisors allows to describe the action of the (topological) monodromy (C. Clemens, N. A'Campo). The fundamental result (the monodromy theorem) says about the eigenvalues and the dimension of Jordan cells of the monodromy operator.

The de Rham cohomologies of non-singular algebraic varieties, from a family, form together a holomorphic vector bundle over the space of non-critical parameters. A similar bundle, the cohomological Milnor bundle, is defined in the local case. The cohomological bundle admits sections, defined by the integer cocycles. This allows to introduce the famous Gauss–Manin connection, such that the integer cocycles represent horizontal sections with respect to it. The holomorphic forms on the ambient space form another class of sections of the cohomological bundle, the geometrical sections. Their integrals along families of integer cycles are holomorphic functions which obey a system of linear differential equations, called the Picard–Fuchs equations. The Picard–Fuchs equations are related with the equations for horizontal sections with respect to the Gauss–Manin connection. They have regular singularities (P. Griffiths, N. Katz). Together with the asymptotic of integrals they constitute invariants of the degeneration (B. Malgrange, V. Arnold, A. Varchenko, J. Steenbrink). The asymptotic of integrals is closely related with the asymptotic and geometry of oscillating integrals appearing in the wave optics and in quantum physics (Arnold, Varchenko).

The integrals of holomorphic forms along integer cycles application in the linearized version of the 16th Hilbert problem, about limit cycles of polynomial planar vector fields. This problem leads to the problem of estimation of the number of zeroes of certain Abelian integral (Arnold). Existence of such estimates and some concrete formulas were obtained by A. Varchenko, A. Khovanski and G. Petrov.

Any compact non-singular projective variety admits a so-called Hodge structure, which says that one can represent the cohomology classes as harmonic forms with their division into the (p,q)-types. P. Deligne proved that non-compact and/or singular variety admits a so-called mixed Hodge structure. It means that there is a weight filtration of the cohomology space by an increasing series of subspaces, such that the quotient spaces are equipped with a Hodge structure (arising from some complete smooth variety obtained after resolution of singularities). J. Steenbrink and W. Schmid proved the existence of a mixed Hodge structure in the case of degeneration of algebraic varieties and Steenbrink constructed such structure in the fibers of the cohomological Milnor bundle. The latter structure is determined from the Jordan cells structure of the monodromy operator and from the asymptotic of geometrical sections (Varchenko).

In the case of degeneration of algebraic varieties the monodromy and the mixed Hodge structure are related with singularities of the period mapping, from the parameter space to the classifying space of Hodge structures (P. Griffiths). This leads to the problems of moduli of algebraic varieties and to theorems of the Torelli type (about injectivity of the period map on the moduli space).

Besides the linear monodromy theory there is its nonlinear part. It is represented by the holonomy groups of some distinguished leaves of holomorphic foliations. This theory is well developed only in two dimensions, where the foliation is defined as a phase portrait of an analytic vector field (with complex 'time'). The singularities were resolved by A. Seidenberg and there remain only foci, nodes, saddles and saddle-nodes. The foci and nodes were classified analytically by Poincaré. The classification of saddles leads to the classification of the holonomy maps associated with loops in one of their separatrices. If the multiplicator of a germ of holomorphic one-dimensional diffeomorphism is resonant, then the situation is like in the case of the Stokes phenomenon. The functional invariants of the analytic classification were found by J. Ecalle and S. Voronin. If the multiplicator is non-resonant, then the analyticity of the formal normal form (which is linear) depends whether the multiplicator satisfies the so-called Briuno condition (A. Briuno, J.-C. Yoccoz). In the case of saddle-node the functional moduli were described by J. Martinet and J.-P. Ramis. Here the main tool of the proof is the sectorial normalization which is proved either by means of some functional analytic methods (M. Hukuhara, T. Kimura, T. Matuda) or by means of the Gevrey expansions.

The holomorphic foliations exist on algebraic surfaces; they are defined by means of polynomial vector fields. J. Jouanolou constructed examples of foliations on the projective plane without algebraic leaves and A. Lins-Neto proved that such foliations are typical. M. Singer proved that, if a polynomial planar vector field has a first integral expressed by quadratures, then it has a simple integrating factor (exponent of integral of a rational 1-form). It turns out that, for a typical foliation with the line at infinity invariant, a generic leaf is dense in the projective plane (M. Hudai-Verenov) and there is infinitely many limit cycles (Yu. Ilyashenko).

The latter two results are proved using the monodromy group of the leaf at infinity. This is a subgroup of the group of germs of one-dimensional diffeomorphisms. The abelian and solvable groups of this type were classified (D. Cerveau, R. Moussu) and the non-solvable groups are rigid, in the sense that their formal or topological equivalence implies their analytical equivalence (J.-P. Ramis, A. Shcherbakov, I. Nakai).

S. Ziglin used the monodromy to prove the non-integrability of certain Hamiltonian systems, e.g., the Poisson–Euler system.

Among modern developments of the classical monodromy theory we underline the generalizations of the Euler hypergeometric integrals to the case with more singularities (P. Deligne, G. Mostow) and to many dimensions (I. Gelfand, A. Varchenko). Here the monodromy group realizes a representation of the fundamental group of complement of the discriminant variety and some classical results (like the theorem of H. Schwarz) were generalized.

The above outlines the history of the monodromy theory. These topics constitute the rough contents of this book.

The monodromy theory can be called a clever bifurcation theory. In the usual bifurcation theory one investigates some objects (functions, varieties, maps, vector fields) depending on real parameter(s) and their changes as the parameter passes through the critical values. For example, the Morse theory describes degeneration of the hypersurface of level of a function as the level tend to the critical value. Usually the objects are defined analytically. In that case the clever investigation relies on observing the transformation of the object as the parameter varies along a loop around the critical value (in the complex parameter space). Therefore the complex analogue of the Morse theory is the Picard–Lefschetz theory. The monodromy approach to the bifurcation problems turns out to be very effective. It allows to obtain results out of reach using the real methods.
There is something mysterious and undefined in the monodromy theory, at least for non-specialists. Often people use it rather loosely, without providing rigorous definitions.

The aim of this monograph is to introduce the reader into the complex of notions and methods used in the monodromy theory. Because these notions and methods involve large parts of the modern mathematics, the book contains a lot of auxiliary mathematical material. It is written to be as self-contained as possible. We strived not to omit the technical parts of the proofs. We have included such elements like the proof of the analytic version of the Hadamard–Perron theorem or the proof of the Thom–Martinet preparation theorem. On the other hand, the results, which are not fundamental and constitute generalizations of simpler results, are treated more loosely. In these cases we present only the ideas and general arguments.

The book touches practically all branches of the monodromy theory. But it does not contain all known results. Many theorems are not even mentioned. Also the literature references is not complete.

The idea of writing this book has appeared in 1996, when the author began to deliver a two-year course in the Warsaw University. The subject was continued in seminars. The lectures were written down and constitute the essential part of the book. Therefore the book is addressed mainly to (graduate) students.

Another reason was the author's self-education. It was great enterprise, which consumed a lot of the author's time and energy. There is hope that this effort was not useless and will help others to learn relatively quickly techniques of the monodromy theory.

At the end of this preface I would like to express my thanks to my students T. Maszczyk, G. Świrszcz, E. Stróżyna, A. Langer, M. Rams, P. Leszczyński, M. Borodzik, Ł. Wiechecki, P. Goldstein and M. Bobienski for their patience during lectures and seminars and for detecting many mistakes. I thank V. Gromak for sending me notes from the lectures of G. Mahoux. I thank to A. Maciejewski for drawing my attention to the works of S. Ziglin, J. Morales-Ruiz and J.-P. Ramis. I thank P. Mormul for showing me some references. I thank F. Loray for giving me preprints of some papers and of his lecture notes. I would like also thank A. Weber, Ś. Gal, P. Pragacz, Yu. Ilyashenko, A. Varchenko and J. Steenbrink for their interest in this book.

During the work on this monograph the author was supported by the Polish KBN Grant No 2 P03A 022 08, No 2 P03A 041 15 and 2 P03A 010 22.

Preface to the Second Edition

About one year ago, Birkhäuser has asked me if I would be willing to work more on my book, by extending it with some new results which have appeared since the first edition. I have agreed and the reader can see the result of my work. Let me describe shortly the additional topics considered in the new version.

I did not make any changes in the initial chapters. In Chapter 6 I have added a result of G. Binyamini, D. Novikov and S. Yakovenko about a double exponential estimate in the weakened Hilbert's 16th problem.

Next, in Chapter 8 I discuss the problem of moduli of analytic classification of deformations of linear differential equations, based on works of A. Glyutsuk and C. Rousseau with collaborators. Also in the same chapter the reader will find an alternative approach to Bolibruch type counterexamples to the Riemann–Hilbert problem due to Yu. Ilyashenko and S. Yakovenko.

The problem of moduli of unfoldings for local diffeomorphisms was investigated by C. Rousseau with coauthors and I discuss it in Chapter 9. There also a new approach to the normal forms of singularities of vector fields and their analytic properties, due to my works with E. Stróżyna is described. It seems to be interesting an association of the normal forms problem with the Invariant Theory. In particular, a new approach to the proof of the finiteness of the ring of polynomial invariants for a linear vector field (the Maurer–Weitzeböck theorem) is presented.

In Chapter 11 I present an application of the Morales-Ruiz–Ramis theory to the problem of (non-)integrability of some Painlevé equations. In particular, I demonstrate this approach in the case of the fourth Painlevé equation based on a work of Ts. Stoyanova. Another approach to this problem was developed in my work with G. Filipuk and I have included it.

In the same chapter I have included some examples of A. Lins-Neto of polynomial vector fields with some special first integrals.

I have added a whole section devoted to the center–focus problem for germs of analytic real vector fields. It contains such subject as algebraic/analytic solvability of the center problem; in particular, a result of N. Medvedeva. The reader will find a result of M. Berthier, D. Cerveau and A. Lins-Neto about neither Liouvillian integrability nor reversibility in the case of a non-elementary singularity. Moreover, I state a conjecture about reduction via a generalized pull-back and present results which support it.

Finally, in Chapter 12 I have added a whole section devoted to generalized hypergeometric functions of one variables. There I included results of a beautiful work of F. Beukers and G. Heckman about the monodromy group of the corresponding equation and results of A. Duval an C. Mitschi about the differential Galois group in the confluent case. I have overlooked those works when writing the first version of my book.

Chapter 1

Analytic Functions and Morse Theory

This chapter is special. Its aim is a quick introduction of the notion of monodromy in applications to multivalued holomorphic function and their Riemann surfaces. The classical theorem about monodromy is a theorem about such functions.

The simplest example of multivalued holomorphic functions is the implicit function defined by means of a quadratic equation in two variables. This leads to the Morse lemma in two (and more) dimensions.

We apply the Morse theory (in the real domain) to calculate the self-intersection index of the cycle generating the homology of a noncritical complex level of a quadratic homogeneous function.

§1 Theorem about Monodromy

1.1. Definition of analytic element and of Riemann surface. By an **analytic element** we mean a pair (D, f), where $D = D_a \subset \mathbb{C}$ is a disc with center at a and $f = f_a$ is a holomorphic function on D, such that the Taylor series of f at a is convergent in D_a. We say that the analytic element (D_a, f_a) has prolongation to an element (D_b, f_b) along a path $\gamma \subset \mathbb{C}$, iff γ can be covered by domains of analytic elements such that the corresponding functions agree at the adjacent intersections. The union of analytic elements obtained from prolongations of (D, f) forms the **Riemann surface** of the (generally multivalued) holomorphic function f.

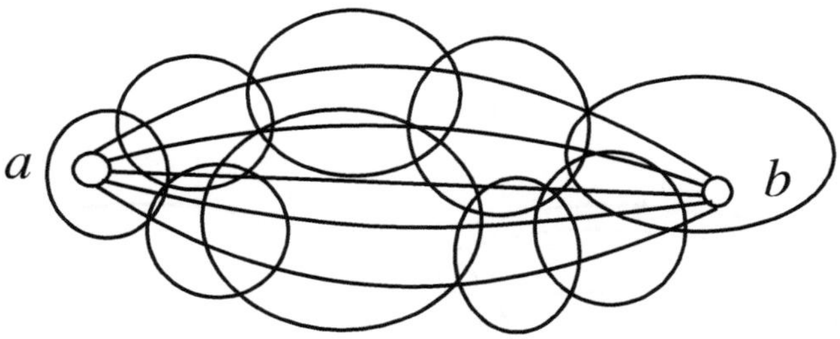

Figure 1.1: Homotopy and analytic elements

1.2. Theorem about monodromy. *If two paths γ_1, γ_2, joining the points a and b, are homotopically equivalent in a domain where the function f is (locally) analytic,*

H. Żołądek, *The Monodromy Group*, Monografie Matematyczne 67,
https://doi.org/10.1007/978-3-031-91270-2_1

then the prolongations of (D_a, f) *to* $(D_{b,i}, f_i)$ *along* γ_i*'s are the same. It means that* $f_1 \equiv f_2$ *in* $D_{b,1} \cap D_{b,2}$.

Proof. Let γ_s be a one-parameter family of paths joining a with b and realizing the homotopy between γ_i. Their union spreads over some compact domain E. We cover this domain by analytic elements, starting from (D_a, f), which agree at intersections. In this way we obtain the Riemann surface of f over E (see Fig. 1.1). It is clear that this Riemann surface is diffeomorphic to E. Thus f is single valued on E. □

1.3. Remark. The reader can see that the theorem about monodromy bears a topological character, it says about coverings. One can formulate it in the following way:

Let $\pi : Y \mapsto X$ *be a covering of topological spaces and let* γ_i, $i = 1, 2$, *be two paths in* X *joining the points* a *and* b. *If the paths are homotopically equivalent, then the two maps* $\pi^{-1}(a) \longmapsto \pi^{-1}(b)$, *defined by lifts of the paths* γ_i *to* Y, *coincide.*

If f is a multivalued function on $U \subset \mathbb{C}$ and M is its Riemann surface, then one has the single valued function $\tilde{f}$ on M,

$$\begin{array}{ccc} M & \stackrel{\tilde{f}}{\longmapsto} & \mathbb{C} \\ \pi \downarrow & & \uparrow f \\ U & = & U \end{array}$$

Example. $f(z) = \sqrt{z}$, $z \neq 0$. The prolongation of this function around 0 does not give the same value, but after two turns around 0 we get the same function. In order to get the Riemann surface of $\sqrt{z}$, we take two copies of the complex plane and cut them along the closed positive axis $z \geq 0$ (see Fig. 1.2).
We put these sheets one above another, turn the upper one along the real axis and glue the boundaries. The Riemann surface of $\sqrt{z}$ is equal $\mathbb{C} \setminus 0$ and we have the diagram

$$\begin{array}{ccc} \mathbb{C}\backslash 0 & \stackrel{x}{\longmapsto} & \mathbb{C} \\ \pi \downarrow & & \uparrow \sqrt{z} \\ \mathbb{C}\backslash 0 & = & \mathbb{C}\backslash 0 \end{array}$$

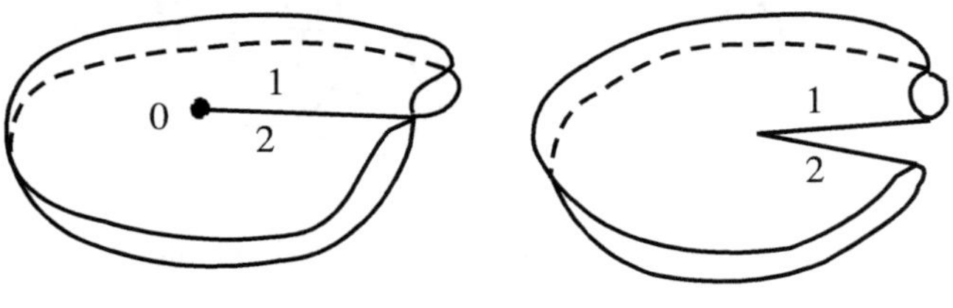

Figure 1.2: Riemann surface of $\sqrt{z}$

We can prolong π to the map on $\mathbb{C}$, $\pi(x) = x^2$. Then we say that π is a **ramified covering**; with one *ramification point* $x = 0$.

We can compactify the complex plane to the projective plane (or the Riemann sphere)

$$\mathbb{C} \cup \infty = \mathbb{CP}^1 = \overline{\mathbb{C}} \simeq \mathbb{S}^2,$$

and we can also compactify the Riemann surface $M \longmapsto M \cup \infty \simeq \mathbb{CP}^1$. The point ∞ is also a ramification point, because after the change of variables we have $1/x \longmapsto 1/z = (1/x)^2$. In Fig. 1.3 the pole ∞ is sent to the pole ∞ and the indicated circles are mapped with degree 2.

We will study multivalued holomorphic functions and their algebraic and topological invariants in Chapter 11.

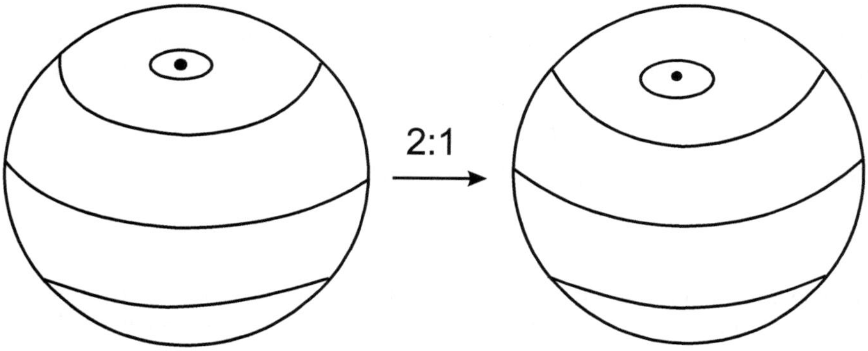

Figure 1.3: Ramified covering

§2 Morse Lemma

Let $U \subset \mathbb{C}$ be a domain containing 0 and let $f : U \longmapsto \mathbb{C}$ be a holomorphic function.

1.4. Definition. We say that the point 0 is **critical** for f iff $f'(0) = 0$. The critical point 0 is called **non-degenerate** iff $f''(0) \neq 0$. The value $f(0)$ is called the **critical value** of f.

Morse lemma in one dimension. *Let* 0 *be non-degenerate critical point of* f. *Then there exists a local holomorphic change* $x = \varphi(y)$, $x(0) = 0$, *such that*

$$f(\varphi(y)) = f(0) + y^2.$$

Proof. We can assume that $f(0) = 0$.

Hadamard's lemma. *If* $f(0)=0$, *then* $f(x)=xg(x)$ *with some analytic function* g.

Proof. We have $f(x)=f(x)-f(0)=\int_0^1\left\{\frac{\mathrm{d}}{\mathrm{d}t}f(tx)\right\}\mathrm{d}t=x\int_0^1 f'(tx)\mathrm{d}t$. □

We have $g(0)=f'(0)=0$. We apply the Hadamard's lemma again and we obtain $g(x)=xh(x)$ and $f(x)=x^2h(x)$, where $h(0)=\frac{1}{2}f''(0)\neq 0$. We put $y=x\sqrt{h(x)}$, where we can choose one of the two unique branches of the square root. □

Consider now the multidimensional case. Let $f:U\longmapsto\mathbb{C}$ be a holomorphic function, $U\subset\mathbb{C}^n$, $0\in U$.

1.5. Definition. The point 0 is **critical** iff $Df(0)=0$. It is **non-degenerate** iff the Hessian matrix $D^2f(0)$ is non-singular, or (equivalently) iff $\det\left(\partial^2 f/\partial x_i\partial x_j\right)(0)$ $\neq 0$. The value $f(0)$ is called the **critical value** of f.

1.6. Morse lemma. *If* 0 *is non-degenerate critical point of the function* f, *then there exists a holomorphic change of variables* $x=\varphi(y)$, $y=(y_1,\ldots,y_n)$, $\varphi(0)=0$, *such that*

$$f(\varphi(y))=f(0)+y_1^2+\ldots+y_n^2.$$

1.7. Remark. In the real case the thesis of the Morse lemma says that

$$f(\varphi(y))=f(0)+y_1^2+\ldots+y_k^2-y_{k+1}^2-\cdots-y_n^2.$$

Using the Morse lemma we shall investigate the level surfaces of holomorphic function in a neighborhood of non-degenerate critical point. Such level surface forms an analytic subvariety in $\mathbb{C}^n$ of complex codimension one, or a codimension two real subvariety in $\mathbb{R}^{2n}$. Let

$$g(y)=y_1^2+\ldots+y_n^2.$$

The case $n=1$. Here $\{g(y)=c\}$ is either one point 0 for $c=0$ or two points $\pm\sqrt{c}$ otherwise.

The case $n=2$. We have $y_2=\sqrt{c-y_1^2}$. The level surface $\{g=c\}$ is the Riemann surface of the function $\sqrt{c-y_1^2}$.
Let $c=0$. Then $y_2=\pm\sqrt{-1}y_1=\pm iy_1$. We get two complex lines joined at one point. Topologically it is diffeomorphic to the cone as in Fig. 1.4.
If $c\neq 0$, then the function $\sqrt{c-y_1^2}$ has two branching points $y_1=\pm\sqrt{c}$. When the variable y_1 varies, turning once around one branching point, then we arrive at the other sheet of the Riemann surface. When y_1 runs around both ramifications, then we arrive to the same place. In order to get the Riemann surface of $y_2(y_1)$, we take two copies of the complex plane cut along the segment joining the ramification points. We put one sheet above another, turn the upper sheet around the line passing through the branching points and glue the two sheets along the cuts (see Fig. 1.5. We obtain a surface diffeomorphic to the infinite cylinder. The image of

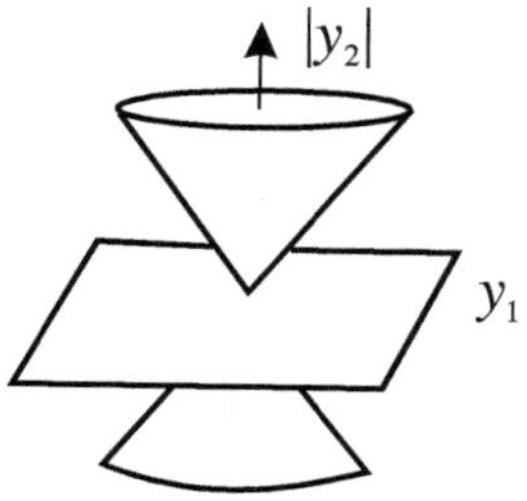

Figure 1.4: Cone for $c = 0$

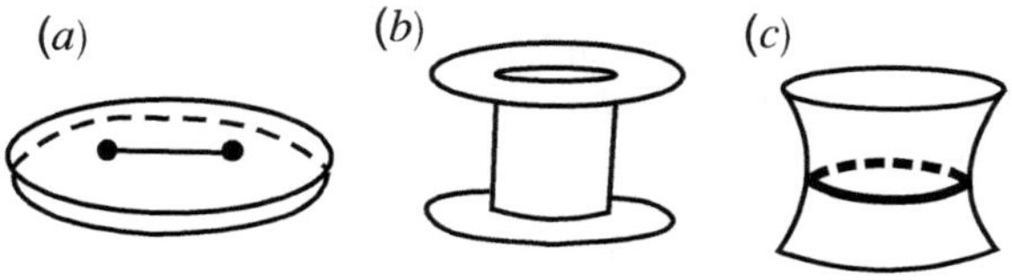

Figure 1.5: Cylinder for $c \neq 0$

the cut forms a closed curve Δ; it is a cycle generating the homology group of this surface in dimension one.

1.8. Proposition (Topology of levels of Morse function).

(a) *If* $c \neq 0$, *then the surface* $\{g = c\}$ *is diffeomorphic to* $T\mathbb{S}^{n-1}$ *(i.e., the tangent bundle to the unit sphere in* $\mathbb{R}^n$*) and the zero section* Δ *of this bundle is a cycle generating the reduced homology groups of this surface.*

(b) *Moreover, the space* $\{x : 0 \leq g(x) \leq 1\}$ *is homotopically equivalent to the space* $\{g(x) = 1\} \cup \mathbb{D}^n$, *where* $\mathbb{D}^n$ *is a ball glued to* $\Delta \subset \{g = 1\}$ *along the boundary. The deformation retraction of* $\{0 \leq g \leq 1\}$ *to* $\{g = 1\} \cup \mathbb{D}^n$ *can be realized in such way that the part of* $\{0 \leq g \leq 1\}$ *outside some neighborhood of* 0 *is sent to the analogous part of* $\{g = 1\}$.

1.9. Remark. In Chapter 3 below we give definitions of the homology groups and other notions from algebraic topology which will be used in the monodromy theory.

Proof of Proposition 1.8. (a) For $n = 1$ this is obvious. For $n = 2$ this follows from Fig. 1.5. Below we present the formulas realizing this diffeomorphism.
Denote $y_1 = u_1 + iv_1$, $y_2 = u_2 + iv_2$, $u = (u_1, u_2)$, $v = (v_1, v_2)$. Assume that $c > 0$. The equation $y_1^2 + y_2^2 = c$ means that

$$u_1^2 + u_2^2 = c + v_1^2 + v_2^2, \; u_1 v_1 + u_2 v_2 = 0.$$

The second equation means that the vectors u and v are orthogonal, $\langle u, v\rangle = 0$. The diffeomorphism is following

$$(u,v) \mapsto \left(u/\sqrt{c+|v|^2}, v\right).$$

(The first component lies in $\mathbb{S}^1$, the second component lies in the linear subspace of $\mathbb{R}^2$ orthogonal to the first component, i.e., tangent to $\mathbb{S}^1$).
If $c = |c|e^{i\theta} \not> 0$, then we apply the transformation

$$(y_1, y_2) \longmapsto (e^{i\theta/2}y_1, e^{i\theta/2}y_2),$$

which is a diffeomorphism, and use the above arguments.

Let $n > 2$. It turns out that the formulas obtained in the previous case are in use in the general situation. Let $y_j = u_j + iv_j$, $u = (u_1, \ldots, u_n)$, $v = (v_1, \ldots, v_n)$. If $c > 0$, then the equation $y_1^2 + \ldots + y_n^2 = c$ means that

$$|u|^2 = c + |v|^2, \ \langle u, v\rangle = 0.$$

The map

$$(u,v) \mapsto \left(u/\sqrt{c+|v|^2}, v\right)$$

transforms the level surface $\{g = c\}$ to $T\mathbb{S}^{n-1}$, the tangent space to the unit $(n-1)$-dimensional sphere in $\mathbb{R}^n$. We treat the case $\arg c \neq 0$ in the same way as before.
Here also we have the $(n-1)$-dimensional cycle Δ, the preimage of the zero section of this tangent bundle. It generates the reduced homology group of the surface $\{g = c\}$ in dimension $n-1$. As $c \to 0$ the cycle Δ tends to the critical point.

(b) We note that the space $T\mathbb{S}^{n-1}$ is contractible, along fibers, to $\mathbb{S}^{n-1}$. Thus the set $\{0 \le g \le 1\}$ is contractible to a disc, the union of spheres $\mathbb{S}_r^{n-1} \subset \{g = r^2\}$ with radii $r \in [0,1]$. This is the ball $\mathbb{D}^n$ from Proposition 1.8(b). Also it is not difficult to construct the deformation retraction as in the thesis of Proposition 1.8(b). □

1.10. Definition. The cycle Δ is called the **vanishing cycle**.

Proof of the Morse lemma. . Here I present the proof suggested to me by T. Maszczyk.
Firstly we need a multidimensional version of the Hadamard's lemma.

Hadamard's lemma. *If $f(x)$, $x \in (\mathbb{C}^n, 0)$, is a germ of analytic function such that $f(0) = 0$, then $f(x) = \sum x_i g_i(x)$ with analytic functions g_i.*

Let 0 be non-degenerate critical point of the function f. We can assume that $f(0) = 0$. We apply the Hadamard's lemma two times, to f and to g_i's, and obtain $f(x) = \sum h_{ij}(x)x_i x_j$, where h_{ij} are analytic functions. Applying the change

$(h_{ij}) \mapsto (\frac{1}{2}(h_{ij} + h_{ji}))$, we can assume that the matrix $[h_{ij}(x)]$ is symmetric. It is non-singular for small x, (because $2[h_{ij}(0)] = D^2 f(0))$.
Consider the following quadratic form

$$\xi \mapsto \Phi(\xi) = \sum h_{ij}(x)\xi_i\xi_j.$$

It is diagonalizable, which means that there exists a linear change $(\xi_i) \longmapsto (\eta_i = \sum \alpha_{ij}(x)\xi_j)$ such that $\Phi(\xi) = \sum \eta_i^2$. Now it is enough to put $y_i = \sum \alpha_{ij}(x)x_j$. □

§3 Morse Theory

Consider the tangent bundle to the sphere $T\mathbb{S}^n$. Let $\Delta = \{(x, v) : v = 0\}$ be its zero section representing a closed n-dimensional cycle. Is it possible to perturb little the cycle Δ, to some cycle Δ_1, in such a way that $\Delta \cap \Delta_1 = \emptyset$?
In order to study this problem we reformulate it. We can consider the cycle Δ as a map $\mathbb{S}^n \longmapsto T\mathbb{S}^n$

$$\Delta : x \longmapsto (x, 0).$$

Its perturbation also will be a map from $\mathbb{S}^n$ to its tangent bundle of the form

$$\Delta_1 : x \longmapsto (y(x), v(x)),$$

where $\sup_x\{|y(x) - x| + |v(x)|\} < \epsilon$. It is clear that we can assume that $y(x) \equiv x$. In such case we have a vector field $\{v(x\}$ on the sphere and the previous problem is following: can we comb the sphere? (Is there a vector field on $\mathbb{S}^n$ non-vanishing at any point?)
If $n = 1$, then the vector field $(x_1, x_2) \longmapsto (x_2, -x_1)$ is the example. Generally, if n is odd, then the answer is positive: $v(x_1, x_2, x_3, x_4, \ldots, x_{2k-1}, x_{2k}) = (x_2, -x_1, x_4, -x_3, \ldots, x_{2k}, -x_{2k-1})$.

If n is even, then the answer is negative.

To show this we need some new notions. Let M be a real n-dimensional manifold and let $\{\boldsymbol{V}(x)\}_{x\in M}$ be a vector field on it, $\boldsymbol{V}(x) \in T_x M$.

1.11. Definition. A point $x_0 \in M$ is called **singular** (or critical or equilibrium) point iff $\boldsymbol{V}(x_0) = 0$. Assume that x_0 is isolated singular point.
Take a small sphere $\mathbb{S}(x_0, \epsilon)$ around x_0 of radius ϵ and consider the map

$$\mathbb{S}(x_0, \epsilon) \ni x \overset{\phi}{\longmapsto} \frac{\boldsymbol{V}(x)}{|\boldsymbol{V}(x)|} \in \mathbb{S}^{n-1}.$$

The degree of the map ϕ is called the **index** of the singular point x_0 and is denoted by $i_{x_0}\boldsymbol{V}$.

1.12. Remark. If a vector field has non-isolated critical points, then it can be perturbed, in the class of differentiable vector fields, to such with only isolated critical points. It is done using the Sard's theorem as follows.

The singular point x_0 is called **degenerate** iff $\det DV(x_0) = 0$. The non-isolated singular points are degenerate. The degenerate singular points for $v(x)$ are the critical points for the maps $x \longmapsto V(x)$. Because the critical values form a set of Lebesque measure zero (the Sard's theorem), then the vector field $V(x) - w$, for suitable small $w \in \mathbb{R}^n$, has no degenerate singular points (in a chart of M diffeomorphic to a subset of $\mathbb{R}^n$).

1.13. Remark. The **degree of a map** f between differentiable oriented manifolds M and N of the same dimensions is defined in Chapter 3 below in the homological terms. Here we give the analytic definition.
If the map is sufficiently regular, then the degree is calculated as follows. Let $y \in N$ and $f^{-1}(y) = \{x_1, \ldots, x_k\}$. Then we have

$$\deg f = \sum_{i=1}^{k} \pm 1,$$

where the sign is $+$, when $Df(x_i)$ preserves the orientation, and is $-$, if it reverses the orientation.
If the map f is not regular, then we approximate it by a regular map f_ϵ, and put $\deg f = \deg f_\epsilon$.

Examples. For the vector field $\dot{x} = x$, $\dot{y} = -y$ (or $x\frac{\partial}{\partial x} - y\frac{\partial}{\partial y}$) the index at 0 is -1 (Fig. 1.6(a). For the vector field $\dot{x} = x + y$, $\dot{y} = -x + y$ the index is 1 (Fig. 1.6(b). The examples of vector fields with index 2 and -2 provide the fields $\dot{z} = z^2$ and $\dot{z} = \bar{z}^2$ respectively (written using the complex variable $z = x + iy$), see. Fig. 1.6(c),(d).

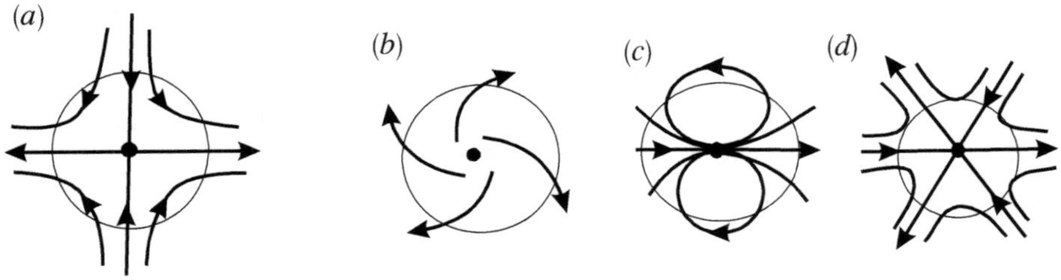

Figure 1.6: Vector fields and their indices

Generally, for a linear vector field, given by the non-singular matrix A, the index at 0 is sign(det A).
Consider the vector field $\dot{z} = 1$ in $\mathbb{C} \simeq \mathbb{R}^2$. It prolongs itself to a vector field in the Riemann sphere with index at ∞ equal to 2, see Fig. 1.7. Indeed, in the variable $\xi = 1/z$ we get $\dot{\xi} = -\xi^2$, which is (up to sign) the same as the vector field from one of the previous examples.

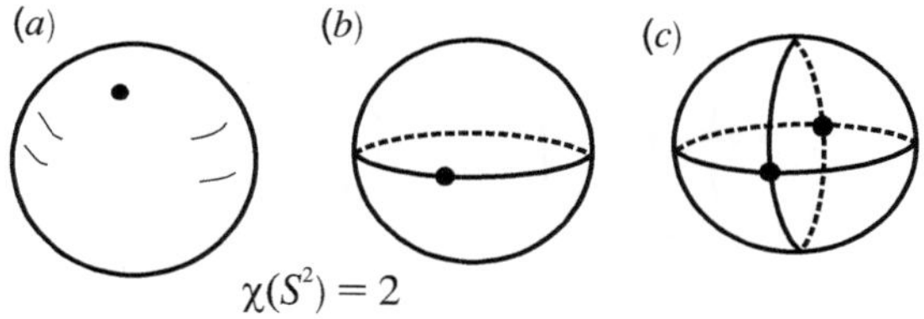

Figure 1.7: Partitions of the 2-sphere

For the vector field $\dot{z} = z$, the index at $z = 0$ is 1 and the index at $z = \infty$ is also 1. Thus the sum of indices is the same for the both vector fields and is equal to 2.

1.14. Poincaré–Hopf theorem. *If the vector field $\boldsymbol{V}(x)$ has only isolated singular points $x_1, \ldots, x_r$, then*

$$\sum i_{x_j} \boldsymbol{V} = \chi(M),$$

where $\chi(M)$ is the Euler characteristic of the manifold M.

1.15. Remark. The **Euler characteristic** is calculated as follows. We take a partition of the manifold M into cells (or simplices) and obtain a complex which consists of m_0 0-dimensional cells, m_1 1-dimensional cells, etc. Then we define

$$\chi(M) = m_0 - m_1 + m_2 - m_3 + \ldots \pm m_n.$$

Proof of Theorem 1.14. We sketch the proof of the Poincaré–Hopf theorem based on the Morse lemma. Let $\mathcal{X}$ be the space of all differentiable vector fields (equipped with some natural topology). Let its subset $\mathcal{X}_0$ consists of vector fields with only **non-degenerate** singular points, i.e., such that $\det D\boldsymbol{V}(x_i) \neq 0$. Note that then $i_{x_i}\boldsymbol{V} = \text{sign}[\det D\boldsymbol{V}(x_i)]$.
The set $\mathcal{X}_0$ is open and dense in $\mathcal{X}$ (see Remark 1.12 above). The function $v \longmapsto \sum$(indices) is locally constant at $\mathcal{X}_0$. It is enough to show that it is the same at the boundary of $\mathcal{X}_0$.
Let $v \in \mathcal{X} \backslash \mathcal{X}_0$ be a vector field with degenerate critical points but isolated and with finite indices. We take some small perturbation $v_\epsilon \in \mathcal{X}_0$ of v. The field v_ϵ has only non-degenerate critical points, where some of them may coalesce as $\boldsymbol{V}_\epsilon \longmapsto \boldsymbol{V}$. From Fig. 1.8(b) it is seen that the sums of indices of $\boldsymbol{V}$ and of its perturbation are the same; the index calculated along the outer cycle is equal to the sum of indices calculated along the inner circles.
Therefore, it remains to calculate the sum of indices of some particular vector field from $\mathcal{X}_0$.

1.16. Definition. A twice differentiable function $f : M \longmapsto \mathbb{R}$ which has only non-degenerate critical points and different critical values is called the **Morse function.**

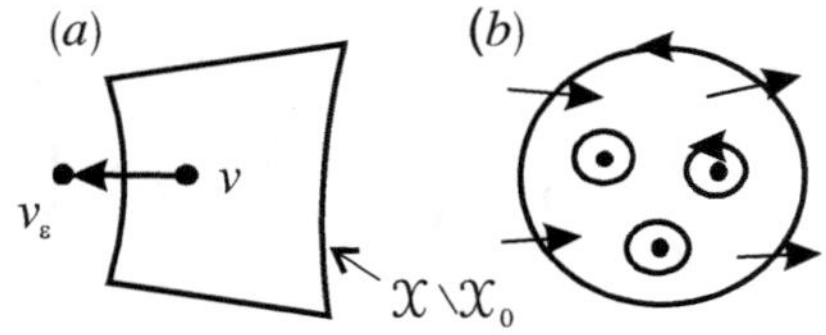

Figure 1.8: Sums of indices ov v and of its perturbation

Assume that M is a Riemannian manifold, i.e., it is equipped with a Riemannian metric $\langle\cdot,\cdot\rangle_x$. (If M is compact and smooth, then using the partition of unity one can always construct such metric.) This metric tensor defines the isomorphism $T_xM \ni \boldsymbol{W} \longmapsto \langle\cdot,\boldsymbol{W}\rangle \in T_x^*M$. If f is a function on M then, applying the inverse of this isomorphism to $\mathrm{d}f(x) \in T_x^*M$, we obtain the *gradient vector field* $\boldsymbol{V}(x) = \mathrm{grad} f(x) = \nabla f(x)$. In local coordinates with the euclidean metric, we have $\dot{x}_i = \partial f/\partial x_i$ and $D\boldsymbol{V} = (\partial^2 f/\partial x_i \partial x_j)$. In particular, the index of the gradient vector field at a critical point x_0 is $\mathrm{sign}[\det(\partial^2 f/\partial x_i\partial x_j)]$. In the case of general metric $\langle\cdot,\cdot\rangle_x = (A(x)\cdot,\cdot)$, we have $\nabla f = A^{-1}\partial f/\partial x$, and the same formula for index holds.
Let $f: M \longmapsto \mathbb{R}$ be a Morse function. As the model vector field, for calculating the sum of indices, we take $\nabla f(x)$.

Now we present the Morse theory about determination of the topology of a manifold using its Morse function (see [**Mil1**]). It main ingredient is the behavior of the level surfaces of the function f in neighborhoods of its critical points.
By the real Morse Lemma it is enough to study the bifurcations of the level surfaces for the function

$$f(x) = f(0) + x_1^2 + \ldots + x_k^2 - x_{k+1}^2 - \ldots - x_n^2,$$

a constant plus a quadratic form. The **Morse index** of this quadratic form is equal to the number of its minuses; we call it the **Morse index** of the critical point of the function f.
We investigate the sets $\{f = c\}$ and $\{f \leq c$, as c varies from its minimal value to its maximal value. Altogether we construct some partition of M into cells.
If $k = n$, then the Morse index of the critical point is 0 and we have a local minimum. The sets $\{f \leq c\}$ are discs. We associate with each critical point x_j of index 0 the 0-dimensional cell $\sigma_j^0 = \{x_j\}$ of the promised cell complex.
If $k = n - 1$, then we observe the following bifurcation. Locally the sets $\{f \leq c\}$, $c < f(0)$, consist of two pieces; they are diffeomorphic to $\mathbb{D}^n \times \mathbb{S}^0$, where $\mathbb{D}^k$ denotes the k-dimensional ball with the boundary $\partial\mathbb{D}^k = \mathbb{S}^{k-1}$. (It is homotopically equivalent to $\mathbb{S}^0$.) After bifurcation, $c > f(0)$, the two components became connected which, up to homotopy, means adding a segment joining the components. With each critical point of Morse index 1 we associate a 1-dimensional cell (a closed

connected curve) σ_j^1, which is adjoined to the 0-dimensional skeleton in the way the bifurcation of passing through this critical value says: $\partial\sigma_j^1 = \mathbb{S}^0 \subset \{f < f(0) - \epsilon\}$. Generally, near any critical point of Morse index $i = n - k$ the sets $\{f \le c\}$, $c < f(0)$ are diffeomorphic to $\mathbb{D}^{n-i+1} \times \mathbb{S}^{i-1} \simeq \mathbb{S}^{i-1}$. The bifurcation is equivalent to adjoining to this set the handle $\mathbb{D}^{n-i} \times \mathbb{D}^i \simeq \mathbb{D}^i$. So, we add to our complex the i-dimensional cell σ_l^i glued along the boundary to the $(i-1)$-dimensional skeleton (see Fig. 1.9).

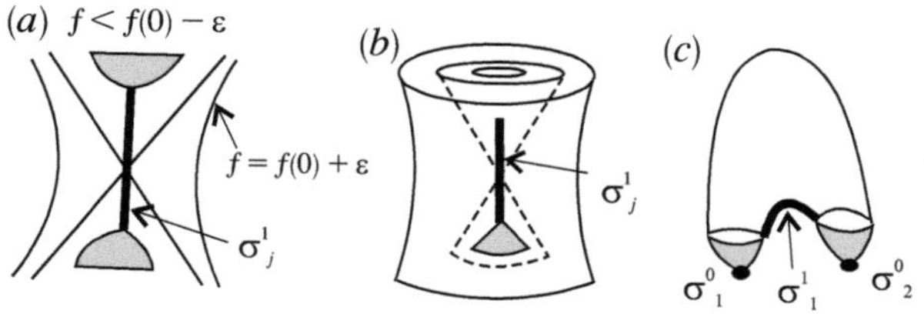

Figure 1.9: Examples of Morse functions and handles

Of course, the Euler characteristic of M is equal to the number of cells associated with critical points of index 0 minus the number of cells associated with points of index 1 plus, etc. This completes the proof of the Poincaré–Hopf theorem. □

1.17. Self-intersection of the cycle Δ. We have $\chi(\mathbb{S}^n) = 0$, if n is odd and $= 2$, if n is even. By the Poincaré–Hopf theorem, this means that the odd-dimensional spheres can have empty intersections with their deformations in their tangent spaces and the even-dimensional spheres do not have this property.
The sum of indices of vector field on $\mathbb{S}^n$ can be treated as the index of self-intersection of this sphere in its tangent bundle. We prove that this number equals to

$$1 + (-1)^n.$$

We shall use these facts in the sequel.

Chapter 2

Normal Forms of Functions

In this chapter we present elements of the theory of singularities of holomorphic functions. We introduce the notions of multiplicity, stability, versal deformation, normal form and describe their main properties. We present also the beginning of the list of normal forms for singularities.
This subject is rather standard and well elaborated in many sources. We follow mainly the first volume of the book of V. Arnold, A. Varchenko and S. Gusein-Zade [**AVG**]

§1 Tougeron Theorem

2.1. Notations and definitions. By $\mathcal{O}_{x_0} = \mathcal{O}_{x_0}(\mathbb{C}^n)$ we denote the **local ring** of **germs** at x_0 of holomorphic functions, i.e., functions holomorphic in some neighborhood of x_0. Two functions f at U and g at V are equivalent iff $f \equiv g$ at $U \cap V$. Usually we put $x_0 = 0$ and write $\mathcal{O}$ or $\mathbb{C}\{x\} = \mathbb{C}\{x_1, \ldots, x_n\}$, instead of $\mathcal{O}_0$. It is used to write $f : (\mathbb{C}^n, 0) \longmapsto \mathbb{C}$.
By $\mathfrak{m}$ we denote the **maximal ideal** of the ring $\mathcal{O}$, $\mathfrak{m} = \{f : f(0) = 0\}$. $\mathfrak{m}$ is generated by $x_1, \ldots, x_n$ (the Hadamard's lemma).
By $j^k f = j^k f(0)$, i.e., the kth **jet** of f, we denote the Taylor series of f up to order k. By J^k we denote the space of k-jets.
The **gradient ideal** of the germ f is generated by $\partial f/\partial x_i$ and is denoted by

$$I_f = (\partial f/\partial x_1, \ldots, \partial f/\partial x_n).$$

The **local algebra** of the germ f is

$$A_f = \mathcal{O}/I_f.$$

The **Milnor number**, or the **multiplicity**, of the germ f is

$$\mu = \dim A_f.$$

Examples. 1. Let $f(x) = x^{n+1}$. Then $I_f = (x^n)$, the set of polynomials with zero first $n-1$ derivatives at $x = 0$. The local algebra is generated by the monomials $1, x, x^2, \ldots, x^{n-1}$ and $\mu(f) = n$. The functions x^{n+1} form the series $\mathbf{A}_n$ of simple singularities (see below).

2. Let $f(x, y) = x^3 + y^4$, i.e., the simple singularity $\mathbf{E}_6$. Its gradient ideal is generated by x^2 and y^3. In order to calculate the local algebra of this function, we present the situation graphically.

H. Żołądek, *The Monodromy Group*, Monografie Matematyczne 67,
https://doi.org/10.1007/978-3-031-91270-2_2

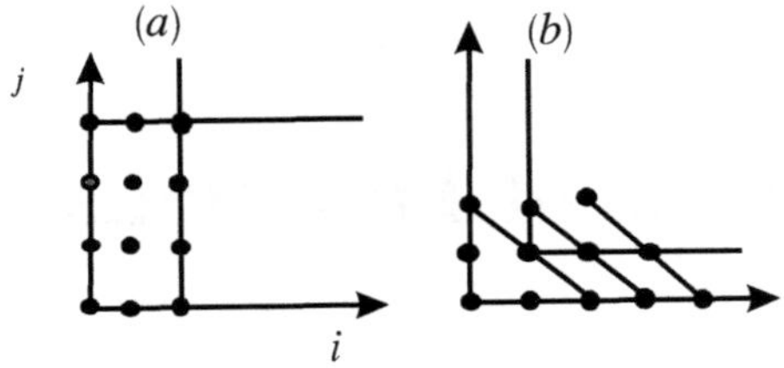

Figure 2.1: Generators of the gradient ideal

At Fig. 2.1(a) we have the lattice $\mathbb{Z}_+^2$, consisting of points of the plane with non-negative integer coordinates (i, j). Each such point represents the monomial $x^i y^j \in \mathcal{O}$. The ideal I_f contains all monomials from the set represented by

$$\left((2,0)+\mathbb{Z}_+^2\right) \cup \left((0,3)+\mathbb{Z}_+^2\right),$$

i.e., we add to the basic points the whole quarters $\mathbb{Z}_+^2$. The remaining points from the lattice represent the basis of the local algebra. Its dimension is 6.

3. Let $f(x, y) = x^2 y + y^3$, i.e., the simple singularity $\mathbf{D}_4$. Then the generators of the gradient ideal $(2xy, x^2 + 3y^2)$ are represented by: the point $(1,1)$ and by two points $(2,0)$, $(0,2)$, which are associated one with another (see Fig. 2.1(b)).
Of course, I_f contains $(1,1)+\mathbb{Z}_+^2$. It is also clear that the monomials represented by $(0,0),(1,0),(0,1)$ are outside I_f and form a part of the basis of the local ring A_f.
The two points $(2,0),(0,2)$ cannot lie simultaneously in the ideal as well as cannot be simultaneously outside it, (they are dependent in A_f). So, we add one of them, e.g., $(0,2)$, to the basis of A_f. Considering the quadratic parts of the Taylor expansions of the functions from our (preliminary) basis and from the ideal I_f, we see that the monomials $(0,0),(1,0),(0,1),(0,2)$ are independent in A_f. The rest is in the gradient ideal, which means that $J = \mathbb{C} + x\mathbb{C} + y\mathbb{C} + y^2\mathbb{C} + I_f = \mathcal{O}$.
To prove this, it is enough to show that the monomials x^i, y^j are in J. But $x^2 = (x^2+3y^2) - 3(y^2) \in I_f + y^2\mathbb{C}$ and $x^i = x^{i-2}(x^2+3y^2) + 3x^{i-3}y(xy) \in I_f$. Similarly we treat the monomials y^j.
Therefore $\mu(f) = 4$.

4. *Problem*: show that $\mu(x^2 y + x^{k-1}) = k$.

2.2. Theorem (Isolated critical point). *The Milnor number $\mu < \infty$ iff $x = 0$ is an isolated critical point of the function f.*

Parallel with the multiplicity of a function one can define the multiplicity for germs of vector fields.
Let $\boldsymbol{F} : (\mathbb{C}^n, 0) \longmapsto (\mathbb{C}^n, 0)$, $\boldsymbol{F} = (f_1, \ldots, f_n)$, be a germ of holomorphic map. Let $A_{\boldsymbol{F}} = \mathcal{O}/(f_1, \ldots, f_n)$ be the local algebra of the germ $\boldsymbol{F}$. Then $\mu(\boldsymbol{F}) = \dim A_{\boldsymbol{F}}$.

2.3. Theorem. *The multiplicity* $\mu = \dim A_{\boldsymbol{F}} < \infty$ *iff* $x = 0$ *is isolated solution of the equation* $\boldsymbol{F} = 0$.

2.4. Theorem (Index and multiplicity). *When we treat* $\boldsymbol{F}$ *as a vector field, then* $\mu = i_0 \boldsymbol{F}$.

The above two theorems will be proved in the next section.

2.5. Theorem of Tougeron. *Let* f *be a germ of holomorphic function such that* $\mu = \mu(f) < \infty$. *Then there exists an analytic change of variables* $y = h(x)$, *such that* $f \circ h = j^{\mu+1} f(0)$.

2.6. Remark. The jet $j^k f$ is called **sufficient** iff any two germs with this k^{th} jet are analytically equivalent. It means stability with respect to high order perturbations. The theorem of Tougeron says that the jet $j^{\mu+1} f$ is sufficient.

Proof of Theorem 2.5. Unfortunately, here we cannot repeat the proof of the Morse lemma. We follow the book of Arnold, Varchenko and Gusein-Zade [**AVG**]. Assume that f has critical point at 0 of multiplicity μ and let $\phi \in \mathfrak{m}^{\mu+2}$. We shall show that $f + \phi \sim f$.

We use the *homotopy method.* Namely, we join the functions f and $f + \phi$ by an arc in the functional space and we seek a one-parameter family of diffeomorphisms realizing the equivalences with f. In other words, we try to solve the equation

$$(f + t\phi) \circ h_t(x) \equiv f(x), \ t \in [0, 1], \tag{2.1}$$

where h_t is unknown.

Introduce the non-autonomous vector field $\boldsymbol{V}_t(x)$ by the formula

$$\mathrm{d}h_t/\mathrm{d}t = \boldsymbol{V}_t(h_t(x)).$$

We shall find the vector field $\boldsymbol{V}_t$ first and then, integrating the latter equation, we shall find the diffeomorphisms h_t.

Differentiating Eq. (2.1) with respect to t, we get the equation $\phi \circ h_t + (f + t\phi)_* \cdot \boldsymbol{V}_t \circ h_t \equiv 0$. Thus we have to solve the equation

$$(f + t\phi)_* \cdot \boldsymbol{V}_t = -\phi \tag{2.2}$$

with respect to $\boldsymbol{V}_t$.

2.7. Lemma. *We have* $\mathfrak{m}^\mu \in (\partial(f + t\phi)/\partial x_1, \ldots, \partial(f + t\phi)/\partial x_n)$, *which means that any monomial of sufficiently high degree lies in the gradient ideal of the function* $f + t\phi$.

Example. For non-degenerate critical point the gradient ideal coincides with the maximal ideal and $\mu = 1$.

From Lemma 2.7 the theorem of Tougeron follows. Indeed, because $\phi \in \mathfrak{m}^{\mu+2}$, Eq. (2.2) has a solution $\boldsymbol{V}_t = \sum v_{t,i} \frac{\partial}{\partial x_i}$. Its components $v_{t,i} \in \mathfrak{m}^2$ and hence $\boldsymbol{V}_t(0) = 0$, $D\boldsymbol{V}_t(0) = 0$. Moreover, $\boldsymbol{V}_t$ depends smoothly on t.

So, in order to find the family of diffeomorphisms h_t, it is sufficient to solve the following initial value problem:

$$\frac{\mathrm{d}}{\mathrm{d}t} h_t = \boldsymbol{V}_t(h_t(x)), \ h_0 = Id.$$

The assumption $\phi \in \mathfrak{m}^{\mu+2}$ is needed to ensure the existence and uniqueness of solutions to the latter problem. Because $\boldsymbol{V}_t(0) = 0$, we get $h_t(0) = 0$. □

Proof of Lemma 2.7. Consider firstly the case $\phi \equiv 0$. Let $\phi_1, \ldots, \phi_\mu \in \mathfrak{m}$. It is enough to show that $\phi_1 \cdot \ldots \cdot \phi_\mu \in I_f$, where I_f is the gradient ideal.
Consider the series of functions: $\phi_0 = 1$, ϕ_1, $\phi_1\phi_2, \ldots, \phi_1 \ldots \phi_\mu$. They are linearly dependent in the local algebra A_f. So, we have

$$c_0 + c_1\phi_1 + c_2\phi_1\phi_2 + \ldots + c_\mu\phi_1 \ldots \phi_\mu \in I_f$$

for some constants c_j. If c_r is the first nonzero coefficient, then $\phi_1 \ldots \phi_r(c_r + \ldots) \in I_f$, or $\phi_1 \ldots \phi_r \in I_f$. Of course, in this case the product of all ϕ_i's also lies in the gradient ideal.
Consider now the general case $\phi \not\equiv 0$. Let $M_1, \ldots, M_r$ be the all homogeneous monomials of degree μ; they form a basis in the space of homogeneous polynomials of degree μ. We know already that $M_j \in I_f$. This means that

$$M_j = \sum \frac{\partial f}{\partial x_i} h_{ij} = \sum \frac{\partial (f + t\phi)}{\partial x_i} h_{ij} - \sum \frac{\partial t\phi}{\partial x_i} h_{ij}.$$

The last sum in the above formulas belongs to $\mathfrak{m}^{\mu+1}$ and can be expressed by means of the monomials M_j (the Hadamard's lemma). We get

$$M_j = \sum \frac{\partial (f + t\phi)}{\partial x_i} h_{ij} - t \sum_k M_k \sum_l x_l a_{kl}(x),$$

where $a_{kl} \in \mathfrak{m}$ (by assumption about ϕ). We can rewrite this system of equations in the matrix form

$$(I - tA)\boldsymbol{M} = \boldsymbol{B},$$

where tA is a small matrix and the components of the vector $\boldsymbol{B}$ belong to the gradient ideal. Because the matrix $I - tA$ is invertible, also the components M_i of the vector $\boldsymbol{M}$ are in this ideal. □

2.8. Corollary. *Any germ of function of finite multiplicity can be replaced by an equivalent polynomial.*

§2 Deformations

We have to introduce some notions concerning actions of infinite-dimensional groups on infinite-dimensional functional spaces. So, firstly we demonstrate them in the finite-dimensional case.

Let M be a manifold (of finite dimension for a while) and let a group G acts on it: $(f,g) \longmapsto gf$, $f \in M$, $g \in G$. Let $f \in M$. We denote its orbit under the action by Gf.

2.9. Definition. A **deformation** of f is a map $F : \Lambda \longmapsto M$, where Λ is the **base** of the deformation with a distinguished point 0 and $F(0) = f$.
Two deformations F, F' are **equivalent**, iff there is a family $g(\lambda) \in G$, $\lambda \in \Lambda$, such that

$$F'(\lambda) = g(\lambda)F(\lambda),$$

i.e., the equivalence along the orbits.
If $\phi : (\Lambda', 0) \longmapsto (\Lambda, 0)$ is a map between the base spaces, then the **induced deformation** (from F by means of ϕ) is

$$\phi^* F(\lambda') = F(\phi(\lambda')),$$

i.e., a change of parameters.
A deformation F (of f) is called **versal** iff any other deformation of f is equivalent to a deformation induced from the deformation F.
A deformation is called **mini-versal** iff it is versal and the dimension of its base is minimal.

We can say that a deformation is versal iff it intersects all orbits near f (see Fig. 2.2). In particular, the deformation with the base M and identity map is versal; but usually is not mini-versal.

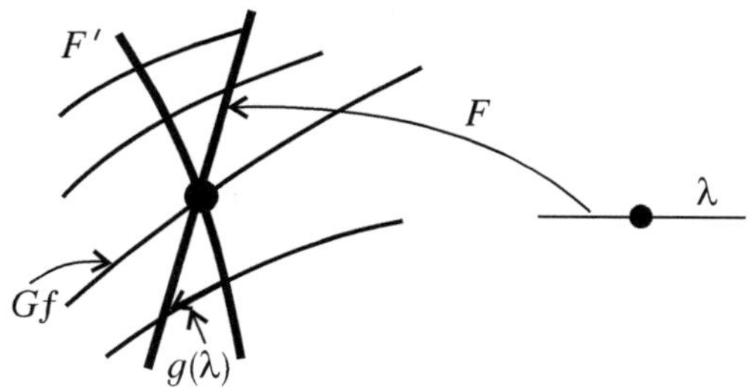

Figure 2.2: Orbits and two deformations

In the singularity theory we deal with the infinite-dimensional situation. The role of the manifold M is played by the space of germs $f = f(x)$ of holomorphic functions and the role of G is played by the group of local analytic diffeomorphisms $h = h(x)$, acting on functions by compositions on the right; it is called the *right equivalence*. However, the definitions from the point 2.9 pass to the infinite-dimensional case unchanged.
The deformation of a germ $f : (\mathbb{C}^n, 0) \longmapsto \mathbb{C}$ is a germ $F : (\mathbb{C}^n \times \mathbb{C}^k, 0) \longmapsto \mathbb{C}$, $F(x, 0) = f(x)$. The equivalence of two deformations is written as $F'(x, \lambda) =$

$F(h(x,\lambda),\lambda)$, where $h(\cdot,\lambda)=h_\lambda$ is a family from G) and the induced deformation is given by $F'(x,\lambda')=F(x,\phi(\lambda')$.

2.10. Definition. We say that a germ f is **stable** iff the orbit of f contains a whole neighborhood of f.
We say that a germ f is **simple** iff a neighborhood of f is covered by finite number of orbits.
If a neighborhood of f is covered by l-parameter families of orbits such that $\max l = m$, then we say that the germ f is m-**modal**; as we shall see below, the normal form of f contains m parameters (sometimes called moduli).
By a **normal form** we mean some (simultaneous) choice of a member from each orbit. This choice is not unique, so one should do it in a way as natural as possible.

2.11. Remarks. (a) In [**AVG**]singularities of other objects are considered: of maps from $\mathbb{C}^n$ to $\mathbb{C}^m$, with the so-called left-right equivalence (when we can make independent changes in the source space and in the target space) and with respect to the so-called V-equivalence (when the change in the target space is linear and depends on x). There analogous definitions, as in the case of functions, are introduced and analogous results are obtained.

(b) The singularity theory is used not only to the local analysis. Usually one has a function on a manifold, where it has finite number of critical points. During deformation of a function the critical points also can move with the parameter. For example $f_\epsilon(x)=x^2-\epsilon x$ has critical point at $\epsilon/2$. Therefore, it is reasonable to keep some neighborhood of a critical point fixed during the deformation.

The notions of stability and versality have their infinitesimal versions. The infinitesimal stability is obtained from differentiation of the equality $(f+t\phi)(x)=f\circ h_t(x)$ with respect to t at $t=0$

$$\phi(x)=\sum\frac{\partial f}{\partial x_i}v_i(x). \tag{2.3}$$

2.12. Definition (infinitesimal stability). The germ f is **infinitesimally stable** iff Eq. (2.3) has solution (v_i) for every ϕ.

In particular, the proof of the theorem of Tougeron is the proof of the implication: infinitesimal stability $\Rightarrow$ stability (i.e., stability with respect to perturbations of high order).
In fact, the notion of stability and of its infinitesimal version has greater application in the theory of maps, e.g., the Whitney singularities of maps from the plane to the plane (see [**AVG**]):
$(x,y)\longmapsto(x^2,y)$ (the **fold**),
$(x,y)\longmapsto(x^3+xy,y)$ (the **cusp**).

Let us differentiate the equation

$$F'(x,\lambda')=F(g(x,\lambda'),\phi(\lambda')),$$

$F' = f(x) + \lambda'\alpha(x)$, $g(x,0) = x$, with respect to $\lambda' \in \mathbb{C}$. We get

$$\alpha(x) = \sum_{i=1}^{n} \frac{\partial F}{\partial x_i} v_i(x) + \sum_{j=1}^{k} \frac{\partial F}{\partial \lambda_j} c_j. \tag{2.4}$$

2.13. Definition (infinitesimal versality). We say that a deformation $F(x,\lambda)$ is **infinitesimally versal** iff the equation (2.4) has a solution $v_i(x) \in \mathcal{O}$, $c_j \in \mathbb{C}$, for any function $\alpha(x) \in \mathcal{O}$.

2.14. Theorem (versal deformations). *Any deformation infinitesimally versal is versal.*

Examples. 1. The deformation $F(x,\lambda) = x^2 + \lambda$ is versal, because the equation $\alpha(x) = 2xv(x) + c$ has the solution $v(x) = (\alpha(x) - \alpha(0))/2x$, $c = \alpha(0)$.
2. Similarly, the deformation $F = x^3 + \lambda_1 x + \lambda_2$ is versal.
3. Generally, if $e_1(x), \ldots, e_\mu(x)$ define a basis of the local algebra A_f of the germ f, then the deformation

$$F(x,\lambda) = f(x) + \lambda_1 e_1(x) + \ldots + \lambda_\mu e_\mu(x)$$

is versal. It is also mini-versal deformation.

2.15. Corollary. *For any germ of finite multiplicity we can choose the function, as well as the mini-versal deformation, in polynomial forms.*

Remark. In the theory of singularities of functions and maps theorems about reductions (to a sufficient jet, or to a normal form) are formulated in the analytic versions. The corresponding changes of variables are analytic. In particular, the formal classification (reduction by means of formal power series) coincide with the analytic classification of singularities.
As the reader will see, this is not the case in the Ordinary Differential Equations Theory and in the Dynamical Systems Theory. Very often power series, which reduce some singularity of a vector field or of a diffeomorphism, diverge.

Proof of Theorem 2.14. This proof relies mostly on local algebra.
Let $F(x,\lambda)$ be the infinitesimal deformation of the germ f and let $F'(x,\lambda')$, $\lambda' \in (\mathbb{C}^{k'},0)$, be other deformation of f.
We apply certain trick, which allows to reduce the problem to the case, when F' is a deformation of F with one parameter. Take the following function:

$$\widetilde{F}(x,\lambda,\lambda') = F(x,\lambda) + F'(x,\lambda') - f(x).$$

It is a deformation of f with parameters (λ,λ'), as well as a deformation of F with the parameter λ'.
Any extension of infinitesimally versal deformation is an infinitesimally versal deformation. Consider the chain

$$\mathbb{C}^k \subset \mathbb{C}^{k+1} \subset \ldots \subset \mathbb{C}^{k+k'}$$

of spaces, which define the chain of deformations with one parameter. Now, step-by-step, we show the equivalence of each of these deformations with some deformations induced from the previous one.

Consider therefore the following special case

$$\Phi(x,\lambda,\mu),\quad \lambda\in\mathbb{C}^l,\ \mu\in\mathbb{C};\quad \Phi(x,\lambda,0)=F(x,\lambda).$$

2.16. Proposition. *The deformation Φ is equivalent to a deformation induced from F.*

Proof. The property that Φ is equivalent to a deformation induced from F can be formulated as follows:

$$\Phi(g_\mu(x,\lambda),\phi_\mu(\lambda),\mu)\equiv F(x,\lambda), \tag{2.5}$$

where $h_\mu(x,\lambda)=(g_\mu(x,\lambda),\phi_\mu(\lambda)$ is a 1-parameter family of local diffeomorphisms. (Apply h_μ^{-1} to (2.5) and you obtain the definition from Definition 2.9.) The family h_μ defines the non-autonomous vector field $\mathrm{d}h_\mu/\mathrm{d}\mu=\boldsymbol{V}_\mu\circ h_\mu$,

$$\boldsymbol{V}_\mu=\sum H_j(x,\lambda,\mu)\frac{\partial}{\partial x_j}+\sum \xi_i(\lambda,\mu)\frac{\partial}{\partial\lambda_i}$$

in $\mathbb{C}^n\times\mathbb{C}^l$, analogously as in the proof of the Tougeron theorem.
Differentiating the identity (2.5) with respect to μ, we get the equation

$$\frac{\partial\Phi}{\partial\mu}+\sum H_j\frac{\partial\Phi}{\partial x_j}+\sum\xi_i\frac{\partial\Phi}{\partial\lambda_i}\equiv 0.$$

As in the proof of the Tougeron theorem, the problem reduces to the problem of solving the equation

$$\alpha(x,\lambda;\mu)=H(x,\lambda;\mu)\cdot\frac{\partial\Phi}{\partial x}+\Xi(\lambda;\mu)\cdot\frac{\partial\Phi}{\partial\lambda} \tag{2.6}$$

for any α.
The assumption of Theorem 2.14, i.e., the infinitesimal versality, ensures existence of a solution of Eq. (2.6) for $\lambda=0,\mu=0$. We need to extend this solution to a solution of Eq. (2.6) in the general case.
In order to pass from the particular solutions to the general solutions, we need some preparation theorems. There are three such theorems: the *Weierstrass Preparation Theorem,* the *Division Theorem* and the *Thom–Martinet Preparation Theorem.* What we need is Thom–Martinet Preparation Theorem for modules over local rings of holomorphic functions.
Let $(x,y)\in\mathbb{C}^n\times\mathbb{C}^k$, $\mathcal{O}_{n+k}=\mathcal{O}_0(\mathbb{C}^n\times\mathbb{C}^k)$, $\mathcal{O}_k=\mathcal{O}_0(\mathbb{C}^k)$, $\mathcal{O}_n=\mathcal{O}_0(\mathbb{C}^n)$. Let $I\subset\mathcal{O}_{n+k}$ be an ideal and denote $I_{x,0}=\{f(x,0):\ f\in I\}$.

2.17. Thom–Martinet preparation theorem. *If some elements $e_1, \ldots, e_r \in \mathcal{O}_{n+k}$ are such that the functions $e_i(x,0)$ generate the module $\mathcal{O}_n/I_{x,0}$ (over $\mathbb{C}$), then the functions e_i generate the module $\mathcal{O}_{n+k}/I$ over $\mathcal{O}_k$.*
In other words, for any $\alpha \in \mathcal{O}_{n+k}$ there exist germs $g_i(y)$ such that

$$\alpha(x,y) = \sum g_i(y) e_i(x,y) \pmod{I}.$$

Finishing the proof of Theorems 2.16 *and* 2.14. We put $y = (\lambda; \mu)$, $I = (\partial\Phi/\partial x_1, \ldots, \partial\Phi/\partial x_n)$, $e_i = \partial\Phi/\partial\lambda_i$ in Theorem 2.17. Its thesis says that Eq. (2.6) has a solution in the class of germs of analytic functions. This gives Proposition 2.16 and then Theorem 2.14. □

Now we make some moves in the direction of the proof of Theorem 2.17. For this we need the other two preparation theorems.

2.18. Weierstrass Preparation Theorem. *Let $f(z_1, \ldots, z_m; w) = f(z,w), w \in \mathbb{C}$, be a germ of holomorphic function such that $f(0,w) = w^n + \ldots$. Then there exist a holomorphic function $h(z,w)$, $h \neq 0$, and holomorphic functions $a_1(z), \ldots, a_n(z)$ such that*

$$f = gh, \quad g(z,w) = w^n + a_1(z) w^{n-1} + \ldots + a_n(z).$$

The function g is called the Weierstrass polynomial.

Proof. If we denote by $b_i(z)$ the zeroes of the function f, then we have the representation $f = h \prod (w - b_i(z))$, $h \neq 0$. The coefficients $a_q(z)$ of the Weierstrass polynomial are symmetric polynomials of the zeroes b_i. Moreover, the ring of symmetric polynomials is generated by the sums of powers of b_i. The latter are given by the formulas

$$b_1^q + \ldots + b_n^q = \frac{1}{2\pi \mathrm{i}} \oint_{|w|=\mathrm{const}} w^q \cdot \frac{\partial f/\partial w}{f} \mathrm{d}w,$$

where the subintegral function is holomorphic in (z,w), if $|w|$ is sufficiently small. Therefore a_q and g are holomorphic functions.
The analyticity of the function h follows from the formula

$$h = \frac{1}{2\pi \mathrm{i}} \oint_{|u|=\mathrm{const}} \frac{h(z,u)\mathrm{d}u}{u-w} = \frac{1}{2\pi \mathrm{i}} \oint_{|u|=\mathrm{const}} \frac{(f/g)\mathrm{d}u}{u-w},$$

where the subintegral function is holomorphic for small $|w|$ and $|z|$. □

2.19. Division Theorem. *Let $f(z,w)$ be as in Theorem 2.18. Then for any germ $\phi(z,w)$ of holomorphic function there exist holomorphic germs $h(z,w)$ and $h_i(z)$, $i = 0, \ldots, n-1$, such that*

$$\phi = hf + \sum_0^{n-1} h_i(z) w^i.$$

Proof. Using the Weierstrass theorem we can assume that f is a Weierstrass polynomial. Define the function

$$h(z,w) = \frac{1}{2\pi \mathrm{i}} \oint_{|u|=\mathrm{const}} \frac{\phi(z,u)}{f(z,u)} \cdot \frac{\mathrm{d}u}{u-w},$$

which is analytic for small $|z|, |w|$. Now the function

$$\phi - fh = \frac{1}{2\pi \mathrm{i}} \oint_{|u|=\mathrm{const}} \frac{\phi(z,u)}{f(z,u)} \cdot \frac{f(z,u)-f(z,w)}{u-w} \mathrm{d}u$$

is a polynomial in w. □

Proof of Theorem 2.17. We follow the S. Łojasiewicz's book [**Loj2**]. Let I be an ideal in $\mathcal{O}_{n+k}$ and let $e_1, \ldots, e_r \in \mathcal{O}_{n+k}$ be such that:

(i) for any $\alpha \in \mathcal{O}_{n+k}$ we have $\alpha(x,0) = \sum a_i e_i(x,0) \pmod{I_{x,0}}$, $a_i \in \mathbb{C}$.

We have to show that:

(ii) for any $\alpha \in \mathcal{O}_{n+k}$ we have $\alpha(x,y) = \sum g_i(y) e_i(x,y) \pmod{I}$, $x \in \mathbb{C}^n$, $y \in \mathbb{C}^k$.

Condition (i) means that

$$\alpha \in I + \sum g_i(y) e_i + \mathfrak{m}_k \mathcal{O}_{n+k},$$

where $\mathfrak{m}_k$ denotes the maximal ideal in $\mathcal{O}_k$. Let $M = \mathcal{O}_{n+k}/I$. It is a module over $\mathcal{O}_{n+k}$ (finitely generated) as well as a module over $\mathcal{O}_k$; it is not yet proven that the latter is finite.
Let $N \subset M$ be the $\mathcal{O}_k$-submodule generated by e_i's. We have

$$M = N + \mathfrak{m}_k M$$

in the class of $\mathcal{O}_k$ moduli. We have to show that $M = N$.

2.20. Nakayama Lemma. *If $\widetilde{M}$ is a finitely generated module over $\mathcal{O}_k$ such that $\widetilde{M} = m\widetilde{M}$, then $\widetilde{M} = 0$.*

Proof. Let $a_1, \ldots, a_s$ be generators of $\widetilde{M}$. By assumption we have $a_i = \sum b_{ij} a_j$, $b_{ij} \in \mathfrak{m}$, or

$$(I - B)\boldsymbol{A} = 0, \quad \boldsymbol{A} = (a_1, \ldots, a_s)^\top, \quad B = (b_{ij}),$$

and the Cramer's formula gives $\det(I-B) \cdot a_i = 0$. But $\det(I-B) = 1 \pmod{\mathfrak{m}}$, which means that the matrix $I - B$ is invertible. Therefore $a_i = 0$. □

Finishing of the proof of Theorem 2.17. If we knew that M is finitely generated over $\mathcal{O}_k$, then we would apply Nakayama Lemma to the module M/N and obtain the desired equality $M = N$.
So, we strive to prove that M is finitely generated. We use induction with respect to the dimension n.

Let $n = 1$, $x \in \mathbb{C}$, $y \in \mathbb{C}^k$. By the assumption (i), M is finite over the ring $\mathcal{S} = \mathcal{O}_k + \mathfrak{m}_k\mathcal{O}_{1+k}$, with the generators $e_1, \ldots, e_r$.
Consider the operator of multiplication by x in the module M over $\mathcal{S}$. If a_i are the generators of M (over $\mathcal{S}$), then

$$xa_i = \sum \gamma_{ij} a_j, \quad \gamma_{ij} \in \mathcal{S}.$$

It means that $\det(x - \Gamma) \cdot a_j = 0$ for $\Gamma = (\gamma_{ij})$. Therefore, there is an element $\eta = x^r + \gamma_1 x^{r-1} + \ldots + \gamma_r$, $\gamma_i \in \mathcal{S}$, annihilating M.
We have $\eta(x, 0) \not\equiv 0$; thus $\eta(x, 0) = c_p x^p + \ldots$ with $c \neq 0$. It means that η satisfies the assumption of the Division Theorem.
We construct now the generators of M over $\mathcal{O}_k$. They are of the form $x^j m_i$, $j = 0, \ldots, p-1$, where $m_1, \ldots, m_s$ are generators of M over $\mathcal{O}_{1+k}$.
Let $u \in M$. We have $u = \sum f_i m_i$, $f_i \in \mathcal{O}_{1+k}$. From the Division Theorem, applied to f_i's, we get

$$f_i = g_i \eta + \sum_0^{p-1} \alpha_{ij} x^j, \quad \alpha_{ij} \in \mathcal{O}_k, \quad g_i \in \mathcal{O}_{1+k}.$$

Because $\eta m_i = 0$, we obtain

$$u = \sum_{i,j} \alpha_{ij} x^j m_i,$$

which ends the first induction step.
The general induction step: Let $n > 1$ and $M = N + \mathfrak{m}_k M$. Then $M = \mathcal{O}_{n-1+k} N + \mathfrak{m}_{n-1+k} M$. We apply the first induction step (with k replaced by $n - k + 1$) to show that the module M is finite over $\mathcal{O}_{n-1+k}$. The induction assumption gives that is finite over $\mathcal{O}_k$. □

Remark. There is one more preparation theorem (not used here).
Note that, if an ideal I is generated by components $f_1(x), \ldots, f_n(x)$ of a holomorphic map $\boldsymbol{F} : (\mathbb{C}^n, 0) \longmapsto (\mathbb{C}^n, 0)$ and $A_{\boldsymbol{F}} = \mathcal{O}/I$ has basis $e_1(x), \ldots, e_\mu(x)$, then any germ α has the representation $\alpha(x) = \sum c_i e_i(x) + \sum \alpha_j(x) f_j(x)$, $c_i \in \mathbb{C}$. Repeating the same for $\alpha_j(x)$, we get

$$\alpha(x) = \sum (c_i + \sum c_{ij} f_j) e_i(x) + \sum \alpha_{ij}(x) f_i(x) f_j(x),$$

etc. Finally, one gets the representation

$$\alpha(x) = \psi_1(f) e_i(x) + \ldots + \psi_\mu(f) e_\mu(x).$$

It is also called the *Weierstrass preparation theorem* (see [**AVG**] and [**Mal1**]).

§3 Proofs of Theorems 2.2 and 2.4

Here we prove Theorems 2.3 and 2.4 and we give another definition of index of vector field.

Recall that we consider a germ $\boldsymbol{F} = (f_1, \ldots, f_n) : (\mathbb{C}^n, 0) \longmapsto (\mathbb{C}^n, 0)$ of holomorphic map (or holomorphic vector field). By $A_{\boldsymbol{F}} = \mathcal{O}/I_{\boldsymbol{F}} = \mathcal{O}/(f_1, \ldots, f_n)$ we denote its local algebra and by multiplicity $\mu(\boldsymbol{F})$ we denote the dimension of $A_{\boldsymbol{F}}$. By $i = i_0\boldsymbol{F}$ we denote the index of the vector field $\boldsymbol{F}$ on $\mathbb{R}^{2n} (\simeq \mathbb{C}^n)$, which is defined as the topological degree of the map $x \longmapsto \frac{\boldsymbol{F}(x)}{|\boldsymbol{F}(x)|} \in \mathbb{S}^{2n-1}$, restricted to a small $(2n-1)$-dimensional sphere around $x = 0$.

Theorem 2.3 says that $\mu < \infty$ iff $\boldsymbol{F}^{-1}(0) = \{0\}$, i.e., iff $x = 0$ is isolated solution of the equation $\boldsymbol{F} = 0$. Theorem 2.4 says that in this case $\mu(\boldsymbol{F}) = i_0\boldsymbol{F}$.

Of course, from Theorem 2.3 Theorem 2.2 (about finite multiplicity of critical point of a function) follows.

2.21. *Proof of the implication*: $\mu < \infty \Rightarrow 0$ *is an isolated point in* $\boldsymbol{F}^{-1}(0)$.
We have shown that $\mathfrak{m}^\mu \subset I_F$ (see Lemma 2.7, where the role of f_i is played by $\partial f/\partial x_i$). Thus we have

$$x_i^\mu = \sum h_{ij} f_j, \quad i = 1, \ldots, n.$$

We see that the zeroes of $\boldsymbol{F}$ are contained in the set of common zeroes of the functions $x_1^\mu, \ldots, x_n^\mu$. □

2.22. *Proof of the implication*: $\boldsymbol{F}^{-1}(0) = \{0\} \Rightarrow \mu < \infty$.
We use one classical result, the Hilbert's theorem about zeroes (Theorem 2.23 below).

From it follows that there exists a positive integer N such that $x_i^N \in I_{\boldsymbol{F}}$, $i = 1, \ldots, n$. This means that any monomial of degree $\geq N$ lies in the ideal I_F and the algebra $\mathcal{O}/I_{\boldsymbol{F}}$ is finite dimensional. □

2.23. Hilbert's theorem about Zeroes. *Let* $V = \{f_1 = \ldots = f_k = 0\} \subset (\mathbb{C}^m, 0)$ *be a germ of analytic set, i.e.,* $f_i \in \mathcal{O}_0(\mathbb{C}^m)$ *are analytic functions. Consider the ideal* $I(V) = \{h : h|_V \equiv 0\} \subset \mathcal{O}_0(\mathbb{C}^m)$ *of functions vanishing on* V.
Then $I(V)$ *is equal to the radical of the ideal* $I_F = (f_1, \ldots, f_k)$*, defined as*

$$r(I_{\boldsymbol{F}}) = \{h : \exists r \ \ h^r \in I_{\boldsymbol{F}}\}.$$

(The same theorem holds when we replace the local ring $\mathcal{O}_0(\mathbb{C}^m)$ *by the ring* $\mathbb{C}[x_1, \ldots, x_m]$ *of polynomials and* V *by an algebraic set.)*

We do not give the proof of this theorem. In fact one can prove it using induction with respect to the dimensions $m - k$ (of V) and m and applying Weierstrass Preparation Theorem (see [**Loj2**]).

2.24. *Proof of Theorem* 2.4. We begin with the presentation of some properties.
(a) Properties of the index:

(i) *If* $\mu = 1$, *then* $i = 1$.

Indeed, if $\mu = 1$, then $I_{\boldsymbol{F}} = \mathfrak{m}$ which implies that $\det_{\mathbb{C}} D\boldsymbol{F}(0) \neq 0$. The index of $\boldsymbol{F}$ is the same as the index of the linear vector field $D\boldsymbol{F}(0)x$. Its index is equal to the sign of its (real) determinant. It is known, from complex analysis, that $\det_{\mathbb{R}} D\boldsymbol{F} = |\det_{\mathbb{C}} D\boldsymbol{F}|^2$.

The above has the following corollaries:

(i) $i > 0$,

(ii) *i is equal to the cardinality of the set $\boldsymbol{F}^{-1}(y)$ for some noncritical value $y \neq 0$.*

Here we use the fact that index is an additive function. The index $i_0\boldsymbol{F}$ (calculated along a fixed sphere) is equal to the sum of indices over singular points of the vector field $\boldsymbol{F}(x) - y$ (calculated along very small spheres around these points, see the point 1.11 in Chapter 1.)

(iv) *If we perturb $\boldsymbol{F}$ and obtain some new singular points, then the index of $\boldsymbol{F}$ is equal to the sum of indices of the perturbed vector fields.*

We notice also the following stability property of the index:

(v) *If the components f_i' of a vector field $\boldsymbol{F}'$ belong to $\mathfrak{m}^{\mu+1}$, then $i_0\boldsymbol{F} = i_0(\boldsymbol{F} + \boldsymbol{F}')$.*

Indeed, we have $f_i' = \sum h_{ij}f_j$, $h_{ij} \in \mathfrak{m}$. Therefore $\boldsymbol{F} + \boldsymbol{F}' = (I + A)\boldsymbol{F}$, where $\det(I + A) > 0$. The map from a small sphere to $\mathbb{S}^{2n-1}$, defined by means of $\boldsymbol{F} + \boldsymbol{F}'$, is a composition of a map defined by means of $\boldsymbol{F}$ and of an (orientation preserving) diffeomorphism defined by the linear vector field $(I + A)x$.

Properties of multiplicity.

(b) **Proposition (Subadditivity of Multiplicity).** *Let $\boldsymbol{F}_\epsilon$ be a deformation of $\boldsymbol{F}$. Then for small $|\epsilon|$*

$$\nu = \#\{\boldsymbol{F}_\epsilon = 0\} \leq \mu.$$

Proof. Using the main result of the previous section, we can assume that the deformation $\boldsymbol{F}_\epsilon(x) = (f_1(x,\epsilon), \ldots, f_n(x,\epsilon))$ is a polynomial deformation and the polynomial algebra $A_{\boldsymbol{F}_0}$ is generated by polynomials $e_1(x), \ldots, e_\mu(x)$.
Recall that the Thom–Martinet preparation theorem states that

$$P(x) = \sum g_j(\epsilon)e_j(x) + \sum \varphi_i(x,\epsilon)f_i(x,\epsilon).$$

The coefficients $g_j(\epsilon)$ and $\varphi_i(x,\epsilon)$ are analytic functions; their domains of analyticity depend on the function P. But, when P is a polynomial, the domains of analyticity of g_j and φ_i can be chosen not depending on P.
Indeed, it is enough to fix these domains in the cases $P(x) = 1, x_1, \ldots, x_n$ and $P(x) = x_ie_j(x)$. Then one can apply induction with respect to $\deg P$.

Therefore we can assume that $x \in U$, $\epsilon \in V$, where U and V are fixed neighborhoods of the origins. Moreover, we assume that all the zeroes of $\boldsymbol{F}_\epsilon$, which bifurcate from $x = 0$, lie in U for $\epsilon \in V$.
If $\mathrm{Hol}(U)$ is the ring of holomorphic functions on U, then, for fixed $\epsilon \in V$, we have the algebra $A_{\boldsymbol{F}_\epsilon}(U) = \mathrm{Hol}(U)/(\boldsymbol{F}_\epsilon)$. This algebra contains the image of the polynomial ring $\mathbb{C}[x]$, we denote it by

$$A_{\boldsymbol{F}_\epsilon}[U].$$

From the above it follows that $\dim_{\mathbb{C}} A_{\boldsymbol{F}_\epsilon}[U] \leq \mu$.
If $a_1, \ldots, a_\nu \in U$ are the zeroes of $\boldsymbol{F}_\epsilon$, then we associate with them the multi-local algebra

$$B = \bigoplus_{i=1}^{\nu} A_{a_i}, \quad A_{a_i} = \mathcal{O}_{a_i}/I_{\boldsymbol{F}_\epsilon},$$

and the natural map

$$\Pi : A_{\boldsymbol{F}_\epsilon}[U] \longmapsto B,$$

which sends a polynomial to its classes in the local algebras at the points a_i. The next lemma completes the proof of Proposition (b). □

Lemma. *The map* Π *is a surjection.*

Proof. It follows from the fact that, having given finite jets at the points a_i, one can always find a polynomial in $\mathbb{C}^n$ having just those jets at a_i. □

(c) From Proposition (b) and its proof we get the following corollaries:

(i) $\sum \mu_{a_i}(\boldsymbol{F}_\epsilon) \leq \mu(\boldsymbol{F})$ *and* $\nu \leq \mu$;

(ii) $i_0\boldsymbol{F} \leq \mu(\boldsymbol{F})$.

The latter property is obtained by application of Proposition (b) to the map $\boldsymbol{F}_\epsilon = \boldsymbol{F} - \epsilon$ (with noncritical ϵ) and the equality $i = \mu = 1$ for the new singular points.

(d) The **Pham's map.** It is defined as

$$\boldsymbol{\Phi} : (x_1, \ldots, x_n) \longmapsto (x_1^{m_1}, \ldots, x_n^{m_n}).$$

(e) **Lemma.** *We have* $i_0\boldsymbol{\Phi} = \mu(\boldsymbol{\Phi}) = m_1 \cdot m_2 \ldots m_n$.

Proof. $i_0\boldsymbol{\Phi} = \#\{\boldsymbol{\Phi} = \epsilon\} = \#\{x_1^{m_1} = \epsilon_1, \ldots, x_n^{m_n} = \epsilon_n\} = \#\{x_1^{m_1} = \epsilon_1\} \cdot \ldots \cdot \#\{x_n^{m_n} = \epsilon_n\}$. On the other hand, the local algebra is generated by the monomials $x_1^{k_1} \ldots x_n^{k_n}$, $0 \leq k_i < m_i$. □

(f) *Continuation of the proof of Theorem* 2.4. Take a map of the form $\boldsymbol{\Phi}_\epsilon = \boldsymbol{\Phi} + \epsilon\boldsymbol{F}$, where $\boldsymbol{\Phi}$ is the Pham's map with high exponents. We have:

$i_0\boldsymbol{\Phi}_\epsilon = i_0\boldsymbol{F}$ for $\epsilon \neq 0$, and

$\mu_0(\boldsymbol{\Phi}_\epsilon) = \mu_0(\boldsymbol{F})$ for $\epsilon \neq 0$.

We shall show that

$$i_0\boldsymbol{\Phi}_\epsilon = \mu_0(\boldsymbol{\Phi}_\epsilon). \tag{2.7}$$

Let $a_1, \ldots, a_\nu$ be the (small) zeroes of $\boldsymbol{\Phi}_\epsilon$. We have:

$$\begin{aligned}
\mu_0(\boldsymbol{\Phi}) &= \mu_0(\boldsymbol{\Phi}_0) \geq \sum \mu_{a_j}(\boldsymbol{\Phi}_\epsilon) \text{ (for } \epsilon \neq 0\text{, by subadditivity)}\\
&\geq \sum i_{a_j}\boldsymbol{\Phi}_\epsilon \text{ (because } i \leq \mu)\\
&= i_0\boldsymbol{\Phi}_0 = i_0\boldsymbol{\Phi} \text{ (by additivity)}\\
&= \mu_0(\boldsymbol{\Phi}) \text{ (it is the Pham's map).}
\end{aligned}$$

Therefore, in the above chain, all inequalities become equalities. In particular, we have $\mu_{a_j}(\boldsymbol{\Phi}_\epsilon) = i_{a_j}\boldsymbol{\Phi}_\epsilon$. Because the point $x = 0$ is among a_j's, the equality (2.7) follows. □

2.25. Index formula in two dimensions and Puiseux expansion.
(a) Let $f : (\mathbb{C}^2, 0) \longmapsto (\mathbb{C}, 0)$ be a germ of irreducible holomorphic function and let $\Gamma = f^{-1}(0)$.

Remark. Here the **irreducibility** of f is understood in any of the following two equivalent ways. Firstly, f cannot be written as a product $f_1 f_2$ of two germs vanishing at 0. Another definition says that the set $\Gamma \backslash \text{sing}(\Gamma)$ is non-empty connected, where $\text{sing}(\Gamma)$ is the set of singular points of Γ, i.e., points where $\mathrm{d}f = 0$. The both definitions work for functions on $(\mathbb{C}^n, 0)$. We say also that the hypersurface Γ is irreducible.
Generally, any germ g (of function) has an unique (up to permutation of factors) factorization $g = g_1^{k_1} \ldots g_r^{k_r}$ with irreducible g_j.
Below we present some arguments, based on the monodromy theory, justifying the above statements.

Let us pass to the curve Γ. Using Weierstrass Preparation Theorem 2.18 (in a suitable linear system of coordinates), we can assume that Γ is a zero set of the Weierstrass polynomial

$$f(x, y) = y^n + a_1(x)y^{n-1} + \ldots + a_n(x), \quad a_j(0) = 0,$$

where a_j are analytic germs, $a_j(0) = 0$. The equation $f = 0$ defines a multivalued function $y = \phi(x)$. ϕ has branching at $x = 0$ and its Riemann surface is isomorphic to the curve Γ (see Chapter 1 and Chapter 12 below).

(b) **Puiseux theorem.** *If the germ* $(\Gamma, 0)$ *is irreducible, then it is topologically equivalent to a disc* $(\mathbb{C}, 0)$.
More precisely, the map $(\mathbb{C}, 0) \longmapsto (\mathbb{C}^2, 0)$

$$t \longmapsto (x, y) = (t^n, \phi(t^n))$$

is a homeomorphism onto $(\Gamma, 0)$ *(analytic outside* 0*).*
Moreover, the function ϕ *takes the form of the Puiseux expansion*

$$\phi = \psi(x^{1/n})$$

with an analytic germ ψ*.*
The above parametrization of $(\Gamma, 0)$ *by the disc is called the* **primitive parametrization.**

Proof. We use monodromy properties of ϕ. Fix a disc $D = \{|x| < \epsilon\}$ as the domain of definition of ϕ and a basic point $a \in D \setminus 0$. Let $\phi_{a,1}(x), \ldots, \phi_{a,n}(x)$ be the germs (sheets) of ϕ near a (roots of $f = 0$ with respect to y).
The variation of the sheets $\phi_{a,j}$ along the loop in $(D \setminus 0, a)$ surrounding the origin results in a permutation of these sheets. Thus the monodromy transformation σ associated with ϕ is an element of the group $S(n)$ of permutations of the n-element set $\{\phi_{a,1}(a), \ldots, \phi_{a,n}(a)\}$.
By the irreducibility, $\Gamma \setminus 0$ is connected. So, any two germs $\phi_{a,i}$ and $\phi_{a,j}$ can be connected by a path in $\Gamma \setminus 0$. The projection of this path onto the x-plane is a loop in $(D \setminus 0, a)$. This implies that σ acts transitively on the set of sheets of ϕ; σ is a cyclic permutation. In particular, $\sigma^n = id$.
Now one can see that Γ is indeed a topological disc. Consider the function $\psi(t) = \phi(t^n)$. Its monodromy around $t = 0$ is the same as $\sigma^n = id$. Thus ψ is analytic outside 0, single-valued and bounded. By the Riemann's theorem about removal of singularities, ψ prolongs itself to an analytic function in D. □

(c) **Remark.** Using the same arguments as in the proof of Puiseux theorem, i.e., Weierstrass Preparation Theorem and monodromy, one can prove the *factorization theorem* $g = g_1^{k_1} \ldots g_r^{k_r}$, g_j – irreducible, and the equivalence of the two definitions of irreducibility.
The factorization theorem can be used in the proof of the Hilbert Theorem about Zeroes (Theorem 2.23). The reader can reconstruct this proof itself.
Of course, these arguments work well only in the 2-dimensional case. The proof of the factorization theorem in general n-dimensional case is more complicated; it relies essentially on Weierstrass Preparation Theorem (see [**Loj2**]).

(d) We pass to the index formula. Let $\boldsymbol{F} = (f_1, f_2) : (\mathbb{C}^2, 0) \longmapsto (\mathbb{C}^2, 0)$ be a germ of holomorphic mapping of finite multiplicity.
Assume firstly, that f_1 is irreducible and let $\alpha : (\mathbb{C}, 0) \longmapsto f_1^{-1}(0)$ be some primitive parametrization of the curve $f_1 = 0$. Take the composition $f_2 \circ \alpha = at^p + \ldots$, $a \neq 0$; (it is not $\equiv 0$ by the finite multiplicity). We define

$$\iota_0 \boldsymbol{F} = \iota_0(f_1, f_2) = p.$$

If $f_1 = g_1^{k_1} \ldots g_r^{k_r}$ with irreducible g_j, then we put

$$\iota_0 \boldsymbol{F} = \sum k_j \iota_0(g_j, f_2).$$

(e) **Theorem.** *We have* $\iota_0 \boldsymbol{F} = i_0 \boldsymbol{F}$*.*

Proof. We use the criterion $i_0\boldsymbol{F} = \#\{\boldsymbol{F}_\delta = 0\}$ for a perturbation $\boldsymbol{F}_\delta$ such that each zero of $\boldsymbol{F}_\delta$ is not degenerate (see the proof of Theorem 2.4 above).
In the case of irreducible f_1 we put $\boldsymbol{F}_\delta = (f_1, f_2 - \delta)$, $\delta > 0$. Thus $\#\{\boldsymbol{F}_\delta = 0\} = \#\{at^p(1+\ldots) = \delta\} = p$. The same perturbation is good in the case $f_1 = g_1 \cdots g_r$, g_j irreducible.
In the cases when f_1 contains a factor g^k (g irreducible), we replace this factor by $(g+\delta h_1)\cdot\ldots\cdot(g+\delta h_k)$, where $h_j \in \mathfrak{m}^N$ (N large) are such that the curves $g + \delta h_j = 0$ are disjoint outside $t = 0$. □

§4 Classification of Singularities

In this section we present only the beginning of the list of normal forms for singularities of functions. We begin with some technical tools. The first result is an immediate consequence of the proof of Morse Lemma.

2.26. Morse Lemma with Parameters. *Let $f(x;\lambda) = f(0;\lambda) + \sum a_{ij}(\lambda)x_i x_j + \ldots$ be such function that the matrix $a_{ij}(0)$ is not degenerate. Then there is a holomorphic change h_λ such that*

$$f(h_\lambda(y);\lambda) = f(0;\lambda) + \sum y_i^2.$$

Consider a functions of the form

$$f(x_1,\ldots,x_k,x_{k+1},\ldots,x_n) = x_1^2 + \ldots + x_k^2 + (\text{terms of degree } \geq 3).$$

By applying a preliminary change of the form $x_1 \longmapsto x_1 + \varphi_1(x_{k+1},\ldots,x_n),\ldots,$ $x_k \longmapsto x_k + \varphi_k(x_{k+1},\ldots,x_n)$, we can assume that we are in a situation from Lemma 2.26 (the critical point with respect to $(x_1,\ldots,x_k)$ is shifted to the origin and $\lambda = (x_{k+1},\ldots,x_n)$). Therefore, there is a change h such that

$$f \circ h = y_1^2 + \ldots y_k^2 + g(y_{k+1},\ldots,y_n),$$

where $g(0) = 0$, $Dg(0) = 0$, $D^2g(0) = 0$. In this way we have separated the quadratic part, depending on one set of variables, from the essential non-quadratic part, depending on the other set of variables.

2.27. Definitions. By the **corank** of the singularity f we mean the corank of the quadratic form $D^2f(0)$, i.e., $n-k$ for the above function.
We say that two germs $f(x_1,\ldots,x_k)$ and $g(y_1,\ldots,y_l)$ are **stably equivalent** iff there is an integer n such that the germs $f(x_1,\ldots,x_k) + x_{k+1}^2 + \ldots + x_n^2$ and $g(y_1,\ldots,y_l) + y_{l+1}^2 + \ldots + y_n^2$ are equivalent (by a change of coordinates).

We study only singularities with corank one and two.

Corank 1. Here we can assume that the functions depend on one variable x.

2.28. Theorem. *If a germ f has singularity of finite multiplicity and of corank 1, then it is stably equivalent to the function x^n.*

Proof. Let

$$f = a_n x^n + a_{n+1} x^{n+1} + \dots,$$

$a_n \neq 0$. (Notice that, if all $a_i = 0$, then $\mu(f) = \infty$). We put

$$y = h(x) = x(a_n + a_{n+1}x + \dots)^{1/n}$$

and we get $f = y^n$. □

2.29. Definition. We say that a singularity has type $\mathbf{A}_k$ iff it is stably equivalent to x^{k+1}.

Corank 2. Here the functions depend on two variables x, y. Consider such a function. Its Taylor series begins from cubic terms.
Firstly, we classify the homogeneous cubic forms of two variables, $ax^3 + bx^2y + cxy^2 + dy^3$; or the holomorphic sections of certain line bundle $\mathcal{O}(3)$ above the projective space $\mathbb{CP}^1$ (see Definitions 8.42 and 10.10). They are associated with cubic the polynomials $P = a + b\lambda + c\lambda^2 + d\lambda^3$.
There are four possibilities:

(a) P has three different zeroes, which can be moved (via an automorphism of $\mathbb{CP}^1$) to $0, \pm i$, and the cubic form is equal to $x^2y + y^3$;

(b) P has one simple zero (at $\lambda = 0$) and one double zero (at $\lambda = \infty$): x^2y;

(c) P has a triple zero: x^3;

(d) $P \equiv 0$.

We study in detail only the first two cases.

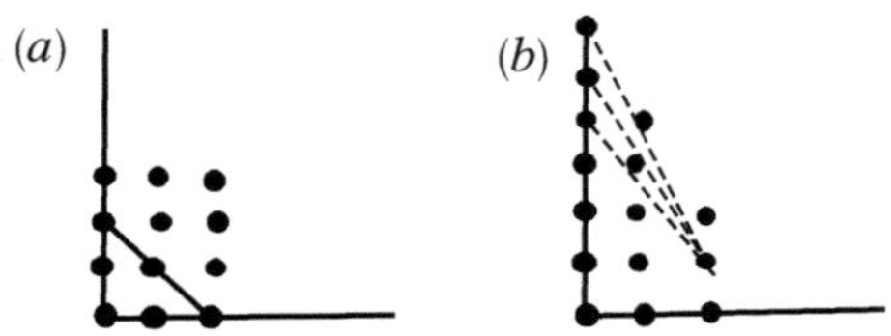

Figure 2.3: Newton diagrams

2.30. Theorem. *Any function $f = x^2y + y^3 + \dots$ is equivalent to $x^2y + y^3$.*

Proof. By applying the substitutions

$$(x, y) \longmapsto (x_1, y_1) = (x + \phi, y + \psi)$$

into the part $x^2y + y^3$ of f, we strive to cancel all higher order terms in f. We have

$$\begin{aligned} x_1^2 y_1 + y_1^3 &= (x + \phi)^2 (y + \psi) + (y + \psi)^3 \\ &= x^2y + y^3 + [2xy\phi + (x^2 + 3y^2)\psi] + \text{h.o.t.} \end{aligned}$$

(with respect to ϕ, ψ). We claim that the term χ in the square bracket contains all terms of degree ≥ 4.
We present this situation in the Newton's diagram (see Fig. 2.3(a)). There we have distinguished the monomials appearing in the gradient ideal $I_{x^2y+y^3}$, i.e., xy and x^2, y^2 (jointed one with another). It is enough to cancel only the terms of degree four and five; because $\mu(f) = 4$ and we can use the Tougeron Theorem.
For the terms of degree four in χ we have $\deg(\phi, \psi) = 2$. With the term x^4 in χ, we associate the term x^2 in ψ and with the term y^4 in χ, we associate the term $y^2/3$ in ψ. The remaining part, from χ as well as from $(x^2 + 3y^2)\psi$, is divisible by xy and can be put into $2xy\phi$.
The terms of degree five are treated in the same way. □

Now we study the next case. Assume that $f = x^2y + \ldots$. The further terms of the expansion of f are ordered by means of the *method of rotating line.*

2.31. Definition. The **Newton support** of function $f = \sum a_{ij}x^iy^j$ is defined as

$$\mathrm{supp}(f) = \{(i,j) : a_{ij} \neq 0\}.$$

The **Newton diagram** of f is the convex hull of the set $\mathrm{supp}(f) + (\mathbb{Z}_+)^2$.

Take a line in $\mathbb{R}^2$ with one its point fixed at $(2,1)$. At the beginning the line passes through the points $(2,1)$ and $(0,3)$. We start to rotate it in the clockwise direction (see Fig. 2.3(b)). We look at the points from the support of f, which are passed through the left part of the line.
There appear in order the terms: y^4; xy^3, y^5; y^6; xy^4, y^7; .... We have two possibilities of the first appearance of these terms in $\mathrm{supp}(f)$:

(i) There appears $(0,k)$ and we have

$$f = x^2y + y^k + \ldots$$

(ii) There appear the points $(1, k+1)$, $(0, 2k+1)$, and $f = x^2y + Axy^{k+1} + By^{2k+1} + \ldots$. Here the first part is of the form y times a quadratic form of (x,z), $z = y^k$; one associates with it the cubic form $z(x^2 + Axz + Bz^2)$ with zeroes at $z = 0$, $x = \lambda_{1,2}z$. Assume that they are different zeroes, i.e., $B(A^2 - 4B) \neq 0$. It is clear that the change $x \longmapsto \lambda x + \mu z$, $z \longmapsto \nu z$, reduces f to

$$x^2y + y^{2k+1} + \ldots$$

(Now the dots mean terms whose support is above the rotating line). Notice that, if $\mu(f) < \infty$, then the rotating line must sooner or later encounter the support of f.

2.32. Theorem. *If $f = x^2y + y^k + \ldots$, then it is equivalent to $x^2y + y^k$.*

Proof. It relies on the fact that the principal part x^2y+y^k is a quasi-homogeneous function.

2.33. Definition. A function $f(x_1,\ldots,x_n)$ is called **quasi-homogeneous** of **degree** d with **exponents** $\alpha_1,\ldots,\alpha_n$ (or weighted homogeneous) iff

$$f(\lambda^{\alpha_1}x_1,\ldots,\lambda^{\alpha_n}x_n)=\lambda^d f(x_1,\ldots,x_n)$$

for any $\lambda\in\mathbb{C}$.

If $f=\sum a_k x^k$, $k=(k_1,\ldots,k_n)$, $x^k=x_1^{k_1}\ldots x_n^{k_n}$, then $\operatorname{supp}(f)\subset\Gamma=\{k:\alpha_1k_1+\ldots+\alpha_nk_n=d\}$. The set Γ is called the **diagonal**. Usually one takes $\alpha_i\in\mathbb{Q}$ and $d=1$.

Example. The function x^2y+y^k is quasi-homogeneous of degree $1+1/(k-1)$ and the exponents $1/2, 1/(k-1)$.

One can define the (quasi-homogeneous) filtration of the ring $\mathcal{O}$. It consists of the decreasing family of ideals $\mathcal{A}_d\subset\mathcal{O}$, $\mathcal{A}_{d'}\subset\mathcal{A}_d$ for $d<d'$. Here $\mathcal{A}_d=\{g:$ degrees of monomials from $\operatorname{supp}(g)$ are $\geq d\}$; the degree is quasi-homogeneous.

When $\alpha_1=\ldots=\alpha_n=1$, this filtration coincides with the usual filtration given by the usual degree.

2.34. Definition. A function f is called **semi-quasi-homogeneous** iff $f=f_0+f_1$, where f_0 is quasi-homogeneous of degree d of finite multiplicity and $f_1\in\mathcal{A}_{d'}$, $d'>d$.

2.35. Theorem. *Let f be a semi-quasi-homogeneous function, $f=f_0+f_1$ with quasi-homogeneous f_0 of finite multiplicity. Then f is equivalent to the function $f_0+\sum c_k e_k(x)$, where $e_1,\ldots,e_s$ are the elements of monomial basis of the local algebra A_{f_0} such that $\deg e_i>d$ and $c_k\in\mathbb{C}$.*

Example. If $f=f_0+f_1$ and $f_0=x^2y+y^k$, then f is equivalent to f_0.

Indeed, the basis of the local algebra $\mathcal{O}/(xy, x^2+ky^{k-1})$ is $1,x,y,y^2,\ldots,y^{k-1}$, and lies below the diagonal Γ. Here $\mu(f_0)=k+1$.

Having Theorem 2.35 we obtain the thesis of Theorem 2.32. □

2.36. Definition. We say that a singularity is of the type $\mathbf{D}_k$ iff it is stably equivalent to x^2y+y^{k-1}.

Proof of Theorem 2.35. As in the proof of Theorem 2.28 we cancel the high order terms using the quasi-homogeneous part f_0 and the changes

$$x_1=x+\phi,\quad y_1=y+\psi,$$

where ϕ,ψ are quasi-homogeneous. However, the degrees of ϕ and ψ are calculated differently.

With such a change one associates the vector field $\phi\frac{\partial}{\partial x}+\psi\frac{\partial}{\partial y}$ (then the change is approximately induced by the flow map generated by this vector field). The

(quasi-homogeneous) degree of $\phi\frac{\partial}{\partial x}$ is equal to $\deg\phi - \deg x$ (because x is in the 'denominator'). Also $\deg(\psi\frac{\partial}{\partial y}) = \deg\psi - \deg y$.
We obtain

$$f_0 + \left[\frac{\partial f_0}{\partial x}\phi + \frac{\partial f_0}{\partial y}\psi\right] + \text{h.o.t.}$$

We use the terms in the square brackets to delete some terms from the power expansion of f. Because the part in the square brackets belongs to the gradient ideal I_{f_0}, we can cancel only the terms from this ideal. □

Recall that a singularity is m-modal iff its neighborhood (in the functional space of germs of functions) is covered by l-parameter families of orbits of the action of the group of equivalences (changes of variables) and $\max l = m$ (see Definition 2.10). The singularity is simple iff $m = 0$. From Theorem 2.33 one gets the formula for the modality.

2.37. Theorem. *The modality of a semi-quasi-homogeneous singularity is equal to the number of monomials e_k, from its mini-versal deformation, which lie on the diagonal and above the diagonal.*

2.38. Theorem. *If f is a simple singularity, then it is stably equivalent to one of the following singularities:*

$$\begin{aligned} \mathbf{A}_k &: x^{k+1}, \\ \mathbf{D}_k &: x^2y + y^{k-1}, \\ \mathbf{E}_6 &: x^3 + y^4, \\ \mathbf{E}_7 &: x^3 + xy^3, \\ \mathbf{E}_8 &: x^3 + y^5. \end{aligned}$$

We do not present the proof of this result. In fact it is not difficult, because we have all the tools: the choice of monomials using the rotating line, Theorem 2.35 and Theorem 2.37.
We shall present some unimodal singularities. (Outside the below list there remain 14 exceptional unimodal singularities, see [**AVG**]).

2.39. Theorem. *The following series of singularities are unimodal:*

$$\begin{array}{ll} \mathbf{P}_8: & x^3 + y^3 + z^3 + axyz, \\ \mathbf{X}_9: & x^4 + y^4 + ax^2y^2, \\ \mathbf{J}_{10}: & x^3 + y^6 + ax^2y^2, \\ \mathbf{T}_{p,q,r}: & x^p + y^q + z^r + axyz, \quad \frac{1}{p} + \frac{1}{q} + \frac{1}{r} < 1. \end{array}$$

The singularities $\mathbf{P}_8, \mathbf{X}_9, \mathbf{J}_{10}$ are called **parabolic** *and $\mathbf{T}_{p,q,r}$ is called* **hyperbolic.**

Chapter 3

Algebraic Topology of Manifolds

This chapter is auxiliary. Here we collect the basic facts from the algebraic topology of manifolds and fibre bundles which will be used in the further chapters. Special attention is focused on the notion of intersection index.
We present also the beginnings of cohomology theory with coefficients in sheaves, e.g., the proof of the de Rham theorem. The other notions from this theory (like the hypercohomology and the spectral sequence) will be introduced in Chapter 7.

§1 Homology and Cohomology Groups

3.1 Definition of homology groups. Let X be a manifold or a CW-complex. It means that X is obtained from cells (diffeomorphic to euclidean balls) glued together in such a way that the lower-dimensional cells lie in the boundaries of higher-dimensional cells. A k-dimensional cell in X is an image of the ball $\mathbb{D}^k = \{x \in \mathbb{R}^k : |x| \leq 1\}$ under a continuous map $\sigma : \mathbb{D}^k \longmapsto X$, we denote this cell also by $\sigma = \sigma^k$. Here the ball $\mathbb{D}^k$ is oriented (by a choice of some orthogonal frame $v_1, \ldots, v_k$ at one of its points) and the cell σ also bears the orientation.
The group of k-dimensional **chains** $C_k(X)$ is a free abelian group generated by all possible k-dimensional *singular cells*, $C_k(X) = \{c = \sum a_\tau \tau : a_\tau \in \mathbb{Z}\}$, where $\tau : \mathbb{D}^k \longmapsto X$ are continuous maps. It is infinitely-generated free abelian group.
A **boundary** $\partial\tau$ of the singular cell τ is the image of the map τ restricted to $\partial\mathbb{D}^k = \mathbb{S}^{k-1}$. The orientation of the boundary $\partial\mathbb{D}^k$ is given by a frame $w_1, \ldots, w_{k-1}$ such that, if n is a vector normal to $\partial\mathbb{D}^k$ and directed outside of $\mathbb{D}^k$, then the frame $n, w_1, \ldots, w_{k-1}$ defines the orientation of $\mathbb{D}^k$. By definition,

$$\partial\left(\sum a_\tau \tau\right) = \sum a_\tau \partial\tau$$

is the **boundary** of the chain c. The operator $\partial = \partial^k$ acts from $C_k(X)$ to $C_{k-1}(X)$.
A chain c is a **cycle** iff $\partial c = 0$.
The group

$$H_k(X) = H_k(X, \mathbb{Z}) = \text{group of cycles/group of boundaries}$$

is called the **group of k-dimensional homologies** of X (or the kth singular homology group or simply the kth homology group). This definition is correct, because we have $\partial \circ \partial = 0$.
The above definition can be modified when, in the definition of the chain, we assume that the coefficients a_τ take values from some other abelian group (or

H. Żołądek, *The Monodromy Group*, Monografie Matematyczne 67,
https://doi.org/10.1007/978-3-031-91270-2_3

ring) R. The corresponding quotient groups are denoted by $H_k(X, R)$. Usually the rings $\mathbb{Z}_p$, $\mathbb{Q}$, $\mathbb{R}$, $\mathbb{C}$ are chosen.
The homologies appear in many situations. So, we describe the general (algebraic) their definition. The series of groups and maps

$$\ldots \longmapsto C_k \overset{\partial_k}{\longmapsto} C_{k-1} \overset{\partial_{k-1}}{\longmapsto} \ldots$$

satisfying the property $\partial_{k-1}\partial_k = 0$, is called the **chain complex** and is denoted by $C_\bullet$. Its homology groups are defined as $\ker \partial_k/\mathrm{Im}\partial_{k+1}$. Such series is called **exact** (or *acyclic*) iff its homology groups are zero.

3.2. Definition (cohomology groups). The group of k-dimensional **cochains** $C^k(X)$ is the abelian group of linear functionals on the group of k-dimensional chains, which take integer values at the singular cells. The **coboundary** of a chain c^* is defined by the formula $\delta c^*(c) = c^*(\partial c)$. A cochain of the form δc^* is called the **coboundary**. A cochain is a **cocycle** iff its coboundary is zero.
The abelian group

$$H^k(X) = H^k(X, \mathbb{Z}) = \text{cocycles/coboundaries}$$

is the **group of k-dimensional cohomologies** of X (or the kth cohomology group). Analogously to the groups $H_k(X, R)$ we define the cohomology groups with coefficients in the group (ring) R, $H^k(X, R)$.
The series

$$\ldots \longmapsto C^k \overset{\delta^k}{\longmapsto} C^{k+1} \longmapsto \ldots$$

satisfying $\delta^{k+1} \circ \delta^k = 0$ is called the **cochain complex.** It is denoted by $C^\bullet$ and its cohomology groups are equal to $\ker \delta^k/\mathrm{Im}\delta^{k-1}$.

Remarks. The formulas $H_k(X, R) = H_k(X, \mathbb{Z}) \otimes R$, $H^k(X, R) = H^k(X, \mathbb{Z}) \otimes R$, $H^k(X, R) = H_k(X, R)^*$ $(= \mathrm{Hom}(H_k(X, R), R))$ do not always hold.
For example, we have $H^k(X, \mathbb{Z}) = \mathrm{Hom}(H_k(X, \mathbb{Z}), \mathbb{Z}) \oplus \mathrm{Ext}(H_{k-1}(X, \mathbb{Z}), \mathbb{Z})$, where $\mathrm{Ext}(H_{k-1}(X, \mathbb{Z}), \mathbb{Z})$ lies in the torsion part of the cohomology group. Recall that an element x of an abelian group is *torsion* iff $mx = 0$ for some $m \in \mathbb{Z} \setminus 0$.
Although the groups of chains and cochains are infinite dimensional, usually the homology and cohomology groups are finitely generated. This holds for the compact manifolds and for the CW-complexes.
The homology groups were introduced by Poincaré at the beginning of the last century. Starting from the middle of the last century a rapid development of the algebraic topology and of the homological algebra began. Now the homology groups are used in every branch of mathematics. Usually, their definitions and calculations are very algebraic. We shall focus on their geometrical meaning.

Below we present a practical method of calculation of the groups $H_k(X)$ and $H^k(X)$. Let X be a CW-complex of dimension n. We associate with it the following series $X_0 \subset X_1 \subset \ldots \subset X_n = X$ of its subcomplexes.

X_0 is a finite set of points, 0-dimensional cells of the complex. X_1 is a 1-dimensional complex obtained from X_0 by adding 1-dimensional cells (arc from X) joined to X_0 at their ends. X_2 consists of 2-dimensional cells with their boundaries in X_1, etc. All cells are oriented.

We have

$$X_k \setminus X_{k-1} = \overset{\circ}{\sigma}{}_1^k \cup \ldots \cup \overset{\circ}{\sigma}{}_s^k,$$

where $\overset{\circ}{\sigma}$ denotes the open cell. The group of k-dimensional chains is replaced by the group

$$\mathbb{Z}\sigma_1^k + \ldots + \mathbb{Z}\sigma_s^k.$$

The boundary of the cell σ_i^k is calculated as follows. If a part of $\partial\mathbb{D}^k$ is sent by means of σ_i^k to σ_j^{k-1}, then the latter cell gives the contribution into the (formal) boundary $\partial\sigma_i^k$ with the coefficient ± 1, depending on the preserving or the reversing of orientation by this map. We have $\partial\sigma_i^k = \sum a_j \sigma_j^{k-1}$ with integer a_j's.

We get a chain complex of finitely generated groups with the boundary maps, whose compositions are zero, and we define the homology groups of this chain complex. By passing to the dual complex, we analogously define its cohomology groups.

3.3. Theorem. ([**Spa**]). *The just defined groups coincide with the groups* $H_k(X,\mathbb{Z})$ *and* $H^k(X,\mathbb{Z})$.

Examples. 1. The circle $\mathbb{S}^1$ consists of the point σ^0 and the arc σ^1 attached to it at the ends (see Fig. 3.1(a)). We have $\partial\sigma^0 = 0$, $\partial\sigma^1 = 0$, and the chain complex takes the form

$$0 \longmapsto \mathbb{Z} \overset{0}{\longmapsto} \mathbb{Z} \longmapsto 0.$$

Thus we have $H_0(\mathbb{S}^1,\mathbb{Z}) = H_1(\mathbb{S}^1,\mathbb{Z}) = \mathbb{Z}$.

2. For the torus $\mathbb{T}^2$ we consider two partitions into cells, presented at Fig. 3.1(b) and 3.1(c). In the case of the first partition we get the complex

$$0 \longmapsto \mathbb{Z}^2 \overset{\partial_2}{\longmapsto} \mathbb{Z}^4 \overset{\partial_1}{\longmapsto} \mathbb{Z}^2 \longmapsto 0,$$

where

$$\begin{array}{ll} \sigma_1^2 \longmapsto \sigma_1^1 + \sigma_2^1, & \sigma_2^2 \longmapsto -\sigma_1^1 - \sigma_2^1, \\ \sigma_1^1 \longmapsto 0, & \sigma_2^1 \longmapsto 0, \\ \sigma_3^1 \longmapsto \sigma_2^0 - \sigma_1^0, & \sigma_4^1 \longmapsto \sigma_2^0 - \sigma_1^0. \end{array}$$

Therefore, $\ker\partial_2 = H_2(\mathbb{T}^2) = \mathbb{Z}$. Next, $\mathrm{Im}\partial_2 = \mathbb{Z}(\sigma_1^1 + \sigma_2^1)$, $\ker\partial_1 = \mathbb{Z}\sigma_1^1 + \mathbb{Z}\sigma_2^1 + \mathbb{Z}(\sigma_3^1 - \sigma_4^1)$ and hence $H_1(\mathbb{T}^2) = \mathbb{Z}\oplus\mathbb{Z}$. Finally, because $\mathrm{Im}\partial_1 = \mathbb{Z}(\sigma_2^0 - \sigma_1^0)$, we have $H_0(\mathbb{T}^2) = \mathbb{Z}$.

For the partition from Fig. 3.1(c), where the edges are identified, as usually in the construction of the torus, the chain complex is much simpler

$$0 \longmapsto \mathbb{Z} \overset{\partial_2}{\longmapsto} \mathbb{Z}^2 \overset{\partial_1}{\longmapsto} \mathbb{Z} \longmapsto 0,$$

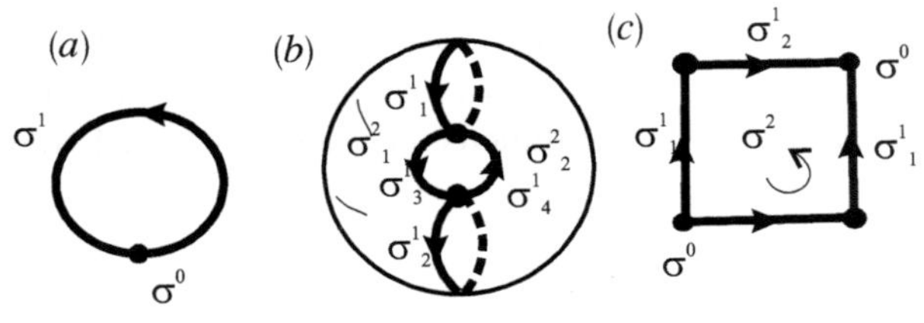

Figure 3.1: Cell complexes for the circle and for the torus

where the boundary maps are zero. Thus the four distinguished cycles σ_i^j generate all the homology groups.

3. The 2-dimensional sphere $\mathbb{S}^2 = \mathbb{CP}^1$ has the partition $\sigma^0 \cup \sigma^2$ inducing the complex

$$0 \longmapsto \mathbb{Z} \longmapsto 0 \longmapsto \mathbb{Z} \longmapsto 0,$$

giving $H_0(\mathbb{S}^2) = H_2(\mathbb{S}^2) = \mathbb{Z},\ H_1(\mathbb{S}^2) = 0$.
Generally, the complex projective space $\mathbb{CP}^n$ has the partition $\sigma^0 \cup \sigma^2 \cup \sigma^4 \cup \ldots \cup \sigma^{2n}$ induced by the chain of immersions $\mathbb{CP}^0 \subset \mathbb{CP}^1 \subset \ldots \subset \mathbb{CP}^n$. All these cells are the cycles generating the even-dimensional homology groups.

Problem. Calculate the groups $H_i(\mathbb{RP}^n, R)$ for $R = \mathbb{Z}$, $\mathbb{Z}_2$, $\mathbb{Q}$.

The homology groups have many important properties.

3.4. If $f : X \longmapsto Y$ is a continuous map, then it induces the homomorphism of homology groups $f_* : H_\bullet(X, R) \longmapsto H_\bullet(Y, R)$, $f_*\sigma = \sigma \circ f$. It induces also the contravariant homomorphism of the cohomology groups $f^* : H^\bullet(Y, R) \longmapsto H^\bullet(X, R)$.

3.5. If two maps f and g are homotopically equivalent, then they induce the same homomorphisms between the homology groups and the cohomology groups. This means that the homology groups of a topological space depend only on its homotopy type. Indeed, if $F : X \times [0, 1] \longmapsto Y$ is the homotopy between f and g and σ is a cycle, then $f_*\sigma - g_*\sigma = \partial(\sigma \circ F)$.

3.6. The *Euler characteristic* of X (see Remark 1.15) is equal to

$$\chi(X) = b_0 - b_1 + b_2 - \ldots,$$

where $b_i = b_i(X) = \dim H_i(X, \mathbb{R})$ are the *Betti numbers*. Indeed, by definition $\chi(X) = \dim C_0 - \dim C_1 + \ldots$. We can write $C_i = \mathrm{Iim}\partial_{i+1} \oplus H_i \oplus \mathrm{coker}\partial_i$, where $\mathrm{coker}\partial_i \simeq \mathrm{Iim}\partial_i$. Thus, after calculating the alternative sum of dimensions of the spaces C_i, the contributions arising from images and cokernels cancel themselves mutually.

3.7. If X is a manifold of dimension n, then $H_n(X, \mathbb{Z})$ is equal either to $\mathbb{Z}$ or to 0. This follows from the partition of X into the simplices (the simplicial partition).

Let σ_i^n, $i = 1, 2, \ldots, r$, be the n-dimensional simplices of the partition. Then we have two possibilities:
either it is possible to choose the orientations of the all σ_i^n's in a compatible way (so that $\partial \sum \sigma_i^n = 0$, see Fig. 3.2), or not.
The first case occurs for **orientable** manifold and the indicated cycle, denoted by $[X]$ and called the *fundamental cycle*, generates $H_n(X, \mathbb{Z})$. In the second case the manifold is not orientable and there is no nonzero n-dimensional cycles.

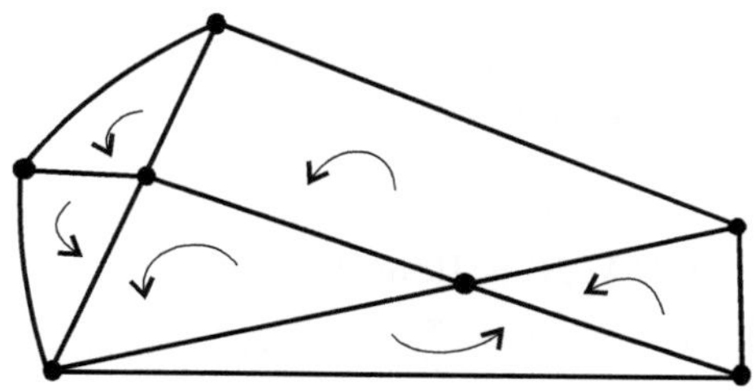

Figure 3.2: Fundamental cycle

3.8. If f is a map between orientable manifolds X and Y of the same dimension, then $f_*[X] = d[Y]$ for some integer d. This number $d = d(f)$ is called the **degree** of the map f (see also Remark 1.13).

3.9. We shall use also the relative homology groups defined as follows. Let $Y \subset X$ be two CW-complexes. The relative group of k-dimensional chains is the quotient group $C_k = C_k(X)/C_k(Y)$. The differential homomorphism acts in usual way and is well defined. Thus we get chain complex and its homology groups are called the **relative homology groups** of the pair (X, Y) and are denoted $H_k(X, Y; R)$. The relative cycles, (generating the relative homologies), are chains c in X such that ∂c is a chain in Y. By passing to the dual complexes, we define the **relative cohomology groups**.
If we take $Y = \{p\}$ (one point), then the groups $H_k(X, \{p\})$ are called the **reduced homology groups** and are denoted by $\widetilde{H}_k(X)$. They are equal to the homology groups of the *completed chain complex*

$$\ldots \longmapsto C_2 \longmapsto C_1 \longmapsto C_0 \longmapsto \mathbb{Z} \longmapsto 0,$$

where the additional boundary operator counts the coefficients, $\sum m_i \tau_i^0 \longmapsto \sum m_i$. Thus the difference between the relative and the non-relative homology groups lies only in the 0-dimensional group. If X is connected, then $H_0(X) = \mathbb{Z}$ and $\widetilde{H}(X) = 0$. We have also $H_k(X, Y) = \widetilde{H}(X/Y)$, where X/Y is obtained from X by identifying the points from Y.

3.10. The relative homology groups appear in the following **long exact sequence of a pair**:

$$\dots \longmapsto H_k(Y) \overset{i_*}{\longmapsto} H_k(X) \overset{j_*}{\longmapsto} H_k(X,Y) \overset{\partial}{\longmapsto} H_{k-1}(Y) \longmapsto \dots .$$

Here the homomorphisms i_*, j_* are induced by the natural maps between the topological spaces. The 'boundary' homomorphism(or connecting homomorphism) ∂ is defined as follows: if c is a relative cycle from its class of relative homology $[c]$, then ∂c represents an (absolute) cycle in Y, and we put $\partial[c] = [\partial c]$.

3.11. If X and Y are such spaces that their homology groups are without torsion, i.e., without nonzero elements of finite order, then

$$\begin{aligned} H_k(X,R) &= H_k(X,\mathbb{Z}) \otimes_{\mathbb{Z}} R, \\ H^k(X,R) &= \mathrm{Hom}\,(H_k(X,\mathbb{Z}),R), \\ H_k(X\times Y,\mathbb{Z}) &= \sum_{i+j=k} H_i(X,\mathbb{Z}) \otimes H_j(Y,\mathbb{Z}), \\ \widetilde{H}_{k+1}(X * Y,\mathbb{Z}) &= \sum_{i+j=k} \widetilde{H}_i(X,\mathbb{Z}) \otimes \widetilde{H}_j(Y,\mathbb{Z}). \end{aligned}$$

Here $X * Y$ is the **joint** of the spaces X and Y defined as $X \times Y \times [0,1]/\sim$, where the subsets $X \times \{q\} \times \{0\}$ and $\{p\} \times Y \times \{1\}$ are squeezed to points and p, q are the base points. In particular, the latter two formulas hold for the homology groups with coefficients in $\mathbb{R}$ and in $\mathbb{C}$.
The first isomorphism, called the **Künneth formula**, is realized by means of the map $(\sigma^i, \delta^j) \longmapsto \sigma^i \times \delta^j \subset X \times Y$, where $\sigma^i \subset X$ and $\delta \subset Y$ are cycles. The second isomorphism is realized by means of the map $(\sigma^i, \delta^j) \longmapsto \sigma^i * \delta^j$.

§2 Index of Intersection

3.12. Definition (index of intersection). (We follow mainly [**GH**]). Let A and B be two cycles in a two-dimensional oriented manifold X represented by closed oriented curves, denoted also by A and B. Assume that A and B intersect transversally one another. If p is such intersection point, then we define the *local index of intersection* of these cycles at p by

$$i_p(A,B) = \pm 1,$$

depending whether the two (ordered) vectors tangent to A and B at p, and defining their orientations, give an orientation of the surface the same as the initial orientation of X or give the opposite orientation (see Fig. 3.3).
The **index of intersection** of the cycles A and B is

$$(A,B) = \sum_p i_p(A,B).$$

If $A = \sum m_i A_i$, $B = \sum n_j B_j$, where A_i, B_j are 1-dimensional cells (arcs) such that $A_i \cap B_j$ intersect only at their interior points and transversally, then

$$(A, B) = \sum_{ij} m_i n_j (A_i, B_j).$$

If the cycles are general, then, in order to compute their intersection, one should choose their representatives (in their homology classes) which satisfy the above requirements. It means that they consist of smooth pieces, which do not intersect one another at their boundaries and the intersections are transversal. It is achieved by small perturbations of the initial cycles.

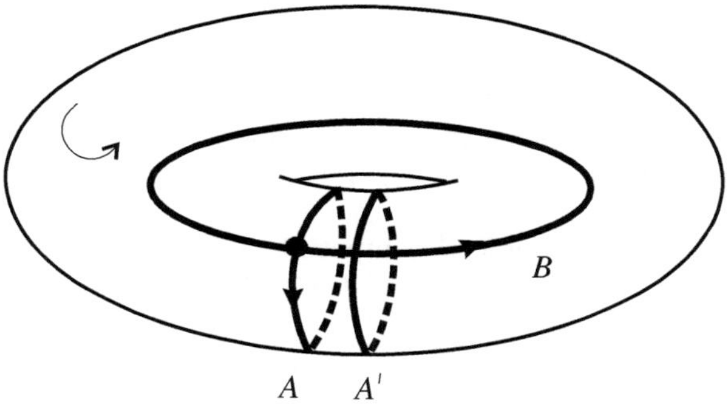

Figure 3.3: Example of interesection index

The intersection index is also denoted by $A \cdot B$, $A \circ B$, and $\langle A, B \rangle$.
At the example from Fig. 3.3 we have $(A, B) = 1$, $(A, A) = 0$, $(B, B) = 0$.

3.13. Lemma. *(A, B) does not depend on the homology classes of the cycles. It means that if $A = \partial C$, then $(A, B) = 0$.*

Proof. Assume that the cycles are represented by the curves A and B satisfying the requirements needed to compute their intersection number. Assume also that $A = \partial C$, where C is a two-dimensional domain in X. The number of points where the curve B enters the domain C is equal to the number of points where B exits C. This follows from the fact that B is closed (as a cycle) and that A divides X into two domains ; here we use the orientability of X. From Fig. 3.4 we see that the orientation, defined by T_pA and T_pB at an income point p, is opposite to the orientation at the outcome point. This shows that $(A, B) = 0$.
From this one can easily complete the proof in the general case. □

If X is an oriented surface, then the intersection index of cycles defines the bilinear map

$$H_1(X, \mathbb{Z}) \times H_1(X, \mathbb{Z}) \longmapsto \mathbb{Z}.$$

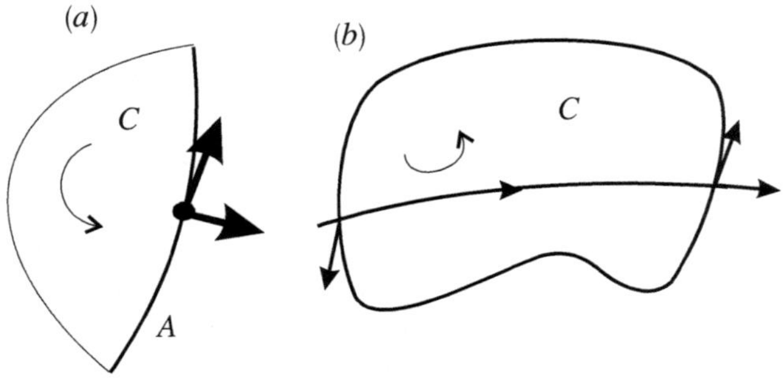

Figure 3.4: Cells and their boundaries

Consider now the situation when X is an n-dimensional manifold and A and B are cycles in X, of dimensions k and $n-k$ respectively. The local index of their intersection is $i_p(A,B) = \pm 1$ and the sign is chosen according to the following rule.
Let $(v_1, \ldots, v_k)$ be the basis of T_pA compatible with the orientation of A and let $(w_1, \ldots, w_{n-k})$ be an analogous basis of T_pB. Then we look at the compatibility of the orientation given by $(v_1, \ldots, v_k, w_1, \ldots, w_{n-k})$ with the orientation of X. As before, we define the index of intersection of the cycles A and B as $(A,B) = \sum_p i_p(A,B)$ and we have the analogue of Lemma 3.13.

Proof of Lemma 3.13 *in the general case.* Let $A = \partial C$ for a $(k+1)$-dimensional CW-complex C and let B be a sum of closed $(n-k)$-dimensional CW-complexes. Assuming that they are in general position, we have that: $A \cap B$ consists of finite number of points of transversal intersection and $C \cap B$ consists of finite number of connected smooth curves γ_α with ends at A. Moreover, the intersections between the cells of A and/or C and of B are transversal; the intersection between cells of maximal dimensions holds along curves, and the intersection holds at points when some cell has dimension smaller by one. It is enough to show that the orientations given by A and B at the endpoints of any such curve γ are different.
One can construct the system of vector fields

$$v_1(t), \ldots, v_k(t), w_1(t), \ldots, w_{n-k-1}(t)$$

along $\gamma : x = x(t)$, satisfying the following properties:
$v_1, \ldots, v_k$ are tangent to C and to A at the endpoints of γ;
$\dot{x}(t), w_1(t), \ldots, w_{n-k-1}(t)$ define the orientation of B (so these vectors are tangent to B) and
$v_1(t), \ldots, v_k(t), \dot{x}(t), w_1(t), \ldots, w_{n-k-1}(t)$ define the orientation of X.
(If we have such system of vectors at one point $x(t_0)$, then we prolong it continuously to the whole curve. Near the endpoints we improve slightly the system (v_i), in order to satisfy the condition of tangency to A.) Therefore, $v_1(t), \ldots, v_k(t), \dot{x}(t)$

define some orientation of C; it can agree with the initial orientation of C or not. The boundary A of C is oriented in such a way that the system (n,(orientation of A)) is the orientation of C, where n is the vector normal to A and directed outside of C. Now it is clear that the orientations $(v_1, \ldots, v_k)$ of A at the income point and at the outcome point of γ are different. □

Thus we have defined the bilinear map, called the **intersection form**

$$H_k(X,\mathbb{Z}) \times H_k(X,\mathbb{Z}) \longmapsto \mathbb{Z}.$$

3.14. Remark. The intersection form vanishes at the torsion part of the homology groups. Indeed, if A is such a cycle that $mA = \partial C$, $m \in \mathbb{Z}$, then $m(A,B) = (mA,B) = (\partial C, B) = 0$ and also $(A,B) = 0$.
We have also the following commutativity relation

$$(A,B) = (-1)^{k(n-k)}(B,A).$$

3.15. Poincaré duality theorem. *Let X be a compact oriented n-dimensional manifold. Then the intersection form is unimodular, which means that it is non-degenerate as a form on the torsion free parts of the homology groups*

$$(\cdot,\cdot) : H_k(X)/\mathrm{Tor} \times H_{n-k}(X)/\mathrm{Tor} \longmapsto \mathbb{Z}.$$

In other words, any functional $H_{n-k}(X)/\mathrm{Tor} \longmapsto \mathbb{Z}$ is an intersection index of some homology class from $H_k(X)$ and, if a class $A \in H_k(X)$ is such that $(A,B) = 0$ for any $B \in H_{n-k}$, then A lies in the torsion part of $H_k(X)$ ($mA = 0$ for some $m > 0$).

Proof. Let $K = \{\sigma^i_\nu\}$ be some simplicial partition of X. Here the upper index i denotes the dimension of a simplex σ^i_μ which is an image in X of the standard (and oriented in the standard way) simplex $\{(x_1, \ldots, x_i) : x_j \geq 0, \sum x_j \leq 1\}$ under a smooth map. They define the chain complex $C_\bullet(K)$ and the cycles, representing elements of the homology groups, are expressed as combinations of the cells from K. We shall construct certain complex dual to the $C_\bullet(K)$.

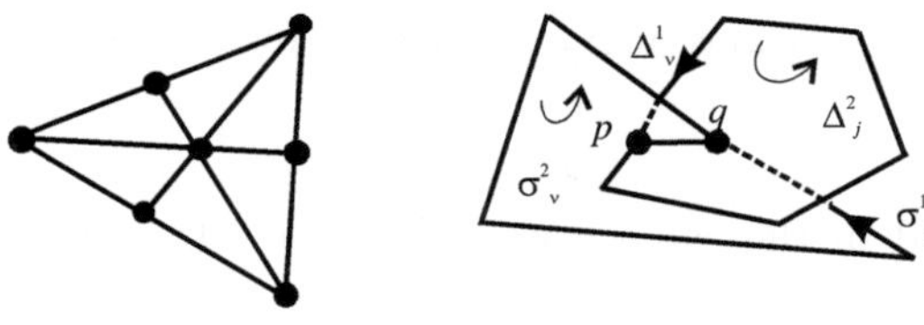

Figure 3.5: Baricentric partitions and dual cells

Let $\{\tau^j_\mu\}$ be the baricentric partition of K (see Fig. 3.5). In the construction of the dual complex, we associate to each simplex from K certain cell consisting of simplices of the baricentric partition of K.

To a vertex $\sigma_\nu^0 \in K$ we associate the n-dimensional cell (the *star* of σ_ν^0)

$$\Delta_\nu^n = *\sigma_\nu^0 = \bigcup \tau_\mu^n,$$

where the sum runs over simplices τ adjacent to the vertex σ_ν^0, $\sigma_\nu^0 \in \tau_\mu^n$.
To a k-dimensional simplex σ_ν^k we associate the $(n-k)$-dimensional cell

$$\Delta_\nu^{n-k} = *\sigma_\nu^k = \bigcap \Delta_\mu^n,$$

where the intersection is over all n-dimensional cells Δ_μ^n associated with the vertices of the simplex σ_ν^k.
The dual cell to σ_ν^n is equal to its baricentric center.
The cell Δ_ν^{n-k} intersects the simplex σ_ν^k transversally at one point, which we denote by p. We define the orientations of the dual cells in such a way that $i_p(\sigma_j, \Delta_j) = 1$.
Next, we introduce the coboundary operation δ on the dual complex $C^\bullet = \{\Delta_\nu^{n-k}\}$.
If $\Delta_\nu^{n-k} = \bigcap_{i=1}^{k+1} \Delta_i^n$, the intersection of $k+1$ n-cells, then

$$\delta(\Delta_\nu^{n-k}) = \sum_j \bigcap_{i\neq j} \Delta_i^n = \sum_j \pm\Delta_j^{n-k+1}$$

(like in the coboundary operator in the complex of cochains defining the Čech cohomology). The sign $\epsilon_j = \pm 1$ before Δ_j is chosen in such a way that the orientation of the cell $\Delta_\nu = \Delta_\nu^{n-k}$ coincides with the orientation of the boundary of $\epsilon_j\Delta_j$, i.e., the orientation of Δ_ν differs by the factor ϵ_j form the orientation of the boundary of Δ_j.
The thesis of the theorem follows from the identity

$$\delta(*\sigma_\nu^k) = (-1)^k * (\partial\sigma_\nu^k),$$

which says that $*$ is a complex homomorphism.
To its proof one considers $\sigma_j = \sigma_j^{k-1}$, the simplices appearing in the boundary of $\sigma_\nu = \sigma_\nu^k$, and $\Delta_j = \Delta_j^{n-k+1} = *\sigma_j$. Assume that σ_j are oriented as the boundaries of σ_ν^k and that the orientations of Δ_j are as described above. It is enough to show that

$$\epsilon_j = (-1)^k.$$

The latter is proved in the same way as Lemma 3.13. One constructs suitable vector fields $v_1(t), \ldots, v_{k-1}(t)$ (tangent to σ_ν) and $w_1(t), \ldots, w_{n-k}(t)$ (tangent to Δ_j) along the segment $[p, q]$ which joins the intersection points $p = \Delta_\nu \cap \sigma_\nu$ and $q = \Delta_j \cap \sigma_j$. One adds to them the vector field $\dot{x}$ tangent to $[p, q]$ and compares induced orientations at the endpoints.
We omit the details, which can be found in [**GH**].

The map $*$ defines an isomorphism between the complex $C_\bullet(K)$ and the complex $C^\bullet(K)$ of cochains of the dual cell partition. Because the cohomology groups of the cochain complex $C^\bullet(K)$ are equal to the cohomology groups of X, we get the isomorphism

$$D:\ H_k(X,\mathbb{Z}) \longmapsto H^{n-k}(X,\mathbb{Z})$$

induced by $*$. This statement is also called the Poincaré duality.
The isomorphism D has the property that

$$\langle D(\gamma),\lambda\rangle = (\gamma,\lambda)$$

for any cycles $\gamma \in H_k(X)$, $\lambda \in H_{n-k}(X)$. This means that the map D/Tor, i.e., the homomorphism defined by $*$ on the torsion-free part, is induced by the intersection form. The invertibility of D/Tor means the unimodularity of the form $(\cdot,\cdot)$.
From this Theorem 3.15 follows. □

Problem: Calculate the intersection form on $H_{2k}(\mathbb{CP}^n) \times H_{2n-2k}(\mathbb{CP}^n)$.

3.16. Remark. The definition of intersection index and of the Poincaré duality can be generalized to the relative situation.
Let $Y = \partial X$ be the boundary of a manifold X. If γ is a relative cycle in the pair (X,Y), i.e., $\gamma \subset X$ and $\partial\gamma \subset Y$, and δ is an absolute cycle in X, then we can define their intersection index (as usual) and obtain the bilinear form

$$H_k(X,Y)/\mathrm{Tor} \times H_{n-k}(X)/\mathrm{Tor} \longmapsto \mathbb{Z},$$

which is non-degenerate (see Fig. 3.6).
Sometimes a chain $\gamma - \partial\gamma \subset X - Y$ is called a *cycle with closed support* in the non-compact manifold $X - Y$ (or a *locally finite cycle*). One also introduces the **homology groups with closed support** (or the *locally finite homology groups*) $H_k^{\mathrm{cl}}(X-Y)$ (or $H_k^{\mathrm{lf}}(X-Y)$ respectively) and the intersection index gives the pairing

$$H_k^{\mathrm{lf}}(X-Y)/\mathrm{Tor} \times H_{n-k}(X)/\mathrm{Tor} \longmapsto \mathbb{Z}.$$

The corresponding dual groups are called the **cohomology groups with compact support**. They are defined by means of the complex of cochains, which are nonzero only on finite number of cells in $X-Y$, and are denoted by $H_{\mathrm{c}}^k(X-Y)$. We have the pairing $H_k^{\mathrm{lf}}(X-Y,\mathbb{R}) \otimes H_{\mathrm{c}}^k(X-Y,\mathbb{R}) \longmapsto \mathbb{R}$.

3.17. The linking number. Let $X = \mathbb{S}^m$ be the m-dimensional sphere and let a and b be two cycles in $\mathbb{S}^m$. The dimension of a is $i < m/2$ and the dimension of b is $m-i-1$ and we assume that $i \le m-i-1$.
Because $H_i(\mathbb{S}^m) = 0$, the cycle a is a boundary, $a = \partial A$. If $i = 0$, then we assume that $a = 0$ in the homology group $H_0(X)$. The intersection number

$$l(a,b) = (A,b)$$

is called the **linking coefficient** of the cycles a and b. It is well defined, because if $a = \partial A'$, then $A - A' = \partial C$, and we use the properties of the intersection index.

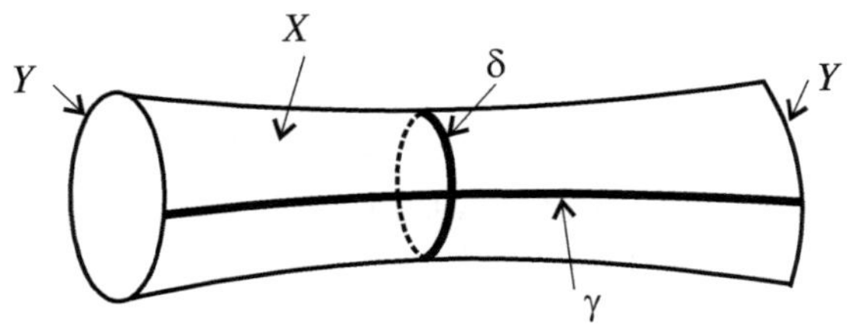

Figure 3.6: Intersection index of absolute and relative cycles

There is another definition of the linking number. Of course, the sphere is a boundary of a ball, $\mathbb{S}^m = \partial\mathbb{D}^{m+1}$. The cycles a and b define the zero classes in the homology groups of $\mathbb{D}^{n+1}$. Thus $a = \partial\widetilde{A}$, $b = \partial\widetilde{B}$, where $\widetilde{A}, \widetilde{B} \subset \mathbb{D}^{m+1}$.

3.18. Lemma. $l(a, b) = (-1)^{i+1}(\widetilde{A}, \widetilde{B})$.

Proof. Proof. Let us introduce the 'polar' coordinates in $\mathbb{D}^{m+1}$: $[0, 1] \times \mathbb{S}^m \ni (r, \theta) \longmapsto r\theta \in \mathbb{D}^{m+1}$.
Assume that a, A and b are represented as smooth submanifolds. Define

$$\begin{aligned}\widetilde{A} &= [1/2, 1] \times a \cup \{1/2\} \times A,\\ \widetilde{B} &= [0, 1] \times b\end{aligned}$$

(see Fig. 3.7). The surfaces $\widetilde{A}$ and $\widetilde{B}$ intersect at $\{1/2\} \times (A \cap b)$. Therefore the

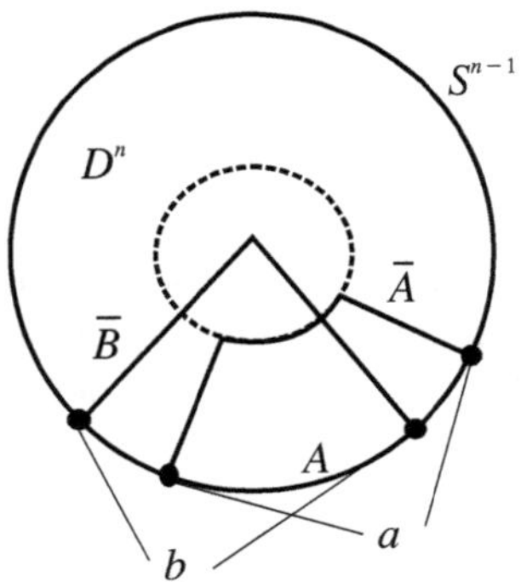

Figure 3.7: Linking number of two cycles a and b

number of intersection points appearing in the both definitions are the same. It remains to control the orientations of the chains.
The orientation of $\widetilde{A}$ is the same as the corresponding orientation of $\{1/2\} \times A = A$.
The orientation of $\widetilde{B}$ is given by $(\partial_r, \textit{orientation of } b)$. Therefore the orientation of

$\mathbb{D}^{m+1}$, given by $\widetilde{A}, \widetilde{B}$, is following: (*orientation of* A, ∂_r,*orientation of* b). Because the standard orientation of $\mathbb{D}^{m+1}$ is (∂_r,*orientation of* $\mathbb{S}^m$), then we must move the radial vector to the beginning. From this we get the factor $(-1)^{i+1}$. □

3.19. Alexander duality theorem. *Let* $K \subset \mathbb{S}^m$ *be a submanifold,* $p \in K$, $q \in \mathbb{S}^m \setminus K$ *be some base points. The linking number of cycles defines the duality of the homology groups* $H_i(K,p)/\mathrm{Tor}$ *and* $H_{m-i-1}(\mathbb{S}^m \setminus K, q)/\mathrm{Tor}$.

Proof. Repeating the proof of the Poincaré duality theorem, we get the duality of the homology groups $H_i(K,p)$ and $H_{m-i}(\mathbb{S}^m \setminus p, \mathbb{S}^m \setminus K)$; (by means of intersection of relative cycles in (K,p) and in $(\mathbb{S}^m \setminus p, \mathbb{S}^m \setminus K)$).
We use the long exact sequence of the pair $(\mathbb{S}^m \setminus p, \mathbb{S}^m \setminus K)$:

$$H_j(\mathbb{S}^m \setminus K, q) \longmapsto H_j(\mathbb{S}^m \setminus p, q) \longmapsto H_j(\mathbb{S}^m \setminus p, \mathbb{S}^m \setminus K) \overset{\partial}{\longmapsto} H_{j-1}(\mathbb{S}^m \setminus K, q) \longmapsto$$

to obtain the isomorphism between $H_j(\mathbb{S}^m \setminus p, \mathbb{S}^m \setminus K)$ and $H_{j-1}(\mathbb{S}^m \setminus K, q)$; because $H_j(\mathbb{S}^m \setminus p, q) = 0$ for every j.
It is also clear that this isomorphism identifies the intersection index with the linking coefficient. □

3.20. Remark. Using the Poincaré duality, one can associate with a cycle A in a manifold X certain cohomology class (with real coefficients). Namely, if the dimension of A is k and $B \in H_{n-k}(X, \mathbb{R})$, then the functional $\phi(B) = (A, B)$ is an element of $H^k(X, \mathbb{R}) = H_k(X, \mathbb{R})^*$.
Note that, in this definition, the coorientation of A is important. Assume that A is a submanifold of X. Its **coorientation** is the compatible choice of a field of frames in the normal spaces $N_pA = T_pX/T_pA$. The manifold X and the submanifold A can be not oriented, but, if A is cooriented, then it defines the cohomology class.

3.21. Example (Maslov index and Morse index). Some cohomology classes are defined just in the way, which we have described above. One of them is the Maslov index.
Let X be an n-dimensional Riemannian manifold. Its cotangent bundle $M = T^*X$ is a **symplectic manifold**. It means that there is a closed and non-degenerate 2-form ω on M: $\mathrm{d}\omega = 0$ and, for each $m \in M$, the bilinear (antisymmetric) form $\omega(\cdot,\cdot)$ on T_mM is non-degenerate. In the case of cotangent bundle the symplectic form is defined as follows. Let $q_1, \ldots, q_n$ be local coordinates in X and let $p_1, \ldots, p_n$ be the associated with them coordinates in $T_q^*X = \{p_1\mathrm{d}q_1 + \ldots + p_n\mathrm{d}q_n\}$. Then $\omega = \sum \mathrm{d}p_i \wedge \mathrm{d}q_i = \mathrm{d}(\sum p_i\mathrm{d}q_i) = \mathrm{d}\alpha$. The 1-form α is called the *Liouville form.*
An n-dimensional submanifold $L \overset{i}{\subset} M$ is called **Lagrangian submanifold** iff

$$i^*\omega = 0,$$

i.e., $\omega|_{T_mL} \equiv 0$.
If $\pi : M \longmapsto X$ is the natural projection map, then the fibers of π are Lagrangian submanifolds; because then $\mathrm{d}q_i = 0$. If $S = S(q)$ is a function on X, then the

set $\{(p,q) : p = \partial S/\partial q\}$ forms a Lagrangian submanifold, because $\alpha = \mathrm{d}S$ and $\mathrm{d}^2S = 0$. If Y is a submanifold in X, then $N^*Y = \{(q,p) : p|_{T_qY} = 0\}$ is also Lagrangian submanifold. Note that for $Y = \{q\}$ we get the first example.
The Maslov index, which we are going to define, is a 1-dimensional cohomology class on the Lagrangian submanifold. It is defined on curves $\gamma \subset L$.
The projection π, restricted to the Lagrangian submanifold L, is a differentiable map between two n-dimensional manifolds. If L is in general position, then $\pi|_L$ is regular at almost every point of L. Define

$$\Sigma = \{z \in L : z \text{ is critical point for } \pi|_L\},$$

the set of points, where L projects badly at X. The image of this set, $\pi(\Sigma)$ is called the **caustics** of the manifold Σ.
If L is in general position, then Σ forms a subset of codimension one in L; given by one equation $\det D(\pi|_L) = 0$. Moreover, typically, Σ is smooth outside some its subset of codimension two (in Σ). We do not prove this property, we demonstrate it in the following example.

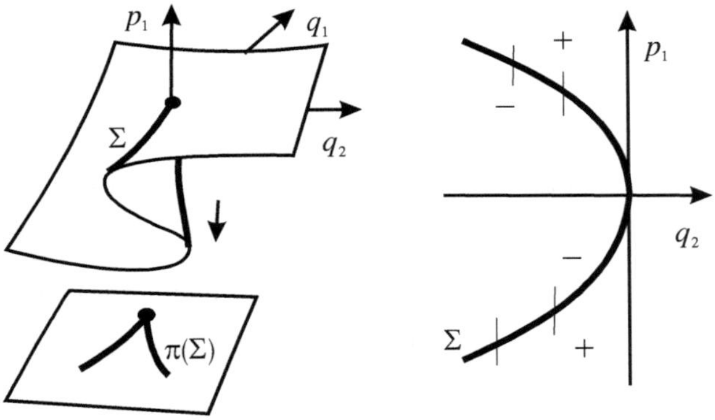

Figure 3.8: Whitney cusp map

Example. (**A**$_3$ *Lagrangian singularity*, see [**Arn1**]). Let $F(p_1, q_2) = p_1^4 + q_2p_1^2$. We define the following Lagrangian manifold in $T^*\mathbb{R}^2$

$$L = \{(q_1, q_2, p_1, p_2) : q_1 = \partial F/\partial p_1,\ p_2 = -\partial F/\partial q_2\}.$$

This manifold can be parameterized by means of the variables p_1, q_2: $(p_1, q_2) \longmapsto (4p_1^3 + 2q_2p_1, q_2, p_1, -p_1^2)$, and the projection $\pi|_L$ is the Whitney cusp map

$$(p_1, q_2) \longmapsto (4p_1^3 + 2q_2p_1, q_2).$$

The set of critical points of the projection is the smooth curve

$$\Sigma = \{12p_1^2 + 2q_2 = 0\}.$$

and its image is the cusp $q_1 = -8p_1^3, q_2 = -6p_1^2$ (see Fig. 3.8).
It turns out that Σ has two sides in L. Take a typical point z from Σ. The kernel of $D\pi(z)$ is 1-dimensional and L is given locally as a graphic of a mapping $(p_1, q_2, \ldots, q_n) \longmapsto (q_1, p_2, \ldots, p_n)$. The critical set $\Sigma = \{\partial q_1/\partial p_1 = 0\}$.
We define the positive side of Σ as that, where $\partial q_1/\partial p_1 > 0$. The negative side is defined by the opposite inequality. This definition is correct. We can see that the sides are well defined at our example. In general situation one can prove that typical codimension two singularities are of the $\mathbf{A}_3$ type.
Hence, Σ is cooriented and defines a 1-dimensional cohomology class. The value of this class on a curve γ is equal to the number of (transversal) passes of γ from the negative side of Σ to the positive side minus the number of reverse passes. This number is called the **Maslov index** of the curve γ.

The Maslov index has applications in asymptotic solutions of partial differential equations and in geometry. Below we present one such application.
Let $q \in X$ and let γ be a geodesic line starting at q and with end at some point q'. (Recall that X is Riemannian).
The **Morse index** of the geodesic γ is the number of points in γ which are conjugate with q. Recall that a point q_1 is **conjugate** with q iff the near geodesics starting at q are infinitesimally focused at q_1. One can also define the multiplicity of the conjugate point.
With this situation one can associate a Lagrangian submanifold $L \subset T^*X$. We identify TX with T^*X by means of the Riemannian metric. Let

$$L_0 = \{(q, v) \in T^*X : |v| = 1\}.$$

It is an $(n-1)$-dimensional submanifold such that $\omega|_{L_0} = 0$.
Let $g^t : (q, v) \longmapsto (q(t), v(t))$ be the geodesic flow; here $v(t) = \dot{q}(t)^*$ is the dual covector to $\dot{q}(t)$. The following n-dimensional manifold

$$L = \bigcup_t g^t(L_0)$$

is Lagrangian.
The curve γ has lift $\delta = (\gamma, \dot{\gamma}^*)$ to L and we can define the Maslov index of δ.

3.22. Theorem. *The Morse index of γ is equal to the Maslov index of δ.*

We do not give the proof of this theorem. We notice only that near the conjugate point on one geodesic near geodesics intersect between themselves. This means that their lifts to the Lagrangian submanifold have bad projections onto the configuration space X. Thus δ passes through the critical surface Σ.

3.23. Definition (de Rham cohomologies). If X is a differentiable manifold, then one defines the **de Rham cohomology groups** of X, $H^k_{\mathrm{dR}}(X, \mathbb{R})$, as the cohomology groups of the following cochain complex

$$0 \longmapsto \Gamma(X, \mathcal{E}^0) \overset{\mathrm{d}}{\longmapsto} \Gamma(X, \mathcal{E}^1) \overset{\mathrm{d}}{\longmapsto} \Gamma(X, \mathcal{E}^2) \longmapsto \ldots \longmapsto \Gamma(X, \mathcal{E}^n) \longmapsto 0$$

Here $\Gamma(X, \mathcal{E}^k)$ are the spaces of all differential forms of degree k defined globally on X. More precisely, $\mathcal{E}^k = \mathcal{E}^k_X$ denotes the sheaf of differential k-forms, defined locally as

$$\omega = \sum a_{i_1,\ldots,i_k} \mathrm{d}x_{i_1} \wedge \ldots \wedge \mathrm{d}x_{i_k}.$$

$\Gamma(X, \mathcal{E}^k)$ denotes the set of global sections of this sheaf, or the set of global sections of the bundle $\wedge^k T^*X$. The operator d is the external derivative

$$\mathrm{d}\omega = \sum \sum_j \frac{a_{i_1,\ldots,i_k}}{\partial x_j} \mathrm{d}x_j \wedge \mathrm{d}x_{i_1} \wedge \ldots \wedge \mathrm{d}x_{i_k}.$$

If the coefficients $a_{i_1,\ldots,i_k}$ of the form ω take complex values, then we have the *complex de Rham cohomology groups* $H^k_{\mathrm{dR}}(X, \mathbb{C})$.
ω is called **closed** iff $\mathrm{d}\omega = 0$ and is called **exact** iff $\omega = \mathrm{d}\eta$ for some $(k+1)$-form η. Thus $H^k_{\mathrm{dR}}(X, \mathbb{R}) = (\text{closed forms})/(\text{exact forms})$.

3.24. De Rham theorem. *We have* $H^k(X, \mathbb{R}) = H^k_{\mathrm{dR}}(X, \mathbb{R})$.

Before giving the proof of de Rham theorem we shall give interpretation of the intersection index in terms of differential forms.
The exterior product of forms ω^k and η^m, $\omega^k \wedge \eta^m$, is compatible with the external derivative, $\mathrm{d}(\omega \wedge \eta) = \mathrm{d}\omega \wedge \eta + (-1)^k \omega \wedge \mathrm{d}\eta$. This defines the bilinear map

$$H^k_{\mathrm{dR}}(X, \mathbb{R}) \times H^m_{\mathrm{dR}}(X, \mathbb{R}) \longmapsto H^{k+m}_{\mathrm{dR}}(X, \mathbb{R}).$$

In this way we obtain the cohomology algebra. In the case of singular cohomologies, this bilinear map coincides with the so called *cup-product*.

Examples. 1. The algebra $H^\bullet(\mathbb{CP}^n, \mathbb{C})$ is isomorphic to $\mathbb{C}[x]/(x^{n+1})$, where x represents the generator of the second cohomology group with integer coefficients.
2. The algebra $H^\bullet(\mathbb{T}^2, \mathbb{R})$ is isomorphic to the exterior algebra $\wedge \mathbb{R}^2$.

The last cohomology group $H^n_{\mathrm{dR}}(X, \mathbb{R})$ is identified with $\mathbb{R}$ by means of the integration along X: $\langle \xi^n, [X] \rangle = \int_X \xi$, the value on the fundamental cycle $[X]$.
Therefore we have $(\omega^k, \eta^{n-k}) \longmapsto \int_X \omega \wedge \eta$. We obtain the pairing

$$H^k_{\mathrm{dR}}(X, \mathbb{R}) \times H^{n-k}_{\mathrm{dR}}(X, \mathbb{R}) \longmapsto \mathbb{R}.$$

3.25. Theorem. *The latter pairing is dual to the intersection index. It means that, if A, B are cycles of complementary dimensions and ω, η are differential forms such that for any cycles C, D, $\int_C \omega = (A, C)$ and $\int_D \eta = (B, D)$, then $\int_X \omega \wedge \eta = (A, B)$.*

Proof. Let $\Delta : X \longmapsto X \times X$ be the diagonal mapping, $x \longmapsto (x, x)$, and let its image also be denoted by Δ. If A, B are cycles in X, then their index of intersection in X is equal to the index of intersection of $A \times B$ and Δ in $X \times X$.
By the Künneth formula $H^n(X \times X, \mathbb{R}) = \bigoplus_k H^k(X, \mathbb{R}) \otimes H^{n-k}(X, \mathbb{R})$ and it is not difficult to see that

$$\Delta^*([\omega] \otimes [\eta]) = [\omega \wedge \eta].$$

Combining these two fact, we get the result. □

3.26. Sheaves. If X and E are manifolds, then a map $p : E \longmapsto X$ is called the **fibre bundle** iff the manifold X is covered by charts U_α such that $p^{-1}(U_\alpha)$ is diffeomorphic to $U_\alpha \times F$, where F is a fixed manifold, a *fiber*. X is the *base* of the bundle. If F is a vector space, then we get a **vector bundle**. If X is complex manifold and $F = \mathbb{C}$, then we have a **line bundle**. If F is a discrete space, then the bundle is a **covering**.

Bundles over X with fixed fiber F are given by identifications (gluing) of the domains $U_\alpha \times F$ and $U_\beta \times F$ above $U_\alpha \cap U_\beta$. It is realized by means of the functions $h_{\alpha,\beta}$ on $U_\alpha \cap U_\beta$ taking values in the group of diffeomorphisms of the fiber. In the case of vector bundles, this is the linear group $\mathrm{Aut}(F)$.

In each chart U_α we can define local sections of the bundle $E \longmapsto X$, as functions on U_α with values in F. If s_α is a section in U_α and s_β is a section in U_β, then they define the section in $U_\alpha \cup U_\beta$ iff $s_\beta = h_{\alpha,\beta} s_\beta$. With each domain $U \subset X$ we can associate its space of sections $\mathcal{F}(U)$. In this way we obtain the **sheaf** $\mathcal{F}$ of sections of the bundle E.

The formal definition of sheaf (of groups on a manifold X) is the following. Firstly, one has a *presheaf*, which associates to any open subset $U \subset X$ an abelian group $\Gamma(U)$ and to any inclusions $V \subset U$ a restriction homomorphisms $\Gamma(U) \longmapsto \Gamma(V)$, $f \longmapsto f|_V$ (with natural properties). A presheaf is a **sheaf** iff it has the additional property: if $f_j \in \Gamma(U_j)$ satisfy $f_i|_{U_i \cap U_j} = f_j|_{U_i \cap U_j}$ then there is unique $f \in \Gamma(\bigcup U_j)$ such that $f|_{U_j} = f_j$.

There are many sheaves associated with some natural bundles: the sheaf $\mathcal{E}_X^0$ of smooth functions (sections of the trivial bundle $X \times \mathbb{R}$), the sheaf $\mathcal{E}_X^k$ of differentiable k-forms on X, the sheaf of vector fields, the constant sheaves (where $\mathcal{F}(U) = \mathbb{Z}$, $\mathbb{C}$, $\mathbb{R}$ consists of locally constant functions).

In algebraic geometry there are more natural bundles and sheaves associated with them. In fact, there the notion of a sheaf is more natural, because it includes some natural situations with singularities, (e.g., when the dimension of a fiber ceases to be constant).

Important are: the sheaf $\mathcal{O}(X) = \mathcal{O}_X$ (of germs of holomorphic functions) and sheaves $\mathcal{F}$ of $\mathcal{O}_X$-moduli; $\mathcal{F}(U)$ is a module over the ring $\mathcal{O}_X(U)$. Such sheaf $\mathcal{F}$ is called **coherent** iff it is locally finitely generated and the relations sheaf is locally finitely generated; for $U \subset X$ there is a surjection $\Gamma(U, \mathcal{O}_X)^m \longmapsto \Gamma(U, \mathcal{F})$ and its kernel defines the sheaf of relations.

In Section 7 we use some natural morphisms of sheaves. If $F : X \longmapsto Y$ is a holomorphic map and $\mathcal{F}$ is a sheaf of $\mathcal{O}_X$-moduli on X, then the *direct image* of $\mathcal{F}$ is the sheaf $F_*\mathcal{F}$ of $\mathcal{O}_X$-moduli on Y (and of $\mathcal{O}_Y$-moduli) such that

$$(F_*\mathcal{F})(V) := \mathcal{F}(F^{-1}(V)).$$

If $\mathcal{G}$ is a sheaf of $\mathcal{O}_Y$-moduli on Y, then the *inverse image* of $\mathcal{G}$ is the sheaf $F^{-1}\mathcal{G}$ of $\mathcal{O}_Y$-moduli on X such that $(F^{-1}\mathcal{G})(U) := \lim_{F(U) \subset V} \mathcal{G}(V)$ (direct limit over V's). One defines also $F^*\mathcal{G} := F^{-1}\mathcal{G} \otimes_{\mathcal{O}_Y} \mathcal{O}_X$, a sheaf of $\mathcal{O}_X$-moduli on X; here $\mathcal{O}_Y$ is treated as a subsheaf of $\mathcal{O}_X$.

3.27. The Čech cohomologies. Let $\mathcal{F}$ be a sheaf on X and $\mathcal{U} = \{U_\alpha\}$ be its locally finite covering. We associate with it the following cochain complex.

- Its groups of k-cochains $C^k(\mathcal{U}, \mathcal{F})$ are the direct sums of the groups $\mathcal{F}(U_{\alpha_0} \cap \ldots \cap U_{\alpha_k})$ (with different indices α_i's); its elements are denoted by $\sigma_{\alpha_0,\ldots,\alpha_k}$ and are anti-symmetric with respect to the indices.

- If $\sigma = \{\sigma_{\alpha_0,\ldots,\alpha_k}\}$, then the coboundary operator acts on it as follows:

$$(\delta\sigma)_{\alpha_0,\ldots,\alpha_{k+1}} = \sum_{j=0}^{k+1} (-1)^j \sigma_{\alpha_0,\ldots,\alpha_{j-1},\alpha_{j+1}\ldots\alpha_{k+1}}.$$

We define $H^k(\mathcal{U}, \mathcal{F})$ as the cohomology groups of this cochain complex. If $\mathcal{V}$ is a subcovering of $\mathcal{U}$, (i.e., each V_β lies in some U_α), then we have a natural map $H^k(\mathcal{U}) \longmapsto H^k(\mathcal{V})$. The direct limit of these groups is the **Čech cohomology group** $H^k(X, \mathcal{F})$.

Examples. 1. The group $H^0(X, \mathcal{F})$ is equal to the group of global sections of the sheaf $\mathcal{F}$. Indeed, if $\sigma^0 = \{\sigma_U\}$, then the condition $(\delta\sigma)_{UV} = \sigma_V - \sigma_U = 0$ means that the sections σ_U and σ_V coincide at $U \cap V$.

2. The vector bundles on X with fiber V are classified accordingly to the elements of the group $H^1(X, \mathcal{G})$, where $\mathcal{G}$ is the sheaf of functions on X with values in the group $\mathrm{Aut}(F)$ (of automorphisms of the fiber).
However, if the group $\mathrm{Aut}(F)$ is non-abelian, then we have the "non-abelian" cohomology group $H^1(X, \mathcal{G})$, whose definition needs some modification. It equals to $\{\sigma^1 : \delta\sigma^1 = e\}/\sim$, where the cochain $\sigma^1 = (\sigma_{UV})$ is a cocycle, iff $(\delta\sigma)_{UVW} = \sigma_{UV}\sigma_{VW}\sigma_{WU} = e$, and $(\sigma_{UV}) \sim (\tau_{UV})$, iff $\sigma_{UV} = \rho_U \tau_{UV} \rho_V^{-1}$ for some 0-cochain $\rho^0 = (\rho_U)$.

3. If $\mathcal{F}$ is a sheaf on X and $F_*\mathcal{F}$ is the direct image of $\mathcal{F}$ defined by a map $F : X \longmapsto Y$, then we have

$$H^j(X, \mathcal{F}) \simeq H^j(Y, F_*\mathcal{F}).$$

Usually the limit in the definition of Čech cohomology is achieved at some sufficiently fine covering. In particular, J. Leray proved that, if the intersections of the elements of the covering have the property that their cohomology groups vanish, then such covering is good (see [**God**]). In this way one obtains the isomorphism of the Čech cohomologies with coefficients in a constant sheaf, denoted by $\check{H}^k$, with the singular cohomologies. The latter fact is proved as follows. Take a simplicial partition K of X. With each vertex s^0_α we associate the open subset U_α equal to the union of n-simplices having s^0_α as a vertex (the *star* with vertex at s^0_α). The system of these stars defines the needed covering of X. The intersection $\bigcap U_{\alpha_i}$ is not empty iff s^0_i are vertices of some simplex in K. The Čech cochain associates integer numbers to such intersections. The latter numbers can be interpreted as singular cochains. In this way we obtain homomorphism of the cochain complexes. If the simplicial partition is sufficiently small, then this gives the isomorphism of the cohomology groups.

Proof of de Rham Theorem 3.24.

Lemma. *We have* $H^k(X, \mathcal{E}^r) = 0$ *for* $k > 0$.

Proof. It relies on existence of a smooth partition of unity for sections of the sheaf of smooth differential forms. The sheaves which admit a partition of unity are called the *flabby sheaves*. (The holomorphic or algebraic sheaves are not flabby). Let $\{U_\alpha\}$ be some covering of X and let ρ_α be some associated with it **partition of unity**: $\operatorname{supp}\rho_\alpha \subset U_\alpha$, $\rho_\alpha \geq 0$, $\sum \rho_\alpha = 1$. If $\sigma = \{\sigma_{\alpha_0,\ldots,\alpha_k}\}$ is a cocycle, i.e., $\delta\sigma = 0$, then we define the cochain $\tau \in C^{k-1}$ by putting

$$\tau_{\alpha_0,\ldots,\alpha_{k-1}} = \sum_\beta \rho_\beta \sigma_{\beta,\alpha_0,\ldots,\alpha_{k-1}}.$$

It turns out that $\delta\tau = \sigma$. For example, in the case $k = 1$, we have $\sigma = \{\sigma_{UV}\}$ such that $\sigma_{UV} + \sigma_{VW} + \sigma_{WU} = 0$ in $U \cap V \cap W$. If $\tau_U = \sum_V \rho_V \sigma_{VU}$, then

$$(\delta\tau)_{UV} = \tau_V - \tau_U = \sum_W \rho_W \sigma_{WV} - \sum_W \rho_W \sigma_{WU} = \sum_W \rho_W \sigma_{UV} = \sigma_{UV}. \qquad \square$$

If ω is a k-form, then it defines a singular cochain: its value on the chain $c = \sum a_i c_i^k$, $a_i \in \mathbb{R}$, is

$$\langle \omega, c \rangle = \sum a_i \int_{c_i} \omega.$$

This homomorphism of cochain complexes commutes with the coboundary operator. The latter fact is equivalent to the **Stokes theorem**

$$\int_{\partial\sigma} \eta = \int_\sigma \mathrm{d}\eta,$$

which is a generalization of the fundamental theorem of analysis (the integral of a derivative is equal to the difference of values of the function at the ends of the interval). In this way we obtain a map from $H^\bullet_{\mathrm{dR}}(X)$ to $H^\bullet(X, \mathbb{R})$. We must show that this map is an isomorphism. Consider the sequence of sheaves

$$0 \longmapsto \mathbb{R} \longmapsto \mathcal{E}^0 \overset{\mathrm{d}}{\longmapsto} \mathcal{E}^1 \overset{\mathrm{d}}{\longmapsto} \mathcal{E}^2 \longmapsto \ldots,$$

where the second arrow is the inclusion.

3.28. Poincaré Lemma. *The above sequence is exact. It means that every closed k-form is locally equal to the differential of some $(k-1)$-form.*

Proof. Because the situation is local, it is enough to consider the case when the closed k-form ω is defined in a neighborhood of 0 in $\mathbb{R}^n$. Denote by $\boldsymbol{E} = \sum x_i\,(\partial/\partial x_i)$ the Euler vector field. If $x \in \mathbb{R}^n$ and $v_1, \dots, v_{k-1}$ are vectors tangent to $\mathbb{R}^n$ at x, then we define the $(k-1)$-form η as follows

$$\langle \eta, (v_1, \dots, v_{k-1})\rangle = \int_0^1 \langle \omega(tx), (\boldsymbol{E}, v_1, \dots, v_{k-1})\rangle \mathrm{d}t.$$

One can check that $\mathrm{d}\eta = \omega$.
(This is generalization of the well-known formula from mechanics: if a field of forces $\boldsymbol{F}(x)$ is such that the work $\int_\gamma \boldsymbol{F}\cdot \mathrm{d}x$ depends only on the ends of the path γ, then the field $\boldsymbol{F}$ is a potential field and the potential energy is given by the formula $U(x) = \int_0^x \boldsymbol{F}\cdot \mathrm{d}s$.) □

Let $\mathcal{Z}^k$ denotes the sheaf of closed k-forms. Poincaré's Lemma says that the short sequences of sheaves

$$0 \longmapsto \mathcal{Z}^k \longmapsto \mathcal{E}^k \overset{\mathrm{d}}{\longmapsto} \mathcal{Z}^{k+1} \longmapsto 0 \tag{3.1}$$

are exact.

3.29. Lemma (long exact sequence). *With each short exact sequence of sheaves*

$$0 \longmapsto \mathcal{E} \overset{\alpha}{\longmapsto} \mathcal{F} \overset{\beta}{\longmapsto} \mathcal{G} \longmapsto 0$$

one associates the following long exact sequence of their Čech cohomologies (analogous to the long exact sequence of the homology groups of a pair):

$$\dots H^k(X,\mathcal{E}) \longmapsto H^k(X,\mathcal{F}) \longmapsto H^k(X,\mathcal{G}) \overset{\delta^*}{\longmapsto} H^{k+1}(X,\mathcal{E}) \longmapsto \dots$$

Proof (sketch). Here the first two homomorphisms are induced by the maps of the sheaves and the coboundary operator δ^* is defined as follows. If σ^k is a Čech cocycle with values in $\mathcal{G}$, then it is image of a cochain τ^k with values in $\mathcal{F}$, $\beta(\tau) = \sigma$. Take the coboundary $\delta\tau$. Because of the commutation relation $\delta\beta = \beta\delta$, $\delta\tau$ belongs to the kernel of β and, by the exactness, to the image of α. Thus $\delta\tau = \alpha(\mu)$, where μ is some $(k+1)$-cocycle with values in $\mathcal{E}$. The cohomology class of μ is the value of the coboundary operator on σ. □

We associate the long exact sequences with the short exact sequences (3.1). Using the fact that $H^q(X,\mathcal{E}^p) = 0$ for $q > 0$, these sequences give us the series of isomorphisms

$$\begin{aligned} \check{H}^k(X,\mathbb{R}) &\simeq H^{k-1}(X,\mathcal{Z}^1) \simeq H^{k-2}(X,\mathcal{Z}^2) \simeq \dots \\ &\simeq H^1(X,\mathcal{Z}^{k-1} \simeq H^0(X,\mathcal{Z}^p)/\mathrm{d}H^0(X,\mathcal{E}^{k-1}). \end{aligned}$$

The group $H^0(X,\mathcal{F})$ is the group of global sections of the sheaf $\mathcal{F}$. Therefore the latter group in the above series of isomorphisms is identified with the group of de Rham cohomologies. □

3.30. Lefschetz fixed point theorem. *Assume that $f : X \longmapsto X$ is a continuous map of a CW-complex. Define its Lefschetz number*

$$L(f) = \sum (-1)^q \mathrm{tr}\left(f_{*q} : H_q(X, \mathbb{Q}) \longmapsto H_q(X, \mathbb{Q})\right).$$

(i) *In the case when X is a simplicial complex and f is simplicial map, the number $L(f)$ can be defined as $\sum (-1)^q \mathrm{tr}\,(f_q : C_q \longmapsto C_q)$ where f_q acts on the space of q-dimensional chains.*

(ii) *In the case when f is a differentiable map of differentiable manifold with isolated fixed points, we have $L(f) = \sum_p i_p(f)$, where the sum is over the set of fixed points and the index $i_p f$ of the map f at p is defined as the index of the vector field $x - f(x)$ (in a local chart $U \subset \mathbb{R}^n$).*

(iii) *If $L(f) \neq 0$, then f has fixed point.*

Proof. (i) The number $L(f)$ has the same properties as the Euler characteristic; and equals to it in the case $f = id$. (ii) This point is the analogue of the Poincaré–Hopf theorem. (iii) This point is proved by means of a fine simplicial approximation of the map. For more details, see [**Spa**]. □

§3 Homotopy Theory

We shall need also little of the homotopy theory and we present here some its fundamental notions. Two continuous maps $f, g : X \longmapsto Y$ are **homotopically equivalent** iff there exists a continuous map $F : X \times [0,1] \longmapsto Y$ such that $F(x,0) = f(x)$, $F(x,1) = g(x)$.

The **fundamental group** $\pi_1(X) = \pi_1(X, x)$, $x \in X$, is the space of homotopy classes of maps $(\mathbb{S}^1, 1) \longmapsto (X, x)$, where $\mathbb{S}^1$ is a circle $\{z \in \mathbb{C} : |z| = 1\}$. This set has a natural structure of a group; multiplication is a composition of loops (after a change of parametrization). Often the fundamental group is not abelian, but we have the following identity

$$H_1(X, \mathbb{Z}) = \pi_1(X)/[\pi_1(X), \pi_1(X)],$$

where $[a, b]$ denotes the commutator $aba^{-1}b^{-1}$.

The set of homotopy equivalence classes of maps $\mathbb{S}^k \longmapsto X$ is denoted by $\pi_k(X)$. It forms an abelian group (for $k > 1$), called the kth **homotopy group** of X.

The space X is called the **Eilenberg–MacLane space** $K(\pi, n)$ iff $\pi_n(X) = \pi$ and $\pi_i(X) = 0$, $i \neq n$. For example, $K(\mathbb{Z}, 1) = \mathbb{S}^1$, $K(\mathbb{Z}_2, 1) = \mathbb{RP}^\infty$, $K(\mathbb{Z}, 2) = \mathbb{CP}^\infty$.

A map $p : X \longmapsto Y$, $Y \subset X$ is called the **retraction** iff $f|_Y = id$. We say that $Y \subset X$ is a **deformation retract** of X iff there is a homotopy $X \times [0,1] \longmapsto X$ which is identity on Y and which joins the identity map on X with some retraction $p : X \longmapsto Y$.

The next lemma, though easy to prove, will be frequently used.

3.31. Lemma (Prolongation of homotopy). *Let $p : E \longmapsto B$ be a fibre bundle with fiber F and let X be some CW-complex. If there are two continuous maps*

$$G : X \times \{0\} \longmapsto E, \ g : X \times [0,1] \longmapsto B,$$

which agree, i.e., $p \circ G(x,0) = g(x,0)$, then there exists a map $\widetilde{G} : X \times [0,1] \longmapsto E$ which prolongs G and agrees with g, $p \circ \widetilde{G} = g$. Moreover, any two such lifts are homotopically equivalent.

3.32. Long sequences associated with a fibration. Let $E \longmapsto B$ be a (locally trivial) fibre bundle with the fiber F. We have the *long exact sequence of the homotopy groups*

$$\dots \pi_k(F) \longmapsto \pi_k(E) \longmapsto \pi_k(B) \longmapsto \pi_{k-1}(F) \longmapsto \dots$$

In general, there is no any analogous sequence of the homology groups. There are such sequences in the cases when:

(i) $F = \mathbb{S}^m$

$$\dots H_{k-m}(B) \longmapsto H_k(E) \longmapsto H_{k-m-1}(B) \longmapsto H_{k-1}(E) \longmapsto \dots,$$

called the **Gysin sequence**, and

(ii) $B = \mathbb{S}^n$

$$\dots H_k(F) \longmapsto H_k(E) \longmapsto H_{k-n}(F) \longmapsto H_{k-1}(F) \longmapsto \dots,$$

called the **Wang sequence**.

Fore more details we refer the reader to [**Spa**].

Chapter 4

Topology and Monodromy of Functions

This chapter is devoted to the topological invariants of analytic and algebraic functions and sets. We begin with description of the homology groups of non-singular levels of germs of holomorphic functions with isolated singularity (the Milnor theorem) and with introduction of the Milnor bundle over a punctured disc. The action of the generator of the fundamental group of the base on a fiber defines the monodromy operator in the homologies of the fiber. Its action is described by the Picard–Lefschetz formula.

In the case of a deformation of singularity one has a Milnor bundle over complement of the discriminant set. The fundamental group of such complement is the braid group. It also acts on the homology of the fiber. We describe the monodromy group arising in this way.

To obtain further invariants of the monodromy, one resolves the singularities, i.e., replaces the singular hypersurface by an union of smooth divisors with normal crossings. This theorem (due to Hironaka) holds in the local as well as in the algebraic situation and in the case of non-isolated singularity. Sometimes (in the case of multiple divisors) one needs to apply certain base change and the normalization of the fiber space. We describe (without proofs) the desingularization process and the semi-stable reduction.

We present the Clemens' construction, i.e., the contraction of the fiber space to the resolved singular fiber. We use it in the proof of the fundamental result, the monodromy theorem (about eigenvalues of the monodromy operator and dimensions of its Jordan cells), and in derivation of the formula for the zeta function of the monodromy.

In the appendix (Section 3) we collect basic properties of such objects like the root systems, semi-simple Lie algebras and finite groups generated by reflections. They have analogies with the monodromy groups.

§1 Topology of a Non-singular Level

Let $f : (\mathbb{C}^n, 0) \longmapsto (\mathbb{C}, 0)$ be a germ of holomorphic function with isolated critical point at 0. We assume that f is a polynomial. We are going to calculate the homology groups of the surfaces $f(x) = z$ intersected with a small ball. Firstly we formulate some preparatory lemmas.

H. Żołądek, *The Monodromy Group*, Monografie Matematyczne 67,
https://doi.org/10.1007/978-3-031-91270-2_4

4.1. Lemma. *There exists a perturbation of f of the form $f_\epsilon = f + \epsilon g$ (with linear g) such that the function f_ϵ has only non-degenerate critical points with different critical values. In other words, f_ϵ is a Morse function.*

Proof. Consider the map $Df : \mathbb{C}^n \longmapsto \mathbb{C}^n$. A point x_0 is critical for Df iff $\det D^2 f(x_0) = 0$. Thus critical points of Df with value 0 are degenerate critical points of f. The Sard's theorem says that almost all values $L \in \mathbb{C}^n$ are noncritical for the map Df. Therefore for the function

$$f_1(x) = f(x) - L \cdot x,$$

with small noncritical L, 0 is not critical value for Df_1. This means that all critical points of f_1 are different and of Morse type.
Now it is enough to change little L in such a way that we remain in the non-degenerate case but with different critical values. □

4.2. Lemma. *There exists $\rho > 0$ such that the spheres $\mathbb{S}_r = \{|x| = r\} \subset \mathbb{C}^n = \mathbb{R}^{2n}$, $0 < r \leq \rho$, intersect the set $f^{-1}(0)$ transversally.*

Proof. The statement of Lemma 4.2 is equivalent to the fact that the function

$$h : x \longmapsto |x|^2,$$

restricted to the set $f^{-1}(0) \setminus 0$, does not have critical points inside a small ball $\mathbb{B}_\rho$ with center at 0 (and radius ρ).
The property that h has critical point at x is a property for implicit extremum in $\mathbb{R}^{2n}$ and can be expressed as the property of vanishing of some minors of certain matrix. More precisely, let $f = f_1 + \mathrm{i} f_2$, $\mathrm{i} = \sqrt{-1}$. Then $f^{-1}(0)$ is given by the equations $f_1 = f_2 = 0$. Let also $y_1, \ldots, y_{2n}$ denote the real coordinates. Then the matrix

$$\begin{bmatrix} \partial h/\partial y_1 & \ldots & \partial h/\partial y_{2n} \\ \partial f_1/\partial y_1 & \ldots & \partial f_1/\partial y_{2n} \\ \partial f_2/\partial y_1 & \ldots & \partial f_2/\partial y_{2n} \end{bmatrix}$$

should have rank less or equal to 2. Because h, f_1, f_2 are polynomials, we obtain a system of algebraic equations defining a real algebraic subset V of $\mathbb{R}^{2n}$.
If one shows that this set consists of finite number of points, then one obtains the thesis of Lemma 4.2. Indeed, one can choose ρ such that the punctured ball $\mathbb{B}_\rho \setminus 0$ does not contain points from V.
The set V is 0-dimensional; otherwise the function h would be constant along some its curve tending to the origin.
If it had a sequence of points accumulating at the origin, then one would choose a subsequence tending to zero with some definite asymptotic direction. We can assume that this asymptotic direction is such that $y_1 > 0, \ldots, y_{2n} > 0$. In particular, the complexification $V^{\mathbb{C}} \subset (\mathbb{R}^{2n})^{\mathbb{C}}$ should have complex dimension ≥ 1.
Suppose firstly that $\dim_{\mathbb{C}} V^{\mathbb{C}} = 1$, i.e., $V^{\mathbb{C}}$ is a complex analytic curve with definite asymptotic direction and having sequence of real points tending to the origin.

We can represent $V^{\mathbb{C}}$ in the form of Puiseux expansion, i.e., the expansion of $y_2, y_3, \ldots$ in powers of $y_1^{1/\nu}$ (see 2.25(b)). Thus, putting $y_1 = t^\nu$, we get $y = (t^\nu, \sum_j a_{2,j}t^j, \ldots, \sum_j a_{2n,j}t^j)$, $t \in (\mathbb{C}, 0)$. $V^{\mathbb{C}}$ contains infinitely many real points; we can assume that these points correspond to real parameters t. We claim that all the coefficients $a_{i,j}$ are real. Indeed, if some $\mathrm{Im} a_{i,j} \neq 0$ with smallest j, then $\mathrm{Im} y_i = \mathrm{Im} a_{i,j} t^j(1 + \ldots) \neq 0$ for small $t > 0$. This means that V is a real 1-dimensional analytic curve.
Suppose now that $d = \dim_{\mathbb{C}} V^{\mathbb{C}} > 1$. We claim that:

There exists a (real) algebraic subset $V_1 \subset V$ with $\dim_{\mathbb{C}} V_1^{\mathbb{C}} < d$ and containing points arbitrarily close to 0.

Indeed, let $g_1, \ldots, g_k$ be the generators of the ideal $I(V)$ of polynomials vanishing on V. Let $\mathrm{Sing}V = \{\mathrm{rank}(\mathrm{d}g_1, \ldots, \mathrm{d}g_k) < d\}$ be the set of singular points of V. It is a proper algebraic subset of V. If the point 0 is not isolated in $\mathrm{Sing}V$, then we put $V_1 = \mathrm{Sing}V$. Assume then the opposite.
Consider the function $h = |y|^2$. We know that $\mathrm{rank}(\mathrm{d}g_1, \ldots, \mathrm{d}g_k) = d$. But it turns out that $\mathrm{rank}(\mathrm{d}g_1, \ldots, \mathrm{d}g_k, \mathrm{d}h) = d + 1$ for $y \neq 0$ and close to 0.
Indeed, the complexification of h on $V^{\mathbb{C}} \setminus 0$ is constant on each component of its critical set and hence $h|_{V\setminus 0}$ has finitely many critical values.
Thus the varieties $V \cap \mathbb{S}_\rho$, $\rho > 0$ and small, are smooth; infinitely many of them are nonempty. Consider the functions $\varphi_i = y_i|_{V\cap\mathbb{S}_\rho}$, $V \cap \mathbb{S}_\rho \neq \varnothing$. Each φ_i has a maximum in the domain $y_i > 0$, i.e., a critical point. Therefore the sets $W_i = \{y \in V : \mathrm{rank}(\mathrm{d}g_1, \ldots, \mathrm{d}g_k, \mathrm{d}|y|^2, \mathrm{d}y_i) \leq d + 1\}$ are nonempty.
If all $W_i = V$, then this would mean that all the differentials $\mathrm{d}y_i$ belong to the space generated by $\mathrm{d}g_j, \mathrm{d}|y|^2$ which has the dimension $d + 1 < 2n$. Thus some $W_i \neq V$ and we put $V_1 = W_i$.
Repeating this, we find a variety $\widetilde{V} \subset V$ with $\dim \widetilde{V}^{\mathbb{C}} = 1$ and containing points arbitrarily close to 0.
More details are in the Milnor's book [**Mil2**]. □

Remark. Note also that the assumption (of Lemma 2) that the function f is a polynomial can be weakened. We can assume that f is analytic. Then one would deal with local real analytic varieties. In the same way one proves that an analytic variety in $\mathbb{R}^m$ has only finitely many local components and that any two points on such a component can be joined by a curve which consists of finitely many smooth analytic pieces. (Here, by an analytic subset of $\mathbb{R}^m$ we mean such subset that near any point $y \in \mathbb{R}^m$ it is given as a common zero of a system of functions analytic near y).

4.3. Corollary.

(a) *If, in the assumptions of the previous lemma, we fix the small sphere $\mathbb{S}_r$ and perturb little the function f to $\tilde{f}$, then also $\tilde{f}^{-1}(z)$ is transversal to $\mathbb{S}_\rho$ for small $|z| \leq \epsilon$, $\epsilon = \epsilon(\rho)$. In particular, the sets $\tilde{f}^{-1}(z) \cap \mathbb{B}_\rho$, z not critical, are regular manifolds with smooth boundary.*

(b) *he independence of* $\mathrm{d}f_1, \mathrm{d}f_2, \mathrm{d}h$ *implies that the sets* $(\mathbb{B}_\rho \backslash \{f=0\}) \cap \{\arg f = \theta\}$, $\theta \in [0, 2\pi)$, *are regular manifolds.*

4.4. Remark. Repeating the proof of Lemma 4.2 one obtains the following result: *The function* $\mathbb{S}_\rho \cap \{f \neq 0, \arg f = \theta\} \ni x \longmapsto |f(x)|$ *does not have critical points for* ρ *and* $|f|$ *small enough and any* $\theta \in [0, 2\pi]$.

Indeed, the proof of this statement leads to the real algebraic variety $\{\mathrm{rank}\,(\mathrm{d}\arg f, \mathrm{d}|f|, \mathrm{d}h) = 2\}$ and to showing that 0 is its isolated point.

4.5. Definition (Milnor bundle). There are two definitions of the Milnor bundle: one as in the second volume of the book of V. Arnold, A. Varchenko and S. Gusein-Zade [**AVG**] and the original Milnor's one as in his book [**Mil2**]. Firstly, we present the Arnold–Varchenko–Gusein-Zade's definition.

Let f be as above, $\mathbb{D}_\epsilon = \{|z| < \epsilon\}$ and $\mathbb{B} = \mathbb{B}_\rho = \{|x| \leq \rho\}$, with ρ like in the thesis of Lemma 4.2. Denote

$$S = \mathbb{D}_\epsilon \setminus 0, \quad V = \mathbb{B} \cap f^{-1}(S).$$

The map

$$V \overset{f}{\longmapsto} S \tag{4.1}$$

is called the **Milnor fibration** associated with the function f. Its fiber over $z \in S$ is denoted by V_z. By Corollary 4.3(a), V_z is a regular $(2n-2)$-dimensional manifold with regular boundary.

The Milnor's definition is following. Take a small sphere $\mathbb{S}_\rho \subset \mathbb{C}^n$ and let $K = \mathbb{S}_\rho \cap f^{-1}(0)$. If ρ is small, then K is a smooth manifold (see Lemma 4.2). Take a tubular neighborhood T of K in $\mathbb{S}_\rho$: $T = \{x \in \mathbb{S}_\rho : |f(x)| < \epsilon_1\}$, $\epsilon_1 << \epsilon$. The map

$$\Phi : \mathbb{S}_\rho \setminus T \longmapsto \mathbb{S}^1, \quad \Phi(x) = \frac{f(x)}{|f(x)|} \tag{4.2}$$

defines a fibration, also called the **Milnor fibration**. By Corollary 4.3(b) the fibers are regular manifolds with boundary.

We shall distinguish the above two fibrations by their numbering, (4.1) or (4.2). We shall work mostly with the fiber bundle (4.1).

4.6. Proposition. *The maps* (4.1) *and* (4.2) *really define fiber bundles. Moreover, they are equivalent in the following sense. For any fiber* V_z *of the bundle* (4.1) *there is a diffeomorphism transforming it to the fiber* $\Phi^{-1}(\theta)$, $\theta = \arg z$, *of the bundle* (4.2). *In other words, the bundle* (4.1) *restricted to some small circle* $\{|z| = \epsilon\} \subset S$ *is isomorphic to the bundle* (4.2).

Proof. We shall prove the local triviality only for the fibration (4.1). By Corollary 4.3(a) the fibers V_z and $V_{z'}$ are homeomorphic for z' close to z. (In the proof of local triviality of the bundle (4.2) one uses Corollary 4.3(b)). However, this observation does not provide a proof because it says nothing about this homeomorphism. The precise proof goes through phase flows of vector fields.

Let us fix a point $z \in S$ and let the points z' belong to a small neighborhood $\mathcal{U}$ of z (in S). We move from the point z to the points z' along trajectories of the vector field $\mathrm{e}^{\mathrm{i}\alpha} \cdot \frac{\partial}{\partial z} = \cos\alpha \cdot \partial/\partial_{z_1} + \sin\alpha \cdot \partial/\partial z_2$. Here we identify $T_{z'}\mathbb{C}$ with $\mathbb{C}_{z_1+\mathrm{i}z_2}$ and with $\mathbb{R}^2_{z_1,z_2}$ and we treat its elements either as complex numbers or as vectors. The angle $\alpha = \arg(z'-z)$. Thus $z' = g^t_{\mathrm{e}^{\mathrm{i}\alpha}\partial/\partial z}$, $t = |z'-z|$ (the flow map after the time t).

We construct a vector field $\boldsymbol{V} = \boldsymbol{V}(x)$ in $f^{-1}(\mathcal{U}) \subset V$ such that $\pi_* \boldsymbol{V} = \frac{\partial}{\partial z}$. More precisely, identifying $T_x V$ with $\mathbb{C}^n$, we represent $\boldsymbol{V}(x)$ as a collection of complex numbers, $\boldsymbol{V}(x) = (v_1(x), \ldots, v_n(x)) \in \mathbb{C}^n$. Then $\mathrm{e}^{\mathrm{i}\alpha}\boldsymbol{V}(x)$ is also a vector from $\mathbb{C}^n$, representing a real vector field in $\mathbb{R}^{2n}$.

The homeomorphism $g : \mathcal{U} \times V_z \longmapsto f^{-1}(\mathcal{U})$ is defined as follows

$$g(z,x) = x; \quad g(z',x) = g^t_{\mathrm{e}^{\mathrm{i}\alpha}\boldsymbol{V}}(x), \ z' \neq z.$$

The field $\boldsymbol{V}(x)$ should satisfy also another condition; it should be tangent to the boundary of V, i.e., to the sphere $\mathbb{S}_\rho$. So, we have the conditions $\dot{f} = 1$, $\dot{h} = 0$ (where $h(x) = x \cdot \bar{x}$, $x \cdot y = \sum x_i y_i$), i.e.,

$$\boldsymbol{V}(x) \cdot \nabla f(x) \equiv 1, \quad \mathrm{Re}(\boldsymbol{V}(x) \cdot \bar{x}) = 0.$$

The second condition is essential only for $n \geq 2$.

Locally, near a given point $x_0 \in V_z$, it is easy to construct $\boldsymbol{V}(x)$ We can assume that $f = u_1$ is a first coordinate from a local holomorphic coordinate system $u_1, \ldots, u_n$. We have $\boldsymbol{V}(u) = (1, v_2(u), \ldots, v_n(u))$ and $\dot{h} = (\partial h/\partial u) \cdot \boldsymbol{V}(u) + (\partial h/\partial \bar{u}) \cdot \overline{\boldsymbol{V}(u)} = 0$ is an additional condition for v. Because $\mathrm{d}h$ is independent on $\mathrm{d}u_1$ and $n \geq 2$, one can always find $v_2, \ldots, v_n$ satisfying it.

Next we take a covering $\{W_\beta\}$ of the set $f^{-1}(\mathcal{U})$ such that in each W_β one has a vector field $\boldsymbol{V}_\beta$ satisfying the needed properties. Take a partition of unity $\{\phi_\beta\}$ associated with this covering. Then the vector field $\boldsymbol{V}(x) = \sum \phi_\beta(x)\boldsymbol{V}_\beta(x)$ is good.

The reader can notice that the field $\boldsymbol{V}(x)$ is not holomorphic; (i.e., not of the form $\sum v_i(x)\partial/\partial_{x_i}$ with holomorphic components $v_i(x)$). Generally, the Milnor bundle is not holomorphic. It is only C^∞-bundle.

Now we give the idea, why the fibers of the bundles (4.1) and (4.2) are diffeomorphic. Note that they have the same (real) dimensions: V_z has complex dimension $n-1$ and $\Phi^{-1}(\theta)$ is a hypersurface in $(2n-1)$-dimensional sphere. Assume that $z = \mathrm{e}^{\mathrm{i}\theta}$.

Here also we use vector fields. We construct a (real) vector field $\boldsymbol{W}(x)$ satisfying the following properties

$$\boldsymbol{W}(x) \cdot \nabla \ln f(x) > 0, \quad \mathrm{Re}(\boldsymbol{W}(x) \cdot \bar{x}) > 0.$$

If $x(t)$ is an integral curve of $\boldsymbol{W}$, i.e., $\dot{x}(t) = \boldsymbol{W}(x(t))$, and $\theta = \arg f(x) = \mathrm{Im} \ln f(x)$, then $\frac{\mathrm{d}}{\mathrm{d}t}(\ln|f|) + \mathrm{i}\dot{\theta} = \frac{\mathrm{d}}{\mathrm{d}t}[\ln f] = \boldsymbol{W} \cdot \nabla \ln f > 0$, i.e.,

$$\dot{\theta} = 0, \quad \frac{\mathrm{d}}{\mathrm{d}t} \ln|f| > 0$$

and
$$\frac{\mathrm{d}}{\mathrm{d}t}(|x|^2) = 2\,\mathrm{Re}(\dot{x}\cdot\bar{x}) > 0.$$

This means that the phase curve $x(t)$ lies in one fiber $\Phi^{-1}(\theta)$ of the map $x \overset{\Phi}{\longmapsto} f(x)/|f(x)|$ and it eventually achieves the sphere S_ρ. In this way we define a diffeomorphism between $f^{-1}(z) \cap \mathbb{B}_\rho$ and $\mathbb{S}_\rho \cap \{\arg f = \theta,\ |f| \geq |z|\} = \Phi^{-1}(\theta) \cap \{|f| \geq |z|\}$.
The vector field $\boldsymbol{W}(x)$ is constructed in an analogous way as the vector field $\boldsymbol{V}(x)$ above. Firstly, we construct it locally and then we use some partition of unity to glue together these local fields to a global field in $\mathbb{B}_\rho \setminus \{f = 0\}$.
It remains to show that the manifolds with boundary $\Phi^{-1}(\theta) \cap \{|f| \geq |z|\}$ and $\Phi^{-1}(\theta)$ are diffeomorphic in the sphere $\mathbb{S}_\rho$; recall that in $\Phi^{-1}(\theta)$ we have $|f(x)| \geq \epsilon_1$. This is also done using vector fields. Because the vector field $\nabla(\ln|f||_{S_\rho})$ is regular in $\Phi^{-1}(\theta)$ (see Remark 4.4), we can use it to the construction of the needed diffeomorphism. □

Remark. The above proof of the local triviality of the Milnor's fibration works in the more general case. If $f : E \longmapsto B$ is a smooth map between real and smooth manifolds, such that the rank of Df is constant, then it defines a locally trivial fibration. This result belongs to C. Ehresmann [**Ehr**].

Let $\tilde{f}$ be a Morse perturbation of f. Let $x_1, \ldots, x_\mu$, $\mu = \mu(f)$, be its set of critical points and let $z_1, \ldots, z_\mu$ denote the corresponding critical values.
Denote $\widetilde{S} = \mathbb{D}_\epsilon \setminus \{z_1, \ldots, z_\mu\}$ and $\widetilde{V} = \mathbb{B} \cap \tilde{f}^{-1}(\widetilde{S})$. The map
$$\tilde{f} : \widetilde{V} \longmapsto \widetilde{S}$$
is also called the **Milnor fibration** associated with the perturbation $\tilde{f}$. The fibers are denoted by $\widetilde{V}_z$. It is also a locally trivial fibration (of class C^∞, but not analytic). (In this section we shall denote by x the arguments of functions and by z their values).

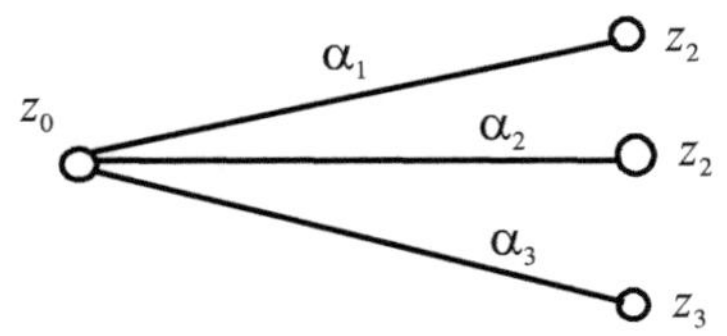

Figure 4.1: Distinguished paths in the target plane

4.7. Definition (Vanishing cycles). Let as before f be a germ of function with isolated critical point, $\tilde{f}$ be its Morse perturbation and $x_1, \ldots, x_\mu$, $\mu = \mu(f)$, be its set of critical points of $\tilde{f}$ with critical values $z_1, \ldots, z_\mu$.

Choose $z_0 \neq z_1, \dots, z_\mu$ and join it with the critical values z_i by means of certain non-intersecting paths $\alpha_i : z = u_i(t)$, $u_i(0) = z_i$, $u_i(1) = z_0$ (see Fig. 4.1).
We want to calculate the homology groups of the space $V_{z_0} = f^{-1}(z_0) \cap \mathbb{B}$. If $\tilde{f}$ is a small perturbation of f, then this space is homeomorphic to the space $\widetilde{V}_{z_0} = \tilde{f}^{-1}(z_0) \cap \mathbb{B}$ (one is a small deformation of the other).
In a neighborhood of any critical point x_i we can express the perturbed function in the form

$$\tilde{f} = z_i + \sum y_j^2.$$

From Chapter 1 we know that there vanishes some $(n-1)$-dimensional cycle. In particular, we get a 1-parameter family $\Delta_j(t)$ of cycles in $\widetilde{V}_{u_j(t)}$ defined as follows. If t is close to 0, then

$$\Delta_j(t) = \sqrt{u_j(t) - z_j} \cdot \mathbb{S}^{n-1},$$

where $\mathbb{S}^{n-1} = \left\{\sum y_j^2 = 1,\ \mathrm{Im} y_j = 0\right\}$. We see that $\Delta_j(0) = \{z_j\}$.
Using Lemma 3.31 we prolong the local homotopy $(\mathbb{R}_+, 0) \times \mathbb{S}^{n-1} \longmapsto V$, $\{t\} \times \mathbb{S}^{n-1} \longmapsto \Delta_j(t)$, to a map $D_j : [0,1] \times \mathbb{S}^{n-1} \longmapsto V$, $D_j(\{t\} \times \mathbb{S}^{n-1}) = \Delta_j(t)$. We denote by Δ_j the cycle $\Delta_j(1)$. It lies in $\widetilde{V}_{z_0}$.
The system $\Delta_1, \dots, \Delta_\mu$ is called the **distinguished system of vanishing cycles**.

4.8. Milnor theorem. *The distinguished system of vanishing cycles generates a basis of the reduced homology group $\widetilde{H}_{n-1}(\widetilde{V}_{z_0}, \mathbb{Z})$. The other groups are zero.*

Proof. The next lemma will be proved later.

Lemma. *The space $X = f^{-1}(\mathbb{D}_\epsilon) \cap \mathbb{B}$ is contractible.*

Let, as before, $\tilde{f}$ be a small Morse perturbation of f. Let $\widetilde{X} = \tilde{f}^{-1}(\mathbb{D}_\epsilon) \cap \mathbb{B}$; it is diffeomorphic to the space X (because it has the "same" boundary).
Let $z_1, \dots, z_\mu$ be the critical values and $\alpha_1, \dots, \alpha_\mu$ be the paths joining the base point z_0 with z_i. Denote

$$A = \bigcup_i \alpha_i.$$

A is a deformation retract of $\mathbb{D}_\epsilon$. Let

$$Y = \tilde{f}^{-1}(A) \cap B.$$

It is a deformation retract of $\widetilde{X}$ and has trivial homotopy type.
We delete from A the critical values, $A' = A \setminus \{z_1, \dots, z_\mu\}$ and we delete from Y the critical fibers, $Y' = Y \setminus \bigcup \widetilde{V}_{z_i}$. Y' over A' is trivial, $Y' \approx \widetilde{V}_{z_0} \times A'$. Thus Y' has the homotopy type of $\widetilde{V}_{z_0}$.
The space Y is homotopically equivalent to the space obtained from $\widetilde{V}_{z_0}$ by gluing the vanishing cycles Δ_i by balls K_i. Indeed, by Proposition 1.8, the part of Y near the critical point z_1 can be continuously deformed to $\widetilde{V}_{u_1(t_0)} \cup \mathbb{D}^n$, where the point

$u_1(t_0) \in \alpha_1$ is close to z_1. This deformation can be prolonged to the remaining part of Y above the arc $\alpha_1' = [z_1, u_1(t_0)] \subset \alpha_1$ with the image in $\widetilde{V}_{u_1(t_0)}$. We do the same above the analogous arcs α_i' near z_i. We obtain that Y is deformed to its part above $\overline{A \setminus \bigcup \alpha_i'}$ and to μ balls $\mathbb{D}^n$ glued along $\Delta_i(t_0)$. Of course, the latter can be squeezed to the fiber above z_0 and to the glued balls, which we denote by $K_1, \ldots, K_\mu$.

The homotopy equivalence map, from the disjoint union $\widetilde{V}_{z_0} \sqcup \bigsqcup_i K_i$ to Y, is identity on the first component. If we represent each K_i as $\bigcup_{t\in[0,1]} t \cdot \mathbb{S}^{n-1}$, then the point $t \cdot a$ is sent to the point $D_j(t,a)$ (under the identification of the cycles $\Delta_j(t)$ with $D(\{t\} \times \mathbb{S}^{n-1}$, see Definition 4.7).

Let us calculate the homology groups of $Y' \approx \widetilde{V}_{z_0}$. We use the long exact sequence of homology groups of the pair (Y, Y'). We have

$$\ldots \longmapsto H_{k+1}(Y) \longmapsto H_{k+1}(Y, Y') \longmapsto \widetilde{H}_k(Y') \longmapsto \widetilde{H}_k(Y) \longmapsto \ldots,$$

where the tildes denote the reduced homology groups.

Here $\widetilde{H}_j(Y) = 0$ (as Y is homotopically trivial). Therefore $\widetilde{H}_k(Y') \approx H_{k+1}(Y, Y')$. But the pair (Y, Y') is homotopically equivalent to the union of pairs $(K_i, \partial K_i) \approx (\mathbb{D}^n, \partial \mathbb{D}^n)$. We know that $H_{k+1}(K_i, \partial K_i) = 0$, if $k \neq n-1$, and $= \mathbb{Z}$ for $k = n-1$. Hence, $\widetilde{H}_k(Y') = 0$, if $k \neq n-1$, and $= \mathbb{Z}^\mu$ for $k = n-1$. The generators of $\widetilde{H}_{n-1}(Y')$ arise from the images of the relative cycles (K_i, Δ_i) under the "boundary" homomorphism, i.e., from the cycles Δ_i. □

Proof of the lemma. The set $V_0 = f^{-1}(0) \cap \mathbb{B}_\rho$ is contractible, because it is transversal to the spheres $\mathbb{S}_r$. For example, we can use the vector field $-\nabla(|x|^2|_{f=0})$. The disc $\mathbb{D}_\epsilon$ is contractible to 0. Therefore, the proof will be finished, when we construct a prolongation of the latter contraction to a contraction $X \longmapsto V_0$. Because the map $X \longmapsto \mathbb{D}_\epsilon$ is not a fibration, we cannot use directly Lemma 3.31. We must work a little more.

Take two sequences $\rho > r_1 > \ldots > 0$, $\epsilon_0 = \epsilon > \epsilon_1 > \ldots > 0$, such that the level surfaces $f = z$ are transversal to the spheres $\mathbb{S}_{r_i}$ for $|z| < \epsilon_i$. We restrict f to $\mathbb{B} - \mathbb{B}_{r_i}$; above $\mathbb{D}_{\epsilon_i}$ they define fibrations $E_i \longmapsto \mathbb{D}_{\epsilon_i}$. Because the bases are contractible, these bundles are trivial. We can assume that the trivializations are compatible, i.e., the maps $E_i \longmapsto \mathbb{D}_{\epsilon_i} \times (V_0 \cap (\mathbb{B} \setminus \mathbb{B}_{r_i}))$ and $E_{i+1} \longmapsto \mathbb{D}_{\epsilon_{i+1}} \times (V_0 \cap (\mathbb{B} - \mathbb{B}_{r_{i+1}}))$ are equal at $E_i \cap E_{i+1}$ above $\mathbb{D}_{\epsilon_{i+1}}$.

The retraction $X \longmapsto V_0$ is constructed as follows. Let $p_1(t)$ be the deformation retraction of $\mathbb{D}_\epsilon$ to $\mathbb{D}_{\epsilon_1}$ (along the rays). We lift it to a deformation retraction $\tilde{p}_1(t)$ of $X = f^{-1}(\mathbb{D}_\epsilon)$ to $f^{-1}(\mathbb{D}_{\epsilon_1})$ in such a way that it coincides with $p_1(t) \times id$ in the fiber space E_1. Next, we lift the deformation retraction of $\mathbb{D}_{\epsilon_1}$ to $\mathbb{D}_{\epsilon_2}$ to a deformation retraction of $f^{-1}(\mathbb{D}_{\epsilon_1})$ to $f^{-1}(\mathbb{D}_{\epsilon_2})$ (naturally in E_2). We repeat this infinitely many times. Every point from $X \setminus V_0$ either eventually falls into one of the bundles E_i, where the retraction is well defined, or it tends to 0.

We compose all these retractions to one deformation retraction of X to V_0. □

4.9. Remarks. (a) The reason to use the reduced homologies is simple. If $f(x) = x^{\mu+1}$, then the multiplicity is μ and the sets $V_z = \{x_i = \zeta^i x_0 : i = 0, 1, \ldots, \mu\}$, $\zeta = e^{2\pi i/(\mu+1)}$, $z = x_0^{\mu+1} \neq 0$, consist of $\mu + 1$ elements. The reduced homology group $\widetilde{H}_0(V_z)$ is identified with $\mathbb{Z}^\mu = \{(m_0, \ldots, m_\mu : \sum m_i = 0\} \subset \mathbb{Z}^{\mu+1}$. The reduced cohomology group $\widetilde{H}^0(V_z)$ is identified with the quotient group $\mathbb{Z}^{\mu+1}/(1, \ldots, 1)\mathbb{Z}$.
(b) If one worked more on the above proof, then one would obtain the homotopy equivalence of the non-singular level surface $\widetilde{V}_{z_0}$ with the bucket $\mathbb{S}^{n-1} \vee \ldots \vee \mathbb{S}^{n-1}$ of μ spheres (joined together at their base points). It is the original result of Milnor. In [**Mil2**] Milnor proves this homotopy equivalence using the Hurewicz theorem (to get an isomorphism of $H_{n-1}(V_z)$ with the homotopy group $\pi_{n-1}(V_z)$ for $n \geq 2$)) and a Whitehead theorem (to show that the map $\mathbb{S}^{n-1} \vee \ldots \vee \mathbb{S}^{n-1} \longmapsto V_z$, induced by the isomorphism of the homotopy and homology groups, is a homotopy equivalence).

§2 Picard–Lefschetz Formula

The Picard–Lefschetz formula describes the change in homologies of the level surface of a holomorphic function induced by variation of the value of the function around some its critical value. Firstly we present this change in two examples (of Morse critical point and of family of elliptic curves).

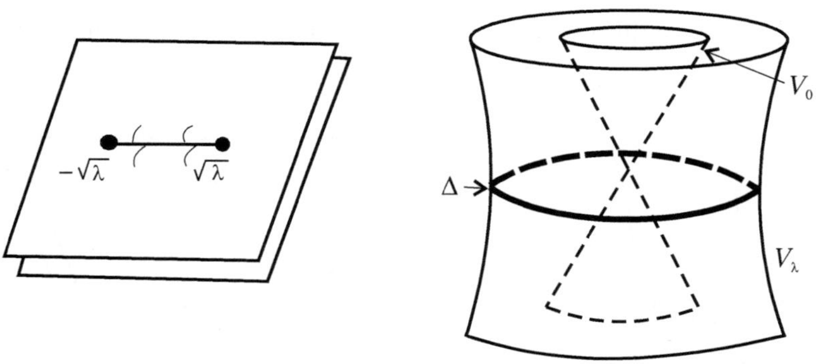

Figure 4.2: Singular and non-singular levels

4.10. Example. Let $f(z, w) = z^2 + w^2$, $z, w \in \mathbb{C}^2$. Let $V_\lambda = \{f(z, w) = \lambda\}$ denote the level surfaces of the function f. V_λ is the Riemann surface of the function $w = \sqrt{\lambda - z^2}$ (see Fig. 4.2).
Let $\lambda(t) = e^{2\pi i t}\alpha$, $0 \leq t < 1$, $\alpha > 0$. We want to see what happens with the surfaces $V_{\lambda(t)}$ in the process of deformation. They are presented at Fig. 4.3. The ramification points rotate with velocity two times smaller than the velocity of rotation of λ.

We construct a family of diffeomorphisms $\Gamma_t : V_{\lambda(0)} \longmapsto V_{\lambda(t)}$. Let $\chi(r)$ be a bump function such that $\chi(r) = 1$ for $0 \le r \le 2\sqrt{\alpha}$ and $\chi(r) = 0$ for $r > 3\sqrt{\alpha}$. We put

$$z(t) = \mathrm{e}^{\pi \mathrm{i} t \chi(|z|)} z.$$

It is the action of the family of diffeomorphisms on the z-variable. The w-variable changes continuously and accordingly to the formula

$$w(t) = \pm\sqrt{\lambda(t) - z^2(t)}.$$

We put $\Gamma_t : (z, w) \longmapsto (z(t), w(t))$. We see that, if z, w are small, then $(z, w)(t) = \mathrm{e}^{\pi \mathrm{i} t}(z, w)(0)$ and, if they are large, then $(z, w)(t) = \text{const}$.

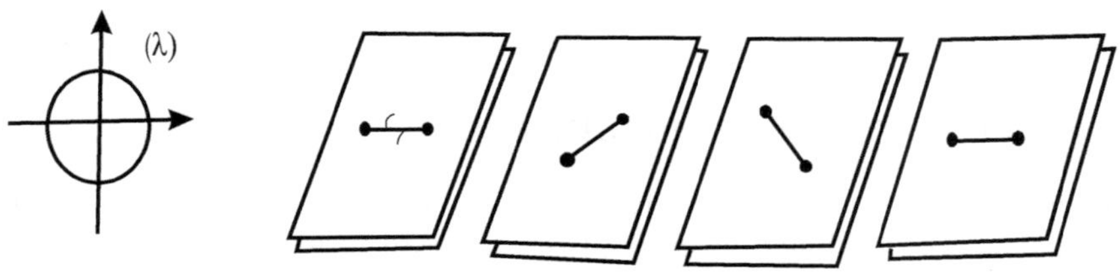

Figure 4.3: Deformation of the non-singular level

The map $h = \Gamma_1$ is a diffeomorphism of V_α; it induces some maps in the homology groups.

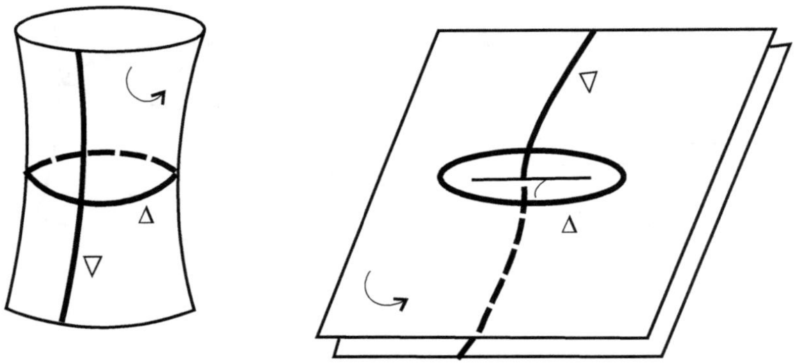

Figure 4.4: Absolute and infinite cycles

We have here two homology groups to deal with: the usual homologies $H_1(V_\alpha, \mathbb{Z}) \approx \mathbb{Z}$ and the homology group with closed support $H_1^{\mathrm{cl}}(V_\alpha, \mathbb{Z}) \approx \mathbb{Z}$. The generator of the first group is the (compact) cycle Δ and the generator of the second group

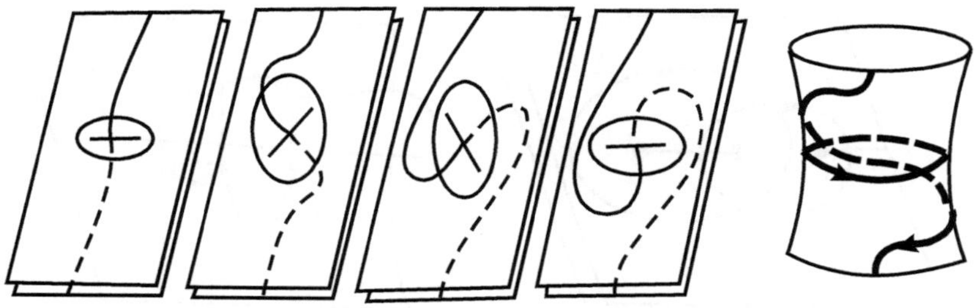

Figure 4.5: Deformation of the cycles

is the infinite cycle ∇ as at Fig. 4.4. We can assume that their index of intersection $(\nabla, \Delta) = 1$. The cycle ∇ can be also treated as a relative cycle in the pair $(\overline{V}_\alpha, \{\infty\})$, where $\overline{V}_\alpha$ is the closure of V_α in $\mathbb{CP}^2$.
The deformations of these cycles under the family of diffeomorphisms Γ_t are presented at Fig. 4.5. We see that $h_*\Delta = \Delta$ and

$$h_*\nabla - \nabla = -\Delta.$$

Generally, if δ is a relative cycle (or a cycle with closed support) then the quantity

$$\mathrm{var}\delta = h_*\delta - \delta$$

is called the *variation of the relative cycle* δ. The values of the operator var lie in the group of absolute cycles.
The *Picard–Lefschetz formula* states here that

$$\mathrm{var}\delta = (\Delta, \delta)\Delta.$$

The formula $h_*\delta = \delta + (\Delta, \delta)\Delta$ holds also for absolute cycles, e.g., for Δ (because $(\Delta, \Delta) = 0$.
Namely, in this way the formula for variation of cycles was introduced in the books of E. Picard and G. Simart [**PiSi**] and of S. Lefschetz [**Lef**].

4.11. Example. Let $f(z, w) = y^2 + x^3 - x$, $V_t = \{f = t\}$.
Here V_t is the Riemann surface of the function $y = \sqrt{t + x - x^3}$. It has (generally) three branching points x_1, x_2, x_3. Fig. 4.6 shows that this Riemann surface is diffeomorphic to a 2-dimensional torus without one point; one has to squeeze the boundary circle to a point.
The first homology group of V_t is generated by two (absolute) cycles: Δ_1 (surrounding x_1 and x_2) and Δ_2 (surrounding x_2 and x_3 as in Fig. 4.6). They represent the two basic cycles in the punctured torus and we have $(\Delta_1, \Delta_2) = 1$.
The cycles $\Delta_{1,2}$ are vanishing cycles. They disappear for those values of t at which two of the branching points x_i coalesce. In order to find them, we solve the system of equations

$$P(x) = t + x - x^3 = 0, \ P'(x) = 1 - 3x^2 = 0,$$

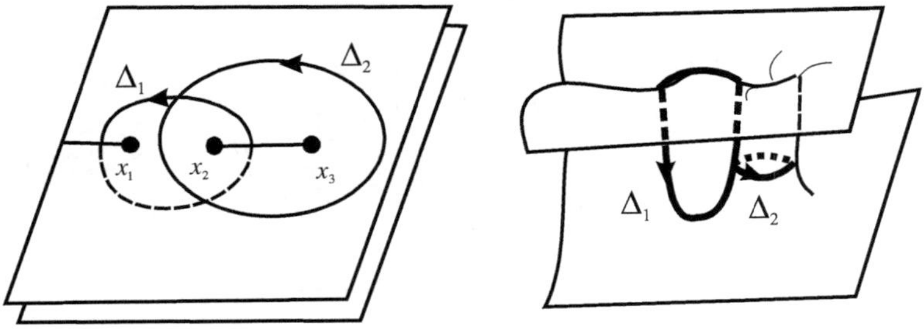

Figure 4.6: Cycles in the Riemann surface from the example

which gives

$$x_{1,2} = \mp\sqrt{1/3},\ t_{1,2} = \pm 2/3\sqrt{3}.$$

Fix some noncritical value t_0. We look at the behavior of the cycles $\Delta_i \subset V_t$ as t varies along two loops starting at t_0 and surrounding one of the critical values t_j. As a result we obtain two monodromy maps in $H_1(V_{t_0})$ which generate the *monodromy group*, a subgroup in $\mathrm{Aut}H_1(V_{t_0})$. The monodromy group is the image of the representation of the fundamental group $\pi_1(\mathbb{C} \setminus \{t_1, t_2\})$ in $\mathrm{Aut}H_1(V_{t_0})$. The action of the monodromy map corresponding to the loop around t_1 is presented at Fig, 4.7. Here the cycle Δ_1 is vanishing and it behaves like the cycle Δ from the previous example. The cycle Δ_2 should be treated as a relative cycle (in neighborhood of x_1) and behaves like the ∇. We get

$$\begin{aligned} &\Delta_1 \longmapsto \Delta_1, \\ &\Delta_2 \longmapsto \Delta_2 + \mathrm{var}\,\Delta_2 = \Delta_2 + (\Delta_1, \Delta_2)\Delta_1 = \Delta_2 + \Delta_1. \end{aligned}$$

The monodromy corresponding to the loop around the point t_2 is the following

$$\begin{aligned} \Delta_1 &\longmapsto \quad \Delta_1 + (\Delta_2, \Delta_1)\Delta_2 = \Delta_1 - \Delta_2, \\ \Delta_2 &\longmapsto \quad \Delta_2. \end{aligned}$$

Thus, the monodromy group is generated by the two matrices

$$\begin{pmatrix} 1 & 1 \\ 0 & 1 \end{pmatrix}, \quad \begin{pmatrix} 1 & 0 \\ -1 & 1 \end{pmatrix}.$$

4.12. Remark (choice of orientation). As we know, the intersection index depends on the choice of orientation of the manifold. In the monodromy theory of vanishing cycles we shall fix the following orientation of $\mathbb{C}^n$:

If $x_1, x_2, \ldots, x_n$, $x_j = u_j + \mathrm{i}v_j$ are the coordinates in $\mathbb{C}^n$, then the canonical orientation is given by the following order of coordinates

$$u_1, v_1, u_2, v_2, \ldots, u_n, v_n.$$

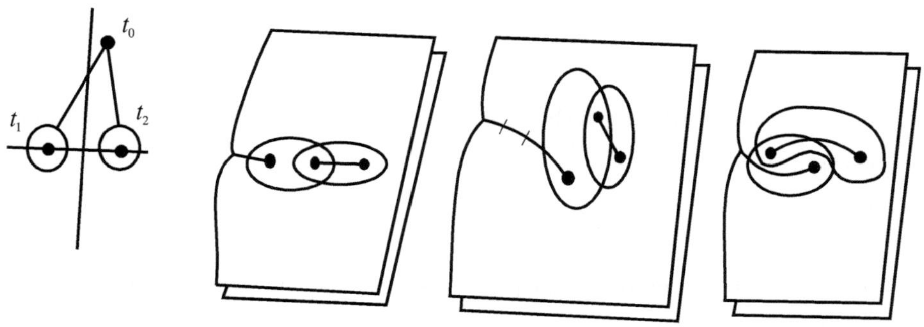

Figure 4.7: Deformation of the cycles from the example

In applications of Morse Lemma (see Chapter 1) we proved the following property

$$\{x_1^2 + \ldots + x_n^2 = z\} \approx T\mathbb{S}^{n-1},$$

where the sphere is identified with the set $S^{n-1} = \{v_1 = \ldots = v_n = 0\}$. However, the natural orientation of $T\mathbb{S}^{n-1}$ is given by

$$u_1, \ldots, u_{n-1}, v_1, \ldots, v_{n-1}. \tag{4.3}$$

The canonical orientation differs from the orientation (4.3) by the factor

$$(-1)^{(n-1)(n-2)/2}$$

(because we need $(n-1) + (n-2) + \ldots + 1$ transpositions).

4.13. Definition (Monodromy operator in the general case). Let $f : (\mathbb{C}^n, 0) \longmapsto (\mathbb{C}, 0)$ be a germ of finite multiplicity and let $\tilde{f}$ be its Morse perturbation with critical points $x_1, \ldots, x_\mu$ and critical values $z_1, \ldots, z_\mu$. As before, we choose the ball $\mathbb{B} = \mathbb{B}_\rho \subset \mathbb{C}^n$ with the boundary $\mathbb{S}_\rho$ and the small disc $\mathbb{D}_\epsilon \subset \mathbb{C}$ with the distinguished (noncritical) point z_0. Let $\widetilde{S} = \mathbb{D}_\epsilon \diagdown \{z_1, \ldots, z_\mu\}$ and $\widetilde{V} = \mathbb{B} \cap \tilde{f}^{-1}(\widetilde{S})$. The map

$$f : \widetilde{V} \longmapsto \widetilde{S}$$

defines a locally trivial fibration with fibers $\widetilde{V}_z$. Its restriction to the boundary $\mathbb{S}_\rho \cap \widetilde{V}$ prolongs itself to a fibration over the disc $\mathbb{D}_\epsilon$ with fibers $\partial\widetilde{V}_z = \tilde{f}^{-1}(z) \cap \mathbb{S}_\rho$ (see Lemma 4.2). Thus the latter (boundary) fibration is trivial (since the base is contractible) and $\mathbb{S}_\rho \cap \widetilde{V}$ is diffeomorphic to the Cartesian product $\partial\widetilde{V}_{z_0} \times \mathbb{D}_\epsilon$. Let $\gamma = \gamma(t)$, $t \in [0, 1]$ be a loop in $\widetilde{S}$ with base point at z_0, $\gamma(0) = \gamma(1) = z_0$. Using the prolongation of homotopy Lemma 3.31, we can construct a 1-parameter family of maps (diffeomorphisms)

$$\Gamma_t : \widetilde{V}_{z_0} \longmapsto \mathbb{B}$$

such that:

1) $\Gamma_0 = \mathrm{id}$;

2) $\Gamma_t(\widetilde{V}_{z_0}) \subset \widetilde{V}_{\gamma(t)}$;

3) $\Gamma_t|_{\partial\widetilde{V}_{z_0}}$ is compatible with the structure of Cartesian product, i.e., $\Gamma_t(z_0, x) = (\gamma(t), x) \in \mathbb{D}_\epsilon \times \partial\widetilde{V}_{z_0}$.

The diffeomorphism

$$h_\gamma = \Gamma_1$$

is called the **monodromy diffeomorphism** induced by the loop γ, and the homomorphism $h_{\gamma *} : \widetilde{H}_*(\widetilde{V}_{z_0}) \longmapsto \widetilde{H}_*(\widetilde{V}_{z_0})$ is called the **monodromy operator** induced by the loop γ. The **operator of relative monodromy**, denoted by $h^{(r)}_{\gamma *}$, acts on the group $H_*(\widetilde{V}_{z_0}, \partial\widetilde{V}_{z_0})$ and is also induced by h_γ. (Note that $h_\gamma|_{\partial\widetilde{V}_{z_0}} = id$).
The **variation operator** induced by the loop γ,

$$\mathrm{var}_\gamma : H_*(\widetilde{V}_{z_0}, \partial\widetilde{V}_{z_0}) \longmapsto \widetilde{H}_*(F_{z_0}),$$

is defined as

$$\Delta \longmapsto h^{(r)}_{\gamma *}\Delta - \Delta.$$

The **monodromy group** associated with the singularity f is the image of the fundamental group $\pi_1(\widetilde{S})$ in Aut $\widetilde{H}_*(\widetilde{V}_{z_0})$ for any Morse perturbation $\widetilde{f}$.
The **operator of classical monodromy** (or the **Picard–Lefschetz transformation**) h_* of the singularity f is the monodromy operator $h_{\gamma_0 *}$ induced by the loop γ_0 in $\widetilde{S}$, which surrounds once and counterclockwise all the critical values z_i. Equivalently, the Picard–Lefschetz transformation of f is defined as the operator in $\widetilde{H}_*(V_{z_0})$ (homologies of the level of the unperturbed function), induced by the loop in $(\mathbb{D}_\epsilon \setminus 0, z_0)$ around $z = 0$.

4.14. Remarks. 1. Let $i_* : \widetilde{H}(\widetilde{V}_{z_0}) \longmapsto H_*(\widetilde{V}_{z_0}, \partial\widetilde{V}_{z_0})$ be the natural map induced by the inclusion. Then we have

$$\begin{aligned} h_{\gamma *} &= I + \mathrm{var}_\gamma \circ i_*, \\ h^{(r)}_{\gamma *} &= I + i_* \circ \mathrm{var}_\gamma. \end{aligned}$$

2. If $\gamma_{1,2}$ are two loops, then $h_{\gamma_1 \cdot \gamma_2 *} = h_{\gamma_1 *} h_{\gamma_2 *}$ and the analogous formula holds in the relative case.

3. $\mathrm{var}_{\gamma_1 \cdot \gamma_2} = \mathrm{var}_{\gamma_1} + \mathrm{var}_{\gamma_2} + \mathrm{var}_{\gamma_2} \cdot i_* \cdot \mathrm{var}_{\gamma_1}$.

4.15. Lemma. *If a, b are two relative cycles (with empty intersection at the boundary $\partial\widetilde{V}_{z_0}$), then*

$$(h_{\gamma *}a, h_{\gamma *}b) = (a, b),$$
$$(\mathrm{var}_\gamma a, \mathrm{var}_\gamma b) + (a, \mathrm{var}_\gamma b) + (\mathrm{var}_\gamma a, b) = 0,$$

where $(\cdot, \cdot)$ is the intersection index.

Proof. The diffeomorphisms Γ_t preserve the orientation (because they are isotopic with the identity). So, if the cycles a, b are such that they have empty intersections at the boundary of $\widetilde{V}_{z_0}$, then this holds along the deformation Γ_t and we have $(\Gamma_{t*}a, \Gamma_{t*}b) = (a, b)$. From this also the second formula follows. □

4.16. Definitions. Let α_i be the paths connecting z_0 with the critical values z_i and let $\Delta_i \in \widetilde{H}_*(\widetilde{V}_{z_0})$ be the cycles vanishing at $z = z_i$ along α_i. The **simple loop** τ_i associated with α_i is the loop as in Fig. 4.8. The operator $h_i = h_{\tau_i *}$ is called the **Picard–Lefschetz transformation** corresponding to the cycle Δ_i.

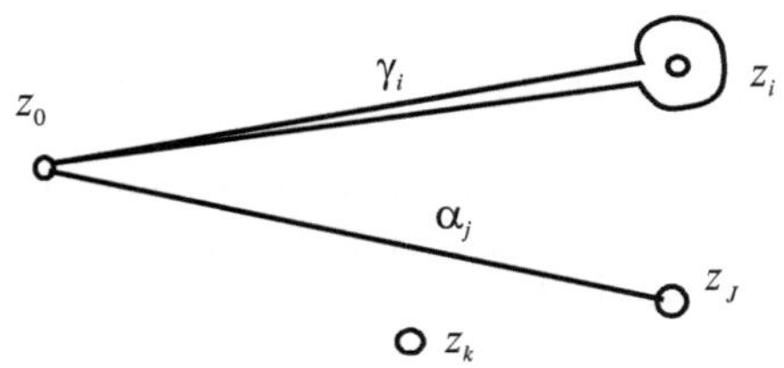

Figure 4.8: Distinguished loops in the target plane

In order to describe the Picard–Lefschetz operator, we consider the situation near a non-degenerate critical point. Assume then that $\widetilde{f} - z_i = y_1^2 + \ldots + y_n^2$, $y_j = u_j + \mathrm{i}v_j$. Take a simple loop (from Definition 4.16) in the form

$$\tau : t \longmapsto z = \mathrm{e}^{2\pi \mathrm{i}t}, \; t \in [0, 1].$$

The fiber V_z is of the form $\{\sum y_i^2 = z, \; \sum |y_i|^2 \leq 4\}$ with the cycle $\Delta = \{(u_1, v_1, \ldots, v_n) : v_1 = \ldots = v_n = 0\}\}$.

4.17. Lemma. *We have*

$$(\Delta, \Delta) = \begin{cases} 0, & n \quad \text{even} \\ 2, & n \equiv 1 (\mathrm{mod}\ 4) \\ -2, & n \equiv -1 (\mathrm{mod}\ 4) \end{cases}$$

Proof. By Remark 4.12 and 1.17, this intersection index equals $\chi(\mathbb{S}^{n-1}) \cdot (-1)^{(n-1)(n-2)/2}$. □

Recall that $\widetilde{H}_{n-1}(V_1)$ is generated by $\Delta = \{(u_1, v_1, \ldots, u_n, v_n) : v_1 = \ldots = v_n = 0\}$ (with the orientation given by the ordered system $u_2, \ldots, u_n$) and, by the Poincaré duality, the group $H_{n-1}(V_z, \partial V_z) = \mathbb{Z}$ and is generated by

$$\nabla = \left\{ u_2 = \ldots = u_n = 0, \; v_1 = 0, \; u_1 = \sqrt{1 + v_2^2 + \ldots + v_n^2} \right\}.$$

We choose the orientation of ∇ in such a way that $(\nabla, \Delta) = 1$.

4.18. Picard–Lefschetz theorem. *In the Morse situation we have the formula*

$$\mathrm{var}_\tau \nabla = (-1)^{n(n+1)/2}\Delta. \tag{4.4}$$

In the general situation for any relative cycle a, we have

$$\mathrm{var}_{\tau_i} a = (-1)^{n(n+1)/2}(a, \Delta_i)\Delta_i,$$
$$h^{(r)}_{\tau_i *} a = a + (-1)^{n(n+1)/2}(a, \Delta_i) i_* \Delta_i,$$

and for any absolute cycle b we have

$$h_{\tau_i *} b = b + (-1)^{n(n+1)/2}(b, \Delta_i)\Delta_i.$$

Proof. It is enough to prove the **Picard–Lefschetz formula** (4.4).
One way to show it is to construct the family of diffeomorphisms Γ_t, appearing in the definition of the monodromy operators. Unfortunately, this construction is very technical and rather difficult. Following [**AVG**], we will avoid these technicalities by a proper use of the properties of the intersection index and of the linking number in the Milnor's version of the Milnor bundle (see (4.2) and Definition 4.6).

The case with odd n.
Take the following lift to the fibration $V \longmapsto S$, i.e., the fibration (4.1), of the loop $\tau(t) = e^{2\pi i t}$:

$$\Omega_t(x) = e^{\pi i t} x.$$

This homotopy does not satisfy the property 3) from definition of Γ_t: it is not identity at the boundary ∂V. Nevertheless we shall use it. The induced operators Ω_{t*} act correctly on the absolute cohomologies $\widetilde{H}_{n-1}$ and this will be sufficient for us.
We shall use the formula $h_{\tau *} = id + \mathrm{var}_\tau \cdot i_*$. We have $\Omega_1(x) = -x$ and then $\Omega_{1*}\Delta = h_{\tau *}\Delta = (-1)^n\Delta$. So it is enough to calculate $i_*\Delta$, i.e., to find the integer m in the identity $i_*\Delta = m\nabla$.
We have $(i_*\Delta, \Delta) = (\Delta, \Delta) = 2(-1)^{(n-1)/2}$ for odd n. Because $(\nabla, \Delta) = 1$ (by the definition of the orientation of ∇), we get $m = 2(-1)^{(n-1)/2}$.
The formula for the action of the monodromy operator gives the relation

$$-\Delta = \Delta + \mathrm{var}_\tau \cdot i_*\Delta = \Delta + 2(-1)^{(n-1)/2}\,\mathrm{var}_\tau \nabla$$

and hence $\mathrm{var}_\tau \nabla = (-1)^{n+1)/2}\Delta$.

The case with even n.
Now $i_*\Delta = 0$ and the above proof does not work. Here we work with the Milnor fibration (4.2) defined as follows. One takes the sphere $\mathbb{S}_\rho \subset \mathbb{C}^n$ and deletes from it a tubular neighborhood T of the submanifold $f = 0$, the fibration projection is $\Phi : \mathbb{S}_\rho \backslash T \longmapsto \mathbb{S}^1$, $\Phi(x) = f(x)/|f(x)|$. We study the monodromy operator (acting in $\widetilde{H}_{n-1}(\Phi^{-1}(1))$) induced by the loop $\tau : t \longmapsto e^{2\pi i t}$ in the base $\mathbb{S}^1$ and by means

of a family of diffeomorphisms $\Gamma_t : \Phi^{-1}(1) \longmapsto \Phi^{-1}(e^{2\pi i t})$ which are identity at the boundary ∂T.

4.19. Definition. The bilinear form $L : \widetilde{H}_{n-1}(\Phi^{-1}(1)) \times \widetilde{H}_{n-1}(\Phi^{-1}(1)) \longmapsto \mathbb{Z}$

$$L(a,b) = l(a, \Gamma_{1/2\,*}b)$$

is called the **Seifert form**. Here $l(\cdot,\cdot)$ is the linking number of cycles in the sphere S_ρ (see 3.17).

By the Alexander duality (Theorem 3.19) the form L is non-degenerate.

We introduce the following matrices. Fix some basis in $\widetilde{H}_{n-1}(\Phi^{-1}(1))$. Let:
L be the matrix of the Seifert form in this basis,
S be the matrix of the intersection form,
H be the matrix of the monodromy operator $\Gamma_{1\,*}$,
$H^{(r)}$ be the matrix of the relative monodromy operator (in $H_*(\Phi^{-1}(1), \partial\Phi^{-1}(1))$).

4.20. Lemma. *We have the following relations:*

(a) $L(\mathrm{var}_\tau a, b) = (a,b)$ *(equivalently,* $L(a,b) = (\mathrm{var}_\tau^{-1} a, b)$;

(b) $(a,b) = -L(a,b) + (-1)^n L(b,a)$ *(or* $S = -L + (-1)^n L^\top$;

(c) $h_{\tau\,*} = (-1)^n \,\mathrm{var}_\tau (var_\tau^{-1})^\top$ *(or* $H = (-1)^n L^{-1} L^\top$;

(d) $h^{(r)}_{\tau\,*} = (-1)^n (\mathrm{var}_\tau^{-1})^\top \,\mathrm{var}_\tau$ *(or* $H^{(r)} = (-1)^n L^\top L^{-1}$*)*.

Proof. (a) Let $a \subset \Phi^{-1}(1)$ represents a relative cycle, i.e., $\partial a \subset \partial\Phi^{-1}(1)$. We denote here the geometrical cycles and their homology classes by the same letters $a, b, \ldots$. We define the map

$$A : [0,1] \times a \longmapsto \mathrm{S}_\rho, \ (t,y) \longmapsto \Gamma_t(y).$$

Its image is an n-dimensional cycle with the boundary consisting of:

$$\Gamma_1(a) - a = \mathrm{var}_\tau a$$

and a part in $\partial\Phi^{-1}(1) \subset \partial T$. The latter part is negligible, because it has too small dimension. Thus we can assume that $\partial A = \mathrm{var}_\tau a$. By definition of the linking number, we have $l(\partial A, \cdot) = (A, \cdot)$
Next, if b is some absolute cycle, then the cycle $A \cap \Gamma_{1/2}(b) = \Gamma_{1/2}(a) \cap \Gamma_{1/2}(b)$ and lies in the fiber $\Phi^{-1}(-1)$. We have then

$$L(\mathrm{var}_\tau a, b) = l(\mathrm{var}_\tau a, \Gamma_{1/2\,*}b) = (A, \Gamma_{1/2\,*}b),$$

where the latter intersection is calculated in the sphere. It is clear that it equals to the following intersection indices (calculated in the fibers $\Phi^{-1}(-1)$ and $\Phi^{-1}(1)$):

$$(\Gamma_{1/2\,*}a, \Gamma_{1/2*}\, b) = (a,b)$$

(because $A\cap\Phi^{-1}(-1)=\Gamma_{1/2}(a)$). So, (a) is proved.
(b) If $a,b\in\widetilde{H}_{n-1}(\Phi^{-1}(1))$, then, due to the non-degeneracy of the variation operator (see the point (a)), there exist relative cycles a',b' such that $a=\mathrm{var}_\tau a'$, $b=\mathrm{var}_\tau b'$.
Then, applying Lemma 4.15, we get

$$(\mathrm{var}_\tau a',\mathrm{var}_\tau b')+(a',\mathrm{var}_\tau b')+(\mathrm{var}_\tau a',b')=0$$

and it is enough to substitute the relations $(\mathrm{var}_\tau a',\mathrm{var}_\tau b')=(a,b)$, $(a',\mathrm{var}_\tau b')=L(a,b)$, $(\mathrm{var}_\tau a',b')=(a,\mathrm{var}_\tau^{-1}b)=(-1)^{n-1}L(b,a)$.
(c) Because $(c,d)=(i_*c,d)$, then from (b) we get

$$i_*=-\mathrm{var}^{-1}+(-1)^n(\mathrm{var}^{-1})^\top$$

and

$$h_{\tau *}=id+\mathrm{var}\cdot i_*=id-id+(-1)^n\mathrm{var}(\mathrm{var}^{-1})^\top.$$

(d) is proved in an analogous way. □

Proof of the Picard–Lefschetz formula for even n. In the case of the function $x_1^2+\dots x_n^2$ we have

$$(\mathrm{var}^{-1}\Delta,\Delta)=L(\Delta,\Delta)=l(\Delta,\Gamma_{1/2*}\Delta)=(-1)^n(\widetilde{A},\widetilde{B}),$$

where $\widetilde{A},\widetilde{B}$ are cells in the disc D^{2n} with boundaries Δ and $\Gamma_{1/2}(\Delta)$ respectively (see Lemma 3.18). The trick relies on the fact that, in order to compute the linking number $l(\Delta,\Gamma_{1/2*}\Delta)$, we do not need the homotopy Γ_t fixed at the boundary (because the cycles are not relative). So, we choose

$$\Gamma_t(x)=\mathrm{e}^{\pi\mathrm{i}t}x.$$

As the n-cells we choose

$$\widetilde{A}=\{v=0\},\quad \widetilde{B}=\{u=0\}$$

with the orientations $u_1,\dots,u_n$ and $v_1,\dots,v_n$ respectively. (Note that $\Gamma_{1/2}$ is the multiplication by $\mathrm{i}=\sqrt{-1}$ and transforms $\widetilde{A}$ to $\widetilde{B}$ with preserved orientations.) We have $(\widetilde{A},\widetilde{B})=(-1)^{n(n-1)/2}$, or $(\mathrm{var}^{-1}\Delta,\Delta)=(-1)^{n(n+1)/2}$, which means that

$$\mathrm{var}^{-1}\Delta=(-1)^{n(n+1)/2}\nabla.$$

This completes the proof of the Picard–Lefschetz theorem. □

Recall that the operator of classical monodromy h_* of the singularity is the monodromy operator $h_{\gamma_0 *}$, induced by a loop in the base $\widetilde{S}$ surrounding all critical values of the perturbation $\tilde{f}$. If τ_i are the simple loops (around critical values z_i) and $h_i=h_{\tau_i *}$ are the corresponding monodromy operators, then we have

$$h_*=h_\mu\cdot\ldots\cdot h_2\cdot h_1.$$

4.21. Theorem. *The matrix H of the operator h_* is determined by the matrix S of the intersection form and vice versa, the intersection form is uniquely determined by the operator of classical monodromy.*

Proof. The first part of this theorem follows from Picard–Lefschetz Theorem. The intersection matrix S is symmetric when n is odd and is anti-symmetric if n is even. In the first case its diagonal elements are equal to ± 2, in the second case they are equal to 0.
Let $\nabla_1, \ldots, \nabla_\mu$ be the basis in $H_*(V_{z_0}, \partial V_{z_0})$, dual to the basis $\{\Delta_i\}$ of vanishing cycles: $(\nabla_i, \Delta_j) = \delta_{ij}$.
Recall that $h_i : a \longmapsto a + (-1)^{n(n+1)/2}(a, \Delta_i)\Delta_i$. Expressing h_i as $id + \text{var}_{\tau_i} \cdot i_*$ and h_* as $id + \text{var}_f \cdot i_*$, we obtain

$$\text{var}_f = \sum_{i_1 < \ldots < i_r, r \leq \mu} \text{var}\tau_{i_1} \cdot i_* \cdot \text{var}_{\tau_{i_2}} \cdot \ldots \cdot i_* \cdot \text{var}_{\tau_{i_r}}.$$

Because $\text{var}_{\tau_i} \nabla_j = (-1)^{n(n+1)/2} \sum \delta_{ij} \Delta_i$ (δ_{ij} – the Kronecker symbol) and $\text{var}_{\tau_i} \cdot i_* \Delta_j = 0$ for $j > i$, we get

$$\text{var}_f \nabla_i = (-1)^{n(n+1)/2} \Delta_i + \sum_{j<i} c_{ij} \Delta_j.$$

It means that the matrix of the operator var_f is upper-triangular, with the same nonzero diagonal entries. By Lemma 4.20 (the points (a) and (c)) the Seifert matrix $L = \text{var}_f^{-1}$ is also upper-triangular and $H = (-1)^n L^{-1} L^\top$ is a product of a lower-triangular matrix and an upper-triangular matrix. We shall show that L is uniquely defined by H.
Above the lower- and upper-triangular matrices have fixed (and the same) diagonal elements. Multiplying L by some constant, we arrive to the following problem:

Let A and B be two upper-triangular matrices with 1s at the diagonals and let $C = A \cdot B^\top$. Are A, B defined uniquely by C?

The answer is yes, because, for any other such representation $C = A_1 \cdot B$, we get the identity $A^{-1} A_1 = B^\top (B_1^\top)^{-1}$ between upper and lower-triangular matrices. □

4.22. Definition (Dynkin diagram of a singularity). Let f be a singularity. We apply to it some special stabilization. We add a sum of squares of new variables, such that the number of variables is

$$n = 4k + 3.$$

The **quadratic form of the singularity** is the quadratic form defined by the intersection index in homologies of a non-singular level surface of the stabilization with the number of variables $n = 3 \pmod 4$.
The **Dynkin diagram of the singularity** is the graph defined as follows:

1) Its vertices are in one-to-one correspondence with elements Δ_i of the distinguished basis of vanishing cycles for the above stabilization.

2) The ith and j^{th} vertices are joined by an edge of multiplicity (Δ_i, Δ_j), if $(\Delta_i, \Delta_j) > 0$, and by a punctured edge of multiplicity $|(\Delta_i, \Delta_j)|$, if $(\Delta_i, \Delta_j) < 0$.

   If the number $|(\Delta_i, \Delta_j)|$ is small, i.e., $= 1, 2, 3$, then one draws as many edges as this intersection number says; otherwise, one draws only one edge and puts a corresponding integer above it.

Remark. The Dynkin diagrams are met in many other situations; e.g., in the classification of the semi-simple Lie algebras, in the classification of the Coxeter groups (see below). Dynkin diagrams provide a very practical way to encode a symmetric matrix with integer entries. Similar procedures are applied to encode integer anti-symmetric matrices.

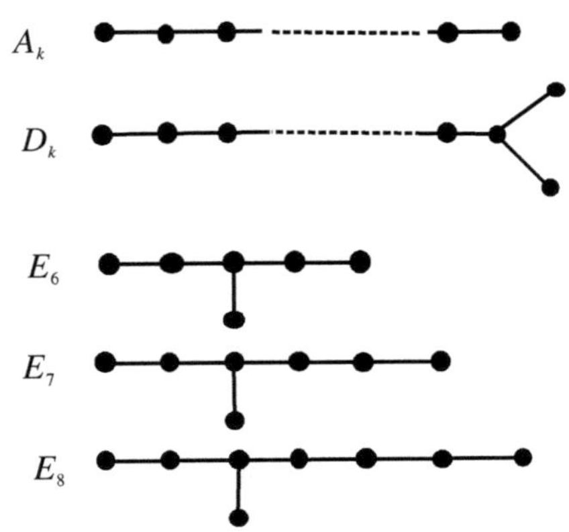

Figure 4.9: Dynkin diagrams of simple singularities

4.23. Theorem (Dynkin diagrams of simple singularities). *The Dynkin diagrams of the simple singularities are the same as at Fig.* 4.9.

4.24. Problem. Show that the quadratic forms of the simple singularities are negatively defined.

As the next theorem shows, the (non-distinguished) unimodal singularities have degenerate their quadratic forms. Let μ_0, μ_+, μ_- denote the number of 0 eigenvalues, the number of positive eigenvalues and the number of negative eigenvalues (respectively) of the quadratic form S of the singularity. The parabolic singularities are semi-definite and the hyperbolic ones are indefinite. This justifies their names. We do not prove this theorem.

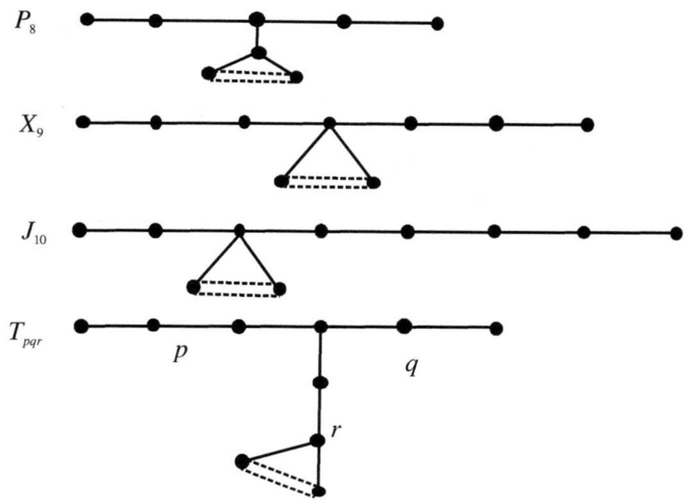

Figure 4.10: Dynkin diagrams of unimodal singularities

4.25. Theorem. *The (non-exceptional) unimodal singularities from Theorem* 2.39 *have the Dynkin diagrams and the signature invariants of their quadratic forms the same as at Fig.* 4.10. *We have* $\mu_0 = 2$, $\mu_+ = 0$ *for* $\mathbf{P}_8$, $\mathbf{X}_8$, $\mathbf{J}_{10}$ *and* $\mu_0 = \mu_+ = 1$ *for* $\mathbf{T}_{pqr}$.

Proof of Theorem 4.23. Because the simple singularities are defined by functions depending on only two variables, we begin the proof with the analysis of the real functions on the plane.
Assume that $f : (\mathbb{C}^2, 0) \longmapsto (\mathbb{C}, 0)$ is a germ of real analytic function. It means that

$$f|_{\mathbb{R}^2} : (\mathbb{R}^2, 0) \longmapsto (\mathbb{R}, 0).$$

Very often occurs that there exists a deformation $\tilde{f}$ which is:

(i) real,

(ii) has only real Morse critical points (local minima, local maxima and saddle points with real critical values) and

(iii) the critical values at the saddle points are all equal to 0 (see Fig. 4.11).

Let $l = \tilde{f}^{-1}(0) \cap \mathbb{R}^2$. It is a curve with self-intersections. With the above real picture we associate certain cycles in the (complex) non-singular level surface of $\tilde{f}$. With the minima we associate the cycles Δ_i^-, which vanish at these points; with the saddle points we associate the cycles Δ_j^0 and with the maxima we associate cycles Δ_k^+. If $\{\tilde{f} = z\}$, $z > 0$, then Δ_k^+ are *ovals*, i.e., components of $\{\tilde{f} = z\}$, surrounding corresponding points of local maximum. The curve l divides the neighborhood of

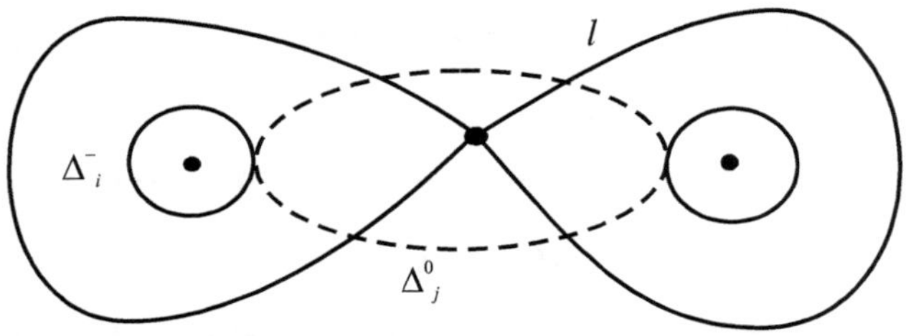

Figure 4.11: Zero level and cycles

the origin into domains (basins) containing exactly one minimum or maximum and an outer domain.

Introduce the numbers:

$n_{0,-}(j,i)$ is the number of times of appearing of the vertex corresponding to Δ^0_j in the boundary of the basin corresponding to Δ^-_i;

$n_{+,0}(k,j)$ is the analogous number associated with the basin of Δ^+_k;

$n_{+,-}(k,i)$ is the number of smooth pieces of the curve l separating the basins corresponding to the cycles Δ^+_k and Δ^-_i.

4.26. Theorem.

(a) *The intersection form of the function $f(x,y)$ is anti-symmetric and defined by the formulas*

$$\begin{aligned}(\Delta^\#_m, \Delta^\#_n) &= 0, \ \# = 0, \pm,\\ (\Delta^0_j, \Delta^-_i) &= n_{0,-}(j,i),\\ (\Delta^+_k, \Delta^0_j) &= n_{+,0}(k,j),\\ (\Delta^-_i, \Delta^+_k) &= n_{+,-}(k,i).\end{aligned}$$

(b) *The intersection form of the function $f(x,y)+t^2$ is a symmetric form given by the formulas*

$$\begin{aligned}(\Delta^\#_m, \Delta^\#_n) &= -2\delta_{mn},\\ (\Delta^0_j, \Delta^-_i) &= n_{0,-}(j,i),\\ (\Delta^+_k, \Delta^0_j) &= n_{+,0}(k,j),\\ (\Delta^-_i, \Delta^+_k) &= -n_{+,-}(k,i).\end{aligned}$$

Proof. (a) Because the cycles $\Delta^\#_m, \Delta^\#_n$ are separated in the Riemann surface $\tilde{f} = z$, they have zero intersection index.

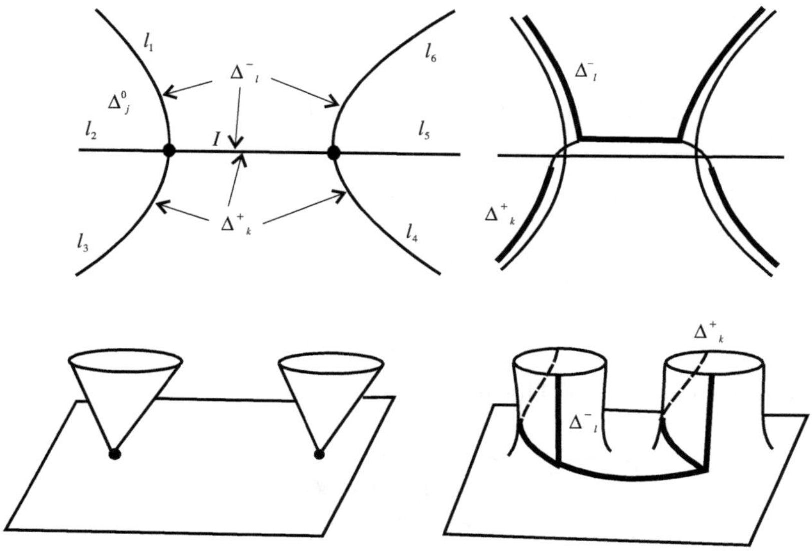

Figure 4.12: Cycles and their intersections

If the value z approaches the critical value 0 from below, then the cycle Δ_i^- (real oval) approaches the boundary of its basin. Near the vertex corresponding to Δ_j^0, the cycle Δ_i^- intersects the cycle Δ_j^0, vanishing at that saddle (see Fig. 4.12. In this way we obtain the next two intersection indices.

If the basins corresponding to the cycles Δ_i^- and Δ_k^+ have common piece I of the curve l, then these cycles intersect each other along this piece at the (singular) level surface $\{\tilde{f} = z\}$, $z = 0$. Assume that this piece ends at saddle points corresponding to the cycles Δ_j^0 and Δ_m^0 with separatrices I, l_1, l_2, l_3 and I, l_4, l_5, l_6 respectively (see Fig. 4.12).

We perturb them at a near level surface, e.g., for $z < 0$. The cycle Δ_i^- is real (it passes near l_1, I, l_6) and the interesting for us part of the cycle Δ_k^+ consists of: the two real pieces close to l_3, l_4, half of the cycle Δ_j^0, half of the cycle Δ_m^0, and the piece of Δ_i^- close to I. The half-cycles of Δ_k^0 and Δ_m^0 are taken simultaneously: if $f = xy$ near such point, then for $z = \epsilon e^{i\theta}$ we take $x = y = \sqrt{\epsilon} e^{i\theta/2}$. Now from Fig. 4.12, it is clear that our cycles intersect themselves in unavoidable way. We cannot destroy this intersection by a perturbation.

From this the fourth formula follows.

(b) The proof of this point relies on investigation, what happens with the vanishing cycles, when one adds a square to a function.

Let $\Delta_1(z), \ldots, \Delta_\mu(z)$ be the distinguished system of vanishing cycles at the level surface $\{f(x, y) = z\} \subset \mathbb{C}^2$. We associate with them the following family of vanishing cycles at the surface $\{f(x,y)+t^2=z\}\subset\mathbb{C}^2\times\mathbb{C}^1$:

$$\widetilde{\Delta_j}=\left(\bigcup_{\tau\in[0,1]}\Delta_j((1-\tau)z)\times\{\sqrt{\tau z}\}\right)\cup\left(\bigcup_{\tau\in[0,1]}\Delta_j((1-\tau)z)\times\{-\sqrt{\tau z}\}\right).$$

In other words, the cycles $\widetilde{\Delta_j}$ are suspensions of the cycles Δ_j; one adds to Δ_j two half-spheres.
If we choose appropriately the orientations of the new cycles, then we get the properties $(\widetilde{\Delta_j},\widetilde{\Delta_j})=-2$ (it is automatic) and

$$(\widetilde{\Delta_i},\widetilde{\Delta_j})=\operatorname{sign}(j-i)(\Delta_i,\Delta_j).$$

This completes the proof of Theorem 4.26. □

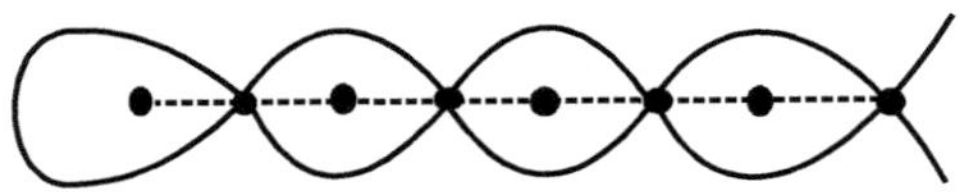

Figure 4.13: Zero level for deformation of the $\mathbf{A}_k$ singularity

We apply the latter theorem to the simple singularities.
$\mathbf{A}_k$. We choose the deformation

$$x^{k+1}+y^2\longmapsto(x-x_1)\dots(x-x_{k+1})+y^2$$

and from Fig. 4.13 we obtain the intersection form and the Dynkin diagram of this singularity.
$\mathbf{D}_k$. The deformation $x^2y+y^{k-1}\longmapsto y(x^2+(y-y_1)\dots(y-y_{k-2}))$ gives the answer (see Fig. 4.14).

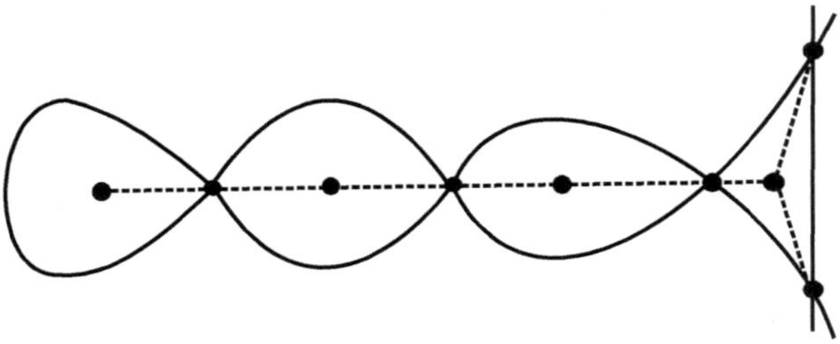

Figure 4.14: Zero level for deformation of the $\mathbf{D}_k$ singularity

$\mathbf{E}_6$. We choose the deformation in the following general form:

$$x^3+y^4\longmapsto x^3-\lambda x+y^4-\mu y^2+\nu,$$

where the parameters λ, μ, ν should be chosen in such a way, that the polynomials $x^3 - \lambda x$ and $y^4 - \mu y + \nu$ have the same absolute values of the critical values. (This ensures that the saddle points are at the zero level). It is not difficult to do it; one can choose the Chebyshev polynomials $T_j(t) = \cos(j \arccos t)$ and normalize them suitably. Similarly we act in the case $\mathbf{E}_8$ below.

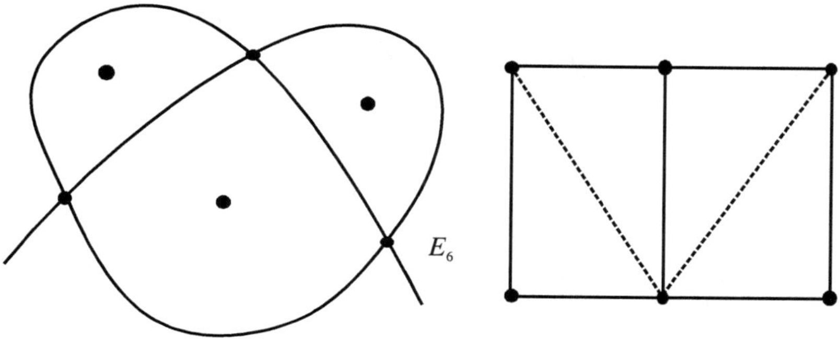

Figure 4.15: Zero level for deformation and the Dynkin diagram for the $\mathbf{E}_6$ singularity

The situation is presented at Fig. 4.15, where the Dynkin diagram differs from the one from Theorem 4.23.
$\mathbf{E}_7$: $x(x^2 + y^3) \longmapsto (x - \lambda)(x^2 + (y - y_1)(y - y_2)(y - y - 3))$.
$\mathbf{E}_8 : x^3 + y^5 \longmapsto x^3 - \lambda x + y^5 - \mu y^3 + \nu y$ (see Fig. 4.16).

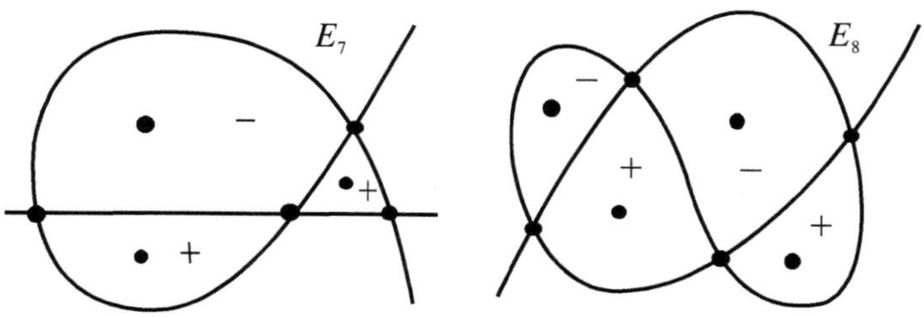

Figure 4.16: Zero levels for deformations of the $\mathbf{E}_7$ and $\mathbf{E}_8$ singularities

Because the Dynkin diagrams for the $\mathbf{E}_k$ singularities do not agree with those from Theorem 4.23, some additional arguments are needed. These arguments rely on deformations of the curve l, representing the 0-level curve of the real deformation of the singularity.
There are two kinds of typical bifurcations, occurring in generic one-parameter families of planar imbedded curves (with possible self-intersections). They are presented at Fig. 4.17.

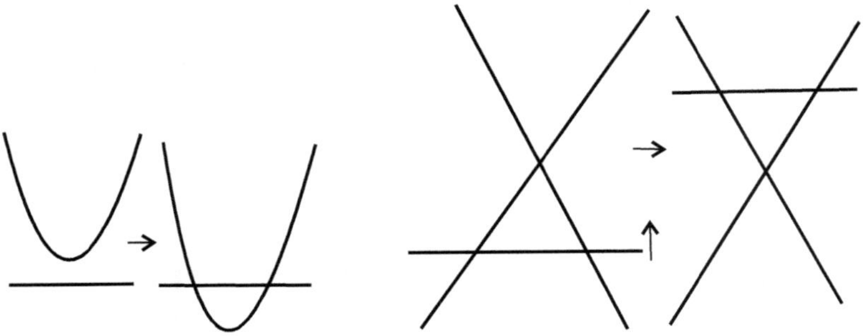

Figure 4.17: Two types of bifurcations of plane curves

We use only deformations containing bifurcations of the second type (i.e., passing through triple self-intersection). We call them the *proper* deformations.

4.27. Proposition. *Any proper deformations of the curve l can be realized in certain real deformation of the singularity. Its influence on the topology of the complex level surface relies on some change of the distinguished basis of vanishing cycles.*

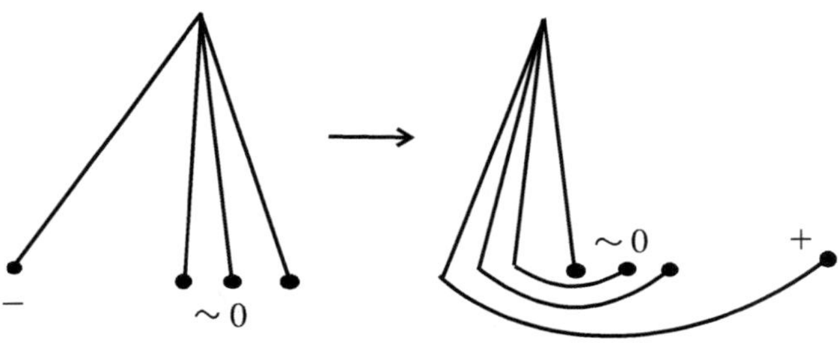

Figure 4.18: Deformation of a system of paths

Proof. Because the space of all possible unfoldings of f is infinite dimensional, the first point of this proposition is rather obvious.
To see the change in topology of level surfaces, we must omit the (real) bifurcating point by passing to the complex domain. We look what happens with the critical values z_i and with the paths α_i (joining them with z_0) near the bifurcation.
The situation is presented at Fig. 4.18, where we get new, deformed paths u_i. Of course, they also define the distinguished basis of homologies. In [**AVG**] there are formulas expressing the matrices of the base changes in this and in other analogous situations. □

Applying the above procedure of deformations of the 0-level curve l to the remaining simple singularities as at Fig. 4.19, we obtain the needed Dynkin diagrams and we complete the proof of Theorem 4.23. □

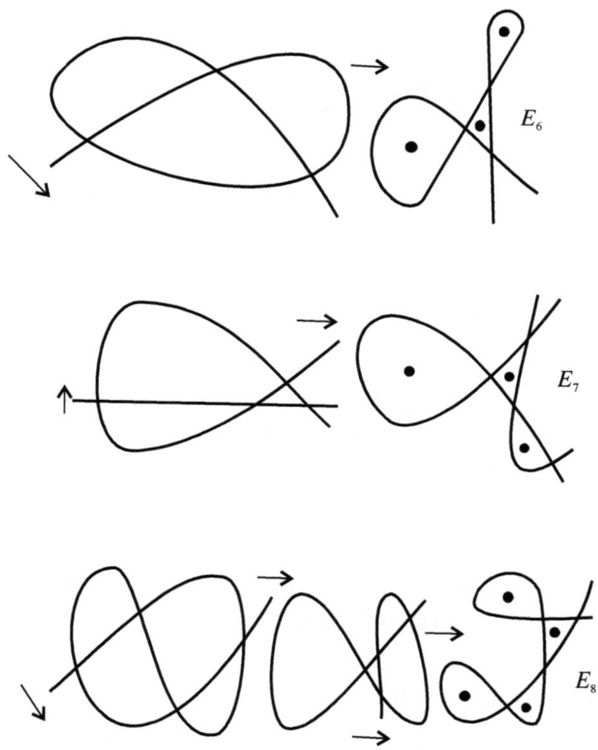

Figure 4.19: Deformations of the zero levels in the cases of exceptional singularities

§3 Root Systems and Coxeter Groups

4.28. Root systems. Let $(\cdot,\cdot)$ denote the scalar product in $\mathbb{R}^n$. If $\beta \in \mathbb{R}^n \setminus 0$, then the map

$$s_\beta : \alpha \longmapsto \alpha - 2\frac{(\alpha,\beta)}{(\beta,\beta)}\beta$$

is a *reflection* with respect to the plane orthogonal to β.
The **root system** $\mathcal{R}$ is a finite subset of $\mathbb{R}^n \setminus 0$ such that

1. $c_{\alpha,\beta} = 2\frac{(\alpha,\beta)}{(\alpha,\alpha)} \in \mathbb{Z}$ for any $\alpha, \beta \in \mathcal{R}$,
2. $s_\beta : \mathcal{R} \longmapsto \mathcal{R}$ for any $\beta \in \mathcal{R}$.

If $\mathcal{R}$ is a root system, then the group generated by the reflections s_α, $\alpha \in \mathcal{R}$ is called the **Weyl group**.

From the above two axioms one obtains the following properties of the root system $\mathcal{R}$.

(i) *If* $\alpha \in R$, *then* $-\alpha \in \mathcal{R}$.

(ii) *If* $\alpha \in R$, $m\alpha \in \mathcal{R}$, *then* $m = \pm 1/2, \pm 1, \pm 2$.

(iii) *If* $\alpha, \beta \in R$, *then* $c_{\alpha,\beta} c_{\beta,\alpha} = 4\cos^2\theta_{\alpha,\beta} \in [0,4]$, *where* $\theta_{\alpha,\beta}$ *is the angle between the vectors* α *and* β. *(If* α, β *are not parallel, then this number takes one of the values* $0, 1, 2, 3$*).*

The system $\mathcal{R}$ is called **reduced** iff for any $\alpha \in \mathcal{R}$ the vector $m\alpha \in \mathcal{R}$ only for $m = \pm 1$.
A subsystem $\mathcal{S} \subset \mathcal{R}$ is called the **basis** of the root system $\mathcal{R}$ (or the system of *simple roots*) iff any $\alpha \in \mathcal{R}$ has unique representation in the form $\sum n_i \alpha_i$, $\alpha_i \in \mathcal{S}$, where all n_i are integers of the same sign. The root α is called *positive* iff all $n_i \geq 0$ and *negative* iff all $n_i \leq 0$.
One can construct the basis $\mathcal{S}$ in the following way. If the linear space, spanned by $\mathcal{R}$, is $\mathbb{R}^n$ with the standard basis $e_1, \ldots, e_n$, then we introduce the lexicographic order in $\mathbb{R}^n$: $\alpha = (\alpha_1, \ldots, \alpha_n) > \beta = (\beta_1, \ldots, \beta_n)$, if $\alpha_1 > \beta_1$, or $\alpha_1 = \beta_1, \alpha_2 > \beta_2$, etc. A root $\alpha > 0$ is called *simple* iff it cannot be represented as a sum of two roots $\beta > 0$ and $\gamma > 0$. It turns out that the system of simple roots is a basis of $\mathcal{R}$.

(iv) *If* $\alpha, \beta \in S$, *then* $(\alpha, \beta) \leq 0$.

If $\mathcal{S} = \{\alpha_1, \ldots, \alpha_r\}$ is the system of simple roots of the reduced root system $\mathcal{R}$, then from (iv) and (v) it follows that the angles θ_{ij} (between α_i and α_j) can take the values 90°, 120°, 135°, 150°, corresponding to $n_{ij} = c_{\alpha_i,\alpha_j} c_{\alpha_j,\alpha_i} = c_{ij} c_{ji} = 0$, 1, 2, 3 respectively. In this case $|\alpha_i|^2 = n_{ij}|\alpha_j|^2$ and the lengths of the (non-orthogonal) roots α_i, α_j are determined up to the order; e.g., if $n_{ij} = 2$, then either $|\alpha_i|^2 = 1$ and $|\alpha_j|^2 = 2$, or $|\alpha_i|^2 = 2$ and $|\alpha_j|^2 = 1$.
The matrix $\{c_{ij}\}$ is called the **Cartan matrix.** It is integer matrix with the entries 2 at the diagonal and with negative entries outside the diagonal.
The above situation is encoded in the **Dynkin diagram** constructed as follows. The vertices correspond to the simple roots $\alpha_i \in \mathcal{S}$. The vertices α_i and α_j are connected by means of n_{ij} edges. If $n_{ij} \neq 0, 1$, then we add an arrow going from a longer root to a shorter root; (sometimes the authors denote the length of the corresponding root by putting the number, the square of the length, above it).
The root system is called **irreducible** iff its Dynkin diagram is connected. Otherwise, $\mathcal{R}$ is a union of two root systems laying in orthogonal subspaces of $\mathbb{R}^n$.
(There exists also the notion of affine Weyl groups (see [**Hum**]). They are infinite groups generated of reflection with respect to hyperplanes which do not necessarily pass through the origin. The **affine Weyl group** W_a, associated with the root system

$\mathcal{R}$ and the finite Weyl group W, is the group generated by reflections with respect to the hyperplanes $\{x : (\alpha, x) = k\}$, $\alpha \in \mathcal{R}$, $k \in \mathbb{Z}$.)

4.29. Theorem. *If $\mathcal{R}$ is an irreducible and reduced root system, then it has the Dynkin diagram of one of the forms at Fig.* 4.20.

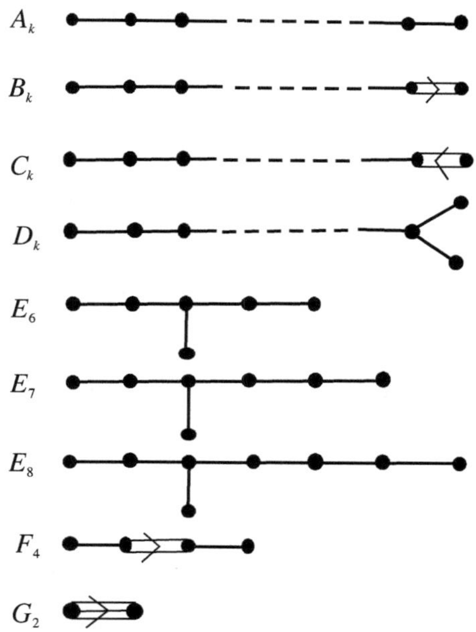

Figure 4.20: Dynkin diagrams of root systems

4.30. Examples. (a) The root system $\mathbf{A}_k$ is equal to

$$\alpha_{ij} = e_i - e_j,\ i \neq j,\ i, j = 1, \ldots, k+1$$

in the lattice $\mathbb{Z}^{k+1}$. It spans the space $\mathbb{R}^k \simeq \{\sum x_i = 0\} \subset \mathbb{R}^{k+1}$. The simple roots are

$$\alpha_1 = e_2 - e_1,\ \alpha_2 = e_3 - e_2, \ldots, \alpha_k = e_{k+1} - e_k.$$

The Weyl group is isomorphic to the group $S(k+1)$ of permutations of the vectors $e_1, \ldots, e_{k+1}$.

(b) The root system $\mathbf{D}_k$ is the set

$$\pm e_i \pm e_j,\ i \neq j,\ i, j = 1, \ldots, k$$

in $\mathbb{Z}^k$. Its system of simple roots is

$$e_1 - e_2, \ldots, e_{k-1} - e_k, e_{k-1} + e_k.$$

The Weyl group is equal to the product of the group of permutations of the vectors e_i and of the group of changes of signs before even number of vectors e_i.

(c) The root system $\mathbf{B}_k$ is following

$$\pm e_i,\ i = 1, \ldots, k; \quad \pm e_i \pm e_j,\ \ i < j.$$

Its Weyl group is the product of the group of permutations of the vectors e_i and of the group of changes of signs before e_i's.

(d) The root system $\mathbf{C}_k$ is

$$\pm 2e_i; \quad \pm e_i \pm e_j.$$

(e) The root system of the type $\mathbf{E}_6$ is the following subset of the space $\{x \in \mathbb{R}^8 : x_6 = x_7 = -x_8\} \subset \mathbb{R}^8$:

$$\pm e_i \pm e_j,\ 1 \le i < j \le 5; \ \pm \frac{1}{2}\left(e_8 - e_7 - e_6 + \sum_1^5 \epsilon_i e_i\right),\ \epsilon_i = \pm 1,\ \prod \epsilon_i = 1.$$

(f) The root system $\mathbf{E}_7$ lies in the subspace of $\mathbb{R}^8$ orthogonal to $e_7 + e_8$ and consists of the vectors

$$\pm e_i \pm e_j,\ 1 \le i < j \le 6; \ \ \pm \frac{1}{2}\left(e_7 - e_8); \ (e_7 - e_8 + \sum_1^6 \epsilon_i e_i\right),\ \prod \epsilon_i = -1.$$

(g) The root system $\mathbf{E}_8$ lies in $\mathbb{R}^8$ and consists of the vectors

$$\pm e_i \pm e_j,\ i < j; \ \frac{1}{2}\left(\sum \epsilon_i e_i\right),\ \prod \epsilon_i = 1.$$

(h) The system $\mathbf{F}_4$ consists of

$$\pm e_i; \ \pm e_i \pm e_j,\ \ i < j; \ \ (\pm e_1 \pm e_2 \pm e_3 \pm e_4)/2$$

in $\mathbb{R}^4$.

(i) The vectors

$$\begin{aligned} &\pm (e_1 - e_2),\ \pm(e_1 - e_3),\ \pm(e_2 - e_3), \\ &\pm (2e_1 - e_2 - e_3),\ \pm(2e_2 - e_1 - e_3),\ \pm(2e_3 - e_1 - e_2) \end{aligned}$$

in the plane $\{x_1 + x_2 + x_3 = 0\} \subset \mathbb{R}^3$ form the system $\mathbf{G}_2$.

Remark. The only non-reduced root system is $\mathbf{BC}_k$ and is defined as the union of the root systems $\mathbf{B}_k$ and $\mathbf{C}_k$.

4.31. Classification of complex semi-simple Lie algebras. (See [**Ser1**]). Let $\mathfrak{g}$ be a complex Lie algebra. It is called *semi-simple* iff the following Cartan–Killing form

$$(x, y) \longmapsto \operatorname{tr} \operatorname{ad}_x \operatorname{ad}_y$$

is non-degenerate. (Here $\mathrm{ad}_x z = [x, z]$).
It turns out that, if $\mathfrak{g}$ is semi-simple, then there exists a maximal abelian subalgebra $\mathfrak{h}$ of $\mathfrak{g}$ (i.e., $[x, y] = 0$ for $x, y \in \mathfrak{h}$ and there is no larger subalgebra with this property containing $\mathfrak{h}$). It is called the **Cartan subalgebra.** (One can choose the Cartan subalgebra as $\{z : \mathrm{ad}_x z = 0\}$ for typical $x \in \mathfrak{g}$). The Cartan form restricted to the Cartan subalgebra is non-degenerate. Due to this, one can equip also the space $\mathfrak{h}^*$, conjugate to $\mathfrak{h}$, with the non-degenerate bilinear form, denoted also by $(\cdot, \cdot)$.
The operators $x \longmapsto [h, x]$, $h \in \mathfrak{h}$, commute and have the same eigenspaces expansions. The eigenvalues are linear functionals on $\mathfrak{h}$, $[h, v] = \alpha(h)v$. Denote by $\mathfrak{g}^\alpha = \{v : [h, v] = \alpha(h)v\}$. Consider the set

$$\Delta = \{\alpha \in \mathfrak{h}^* \setminus 0 : \mathfrak{g}^\alpha \neq 0\}.$$

Its elements are called the **roots**.

4.32. Theorem. *The system Δ is a reduced root system. Moreover, any reduced root system is a system of roots associated with some semi-simple complex Lie algebra and some its Cartan subalgebra; this correspondence is one-to-one.*
Such algebras are classified by means of the Dynkin diagrams from Theorem 4.30.

One can find the proof of this result in the book of Serre [**Ser1**]. Because the classification of the root systems was completed, in this way the complete classification of semi-simple Lie algebras was achieved. (Note that the classification of solvable Lie algebras is not finished yet, see [**Kir**]).

4.33. Examples. (a) The Lie algebra corresponding to the root system $\mathbf{A}_k$ is $\mathfrak{g} = sl(k + 1, \mathbb{C})$, the space of traceless $(k + 1) \times (k + 1)$ matrices. Here, (and in the next example), the Cartan form is proportional to $(A, B) = \mathrm{tr}\, AB^\top$. The Cartan subalgebra consists of diagonal matrices

$$\mathfrak{h} = \{\mathrm{diag}\,(\lambda_1, \ldots, \lambda_{k+1}) : \textstyle\sum \lambda_i = 0\}.$$

The roots are $\alpha_{ij}(\lambda) = \lambda_j - \lambda_i$ and the corresponding eigenspaces $\mathfrak{g}^{\alpha_{ij}}$ are one-dimensional (it is general property) and are generated by the matrices $E_{ij} = \{a_{kl}\}$ with $a_{ij} = 1$ and $a_{kl} = 0$ otherwise.

(b) The Lie algebra with the Dynkin diagram $\mathbf{D}_k$ is $so(2k, \mathbb{C})$, i.e., the Lie algebra of the group of $(2k) \times (2k)$matrices preserving some non-degenerate symmetric quadratic form. (Note that over $\mathbb{C}$ any two such forms are equivalent).
We choose this quadratic form given by the symmetric matrix $\begin{pmatrix} 0 & E \\ E & 0 \end{pmatrix}$, where E is the unit matrix. Then $so(2k)$ consists of matrices $\begin{pmatrix} X & Y \\ Z & U \end{pmatrix}$ such that $U = -X^\top$, $Y^\top = -Y$, $Z^\top = -Z$. The Cartan subalgebra is equal to $\mathfrak{h} = \{\mathrm{diag}(\lambda_1, \ldots, \lambda_k, -\lambda_1, \ldots, -\lambda_k)\}$. The roots and the generators of the eigenspaces

are following: $\alpha_{ij}(\lambda) = \lambda_i - \lambda_j$, $\begin{pmatrix} E_{ij} & 0 \\ 0 & E_{ji} \end{pmatrix}$, $i \neq j$, where E_{ij} is the matrix from the previous example, and $\beta_{ij} = \lambda_i + \lambda_j$, $\begin{pmatrix} 0 & E_{ij} - E_{ji} \\ 0 & 0 \end{pmatrix}$, $i < j$; $-\beta_{ij}$, $\begin{pmatrix} 0 & 0 \\ E_{ij} - E_{ji} & 0 \end{pmatrix}$.

(c) The Lie algebra of the type $\mathbf{B}_k$ is $so(2k+1, \mathbb{C})$. The algebra $\mathbf{C}_k$ is equal to $sp(k, \mathbb{C})$, i.e., the algebra of the group of $(2k) \times (2k)$ matrices preserving some non-degenerate anti-symmetric quadratic form.

4.34. Coxeter groups. (See [**Bou**] and [**Hum**]). The **Coxeter group** W is formally defined by means of (finitely many) generators $s \in S$ and relations

$$s^2 = e, \quad (ss')^{m(s,s')} = e.$$

We have $m(s, s') \geq 2$ for $s \neq s'$. The matrix $M = \{m(s, s')\}_{s,s' \in S}$ is called the *Coxeter matrix*. It is symmetric, integer, with the entries 1 at the diagonal and the entries ≥ 2 outside the diagonal.

One can associate with the Coxeter matrix the following bilinear form on $\mathbb{R}^S$ (with the canonical basis (e_s)):

$$B(e_s, e_{s'}) = -\cos(\pi/m(s, s')).$$

The map

$$\sigma_s : x \longmapsto x - 2B(e_s, x)e_s$$

is called *pseudo-reflection*. It preserves the form B and, when the matrix M is positive and non-degenerate, we have the realization of the Coxeter group as a group generated by reflections.

It turns out that the form B is positive and non-degenerate if and only if the Coxeter group W is finite (see Theorem 2 in [**Bou**], Ch. V, §4).

One encodes the Coxeter group by means of the **Coxeter graph** defined as follows. Its vertices correspond to the generators S. Vertices s and s' are connected by means of an edge without index if $m(s, s') = 3$ and by an edge with the index $m = m(s, s')$ if $m > 3$. (If $m(s, s') = 2$, then there is no edge).

The Coxeter group is called *irreducible* iff its Coxeter graph is connected.

A group generated by reflections in $\mathbb{R}^n$ is called **crystallographic** iff it preserves some lattice $\Lambda \subset \mathbb{R}^n$, $\Lambda \simeq \mathbb{Z}^n$.

4.35. Theorem. *If W is a finite Coxeter group, then it has the Coxeter graph of one of the types presented at Fig.* 4.21.

4.36. Examples. (a) The Coxeter groups of the types $\mathbf{A}_k, \mathbf{B}_k, \mathbf{D}_k, \mathbf{E}_k, \mathbf{F}_4, \mathbf{G}_2 \simeq \mathbf{I}_2(6)$ are equal to the Weyl groups of the corresponding root systems.

Note that there is no the case $\mathbf{C}_k$ in the latter theorem. The reason is that the Weyl group of this root system is isomorphic to the Weyl group of the system $\mathbf{B}_k$.

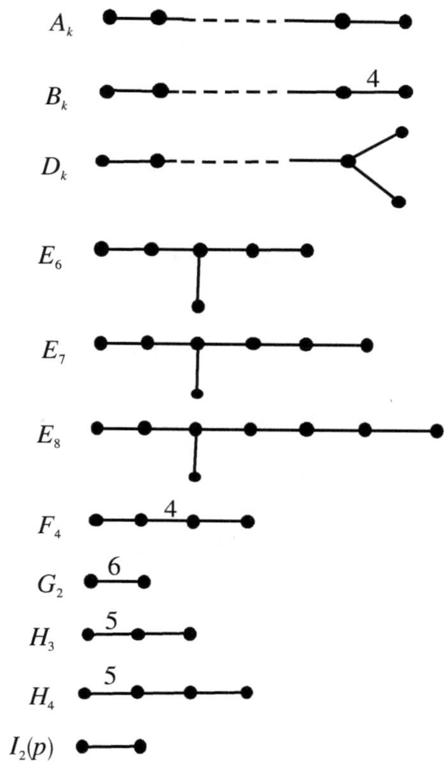

Figure 4.21: Coxeter graphs of Coxeter groups

The map $\alpha \longmapsto \alpha^\vee = 2\alpha/(\alpha, \alpha)$ transforms the root system of the $\mathbf{B}_k$ type to the system of the $\mathbf{C}_k$ type.

(b) The group $\mathbf{H}_3$ is the group of isometries of the icosahedron (and of its dual, the dodecahedron).

(c) The group $\mathbf{H}_4$ is the group of isometries of the regular 120-cell in $\mathbb{R}^4$ (with 120 3-dimensional cells, 720 faces, 1200 edges and 600 vertices). The groups $\mathbf{H}_3$ and $\mathbf{H}_4$ are not crystallographic.

(d) The group $\mathbf{I}_2(p)$ is the *dihedral group* of isometries of the regular p-gon imbedded in the space: rotations by the angles $2\pi j/p$ and reflections along p lines. Often this group is denoted by $\mathcal{D}_p$; (however it is not the Weyl group of the root system $\mathbf{D}_p$).
The groups $\mathbf{I}_2(p)$, $p = 5$ (denoted also by $\mathbf{H}_2$), or $p \geq 7$, are also not crystallographic. All other groups from Fig. 4.21 are crystallographic.

From the previous subsection we get the following result.

4.37. Theorem. *The monodromy groups of the simple singularities* $\mathbf{A}_k$, $\mathbf{D}_k$, $\mathbf{E}_6$, $\mathbf{E}_7$, $\mathbf{E}_8$ *of holomorphic functions are isomorphic to the corresponding Weyl groups.*

4.38. Remark. Already all other finite Coxeter groups have found applications in the singularity theory.
The Dynkin diagrams of the type $\mathbf{B}_k$, $\mathbf{C}_k$ and $\mathbf{F}_4$ appear in the analysis of singularities of functions on manifolds $(\mathbb{C}^n, 0)$ with boundary $(x_1 = 0)$: x_1^k, $x_1x_2 + x_2^k$ and $x_1^2 + x_2^3$. (One treats the manifold with boundary as the result of the 2-fold covering of $(\mathbb{C}^n, 0) \longmapsto (\mathbb{C}^n, 0)$ ramified along $\mathbb{C}^{n-1}$. If we pull-back the function f to the covering space, i.e., $\hat{f}$, then it becomes $\mathbb{Z}_2$-invariant. The corresponding space of vanishing homologies is divided into two different representations of $\mathbb{Z}_2$: the invariant cycles and the anti-invariant cycles. The Dynkin diagrams of $\mathbf{B}_k$, $\mathbf{C}_k$, $\mathbf{F}_4$ are associated with the anti-invariant component, see [**AVG**].)
The groups $\mathbf{G}_2$, $\mathbf{I}_2(p)$ and $\mathbf{H}_3$ have found application in symmetric singularities $x_1^3 + x_2^3$, $x_1^p + x_2^p + x_1^2x_2^2$ and $x_1^5 + x_2^5$; as monodromy groups in some components (character eigenspaces) of representation of the symmetry group in the space of vanishing homologies (see [**VC**]).
The Coxeter graphs of the type $\mathbf{H}_3$ and $\mathbf{H}_4$ appear in certain variational problems (see [**Arn6**]).
Below the reader will find another connection between the Weyl groups of root systems and the singularities.

§4 Bifurcational Diagrams

Let

$$F : (\mathbb{C}^n \times \mathbb{C}^\mu, 0) \longmapsto (\mathbb{C}, 0)$$

be a mini-versal deformation of a singularity $f(x) = F(x, 0)$. One can associate with it several bifurcational diagrams.

4.39. Definition. The **bifurcational diagram of zeroes** of the deformation F is

$$\Sigma = \{\lambda : 0 \text{ is critical value of } F(\cdot, \lambda)\}.$$

Examples. 1. The bifurcational diagram of zeroes of the singularity $\mathbf{A}_2 : x^2 + \lambda$ consists of one point $\lambda = 0$.

2. $\mathbf{A}_3 : F = x^3 + \lambda_1 x + \lambda_2$. Here Σ consists of those λ's, for which the above cubic polynomial has double root, i.e., it is the discriminant of this polynomial.
We have $x^3 + \lambda_1 x + \lambda_2 = 0$ and $F_x' = 3x^2 + \lambda_1 = 0$. Eliminating x, we get the equation of the *cusp*, whose real part is presented at Fig. 4.22(a).

$$4\lambda_1^3 + 27\lambda_2^2 = 0.$$

3. $\mathbf{A}_4$. We have $x^4+\lambda_1 x^2+\lambda_2 x+\lambda_3 = 4x^3+2\lambda_1 x+\lambda_2 = 0$. Substituting λ_2 from the second equation to the first, we obtain the parametric representation of the diagram Σ (by means of x, λ_1):

$$\lambda_2 = -2\lambda_1 x - 4x^3, \ \lambda_3 = \lambda_1 x^2 + 3x^4.$$

We shall study the sections of the surface Σ by means of the planes $\lambda_1 = \text{const}$. Moreover, due to symmetry, it is enough to study the behavior of the obtained curves only for $x \geq 0$. Depending on λ_1 we have three cases.
If $\lambda_1 = 0$, then we get the 'parabola' with the exponent 4/3: $\lambda_3 = \text{const} \cdot \lambda_2^{4/3}$.
If $\lambda_1 > 0$, then the curve $\lambda_{2,3}(x)$, $x > 0$, is regular, lies in the first quadrant, and the coordinates grow with x.
If $\lambda_1 < 0$, then, for small $x > 0$, our curve lies in the fourth quadrant ($\lambda_3 < 0 < \lambda_2$), and, for large x, it tends to infinity in the second quadrant ($\lambda_2 < 0 < \lambda_3$). Differentiating $\lambda_{2,3}(x)$: $\lambda_2' = -2(\lambda_1 + 6x^2)$, $\lambda_3' = 2x(\lambda_1 + 6x^2)$, we find that the curve is singular for $x = \sqrt{-\lambda_1/6}$ at $\lambda_{2,0} = 8(-\lambda_1/6)^{3/2}$, $\lambda_{3,0} = -\lambda_1/12$.
The further differentiations give $\lambda_2'' = -24\sqrt{-\lambda_1/6}$, $\lambda_3'' = -4\lambda_1 = -\sqrt{-\lambda_1/6}\cdot\lambda_2''$, $\lambda_2''' = -24$, $\lambda_3''' = 72\sqrt{-\lambda_1/6}$ at the singular point. This shows that this singular point is of the cusp type: $\lambda_2 - \lambda_{2,0} = \text{const}\cdot x^2 + \ldots$, $\sqrt{-\lambda_1/6}(\lambda_2 - \lambda_{2,0}) + (\lambda_3 - \lambda_{3,0}) = \text{const}\cdot x^3 + \ldots$.
The real part of this bifurcational diagram, called the *swallow tail*, is presented at Fig. 4.22(b).

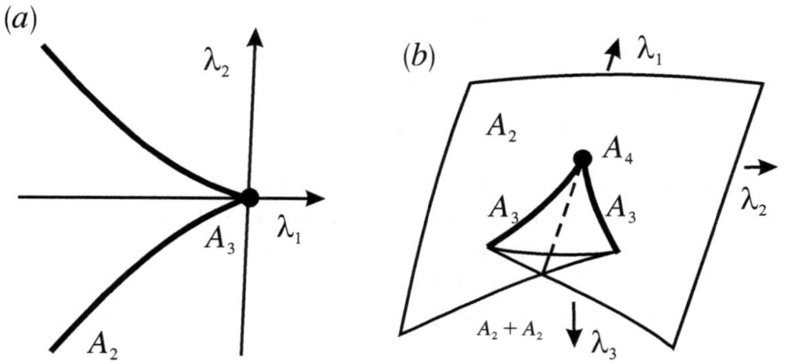

Figure 4.22: Cusp and swallow tail for the $\mathbf{A}_3$ and $\mathbf{A}_4$ singularities

4.40. Definition (Caustic). The set

$$\{\lambda : F(\cdot, \lambda) \text{ has degenerate critical point}\},$$

called the **caustics,** appears in the *catastrophe theory* in the following situations. Consider the conservative Newton system $\ddot{x} = -\nabla V(x)$, $x \in \mathbb{R}^n$. This system is in an equilibrium state iff its potential energy takes locally minimal value, i.e., $\dot{x} = 0$

and x equals to certain critical point of V. If $V = F(x, \lambda)$ varies with parameters, then the critical points move and can coalesce for λ at the **caustics**.
Because the physics of the system does not depend on the free constant in the definition of the potential energy, we can consider only the *restricted mini-versal deformation* $F_0(x, \lambda) = f(x) + \sum_{i=1}^{\mu-1} \lambda_i \phi_i(x)$, where there is no constant term; (ϕ_i form the basis of $\mathfrak{m}/I_f$). Moreover, because our space-time is four-dimensional, one should study only the singularities of codimension ≤ 5. There are seven such singularities:

$$\mathbf{A}_2,\ \mathbf{A}_3,\ \mathbf{A}_4,\ \mathbf{A}_5,\ \mathbf{D}_4^+,\ \mathbf{D}_4^-,\ \mathbf{D}_5.$$

(Here the singularities $\mathbf{D}_4^{\pm}$ have the form $x^3y \pm y^3$ and are different, because the space is real). These are the seven *elementary catastrophes* of R. Thom. In his book [**Tho**] one can also find the complete bifurcational diagrams (caustics) of these singularities.
Later it turned out that the bifurcations of the potential energy are not sufficient to describe the changes in the qualitative behavior of the corresponding Newton system (J. Guckenheimer). Because the gradient is defined by means of the metric, one should study altogether the bifurcations of the potential as well as the metric tensor (see [**Arn7**]).

The origin of the notion of caustic comes from the wave and geometrical optics. If $S \subset \mathbb{R}^3$ is a surface which is a source of the light (or of the electromagnetic waves), then some points of the space are more intensely lighted than others. These are the points where the rays starting at the surface S are focused. If we denote by x the points in S, by λ the points in the surrounding space and by $F(x, \lambda)$ the optical length of the light ray from x to λ, then the optical caustics coincides with the corresponding bifurcational diagram of the unfolding F:

$$\{\lambda : \exists x \ \ D_x F(x, \lambda) = 0, \ \ \det D_{xx}^2 F(x, \lambda) = 0\}.$$

In [**AVG**] the authors introduce also the *bifurcational diagram of function* as the set of those parameters of the restricted mini-versal deformation, for which the function $F(\cdot, \lambda)$ is not a Morse function. The bifurcational diagram of function is obtained from the caustics, by adding the bifurcational surfaces, for which the function has several Morse critical points with the same value. The latter are called the *Maxwell strata*.

4.41. Examples. (a) The singularity $\mathbf{A}_1$ has no caustic. The caustic of $\mathbf{A}_k$ is obtained by the equations $F'_x(x, \lambda) = F''_{xx}(x, \lambda) = 0$ for $F = x^{k+1} + \lambda_1 x^{k-1} + \ldots + \lambda_{k-1} x$. In the case of $\mathbf{A}_2$ it is a point, in the case $\mathbf{A}_3$ it is the cusp and in the case $\mathbf{A}_4$ it is the swallow tail (see Fig. 4.22).
(b) Consider the restricted mini-versal deformation of the singularity $\mathbf{D}_4^{\pm}$:

$$F = x^2y \pm y^3/3 + \lambda_1 y^2 + \lambda_2 x + \lambda_3 y.$$

The condition for the partial derivatives gives

$$2xy + \lambda_2 = 0, \tag{4.5}$$

$$x^2 \pm y^2 + 2\lambda_1 y + \lambda_3 = 0. \tag{4.6}$$

The condition for vanishing of the determinant of the Hessian gives the third equation

$$x^2 \mp y^2 - \lambda_1 y = 0. \tag{4.7}$$

Let us concentrate on the case $\mathbf{D}_4^-$. As in the case of swallow tail we draw the sections of the caustics by the planes $\lambda_1 = \text{const}$; so, we fix λ_1.
From the equation (4.7) we get

$$x = \pm\sqrt{y(\lambda_1 - y)}$$

and the equations (4.5) and (4.6) allow us to give the parametric representation of the caustic by means of the 'parameters' λ_1, y:

$$\lambda_2 = \mp 2y\sqrt{y(\lambda_1 - y)}, \quad \lambda_3 = y(2y - 3\lambda_1).$$

If $\lambda_1 = 0$, then $y = 0$ (because of $\sqrt{-y^2}$) and thus we get one point $\lambda_2 = \lambda_3 = 0$. If $\lambda_1 < 0$, then from the formula for λ_2 it follows that y can take values between λ_1 and 0. Near $y = 0$ we get the cusp type singularity: $\lambda_2 \sim (-y)^{3/2}, \lambda_3 \sim y$ (here $\lambda_3 < 0$). Near $y = \lambda_1$, λ_2 is close to 0, λ_3 is close to $-\lambda_1^2$ and the curve is smooth. Differentiation with respect to y gives

$$(\lambda_2^2/4)' = y^2(3\lambda_1 - 4y), \quad \lambda_3' = 4y - 3\lambda_1.$$

Thus the point corresponding to $y = 3\lambda_1/4$ is singular. The calculations of the next derivatives show that it is of the cusp type: $(\lambda_2^2/4)'' = -(3\lambda_1/2)^2$, $\lambda_3'' = 4$, $(\lambda_2^2/4)''' = -12\lambda_1 \neq 0$, $\lambda_3''' = 0$.
This allows to draw the complete curve (symmetric with respect to the λ_3-axis, see Fig. 4.23(b)).
If $\lambda_1 > 0$, then the change $y \longmapsto -y$, $\lambda_1 \longmapsto -\lambda_1$, $\lambda_2 \longmapsto -\lambda_2$ returns us to the previous case.
The complete caustics for $\mathbf{D}_4^-$, the so-called the *pyramid*, is presented in Fig. 4.23(b).
The analogous analysis can be performed in the case $\mathbf{D}_4^+$. The corresponding bifurcational diagram, called the *basket*, is presented in Fig. 4.23(a).

4.42. Milnor bundle associated with a deformation. Let $F(x, \lambda)$ be a mini-versal deformation of $f = F(\cdot, 0)$ and let Σ be its bifurcational diagram of zeroes. Let $\widetilde{\mathbb{D}}_\epsilon$ be the small ball in the space of parameters λ, of radius ϵ, and let $\mathbb{B} = \mathbb{B}_\rho$ be the small ball in the space of variables x, of radius ρ. Denote $\Sigma_\epsilon = \Sigma \cap \widetilde{\mathbb{D}}_\epsilon$, $V = \{F(x, \lambda) = 0, \lambda \notin \Sigma\} \cap \mathbb{B} \times \widetilde{\mathbb{D}}_\epsilon$.
The space $\widetilde{\mathbb{D}}_\epsilon \setminus \Sigma$ forms a base of the following locally trivial fibration

$$V \stackrel{F}{\longmapsto} \widetilde{\mathbb{D}}_\epsilon \setminus \Sigma \tag{4.8}$$

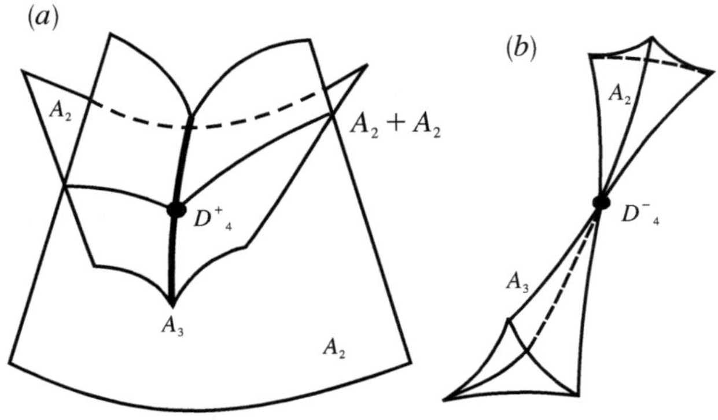

Figure 4.23: Basket and Pyramid for $\mathbf{D}_4^{\pm}$ singularities

with the fiber V_λ diffeomorphic to the non-singular level surface of f. It can be also called the *Milnor fibration*.
In this section we study action of the fundamental group of the base of this fibration $\pi_1(\widetilde{\mathbb{D}}_\epsilon \setminus \Sigma, \lambda_*)$ on the homology group of the distinguished fiber, $H_{n-1}(V_{\lambda_*})$ (λ_* is a distinguished parameter). It is defined as follows. Any loop in $\widetilde{\mathbb{D}}_\epsilon$ around Σ is lifted to a deformation of fibers and defines a diffeomorphism of the distinguished fiber V_{λ_*} inducing the operator in the homology group. We have the following well defined homomorphism of groups:

$$\pi_1(\widetilde{\mathbb{D}}_\epsilon \setminus \Sigma) \longmapsto \operatorname{Aut} H_{n-1}(V_{\lambda_*}).$$

Recall also that the monodromy group of the germ f is the image in $\operatorname{Aut} H_{n-1}(\widetilde{V}_{z_0})$ of the fundamental group $\pi_1(\mathbb{D}_\epsilon \setminus \{z_1, \dots, z_\mu\}, z_0)$ of the base of the Milnor fiber bundle $\tilde{f} : \widetilde{V} \longmapsto \mathbb{D}_\varepsilon \setminus \{z_1, \dots, z_\mu\}$ (see (4.8)); (here z_i are the critical values of the Morse perturbation $\tilde{f}$ of f).

4.43. Theorem. *The image of the group* $\pi_1(\widetilde{\mathbb{D}}_\epsilon \setminus \Sigma)$ *in* $\operatorname{Aut} H_{n-1}(V_{\lambda_*})$ *is isomorphic to the monodromy group of the germ* f.

Proof. One can choose the mini-versal deformation of the germ f in the following form

$$F_0(x, \lambda') - \lambda_0, \quad \lambda' \in \mathbb{C}^{\mu-1},$$

where F_0 is the restricted mini-versal deformation (without the constant term). As a Morse perturbation of f we choose

$$\tilde{f} = F_0(x, \lambda'_*),$$

where λ'_* is some properly chosen parameter in $\mathbb{C}^{\mu-1}$.

Let $\pi : (\lambda_0, \lambda') \longmapsto \lambda'$ be the projection and define the 1-dimensional disc $\mathbb{D} = \pi^{-1}(\lambda'_*)$. The intersection $\mathbb{D} \cap \Sigma$ consists of the points $(\lambda_{0,i}, \lambda'_*)$ such that $\lambda_{0,i}$ is the critical value of $\tilde{f}$. Let $\lambda_{0,*}$ be noncritical value of $\tilde{f}$ and denote $\lambda_* = (\lambda_{0,*}, \lambda'_*)$. We have the following commutative diagram

$$\begin{array}{ccc} \pi_1(\mathbb{D} \setminus \Sigma) & \longmapsto & \operatorname{Aut} H_{n-1}(V_{\lambda_*}) \\ \downarrow i_* & & \uparrow \\ \pi_1(\widetilde{\mathbb{D}}_\epsilon \setminus \Sigma) & = & \pi_1(\widetilde{\mathbb{D}}_\epsilon \setminus \Sigma) \end{array}$$

where i_* is the homomorphism induced by the inclusion $i : \mathbb{D} \setminus \Sigma \longmapsto \widetilde{\mathbb{D}}_\epsilon \setminus \Sigma$.
We see that the thesis of Theorem 4.43 is equivalent to the following:

$$i_* \text{ is epimorphism.}$$

(Then the images of the two fundamental groups in Aut coincide).
We know the structure of the fundamental group of the punctured disc $\mathbb{D} \setminus \{\lambda_{0,1}, \ldots, \lambda_{0,\mu}\}$. It is the free group generated by μ simple loops surrounding the punctures.
The structure of the second fundamental group is much more complicated. Using the results of the Chapter 2, we can restrict ourselves to the situation when everything is algebraic. In particular, the bifurcational diagram of zeroes Σ is an affine algebraic subvariety in $\mathbb{C}^\mu$. So, we want to know the structure of the fundamental group of complement of an affine algebraic hypersurface. It is partially described by the theorem of Zariski and van Kampen formulated below (see [**Zar**]).
Let $M^{n-1} \subset \mathbb{C}^n$ be an algebraic hypersurface. If $L \subset \mathbb{C}^n$ is a line in general position, then it intersects M transversally at a finite set $L \cap M = \{p_1, \ldots, p_m\}$. If $i : L \setminus M \longmapsto \mathbb{C}^n \setminus M$ is the inclusion, then it induces the homomorphism $i_* : \pi_1(L \setminus M) \longmapsto \pi_1(\mathbb{C}^n \setminus M)$. The Zariski–van Kampen theorem says that i_* is a surjection and also describes generators and relations in the group $\pi_1(\mathbb{C}^n \setminus M)$. Let $p_0 \in L \setminus M$. We choose loops $\tau_i \subset L \setminus M$ starting at p_0 and surrounding just one point p_i, e.g., the simple loops as in the definition of the monodromy group. The loops $i_*\tau_i$ are the generators of $\pi_1(\mathbb{C}^n \setminus M)$. Below we describe the relations satisfied by these generators.
Let $\pi : \mathbb{C}^n \longmapsto \mathbb{C}^{n-1}$ be the projection along the line L, $\pi(x, y) = x$. Its fibers are the lines L_x and $L_x \cap M = \{p_1(x), \ldots, p_\mu(x)\}$. We take the restriction $\pi|_M$ and consider the set of critical points of the latter map. Its image under $\pi|_M$, i.e., the set of critical values of $\pi|_M$, is an algebraic subvariety N in $\mathbb{C}^{n-1}$. If L is generic, then N is a hypersurface. The surface N consists of those x at which some of the points $p_i = p_i(x)$ coalesce (see Fig. 4.24).
Choose a generic point x_0 from N, e.g., non-singular. Let $\widetilde{L} \subset \mathbb{C}^{n-1}$ be a generic line passing through x_0 and transversal to N. If (for example) for $x = x_0$ only two points $p_i = p_i(x)$ and $p_j = p_j(x)$ coalesce and the point $x \in \widetilde{L}$ runs around x_0, along a loop σ, then the points p_i and p_j exchange their positions. This means that σ induces an automorphism T_σ of the fundamental group of the punctured line $L \setminus M$.

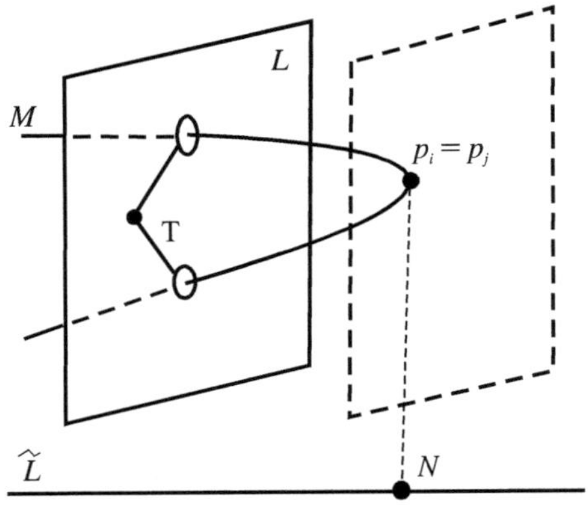

Figure 4.24: The projection π

Generally, let $\widetilde{L} \cap N = \{q_1, \ldots, q_k\}$ and let $\sigma : x = x(t),\ t \in [0,1]$ be a loop in $\widetilde{L} \setminus N$. It turns out that

$$i_*\tau_l = i_*T_\sigma\tau_l, \ \ l = 1, \ldots, m \tag{4.9}$$

in $\pi_1(\mathbb{C}^n \setminus M)$.
Indeed, if we fix the base points $p_0(t) = \{y = y_0\}$ in $L_{x(t)}$ away from the points $p_i(t) = p_i(x(t))$, then the loops τ_i are deformed to a family of loops $\tau_{i,t} \in L_{x(t)}$; the final loops $\tau_{i,1}$ are equal to $T_\sigma\tau_i$. These deformations do not provide the homotopy between $i_*\tau_i$ and $i_*T_\sigma\tau_i$ because the base points in the loops $\tau_{i,t}$ are not constant. However, it is not difficult to improve this: the new deformed loops $\widetilde{\tau}_{i,t}$ start at $p_0 \in L$, do not lie completely in the lines $L_{x(t)}$ and form small perturbations of the curves $s \longmapsto (x(st), y_t(s))$, where $\{y_t(s)\}_{s\in[0,1]}$ defines $\tau_{i,t}$.
All this is summarized in the next result, which completes the proof of Theorem 4.43. □

4.44. Theorem of Zariski–van Kampen. *The fundamental group of the complement of the affine hypersurface $M \subset \mathbb{C}^n$ has the loops $i_*\tau_l$ as generators, which are subject to the relations (4.9). In particular, it has m generators $i_*\tau_l$ and mk relations $i_*\tau_l = i_*T_{\sigma_j}\tau_l$ (where σ_j generate $\pi_1(\widetilde{L} \setminus N)$, and the homomorphism i_* is an epimorphism).*

4.45. Examples. (a) Assume that the hypersurface M is smooth. We can assume that the lines L, $\widetilde{L}$ and the hypersurface N are such that at each point $q_k \in \widetilde{L} \cap N$ only two points p_i and p_j coalesce. Now we look at Fig. 4.25. We have $i_*\tau_i = i_*T\tau_i = i_*\tau_j$ and, therefore, all the generators are equal. The other relation $i_*\tau_j = i_*T\tau_j = i_*\tau_i\tau_j\tau_i^{-1}$ is now trivial.
Generally, if the points p_i and p_j coalesce at a smooth point of M with first order of tangency of the corresponding line L_x, then the generators defined by means of

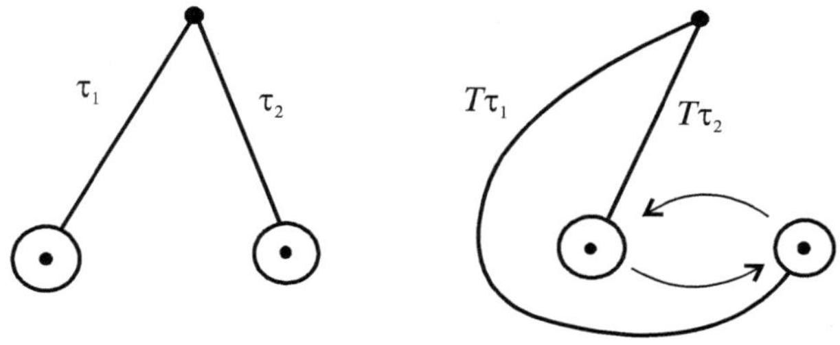

Figure 4.25: Application of Th. 4.44 in the smooth case

them coincide, $i_*\tau_i = i_*\tau_j$.
Unfortunately, this phenomenon is only local and one cannot make definite conclusions about the group $\pi_1(\mathbb{C}^n \setminus M)$. In the first version of this monograph it was claimed that this group is $\mathbb{Z}$, but an analysis done by the author with Z. Jelonek yielded that it can be not true in some cases.

(b) If p_i and p_j coalesce at some singular point of M of the type of transversal intersection of two smooth local components (double point), then, for suitably chosen lines L, $\widetilde{L}$, we have $M \approx \{x^2 = y^2\}$, where x is a coordinate in $\widetilde{L}$ and y is a coordinate in L. The action of the loop $\sigma : x = \epsilon e^{2\pi i t}$ results in full turn of the points $y_{1,2} = \pm x$. From Fig. 4.26 we get $i_*T\tau_1 = i_*\tau_2\tau_1\tau_2^{-1}$, which means that the generators $i_*\tau_1$ and $i_*\tau_2$ commute. (The second identity $i_*T\tau_2 = i_*\tau_2\tau_1\tau_2\tau_1^{-1}\tau_2^{-1}$ gives nothing new).
The above commutativity is a consequence of the fact that $\pi_1(X \times Y) = \pi_1(X) \times \pi_1(Y)$ applied to $X \approx Y \approx \mathbb{C} \setminus 0$.

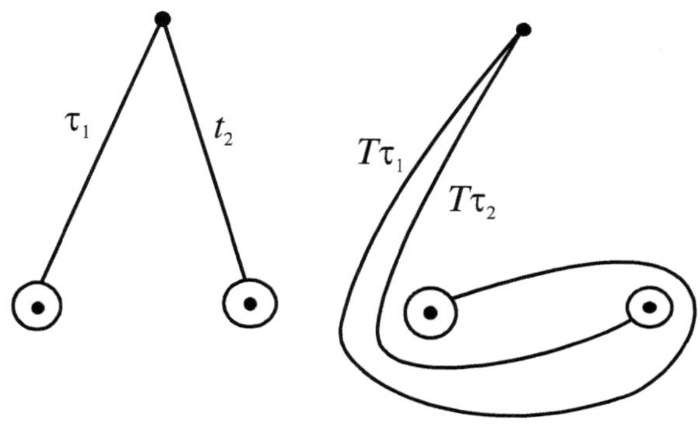

Figure 4.26: Application of Th. 4.44 in the case of node singularity

(c) Consider the curve $M = \{x^2 = y^3\}$ with the cusp singularity. Here we can choose the lines as $L_x = \{x = \text{const}\}$ and the fundamental group $\pi_1(\mathbb{C}^2 \setminus M)$ is generated by the loops i_*a, i_*b, i_*c surrounding the points $y_{1,2,3} = x^{2/3}$ (see Fig. 4.27). The action of the loop $\sigma : x = \epsilon e^{2\pi i t}$ results in the rotation of the points y_i by the angle $4\pi/3$. From Fig. 4.27 we get the action T of the loop σ on the generators a, b, c:

$$\begin{aligned} i_*a &= i_*Ta = i_*cbc^{-1}, \\ i_*b &= i_*Tb = i_*(cb)c(cb)^{-1}, \\ i_*c &= i_*Tc = i_*(cb)a(cb)^{-1}. \end{aligned}$$

The first from the above identities allows us to express i_*a by means of i_*b, i_*c (so we have two generators). The second identity means that

$$i_*b \cdot i_*c \cdot i_*b = i_*c \cdot i_*b \cdot i_*c. \tag{4.10}$$

The reader can check that the third relation follows from the first two.
We shall generalize this example in Theorem 4.46 below.

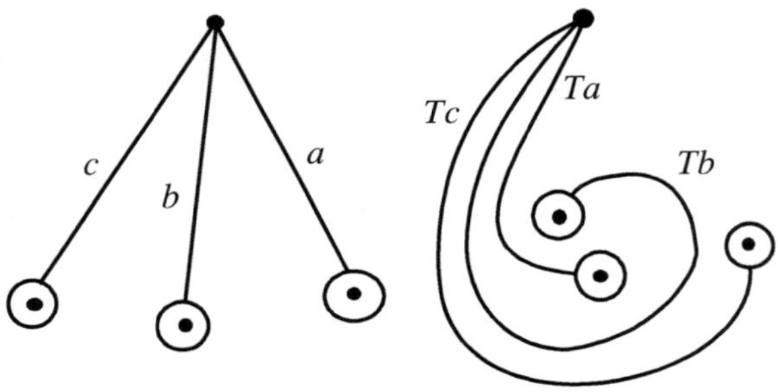

Figure 4.27: Application of Th. 4.44 in the case of the cusp singularity

(d) The above construction (with the lines L, $\widetilde{L}$) can be generalized to the projective case. Let M be an algebraic hypersurface in $\mathbb{CP}^n$. Take a generic point $z \in \mathbb{CP}^n \setminus M$ and a generic hyperplane $H \subset \mathbb{CP}^n \setminus z$. They define the projection $\pi : \mathbb{CP}^n \setminus z \longmapsto H$; $\pi(p)$ is the intersection of the projective line, passing through z and p, with the hyperplane H. If H is parameterized by x, then the projective lines $L_x = \pi^{-1}(x)$ play the role of the affine lines in the previous construction. Let L be some distinguished (but generic) projective line from this family, let $L \cap M = \{p_1, \dots, p_m\}$ and let $i : L \setminus M \longmapsto \mathbb{CP}^n \setminus M$ denotes the embedding. We denote the set of critical values of the restriction of the projection π to M by N, it is an algebraic hypersurface in the projective space $\mathbb{CP}^{n-1} = H$. We choose a general projective line $\widetilde{L} \subset H$ intersecting N transversally at smooth points, say at $q_1, \dots, q_k$.
As before, we choose the generators $i_*\tau_i$, $i = 1, \dots, m$. They are subject to the relations $i_*T_\sigma\tau_i = i_*\tau_i$ for loops $\sigma \in \pi_1(\widetilde{L} \setminus N)$.

However, there is one more relation. Note that the composed loop $\tau_m \tau_{m-1} \dots \tau_1$ is contractible in the punctured sphere $L \setminus M \simeq \mathbb{S}^2 \setminus \{p_1, \dots, p_m\}$. Thus, the product of the generators is the neutral element of our fundamental group.
Combining this with the previous examples, we get the following general result (see [**Del1**], [**Zar**]).

4.46. Definition. We say that the analytic hypersurface $S \subset \mathbb{C}^n$ has the **normal crossings singularities** iff near each point $p \in S$ there is a local analytic system of coordinates $x_1, \dots, x_n$ such that we have

$$S = \{x_1 x_2 \cdots x_k = 0\},$$

for some k between 1 and n.

4.47. Theorem of Zariski–Fulton–Deligne. *If an algebraic hypersurface $M \subset \mathbb{CP}^n$ has only normal crossings singularities, then $\pi_1(\mathbb{CP}^n \setminus M)$ is abelian. If, additionally, $M = M_1 \cup M_2 \cup \dots \cup M_s$, where M_i are irreducible components of degree d_i, then the above (abelian) fundamental group has s generators $\tau_1, \dots, \tau_s$ (each representing a loop surrounding just one M_i) which are subject to the relation*

$$\tau_1^{d_1} \tau_2^{d_2} \dots \tau_s^{d_s} = e.$$

4.48. The braid group and the colored braid group

(a) **Definition of the braid group.** The **braid group** $B(n)$ has $n-1$ generators $g_1, \dots, g_{n-1}$ with the relations

$$\begin{aligned} g_i g_{i+1} g_i &= g_{i+1} g_i g_{i+1}, \\ g_i g_j &= g_j g_i, \qquad\qquad |i-j| > 1; \end{aligned}$$

(note that the first from these relations coincides with the relation (4.10) in Example 4.45(c)). This formal definition is not very useful. Below we present the geometrical definition.

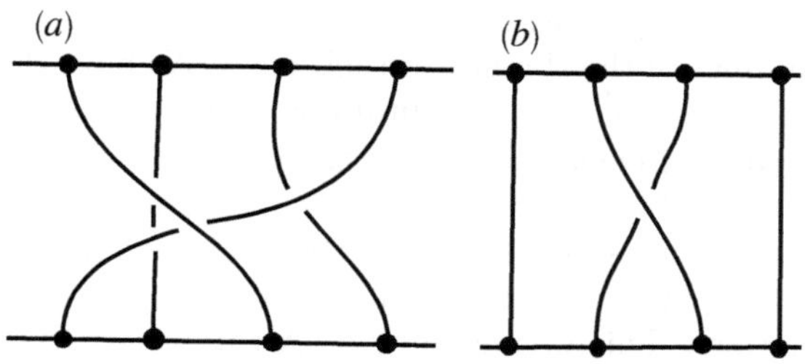

Figure 4.28: Briads: (a) a typical braid, (b) a generator of the braid group

The **braid** consists of n non-intersecting threads $(x_i(t), t)$, $t \in [0,1]$ in $\mathbb{R}^3$ (which we identify with $\mathbb{C} \times \mathbb{R}$) connecting the n fixed points, e.g., $1, 2, \dots, n \in \mathbb{C}$, at the lower base $\mathbb{C}\times\{0\}$ with the same points at the upper base $\mathbb{C}\times\{1\}$ (see Fig. 4.28). Two braids are treated as equivalent iff one can deform one to the other in such a way that the monotonicity of the threads and non-intersection is preserved during the deformation. The braids can be multiplied by adjoining one above the other. They form a group, called the **braid group** $B(n)$, whose unit element is represented by the non-knotted braid, consisting of vertical intervals. The reverse braid to the given one is obtained from it by reflection with respect to some horizontal plane. It is not difficult to see that the group $B(n)$ is generated by $n-1$ elementary braids $g_1,\dots,g_{n-1}$, where g_i is presented at Fig. 4.28(b). Moreover, Fig. 4.29 shows that the elements g_i satisfy the relations appearing in the formal definition of the braid group.

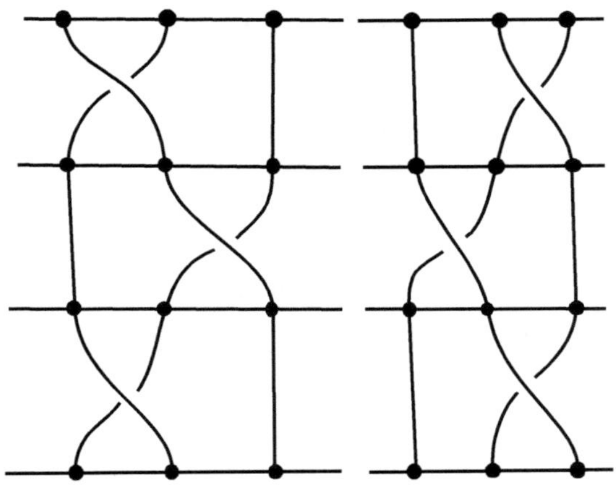

Figure 4.29: A relation between generators of the braid group

(b) **The colored braid group.** With every braid a permutation of the set $\{1,\dots,n\}$ is associated: beginnings of the threads to their ends. We have the homomorphism from $B(n)$ to $S(n)$, the group of permutations of the n-element set. Its kernel $\widehat{B}(n)$ is called the **colored braid group** (or the *dyed braid group* or the *pure braid group*). A braid is colored iff it returns to the same point.

In the further chapters we shall deal with the braid group as well as with the colored braid group. For this reason it is desirable to represent $\widehat{B}(n)$ by means of generators and relations. This was done by E. Artin in [**Art**] (see also [**Mos1**]). The generators of $\widehat{B}(n)$ are the braids σ_{ij}, $i<j$, presented in Fig. 4.30. The relations are following:

$$\begin{array}{lll} r<s<i<j & \Rightarrow & \sigma_{rs}^{-1}\sigma_{ij}\sigma_{rs}=\sigma_{ij}, \\ i<r<s<j & \Rightarrow & \sigma_{rs}^{-1}\sigma_{ir}\sigma_{rs}=\sigma_{is}\sigma_{ir}\sigma_{is}^{-1}, \quad \sigma_{rs}^{-1}\sigma_{is}\sigma_{rs}=\sigma_{is}\sigma_{ir}\sigma_{is}(\sigma_{is}\sigma_{ir})^{-1}, \\ i<r<j<s & \Rightarrow & \sigma_{rs}^{-1}\sigma_{ij}\sigma_{rs}=(\sigma_{ir}\sigma_{is}\sigma_{ir}^{-1}\sigma_{is}^{-1})\sigma_{ij}(\sigma_{ir}\sigma_{is}\sigma_{ir}^{-1}\sigma_{is}^{-1})^{-1}. \end{array}$$

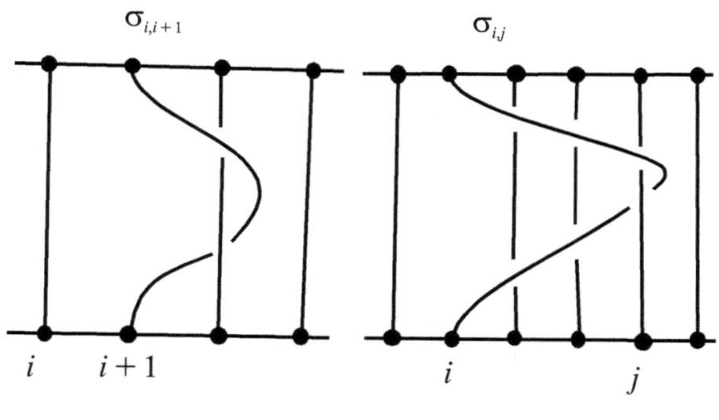

Figure 4.30: Generators of the colored braid group

(c) **Problem**: prove these relations.

(d) **Lemma.** *Let $\Delta \subset \mathbb{C}^n$ consists of points with some coordinates coinciding and let $\widetilde{\Delta} = \Delta/S(n)$, where the symmetric group $S(n)$ acts on $\mathbb{C}^n$ by permutation of the coordinates.*
The braid group $B(n)$ coincides with the fundamental group of $\mathbb{C}^n/S(n) \setminus \widetilde{\Delta}$ and the colored braid group is isomorphic with the group $\pi_1(\mathbb{C}^n \setminus \Delta)$.

Proof. The space of subsets $\{x_1, \ldots, x_n\}$ of $\mathbb{C}$ is equal to $\mathbb{C}^n/S(n)$. The space of n-element subsets is equal to $(\mathbb{C}^n \setminus \Delta)/S(n)$ and the space of colored subsets is equal to $\mathbb{C}^n \setminus \Delta$.
If $\{(x_i(t), t),\ t \in [0, 1]\}$ is the braid, then the map $t \longmapsto \{x_1(t), \ldots, x_n(t)\}$ defines a path in $\mathbb{C}^n/S(n) \setminus \widetilde{\Delta}$, with the begin and the end at the set $\{1, \ldots, n\}$. If, additionally, $x_i(1) = x_i(0) = i$, then we obtain a loop in $\mathbb{C}^n \setminus \Delta$.
The braids σ_{ij}, $i < j$, correspond to loops in $\mathbb{C}^n \setminus \Delta$ which run around exactly one hyperplane $x_i = x_j$. When we represent a point in $\mathbb{C}^n \setminus \Delta$ as the colored (ordered) set $\{x_1, \ldots, x_n\} \subset \mathbb{C}$, then the loop σ_{ij} defines the family $\{x_1(t), \ldots, x_n(t)\}$, $t \in [0, 1]$, such that the point $x_j(t)$ overruns the constant point $x_i(t) \equiv x_i(0)$ and other points do not move, $x_k(t) \equiv$ const. □

(e) The Fundamental Theorem of Algebra allows to construct a diffeomorphism between the spaces $\mathbb{C}^n$ and $\mathbb{C}^n/S(n)$, where the first space is identified with the space of polynomials $P_a(x) = x^n + a_1 x^{n-1} + \ldots + a_n$. One associates with the polynomial $P_a(x)$ its set of roots. The coefficients a_i form the elementary symmetric polynomials of the roots x_j: $a_1 = x_1 + \ldots + x_n, \ldots, a_n = x_1 \cdots x_n$.
Note that, in this identification, the polynomials with double or multiple roots correspond to elements of $\widetilde{\Delta}$. The set of $(a_1, \ldots, a_n)$'s corresponding to polynomials with multiple roots, i.e., the *discriminant set*, is just the bifurcational diagram $\widetilde{\Sigma}$ of zeroes of the deformation $P_a(x)$ of the function x^n, i.e., the $\mathbf{A}_{n-1}$ singularity.

The deformation $P_a(x)$ is versal, but it is not a mini-versal deformation of x^n; in the mini-versal deformation the coefficient a_1 is equal to zero. The discriminant set $\widetilde{\Sigma}$ is diffeomorphic to $\mathbb{C}\times\Sigma$, where Σ is the bifurcational diagram of zeroes of the standard unfolding $\lambda_0+\ldots+\lambda_{n-2}x^{n-2}+x^n$ of the $\mathbf{A}_{n-1}$ singularity.
This is related with the fact that the action the group of permutations $S(n)$ in $\mathbb{C}^n$ is reducible. It is decomposed into the action on the $(n-1)$-dimensional subspace $\{x_1+\ldots+x_n=0\}$ and into the trivial action on the 1-dimensional subspace $\{x_1=\ldots=x_n\}$.

4.49. Theorem (Complement of the discriminant). *The fundamental group of the complement of the bifurcational diagram of zeroes of the mini-versal deformation of the $\mathbf{A}_{n-1}$ singularity is equal to the braid group $B(n)$.*
Moreover, the space $\mathbb{C}^{n-1}\setminus\Sigma$ is the Eilenberg–MacLane space $K(\pi,1)$, where $\pi=B(n)$.

Proof. The first part of this theorem has been just proven.
Recall that the Eilenberg–MacLane space $X=K(\pi,k)$ is characterized by the property that $\pi_k(X)=\pi$ and $\pi_j(X)=0$ for $j\neq k$ (see Chapter 3, Section 3). To show the second part of Theorem 4.49, we notice that the space $\mathbb{C}^n\setminus\Delta$ forms the $n!$-fold covering of the space $\mathbb{C}^n/S(n)\setminus\widetilde{\Delta}$. The fundamental group of the first space is the group $\widehat{B}(n)$ of colored braids and the fundamental group of the second space is the group $B(n)$ of braids. The higher homotopy groups of these two spaces coincide. (This follows from the long exact sequence of the homotopy groups of a fibration, see 3.32).
On the other hand, using the fibrations $\mathbb{C}^n\setminus\Delta\longmapsto\mathbb{C}^{n-1}\setminus\Delta:\ (x_1,\ldots,x_n)\longmapsto(x_1,\ldots,x_{n-1})$ with the fiber $\mathbb{C}\setminus\{x_1,\ldots,x_{n-1}\}$ (which has trivial higher homotopy groups), it is easy to show that the space $\mathbb{C}^n\setminus\Delta$ is of the type $K(\widehat{B}(n),1)$. (Here again we use the long exact sequence of homotopy groups). □

There is another approach to topology of the bifurcational diagram of zeroes Σ. We recall that the group $S(n)$ is the Weyl group W of the root system of the type $\mathbf{A}_{n-1}$; it is also the Coxeter group generated by reflections and acting in the space $\mathbb{C}^{n-1}=\{x_1+\ldots+x_n=0\}\subset\mathbb{C}^n$. The reflections are of the form $s_\alpha: x\longmapsto x-2[(x,e_i-e_j)/|e_i-e_j|^2](e_i-e_j)=x-(x_i-x_j)(e_i-e_j)$ or $(x_1,\ldots,x_i\ldots,x_j,\ldots,x_n)\longmapsto(x_1,\ldots,x_j\ldots,x_i,\ldots,x_n)$ (the transpositions). The surfaces of fixed points of the reflections s_α, $\alpha\in\mathcal{R}$ (the root system), are called the *mirrors*. Here $\alpha=e_i-e_j$ and the mirrors are equal to $\{x_i=x_j\}$. The set Δ is an union of the mirrors of the Weyl group $W=S(n)$.
In particular, from the above it follows that the pair $(\widetilde{\mathbb{D}}_\epsilon,\Sigma_\epsilon)$ (neighborhood of zero in the parameter space and the germ of the bifurcational set) is isomorphic to the pair $(\mathbb{C}^n/W,\Delta/W)$. V. Arnold proved the following result (see [**AVG**]).

4.50. Theorem. *For the simple singularities $\mathbf{A}_k$, $\mathbf{D}_k$, $\mathbf{E}_k$ the pair $(\widetilde{\mathbb{D}}_\epsilon,\Sigma_\epsilon)$ is isomorphic to the pair $(\mathbb{C}^k/W,\Delta/W)$, where W is the Weyl group of the corresponding root system and Δ is the union of its mirrors.*

In [**AVG**] the generalized braid group $B_W(n) = \pi_1(\mathbb{C}^n/W \setminus \Delta/W)$ is introduced and the analogous result about $K(B_W(n), 1)$ property is proved. These results are generalized to the case of boundary singularities $\mathbf{B}_n$.

Below we present some general properties of the bifurcational diagrams, which will be used by us later.

4.51. Theorem. *Let $F(x, \lambda)$ be a mini-versal deformation of a singularity f of finite multiplicity. The local bifurcational diagram of zeroes Σ_ϵ is an irreducible analytic variety. More precisely, there is a germ of holomorphic map $(\mathbb{C}^{\mu-1}, 0) \longmapsto (\mathbb{C}^\mu, 0)$ whose image is $(\Sigma, 0)$.*

Proof. Recall that an analytic set S is called irreducible iff the set $S \setminus$ (singular points of S) is connected. For example, the set $\{xy = 0\} \subset \mathbb{C}^2$ is reducible. Of course, the image of a connected set by means of an analytic map is irreducible. Consider the set

$$\mathcal{A} = \{g : (\mathbb{C}^n, 0) \longmapsto (\mathbb{C}, 0) | \quad dg(0) = 0\}.$$

In particular, 0 is a critical value for germs from $\mathcal{A}$. Let $\mathcal{G}$ be the group of germs of analytic diffeomorphisms of $(\mathbb{C}^n, 0)$, i.e., preserving $x = 0$. The group $\mathcal{G}$ acts on the space $\mathcal{A}$. The orbit $\mathcal{G}f$ forms a subset of $\mathcal{A}$ of codimension $\mu - 1$; because the orbit $\mathcal{G}f$ in $\mathcal{O}_0(\mathbb{C}^n)$ has codimension μ and $\mathcal{A}$ lies in the hypersurface $\{f : f(0) = 0\}$. Let us choose a $(\mu - 1)$-parameter holomorphic transversal (at f) to $\mathcal{G}f$ in $\mathcal{A}$. It is a deformation of f, which is induced from the mini-versal deformation. The map realizing this induction is the needed map onto Σ_ϵ. Indeed, because all functions from this deformation have 0 as critical value, the whole neighborhood of $0 \in \mathbb{C}^{\mu-1}$ is sent to Σ_ϵ. □

This result has interesting consequences.

4.52. Theorem. *The monodromy group of an isolated singularity of a function acts transitively on the set of some system of vanishing cycle of its non-singular hypersurface level surface (associated with a system of simple loops, see Definition 4.7).*

Proof. We have to show that for any two vanishing cycles Δ_1, Δ_2 there is a monodromy map sending Δ_1 to Δ_2 or to $-\Delta_2$.
Let $F(x, \lambda) = F_0(x, \lambda') - \lambda_0$, where $F_0(x, \lambda')$ is the restricted deformation (without the constant term). We choose the Morse perturbation of f in the form $\tilde{f} = F_0(x, \lambda'_*)$ for some general fixed $\lambda'_* \in \mathbb{C}^{\mu-1}$. Let $L = \{(z, \lambda'_*) : z \in \mathbb{C}\}$ be the line in the space of parameters; (here z plays the role of λ_0).
We define the vanishing cycles Δ_i, $i = 1, 2$, by means of paths $\alpha_i \subset L$, joining the fixed point z_0 with the corresponding critical values z_i of $\tilde{f}$.
Let $\pi : (z, \lambda') \longmapsto \lambda'$ be the projection. The set of critical values of the map π, restricted to the smooth part of Σ, forms a subset of complex codimension one. Because Σ is irreducible, then $\Sigma \setminus$ (set iff its singular points)$\setminus$(set of critical points

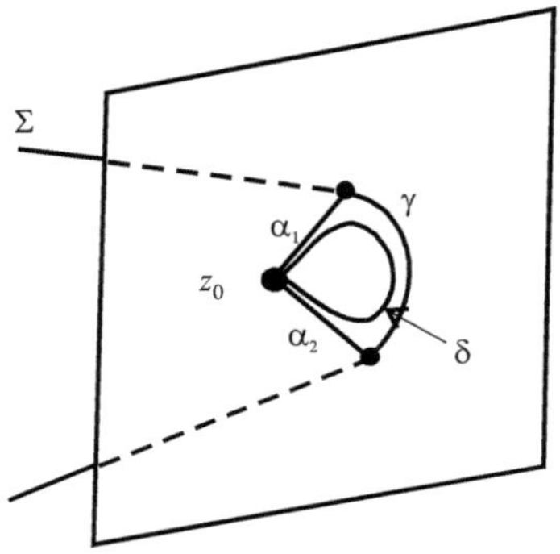

Figure 4.31: Paths and loops used in the proof of Th. 4.52

of $\pi|_\Sigma$) is connected. We choose a path γ lying in Σ, joining the points (z_1, λ'_*) with (z_2, λ'_*) and omitting the 'bad' points (singular and critical for π).
We define a loop $\delta \subset \mathbb{C}^\mu \setminus \Sigma$: firstly we move along the path α_1, then we move near the path γ and finally we follow the path α_2 in the reverse direction (see Fig. 4.31).
Along α_1, the cycle Δ_1 vanishes (tends to one point), along γ, it is close to some varying critical point of $F_0(\cdot, \lambda')$ and passes to a small cycle vanishing at the critical value z_2. Finally, along α_2, it passes to $\pm\Delta_2$. □

4.53. Problems. (a) (A. Gabrielov, F. Lazzeri). Show that the Dynkin diagram of an isolated singularity is connected.

(b) Let f_t be some deformation with the critical points $p_1(t), \ldots, p_k(t)$. Show that, if $f_t(p_i(t)) \equiv f_t(p_j(t))$, then $k = 1$.

4.54. Remarks. In this chapter we have omitted many other results concerning the topology of singularities. It turns out that for the *complete intersections*, i.e., singularities of local maps $f : (\mathbb{C}^n, 0) \longmapsto (\mathbb{C}^m, 0)$, an analogue of the Milnor theorem, about the homology groups of the non-singular level surfaces $f = \text{const}$, holds (see [**AVG**]). It means that one can define the vanishing cycles and the Milnor number.

V. Goryunov developed theory of singularities of projections $E \overset{i}{\subset} \mathbb{C}^n, 0) \overset{\pi}{\longmapsto} (\mathbb{C}, 0)$, where E is the complete intersection and π is the projection. There also one can define vanishing cycles.
The analysis of singularities of meromorphic functions leads to classification of pairs of functions (under some restrictions). This direction was developed in [**Arn9**].

Some work was done in the classifications and the monodromy theory of non-isolated singularities (see [**AVGL**]). This subject seems to be important because in applications one encounters such situations, e.g., Hamiltonian planar vector fields with polynomial Hamilton function. The French school of algebraic geometry (A. Grothendieck, Deligne and others) developed theory which solves these problems. We deal with these topics in the next section and in Chapter 7.

§5 Resolution and Normalization

Let $f : (\mathbb{C}^n, 0) \longmapsto (\mathbb{C}, 0)$ be a germ of holomorphic function with an isolated critical point at 0. Denote by $X_0 = f^{-1}(0)$ the germ of singular hypersurface in $\mathbb{C}^n$.

4.55. Definition (resolution of singularity). By a **resolution of singularity of the function** f (or of the hypersurface X_0) we mean an analytic map of complex manifolds

$$\pi : (Y, Y_0) \longmapsto (\mathbb{C}^n, 0)$$

satisfying the following conditions:

(i) $\pi|_{Y \setminus Y_0}$ is an analytic isomorphism between $Y \setminus Y_0$ and $\mathbb{C}^n \setminus 0$, i.e., between a neighborhood of Y_0 in Y (minus Y_0) and a punctured neighborhood of 0;

(ii) the sets $Y_0 = \pi^{-1}(0)$ and $\pi^{-1}(X_0)$ are unions of smooth analytic hypersurfaces in Y with normal crossings;

(iii) near any point in Y_0 there exists such a local system of analytic coordinates $y_1, \ldots, y_n$ and non-negative integers $k_1, \ldots, k_n, m_1, \ldots, m_n$ that

$$f \circ \pi(y) = y_1^{k_1} \cdot \ldots \cdot y_n^{k_n},$$
$$\det(\pi_*)(y) = g(y) y_1^{m_1} \cdot \ldots \cdot y_n^{m_n}, \quad g(0) \neq 0.$$

4.56. Theorem of Hironaka. ([**Hir**]). *Any singularity of complex hypersurface $X_0 = f^{-1}(0)$ has its resolution.*

The proof of this theorem is very long and technical. E. Briestone and P. Milman [**BM**] significantly simplified some parts of it. For elementary presentation of this proof we refer the reader to the article [**Hau**] by H. Hauser. Most recent proof is due to J. Włodarczyk [**Wl**].

We note also that, when one uses so-called *alterations*, i.e., typically finite-to-one maps (instead of blowing-ups) , then the analogous theorem about resolution of singularities becomes very short and simple (see [**Oor**]).

In fact, the statement of the Hironaka's theorem holds also in the case of real hypersurfaces. He proved it for algebraic varieties, defined over fields of characteristic zero.

The general idea of the resolution can be explained in the two dimensional case.

4.57. Elementary blowing-up. The (elementary) **blowing-up** (or *σ-process*) in $\mathbb{C}^2$ relies on replacing the point $x = y = 0$ by a projective line $\mathbb{CP}^1$. Namely, one

performs a partial compactification of the punctured plane $\mathbb{C}^2 \setminus 0$ by adding to it the directions of lines passing through $0 \in \mathbb{C}^2$.
We can write it in coordinates. The map $U_1 = \mathbb{C}^2 \longmapsto \mathbb{C}^2$

$$(x, u) \longmapsto (x, y) = (x, ux)$$

is a diffeomorphism between $U_1 \setminus \{x = 0\}$ and $\mathbb{C}^2 \setminus \{x = 0\}$ and sends the line $\{x = 0\}$ to the point $(0, 0)$. Similarly, the map $U_2 = \mathbb{C}^2 \longmapsto \mathbb{C}^2$

$$(y, v) \longmapsto (x, y) = (yv, y)$$

realizes a diffeomorphism between $U_2 \setminus \{y = 0\}$ and $\mathbb{C}^2 \setminus \{y = 0\}$ and sends the line $\{y = 0\}$ to the origin. The open sets U_1 and U_2 form an atlas of the manifold Y, with the gluing diffeomorphism $y = ux, v = 1/u$ in $U_1 \cap U_2 = U_1 \setminus \{u = 0\}$. We see that $\pi^{-1}(0, 0) = \mathbb{CP}^1$, parameterized by $(u : 1) = (1 : v)$.
Moreover, the self-intersection index of the curve $Y_0 = \pi^{-1}(0, 0)$ in the surface Y is equal to -1, which means that Y is a nontrivial surface. This follows from the fact that one can choose a section of the normal bundle $NY_0 = T_{Y_0}Y/T_{Y_0}Y_0$ in form of a meromorphic function on $\mathbb{CP}^1$, which has no zeroes and one simple pole: in U_1 the section is given by $x = x(u) \equiv 1$ and in U_2 we have $y = y(v) = u = 1/v$. Such section corresponds to the meromorphic vector field $\dot{z} = z^{-1}$ on $\mathbb{CP}^1 \cap U_2$ (in the trivialization $U_2 \approx \mathbb{C} \times \mathbb{C} \approx T\mathbb{C}$). The latter vector field is replaced by the continuous vector field $\dot{z} = \bar{z}$ with the index $i_0 = -1$.
(One can explain the negativity of $Y_0 \cdot Y_0$ more geometrically. The positivity of self-intersection of a complex submanifold means that it can be slightly deformed. If $Y_1 \neq Y_0$ were such a deformation then, after applying to it the blowing-down map π, we would obtain a compact analytic variety wholly located in a small neighborhood of a point in $\mathbb{C}^2$.)
In the two-dimensional case ($n = 2$) the resolution of singularity consists of a finite sequence of elementary blowing-ups. In Chapter 9 (Theorem 9.18) below we present the proof of resolution of a singular point of a planar analytic vector field. Applying it to the Hamiltonian vector field, with the Hamilton function f, we get the proof of the two-dimensional Hironaka theorem.
If $n > 2$, then the *elementary blowing-up* is given by the system of local maps

$$((u_1 : \ldots : \overset{j}{1} : \ldots : u_{n-1}), x_j) \longmapsto (x_j u_1, \ldots, x_j, \ldots, x_j u_n)$$

and has the property $\pi^{-1}(0) = \mathbb{CP}^{n-1}$. One adds to $\mathbb{C}^n \setminus 0$ the directions of all complex lines passing through 0.
Here, like in the 2-dimensional case, the *exceptional divisor* $\pi^{-1}(0)$ is not movable in the surface $Y = \pi^{-1}(\mathbb{C}^n)$. However, here we cannot prove it so simply, because we cannot use the index of self-intersection.

4.58. Example (Cusp singularity)

$$y^2 = x^3.$$

Its resolution is presented at Fig. 4.32. We see that $\pi^{-1}(0) = E_1 \cup E_2 \cup E_3$, where each $E_j = \mathbb{CP}^1 (\approx \mathbb{S}^2)$ and they intersect one another transversally. Moreover, $\pi^{-1}(X_0) = (f \circ \pi)^{-1}(0) = \Gamma \cup E_1 \cup E_2 \cup E_3$ with transversal intersections. Here Γ is called the *proper preimage of* X_0 (or the *strict transform of* X_0).

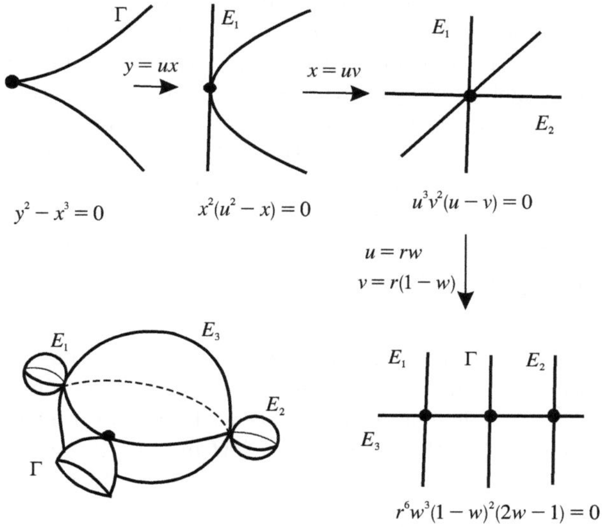

Figure 4.32: Resolution of the cusp singularity

It turns out that only the blowing-ups are not sufficient to prove the Hironaka theorem. One needs the following construction which generalizes the blowing-up construction. If $S \subset (\mathbb{C}^n, 0)$ is a smooth analytic surface of codimension k, then locally its neighborhood is isomorphic to $S \times \mathbb{D}^k$, where $\mathbb{D}^k$ is a ball. The *blowing-up along* S means replacing $S \times \mathbb{D}^k$ by

$$S \times (\mathbb{D}^k \setminus 0) \ \cup \ S \times \mathbb{CP}^{k-1},$$

i.e., we add to $S \times (\mathbb{D}^k \setminus 0)$ the directions of lines in the normal bundle NS passing through its zero section. This construction is called the *blowing-up along the subvariety* S.

After resolution we get $\pi^{-1}(0) = E_1 \cup \ldots \cup E_r$, where E_j are divisors (i.e., hypersurfaces) with normal intersections and $\pi^{-1}(X_0) = \Gamma_1 \cup \ldots \Gamma_s \cup E_1 \cup \ldots \cup E_r$. The set $\Gamma_1 \cup \ldots \Gamma_s$ is called the *proper preimage of* X_0 or the *strict transform of* X.

The Hironaka theorem has many applications. We list some of them.

4.59. Remark. It is not very difficult to show that, if X_0 is a singular algebraic hypersurface in a projective smooth variety X, then the resolution of singularities of X_0 gives a projective algebraic variety Y with algebraic map π satisfying properties analogous to those from the above definition of resolution. In particular, $\pi^{-1}(X_0)$ has only singularities in the form of normal intersections.

One can obtain even stronger resolution here. If, for example, an irreducible component Z of X_0 has self-intersection singularity, then, resolving this self-intersection, we obtain smooth strict transform of Z and some pasted divisors. Thus, we are able to obtain a resolution such that $(f \circ \pi)^{-1}(0) = T_1 \cup T_2 \cup \ldots \cup T_N$ is an union of *smooth* algebraic hypersurfaces $T_j \subset Y$ with normal intersections.

Moreover, the assumption that the singular point of the function f, defining $X_0 = f^{-1}(0)$, is isolated is not necessary. The Hironaka's theorem holds in the general case, when X, X_0 are analytic varieties and f is an analytic map from (X, X_0) to $(\mathbb{C}, 0)$.

The fibers $X_z = f^{-1}(z)$ can be compact projective algebraic varieties. This situation, called the *degeneration of algebraic manifolds*, will be subject of study below in this chapter and in Chapter 7.

4.60. Remark. If $M \subset \mathbb{CP}^N$ is an open smooth quasi-projective variety, i.e., given by analytic equations and inequalities, then there is its algebraic compactification $\overline{M}$ such that $N = \overline{M} \setminus M$ is a sum of divisors with normal intersections.

To show this, we notice that the closure M_1 of M in $\mathbb{CP}^N$ is a closed algebraic set. Next, one performs resolution of singular points in $M_1 \setminus M$.

4.61. Remark. One can realize the resolution of a singularity of a function on the basis of its Newton diagram. This leads to a beautiful construction of a toroidal embedding. The divisors E_j, Γ_k and integers k_j, m_j are computed from the Newton diagram. This is done in and [**Mum2**] and [**AVG**]. This construction works well only for germs, which are generic among functions with fixed Newton's diagram.

4.62. Resolution and normalization in algebraic geometry. Sometimes the resolution of singularity of a hypersurface is not sufficient in its analysis. Note that $Z = \pi^{-1}(X_0)$ is a manifold with singularities. These singularities are of two types. Firstly, as a topological space Z has singular points, intersections of algebraic components. The second type of singularity deals more with the language of modern algebraic geometry.

In algebraic geometry, the algebraic manifolds cannot be treated as sets of geometrical points. They are ringed topological spaces. This means that locally, in any affine part of $\mathbb{CP}^N$, we have a set $Z \subset \mathbb{C}^N$ and a ring $\mathcal{O}_Z$ of regular analytic functions on it.

For example, if $I \subset \mathbb{C}[x_1, \ldots, x_N] = \mathbb{C}[x]$ is an ideal, then it defines the affine algebraic variety $(Z, \mathcal{O}_Z) = (V(I), \mathbb{C}[x]/I)$, $V(I) = \{p : f(p) = 0,\ f \in I\}$. If the ideal I is *radical*, i.e., $r(I) = I(V(I)) = I$ (where $I(A) = \{f \in \mathbb{C}[x] : f|_A = 0\}$), then the ring $\mathcal{O}_Z$ coincides with the standard meaning of analytic function on Z. However, if the radical $r(I) = \{h : \exists_n\ h^n \in I\}$ of I is different from I, then $\mathbb{C}[x]/I$ contains functions, which are divisors of zero. It means that there are nonzero analytic functions on Z, which vanish at any point of Z; because the

value of f at a point p is the residue class in the field $\mathcal{O}_{p,Z}/\mathfrak{m}_p$. (Here the local ring $\mathcal{O}_{p,Z}$ consists of meromorphic functions on Z (quotients of elements of $\mathcal{O}_Z$), which are analytic at p and $\mathfrak{m}_p$ is its maximal ideal of functions vanishing at p). If the ideal I is radical, then the ringed algebraic variety (associated with it) is called *reduced*. If $(Z, \mathcal{O}_Z) = (V(I), \mathbb{C}[x]/I)$, then by its reduced version one means $(Z_{\text{red}}, \mathcal{O}_{\text{red}}) = (V(I), \mathbb{C}[x]/r(I))$. This ideology forms a basis of the Grothendieck's theory of algebraic schemes.
The second type of singularity of Z relies on the fact that $Z \neq Z_{\text{red}}$.
In order to avoid the difficulties associated with the above type singularities, one introduces the notion of normalization.

4.63. Definition (Normality). Let X be an analytic variety with the subset $\operatorname{Sing} X$ of its singular points. We say that X is **normal** at $x \in X$ iff any holomorphic and bounded function defined on $U \setminus \operatorname{Sing} X$, where U is a neighborhood of x, prolongs itself to a holomorphic function in U.
The space X is called **normal** iff it is normal in any of its points.

It can be shown that X is normal at x iff the local ring $\mathcal{O}_{x,X}$ is *integrally closed* in its field of quotients. This means that, if $f = g/h$, $g, h \in \mathcal{O}_{x,X}$, satisfies a *monic* algebraic equation

$$f^n + a_{n-1}f^{n-1} + \ldots + a_0 = 0, \quad a_j \in \mathcal{O}_{x,X},$$

then $f \in \mathcal{O}_{x,X}$; note that the leading coefficient equals 1. In the case when X is an algebraic variety, its normality is equivalent to the integral closeness of their affine rings $\mathbb{C}[X_i]$ of regular (polynomial) functions on the affine parts $X_i \subset \mathbb{C}^N$ (see [**Sha**]).
Moreover, if X is normal, then its set of singular points $\operatorname{Sing} X$ has codimension ≥ 2.

4.64. Examples. (a) If X is smooth in x, then it is also normal at x.

(b) If $X = \{x^2 = y^2\} \subset \mathbb{C}^2$, then the function $f(x, y) = x/y$ is equal to ± 1 in $X \setminus 0$ but is not continuous at 0.

(c) If $(X, \mathcal{O}_X) = (\{x = 0, 1\}, \mathbb{C}[x]/(x^2(x-1)))$, then the function $f = 1/x$ is equal to 1 at $X - \operatorname{Sing} X = \{x = 1\}$ but f does not belong to $\mathcal{O}_X$.

(d) If X is a normal affine algebraic variety and G is a finite group of automorphisms of X, then the quotient algebraic variety $Y = X/G$ is also normal.
If $\mathbb{C}[X]$ is the ring of regular functions on X, then the ring of invariants $\mathbb{C}[X]^G = \{f \in \mathbb{C}[X] : f \circ g = f,\ g \in G\}$ can be identified with the ring of regular functions on Y. In [**Sha**] it is shown that $\mathbb{C}[X]^G$ is finitely generated, does not contain divisors of zero and forms a ring of regular functions on its spectrum $\operatorname{Spec}_{\mathbb{C}}\mathbb{C}[X]^G$, which equals (as a set) to the set of maximal ideals of this ring.
The fields of quotients of the above two rings $\mathbb{C}(X)$, $\mathbb{C}(Y)$ form the fields of meromorphic functions on X and Y respectively. We have $\mathbb{C}(Y) = \mathbb{C}(X)^G$.

Let $f \in \mathbb{C}(Y)$ be an integer element over $\mathbb{C}[Y] \subset \mathbb{C}[X]$, i.e., satisfies a monic algebraic equation with coefficients in $\mathbb{C}[Y]$. So, f is integer also over $\mathbb{C}[X]$ and, by the normality of X, we have $f \in \mathbb{C}[X]$. Because f is G-invariant we have $f \in \mathbb{C}[Y]$.
For example, if $G = \mathbb{Z}_m$ with the generator acting on $X = \mathbb{C}^1$ as rotation by the angle $2\pi/m$, then $Y = X/G \simeq \mathbb{C}^1$ and the quotient map is $z \longmapsto z^m$.
If $G = \mathbb{Z}_2$ with the generator acting on $X = \mathbb{C}^2$ as the central symmetry $(u, v) \longmapsto (-u, -v)$, then $X/\mathbb{Z}_2$ is the cone $\{(x, y, z) : xy = z^2\}$ (because $\mathbb{C}[u, v]^G = \mathbb{C}[u^2, v^2, uv]$). By the above, the quadratic cone is normal. Note that its singular set is the vertex (of codimension 2).

4.65. Normalization theorem. *For any analytic manifold X there exists a unique (up to isomorphism) normal analytic manifold $\widetilde{X}$ and a holomorphic map $\tilde{\pi} : \widetilde{X} \longmapsto X$, such that:*

(i) *$\tilde{\pi}$ is an analytic diffeomorphism between $\widetilde{X} \setminus \tilde{\pi}^{-1}(\operatorname{Sing} X)$,*

(ii) *$X \setminus \operatorname{Sing} X$, $\sharp\tilde{\pi}^{-1}(point) < \infty$, and*

(iii) *$\tilde{\pi}^{-1}(X \setminus \operatorname{Sing} X)$ is dense in $\widetilde{X}$.*

The pair $(\widetilde{X}, \tilde{\pi})$ is called the **normalization** of X.

We do not prove this theorem; for the proof we refer to [**Sha**] and [**Loj2**]. We discuss only the normalization in the above examples.
In the example (b) $\widetilde{X}$ consists of two non-intersecting lines, $\widetilde{X} = X_1 \cup X_2 \subset \mathbb{C}^2$, $X_{1,2} = \{(u, v) : u = \pm 1\}$, $\tilde{\pi}|_{X_1}(u, v) = (v, v)$, $\tilde{\pi}|_{X_2}(u, v) = (v, -v)$.
In the example (c) $(\widetilde{X}, \mathcal{O}_{\widetilde{X}}) = (\{y = 0, \pm 1\}, \mathbb{C}[y]/(y(y^2 - 1)))$, $\tilde{\pi}(y) = y(y + 1)/2$. This normalization separates the twofold point $x = 0$, replaces it by the two points $y = 0, -1$.
The general difference between desingularization and normalization lies in the fact that, in the desingularization, we replace the hypersurface together with its neighborhood, whereas, in the normalization, we replace only the hypersurface itself.

4.66. Semi-stable reduction. Assume that, after resolution of a singularity and the change the target $(\mathbb{D}_\epsilon, 0)$ to $(\mathbb{D}, 0)$, $\mathbb{D} = \{|t| < 1\}$, we get a function $g = f \circ \pi : Y \longmapsto \mathbb{D}$ with local representations $g = y_1^{k_1} \ldots y_n^{k_n}$. When all $k_i = 0, 1$, then we say that the morphism g is **semi-stable**.
(Note that here we have changed the notation of the target variable, from $z \in (\mathbb{C}, 0)$ to $t \in \mathbb{D}$. This notation agrees with the notation used in other sources. We will keep it in the sequel.)
The **semi-stable reduction** means replacing each multiple divisor $E_j = \{y_j = 0\}$ by several copies of single divisors $E_{j,1}, \ldots, E_{j,l_j}$ each of them is a ramified covering of E_j (separation of multiple divisors).
The semi-stable reduction is realized in two steps. Firstly, one applies the base change $\alpha : s \longmapsto t = s^l$, where l is the least common multiplier (lcm) of all k_i's

appearing in local singularities of $Z = g^{-1}(0)$, $l = \operatorname{lcm}(k_1, \ldots, k_n)$. One takes $Y_\alpha = Y \times_{\mathbb{D}} \mathbb{D} = \{(y, s) : g(y) = \alpha(s)\}$ and the projection $g_\alpha : Y_\alpha \longmapsto \mathbb{D}$, $(y, s) \longmapsto s$. Usually, the space Y_α is not normal. So, the second step is the replacement of Y_α by its normalization $\widetilde{Y}$ (with the normalization map $\tilde{\pi} : \widetilde{Y} \longmapsto Y_\alpha$).
The map $\tilde{g} = g_\alpha \circ \tilde{\pi} : \widetilde{Y} \longmapsto \widetilde{\mathbb{D}} = \mathbb{D}$ realizes the semi-stable reduction. We have the following diagram:

$$\begin{array}{ccccc} \widetilde{Y} & \overset{\tilde{\pi}}{\longmapsto} & Y_\alpha & \longmapsto & Y \\ \downarrow \tilde{g} & & \downarrow g_\alpha & & \downarrow g \\ \widetilde{\mathbb{D}} & = & \widetilde{\mathbb{D}} & \overset{\alpha}{\longmapsto} & \mathbb{D} \end{array}$$

We have the following result (see [**Mum2**]).

Theorem of Mumford (Semi-stable reduction). *If $f : X \longmapsto \mathbb{D}$ is an analytic map from a smooth projective algebraic manifold X with unique critical value $t = 0$, then there exists a base change $\alpha : s \longmapsto t = s^l$ and a resolution $\widetilde{Y} \longmapsto X \times_{\mathbb{D}} \mathbb{D}$ such that the induced map $\tilde{g} : \widetilde{Y} \longmapsto \mathbb{D}$ is semi-stable.*

4.67. Formulas for the semi-stable reduction. (We follow [**Ste2**]). If, near a point $y_0 \in Y$, we have $f = y_1^{k_1} \ldots y_m^{k_m}$, $m < n$, $k_i > 0$ and $\alpha : s \longmapsto s^l$, $k_i | l$, then locally we have $Y_\alpha = \{(y, s) : y_1^{k_1} \ldots y_m^{k_m} = s^l\}$. If the greatest common divisor of the k_i's is greater than one, then Y_α has several components; (note that $\gcd(k_1, \ldots, k_m) | l$). Thus, in the first step of the construction of $\widetilde{Y}$, one separates these components. Assume now that $\gcd(k_1, \ldots, k_m) = 1$. Put $l_i = l/k_i$, $i = 1, \ldots, n$, where $k_{m+1} = \ldots = 1$. We define the map from $\mathbb{C}_z^n$ (with coordinates z_i) to Y_α by the formula $y_i = z_i^{l_i}$, $s = z_1 \cdots z_n$. Of course, the space $\mathbb{C}_z^n$ is normal but it is not our normalization; it is too big.
One divides $\mathbb{C}_z^n$ by the group $G \subset \mathbb{Z}_{l_1} \times \ldots \times \mathbb{Z}_{l_n}$ (acting diagonally by means of rotations) and consisting of those rotations which preserve $s = z_1 \cdots z_n$. By Example 4.64(d), $\widetilde{Y} = \mathbb{C}_z^n / G$ is normal.
Moreover, the quotient group $\mathbb{Z}_{l_1} \times \ldots \times \mathbb{Z}_{l_n} / G$ is cyclic, equal to $\mathbb{Z}_l$, and it acts on $\widetilde{Y}$. The quotient $\widetilde{Y} / \mathbb{Z}_l$ is equal to $X = \mathbb{C}_y \times \mathbb{D}$.
These formulas hold locally in suitable (local) charts. In general case one covers X with such charts and performs the normalizations successively in each chart.

4.68. Examples. (a) $f(y_1, y_2) = y_1^2 y_2^2$. Then $Y_\alpha = \{(y_1 y_2)^2 = s^2\}$ consists of two intersecting components and the normalization $\widetilde{Y}$ consists of two separated smooth components $y_1 y_2 = \pm s$.

(b) $f = y_1^2 y_2$. Then $Y_\alpha = \{y_1^2 y_2 = s^2\}$ is not normal; because the function s/y_1, smooth on $Y_\alpha - \operatorname{Sing} X$, $\operatorname{Sing} X = \{y_1 y_2 = s = 0\}$, is not analytic in the whole Y_α. The normalization map is $y_1 = z_1$, $y_2 = z_2^2$, $s = z_1 z_2$. Here the group $G = \{e\}$. The divisor $E_1 = \{y_1 = 0\}$ is replaced by the divisor $\widetilde{E}_1 = \{z_1 = 0\}$, with 2-fold covering over E_1, and $E_2 = \{y_2 = 0\}$ is replaced by $\widetilde{E}_2 \approx E_2$ (see Fig. 4.33).

(c) *Semi-stable reduction of the cusp resolution* (see [**Ste2**]). Let $X \subset \mathbb{CP}^2 \times \mathbb{D}$ be given by $x_0 x_2^2 - x_1^3 = t x_0^3$ and let $f = t$. In the affine part we get $y^2 - x^3 = t$

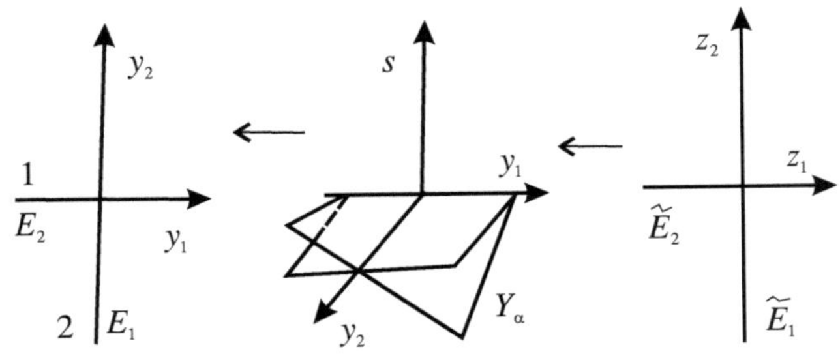

Figure 4.33: Semi-stable reduction in Ex. 4.68(b)

with the cusp singularity of X_0. After resolution (see Example 4.58) we obtain the surface Y with the map $g = f \circ \pi : Y \longmapsto \mathbb{D}$ and $g^{-1}(0) = \Gamma + 2E_1 + 3E_2 + 6E_3$ (with multiple divisors). Here $\alpha : s \longmapsto t = s^l$, $l = 6$, and the surface Y_α is not normal.

Applying the normalization to each of the intersections of divisors, we obtain the following.

The normalization near $\Gamma \cap E_3$: $r^6(2w-1) = s^6$, is the map $\tilde{\pi} : (r,z) \longmapsto (r, 2w-1) = (r, z^6)$, i.e., we have local branched covering of E_3 with ramification index 6 (see the discussion in 11.31 below).

The normalization near $E_1 \cap E_3 : r^6(w-1)^2 = s^6$ gives two copies of the plane $\mathbb{C}^2$ with the normalization map $(r,z) \longmapsto (r, w-1) = (r, z^3)$.

The normalization near $E_2 \cap E_3$ gives three copies of $\mathbb{C}^2$ with the map $(r,z) \longmapsto (r,w) = (r,z^2)$.

Thus $\tilde{\pi}^{-1}(\Gamma) = \widetilde{\Gamma}$ is one divisor $\mathbb{CP}^1$, $\tilde{\pi}^{-1}(E_1) = E_{11} \cup E_{12}$, $\tilde{\pi}^{-1}(E_2) = E_{21} \cup E_{22} \cup E_{23}$. Finally, $\widetilde{E}_3 = \tilde{\pi}^{-1}(E_3)$ is a smooth projective variety and $\tilde{\pi}$ realizes the ramified covering $\widetilde{E}_3 \longmapsto E_3 = \mathbb{CP}^1$ of degree 6; one can see that generically $\tilde{\pi}$ is 6 to 1. Using the Riemann–Hurwitz formula $6\chi(E_3) = \chi(\widetilde{E}_3) + \sum(\nu(p) - 1)$, (connecting the Euler characteristics and the ramification indices $\nu(p)$, see Theorem 11.32 below), we find that $\widetilde{E}_3$ is a topological torus, i.e., an elliptic curve. (See also Fig. 4.34.)

4.69. Remark (orbifold character of the space Y). The normalization map $\tilde{\pi} : \widetilde{Y} \longmapsto Y$ is finite-to-one. More precisely, from the formulas in the point 4.67, it follow that $\widetilde{Y}$ can be covered by charts of the form $\mathbb{C}^n/G$, where $G \subset GL(n, \mathbb{C})$ is a finite group (of rotations in coordinates).

Such manifolds are called *orbifolds* (or *V-manifolds*, see [**Ste2**]).

It turns out that most of the cohomological theory on smooth algebraic varieties can be applied to algebraic orbifolds. In particular, one can define the analogues of the de Rham sheaves, holomorphic de Rham sheaves, hypercohomology etc. (see Chapter 7).

We will use the above constructions (resolution and normalization) in Chapter 7.

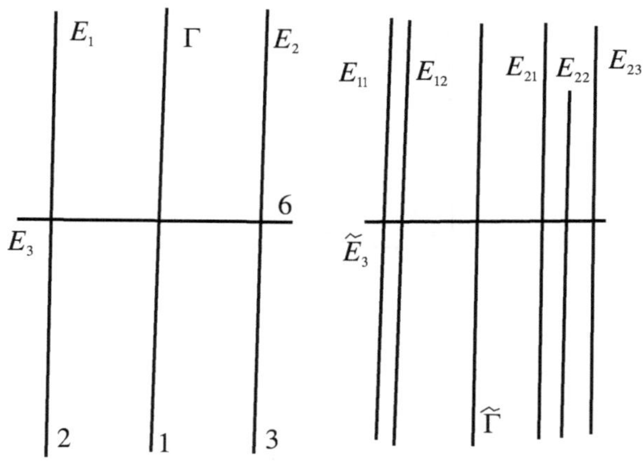

Figure 4.34: Semi-stable reduction in Ex. 4.68(c)

4.70. The Clemens contraction. (a) Let $f : X \longmapsto \mathbb{D}$ be a holomorphic mapping with unique critical value $t = 0$ and such that locally $f = x_1^{k_1} \dots y_m^{k_m}$ for some analytic coordinates $x_1, \dots, x_n$, i.e., with resolved singularities. $X_t = f^{-1}(t)$ are projective algebraic varieties; smooth for $t \neq 0$ and with normal intersection singularities for $t = 0$. Assume also that $X = f^{-1}(\mathbb{D})$. The map $f : X - X_0 \longmapsto \mathbb{D}^* = \mathbb{D} \setminus 0$ defines a fibre bundle, the algebro-geometrical analogue of the Milnor bundle. Here the fibers X_t have the same homology groups $H_*(X_t, \mathbb{C})$; (in contrast with the Milnor's case they have many non-zero components in general). One also defines the monodromy diffeomorphism $h_t : X_t \longmapsto X_t$ and the **monodromy operator** (or the **Picard–Lefschetz transformation)** $M = h_{t*} : H_*(X_t) \longmapsto H_*(X_t)$.
C. Clemens [**Cle**] constructed a continuous mapping $p : X \longmapsto X_0$ with some special properties.
We explain his construction on the special case $f(x_1, x_2) = x_1 x_2$ (or $X = \{x_1 x_2 = t x_0^2\} \subset \mathbb{CP}^2 \times \mathbb{D}$ and $f = t$).
Let $U = \{|x_{1,2}| < 1\}$ be the neighborhood of the origin. We define a certain family of maps $p_t = p|_{X_t} : X_t \cap U \longmapsto X_0$. We divide $X_t \cap U$ into three parts: $A_1 = X_t \cap \{|t|^{1/2} < |x_2| < 1\}$, $A_2 = X_t \cap \{|t|^{1/2} < |x_1| < 1\}$ and $A_3 = \{|x_1| = |x_2| = |t|^{1/2}\}$.
On A_1 the map p_t is defined as composition of the projection $(x_1, x_2) \longmapsto (0, x_2)$ and the stretching

$$(0, x_2) \longmapsto (0, [(|x_2| - |t|^{1/2})/(1 - |t|^{1/2})] x_2)$$

(with the punctured disc $0 \times \mathbb{D}^*$ as the image). Analogously p_t on A_2 is defined (with the image $\mathbb{D}^* \times 0$). Finally, $p(A_3) = 0$ (see Fig. 4.35).
On the remaining part $X_t \setminus U \subset \mathbb{CP}^2$ one defines p_t as the projection onto the nearest component of X_0 along some affine coordinate.

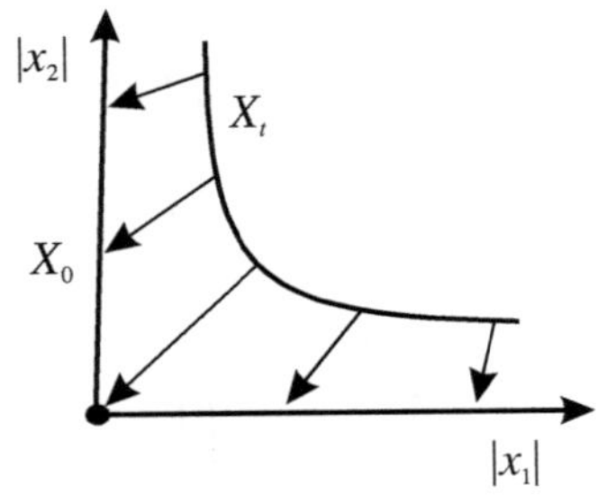

Figure 4.35: Clemens contraction

The maps p_t organize themselves into a contraction (or retraction) of X to X_0, i.e., with identity on X_0. Moreover, if $x_0 \in X_0$ is smooth, then $p_t^{-1}(x_0)$ is one point and, if x_0 is singular, then $p_t^{-1}(x_0)$ is a topological 1-dimensional torus.
Note also that this construction can be generalized to the case when $f = x_1^{k_1} x_2^{k_2}$. Then $p_t^{-1}(x_0)$ consists of k_1 (or of k_2 points) for smooth point x_0.
The general result is the following.

(b) **Proposition–Definition.** ([**Cle**]) *Let $f : X \longmapsto \mathbb{D}$ be a holomorphic map with only critical value 0 and with resolved singularities. Then there exists a continuous family of maps $p_t : X_t \longmapsto X_0$ such that the total map $p = (p_t)$ is a contraction of $f^{-1}(\mathbb{D})$ to X_0 forming a prolongation of the contraction of $\mathbb{D}$ to 0 along the radii.*
If, near $x_0 \in X_0$, we have $f = x_1^l$ then $p_t^{-1}(x_0)$ is an l-element set. If, near x_0, we have $f = x_1^{k_1} \dots x_m^{k_m}$, $m > 1$, then $p_t^{-1}(x_0)$ is a topological $(m-1)$-dimensional torus $\mathbb{T}^{m-1} = \{(\phi_1, \dots, \phi_m) \pmod{2\pi} : \sum k_i \phi_i = \arg t\}$.
We call the map p the **Clemens contraction.**

(c) Unfortunately, the Clemens contraction p_t is not compatible with the action of the monodromy map $h_t : X_t \longmapsto X_t$.
Indeed, if $p_t^{-1}(x_0) = \{y_0, \dots, y_{l-1}\}$, $y_j = (|t|^{1/l} e^{i(\arg t + 2\pi j)/l}, 0, \dots, 0)$, then the natural lifts (to the fibers X_t's) of the path $\theta \longmapsto t(\theta) = e^{i\theta} t$, $\theta \in [0, 2\pi\}$, would be $y_j e^{i\theta/l}$. Thus h_t should realize a cyclic permutation.
The action of h_t on the torus $\mathbb{T}^{m-1}$ should be translation $(\phi_1, \dots, \phi_m) \longmapsto (\phi_1 + 2\pi/(k_1 m), \dots, \phi_m + 2\pi/(k_m m))$.
We see that the above 'action' cannot be continuous, with discontinuity at the tori.
So, there is necessity to improve the Clemens contraction in such a way that it be compatible with the action of monodromy.
The idea of this construction is following. We replace the target space X_0 by another target space $\widehat{X}_0$, which is obtained from X_0 by replacing each point x_0 of an m-fold intersection by $(m-1)$-dimensional simplex Δ^{m-1} (with baricentric coordinates $(a_1, \dots, a_m)$, $\sum a_j = 1$). Also the torus $\mathbb{T}^{m-1} = p_t^{-1}(x_0)$ is replaced by

$\mathbb{T}^{m-1} \times \Delta^{m-1}$. We get a new space $\widehat{X}_t$. Of course, this $\widehat{X}_t$ has the same homotopy type as X_t (see Fig. 4.36).
The action of the monodromy $\hat{h}_t$ on the thick torus $\mathbb{T}^{m-1} \times \Delta^{m-1}$ is following

$$(\phi_1, \dots, \phi_m; a_1, \dots, a_m) \longmapsto (\phi_1 + 2\pi a_1/k_1, \dots, \phi_m + 2\pi a_m/k_m; a_1 \dots, a_m).$$

These local transformations glue themselves together to a continuous homeomorphism $\hat{h}_t$ of $\widehat{X}_t$. It induces the same homomorphism in homologies as h_{t*}.
We call the corresponding map $\hat{p}_t$ the **thick Clemens contraction** and the homeomorphism $\hat{h}_t$ the **thick monodromy transformation**.
The thick Clemens contraction is compatible with the thick monodromy transformation: $\hat{p}_t \circ \hat{h}_t = \hat{p}_t$.
This construction was used by Clemens in his direct proof of the following fundamental theorem in the monodromy theory.

4.71. Monodromy theorem. *For any k the induced homology homomorphism $(h_t)_{*k} : H_k(X_t, \mathbb{C}) \longmapsto H_k(X_t, \mathbb{C})$ is quasi-unipotent. It means that all its eigenvalues are roots of unity, $(h_*^A - I)^B = 0$ for some natural A, B. Moreover, $B \leq n - |k - n + 1|$ where $n = \dim X$.*

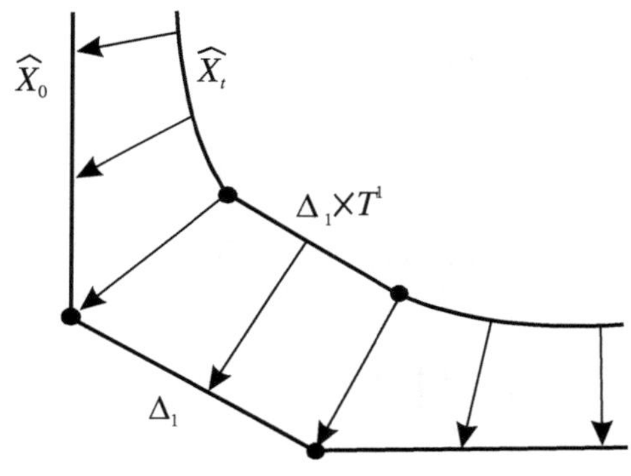

Figure 4.36: Thick Clemens contraction

The Monodromy Theorem was first proved by A. Landman in his Berkeley Thesis in 1966 (see [**Lan**] and the point 7.26 in Chapter 7). Besides the Landman's and the below Clemens' [**Cle**] proofs there are three more proofs. One is algebraic by A. Grothendieck (see [**Gro2**] and [**DK**]), there is a proof by E. Brieskorn (using the solution of the 7th Hilbert's problem, see [**Brie**] and 7.37(i)) and A. Borel [**BN**] gave another proof using the period mapping and hyperbolic geometry (see 7.56(d)).

Clemens' proof of the Monodromy Theorem. Denote by A the greatest common multiplier of all the multiplicities of the divisors in X_0. It will be the integer A

from the thesis of Monodromy Theorem. Denote $M = (h_t)_{*k}$. We have to show that the operator $M^A - I$ is nilpotent with order of nilpotency $\leq n - |k - n + 1|$. Of course, we shall work with the map $\hat{h}_t$ acting on $\widehat{X}_t$. Let the subset $Z_j \subset \widehat{X}_t$ consist of points corresponding to $(\geq j)$-fold intersections in X_0. Thus we have $\widehat{X}_t = Z_1 \supset Z_2 \supset \ldots \supset Z_{n+1} = \emptyset$, $n = \dim X$.
If σ is a (geometrical) cycle in Z_1, then the cycle $\hat{h}_t^A(\sigma) - \sigma$ lies in Z_2; (because $\hat{h}_t$ acts periodically on $Z_1 \setminus Z_2$).
Let us look at the action of $\hat{h}_t$ on $Z_2 \setminus Z_3$. Consider a component $\mathbb{T}^1 \times \Delta^1 \times E_i^{(2)}$, where $E_i^{(2)}$ is a $(n-2)$-dimensional variety, locally of the form $x_1 = x_2 = 0$ and with $\{f = x_1^{k_1} x_2^{k_2}\}$. Here the action of $\hat{h}_t$ is homotopically equivalent to the action in one endpoint of the interval $\Delta^1 : (\phi_1, \phi_2; 1, 0; z) \longmapsto (\phi_1 + 2\pi/k_1, \phi_2; 1, 0; z)$.
If σ is a (geometrical) cycle in Z_2, then we can move it homotopically, in any component of $Z_2 - Z_3$, to a part of the boundary corresponding to one vertex of the simplex Δ^1. It is not difficult to see that this can be done in continuous way. The new cycle σ' represents the same homology class as σ.
Because the action of $\widehat{h_t}$ at the boundary of $\mathbb{T}^1 \times \Delta^1 \times E_i^{(2)}$ is periodic, the cycle $\hat{h}^A\sigma' - \sigma'$ lies in Z_3.
The same construction can be repeated for cycles in Z_j, $j > 2$. The action of $\hat{h}_t^A - id$ gives a cycle homologically equivalent to a cycle lying in Z_{j+1}.
Consider the case $k > n-1$. Because the subsets Z_j are homotopically equivalent to spaces of (real) dimension $2(n-1)-j$, then a k-dimensional cycle is sent to zero after $2n-1-k$ times; thus $B \leq 2(n-1)-k+1$. In the case $k \leq n-1$ one should use the Poincaré duality which, together with the above, gives $B \leq k+1$. □

4.72. Zeta function of the monodromy. If $F : X \longmapsto X$ is a diffeomorphism of a differentiable manifold X, then the **ζ-function** *of* F is defined as

$$\zeta_F(s) = \prod_{q \geq 0} [\det I - sF_{*q})]^{(-1)^{q+1}},$$

where $F_{*q} : H_q(X, \mathbb{C}) \longmapsto H_q(X, \mathbb{C})$ are the induced homology homomorphisms. ζ_F is a rational function, which can be expressed in terms of the structure of its set of periodic points. Namely, with this aim it was introduced in the theory of dynamical systems.
Indeed, let

$$\Lambda(F^k) = \sum_q (-1)^q \operatorname{tr}(F_{*q})^k$$

be the Lefschetz number of the map $F^k = F \circ \ldots \circ F$ (k times). Let the positive integers $r_1, r_2, \ldots$ be defined inductively by the relations

$$\Lambda(F^k) = \sum_{i|k} r_i,$$

e.g., $r_1 = \Lambda(F)$, $r_2 = \Lambda(F^2) - r_1$, $r_4 = \Lambda(F^4) - r_1 - r_2$. Applying the identity $\det(I - sA) = e^{\operatorname{tr}\ln(I-sA)} = \exp(-\sum(s^i/i)\operatorname{tr} A^i)$ to each factor $\det(I - sF_{*q})$, we obtain the formula

$$\zeta_F(s) = \prod(1 - s^i)^{-r_i/i}.$$

Recall that $\Lambda(F^k)$ counts the number (with multiplicities) N_k of periodic points of F with period divisible by k (see the Lefschetz theorem 3.30). We get $\zeta_F = \exp(-\sum(N_k/k)s^k)$.

We apply this to the monodromy diffeomorphism $F = h : X_t \longmapsto X_t$ of a fibre of the cohomological bundle associated with a *degeneration of analytic manifolds* $g : (X, X_0) \longmapsto (\mathbb{D}, 0)$, where $X_t = g^{-1}(t)$ are smooth analytic manifolds and X_0 is a sum of smooth submanifolds with normal crossings and near any point of X_0 there is a system of coordinates such that $g = x_1^{k_1} \dots x_n^{k_n}$. The map $X^* = X \setminus X_0 \longmapsto \mathbb{D}^* = \mathbb{D} \setminus 0$ is a locally trivial topological bundle, the analogue of the Milnor bundle. The Picard–Lefschetz transformation acting on $H_*(X_{t_0}, \mathbb{C})$ is induced by the monodromy diffeomorphism $h = h_{t_0}$ of a fixed fiber X_{t_0} (see above). A'Campo [**AC2**] has found a beautiful formula for the ζ_h.

The special case of this situation appears when we have a singularity $f : (\mathbb{C}^n, 0) \longmapsto (\mathbb{C}, 0)$ of multiplicity μ; $X_t = f^{-1}(t) \cap \mathbb{B}_\rho$. Thus, we can assume that f is a polynomial. Let $\pi : Y \longmapsto \mathbb{B}_\rho \subset \mathbb{C}^n$ be a resolution of f, $g = f \circ \pi$, $X_0 = g^{-1}(0)$. We have $g^{-1}(t) \simeq V_t$ for $t \neq 0$. Because $h_{*0} = I$ on $H_0(V_t) = \mathbb{C}$ and $H_q(V_t) = 0$ for $q \neq 0, n-1$, we can express the zeta function of h by means of the characteristic polynomial of the monodromy operator h_*, $P(\lambda) = \det(h_* - \lambda I)$. We have $\zeta_h(s) = (1-s)^{-1}\det(I - sh_*)^{(-1)^n} = (1-s)^{-1}\left[(-s)^\mu P(1/s)\right]^{(-1)^n}$.

For any natural m we put

$$S_m = \{y \in X_0 : g = x_1^m \text{ near } y\}.$$

Denote by $\chi(S_m)$ the Euler characteristic of S_m.

4.73. Theorem of A'Campo ([**AC1**], [**AC2**])

(a) *We have*

$$\Lambda(h^0) = \chi(V_t) = \sum_m m\chi(S_m),$$

$$\Lambda(h^k) = \sum_{m|k} m\chi(S_m).$$

In particular,

$$\zeta_h(s) = \prod_m (1 - s^m)^{-\chi(S_m)},$$

$$P(s) = (-1)^\mu \left[(s-1)\prod_m (s^m - 1)^{-\chi(S_m)}\right]^{(-1)^n},$$

$$\mu = (-1)^n[1 - \sum_m m\chi(S_m)].$$

(b) *If h is a monodromy diffeomorphism of the Milnor bundle associated with critical point of a function with multiplicity* $0 < \mu < \infty$, *then*

$$\Lambda(h) = \chi(S_1) = 0.$$

4.74. Examples. (a) For the cusp singularity we have: $S_6 = E_3 \setminus \{3 \text{ pts}\} = \mathbb{S}^2 \setminus \{3 \text{ pts}\}$, $\chi(S_6) = -1$; $S_2 \simeq S_3 = S^2 - \{1 \text{ pt}\}$, $\chi(S_2) = \chi(S_3) = 1$; $S_1 = \mathbb{D}^*, \chi(S_1) = 0$. Thus $\mu = 1-2-3+6 = 2$, $\zeta_h = (1-s^2)^{-1}(1-s^3)^{-1}(1-s^6)$, $P(s) = s^2 - s + 1$.

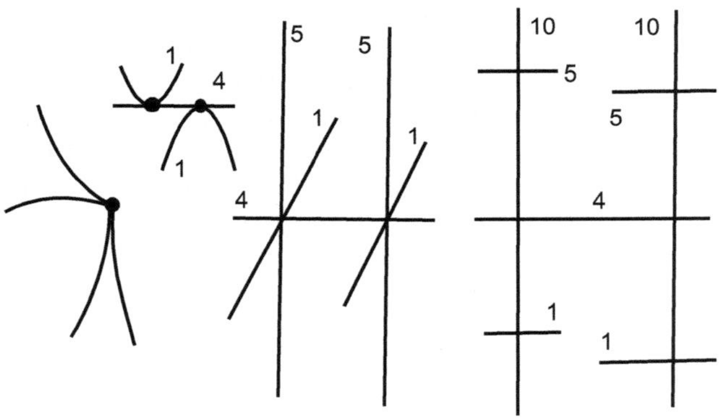

Figure 4.37: Resolution of the singularity from Ex. 4.74(b)

(b) For $f = (x^3 + y^2)(x^2 + y^3)$ we have the resolution process presented in Fig. 4.37.
Here $\chi(S_{10}) = 2\cdot(-1)$, $\chi(S_5) = 2\cdot 1$, $\chi(S_4) = \chi(S_1) = 0$. This gives $\mu = 11$, $\zeta_h = (s^5+1)^2$, $P(s) = -(s-1)(s^5+1)^2$. This example is considered in [**AVG**] (v. 2).
(c) The reader can make calculations for $f = (x^2+y^3)(x^2y^2+x^6+y^6)$ from [**AC2**].
(d) For the Pham singularity $f = x_1^d + \ldots + x_n^d$ we have $\mu = (d-1)^n$ and after blowing-up we obtain one divisor $\mathbb{CP}^{n-1}$ and only two sets S_m, i.e., $S_1 \simeq X_0 \setminus 0$ and $S_d = \mathbb{CP}^{n-1} \setminus Q$, where Q is a smooth hypersurface of degree d.

The surface $S_1 \subset \mathbb{C}^n \setminus 0$ can be deformed along the radii to a submanifold $\widetilde{S} \subset \mathbb{S}^{2n-1}$. But $\widetilde{S}$ admits an action of the circle $\mathbb{S}^1$. So, there is a vector field on $\widetilde{S}$ without singularities and (by the Poincaré–Hopf theorem) $\chi(\widetilde{S}) = \chi(S_1) = 0$, which agrees with the point (b) of the A'Campo theorem.
Next, by additivity of the Euler characteristic, we have $\chi(S_d)+\chi(Q)=\chi(\mathbb{CP}^{n-1})= n$. Therefore we obtain the following formula for the Euler characteristic of a general hypersurface in projective space:

$$\chi(Q) = n + [(1-d)^n - 1]/d.$$

4.75. *Proof of Theorem* 4.70. (a) (This part was proved in [**AC2**].) We use the thick Clemens map $\hat{p}_t$ and notations from the proof of Theorem 4.71. One divides the variety $\widehat{X}_t$ into the following subsets invariant with respect to the monodromy map $\hat{h}_t$. Let $\widehat{X}^m \simeq p_t^{-1}(S_m)$, $m = 1, 2, \ldots$; thus the set $Z_1 - Z_2 = \widehat{X}^1 \cup \widehat{X}^2 \cup \ldots$. The set $Z_2 - Z_3$ consists of components of the form $\mathbb{T}^1 \times \Delta^1 \times E_i^{(2)}$. Generally $Z_j - Z_{j+1}$ is composed of sets of the form $\mathbb{T}^{j-1} \times \Delta^{j-1} \times E^{(j)}$, where the subsets $E_i^{(j)}$ are the connected components of the set of points from X_0, at which exactly j divisors intersect themselves.
If $x \in S_m$, then $p_t^{-1}(x)$ is an m-point set, at which the monodromy map acts as a cyclic permutation. Thus, $(h|_{p_t^{-1}(x)})^k = id$ if $m|k$, and it does not have fixed points otherwise. So, $\Lambda((h|_{p_t^{-1}(x)})^k) = \chi(p_t^{-1}(x)) = m$ or $= 0$, depending whether $m|k$ or not.
At each set $V = \mathbb{T}^{j-1} \times \Delta^{j-1} \times E_i^{(j)}$ the action of $\hat{h}$ is homotopically equivalent to a rotation by angles commensurable with 2π. Thus $(h|_{\mathbb{T}^{j-1}})^k = id$ or it does not have fixed points. Because $\chi(\mathbb{T}^r) = 0$ we have $\Lambda((h_{\mathbb{T}^{j-1}})^k) = 0$.
Next, we use the standard properties of the Lefschetz numbers (analogous to the properties of the Euler characteristic): (i) $\Lambda(F_1 \times F_2) = \Lambda(F_1) \cdot \Lambda(F_2)$ for $F_1 \times F_2$ acting diagonally on $X_1 \times X_2$ and (ii) $\Lambda(F) = \Lambda(F_1) + \Lambda(F_2)$, if $F_{1,2}$ act on two subsets $X_{1,2}$ of a partition of X.
We obtain $\Lambda(h^k) = \sum_{m|k} m\chi(S_m)$, which implies the other formulas from the first part of the theorem.
(b) (This part was proved in [**AC1**].) The proper image of $\Gamma = \pi^{-1}(f^{-1}(0) \setminus 0)$ of the set $f^{-1}(0) \setminus 0$ lies in S_1 (by assumption that 0 is isolated critical point). Any other divisor, i.e., from $\pi^{-1}(0)$, belongs (almost completely) to some S_m, $m > 1$. It follows from the process of resolution (see the examples). Because $f^{-1}(0) \cap B_\epsilon$ is contractible to 0, one can use this contraction and the Clemens map to obtain a contraction of $\widehat{X}_t$ to the subset $p_t^{-1}(\widehat{X}_0 \setminus S_1)$.
On the latter set we repeat the analysis from the point (a). On the sets $\widehat{X}^m$, $m > 1$, the monodromy map does not have fixed points. The remaining part has zero Euler characteristic. □

Chapter 5

Integrals along Vanishing Cycles

The previous chapter was devoted to the topology of critical point of a holomorphic function. This chapter is devoted to its analysis.
The principal objects of investigation are integrals of holomorphic forms (defined in a neighborhood of the critical point) along cycles lying in a level surface of the function and vanishing at the critical point. It turns out that these integrals, treated as a functions of the (noncritical) value of the function, form holomorphic and multivalued functions with ramification at the critical value. They satisfy a system of differential equations, the Picard–Fuchs equations.
The holomorphic forms define sections (the geometrical sections) of a certain holomorphic vector bundle above the set of noncritical values of the function. This is the cohomological Milnor bundle, with fibers equal to the cohomology groups with complex coefficients of the local level of the function. Any such fiber contains an integer lattice consisting of integer cocycles. These integer cocycles extend themselves into sections of the cohomological Milnor bundle. They are sections horizontal with respect to the so-called Gauss–Manin connection. The Picard–Fuchs equations are related to the equations for horizontal sections (with respect to the Gauss–Manin connection), expressed in a basis consisting of geometrical sections.
The asymptotic behavior of integrals of holomorphic forms along vanishing cycles, as the value tends to the critical value, gives us very important information about geometry of the neighborhood of the critical point.
The asymptotic of the integrals along vanishing cycles is used in analysis of asymptotic of oscillating integrals, which appear in the geometrical optics.
In this chapter the values of the function will be denoted by t. The reason is that we shall differentiate the sections of the cohomological bundle with respect to the 'time' t.

§1 Analytic Properties of Integrals

5.1. Example (Complete elliptic integrals) Let

$$V_t : y^2 + x^3 - x = t$$

be a family of elliptic curves in $\mathbb{C}^2$ (see also Example 4.11). If t is typical (different from the critical values $t_{1,2} = \pm 2/3\sqrt{3}$, corresponding to the saddle point and the local minimum), then the corresponding Riemann surface is diffeomorphic to the punctured torus $\mathbb{T}^2 \setminus \{p\}$. Its first homology group is generated by two cycles

H. Żołądek, *The Monodromy Group*, Monografie Matematyczne 67,
https://doi.org/10.1007/978-3-031-91270-2_5

$\gamma = \gamma(t)$ and $\delta = \delta(t)$, where γ is represented as a real oval of this elliptic curve (see Fig. 5.1). The cycle δ cannot be deformed to the real part of the curve. We assume for a while that the geometrical positions of the cycles are fixed. (In notations from Example 4.11 we have $\gamma = \Delta_2$ and $\delta = \Delta_1$.)

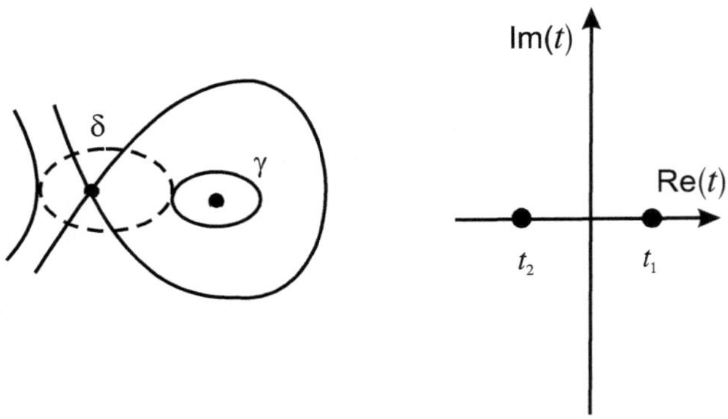

Figure 5.1: Critical points, critical values and cycles in the example

Consider the following functions:

$$I_0(t) = \int_{\gamma(t)} y\mathrm{d}x, \quad I_1(t) = \int_{\gamma(t)} xy\mathrm{d}x.$$

The integral I_0 is equal to the area of the domain bounded by the oval γ and I_1 is equal to a moment of the Lebesque measure in this domain. In particular, I_1/I_0 defines the position of the mass center of the domain $\{y^2 + x^3 - x \leq t\}$.

5.2. Proposition. *The functions I_0, I_1 can be prolonged to holomorphic functions outside the critical value $t_1 = 2/3\sqrt{3}$, corresponding to the saddle point.*

Proof. The cycle $\gamma(t)$ is well defined for $t \in [t_2, t_1]$. The functions $I_{0,1}(t)$ are differentiable in this interval. Indeed, they can be written in the form

$$I_0(t) = 2\int_{x_2(t)}^{x_3(t)} y(x,t)\mathrm{d}x, \quad I_1(t) = 2\int_{x_2(t)}^{x_3(t)} xy(x,t)\mathrm{d}x,$$

where $x_{2,3}(t)$ are the points of intersection of the cycle $\gamma(t)$ with the x-axis and $y(x,t) = \sqrt{t + x - x^3}$ (see Fig. 5.2(a)). We can differentiate these functions because $y(x_{2,3}, t) = 0$.
Unfortunately that approach is wrong, because we do not know how to perform the further differentiations. Above, the subintegral function, as well as the contour of integration, depend on t. The right idea is to transform this problem to a problem, where the contour of integration is fixed, i.e., does not depend on t, and the subintegral 1-form is analytic with respect to t. In order to do this we must

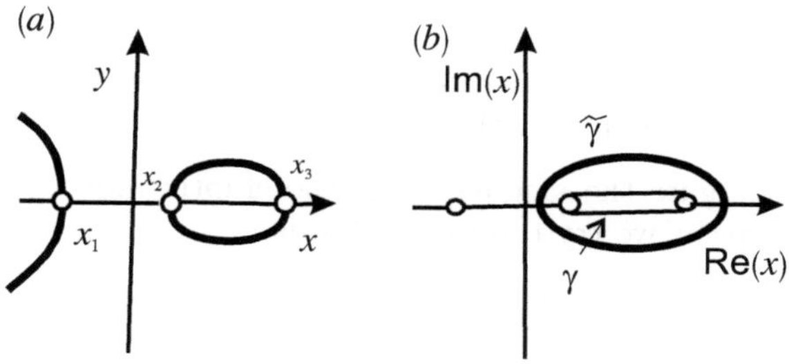

Figure 5.2: Cycles: in the real domain (a) and in the complex domain (b)

deform the cycle $\gamma(t)$. The first problem, which we encounter, is the dependence of the integral (of the form along varying cycle) on the deformation.

5.3. Lemma. *If two* 1*-dimensional cycles* σ *and* σ'*, lying in a Riemann surface* V*, are homologous one to another and a* 1*-form* ω *is holomorphic, then* $\int_\sigma \omega = \int_{\sigma'} \omega$.

Proof. Let z be a local holomorphic parameter in the Riemann surface V. The form ω is called **holomorphic** iff its local representation is of the form $f(z)\mathrm{d}z$, where f is a holomorphic function. Such form is obviously closed, $\mathrm{d}\omega = 0$, and represents an element of the first cohomology group $H^1(V, \mathbb{C})$ (de Rham Theorem). If $\sigma - \sigma' = \partial A$, then the values of the class $[\omega]$ on the two cycles are equal (the Stokes theorem). □

Continuation of the proof of Proposition 5.2. We choose a loop $\hat{\gamma}$, homologically equivalent to γ, as the lift to the Riemann surface of the function $x \mapsto y(x, t)$ of certain loop $\tilde{\gamma}$ in the complex x-plane, surrounding just two ramification points x_2, x_3 of the algebraic function $y(x, t)$. We choose the following branch of $y(x, t)$ along $\tilde{\gamma}$ (one of two): it is such that $\lim y(x + \mathrm{i}b, t) > 0$ for $b \to 0^+$, $x \in (x_2, x_3)$ and $t \in [t_1, t_2]$.
Now the subintegral form $y(x, t)\mathrm{d}x$ (or $xy(x, t)\mathrm{d}x$) is single-valued (along $\tilde{\gamma}$) and holomorphic with respect to t. If the loop $\tilde{\gamma}$ is fixed, then the corresponding integrals are locally holomorphic functions of t.
If t moves outside the interval $[t_2, t_1]$, then the roots $x_{2,3}(t)$ also move in the complex plane but the analyticity property holds as long as any of the points $x_{2,3}$ do not coalesce with the third branching point $x_1(t)$. The latter possibility occurs only when $t = t_1$.
This proves Proposition 5.2. □

5.4. Corollary. *The functions* $I_j(t)$ *are analytic near the critical value* $t_2 = -2/3\sqrt{3}$*; they are equal to* 0 *at* t_2 *and this is a simple zero.*

Proof. The latter statement follows from the fact that, for $t-t_1$ positive and small, the functions $I_j(t)$ are proportional to the area of the domain $\{y^2+x^3-x<t\}$, which is an approximate ellipse with the axes of length about $\sqrt{t-t_2}$. □

Now we shall investigate the singularity of the elliptic integrals I_j at the other critical value t_1. Firstly, we try the real methods.

Lemma. *Each of the functions $I_j(t)$ has the following behavior as $t\to t_1^-$:*

$$c_1+c_2(t-t_2)\ln(t-t_1)+\cdots.$$

Proof. We calculate the singularity of the derivatives of the elliptic integrals. We have

$$I_j'(t)=\frac{1}{2}\int_{\gamma(t)}\frac{x^j\,\mathrm{d}x}{y}=\int_{x_2}^{x_3}\frac{x^j\,\mathrm{d}x}{\sqrt{t+x^3-x}}.$$

The approximate local equation of the curve $\gamma(t)$, near the critical point $(x_0,0)$, $x_0<0$, and for $t\approx t_1$, $t<t_1$, is: $y^2-c(x-x_0)^2\approx t-t_1$, $c>0$. Thus we get $x_2\approx x_0+\sqrt{(t_1-t)/c}$ and $I_j'\sim\text{const}\cdot\int_{x_2}^{0}\left[c(x-x_0)^2-(t_1-t)\right]^{-1/2}\sim\text{const}\cdot\ln(t_1-t)$. □

Using the monodromy we can investigate this integral much better. From the analysis performed in Example 4.11 it follows that, as the argument t turns around t_1, the cycle γ changes to $\gamma+\delta$ (Example 4.11). Thus

$$I(t)=\int_\gamma\omega\mapsto I(t_2+(t-t_1)\mathrm{e}^{2\pi\mathrm{i}})=\int_\gamma\omega+\int_\delta\omega.$$

The function $J(t)=\int_{\delta(t)}\omega$ is holomorphic near $t=t_1$ (by the same reason as the function $I(t)$ is analytic near $t=t_2$).
Consider the function $f(t)=I(t)-\frac{1}{2\pi\mathrm{i}}J(t)\ln(t-t_1)$. One checks that it is univalent and holomorphic in a small punctured disc around t_1. It is bounded there and (by the Riemann theorem) it prolongs itself analytically to the whole neighborhood of t_1. This means that the only singularity of the elliptic integrals is of the logarithmic type. We have proved the following.

5.5. Proposition. *The functions $I_j(t)$ have following representations near $t=t_1$:*

$$I_j(t)=f(t)+g(t)\ln(t-t_1),$$

where f and g are analytic near t_1 functions.

Now we study the asymptotic behavior of the integrals $I_j(t)$ for $t\to\infty$. Here the three branching points of the algebraic function $y(x,t)$ are approximately equal to three cubic roots of t, $x_k\sim\mathrm{e}^{2\pi\mathrm{i}k/3}t^{1/3}$ (see Fig. 5.3). Therefore, in the subintegral function we have

$$x\sim t^{1/3},\ y\sim\sqrt{t-x^3}\sim t^{1/2},\ \mathrm{d}x\sim t^{1/3}.$$

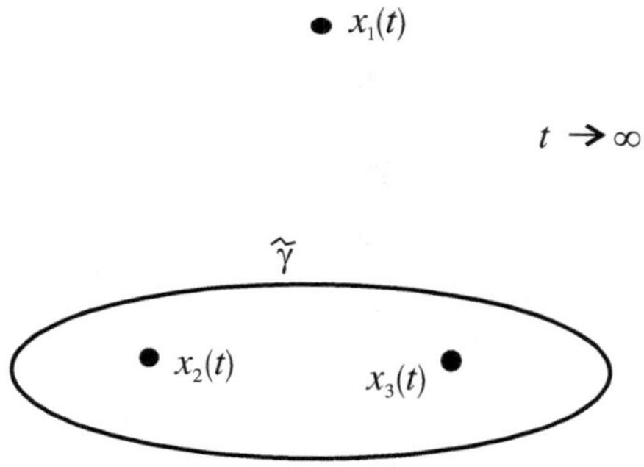

Figure 5.3: Ramification points and the cycle as $t \to \infty$

This gives:

5.6. Lemma. $I_0(t) \sim t^{5/6}$ and $I_1(t) \sim t^{7/6}$ *as* $t \to \infty$.

Let us look now at the action of the monodromy map induced by the variation of t along a loop around infinity. This map is the composition of two maps $h_{\tau_1 *} h_{\tau_2 *}$, where $\tau_{1,2}$ are the two loops surrounding the critical values $t_{1,2}$. From Example 4.11 we get that it has the matrix form $\begin{pmatrix} 0 & 1 \\ -1 & 1 \end{pmatrix}$, or $(\gamma, \delta) \mapsto (\delta, -\gamma + \delta)$. The characteristic polynomial is $\lambda^2 - \lambda + 1$ and the eigenvalues are $\mathrm{e}^{\pm \pi \mathrm{i}/3}$ (the roots of unity of order 6).
If $I = \int_\gamma \omega$, $J = \int_\delta \omega$, then we get

$$(I, J)(t\mathrm{e}^{2\pi\mathrm{i}}) = (J, -I + J)(t).$$

The functions $I(t)$, $J(t)$ generate a 2-dimensional complex vector space. The monodromy operator acts on this space. Let $F(t)$ and $G(t)$ be the eigenvectors of the monodromy operator: $F(t\mathrm{e}^{2\pi\mathrm{i}}) = \mathrm{e}^{\pi\mathrm{i}/3} F(t)$, $G(t\mathrm{e}^{2\pi\mathrm{i}}) = \mathrm{e}^{-\pi\mathrm{i}/3} G(t)$. We find that $F(t) = t^{1/6} \widetilde{F}(t)$, $G(t) = t^{-1/6} \widetilde{G}(t)$, where $\widetilde{F}$, $\widetilde{G}$ are univalent near $1/t = 0$.
Using Lemma 5.6 we obtain the following result.

5.7. Proposition. *The functions* $I_j(t)$ *have following representations near* $t = \infty$:

$$I_j = t^{5/6} f(1/t) + t^{7/6} g(1/t),$$

where f, g *are analytic functions near* $1/t = 0$.

Now we consider the general situation. Let $f : (\mathbb{C}^n, 0) \mapsto \mathbb{C}$ be a germ of holomorphic function and let t be a noncritical value for f. Let $\Delta(t)$ be some $(n-1)$-dimensional cycle in the surface $f^{-1}(t)$, depending continuously on t. Let

$$\omega = \sum h_i(x) \mathrm{d}x_1 \wedge .\overset{i}{\vee}. \wedge \mathrm{d}x_n$$

be a holomorphic $(n-1)$-form on $\mathbb{C}^n$. (Here the name *holomorphic* means that the functions $h_i(x)$ are holomorphic and there are no differentials $\mathrm{d}\bar{x}_k$.)
Define the function

$$I(t) = \int_{\Delta(t)} \omega.$$

Because ω, restricted to the surface $f^{-1}(t)$, is closed, the definition of the function $I(t)$ does not depend on the choice of the cycle Δ in its homology class (see the proof of Lemma 5.3 above).

5.8. Theorem. *The function $I(t)$ depends holomorphically on t.*

Proof. As in the proof of Proposition 5.2 we want to replace the integration of the fixed $(n-1)$-form along the varying cycle by integration of a varying form along a fixed cycle. The tool for this construction provides us the Leray coboundary operator.

5.9. Definition. Let M be a complex manifold and let $N \subset M$ be its complex submanifold of codimension 1. Take a tubular neighborhood T of N, take its boundary ∂T and the projection $\pi : \partial T \mapsto N$. This is a fibration with the fiber $\mathbb{S}^1$.
If η is some l-dimensional cycle in N, then $\pi^{-1}(\eta)$ is a $(l+1)$-dimensional cycle in $M \setminus N$. The operator $\delta : H_l(N) \mapsto H_{l+1}(M \setminus N)$:

$$[\eta] \mapsto [\pi^{-1}(\eta)]$$

is called the **Leray coboundary operator**.

Fix a noncritical value t_0. We take a tubular neighborhood of $f^{-1}(t_0)$ in $\mathbb{C}^n$ (included completely in the set of noncritical points for f). Let $\delta\Delta$ be the image of the cycle $\Delta(t_0)$ under the Leray coboundary operator. If t is close to t_0, then the cycle $\delta\Delta$ also plays the role of the value of the coboundary operator for the cycle $\Delta(t)$.

Lemma. *We have*

$$I(t) = \frac{1}{2\pi \mathrm{i}} \int_{\delta\Delta} \frac{\mathrm{d}f \wedge \omega}{f-t}.$$

Finishing of the proof of Theorem 5.8. In the latter formula the n-cycle is fixed (does not depend on t) and the subintegral n-form is analytic with respect to t, for t from a small neighborhood of t_0. From this the analyticity of $I(t)$ follows. □

Proof of the lemma. The Leray coboundary operator applied to the family of cycles $\Delta(t)$ has very simple construction. For fixed t consider the small circle $\gamma = \{s \in \mathbb{C} : |s-t| = \epsilon\}$. Then we can choose $\delta\Delta(t) = \bigcup_{s\in\gamma} \Delta(s)$. Therefore we get

$$\frac{1}{2\pi \mathrm{i}}\int_{\delta\Delta(t)}\frac{\mathrm{d}f\wedge\omega}{f-t}=\frac{1}{2\pi \mathrm{i}}\int_{\gamma}\left(\int_{\Delta(s)}\omega\right)\frac{\mathrm{d}s}{s-t}$$
$$=\frac{1}{2\pi \mathrm{i}}\int_{\gamma}\left(\int_{\Delta(t)}\omega\right)\frac{\mathrm{d}s}{s-t}+\frac{1}{2\pi \mathrm{i}}\int_{\gamma}\left(\int_{\Delta(s)-\Delta(t)}\omega\right)\frac{\mathrm{d}s}{s-t}. \tag{5.1}$$

The first integral in Eq. (5.1) equals to $\int_{\Delta(t)}\omega$. The function $\left(\int_{\Delta(s)-\Delta(t)}\omega\right)/(s-t)$ is bounded and, because γ tends to a point as $\epsilon\to 0$, the second integral in (5.1) is equal to 0. □

5.10. Remark. If the function f depends analytically on some parameter y, $f=F(\cdot,y)$, and the form $\omega=\omega(x,y)$ also depends analytically on y then, repeating the proof of Theorem 5.8, we show that the function

$$I(t,y)=\int_{\Delta(t,y)}\omega$$

depends analytically on all variables. Here $\Delta(t,y)$ is some continuous family of $(n-1)$-cycles in $\{x:F(x,y)=t\}$.

Let $f:\mathbb{C}^n\mapsto\mathbb{C}$ be a holomorphic function and let ω be a holomorphic n-form on $\mathbb{C}^n$.

5.11. Lemma and definition. *Let x_0 be such a point that $\mathrm{d}f(x_0)\neq 0$. There exists an $(n-1)$-form η in a neighborhood of x_0 such that the equation*

$$\mathrm{d}f\wedge\psi=\omega$$

*holds. The form η restricted to the surface $f=\mathrm{const}$ is defined uniquely.
It is called the* **Gelfand–Leray form** *and is denoted by*

$$\omega/\mathrm{d}f.$$

Proof. After applying some holomorphic change of coordinates, we can assume that f is one of the coordinates, $f=x_1$. Let $\omega=h(x)\mathrm{d}x_1\wedge\ldots\wedge\ldots\mathrm{d}x_n$. Then we can choose

$$\omega/\mathrm{d}f=h(x)\mathrm{d}x_2\wedge\ldots\wedge\mathrm{d}x_n$$

It is also clear that the non-uniqueness in the choice of $\omega/\mathrm{d}f$ lies in the space of forms of the type $\rho\wedge\mathrm{d}x_1=\rho\wedge\mathrm{d}f$. □

Example. If $f=y^2-x^3+x$, $\omega=\mathrm{d}x\wedge\mathrm{d}y$, then

$$\frac{\omega}{\mathrm{d}f}=-\frac{\mathrm{d}x}{2y}=\frac{-\mathrm{d}x}{2\sqrt{x^3-x+t}}.$$

It is the 1-form appearing in the derivative of the elliptic integral I_0 (see Example 5.1).
The Gelfand–Leray form can be defined in the real domain. The definition is the same but now the variables x_i are real, f is real function on $\mathbb{R}^n$ and ω is real n-form. The notion of the Gelfand–Leray form is very useful here. It is demonstrated in the following lemma.

5.12. Lemma. *We have the following formulas*

$$\int_{\mathbb{R}^n} \omega = \int_{-\infty}^{\infty} \left(\int_{f=t} \frac{\omega}{\mathrm{d}f} \right) \mathrm{d}t, \tag{5.2}$$

if the support $\operatorname{supp}\omega$ *of* ω *is disjoint with the set* $\{\mathrm{d}f = 0\}$, *and*

$$\frac{\mathrm{d}}{\mathrm{d}t} \left(\int_{f=t} \eta \right) = \int_{f=t} \frac{\mathrm{d}\eta}{\mathrm{d}f}. \tag{5.3}$$

This means that we can differentiate the integrals by means of the formal differentiation of the subintegral form.

Proof. The identity (5.2) follows from the Fubini theorem. Next, we have

$$\int_{f=t+h} \eta - \int_{f=t} \eta = \int_{t \le f \le t+h} \mathrm{d}\eta = \int_t^{t+h} \mathrm{d}s \left(\int_{f=s} \frac{\mathrm{d}\eta}{\mathrm{d}f} \right) \qquad \square$$

The identity (5.3) holds also in the complex case. Indeed, we have

$$\begin{aligned} \frac{\mathrm{d}}{\mathrm{d}t} \int_{\Delta(t)} \omega &= \frac{\mathrm{d}}{\mathrm{d}t} \frac{1}{2\pi \mathrm{i}} \int_{\delta\Delta} \frac{\mathrm{d}f \wedge \omega}{f - t} = \frac{1}{2\pi \mathrm{i}} \int_{\delta\Delta} \frac{\mathrm{d}f \wedge \omega}{(f-t)^2} = \frac{1}{2\pi \mathrm{i}} \int_{\delta\Delta} \mathrm{d} \left(\frac{-1}{f-t} \right) \wedge \omega \\ &= \frac{1}{2\pi \mathrm{i}} \int_{\delta\Delta} \frac{\mathrm{d}\omega}{f-t} = \frac{1}{2\pi \mathrm{i}} \int_{\delta\Delta} \frac{\mathrm{d}f \wedge (\mathrm{d}\omega/\mathrm{d}f)}{f-t} = \int_{\Delta(t)} \frac{\mathrm{d}\omega}{\mathrm{d}f}. \end{aligned}$$

In what follows we shall investigate the integrals of Gelfand–Leray forms along vanishing cycles. If $\eta = h(x)\mathrm{d}x_1 \wedge \ldots \wedge \mathrm{d}x_n$ is a holomorphic n-form and $\Delta(t)$ is a family of cycles, then we define the function

$$\tilde{I}(t) = \int_{\Delta(t)} \frac{\eta}{\mathrm{d}f}.$$

The next theorem is analogous to Theorem 5.8.

5.13. Theorem. $\tilde{I}(t)$ *is a holomorphic function of* t *and of eventual parameters.*

Proof. We have the formula

$$\tilde{I} = \frac{1}{2\pi \mathrm{i}} \int_{\delta\Delta} \frac{\eta}{f - t},$$

where the Leray coboundary cycle is fixed and the subintegral form is holomorphic. $\square$

§2 Singularities and Branching of Integrals

Let $f:(\mathbb{C}^n,0)\mapsto(\mathbb{C},0)$ be a germ of holomorphic function with the isolated critical point $x=0$ (and the critical value $t=0$) and let $\Delta_1(t),\ldots,\Delta_\mu(t)$ be a family of $(n-1)$-cycles at the level surfaces $V_t=\{f=t\}\cap\mathbb{B}_\rho$, vanishing at $x=0$ and forming a basis of the space $\widetilde{H}_{n-1}(V_t,\mathbb{C})$. Let ω be a holomorphic $(n-1)$-form and η be a holomorphic n-form. Consider the vector-valued functions

$$I(t)=\left(\int_{\Delta_1(t)}\omega,\ldots,\int_{\Delta_\mu(t)}\omega\right),$$

$$\tilde{I}(t)=\left(\int_{\Delta_1(t)}\eta/\mathrm{d}f,\ldots,\int_{\Delta_\mu(t)}\eta/\mathrm{d}f\right).$$

5.14. Theorem. *There exists an expansion of the function $I(t)$ into the series*

$$I(t)=\sum_{\alpha,k}a_{k,\alpha}t^\alpha(\ln t)^k,$$

where $a_{k,\alpha}\in\mathbb{C}^\mu$, the α's are non-negative rational numbers, such that $\mathrm{e}^{2\pi\mathrm{i}\alpha}$ form eigenvalues of the monodromy operator, and the k's are positive integers $\leq$ maximal dimension of the Jordan cells, corresponding to the eigenvalue $\mathrm{e}^{2\pi\mathrm{i}\alpha}$.
This series is convergent for $|t|$ sufficiently small and for $\arg t$ lying in some fixed bounded sector.
An analogous expansion holds also for the function $\tilde{I}$, only the summation runs over rational α's which are >-1.

Proof. We need the following upper bound.

5.15. Lemma. *If $|t|$ is small and $\arg t$ is bounded, the form ω is holomorphic and $\Delta(t)$ is a family of cycles in $f=t$ vanishing at $t=0$, then there exists a constant C such that*

$$\left|\int_{\Delta(t)}\omega\right|<C.$$

Proof. Because the form is are locally bounded, the estimates for the integrals follow from the construction of vanishing cycles. Recall that it uses the mini-versal deformation of f of the form $F_0(x,\lambda')-t$, where F_0 is the restricted deformation (without the constant term). One takes a generic parameter λ'_* and puts $\tilde{f}=F_0(\cdot,\lambda'_*)$ as the Morse perturbation of f with critical values $t_1,\ldots,t_\mu$. With the paths $\alpha_i=\alpha_{\lambda'_*,i}$, joining some distinguished noncritical value t with t_i, one associates the cycles $\Delta_i=\Delta_{\lambda',i}$ vanishing along α_i. We see that, when the argument $\arg t$ is bounded, then, passing with (good) λ'_* to 0, we obtain the cycles $\Delta_i=\Delta_{0,i}$ vanishing along $\alpha_{0,i}$. The cycles Δ_i are uniformly bounded as far as t varies within a sector. □

5.16. Remark. An analogous estimate

$$\left|\int_{\Delta(t)} \omega\right| < C|t|^{-N}$$

for some integer N holds in the case when f is a polynomial, ω is a polynomial $(n-1)$-form, $\Delta(t)$ is a family of $(n-1)$-cycles in the level surface $f = t$ and $t = 0$ is the critical value for f in some generalized sense; $t = 0$ is an *atypical value*. The latter means that the map $f : \mathbb{C}^n \mapsto \mathbb{C}$ is not a local fibration above a neighborhood of $t = 0$. In particular, the function f can have bad behavior at infinity or it can have non-isolated critical points.
In this case it is possible that the cycle $\Delta(t)$ grows to infinity. However, this growth is not too fast. It grows polynomially and, because the form is also polynomial, the estimate follows.
The results of these kind are known in algebraic geometry. The book of Lefschetz [**Lef**] contains description of cycles in complex algebraic varieties in a form of real semi-algebraic sets. The equations and inequalities (involved in the definitions of the cycles) depend algebraically on t and grow at most polynomially at infinity.

The next lemma is from the linear algebra.

5.17. Lemma. *Let A be $\mu \times \mu$ matrix such that* $\det A \neq 0$. *Then there exists* $\ln A$, *i.e., such matrix B that* $\mathrm{e}^B = A$.

Proof. Take the Jordan representation of A: $A = A_s + A_n$, where A_s is the *semi-simple* part (diagonal in a suitable basis) and A_n is the *nilpotent* part (upper-triangular in that basis). Moreover, the matrices A_s and A_n commute. From this the representation (called the *Chevalley decomposition)*

$$A = A_u A_s$$

follows. Here $A_u = I + A_n A_s^{-1}$ is the unipotent part of A and also commutes with A_s. (An operator C is called *unipotent* iff $C - I$ is nilpotent).
Now $\ln A = \ln A_u + \ln A_s$. For $\ln A_u$ we get the finite series

$$\ln A_u = \ln(I + (A_u - I)) = (A_u - I) - (A_u - I)^2/2 + (A_u - I)^3/3 - \cdots .$$

The operator A_s is the direct sum of its components in the eigenspaces of A_s. If $C = \lambda \cdot I$, then $\ln C = \ln \lambda \cdot I$. □

Proof of Theorem 5.14. Let us apply the monodromy to the vector-valued function $I(t) = \left(\int_{\Delta_1(t)} \omega, \ldots, \int_{\Delta_\mu(t)} \omega\right)$. As the t surrounds the origin, the function $I(t)$ undergoes the transformation

$$I(t) \mapsto I(t) \cdot M,$$

where M is the matrix of the monodromy operator in the basis $\{\Delta_i\}$.

Consider the matrix

$$F(t) = t^{-\ln M/2\pi \mathrm{i}} = \exp\left[-\ln t \frac{\ln M}{2\pi \mathrm{i}}\right].$$

The monodromy of the matrix-valued function $J(t)$ is the following

$$F(t) \mapsto M^{-1}F(t).$$

Therefore, the vector-valued function $t \mapsto \Phi(t) = I(t)F(t)$ is univalent. We shall show that it is meromorphic.

For this it is enough to show that it has at most polynomial growth as $t \to 0$; then, after multiplying by some power of t, we can use the Riemann theorem. Lemma 5.15 says that I is bounded. On the other hand, the matrix-function $F(t)$ has the following finite expansion

$$F(t) = \sum t^{-\alpha} P_\alpha(\ln t), \tag{5.4}$$

where $\mathrm{e}^{2\pi \mathrm{i}\alpha}$ are eigenvalues of M and $\deg P_\alpha$ is the maximal dimension of any Jordan cell with this eigenvalue. By Monodromy Theorem 4.71 (Chapter 4) the exponents α are rational numbers. An analogous expansion (with $t^\alpha Q_\alpha(\ln t)$) holds for $F^{-1}(t)$. We see that $|F(t)| < C|t|^{-N}$ for some integer N and constant C.

Thus we have the convergent Laurent expansion $\Phi(t) = \sum_{j=-m}^{\infty} \phi_j t^j$. From this, Eq. (5.4) for F^{-1} and the formula $I(t) = \Phi(t)F^{-1}(t)$ one gets the expansion $\sum a_{k,\alpha} t^\alpha \ln^k t$ from thesis of Theorem 5.14.

In the same way we obtain the same general expansion for the function $\tilde{I}(t)$ defined by means of the integrals of the Gelfand–Leray form $\eta/\mathrm{d}f$.

To obtain a bound for the exponents α we notice that, by Lemma 5.15, the function $I(t)$ is bounded and vanishes at $t = 0$. Therefore $\alpha > 0$.

The functions $\tilde{I}_j = \int_{\Delta_j(t)} \eta/\mathrm{d}f$ are derivatives of some functions of the type $\int_{\Delta_j(t)} \omega$ for ω such that $\eta = \mathrm{d}\omega$. (It follows from the *holomorphic Poincaré Lemma*, which is proved in the same way as the usual Poincaré Lemma 3.28.) This gives $\alpha > -1$ in the expansion of $\tilde{I}$.

Theorem 5.14 is complete. □

5.18. Example. Let $f = x_1^2 + \ldots + x_n^2$ and let ω be a holomorphic $(n-1)$-form. Here we have only one vanishing cycle $\Delta(t)$. The monodromy operator acts on it as follows: $\Delta(t) \mapsto (-1)^n \Delta(t)$.

If $I(t) = \int_{\Delta(t)} \omega$, then we have

$$I = \sum_{j=0}^{\infty} a_j t^j,$$

for even n, and

$$I = \sum_{j=0} a_{j+1/2} t^{j+1/2},$$

for odd n.
If we take the expansion $\omega = \omega_{n-1} + \omega_n + \dots$ into components homogeneous with respect to the dilations $x \mapsto \lambda x$, then we get $\int \omega_p = 0$ for $p - n$ odd and $= a_{p/2} t^{p/2}$ otherwise.
Consider the special case $\omega_n = \sum (-1)^j A_j x_j \mathrm{d}x_1 \wedge .\overset{j}{\cdot}. \wedge \mathrm{d}x_n$. Using the formula $\int_{\Delta(t)} \omega_n = \int_{D(t)} \mathrm{d}\omega_n$, where $D(t)$ is the n-ball with the boundary $\Delta(t)$, we get $a_{n/2} t^{n/2} = \int_{x_1^2 + \dots + x_n^2 \le t} (\sum A_j) \mathrm{d}x_1 \wedge \dots \wedge \mathrm{d}x_n$. This gives $a_{n/2} = (\sum A_j) \times$ (volume of the n-dimensional unit ball).

§3 Picard–Fuchs Equations

Before formulating the next theorems, we proceed the study of the elliptic integrals from 5.1.

5.19. Picard–Fuchs equations for elliptic integrals. As in 5.1 we consider the family

$$V_t : f(x, y) = y^2 + x^3 - x = t$$

of elliptic curves. We take the 1-forms $\omega_0 = y\mathrm{d}x$, $\omega_1 = xy\mathrm{d}x$ and consider the abelian integrals

$$I_{0,1} = \int_{\gamma(t)} \omega_{0,1}, \quad J_{0,1} = \int_{\delta(t)} \omega_{0,1},$$

where $\gamma(t)$, $\delta(t)$ are the vanishing cycles; the first vanishes at the local minimum and the second vanishes at the saddle point. They are periods of the basic cycles in the elliptic curve V_t.
Let $\omega = A(x, y)\mathrm{d}x + B(x, y)\mathrm{d}y$ be a polynomial 1-form of the degree $\deg \omega = \max(\deg A, \deg B) \le n$. Our first task is to express the integrals

$$I_\omega(t) = \int_\gamma \omega, \quad J_\omega(t) = \int_\delta \omega$$

by means of the integrals $I_{0,1}$ (or $J_{0,1}$ respectively). The next lemma was firstly formulated by G. Petrov [**Pet1**].

5.20. Lemma. *There exist two polynomials $P_0(t)$ and $P_1(t)$ of degrees $\le [(n-1)/2]$ and $\le [n/2] - 1$ (respectively) and such that*

$$I_\omega = P_0(t) I_0 + P_1(t) I_1,$$
$$J_\omega = P_0(t) J_0 + P_1(t) J_1.$$

More precisely, if we denote by V_n the linear space of I_ω's with $\deg \omega \le n$, *then it coincides with the space of pairs of polynomials P_0, P_1 of the above degrees. In particular,* $\dim V_n = n$.

Proof. We use three ingredients: (i) the integration by parts $\int \mathrm{d}R = 0$, (ii) the property $\mathrm{d}(y^2 + x^3 - x) = 0$ and (iii) the substitution $y^2 = t + x - x^3$.
Thus we have

$$\int x^i y^j \mathrm{d}y = (-i/(j+1)) \int x^{i-1} y^{j+1} \mathrm{d}x,$$

$$\int x^i y^{2k} \mathrm{d}x = \int x^i (t + x - x^3)^k \mathrm{d}x = 0,$$

$$\int x^i y^{2k+1} \mathrm{d}x = \int x^i (t + x - x^3)^k y \mathrm{d}x = \sum_{j=0}^{3k+i} Q_j(t) I_j.$$

where $I_j = \int x^j y \mathrm{d}x$ and Q_j are polynomials.
In order to analyze I_j we use the property (ii): $3x^2 \mathrm{d}x = \mathrm{d}x - 2y\mathrm{d}y$. Hence

$$\begin{aligned} 3I_j &= \int x^{j-2} y \mathrm{d}x - 2 \int x^{j-2} y^2 \mathrm{d}y \\ &= I_{i-2} + \frac{2(j-2)}{3} \int x^{j-3} y^3 \mathrm{d}x \\ &= I_{j-2} + \frac{2(j-2)}{3} \int x^{j-3} (t + x - x^3) y \mathrm{d}x \\ &= I_{j-2} + \frac{2j-4}{3} t I_{j-3} + \frac{2j-4}{3} I_{j-2} - \frac{2j-4}{3} I_j. \end{aligned}$$

Calculating from this I_j (with nonzero coefficient), we get the representation of I_j as a combination of I_{j-2} and tI_{j-3}. This shows that the representation from Lemma 5.20 with some $P_{0,1}$ holds. In particular, we have $I_2 = I_0$, $I_3 = c_1 t I_0 + c_2 I_1$, etc.
Lemma 5.6 says that $I_0 \sim t^{5/6}$ and $I_1 \sim t^{7/6}$ as $t \to \infty$. This shows that the functions $t^i I_0$ and $t^j I_1$ are linearly independent and are good candidates for a basis of the space V_n. It remains to estimate the degrees of P_i's and show that the separate monomials in P_i's are obtained from certain 1-forms of degree $\leq n$. We will do it also using the asymptotic behavior of the integrals as t tends to infinity.
Because $\int y^n \mathrm{d}y = 0$ and because of (i) and (ii), it is enough to consider the integrals of the forms $x^k y^l \mathrm{d}x$ with $k + l = n$ and with odd l. We have $\int x^k y^l \mathrm{d}x = O(t^{l/2+(k+1)/3})$.
If n is even, then the integral $\int x y^{n-1} \mathrm{d}x = O(t^{(n-1)/2+2/3})$ is dominating and gives the term $\sim t^{n/2-1} I_1$. If n is odd, then the integral $\int y^n \mathrm{d}x = O(t^{n/2+1/3})$ is dominating and gives the term $\sim t^{(n-1)/2} I_0$. □

5.21. Remark. During the demonstration of Lemma 5.20 we proved the following fact:

The integral $I_\omega(t) \equiv 0$ iff the form is of the type $\omega = g\mathrm{d}f + \mathrm{d}R$ with polynomial functions g and R. This means that ω restricted to the surface $f =$ const is exact.

Such property is characteristic for other functions f: x^2+y^2 and for generic polynomials f.

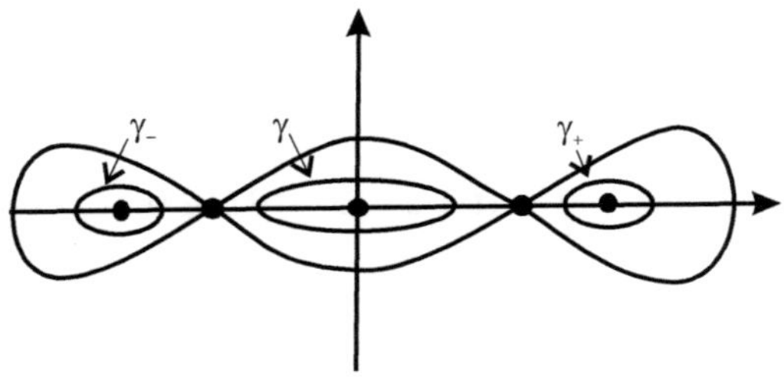

Figure 5.4: Cycles for $f = y^2 + x^2(x^2-1)^2$

But the polynomial $f = y^2 + x^2(x^2-1)^2$ and the form $xy\mathrm{d}x$ provide an example, where the above property fails. The surface $f=t$ contains one symmetric cycle γ vanishing at $x=y=0$ and two cycles $\gamma_\pm$ vanishing at the points $x=\pm 1, y=0$ (see Fig. 5.4). The integral of $xy\mathrm{d}x$ along γ is equal to zero but the analogous integrals along $\gamma_\pm$ are nonzero.

Problem. Generalize the result of Lemma 5.20 to the case of hyperelliptic curve $y^2 + x^n + \lambda_{n-2}x^{n-2} + \ldots + \lambda_1 x = t$.

Using the Gelfand–Leray form we get the formulas for the derivatives of the elliptic integrals

$$\frac{\mathrm{d}I_j}{\mathrm{d}t} = I_j' = \frac{1}{2}\int_\gamma \frac{x^j \mathrm{d}x}{y}.$$

5.22. Lemma (Picard–Fuchs equations). *We have*

$$5I_0 = 6tI_0' + 4I_1',$$
$$21I_1 = 4I_0' + 18tI_1'.$$

Proof. We have

$$I_0 = \int y^2 \mathrm{d}x/y = \int (t + x - x^3)\mathrm{d}x/y = 2tI_0' + 2I_1' - \int x^3 \mathrm{d}x/y.$$

Next, $\int x^3 \mathrm{d}x/y = \frac{1}{3}\int x\mathrm{d}x/y - \frac{2}{3}\int x\mathrm{d}y = \frac{2}{3}(I_1' + I_0)$. From this we get I_0 expressed by means of I_j''s.
The second equation is proved analogously. □

The reader can observe that the determinant of the matrix of the right-hand-sides of the Picard–Fuchs equations is equal $4(27t^2-4)$ and vanishes exactly at the two critical values of the function f.
Solving these equations, one can express the derivatives of integrals as linear functions of the integrals, i.e., one obtains a non-autonomous linear differential system.

Any linear 2-dimensional linear differential equation is equivalent to some linear second order differential equation. Indeed, after some transformations we obtain the following result.

5.23. Lemma.

(a) $4(27t^2 - 4)I_0'' = -15I_0$;

(b) $4(27t^2 - 4)I_1'' = 21I_1$.

From this one can conclude that any integral I_ω satisfies some second order linear equation (with rational coefficients).
Note also that all the above (i.e., the representation and the differential equations) holds also for the integrals J_ω along the second cycle δ. In particular, we have two solutions of the Picard–Fuchs system from Lemma 5.22: $(I_0, I_1)^\top$ and $(J_0, J_1)^\top$. It is natural to investigate the determinant of the matrix formed by these solutions

$$W(t) = \det \begin{pmatrix} I_0 & J_0 \\ I_1 & J_1 \end{pmatrix},$$

i.e., the Wronskian of the corresponding fundamental matrix.

5.24. Lemma. *We have* $W(t) = \frac{8}{315} \cdot \pi \cdot \mathrm{i} \cdot (27t^2 - 4)$.

Proof. Firstly, we show that $W(t)$ is analytic in the whole plane $\mathbb{C}$. For this it is enough to consider neighborhoods of branching points of the elliptic integrals $t = \pm 2/3\sqrt{3}$. However, from the Picard–Fuchs formula it follows that, as t surrounds $2/3\sqrt{3}$, the fundamental matrix undergoes the transformation $\begin{pmatrix} I_0 & J_0 \\ I_1 & J_1 \end{pmatrix} \mapsto \begin{pmatrix} I_0 + J_0 & J_0 \\ I_1 + J_1 & J_1 \end{pmatrix}$. Thus $W(t)$ is single valued near $t = 2/3\sqrt{3}$. The case $t = -2/3\sqrt{3}$ is analogous.
From the asymptotic behavior of the integrals for large t (see Lemma 5.6) it follows that $W(t)$ grows like t^2 and is a quadratic polynomial.
Because one of the integrals vanishes at critical value and the other integral is bounded there, the Wronskian must be equal zero at this value. Thus $W(t) = K \cdot (27t^2 - 4)$. (One can obtain the latter formula directly from the Picard–Fuchs system 5.22.)
It remains to calculate the constant K. We use the asymptotic behavior at infinity. Let $x_j = \mathrm{e}^{2\pi \mathrm{i} j/3}$, $j = 1, 2, 3$, be the roots of the equation $1 - x^3 = 0$ and let $A_i = 2\int_{x_2}^{x_3} \sqrt{1 - x^3} x^i \mathrm{d}x$, $B_i = 2\int_{x_1}^{x_2} \sqrt{1 - x^3} x^i \mathrm{d}x$, $i = 0, 1$. Then we have $27K = A_0 B_1 - A_1 B_0$. We represent $A_i/2$ as $\int_0^{x_3} - \int_0^{x_2}$ (along the rays) and similarly we represent $B_i/2$. Note that, passing from one ray to another, means a change of the argument in x and in $y = \sqrt{1 - x^3}$: for $0 < x < x_3$ we have $\arg x = \arg y = 0$, for $x \in [0, x_2]$ we have $\arg x = -2\pi/3$ and $\arg y = \pi$, and for $x \in [0, x_1]$ we have $\arg x = 2\pi/3$ and $\arg y = 0$.

We get the result in terms of the Euler Beta-function $B(\alpha,\beta) = \int_0^1 \tau^{\alpha-1}(1-\tau)^{\beta-1}\mathrm{d}\tau = \Gamma(\alpha)\Gamma(\beta)/\Gamma(\alpha+\beta)$. Namely $27K = \frac{4}{9}\cdot[(1+\mathrm{e}^{-2\pi\mathrm{i}/3})(-\mathrm{e}^{-4\pi\mathrm{i}/3}-\mathrm{e}^{4\pi\mathrm{i}/3}) - (1+\mathrm{e}^{-4\pi\mathrm{i}/3})\,(-\mathrm{e}^{-2\pi\mathrm{i}/3}-\mathrm{e}^{2\pi\mathrm{i}/3})]\cdot B(\frac{2}{3},\frac{3}{2})\cdot B(\frac{1}{3},\frac{3}{2})$. Using the identity $\Gamma(z)\Gamma(1-z) = \frac{\pi}{\sin(\pi z)}$, after some transformations, we get the result. □

When we add a parameter in the above analysis, i.e., when we consider the curves $y^2+x^3-\lambda x = t$, and we pass with λ to zero, then we obtain the formula $\det(\int_{\Delta_j}\mathrm{d}\omega_i/\mathrm{d}f) = \text{const}$. (Here $\Delta_1=\delta$, $\Delta_2=\gamma$). This property is generalized in the next theorem.

Let $f:(\mathbb{C}^n,0)\mapsto(\mathbb{C},0)$ be a germ with isolated critical point.
Let $\Delta_1(t),\ldots,\Delta_\mu(t)$ be a system of cycles vanishing at $t=0$ and forming a basis of the $(n-1)^{\text{th}}$ reduced homology group of the local level surface $f=t$.
Let $\omega_1,\ldots,\omega_\mu$ be a system of germs of holomorphic n-forms.
The matrix

$$\left(\int_{\Delta_j(t)}\omega_i/\mathrm{d}f\right)_{i,j=1,\ldots,\mu}$$

is called the **period matrix**. We study the following function

$$(\det)^2\, t\mapsto(\det)^2\left(\int_{\Delta_j(t)}\omega_i/\mathrm{d}f\right).$$

5.25. Theorem (Determinant of the period matrix).

(a) *The function* $(\det)^2(t)$ *is univalent near* $t=0$.

(b) *It is of the form* $t^{\mu(n-2)}\cdot g(t)$, *where* g *is analytic near* 0.

(c) *If the system* (ω_i) *of forms is general, then* $g(0)\neq 0$.

The proof will be given later.

5.26. Definition. The system $\omega_1,\ldots,\omega_\mu$ is called the **trivialization** iff $(\det)^2\not\equiv 0$. It is called the **basic trivialization** iff $g(0)\neq 0$.

(The name 'trivialization' will be justified in the next section. Namely, the de Rham classes of the forms $\omega_i/\mathrm{d}f|_{f=t}$ are sections of the cohomological Milnor bundle and define local trivializations of this bundle.)

5.27. Examples. (a) For the function $f(x)=x^k$ the level surface $V_t=\{f=t\}$, $t\neq 0$, consists of k points $x_j=\zeta^j t^{1/k}$, $\zeta=\mathrm{e}^{2\pi\mathrm{i}/k}$ $j=0,\ldots,k-1$. The reduced 0th homology group $\widetilde{H}_0(V_t,\mathbb{C})$ is $(k-1)$-dimensional and is generated by the cycles $\Delta_j=x_{j+1}-x_j$, $j=0,\ldots,k-2$. If $\omega_i=x^i\mathrm{d}x$, $i=0,\ldots,k-2$, then $\omega_i/\mathrm{d}f=(1/k)x^{i-k+1}$ and, after some calculations, we get

$$\begin{aligned}\det(t) &= k^{1-k}\cdot(t^{-1/k})^{1+2+\ldots+(k-1)}\cdot(\zeta^{-1}-1)(\zeta^{-2}-1)\ldots(\zeta^{1-k}-1)\\ &\quad\times \mathrm{Vand}\,(\zeta^{1-k},\zeta^{2-k},\ldots,\zeta^{-1}),\end{aligned}$$

where $\mathrm{Vand}(y_1, \ldots, y_n)$ is the van der Monde determinant, equal to $\prod_{i<j}(y_i - y_j)$. Therefore $(\det)^2 = \mathrm{const} \cdot t^{\mu \cdot (n-2)}$, $\mu = k-1$, $n = 1$, and the system (ω_i) forms the basic trivialization.

(b) For the function $y^2 + x^3$ the forms $dx \wedge dy$ and $xdx \wedge dy$ form the basic trivialization.

5.28. Corollary

(a) *If $\omega_1, \ldots, \omega_\mu$ is a trivialization, then the forms $\omega_i/\mathrm{d}f$ form a basis of the reduced cohomology group $\widetilde{H}^{n-1}(V_t, \mathbb{C})$ for small $t \neq 0$.*

(b) *If $\omega_1, \ldots, \omega_\mu$ is a basic trivialization and ω is a holomorphic n-form, then there exist holomorphic functions $p_1(t), \ldots, p_\mu(t)$ such that for any vanishing cycle $\Delta(t)$ we have*

$$\int_\Delta \omega/\mathrm{d}f = \sum_j p_j(t) \int_\Delta \omega_j/\mathrm{d}f.$$

Proof. (a) follows from the fact that $\det(t)$ is nonzero for small $t \neq 0$.
For fixed small $t \neq 0$ consider μ vectors $I_i = \left(\int_{\Delta_1} \omega_i \mathrm{d}f \ldots, \int_{\Delta_\mu} \omega_i/\mathrm{d}f\right)$. By the above, they form a basis in $\mathbb{C}^\mu$. Thus the vector $I = \left(\int_{\Delta_1} \omega/\mathrm{d}f, \ldots, \int_{\Delta_\mu} \omega/\mathrm{d}f\right)$ is a combination of the vectors I_i, $p_i(t)$ being the coefficients. Moreover, this combination is unique. So, it remains to show that these coefficients are analytic. Firstly, we notice that the monodromy around $t = 0$ acts in the same way onto I and onto I_i: $I_i(t\mathrm{e}^{2\pi \mathrm{i}}) = I_i(t)M$. So, we have $I(t) = \sum p_i(t\mathrm{e}^{2\pi \mathrm{i}}) I_i(t)$ and, by the uniqueness, the functions p_i are univalent.
Next, we use the expression of $p_i(t)$ as the ratios of two determinants, the denominator is the determinant of the matrix formed by the vectors $I_1, \ldots, I_\mu$ and the numerator is the determinant of the matrix obtained from the previous one by replacing the vector I_i with I (the Cramer's formula). Because the numerator has $t = 0$ as zero of order not smaller than the order of zero of the denominator, the functions p_i can be prolonged analytically to 0. □

We see that (b) of the previous corollary forms a local analogue of Lemma 5.20. The next theorem is the local analogue of Lemmas 5.22 and 5.23.

5.29. Theorem.

(a) *Let $I_\omega(t) = \int_{\Delta(t)} \omega/\mathrm{d}f$ be the integral of holomorphic Gelfand–Leray form along vanishing cycle. Then there exist functions $q_1(t), \ldots, q_l(t)$, $l \leq \mu$, meromorphic near $t = 0$ and such that the following* **Picard Fuchs equation** *holds:*

$$\frac{\mathrm{d}^l}{\mathrm{d}t^l} I_\omega + q_1(t) \frac{\mathrm{d}^{l-1}}{\mathrm{d}t^{l-1}} I_\omega + \ldots + q_l(t) I_\omega = 0. \tag{5.5}$$

Moreover, any solution to this equation is a linear combination of the integrals $\int_{\Gamma(t)} \omega/\mathrm{d}f$ along some family of vanishing cycles $\Gamma(t)$.

(b) *If* $\omega_1, \ldots, \omega_\mu$ *is a trivialization, then the vector-valued function*

$$J(t) = \left(\int_{\Delta(t)} \omega_1 / \mathrm{d}f, \ldots, \int_{\Delta(t)} \omega_\mu / \mathrm{d}f \right)^\top$$

satisfies the differential Picard–Fuchs equation

$$\frac{\mathrm{d}J}{\mathrm{d}t} = A(t)J,$$

where $A(t)$ *is a matrix-valued meromorphic function. Moreover, the space of solutions of this equation is spanned by* J*'s of the above type (integrals along a family of vanishing cycles).*

Proof. (a) Let $\Delta_1, \ldots, \Delta_\mu$ be the basis of the lattice of vanishing cycles. Consider the vector-valued function $I(t) = \left(\int_{\Delta_1} \omega / \mathrm{d}f, \ldots, \int_{\Delta_\mu} \omega / \mathrm{d}f \right)$ and its successive derivatives $I'(t)$, $I''(t)$, $I'''(t), \ldots$. For fixed $t \neq 0$ we consider the sequence of subspaces $L_k(t) \subset \mathbb{C}^\mu$ spanned by first k of the vectors $I^{(j)}$.
From the expansion into the asymptotic series (see Theorem 5.14) it follows that the dimensions of the subspaces $L_k(t)$ are constant for small $t \neq 0$. Indeed, the minors of the matrix composed by the vectors $I, I', \ldots, I^{(k-1)}$ also can be expanded into series with powers of t and $\ln t$. Such series is either identically equal to zero or has some nonzero dominating term.
Let l be the maximal dimension of L_k's. Then $I^{(l)}$ is expressed as a linear combination of the lower order derivatives of I, $I^{(l)} = -\sum q_i(t) I^{(l-i)}$. Moreover, this representation is unique. From this the univalency of the functions q_i follows.
Indeed, the monodromy around $t = 0$ acts in the same way onto $I^{(l)}$ and onto $I^{(i)}$ and (by the uniqueness) the functions q_i do not change themselves (see the proof of Corollary 5.28).
Because the integrals have polynomial growth as $t \to 0$ ($< C|t|^{-N}$ within some sector), then also q_i have polynomial growth. So, they are meromorphic.
By construction, each component of the vector I is a solution of Eq. (5.5). Their linear combinations generate the l-dimensional space of solutions.
(b) Let $J_i(t) \in \mathbb{C}^\mu$ be the vectors defined like $J(t)$ but with $\Delta(t) = \Delta_i(t)$. For each small $t \neq 0$ any vector $J_i'(t)$ can be written as a linear combination of the vectors $J_k(t)$. The coefficients of these combinations form the matrix elements of the matrix $A(t)$ from thesis of the point (b). Its univalency and polynomial growth is proved in the same way as in the point (a). □

5.30. Proof of Theorem 5.25. Part (a) is a consequence of the following:

1. **Lemma.** *The function* $(\det)^2$ *is single valued.*

Proof. If M is the matrix of the monodromy operator in the basis $\Delta_1, \ldots, \Delta_\mu$, then the monodromy of the function $(\det)^2(t)$ is following: $(\det)^2 \mapsto (\det M)^2 (\det)^2$.

Notice now that M is integer matrix and its determinant is an integer number. Also M^{-1} is an integer matrix (it corresponds to surrounding $t = 0$ in the reverse direction) with integer determinant. This implies that $\det M = \pm 1$. $\square$

Consider a mini-versal deformation $F = F_0(x, \lambda') - \lambda_0 : \mathbb{C}^n \times \mathbb{C}^\mu$ of f with the bifurcational diagram of zeroes $\Sigma = \{h(\lambda) = 0\}$; here $\lambda_0 = t$. Let $\Delta_j = \Delta_j(\lambda)$ be the family of cycles in $F^{-1}(0) \cap \{\lambda = \text{const}\}$, prolongations of the cycles $\Delta_j(t)$. Define the functions $\lambda \mapsto \int_{\Delta_j(\lambda)} \omega_i(x)/\mathrm{d}_x F$, where d_x denotes the exterior derivative with respect to x (λ fixed). As above, we define the function $(\det)^2(\lambda)$.

2. **Lemma.** *We have* $\det^2(\lambda) = h^{n-2} \cdot g(\lambda)$ *for some holomorphic function* g.

From this the point (b) of Theorem 5.25 (i.e., that $(\det)^2$ is divisible by $t^{\mu(n-2)}$) follows. Indeed, if λ' is fixed and generic, then the intersection of Σ with the plane $\lambda' = \text{const}$, consists of μ points $t_1, \ldots, t_\mu$. Because $\det^2(\lambda', t) = (\prod(t - t_i))^{n-2} \times g(\lambda', t)$, then passing with λ' to zero we get the result. $\square$

Proof of Lemma. 2. We use the following general rule from analytic geometry:

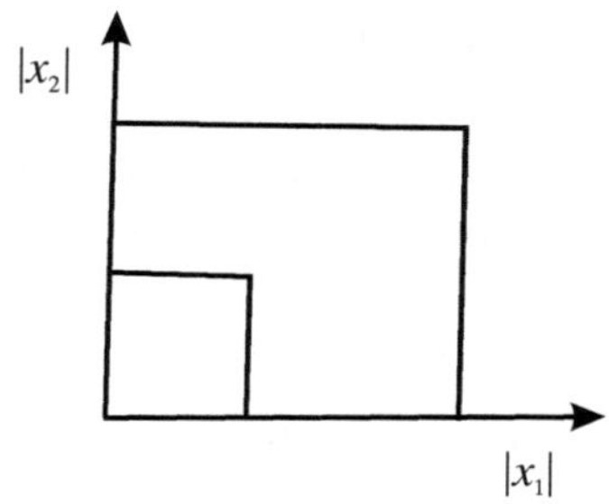

Figure 5.5: Hartog's theorem

3. **Theorem.** *If a function f, defined outside an analytic subset Z of a domain $\Omega \subset \mathbb{C}^n$ of complex codimension ≥ 2, is analytic (in $\Omega \setminus Z$), then it prolongs itself to a function holomorphic in Ω.*

(Its proof reduces itself to the proof of the Hartogs' theorem, see Fig. 5.5. The prolongation is given by the formula $f(z_1, z_2) = \frac{1}{2\pi i} \int_{|\zeta|=\text{const}} f(\zeta, z_2)/(\zeta - z_1)$.)

Using the above principle, we can study analytic properties of the function $(\det)^2$ near those points of the bifurcational diagram Σ, where exactly one cycle vanishes at a point which is of Morse type. Other points form a subset of complex codimension ≥ 1 in Σ and ≥ 2 in $\mathbb{C}^\mu$. Assume that the vanishing cycle is Δ_1, that it vanishes at $x = 0$, and that $F_0(\cdot, \lambda') = x_1^2 + \ldots + x_n^2$ near $x = 0$. The cases of different parity of n are treated separately.

If n is odd, then the Picard–Lefschetz formula says that the monodromy operator, generated by the loop around $t = 0$, acts as the reflection: $\Delta_1 \mapsto -\Delta_1$, $\Delta_j \mapsto \Delta_j$,

where we can assume that Δ_j, $j > 1$, lie in the subspace of fixed points of the reflection. Thus the functions $\int_{\Delta_j} \omega_i/\mathrm{d}_x F$, $j > 1$ are holomorphic.
The integral $\int_{\Delta_1} \omega_i/\mathrm{d}_x F = \int \phi(x)\mathrm{d}x_1 \ldots \mathrm{d}x_n/\mathrm{d}(x_1^2 + \ldots + x_n^2)$ is of the form $\sqrt{t} \times$ (univalent function). The change $x = \sqrt{t}y$ shows that it is of order at least $t^{n/2-1}$. So, also $\det(\lambda', t)$ is divisible by $t^{n/2-1}$.
If n is even, then all the eigenvalues of the monodromy operator are equal to 1: $\Delta_1 \mapsto \Delta_1$, $\Delta_j \mapsto \Delta_j + (\Delta_j, \Delta_1)\Delta_1$. The functions $\int_{\Delta_1} \omega_i/\mathrm{d}_x F$ are holomorphic and of order at least $t^{n/2-1}$.
The functions $\int_{\Delta_j}$ are expressed as sums of holomorphic parts and of $\frac{\ln t}{2\pi \mathrm{i}} \int_{\Delta_1}$. In calculation of the determinant the parts with logarithm disappear. We get $\det(\lambda', t) \sim t^{n/2-1}$.
Therefore Theorem 5.25(c) is complete. □

Before proving (c) (i.e., the existence of basic trivializations) we present such trivialization in the case of Pham's singularity.

4. **Example (Pham's singularity).** Recall that this singularity is of the form

$$f = x_1^{m_1} + \ldots + x_n^{m_n}.$$

Firstly, we compute the basis of vanishing cycles following the Milnor's book [**Mil2**] We have the Milnor fibration (the Milnor's version)

$$\mathbb{S}^{2n-1} \setminus \{f = 0\} \overset{\Phi}{\mapsto} \mathbb{S}^1, \; \Phi(x) = f(x)/|f(x)|$$

and we study the homotopy type of the fiber $\Phi^{-1}(\mathrm{e}^{\mathrm{i}\theta})$.
The diffeomorphisms $h_t(x_1, \ldots, x_n) = (\mathrm{e}^{\mathrm{i}t/m_1}x_1, \ldots, \mathrm{e}^{\mathrm{i}t/m_n}x_n)$ transform the fiber $\Phi^{-1}(\mathrm{e}^{\mathrm{i}\theta})$ to the fibers $\Phi^{-1}(\mathrm{e}^{\mathrm{i}(\theta+t)})$.
Consider the map $(x, r) \mapsto (\mathrm{e}^{r/m_1}x_1, \ldots, \mathrm{e}^{r/m_n}x_n)$ from $(\mathbb{S}^{2n-1} \setminus \{f = 0\}) \times \mathbb{R}$ to $\mathbb{C}^n \setminus \{f = 0\}$. It is easy to see that this map realizes the homotopy equivalence between $\Phi^{-1}(\mathrm{e}^{\mathrm{i}\theta}) \times \mathbb{R}$ and $\Phi^{-1}(\mathrm{e}^{\mathrm{i}\theta})$. Therefore, we shall concentrate ourselves on the fibration $\mathbb{C}^n \setminus \{f = 0\} \ni x \overset{\Psi}{\mapsto} f(x)/|f(x)|$.
Denote by Ω_m the set of mth degree roots of unity, $\Omega_m = \{\mathrm{e}^{2\pi \mathrm{i}j/m},\ j = 0, \ldots, m-1\}$. Let $\Omega = \Omega_{m_1} * \Omega_{m_1} * \ldots * \Omega_{m_n}$ be the joint of the 0-dimensional spaces Ω_{m_j}. It is a $(n-1)$-dimensional CW-complex and can be embedded into $\mathbb{C}^n$ as follows (see Fig. 5.6)

$$\Omega = \{(t_1\omega_1, t_2\omega_2, \ldots, t_n\omega_n) : t_j \geq 0,\ \ t_1 + \ldots + t_n = 1,\ \ \omega_j \in \Omega_{m_j}\}.$$

1. **Lemma (Pham).** *The set Ω is a deformation retract of the fiber $\Psi^{-1}(1)$.*

Proof. If $x \in \Psi^{-1}(1)$, then we vary this point along a curve $x(t)$ in such a way that the powers of coordinates $x_j(t)^{m_j}$ move along the intervals orthogonal to the real axis and tend to this axis. After the first step of deformation, we arrive at a point x' such that its coordinates satisfy the property $(x_j')^{m_j} \in \mathbb{R}$. Note that

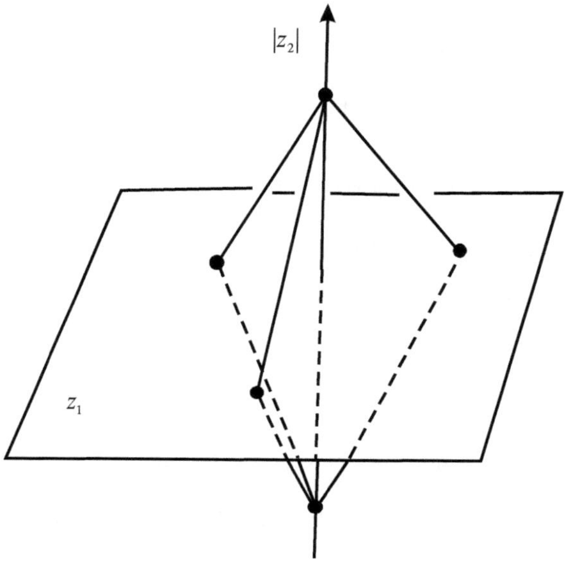

Figure 5.6: The set Ω

during the deformation $\mathrm{Re} x_j(t)^{m_j} = \text{const}$, which means that $f(x(t)) = \text{const}$. Thus the deformation takes place in the fiber $\Psi^{-1}(1)$.
The next step of the deformation relies on a contraction of the coordinates x'_j such that the quantities $(x'_j)^{m_j}$ which are < 0 tend to zero. The other coordinates remain unchanged. This deformation is also realized within $\Phi^{-1}(1)$ and we arrive at a point x'' such that $(x''_j)^{m_j} \geq 0$.
We can write $x''_j = t_j \omega_j$ for some root $\omega_j \in \Omega_j$. The final deformation transforms x'' to $x''/(t_1 + \ldots + t_n)$. □

6. Recall that, if $\widetilde{H}_*(A)$ and $\widetilde{H}_*(B)$ are without torsion, then $\widetilde{H}_{k+1}(A * B) = \bigoplus_{i+j=k} \widetilde{H}_i(A) \otimes \widetilde{H}_j(B)$ (see 3.11 in Chapter 3). Therefore

$$\widetilde{H}_{n-1}(\Phi^{-1}(1)) = \widetilde{H}_0(\Omega_{m_1}) \otimes \ldots \otimes \widetilde{H}_0(\Omega_{m_n}).$$

Note that each $\widetilde{H}_0(\Omega_{m_j})$ is generated by the $(m_j - 1)$ cycles $\Delta_{jk} = \mathrm{e}^{2\pi \mathrm{i}(k+1)/m_j} - \mathrm{e}^{2\pi \mathrm{i} k/m_j}$, $k = 0, \ldots, m_j - 2$.
The monodromy map $h : x \mapsto (\mathrm{e}^{2\pi \mathrm{i}/m_1} x_1, \ldots, \mathrm{e}^{2\pi \mathrm{i}/m_n} x_n)$ induces the homeomorphism $h_1 * \ldots * h_m$ of the complex Ω: $h_j \omega_j = \mathrm{e}^{2\pi \mathrm{i}/m_j} \omega_j$. Therefore the monodromy operator has the form

$$h_* = h_{1*} \otimes \ldots \otimes h_{n*}$$

Because each h_{j*} is isomorphic to the operator of cyclic permutation of coordinates $z_1, \ldots, z_{m_j}$ in the space $\{\sum z_k = 0\} \subset \mathbb{C}^{m_j}$, the monodromy operator h_* is diagonalizable with the eigenvalues $\omega_1\omega_2\ldots\omega_n$, $\omega_j \in \Omega_{m_j} \setminus 1$.
This result will be generalized in Theorem 5.31 below.

7. Now we translate the above construction of generators of the homology group of a fiber of the Milnor's version of the Milnor bundle to the construction of vanishing cycles in a fiber of the Arnold–Varchenko–Gusein-Zade's version of the Milnor bundle. We use the induction with respect to the dimension n.

Let $t > 0$. At the surface $f = t$ (where, as before, $f = x_1^{m_1} + \ldots + x_n^{m_n}$) we have

$$x_n = \left(t - (x_1^{m_1} + \ldots + x_{n-1}^{m_{n-1}})\right)^{1/m_n}.$$

Let $\Delta(t_{n-1}) \subset \mathbb{C}^{n-1}$ be some family of $(n-2)$-dimensional cycles in $x_1^{m_1} + \ldots + x_{n-1}^{m_{n-1}} = t_{n-1}$, $0 \le t_{n-1} \le t$. With the cycle $\Delta(t_{n-1})$, which is diffeomorphic to $(n-2)$-dimensional sphere, we associate its suspension $\Delta(t) \approx S\Delta(t_{n-1}) \subset \mathbb{C}^{n-1} \times \mathbb{C}$. (If X is a topological space, then its *suspension* SX is defined as $X \times [0,1]/\sim$, where we identify $X \times \{0\}$ and $X \times \{1\}$ with points.) Fix an integer $k = k_n$ between 0 and $m_n - 2$. Then

$$\Delta(t) = \bigcup_{t_{n-1}=0}^{t} \Delta(t_{n-1}) \times \left\{\mathrm{e}^{2\pi \mathrm{i}k/m_n} (t - t_{n-1})^{1/m_n}\right\}$$
$$\cup \bigcup_{t_{n-1}=0}^{t} \Delta(t_{n-1}) \times \left\{\mathrm{e}^{2\pi \mathrm{i}(k+1)/m_n} (t - t_{n-1})^{1/m_n}\right\}$$

as a topological space. As an $(n-1)$-dimensional cycle, it is equipped also with the corresponding orientation. The projection of $\Delta(t)$ onto the x_n-plane consists of two intervals $\left[0, \mathrm{e}^{2\pi\mathrm{i}(k+1)/m_n} t^{1/m_n}\right]$ and $\left[0, \mathrm{e}^{2\pi\mathrm{i}k/m_n} t^{1/m_n}\right]$ (with opposite orientations).
The $(n-2)$-cycle $\Delta(t_{n-1})$ is also defined as some suspension of a certain $(n-3)$-cycle, etc. We see that the choice of an element of the basis of $\widetilde{H}_{n-1}(\{f = t\})$ relies on choosing the system $K = (k_1, \ldots, k_n)$ of integers $k_1 \in \{0, \ldots, m_1 - 2\}, \ldots, k_n \in \{0, \ldots, m_n - 2\}$. The corresponding cycles are denoted by $\Delta_K(t)$.

8. **Lemma.** *The system*

$$\omega_I = x_1^{i_1} \ldots x_n^{i_n} \mathrm{d}x_1 \wedge \ldots \wedge \mathrm{d}x_n, \; I = (i_1, \ldots, i_n), \; 0 \le i_j \le m_n - 2,$$

defines a basic trivialization of the Pham singularity.

Proof. Let us calculate the integral $I = \int_{\Delta_K(t)} \omega_I/\mathrm{d}f$. It is a homogeneous function of t. If we introduce the new variables by means of the formula $x_i = t^{1/m_i} X_i$, then ω_i is proportional to $t^{(i_1+1)/m_1} \ldots t^{(i_n+1)/m_n}$, $\mathrm{d}f$ is of order t and the integral equals

$$I = C \cdot t^{\alpha_1 + \ldots + \alpha_n - 1},$$

where $\alpha_j = (i_j + 1)/m_j$. We have to calculate the constant C in this formula. Let $\zeta_j = \mathrm{e}^{2\pi \mathrm{i}/m_j}$. Representing $\mathrm{d}f$ as $m_n x_n^{m_n-1}\mathrm{d}x_n + \ldots$, we get

$$C = (1/m_n)\cdot\left[\zeta_n^{(k_n+1)(i_n+1)} - \zeta_n^{k_n(i_n+1)}\right]$$
$$\times \int_0^1 (1-t_{n-1})^{\alpha_n - 1}\mathrm{d}t_{n-1}\int_{\Delta(t_{n-1})} x_1^{i_1}\ldots x_{n-1}^{i_{n-1}}\mathrm{d}x_1\ldots \mathrm{d}x_{n-1}/\mathrm{d}f_{n-1},$$

where $f_{n-1} = x_1^{m_1} + \ldots + x_{n-1}^{m_{n-1}}$ and $\int_{\Delta(t_{n-1})} \sim \mathrm{const}\cdot t_{n-1}^{\alpha_{n-1}+\ldots+\alpha_1 - 1}$. Next, we act by induction with respect to the dimension. We obtain

$$C = \left(\prod_j \frac{\zeta_j^{k_j+1} - \zeta_j^{k_j}}{m_j}\right)$$
$$\times B(\alpha_n, \alpha_{n-1} + \ldots + \alpha_1)\cdot B(\alpha_{n-1}, \alpha_{n-2} + \ldots + \alpha_1)\cdot \ldots \cdot B(\alpha_2, \alpha_1).$$

Here, the above product of the Beta-functions is equal to $B(\alpha_1, \ldots, \alpha_n) = \Gamma(\alpha_1)\cdots\Gamma(\alpha_n)/\Gamma(\alpha_1 + \ldots + \alpha_n)$.
Now we are ready to calculate the determinant $\det(\int_{\Delta_K}\omega_I/\mathrm{d}f)$. It is equal to $C_1 \cdot t^\nu$, where

$$\begin{aligned}\nu &= \sum_{i_1, i_2, \ldots, i_n}\left(\frac{i_1+1}{m_1} + \ldots + \frac{i_n+1}{m_n} - 1\right)\\ &= \left[\frac{1}{m_1}\cdot\frac{(m_1-1)m_1}{2}\cdot(m_2-1)\ldots(m_n-1)\right] + \ldots\\ &\quad + \left[\frac{1}{m_n}\cdot(m_1-1)\ldots(m_{n-1}-1)\cdot\frac{(m_n-1)m_n}{2}\right] - \mu\\ &= \mu(n/2-1),\end{aligned}$$

where $\mu = \prod(m_j - 1)$. This power agrees with the statement of Theorem 5.25. The constant C_1 is proportional to the product of Beta-functions, which are non-zero, and of the determinant of the matrix with the entries $\prod_j\left(\zeta_j^{(k_j+1)(i_j+1)} - \zeta_j^{k_j(i_j+1)}\right)$. This matrix is the matrix of the tensor product of the n matrices associated with one-dimensional singularity. The latter was studied in Example 5.27(a), where it was shown that the corresponding determinant is nonzero. Therefore, $C_1 \neq 0$, which shows that the system $\{\omega_I\}$ is a basic trivialization. □

9. *Proof of Theorem* 5.25(c).

We can assume that the germ f is a polynomial. Take the following deformation of f

$$P(x, \rho, \epsilon) = f(\rho x) + \sum(1+\epsilon_j)x_j^N,$$

where $N \geq \mu + 2$ is a large positive integer and ρ, ϵ_j are parameters. For fixed $\rho \neq 0$ and ϵ_j this function is equivalent to f and the hypersurface $P(\cdot, \rho, \epsilon)$ is non-singular outside the origin.
P can be treated as a deformation of the Pham's function $\sum x_j^N$. It is induced from the mini-versal deformation $Q(x, \lambda) = Q_0(x, \lambda') - \lambda_0$ of the Pham's singularity. Let $\tilde{\mu} = (N-1)^n$ be the Milnor number of the Pham singularity.
The space of parameters of the deformation of the Pham's singularity contains parameters λ, for which:

(i) the function $Q(\cdot, \lambda)$ has critical point with critical value 0, which is equivalent to f, and

(ii) $Q(\cdot, \lambda)$ has no other critical points with critical value 0.

Fix such parameter $(\lambda'_*, \lambda_{0*})$ and let $s = \lambda_* - \lambda_{0*}$. Let $\Delta_1(s), \ldots, \Delta_\mu(s)$ be the basis of cycles vanishing at this distinguished critical point. We complete this system to a basis of vanishing cycles for the Pham's singularity. It is enough to show that:

There is a system of μ Gelfand–Leray forms, among the forms defining the basic trivialization for the Pham's singularity, such that the determinant of the period matrix of integrals of these forms along Δ_j has the exact order $s^{(n-2)\mu/2}$.

Let $D(s)$ be the determinant of the period matrix for the whole Pham's singularity. We proved that it has the exact order $h^{(n-2/2} = s^{(n-2)\mu/2}$, where $h(\lambda) = 0$ is the equation for the bifurcational diagram (see Lemmas 2 and 8 above).
On the other hand, we have the formula

$$D = \sum_I D^I \overline{D}^I,$$

where I are subsets of μ elements from the set $\{1, 2, \ldots, \tilde{\mu}\}$, D^I is the $(\mu \times \mu)$ minor of the period matrix, lying in the intersection of the first μ rows and columns with numbers from I, and $\overline{D}^I$ is the algebraic completion of the minor D^I. (This is a generalization of the well-known formula from the linear algebra).
Because each minor $D^I(s)$ has order $\geq (n-2)\mu/2$ and $\overline{D}^I$ has order ≥ 0, (as expressed by integrals along non-vanishing cycles), then there exists D^{I_0} with the order equal exactly to $(n-2)\mu/2$.
The Theorem 5.25(c) is complete. □

Now we present a generalization of the results proven in the points 5 and 6 of the above proof.

5.31. Theorem of Thom and Sebastiani. ([**TS**]). *Let $f(x, y) = g(x) + h(y)$, $g : (\mathbb{C}^m, 0) \mapsto (\mathbb{C}, 0)$, $h : (\mathbb{C}^k, 0) \mapsto (\mathbb{C}, 0)$ with isolated critical points. Then we have:*

(i) *the generic fiber $V_t^f = f^{-1}(t) \cap (\text{small ball})$ of the Milnor fibration is diffeomorphic to the joint $V_u^g * V_v^h$, $u + v = t$ (of non-singular fibers) and hence $\widetilde{H}^{m+k-1}(V_t^f) = \widetilde{H}^{m-1}(V_u^g) \otimes \widetilde{H}^{k-1}(V_v^h)$;*

(ii) *the monodromy operator* M_f *associated with the function* f *equals to* $M_g \otimes M_h$.

Proof. This theorem is a generalization of the Pham's lemma (the point 5 of the proof of Theorem 5.25) and of the succeeding description of monodromy for the Pham's singularity (the point 6). The proof in the general case can be easily reconstructed from the proof in the particular case. □

§4 Gauss–Manin Connection

With any singularity $f(x)$, or with its versal deformation $F(x,\lambda)$, the Milnor fibration: $f : V = \mathbb{B}^n \cap f^{-1}(\mathbb{D}\setminus 0) \mapsto \mathbb{D}\setminus 0$ (or $\mathbb{B}^n \times \mathbb{B}^\mu \setminus \pi^{-1}(\Sigma) \mapsto B^\mu \setminus \Sigma$) is associated; here $\mathbb{B}^k$ denotes a small ball in $\mathbb{C}^k$ and $\mathbb{D}$ denotes a small disc. In this subsection we restrict ourselves to the first bundle. We shall use the notation $\mathbb{D}^* = \mathbb{D}\setminus 0$.
With this Milnor bundle two other (linear) bundles are associated.

5.32 Definition. The vector bundle $\pi : \mathcal{H}_{n-1} \mapsto \mathbb{D}^*$, with the fiber at a point t equal to $H_{n-1,t} = \widetilde{H}_{n-1}(V_t,\mathbb{C})$, is called the **homological Milnor bundle**.
The analogous vector bundle $\mathcal{H}^{n-1} = \mathcal{H}_{n-1}^\vee$ with the fibers $H_t^{n-1} = (H_{n-1,t})^\vee = \widetilde{H}^{n-1}(V_t,\mathbb{C})$ is called the **cohomological Milnor bundle**.
(Here we use the notation $\vee$ for dual complex space or a dual complex vector bundle. This notation is widely used in algebraic geometry and it allows to distinguish it from a dual to Hermitian space, denoted by $*$.)

The fiber $H = H_t = H_{n-1,t}$ of the homological Milnor bundle contains the lattice $H_\mathbb{Z} = \widetilde{H}_{n-1}(V_t,\mathbb{Z})$ (of integer cycles), isomorphic to $\mathbb{Z}^\mu$ and generated by the vanishing cycles. Similarly $H^\vee = H_t^{n-1}$ contains the integer lattice $H_\mathbb{Z}^\vee = \{w \in H^\vee : w(H_\mathbb{Z}) \subset \mathbb{Z}\}$ (of integer cocycles).
Thse lattices have the following properties:

Lemma.

(a) *The basis of* $H_\mathbb{Z}$ *is a basis for* H *(and similarly for the conjugate space). The complex space can be written as* $H = H_\mathbb{Z} \otimes_\mathbb{Z} \mathbb{C}$.

(b) *The monodromy operator preserves the lattice.*
This means that $MH_\mathbb{Z} = M^{-1}H_\mathbb{Z} = H_\mathbb{Z}$.

Due to the property (a) we can define the real analogue $H_\mathbb{R} = H_\mathbb{Z} \otimes_\mathbb{Z} \mathbb{R}$ of the complex space H. We have $H = H_\mathbb{R} \oplus \sqrt{-1}H_\mathbb{R}$.
The linear fibrations defined above have one very important property.

5.33. Lemma. *Their transition functions between local trivializations can be chosen constant. This means that if* $\pi^{-1}(U_\alpha) \approx U_\alpha \times H$, $\pi^{-1}(U_\beta) \approx U_\beta \times H$ *and the map* $(t,v) \mapsto (t, h_{\alpha,\beta}(t)v)$ *is their isomorphism over* $U_\alpha \cap U_\beta$, *then* $h_{\alpha,\beta} \equiv \mathrm{const}$.

Proof. It follows from the fact that the operators $h_{\alpha,\beta}$ must preserve the lattice $H_{\mathbb{Z}}$. □

Vector bundles with the above property are called the **local systems** and will be investigated later. In particular, they are holomorphic vector bundles (i.e., with holomorphic transition functions).
The algebraists associate with the cohomological bundle $\mathcal{H}^{n-1}$ its sheaf of sections. They denote it by

$$R^{n-1}f_*(\mathbb{C}),$$

and call it the $(n-1)^{\text{th}}$ *direct image of the constant sheaf* $\mathbb{C}$, or the $(n-1)^{\text{th}}$ *Leray sheaf*. In the language of sheaves the sheaf $R^\bullet f_*(\mathbb{C})$ is treated as a sheaf in the whole neighborhood $\mathbb{D}$ of $0 \in \mathbb{C}^1$ and is associated with the presheaf $U \mapsto H^\bullet(f^{-1}(U), \mathbb{C})$, $U \subset \mathbb{D}$.
In the completely algebraic situation $f : X \mapsto S$, where X and S are algebraic projective manifolds (or schemes) and f is an algebraic morphism, the Leray sheaf is defined on the whole S.

If the transition functions of the homological fibration are locally constant, then we can easily define the parallel transport of vectors from one fiber to another along paths in the base.
Assume that $\gamma : \tau \mapsto t(\tau) \in \mathbb{D}^*$ is a path joining $t(0)$ with $t(1)$ and that $\Delta \in H_{t(0)}$. Let us represent Δ as $\sum a_i \Delta_i(0)$, where $a_i \in \mathbb{C}$ and $\Delta_i(0)$ form a basis of $H_{t(0)}$ (e.g., vanishing integer cycles) and let us prolong the latter cycles to the cycles $\Delta_i(\tau)$ along the path γ. Then the expression

$$\Delta(\tau) = \sum a_i \Delta_i(\tau)$$

defines the family of parallel vectors in $H_{t(\tau)}$.
Analogously one defines the parallel transport of vectors in the cohomological Milnor bundle.

5.34. Definition. The parallel transport in the (co-)homological bundle is called the **Gauss–Manin connection.**

Remark. The connection, or the parallel transport, in vector bundle $p : E \mapsto B$ is a very important notion from differential geometry. It allows to compare vectors at different fibers (see Fig. 5.7).
The connection can be described in the infinitesimal form by means of differential calculus. If $\gamma = \gamma(\tau)$ is a curve in the base B and $v = v(\tau)$ is a family of parallel vectors in the fibers $E_\tau = p^{-1}(\tau)$, then $v(\tau + d\tau) = v + (\nabla_{\dot\gamma} v)d\tau$, where the operator ∇ is linear in $\dot\gamma = \frac{\mathrm{d}\gamma}{\mathrm{d}\tau}$ and in v. One can say that it is a functional on the space of vectors tangent to B taking values in the group of automorphisms of the fiber. If $b_1, \ldots, b_k$ are local coordinates in B, then we write the connection ∇ as the 1-form

$$\theta = \sum A^j(b)\mathrm{d}b_j,$$

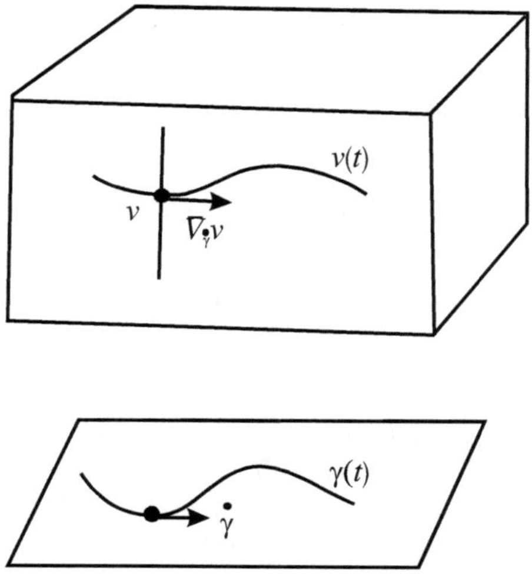

Figure 5.7: Vector bundle and connection

where $A_j(b)$ are matrix-valued functions. In the language of sheaves the connection is denoted as $\nabla : \mathcal{O}_B(E) \mapsto \mathcal{E}^1_B(E)$, i.e., as a morphism from the sheaf of sections of the bundle E to the sheaf of 1-forms on B with values in the fiber.
One can say also that at each point $v \in E$ above $b \in B$ we have distinguished a *horizontal subspace* Z_v of T_vE, which projects isomorphically onto T_bB. The subspace Z_v is transversal to the fiber and is generated by *vectors horizontal with respect to the connection*, i.e., the vectors $\dot{v}$. This family of subspaces forms the so-called distribution in TE and is given as the family of kernels of the systems of 1-forms $\mathrm{d}v - \sum(A^j(b)v)\mathrm{d}b_j$.
In linear coordinates $v_1, \ldots, v_m$ the system of 1-forms, defining the horizontal distribution Z_v, is $\mathrm{d}v_i - \sum_{j,l} A^j_{il} v_l \mathrm{d}b_j$. The horizontal family $v(\tau)$ satisfies the system of linear differential equations $\dot{v}_i = \sum_{j,l} A^j_{il}(b(\tau))\dot{b}_j v_l$. In the particular case of the tangent bundle $E = TB$ and the *Levi-Civita connection,* we have the equation for the geodesics $\ddot{b}_i = \sum \Gamma^j_{il}\dot{b}_j\dot{b}_l$, $A^j_{il} = \Gamma^j_{il}$.
Sometimes people write $(\mathrm{d} - \theta)v = 0$, with the operator

$$\mathrm{d} - \theta = \sum \left[\partial_j - A^j(b)\right] \mathrm{d}b_j.$$

This equation is thought as an equation for a section $v = v(b)$ of the bundle which would consist of parallel vectors, i.e., for section horizontal with respect to the connection. However the latter equation can have no solution if the dimension of the base is greater than one. The obstacle to integrability of such equation lies in the *curvature*. Namely, when we transport some vector v along the boundary of a small parallelogram in B with sides ϵu and ϵw, then, at the end, we arrive to the vector $v + \epsilon^2 R(u,w)v + \dots$. Here R is the curvature, which can be treated as a 2-form with values in the group of automorphism of the fiber. The classical *Cartan structure equation* says that $R = -(\mathrm{d}-\theta)\circ(\mathrm{d}-\theta) = \mathrm{d}\theta - \theta\wedge\theta$. Indeed, we have

$$\begin{aligned}(\mathrm{d}-\theta)^2 v &= \sum_{k,j}\left[-\partial_k\left(A^j\right)v - A^j\partial_k v - A^k\partial_j v + A^k A^j\right]\mathrm{d}b_k\wedge\mathrm{d}b_j \\ &= -\sum_{k<j}\left[\partial_k A^j - \partial_j A^k - A^k A^j + A^j A^k\right]\mathrm{d}b_k\wedge\mathrm{d}b_j.\end{aligned}$$

(It is equivalent to the Frobenius condition for the integrability of the distribution Z_v in terms of the 1-forms defining it (see Theorem 9.2 in Chapter 9). If the connection is *flat*, i.e., with curvature $R = 0$, then the horizontal distribution is integrable in the sense that through every point in E a k-dimensional surface tangent everywhere to the distribution passes. These surfaces are of the form $v = v(b)$ and consist of parallel vectors.
The above theory is formulated for the real vector bundles over real manifolds as well as for the complex ones. We are dealing with the complex case (like in [**GH**]and [**Wel**]).
If the dimension of the base is equal to 1, then the curvature of any connection is equal to zero (no nontrivial 2-forms). In the case of the Gauss–Manin connection the curvature is also equal to zero, because the formula $\Delta(b) = \sum a_i\Delta_i(b)$, $a_i =$ const, gives the horizontal section. (This specially concerns the case with multi-dimensional base, i.e., the homological bundle associated with the fibration $\mathbb{B}^n\setminus\pi^{-1}(\Sigma)\mapsto\mathbb{B}^\mu\setminus\Sigma$).

Let us write the differential equations for the horizontal sections of the (co-)homological bundle with respect to the Gauss–Manin connections. They turn out to be related with the Picard–Fuchs equations.
We know already some horizontal sections of the homological fibration. They are of the form $\Delta(t) = \sum a_i(t)\Delta_i(t)$, where Δ_i form the basis of vanishing cycles and a_i satisfy the equations $\dot a_i(t)\equiv 0$. Similarly, the sections $\Delta^\vee(t) = \sum a_i\Delta_i^\vee$, $\dot a_i\equiv 0$ (with $\{\Delta_i^\vee(t)\}$ as the dual basis to $\{\Delta_i(t)\}$), are *horizontal sections* of the cohomological bundle.
However there are some other natural sections of the cohomological bundle.

5.35. Definition. Let $\omega(x)$ be a holomorphic n-form in a neighborhood of the origin and let for each $t\neq 0$ the symbol $[\omega/\mathrm{d}f|_{V_t}]$ denote the cohomology class of the Gelfand–Leray $(n-1)$-form in $\widetilde{H}^{n-1}(V_t,\mathbb{C})$. This family of cohomological classes defines a section of the cohomological Milnor bundle and is called the **geometrical section corresponding to the form** ω. This section is denoted by $s[\omega]$ and its value on a family $\Delta(t)$ of vanishing cycles is equal to

$$\langle s[\omega],\Delta\rangle = \int_\Delta \omega/\mathrm{d}f.$$

The latter represents some multivalued analytic function.

The geometrical sections of the cohomological bundle are not parallel (i.e., not horizontal) with respect to the Gauss–Manin connection. Indeed, if $\Delta(t)$ is a family of vanishing cycles (which represents a horizontal section of the homological bundle), then $\langle s[\omega], \Delta(t)\rangle$ is generally non-constant function; thus the condition $\langle s[\omega], \Delta\rangle \in \mathbb{Z}$ does not hold.
Let $\omega_1, \ldots, \omega_\mu$ be n-forms defining a trivialization of the singularity. It means that their geometrical sections $s[\omega_j]$ form a basis in each fiber of the cohomological bundle. We look for the combinations

$$s = \sum F_i(t) \cdot s[\omega_i],$$

such that s is horizontal with respect to the Gauss–Manin connection.
Recall that in Theorem 5.29(b) we have defined the vector-valued functions $J(t) = \left(\int_{\Delta(t)} \omega_1/\mathrm{d}f, \ldots, \int_{\Delta(t)} \omega_\mu/\mathrm{d}f\right)^\top$ for a family of vanishing cycles. It was proved that $J(t)$ satisfies the Picard–Fuchs equation $\dot{J} = A(t)J$ and that any solution of the Picard–Fuchs equations is of this form.
The condition for horizontality of the section s means that for any family $\Delta(t)$ the derivative $\frac{\mathrm{d}}{\mathrm{d}t}(\sum F_i(t) \int_{\Delta(t)} \omega_i/\mathrm{d}f) \equiv 0$. If $F = (F_1, \ldots, F_\mu)^\top$ is a vertical vector, then we get $0 = \frac{\mathrm{d}}{\mathrm{d}t}\langle F, J\rangle = \langle \dot{F}, J\rangle + \langle F, \dot{J}\rangle = \langle \dot{F} + A^\top F, J\rangle$. We have proved the following result.

5.36. Lemma. *The section $s = \sum F_i(t) \cdot s[\omega_i]$ of the cohomological Milnor fibration is horizontal with respect to the Gauss–Manin connection iff the vector-function $F = (F_1, \ldots, F_\mu)^\top$ satisfies the non-autonomous linear equation*

$$\dot{F} = -A^\top(t)F, \tag{5.6}$$

where $A(t)$ is the matrix from the Picard–Fuchs equation $\dot{J} = A(t)J$.

Above we have defined the Picard–Fuchs equation and the equation (5.6) for horizontal sections. There appear natural questions about the dependence of these equations on the choice of the ingredients defining them. These ingredients are: the basis of the lattice of vanishing cycles and the trivialization.
The change of the basis $\{\Delta_i\}$ is realized by means of a constant matrix R. We get the change $A(t) \mapsto RA(t)R^{-1}$.
The choice of the trivialization $\omega_1, \ldots, \omega_\mu$ is significant, but not very much. Let $\omega_1', \ldots, \omega_\mu'$ be another trivialization. If $J'(t) = (\int_{\Delta(t)} \omega'/\mathrm{d}f, \ldots, \int_{\Delta(t)} \omega_\mu'/\mathrm{d}f)^\top$ is the associated with it vector-valued integral, then we have $J' = QJ$ for some matrix-valued function $Q = Q(t)$, which is meromorphic in a neighborhood of the origin and invertible for $t \neq 0$. Differentiating it we get $\dot{J}' = \dot{Q}J + Q\dot{J} = (\dot{Q}Q^{-1} + QAQ^{-1})J'$. So, we have the following result.

5.37. Lemma. *The change of trivialization in the Picard–Fuchs equation results in the application of the gauge transformation*

$$A(t) \mapsto A'(t) = QAQ^{-1} + \dot{Q}Q^{-1}.$$

Remark. The terminology gauge transformation is taken from physics. The Yang–Mills fields are connections in some vector bundles over the Minkowski space-time. There, quantities not depending on the application of the gauge transformation are physically important. For example, in the electrodynamics the connection is given by a 4-potential, which is not uniquely defined, but its curvature (describing the tensor of the electromagnetic field) is gauge invariant.

The Picard–Fuchs equation is characterized by the triple

$$(A, W, W_{\mathbb{Z}}),$$

where $A = A(t)$ is the matrix, W is the space of solutions and $W_{\mathbb{Z}}$ is the lattice generated by integrals over the basic vanishing cycles. The application of the gauge transformation from Lemma 5.36 defines the equivalence relation between such triples: $(A, W, W_{\mathbb{Z}}) \sim (A', W', W'_{\mathbb{Z}})$ iff $W' = QW$, $W'_{\mathbb{Z}} = QW_{\mathbb{Z}}$ and $A' = QAQ^{-1} + \dot{Q}Q^{-1}$.

5.38. Definition. The equivalence class of the triple $(A, W, W_{\mathbb{Z}})$ is called the **Picard–Fuchs singularity** of the critical point.

In Chapter 8 below we will study the non-autonomous systems of differential equations $\dot{z} = A(t)z$, where $x \in \mathbb{C}^m$ and $A(t)$ is a germ of meromorphic matrix-valued function in $(\mathbb{C}, 0)$. We shall classify them with respect to the gauge transformations by means of analytic (or meromorphic) matrices $Q(t)$.
Important problem is to determine the class of meromorphic linear systems, which form a Picard–Fuchs singularity of critical point of some analytic function. The following definition will appear again in Chapter 8.

5.39. Definition. The point $t = 0$ of the equation $\dot{x} = A(t)x$ is called **regular** iff any its solution satisfies the estimate

$$|x(t)| < C|t|^{-N},$$

for some constants N, C, as t tends to zero within some fixed sector.

5.40. Theorem (Regularity of the Gauss–Manin connection). *The singular point $t = 0$ of the Picard–Fuchs equation associated with isolated critical point of a holomorphic function is regular.*
Moreover, there is a germ of meromorphic matrix function $Q(t)$ (with only pole at) $t = 0$ such that the corresponding gauge transformation sends A to $A' = \frac{\ln M}{2\pi i t}$.

Proof. The first part follows from the expansion of the integrals along vanishing cycles into asymptotic series containing rational powers of t and powers of logarithms (see Theorem 5.14). The fundamental matrix of solution of the Picard–Fuchs system is $\mathcal{F}(t) = \left(\int_{\Delta_j} \omega_i/\mathrm{d}f\right)$ and it has the property $\mathcal{F}(t\mathrm{e}^{2\pi\mathrm{i}}) = \mathcal{F}(t) \cdot M$, where M is the monodromy operator (acting on the fiber $\mathcal{H}_{t_0}$ of the homological fibration). As in the proof of Theorem 5.14 we have

$$\mathcal{F}(t) = Q(t)t^{\ln M/(2\pi\mathrm{i})},$$

where $Q(t)$ is a meromorphic matrix function (by the regularity of $\mathcal{F}(t)$ and of $t^{\ln M/(2\pi\mathrm{i})}$). Any solution of the Picard–Fuchs equation is of the form $J(t) = \mathcal{F}(t)v$, where v is a constant vector. We define J' by the formula $J = QJ'$.
The vector function J' satisfies the equation $\frac{\mathrm{d}}{\mathrm{d}t}J' = \frac{\ln M}{2\pi\mathrm{i}t}J'$. □

Remark. As we have noticed in Remark 5.16, the regularity of the Picard–Fuchs equation holds not only in the local case of isolated singularity.
We have seen that the elliptic integrals satisfy certain differential equations defined in the whole plane (with regular singular points, see Lemma 5.22).
Such equations can be obtained also in the case of non-isolated critical points, or in the case of critical values associated with bifurcations at infinity; (generally the Picard–Fuchs equations are associated with the Leray sheaf induced by any algebraic morphism $f : X \mapsto S$). The regularity of singularities of such equations are associated with polynomial growth of the integrals along cycles which undergo bifurcations (vanishing or escaping to infinity, etc.). We will return to this subject in Chapters 7 and 8.

If ω is a holomorphic n-form and $s[\omega]$ is the corresponding geometrical section, then the asymptotic expansion $\int_{\Delta(t)} \omega/\mathrm{d}f = \sum_{\alpha,k} (a_{\alpha,k}/k!) \cdot t^\alpha \cdot (\ln t)^k$ defines some natural cohomological classes. Notice that the map $\Delta \mapsto a_{\alpha,k}$ is a linear functional and can be treated as a section of the cohomological Milnor bundle. We denote this section by $A_{\alpha,k} = A^\omega_{\alpha,k} = A_{\alpha,k}(t)$:

$$a_{\alpha,k} = \langle A_{\alpha,k}, \Delta \rangle.$$

We can also write

$$s[\omega] = \sum_{\alpha,k} \frac{A_{\alpha,k}}{k!} t^\alpha (\ln t)^k.$$

5.41. Lemma.

(a) *The classes $A_{\alpha,k}(t) \in \widetilde{H}^{n-1}(f^{-1}(t))$ and define the horizontal sections with respect to the Gauss–Manin connection.*

(b) *They belong to the eigenspaces of the monodromy operator M^{cohom}, corresponding to the eigenvalue $\mathrm{e}^{-2\pi\mathrm{i}\alpha}$ of the monodromy operator $M = M_{\mathrm{hom}}$ in homologies.*

(c) *We have $A_{\alpha,k} = \left(-\ln M_u^\top/2\pi\mathrm{i}\right)^k A_{\alpha,0}$, where M_u is the unipotent factor of the (homological) monodromy operator.*

Proof. (a) is a consequence of the fact that $\langle A_{\alpha,k}, \Delta(t)\rangle \equiv \text{const.}$

(b) and (c) follow from the action of the monodromy operator onto the asymptotic expansion.

The monodromy acts on the fibers of the homological bundle and on the fibers of the cohomological bundle. If $M = M_{\text{hom}}$ is the matrix of the monodromy, expressed in some basis $\{\Delta_i(t_0)\}$ (of integer cycles) of $\mathcal{H}_{t_0}$, and M^{cohom} is the analogous matrix expressed in the dual basis $\{\Delta_i^\vee(t_0)$, then we have $M_{\text{hom}}^\vee M^{\text{cohom}} = I$. It is because $\langle\Delta_i^\vee(t), \Delta_j(t)\rangle \equiv \text{const.}$ So, we have $M^{\text{cohom}} = (M^\vee)^{-1} = (M^\top)^{-1}$. The same property holds, when we express M^{cohom} in a basis consisting of horizontal sections (like $A_{\alpha,k}$).

We have

$$\begin{aligned}\langle (M^\top)^{-1}s[\omega], \cdot\rangle &= \sum t^\alpha (\ln t)^k (M^\top)^{-1} A_{\alpha,k}/k!\\ &= \sum (te^{2\pi \mathrm{i}})^\alpha (\ln t + 2\pi\mathrm{i})^k A_{\alpha,k}/k!\\ &= \sum_{\alpha,l} t^\alpha (\ln t)^l \cdot \mathrm{e}^{2\pi\mathrm{i}\alpha} \frac{1}{l!} \sum_m \frac{(2\pi\mathrm{i})^m}{m!} A_{\alpha,l+m}.\end{aligned}$$

So $(M^\top)^{-1}A_{\alpha,k} = \left(e^{-2\pi\mathrm{i}\alpha}\right)^{-1} \cdot (M_u^\top)^{-1}A_{\alpha,k}$, where

$$(M_u^\top)^{-1}A_{\alpha,k} = \sum_m \left((2\pi\mathrm{i})^m/m!\right) A_{\alpha,k+m} = e^{2\pi\mathrm{i}N}A_{\alpha,k}.$$

We see that the action of M^{cohom} consists of multiplying by the eigenvalue (i.e., the action of M_s^{-1}) and of the action of the exponent $\exp(2\pi\mathrm{i}N)$ of the nilpotent Jordan cell $N = -\ln(M_u)^\top$. □

5.42. Definition. Let ω be a holomorphic form. The smallest number α such that the coefficient $A^\omega_{\alpha,0}$ is nonzero is called the **order of the form** ω and is denoted by

$$\alpha(\omega).$$

The **principal part of the geometrical section defined by** ω is the section

$$s_{\max}[\omega] = t^{\alpha(\omega)}\left(A^\omega_{\alpha(\omega),0} + \ldots + (\ln t)^{n-1}A^\omega_{\alpha(\omega),n-1}\right).$$

(Note that, due to Lemma 5.41(c), all terms of the above expression can be expressed by means of one of them, e.g., by $A_{\alpha(\omega),0}$).

For the fixed germ f of holomorphic function we denote by

$$\alpha_{\min}$$

the smallest of all orders $\alpha(\omega)$ of germs of holomorphic forms. The number

$$-(1+\alpha_{\min})$$

is called the **complex index of oscillation** of the critical point of f.
The number

$$n/2 - (1 + \alpha_{\min})$$

is called the **complex index of singularity** of the germ f.
The system

$$\alpha_1 \leq \alpha_2 \leq \ldots \leq \alpha_\mu,$$

obtained as the minimal orders of holomorphic forms generating trivialization, is called the **spectrum of critical point** of f. It means that $\alpha_1 = \alpha_{\min} = \alpha(\omega_1)$, $\alpha_2 = \min\{\alpha(\omega_2) : s[\omega_2]$ is independent of $s[\omega_1]\}$, etc.

Example. Let f be a quasi-homogeneous polynomial and let ω be a quasi-homogeneous n-form. It means that $f(x) = \lambda^\beta f_0(y)$, $\omega(x) = \lambda^\gamma \omega_0(y)$, where (λ, y) are such coordinates that $x_i = \lambda^{\nu_i} y_i$ and $f|_{\lambda=1} = 1$; (the quasi-homogeneous blowing-up).
After this change, the integral $\int_{\Delta(t)} \omega/\mathrm{d}f$ becomes equal to $\lambda^{\gamma-\beta} \int_\Delta \omega_0/\mathrm{d}f_0$, where Δ is some fixed cycle. Because $t = \lambda^\beta$, we get

$$\alpha(\omega) = \gamma/\beta - 1.$$

We see also that the principal part of the geometrical section is equal to the whole section, $s_{\max}[\omega] = s[\omega]$.

The constants β and γ can be calculated from the Newton diagrams in $\mathbb{R}^2_+ = \{(k_1, \ldots, k_n)\}$ of the germ f and of the form ω. Let $\omega = g(x)\mathrm{d}x_1 \wedge \ldots \wedge \mathrm{d}x_n$. Let Δ be the Newton polyhedron of the function f and Δ_1 be the Newton polyhedron of the function $x_1 \ldots x_n g$.
(Recall that the Newton polyhedron Δ of f is the convex hull of the set $\mathrm{supp} f + \mathbb{R}^n_+$ and its Newton diagram Γ is the part of the boundary of the Newton polyhedron not contained in the coordinate hyperplanes. In this example Δ is the first orthant cut by the hyperplane $\sum \nu_i k_i = \beta$).
Let ν be the smallest number such that $\nu\Delta_1 \subset \Delta$. It is called the *coefficient of embedding of* Δ_1 *into* Δ. The number $-1/\nu$ is called the *distance of polyhedrons of* f *and of* ω.
We have $\nu = \beta/\gamma = \mathrm{ord}\, f/\mathrm{ord}\, \omega$ and $\alpha(\omega) = 1/\nu - 1$. In particular, the number $-(\alpha(\omega) + 1)$ is equal to the distance of diagrams of the function and of the form. Because the form with minimal order is obtained by putting $g = 1$, the above allows us to compute the index of oscillation of f and its index of singularity.
Take the intersection of the diagonal line $k_1 = k_2 = \ldots = k_n$ with the diagram Γ_1. It is one point $(\tilde{\nu}, \ldots, \tilde{\nu})$ and we have $\alpha_{\min} = 1/\tilde{\nu} - 1$. Thus the complex index of oscillation is equal to $1/\tilde{\nu}$; here $\tilde{\nu} = \nu = \beta/\gamma$.
It turns out that the above calculations can be generalized.

5.43. Definition. Fix the Newton polyhedron Δ and its Newton diagram Γ such that $\mathbb{R}^n_+ \setminus \Delta$ has finite volume. Γ consists of finite number of $(n-1)$-dimensional

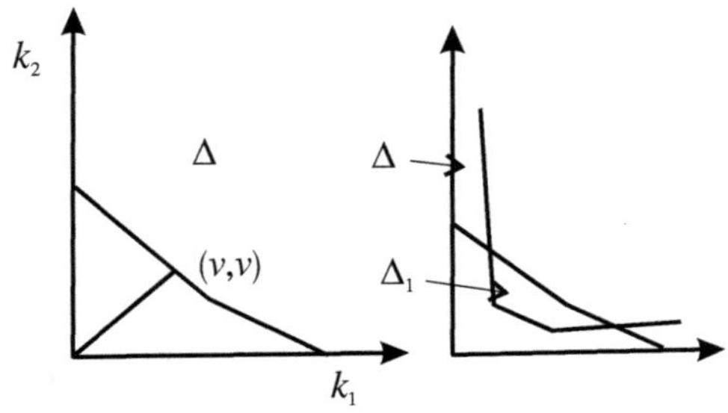

Figure 5.8: Newton polyhedra

faces, of $(n-2)$-dimensional faces, etc. Let $(\tilde{\nu}, \dots, \tilde{\nu})$ be the point of intersection of Γ with the diagonal $k_1 = \dots = k_n$. The number $-1/\tilde{\nu}$ is called the **distance of the Newton diagram** Γ (see Fig. 5.8).
Let $\omega = g(x)\mathrm{d}x_1 \dots \mathrm{d}x_n$ be a germ of holomorphic n-form and let Δ_1 be the Newton polyhedron of the function $x_1 \dots x_n g$. The smallest number ν such that $\nu\Delta_1 \subset \Delta$ defines the **distance of the polyhedrons** Δ **and** Δ_1 as the number $-1/\nu$.

5.44. Theorem. ([**AVG**]) *Consider the space of germs f with Δ and Γ as their Newton polyhedron and Newton diagram. This set contains an open and dense subset of germs, for which the following properties hold:*

(i) *The complex index of oscillation of f does not exceed the distance of Γ. If this distance is > -1 , i.e., the diagram is distant, then the index of oscillation and the distance coincide.*

(ii) *If ω is a germ of a form, then the number $-(\alpha(\omega)+1)$ does not exceed the distance of Δ and Δ_1. These numbers are equal if the distance is > -1.*

The following conjecture is associated with the spectrum.

5.45. Semi-continuity of the spectrum conjecture (Arnold–Malgrange ([AVG], [Mal2])). *The spectrum of an isolated critical point of a holomorphic function is semi-continuous in the following sense: if the critical point P with the spectrum $(\alpha_1, \dots, \alpha_\mu)$ is deformed to a simpler point P' with the spectrum $\alpha'_1, \dots, \alpha'_{\mu'}$, $\mu' < \mu$, then $\alpha_k \le \alpha'_k$.*

The reader interested in more information about the results concerning this conjecture is referred to [**AVG**], [**Ste2**] and [**Kul**].
In the next section and in the next chapter we present applications of the theory developed in this chapter to the asymptotic of oscillating integrals and to the qualitative theory of planar vector fields. Later we will return to the asymptotic of integrals in order to define a mixed Hodge structure in the Milnor cohomological bundle.

§5 Oscillating Integrals

5.46. Definition of the oscillating integral. Let S be a surface (or a curve or a finite set of points) in the space $\mathbb{R}^3$. Assume that each point of S is a source of light with fixed frequency ω and intensity $\rho(x)$, depending on the point $x \in S$. We look at the amplitude of the light at the points y in the space (outside S) (see Fig. 5.9).

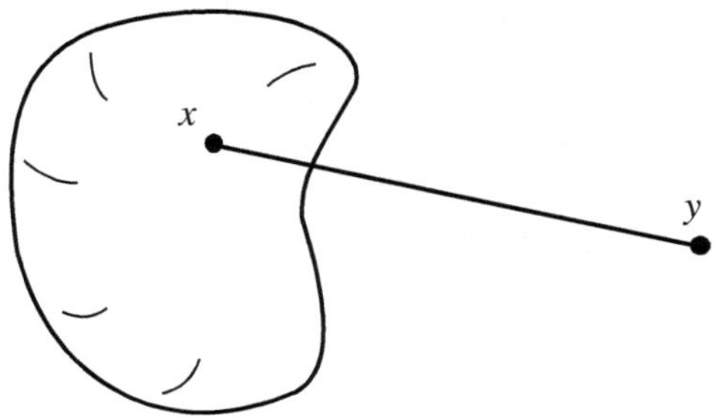

Figure 5.9: Intensity of light at y comes from contributions at points x in the source surface

Using the fundamental solution of the wave equation

$$\partial_{tt}^2 u = c^2 \Delta u,$$

with suitable boundary conditions, we get the formula

$$u(y,t) = \int_S \frac{\mathrm{e}^{2\pi \mathrm{i}(\omega t - k|x-y|)}}{4\pi|x-y|} \rho(x) \mathrm{d}x$$

for some component of the corresponding strength of the electromagnetic field. Here c is the light velocity, $k = \omega/c$ is the length of the wave vector, and $|\cdot|$ is the euclidean norm.

The amplitude of the light at y is the absolute value of the above expression and equals to the absolute value of the following **oscillating integral**:

$$I(\tau, y) = \int \mathrm{e}^{\mathrm{i}\tau F(x,y)} \phi(x,y) \mathrm{d}x, \tag{5.7}$$

with $\tau = -2\pi k$, $F = |x-y|$, $\phi = \rho(x)/(4\pi|x-y|)$. In the case of light the number τ is large and the main problem is to investigate the asymptotic behavior of the function $I(\tau, y)$ as $\tau \to \infty$. Here y is treated as a parameter. The function $\tau F(x,y)$ is called the **phase**.

If the function $F(\cdot, y)$ were regular, i.e., noncritical at S, then we would obtain that $|I| < C_N/\tau^N$ for any N (with suitable constants C_N); we prove it below. This would mean that the amplitude is very small (it decreases faster than any power) and we could treat the source S as very weak. However, usually S is a compact surface and the function $|x-y|$ has critical points on it (e.g., minima and maxima). If these critical points are of Morse type, which holds for typical y's, then $I \sim \tau^{-n/2}$, $n = \dim S$. But when y varies, then the function $|\cdot - y|$ can have more degenerate singularities on S. In this case the function I decreases more slowly. The latter points form the so called **caustic**.

In what follows we investigate the function (5.7) with real-analytic $F(x,y)$, $x \in \mathbb{R}^n$, $y \in \mathbb{R}^m$, and with the function ϕ smooth with compact support. Thus, F is a deformation of a singularity of some real function. (Using the partition of unity, we can always reduce any oscillating integral to a sum of integrals satisfying the latter assumptions).

If $m = 0$, i.e., there are no parameters, then we have the integral

$$I(\tau) = \int \mathrm{e}^{\mathrm{i}\tau f(x)} \phi(x) \mathrm{d}x. \tag{5.8}$$

5.47. Proposition. *If the support of $\phi(x)$ is compact and $f(x)$ has no critical points in this support, then the integral I decreases to 0 faster than any power of $1/\tau$.*

Proof. It is enough to consider the case $n = 1$. We have

$$\int_{-\infty}^{\infty} \mathrm{e}^{\mathrm{i}\tau f(x)} \phi(x) \mathrm{d}x = \frac{1}{\mathrm{i}\tau} \int \mathrm{d}(\mathrm{e}^{\mathrm{i}\tau f})(\phi/f') = \frac{-1}{\mathrm{i}\tau} \int \mathrm{e}^{\mathrm{i}\tau f} \left(\phi/f'\right)' \mathrm{d}x,$$

where $f' \neq 0$. Repeating this transformation we get the result. □

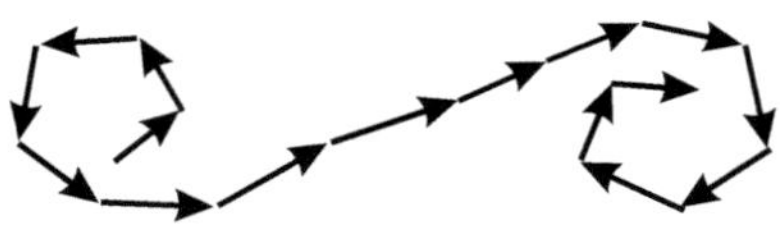

Figure 5.10: The dominant contribution comes from neighborhoods of critical points of the phase

The conclusion of this theorem is that the main contribution to integral (5.8) comes from small neighborhoods of the critical points of the phase function. This can be seen geometrically, when we replace the (one-dimensional) integral by its Riemann sum (see Fig. 5.10).

Next to consider is the case with Morse critical point. The particular example of such integral is the *Fresnel integral* $\int \cos(\tau x^2)\mathrm{d}x$, where from Fig. 5.11 it is seen that the main contribution comes from the first bump and is of order $C/\sqrt{\tau}$. The value of the constant C can be computed in the following way. The Fresnel integral is the limit as $\sigma \to 0^+$ of $\mathrm{Re} \int \mathrm{e}^{(\mathrm{i}\tau - \sigma)x^2} \mathrm{d}x = \sqrt{\pi/(\sigma - \mathrm{i}\tau)}$; thus $C = \sqrt{\pi/2}$.

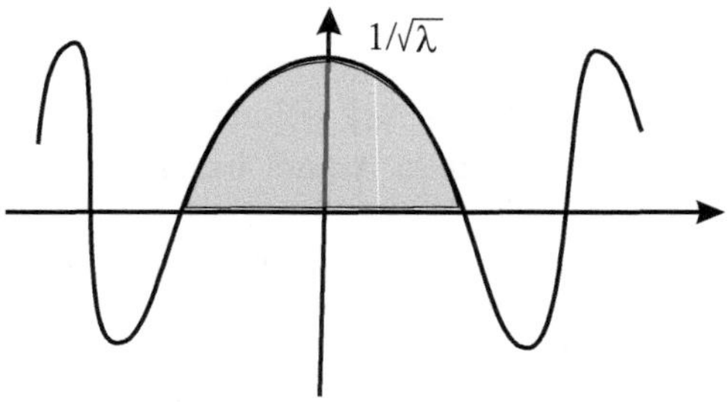

Figure 5.11: Dominating part of the oscillating integral comes from the shaded domain

5.48. Theorem (Stationary phase formula). *Assume that $\phi(x)$, $\phi(0)\neq 0$, has compact support containing only one critical point 0 of $f(x)$ which is non-degenerate. Then the integral (5.8) has the asymptotic*

$$\phi(0)\cdot \mathrm{e}^{\mathrm{i}\tau f(0)}\cdot \left(\frac{2\pi}{\tau}\right)^{n/2}\cdot |\det D^2 f|^{-1/2}\cdot \mathrm{e}^{\mathrm{i}(\pi/4)\,\mathrm{sign} D^2 f},$$

where $\mathrm{sign} D^2 f(0)$ *is the inertia signature of the quadratic form defined by the second derivative of f (the number of pluses minus the number of minuses).*

Proof. Using the Morse lemma we reduce the problem to the case with $f = z_1^2 + \ldots + z_k^2 - z_{k+1}^2 - \ldots - z_n^2$ with the signature $k-(n-k)$. The coefficient $1/\sqrt{|\det D^2 f|}$ arises from the change of variables.
We approximate the exponent $\mathrm{e}^{\mathrm{i}\tau f}$ by $\exp(-\sum a_j Z_j^2)$ with $\mathrm{Re} a_j > 0$ and $a_j \to \pm \mathrm{i}$, $Z_j = \sqrt{\tau} z_j$. Next we expand $\phi(x)$ as $\phi(0)+\phi_1(x)$. Then it is easy to see that the leading part of the integral, with the modified exponent, comes from the integral with $\phi(0)$. The latter is a product of the line integrals $\int \exp(-a_j Z_j^2)\mathrm{d}Z_j = \sqrt{\pi/a_j}$, which gives the formula from Theorem 5.48. □

5.49. Remark. The number $\mathrm{sign} D^2 f$ appearing in the last term of the asymptotic formula from Theorem 5.48 is equal to the *Maslov index*, the same as in 3.21 (Chapter 3). In terms of the problem from Fig. 5.9 this Maslov index is the Maslov index associated to a lift (to some Lagrangian subvariety) of the ray δ passing from the critical point $x\in S$ to y.
Namely, the light particles (with velocity $c=1$) move along straight lines. They satisfy the system of Hamilton equations $\dot{y}=p$, $\dot{p}=0$, with the Hamilton function $H(y,p)=|p|^2$, and lie in the level hypersurface $H=1$. With the surface S the set of initial conditions for this system is associated: $y(0)\in S$ and $p(0)$ orthogonal to S.
They form the normal bundle NS to S in $T\mathbb{R}^3$, where the latter is identified with the cotangent bundle (by means of the euclidean metric). The 2-dimensional

surface $\Lambda_0 = NS$ has the property that the symplectic form $dydp$ restricted to it is equal zero. Λ_0 is not a Lagrangian submanifold (lack of dimension). But the set $\Lambda = \bigcup_t \{(y(t), p(t)); (y(0), p(0)) \in \Lambda_0\}$ is Lagrangian. The ray δ is lifted to a curve $\tilde{\delta}$ in the Lagrangian submanifold Λ and, using the definition from 3.21, we can define its Maslov index.
If the interval δ, joining x and y realizes, the local minimum of the distance from y to S, then its Maslov index is 0.
If x is a saddle point of the distance function $x \to |x - y|$, then the Maslov index of the corresponding ray is 1. One can see that this ray must pass through the envelope of the system of rays starting orthogonally from S. This is the *caustic* of S and forms the set of critical values of the projection of the Lagrangian manifold to the configuration space.
The Maslov index of a ray starting from a local maximum is 2.
Analogous formulas, with oscillating integral and stationary phase formula (with Maslov index), appear also in the quasi-classical approximations of solutions of the Schrödinger equation. There, the role of a large parameter plays the inverse of the Planck constant $\tau \sim h^{-1}$ (see [**Arn1**]).

5.50. Caustics in $\mathbb{R}^3$. Recall that for the deformation $F(x, y)$ of $F(x, 0)$ the caustics is defined as the set of those parameters y for which the function $F(\cdot, y)$ has degenerate critical point or

$$\Sigma = \{y : \exists_x \ D_x F = 0, \ D^2_{xx} F = 0\}.$$

If y is three-dimensional, which corresponds to the real space, then we obtain the five generic singularities of caustics presented at Fig. 5.12 (see [**AVG**]).

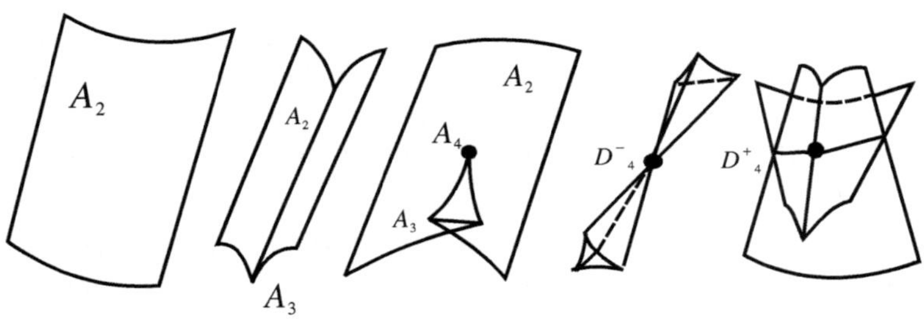

Figure 5.12: Caustics

In [**AVG**] the generic caustics in the space-time $\mathbb{R}^4$ are classified.

The next result says about the whole expansion of the oscillating integral. This expansion turns out to be not convergent, it is only asymptotic.

5.51. Definition. If $f(t)$ is a function, defined in a neighborhood of 0 in $\mathbb{R}$ (or in $(\mathbb{R}_+, 0)$ or in $(\mathbb{C}, 0)$ or in a sector in $\mathbb{C}$ with vertex at 0), then we say that f has

asymptotic expansion $\sum_j f_j(t)$, with well defined 'model functions' f_j (such that $|f_{j+1}| < |f_j| \to 0$ as $j \to \infty$), iff $|f(t) - \sum_{j=1}^N f_j(t)| < g_N(t)$ with $g_N(t) \to 0$ as $N \to \infty$. We write

$$f(t) \sim \sum_j f_j(t).$$

There is an analogous definition in the case $t \to \infty$.

We shall meet functions which have asymptotic expansion in some sector S in the complex t-plane with vertex at 0. Often the function is analytic in such sector. For example, $f(t) = \int_{-\infty}^{\infty} \exp[-s^2 - ts^4]\mathrm{d}t \sim \sum \Gamma(2n - 1/2)(-t)^n$ is analytic only in the half-plane $\mathrm{Re}\, t > 0$. It turns out that also the reverse statement is true.

5.52. The Borel–Ritt theorem. *For any formal power series and any sector, this series is an asymptotic series of a certain holomorphic function defined in this sector.*

Proof. If $\sum a_n t^n$ is this series, then this new function is

$$\sum a_n(1 - \exp[-b_n t^{-\beta}])t^n. \tag{5.9}$$

The exponent β depends on the sector S, which we assume to be symmetric with respect to the real positive semi-axis and is such that $\mathrm{Re}\, t^{-\beta} > 0$ in S. The positive coefficients b_n are chosen in order to make series (5.9) convergent: because $|1 - e^z| < |z|$ for $\mathrm{Re}\, z < 0$, series (5.9) is estimated by $\sum |a_n| \cdot b_n \cdot |t|^{n-\beta}$ and is convergent for $b_n < |a_n|^{-1}$. Finally, all derivatives of $1 - \exp[-b_n t^{-\beta}]$ are zero at $t = 0$. □

5.53. Theorem (Asymptotic expansion of oscillating integrals). *If the support of the function $\phi(x)$ contains only one critical point 0 of $f(x)$, with finite Milnor number, then the integral* (5.2) *has the asymptotic expansion*

$$\mathrm{e}^{\mathrm{i}\tau f(0)} \sum_{\alpha,k} a_{\alpha,k} \tau^{\alpha} (\ln \tau)^k,$$

where the index α runs over finite set of decreasing arithmetic sequences of rational numbers and k are integers between 0 and $n-1$. The coefficients $a_{\alpha,k} = a_{\alpha,k}(\phi)$ are distributions (generalized functions) with support in 0.

Proof. From the Gelfand–Leray formula (see Lemma 5.12) we get

$$\int_{\mathbb{R}^n} \mathrm{e}^{\mathrm{i}\tau f(x)} \omega = \int_{-\infty}^{\infty} \mathrm{e}^{\mathrm{i}\tau t} \left(\int_{f=t} \omega / \mathrm{d}f \right) \mathrm{d}t,$$

where $\omega = \phi \mathrm{d}x_1 \ldots \mathrm{d}x_n$.

5.54. Theorem. *Let $f(0) = 0$. We have*

$$\int_{f=t} \omega/\mathrm{d}f \sim \sum_{\beta,k} b^{\pm}_{\beta,k} t^{\beta} (\ln t)^k,$$

as $t \to 0^{\pm}$. Here the set of indices β runs over finite set of growing arithmetic sequences of rational numbers.

From this theorem, Theorem 5.53 easily follows. We have

$$\int_0^\infty e^{i\tau t} t^\beta dt = \Gamma(\beta+1)/(-i\tau)^{\beta+1},$$

$$\int_0^\infty e^{i\tau t} t^\beta (\ln t)^k dt = \frac{d^k}{d\beta^k}(\Gamma(\beta+1)/(-i\tau)^{\beta+1}).$$

In our case, we replace the integral from the Gelfand–Leray formula by the sum of two integrals: one over positive t's and the other over negative t's. Moreover, if $|t|$ is sufficiently large, then the function $\int_{f=t}$ is zero (because ω has compact support). Thus we should consider integrals of the form $\int_0^\infty e^{i\tau t} t^\beta (\ln t)^k \chi(t) dt$, where $\chi(t)$ is a function with compact support and equals identically to 1 near $t = 0$. For the latter integral the above formulas hold, but are not exact; nevertheless, the differences are flat functions. □

Proof of Theorem 5.54. In [**Jea**] this theorem is proved using the resolution of the singularity of the phase function f (see Chapter 4 above). Here we present a proof which is suggested in [**AVG**]. We show only the expansion and omit the localization of the exponents β.

Firstly, using the theory developed in Chapter 2, we can assume that the phase function f is a polynomial such that $x = 0$ is its only critical point with the critical value 0. Assume that the support of ω lies in a small ball $\mathbb{B}_\rho$.

Consider the surfaces $\{f = t\} \cap \mathbb{R}^n$ for $t \to 0^{\pm}$. It may consist of several pieces. Some of these pieces tend uniformly to 0 and represent vanishing cycles of the complex hypersurfaces $\{f = t\} \subset \mathbb{C}^n$; they are elements of $H_{n-1}(\{f = t\} \cap \mathbb{B}_\rho)$. If we approximate the form ω by a polynomial form, then we obtain an integral of a holomorphic Gelfand–Leray form along vanishing cycle. Then one can use the expansion from Theorem 5.14. In this sense, the theory of oscillating integrals is connected with the monodromy theory. Other pieces of the real hypersurface $\{f = t\} \cap \mathbb{B}_\rho$ can be treated as relative cycles.

Choose one component γ_t of $\{f = t\} \cap \mathbb{B}_\rho$, where we can assume that $t > 0$. The union $\bigcup_{0 \le s \le t} \gamma_s$ forms a closed n-dimensional set A_t, whose boundary consists of: parts of the hypersurfaces $\{f = 0\}$ and $\{f = t\}$ and a part of $\partial \mathbb{B}_\rho$.

By Lemma 5.12 we have $\int_{\gamma_t} \omega/df = \frac{d}{dt}\int_{A_t} \omega$. We will show that the integral $J(t) = \int_{A_t} \omega$ has the expansion

$$J(t) \sim \sum_{\beta,k} c_{\beta,k} t^\beta (\ln t)^k,$$

where all $\beta > 0$. The proof of the latter statements uses induction with respect to the dimension n.

If $n = 1$, then $\{f = 0\} = \{0\}$ and γ_t is one of the roots of the equation $f(x) = t$, (it has the Puiseux expansion $\gamma_t = \sum c_j t^{j/k}$), and A_t is the interval joining 0 and γ_t. Here the result follows from the Taylor expansion formula. (Note that, if $n = 1$, then the expansion of $J(t)$ does not contain logarithms.)
In the induction step we get the integral

$$\int_{\mu(t)}^{\nu(t)} \mathrm{d}x_n \int_{A(t,x_n)} \eta,$$

where $A(t, x_n)$ is the intersection of A_t with the hyperplane $x_n = \text{const}$. By the induction assumption, for any fixed x_n, the integral $\int_{A(t,x_n)} \eta$ has an expansion as expected. Next, one divides the interval $[\mu(t), \nu(t)]$ into subintervals in such a way that, for x_n from any of these intervals, the exponents in the expansion of $\int_A \eta$ are constant and the coefficients are smooth functions with algebraic singularities at the ends of the interval; moreover, the points of partition have Puiseux expansions in t. From this the thesis of Theorem 5.54 easily follows.
The existence of such partition follows from the stratification (a kind of partition) of the boundary of the real semi-analytic set A_t into strata. Firstly, one distinguishes the components of the smooth part of ∂A_t. Next, one takes the set $\mathrm{Sing}(\partial A_t)$ of singular point of ∂A_t and separates the smooth components of $\mathrm{Sing}(\partial A_t)$, etc. We project all the strata onto the x_n axis. These projections have their sets of critical points (semi-analytic subsets); we stratify them too. The projections of the strata of the new stratification give the set of intervals in the real line.

In order to show that α's from Theorem 5.53 and β's from Theorem 5.54 belong to finite set of arithmetic rational sequences one must apply the resolution of singularity theorem (see Theorem 4.56 in Chapter 4). There the problem reduces itself to the expansion of the integrals

$$\int \exp[\mathrm{i}\tau x_1^{k_1} \dots x_n^{k_n}] \cdot x_1^{m_1} \dots x_n^{m_n} \cdot \phi(x)\mathrm{d}x.$$

The results of Theorems 5.53 and 5.54 hold also in the case, when the analytic phase function $f(x)$ has critical point of infinite codimension, e.g., is non-isolated. (Note that, in the real case, a critical point can be isolated but have infinite Milnor number; example: $(x^2 + y^2)^2$). □

5.55. The index of oscillation and the index of singularity. From now on we assume that $\phi(0) \neq 0$.
The leading term in the asymptotic expansion of the oscillating integral is of the form

$$\tau^{-n/2+\beta}(\ln t)^k,$$

where $-n/2 + \beta$ is called the **index of oscillation** and β is called the **index of singularity** (see the definition of the complex index of oscillation in Definition

5.41). For the simple singularities the index of singularity is easy to compute and its values are presented in the following table:

Sing.	$\mathbf{A}_k$	$\mathbf{D}_k$	$\mathbf{E}_6$	$\mathbf{E}_7$	$\mathbf{E}_8$
β	$\frac{k-1}{2k+2}$	$\frac{k-2}{2k-2}$	$\frac{5}{12}$	$\frac{4}{9}$	$\frac{7}{5}$

If f is quasi-homogeneous, then its index of oscillation is calculated as follows. Take the Newton's diagram Γ of f. It is a part of the hyperplane $\sum \nu_i k_i = d$, restricted to the first orthant $k_i \geq 0$. Here ν_i are indices of the quasi-homogeneity and d is the degree: $f(\lambda^{\nu_1} x_1, \ldots, \lambda^{\nu_n} x_n) = \lambda^d f(x)$. If the point $(\nu, \ldots, \nu)$ is the intersection point of the diagonal $\{k_i = k_j\}$ with Γ, then the number $-1/\nu$ is the distance of Γ (see Definition 5.42).

Proposition. *If f is quasi-homogeneous, then the index of oscillation is equal to the distance of the Newton diagram.*

Proof. We have

$$\int e^{i\tau f} \sim \int_0^\infty e^{i\tau t} t^{\gamma-1} dt \sim (i\tau)^{-\gamma},$$

where $\gamma = (\sum \nu_i)/d$.
On the other hand, for the diagonal point $(\nu, \ldots, \nu)$ we have

$$\nu \cdot \sum \nu_i = d.$$

Thus $\gamma = 1/\nu$. □

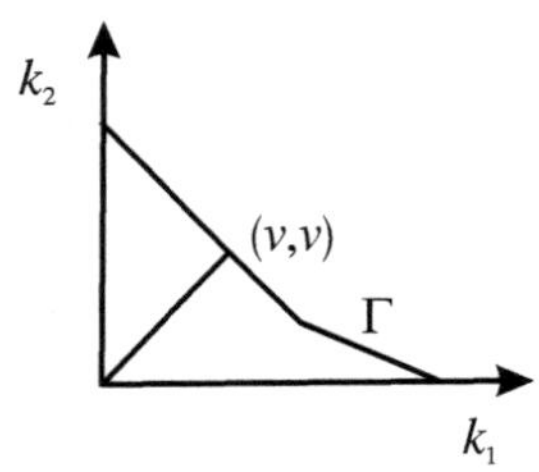

Figure 5.13: Newton diagram and its distance

Using the resolution of singularity the authors of [**AVG**] proved the following result, which generalizes the above proposition and is an analogue of Theorem 5.43. If Γ is a Newton's diagram of some function, then its distance is $-1/\tilde{\nu}$, where $(\tilde{\nu}, \ldots, \tilde{\nu})$ is the intersection point of the diagonal with the diagram. The *multiplicity* of this point is equal to the codimension of the face of Γ which contains it (see Fig. 5.13).

5.56. Theorem. *If the germ f is typical among the set of germs with Γ as their Newton's diagram, then:*

(i) *the index of oscillation of* f *is equal to the distance of* Γ;

(ii) *the degree of* $\ln t$, *in the first term of the asymptotic expansion from Theorem 5.54, is equal to the multiplicity of the diagonal point from* Γ.

5.57. Laplace method. Here we deal with the integrals of the Laplace type

$$\int e^{-\tau f(x)} \phi(x) dx,$$

where $\tau \to +\infty$ and f has local minimum at $x = 0$. For this integral the asymptotic expansion formula, analogous to that from Theorem 5.53, holds. We do not formulate the theorem and restrict ourselves to the following example.

5.58. Stirling formula. The integral defining the Gamma-function $\Gamma(\alpha + 1)$ can be treated as the Laplace integral

$$\int e^{-f(t,\alpha)} dt,$$

where the 'phase' $f(t, \alpha) = t - \alpha \ln t$ has the critical point at $t \sim \alpha$ with the second derivative $1/\alpha$. Thus

$$f \approx -\alpha \ln \alpha + \alpha + \frac{1}{2\alpha}(t - \alpha)^2$$

and we get $\Gamma(\alpha + 1) \sim \alpha^\alpha e^{-\alpha} \sqrt{2\pi\alpha}$. If $\alpha = n$, then we get the Stirling formula for $n!$.

Chapter 6

Vector Fields and Abelian Integrals

§1 Phase Portraits of Vector Fields

If M is a smooth manifold and $\boldsymbol{V} \in \Gamma(M, TM)$ is a vector field on M (a global section of the tangent bundle), then it defines the differential equation on M

$$\dot{x} = \boldsymbol{V}(x). \tag{6.1}$$

Here the dot means the derivative with respect to (physical) time.

6.1. Examples.

(a) The Newton equations: $m_i \ddot{\vec{x}}_i = \boldsymbol{F}_i(\vec{x}_1, \ldots, \vec{x}_n)$.

(b) Population of bacteria: $\dot{x} = x^2$.

(c) Population of one species: $\dot{x} = x(1 - ax)$. Here the factor $1 - ax$ denotes the amount of food.

(d) The Lotka–Volterra system (populations of predators and preys): $\dot{x} = ax(1 - by)$, $\dot{y} = cy(1 - dx)$.

(e) The *Hamiltonian system*

$$\dot{x} = \partial H/\partial y, \quad \dot{x} = -\partial H/\partial x,$$

where H is the Hamilton function. The conservative Newton system $\dot{x} = y$, $\dot{y} = -x + x^2$ is a particular case of Hamiltonian system.

Below we present some elements of the qualitative theory of Ordinary Differential Equations; for more informations we refer the reader to the classical monograph [**Arn5**] and more recent book [**ZM**].

The **phase portrait** of the vector field $\boldsymbol{V}$ is the partition of the phase space M into the phase curves of the vector field. The phase curves are the images of the solutions $\mathbb{R} \ni t \to x(t)$ of equation (6.1).

There are three types of phase curves:

- the equilibrium states, i.e., points x_i at which $\boldsymbol{V}(x_i) = 0$. They are called also singular points, critical points or stationary points.
- closed curves corresponding to periodic solutions of (6.1): there exists $T > 0$ such that $x(t + T) = x(t)$ for any t. Minimal such $T > 0$ is called the period of this trajectory (or of the phase curve).

H. Żołądek, *The Monodromy Group*, Monografie Matematyczne 67,
https://doi.org/10.1007/978-3-031-91270-2_6

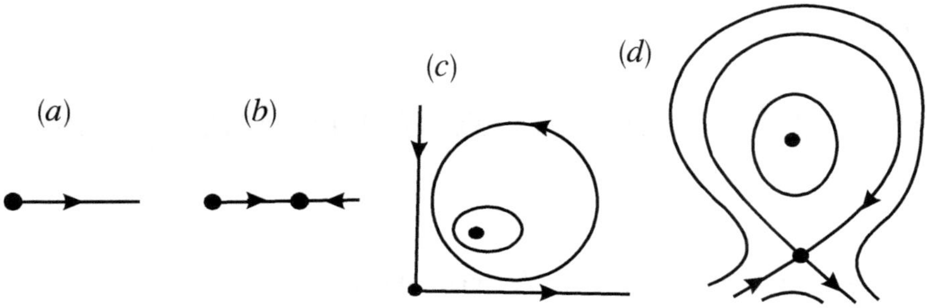

Figure 6.1: Phase portraits of some vector fields

– immersed embeddings of the real line.

The phase portraits of the vector fields from example (b), (c), (d), (e) are presented at Fig. 6.1.
Singular points are the simplest elements of the phase portrait to study. Near such point $x = 0$ we have

$$\dot{x} = Ax + O(|x|)^2, \; x \in \mathbb{R}^n,$$

where A is a constant matrix.

6.2. Definition. The point $x = 0$ is called **hyperbolic** iff $\mathrm{Re}\lambda_i \neq 0$, where λ_i are the eigenvalues of the matrix A. Otherwise the point $x = 0$ is called non-hyperbolic.

6.3. Grobman–Hartman theorem. *In a neighborhood of a hyperbolic critical point there exists a homeomorphism $y = y(x)$, which transforms the phase portrait of the system* (6.1) *to the phase portrait of the system*

$$\dot{y}_1 = y_1, \; \dot{y}_2 = -y_2,$$

where $y_1 \in \mathbb{R}^k$, $y_2 \in \mathbb{R}^{n-k}$, and k is the number of positive $\mathrm{Re}\lambda_i$*'s.*

In particular, if all $\mathrm{Re}\lambda_i < 0$, then the equilibrium point $x = 0$ is *asymptotically stable*; all trajectories starting near 0 remain there for all $t > 0$ and tend to 0 as $t \to \infty$.

6.4. Bifurcations of critical points. If we chose randomly a vector field, then almost surely it will have only hyperbolic critical points. The situations with non-hyperbolic critical points occur inevitably when we have a family V_μ of vector fields depending on a parameter(s). For typical values of μ, $\boldsymbol{V}_\mu$ has only hyperbolic stationary points but, for some *bifurcational* values of the parameter, $\boldsymbol{V}_\mu$ encounters non-hyperbolic singularities. The theory describing the change of the phase portrait near such situations is called the *bifurcation theory.*
In one-parameter typical families we meet only two types of non-hyperbolicity:

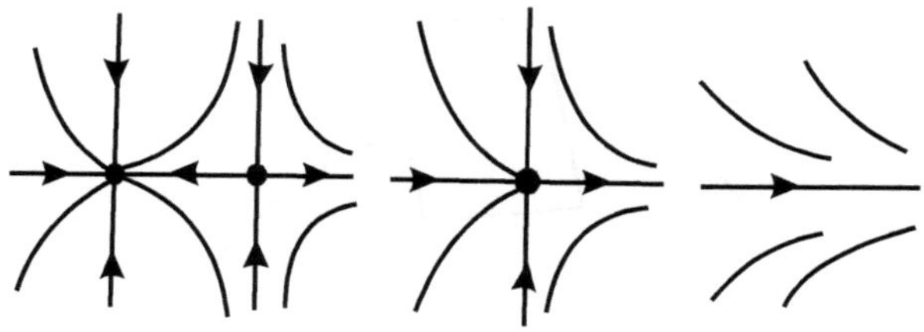

Figure 6.2: Saddle-node bifurcation

(a) $\lambda_1 = 0$, $\mathrm{Re}\lambda_j \neq 0$, or

$$\dot{x} = \mu + x^2, \ \dot{y} = \pm y$$

(for $n = 2$). It is the **saddle-node bifurcation** (see Fig. 6.2).

(b) $\mathrm{Re}\lambda_{1,2} = 0 \neq \mathrm{Re}\lambda_j$, or

$$\dot{r} = r(\mu - r^2), \ \dot{\varphi} = 1$$

(in the polar coordinates in the plane). This is the well-known **Andronov–Hopf bifurcation** (see Fig. 6.3).

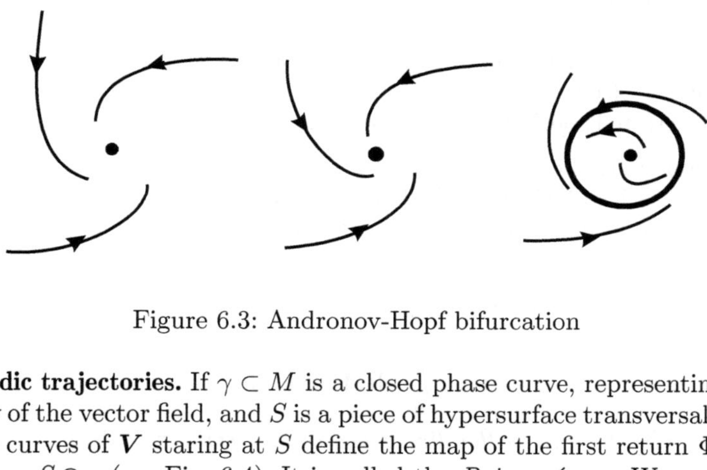

Figure 6.3: Andronov-Hopf bifurcation

6.5. Periodic trajectories. If $\gamma \subset M$ is a closed phase curve, representing periodic trajectory of the vector field, and S is a piece of hypersurface transversal to γ, then the phase curves of $\boldsymbol{V}$ staring at S define the map of the first return $\Phi : S \to S$, defined near $S \cap \gamma$ (see Fig. 6.4). It is called the *Poincaré map*.We parameterize S by $y \in \mathbb{R}^{n-1}$ and take its linearization at $\{y = 0\} = S \cap \gamma$,

$$\Phi(y) = By + O(|y|^2).$$

Here the point $x = 0$ is the *fixed point* for the map Φ.

6.6. Definition. The fixed point $y = 0$ is called **hyperbolic** iff all $|\lambda_i| \neq 1$, where λ_i are the eigenvalues of the matrix B.

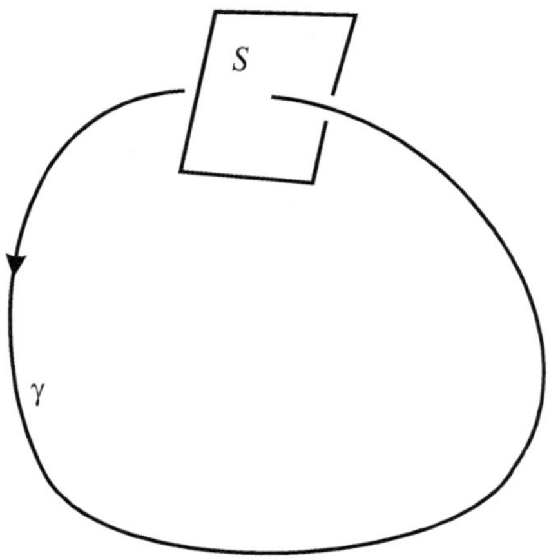

Figure 6.4: Poincaré map

In such case the trajectory γ is called **hyperbolic**. The numbers $\nu_i = \ln|\lambda_i|$ are called the *characteristic multipliers*.

There is an analogue of the Grobman–Hartman theorem for hyperbolic fixed point of local diffeomorphisms; we do not provide its formulation. In particular, if γ is hyperbolic, then the phase portrait near γ is stable with respect to perturbations of the vector field (it is *structurally stable*).
We will be interested specially in the case of planar vector fields, i.e., with $M = \mathbb{R}^2$.

6.6. Definition. Periodic trajectories in $\mathbb{R}^2$, which are isolated among the set of periodic trajectories of $\boldsymbol{V}$, are called the **limit cycles**. The singular point, which has a neighborhood filled by periodic trajectories of $\boldsymbol{V}$, is called the **center**.

6.8. Theorem (Dulac criterion). *If $\gamma : x = x(t)$ is a periodic trajectory (with period T) of a planar vector field $\boldsymbol{V} = \sum v_i(x)\partial/\partial x_i$, then its (unique) characteristic multiplier is equal to*

$$\nu = \int_0^T \mathrm{div}\boldsymbol{V}(x(t))\mathrm{d}t,$$

where $\mathrm{div}\boldsymbol{V} = \sum \partial v_i/\partial x_i$ *is the divergence of the vector field* $\boldsymbol{V}$.

Problem. Prove it using the Liouville's theorem $\det \mathrm{e}^{At} = \mathrm{e}^{\mathrm{tr}At}$.

6.9. Bifurcations of periodic orbits. There are three generic one-parameter bifurcations of local diffeomorphisms, corresponding to bifurcations of periodic trajectories:

(a) $\lambda_1 = 1, \ |\lambda_j| \neq 1$ or

$$y \to y + \mu + y^2.$$

It is the *saddle-node bifurcation* for diffeomorphisms.

(b) $\lambda_1 = -1$, $|\lambda_j| \neq 1$ or
$$\Phi^2(y) = (1+\mu)y - y^3.$$
This is called the *pitchfork bifurcation*, or the *period doubling bifurcation*. The initial periodic orbit γ lies (as the equator) in a Möbius band and the bifurcation relies on creating a periodic trajectory covering γ two times.

(c) $\lambda_{1,2} = e^{\pm 2\pi i \alpha}$, $|\lambda_j| \neq 1$. It is the *Andronov–Hopf bifurcation for diffeomorphisms*. The resonant cases, i.e., with rational α, are qualitatively different from the non-resonant cases. Here periodic orbits of very long period and many other features, characteristic for the general theory of dynamical systems, can be observed. (We shall discuss the resonant case below).

6.10. Separatrix connection. In the qualitative analysis of planar vector fields one encounters also the (one-parameter) bifurcations of **separatrix connection** and **separatrix loop** presented at Fig. 6.5. The *separatrices* of a singular point of a vector field are the phase curves which tend to critical point with definite limit direction (as time goes to $+\infty$ or to $-\infty$) and are 'distinguished' among such trajectories.

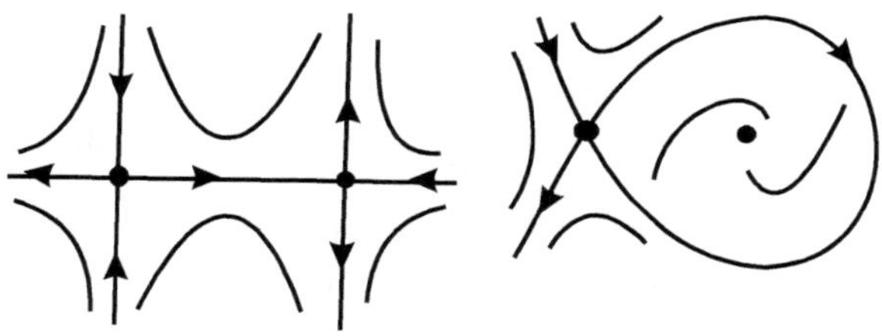

Figure 6.5: Saddle connection and saddle loop

In the analysis of phase portraits of planar vector fields (or vector fields on two-dimensional manifolds) important is the analysis of: singular points, periodic solutions (limit cycles) and positions of separatrices. The singular points can be analyzed using purely algebraic methods, but the analysis of limit cycles and separatrices needs application of transcendental methods and is difficult in general. Recall that one of the Hilbert's problems deals with limit cycles.

6.11. Second part of the 16th Hilbert Problem. *Find the estimate from above $H(n)$ for the maximum number of limit cycles of any planar polynomial vector field of degree n.*

Despite its simple formulation the progress in the solution of this problem is very slow. The history of its investigations is full of errors. For example, in the 1920s H. Dulac published the mémoire [**Dul1**] with the 'proof' of finiteness of the number of limit cycles of an individual polynomial vector field. This proof remained 'correct' until the end of 70-ties, where a significant gap in the Dulac's arguments was discovered. This gap was filled independently by Yu. Ilyashenko [**Il4**] and by J. Ecalle [**Ec3**].

6.12. Theorem (Finiteness of the number of limit cycles). *If* $\boldsymbol{V}$ *is a polynomial vector field on a plane, then it has a finite number of limit cycles.*

It is not known whether the number $H(n)$ is finite. The problem lies in proving a locally uniform bound for the number of limit cycles in a local family of polynomial vector fields (see [**IlY2**], **Rou**] and [**Il7**])).
Even the number $H(2)$ is not known.
There are many special results in the problem 6.11. One case, where much was done, concerns perturbations of the integrable systems.

§2 Method of Abelian Integrals

6.13. Perturbations of Hamiltonian systems. These are the systems $\boldsymbol{V}_\epsilon$ of the type

$$\dot{x} = \frac{\partial H}{\partial y} + \epsilon P(x,y;\epsilon),\ \dot{y} = -\frac{\partial H}{\partial y} + \epsilon Q(x,y;\epsilon),$$

where ϵ is a small parameter.
Before perturbation, i.e., for $\epsilon = 0$, the phase space contains domains filled completely with the *ovals* of levels the Hamilton function , i.e., the connected components of the (real) curves $\{H(x,y) = h\}$.
After perturbation usually there remain only finite number of closed phase curves. The problem is to count their number. More precisely, if γ_ϵ is a limit cycle for V_ϵ such that $\gamma_\epsilon \to \gamma(h_i) \subset \{H = h_i\}$, then we say that the oval $H = h_i$ generates a limit cycle. We calculate the number of such h_i's.
This problem can be treated as the linearization of the Hilbert 16th problem in a Hamiltonian vector field.
Such questions appear naturally in the bifurcation theory, as the below examples show.

6.14. Example (Bogdanov–Takens bifurcation). This is the two-parameter bifurcation corresponding to the nilpotent Jordan cell with zero eigenvalues

$$\dot{x} = y, \quad \dot{y} = -\mu_1 + \mu_2 y + x^2 + xy.$$

Here the singular points are $y = 0,\ x_{1,2} = \pm\sqrt{\mu_1},\ \mu_1 \geq 0$ with the linear parts $\begin{pmatrix} 0 & 1 \\ 2x & \mu_2 \end{pmatrix}$.
We see that the line $\mu_1 = 0$ is bifurcational, with the saddle-node bifurcation (two singular points disappear).
Also the line $\mu_2 = 0$ ($\mathrm{tr} = 0$) is bifurcational, with the Andronov–Hopf bifurcation. Thus, after passing through this line in the direction of growing μ_2, a limit cycle is born (see Fig. 6.6). The problem is what happens with this cycle, when the point

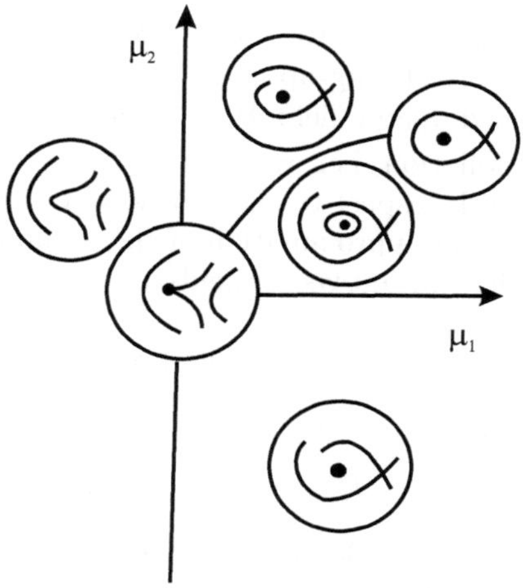

Figure 6.6: Bogdanov-Takens bifurcation

in the parameter space is away from the line $\mu_2 = 0$. It turns out that the system can be reduced to a perturbation of a Hamiltonian system. If the terms $\mu_2 y$ and xy are negligibly small with respect to μ_1, then the system is Hamiltonian with the Hamilton function $\frac{1}{2}y^2 + \mu_1 x - \frac{1}{3}x^3$. Here $x \sim \sqrt{\mu_1}$, $y \sim \mu_1^{3/4}$. This implies $xy \sim \mu_1^{5/4} << \mu_1$ and $\mu_2 y << \mu_1$ for $\mu_2 << \mu_1^{1/4}$.
After normalization $x = \mu_1^{1/2}X$, $y = \mu_1^{3/4}Y$, and division of the vector field by $\mu_1^{1/4}$, we get

$$\dot{X} = Y, \quad \dot{Y} = -1 + X^2 + \epsilon(\nu + X)Y,$$

where $\epsilon = \mu_1^{1/4}$, $\nu = \mu_2/\sqrt{\mu_1}$.
This bifurcation was firstly completely investigated by R. Bogdanov in [**Bog**], who proved that this vector field has at most one limit cycle and its bifurcations are as in Fig. 6.6 (see also [**Arn5**]).

6.15. Example (Resonant periodic trajectory in a space). Let $\gamma \subset \mathbb{R}^3$ be a closed trajectory of an unperturbed vector field $\boldsymbol{V}_0$. Assume additionally that the eigenvalues of the linearization of the Poincaré map lie in the unit circle $\lambda_{1,2} = \mathrm{e}^{\pm 2\pi i\alpha}$. If α is irrational, then we have a singularity of codimension one (in principle). In the resonant case $\alpha = \frac{p}{q}$ one should consider a two-parameter deformation of this situation.
The section S transversal to γ can be parameterized by points from the complex plane $z \in \mathbb{C}$. We choose such neighborhood of γ, parameterized by (φ (mod 2π), z), that the linear parts of the natural correspondence maps $\{\varphi = \varphi_1\} \to \{\varphi = \varphi_2\}$ (defined by trajectories of V_0) are the homogeneous rotations $z \to \mathrm{e}^{\mathrm{i}(\varphi_2 - \varphi_1)p/q} z$.

This system of linear maps defines the *Seifert foliation* near γ. Its generic leaf makes q turns along γ before closing-up.
Now we take the deformed vector field $\boldsymbol{V}_\mu$ and average its z-component along the leaves of the Seifert foliation, i.e., we take $\int_0^{2\pi q} \dot{z}$. We obtain a planar vector field, which is invariant with respect to the rotation by the angle $2\pi/q$ and whose dynamics gives a rather good approximation of the dynamics of $\boldsymbol{V}_\mu$.
The versal families of such invariant vector fields are given in the following formulas (see [**Arn5**]):

$$\dot{x} = y,\ \dot{y} = -\mu_1 + \mu_2 y + x^2 + xy, \quad (q = 1),$$

$$\dot{x} = y,\ \dot{y} = \mu_1 x + \mu_2 y + ax^3 + bx^2 y, \quad (q = 2),$$

$$\dot{z} = \mu z + A|z|^2 + B\bar{z}^{q-1}, \quad \mu = \mu_1 + \mathrm{i}\mu_2, \quad (q \geq 3).$$

We see that the case $q = 1$ is the Bogdanov–Takens bifurcation. If $q = 2$, then for $\mu_2 = b = 0$ the system is Hamiltonian with the Hamilton function $(2y^2 - 2\mu_1 x^2 - ax^4)/4$. If $q \geq 3$, then for $\mu_1 = \mathrm{Re}A = 0$ the system is Hamiltonian. It follows from the formula $\mathrm{div}P = 2\mathrm{Re}\partial P/\partial z$ for a vector field $\dot{z} = P(z, \bar{z})$.

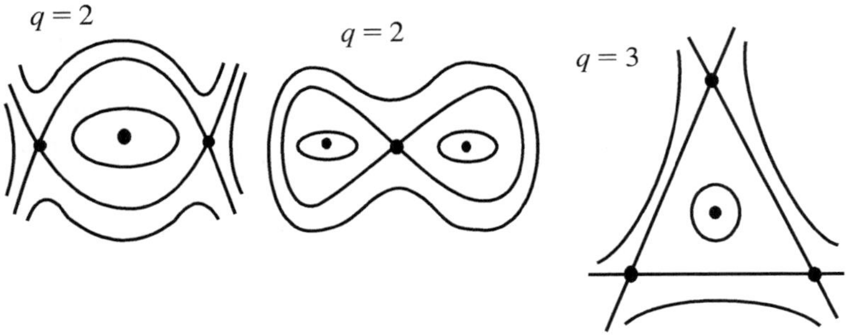

Figure 6.7: Levels of Hamiltonian functions

In the cases $q = 2, 3$ the analysis of phase portraits is reduced to analysis of limit cycles in perturbation of the Hamiltonian system; it was done by E. Horozov [**Hor**] and by Yu. Ilyashenko [**Il1**] (see Fig. 6.7). The cases $q \geq 5$ are called weak resonances and are simple to investigate (see [**Arn5**]).
The case $q = 4$ is still not finished. The Abelian integrals were studied by A. Neishtadt in [**Nei**], by F. Berezovskaya and A. Khibnik [**BeKh**] and by B. Krauskopf [**Kr**].

6.16. Example (One zero and a pair of imaginary eigenvalues). Assume that an unperturbed system in $\mathbb{R}^3$ has a singular point with these eigenvalues of the linear part. It is a codimension two phenomenon.
Here one performs the averaging along the trajectories of the linear system (circles) and obtains the following two-dimensional vector field (where one variable is the

amplitude of oscillations):

$$\dot{x} = \mu_1 + \mu_2 x + ax^2 \pm y^2 + by^3, \; \dot{y} = -2xy$$

If $\mu_2 = b = 0$, then the system has center. It is not Hamiltonian. but it has the first integral

$$y^a \left(x^2 \pm y^2/(a+2) + \mu_1/a\right)$$

(see Fig. 6.8). This bifurcation was analyzed in [**Zo1**] (see also [**KoZe**]).

6.17. Example (Two pairs of imaginary eigenvalues). This case, after averaging along the 2-tori corresponding to the two independent rotations (of the linear part), gives rise to the following generalized Lotka–Volterra system

$$\dot{x} = x(\mu_1 + ax + by), \quad \dot{y} = y(\mu_2 + cx + dy + ex^2).$$

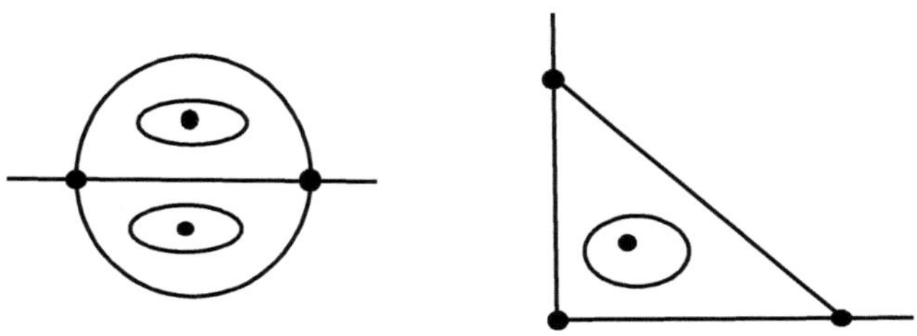

Figure 6.8: Levels of first integrals in the cases of zero-Hopf and Hopf-Hopf bifurcations

Here also we obtain situation with perturbation of a system with the first integral

$$x^\alpha y^\beta (1 + kx + ly)$$

(see Fig. 6.8). The bifurcations and corresponding Abelian integrals were studied in [**Zo2**] (see also [**KoZe**]).

6.18. Reduction to zeroes of Abelian integrals. Assume that we have the situation as in 6.13. Take a section (interval) S transversal to the family of closed curves $H(x,y) = t$ (see Fig. 6.9). We parameterize it by the function H restricted to it, $t = H|_S$.
Beginning from here we denote the values of the Hamilton function by t (not by h). This notation agrees with the notation used in Chapter 5.
We compute the first approximation of the Poincaré map. If $P \in S$, $H(P) = t$, is an initial point of the positive trajectory Γ of the perturbed system, then the first intersection of Γ with S is the value of the *return map*, $Q = \Phi(P)$ (see Fig. 6.9). We calculate the increment $\Delta H = H(Q) - H(P)$ of the Hamilton function along

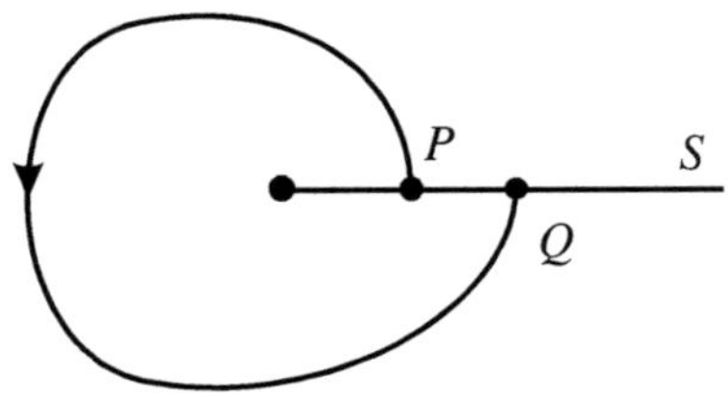

Figure 6.9: Return map

Γ. Note that Γ is periodic iff $\Delta H = 0$ and it is hyperbolic stable (respectively, unstable) limit cycle iff, additionally, $(\Delta H)'(P) < 0$ (respectively, > 0).
Using the representations $H'_x = -\dot{y} + \epsilon Q$, $H'_y = \dot{x} - \epsilon P$, we get

$$\Delta H = \int_0^T \dot{H} \mathrm{d}t = \int (H'_x \dot{x} + H'_y \dot{y}) \mathrm{d}t = \epsilon \int (H'_x P + H'_y Q) \mathrm{d}t = \epsilon \int (Q\dot{x} - P\dot{y}) \mathrm{d}t$$
$$= \epsilon \int_\Gamma Q \mathrm{d}x - P \mathrm{d}y.$$

Because the phase curve Γ is close to the oval of $H = t$ (up to the order $O(\epsilon)$), then we get

$$\Delta H = \epsilon I(t) + O(\epsilon^2),$$

where

$$I(t) = \int_{H=t} Q \mathrm{d}x - P \mathrm{d}y$$

is the **Abelian integral**. In fact, the path of integration is some real oval $\gamma(t)$ of the curve $H = t$ and the function I is defined in an interval $(t_{\min}, t_{\max})$ of t's, for which the ovals $\gamma(t)$ are compact and smooth.
(Probably this integral first appeared in the work [**Pon**] of L. Pontryagin. It was used intensively by V. Melnikov [**Mel**] as a tool for detecting subharmonic solutions in some periodic non-autonomous Hamiltonian systems. Some people (e.g., Arnold) claim that it was certainly known already to Poincaré. Therefore in the literature it appears under different names: *Pontryagin integral, Poincaré–Pontryagin integral, Melnikov integral, Pontryagin–Melnikov integral, generating function.*)
We see that:

The necessary condition for existence of limit cycle $\gamma_\epsilon \to \delta(t_i)$ is the equality $I(t_i) = 0$.

Under some generic assumptions it is also a sufficient condition.

6.19. Weakened 16th Hilbert problem. *Consider the space of integrals $I(t)$ with P, Q, H polynomials of degree $\leq n$ and defined in the intervals $(t_{\min}, t_{\max})$. Find* *an estimate $C(n)$ for the number of zeroes of $I(t)$, uniform with respect to the polynomials P, Q, H.*

This problem (stated by V. Arnold [**Arn5**]) is also not solved completely, but there are many nice results concerning it.

6.20. Results. Firstly, A. Varchenko [**Var3**] and A. Khovanski [**Kh2**] have proved that

$$C(n) < \infty,$$

i.e., the existence of an uniform estimate. However they do not give any formula for $C(n)$. The proof of Varchenko bases on the methods developed in the book [**AVG**], (asymptotic expansions of integrals along cycles in complex algebraic curves), and some finiteness results from the real analytic geometry. Khovanski observed that Abelian integrals belong to his class of Pfaff functions and applied his theory of fewnomials. Below we present some of the Varchenko–Khovanski's arguments.
Concrete estimates are given with some restrictions onto the Hamilton function.
In the case of elliptic Hamiltonian $y^2 + x^3 - x$ G. Petrov [**Pet2**] proved the Chebyshev property of Abelian integrals. We present his beautiful proof below.
In the case of hyperelliptic Hamiltonian $y^2+R(x)$ (with fixed polynomial R) Petrov [**Pet3**] proved the linear estimate $\leq a \cdot n + b$ for the number of zeroes of any form $Q\mathrm{d}x - P\mathrm{d}y$ of degree n.
For cubic H and quadratic P and Q L. Gavrilov [**Gav1**] proved that the number of zeroes is ≤ 2 (also around two foci).
Another general estimates were obtained by Yu. Ilyashenko and S. Yakovenko ($\leq 2^{2^{cn}}$, $c = c(H)$ in [**IlY1**]) for generic Hamiltonians, by D. Novikov and Yakovenko ($\leq 2^{cn}$ in [**NY**]) and by A. Glyutsuk and Yu. Ilyashenko ($\leq e^{2500n^2}$ for Hamiltonians with critical points of absolute value ≤ 1 in [**GY**]).
In 1996 A. Khovanski and G. Petrov announced the estimate

$$\leq a \cdot n + b,$$

where $a = a(\deg H)$ and $b = b(\deg H)$ depend only on degree of the Hamiltonian (without an explicit formula). No restriction on H is made. This result is not yet published, but we present this proof below.

After the first version of this book was published, G. Binyamini, D. Novikov and S. Yakovenko [**BNY**] proved the estimate

$$\leq 2^{2^{\mathrm{Poly}(n)}},$$

(where $\mathrm{Poly}(n) = O(n^p)$, $p \leq 61$, is some explicit polynomial), without any restrictions. We present the main ideas behind its proof below.

Below we present some estimates, with proofs, for the number of zeroes of $I(t) = I_\omega(t) = \int_{H=t} \omega$ in the case of polynomial 1-form

$$\omega = A(x, y)\mathrm{d}x + B(x, y)\mathrm{d}y$$

of degree $\deg \omega = \max(\deg A, \deg B)$ and with concrete H's. We begin with the quadratic Morse Hamiltonian.

6.21. Proposition. *If $H = x^2+y^2$, then $I_\omega(t)$ is a polynomial of degree $\le (n+1)/2$. It has at most $[(n+1)/2]-1$ positive zeroes corresponding to eventual limit cycles.*

Proof. Consider the case when ω is homogeneous of degree j. Then, putting $x = \sqrt{h}\cos\theta,\ y = \sqrt{h}\sin\theta$, we get the trigonometric integral

$$I = t^{(j+1)/2}\int_0^{2\pi} R(\cos\theta, \sin\theta)\mathrm{d}\theta,$$

where R is homogeneous polynomial of degree $j+1$. This integral vanishes for odd $j+1$. □

Next is the case of elliptic Hamiltonian

$$H = y^2 + x^3 - x$$

studied before. We integrate the *real* 1-form ω along the *real* oval $\gamma(t)$. The latter represents one of the two generators of the first homology group of the complex elliptic curve $\{H = t\} \subset \mathbb{CP}^2$. It vanishes at the critical point $x = 1/\sqrt{3}, y = 0$, with the critical value $t = -2/3\sqrt{3}$. The other generator is $\delta(t)$ and vanishes at $(-1/\sqrt{3}, 0)$, with the critical value $t = 2/3\sqrt{3}$ (see 5.19–5.24 and Fig. 6.10).

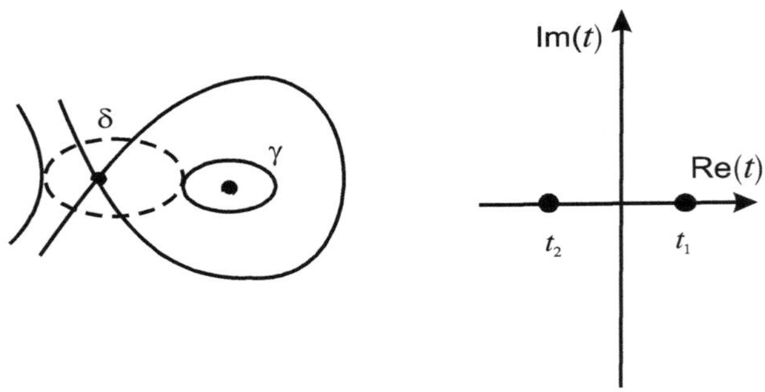

Figure 6.10: Critical points, critical values and cycles in the case of elliptic Hamiltonian

The functions $I_\omega(t)$ have analytic prolongation to the complex arguments t. They are multivalued function with unique branching point at $t = 2/3\sqrt{3}$. Thus, in the complex plane cut along the half-line $\{t \ge 2/3\sqrt{3}\}$, the functions I_ω are analytic and univalent. We denote by Ω the set

$$\mathbb{C} \setminus \{-2/3\sqrt{3}\} \setminus [2/3\sqrt{3}, \infty)$$

and consider I_ω as functions on Ω. We further note hat Ω contains the interval $(-2/3\sqrt{3}, 2/3\sqrt{3})$, which is of interest for us. All these functions with $\deg\omega \leq n$ form a finite dimensional vector space W_n.

6.22. Definition. A linear space W of functions defined on a set A, $(A \subset \mathbb{R}, \mathbb{C})$, is called **Chebyshev** iff any nonzero function from W has at most $\dim W - 1$ zeroes in A.
W is **Chebyshev with accuracy** k iff this number of zeroes is $\leq \dim W - 1 + k$.

The reader can prove that in a k-dimensional linear space W of functions on A one can always choose a nonzero function vanishing at any $k-1$ previously chosen points.

6.23. Theorem of Petrov. ([**Pet2**]) *The space* $W_n = \{\int_{\gamma(t)} \omega : \deg\omega \leq n,\, t \in \Omega\}$ *is Chebyshev.*

Proof. In 5.19–5.24 the following properties of the elliptic integrals were proved.

1. $I_\omega = P_0(t)I_0 + P_1(t)I_1$, *where* $I_j = \int_{\gamma(t)} x^i y\mathrm{d}x$ *and* $P_{0,1}$ *are polynomials of degrees* $\leq [(n-1)/2]$ *and* $\leq [n/2] - 1$ *respectively. The space* W_n *has dimension equal to* n

2. $5I_0 = 6tI_0' + 4I_1'$, $21I_1 = 4I_0' + 18hI_1'$.

3. $4(27t^2 - 4)I_1'' = 21I_1$.

4. $I_0 \sim t^{5/6}$, $I_1 \sim t^{7/6}$ *as* $t \to \infty$

Below we prove additional properties of the elliptic integrals.

5. **Lemma.** *We have* $\mathrm{Im}I_1(t) \neq 0$ *for* $t > 2/3\sqrt{3}$.

Proof. In order to understand properly the statement of Lemma 5, we must recall the definition of I_1 as the analytic prolongation of an integral of a holomorphic form along the cycle $\gamma(t)$. At the point $t = 2/3\sqrt{3}$ this function has ramification, but the limit values of I_1 along the upper and the lower ridges of the cut $[2/3\sqrt{3}, \infty)$ are well defined. Thus the $\mathrm{Im}I_1$ is the imaginary value at the upper ridge.
Because the initial function (i.e., for $|t| < 2/3\sqrt{3}$) was real, then I_1 behaves well under conjugation of the argument. In particular, the value of $\mathrm{Im}I_1$ at the lower ridge is equal to minus its value at the upper ridge. The function $\mathrm{Re}I_1$ is the same at both ridges.
The difference between values of I_1 at the upper ridge and lower ridge, i.e., $2\mathrm{Im}\, I_1$, is equal to the variation of the integral as the value t varies around the critical value $2/3\sqrt{3}$. By the Picard–Lefschetz formula, this variation is equal to the value of the form $xy\mathrm{d}x$ at the other generator of the first homology group $\delta(t)$. These arguments show that

$$\mathrm{Im}I_\omega(t) = \frac{1}{2}\int_{\delta(t)} \omega,\ t \geq 2/3\sqrt{3},$$

for any real form ω.
Therefore, $z(t) = \mathrm{Im} I_1$ is expressed by means of integrals along cycles. In particular, it satisfies the Picard–Fuchs equation 3, i.e., $4(27t^2 - 4)z'' = 21z$. Moreover, $z(2/3\sqrt{3}) = 0$ (because δ vanishes there).
Because the factor $27t^2 - 4 > 0$, we have $z'' > 0$, if $z > 0$, and $z'' < 0$, if $z < 0$. Thus z is either positive and convex, or negative and concave. In any case it cannot have zeroes. □

6. **Lemma.** *$I_1(t)$, $t < 2/3\sqrt{3}$, vanishes only at the point $t = -2/3\sqrt{3}$ and this is a simple zero.*

Proof. Of course, $I_1(t)$ is real in this half-line and vanishes at $-2/3\sqrt{3}$.
Moreover, because $I_1 = \int\int_{H\leq t} x dx dy$ and the domain $H \leq t$ is an approximate ellipse around the point $(1, 0)$ with the semi-axes $\sim \sqrt{t + 2/3\sqrt{3}}$, then $I_1 \sim (t + 2/3\sqrt{3})$ as $t \to -2/3\sqrt{3}$. This gives the simplicity of the zero.
The negativity of I_1 at the half-line $t < -2/3\sqrt{3}$ is proved in the same way as in the proof of Lemma 5.
The positivity of I_1 at the interval $(-2/3\sqrt{3}, 2/3\sqrt{3})$ needs application of some geometrical arguments. Note that the integral I_1 is proportional to the center of mass of the domain $H \leq t$. It is seen from Fig. 6.10 that this center of mass lies in the right half of the plane. Also it is not difficult to show it analytically. □

7. **Lemma.** *$I_1(t)$ has only one zero in the complex plane cut along $[2/3\sqrt{3}, \infty)$.*

Proof. Because I_1 is real on $\mathbb{R}$ and has only one zero on the real part of the domain from Lemma 7, the number of its zeroes is odd. It is enough to show that this number is < 3.
We use the *argument principle* which says that:

If a contour Γ bounds some region of analyticity of a function g and $g|_\Gamma \neq 0$, then the number of zeroes of g in this domain is equal to the increment of the argument of g along Γ (divided by 2π).

We take the contour as in Fig. 6.11: a small loop around $t = 2/3\sqrt{3}$, a large loop around infinity and along the ridges of the cut.
The increment of $\arg I_1$ along the large circle is defined by the asymptotic of the integral at infinity (see (4)) and equals to $7/6$. Because $\mathrm{Im} I_1$ does not vanish along the cut, then the increment of $\arg I_1$ along the ridges and the small circle is ≤ 1. Thus $\Delta_\Gamma(\arg I_1) \leq 13/6 < 3$. □

8. **Lemma.** *The function $(I_0/I_1)(t)$, $t > 2/3\sqrt{3}$, does not take real values.*

Proof. If that happens for $t = t_0$, then the vectors $(\mathrm{Im} I_0, \mathrm{Im} I_1)^\top$ and $(\mathrm{Re} I_0, \mathrm{Re} I_1)^\top$ would be parallel. However, both satisfy the same system of linear differential equations 2; they form two solutions of it. The Wronskian of these solutions would vanish at t_0 and then it should vanish on the whole interval. This would mean

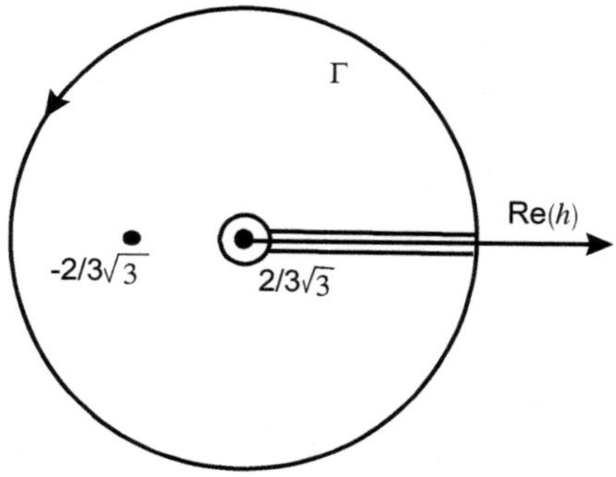

Figure 6.11: Contour for calculating the argument's increment

that $(I_0/I_1)(t) \in \mathbb{R}$ for all $t > 2/3\sqrt{3}$. By the Schwarz reflection principle, I_0/I_1 could be prolonged to analytic function in the plane outside $\{2/3\sqrt{3}\}$. Because it is bounded near $2/3\sqrt{3}$, then this singularity would be removable and I_0/I_1 would be an entire function.
However, its exact asymptotic $\sim t^{-1/3}$ at infinity (see the point 4) contradicts the above. □

9. *Finishing of the proof of Theorem* 6.23.

Because $I_\omega(-2/3\sqrt{3}) = 0$ and I_1 has the only simple zero just in this point, then the number of zeroes of I_ω in the domain Ω is the same as the number of zeroes of the holomorphic function

$$I_\omega/I_1 = P_0(t) \cdot (I_0/I_1) + P_1(t)$$

in the plane cut along $[2/3\sqrt{3}, \infty)$.
Again we use the argument principle with the same contour Γ.
The increment of the argument along the large circle is $\sim$ max(deg $P_0 - 1/3$, deg P_1). Along the ridges of the cut, the number of turns of I_ω/I_1 around 0 is bounded by the number of zeroes of the imaginary part of the function at the cut plus 1; (here we use the reality of the form ω). This gives $\leq \deg P_0 + 1$.
Summing it up (using the 1) we get the increment $\leq (n-1)$. □

6.24. Application to bifurcations. In the case of the Bogdanov–Takens bifurcation we have the elliptic integral

$$\int (\alpha x + \beta) y \mathrm{d}x = \alpha I_0 + \beta I_1.$$

Thus $\deg P_{0,1} = 0$, the space of integrals is 2-dimensional and this elliptic integral has at most one zero corresponding to the unique limit cycle.

Below we present calculations of the number of zeroes of elliptic integrals appearing in bifurcations of periodic orbits with 1 : 2 resonance ($q = 2$ in Example 6.15). In this case we show the divergence from the Chebyshev property (the accuracy grows with the degree). We also present applications of some real methods. We follow the work [**RZ**].

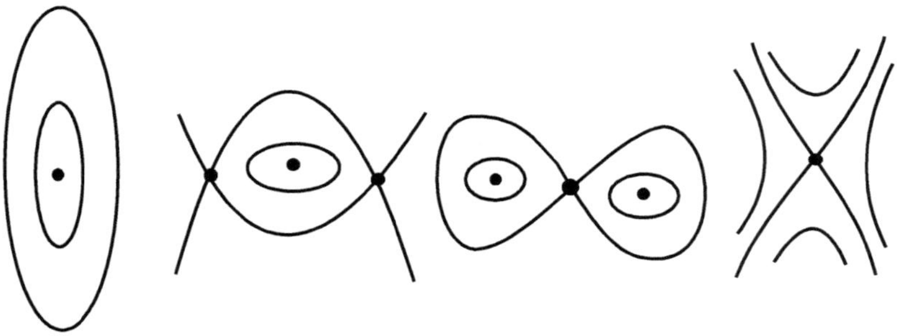

Figure 6.12: Level curves for the Hamiltonian $H = y^2 \pm x^4 \pm 2x^2$

The Hamiltonian is also elliptic (see Fig. 6.12)

$$H = y^2 \pm x^4 \pm 2x^2.$$

In the cases (a) and (b) the corresponding spaces of Abelian integrals are Chebyshev (Petrov).
We shall consider the case (c) with the symmetric form $\omega = A\mathrm{d}x + B\mathrm{d}y$, i.e., the polynomials A, B contain only monomials of odd degree.
The Hamilton function $H = y^2 + x^4 - 2x^2$ has three critical points: $(\pm 1, 0)$ with the critical value $t = -1$ and $(0,0)$ with the critical value 0. At $(\pm 1, 0)$ two cycles γ, γ' vanish. We study the integral

$$I_\omega(t) = \int_\gamma \omega + \int_{\gamma'} \omega = 2\int_\gamma \omega, \qquad -1 \le h < 0,$$

$$I_\omega(t) = \int_{\gamma''} \omega, \qquad h > 0,$$

where $\gamma'' = \gamma + \gamma'$ is the cycle presented at Fig. 6.12(c).
Let $W_n = \{I_\omega : \deg \omega \le n, \omega \text{ is symmetric}\}$.

6.25. Theorem. ([**RZ**]) *We have* $\dim W_n = 2[(n-1)/2]+1$ *and the maximal number of zeroes of a nonzero function from* W_n *in the interval* $(-1, \infty)$ *is* $3[(n-1)/2]$.

Remark. We see that the accuracy of the Chebyshev property of W_n is about $n/2$ and grows. However, when one restricts the integrals to the interval $(0, \infty)$, then the Chebyshev property with the accuracy 1 holds (see [**RZ**]).

Proof of Theorem 6.25. 1. Define the integrals for $t \geq -1$

$$I_0(t) = \int_{H=t} y\mathrm{d}x, \quad I_1(t) = \int_{H=t} x^2 y\mathrm{d}x.$$

We have the following properties (proved like the analogous properties of the Petrov's elliptic integrals).

2. $I_\omega = P_0(t)I_0 + P_1(t)I_1$, *where* $P_{0,1}$ *are polynomials of degrees* $\leq [(n-1)/2]$ *and* $\leq [(n-1)/2] - 1$ *respectively, thus* $\dim W_n = 2[(n-1)/2] + 1$.

3. $3I_0 = 4tI_0' + 4I_1'$, $15I_1 = 4tI_0' + (12t + 16)I_1'$ (Picard–Fuchs equations).

4. $I_0(t) = c_1 + c_2 t \ln|t|^{-1} + \dots$, $I_1(t) = d_1 + d_2 t \ln|t|^{-1} + \dots$ *as* $t \to 0$, (*with* $c_{1,2}, d_{1,2} > 0$).

Denote

$$Q = I_1/I_0.$$

The properties 3 and 4 imply the following.

5. $Q(t) = e_1 + e_2 t \ln|t|^{-1} + \dots$ *as* $t \to 0$.

6. $4h(t+1)Q' = 5Q^2 + 2tQ - 4Q - t$ *and the graphic of the function* $t \to Q(t)$ *consists of the phase curves of the vector field*

$$\dot{t} = 4t(t+1), \ \dot{Q} = 5Q^2 + 2tQ - 4Q - t \tag{6.2}$$

indicated at Fig. 6.13.

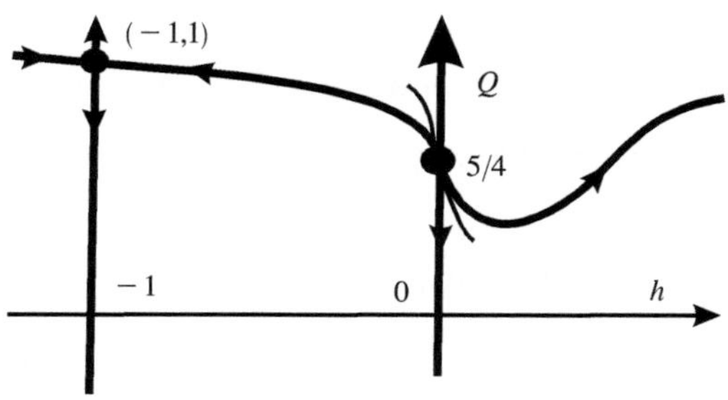

Figure 6.13: Graph of the function Q

The linear part of the vector field (6.2) at the singular point $(0, 4/5)$ is the Jordan cell $\begin{pmatrix} 4 & 0 \\ 3/5 & 4 \end{pmatrix}$ and all trajectories near it have the form $Q = 4/5 + ct\ln|t| + \dots$.

The singular point $(-1, 1)$ is a saddle with the linear part $\begin{pmatrix} -4 & 0 \\ 1 & 6 \end{pmatrix}$ and our graphic lies in two separatrices of this saddle.

7. $I_0 \sim t^{3/4}$, $I_1 \sim t^{5/4}$, $Q \sim t^{1/2}$ *as* $t \to \infty$.

8. $Q(t)$ *is decreasing for* $t < 0$.

Proof. We have $Q'(-1) < 0$, $Q'(0) = -\infty$, $Q''(0) = -\infty$, and $Q'(t) < 0$, $Q''(t) < 0$ as $t \to -\infty$.
On the other hand, $(4t(t+1)Q')'|_{Q'=0} = 2Q - 1$. But $\dot{Q}|_{Q=1/2} = -3/4 < 0$, which implies $Q(t) > 1/2$. This means that the function $Q(t)$ would be concave at each its critical point in the interval $(-1, 0)$ and convex at critical points in $(-\infty, -1)$. This, combined with the behavior at the endpoints, gives the result. □

9. *$Q(t)$ has an unique minimum at $t_* > 0$.*

Proof. From the asymptotic behaviors at $t = 0$ and at $t = \infty$ it follows that such minimum exists. The condition $Q' = Q'' = 0$ means intersection of the line $2Q = 1$ and the hyperbola $2Q^2 + 2tQ - 4Q - t = 0$. We have shown that this intersection is empty, which means that Q'' has the same sign at each critical point of Q. □

Let

$$g(t) = \frac{I_1}{I_0} + \frac{P_0}{P_1} = Q + R.$$

Its zeroes are the zeroes of I_ω, which are > -1.

10. *We have $4P_1^2 t(t+1)g'|_{g=0} = S(t)$, where $S(t)$ is a polynomial of degree $\leq 2[(n-1)/2]$.*

11. *The upper bound for the number of zeroes.* Here we use the idea of Petrov from [**Pet1**].
We divide the interval $(-1, \infty)$ into the subintervals of continuity of g. The number of such subintervals is $\leq \deg P_1 + 1$.
In each such subinterval we apply the *Rolle principle* to the function g:

Between any two zeroes of a function a zero of the derivative lies.

Thus, by 10, between two zeroes of g there is either a zero of $S(t)$ or the point $t = 0$.
On the other hand, the zeroes of $g'|_{g=0}$ are the points of contact of the vector field (6.2) with the curve $Q = -R(t)$. There is such a contact point between the line $t = -1$ and the first zero of the function g.
We have then:

$$\#\{g = 0\} \leq [(\#\{S = 0\} - 1) + 1] + \#(\text{intervals}) \leq 3[(n-1)/2].$$

12. *The lower bound.* The function I_ω has the following expansion in a neighborhood of the point $t = 0$

$$I_\omega(t) = b_0 + a_1 t \ln|t|^{-1} + b_1 t + a_2 t^2 \ln|t|^{-1} + b_2 t^2 + \dots$$

If $t > 0$, then the t^i and $t^i \ln|t|^{-1}$ are positive. We have also

$$I_\omega(-t) = b_0 - a_1 t \ln|t|^{-1} - b_1 t + a_2 t^2 \ln|t|^{-1} + b_2 t^2 - \dots$$

We use the *Descartes principle*:

The number of positive zeroes of a polynomial is bounded by the number of sign changes of its coefficients. Moreover, this bound is achieved for a suitable choice of the absolute values of the coefficients.

In fact, this principle holds also in the case when the polynomial is replaced by the finite sum of functions like t^i and $t^i \ln |t|^{-1}$ and their small perturbations.
We apply this principle to the cases of functions $I_\omega(t)$ and $I_\omega(-t)$, where the coefficients are chosen in a way to get the maximal number of zeroes. Thus we have the leading coefficient $a_m = O(1) \neq 0$ (or $b_m = O(1) \neq 0$) and the absolute values are chosen in a way to give the number of zeroes in the intervals $t < 0$ and $t > 0$ prescribed by the sign changes, i.e.,

$$0 < |b_0| << |a_1| << |b_1| << |a_2| << \ldots << |a_m|.$$

Note the following property:

The sign change between a_i and b_i in $I(t)$ leads to the corresponding sign change in $I(-t)$ and the sign change between b_i and a_{i+1} implies no sign change in $I(-t)$.

Let k_1 be the number of sign changes in $I(t)$ implying the sign changes in $I(-t)$ and k_2 be the number of remaining sign changes. The number of positive zeroes of I_ω is $k_1 + k_2$
If $a_m \neq 0$ is the leading coefficient, then the number of sign changes in $I(-t)$ is $k_1 + (m - k_2)$. The total number of small zeroes of I is $2k_1 + m$, where $k_1 \leq m - 1$. This gives the maximal number $3m - 2$ of zeroes.
If $b_m \neq 0$ is the leading coefficient, then the same arguments give the maximal number $3m$ of zeroes.
It remains to show that the coefficients a_i, b_i in the expansion of the integral are controlled by the coefficients of the polynomials $P_{0,1}$ in the expansion 2. Indeed, using the properties 4, we see that, for $P_0 = at^m$, $m = [(n-1)/2]$ and $P_1 \equiv 0$, the first nonzero coefficient in the asymptotic expansion of I is b_m. Taking a suitable $P_0 = at^{m-1}$ and $P_1 = bt^{m-1}$, we get a_m as leading. Other coefficients are controlled in the same way.
Theorem 6.25 is complete. □

Below we present the promised proof of the linear bound, which is not published yet. Initial claim of Petrov and Khovanski was the estimate $an + b$, where the constant a can be chosen universal, not depending on n and H, and b depends on H). The below proof is based on the conception of rational envelopes of Abelian integrals developed in [**Yak**] and [**IlY1**], on the lecture of Petrov in the Banach Center in Warsaw (1995) and on estimates in the style of Khovanski [**Kh2**] and Varchenko [**Var3**].

6.26. Theorem of Petrov and Khovanski. *The number of zeroes of any Abelian integral I_ω along a family of real ovals of the curve $H = t$ with* $\deg \omega = n$ *is bounded by*

$$a \cdot n,$$

where $a = a(\deg H)$ *is a constant depending only on* $\deg H$.

Proof. 1. We begin with a representation of I_ω as a combination of given Abelian integrals with rational coefficients. This proposition (in a restricted form) was proved by S. Yakovenko in [**Yak**] . His proof uses the theorem of Röhrl and Plemelj (Theorem 8.37 in Chapter 8) about realization of a given monodromy group by means of a linear rational differential system. Our proof is different and without restrictions, i.e., the critical points of H can be non-isolated and may lie at infinity.

2. **Lemma (Rational envelopes).** *There exist real polynomial holomorphic* 1*-forms* $\omega_1, \ldots, \omega_k$ *such that, for any real polynomial holomorphic* 1*-form* ω *of degree* n*, the following representation holds:*

$$P_0(t)^K \cdot I_\omega = P_1(t) I_{\omega_1} + \ldots + P_k(t) I_{\omega_k}.$$

Here the integer $k = k(d)$ *depends only on* $d = \deg H$*,* P_i *are real polynomials such that* P_0 *depends only on* H*,* $\deg P_i \leq c{\cdot}n$*,* $i > 0$*, the positive integer* $K \leq c{\cdot}n$ *and the constant* c *depends only on* $d = \deg H$*.*

Proof. (a) Fix a Hamilton function H. We take the family of complex level curves $\{H = t\} \subset \mathbb{C}^2$. They form the locally trivial fibration (an analogue of the Milnor fibration)

$$\mathbb{C}^2 \setminus H^{-1}(\text{atypical values}) \xmapsto{H} \mathbb{C} \setminus (\text{atypical values}).$$

Here by the atypical values we mean the usual values of H at the critical points and the values corresponding to the 'bad' behavior of the family of levels at infinity. The generic fibre of this fibration is an open Riemann surface, whose homotopical type is a bucket of circles. Its first homology group is generated by concrete geometric cycles $\delta_1(t), \ldots, \delta_k(t)$.

Fix a noncritical value t_0 and the cycles $\delta_i(t_0)$. We choose real polynomial forms $\omega_1, \ldots, \omega_k$ such that their restrictions to the curve $H = t_0$ generate $H^1(\{H = t_0\})$. It is done by a suitable approximation of holomorphic generators of the first de Rham cohomology group in such a way that $\det\left(\int_{\delta_i(t_0)} \omega_j\right) \neq 0$. Later we will impose other restrictions for the forms ω_i.

(Such polynomial generators of the de Rham cohomology of an affine algebraic variety exist in general case (A. Grothendieck). This follows from the quasi-equivalence between the holomorphic de Rham complex $\Omega^\bullet$ and the complex $\Omega^\bullet_{\text{alg}}$ of holomorphic forms with regular behavior at infinity (see 7.33(e) in Chapter 7 and [**GH**]).)

Repeating the proof of Corollary 5.28(b), we obtain the locally unique representation

$$\int_{\delta(t)} \omega = \sum_j p_j(t) \int_{\delta(t)} \omega_j \tag{6.3}$$

for any polynomial holomorphic form ω and any family $\delta(t)$ of integer cycles in $H = t$. The functions $p_j(t)$ do not depend on δ and are prolonged to holomorphic and single-valued functions in

$$\mathbb{C}\setminus\{t_1,\ldots,t_r\},$$

where t_j are atyptical values. Each function $p_j(t)$ is expressed as a ratio of two determinants, with the denominator equal det $\left(\int_{\delta_i}\omega_j\right)$ (the same for all p_j) and with the numerator equal to the determinant of the matrix $\int_{\delta_i}\tilde{\omega}_j$ obtained from the previous one by replacing the jth column by $\int_{\delta_i}\omega$; $\tilde{\omega}_j=\omega,\ \tilde{\omega}_l=\omega_l,\ l\neq j$.
The atyptical values t_i are of the four types:

(i) the critical values of isolated critical points of H in the finite plane $\mathbb{C}^2$;

(ii) the critical values of non-isolated critical points of H in $\mathbb{C}^2$;

(iii) the critical values of critical points at infinity (in $\mathbb{CP}^2$);

(iv) the zeroes of the determinant det $\int_{\delta_i}\omega_j$.

In Chapter 5 it was shown that the integrals are regular near the critical points of the type (i). Thus $p_j(t)$ are meromorphic near them, $p_j=q_j(t)/(t-t_j)^{K_j}$, where the power K_j depends on H and on the forms ω_i and q_j is holomorphic.
The same statement is true near the points of the type (iv).
For the points of the type (ii) we apply the resolution of the (non-isolated) singularities in the finite plane. In some local analytic coordinates $\tilde{x},\tilde{y}$, we obtain $H-t_j=\tilde{x}^p\tilde{y}^q$. If $(p,q)=gcd(p,q)=1$, then the Riemann surface $\tilde{x}^p\tilde{y}^q=t-t_j$ contains the vanishing cycle $\tau(t):\ \tilde{x}(\theta)=\epsilon e^{iq\theta},\ \tilde{y}=\epsilon e^{-ip\theta},\ 0\leq\theta\leq 2\pi,\ \epsilon=(t-t_j)^{1/(p+q)}$. It is easy to estimate the integrals along $\tau(t)$. If p and q are not relatively prime, then several such cycles vanish.
Near a smooth part E of the set of non-isolated critical points, i.e., where $H-t_j=\tilde{x}^p$, we do not have vanishing cycles. Here p local components of $H=t$ approach E and some part of a cycle δ_i may pass near E. The corresponding contributions to the integrals are easy to estimate.
As the example of the function $x(1-x(y^2+1))$ shows, the critical points of H may lie at the line $L_\infty=\{(x:y:0)\}\subset\mathbb{CP}^2$ (at infinity). We include also the value $t=\infty$ into the part (iii) of the set of critical values. The foliation of $\mathbb{C}^2$ into the level curves $H(x,y)=t$ is a holomorphic foliation which can be prolonged to a holomorphic foliation $\mathcal{F}$ in $\mathbb{CP}^2$ (see Chapter 9 below). Near a critical point $(x:y:z)=(x_0:y_0:0)$ at infinity, the function H is rational, equal to $F_m(u,z)/z^m,\ m=\deg H$ (u and z) are local projective coordinates). If the curves $H=t$ have only isolated (transversal) intersections with L_∞, then these points are singular points of the foliation $\mathcal{F}$ of the node type (see Chapter 9 below).
For the points with bad behavior of the $\mathcal{F}$ we apply the resolution process: by Theorem 9.18 in Chapter 9, such resolution exists. We obtain either $H-t_j=\tilde{x}^p\tilde{y}^q$ (a saddle point) or $H-t_j=\tilde{x}^pz^{-m}$ (a node). In the first case certain cycle vanishes. In the second case the local Riemann surfaces $H=t$ are punctured discs, with

one cycle $\sigma(t)$ generating its local first homology group. However, the cycle $\sigma(t)$ does not intersect the cycles which have nontrivial monodromy and are in some way associated with the real ovals of $H = t$. σ is monodromy invariant and the integral $\int_\sigma \eta$ equals to the residuum of the form η at the point $\tilde{x} = z = 0$.
The above analysis shows that we have to study the asymptotic of the integrals along those cycles $\delta(t)$, which become large as $t \to t_j$ or as $t \to \infty$. Here we can either use the general result about regularity of the Gauss–Manin connection (see Remark 5.16 above or Remark 8.19 below) or apply the a priori estimates (like in [**Yak**]). We recall the Yakovenko's arguments.
The cycles $\delta_i(t)$ can be chosen as lifts to the Riemann surface of the algebraic function $y(x)$ (defined by $H(x,y) = t$) of some loops $\gamma_i(t)$ in the x-plane, deprived of the ramification points x_k of the function $y(x)$. The points $x_k = x_k(t)$ vary regularly, with power type escape to infinity. This implies that the absolute values of the coordinates along the cycles $\delta_i(t)$ also have regular growth.
Because the forms ω and ω_i are polynomial, their integrals along the cycles $\delta_i(t)$ also have polynomial growth. The coefficient functions $p_i(t)$ (from the expansion $\int \omega = \sum p_i \int \omega_i$ in Eq. (6.3)) are expressed as ratios of determinants, which are regular. This implies that $p_i(t) = P_i(t)/P_0(t)^K$, where P_i are polynomials and $P_0 = \prod(t - t_i)$. It is also clear that the degrees of $P_1, P_2, \dots$ grow linearly with $n = \deg \omega$ and the exponent K is linear in n.

(b) So, the proposition is proved, but with constants depending on the Hamilton function H. We need some arguments, which would show uniformity of these constants.
The space of real polynomial Hamiltonians of degree $\leq d$ (and without the constant term) can be identified with a sphere $\mathcal{H} \subset \mathbb{R}^N$, $\mathcal{H} = S^{N-1}$, $N = (d+1)(d+2)/2-1$: the Hamilton functions H and λH lead to the same Abelian integrals. So, it is enough to show that the estimates are uniform with respect to H, locally in $\mathcal{H}$.
$\mathcal{H}$ is a real algebraic variety; its complex variant $\mathcal{H}^{\mathbb{C}}$ is a quasi-projective variety (a quadric). Consider the space $\mathcal{H}^{\mathbb{C}} \times \mathbb{C}$; (its points (H,t) correspond to algebraic curves $H(x,y) = t$). Let $\mathcal{M}^{\mathbb{C}} = \overline{\mathcal{H}^{\mathbb{C}}} \times \mathbb{CP}^1$ be the closure of this set; (here $\overline{\mathcal{H}^{\mathbb{C}}}$ is the projective closure). Denote $\Sigma_1^{\mathbb{C}} = \mathcal{M}^{\mathbb{C}} \setminus \mathcal{H}^{\mathbb{C}} \times \mathbb{C}$; it is a hypersurface in $\mathcal{M}^{\mathbb{C}}$. (We agree to denote complex varieties by the upper index $\mathbb{C}$, their real parts are denoted without this index).
The degenerate curves $H = t$ correspond to points from a bifurcational subset $\Sigma_2^{\mathbb{C}} \subset \mathcal{H}^{\mathbb{C}} \times \mathbb{C}$. $\Sigma_2^{\mathbb{C}}$ is a proper algebraic hypersurface. The restriction of the projection $\mathbb{C}^2 \times (\mathcal{H}^{\mathbb{C}} \times \mathbb{C}) \to \mathcal{H}^{\mathbb{C}} \times \mathbb{C}$ to $\{(x,y,H,t) : H(x,y) = t,\ (H,t) \notin \Sigma_2^{\mathbb{C}}\}$ is a locally trivial (Milnor) fibration over $\mathcal{M}^{\mathbb{C}} \setminus (\Sigma_1^{\mathbb{C}} \cup \Sigma_2^{\mathbb{C}})$.
The polynomial forms ω_i, $i = 1, \dots, k$, are defined as the same for all $H \in \mathcal{H}^{\mathbb{C}}$. For generic $(H,t) \in \mathcal{M}^{\mathbb{C}} \setminus (\Sigma_1^{\mathbb{C}} \cup \Sigma_2^{\mathbb{C}})$ the cohomology classes of the forms ω_i in $H^1(\{H = t\})$ form the basis of this space. The integrals $\int_{\delta_i(H,t)} \omega_j$ and $\int_{\delta_i(H,t)} \omega$ are well defined for such (H,t); because the cycles $\delta_i = \delta_i(H,t) \subset \{H = t\}$ are well defined there. Denote by $\Sigma_3^{\mathbb{C}} \subset \mathcal{M}^{\mathbb{C}} \setminus (\Sigma_1^{\mathbb{C}} \cup \Sigma_2^{\mathbb{C}})$ the set of those (H,t) for which $\det \int_{\delta_i(H,t)} \omega_j = 0$. It is analytic subset; (as we shall see, it is also algebraic

subset of $\mathcal{M}^{\mathbb{C}}$). Denote also $\Sigma^{\mathbb{C}} = \Sigma_1^{\mathbb{C}} \cup \Sigma_2^{\mathbb{C}} \cup \Sigma_3^{\mathbb{C}}$.
The integrals, treated as functions of (H, t), have singularities along $\Sigma_1^{\mathbb{C}} \cup \Sigma_2^{\mathbb{C}}$. These singularities are investigated using the resolution of singularities of this variety.
Let $\pi : \widetilde{\mathcal{M}}^{\mathbb{C}} \to \mathcal{M}^{\mathbb{C}}$ be the resolution of singularities of $\Sigma_1^{\mathbb{C}} \cup \Sigma_2^{\mathbb{C}}$.
Near any point of $\pi^{-1}(\Sigma_1^{\mathbb{C}} \cup \Sigma_2^{\mathbb{C}})$ there exists a local system of coordinates $z_1, \ldots, z_N$ such that $\Sigma_1^{\mathbb{C}} \cup \Sigma_2^{\mathbb{C}} = \{z_1 \cdot z_2 \cdot \ldots \cdot z_m = 0\}$. The integrals are treated as functions of $z = (z_1, \ldots, z_N)$.
They admit the following finite expansions

$$\sum_{\vec{k},\vec{\alpha}} a_{\vec{k},\vec{\alpha}}(z) \cdot z_1^{\alpha_1} (\ln z_1)^{k_1} \cdot \ldots \cdot z_m^{\alpha_m} (\ln z_m)^{k_m}, \tag{6.4}$$

where the (finite) sum runs over integer vectors $\vec{k} = (k_1, \ldots, k_m)$ and rational vectors $\vec{\alpha} = (\alpha_1, \ldots, \alpha_m)$. The exponents α_l are such that $\mathrm{e}^{2\pi \mathrm{i} \alpha_l}$ are eigenvalues of the monodromy operators M_l associated with the loops $\{z_l = \epsilon \mathrm{e}^{\mathrm{i}\theta},\ z_j = \text{const}\ (j \neq l)\}$ (around the hypersurfaces $\{z_l = 0\}$). The integers k_l take values 0 or 1. The coefficients $a_{\vec{k},\vec{\alpha}}(z)$ are analytic functions. The proof of this expansion is the same as the proof of Theorem 5.14 (with the use of the regularity of integrals).
The determinants $\det\left(\int_{\delta_i(t,H)} \omega_j\right)$ and $\det\left(\int_{\delta_i(t,H)} \tilde{\omega}_j\right)$ (where $\tilde{\omega_j} = \omega_j$ or $= \omega$) have simple monodromy properties. They remain fixed or change sign when the z turns around the divisor $z_l = 0$ (like in the proof of Theorem 5.25). They take the form $\sqrt{z_{l_1} \ldots z_{l_r}} \cdot \phi(z)$, where ϕ is meromorphic. This shows that $\Sigma_3^{\mathbb{C}} = \left\{\det\left(\int_{\delta_i(t,H)} \omega_j\right) = 0\right\}$ has algebraic singularities, and hence is algebraic.
The formula (6.3) is generalized as follows

$$\int_{\delta(t,H)} \omega = \sum p_j(t, H) \int_{\delta(t,H)} \omega_j$$

for $(H, t) \in \mathcal{M}^{\mathbb{C}} \setminus \Sigma^{\mathbb{C}}$. The functions $p_j(t, H)$ are holomorphic, single valued and regular. Thus, they are restrictions of rational functions on $\mathcal{M}^{\mathbb{C}}$, with poles at $\Sigma^{\mathbb{C}}$.
Take one function p_j. We restrict it to the real lines $L_H = \{(H, t),\ t \in \mathbb{R}\}$, $p_j|_{L_H} = p_j(\cdot, H)$. There are three possibilities:

(i) $p_j(\cdot, H) \equiv 0$,

(ii) $p_j(\cdot, H) \equiv \infty$,

(iii) $p_j(\cdot, H) = P_j(t)/Q_j(t)$ is a nontrivial rational function.

We must eliminate the case (ii). It corresponds to the situation when $\det\left(\int_{\delta_i} \omega_j\right)|_{L_H} \equiv 0$. We achieve the goal by imposing additional restrictions onto the forms ω_i.
We deal with two situations: (α) the line $L_H \not\subset \Sigma_2$ (the real part of $\Sigma_2^{\mathbb{C}}$) and (β) $L_H \subset \Sigma_2$.

In the case (α) we deal with irregular position of the component Σ_3 with respect to the fibers of the projection $\mathcal{H}\times\mathbb{R}\to\mathcal{H}:\ L_H\subset\Sigma_3$. It is clear that the opposite property, $L_H\not\subset\Sigma_3$, is an open condition with respect to the forms $\omega_1,\ldots,\omega_k$.
In the case (β) all the curves $\{H(x,y)=t,\ t\in\mathbb{R}\}$ are non-generic; (e.g., when $\deg H<d$). It is not surprising that there $\det\int_{\delta_i}\omega_j|_{L_H}\equiv 0$, but then also $\det\int_{\delta_i}\tilde{\omega_j}|_{L_H}\equiv 0$. Here $p_j(t,H)$ is understood as a limit of the ratio of determinants $\det\int_{\delta_i}\tilde{\omega}_j/\det\int_{\delta_i}\omega_j$ at points $(H',t')\in\mathcal{H}\times\mathbb{R}-(\Sigma_2\cup\Sigma_3)$, $(H',t')\to(H,t)$. This shows that we have to compare the asymptotic expansions of two determinants. The condition that the expansion of the $\det\int_{\delta_i}\omega_j$ is optimal (i.e., with smallest exponents) at a general point of the line $L_H\subset\Sigma_2$ is an open condition for $\omega_1,\ldots,\omega_k$.
Using the openness of the above conditions for the 1-forms and the compactness of the set of lines L_H, $H\in\mathcal{H}$, we see that there exist forms $\omega_1,\ldots,\omega_k$ such that $p_j|_{L_H}\neq\infty$.

(c) Let us pass to the estimation of the constant c from the proposition (in $\deg P_i\le cn$ and $Q=P_0^K$, $K\le cn$). It is clear that it is enough to get a local uniform estimate for the asymptotic as $n=\deg\omega\to\infty$ of the integral $\int_{\delta(t,H)}\omega$ with respect to ω and H. Here the forms ω belong to the compact space $\Omega=\{\sum_{i+j\le n}x^iy^j(a_{ij}\mathrm{d}x+b_{ij}\mathrm{d}y):a_{ij},b_{ij}\in\mathbb{R},\ \sum a_{ij}^2+b_{ij}^2=1\}$ and the Hamiltonians H belong to the compact space $\mathcal{H}$.
This uniform estimate follows from the expansion (6.4) (at points from $\Sigma_1\cup\Sigma_2$), where the exponents α_j are of order $O(n)$ and the coefficients $a_{\vec{k},\vec{\alpha}}(z)$ are uniformly bounded, locally with respect to z and globally with respect to $\omega\in\Omega$. The bound $|\alpha_j|<\mathrm{const}\cdot n$ is proved using the arguments from the point (a) above. We represent the cycles $\delta_i(t,H)$ as lifts (to the Riemann surface of the algebraic function $y(x)$ defined by $H(x,y)=t$) of loops $\gamma_i(t,H)$ in the x-plane deprived of the ramification points $x_j=x_j(t,H)$. The points $x_j=x_j(z)$ escape to infinity in a regular way as $z\to 0$. □

3. **The linear estimate when H has only real critical values.** It is enough to study the zeroes of the real polynomial envelope $P_1I_1+\ldots P_kI_k$, or of

$$g(t)=P_1+P_2\frac{I_2}{I_1}+\ldots+P_k\frac{I_k}{I_1}.$$

We apply the argument principle to the contour presented at Fig. 6.14. The increment of the argument along the large circle is bounded by $\mathrm{const}\cdot n$. Along the small semi-circles around the singular points t_i the increment of the argument is uniformly bounded; they are negatively oriented. The increment along the cuts is bounded by the number of zeroes of the imaginary part of g at the cuts.
The latter imaginary parts are proportional to the combinations

$$P_2K_2+\ldots+P_kK_k,$$

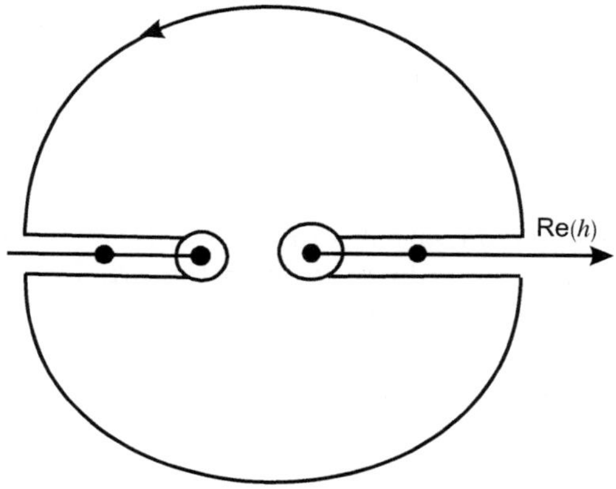

Figure 6.14: Contour for evaluation of the argument's increment

where K_j are expressed as some simple functions of integrals of ω_i along the cycles δ_j (i.e., we can put $K_j = \int_{\mathrm{var}\,\delta}\omega_j \cdot \int_\delta \omega_1 - \int_\delta \omega_j \cdot \int_{\mathrm{var}\,\delta}\omega_1$). This is done like in the proof of Theorem 6.24. Here we use the reality of I_j's.

Therefore, the problem is reduced to the estimation of the number of zeroes of a polynomial envelope of $k-1$ functions K_i which have regular singularities at the critical points of H and are real at an interval of the real axis.

We take the function $P_2 + P_3(K_3/K_2) + \ldots + P_k(K_k/K_2)$ and repeat the above estimation of its argument along contours as in Fig. 6.14.

After finite number of such steps we obtain the estimate $a(d)\cdot n + b(H)$ for the number of zeroes of the Abelian integral. Here the constant $b = b(H)$ is approximately equal to the number of zeroes (in $\mathbb{R}\setminus\{t_1, \ldots, t_r\}$) of an expression $A = A(t)$, which depends on the integrals $\int_{\delta_i}\omega_j$. A is a combination of products of such integrals. Here we find that b depends on H; in the points 5, 6, 7 below we shall show that b depends only on $d = \deg H$. $\square$

4. **The linear estimate in the general case.** Assume that H has non-real critical values.

We construct a polynomial F such that the composition $F \circ H$ has only real critical values. The function F is a composition of functions of the form $t \to (t-a_j)^2$, where $a_j \pm \mathrm{i}b_j$, $b_j \neq 0$, is a critical value of the function obtained in the previous step. Then the composed function acquires a real critical value $-b_j^2$. After finite number of steps we eliminate all non-real critical values.

Thus $F = F_m \circ \ldots \circ F_1$, $F_i(t) = (t-a_i)^2$. The values of the composition $F \circ H$ are denoted by s.

If $m = 1$, then we have $t = a_1 + \sqrt{s}$. If $m = 2$, then $t = a_1 + \sqrt{a_2 + \sqrt{s}}$. Generally, $t = a_1 + \sqrt{a_2 + \sqrt{a_3 + \ldots + \sqrt{s}}}$.

Denote $\Phi_m(s) = \sqrt{s}$, $\Phi_{m-1}(s) = \sqrt{a_{m-1} + \sqrt{s}}, \ldots, \Phi_1 = \sqrt{a_2 + \sqrt{a_3 \ldots + \sqrt{s}}}$ and $\tilde{I}_j(s) = I_j(t)$. The functions Φ_k and $\tilde{I}_j$ are multivalued holomorphic functions with singularities at the set of critical values of $F \circ H$. These singularities are regular (of power type).
Next, the polynomials P_j (from the proposition about rational envelopes) are expressed as polynomial combinations of products of the functions Φ_j.
This shows that the problem reduces itself to the problem of estimation of the number of zeroes of the combination

$$Q_1(s)J_1 + \ldots + Q_p(s)J_p,$$

where Q_j are polynomials of degrees $\leq$ const$\cdot n$ and J_i are multivalued holomorphic functions with regular real singularities.
For such combination one can apply the proof from the previous case.

5. **Towards an estimate of the constant** $b(H)$**.** Because any form of degree $n = 0$ is exact, its integral vanishes and $\#\{\text{zeroes}\}|_{n=0} = 0$. This means that we should show that $b(H)$ is uniformly bounded with respect to $H \in \mathcal{H}$, where $\mathcal{H}$ is the space of Hamiltonians of degree $\leq d$.
Here we follow the proof of existence of a general estimate $\#\{I_\omega = 0\} \leq C(d, n)$ given by A. Varchenko in [**Var3**]. The ingredients of the proof are following: the asymptotic expansion (6.4) (in its real form, corresponding the real resolution of a real algebraic variety), the Khovanski's theory [**Kh2**, **Kh3**] (of estimation of the number of solutions of a Pfaff system by the number of solutions of an analytic system) and the local uniform estimation of the number of solutions of analytic systems depending on parameters (theorem of Gabrielov [**Gab**]).
Recall that we have to estimate the number of zeroes of a function $A(s) = A(s; H)$, which is a combination of products of Abelian integrals (of forms ω_i) and of the functions Φ_j. Here the argument is s, where $s = (\ldots((t-a_1)^2 - a_2)^2 \ldots - a_m)^2 = F(t)$. This suggests that we have to replace the space $\{(H,t)\} = \mathcal{H} \times \mathbb{R}$ by the space $\mathcal{N}$ of pairs (H, s). So, we have the algebraic mapping $\Psi : \mathcal{M} = \mathcal{H} \times \mathbb{RP}^1 \to \mathcal{N}$ and a new "bifurcational" subset $\Lambda \subset \mathcal{N}$. Λ is the union of $\Psi(\Sigma_1 \cup \Sigma_2)$ and of the set of critical values of Ψ. The mapping Ψ can be extended to a complex holomorphic mapping from a neighborhood of $\mathcal{M}$ in $\mathcal{M}^{\mathbb{C}}$ to a neighborhood of $\mathcal{N}$ in its complex analogue $\mathcal{N}^{\mathbb{C}}$.
We apply the resolution of singularities of the real hypersurface Λ. Thus, we have a real manifold $\widetilde{\mathcal{N}}$ with a resolution mapping $\widetilde{\mathcal{N}} \to \mathcal{N}$. The function A can be treated as a function on $\widetilde{\mathcal{N}}$. Near points from $\widetilde{\Lambda}$ (the inverse images of Λ) A admits asymptotic expansion of the type (6.4).
We cover the (compact) variety $\widetilde{\mathcal{N}}$ by finitely many charts U_i, defined by analytic inequalities, such that either (i) $U_i \cap \widetilde{\Lambda} = \emptyset$, or (ii) A admits expansion (6.4) in U_i.
In charts of the type (i) the function $A(s; H)$ is real analytic.
In charts of the type (ii) we have (after eventual multiplication by a nonzero

function)

$$A = A_0(z_1, \ldots, z_N; (-\ln z_1)^{-1}, \ldots, (-\ln z_m)^{-1}; z_1^{\alpha_{11}}, \ldots, z_1^{\alpha_{1r_1}}; z_2^{\alpha_{21}}, \ldots, z_m^{\alpha_{mr_m}}).$$

Here the function $A_0(z_1, \ldots, z_N; v_1, \ldots, v_m; u_{11}, \ldots, u_{mr_m}) = A_0(z, v, u)$ is analytic in all variables for $z_1, \ldots, z_m > 0$ and small and for $z_{m+1}, \ldots, z_N$ small. Note that the quantities $v_i = (-\ln z_i)^{-1}$ and $u_{ij} = z_i^{\alpha_{ij}}$ are also positive and small; because α_{ij} are rational and > 0 (we can assume it). We will assume (for definiteness) that $z_i, v_i, u_{ij} \in (0, 1)$.
We have the analytic map $\Pi : U_i \to \mathcal{H}$. Our task is to estimate the cardinality of $\{A = 0\} \cap \Pi^{-1}(H)$ locally uniformly with respect to H, e.g., for $H \in V_i$, where $V_i = \{H \in \mathcal{H} : |H - H_0|^2 \leq \theta\}$ is a ball in $\mathcal{H}$. For the charts of the type (i) we have a problem of estimation of local solutions of an analytic equation depending analytically on parameters. In the next point we show that also in the charts of the type (ii) the problem can be reduced to an analytic problem of the same type.

6. **Separating solutions of Pfaff systems.** In this point the equation $A(s; H) = 0$ (shortly $A(s) = 0$) is treated as an equation for s depending on the parameter H. Also the variables z_i are treated as functions of s. The exterior derivative, denoted by d, means the derivative with respect to v, u, s with H fixed.
The equation $A(s) = 0$ can be rewritten in the form

$$A_0|_\Gamma = 0,$$

where Γ is a separating solution of the following Pfaff system (of M equations),

$$\begin{aligned} z_i \mathrm{d}v_i &= v_i^2 \mathrm{d}z_i, \\ z_i \mathrm{d}u_{ij} &= \alpha_{ij} u_{ij} \mathrm{d}z_i, \end{aligned} \tag{6.5}$$

(recall that $A(s) = A_0(z(s), v(s), u(s))$).
According to Khovanski, a **separating solution** of a Pfaff equation $\eta = 0$ in a real manifold U (where η is a 1-form) is a smooth hypersurface Γ (with the embedding $i : \Gamma \longmapsto U$) such that:

(a) $i^*\eta \equiv 0$ (i.e., Γ is an integral surface of the Pfaff equation);

(b) Γ does not pass through singular points of η (i.e., $\eta(x) \neq 0$ for $x \in \Gamma$) and

(c) Γ is the boundary of a domain $D \subset U$ and the coorientation of Γ by means of η coincides with its coorientation as the boundary ∂D.

A smooth submanifold $\Gamma \subset U$ of codimension l is a **separating solution** of a Pfaff system $\eta_1 = \ldots = \eta_l = 0$ iff there is a sequence of submanifolds $\Gamma = \Gamma_0 \subset \Gamma_1 \subset \ldots \subset \Gamma_l = U$ such that each Γ_{j-1} is a separating solution of $\eta_j|_{\Gamma_j}$ in Γ_j.
There holds the following fundamental statement:

Khovanski–Rolle lemma. *Let $\Gamma \subset U$ be a separating solution of a Pfaff equation $\eta = 0$ and let $\gamma \subset U$ be an oriented curve. Then between any two consecutive intersection points of γ with Γ there exists a point (in γ), where the tangent vector γ_* (of γ) lies in the hyperplane $\eta = 0$*

Proof. Assume that the intersections are transversal; if not, then $\langle \eta, \gamma_* \rangle = 0$ at one of the intersection points. In the consecutive points of $\gamma \cap \Gamma$ the values of η at γ_* have different signs. This implies that $\langle \eta, \gamma_* \rangle$ vanishes at some intermediary point. □

The Khovanski–Rolle lemma allows us to replace the non-analytic equation $A(s) = 0$ by a collection of systems of analytic equations of the type $A_0(v,u,s) = A_1(v,u,s) = \ldots = A_M(v,u,s) = 0$ in $U = (0,1)^M \times \{t : z_i(t) \in (0,1),\ i = 1,\ldots,N\}$, depending analytically on H.
Indeed, the formulas $v_i = (-\ln z_i)^{-1}$ and $u_{ij} = z_i^{\alpha_{ij}}$ define a 1-dimensional submanifold $\Gamma_0 \subset U$. It is a separating solution of the Pfaff system (6.5). To see this, it is enough to introduce some order in the system (6.5) and observe that each hypersurface $v_i = (-\ln z_i)^{-1}$ (or $u_{ij} = z_i^{\alpha_{ij}}$) is a boundary. Thus we have the sequence $\Gamma_0 \subset \Gamma_1 \subset \ldots \subset \Gamma_M = U$.
Take the surface Γ_1. It contains the separating solution Γ_0 of the Pfaff equation $\eta_1|_{\Gamma_1} = 0$ and the curve $\gamma_0 = \{A_0|_{\Gamma_1} = 0\}$. We are interested in estimation of the number (multiplicity counting) of intersection points $\gamma_0 \cap \Gamma_0$. On each compact connected component of γ_0 this number is equal to the number of contact points of this component with the field of directions $\eta_1 = 0$ (by Khovanski–Rolle Lemma). On each non-compact connected component of γ_0 this number is equal to 1 plus the number of contact points of this component with the field of directions $\eta_1 = 0$. The contact points are defined by the system of analytic equations in Γ_1: $A_0 = 0$, $\mathrm{d}A_0 \wedge \eta_1 = 0$ (or $A_0 = A_1 = 0$). We find that

$$\#\{A = 0\} = \#\{\text{non-compact components of } \gamma_0\} + \#\{A_0 = A_1 = 0\}.$$

The number of non-compact components can be estimated as follows. Following [**Kh2**] we introduce the *special bump function* $\chi(s,v,u)$ which is positive, analytic and vanishes at the boundary of U, e.g., $\chi = \prod z_i(1-z_i) \prod v_i(1-v_i) \prod u_{ij}(1-u_{ij})$. Take a small positive value ϵ_1, non-critical for χ. We have $\#(\Gamma_1 \cap \{\chi - \epsilon_1 = 0\}) \geq 2 \cdot \#\{\text{non-compact components}\}$.
Consider the 3-dimensional manifold Γ_2. It contains the separating solution Γ_1 of $\eta_2 = 0$ and the curves $A_0 = A_1 = 0$ and $A_0 = \chi - \epsilon_1 = 0$. As in the previous case we obtain that $\#\Gamma_1 \cap \{\text{curve}\} = \#\{\text{non-compact components of curve}\} + \#\{\text{contact points of curve}\}$. The number of contact points leads to a system of three analytic equations. The number of non-compact components is estimated using the special bump function χ and some non-critical value ϵ_2.
Repeating this algorithm finite number of times, we reduce the problem of estimation of the number of solutions of $A(s) = 0$ to the problem of estimations of the numbers of solutions of 2^M systems of analytic equations in U.
Recall also that the estimate should be locally uniform with respect to the parameter H, i.e., for $H \in V = V_i \subset \mathcal{H}$. It may occur (and occurs) that, for some H, the system has infinitely many solutions, the set of solutions becomes a variety of positive dimension. Because (in the Hilbert's problem) we are interested in the

isolated periodic trajectories of a vector field, here we are interested in the *isolated* solutions of the systems of analytic equations.
We can formulate the problem as follows. Consider a real semi-analytic set $W = \{((z,v,u);H) : (z,v,u) \in U,\ H \in V,\ z = z(s,H),\ A_0(z,v,u) = A_1 = \ldots = A_M = 0\} \subset U \times V$. We have the projection $\Pi : W \to V$, the restriction of the projection onto the second factor. The problem is:

Show that there exists a constant C such that the number of isolated points in $\Pi^{-1}(H)$ is $\leq C$ for any $H \in V$.

Its positive answer, given in Theorem 6.28 below, completes the proof of Theorem 6.26. □

6.27. Theorem of Binyamini–Novikov–Yakovenko (Double exponential bound). *For any polynomial H of degree $\leq n+1$ and any polynomial 1-form ω of degree $\leq n$, the sum $N(H,\omega)$ of the numbers of isolated zeroes (with multiplicities) of the Abelian integrals $\int \omega$ over all ovals γ_t of the curve $\{H = t\}$ satisfies*

$$N(H,\omega) \leq 2^{2^{\mathrm{Poly}(n)}},$$

where $\mathrm{Poly}(n) = O(n^p)$ *is an explicit polynomial of degree $p \leq 61$.*

Sketch of the proof of the Binyamini–Novikov–Yakovenko theorem. [**BNY**] Here we present some tools used in this proof.

(a) Recall that, in the proof of the Khovanski–Petrov theorem 6.26, Lemma 2 about rational envelopes allowed to represent the integral $I_\omega(t) = \int \omega$, along an oval γ, as a finite combination of basic integrals I_{ω_j} with coefficients being rational functions $P_j(t)/P_0(t)^K$ of t. (The forms ω_j constitute a basis of the so-called Petrov module.) Then one estimated the number of zeroes of $P_1 I_{\omega_1} + \ldots + P_k I_{\omega_k}$ in suitable real interval in $\mathbb{R}$, or rather of $g(t) = P_1 + P_2 I_{\omega_2}/I_{\omega_1} + \ldots + P_k I_{\omega_k}/I_{\omega_1}$, by estimating the increment of the argument of the function $g(t)$ along a contour presented in Fig. 6.14 (the Petrov's trick). This allows to reduce the number k of summands, by replacing the ratios $I_{\omega_j}/I_{\omega_1}$ by some more complicated functions.
In [**BNY**] also the increment of argument $\Delta \arg$ is used, but to a solution $y(t)$ of differential equation

$$a_0(t,\lambda)y^{(k)} + \ldots + a_k(t,\lambda)y = 0,\ t \in \mathbb{C},\ \lambda \in \mathbb{C}^m. \tag{6.6}$$

The classical de la Valleé Poussin theorem [**dlVP**] asserts that, when $a_0 \equiv 1$and a_j, $j > 0$, are analytic, then

$$\Delta_\Gamma \arg y(t),$$

Γ – a circular arc, is estimated in terms of the length $|\Gamma|$ and $\sup_\Gamma |a_j|$. In the case of Eq. (6.6) the situation is more complicated.

(b) Before deriving Eq. (6.6) we take a broader look at the situation. We consider complex curves $\Gamma = \Gamma_\lambda \subset \mathbb{C}^2$ defined by

$$H(x,y;\lambda) = 0,$$

where H is a polynomial in x, y of degree $\leq n+1$,with $\lambda \in \mathbb{CP}^m$ as the vector of the coefficients.
For generic λ the first homology group of Γ_λ has $\ell = n^2$ generators $\delta_j = \delta_j(\lambda)$ and the 1-forms $\eta_\alpha = x^{\alpha_1+1}y^{\alpha_2}\mathrm{d}y$, $0 \leq \alpha_{1,2} \leq n-1$ generate the first cohomology group of Γ_λ (denote them by ω_k). We have the period matrix

$$X = X(\lambda) = \left(\int_{\delta_j} \omega_k \right).$$

It satisfied the Pfaff system (an analogue of the Picard–Fuchs equations)

$$\mathrm{d}X = \Omega X, \tag{6.7}$$

where $\Omega(\lambda)$ is a $\ell \times \ell-$ matrix with entries being 1-forms. It is well defined in $\mathbb{CP}^m \diagdown \Sigma$, where Σ is a bifurcational hypersurface. The proof (which we skip) uses the Gelfand–Leray formula for derivative $(\partial/\partial\lambda_\alpha)\int \eta_\beta = \int x^{\alpha_1}y^{\alpha_2}\mathrm{d}\eta_\beta/\mathrm{d}H$ and the representation

$$x^{\alpha_1}y^{\alpha_2}\mathrm{d}\eta_\beta = \sum_\gamma R^\gamma_{\alpha,\beta}(H)\eta_\gamma + u_{\alpha,\beta}\mathrm{d}H + \mathrm{d}v_{\alpha,\beta},$$

where $R^\gamma_{\alpha,\beta}$ and $u_{\alpha,\beta}(x,y)$, $v_{\alpha,\beta}(x,y)$ are suitable polynomials.
The matrix Ω has some important properties:

- The system (6.7) is integrable, in the sense that curvature of the corresponding connection $\mathrm{d}-\Omega$ vanishes, $\mathrm{d}\Omega - \Omega\wedge\Omega = 0$. This is the general property of the Gauss–Manin connection; compare Definition 5.34 and the remark after it.
- Ω is regular.
- Ω is *quasi-unipotent* in the sense that the eigenvalues of corresponding monodromy matrices are roots of 1; these eigenvalues are ± 1, in fact.
- Ω is defined over $\mathbb{Q}$ with an estimate for the so-called *size*, $s \leq 2^{\mathrm{Poly}(n)}$.

The *size* of a rational function P/Q, $P, Q \in \mathbb{Z}[z_1, \ldots, z_r]$, is roughly $\|P\| + \|Q\|$, where $\|P\|$ is the sum of absolute values of tits coefficients. This allows to define the size of Ω (we skip the detailed definition).

(c) For a given line $\mathbb{CP}^1 \subset \mathbb{CP}^m$ one has a birational equivalence of $\mathbb{CP}^m$ with $\mathbb{CP}^1 \times \mathbb{CP}^{m-1} = \{(t, \hat{\lambda})\}$. Restriction of the system (6.7) to $\mathbb{CP}^1 \times \{\hat{\lambda}\}$ gives a system of linear differential equations $\frac{\mathrm{d}}{\mathrm{d}t}X = A(t;\hat{\lambda})X$, which leads to the equation (6.6).
The problem with Eq. (6.6) is that the leading coefficient can vanish; not only for isolated values of t, but also identically for some values of $\hat{\lambda} \in \mathcal{S}$ (like in the so-called singular perturbation).

Below we will use the notation λ, instead of $\hat{\lambda}$.
As the authors of [BNY] write: 'Somewhat miraculously, the singular perturbation that occurs in the reduction of a regular system to the derived equation, is only apparent: together with the leading coefficient, all other coefficients of the equation (6.6) necessarily vanish for the same values of the parameter'.

(d) One defines the so-called *slope*

$$\max_j \sup_{\lambda \notin \mathcal{S}} \|a_j(\cdot;\lambda)\| \,/\, \|a_0(\cdot;\lambda)\|\,,$$

where the norm $\|\cdot\|$ is the same as in the above definition of the size. We note that the finiteness of this uniform bound depends crucially on the regularity of the original system.
The de la Valleé Poussin technique provide a uniform bound for the number of zeros of solutions long as the singular points remain well apart. A substantial difficulty which needs to be addressed (and cannot be circumvented for general Fuchsian systems) is the study of colliding singular points. However, under the regularity and integrability assumptions, this problem can be resolved. It turns out that for Fuchsian equations (operators) the slope remains uniformly bounded from above. This fact, somewhat surprising, means that the collision of singularities can be treated by a proper resolution of singularities.
The authors introduce the notion of the invariant slope of a linear operator, and subsequently show that it is finite and uniformly bounded.
Next, the variation of argument of any solution can be explicitly majorized along any arc (circular or rectilinear), which does not pass near the singular points where the leading coefficient vanishes. The meaning of this dangerous proximity can be made precise using the scale invariance of the invariant slope: the upper bounds for the variation of arguments of an arbitrary solution of the differential equation are possible for arcs whose length is not very large relative to the distance to the singular locus. This scale invariance is one of the key tools in the subsequent construction. For brevity they refer to such arcs as "short arcs". For instance, any sufficiently small circular arc centered at an isolated singular point, is "short" in this sense.
The variation of argument of any nonzero solution of the equation (6.6) along an arc or segment γ is explicitly bounded,

$$\Delta_\gamma \arg y(t) \leq k L^{O(d)} \cdot \measuredangle D,$$

where k is the order of the equation, $L \sim |\gamma| / \mathrm{dist}(\gamma, \mathcal{S})$ is the normalized length of the arc and $\measuredangle D$ is the so-called invariant slope of the corresponding differential operator.

(e) One has the bounds

$$\measuredangle D \leq s^{(d\ell)^{O(m)}}$$

and

$$L \leq s^{2^{O(\ell^4 dm)5}},$$

where s is the size s of Ω (as roughly defined above), m is the number of parameters and ℓ is the dimension of the Picard–Fuchs system. Moreover, $k = \ell$.
They obtained following bounds for s, d, m and ℓ:

$$s \leq 2^{\mathrm{Poly}(n)}, \; d \leq O(n^2), \; m \leq O(n^2), \; \ell = n^2.$$

Because the number of zeroes of the Abelian integral is estimated by the variation of argument of solution of suitable equation (6.6), the bound from Theorem 6.27 follows. □

6.28. Theorem of Gabrielov. ([**Gab**]) *Let* $U = [0,1]^m \subset \mathbb{R}^m$ *and* $V = [0,1]^n \subset \mathbb{R}^n$, *let* $W \subset U \times V$ *be a semi-algebraic set and let* $\Pi : W \to V$ *be the restriction to* W *of the projection onto the second factor. Then there exists a constant* C *such that, for any* $y \in V$, *the number of connected components of* $\Pi^{-1}(y)$ *is bounded by* C.

6.29. Remarks. The theorem is a result from the real analytic geometry. Recall that, by definition, a subset $W \subset \mathbb{R}^k$ is (real) **analytic** iff near any point $z_0 \in \mathbb{R}^k$ (not only in W!) it is defined by a system of equations analytic in a neighborhood of z_0. W is (real) **semi-analytic** iff near any point $z_0 \in \mathbb{R}^n$ it is a finite union of subsets defined by finite systems of analytic equations and analytic inequalities. Thus, the above sets U, V, W are real semi-analytic. When the equations (or/and inequalities) are algebraic, then we have *real (semi-)algebraic sets.* The real semi-algebraic sets form real analogies of the complex quasi-projective varieties (i.e., Zariski open subsets of projective algebraic varieties) and of the complex constructible sets (i.e., finite unions of sets which are Zariski open in their closures).
The complex analytic, algebraic and quasi-projective varieties have some nice natural properties. Complex analytic variety has locally finitely many components. The image $f(W)$ of a complex analytic variety $W \subset \mathbb{C}^k$, under a *proper* analytic map f, is an analytic set (theorem of Remmert, see [**GH**]). The equivalent formulation of this statement says that a proper projection of a complex analytic set is an analytic set. Here the assumption that f is proper (i.e., that the inverse images of compact subsets are compact) is essential: for example, the image of the hyperbola $xy = 1$ onto the y-axis equals to $\mathbb{C}^*$. Also the existence of a local uniform bound for the number of connected components of $f^{-1}(y)$ is easy in the complex case.
In the real case the situation is not that clear. Real (semi-)analytic subsets $W \subset \mathbb{R}^k$ have finitely many local (e.g., in a cube) connected components (see [**Loj1**]). In fact, one can prove this property following the lines of the proof of Lemma 4.2 in Chapter 4. It was shown there that, if a real algebraic variety (or real analytic variety) has a sequence of points accumulating at z_0, then it contains a semi-analytic curve through z_0. Real semi-analytic sets can be stratified; they form a CW-complexes with some additional differential properties.

A. Seidenbergand A. Tarski proved that a projection (or equivalently an image under real algebraic map) of a semi-algebraic set is semi-algebraic (see [**Loj1**]). In the proof they used methods from the mathematical logic, but there is an analytic proof based on algebraic functions.
The Bezout theorem (about estimation of the number of solutions of a system of complex algebraic equations by the product of degrees) does not hold in the real case; here is the example:

$$x = y = x^2 + y^2 + [z(z^2-1)]^2 + [t(t^2-1)]^2 = 0.$$

The example with the analytic set

$$W = \{(x,y,z,u,s,t) : x^2+y^2+z^2+u^2 = 1,\ s^2+t^2 = 1,\\ z = x \cdot \exp\left[s^2/(s^2+2t^2)\right], tx = sy\} \subset \mathbb{R}^4 \times \mathbb{R}^2$$

with its projection f onto $\mathbb{R}^4$, equal to $f(W) = \{z = x \cdot \exp\left[x^2/(x^2+2y^2)\right],\ x^2 + y^2 + z^2 + u^2 = 1\}$ (which is not analytic at $(0,0,0,1)$), shows that proper projections of semi-analytic sets can be not semi-analytic. This may occur only when $\dim f(W) > 2$. The subsets of $\mathbb{R}^n$, which are projections of relatively compact semi-analytic sets are called the *$\mathcal{P}$-sets,* or the *sub-analytic sets*. Here the assumption of relative compactness replaces the assumption of properness. It is useful here to have in mind the following example, with the (not relatively compact) analytic set $W = \{(x,y,z) : z(x^2+y^2) = 1,\ |x+\mathrm{i}y| = \arg(x+\mathrm{i}y)\} \subset \mathbb{R}^3 = \mathbb{R}^2 \times \mathbb{R}$, whose projection is the logarithmic spiral.
The sub-analytic sets have following properties. Their complements are sub-analytic sets. Near any their point they are defined by a system of analytic equations and inequalities. They are locally arcwise connected (by means of semi-analytic curves). There holds the sub-analytic analogue of the Gabrielov's theorem (it is in fact the original theorem of Gabrielov): if $Z \subset [0,1]^m \times [0,1]^n$ is a sub-analytic set and Π is the restriction (to Z) of the projection onto the second factor, then the numbers of connected components of the fibers $\Pi^{-1}(y)$ are uniformly bounded.

6.30. Sketch of the proof of the Gabrielov's theorem. Because the original proof Gabrielov's proof concerns the general case of projection of a $\mathcal{P}$-set and is rather technical, we provide an independent proof. We will deal with restricted assumptions (for simplicity):
We have an analytic subset W of an open bounded domain $U \subset \mathbb{R}^{m+n} = \{(x,y)\}$ such that the restriction Π of the projection onto $\mathbb{R}^n = \{y\}$ has the fibers: either finite, or ≥ 1-dimensional; but the typical fiber is finite (maybe empty). We assume also that the variable x is 1-dimensional (i.e., $m = 1$); later we will say how work in the general case. We denote the fibers by $W_y = \Pi^{-1}(y)$.
Assume that the thesis of the Gabrielov's theorem does not hold. It means that there exists a sequence of points $y_k \to y_*$ (which we put $= 0$) such that the fibers W_{y_k} contain growing number of bounded isolated points x_{kj}. Assume that a point (x_*, y_*) (which we put $= (0,0)$) is an accumulation point of the set $\{(x_{kj}, y_k)\}$.

Let $f_1(x,y),\ldots,f_r(x,y)$ be the generators of the ideal (in the local ring $\mathcal{O}_0(\mathbb{R}^{n+1})$) of germs of functions vanishing at W. If some of the functions f_j is such that its restriction to the line $y=0$ is of finite multiplicity, $f_j(x,0)=ax^d+\ldots$, then any fiber W_y does not contain more than d points (by the Weierstrass theorem).
Assume then that all $f_j(x,0)\equiv 0$; in other words, the line $\{y=0\}\subset W$. Here the essence of the theorem lies. When we treat f_j as functions of x depending on the parameter y, then for $y=0$ the point $x=0$ is singular of infinite codimension. In the smooth (i.e., C^∞) case unbounded number of singular points could be born after perturbation, i.e., for $y\neq 0$. In the analytic case this number turns out to be bounded. We shall meet this phenomenon in the next section (see also [**FY**]).
Let $Z=\Pi(W)\subset\mathbb{R}^n$. It is a sub-analytic set (maybe not semi-analytic) containing the origin $y=0$. We have a mapping $\Pi: W\to Z$ between sets of the same dimension. The cardinality of the fibers changes. The changes of the cardinality of $\Pi^{-1}(y)$ occur due to bifurcations. Such bifurcation points are the critical points the restriction of the projection Π to the corresponding stratum of the stratification of W (the fold singularity).
These critical points are obtained by differentiation of the generators f_j with respect to x. Indeed, if (locally outside $y=0$) W is represented as $F(x;y)=0$, $F:\mathbb{R}\times Z\to\mathbb{R}$, then the Rolle principle says that we should calculate the zeroes of $\partial_x F$.
The system of equations $F=\partial_x F=0$ defines a subvariety $W_1\subset W$, which can be defined in an analytic way. W_1 is an analytic variety. It contains the line $y=0$. Let $Z_1=\Pi(W_1)\subset\mathbb{R}^n$ and $\widetilde{W}_1=\Pi^{-1}(Z_1)\subset W$. The latter are sub-analytic sets (in general). The restriction $\Pi_1=\Pi|_{\widetilde{W}_1}:\widetilde{W}_1\to Z_1$ is a mapping with the same properties as $\Pi: W\to Z$. The cardinalities of fibers are unbounded.
We proceed as before, we look for the critical points of Π_1. It is not difficult to see that they form an analytic subset $W_2\subset W_1$, defined by means of $f_j,\partial_x f_j$ and $\partial_x^2 f_j$. We put $Z_2=\Pi(W_2)$, $\widetilde{W}_2=\Pi^{-1}(Z_2)$. The mapping $\Pi|_{\widetilde{W}_2}$ has the same properties as $\Pi|_{\widetilde{W}_1}$ etc.
Due to the analyticity of W, the sequence of varieties W_j stabilizes, $W_p=W_{p+1}=\ldots=W_\infty$ (because of the dimension argument); assume that p is minimal such integer. The generators of the ideal of functions vanishing at W_∞ have the property that all their derivatives with respect to x are equal to zero; they depend only on y. Thus $W_\infty=\mathbb{R}\times Z_\infty$.
For any $y\in Z\setminus Z_\infty$ the maximal order of zero of the function $F(\cdot,y)$ (where $W=\{F=0\}$) is $p-1$. Thus $\partial_x^{p-1}F\neq 0$ near $x=0$ and there are no more than $p-1$ isolated zeroes of F.
For example, if $n=2$ and $W=\{F=y_1g_1(x,y)+y_2g_2(x,y)=0\}$, then we get $W_1=\{y=0\}\cup\{F=\Delta=0\}$, $\Delta=g_1g_{2x}-g_2g_{1x}$. Assume that the surface $\Delta=0$ intersects the line $y=0$ transversally; then we have $W_\infty=\{y=0\}$ and the curve $\{F=\Delta=0\}$ is the curve of non-degenerate fold points. Any fiber W_y, $y\neq 0$, contains at most two points.

The general case of multidimensional $x=(x_1,\ldots,x_m)$ can be treated (in principle) in the same way. We take the generators $f_1, \ldots, f_r$, $r \geq m$, of the ideal of functions vanishing on W. If the restrictions to the plane $y = 0$ of some m of the f_j's form a vector field of finite multiplicity, then we get the bound for cardinalities of the fibers. Otherwise, we look for the critical points of the map $\Pi : W \to Z = \Pi(W)$. These critical points form an analytic subset W_1, etc. □

§3 Quadratic Centers and Bautin's Theorem

The correspondence between zeroes of Abelian integrals and limit cycles of perturbations of Hamiltonian system is not one-to-one. In particular, existence of a uniform bound for the number of zeroes of Abelian integrals does not lead automatically to a uniform bound for the number of limit cycles. As an example, where there can be more limit cycles than zeroes of the Abelian integrals, we consider the case of quadratic perturbation of the linear center.

Consider the perturbation

$$\dot{x} = -y + \epsilon P, \ \dot{y} = x + \epsilon Q,$$

where P and Q are quadratic polynomials. Without loss of generality we can assume that the point $(0,0)$ remains fixed during the perturbation and that the linear part has the canonical form with the complex eigenvalues $\lambda \pm \mathrm{i}$, $\lambda = \mathrm{const} \cdot \epsilon$. Then the corresponding Abelian integral reduces to

$$\epsilon I = \lambda \int y \mathrm{d}x - x \mathrm{d}y = 2\lambda \times \text{area}$$

and, of course, $I \neq 0$. However, we have the following result.

6.31. Theorem of Bautin. ([**Bau**]) *The maximal number of limit cycles appearing after the above perturbation is equal to* 3.

Proof. In order to calculate the number of all limit cycles, we have to calculate the Poincaré return map up to a sufficiently high exactness

$$\Delta H = \epsilon I + \epsilon^2 I_2 + \ldots$$

In order to free ourselves of the parameter ϵ, we make the normalization $\epsilon z \to z$. Then the cycles of order $O(1)$ become small cycles, of order $O(\epsilon)$ and the system written in the complex form becomes following

$$\dot{z} = (\mathrm{i} + \lambda)z + Az^2 + Bz\bar{z} + C\bar{z}^2, \quad z = x + \mathrm{i}y,$$

where A, B, C are complex parameters.
Because the real parameter λ controls the Abelian integral, in the further calculations of the return map we put $\lambda = 0$.

When $\Delta H \equiv 0$, then all the trajectories of the system near $z = 0$ are closed. In particular, the system has first integral. As H. Poincaré and A. Lyapunov have shown (using a priori estimates) this first integral can be chosen as an analytic function of x, y (or $z, \bar{z}$); see also Theorem 11.51(a) below. We try to find this integral.
Assume that it is of the form

$$F = z\bar{z} + F_3(z, \bar{z}) + \ldots = \sum a_{mn} z^n \bar{z}^n,$$

where F_j are homogeneous parts of degree j. We solve step-by-step the equation

$$\dot{F} = \frac{\partial F}{\partial z}(\mathrm{i}z + Az^2 + Bz\bar{z} + C\bar{z}^2) + \frac{\partial F}{\partial \bar{z}}(-\mathrm{i}\bar{z} + \overline{A}\bar{z}^2 + \overline{B}z\bar{z} + \overline{C}z^2) = 0$$

which, rewritten in the homogeneous parts, gives

$$\mathrm{i}(z\partial_z - \bar{z}\partial_{\bar{z}})F_k + [\frac{\partial F_{k-1}}{\partial z}(Az^2 + Bz\bar{z} + C\bar{z}^2) + \text{conjugate}] = 0.$$

The latter equation gives the recursive equations for the coefficients

$$\mathrm{i}(m - n)a_{mn} + (\text{terms calculated earlier}) = 0.$$

We see the obstacle when $m = n$. The terms $(z\bar{z})^m$ cannot be cancelled. This means that, if $\lambda = 0$, then

$$\dot{F} = g_1|z|^4 + g_2|z|^6 + \ldots$$

6.32. Definition. The numbers g_j are called the **Poincaré–Lyapunov focus numbers** and the function F is called the **Lyapunov function.**

The quantities g_j are polynomials of $A, \overline{A}, B, \overline{B}, C, \overline{C}$ of degree j. Moreover, they are invariant with respect to the action of the rotations group $A \to e^{\mathrm{i}\theta}A$, $B \to e^{-\mathrm{i}\theta}B$, $C \to e^{-3\mathrm{i}\theta}C$, induced by the rotations $z \to e^{\mathrm{i}\theta}z$. Because the first integral (or the Lyapunov function) is not defined uniquely (one can add cF^k to F) also the focus quantities are not uniquely defined (one can add fg_k, $k < j$, to g_j). Well defined is the series of ideals (g_2), $(g_2, g_4), \ldots$ in the ring $\mathbb{C}[A, \overline{A}, B, \ldots]$. This series stabilizes (the Hilbert's theorem on basis) and the final ideal is called the **Bautin ideal.**
If F is not first integral, then it can be non-analytic.
The focus quantities are used to express the Taylor expansion of the Poincaré return map $\mathcal{P} : S \to S$, where S is the positive half-line parameterized by r. We can also parameterize S by some high order jet of the Lyapunov function, $\Phi = j^N F = r^2 + \ldots$. We have

$$\Delta r \approx \frac{\mathrm{d}r}{\mathrm{d}\Phi}\Delta\Phi \approx \frac{1}{2r}\int_0^{2\pi} \dot{F}\mathrm{d}t \approx \pi g_k r^{2k+1},$$

if g_k is the first nonzero focus quantity.
In the general situation, the condition for limit cycles is following

$$\frac{1}{\pi}(\mathcal{P}(r)-r)=g_0 r(1+\ldots)+g_1 r^3(1+\ldots)+\ldots+g_M r^{2M+1}(1+\ldots)=0,$$

where $g_0=(\mathrm{e}^{2\pi\lambda}-1)/\pi$ and $g_1,\ldots,g_M$ generate the Bautin ideal.
If all $g_j=0$, then the point $=0$ is a center. Thus the system of algebraic equations $\lambda=g_1=\ldots=g_M=0$ defines an algebraic variety, called the **center variety**.

6.33. Theorem of Dulac and Kapteyn. ([**Dul2**], [**Kapt**]) *The Bautin ideal is generated by the first three focus quantities which take the form*

$$\begin{aligned}
g_1 &= -2\mathrm{Im}AB,\\
g_2 &= -(2/3)\mathrm{Im}[(2A+\overline{B})(A-2\overline{B})\overline{B}C],\\
g_3 &= -(5/4)\mathrm{Im}[(|B|^2-|C|^2)(2A+\overline{B})\overline{B}^2C],
\end{aligned}$$

and the center variety consists of the following four irreducible components

$$\begin{aligned}
Q_3^{\mathrm{H}} &: 2A+\overline{B}=0,\\
Q_3^{\mathrm{R}} &: \mathrm{Im}AB=\mathrm{Im}\overline{B}^3C=\mathrm{Im}A^3C=0,\\
Q_3^{\mathrm{LV}} &: B=0,\\
Q_4 &: A-2\overline{B}=|B|-|C|=0.
\end{aligned}$$

Proof of sufficiency of the center conditions. We shall not calculate the focus quantities. We shall restrict ourselves to showing that they really describe systems with center at $z=0$. In the notations $Q_i^\sharp$ (of the components of the center variety) the lower index denotes their codimension in the space of all quadratic systems.
The Hamiltonian systems. Because $\mathrm{Re}\partial\dot{z}/\partial z=\mathrm{Re}(2A+\overline{B})z^2$, the component Q_3^H consists of Hamiltonian systems.
The Lotka–Volterra systems. One can check that in the case Q_3^{LV} the system has three invariant lines $l_1(z,\bar{z})=0, l_2=0, l_3=0$, not passing through the origin. This means that $\dot{l}_i=l_i g_i$, where g_i are linear homogeneous functions. These g_i are linearly dependent, $\alpha_1 g_1+\alpha_2 g_2+\alpha_3 g_3=0$. This implies that the following Darbouxian function $l_1^{\alpha_1}l_2^{\alpha_2}l_3^{\alpha_3}$ is the first integral.
The reversible systems. In the case Q_3^{R} we can apply a rotation of the variable z such that the coefficients A,B,C become real. Rewriting the differential equation in the real variables, we obtain

$$\dot{x}=-y+ax^2+by^2,\quad \dot{y}=x+cxy.$$

This system is invariant with respect to the transformation : $(x,y,t)\to(-x,y,-t)$ (it is *time-reversible*). This implies existence of a center.

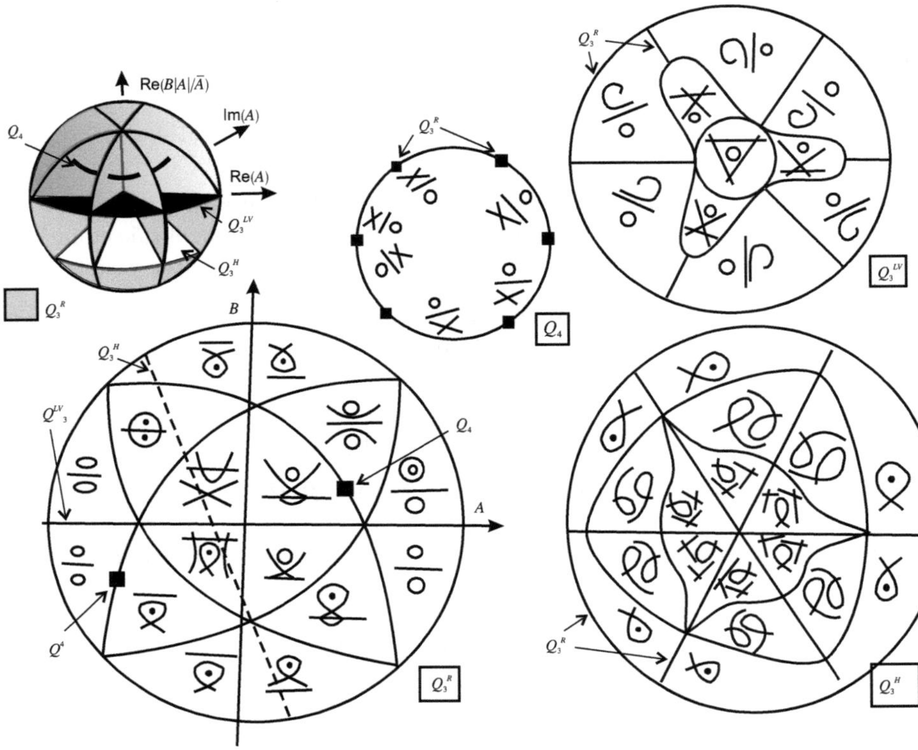

Figure 6.15: Bifurcation diagram for quadratic centers

The codimension four center. In the case Q_4 one can reduce the equation to the following one

$$\dot{z} = \mathrm{i}z + 2z^2 + |z|^2 + \mathrm{e}^{\mathrm{i}\xi}\bar{z}^2.$$

One can check that this system has the invariant conic $f_2 = x_1^2 + 4y + 1 = 0,\ x_1 = 2\mathrm{Im}(\mathrm{e}^{-\mathrm{i}\xi/2}z)$, the invariant cubic $f_3 = \cos(\xi/2)(x_1^3+6x_1y)+6y+1$ and the following first integral f_2^3/f_3^2. The bifurcational diagram and phase portraits are presented in Fig. 6.15.
For more details see [**Zo3**]. □

Proof of Bautin's theorem. Recall that the focus quantities g_j are polynomials of the coefficients and are invariant with respect to the action of the group $\mathbb{S}^1$. We consider the ring $\mathcal{R}$ of real polynomials of $A, \overline{A}, \ldots, \overline{C}$, which are invariant with respect to the action of $\mathbb{S}^1$. The following algebraic result is proved in [**Zo3**].

6.34. Proposition. *If $f \in \mathcal{R}$ vanishes at the center variety, then it takes the form $f = f_1g_1 + f_2g_2 + f_3g_3$. In other words, the Bautin ideal is radical in the ring $\mathcal{R}$.*

Because all the focus numbers vanish at the center variety, then, by Proposition 6.31, they all belong to the ideal generated by first three of them. This allows to write down the Poincaré map in the finite expansion

$$\Delta r = g_0 r(1+\ldots) + \ldots + g_3 r^7(1+\ldots).$$

Next we apply the derivation-division algorithm of Bautin. We divide Δr by $r(1+\ldots)$ and apply the Rolle's principle to it; we need to estimate the number of zeroes of the derivative, equal to $2g_1 r(1+\ldots)+\ldots$. We divide the latter expression by $r(1+\ldots)$ and again apply the Rolle's principle. At the end we arrive to the estimation the number of zeroes of $48g_3+\ldots$.
We see that we get at most 3 small zeroes. Moreover, because the coefficients $g_{0,1,2,3}$ are independent functions of parameters, they can vary independently and we can construct an example with three zeroes of Δr. □

6.35. Remark. The analogues of the Dulac–Kapteyn and Bautin theorems hold in the case of systems with cubic homogeneous nonlinearity

$$\dot z = \mathrm{i}z + Dz^3 + Ez^2\bar z + Fz\bar z^2 + G\bar z^3.$$

Here we have there cases of center: $\mathrm{Re}E = 3D + \overline{F} = 0$ (the Hamiltonian case), $\mathrm{Re}E = \mathrm{Im}DF = \mathrm{Re}D^2G = 0$ (the reversible case) and $E = D - 3\bar F = |G| - 2|F| = 0$ (the Malkin's case). The maximal number of small limit cycles is 5. These results were proved by K. Sibirski [**Sibi**] (see also [**Zo4**]).
In the general case of polynomial vector fields of degree n, the problem of center and the problem of small limit cycles are not solved.

6.36. Generalizations of Abelian integrals. 1. When the Abelian integral $I_\omega(t) = \int_{\gamma(t)}\omega$ vanishes identically (as a function of t), but the return map is not the identity, then we have

$$\Delta(t) = \varepsilon^k M_k(t) + O(\varepsilon^{k+1}),$$

where $k > 1$ and $M_k \not\equiv 0$. The function $M_k(t)$ is called the *higher order Melnikov integral* and can be written in form of a multiple integral of some polynomial k-form.
Often M_k can be reduced to the usual Abelian integral. This holds when

$$\omega = f\mathrm{d}H + \mathrm{d}g \quad \text{whenever} \quad I_\omega \equiv 0;$$

here f and g are some polynomials. Such situation takes place for a generic Hamiltonian and for the elliptic Hamiltonian (see also Remark 5.21). Then the Pfaff equation $\mathrm{d}H + \varepsilon\omega = 0$ (for the phase curves) reads as $0 = (1+\varepsilon f)^{-1}[\mathrm{d}H(1+\varepsilon f) + \varepsilon \mathrm{d}g] = \mathrm{d}(H+\varepsilon g) - \varepsilon^2 f\mathrm{d}g + O(\varepsilon^3)$. It follows that

$$\Delta H = \varepsilon^2 \int_{\gamma(t)} (-f)\mathrm{d}g + \ldots$$

If $M_2 \equiv 0$, then $f\mathrm{d}g = f_1\mathrm{d}H + \mathrm{d}g_1$ and $M_3 = \int f_1\mathrm{d}g_1$, etc. This algorithm, called now the *Françoise algorithm*, was introduced by J.-P. Françoise [**Fr**]. L. Gavrilov [**Gav2**] proved that all higher order Melnikov integrals satisfy some linear Picard–Fuchs equation with regular singularities.

2. For perturbations of vector fields with an integrating multiplier, i.e.,

$$\dot{x} = M^{-1}H'_y + \epsilon P, \ \dot{y} = -M^{-1}H'_x + \epsilon Q,$$

the corresponding Melnikov integral takes the form

$$\Delta H = \varepsilon \int M(Qdx - Pdy) + O(\varepsilon^2).$$

Such situations occur in the cases of perturbations of quadratic cases of center: Q_3^{LV}, Q_3^{R} and Q_4. The corresponding theory is not as developed as the infinitesimal Hilbert's problem theory (see [**Zo1–Zo3**]).

3. There is a multi-dimensional generalization of the infinitesimal Hilbert problem, developed by the author with P. Leszczyński and M. Bobieński [**LZ**], [**BZ1**], [**BZ2**] (see also [**ZM**]). One deals with the system

$$\dot{x} = H'_y + zR + \varepsilon P, \quad \dot{y} = -H'_x + zS + \varepsilon Q, \quad \dot{z} = Az + \varepsilon B.$$

Here H, P, Q, R, S, A, B are polynomials in (x, y). For $\varepsilon = 0$ the surface $z = 0$ is invariant, with a Hamiltonian system in it. Under some natural assumptions, after perturbation, there remains invariant surface of the form $z = \varepsilon g(x, y) + \dots$, where g satisfies the equation $\dot{g} = Ag + B$. The limit cycles on the latter surface are generated by zeroes of $I(t) + J(t)$, where I is the standard Abelian integral and J is the following *generalized Melnikov integral*

$$J(t) = \int_{\gamma(t)} g \cdot (S\mathrm{d}x - R\mathrm{d}y).$$

Here g can be expressed in an integral form (with an exponent), so J is a double integral. Under some assumptions the estimate $\leq \mathrm{const} \cdot n$, $n = \deg(P, Q, R, S, B)$, for the number of zeroes of $I + J$ was proved.
In [**LZ1, LZ2**] perturbations of the so-called Hess–Appelrot case in the rigid body dynamics (see the Euler–Poisson equations in the point 14.40 in Chapter 10) were treated in this way.

Chapter 7

Hodge Structures and Period Map

The Hodge structure is defined in the cohomology ring of any smooth compact projective algebraic complex manifold. It presents deep a connection between topological and analytical properties of algebraic varieties. It provides us with new invariants of algebraic varieties which are functorial; it means that are preserved by algebraic morphisms.

If the variety is not compact, i.e., it is a *quasi-projective variety* (Zariski open subset of a closed projective variety), or it has singularities, then its cohomology group may not admit any Hodge structure. However, P. Deligne proved that such manifold admits some structure which is functorial and forms a generalization of the Hodge structure. He called it the mixed Hodge structure.

The mixed Hodge structure appears also in the situation, when we have an analytic family of algebraic hypersurfaces $N_t \subset M$, $t \in \mathbb{D}$, with exactly one singular fiber N_0. Each N_t admits a pure Hodge structure. In the limit $t \to 0$ these Hodge structures degenerate to certain limit mixed Hodge structure, introduced independently by W. Schmid and J. Steenbrink.

Steenbrink has gone further and, using the construction of the limit mixed Hodge structure, defined certain mixed Hodge structure in the cohomological Milnor fibration (of an isolated critical point of a holomorphic function). A. Varchenko proved that the Steenbrink's mixed Hodge structure can be defined by means of the asymptotic expansions of the integrals along vanishing cycles and by the monodromy operator.

The situation with degeneration of a family N_t of hypersurfaces leads to investigation of the so-called period map $t \to (\int_{\delta_j(t)} \omega_i)$ (integrals of holomorphic forms along basic cycles in N_t) with values in a period matrix space D (certain homogeneous space). In the case of curves N_t, the period mapping is an embedding (Torelli theorem). We have here the Gauss–Manin connection and the monodromy operator inducing an automorphism of D. P. Griffiths investigated the singularities of the period mapping as $t \to 0$. In some cases the limit turns out to lie in certain natural partial compactification of the homogeneous space D.

We are going to describe general ideas lying in foundations of this theory. We cannot present all the details, because they are very complicated and need advanced knowledge of the homological algebra and sheaf theory. Even without these details this chapter is decidedly more difficult than the others.

7.1. Agreement. By an algebraic variety M we will mean a *Zariski open subset* of a projective complex variety. It means that M is defined by a finite system of algebraic equations $f_1(x) = \ldots = f_k(x) = 0$, $x = (x_0, x_1, \ldots, x_N) \in \mathbb{C}^{N+1}$, (with

H. Żołądek, *The Monodromy Group*, Monografie Matematyczne 67,
https://doi.org/10.1007/978-3-031-91270-2_7

homogeneous f_j) and by finite system of algebraic inequalities $g_1(x) \cdot g_2(x) \ldots \neq 0$ (g_j also homogeneous). Such variety is called **quasi-projective**. If there are no inequalities, then the subset (of $\mathbb{CP}^N$) is called *Zariski closed*.
In the affine part $\mathbb{C}^N \subset \mathbb{CP}^N$ defined by $\{x_0 \neq 0\}$, with the affine coordinates $y_i = x_i/x_0$, we have a variety given by algebraic (non-homogeneous) equations and inequalities $f_i(1, y_1, \ldots, y_N) = 0 \neq g(1, y_1, \ldots, y_N)$. Similarly the other affine parts in the hyperplanes $\{x_i \neq 0\}$ are defined.
If the definition of M does not contain inequalities, then such variety is called *closed*.
M is called *smooth* iff it is of the class C^∞ at each its point as a real manifold. Equivalently, iff the rank of the matrix $(\mathrm{d}f_1, \ldots, \mathrm{d}f_k)$ is constant at each affine part of M.

§1 Hodge Structure on Algebraic Manifolds

7.2. Harmonic forms on Riemannian manifolds. Let M be a real Riemannian oriented manifold, i.e., with a scalar product $\mathrm{d}s^2 = \langle \cdot, \cdot \rangle_x$, defined in each tangent space $T_x M$ and depending smoothly on x. This product defines the scalar product in the cotangent space $T_x^* M$ and in its exterior products $\bigwedge^k T_x^* M$. Moreover, there is defined the *volume form* $\mathrm{VOL} \in \Gamma(M, \mathcal{E}^n)$, where $n = \dim M$ and $\mathcal{E}^k$ denotes the sheaf of smooth k-forms: $\langle \mathrm{VOL}, (v_1, \ldots, v_n) \rangle = \pm(\det(\langle v_i, v_j \rangle))^{1/2}$. Here the proper choice of the sign $\pm$ is possible due to the orientability of M.
Locally one can write $\mathrm{VOL} = \phi_1 \wedge \ldots \wedge \phi_n$, where $\phi_i = \phi_i(x) \in T_x^* M$ are 1-forms diagonalizing the metric: $\mathrm{d}s^2 = \sum \mathrm{d}\phi_j^2$.
We have also the so-called *star operation* $* : \mathcal{E}^k \to \mathcal{E}^{n-k}$ defined by the equality

$$\omega(x) \wedge *\eta(x) = \langle \omega, \eta \rangle_x \cdot \mathrm{VOL}(x).$$

The metric structure on M allows to define the scalar product on the spaces of differential forms on the manifold $\Gamma(\mathcal{E}^k) = \Gamma(M, \mathcal{E}^k)$: $(\omega, \eta) = \int_M \langle \omega, \eta \rangle_x \mathrm{VOL}(x)$. Thus $\Gamma(M, \mathcal{E}^k)$ becomes pre-Hilbert space. We complete it to the Hilbert space $L = L^k$.
Moreover, if $\mathrm{d} : \Gamma(\mathcal{E}^k) \to \Gamma(\mathcal{E}^{k+1})$ is the exterior derivative, then we have the conjugate operator $\mathrm{d}^* : \Gamma(\mathcal{E}^{k+1}) \to \Gamma(\mathcal{E}^k)$. One can check that

$$\mathrm{d}^* = - * \mathrm{d} * .$$

We define the *laplacian* (or the *Laplace–Beltrami operator*) of M as

$$\Delta = \mathrm{d}\mathrm{d}^* + \mathrm{d}^*\mathrm{d}.$$

Sometimes this laplacian is denoted by Δ_{d}, as associated with the exterior derivative. Note that it depends on the metric.

The forms satisfying the equality $\Delta\omega = 0$ are called the *harmonic forms*. Because $(\Delta\omega, \omega) = ||\mathrm{d}\omega||^2 + ||\mathrm{d}^*\omega||^2$, the harmonic forms have the property that $\mathrm{d}\omega = \mathrm{d}^*\omega = 0$.
Recall that, by de Rham Theorem 3.24, the cohomology group $H^k(M, \mathbb{R})$ is isomorphic to the group of de Rham cohomologies. It is the quotient space of closed k-forms modulo differentials of $(k-1)$-forms. It is desirable to have some natural way of choosing a form from each cohomology class. In the case of Riemannian manifold such choice is given in the following theorem.

7.3. Theorem. *The group $H^k_{\mathrm{dR}}(M, \mathbb{R})$ is isomorphic to the space of smooth harmonic k-forms on M.*

Proof. From each class $[\omega] = \{\omega + \mathrm{d}\eta : \eta \in \Gamma(\mathcal{E}^{k-1})\}$ we take a form with minimal $||\cdot||$-norm. Because $\frac{\mathrm{d}}{\mathrm{d}t}||\omega + t\mathrm{d}\eta||^2|_{t=0} = 2(\omega, \mathrm{d}\eta) = 2(\mathrm{d}^*\omega, \eta) = 0$ for any η, the minimality of ω implies $\mathrm{d}^*\omega = 0$. On the other hand, if $\mathrm{d}^*\omega = 0$, then $||\omega + \mathrm{d}\eta||^2 = ||\omega||^2 + ||\mathrm{d}\eta||^2$, which shows that ω is minimal.
These arguments are only heuristic. Note that the minimal element can be found in the closure of the class $[\omega]$ in the Hilbert space $L = L^k$. We obtain harmonic forms in weak sense, as harmonic generalized functions. To complete the proof we use the partial differential equations methods.
The operator Δ is non-negatively definite, $(\Delta\omega, \omega) \geq 0$; so, the operator $I + \Delta$ is strictly positive. One shows, (using the *Rellich lemma* and the *Gårding inequality)*, that:

The operator $(I + \Delta)^{-1}$ is bounded and compact in L.

(*Proof*: $I + \Delta$ is an isomorphism between the Sobolev space L_1 (the completion of the space of smooth forms in the norm $(||\omega||^2 + ||\mathrm{d}\omega||^2)^{1/2}$) and the Hilbert space L and the inclusion $L_1 \subset L$ is compact operator (by the uniform continuity of functions from the ball in L_1).)

Thus $(I + \Delta)^{-1}$ has discrete eigenvalues of finite multiplicity. This applies to the eigenspace corresponding to the eigenvalue $\lambda = 1$, which is equal to the space of harmonic forms. This shows that the de Rham cohomology groups are finite dimensional (we know it also from topological arguments).
Next, one applies the *Sobolev lemma* to show that harmonic forms are smooth:

If $(I + \Delta)\omega = \eta$ and η is smooth, then ω is smooth.

(See [**GH**] for details of the proof). □

7.4. Dolbeault cohomologies of complex manifolds. Now we pass to the case when M is a smooth complex manifold.
The sheaf of complex differential forms on M can be decomposed into sum of the sheaves $\mathcal{E}^{p,q}_M$. If $z_1 = x_1 + \mathrm{i}y_1, \ldots, z_n = x_n + \mathrm{i}y_i$ are local coordinates in M, then the *forms of the type* (p, q) (or the (p, q)-forms) have the representation $\omega = \sum_{IJ} a_{IJ} \mathrm{d}z^I \wedge \mathrm{d}\bar{z}^J$, where the sum runs over p-element subsets I and q-element subsets J of the set of indices and $a_{IJ} \in C^\infty(M, \mathbb{C})$.

There is also the sheaf of holomorphic p-forms on M which are of the type $(p,0)$ with holomorphic coefficients (see the definition before Theorem 5.8). It is denoted by $\Omega^p = \Omega^p_M$.
The **Dolbeault cohomology groups** are the Čech cohomology groups (see 3.26)

$$H^q(M,\Omega^p).$$

As we shall see, they play a role analogous to the Čech cohomology groups $H^k(M,\mathbb{C})$, i.e., with coefficients in the constant sheaf $\mathbb{C}$.
Recall that the latter groups are calculated by means of (complex) differential forms (see the proof of the de Rham's theorem). Analogously, the Dolbeault cohomology groups are calculated using the differential forms.
The role of the sequence $0 \longmapsto \mathbb{C} \longmapsto \mathcal{E}^0 \overset{\mathrm{d}}{\longmapsto} \mathcal{E}^1 \longmapsto \ldots$ which is exact (Poincaré Lemma) play the *Dolbeault complexes*

$$0 \longmapsto \Omega^p \longmapsto \mathcal{E}^{p,0} \overset{\bar\partial}{\longmapsto} \mathcal{E}^{p,1} \ldots,$$

where $\bar\partial$ is the exterior derivative only with respect to the variables $\bar z_j$. The analogue of the Poincaré Lemma, the **$\bar\partial$-Poincaré Lemma,** says that:

The latter sequences are exact. It means that for any local (p,q)-form ω, such that $\bar\partial\omega = 0$, there is a local $(p,q-1)$-form η such that $\bar\partial\eta = \omega$.

(In one dimension we have the formula $f(z) = \frac{1}{2\pi \mathrm{i}}\int_{\mathbb{D}}[g(w)/(w-z)]\mathrm{d}w \wedge \mathrm{d}\bar w$ for the solution of the equation $\partial f/\partial \bar z = g$; the inductive proof in the general case is given in [**GH**]).
Repeating, word-by-word, the proof of the de Rham theorem we get the following.

7.5. Dolbeault's Theorem. *We have*

$$H^q(M,\Omega^p) = H^{p,q}_{\bar\partial},$$

where the latter groups are the groups of global $\bar\partial$-closed (p,q)-forms on M modulo the subgroups of global $\bar\partial$-exact forms.

7.6. Dolbeault cohomology groups of Hermitian varieties. Assume that M is a complex smooth manifold equipped with a Hermitian product $\langle\cdot,\cdot\rangle_z$ in its tangent bundle.
We can assume that in each tangent plane the Hermitian form is diagonal and given by

$$\mathrm{d}s^2 = \sum \phi_j \otimes \bar\phi_j,$$

where $\phi_j = \phi_j(z) \in T^*_z M$ are local $(1,0)$-forms. With this metric one associates the following $(1,1)$-form: $\omega = \frac{\mathrm{i}}{2}\sum \phi_j \wedge \bar\phi_j$. (If $\mathrm{d}s^2 = \sum h_{ij}(z)\mathrm{d}z_i \otimes \mathrm{d}\bar z_j$, then $\omega = \frac{\mathrm{i}}{2}\sum h_{ij}\mathrm{d}z_i \wedge \mathrm{d}\bar z_j$). One can say that the real part of the Hermitian metric, which is a symmetric tensor, represents the Riemannian structure of M (treated

as real $2n$-dimensional manifold) and the imaginary part of the Hermitian metric is just the anti-symmetric form ω.
One defines the induced Hermitian product on the spaces $T_z^{*(p,q)}M \subset \bigwedge^{p+q} T_z^* M$ (of (p,q)-forms at z) and the volume form as $\mathrm{VOL} = \omega^n/n!$
Repeating the analysis from the point 7.2, we equip the spaces $\Gamma(M, \mathcal{E}^{p,q})$ with Hermitian products and transform them into complex pre-Hilbert spaces.
One defines the *Hodge star operation* $* : \mathcal{E}^{p,q} \longmapsto \mathcal{E}^{n-p,n-q}$ by $\xi(z) \wedge *\eta(z) = \langle \xi, \eta \rangle_z \cdot \mathrm{VOL}(z)$.
It turns out that

$$
\begin{aligned}
** &= (-1)^{p+q}, \\
\bar{\partial}^* &= - * \bar{\partial} *,
\end{aligned}
$$

where $\bar{\partial}^*$ is the operator conjugate to $\bar{\partial}$.
One introduces the *$\bar{\partial}$-laplacian*

$$\Delta_{\bar{\partial}} = \bar{\partial}\bar{\partial}^* + \bar{\partial}^*\bar{\partial}$$

and defines the *harmonic (p,q)-forms* as those which satisfy the equation $\Delta_{\bar{\partial}}\eta = 0$. They satisfy also the equations $\bar{\partial}\eta = \bar{\partial}^*\eta = 0$. The space of harmonic (p,q)-forms is denoted by $\mathcal{H}^{p,q}$.
As in the point 7.2, one has the formal isomorphism

$$H_{\bar{\partial}}^{p,q}(M) = \mathcal{H}^{p,q}, \tag{7.1}$$

where one chooses from each cohomology class an element with minimal norm.
The next result is an analogue of Theorem 7.3 (with the same proof, which we do not repeat here).

7.7. Hodge's Theorem. *The isomorphism* (7.1) *is rigorous. This means that the space $H_{\bar{\partial}}^{p,q}(M)$ is isomorphic with the space of smooth harmonic forms. Moreover, the spaces $\mathcal{H}^{p,q}$ are finite dimensional.*

7.8. Remark. The reader can notice that the above spaces $\mathcal{H}^{p,q}$ are not chosen canonically, they depend on the Hermitian metric.

7.9. Kähler manifolds. A complex manifold M is called **Kähler** iff it admits a Hermitian metric $\mathrm{d}s^2 = \sum \phi_j \otimes \bar{\phi}_j$ such that the associated with it $(1,1)$-form

$$\omega = \frac{\mathrm{i}}{2} \sum \phi_j \wedge \bar{\phi}_j$$

is symplectic. In other words, $\mathrm{d}\omega = 0$

7.10. Example. *Smooth closed projective algebraic varieties are Kähler.*
Firstly, we introduce certain Kähler structure in the projective space $\mathbb{CP}^N$. It is given by the *Fubini–Study form*

$$\omega = \omega_{\mathbb{CP}^N} = \frac{\mathrm{i}}{2\pi} \partial\bar{\partial} \log |Z|^2,$$

where $Z : U \to \mathbb{C}^{N+1}\setminus 0$, $U \subset \mathbb{CP}^N$, is a local holomorphic section of the tautological bundle, i.e., the line $Z(x)\mathbb{C}$ represents the point x. It is easy to see that this definition does not depend on the choices of Z in its class and that ω is closed. Expressing it in the affine coordinates, we show that it is non-degenerate.
The Fubini–Study form ω represents certain class $[\omega]$ in the cohomology group $H^2(\mathbb{CP}^N, \mathbb{C})$. It turns out that $[\omega] \in H^2(\mathbb{CP}^N, \mathbb{Z})$; it takes integer values at the integer 2-dimensional cycles. To see this it is enough to calculate the value of the class $[\omega]$ on the generator of $H_2(\mathbb{CP}^N, \mathbb{Z})$, which can be represented as any projective line $\mathbb{CP}^1 \subset \mathbb{CP}^N$. But ω restricted to the affine plane $\mathbb{C}^1 \subset \mathbb{CP}^1$ equals to $\mathrm{i} \cdot \mathrm{d}z \wedge \mathrm{d}\bar{z}/(1+|z|^2)^2$ and its integral is 1. The class $[\omega]$ is the coorientation class of (any) hyperplane $\mathbb{CP}^{N-1} \subset \mathbb{CP}^N$.
If $M \overset{i}{\subset} \mathbb{CP}^N$ is algebraic, then the restriction of the Fubini–Study metric to M defines the Kähler metric on M. Indeed, $\mathrm{d}s^2|_{TM}$ is a metric (hence non-degenerate) and the corresponding $(1,1)$-form $\omega_M = i^*\omega_{\mathbb{CP}^N} = i^*\omega$ is closed. Moreover, the Kähler form $i^*\omega$ represents integer cohomology class of M. The class $i^*[\omega]$ is the coorientation class of intersection of M with a generic projective hyperplane of $\mathbb{CP}^N$ (hyperplane section).

The Kähler manifolds have the important property

$$\Delta_{\bar{\partial}} = (1/2)\Delta_{\mathrm{d}}.$$

It means that the forms, harmonic in $\bar{\partial}$-sense, are also harmonic in the usual sense. To show this equality one introduces the operators $L : \eta \to \eta\wedge\omega$ and $\mathrm{d_c} = -\mathrm{i}(\partial - \bar{\partial})$ and the notations: $[A, B] = AB - BA$ (for the commutator) and $\{A, B\} = AB + BA$ (for the anti-commutator). Note that $\mathrm{d} = \partial + \bar{\partial}$. Next, one has the following simple identities which are consequences of the closeness of ω:

$$[L, \mathrm{d}] = [L^*, \mathrm{d}^*] = 0.$$

The next two identities are more complicated in proofs,

$$[L, \mathrm{d}^*] = \mathrm{d_c}, \quad [L^*, \mathrm{d}] = \mathrm{d_c^*};$$

they imply $\{\mathrm{d_c}, \mathrm{d}^*\} = \{\mathrm{d_c^*}, \mathrm{d}\} = \{\partial, \bar{\partial}^*\} = \{\bar{\partial}, \partial^*\} = 0$.
(In a local chart, with the metric $\mathrm{d}s^2 = \sum \mathrm{d}z_k \otimes \mathrm{d}\bar{z}_k$ (where $\|\mathrm{d}z_k\|^2 = 2$), one introduces the operators $e_k(\cdot) = \mathrm{d}z_k \wedge (\cdot)$, $\bar{e}_k = \mathrm{d}\bar{z}_k \wedge (\cdot)$, acting on forms with compact support. The conjugate e_k^* acts as follows: $e_k^*(\mathrm{d}z_I \wedge \mathrm{d}\bar{z}_J) = 0$, if $k \notin I$, and $e_k^*(\mathrm{d}z_k \wedge \mathrm{d}z_{I'} \wedge \mathrm{d}\bar{z}_J) = 2\mathrm{d}z_{I'} \wedge \mathrm{d}\bar{z}_J$; analogously acts $\bar{e}_k^*$. There are relations like in the Clifford algebra: $\{e_k, e_k^*\} = \{\bar{e}_k, \bar{e}_k^*\} = 2$ and other anti-commutators are zero. We have $\partial = \sum e_k \partial_k = \sum \partial_k e_k$, $\bar{\partial} = \sum \bar{e}_k \bar{\partial}_k = \sum \bar{\partial}_k \bar{e}_k$, $\partial^* = -\sum \bar{\partial}_k e_k^*$, $\bar{\partial}^* = -\sum \partial_k \bar{e}_k^*$ (integration by parts) and $L = \frac{\mathrm{i}}{2}\sum e_k \bar{e}_k$. Now direct calculations show that $[L^*, \partial] = i\bar{\partial}^*$ and $[L^*, \bar{\partial}] = -i\partial^*$. See also [**GH**] and [**Wel**].)
Finally we get

$$4\Delta_{\bar{\partial}} = \{\mathrm{d} - i\mathrm{d_c}, \mathrm{d}^* + i\mathrm{d_c^*}\} = \Delta_{\mathrm{d}} + \{\mathrm{d_c}, \mathrm{d_c^*}\} + \mathrm{i}\{\mathrm{d}, \mathrm{d_c^*}\} - \mathrm{i}\{\mathrm{d_c}, \mathrm{d}^*\} = 2\Delta_{\mathrm{d}}$$

(because $\Delta_{\mathrm{d}} - \{\mathrm{d}_c, \mathrm{d}_c^*\} = 2\{\partial, \bar\partial^*\} + 2\{\bar\partial, \partial^*\}$).
This gives the following:

7.11. Hodge expansion theorem. *Let M be a Kähler manifold. We have*

$$H^k_{\mathrm{dR}}(M, \mathbb{C}) = \bigoplus_{p+q=k} H^{p,q}(M),$$

where $H^{p,q}(M) = \{closed\ (p,q)\text{-}forms\}/\{exact\ (p,q)\text{-}forms\}$. These groups are isomorphic to the groups $\mathcal{H}^{p,q}$, $H^q(M, \Omega^p)$ and $H^{p,q}_{\bar\partial}$.
In particular, the space $H^{p,0} \approx H^0(M, \Omega^p) = \Gamma(M, \Omega^p)$ consists of global holomorphic n-forms.
Moreover, we have the following isomorphisms (with $n = \dim_{\mathbb{C}} M$):

$$\overline{H^{p,q}} = H^{q,p},$$
$$H^{p,q} \approx H^{n-p,n-q}.$$

The latter isomorphism is given by the Hodge star $*$, follows from the Poincaré duality and is called the *Kodaira–Serre duality*.

Kähler manifolds admit also another expansion. Introduce the *Lefschetz operator* $L : H^{p,q} \longmapsto H^{p+1,q+1}$ defined by (see above)

$$L : [\eta] \longmapsto [\eta \wedge \omega];$$

(it is correct because ω is closed). Moreover, studying the Lie algebra generated by L and L^*, one obtains the following.

7.12. Strong Lefschetz theorem. *The maps $L^k : H^{n-k} \longmapsto H^{n+k}$, $n = \dim_{\mathbb{C}} M$ are isomorphisms.*
If we define the subspace of primitive vectors

$$P^{n-k} = \ker L^{k+1}|_{H^{n-k}} = \ker L^*|_{H^{n-k}},$$

then we have the following Lefschetz expansion

$$H^k = \bigoplus_j L^j P^{k-2j},$$
$$P^l = \bigoplus_{p+q=l} P^{p,q}.$$

7.13. Example. Let $X \subset \mathbb{CP}^n$ be a smooth closed hypersurface of degree d. If $n = 2k$ is even, then we have:

$$H^0(X) = P^0 = \mathbb{C}, \quad H^1(X) = 0, \quad H^2(X) = LP^0 = \mathbb{C},$$
$$H^3(X) = 0, \ldots, H^{2k-2}(X) = L^{k-1}P^0, H^{2k-1}(X) = P^{2k-1},$$
$$H^{2k}(X) = L^k P^0, \quad H^{2k+1}(X) = 0, \ldots, H^{4k-2}(X) = L^{2k-1}P^0.$$

If $n = 2k+1$ is odd, then

$$\begin{gathered} H^0(X) = P^0 = \mathbb{C}, \quad H^1(X) = 0, \quad H^2(X) = LP^0 = \mathbb{C}, \\ H^3(X) = 0, \dots, H^{2k-1}(X) = 0, \quad H^{2k}(X) = L^k P^0 \oplus P^{2k}, \\ H^{2k+1}(X) = 0, \quad H^{2k+2}(X) = L^{k+1} P^0, \dots, H^{4k}(X) = L^{2k} P^0. \end{gathered}$$

Here $\dim P^{p,q}$, $p+q = n-1$, is equal to the number of integer points in the hypercube $[1, d-1]^n$ between the hyperplanes $|k| = qd$ and $|k| = (q+1)d$, $|k| = \sum k_i$, i.e., $\dim P^{p,q} = \#\{k \in \mathbb{Z}^n : 0 < k_i < d,\ qd < |k| < (q+1)d\}$. See also 7.44(b) below.

This is not end of the story. The cohomology spaces of Kähler manifolds are also equipped with certain bilinear form, called the **polarization** $Q : H^{n-k} \otimes H^{n-k} \to \mathbb{C}$ and defined as follows

$$Q(\xi, \eta) = \int_M \xi \wedge \eta \wedge \omega^k.$$

7.14. Theorem. *The polarization form has the following properties, called the bilinear Riemann–Hodge relations:*

$$\begin{array}{lrcll} \text{(i)} & Q(H^{p,q}, H^{p',q'}) & = & 0 & \text{unless } p = q',\ q = p', \\ \text{(ii)} & i^{p-q}(-1)^{m(m-1)/2} Q(\xi, \bar{\xi}) & > & 0 & \text{for } \xi \in P^{p,q},\ m = p+q. \end{array}$$

The expansion of these properties to H^m and the identity $Q(L\xi, L\eta) = Q(\xi, \eta)$ shows that Q extends itself to the whole H^* and is non-degenerate.

The introduced structures (the Hodge decomposition, the Lefschetz decomposition, the polarization) are functorial. It means that, if M and N are smooth compact algebraic varieties and $f : M \longmapsto N$ is a smooth algebraic morphism, then the induced map f^* transforms the above structures from N to the structures in M.

7.15. Example (Hodge structure on the elliptic curve). Let $M^0 = \{y^2 = P_3(x)\}$, where P_3 is a cubic polynomial with three distinct zeroes x_1, x_2, x_3. After adding the point $[1 : 0 : 0]$ at infinity, one obtains a non-singular algebraic curve M in $\mathbb{CP}^2$. As a Riemann surface it is a torus (with genus 1). Its cohomology groups are

$$H^0 = \mathbb{C}, \quad H^1 = \mathbb{C}^2, \quad H^2 = \mathbb{C}.$$

These spaces contain also the integer lattices $H^k(M, \mathbb{Z})$ and real subspaces $H^k(M, \mathbb{R})$. We investigate the influence of the complex structure on M on the cohomology ring.

Consider the form

$$\eta = \mathrm{d}x/y$$

restricted to M. It is holomorphic and nonzero in the affine part of M. Indeed, near the points where $y = 0$ (i.e., $(x_i, 0)$, $P_3(x_i) = 0$) we have $x = x_i + cy^2 + \dots$

and $\mathrm{d}x = (2cy + \dots)\mathrm{d}y$. Let us look at the behavior of η near the intersection of M with the line at infinity. We apply the change $z = 1/y$, $u = x/y$. Then $M = \{z = \tilde{P}_3(u, z)\}$ is smooth near $u = z = 0$ and can be parameterized by u: $z = au^3 + \dots$. After simple transformations, we obtain $\eta = (-2 + \dots)\mathrm{d}u$. This shows that η is holomorphic and nonzero in the whole M.

Of course, η represents an element (nontrivial as we shall see) of the first cohomology group, $[\eta] \in H^1(M, \mathbb{C})$.

If η' is another holomorphic 1-form on M, then the ratio η'/η is a global holomorphic function on M. Such function is bounded and, by an analogue of the Liouville theorem (the image of a non-constant holomorphic function is open), it is constant. This means that the space of global holomorphic 1-forms on the elliptic curve is one dimensional.

We have the following property

$$\mathrm{i}\int_M \eta \wedge \bar{\eta} > 0. \tag{7.2}$$

This follows from the local analysis. If locally $\eta = f(z)\mathrm{d}z$, $z = x + \mathrm{i}y$, then $\eta \wedge \bar{\eta} = |f|^2 \mathrm{d}z \wedge \mathrm{d}\bar{z} = -\mathrm{i}|f|^2 \mathrm{d}x \wedge \mathrm{d}y$.

This means that the class $[\eta]$ is nonzero in H^1 and generates the subspace $H^{1,0}$. We get the following Hodge decomposition of the cohomology space of M:

$$H^{0,0} = \mathbb{C}, \quad H^{1,0} = \mathbb{C} \cdot [\eta], \quad H^{0,1} = \overline{H^{1,0}} = \mathbb{C} \cdot [\bar{\eta}], \quad H^{1,1} = \mathbb{C} \cdot [\eta \wedge \bar{\eta}].$$

Other spaces are equal to zero.

We see also that the property (7.2) defines the polarization form on H^1. It is non-degenerate and anti-symmetric.

7.16. Remark. Generally, if X is an algebraic curve, then the dimension of the space of its holomorphic forms, i.e., $\dim H^0(X, \Omega^1) = \dim \Gamma(X, \Omega^1)$, is called the *algebraic genus* of X. For example, let $X = \mathbb{CP}^1$ and let η be a holomorphic form. Then $\eta|_{\mathbb{C}} = f(x)\mathrm{d}x$ with holomorphic f. Near infinity we have $\eta = -f(1/y)\mathrm{d}y/y^2$, $y = 1/x$ and, in order to have holomorphic η, we should have $f(\infty) = 0$, which implies $f = 0$. Thus the algebraic genus of the projective plane is zero.

One of fundamental facts from the theory of algebraic curves says that the algebraic genus coincides with the *topological genus*, i.e., with the number of handles of the corresponding Riemann surface.

If an algebraic manifold is singular and/or not complete, then usually its cohomological ring does not have Hodge structure, such as the presented above. It has turned out that the cohomologies of such manifolds admit the so-called mixed Hodge structures. Before giving the precise definition of the mixed Hodge structure, we present some additional cohomological tools and examples.

§2 Hypercohomologies and Spectral Sequences

7.17. Preliminary remarks. Recall the proof of the de Rham and Dolbeault theorems.

In order to calculate the Čech cohomology groups with coefficients in some sheaf $\mathcal{F}$, one introduces its **resolvent** (denoted also by $\mathcal{F}^\bullet$)

$$0 \longmapsto \mathcal{F}^0 \overset{\mathrm{d}^0}{\longmapsto} \mathcal{F}^1 \overset{\mathrm{d}^1}{\longmapsto} \mathcal{F}^2 \ldots$$

This is a cochain complex ($\mathrm{d}^{j+1} \circ \mathrm{d}^j = 0$) of *flabby* sheaves, which is exact in the terms $\mathcal{F}^1$, $\mathcal{F}^2, \ldots$ ($\ker \mathrm{d}^j|_{\mathcal{F}^j(U)} = \mathrm{Im}\ \mathrm{d}^{j-1}|_{\mathcal{F}^{j-1}(U)}$) and the sheaf $\mathcal{F}$ is included into $\mathcal{F}^0$ as $\ker \mathrm{d}^0$. One uses sometimes the notation $0 \longmapsto \mathcal{F} \longmapsto \mathcal{F}^\bullet$.

The flabby sheaves admit a partition of unity, which implies $H^q(M, \mathcal{F}^j) = 0$ for $q > 0$. Then, using the short exact sequences of sheaves $0 \longmapsto \ker \mathrm{d}^j \longmapsto \mathcal{F}^j \longmapsto \ker \mathrm{d}^{j+1} \longmapsto 0$ and associated with them the long exact sequences of Čech cohomologies, one obtains the successive reductions $H^n(M, \mathcal{F}) = H^n(M, \ker\ \mathrm{d}^0) \approx H^{n-1}(M, \ker \mathrm{d}^1) \ldots$ and, finally, one gets the group

$$\begin{aligned} &H^0(M, \ker \mathrm{d}^n)/\mathrm{d}^{n-1} H^0(M, \mathcal{F}^{n-1}) \\ &= (\text{global closed sections of } \mathcal{F}^n)/(\text{global exact sections}). \end{aligned}$$

Note that the sum of indices in $H^i(M, \ker \mathrm{d}^j)$ is constant and equal n; this suggests a possible generalization of these theorems.

One can ask, what happens, when the considered resolvent is not flabby. For example, the *holomorphic de Rham complex*

$$\Omega^0 \overset{\mathrm{d}}{\longmapsto} \Omega^1 \overset{\mathrm{d}}{\longmapsto} \ldots$$

is exact (by the holomorphic Poincaré lemma), but is not flabby.

It turns out that in such case the groups $H^n(M, \mathcal{F})$ can be calculated using the so-called hypercohomologies defined as follows.

7.18. Definition. Let $\mathcal{F}^\bullet : 0 \longmapsto \mathcal{F}^0 \overset{\mathrm{d}^0}{\longmapsto} \mathcal{F}^1 \overset{\mathrm{d}^1}{\longmapsto} \mathcal{F}^2 \ldots$ be a complex of sheaves on M on M and let $\mathcal{U} = \{U_\alpha\}$ be a covering of M. One considers the double complex $(K^{\bullet,\bullet}(\mathcal{U}), D', D'')$, where

$$K^{p,q}(\mathcal{U}) = K^{p,q} = C^p(\mathcal{U}, \mathcal{F}^q)$$

are the Čech p-cochains with values in $\mathcal{F}^q$,

$$D' = \delta : K^{p,q} \longmapsto K^{p+1,q}$$

is the Čech coboundary operator, and

$$D'' = (-1)^p \mathrm{d}^q : K^{p,q} \longmapsto K^{p,q+1}.$$

We have

$$\begin{array}{ccccc} K^{0,0} & \stackrel{D'=\delta^0}{\longmapsto} & K^{1,0} & \longmapsto & \dots \\ \mathrm{d}^0 \downarrow D'' & & \mathrm{d}^1 \downarrow D'' & & \\ K^{0,1} & \stackrel{D'=\delta^0}{\longmapsto} & K^{1,1} & \longmapsto & \dots \\ \downarrow & & \downarrow & & \\ \vdots & & \vdots & & \end{array}$$

With the double complex one associates the cochain complex $(K^\bullet(\mathcal{U}, D)$, where

$$K^n(\mathcal{U}) = \bigoplus_{p+q=n} K^{p,q}, \quad D = D' + D'';$$

(here one checks that $D \circ D = 0$).
Let $H^n(\mathcal{U}, \mathcal{F}^\bullet)$ be the cohomology groups of the latter cochain complex. Their direct limit, with respect to the system of coverings $\mathcal{U}$, defines the **hypercohomology groups** denoted by

$$\mathbb{H}^n(M, \mathcal{F}^\bullet) = \lim_{\rightarrow} H^n(\mathcal{U}, \mathcal{F}^\bullet).$$

The next result is the analogue of the de Rham theorem (see [**God**], [**GH**]).

7.19. Theorem (Hypercohomology). *If the complex of sheaves $F^\bullet$ is a resolvent of F, then*

$$H^n(M, \mathcal{F}) \approx \mathbb{H}^n(M, \mathcal{F}^\bullet).$$

We will prove this theorem later.

In order to compute the groups $\mathbb{H}^n$, one uses the spectral sequences. Their idea is to compute the hypercohomology groups, not at once, but successively approximating step-by-step the groups $\mathbb{H}^n$.
The complex $K^\bullet = K^\bullet(\mathcal{U})$ has the natural decreasing filtration $K^\bullet = F^0 \supset F^1 \supset F^2 \supset \dots$, where

$$F^p = K^{n,0} \oplus K^{n-1,1} \oplus \dots \oplus K^{p,n-p},$$

when restricted to K^n. The differential (the coboundary operator) D is compatible with this filtration.
At the first step of calculations, one restricts the coboundary operator D to F^p and neglects the terms modulo F^{p+1}. More precisely, one considers the following homomorphism, induced by D (see Fig. 7.1),

$$\mathrm{d}_0 : F^p/F^{p+1} \longmapsto F^p/F^{p+1}.$$

If one denotes $E_0^{p,q} = (F^p/F^{p+1})|_{K^{p+q}}$ (which equals to $K^{p,q}$), then we have $\mathrm{d}_0 : E_0^{p,q} \longmapsto E_0^{p,q+1}$. Because D is a differential, we have $\mathrm{d}_0 \circ \mathrm{d}_0 = 0$. Here a

chain $a \in F^p$ is a cocycle with respect to d_0, iff $Da = 0 \pmod{F^{p+1}}$; so Da can be nonzero, but must lie in F^{p+1}.
One has the cohomology groups associated with the differential d_0 and denotes them by

$$E_1^{p,q} = \ker \mathrm{d}_0 / \mathrm{Im}\ \mathrm{d}_0$$

on $E_0^{p,q}$. The group $\bigoplus_p E_1^{p,n-p}$ is the first approximation of $\mathbb{H}^n$.
Of course, in our situation (from Definition 7.18), the differential d_0 is equal D''. We have $E_1^{p,q} = H^q_{D''}(K^{p,\bullet})$, the homology groups of the vertical chain complexes. Thus $E_1^{p,q}$ can be associated with the block $K^{p,q}$; it is a quotient of some subspace of $K^{p,q}$ by another smaller subspace.
In the second approximation one also uses the operator D and its action on terms from $E_1^{\bullet,\bullet}$, (i.e., on cocycles in the first approximation), but now one neglects the terms from F^p which lie in F^{p+2}. This gives certain coboundary operator

$$\mathrm{d}_1 : E_1^{p,q} \longmapsto E_1^{p+1,q}.$$

If $a \in F^p$ is a representative of a class $[a]$ from $E_1^{p,q}$, then $Da \in F^{p+1}$ (as it $= 0$ $\mathrm{mod} F^{p+1}$), and we take class of Da in F^{p+1} (mod F^{p+2}), i.e., the coset $Da + F^{p+2}$. Because the degree of Da is greater by 1 than the degree of a and because p grows by 1, we have $[Da] = \mathrm{d}_1[a] \in E_1^{p+1,q}$.
As before we define

$$E_2^{p,q} = \ker \mathrm{d}_1 / \mathrm{Im} \mathrm{d}_1.$$

One can see that $E_2^{p,q} = H^p_{D'}(H^q_{D''}(K^{\bullet,\bullet}))$. Here also the term $E_2^{p,q}$ is, more or less directly, associated with the block $K^{p,q}$.
Before making a generalization, let us look how the differential d_2 is constructed. We have $\mathrm{d}_2 : E_2^{p,q} \longmapsto E_2^{p+2,q-1}$ and $\mathrm{d}_2[a] = \mathrm{d}a \pmod{F^{p+3}}$. Here $a \in F^p$ is such element that $Da = 0 \pmod{F^{p+2}}$.
In general, we cannot identify the corresponding class $[a] \in E_2^{p,q}$ with some element a_p from the space $F^pK^n / F^{p+1}K^n = K^{p,q}$, $p + q = n$. In fact, we have $a = a_p + a_{p+1} + \ldots$, $a_j \in K^{j,n-j}$, and $Da = (D''a_p) + (D'a_p + D''a_{p+1}) + (D'a_{p+1} + D''a_{p+2}) + \ldots$, where the summands are in F^p, $F^{p+1}, \ldots$ respectively. Because $\mathrm{d}_0[a] = \mathrm{d}_1[a] = 0$, we find $D'a_p = 0$ and $D''a_p = -D'a_{p+1}$. Thus $\mathrm{d}_2[a] = [D''a_{p+1} + D'a_{p+2}]$ is identified with an element from $K^{p+2,q-1}$. We see that there is no direct link between $E_2^{p,q}$ and $K^{p,q}$.
Repeating the above procedure, we obtain the triply indexed groups $E_r^{p,q}$ and differentials $\mathrm{d}_r : E_r^{p,q} \longmapsto E_r^{p+r,q-r+1}$. Note that the index p grows by r (Da takes value in F^{p+r}), but the sum of indices (degree) grows only by 1 (like for D as expected).

The above process stabilizes at some moment, i.e., $E_r^{p,q} = E_{r+1}^{p,q} = \ldots = E_\infty^{p,q}$ for $r > r_0$ and we have

$$\mathbb{H}^n = \bigoplus_p E_\infty^{p,n-p}.$$

In fact, $r_0 \le 2 \cdot \dim M$ in our situation.

7.20. Definition. The system $(E_r^{p,q}, \mathrm{d}_r)$ is called the **spectral sequence of hypercohomology groups.**

The complex $K^\bullet$ admits also the reverse filtration: $\widetilde{F}^p = K^{0,n} \oplus K^{1,n-1} \oplus \ldots \oplus K^{n-p,p}$. With this filtration one can associate the corresponding spectral sequence $\widetilde{E}_r^{p,q}$. Here one gets $\widetilde{E}_1^{p,q} = H_{D'}^p(K^{\bullet,q})$ and $\widetilde{E}_2^{p,q} = H_{D''}^q(H_{D'}^p(K^{\bullet,\bullet})$. Therefore the two spectral sequences converge to the hypercohomology groups.

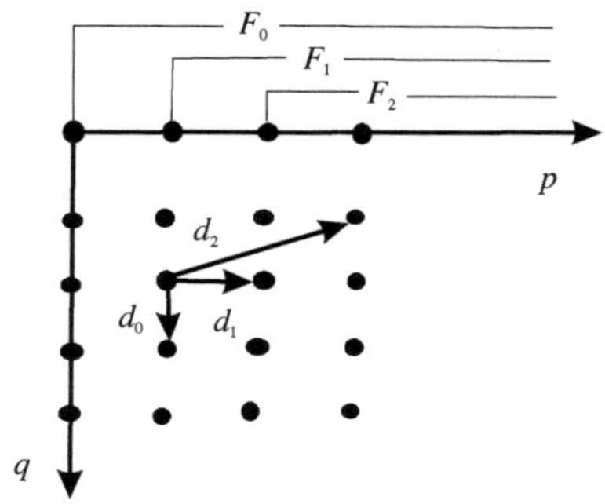

Figure 7.1: Filtration and differentials for the spectral sequence

The reader can also note that the spectral sequence can be introduced in each case, when a cochain complex admits a filtration compatible with the differentials. Using the spectral sequence $\widetilde{E}_\bullet^{\bullet,\bullet}$, one obtains the following useful result about independence of the hypercohomology groups on the complex $\mathcal{F}^\bullet$.

7.21. Definition. A homomorphism of cochain complexes of sheaves $j^\bullet : \mathcal{F}^\bullet \longmapsto \mathcal{G}^\bullet$, i.e., a system of maps $j^k : \mathcal{F}^k \longmapsto \mathcal{G}^k$ commuting with the differentials, is called the **quasi-isomorphism** iff it induces isomorphisms of their cohomological sheaves

$$j_* : H^q(\mathcal{F}^\bullet(U)) \approx H^q(\mathcal{G}^\bullet(U)).$$

7.22. Proposition. *The quasi-isomorphism j induces isomorphism of the hypercohomologies $j_* : \mathbb{H}^*(M, \mathcal{F}^\bullet) \approx \mathbb{H}^*(M, \mathcal{G}^\bullet)$.*

Proof. j induces isomorphism at the first level of the second spectral sequence $E_1^{p,q} = H_{D''}^q(K^{p,0}(\mathcal{U})) = C^q(\mathcal{U}, \ker \mathrm{d}^p / \mathrm{Imd}^{p-1})$. □

Proof of Theorem 7.19. Consider the two complexes of sheaves: $\mathcal{G}^\bullet : 0 \longmapsto \mathcal{F} \longmapsto 0 \longmapsto \ldots$ and $\mathcal{F}^\bullet : 0 \longmapsto \mathcal{F}^0 \longmapsto \mathcal{F}^1 \longmapsto \ldots$.
They are quasi-isomorphic, because they have the same cohomology sheaves: equal $\mathcal{F}$ in the 0th term and 0 in the next terms. So, $\mathbb{H}^*(M, \mathcal{F}^\bullet) = \mathbb{H}^*(M, \mathcal{G}^\bullet)$.
The latter hypercohomology group equals to $H^*(M, \mathcal{F})$. □

7.23. Examples. (a) In the proof of the de Rham theorem the two complexes of sheaves $\mathbb{R}^\bullet : 0 \longmapsto \mathbb{R} \longmapsto 0 \longmapsto 0 \longmapsto \ldots$ and $\mathcal{E}^\bullet : 0 \longmapsto \mathcal{E}^0 \longmapsto \ldots$ are quasi-isomorphic (because both are exact in the right of $\mathbb{R}$). So they have isomorphic hypercohomology groups.
The second spectral sequence $\widetilde{E}_\bullet^{\bullet,\bullet}$, applied to the complex $\mathbb{R}^\bullet$, gives $\widetilde{E}_2^{p,q} = H^p(M,\mathbb{R})$, if $q=0$, and $=0$, otherwise; thus $\widetilde{E}_\infty = \mathbb{H}^* = H^*(M)$.
The first spectral sequence $E_\bullet^{\bullet,\bullet}$, applied to the complex $\mathcal{E}^\bullet$, gives $E_2^{p,q} = H^p_{\mathrm{dR}}(M)$, if $q=0$, and $=0$, otherwise; thus $E_\infty^{n,0} = \mathbb{H}^n = H^n_{\mathrm{dR}}(M)$. This is the de Rham theorem.

(b) The proof of the Dolbeault theorem runs in the same way but with the complexes $0 \longmapsto \Omega^p \longmapsto 0 \longmapsto 0\ldots$ and the Dolbeault complex $0 \longmapsto \mathcal{E}^{p,0} \longmapsto \ldots$.

(c) The quasi-isomorphic complexes: trivial

$$0 \longmapsto \mathbb{C} \longmapsto 0 \longmapsto \ldots$$

and holomorphic de Rham

$$\Omega^\bullet : 0 \longmapsto \Omega^0 \overset{\mathrm{d}}{\longmapsto} \Omega^1 \longmapsto \ldots$$

give the isomorphism $H^*(M,\mathbb{C}) \approx \mathbb{H}^*(M,\Omega^\bullet)$.
In the case of Kähler manifold one obtains additionally that the second spectral sequence, associated with the resolvent $\Omega^\bullet$, stabilizes in the second term $\widetilde{E}_2 = \widetilde{E}_\infty$, which gives the Hodge expansion $\mathbb{H}^n(M,\Omega^\bullet) \approx \bigoplus H^q(M,\Omega^{n-q})$.

7.24. Leray spectral sequence. This is a sequence used in computation of the cohomology groups of a fibre bundle.
Let $\pi : E \longmapsto B$ be a smooth fibre bundle, i.e., the sets $\pi^{-1}(U_\alpha)$ are diffeomorphic to $U_\alpha \times C$, where C is a smooth fiber and (U_α) is a covering of the base B.
The cohomology ring of E is calculated using the de Rham complex $\mathcal{E}^\bullet : 0 \longmapsto \mathcal{E}^0 \longmapsto \mathcal{E}^1 \ldots$. We introduce the following decreasing filtration $F^0 \supset F^1 \supset \ldots$ of the de Rham complex (and of the complex of global sections). If locally $\pi : (x,y) \longmapsto x$, then $F^p\mathcal{E}^n$ consists of forms $\sum a_{IJ}\mathrm{d}x_I \wedge \mathrm{d}y_J$ with $|I| \geq p$.
The **Leray spectral sequence** is the spectral sequence of the de Rham complex associated with the filtration $F^0 \supset F^1 \supset \ldots$.
Obviously

$$E_0^{p,q} = F^p\mathcal{E}^{p+q}/F^{p+1}\mathcal{E}^{p+q}$$

and consists of the forms $\eta = \sum \eta_I \wedge \mathrm{d}x_I$, with $|I| = p$ and η_I contain only $\mathrm{d}y_j$'s. It is also clear that

$$\mathrm{d}_0\eta = \sum \mathrm{d}_y\eta_I \wedge \mathrm{d}x_I.$$

This implies that

$$E_1^{p,q} = \ker \mathrm{d}_y/\mathrm{Im}\mathrm{d}_y = \mathcal{E}^p_B(H^q_{\mathrm{dR}}(E_x))$$

and consists of p-forms on B with values in the q-cohomologies of the fiber $E_x = \pi^{-1}(x)$. More precisely, the groups $H^q_{\mathrm{dR}}(E_x)$ are organized into the cohomological bundle, called the **Leray sheaf** $R^q\pi_*\mathbb{C}$ (denoted also $\mathcal{H}^q$). We have $\mathcal{H}^q(U) = H^q(\pi^{-1}(U)) \simeq U \times H^q(C)$. (This sheaf is the analogue of the cohomological Milnor bundle from Chapter 5). $\mathcal{H}^q$ is a local system with the Gauss–Manin connection induced by the lattices $H^q(E_x, \mathbb{Z})$.
The second term in the Leray spectral sequence is also easy to compute. If $\eta = \sum \eta_I \wedge \mathrm{d}x_I$ represents an element from $E_1^{p,q}$, i.e., $\mathrm{d}_y\eta = 0$, then

$$\mathrm{d}_1\eta = \mathrm{d}_x\eta.$$

This shows that $E_2^{p,q} = \ker \mathrm{d}_x/\mathrm{Im}\mathrm{d}_x$ can be identified with

$$H^p_{\mathrm{dR}}(B, \mathcal{H}^q).$$

In general situations, the Leray spectral sequence does not stabilize here. In [**GH**] it is shown that $E_2 \neq E_\infty$ in the case of the Hopf bundle $\mathbb{S}^{2n+1} \longmapsto \mathbb{CP}^n$.

7.25. Theorem (Leray spectral sequence). ([**Del1**]) *If the spaces E and B are Kähler manifolds, then the Leray spectral sequence degenerates at E_2. In particular*

$$H^n(E) \simeq \bigoplus_{p+q=n} H^p(B, \mathcal{H}^q).$$

Proof. We shall use the strong Lefschetz theorem 7.12.
Let ω be the Kähler $(1,1)$-form on E, whose restriction ω_x to each fiber E_x is also Kähler. Let $L : \eta \longmapsto \eta \wedge \omega_x$ be the Lefschetz operator. The action of L is compatible with the Leray spectral sequence, $L : E_r^{p,q} \longmapsto E_r^{p,q+2}$, and L commutes with the differentials d_r.
By Theorem 7.12: (i) $L^k : \mathcal{H}^{n-k} \longmapsto \mathcal{H}^{n+k}$, $n = \dim_{\mathbb{C}} E_x$, are isomorphisms, and (ii) $\mathcal{H}^q = \bigoplus_k L^k\mathcal{P}^{q-2k}$, where $\mathcal{P}^{n-i}$ are the sheaves of primitive cohomologies of fibers (kernels of L^{i+1}). We have then $E_2^{p,q} = \bigoplus_k L^k H^p(B, \mathcal{P}^{q-2k})$
In order to show that $\mathrm{d}_2 = 0$, it is enough to show that $\mathrm{d}_2 = 0$ on $H^p(B, \mathcal{P}^{n-i})$. Of course, $\mathrm{d}_2 : H^p(B, \mathcal{P}^{n-i}) \longmapsto H^{p+2}(B, \mathcal{H}^{n-i-1})$. Because $L^{i+1} : H^{p+2}(B, \mathcal{H}^{n-i-1}) \longmapsto H^{p+2}(B, \mathcal{H}^{n+i+1})$ is isomorphism, we have to show that $L^{i+1} \circ \mathrm{d}_2 = 0$. On the other hand, $L^{i+1} \circ \mathrm{d}_2 = \mathrm{d}_2 \circ L^{i+1}$, where $L^{i+1} = 0$ on $H^p(B, \mathcal{P}^{n-i})$.
In the same way one shows that other $\mathrm{d}_r = 0$. □

7.26. Landman's proof of the monodromy theorem using the Leray spectral sequence. The Leray spectral sequence is defined also in the situations, when the map $\pi : E \longmapsto B$ is only continuous (not a locally trivial bundle). For this one uses the corresponding filtration in the complex of singular cochains.
The second term of this sequence consists of the groups $E_2^{p,q} = H^q(B, R^p\pi_*\mathbb{C})$, where $R^p\pi_*\mathbb{C}(U) = \mathcal{O}(\mathcal{H}^p)(U) = H^p(\pi^{-1}(U))$, $U \subset B$, defines the Leray sheaf (the 'cohomological bundle').

A. Landman in his thesis in 1966 (see [**Lan**]) applied the Leray spectral sequence for the case of Clemens contraction map to the proof of the quasi-unipotency of the monodromy transformation (see Theorem 4.71). His idea of the proof is following. Recall the definition of the Clemens' contraction from the point 4.70. We have a holomorphic map $f : (X, X_0) \longmapsto (\mathbb{C}, 0)$ such that near any point of X_0 there is a system of holomorphic coordinates such that $f = x_1^{k_1} \dots x_r^{k_r}$. The Clemens map p_t sends a non-singular hypersurface $X_t = f^{-1}(t)$ to X_0 such that the preimage of a smooth point x_0 is finite (equal to multiplicity of the corresponding divisor) and a preimage of a point of r-fold intersection is a $(r-1)$-dimensional torus. The monodromy transformation $h : X_t \longmapsto X_t$ is "homologically" compatible with the map p_t and acts either by cyclic permutation (if $p_t^{-1}(x_0)$ is finite) or as a periodic translation on the corresponding torus.

This implies that the monodromy transformation acts diagonally upon the stalks of the Leray sheaf (with roots of unity as eigenvalues). Therefore, it induces a diagonal and periodic homomorphism h^* on the second terms $E_2^{p,q}$ of the Leray spectral sequence and, in consequence, on the final terms $E_\infty^{p,q} = \mathrm{Gr}_p H^{p+q}(X_t)$. Thus, the eigenvalues of the monodromy are roots of unity.

If $(M_s)^k = (\mathrm{Gr}\, h^*)^k = I$, then the only nonzero entries of the matrix $M^k - I$ are upper-triangular. (Here (M_s is the semi-simple part of the monodromy map and equals to the graded operator $\mathrm{Gr}\, h^* = \bigoplus^p \mathrm{Gr}_p h^*$, where $\mathrm{Gr}_p h^* = h^*|_{F^p}$ (mod F^{p+1})). This shows that the dimension of any Jordan cell is $\leq p+q+1$.

To obtain the Clemens' estimate (by $n - |p+q-n+1|$ where $n = \dim X$) one should apply the Poincaré duality. □

§3 Mixed Hodge Structures

7.27. Examples. (a) Let X be a closed algebraic curve with singularities in $S \subset X$, where we assume that the singular points are the double points (locally given by $xy = 0$).

The singular points can be resolved by means of the *normalization* (see the points 4.63, 4.64 and 4.65). This means that there is a smooth complete algebraic curve $\widetilde{X}$ and a morphism $\pi : \widetilde{X} \longmapsto X$ such that π is a diffeomorphism between $\widetilde{X} \setminus \pi^{-1}(S)$ and $X \setminus S$ and $\pi^{-1}(s)$ is a two-point set $\{x_1, x_2\}$ for each $s \in S$ (see Fig. 7.2).

Consider the constant sheaf $\mathbb{C}_X$ on X (see 3.26); it means that $\mathbb{C}_X(U) \approx \mathbb{C}$ for each open connected $U \subset X$. Analogously, we consider the sheaves $\mathbb{C}_{\widetilde{X}}$ on $\widetilde{X}$ and $\pi_*\mathbb{C}_{\widetilde{X}}$ on X. The latter is defined by means of $\pi_*\mathbb{C}_{\widetilde{X}}(U) = \mathbb{C}_{\widetilde{X}}(\pi^{-1}(U))$: if $U \cap S = \emptyset$, then $\pi_*\mathbb{C}_{\widetilde{X}}(U) \approx \mathbb{C}$, and, if U contains exactly one singular point $s \in S$, then $\pi_*\mathbb{C}_{\widetilde{X}}(U) \approx \mathbb{C} \oplus \mathbb{C}$. Let $\mathbb{C}_S$ be a constant sheaf on X supported in S: $\mathbb{C}_S(U) = 0$, if $U \cap S = \emptyset$, and $\mathbb{C}_S(U) = \mathbb{C}$, if $U \cap S = \{s\}$.

We have the exact sequence of sheaves on X

$$0 \longmapsto \mathbb{C}_X \longmapsto \pi_*\mathbb{C}_{\widetilde{X}} \longmapsto \mathbb{C}_S \longmapsto 0,$$

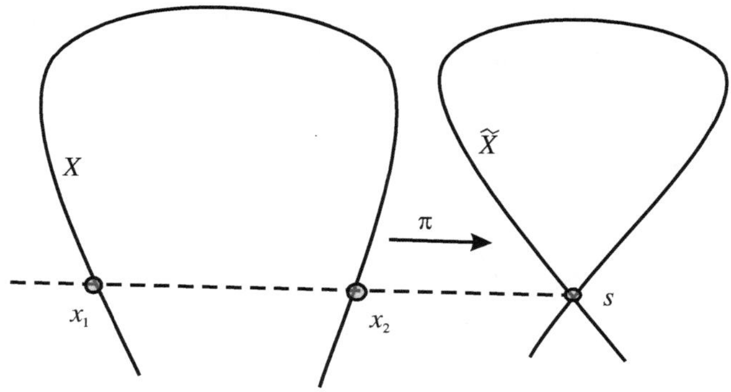

Figure 7.2: Normalization of the node singularity of a curve

(Outside S we have $0 \longmapsto \mathbb{C} \longmapsto \mathbb{C} \longmapsto 0 \longmapsto 0$ and at $s \in S$ we get $0 \longmapsto \mathbb{C} \longmapsto \mathbb{C} \oplus \mathbb{C} \longmapsto \mathbb{C} \longmapsto 0$, where the first arrow is the diagonal embedding and the third arrow is the difference).
As we know from Chapter 3 (see Lemma 3.29) this sequence induces the long exact sequence of the Čech cohomology groups

$$H^0(X, \pi_*\mathbb{C}_{\widetilde{X}}) \stackrel{0}{\longmapsto} H^0(X, \mathbb{C}_S) \longmapsto H^1(X, \mathbb{C}_X) \longmapsto H^1(X, \pi_*\mathbb{C}_{\widetilde{X}}) \longmapsto H^1(X, \mathbb{C}_S).$$

Here we have $H^k(X, \pi_*\mathbb{C}_{\widetilde{X}}) = H^k(\widetilde{X}, \mathbb{C})$ (see Example 3 in 3.26) and $H^k(X, \mathbb{C}_S) = H^k(S, \mathbb{C})$; thus the last group is zero. The first homomorphism is zero, because any element of $H^0(\widetilde{X}, \mathbb{C})$ is a constant function and the difference of its values at $x_{1,2}$ (above s) is equal to 0.
We see that the group $H^1(X, \mathbb{C})$ contains $H^0(S, \mathbb{C})$ as a subgroup and the quotient group $H^1(X, \mathbb{C})/H^0(S, \mathbb{C})$ is isomorphic to $H^1(\widetilde{X}, \mathbb{C})$. Because S and $\widetilde{X}$ are non-singular compact algebraic manifolds, they admit the natural Hodge structures.
One can introduce the filtration

$$0 = W_{-1} \subset W_0 \subset W_1 = H^1(X, \mathbb{C}),$$

where $W_0 = H^0(S, \mathbb{C})$. It has the property that the quotient groups W_n/W_{n-1} admit the Hodge structures

$$W_n/W_{n-1} = \bigoplus_{p+q=n} H^{p,q}.$$

For example, if $X = \{y^2 z = x^2(x - z)\}$ is a singular elliptic curve, then its normalization is a rational curve; $\widetilde{X} = \mathbb{CP}^1$,

$$\pi : t \longmapsto (x, y) = (1 + t^2, t(1 + t^2)).$$

Here $H^1(X, \mathbb{C}) = \mathbb{C}$, $H^1(\widetilde{X}, \mathbb{C}) = 0$, $H^0(S, \mathbb{C}) = \mathbb{C}$ (see Fig. 7.3). Thus $W_0 \simeq W_1 \approx H^1(X, \mathbb{C}) = H^{0,0}$, which means that the first cohomology group of X has the pure Hodge structure of weight 0.

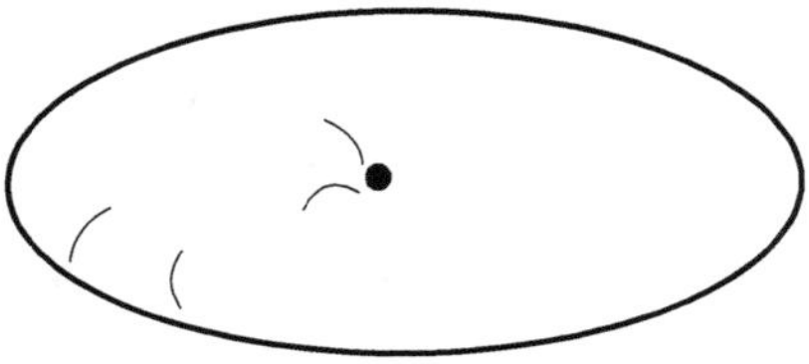

Figure 7.3: Singular elliptic curve

Of course, $H^1(X)$ cannot admit the pure Hodge structure of weight 1, because then it should be of the form $H^{1,0} \oplus H^{0,1}$ and should have even dimension.

(b) Let $X = \mathbb{C} \setminus \{0\}$ be the punctured complex line. Of course, there is no pure Hodge structure on $H^1(X, \mathbb{C}) = \mathbb{C}$. However, one can associate the generator α of $H^1(X, \mathbb{Z})$ with the point $Y = \{0\}$ in the completion $\overline{X}$ of X. α is the generator of $W_0 \approx H^0(Y, \mathbb{C})$. In this (artificial) sense one can introduce the Hodge structure on $W_0 = W_0/W_{-1} = H^{0,0}$.

(c) Let X be a curve consisting of two smooth (algebraic) components X_1 and X_2, intersecting transversally one another in two points Q_1 and Q_2. Let, additionally, X_i be not closed, they are obtained from the smooth closed curves $\overline{X}_i$ by deleting some points $P_1, \ldots, P_N$ (see Fig. 7.4).

We distinguish three kinds of generators of the group $H_1(X)$:

- the cycles α_i, surrounding the points P_i;
- the cycles β_j arising from $H_1(\overline{X}_i)$ (they are not defined uniquely, only modulo elements from the first group);
- the cycle γ which is defined uniquely modulo elements of the first two types.

This argumentation leads to the idea that there should exist a filtration

$$0 \subset W_0 \subset W_1 \subset W_2 = H_1(X),$$

such that the factors W_n/W_{n-1} arise from homologies of smooth complete manifolds.
Analogous filtrations should exist in the dual space $H^1(X)$. Due to this we can introduce the Hodge structure on W_n/W_{n-1}.

7.27. Definition (Mixed Hodge structure of a finite dimensional space). Let $H_{\mathbb{Q}}$ be a finite dimensional vector space over the field $\mathbb{Q}$ of rational numbers. By H we denote the complex space $H_{\mathbb{Q}} \otimes_{\mathbb{Q}} \mathbb{C}$.
A mixed Hodge structure on $H_{\mathbb{Q}}$ consists of:

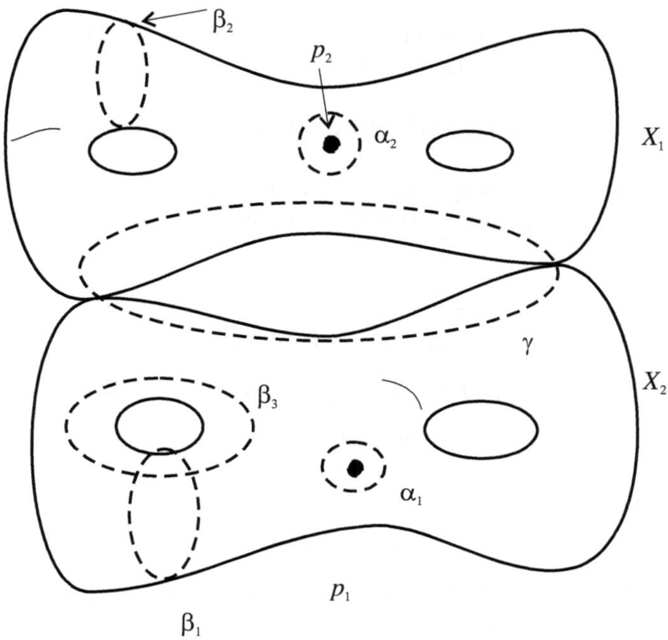

Figure 7.4: Cycles for a singular curve with punctures

(i) An increasing **weight filtration** $W^{\mathbb{Q}}$ on $H_{\mathbb{Q}}$:

$$0 \subset \ldots \subset W_{n-1}^{\mathbb{Q}} \subset W_n^{\mathbb{Q}} \subset \ldots \subset H_{\mathbb{Q}}.$$

We denote $W_j = W_j^{\mathbb{Q}} \otimes \mathbb{C} \subset H$ (see Fig. 7.5).

(ii) A decreasing **Hodge filtration** F on H :

$$H = F^0 \supset F^1 \supset \ldots \supset 0.$$

Let

$$\mathrm{Gr}_k = \mathrm{Gr}_k H = \mathrm{Gr}_k^W H = W_k / W_{k-1}$$

be the graded spaces. One introduces the following filtrations on the spaces $\mathrm{Gr}_k H$, induced by F

$$F^p \mathrm{Gr}_k H = (F^p \cap W_k + W_{k-1})/W_{k-1}.$$

The fundamental condition on the filtrations W, F is the

$$\mathrm{Gr}_k H = F^p \mathrm{Gr}_k H \oplus \overline{F^{k-p-1} \mathrm{Gr}_k H} \tag{7.3}$$

for any p, k. The conjugation operation is well defined due to existence of the canonical real subspace $H_{\mathbb{R}} = H_{\mathbb{Q}} \otimes \mathbb{R}$. If $e_1, \ldots, e_m$ is a basis of $H_{\mathbb{R}}$, then $\overline{\sum c_i e_i} = \sum \bar{c}_i e_i$ for vectors in $H_{\mathbb{C}}$.
If (7.3) holds, then we say that the filtrations W and F define the **mixed Hodge structure on $H_{\mathbb{Q}}$.**
The **morphism of mixed Hodge structures of weight** $2l$ is a $\mathbb{Q}$-linear map $\varphi : H_1 \longmapsto H_2$ of spaces with mixed Hodge structures which is compatible with the filtrations in a way that

$$\varphi(W_k H_1) \subset W_{k+2l} H_2,$$
$$\varphi(F^p H_1) \subset F^{p+l} H_2.$$

7.29. Remark. The filtration $F^\bullet = F^\bullet \mathrm{Gr}_n H$ defines the spaces

$$H^{p,q} = F^p \cap \overline{F^{k-p}}, \quad p + q = k,$$

and condition (7.3) is equivalent to

$$H^{p,q} = \overline{H^{q,p}}.$$

This gives the expansion $\mathrm{Gr}_k^W H = \bigoplus_{p+q=k} H^{p,q}$ and we say that the space $\mathrm{Gr}_k H$ admits the *pure Hodge structure of weight* k. The dimensions $h^{p,q} = \dim H^{p,q}$ are called the *Hodge numbers*.

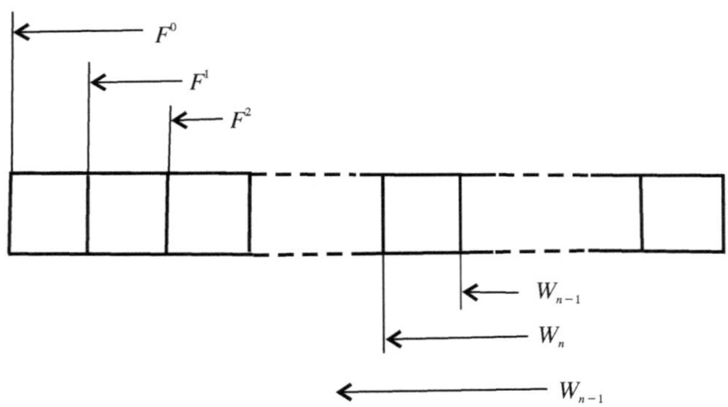

Figure 7.5: Hodge and weight filtrations

7.30. The mixed Hodge structure on the nth cohomology group of a complete Kähler manifold. If $H_{\mathbb{Q}} = H^n(M, \mathbb{Q})$ for a smooth complete algebraic variety M, then the filtrations: weight

$$0 = W_{n-1}^{\mathbb{Q}} \subset W_n^{\mathbb{Q}} = H_{\mathbb{Q}}$$

and Hodge

$$F^0 \supset F^1 \supset \ldots \supset F^n,$$

with $F^p = H^{n,0} \oplus H^{n-1,1} \oplus \ldots \oplus H^{p,n-p}$, define the mixed Hodge structure. Moreover, the Lefschetz operator $L : \eta \longmapsto \eta \wedge \omega$ is a morphism of mixed Hodge structures of weight 2.

In this example the Hodge filtration F is induced from certain filtration of the double complex $K^{\bullet,\bullet}$ (which leads to the expression of $H^*(M, \mathbb{C})$ as the hypercohomology groups of suitable complex of sheaves, see the previous section).
For example, in the case of smooth de Rham resolvent $0 \longmapsto \mathcal{E}^0 \longmapsto \ldots$, one introduces the filtration associated with the expansion of $\mathcal{E}^m$ into the sum of $\mathcal{E}^{p,q}$ (of forms of (p,q)-type). We call this filtration the *standard Hodge filtration*. Thus

$$F^p\mathcal{E}^\bullet = \bigoplus_{r \geq p, q} \mathcal{E}^{r,q}$$

and the induced filtration on $K^{\bullet,\bullet}$, $K^{r,s} = C^r(\mathcal{U}, \mathcal{E}^s)$ (the Čech r-cochains associated with a covering $\mathcal{U}$), is

$$F^pK^{\bullet,\bullet} = \bigoplus_{s,t} C^s(\mathcal{U}, F^p\mathcal{E}^t).$$

In the case of the holomorphic de Rham complex $0 \longmapsto \Omega^0 \longmapsto \ldots$, and $K^{r,s} = C^r(\mathcal{U}, \Omega^s)$, one puts the so called the *filtration bête* (or the *stupid filtration*)

$$F^pK^{\bullet,\bullet} = \bigoplus_{r; s \geq p} C^r(\mathcal{U}, \Omega^s).$$

By Theorem 7.11 the both initial filtrations lead to the same Hodge filtration in $H^n(M, \mathbb{C})$. The same alternative in the choice of the Hodge filtration we shall meet further.

7.31. Theorem of Deligne. ([**Del3**]) *If X is a quasi-projective variety, then its cohomology groups $H^n(X, \mathbb{Q})$ admit a natural mixed Hodge structures. If $f : X \longmapsto Y$ is a morphism of quasi-projective manifolds, then the homomorphism $f^* : H^n(Y, \mathbb{Q}) \longmapsto H^n(X, \mathbb{Q})$ is a morphism of mixed Hodge structures of weight 0. If X is complete and smooth, then the mixed Hodge structure coincides with the classical Hodge structure.*

In fact, P. Deligne proved existence of mixed Hodge structures in more general assumption: X is a scheme of finite type.
One uses the cohomologies with rational coefficients in definition of the weight filtration for the following reason. Firstly, one has the cohomology groups with integer coefficients, which have some torsion part. After passing to the complex coefficients, the torsion part is killed, $H^n(X, \mathbb{C}) = (H^n(X, \mathbb{Z})/\mathrm{Tor}) \otimes \mathbb{C}$ and we have a lattice $H^n(X, \mathbb{Z})/\mathrm{Tor} \subset H^n(X, \mathbb{C})$. The weight filtration should be a filtration of this lattice. The cohomologies with rational coefficients kill the torsion part and also contain an integer lattice. Sometimes one considers the weight filtration on $H^n(X, \mathbb{Z})/\mathrm{Tor}$.

In 7.32 and 7.33 we present the ideas lying behind the proof of this theorem. They will be used in the definition of the mixed Hodge structure associated with a degenerating family of algebraic manifolds. Finally, we introduce the mixed Hodge structure in fibers of the cohomological Milnor fibration.

7.32. Mixed Hodge structure on a semi-smooth variety. (a) Here we are dealing with a d-dimensional manifold of the form

$$M = M_1 \cup \ldots \cup M_m,$$

such that any point of M has neighborhood of the type

$$\{(z_1, \ldots, z_{d+1}) \in \mathbb{C}^{d+1} : \ z_1 \cdot z_2 \ldots z_k = 0, \ \ |z_i| < \epsilon\},$$

and M_i are smooth and compact projective varieties. We call such M a *semi-smooth variety.* Its singularities do not need resolution, the are normal crossing singularities.

(b) **Example.** Let $M = M_1 \cup M_2$, where M_i are smooth, compact and intersect one another transversally.
The exact sequence of sheaves

$$0 \longmapsto \mathbb{C}_M \longmapsto \mathbb{C}_{M_1} \oplus \mathbb{C}_{M_2} \longmapsto \mathbb{C}_{M_1 \cap M_2} \longmapsto 0$$

gives the *Mayer–Vietoris long exact sequence*

$$\ldots \longmapsto H^{n-1}(M_1 \cap M_2) \overset{\gamma_{n-1}}{\longmapsto} H^n(M) \overset{\alpha_n}{\longmapsto} H^n(M_1) \oplus H^n(M_2) \overset{\beta_n}{\longmapsto} H^n(M_1 \cap M_2) \longmapsto \ldots,$$

where $\beta_n(w_1 \oplus w_2) = i_1^* w_1 - i_2^* w_2$, $i_{1,2} : M_1 \cap M_2 \longmapsto M_{1,2}$.
Let $W_{n-1} = \mathrm{Im}\gamma_{n-1} = \mathrm{coker}\,\beta_{n-1}$, $W_n = H^n(M)$. Then $W_n/W_{n-1} \approx \mathrm{Im}\alpha_n \approx \ker \beta_n$. Because the homomorphisms γ_{n-1} and β_n are functorial, they preserve the pure Hodge structures.
Thus, one can introduce the pure Hodge structures on $W_{n-1}/W_{n-2} = W_{n-1}$ (of weight $n-1$ induced from $H^{n-1}(M_1 \cap M_2)$) and on W_n/W_{n-1} (of weight n induced from $H^n(M_1) \oplus H^n(M_2)$).

(c) Let us pass to the general case. One denotes the kth *skeleton* $M^{(k)}$ as the disjoint union of the subvarieties $M_{i_1} \cap \ldots \cap M_{i_k}$, $i_1 < \ldots < i_k$.
On $M^{(k)}$ there are the sheaves $\mathcal{E}^p_{M^{(k)}}$ of differentiable forms. We identify them with sheaves on M, i.e., as sheaves of forms with support in $M^{(k)}$, $i_* \mathcal{E}^p_{M^{(k)}}$, $i : M^{(k)} \longmapsto M$ is the inclusion.
One introduces the following double complex of sheaves

$$\mathcal{K}^{p,q} = i_* \mathcal{E}^p_{M^{(q+1)}},$$

with the differentials

$$D' = \mathrm{d} : \mathcal{K}^{p,q} \longmapsto \mathcal{K}^{p+1,q}$$

equal to the external derivative, and

$$D'' = (-1)^p \sum_{j=0}^{q+1} (-1)^j \delta_j^* : \mathcal{K}^{p,q} \longmapsto \mathcal{K}^{p,q+1},$$

where $\delta_j : M_{i_0} \cap \ldots \cap M_{i_q} \longmapsto M_{i_1} \cap \ldots \cap M_{i_{j-1}} \cap M_{i_{j+1}} \cap \ldots \cap M_{i_q}$ (combinatorial differential like in the Čech complex).
The double complex $(\mathcal{K}^{\bullet,\bullet}, D', D'')$ defines the cochain complex $(\mathcal{K}^\bullet, D)$, where

$$\mathcal{K}^k = \bigoplus_{p+q=k} \mathcal{K}^{p,q}, \quad D = D' + D''.$$

(Here one checks that $D^2 = 0$). Moreover, using Poincaré Lemma, one shows that this cochain complex is exact (as a complex of sheaves). Therefore, it forms a resolvent of the constant sheaf

$$0 \longmapsto \mathbb{C}_M \longmapsto \mathcal{K}^\bullet.$$

(d) Using the theorem about hypercohomology 7.19, we obtain that the group $H^n(M, \mathbb{C})$ is equal to the nth hypercohomology group of the complex $\mathcal{K}^\bullet$. However, the sheaves $\mathcal{K}^j$ are flabby sheaves, admitting partition of unity and the higher cohomologies with values in these sheaves vanish (see the proof of de Rham Theorem). Thus, we are in the same situation as in the case of calculation de Rham cohomology. The hypercohomology group of $\mathcal{K}^\bullet$ is equal to the homology group of the complex of global sections of $\mathcal{K}^\bullet$

$$0 \longmapsto \Gamma(M, \mathcal{K}^0) \longmapsto \Gamma(M, \mathcal{K}^1) \longmapsto \Gamma(M, \mathcal{K}^2) \longmapsto \ldots$$

Thus $H^n(M, \mathbb{C}) = H^n_D(\Gamma(M, \mathcal{K}^\bullet))$.

(e) Let us fix n. One defines three filtrations F, $\widehat{W}$ and W on $\mathcal{K}^\bullet$ as follows. We put

$$F^r = \bigoplus_{p+q=\bullet} F^r \mathcal{K}^{p,q},$$

$$W_k = \bigoplus_{p;q \leq n-k} \mathcal{K}^{p,q}.$$

where $F^\bullet$ is the natural Hodge filtration in the space of forms on a complex variety. The filtration F is called the *Hodge filtration* and the filtration W is called the *weight filtration*.
The Hodge and weight filtrations on $\mathcal{K}$ induce the Hodge and weight filtrations on the space $H = H^n_D(\Gamma(M, \mathcal{K}^\bullet))$.

(f) Let us calculate the induced weight filtration on $H^n(M)$. We have here the situation with filtered cochain complex. The calculations of the cohomology groups of filtered complexes are performed using the spectral sequences. In fact, the weight filtration is increasing, contrary to the case in definition of spectral sequence. So, we use the *reverse weight filtration* $\widehat{W}^\bullet$, $\widehat{W}^j = W_{n-j} = \oplus_{p,q\geq j}\mathcal{K}^{p,q}$. We call the spectral sequence associated with the reverse weight filtration the *weight spectral sequence* and denote it by ${}_W E_r^{p,q} = E_r^{p,q}$.
We have $E_0^{p,q} = \Gamma(M, \mathrm{Gr}^p_{\widehat{W}}\mathcal{K}^{p+q}) = \Gamma(M, \mathcal{K}^{q,p}) = \Gamma(M^{(p+1)}, \mathcal{E}^q)$ (transposed indices!) and the differential d_0 is induced by $D' = \mathrm{d}$. We see that $E_1^{p,q} = H^q(M^{(p+1)})$.
Since $\mathrm{Gr}^p_{\widehat{W}}\mathcal{K}^{p+q} = \mathrm{Gr}^W_q\mathcal{K}^{p+q}$, $p+q=n$, the index q corresponds exactly the index of the weight filtration $W_\bullet$, as well as the degree of the cohomology group $H^q(M^{(p+1)}) \simeq_W E_r^{p,q}$. Thus the terms ${}_W E_1^{p,q}$ admit pure Hodge structure of weight q.
The second term $E_2^{p,q}$ of the weight spectral sequence is nothing else as the cohomology group of the complex $\ldots H^q(M^{(p)}) \longmapsto H^q(M^{(p+1)}) \longmapsto H^q(M^{(p+2)}) \longmapsto \ldots$, where the differentials d_1 are induced by D'', which are composed of the restriction maps for components $A \in M^{(j)}$, $B \in M^{(j+1)}$, $A \subset B$.
Of course, the differentials d_1 are morphisms of pure Hodge structures (of weight 0) and the terms $E_2^{p,q}$ admit pure Hodge structures of weight $n-p$. The next proposition shows that this spectral sequence degenerates here, i.e., $H^n(M) = \bigoplus_q {}_W E_2^{n-q,q}$. This expansion gives the weight filtration of $H^n(M) : \mathrm{Gr}^W_q H^n(M) =_W E_2^{n-q,q}$ with pure Hodge structure of weight q.

(g) **Proposition.** *The differentials* $\mathrm{d}_2 = \mathrm{d}_3 = \ldots = 0$ *in the weight spectral sequence.*

Proof. (We follow [**KK**]). An element from $E_2^{p,q} =_W E_2^{p,q}$ is represented by a harmonic q-form ω_p on $M^{(p+1)}$ such that $D''[\omega_p] = 0$ in $H^q(M^{(p+2)})$, i.e., $D''\omega_p$ is exact form on $M^{(p+2)}$.
The class $[\omega_p]$ is defined modulo $\widehat{W}_{p+1}$. Thus, in order to show that $\mathrm{d}_2[\omega_p] = 0$, it is enough to find a form $\omega_{p+1} \in \widehat{W}_{p+1}$ such that $D(\omega_p + \omega_{p+1}) = 0 \mod \widehat{W}_{p+3}$. Generally, if one finds $\omega_{p+1} \in \widehat{W}_{p+1}, \omega_{p+2} \in \widehat{W}_{p+2}, \ldots, \omega_{p+q} \in \widehat{W}_{p+q}$ such that $D(\omega_p + \ldots + \omega_{p+q}) = 0$, then one will have all $\mathrm{d}_j = 0$. We concentrate on showing that $\mathrm{d}_2 = 0$.
The construction of ω_{p+1}'s can be done separately in each (r,s)-harmonic summands. Thus we, assume that ω_p is a harmonic (r,s)-form.
Because each summand in the skeleton $M^{(p+2)}$ is a smooth projective variety, we can apply the differential calculus on Kähler varieties (see the points 7.10 and 7.11). Thus, $\Delta_{\mathrm{d}} = 2\Delta_\partial = 4\partial\partial^* = 4\partial^*\partial = 2\Delta_{\bar\partial} = 4\bar\partial\bar\partial^* = 4\bar\partial^*\bar\partial$.
Because the form $\eta = D''\omega_p$ is exact on $M^{(p+1)}$, it represents zero cohomology class and has zero harmonic component. Thus $\rho = 4\Delta_{\bar\partial}^{-1}\eta$ is well defined and satisfies $\eta = \partial\partial^*\rho$. Next, as $\mathrm{Im}\partial^* \perp \ker\Delta_\partial$, $\partial^*\rho$ has also zero harmonic component and $\sigma = 4\Delta_{\bar\partial}^{-1}(\partial^*\rho)$ is well defined and satisfies $\partial^*\rho = \bar\partial\bar\partial^*\sigma$. From this and from

the previous we obtain the representation $\eta = \partial\bar{\partial}\gamma$, $\gamma = \bar{\partial}^*\sigma$ (the $\partial\bar{\partial}$-*Lemma*, see [**KK**] and [**GH**]).
We put $\omega_{p+1} = \partial\gamma$. The direct calculations show that $D(\omega_p + \omega_{p+1}) = D''\omega_{p+1} \in \widehat{W}_{p+2}$. But $D''\omega_{p+1}$ represents zero ∂-de Rham cohomology class (mod $\widehat{W}_{p+3}$). Hence $\mathrm{d}_2 = 0$.
The proof of the identity $\mathrm{d}_3 = 0$ uses the fact that $D''\omega_{p+1}$ is exact. One can again apply to it the $\partial\bar{\partial}$-lemma and repeat the above analysis, etc. □

(h) Above, the filtration W was defined only on the complex spaces $H^n(M, \mathbb{C})$; the definition of mixed Hodge structure needs its definition on the rational space $H^n(M, \mathbb{Q})$. We have seen that the weight filtration and the weight spectral sequence are associated with the cohomologies of smooth varieties $M^{(j)}$ and continuous mas between them. They are well defined over $\mathbb{Q}$.
The rigorous proof uses the spectral sequence of the hypercohomology associated with the resolution $0 \longmapsto \mathbb{Q}_M \longmapsto i_*\mathbb{Q}_{M^{(1)}} \stackrel{D''}{\longmapsto} \ldots$. Namely, one considers the double complex whose entries are the Čech l-cochains with values in the sheaves $i_*\mathbb{Q}_{M^{(k)}}$. We do not present the details and refer the reader to the works of P. Deligne [**Del3**] and to the paper of J. Steenbrink [**Ste1**] (see also [**KK**]). The final result follows.

(i) **Theorem (Mixed Hodge structure on semi-smooth variety).** *The space* $H_{\mathbb{Q}} = H^n(M, \mathbb{Q})$ *admits a mixed Hodge structure such that the spaces* $\mathrm{Gr}_k H$ *have pure Hodge structure of weight* k.

(j) **Example (b) revisited.** Let $M = M_1 \cup M_2$ with smooth algebraic closed curves $M_{1,2}$ and $M_1 \cap M_2 = \{p_1, \ldots, p_k\}$. Then the group $\mathrm{Gr}_1^W H^1(M)$ is equal to the homology group of the complex

$$0 \longmapsto H^1(M_1) \oplus H^1(M_2) \longmapsto H^1(M_1 \cap M_2)$$

and $\mathrm{Gr}_0^W H^1(M)$ is calculated from the sequence

$$H^0(M_1) \oplus H^0(M_2) \longmapsto H^0(M_1 \cap M_2) \longmapsto 0.$$

Thus $\mathrm{Gr}_1 H^1(M) = H^1(M_1) \oplus H^1(M_2)$ (with $h^{1,0}(M) = h^{1,0}(M_1) + h^{1,0}(M_2)$) and $\mathrm{Gr}_0 H^1(M) = H^{0,0} = \mathbb{C}^k/\mathbb{C}(1, \ldots, 1) = \mathbb{C}^{k-1}$ (with $h^{0,0}(M) = k - 1$).
Similarly, $\mathrm{Gr}_2 H^2(M) = H^2(M_1) \oplus H^2(M_2) = H^{1,1} = \mathbb{C}^2$ and $\mathrm{Gr}_1 H^2(M) = \mathrm{Gr}_0 H^2(M) = 0$.

(k) **Remark.** There is another possibility to choose the building blocks for the resolution complex $\mathcal{K}^\bullet$. Instead of the flabby sheaves of smooth differential forms on $M^{(k)}$, one can use the sheaves of holomorphic forms

$$\mathcal{K}_{\mathrm{hol}}^{p,q} = i_*\Omega^p_{M^{(q+1)}}.$$

The corresponding complex $0 \longmapsto \mathbb{C} \longmapsto \mathcal{K}_{\mathrm{hol}}^\bullet$ is exact (by the holomorphic Poincaré lemma). Thus $H^n(M) = \mathbb{H}^n(M, \mathcal{K}^\bullet)$. The preliminary weight filtration is the same as in the smooth case, but as the preliminary Hodge filtration we use the filtration bête $F^p\mathcal{K}^{\bullet,\bullet}_{\text{hol}} = \bigoplus_{r\geq p,q}\mathcal{K}^{r,q}_{\text{hol}}$. This approach is preferred by Deligne. In the sequel, we shall introduce various mixed Hodge structures in different situations. In all these cases there is following alternative: either to use the sheaves of smooth forms with their standard Hodge filtration or to use the sheaves of holomorphic forms with their stupid filtration. We shall use preferably the holomorphic sheaves.

7.33. Mixed Hodge structure on smooth incomplete manifold. (a) We assume here that M^* is an open subset of a smooth closed d-dimensional algebraic variety M such that

$$M^* = M\diagdown N,$$

where M is closed smooth variety and N is of the form

$$N_1 \cup \ldots \cup N_m,$$

where N_i are smooth closed hypersurfaces with normal intersections (like M_i from the previous point).
The main tools which are used in the analysis of this case are:

- the sheaf of holomorphic forms with logarithmic singularities along N,
- the Poincaré residuum.

(b) **Definition**. The sheaf on M

$$\Omega^n_M(\log N) = \Omega^n(\log)$$

of *holomorphic n-forms with logarithmic singularities along* N is locally generated by the forms

$$\frac{\mathrm{d}z_1}{z_1}, \ldots, \frac{\mathrm{d}z_k}{z_k}, \mathrm{d}z_{k+1}, \ldots, \mathrm{d}z_n,$$

where $z_1 \cdot \ldots \cdot z_k = 0$ is the local equation for N in M. The coefficients are holomorphic functions on M.

(c) This logarithmic sheaf admits the following *preliminary weight filtration* (or the *filtration by the order of poles*)

$$0 = \widetilde{W}_{-1} \subset \widetilde{W}_0 \subset \widetilde{W}_1 \subset \ldots \subset \widetilde{W}_n,$$

where $\widetilde{W}_k$ consists of forms of the form $\eta = \sum \eta_I \wedge (\mathrm{d}z_I/z_I)$ such that $|I| \leq k$. Here the sum runs over multi-indices and $\mathrm{d}z_I/z_I = (\mathrm{d}z_{i_1}/z_{i_1})\wedge\ldots\wedge(\mathrm{d}z_{i_l}/z_{i_l})$, $I = (i_1, \ldots, i_l)$, and η_I are holomorphic forms. Thus $\widetilde{W}_0 = \Omega^n_M$, $\widetilde{W}_1 \approx \{\sum \eta_i \wedge (\mathrm{d}z_i/z_i)\}$, etc.

(d) **Definition**. The **Poincaré residuum**

$$R^l : \widetilde{W}_l \longmapsto \Omega^{n-l}_{N^{(l)}}$$

is defined by means of the formula

$$R^l(\omega \wedge \mathrm{d}z_I/z_I) = (2\pi \mathrm{i})^l \omega|_{N_{i_1}\cap\ldots\cap N_{i_l}}$$

near points from $N_{i_1} \cap \ldots \cap N_{i_l}$. In other words, we integrate the form $\omega \wedge \mathrm{d}z_I/z_I$ along the l-dimensional cycle $\{|z_{i_j}| = \epsilon\}$ in $M^* = M \setminus N$.
Here we put $N^{(0)} = M$; thus R^0 is the identity.
The operator R^l commutes with the differential d and is equal to zero at $\widetilde{W}_{l-1}$.
The holomorphic logarithmic complex is used to compute the cohomologies of M^*. Namely we have the following.

(e) **Proposition.** *$H^n(M^*, \mathbb{C}) \approx \mathbb{H}^n(M, \Omega^\bullet(\log))$, where $\mathbb{H}$ denotes the hypercohomology groups introduced in Section 2.*

Proof. It is known from 7.21 that the cohomology groups of M^* are isomorphic to the hypercohomology groups associated with the complex of holomorphic forms, $H^n(M^*, \mathbb{C}) \approx \mathbb{H}^n(M^*, \Omega^\bullet_{M^*})$. The sheaves $\Omega^k_{M^*}$ induce the sheaves $j_*\Omega^k_{M^*}$ on M, where $j : M^* \longmapsto M$; they are meromorphic forms with poles along the subvariety N. We have $\mathbb{H}^*(M^*, \Omega^\bullet_{M^*}) = \mathbb{H}^*(M, j_*\Omega^\bullet_{M^*})$ (see Example 3 in 3.26).
The isomorphism $H^n(M^*, \mathbb{C}) \simeq \mathbb{H}^n(M, \Omega^\bullet(\log))$ follows now from the following fact:

The complexes of sheaves $j_*\Omega^\bullet_{M^*}$ and $\Omega^\bullet_M(\log)$ on M are quasi-isomorphic (see Proposition 7.22 about quasi-isomorphism).

To show the above quasi-isomorphism, it is enough to show that the complexes (of local sections) $\Gamma(U, \Omega^\bullet_M(\log))$ and $\Gamma(U, j_*\Omega^\bullet_{M^*})$ have the same homologies for $U = \mathbb{D}^l \times \mathbb{D}^{d-l}$ near a point where $N : x_1 \ldots x_l = 0$. The cohomology groups $H^m_{\mathrm{dR}}(U \cap M^*, \mathbb{C})$ are generated by the forms $\mathrm{d}x_j/(2\pi \mathrm{i} x_j)$. Such are also the generators of $H^m(\Gamma(U, j_*\Omega^\bullet_{M^*}))$ and of $H^m(\Gamma(U, \Omega^\bullet(\log)))$. This induces the desired quasi-isomorphism. □

(f) **Remark.** In calculations of the Čech cohomologies one usually uses a covering $\mathcal{U}$ such that $\check{H}^q(U_{i_1} \cap \ldots \cap U_{i_r}, \mathcal{F}) = 0$ for $q \geq 1$ and $U_i \in \mathcal{U}$. Then by the theorem of Leray about calculation of Čech cohomology (see [**God**]) $\check{H}(X, \mathcal{F}) = H^q(\mathcal{U}, \mathcal{F})$. In all cases considered above one could choose the covering of manifold by means of polydisc $U \simeq \mathbb{D}^d$. In the case of the sheaf $j_*\Omega^m_M$ one should consider also subsets of the form $\mathbb{D}^l \times \mathbb{D}^{d-l}$ around the points where $N = \{x_1 \ldots x_l = 0\}$. We have $\check{H}^q(U, j_*\Omega^m_{M^*}) = H^q((\mathbb{D}^*)^l \times \mathbb{D}^{d-l}, \Omega^m_M)$. The vanishing of the latter group for $q \geq 1$ is a consequence of the Cartan Lemma B on Stein manifolds, proved in [**GuRo**]. (By the *Stein manifold* one means an analytic submanifold of an affine complex space). Similar results are proved in the next chapter (in 8.37 and 8.38).)

(g) We introduce the *weight* and *Hodge filtrations* on the sheaf complex $\Omega^\bullet(\log)$:

$$W_{n+r} = \widetilde{W}_r \Omega^\bullet(\log)$$

(i.e., the shift of indices) and

$$F^p\Omega^\bullet(\log) = \bigoplus_{k\geq p} \Omega^k(\log).$$

(h) **Theorem (Mixed Hodge structure on smooth incomplete variety).** *The above two filtrations induce a mixed Hodge structure on $H^n(M^*, \mathbb{C})$. Moreover, this mixed Hodge structure does not depend on the specific compactification M.*

Proof. The proof that the above two filtrations define the mixed Hodge structure relies on the fact that the Poincaré residuum operator realizes a quasi-isomorphism of the complexes

$$\mathrm{Gr}^W_{n+r} = \mathrm{Gr}^{\widetilde{W}}_r \Omega^\bullet(\log) \longmapsto i_*\Omega^{\bullet-r}_{N^{(r)}}.$$

Recall that $N^{(0)} = M$.
In literature the shift of indices, like $i_*\Omega^{\bullet-r}_{N^{(r)}}$, is often denoted by adding $[-r]$ without changing other indices; e.g., $i_*\Omega^\bullet_{N^{(r)}}[-r]$. We shall prefer indication of a shift, like above.
As in the previous point, one has to introduce the *weight spectral sequence*. For this, one introduces the *reverse weight filtration* $\widehat{W}_k = \widetilde{W}_{-k}$, $k = -d, -d+1, \ldots, 0$ of the complex $\Omega^\bullet(\log)$, reverse to the preliminary weight filtration; (thus $\mathrm{Gr}^{\widehat{W}}_{-r} = \mathrm{Gr}^W_{n+r}$).
The first term of the *weight spectral sequence* ${}_WE_1^{p,q}$, $p = -r$, $p + q = n$, is equal to

$$\begin{aligned} {}_WE_1^{p,q} &= \mathbb{H}^n(M, \mathrm{Gr}^{\widehat{W}}_p \Omega^\bullet(\log)) \\ &= \mathbb{H}^n(N^{(r)}, \Omega^{\bullet-r}) \\ &= H^{n-r}(N^{(r)}) \\ &= H^{q+2p}(N^{(-p)}). \end{aligned}$$

As the Poincaré residuum shifts the Hodge filtration, $R^l : F^s\Omega^\bullet \longmapsto F^{s-l}\Omega^{\bullet-l}$, we have $F^s\mathbb{H}^n(M, \mathrm{Gr}^{\widehat{W}}_{-r}) = \widetilde{F}^{s-r}H^{n-r}(N^{(r)})$, where $\widetilde{F}$ denotes the natural Hodge filtration on $H^{n-r}(N^{(r)})$ of weight $n - r$. We obtain

$$\begin{aligned} \overline{F^s\mathbb{H}^n(M, \mathrm{Gr}^{\widehat{W}}_{-r})} &= \overline{\widetilde{F}^{s-r}H^{n-r}(N^{(r)})} \\ &= \widetilde{F}^{n-s}H^{n-r}(N^{(r)}) \\ &= F^{(n+r)-s}\mathbb{H}^n(M, \mathrm{Gr}^{\widehat{W}}_{-r}). \end{aligned}$$

This shows that

The first terms $E_1^{-r,n+r}$ of the weight spectral sequence admit pure Hodge structures of weight $n + r$.

This agrees with our introduction of the weight filtration: $\mathrm{Gr}^W_{n+r} = \mathrm{Gr}^{\widehat{W}}_{-r}$.
Now, we should describe the differential $\mathrm{d}_1 : E_1^{p,q} \longmapsto E_1^{p+1,q}$ and show that it is a morphism of pure Hodge structures.
d_1 acts from $H^{n-2r}(N^{(r)})$ to $H^{n-2r+2}(N^{(r-1)})$ and its description is more or less the following. Assume that a class from $H^{n-2r}(A)$, A – a component from $N^{(r)}$, is represented by a form η. Let also A be a hypersurface of B, where B is a summand of $N^{(r-1)}$. Locally $A = \{z = 0\} \subset B$. We take $(2\pi\mathrm{i})^{-1}\mathrm{d}(\eta \wedge \mathrm{d}\ln z)$ as the "local" representative of the class $\mathrm{d}_1[\eta]$. In order to improve this definition globally, one replaces the form $\mathrm{d}\ln z$ by $\partial \ln |s|^2$, where s is a global section of certain 1-dimensional holomorphic bundle on B, equipped with some Hermitian metric $|\cdot|$. This bundle is the bundle $[A]$ associated with the divisor A and the section s has first order zero on A (see the point 10.10 in Chapter 10 below).
The map d_1 is known as the *Gysin map.* It can also be defined as the composition $\pi_B \circ j_* \circ \pi_A^{-1}$ of homomorphisms, where $\pi_Z : H^i(Z) \longmapsto H_{2\dim Z - i}(Z)$ is the Poincaré duality and j_* is induced by the inclusion.
It is not difficult to see that the Gysin map is a morphism of pure Hodge structures of weight 2. For example, the homology spaces $H_m(Z)$ are equipped with pure Hodge structures of weight $-m$, as dual to the spaces $H^m(Z)$, and π_Z is a morphism of spaces with Hodge structures.

The weight spectral sequence degenerates in the second term, i.e., $\mathrm{d}_2 = \mathrm{d}_3 = \ldots = 0$ *and* $\mathrm{Gr}_{n+r}H^n(M^*) = E_2^{-r,n+r}$ *is equal to the homology of the complex*

$$H^{n-2r-2}(N^{(r+1)}) \longmapsto H^{n-2r}(N^{(r)}) \longmapsto H^{n-2r+2}(M^{(r-1)}).$$

This fact can be proved in similar way as in the previous point (using smooth de Rham complexes), but we present another proof (from [**Del3**]).
The Hodge filtration is the filtration bête on the sheaf $\Omega^\bullet(\log)$. This filtration induces Hodge filtrations $F^\bullet$ on ${}_W E_1^{p,q}$ (and on ${}_W E_r^{p,q}$, $r > 1$). One can see that the differentials $\mathrm{d}_1, \mathrm{d}_2, \ldots$ are compatible with $F^\bullet$; $\mathrm{d}_r F^a \subset F^a$ and $\mathrm{d}_r \overline{F}^b \subset \overline{F}^b$. We have $E_r^{p,q} = \bigoplus_{a+b=q} F^a(E_r^{p,q}) \cap \overline{F}^b(E_r^{p,q}) = \bigoplus F^a \cap \overline{F}^b$. But then $\mathrm{d}_r\left(F^a \cap \overline{F}^b\right) \subset F^a \cap \overline{F}^b$ in $E_r^{p+r,q-r+1}$. It is clear that the latter intersection is zero.
Because the Poincaré duality and the Gysin map are well defined over $\mathbb{Q}$, the weight filtration is induced from a filtration of $H^n(M^*, \mathbb{Q})$. We omit the rigorous proof of this fact.
The second statement of Theorem (h) states that: if we take another embedding $M^* \subset \widetilde{M}$, with smooth $\widetilde{M}$ and with semi-smooth complement $\widetilde{M} - M^* = \widetilde{N}$, then the corresponding filtrations on the sheaf complex $\Omega^\bullet_{\widetilde{M}}(\log \widetilde{N})$ lead to the same mixed Hodge structure as defined above. We omit the proof of this property. □

(i) **Remark.** The weight filtration has the form

$$0 = W_{n-1} \subset W_n \subset \ldots \subset W_{2n} = H^n(M)$$

and j^* maps $H^n(M)$ onto W_n.

(j) **Example.** Let $M^* = M\setminus\{p_1, \ldots, p_k\}$, where M is a smooth compact curve. Then $\mathrm{Gr}_1 H^1(M^*)$ is calculated from the sequence

$$0 \longmapsto H^1(M) \longmapsto 0$$

and $\mathrm{Gr}_2 H^1(M^*)$ from

$$0 \longmapsto \bigoplus H^0(p_i) \longmapsto H^2(M).$$

We find $\mathrm{Gr}_1 H^1(M^*) = H^1(M)$ with $h^{1,0}(M^*) = h^{1,0}(M)$ and $\mathrm{Gr}_2 H^1(M^*) = H^{1,1} = \mathbb{C}^{k-1}$.

(k) **Remark.** Instead of dealing with the holomorphic logarithmic complex with its stupid filtration, we could deal with the smooth logarithmic de Rham complex $\mathcal{E}^n(\log) = \bigoplus \mathcal{E}^{p,q}(\log) = \bigoplus \Omega^p(\log) \otimes \mathcal{E}^{0,q}$ (of smooth forms containing $\mathrm{d}z_i/z_i$, $\mathrm{d}z_j$ and $\mathrm{d}\bar{z}_k$). The preliminary weight filtration is the same as in the holomorphic case and the Poincaré residuum identifies the graded sheaves with the smooth de Rham sheaves on $N^{(k)}$'s. One can show that $H^n(M^*)$ is equal to the nth cohomology group of the complex of global forms from $\mathcal{E}^\bullet(\log)$ with the external derivative as differential, i.e., to the logarithmic de Rham cohomology group. The natural weight filtration and the standard Hodge filtration associated with the expansion into (p,q)-components induce the mixed Hodge filtrations on the logarithmic de Rham cohomology group. The latter coincides with the above mixed Hodge filtrations of the hypercohomology group of the holomorphic logarithmic complex.

§4 Mixed Hodge Structures and Monodromy

7.34. Degeneration of algebraic manifolds and limits of Hodge structures. Consider a family X_t, $t \in \mathbb{D} = \{|t| < 1\}$, of algebraic closed manifolds degenerating at $t = 0$. More precisely, we have a holomorphic map $f : X \longmapsto \mathbb{D}$ with smooth compact algebraic fibers $X_t = f^{-1}(t)$, $t \neq 0$, and such that the non-smooth X_0 does not need desingularization (union of smooth divisors with normal crossings). Thus we have a degeneration of algebraic manifolds (semi-stable or not).
For each $t \neq 0$ and any integer n, the group $H^n(X_t, \mathbb{C})$ (with the rational subgroup $H^n(X_t, \mathbb{Q})$) admits a pure Hodge structure. Although the groups $H^n(X_t)$ do not change with t, the Hodge structure depends on t. The Hodge filtration $F_t^0 \supset F_t^1 \supset \ldots$ depends on t.
The spaces $H^n(X_t, \mathbb{C})$, $t \neq 0$, organize themselves into the cohomological bundle $\mathcal{H}^n$ above the punctured disc $\mathbb{D}^* = \mathbb{D}\setminus 0$; it is the local system, called also the nth direct image of constant sheaf and denoted $R^n f_* \mathbb{C}_{X - X_0}$. On each fiber $H^n(X_t, \mathbb{C})$ (of the bundle $\mathcal{H}^n$) the monodromy operator M associated with the simple loop in $\mathbb{D}^*$ acts. The monodromy operator preserves the integer lattice $H^n(X_t, \mathbb{Z})/\mathrm{Tor}$ and the rational subspace $H^n(X_t, \mathbb{Q})$. The bundle $\mathcal{H}^*$ is equipped

with the Gauss–Manin connection ∇. The Hodge subspaces F_t^p also organize themselves into bundles $\mathcal{F}^p$ on $\mathbb{D}^*$.

One of the aims of this section is to study the limit of the Hodge filtrations in $H^n(X_t)$ as $t \to 0$. It turns out that the limit filtration ceases to generate a Hodge structure. But one can introduce certain weight filtration so that the both filtrations together define a mixed Hodge structure called the *limit mixed Hodge structure.*

Of course, in order to compare the Hodge filtrations (for different t's) one has to fix some (complex) space H_∞ with isomorphisms $H^n(X_t) \longmapsto H_\infty$. The Hodge filtrations in $H^n(X_t)$ induce a family of Hodge filtrations in H_∞ and the weight filtration is defined in H_∞.

In the W. Schmid's and J. Steenbrink's approaches the choices of the space H_∞ are different and the weight filtrations are defined differently. However, their mixed Hodge structures turn out to be isomorphic.

In the whole section the integers $d = \dim X$ and n (for $H^n(X_t)$) are fixed.

7.35. The Schmid's limit mixed Hodge structure. We sketch briefly the Schmid's approach to this problem (see [**Sch**]).

This approach is based on the *period mapping* $\Phi : \mathbb{D}^* \longmapsto D$ introduced in the next section. Here D is a homogeneous space; it consists of all Hodge structures (with polarization) on a fixed vector space $H = H_\infty$ modulo action of some algebraic subgroup of $GL(H)$ (like the Grassmann variety). The period mapping is a multivalued holomorphic mapping.

We get a family of Hodge filtrations $H = \widetilde{F}_t^0 \supset \widetilde{F}_t^1 \supset \dots \ \widetilde{F}_t^p = \Phi(F_t^p)$. Because Φ is multivalued, the limit of $\widetilde{F}_t^p$, $t \to 0$, may not exist. However, after small modification, the spaces $\Phi(t^{-\log M/2\pi\mathrm{i}} F_t^p) = \Phi(t^{-\log M_u/2\pi\mathrm{i}} F_t^p)$, where M_u is the unipotent factor of the monodromy operator M restricted to $H^n(X_t)$, tend to limits F_∞^p. Here the Hodge filtration is invariant with respect to the semi-simple factor M_s of M.

The limit filtration does not generate a Hodge structure in H, but it induces a mixed Hodge structure, when one introduces the *weight filtration* in H as follows. Let $N = \log M_u$, which we treat as acting on H (see Fig. 7.6). It is a nilpotent operator and there exists m such that $N^m \neq 0 = N^{m+1}$. The weight filtration is of the following form (of length $2m-1$ and with W_n in the central place)

$$0 \subset W_{n-m} \subset \dots \subset W_n \subset \dots \subset W_{n+m} = H.$$

We put $W_{n-m} = \mathrm{Im} N^m$ and $W_{n+(m-1)} = \ker N^m$; thus N^m realizes isomorphism between $\mathrm{Gr}_{n+m} = \frac{W_{n+m}}{W_{n+(m-1)}}$ and $\mathrm{Gr}_{n-m} = W_{n-m}$. Next, we define the following filtration on $\frac{W_{n+(m-1)}}{W_{n-m}}$: one puts $\frac{W_{n-(m-1)}}{W_{n-m}} = \mathrm{Im}\left(N^{m-1}|\frac{W_{n-(m-1)}}{W_{n-m}}\right)$ and $\frac{W_{n+(m-2)}}{W_{n-m}} = \ker\left(N^{m-1}|\frac{W_{n+(m-1)}}{W_{n-m}}\right)$. The same is done with $\frac{W_{n+(m-2)}}{W_{n-(m-1)}}$: $\frac{W_{n-(m-2)}}{W_{n-(m-1)}} = \mathrm{Im}\left(N^{m-2}|\frac{W_{n+(m-2)}}{W_{n-(m-1)}}\right)$ and $\frac{W_{n+(m-3)}}{W_{n-(m-1)}} = \ker\left(N^{m-2}|\frac{W_{n+(m-2)}}{W_{n-(m-1)}}\right)$, etc. In this way we define the whole weight filtration.

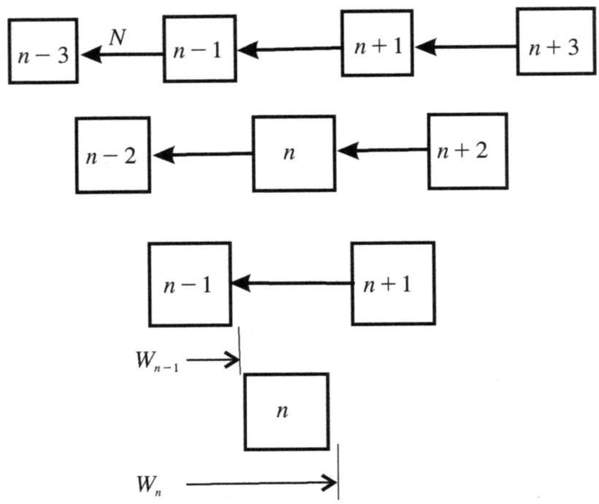

Figure 7.6: Action of the 'nilpotent' part N of the monodromy operator

It has the following properties (see Fig. 7.6):

$$N(W_j) \subset W_{j-2},$$
$$N^k : \mathrm{Gr}_{n+k} \longmapsto \mathrm{Gr}_{n-k} \text{ is an isomorphism.}$$

Thus Schmid's weight filtration is determined completely by the monodromy operator. We call this construction the **monodromy weight filtration with central index** n. It is defined over $\mathbb{Q}$.

The most difficult part of the Schmid's approach is the proof that the filtrations $\{F_\infty^p\}$ and $\{W_j\}$ define a mixed Hodge structure. That proof is essentially algebraic; it uses certain Deligne's classification of pure Hodge structures invariant with respect to an action of the Lie algebra $sl(2, \mathbb{C})$.
We do not present this proof and we shall concentrate on the Steenbrink's geometrical approach.

7.36. The Steenbrink's limit mixed Hodge structure. We follow [**Ste1**], [**Ste2**].

(a) Firstly we define the fixed space H, where the varying Hodge filtration achieves its limit.
Let $\mathbb{H} = \{\mathrm{Im}\, z > 0\}$ be the upper half-plane which is the universal covering of the punctured disc $\mathbb{D}^*$: $z \longmapsto t = \mathrm{e}^{2\pi \mathrm{i} z}$. We put

$$X_\infty = X \times_{\mathbb{D}^*} \mathbb{H} = \{(x, z) : f(x) = \mathrm{e}^{2\pi \mathrm{i} z}\}$$

(i.e., a covering of $X - X_0$). We denote $\pi : X_\infty \longmapsto X$ the projection onto the first factor.

The fixed space H is defined as

$$H = H^n(X_\infty, \mathbb{C}).$$

Of course, the space X_∞ is homotopically equivalent to a non-singular fiber X_t and $H \simeq H^n(X_t, \mathbb{C})$. We can treat H as $\lim_{z\to i\infty} H^n(X_t)$.

(b) The limit mixed Hodge structure is introduced in $H^n(X_\infty)$ in a way to agree with the mixed Hodge structure on the singular fiber X_0 (see the point 7.32). Because X_0 is a deformation retract of X, we have the isomorphism $H^n(X_0) \longmapsto H^n(X)$. On the other hand, the inclusion $X_t \longmapsto X$ induces the homomorphism $H^n(X) \longmapsto H^n(X_t)$. Composing these two homomorphisms and passing to the limit $t \to 0$, we obtain a homomorphism $H^n(X_0) \longmapsto H^n(X_\infty)$. The latter should be a morphism of spaces with mixed Hodge structures.

(c) Define the following **sheaf of relative logarithmic holomorphic forms**

$$\Omega^p_{X/\mathbb{D}}(\log X_0) = \Omega^p_{X/\mathbb{D}}(\log)$$

as the sheaf of logarithmic holomorphic forms modulo $\mathrm{d}t = \mathrm{d}f$. It means that $\Omega^p_{X/\mathbb{D}}(\log) = \Omega^p_X(\log)/f^*\Omega^1_{\mathbb{D}}(\log 0) \wedge \Omega^{p-1}_X(\log)$.
The restriction of the complex of sheaves $\Omega^\bullet_{X/\mathbb{D}}(\log)$ to any non-singular fiber X_t is the same as the usual complex $\Omega^\bullet_{X_t}$ of holomorphic forms on X_t. (The restriction of a sheaf $\mathcal{F}$ on X to an analytic subset Y is denoted usually by $\mathcal{F}\otimes_{\mathcal{O}_X}\mathcal{O}_Y = \mathcal{F}\otimes\mathcal{O}_Y$). For $t = 0$ one also has the restriction to X_0, i.e., $\Omega^\bullet_{X/\mathbb{D}}(\log) \times \mathcal{O}_{X_0}$. However, here the sheaf $\mathcal{O}_{X_0}$ of germs of regular functions on X_0 does not consist of reduced rings $\mathcal{O}_{X_0}(U)$ (they contain nilpotents). For example, near a point $x_0 \in X_0$, where $f = x_1^{k_1}$, $k_1 > 1$, we have $\mathcal{O}_{x_0,X_0} \simeq \mathcal{O}_0(\mathbb{C}^d)/(x_1^{k_1})$ and the function x_1 is a nilpotent element of the local ring.
In order to avoid excessive notations we shall sometimes avoid the tensoring by $\mathcal{O}_{X_t}$.
The next proposition is an analogue of Proposition (e) from 7.33.

(d) **Proposition.** *We have*

$$H^n(X_\infty, \mathbb{C}) \simeq \mathbb{H}^n(X_0, \Omega^\bullet_{X/\mathbb{D}}(\log)).$$

In particular, the groups $\mathbb{H}^n(X_t, \Omega^\bullet_{X/\mathbb{D}}(\log))$ *have constant dimension on* $\mathbb{D}$ *and define a prolongation of the cohomological bundle* H *from* $\mathbb{D}^*$ *to* $\mathbb{D}$.

Proof. The second statement follows from the fact that, for $t \neq 0$, $\dim H^n(X_t) = \dim H^n(X_\infty)$.
Of course, we have $H^n(X_\infty, \mathbb{C}) \simeq \mathbb{H}^n(X_\infty, \Omega^\bullet_{X_\infty})$ (the holomorphic de Rham theorem). The latter hypercohomology group of X_∞ is the same as the hypercohomology group of X, but with values in the direct image under the map π_* of the sheaf of holomorphic forms on X_∞: $\mathbb{H}^n(X, \pi_*\Omega^\bullet_{X_\infty})$, where $\pi : X_\infty \longmapsto X$ is the projection (see Example 3 in 3.26).

Next, X is contractible to X_0. Thus the group $\mathbb{H}^n(X, \pi_*\Omega^\bullet_{X_\infty})$ is the same as the limit $\lim_\to \mathbb{H}^n(V_\epsilon, \pi_*\Omega^\bullet_{X_\infty})$, where the sets $V_\epsilon = \{|f(x)| < \epsilon\}$ form a system of neighborhoods of X_0. The latter limit group can be interpreted as the hypercohomology group of X_0 with coefficients in the sheaf complex, whose groups of local sections are germs of $\pi_*\Omega^\bullet_{X_\infty}$ with center at X_0, treated as $\mathcal{O}_{X_0}$-moduli. The latter sheaf is denoted in literature by $i^\bullet\pi_*\Omega^\bullet_{X_\infty}$, where $i : X_0 \longmapsto X$. (If $g : M \longmapsto N$ and $\mathcal{F}$ is a sheaf on N, then $g^\bullet\mathcal{F}(U) = \lim_\to \mathcal{F}(V)$, $g(U) \subset V$.)
Proposition (d) is derived from the following. □

(e) **Lemma.** *The two sheaves on X_0, $i^\bullet\pi_*\Omega^\bullet_{X_\infty}$ and $\Omega^\bullet_{X/\mathbb{D}}(\log)) \otimes O_{X_0}$, are quasi-isomorphic.*
More precisely, let $x_0 \in X_0$ be such a point that $f = x_1^{k_1} \dots x_l^{k_l}$ near it. Let $k = \gcd(k_1, \dots, k_l)$ and $k_i = m_i k$. Then the stalks at x_0 of the qth local cohomology groups (i.e., the direct limits of the system of groups $H^q(\Gamma(U, F^\bullet))$ over neighborhoods U of x_0) of the above two complexes are generated by the forms:

$$t^{-a/k}\left(\prod x_i^{m_i}\right)^a \mathrm{d}\ln x_{i_1} \wedge \dots \mathrm{d}\ln x_{i_q}, \quad t = \mathrm{e}^{2\pi \mathrm{i} z},$$

and by

$$\left(\prod x_i^{m_i}\right)^a \mathrm{d}\ln x_{i_1} \wedge \dots \mathrm{d}\ln x_{i_q}$$

respectively. Here $a = 1, \dots, k-1$, $1 \le i_1 < i_2 < \dots < i_q \le l$, the rational forms $\mathrm{d}\ln x_i = \mathrm{d}x_i/x_i$ are subject to the relation

$$\sum k_i \cdot \mathrm{d}\ln x_i = 0$$

and the multi-valued function $t^{-a/k}$ is univalent on X_∞.

Proof. (i) The system of open sets $U = V_\epsilon \cap \{|x| < \eta\}$ forms a fundamental system of neighborhoods of x_0. We calculate

$$\lim H^q(\Gamma(U, \pi_*\Omega^\bullet_{X_\infty})) = \lim H^q(\Gamma(\pi^{-1}(U), \Omega^\bullet_{X_\infty})).$$

We have

$$\pi^{-1}(U) = \{(x, z) : \prod x_i^{k_i} = \mathrm{e}^{2\pi \mathrm{i} z},\ |x| < \eta,\ \mathrm{Im} z > C\}.$$

This set is homotopically equivalent to

$$F = \{y \in \mathbb{C}^l | \prod y_i^{k_i} = 1\}.$$

The set F forms disjoint union of k components each of which isomorphic to $(\mathbb{C}^*)^{l-1}$.
The function $\tau = y_1^{m_1} \dots y_l^{m_l}$ generates $H^0(F, \mathbb{C}) = \mathbb{C}^k$, treated as $\mathbb{C}$-algebra. The forms $\mathrm{d}\ln y_j \in H^1(F, \mathbb{C})$ generate $H^*(F, \mathbb{C})$, as the Grassmann $H^0(F)$-algebra. These generators are subject to the relations $\tau^k = 1$, $\sum k_i \cdot \mathrm{d}\ln y_i = 0$.

In $\pi^{-1}(U)$ the function τ becomes $t^{-1/k}\prod x_i^{m_i}$ and the forms $\mathrm{d}\ln y_i$ pass to the forms $\mathrm{d}\ln x_i$.

(ii) Recall the definition of the **Koszul complex** on a $\mathbb{C}$-algebra A with operators $D_1, \ldots, D_k$ (acting on A). It is the complex

$$0 \longmapsto A^k \longmapsto \bigwedge^2 A^k \longmapsto \ldots$$

where $\mathrm{d}(fe_{i_1} \wedge \ldots \wedge e_{i_p}) = \sum D_i(f)e_i \wedge e_{i_1} \wedge \ldots \wedge e_{i_p}$ are the differentials. Here $f \in A$ and $e_1, \ldots, e_k$ is the canonical basis in A^k. It is easy fact that:

If at least one of the operators D_i is bijective then the Koszul complex is exact, i.e., it has zero cohomology groups (is acyclic).

Let us calculate, for example, the first cohomology group. If $\sum f_ie_i$ lies in the kernel of d, then $D_if_j = D_jf_i$. Assuming that D_1 is invertible, we find $f_j = D_j(D_1^{-1}f_1) = D_jg$; thus $\sum f_ie_i = \mathrm{d}g$.
Let us pass to calculation of the local homology groups of the sheaf complex $\Omega^\bullet_{X/\mathbb{D}}(\log) \otimes \mathcal{O}_{X_0}$ near a point x_0 as above, where $f = x_1^{k_1} \ldots x_l^{k_l}$. We can also assume that $l = \dim X$ (because we can contract the neighborhood $U \cap X_0$ along the eventual additional directions).
It is easy to see that the complex $\Gamma(U \cap X_0, \Omega^\bullet_{X/\mathbb{D}}(\log))$ is isomorphic to the Koszul complex on $A = \mathbb{C}\{x_1, \ldots, x_l\} = \mathcal{O}_0(\mathbb{C}^l)$ (the local ring of germs of analytic functions) with the operators $D_i = x_i\partial_{x_i} - (k_i/k_l)x_l\partial_{x_l}$, $i = 1, \ldots, l-1$. The cohomology of this Koszul complex are computed monomial by monomial, because D_i are homogeneous. One gets a nonzero contribution only from those monomials on which D_i are all zero. Because $D_i(x_1^{a_1} \ldots x_l^{a_l}) = (a_i - a_lk_i/k_l)x_1^{a_1} \ldots x_l^{a_l}$ this amounts to saying that $x_1^{a_1} \ldots x_l^{a_l}$ is a power of $x_1^{m_1} \ldots x_l^{m_l}$. □

(f) *The case of semi-stable degeneration.* We introduce the limit mixed Hodge structure firstly under the assumption of semi-stability. We know from Clemens' proof of the monodromy theorem 4.71 (in Chapter 4) that the monodromy operator is unipotent in this case. In **[Ste1]** the limit mixed Hodge structure is introduced under the assumption of unipotency of M.
The semi-stability assumption says that locally $f = x_1 \ldots x_l$ and the (ringed) variety $(X_0, \mathcal{O}_{X_0})$ is reduced. Thus we avoid some algebraic complications.
It is easy to introduce the Hodge filtration on $H^n(X_\infty, \mathbb{C})$, it arises from the filtration bête of the relative holomorphic logarithmic complex.
The weight filtration should arise from the intersections of components of X_0. Assume that

$$X_0 = Y_1 \cup \ldots \cup Y_N$$

(with normal intersections) and denote $Y^{(k)}$ the kth skeleton of X_0 as the disjoint union of k-fold intersections $Y_{i_1} \cap \ldots \cap Y_{i_k}$ (see 7.32 above).

(g) *The complex* $A^\bullet$. We want to replace the complex of relative forms (where we divide by $\mathrm{d}\ln t = \sum \mathrm{d}\ln x_i$) by some more natural complex. Moreover, this new complex should be quasi-isomorphic with $\Omega^\bullet_{X/\mathbb{D}}(\log)$.
The mapping $\omega \longmapsto \omega \wedge \mathrm{d}\ln t$ defines a homomorphism of sheaves (restricted to X_0) $\theta : \Omega^\bullet_{X/\mathbb{D}}(\log) \longmapsto \Omega^{\bullet+1}(\log)/\widetilde{W}_0\Omega^{\bullet+1}(\log)$ (forms with at least one pole); here $\widetilde{W}_\bullet$ is the preliminary weight filtration of the holomorphic logarithmic complex (by order of poles, see the point 7.33(c)). But θ is not a quasi-isomorphism. In order to get a quasi-isomorphism one should be able to associate with a local closed form η on $U \subset X$, a closed form from $\Gamma(U, \Omega^\bullet_{X/\mathbb{D}}(\log))$; for this one needs $\mathrm{d}\ln t \wedge \eta = 0$. So, a natural differential should be $D = \mathrm{d}' + \mathrm{d}''$, where $\mathrm{d}' = \mathrm{d}$ and $\mathrm{d}'' = \mathrm{d}\ln t \wedge (\cdot)$. But the image of d'' lies in $\Omega^{\bullet+1}(\log)/\widetilde{W}_1\Omega^{\bullet+1}(\log)$ (forms with at least two poles). This suggests introduction of the following complex

$$\mathcal{A}^m = \bigoplus_{q=0}^{m} \Omega^{m+1}(\log)/\widetilde{W}_q\Omega^{m+1}(\log) = \bigoplus \mathcal{A}^{m-q,q},$$

with the differential $D = \mathrm{d}' + \mathrm{d}''$ as above.
The map $\theta : \Omega^\bullet_{X/\mathbb{D}}(\log) \longmapsto \mathcal{A}^{\bullet,0}$ induces a homomorphism into $\mathcal{A}^\bullet$. It is not difficult to show that θ is a quasi-isomorphism between $\Omega^\bullet_{X/\mathbb{D}}(\log)$ and $\mathcal{A}^\bullet$ (see [**Ste1**] for details).

(h) **Definition of the limit mixed Hodge structure.** Let us fix n. Using the quasi-isomorphism θ and Proposition (d) we obtain $H^n(X_\infty, \mathbb{C}) \simeq \mathbb{H}^n(X_0, \mathcal{A}^\bullet)$. The mixed Hodge structure on this space is induced from the following two filtrations on the double complex $\mathcal{A}^{\bullet,\bullet}$.
The (decreasing) *Hodge filtration* is equal to $F^r\mathcal{A}^m = \bigoplus_{p\geq r} \mathcal{A}^{p,m-p}$ Notice that $F(\mathcal{A}^\bullet)$ is the stupid filtration induced from the filtration bête on $\Omega^\bullet_{X/\mathbb{D}}(\log)$. Later we will interpret the filtration $F^\bullet H^n(X_\infty)$ as a limit of the filtrations $F^\bullet H^n(X_t)$ (see the point 7.37(h) below).
The (increasing) *weight filtration* is equal to $W_{n+r}\mathcal{A}^{m-q,q} = (\widetilde{W}_{2q+r+1}/\widetilde{W}_q)$ on $\Omega^{m+1}(\log)$; here r can be positive and negative as well, but $q + r \geq 0$.

(i) **Theorem (Limit mixed Hodge structure).** *The above filtrations induce a mixed Hodge structure in* $H^n(X_\infty)$.

Proof. We have $W_{n+r}/W_{n+r1} = \widetilde{W}_{2q+r+1}/\widetilde{W}_{2q+r}$ (on $\Omega^{m+1}(\log)$) (see Fig. 7.7). By Poincaré residuum. this sheaf is isomorphic to $i_*\Omega^{m-2q-r}_{Y(2q+r+1)}$ (i – embedding). Therefore $\mathrm{Gr}^W_{n+r}\mathcal{A}^m = \bigoplus_{q\geq 0,-r} i_*\Omega^{m-2q-r}_{Y(2q+r+1)}$, or

$$\mathrm{Gr}^W_{n+r}\mathcal{A}^\bullet = \bigoplus_{q\geq 0,-r} i_*\Omega^{\bullet-2q-r}_{Y(2q+r+1)}.$$

Calculating the hypercohomology of this, we get

$$\mathbb{H}^n(X_0, \mathrm{Gr}_{n+r}\mathcal{A}^\bullet) = \bigoplus_{q\geq 0,-r} H^{n-2q-r}(Y^{(2q+r+1)}) = \bigoplus_q H_{r,q}.$$

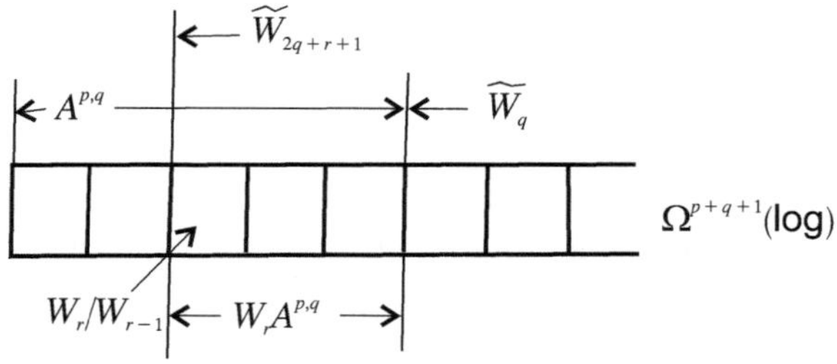

Figure 7.7: Hodge and weight filtrations for the limit mixed Hodge structure

Each component $H_{r,q}$ admits pure Hodge structure as the cohomology group of a closed non-singular algebraic variety.
The weight of this pure Hodge structure is calculated as follows. Let $\widetilde{F}$ be the filtration bête of $\Omega^\bullet(\log)$. We have

$$F^p \mathrm{Gr}^W_{n+r} \mathcal{A}^\bullet = \bigoplus_q \widetilde{F}^{p+q+1} \mathrm{Gr}^{\widetilde{W}}_{2q+r+1} \Omega^{\bullet+1}(\log)$$
$$= \bigoplus_q i_* \widetilde{F}^{p-q-r} \Omega^{\bullet-2q-r}_{Y^{(2q+r+1)}}$$

and hence $F^p H_{r,q} = \widetilde{F}^{p-q-r} H_{r,q}$. Because

$$F^{(n+r)-p} H_{r,q} = \widetilde{F}^{n-p-q} H^{n-2q-r}(Y^{(2q+r+1)})$$
$$= \overline{\widetilde{F}^{p-q-r} H_{r,q}}$$
$$= \overline{F^p H_{r,q}},$$

the pure weight is $n+r$.
The groups $\bigoplus_q H_{r,q}$ form the first terms ${}_W E_1^{-r,n+r} = E_1^{-r,s}$ of the weight spectral sequence of hypercohomology of $\mathcal{A}^\bullet$ associated with suitable reverse weight filtration. The differential d_1 (of this spectral sequence) acts differently on different components $H_{r,q}$. We have the maps

$$\mathrm{d}_1 : H^{s-2q-2r}(Y^{(2q+r+1)}) \longmapsto H^{s-2q'-2r+2}(Y^{(2q'+r)}), \quad q \geq 0, -r; \;\; q' \geq 0, -r+1;$$

they can be induced by inclusions of components, by Gysin maps between components and by compositions of these. This suggests that d_1 consists of morphisms of pure Hodge structures.
The weight spectral sequence degenerates at the second term

$$H^n(X_\infty) \simeq \bigoplus {}_W E_2^{-r,n+r},$$

and ${}_W E_2^{-r,n+r}$ is the rth homology group of the complex $\ldots \longmapsto \bigoplus_{q\geq 0,-r} H_{r,q} \longmapsto \bigoplus_{q\geq 0,-r+1} H_{r-1,q} \longmapsto \ldots$. Proof of this fact is the same as the proof of analogous statement in the case of incomplete manifold (see the point 7.33(i)): it uses the fact that the differentials $d_1, d_2, \ldots$ are compatible with the filtrations $F^\bullet$ and $\overline{F}^\bullet$. □

(j) *The case of not semi-stable degeneration.* In this case the monodromy operator can be unipotent as well as not unipotent. Steenbrink treats the general (not unipotent) case in [**Ste2**].
As in the points 4.66, 4.67 in Chapter 4 we apply firstly the base change $\widetilde{\mathbb{D}} = \mathbb{D} \longmapsto \mathbb{D}$, $s \longmapsto t = s^l$, and replace X by $X \times_{\mathbb{D}} \widetilde{\mathbb{D}}$ with the projection $\hat{f} : X \times_{\mathbb{D}} \widetilde{\mathbb{D}} \longmapsto \widetilde{\mathbb{D}}$. Because $X \times_{\mathbb{D}} \widetilde{\mathbb{D}}$ is usually not normal space, one introduces the normalization $n : \widetilde{X} \longmapsto X \times_{\mathbb{D}} \widetilde{\mathbb{D}}$. Let $\tilde{f} = \hat{f} \circ n : \widetilde{X} \longmapsto \widetilde{\mathbb{D}}$, it plays the role of semi-stable degeneration. We have the diagram

$$\begin{array}{ccccc} \widetilde{X} & \stackrel{n}{\longmapsto} & X \times_{\mathbb{D}} \widetilde{\mathbb{D}} & \longmapsto & X \\ \downarrow \tilde{f} & & \downarrow \hat{f} & & \downarrow f \\ \widetilde{\mathbb{D}} & = & \widetilde{\mathbb{D}} & \longmapsto & \mathbb{D} \end{array}$$

The variety $\widetilde{X}$ is an orbifold (or V-manifold); it is covered by subsets of the form U/G, where $U \approx \mathbb{C}^d$ and G is a finite subgroup of $GL(d, \mathbb{C})$.
The V-manifold has singularities of codimension ≥ 2. (The elements of G are divided into: rotations with smooth quotient and other elements with at least two eigenvalues different from 1.)
The latter fact allows to extend the sheaf and Hodge theories to any V-manifold Z (see Theorem 3 in 5.30). In particular, the complex $\widetilde{\Omega}^\bullet = j_* \Omega^\bullet_{Z - \mathrm{Sing}(Z)}$ (j – embedding) is a coherent resolution of the constant sheaf and defines the Hodge structure $H^{p,q}(Z, \mathbb{C}) = H^q(Z, \widetilde{\Omega}^p) \subset \mathbb{H}^{p+q}(Z, \widetilde{\Omega}^\bullet)$.
If Z is projective, then the Kähler form (induced from the Fubini–Study metric) defines an integer cohomology class $\omega \in H^2(Z, \mathbb{Z})$ (like in 7.10). Also the notions of primitive cohomologies and polarization have the same meaning as in the smooth projective case (see 7.11 and 7.12). For the proofs we refer the reader to the work of Steenbrink.
Let us return to the map $\tilde{f} : \widetilde{X} \longmapsto \widetilde{\mathbb{D}}$ with the fibers $\widetilde{X}_s = \tilde{f}^{-1}(s)$. As in the semi-stable case, one introduces the space $X_\infty = X \times_{\mathbb{D}} \mathbb{H}$, which is the same as $\widetilde{X} \times_{\widetilde{\mathbb{D}}} \mathbb{H}$. Also, analogously, one constructs the holomorphic logarithmic sheaves $\widetilde{\Omega}^p(\log \widetilde{X}_0) = \widetilde{\Omega}^p(\log)$ and the holomorphic logarithmic relative sheaves $\widetilde{\Omega}^p_{\widetilde{X}/\widetilde{\mathbb{D}}}(\log)$.
One has $H^n(X_\infty) = \mathbb{H}^n(\widetilde{X}_0, \widetilde{\Omega}^\bullet_{\widetilde{X}/\widetilde{\mathbb{D}}}(\log))$. The corresponding complex $\mathcal{A}$, with Hodge and weight filtrations, is defined in the same way. Thus, the mixed Hodge structure in $H^n(X_\infty)$ is defined.

(k) *The cohomological bundles on* $\widetilde{\mathbb{D}}$ *and* $\mathbb{D}$. The system of spaces

$$\widetilde{H}^n_s = \mathbb{H}^n(\widetilde{X}_s, \widetilde{\Omega}^\bullet_{\widetilde{X}/\widetilde{\mathbb{D}}}(\log))$$

(of constant dimension) is glued together to a vector fibre bundle $\widetilde{\mathcal{H}}^n \longmapsto \widetilde{\mathbb{D}}$ (the cohomological bundle). This bundle admits the Hodge filtration $\widetilde{F}_s^p/\widetilde{F}_s^{p+1} = H^{n-p}(\widetilde{X}_s, \widetilde{\Omega}^p_{\widetilde{X}/\widetilde{\mathbb{D}}}(\log))$. Using the upper semi-continuity of $\dim \widetilde{F}_s^p$ with respect to s (see [**HaR**]) and the (topological) triviality of $\widetilde{\mathcal{H}}^n$, one gets that the corresponding fibrations $\widetilde{\mathcal{F}}^p$ are also (topologically) trivial; (the sum of dimensions is constant). The sheaf of local sections of $\widetilde{\mathcal{H}}^n$ is the hypercohomological sheaf: $\mathcal{O}(\widetilde{\mathcal{H}}^n) = R^n \tilde{f}_* \widetilde{\Omega}_{\widetilde{X}/\widetilde{\mathbb{D}}}(\log)$.

Applying the finite-to-one maps $\widetilde{X} \longmapsto X$, $\alpha : \widetilde{\mathbb{D}} \longmapsto \mathbb{D}$, we transform the bundle $\widetilde{\mathcal{H}}^n$ to bundle $\mathcal{H}^n$ above $\mathbb{D}$ with the fibers $\mathbb{H}^n(X_t, \Omega^\bullet_{X/\mathbb{D}}(\log))$ and with the analogous Hodge subbundles $\mathcal{F}^p$. The bundle $\mathcal{H}^n$ and its Hodge subbundles $\mathcal{F}^p$ are trivial, i.e., the corresponding sheaves of sections are locally free of finite rank on $\mathbb{D}$.

7.37. Limit mixed Hodge structure and monodromy. (a) The cohomological bundle above the punctured disc $\mathbb{D}^*$, $\mathcal{H}^n|_{X^*} \longmapsto \mathbb{D}^*$ admits the Gauss–Manin connection. It is defined by the condition that sections represented by continuous families of cocycles ϕ_t, taking values in the integer lattices $H^n(X_t, \mathbb{Z})/\mathrm{Tor}$, are horizontal with respect to this connection, $\nabla_{\partial/\partial t}\phi_t \equiv 0$.

The Gauss–Manin connection is defined also as a morphism of sheaves on $\mathbb{D}^*$

$$\nabla : \mathcal{O}(\mathcal{H}^n) \longmapsto \Omega^1_{\mathbb{D}^*} \otimes \mathcal{O}(\mathcal{H}^n) = \Omega^1(\mathcal{H}^n)$$

(thus, $\nabla_{a(t)\partial/\partial t}$ is a section of the sheaf of endomorphisms of $\mathcal{H}^n$).

The Gauss–Manin connection can be defined cohomologically. Note that we have the exact sequence of de Rham complexes on $X^* = X \setminus X_0$

$$0 \longmapsto \Omega^1_{\mathbb{D}^*} \otimes \mathcal{E}^{\bullet-1}_{X^*/\mathbb{D}^*} \overset{\wedge}{\longmapsto} \mathcal{E}^\bullet_{X^*} \longmapsto \mathcal{E}^\bullet_{X^*/\mathbb{D}^*} \longmapsto 0, \tag{7.4}$$

where $\mathcal{E}^p_{X^*/\mathbb{D}^*} = \mathcal{E}^p_{X^*}/\sim$ (with the equivalence $\mathrm{d}t \wedge \omega^{p-1} \sim 0$) are the relative de Rham sheaves. We know also that $\mathcal{H}^n_t = H^n_{\mathrm{d}}(\Gamma(X_t, \mathcal{E}^\bullet_{X^*/\mathbb{D}^*}))$, cohomologies of the complex of global sections of the relative de Rham sheaves.

The exact sequence (7.4) induces the long exact sequence of cohomologies

$$\longmapsto H^n(\Gamma(X_t, \mathcal{E}^\bullet_{X^*})) \longmapsto H^n(\Gamma(X_t, \mathcal{E}^\bullet_{X^*/\mathbb{D}^*})) \overset{\delta}{\longmapsto} H^{n+1}(\Gamma(X_t, \Omega^1_{\mathbb{D}^*} \otimes \mathcal{E}^\bullet_{X^*/\mathbb{D}^*})) \longmapsto \dots$$

Let us calculate the connecting homomorphism δ. If ω_t is a family of closed forms on X_t (i.e., it defines a closed section of $\mathcal{E}^n_{X^*/\mathbb{D}^*}$), then it arises from a form $\tilde{\omega}$ on X^* as restriction, $\omega_t = \tilde{\omega}|_{X_t}$. Next, one takes $\mathrm{d}\tilde{\omega}$; it belongs to the image of the map $\Omega^1 \otimes \mathcal{E}^n \longmapsto \mathcal{E}^{n+1}$. Thus $\mathrm{d}\tilde{\omega} = \mathrm{d}t \wedge \eta$, where $\eta = \mathrm{d}\tilde{\omega}/\mathrm{d}t$ is the Gelfand–Leray form. One puts $\delta[\omega_t] = [\eta]$.

If $\Delta(t)$ is a family of cycles, which are horizontal with respect to the Gauss–Manin connection, then we have $\frac{\mathrm{d}}{\mathrm{d}t}\langle \tilde{\omega}, \Delta(t) \rangle = \langle \nabla_{\partial/\partial t}\tilde{\omega}, \Delta(t) \rangle$. But, by Lemma 5.12 and

remarks after it (see Chapter 5), the above derivative equals to $\langle \mathrm{d}\tilde{\omega}/\mathrm{d}t, \Delta(t)\rangle$. Therefore the class $[\eta] = \delta[\omega]$ is equal to $\nabla_{\partial/\partial t}[\omega]$, i.e., we have $\nabla = \delta$.
The above can be repeated in the case when the smooth de Rham complexes are replaced by complexes of sheaves of holomorphic forms $\Omega^\bullet_{X^*}$, $\Omega^\bullet_{X^*/\mathbb{D}^*}$, and the de Rham cohomology groups are replaced by the hypercohomology groups. The Gauss–Manin connection becomes the connecting homomorphisms in the long exact sequence of hypercohomologies.
The latter (holomorphic) construction has extension to the cohomological bundle above the full disc. One has to replace the sheaves of holomorphic forms by the logarithmic holomorphic de Rham sheaves, More precisely, the following is true.

(b) **Proposition.** *The Gauss–Manin connection* $\nabla : \mathcal{O}(\mathcal{H}^n) \longmapsto \Omega^1(\mathcal{H}^n)$ *is equal to the connecting homomorphism in the long exact sequence of hypercohomologies, i.e.,*

$$\begin{aligned}\mathbb{H}^n(X_t, \Omega^\bullet_{X/\mathbb{D}}(\log)) \longmapsto \mathbb{H}^{n+1}&(X_t, \Omega^1_{\mathbb{D}}(\log 0) \otimes \Omega^\bullet_{X/\mathbb{D}}(\log)) \\ &\simeq \mathbb{H}^n(X_t, \Omega^\bullet_{X/\mathbb{D}}(\log)),\end{aligned}$$

associated with the short exact sequence

$$0 \longmapsto \Omega^1_{\mathbb{D}}(\log 0) \otimes \Omega^{\bullet-1}_{X/\mathbb{D}}(\log) \longmapsto \Omega^\bullet_X(\log) \longmapsto \Omega^\bullet_{X/\mathbb{D}}(\log) \longmapsto 0.$$

(c) The *monodromy operator*, called also the *Picard–Lefschetz transformation,* acts on the fibers of the cohomological bundle over $\mathbb{D}^*$, $M_t : \mathcal{H}^n_t \longmapsto \mathcal{H}^n_t$. Any two of them are conjugate.
Because the cohomological bundle extends to the whole $\mathbb{D}$, we have the limit operator $M = M_0 : \mathcal{H}^n_0 \longmapsto \mathcal{H}^n_0$. The latter can be defined by means of the Gauss–Manin connection.
Define the *residuum of the Gauss–Manin connection* $\mathrm{Res}\nabla : \mathcal{H}^n_0 \longmapsto \mathcal{H}^n_0$ as the composition $\mathrm{res} \circ \nabla$, where $\mathrm{res} : a(t)\mathrm{d}\ln t \otimes \eta \longmapsto a(0) \cdot \eta$.
We have $t\mathrm{d}\omega/\mathrm{d}t = \mathrm{Res}\nabla(\omega) + O(t)$ and, after performing calculations of the monodromy operator with $t = \epsilon e^{\mathrm{i}\theta}$, $\epsilon \to 0$, $\theta \in [0, 2\pi]$, we get the following.

(d) **Proposition.** $M_0 = \exp(2\pi\mathrm{i} \cdot \mathrm{Res}\nabla)$.

(e) Let us express the action of $M = M_0$ on the limit mixed Hodge structure introduced in the point 7.36. Assume firstly that the degeneration is semi-stable. Then X_0 is reduced and M is unipotent. Denote

$$N = \log M = 2\pi\mathrm{i} \cdot \mathrm{Res}\nabla.$$

Because the mixed Hodge structure is introduced by means of the auxiliary complex $\mathcal{A}^\bullet$ and ∇ has cohomological interpretation, we introduce an auxiliary short exact sequence of coherent holomorphic sheaves containing $\mathcal{A}^\bullet$. Namely, there exists some complex $\mathcal{B}^\bullet$ and the following commutative diagram with exact rows and vertical quasi-isomorphisms

$$\begin{array}{ccccccccc} 0 & \longmapsto & \Omega^{\bullet-1}_{X/\mathbb{D}}(\log) & \stackrel{\mathrm{d}\ln t\wedge}{\longmapsto} & \Omega^\bullet_X(\log) & \longmapsto & \Omega^\bullet_X(\log) & \longmapsto & 0 \\ & & \downarrow \mathrm{d}\ln t\wedge & & \downarrow \sigma & & \downarrow \mathrm{d}\ln t\wedge & & \\ 0 & \longmapsto & \mathcal{A}^{\bullet-1} & \longmapsto & \mathcal{B}^\bullet & \longmapsto & \mathcal{A}^\bullet & \longmapsto & 0 \end{array}$$

Thus $\mathrm{Res}\nabla$ is the connecting homomorphism $\mathbb{H}^n(X_0,\mathcal{A}^\bullet)\longmapsto\mathbb{H}^n(X_0,\mathcal{A}^\bullet)$ The complex $\mathcal{B}^\bullet$ consists of the blocks

$$\mathcal{B}^{m-q,q}=\mathcal{A}^{m-q-1,q}\oplus\mathcal{A}^{m-q,q}$$

and the homomorphism $\sigma:\Omega^m_X(\log)\longmapsto\mathcal{B}^m$ is defined as $\omega\longmapsto(\omega,\omega\wedge\mathrm{d}\ln t)\in\mathcal{A}^{m-1,0}\oplus\mathcal{A}^{m,0}$; recall that $\mathcal{A}^{m-q,q}=\Omega^{m+1}(\log)/\widetilde{W}_q\Omega^{m+1}(\log)$. The differential $\mathrm{d}_B=\mathrm{d}'_B+\mathrm{d}''_B$, $\mathrm{d}'_B:\mathcal{B}^{m-q,q}\longmapsto\mathcal{B}^{m-q+1,q}$, $\mathrm{d}''_B:\mathcal{B}^{m-q,q}\longmapsto\mathcal{B}^{m-q,q+1}$, should be chosen to satisfy $\sigma(D\omega)=\mathrm{d}_B\sigma(\omega)$. Therefore, $\mathrm{d}_B(\omega,\omega\wedge\mathrm{d}\ln t)=(\mathrm{d}\omega,\mathrm{d}\omega\wedge\mathrm{d}\ln t)\in\mathcal{A}^{m-q,0}\oplus\mathcal{A}^{m-q+1,0}$. The following choice is natural:

$$\begin{aligned}\mathrm{d}'_B(\omega_1,\omega_2)&=(\mathrm{d}\omega_1,\mathrm{d}\omega_2),\\ \mathrm{d}''_B(\omega_1,\omega_2)&=(\mathrm{d}\ln t\wedge\omega_1+\nu(\omega_2),\mathrm{d}\ln t\wedge\omega_2),\end{aligned}$$

where the operator

$$\nu:\mathcal{A}^{m-q,q}\longmapsto\mathcal{A}^{m-q-1,q+1}$$

equals $(-1)^m$ times the canonical projection $\omega_2(\mathrm{mod}\ \widetilde{W}_q)\longmapsto\varphi_2(\mathrm{mod}\ \widetilde{W}_{q+1})$ (see the point 7.36(g)).
One checks that:

The connecting homomorphism of hypercohomologies is induced by the endomorphism ν, i.e., $\mathrm{Res}\nabla=\nu$.

We see also that $\nu:W_j\mathcal{A}^\bullet\longmapsto W_{j-2}\mathcal{A}^\bullet$ and that $\nu^r:\mathrm{Gr}_{n+r}\mathcal{A}^\bullet\longmapsto\mathrm{Gr}_{n-r}\mathcal{A}^\bullet$ is an isomorphism. Moreover $\nu:F^p\mathcal{A}^\bullet\longmapsto F^{p-1}\mathcal{A}^\bullet$.
Because the endomorphism $\nu=\mathrm{Res}\nabla$ of $\mathcal{H}^n_0$ is conjugate to N, also the operator N possesses the above properties.
In general situation we have the Chevalley decomposition $M=M_sM_u$, M_s – the semi-simple part and M_u – the unipotent part. One has $M_s^l=id$, where l is the order of the ramified covering $\alpha:\widetilde{\mathbb{D}}\longmapsto\mathbb{D}$. We put $N=\log M^l=l\cdot\log M_u$. The general result is as follows.

(f) **Theorem (Action of monodromy on the limit mixed Hodge structure).**

(i) *The operator N satisfies the properties*

$$\begin{aligned}N&:W_j\longmapsto W_{j+1},\\ N&:F^p\longmapsto F^{p-1},\end{aligned}$$

i.e., it is a morphism of the limit mixed Hodge structure of weight -2.

(ii) *The map* N^r *realizes isomorphism between* Gr^W_{n+r} *and* Gr^W_{n-r}. *In particular, the weight filtration is completely determined by the unipotent part of monodromy and is defined on the rational cohomology* $H^n(X_\infty, \mathbb{Q})$.

(iii) *The operator* M_s *is an automorphism of the limit mixed Hodge structure.*

Proof. We have $M^l = \exp(2\pi\mathrm{i}\cdot\mathrm{Res}\widetilde{\nabla})$, where $\widetilde{\nabla}$ is the Gauss–Manin connection of the cohomological fibration $\widetilde{\mathcal{H}}$ associated with the degener ration $\widetilde{X} \longmapsto \widetilde{\mathbb{D}}$ (see the point 7.36(j)). Thus the proofs of (i) and (ii) are the same as in the semi-stable case.
To prove (iii) one must notice that the semi-simple (i.e., diagonal) part of the monodromy is the same as the action of an automorphism Λ^* on $\mathbb{H}^n(\widetilde{X}_0, \widetilde{\Omega}^\bullet_{\widetilde{X}/\widetilde{\mathbb{D}}}(\log)$ induced by the diffeomorphism Λ of $\widetilde{X}$ defined as the lift of the rotation $s \longmapsto \mathrm{e}^{2\pi\mathrm{i}/l}s$ of the base $\widetilde{\mathbb{D}}$. This is shown using local calculations like in the proof of Lemma 7.36(e)).
The fact that Λ^* is semi-simple and that preserves the mixed Hodge filtrations follows from the fact that it acts diagonally at the sheaves $\Omega^p_{\widetilde{X}/\widetilde{\mathbb{D}}}(\log)\otimes\mathcal{O}_{\widetilde{X}_0}$ and preserves the skeletons $Y^{(j)}$. We see also that Λ^* is periodic with period l, which implies that the eigenvalues of M_s are roots of unity of order l. □

(g) **Corollary (Monodromy theorem).** *We have* $(M^l - I)^{n+1} = 0$.

(h) *The Hodge filtration on* $H^n(X_\infty)$ *as limit of Hodge filtrations on* $H^n(X_t)$. We consider the case of semi-stable degeneration; in general situation one must change notations by adding the tildes.
The Steenbrink's Hodge filtration on $H^n(X_\infty) = \mathcal{H}^n_0$ is defined as $\mathcal{F}^\bullet_0 = \mathcal{F}^\bullet|_{t=0}$, where $\mathcal{F}^p \subset \mathcal{H}^n$ are holomorphic subbundles. The fibers of the corresponding graded bundles are $H^{n-p}(X_t, \Omega^p_{X/\mathbb{D}}(\log)) = \mathrm{Gr}^p_F H^n(X_t)$. In this sense the Steenbrink's Hodge filtration is a limit of the Hodge filtrations arising from smooth closed varieties.
Schmid introduced the limit Hodge filtration in another way. If $z \in \mathbb{H}$ (upper half-plane) and $t = \mathrm{e}^{2\pi\mathrm{i}z} \in \mathbb{D}^*$, then the inclusion $X_t \longmapsto X_\infty = X \times_{\mathbb{D}} \mathbb{H}$ induces varying Hodge filtration in $H^n(X_\infty)$, which we denote by $F^\bullet_t$. This family is multivalued, but the switched family $\widehat{F}^\bullet_t = \mathrm{e}^{-zN_t}F^\bullet_t = t^{-N/2\pi\mathrm{i}}F^\bullet_t$, $N_t = \log M_t$, defines a single valued map from $\mathbb{D}^*$ to the corresponding variety of flags of subspaces of $H^n(X_\infty)$. Schmid proved that the limit $F^\bullet_\infty = \lim_{t\to 0}\widehat{F}^\bullet_t$ exists.
Note that the Steenbrink's filtration $\mathcal{F}^\bullet_0$ is invariant with respect to the action of the operator $\mathrm{e}^{-zN} = I - zN + \ldots + (-zN)^d/d!$, $N = \log M_0$: $\mathrm{e}^{-uN}\mathcal{F}^p_0 = \mathcal{F}^p_0$. It is because $\mathrm{e}^{-zN} = M^{-z}$ and M preserves $\mathcal{F}^\bullet$.
We compare the two families of filtrations of $H^n(X_\infty)$: $t^{-N/2\pi\mathrm{i}}F^\bullet_0$ and $t^{-N_t/2\pi\mathrm{i}}F^\bullet_t$. The distance between $\mathcal{F}^\bullet_0 = F^\bullet_0$ and $F^\bullet_t$ is of order $O(|t|)$; also $||N - N_t|| = O(|t|)$. Because $||t^{-N_t/2\pi\mathrm{i}}|| = O(\ln^d |t|)$, we get that $\mathrm{dist}(\widehat{F}^\bullet_t, F^\bullet_0) \to 0$, i.e., $F^\bullet_\infty = \mathcal{F}^\bullet_0$.
By Theorem (f) the Steenbrink's limit weight filtration is the monodromy weight filtration with the central index n. It coincides with the Schmid's limit weight filtration (see its definition in 7.35). Therefore:

The Steenbrink's limit mixed Hodge structure coincides with the Schmid's limit mixed Hodge structure.

(i) **Brieskorn's proof of the first part of the monodromy theorem.** We use here the opportunity to present a clever proof by E. Brieskorn [**Brie**] of the fact that eigenvalues of M are roots of unity (the *first part of the monodromy theorem*). Its ingredients are: the positive solution (by O. Gelfond [**Gel**] and T. Schneider [**Schn**]) of the VIIth Hilbert's problem (if α and $\mathrm{e}^{2\pi\mathrm{i}\alpha}$ are both algebraic numbers, then $\alpha \in \mathbb{Q}$) and the algebraicity of the eigenvalues of residuum of the Gauss–Manin connection.

One has to prove algebraicity of the eigenvalues α_j of the operator $\operatorname{Res}\nabla$ (see the points (a), (b), (c) and (d) above). It is done as follows.

Let $\sigma : \mathbb{C} \longmapsto \mathbb{C}$ be an automorphism of the number field $\mathbb{C}$. Applying σ to coefficients of the Taylor expansions of functions and of differential forms, we obtain the extension of σ to the sheaves $\mathcal{O}_X$, $\Omega^1_{\mathbb{D}}(\log 0)$, $\Omega^p_{X/\mathbb{D}}(\log)$ and to the Leray sheaves $\mathcal{O}(\mathcal{H}^n)$, $\mathcal{O}(\mathcal{H}^{p,q}) = R^q f_* \Omega^p_{X/S}(\log)$. The automorphism σ commutes with the action of ∇. In particular, it permutes the eigenvalues of $\operatorname{Res}\nabla$.

If some eigenvalue α_j were not algebraic, then for any other non-algebraic number β there would exist an automorphism σ sending α_j to it. One can choose β such that $\mathrm{e}^{2\pi\mathrm{i}\beta}$ is non-algebraic. However, we know that $\mathrm{e}^{2\pi\mathrm{i}\alpha_j}$ is algebraic, hence the number $\sigma(\mathrm{e}^{2\pi\mathrm{i}\alpha_j})$ is algebraic too. We have a contradiction. □

7.37. Example (Morse degeneration). Consider a degeneration $g : Z \longmapsto \mathbb{D}$ such that the $Z_t = g^{-1}(t)$, $t \neq 0$, are smooth closed algebraic varieties and Z_0 has unique Morse singularity; i.e., locally $g = z_1^2 + \ldots + z_d^2$. By applying the blowing-up (once) one obtains a degeneration $f : X \longmapsto \mathbb{D}$ satisfying the assumptions of this section. One has $X_0 = \widetilde{Z}_0 + E$, where $\widetilde{Z}_0$ is the smooth strict transform of Z_0 and $E = \mathbb{CP}^{d-1}$ is the exceptional divisor with multiplicity 2: locally $f = z^2(u_1^2 + \ldots + u_d^2)$, $z \in (\mathbb{C}, 0)$, $u \in \mathbb{CP}^{d-1}$. $\widetilde{Z}_0 \cap E$ is a quadric in $\mathbb{CP}^{d-1}$. Application of the base change $t = s^2$ gives $X \times_{\mathbb{D}} \widetilde{\mathbb{D}} = \{z^2(u_1^2 + \ldots + u_d^2) = s^2\}$. The normalization of the latter space is a smooth variety $\widetilde{X} = \{zv = s, v^2 = u_1^2 + \ldots + u_d^2\}$ with $\tilde{f} = zv$, $\widetilde{X}_0 = Y_1 + Y_2$, $Y_1 \simeq \widetilde{Z}_0$ and $Y_2 = \{z = 0, v^2 = u_1^2 + \ldots + u_d^2\}$ a 2-fold covering of E ramified along $\widetilde{Z}_0 \cap E = Y_1 \cap Y_2$.

In calculations of the weight filtrations on $H^n(X_\infty)$, one uses the cohomologies of $Y^{(j)}$, where $Y^{(1)} = Y_1 \sqcup Y_2$ (disjoint union) and $Y^{(2)} = Y_1 \cap Y_2$. The components of the first term of the weight spectral sequence of $\mathbb{H}^n(\widetilde{X}_0, \mathcal{A}^\bullet)$ (associated with the weight filtration) are $E_1^{-r,n+r} = \bigcup_{q \geq 0,-r} H^{n-r-2q}(Y^{(2q+r+1)})$ (see the points (h) and (i) in 7.36). One gets the table (a part of the weight spectral sequence)

$$
\begin{array}{ccccccccc}
0 & \longmapsto & E_1^{-1,n-1} & \longmapsto & H^{n-1}(Y^{(1)}) & \longmapsto & H^{n-1}(Y^{(2)}) & \longmapsto & 0, \\
0 & \longmapsto & H^{n-2}(Y^{(2)}) & \longmapsto & H^{n}(Y^{(1)}) & \longmapsto & H^{n}(Y^{(2)}) & \longmapsto & 0, \\
0 & \longmapsto & H^{n-1}(Y^{(2)}) & \longmapsto & H^{n+1}(Y^{(1)}) & \longmapsto & E_1^{1,n+1} & \longmapsto & 0.
\end{array}
$$

Here the horizontal arrows denote the differential d_1 of the spectral sequence. In particular, $\mathrm{d}_1 : H^j(Y^{(1)}) = H^j(Y_1) \oplus H^j(Y_2) \longmapsto H^j(Y_1 \cap Y_2)$ acts as the difference of restrictions of cohomology classes. Other differentials are the Gysin maps. Because this spectral sequence degenerates at $E_2^{\bullet,\bullet}$, the weight gradation groups $\mathrm{Gr}_{n-1}H^n(X_\infty)$, $\mathrm{Gr}_n H^n(X_\infty)$, $\mathrm{Gr}_{n+1}H^n(X_\infty)$ are the homology groups in the diagonal terms (with the sum of indices equal to $n+1$).

Assume firstly that $d = 2$. Then Z_t are curves, Z_0 has double point, $E = \overline{\mathbb{C}}$, $\widetilde{Z} \cap E = \{p_1, p_2\}$, $Y_2 = \overline{\mathbb{C}}$ with double covering over E.
The spaces $H^0(X_\infty)$ and $H^2(X_\infty)$ have pure weights 0 and 2 respectively.
The interesting happens with $H^1(X_\infty)$. We have $\mathrm{Gr}_0 H^1(X_\infty) = \mathrm{Coker}\,(H^0(Y_1) \oplus H^0(Y_2) \longmapsto H^0(Y_1 \cap Y_2)) = \mathbb{C}$ (because the cycle $[p_1]-[p_2] \not\sim 0$ in $Y_1 \cap Y_2$, but is ~ 0 in Y_1 as well as in Y_2). Next, $\mathrm{Gr}_1 H^1(X_\infty) = H(0 \longmapsto H^1(\widetilde{Z}_0) \longmapsto 0) = H^1(\widetilde{Z}_0)$. Finally, $\mathrm{Gr}_2 H^1(X_\infty) = \mathbb{C}$, because of the isomorphism $N : \mathrm{Gr}_2 \longmapsto \mathrm{Gr}_0$.
Here the monodromy operator has all eigenvalues equal to 1, but the monodromy is unipotent and nontrivial $M - I \neq 0 = (M - I)^2$.
Note that, in this degeneration, not one but two cycles from $H_1(Z_t)$ disappear at $H_1(\widetilde{Z}_0)$. One is the vanishing cycle and the other is removed after desingularization. For example, an elliptic curve, acquiring a double point, is rationally equivalent to the projective line.
The same holds in the case of general even dimension d.

If d is odd, then all the spaces $H^n(X_\infty)$ have pure weight n. If $n = d - 1$, then exactly one of the eigenvalues of the monodromy operator is equal to -1 (at the vanishing cycle). The reader can check these properties in the trivial case $d = 1$. Here $\widetilde{Z}_0$ has the same cohomologies as Z_t.

More informations about degenerations of curves and surfaces the reader can find in [**KK**].

7.39. Relation between the limit mixed Hodge structure and the mixed Hodge structures on X_0 and on $X \setminus X_0$. The mixed Hodge structures on cohomology groups of complete intersections (e.g., X_0) and of open smooth varieties (e.g., $X^* = X \setminus X_0$) were introduced in the points 7.32 and 7.33 respectively. It turns out that all three mixed Hodge structures are related. In order to avoid complications, we assume the semi-stable case.
One has the exact *Wang sequence* (see the point 3.32)

$$\ldots \longmapsto H^n(X^*) \longmapsto H^n(X_t) \overset{M-I}{\longmapsto} H^n(X_t) \longmapsto \ldots, \quad t \in \mathbb{D}^*,$$

(here $M = M_t$). This sequence is obtained from the long exact sequence of the pair $(X_t \times [0,1], X_t \times 0 \cup X_t \times 1)$, when we identify (homotopically) X^* with the quotient space $X_t \times [0,1]/\sim$, where $(x,0) \sim (h(x),1)$ and h is the monodromy diffeomorphism of the fiber.
When we replace X_t by X_∞, then the corresponding Wang sequence is still exact. We have $H^n(X^*) = \mathbb{H}^n(X, \Omega^\bullet(\log)) = \mathbb{H}^n(X, \mathcal{B}^\bullet)$ (see the points 7.32(e) and

7.37(e)). The complex of sheaves $\Omega^\bullet(\log)$ is equipped with the Hodge and weight filtrations, defined in the point 7.33(i), analogous filtrations are introduced in the complex $\mathcal{B}$; they are analogous to the filtration in the complex $\mathcal{A}$ from 7.36(h). The final result follows.

(a) **Proposition.** *The limit Wang sequence* $\ldots \longmapsto H^n(X^*) \longmapsto H^n(X_\infty) \overset{M-I}{\longmapsto} H^n(X_\infty) \longmapsto \ldots$ *is exact sequence of spaces with mixed Hodge structures.*

If M is unipotent, then the operator $M - I$ can be replaced by $N = \log M$. The operator N in the Wang sequence is induced by the operator $\nu : \mathcal{A} \longmapsto \mathcal{A}$ which, when restricted to $\mathcal{A}^{m-q,q}$, is equal (up to a sign) to the projection $\Omega^{m+1}(\log)\widetilde{W}_q \longmapsto \Omega^{m+1}(\log)/\widetilde{W}_{q+1} = \mathcal{A}^{p-1,q+1}$. The kernel of ν consists of the sheaves $\mathrm{Gr}_{q+1}\Omega^{m+1}(\log) \simeq i_*\Omega^{m-q}_{Y(q+1)}$. The latter sheaves form the building blocks $\mathcal{K}^{p,q+1}$ used in the point 7.32(c) in construction of the mixed Hodge filtrations in $H^n(X_0)$. We have the following result.

(b) **Proposition.** *The composition of maps* $H^n(X_0) \longmapsto H^n(X) \longmapsto H^n(X_\infty)$ *is a homomorphism of spaces with mixed Hodge structures.*

The first map above $H^n(X_0) \longmapsto H^n(X)$ is the map induced by the Clemens' contraction. Identifying X_∞ with X_t one gets the next result.

(c) **Invariant cycle theorem.** *Let* $f : X \longmapsto \mathbb{D}$ *be a semi-stable degeneration. Then* $\ker(M_t - I) = \ker N = \mathrm{Im}(H^n(X_0))$ *in* $H^n(X_t)$, *i.e., the cocycles on* X_t, *upon which the monodromy acts trivially, arise from cocycles in the singular fiber* X_0 *under the Clemens contraction.*
In the case of general family $X_t = f^{-1}(t)$ *with degeneration (not semi-stable), we have* $\ker N = \mathrm{Im}(H^n(\widetilde{X}_0)$ *in* $H^n(X_t)$, *where* X_t *are identified with the fibers* $\tilde{f}^{-1}(s)$, $t = s^l$, *of the semi-stable reduction. Thus the cycles in* $H_n(X_t)$, *upon which* M_t *acts semi-simply, are identified with certain cycles in the reduced fiber* $\widetilde{X}_0$.

This result was firstly proved by Deligne [**Del3**] in the case when we have a degeneration $F : X \longmapsto S$ over a general algebraic manifold (scheme) S as a base. Here the degeneration occurs along a subvariety $S_0 \subset S$ and we have the action of the fundamental group $\pi_1(S \setminus S_0)$ on X_t. Deligne also proved that the action of $\pi_1(S \setminus S_0)$ on $H^n(X_t)$ is *completely reducible* (i.e., any invariant subspace has complementary invariant subspace). It is a generalization of the invariant cycle theorem and its proof also uses mixed Hodge structures.

7.40. Mixed Hodge structure on the cohomological Milnor bundle. (a) *The Milnor fibration and degeneration of algebraic varieties.* Let

$$g : (\mathbb{C}^{n+1}, 0) \longmapsto (\mathbb{C}, 0)$$

be a germ of holomorphic function with isolated critical point. For a small disc $\mathbb{D}_\epsilon$ (with center 0 and radius ϵ) in the image, we choose a small ball $\mathbb{B}_\rho$ in the preimage such that the varieties $g^{-1}(t)$ are transversal to the sphere $\partial\mathbb{B}_\rho$. Thus the family

$V_t = \mathbb{B}_\rho \cap g^{-1}(t)$, $t \in \mathbb{D}_\epsilon^*$, organize themselves into the locally trivial topological bundle V above the punctured disc, called the Milnor bundle (see Chapter 4). The spaces $H^n(V_t, \mathbb{C})$ are fibers $\mathcal{H}_t^n$ of the (holomorphic) cohomological Milnor bundle $\mathcal{H}^n$ over $\mathbb{D}_\epsilon^*$ (see Chapter 5). Our aim is to introduce a mixed Hodge structure on the fibers $\mathcal{H}_t^n$.
Using the results of Chapter 2 (Tougeron theorem and its corollaries) we can assume that: (i) g is a polynomial (of degree N); (ii) $\mathbb{D}_\epsilon = \mathbb{D}$; (iii) $t = 0$ is the only critical value of g and the projective closure (in $\mathbb{CP}^{n+1}$) of $g^{-1}(0)$ has only one critical point 0.
Thus we take the variety $\widehat{Y} = \{(z,t) \in \mathbb{CP}^{n+1} \times \mathbb{D} : \tilde{g}(z) - tz_{n+1}^N = 0\}$, where $\tilde{g}$ is the homogenization of g. We have the map $g_1 : \widehat{Y} \longmapsto \mathbb{D}$, defined as the projection $g_1(z,t) = t$. The varieties $\widehat{Y}_t = g_1^{-1}(t)$ intersect each other at $\widehat{Y}^\infty = \{g(z) = z_{N+1} = 0\}$ at infinity. So, we blow-up the set $\widehat{Y}^\infty$. Denote by Y the obtained space and by g_2 the morphism to $\mathbb{D}$ induced by g_1. We have $V_t = V \cap Y_t$, $Y_t = g_2^{-1}(t)$. There is a contraction $Y \longmapsto Y_0$, the prolongation of the radial contraction $\mathbb{D} \longmapsto 0$ (see Chapter 4). Next, the singular fiber V_0 (of the Milnor bundle) can be contracted to the singular point; we prolong it to a map $Y_0 \longmapsto Y_0$ such that $V_0 \longmapsto 0$. The composition of these two contractions gives the following.

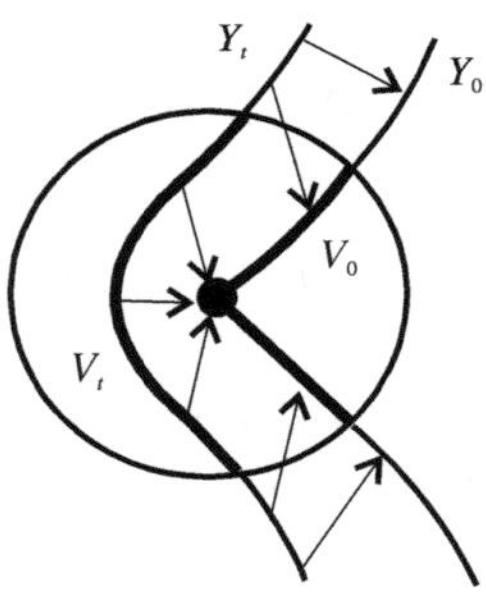

Figure 7.8: Contraction to the zero level

(b) **Contraction lemma.** *There are contraction $Y_t \longmapsto Y_0$ such that $V_t \longmapsto 0$. Thus $Y_0 \simeq Y_t/V_t$. Moreover, these contractions are compatible with the action of the monodromy transformation.*

(c) **Exact sequence.** The long exact sequence of cohomology groups of the pair (Y_t, V_t) gives the following exact sequence

$$0 \longmapsto \widetilde{H}^n(Y_0) \longmapsto H^n(Y_t) \longmapsto H^n(V_t) \longmapsto H^{n+1}(Y_0) \longmapsto H^{n+1}(Y_t) \longmapsto 0,$$

(as the only nonzero reduced cohomology group of the Milnor fiber is $\widetilde{H}^n(V_t)$).
We apply the resolution $\sigma : X \longmapsto Y$ of the singularity of Y_0. We put $f = g_2 \circ \sigma : X \longmapsto \mathbb{D}$. Thus, $f^{-1}(0) = X_0 = E_0 + E_1 + \ldots + E_r$, with smooth divisors E_j

intersecting themselves normally. Here the strict transform (desingularization) of Y_0 is the divisor denoted by E_0 and $E_1 + \ldots + E_r = \sigma^{-1}(0)$. Of course, $X_t = f^{-1}(t) \simeq Y_t$ for $t \neq 0$.
Therefore, we are in the situation of degeneration of a family of algebraic manifolds, as in the previous several points. We define the limit spaces $X_\infty = X \times_{\mathbb{D}} \mathbb{H}$ and $V_\infty = V \times_{\mathbb{D}} \mathbb{H}$, where $\mathbb{H} \longmapsto \mathbb{D}$ is the exponential map $t = e^{2\pi i u}$. They are homotopically equivalent to X_t and to V_t respectively.
Passing to the 'limit' $t \to 0$ in the above exact sequences (with Y_t replaced by X_t) we get the following exact sequence

$$0 \longmapsto \widetilde{H}^n(Y_0) \longmapsto H^n(X_\infty) \longmapsto H^n(V_\infty) \longmapsto H^{n+1}(Y_0) \longmapsto H^{n+1}(X_\infty) \longmapsto 0. \tag{7.5}$$

This sequence is monodromy invariant.
We shall introduce mixed Hodge structures in each entry from this complex.

(d) **Mixed Hodge structures on $H^j(X_\infty)$ and on $H^j(Y_0)$.** The limit mixed Hodge structure on $H^j(X_\infty)$ was introduced in 7.36. We recall shortly its basic ingredients.
We deal with the semi-stable reduction

$$\begin{array}{ccccccc} (V,V_0) & \longmapsto & (Y,Y_0) & \leftarrow & (X,X_0) & \leftarrow & (\widetilde{X},\widetilde{X}_0) \\ \downarrow g & & \downarrow g_2 & & \downarrow f & & \downarrow \tilde{f} \\ (\mathbb{D},0) & \longmapsto & (\mathbb{D},0) & = & (\mathbb{D},0) & \leftarrow & (\widetilde{\mathbb{D}},0) \end{array}$$

Here one applies the base change $\widetilde{\mathbb{D}} \longmapsto \mathbb{D}$, $t = s^l$, and introduces the space $\widetilde{X}$ (normalization of $X \times_{\mathbb{D}} \widetilde{\mathbb{D}}$) with the map $\tilde{f} : \widetilde{X} \longmapsto \mathbb{D}$. The singular fiber $\widetilde{X}_0 = D_0 + C_1 + \ldots + C_s$, where $D_0 \simeq E_0$ (desingularization of Y_0) and the divisors C_j have multiplicity 1 and are sent to some divisors E_i by the map $\widetilde{X} \longmapsto X$. (We use the notation $C_1, C_2, C_3, \ldots$ in order to distinguish them from D_0, but below we will also use the notation $D_i = C_i$, $i = 1, 2, \ldots$.)
The cohomology group $H^j(X_\infty)$ is the same as the jth hypercohomology group of the sheaf complex $\Omega^\bullet_{\widetilde{X}/\widetilde{\mathbb{D}}}(\log)$ on $\widetilde{X}_0$. This complex is quasi-isomorphic to the complex $\mathcal{A}^\bullet$ composed of the building blocks $\mathcal{A}^{p,q} = \Omega^{p+q+1}_{\widetilde{X}/\widetilde{\mathbb{D}}}(\log)/\widetilde{W}_q$, with the preliminary filtration $\widetilde{W}$ by the order of poles. The Hodge filtration $F^r = \bigoplus_{p \geq r, q} \mathcal{A}^{p,q}$ and the weight filtration $W_{j+r} = \widetilde{W}_{2q+r+1}/\widetilde{W}_q$ induce the limit mixed Hodge structure on $H^i(X_\infty)$.

Our next aim is to introduce a sheaf complex $\mathcal{A}^\bullet(Y_0)$, with Hodge and weight filtrations, such that it forms a subcomplex of $\mathcal{A}^\bullet$ and its jth hypercohomology group is equal to $\widetilde{H}^j(Y_0)$.
The reduced cohomology group $\widetilde{H}^j(Y_0) = H^j(Y_0, 0)$ is the same as the relative cohomology group $H^j(\widetilde{X}_0, C)$, where $C = C_1 + C_2 + \ldots$ is the preimage of 0 under the map $\widetilde{X} \longmapsto Y$. The both spaces $\widetilde{X}_0$ and C are semi-smooth varieties. In 7.32

the double complex $\mathcal{K}^{\bullet,\bullet}$, to calculate the cohomology of such semi-smooth variety, was introduced. The above cohomology is equal to the hypercohomology of the corresponding simple complex $\mathcal{K}^{\bullet}$. On $\widetilde{X}_0$ we have $\mathcal{K}^{p,q}(\widetilde{X}_0) = i_*\Omega^p_{D^{(q+1)}}$, where we use the notations $D_i = C_i$, $i = 1, 2, \ldots$. Analogously, on C we have the double complex $\mathcal{K}^{p,q}(C) = i_*\Omega^p_{C^{(q+1)}}$. The second sheaf is a direct factor of the first sheaf. We define the *mixed Hodge complex on* Y_0 by means of the building blocks

$$\mathcal{A}^{p,q}(Y_0) = \mathcal{K}^{p,q}(\widetilde{X}_0)/\mathcal{K}^{p,q}(C) = \begin{cases} i_*\Omega^p_{D_0}, & (q=0), \\ i_*\Omega^p_{D_0\cap C^{(q)}}, & (q>0). \end{cases}$$

The Hodge and weight filtrations on $\mathcal{A}^{\bullet,\bullet}(Y_0)$ are given by

$$F^r = \bigoplus_{p\geq r,q} \mathcal{A}^{p,q}(Y_0),$$
$$W_{n+r} = \bigoplus_{p,q\geq -r} \mathcal{A}^{p,q}(Y_0).$$

Because the corresponding simple complex $\mathcal{A}^{\bullet}(Y_0)$ is a resolvent of constant sheaf, we have $\widetilde{H}^n(Y_0) = \mathbb{H}^n(\widetilde{X}_0, \mathcal{A}^{\bullet}(Y_0)$. As in the case of mixed Hodge structure of a semi-smooth variety, the weight spectral sequence of this hypercohomology degenerates at the second term, ${}_WE_2^{p,q} = \mathrm{Gr}_q^W(H^{p+q}(Y_0))$.

(e) **Definition (mixed Hodge structure on $H^n(V_\infty)$.** The sheaves $\mathcal{A}^{p,q}(Y_0)$ are subsheaves of the sheaves $i_*\Omega^p_{D^{(q+1)}}$. The latter sheaves are isomorphic to

$$\mathrm{Gr}^{\widetilde{W}}_{q+1}\Omega^{p+q+1}_{\widetilde{X}}(\log) = \widetilde{W}_{q+1}/\widetilde{W}_q$$

(via the Poincaré residuum), which in turn are direct factors of the sheaves $\mathcal{A}^{p,q} = \widetilde{W}_{p+q+1}/\widetilde{W}_q$. Therefore, there is an inclusion $\mathcal{A}^{p,q}(Y_0) \subset \mathcal{A}^{p,q}$.
The *mixed Hodge complex on the limit Milnor fiber* is given by

$$\mathcal{A}^{\bullet}(V_\infty) = \mathcal{A}^{\bullet}/\mathcal{A}^{\bullet}(Y_0)$$

with the Hodge and weight filtrations induced from the filtrations on $\mathcal{A}$ and on $\mathcal{A}(Y_0)$.

(f) **Theorem (Mixed Hodge structure on the limit Milnor fiber).** *We have* $H^n(V_\infty) = \mathbb{H}^n(\widetilde{X}_0, \mathcal{A}^{\bullet}(V_\infty))$, *and the Hodge and weight filtrations of the complex* $\mathcal{A}^{\bullet}(V_\infty)$ *induce a mixed Hodge structure on* $H^n(V_\infty)$, *such that the exact sequence* (7.5) *(see the point* (c) *above) is the sequence of spaces with mixed Hodge structures and their morphisms.*
Moreover, the weight spectral sequence of hypercohomology of $\mathcal{A}^{\bullet}(Y_0)$ *degenerates at the second term.*

Proof. The fact that $H^n(V_\infty)$ is the same as the hypercohomology of $\mathcal{A}^\bullet(Y_0)$ and the statement about the sequence (7.5) are consequences of the short exact sequence of the sheaf complexes

$$0 \longmapsto \mathcal{A}^\bullet(Y_0) \longmapsto \mathcal{A}^\bullet \longmapsto \mathcal{A}(V_\infty) \longmapsto 0.$$

The sequence (7.5) is the long exact sequence of hypercohomology associated with this short sequence.
Also the statement about weight spectral sequence can be obtained from this and from the fact, that the weight spectral sequences associated with complexes $\mathcal{A}^\bullet$ and $\mathcal{A}^\bullet(Y_0)$ degenerate at the second terms. □

(g) **The first terms of the weight spectral sequence.** The weight spectral sequence of hypercohomology is associated with the reverse weight filtration of the complex $\mathcal{A}^\bullet(V_\infty)$. Its first terms are equal to ${}_W E_1^{-r,n+r} = \mathbb{H}^n(\widetilde{X}_0, \mathrm{Gr}^W_{n+r}\mathcal{A}^\bullet(V_\infty))$. Here

$$\begin{aligned}\mathrm{Gr}^W_{n+r}\mathcal{A}^\bullet(V_\infty) &= \mathrm{Gr}^W_{n+r}\mathcal{A}^\bullet/\mathrm{Gr}^W_{n+r}\mathcal{A}^\bullet(Y_0)\\ &= \begin{cases}\bigoplus_{q\geq 0} i_*\Omega^{\bullet-2q-r}_{D(2q+r+1)}, & (r>0),\\ i_*\Omega^{\bullet+r}_{C(-r+1)} \oplus \bigoplus_{q>-r} i_*\Omega^{\bullet-2q-r}_{D(2q+r+1)}, & (r\leq 0).\end{cases}\end{aligned}$$

Therefore,

$${}_W E_1^{-r,n+r} = \begin{cases}\bigoplus_{q\geq 0} H^{n-2q-r}(D^{(2q+r+1)}), & (r>0),\\ H^{n+r}(C^{(-r+1)}) \oplus \bigoplus_{q>-r} H^{n-2q-r}(D^{(2q+r+1)}), & (r\leq 0),\end{cases}$$

(where $D^{(j)}$ consist of intersections of all divisors $D_i = C_i$, $i = 1, 2, \ldots$, as well as of D_0). In other notations,

$${}_W E_1^{u,v} = \begin{cases}\bigoplus_{q\geq 0} H^{2n-v-2q}(D^{(2q-u+1)}), & (u<0),\\ H^{v}(C^{(u+1)}) \oplus \bigoplus_{q>u} H^{2n-v-2q}(D^{(2q-u+1)}), & (u\geq 0).\end{cases}$$

The weight of the factor $H^{n+r}(C^{(-r+1)})$ is $n+r$, and, from the proof of Theorem 7.36(i), it follows that the weight of the other factors are also equal to $n+r$. So, the weight of ${}_W E_1^{-r,n+r}$, and of ${}_W E_2^{-r,n+r}$ is $n+r$.

(h) **Example (Homogeneous singularity).** (We use Example 3.12 from [**Ste2**]). Let $g = P$, where P is a homogeneous polynomial on $\mathbb{C}^{n+1}$ of degree d with isolated singularity. The blowing-up of the point $0 \in \mathbb{C}^{n+1}$ gives a manifold X with holomorphic map $f : X \longmapsto \mathbb{C}$; (in an affine chart $\mathbb{C} \times \mathbb{CP}^n \ni (y,u) \longmapsto x = yu$, $f = y^d P(u:1)$). The singular fiber X_0 consists of two components: E_0 (the normalization of $P^{-1}(0)$) and $E_1 \simeq \mathbb{CP}^n$ of multiplicity d and $E_0 \cap E_1$ is a hypersurface of degree d in $\mathbb{CP}^n$. The semi-stable reduction gives $\widetilde{X}$ with $\widetilde{f} : \widetilde{X} \longmapsto \widetilde{\mathbb{D}}$ and $\widetilde{X}_0 = D_0 + C_1$, where $D_0 \simeq E_0$ and $C_1 = D_1$ is a d-fold covering of E_1 with ramification along $E_0 \cap E_1$; here $X \otimes_{\mathbb{D}} \widetilde{\mathbb{D}} = \{y^d P(u:1) = s^d\}$ and $\widetilde{X} = \{yv = s, P(u:1) = v^d\}$

is smooth with $D_0 = \{v = 0\}$, $D_1 = \{y = 0, P(u:1) = v^d\}$. In the Steenbrink's construction, X should be chosen compact, but it is not essential.
We have

$$\mathcal{A}^{p,0} = \Omega^{p+0+1}(\log)/\widetilde{W}_0 \simeq (\Omega^p_{D_0} \oplus \Omega^p_{D_1}) \oplus \Omega^{p-1}_{D_0\cap D_1}, \ \mathcal{A}^{p,1} \simeq \Omega^p_{D_0\cap D_1}$$

and other blocks vanish. Next,

$$\mathcal{A}^{p,0}(Y_0) \simeq \Omega^p_{D_0} \text{ and } \mathrm{d}\mathcal{A}^{p,1}(Y_0) \simeq \Omega^p_{D_0\cap D_1}.$$

Hence $\mathcal{A}^{p,0}(V_\infty) \simeq \Omega^p_{D_1} \oplus \Omega^{p-1}_{D_0\cap D_1}$ and $\mathcal{A}^{p,1}(V_\infty) = 0$.
This gives $\mathcal{A}^\bullet(V_\infty) \simeq \Omega^\bullet_{D_1} \oplus \Omega^{\bullet-1}_{D_0\cap D_1}$ with the weight filtration $\mathrm{Gr}^W_{n+1}\mathcal{A}^\bullet(V_\infty) \simeq \Omega^{\bullet-1}_{D_0\cap D_1}$, $\mathrm{Gr}^W_n \mathcal{A}^\bullet(V_\infty) \simeq \Omega^\bullet_{D_1}$.
One can check that this mixed Hodge complex coincides with the mixed Hodge complex used by Deligne to calculate $H^n(D_1 \setminus D_0)$, where $D_1 \setminus D_0 = \{P(u) = 1\} \subset \mathbb{C}^{n+1}$ is an affine (open) variety.

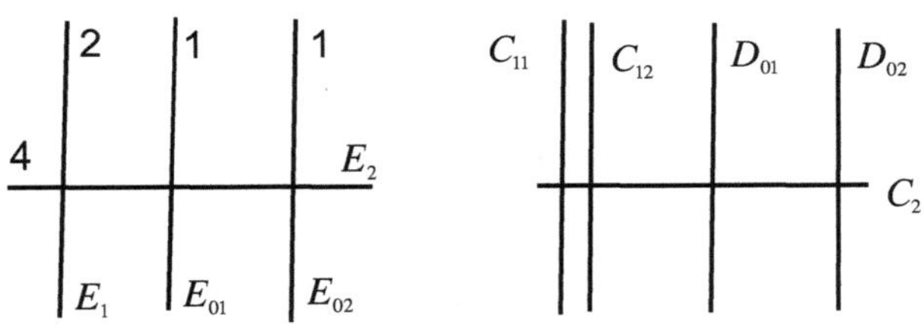

Figure 7.9: Resolution and the semi-stable reduction for the $\mathbf{A}_3$ singularity

(i) **Example.** Let $g = x^4 + y^2$ be the $\mathbf{A}_3$ singularity (see Fig. 7.9). The resolution consists of two blowing-ups: $x = z\tilde{x}$, $y = z\tilde{y}$, $(\tilde{x} : \tilde{y}) \in \mathbb{CP}^1$, $g = z^2(z^2\tilde{x}^4 + \tilde{y}^2)$, and the blowing-up of $z = 0, \tilde{x} = 1, \tilde{y} = 0$ by means of $z = rZ, \tilde{y} = rY$, $g = r^4 Z^2(Z + \mathrm{i}Y + \ldots)(Z - \mathrm{i}Y + \ldots))$. We find the divisors: E_{01}, E_{02} (normalizations of $g^{-1}(0)$), $E_1 \simeq \mathbb{CP}^1$ (of multiplicity 2) and $E_2 \simeq \mathbb{CP}^1$ (of multiplicity 4). Next, the variety $X \times_{\mathbb{D}} \widetilde{\mathbb{D}} = \{r^4 Z^2(Z^2 + Y^2 + \ldots) = s^4\}$ is normalized. The neighborhood of $r = Z = 0$ has two preimages, each of the form $r\widetilde{Z} = s(a + \ldots), Z = \widetilde{Z}^2$. The preimage of neighborhood of any point $E_2 \cap E_{0j}$ has one component of the form $rV = s(a + \ldots), Z \pm \mathrm{i}Y + \ldots = V^4$. This shows that $\widetilde{X}_0 = D_{01} + D_{02} + C_{11} + C_{12} + C_2$, where $D_{0,j} \simeq E_{0j}$, $C_{1j} = D_{1j} \simeq \mathbb{CP}^1$ (sent to E_1) and $C_2 = D_2$ is a 4-fold covering of E_2 with branching points at $D_1 \cap D_{1j}$ (of index 2) and at $D_2 \cap D_{0j}$ (of index 4). The Riemann–Hurwitz formula (Theorem 11.32) shows that C_2 is an elliptic curve.
We are interested in $H^1(V_\infty)$. We find part of the weight spectral sequence in the

form

$$\begin{array}{ccccccc} & & E_1^{-1,0} & \longmapsto & H^0(C^{(1)}) & \longmapsto & H^0(C^{(2)}) & \longmapsto & 0, \\ & & 0 & \longmapsto & H^1(C^{(1)}) & \longmapsto & 0, & & \\ 0 & \longmapsto & H^0(D^{(2)}) & \longmapsto & H^2(C^{(1)}) & \longmapsto & E_1^{1,2}. & & \end{array}$$

We see that $\mathrm{Gr}_0^W H^1(V_\infty) = H^0(C^{(2)})/\mathrm{Im}\, H^0(C^{(1)}) = 0$ (it equals to $\mathbb{C}^2/\mathrm{Im}\mathbb{C}^3$), $\mathrm{Gr}_1^W = H^1(C^{(1)}) = H^1(C_2) = H^{1,0} \oplus H^{0,1} \approx \mathbb{C}^2$ and $\mathrm{Gr}_2^W = \ker(H^0(D^{(2)}) \longmapsto H^0(C^{(1)})) = \ker(\mathbb{C}^4 \longmapsto \mathbb{C}^3) = \mathbb{C} = H^{1,1}$.
In [**Ste2**] (Example 3.13) the reader can find calculations for general singularity $(\mathbb{C}^2, 0) \longmapsto (\mathbb{C}, 0)$.

7.41. Mixed Hodge structure on Milnor bundle and monodromy. (a) The cohomological Milnor bundle over the punctured disc $\mathbb{D}^*$ is prolonged to a bundle $\mathcal{H}^n$ over $\mathbb{D}$ with the central fiber $\mathcal{H}_0^n$ equal to $H^n(V_\infty)$. The monodromy action $M_t : \mathcal{H}_t^n \longmapsto \mathcal{H}_t^n$ has limit M_0.
One expects that the action of the monodromy operator $M = M_0$ on $H^n(V_\infty)$ should be analogous as the action of the monodromy on the limit mixed Hodge structure on $H^n(X_\infty)$. As the reader will see, this is not completely true.
Indeed, by the Thom–Sebastiani theorem 5.31, the monodromy of the singularity $g = x^4 + y^2$ is equal to the tensor product of monodromies of the component functions:

$$M_g = M_{x^4} \otimes M_{x^2} = \begin{pmatrix} 0 & 0 & -1 \\ 1 & 0 & -1 \\ 0 & 1 & -1 \end{pmatrix} \otimes (-1)$$

with the eigenvalues $1, \pm \mathrm{i}$. Thus $M = M_s$ (is semi-simple) and $N = \log M_u = 0$. We should expect that $N^r : \mathrm{Gr}_{1+r}^W \longmapsto \mathrm{Gr}_{1-r}^W$ are isomorphisms. But $N^1 = 0$, $\mathrm{Gr}_2^W \neq 0 = \mathrm{Gr}_0^W$.

(b) If a root of unity λ is an eigenvalue of an operator M acting on a space H, then we denote by H_λ the invariant subspace corresponding to this eigenvalue. By $H_{\neq 1}$ we denote $\bigoplus_{\lambda \neq 1} H_\lambda$.
Recall the exact sequence (7.5):

$$0 \longmapsto \widetilde{H}^n(Y_0) \longmapsto H^n(X_\infty) \longmapsto H^n(V_\infty) \longmapsto H^{n+1}(Y_0) \longmapsto H^{n+1}(X_\infty) \longmapsto 0.$$

We have $H^j(Y_0) = H^j(Y_0)_1$. This shows that

$$H^n(V_\infty)_\lambda \simeq H^n(X_\infty)_\lambda, \quad \lambda \neq 1,$$

and that M acts on $H^n(V_\infty)_{\neq 1}$ in the same way as it acts on $H^n(X_\infty)_{\neq 1}$.

(c) Consider the restriction of the sequence (7.5) to the subspaces corresponding to the eigenvalue $\lambda = 1$. As usually, we denote $N = \log M_u$, where M_u is the unipotent part.

We can represent $H^n(V_\infty)_1$ as

$$(H^n(X_\infty)_1/\mathrm{Im}H^n(Y_0)_1) \oplus \mathrm{Im}\left(H^n(V_\infty)_1 \longmapsto H^{n+1}(Y_0)_1\right).$$

(d) We have $H^j(Y_0) \subset \ker N|H^j(V_\infty)_1$. The invariant cycle theorem 7.39(c) says even more, $\ker N$ is equal to the image of $H^j(\widetilde{X}_0)$, where $\widetilde{X}_0$ is the reduced semi-smooth variety associated with X_0 (via base change and normalization). But X_0 consists of divisors E_0 (with multiplicity 1) and $E_1, E_2, \ldots$ (with multiplicities >1); respectively $\widetilde{X}_0$ consists of divisors $D_0 = E_0$ and C_{ij} (preimages of E_j, $j > 0$). The cycles from $\bigcup C_{ij}$ lie in subspaces corresponding to eigenvalues $\neq 1$ (see also the proof of the monodromy theorem). This leads to the equalities

$$H^j(Y_0) = \ker N|H^n(V_\infty)_1,$$
$$H^n(X_\infty)_1/\mathrm{Im}H^n(Y_0)_1 = H^n(X_\infty)_1/\ker N.$$

The space $H^n(X_\infty)_1$ has weight filtration, being the monodromy weight filtration with central index n; i.e., defined by $N : W_j \longmapsto W_{j-2}$ and the operators N^r : $\mathrm{Gr}^W_{n+r} \longmapsto \mathrm{Gr}^W_{n-r}$ are isomorphisms. In the quotient space $H^n(X_\infty)/\ker N$ we delete all blocks Gr^W_{n-r} at which $N = 0$. It means that the central index is shifted to $n+1$ (see Fig. 7.10).

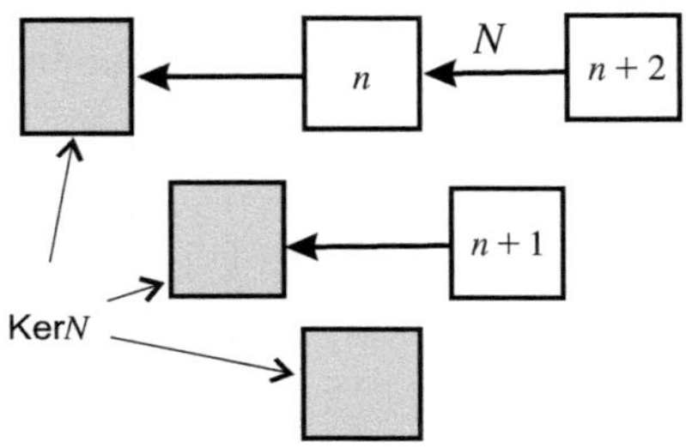

Figure 7.10: Action of the 'nilpotent' part N of the monodromy

(e) Let us study the weight filtration on $\mathrm{Im}\left(H^n(V_\infty)_1 \longmapsto H^{n+1}(Y_0)_1\right)$. One can avoid this term by suitable choice of the polynomial $g : \mathbb{C}^{n+1} \longmapsto \mathbb{C}$ defining the singularity. In **[SS]** it is argued that, if the degree of g is sufficiently high, then the map $H^n(X_t) \longmapsto H^n(V_t)$ is surjective; (equivalently: any n-dimensional cycle in V_t is not a boundary in X_t).
(In **[Ste2]** it is proved that, in the general case, the space $H^{n+1}(Y_0)$ (and $\mathrm{Im}\left(H^n(V_\infty)_1 \longmapsto H^{n+1}(Y_0)_1\right)$) has pure weight $n+1$. The proof is based on the exact sequence $\ldots \longmapsto H^n(E_1 \cup E_2 \cup \ldots) \longmapsto H^{n+1}(Y_0) \longmapsto H^{n+1}(X_0) \longmapsto \ldots$, the fact that the combinatorial complexes $\ldots H^p(E^{(q)}) \longmapsto H^p(E^{(q+1)}) \longmapsto \ldots$ are exact in terms $q > 1$ (as E_j appear in successive blowing-ups), showing that $H^p(E_1 \cup \ldots)$ have pure weight p, and a similar proof that $H^{n+1}(X_0)$ has pure weight $n+1$.)

The above gives the following result.

(f) **Theorem (Action of monodromy on mixed Hodge structure in vanishing cohomology).**

(i) *M_s is an automorphism of the mixed Hodge structure.*

(ii) *The maps*

$$\begin{aligned} N^r: \quad & \mathrm{Gr}^W_{n+r} H^n(V_\infty)_{\neq 1} \longmapsto \mathrm{Gr}^W_{n-r} H^n(V_\infty)_{\neq 1}, \\ N^r: \quad & \mathrm{Gr}^W_{n+1+r} H^n(V_\infty)_1 \longmapsto \mathrm{Gr}^W_{n+1-r} H^n(V_\infty)_1 \end{aligned}$$

are isomorphisms of weight $-2r$.

(g) **Corollary (symmetries of the Hodge numbers).** *We have*

$$\begin{aligned} h^{p,q}_\lambda &= h^{q,p}_{\bar\lambda}, \\ h^{p,q}_\lambda &= h^{n-p,n-q}_\lambda, \quad \lambda \neq 1, \\ h^{p,q}_1 &= h^{n+1-p,n+1-q}_1, \end{aligned}$$

where $h^{p,q}_\lambda = \dim H^{p,q}_\lambda$.

(h) **Corollary.** *The dimension of a Jordan cell of M does not exceed the dimension $n+1$ of the ambient space in the case of eigenvalue different from 1, and is $\leq n$ for eigenvalue 1.*

7.42. Mixed Hodge structure and intersection form for a quasi-homogeneous singularity. (We follow mainly [**Ste3**] and [**Var1**]).
Assume that g is a quasi-homogeneous polynomial, i.e., $g(\lambda^{\alpha_0}x_0, \ldots, \lambda^{\alpha_n}x_n) = \lambda g(x_0, \ldots, x_n)$, where α_j are rational numbers.
Let $\{x^k : k = (k_0, \ldots, k_n) \in I\}$, $I \subset (\mathbb{Z}_+)^{n+1}$, be the set of monomials, whose classes form a basis of the local algebra $\mathcal{O}_0(\mathbb{C}^{n+1})/(\partial f/\partial x)$. For $k \in I$ we put $l_k = \sum_{i=0}^n (k_i + 1)\alpha_i$ and associate with it the form

$$\omega_k = x^k \mathrm{d}x_0 \wedge \ldots \wedge \mathrm{d}x_n.$$

The next theorem of Steenbrink solves a conjecture of V. Arnold about intersection form for quasi-homogeneous singularities [**Arn4**].

Theorem of Steenbrink (quasi-homogeneous singularities). ([**Ste3**])

(i) *With the system of forms ω_k, $k \in I$, one can associate a basis in $H^n(V_t)$ consisting of eigenvectors of the monodromy with eigenvalues $\mathrm{e}^{2\pi \mathrm{i} l_k}$.*

(ii) *The Hodge numbers are equal to*

$$\begin{aligned} h^{p,q} &= \#\{k : q < l_k < q+1\}, \quad p+q = n, \\ h^{p,q} &= \#\{k : l_k = q\}, \quad p+q = n+1. \end{aligned}$$

(iii) *The zero space of the intersection form on* $H^n(V_t)$ *has dimension*

$$\mu_0 = \#\{k \in I : l_k \in \mathbb{Z}\},$$

and, if n *is even, then the numbers of pluses and of minuses in the canonical representation of the intersection form are equal to*

$$\mu_+ = \#\{k \in I : l_k \notin \mathbb{Z}, [l_k] \text{ even}\},$$
$$\mu_- = \#\{k \in I : l_k \notin \mathbb{Z}, [l_k] \text{ odd}\}.$$

Proof. (α) We introduce the notion of weighted projective space. Let the weights be $\alpha_j = a_j/b_j$. We put $b = \operatorname{lcm}(b_0, \dots, b_n)$, $c_i = \alpha_i b$ (integers).
Let G be a subgroup of $PGL(n+1, \mathbb{C})$ consisting of elements represented by the diagonal matrices $\operatorname{diag}(e^{2\pi i m_0/c_0}, \dots, e^{2\pi i m_n/c_n})$, $m_j \in \mathbb{Z}$. The **weighted projective space** is the space $M = \mathbb{CP}^n/G$. Its structural ring is $\mathbb{C}[z_0, \dots, z_n]^G$ and consists of quasi-homogeneous polynomials. The space M is a V-manifold and is a normal space (by Example 4.64(d)).

(β) As in Example 7.39(h), one shows that $H^n(Z_\infty) \simeq H^n(Z^*)$, where $Z^* = \{g(x) = 1\} \subset \mathbb{C}^{n+1}$. We treat Z^* as a subvariety in its weighted-projective closure $Z \subset M$. Thus $Z^* = Z \setminus Z_\infty$, where Z_∞ lies at infinity. Because Z^* is diffeomorphic with any nonzero level $g^{-1}(t)$, $t \neq 0$, (and is homotopically equivalent to the fiber Z_t of the Milnor fibration), we shall study the mixed Hodge structure of the fiber $g = t$.
The mixed Hodge structure of $H^n(Z^*)$ is defined as follows (with the Gysin homomorphisms):

$$\operatorname{Gr}_n^W H^n(Z^*) = \operatorname{coker}\left[H^{n-2}(Z_\infty) \longmapsto H^n(Z)\right],$$
$$\operatorname{Gr}_{n+1}^W H^n(Z^*) = \ker\left[H^{n-1}(Z_\infty) \longmapsto H^{n+1}(Z)\right].$$

One the other hand, we know that $\operatorname{Gr}_n^W H^n(Z_\infty) = H^n(V_\infty)_{\neq 1}$, $\operatorname{Gr}_{n+1}^W H^n(Z_\infty) = H^n(V_\infty)_1$.

(γ) Let us pass to the description of the Hodge filtration. Its definition by means of the filtration bête of the complex $\Omega_Z^\bullet(\log Z_\infty)$ is not useful in calculations. The same is with the stupid filtration of the complex $\mathcal{A}^\bullet(Z_\infty)$.
In 7.45 below the Hodge filtration on $H^n(g^{-1}(t))$ is introduced in another way. So, we use that filtration, leaving aside (for a while) the explanation of its equivalence to the standard Hodge filtration.
If ω is a quasi-homogeneous holomorphic $(n+1)$-form and $l > 0$ is an integer, then $\omega/(g-t)^l$ is a meromorphic quasi-homogeneous form with pole at the hypersurface $g = t$. We associate with $\omega/(g-t)^l$ its *residuum* $\eta = \operatorname{Res}\omega/(g-t)^l$, an n-form on $g = t$, as follows. Let U be a tubular neighborhood of $g = t$ and ∂U its boundary with projection $\pi : \partial U \longmapsto \{g = t\}$. If Δ is an n-dimensional cycle in $\{g = t\}$,

then $\pi^{-1}(\Delta)$ is $(n+1)$-dimensional cycle in $\partial U \subset \mathbb{C}^{n+1}\setminus\{g=t\}$; it is the Leray coboundary $\delta\Delta$ (see Definition 5.9). We put

$$\int_{\Delta} \operatorname{Res} \omega/(g-t)^l = \int_{\pi^{-1}(\Delta)} \omega/(g-t)^l.$$

One can easily check that $\operatorname{Res}\omega/(g-t)^l = \frac{2\pi \mathrm{i}}{(l-1)!}(\nabla_{\partial/\partial t})^{l-1} s[\omega](t)$, where $s[\omega]$ is the geometrical section (of the cohomological Milnor bundle) defined by means of the Gelfand–Leray form $\omega/\mathrm{d}g$ and $\nabla_{\partial/\partial t}$ is the Gauss–Manin connection. The map Res is the extension of the Poincaré residuum homomorphism to meromorphic forms with high order poles.

With any form $\omega_k = x^k \mathrm{d}x_0 \wedge \ldots \wedge \mathrm{d}x_n$, such that x^k is an element of the monomial basis of the local algebra $\mathbb{C}[x]/(\partial g/\partial x)$, we associate the element $\eta_k = \operatorname{Res}\omega_k(g-t)^{[-l_k]-1}$, $l_k = \sum \alpha_i(k_i+1)$. One shows that the geometrical sections satisfy $s[\omega_k] = t^{l_k} A_k$, where A_k are horizontal sections of $\mathcal{H}^n$ (by quasi-homogeneity). They form a basis of $H^n(g^{-1}(t))$ (see Chapter 5). Therefore, the classes

$$\eta_k = \text{const} \cdot t^{l_k + [-l_k]} A_k, \; k \in I,$$

also form a basis.

In 7.45 below it is shown that the Hodge filtration on the space generated by classes η_k is the same as the filtration by the order of pole $-[-l_k]+1$ of the space generated by forms $\omega_k(g-t)^{[-l_k]-1}$. Thus $F^p = \operatorname{span}[\eta_k : \, l_k \le n-p]$.

Next, the elements η_k are eigenvectors of the monodromy with eigenvalues $\mathrm{e}^{2\pi \mathrm{i} l_k}$ and hence $W_n = \operatorname{span}[\eta_k : l_k \notin \mathbb{Z}]$ (corresponding to $\mathrm{e}^{2\pi \mathrm{i} l_k} \neq 1$).

In this way the mixed Hodge structure. on $H^n(g^{-1}(t))$ is defined. The Hodge numbers are the same as in the theorem.

(δ) In order to calculate the invariants of the intersection form on $H^n(Z^*)$, we use the cohomologies with compact support $H^n_{\mathrm{c}}(Z^*)$. $H^n_{\mathrm{c}}(Z^*)$ is the dual to $H^n(Z^*)$ and is equipped with a mixed Hodge structure (with $W_{n-1} \subset W_n$) such that the pairing morphism $H^n_{\mathrm{c}}(Z^*) \otimes H^n(Z^*) \longmapsto \mathbb{C} = H^{n,n}$ is a morphism of mixed Hodge structures. We have $W_{n-1}H^n_{\mathrm{c}} = \{\omega : \langle \omega, \eta\rangle = 0 \; \forall \eta \in W_n H^n\}$. Consider the commutative diagram, where $i : Z^* \longmapsto Z$ is the inclusion and all arrows are morphisms of mixed Hodge structures:

$$\begin{array}{ccc} H^n_{\mathrm{c}}(Z^*) & \overset{i_*}{\longmapsto} & H^n_{\mathrm{c}}(Z) \\ \downarrow j & & \downarrow = \\ H^n(Z^*) & \overset{i^*}{\leftarrow} & H^n(Z) \end{array}$$

Let S be the bilinear form on $H^n_{\mathrm{c}}(Z^*)$, given by $S(\alpha,\beta) = \langle \alpha, j(\beta)\rangle = \int \alpha \wedge \beta$. We get $S(\alpha,\beta) = 0$, if $\alpha \in W_{n-1}H^n_{\mathrm{c}}$ (or, if $\beta \in W_{n-1}H^n_{\mathrm{c}}$); because $j|_{W_{n-1}} = 0 : W_{n-1}H^n_{\mathrm{c}} \longmapsto W_{n-1}H^n = 0$. This shows that $W_{n-1}H^n_{\mathrm{c}}$ is the zero space of the intersection form.

Next, i_* identifies $\operatorname{Gr}^W_n H^n_{\mathrm{c}}(Z^*)$ with $P^n(Z)$, the primitive part of the weighted-projective variety $Z \subset \mathbb{CP}^n/G$. Recall that the primitive cohomology classes are

those classes which have zero intersection with suitable power of the Kähler class, dual to a hyperplane section $[H \cap Z]$ (e.g., at infinity, see Theorem 7.12). In particular, the forms with compact support do not touch infinity and hence represent primitive classes, $H_c^n(Z^*) \subset P^n(Z)$. By the point (ii) (of the theorem under proof), $\mathrm{Gr}_n H^n(Z^*) = \mathrm{Coker}\left(H^{n-2}(Z_\infty) \longmapsto H^n(Z)\right)$ and the arrow is the Gysin homomorphism. Using the Poincaré duality, we identify it with the space of cycles in Z modulo cycles in Z_∞. Because $\mathrm{Gr}_n H_c^n(Z^*)$ is dual to $\mathrm{Gr}_n H^n(Z^*)$, we find that it equals to $P^n(Z)$.
Now we use Theorem 7.14. The form S is non-degenerate on P^n, which shows that the dimension of the zero subspace $\mu_0 = \dim \mathrm{Gr}_{n+1}^W H^n(Z^*)$ (by duality). If n is odd, then the form S is symplectic on $P^n(Z)$. If n is even, then we have the Hodge–Riemann relations on $P^n(Z) = \bigoplus H^{p,q}$: (i) $S(\alpha, \beta) = 0$, if $\alpha \in H^{p,q}$, $\beta \in H^{r,s}$, $(p,q) \neq (s,r)$, and (ii) $(-1)^{n(n-1)/2} \mathrm{i}^{p-q} S(x, \bar{x}) > 0$ for $x \in H^{p,q} \setminus 0$.
If n is even and we consider cohomologies with real coefficients, then the form $S|_{P^n(Z,\mathbb{R})}$ is diagonalizable with $\mu_+ = \sum_{q \text{ even}} h_{\neq 1}^{p,q}$ pluses and with $\mu_- = \sum_{q \text{ odd}} h_{\neq 1}^{p,q}$ minuses □

7.43. Remarks. (a) Steenbrink in [**Ste3**] introduced the Hodge filtration by means the identity $F^p H^n(Z) \approx \Gamma(M, \Omega_M^{n+1}(n+1-p)Z))/\mathrm{d}\Gamma(M, \Omega_M^n((n-p)Z))$ and the analogous identity for $F^p H^{n-1}(W)$. Here $\Omega_M^m(kZ)$ is the sheaf of meromorphic k-forms with poles on Z of order k (see below). In order to prove this, he uses a theorem of Bott about vanishing of Čech cohomologies of the projective space.

(b) The filtration $F^p = \mathrm{span}\,[\eta_k : l_k \leq n-p]$, from the proof of the theorem of Steenbrink, is equivalent to the following filtration, called the *asymptotic Hodge filtration*, in the space spanned by the geometrical sections $s[\omega_k]$, $k \in I$:

$$F_{\mathrm{as}}^p = \mathrm{span}[s[\omega_k] : l_k \leq n-p].$$

Because $s[\omega_k] \sim t^{l_k}$ this is the filtration determined by the asymptotic behavior of the geometrical sections.

(c) In [**Ste2**] Steenbrink has calculated the invariants $\mu_0, \mu_\pm$ in general situation. We have

$$\mu_0 = \sum_{p+q=n+1,n+2} h_1^{p,q}, \quad \mu_\pm = \sum_{p+q=n+2}{}' h_1^{p,q} + 2 \sum_{p+q \geq n+3}{}' h_1^{p,q} + \sum{}' h_{\neq 1}^{p,q},$$

where $\sum'$ runs over even (respectively, odd) q's for even n (respectively, odd n). From this it follows in particular that: if n is odd, then $\mu - \mu_0$ is even; if $n = 2 \pmod 4$, then $\mu - \mu_-$ is even and if $n = 0 \pmod 4$, then $\mu - \mu_+$ is even.

7.44. Applications of the mixed Hodge structure on the Milnor bundle. (a) *The Petrovski–Oleinik inequalities in real algebraic geometry.* V. Arnold [**Arn4**] observed that certain classical inequalities obtained by I. Petrovski and O. Oleinik [**PO**] can be interpreted in terms of the mixed Hodge structure on vanishing cohomology.

The **Petrovski–Oleinik inequalities** are the following:

$$|\chi(A) - 1| \leq \Pi_n(d), \quad n \text{ even},$$
$$|\chi(B_+) - \chi(B_-)| \leq \Pi_n(d), \quad n \text{ odd}, \ d \text{ even},$$

where $A \subset \mathbb{RP}^{n-1}$ is a projective algebraic hypersurface given as $f = 0$, $B_\pm = \{\pm f \geq 0\}$, f is a homogeneous polynomial, χ is the Euler characteristic and $\Pi_n(d) = \#\{k : \sum k_i = dn/2\}$.
The same inequalities in the Arnold's interpretation take the form

$$|\text{ind}| < h_1^{n/2,n/2}, \quad |\text{ind}| < h_1^{[n/2],[n/2]},$$

where ind $= i_0 \nabla f$ is the index of the gradient vector field and $h_1^{k,k}$ are the Hodge numbers of the mixed Hodge structure. in the vanishing cohomology defined by $f(x)$ (n even) and by $f(x) + y^2$ in $\mathbb{C}^{n+1}$ (n odd).
The proof uses the theorem of Steenbrink about quasi-homogeneous singularities (from the point 7.42).
Arnold conjectures that the mixed Hodge structure should play important role in topology of real algebraic manifolds.

(b) *Hodge numbers for hypersurfaces in projective and affine spaces.* Using the Lefschetz theorem about hyperplane sections and the Steenbrink's theorem about quasi-homogeneous singularities, applied to a hypersurface $Z \subset \mathbb{CP}^{n+1}$ defined by generic homogeneous polynomial equation $\tilde{g} = 0$ (e.g., $\tilde{g}(x) = g(x) - x_{n+1}^d$), one can show that $P^n(Z)$ coincides with $\text{Gr}_{n-1}^W(Z_\infty)$. In this way one computes the Hodge structure on Z (see Example 7.13 above).
The *Lefschetz theorem about hyperplane section* says that:

If $X \subset \mathbb{CP}^n$ is smooth compact variety of (complex) dimension m and $Y = X \cap H$ is smooth, where $H \subset \mathbb{CP}^n$ is a projective hyperplane, then the restriction homomorphism $H^j(X) \longmapsto H^j(Y)$ is isomorphism for $j < m-1$ and is embedding for $j = m-1$.
The dual version of this statement says that the Gysin homomorphism $H^j(Y) \longmapsto H^{j+2}(X)$ is isomorphism for $j > m-1$ and is epimorphism for $j = m-1$.

From this it follows also that:

The affine variety $Z^* = \{g(x) = 1\} \subset \mathbb{C}^{n+1}$ has the only nontrivial reduced cohomologies in dimension n.

It is generalization of the Milnor theorem to an algebraic situation.

7.45. Asymptotic mixed Hodge structure. A. Varchenko [**Var1**] defined the following **asymptotic mixed Hodge structure** on a fiber $\mathcal{H}_t = H^n(V_t)$ of the Milnor cohomological fibration, associated with a singularity $g : (\mathbb{C}^{n+1}, 0) \longmapsto (\mathbb{C}, 0)$
The **asymptotic weight filtration** $W_\bullet$ is the same as the Steenbrink's weight filtration; it is the monodromy filtration with central index n on $H^n(V_t)_{\neq 1}$, and with central index $n+1$ on $H^n(V_t)_1$.

The **asymptotic Hodge filtration** $F^\bullet$ is defined by means of the geometrical sections $s[\omega] = \langle \cdot, \omega/\mathrm{d}f\rangle$ of the cohomological Milnor bundle $\mathcal{H}^n$. We have

$$F^p_{\mathrm{as}} = F^p H^n(V_t) = \operatorname{span}\{s[\omega] : \omega \text{ of order } \leq n-p\}.$$

It means that, if $s[\omega] \in F^p$ and $\int_{\Delta(t)} \omega/\mathrm{d}f = \sum_{\alpha,k} A_{k,\alpha}(\Delta)t^\alpha(\ln t)^k$, then there exists $\alpha_0 \leq n-p$, k such that $A_{\alpha_0,k} \neq 0$.

Theorem. ([**Var1**], [**SS**]) *The above two asymptotic filtrations define a mixed Hodge structure equal to the Steenbrink's mixed Hodge structure.*

Varchenko [**Var1**] proved this result in special cases (semi-quasi-homogeneous, two variables, $f(x)+g(y)$). J. Scherk and J. Steenbrink [**SS**] gave a proof of this theorem using F. Pham's [**Ph3**] description of so-called Gauss–Manin moduli (via filtrations on some $\mathcal{D}$–moduli). V. Kulikov in [**Kul**] translated the proof from [**SS**] into the more standard language of sheaves and hypercohomologies.
We will see that the asymptotic Hodge filtration is analogous to the asymptotic Hodge filtration, which we have met in the quasi-homogeneous case.

The idea of the proof of this theorem is to equip the hypercohomologies of a complement $X\setminus Y$ of a hypersurface $Y \subset X$ with a filtration which arises from a filtration of some sheaf complex, but is different from the standard filtration into forms of (p,q)-type (when we use $H^n(X\setminus Y) = \mathbb{H}^n(X, \mathcal{E}^\bullet(\log)))$ and is different from the stupid filtration (when $H^n(X\setminus Y) = \mathbb{H}^n(X, \Omega^\bullet(\log))$. This idea belongs to P. Griffiths [**Gri3**].
We have the short exact sequence

$$0 \longmapsto \Omega^\bullet_X \longmapsto \Omega^\bullet_X(\log Y) \overset{R}{\longmapsto} i_*\Omega_Y^{\bullet-1} \longmapsto 0$$

(R – the Poincaré residuum), which shows that one can calculate the cohomologies of $X\setminus Y$ by means of hypercohomologies of the holomorphic logarithmic complex on X. The complex $\Omega^\bullet(\log Y)$ is quasi-isomorphic to the complex $\Omega(*Y)$ of meromorphic forms with finite order poles on Y. The above short exact sequence becomes quasi-isomorphic to the sequence

$$0 \longmapsto \Omega^\bullet_X \longmapsto \Omega^\bullet_X(*Y) \overset{R}{\longmapsto} DR^{\bullet-1}(B_{[Y]X}) \longmapsto 0,$$

where $DR^\bullet(B_{[Y]X}) = \Omega^\bullet_X \otimes B_{[Y]X}$ is the *de Rham complex of the ring* $B_{[Y]X} = \mathcal{O}_X(*Y)/\mathcal{O}_X$ and consist of classes of meromorphic forms with poles at Y modulo holomorphic forms. Here, by $\mathcal{O}_X(kY)$ we denote the sheaf of meromorphic functions on X with poles at Y of order at most k. $\mathcal{O}_X(*Y)$ is the union of such rings, $\mathcal{O}_X \subset \mathcal{O}_X(Y) \subset \mathcal{O}_X(2Y) \subset \ldots \mathcal{O}_X(*Y)$.
The complexes $\Omega^\bullet_X$, $\Omega^\bullet_X(\log Y)$ and $i_*\Omega_Y^{\bullet-1}$ are equipped with the filtration bête $F^\bullet$. For example, we have $F^p\Omega^m_X(\log Y) = \Omega^m(\log)$, if $m \geq p$, and $=0$, otherwise.
We want to introduce a filtration $P^\bullet$ on $\Omega^\bullet(*Y)$ (and on $DR^{\bullet-1}(B_{[Y]X})$) such that the inclusion map $a : \Omega^\bullet(\log) \longmapsto \Omega^\bullet(*Y)$ is a filtered quasi-isomorphism

(and the maps $i_*\Omega_Y^{\bullet -1} \longmapsto DR^\bullet(B_{[Y]X})$ and $\Omega_X^\bullet(*Y) \longmapsto DR^{\bullet -1}(B_{[Y]X})$ are filtration preserving).
For fixed n we put the *filtration by the order of pole* $P^\bullet$:

$$\begin{aligned} P^p\Omega^m(*Y) &= \Omega^m((n+1-p)Y), \quad p=1,\ldots,n, \\ &= \Omega^m(*Y), \quad p=0. \end{aligned}$$

Then we have $a(F^p\Omega^m(\log)) \subset P^p\Omega^m(*Y)$, $p = 0,1,\ldots n$. The filtration $P^\bullet$ on the complex $DR^\bullet(B_{[Y]X})$ is defined analogously.
It follows that the filtration(s) by order of pole induce the Hodge filtration on $\mathbb{H}^n(X,\Omega^\bullet(*Y))$ and on $\mathbb{H}^n(X,DR^{\bullet-1}(B_{[Y]X}))$, which coincide with the Deligne's Hodge filtrations on $H^n(X\setminus Y)$ and $H^{n-1}(Y)$ respectively.
(In [**Ste3**] the above filtered quasi-isomorphism is derived directly in the case X is a weighted projective space and Y is is (quasi-homogeneous) hypersurface. Steenbrink used long exact sequences, associated with short sequences $0 \longmapsto Z(\Omega^{q-1}(kY)) \longmapsto \Omega^{q-1}(kY) \longmapsto Z(\Omega^q((k+1)Y)) \longmapsto 0$ and $0 \longmapsto Z(\Omega_X^q) \longmapsto Z(\Omega^q(Y)) \longmapsto Z(\Omega_Y^{q-1}) \longmapsto 0$, and vanishing of some cohomology groups $H^p(X,\Omega_X^q \otimes \mathcal{O}(k))$ (Bott theorem).)
One applies this construction to the cases $X = V$ (small ball around the origin in $\mathbb{C}^{n+1}$) and $Y = V_t$, $t \neq 0$). Due to the quasi-isomorphism between $DR^\bullet(B_{[Y]X})$ and $\Omega_Y^\bullet$, which is analogous to the map Res from the proof of the theorem of Steenbrink in the point 7.42, the induced filtration on $H^n(Y)$ becomes the asymptotic Hodge filtration.
The detailed proof is much more complicated. We refer the reader to the book [**Kul**]. □

The asymptotic mixed Hodge structure is closely related with the spectrum of the singularity and the conjecture about its semi-continuity (see the point 5.45, [**AVG**], [**Ste2**] and [**Kul**]).
Varchenko in [**Var2**] proved that the operator $N = \log M_u$ acting on $H^n(V_t) \simeq \mathbb{C}^\mu$ has the same structure as the operator of multiplication by f in the local algebra $\mathbb{C}[x]/(\partial f/\partial x) \simeq \mathbb{C}^\mu$. It follows from the proof of the previous theorem (see [**SS**] and [**Kul**]). In particular, $f^{n+1} \in (\partial f/\partial x)$, where $n+1$ is the dimension of the ambient space. The later property was firstly proven by J. Briançon and H. Skoda [**BS**].

§5 Period Mapping in Algebraic Geometry

7.46. Example (moduli space of elliptic curves). An elliptic curve E can be identified with the quotient $\mathbb{C}/\Lambda$, where $\Lambda = \{mz_1 + nz_2 : m,n \in \mathbb{Z}\} = \mathbb{Z}\cdot z_1 + \mathbb{Z}\cdot z_2$ is a lattice and z_1 and z_2 are **periods**. The automorphisms of $\mathbb{C}$, i.e., $z \longmapsto \lambda z$, define isomorphisms between lattices $\mathbb{C}/\Lambda$ and $\mathbb{C}/\lambda\Lambda$. Thus one can assume that $z_1 = 1$, $\operatorname{Im} z_2 > 0$. Usually z_2 is here denoted by τ.

The generators of $\Lambda \simeq \mathbb{Z}^2$ are not defined uniquely. One can apply an automorphism of $\mathbb{Z}^2$, i.e., an element of the group $SL(2,\mathbb{Z})$ to them. Because $\tau = z_1/z_2$ is the coordinate in the projectivization of the complex plane generated by z_1, z_2, the action of $\begin{pmatrix} a & b \\ c & d \end{pmatrix} \in SL(2,\mathbb{Z})$ is equal to its projective representation: $\tau \longmapsto \frac{a\tau+b}{c\tau+d}$.
Therefore, the space of all elliptic curves is identified with $\mathbb{H}/PSL(2,\mathbb{Z})$, where $\mathbb{H}$ is the upper half-plane. There is one-to-one correspondence between the space $\mathcal{M}_1$ of moduli of elliptic curves and $\mathbb{H}/PSL(2,\mathbb{Z})$.
The map $E \longmapsto \tau(E)$ is called the **period map**. It can be defined by means of integrals of holomorphic forms along cycles in E. The space of holomorphic forms is 1-dimensional, $H^0(E,\Omega^1) = H^{1,0}(E) \simeq \mathbb{C}$ and is generated by $\omega = \mathrm{d}z$. The integer homologies $H_1(E,\mathbb{Z})$ are generated by the cycles $\gamma = \mathbb{R}/\Lambda$ and $\delta = \tau\mathbb{R}/\Lambda$ with the index of intersection $\langle \gamma, \delta \rangle = 1$. The intersection matrix is $\begin{pmatrix} 0 & 1 \\ -1 & 0 \end{pmatrix}$.
The period mapping is identified with the map $E \longmapsto (\int_\gamma \omega, \int_\delta \omega) = (1, \tau)$.
The intersection form of cycles is dual to the external product on de Rham cohomologies, $Q(\phi, \psi) = \int_E \phi \wedge \psi$. In the basis (η, μ) of $H^1(E,\mathbb{Z})$, dual to (γ, δ), we have $Q = \begin{pmatrix} 0 & 1 \\ -1 & 0 \end{pmatrix}$. Here $\eta = \mathrm{d}x - (\tau_1/\tau_2)\mathrm{d}y$, $\mu = \mathrm{d}y/\tau_2$, where $z = x + \mathrm{i}y$, $\tau = \tau_1 + \mathrm{i}\tau_2$.
Moreover, $\omega = \mathrm{d}x + i\mathrm{d}y = \eta + \tau\mu = (\eta, \mu)\Omega^\top$, $\Omega = (1, \tau)$.

If a curve $C \subset \mathbb{CP}^2$ is given by a quartic equation $y^2 = P_4(x)$ (in the affine part), then C is a topological torus (elliptic curve). One can choose a basis (γ, δ) of $H_1(C,\mathbb{Z})$, with $\langle \gamma, \delta \rangle = 1$, and a 1-form $\omega = c \cdot \mathrm{d}x/y \in H^0(C, \Omega^1)$ such that $\int_\gamma \omega = 1$ and $\operatorname{Im} \int_\delta \omega > 0$. Thus the period mapping $(\int_\gamma \omega, \int_\delta \omega) = (1, \tau)$ coincides in the both models of elliptic curve.
There is a holomorphic diffeomorphism between C and E. It is the *Abel–Jacobi map* $p \longmapsto \int_{p_0}^p \omega \pmod{\Lambda}$, where p_0 is a fixed point. The curve E is called the *jacobian of the curve* C.

7.47. The period mapping for general algebraic curves. Let C be a smooth closed projective curve of genus g.
For example, the curve defined by $Q(x,y) = 0$ (in $\mathbb{C}^2$), where Q is a generic polynomial of degree d, has smooth compactification in $\mathbb{CP}^2$ to a Riemann surface of genus $g = (d-1)(d-2)/2$. The compactification of the affine hyperelliptic curve $y^2 = P_{2g+2}(x)$ (P_{2g+2} – a generic polynomial of degree $2g+2 > 4$) is singular at one point at infinity. The resolution of this singular point gives a general *hyperelliptic curve* of genus g. Here the hyperellipticity means the existence of a 2-fold ramified covering above the projective line.
Then $H^1(C,\mathbb{C}) = H^{0,1} \oplus H^{1,0}$, where $h^{0,1} = h^{1,0} = g$, and $H^{1,0}$ is identified with the space of global holomorphic 1-forms.
Let $Q(\phi,\psi) = \int_C \phi \wedge \psi$ be bilinear anti-symmetric form on $H^1(C)$. It has the

property

$$\mathrm{i}Q(\alpha,\overline{\alpha}) > 0, \ \alpha \in H^{1,0} \setminus 0, \tag{7.6}$$

(because locally $\alpha \wedge \overline{\alpha} = -2\mathrm{i}|f|^2 \mathrm{d}x \wedge \mathrm{d}y$ for $\alpha = f(z)\mathrm{d}z$).
There exists a basis $\eta_1, \ldots, \eta_g, \mu_1, \ldots, \mu_g$ of $H^1(C,\mathbb{Z})$ such that

$$Q = \begin{pmatrix} 0 & I_g \\ -I_g & 0 \end{pmatrix} \tag{7.7}$$

in this basis. (I_g is the identity matrix of dimension g). This basis is dual to a basis $\gamma_1, \ldots, \gamma_g, \delta_1, \ldots, \delta_g$ of the lattice of integer cocycles. We choose a basis of holomorphic forms $\omega_1, \ldots, \omega_g \in H^0(C, \Omega^1)$ such that

$$\int_{\delta_i} \omega_j = \delta_{ij}.$$

The expansion of the forms ω_i in the basis $\eta_1, \ldots, \mu_g$ has the form

$$(\omega_1, \ldots, \omega_g) = (\eta_1, \ldots, \eta_g | \mu_1, \ldots, \mu_g)\Omega^\top$$

where $\Omega = (I_g|Z)$ and $Z = X + \mathrm{i}Y$ is a complex $g \times g$ matrix (X, Y real). Ω is called the **period matrix**: $\Omega = (\int_{\gamma_k} \omega_j | \int_{\delta_k} \omega_j)$.
The form $Q|_{H^{1,0}}$, in the basis ω_j, is equal to $\Omega Q \Omega^\top = Z^\top - Z$, where Q is in the form (7.7). Because $Q|_{H^{1,0}} \equiv 0$, we get the *first Hodge–Riemann relation:*

(i) *The matrix Z is symmetric, $Z = Z^\top$.*

The Hermitian form $(\phi,\psi) \longmapsto \mathrm{i}Q(\phi,\bar{\psi})$ restricted to $H^{1,0}$ is equal to $\mathrm{i}\Omega Q \overline{\Omega}^\top = 2\mathrm{Im}\, Z = 2Y$. Because of (7.6), we have the *second Hodge–Riemann relation*:

(ii) *The matrix Y is positive definite, $Y > 0$.*

The space

$$\mathbb{H}_g = \{Z \in gl(g,\mathbb{C}) : \ Z = Z^\top, \ \mathrm{Im} Z > 0\}$$

is called **Siegel upper half-plane of genus** g.
Here the Siegel upper half-plane is identified with the *period matrix space,* or with the *classifying space of Hodge structures* (see below).
The (arithmetic) *Siegel modular group* $Sp(g,\mathbb{Z})$ of matrices $G = \begin{pmatrix} A & B \\ C & D \end{pmatrix} \in SL(2g,\mathbb{Z})$, which preserve the (symplectic) form Q in the basis η_k, μ_l ($GQG^\top = Q$) acts on the space $\mathbb{H}_g$:

$$G(Z) = (AZ + B)(CZ + D)^{-1}.$$

The space $\mathbb{H}_g / Sp(2g,\mathbb{Z})$ is the proper space for values of the period mapping; to be correct, one should write $Sp(2g,\mathbb{Z}) \setminus \mathbb{H}_g$, as we have the left action of the modular group. However, it is more useful to work with the space $\mathbb{H}_g$.

With the curve C one can associate certain torus $J(C)$, called the *jacobian of* C and defined as $H^{1,0}/\pi H^1(C,\mathbb{Z})$, where $\pi : H^1 \longmapsto H^{1,0}$ is the projection along $H^{0,1}$. Thus $J(C)$ is a $2g$ dimensional real torus. The *Abel–Jacobi mapping*, from C to $J(C)$, is defined as $p \longmapsto (\int_{p_0}^p \omega_1, \ldots, \int_{p_0}^p \omega_g)^\top \pmod{\Lambda}$, where the lattice $\Lambda \simeq \mathbb{Z}^{2g}$ is generated by periods, the columns of the matrix Ω.
(The jacobian $J(C)$ turns out to be an algebraic variety, i.e., *abelian variety* with Kähler $(1,1)$-form (polarization). It has other interesting invariants reflecting the geometry of the curve C. We refer the reader to any book from algebraic geometry to get more informations about this interesting subject.)
Denote by $\mathcal{M}_g$ the *space of moduli of algebraic curves of genus* g. Two such curves are treated equivalent iff there is an analytic diffeomorphism between them.
We have defined the map

$$\Phi_g : \mathcal{M}_g \longmapsto \mathbb{H}_g/Sp(2g,\mathbb{Z}).$$

7.47. Torelli theorem for curves. ([**Tor**]) *The map* Φ_g *is embedding. It means that, if two curves have the same period matrix modulo the Siegel modular group, then they are analytically isomorphic.*

The proof of this result needs introduction of some additional algebro-geometrical notions (canonical curve, Θ-divisor) (see [**GH**] for example). So, we omit this proof.

At this moment it is worth to cite some results of D. Mumford and P. Deligne [**DeMu**]. It turns out that the space $\mathcal{M}_g$, $g > 1$, admits a structure of irreducible quasi-projective algebraic variety of dimension $3g-3$. Its closure $\overline{\mathcal{M}_g}$ is a smooth variety consisting of classes of curves admitting at worst double point singularities. The same concerns the space $\mathcal{M}_{g,n}$ (of moduli of curves of genus g with n punctures) and its closure (which forms a so-called algebraic stack). The intersection numbers of certain cycles in $\overline{\mathcal{M}_{g,n}}$ (the Gromov–Witten invariants) generate the partition function of conformal 2-dimensional field theory (see [**Dub**], [**Voi**]). These topics are in the frontiers of modern mathematics. They reveal deep connections between Algebraic Geometry and Mathematical Physics.
We leave these interesting theme aside and we pass to the definition of period matrix space in general situation.

7.49. Definition (polarized Hodge structure). Let $H_{\mathbb{Z}}$ be an integer lattice in its complexification $H = H_{\mathbb{Z}} \otimes \mathbb{C}$. Assume that H admits a pure Hodge structure of weight n, i.e., $H = \bigoplus_{p+q=n} H^{p,q}$ with $H^{p,q} = \overline{H^{q,p}}$. Let $Q : H_{\mathbb{Z}} \otimes H_{\mathbb{Z}} \longmapsto \mathbb{Z}$ be a non-degenerate bilinear form satisfying the *Hodge–Riemann relations*

$$\begin{aligned} Q(\phi,\psi) &= (-1)^n Q(\psi,\phi), \\ Q(\phi,\psi) &= 0, \quad \phi \in H^{p,q},\ \psi \in H^{r,s},\ p \neq s, \\ \mathrm{i}^{p-q} Q(\phi,\bar{\phi}) &> 0, \quad \phi \in H^{p,q} \setminus 0. \end{aligned}$$

The triple $(H_{\mathbb{Z}}, H^{p,q}, Q)$ is called the **polarized Hodge structure of weight** n. We introduce the Hodge numbers $h^{p,q} = \dim H^{p,q}$ and the *Weil operator* $C : H \longmapsto H$, $C|_{H^{p,q}} = \mathrm{i}^{p-q}$.

Example. Let X be a smooth complete projective variety of dimension d. Let $[\omega] \in H^2(X,\mathbb{Z})$ be its Kähler class and let $L : H^j(X) \longmapsto H^{j+2}(X),\ [\eta] \longmapsto [\omega\wedge\eta]$, be the Lefschetz operator.
We define $H_{\mathbb{Z}}$ as $P^n(X,\mathbb{Z}) = \ker\left[L^{d-n+1} : H^n \longmapsto H^{2d-n+2}\right],\ n \leq d$, the subspace of primitive vectors (see Theorem 7.12). The spaces $H^{p,q}$ are $P^n(X,\mathbb{C}) \cap H^{p,q}(X)$ and the form $Q(\phi,\psi)$ equals to $(-1)^{n(n+1)/2}\int_X \phi\wedge\psi\wedge\omega^{d-n}$. Due to Theorem 7.14, these data define a polarized Hodge structure of weight n.

7.50. Definition (classifying space of polarized Hodge structures). Let an integer n, a lattice $H_{\mathbb{Z}}$ and a bilinear form Q (on the lattice) and positive integers $h^{p,q}$ (equal to the dimensions of $H^{p,q}$) be given. Define the space D as the space of all polarized Hodge structures $H_{\mathbb{Z}}$, $H^{p,q}$, Q, with $\dim H^{p,q} = h^{p,q}$, and call it the **period matrix space**.
The space D can be constructed explicitly. The space $\mathbf{F}$ of all filtrations $F^0 \supset F^1 \supset \ldots$ with fixed dimensions $f^j = \dim F^j$ $(= h^{n,0} + h^{n-1,1} + \ldots + h^{j,n-j})$ is a *flag variety* and forms a subvariety of a product of *Grassmann varieties* (i.e., spaces of all subspaces of H of fixed dimension) with the natural complex structure.
The first Hodge–Riemann condition

$$\check{D}:\quad Q(F^p, F^{n-p+1}) = 0$$

defines a closed complex analytic subvariety $\check{D}$ of the flag variety $\mathbf{F}$. The second Hodge–Riemann condition

$$D:\quad Q(C\phi,\bar{\phi}) > 0,\ \ \phi \neq 0,$$

defines the period matrix space D as an open subvariety of $\check{D}$.
The spaces $\check{D}$ and D have also another algebraic descriptions. Let $G_{\mathbb{C}} = \mathrm{Aut}\,(H,Q)$ be the group of linear automorphisms of the space H preserving the form Q. $G_{\mathbb{C}}$ acts on the flag space $\mathbf{F}$ and the subset $\check{D}$ is invariant subset with respect to this action. The action of $G_{\mathbb{C}}$ on $\check{D}$ is transitive; for any a, b we have $b = g(a),\ g \in G_{\mathbb{C}}$ (there is only one orbit). This shows that $\check{D}$ is a homogeneous space and can be represented in the form

$$\check{D} = G_{\mathbb{C}}/B,$$

where B is a stationary subgroup of one point, of a fixed flag $F^0_* \supset F^1_* \supset \ldots$. In a suitable basis of H attached to this flag, the elements of the group B consist of block-triangular matrices. Such group is called *parabolic subgroup*. (The definition of parabolicity of an algebraic subgroup of a linear affine algebraic group requires that the quotient space be a compact projective variety). This shows that the space $\check{D}$ is smooth.
In order to give a similar description of the space D, one takes the subgroup $G_{\mathbb{R}} = \mathrm{Aut}\,(H_{\mathbb{R}},Q) \subset G_{\mathbb{C}}$ (the real variant of $G_{\mathbb{C}}$) and its subgroup $K = B \cap G_{\mathbb{R}}$. K turns out to be compact (because its restrictions to the quotient spaces F^p_*/F^{p+1}_* are orthogonal groups). We have

$$D = G_{\mathbb{R}}/K.$$

One has also the discrete *arithmetic* subgroup $G_{\mathbb{Z}} = \mathrm{Aut}\,(H_{\mathbb{Z}}, Q) \subset G_{\mathbb{R}}$. We treat this subgroup as acting on D from the left (induced from left multiplication in $G_{\mathbb{R}}$), whereas the cosets in $G_{\mathbb{R}}/K$ are cosets of the right multiplication of elements from K on $G_{\mathbb{R}}$. So, one has the left quotient $G_{\mathbb{Z}} \setminus D$ as the proper classifying space of Hodge structures. Because analytic description of the latter space is not explicit, one prefers to deal with D. However, when one has a family of Hodge structures depending on a parameter $s \in S$, then one has a map from S to $G_{\mathbb{Z}} \diagdown D$, which can be multivalued. In order to get single-valued mapping one uses the *period matrix space* in the form

$$\Gamma \setminus D,$$

where $\Gamma \subset G_{\mathbb{Z}}$ is the *monodromy group*, the image in $G_{\mathbb{Z}}$ of the fundamental group of S.

Let us demonstrate the construction of the space D in the case of curves of genus g. Each such curve is C^{∞} diffeomorphic to a fixed curve X_0. The lattice $H_{\mathbb{Z}}$ is the space $H^1(X_0, \mathbb{Z}) \simeq \mathbb{Z}^{2g}$, with the basis $\eta_1, \dots, \eta_g, \mu_1, \dots, \mu_g$ and with the standard symplectic form Q. The Hodge structures on $H = \mathbb{C}^{2g}$ are defined by choices of the subspaces $F^1 = H^{1,0}(X, \mathbb{C})$. The subspaces F^1 form graphs $F^1 = \{(\eta, \mu) : \mu = Z^{\top}\eta\}$, $\eta = \sum a_j \eta_j$, $\mu = \sum b_j \mu_j$. This means that the space of flags $H \supset F^1$ of this form is the same as the space of matrices Z. This space forms an affine part of the Grassmann variety $G(2g, g)$ of all flags. The space $\{Z : \ Z^{\top} = Z\}$ is the affine part of $\check{D}$ and $\{Z \in \check{D} : \ \mathrm{Im}\, Z > 0\}$ is the affine part of D.

7.51. General period mapping. Consider an analytic morphism $f : X \longmapsto S$ of complex analytic manifolds and such that $X \subset \mathbb{CP}^N \times S$, f is the projection and the fibers $X_s = f^{-1}(s) \subset \mathbb{CP}^N \times \{s\}$ are compact smooth projective varieties. Fix a positive integer n. For each s we have the data $(H_{\mathbb{Z}})_s$, $(H^{p,q})_s$, Q_s.
Fix s_0 and $H_{\mathbb{Z}} = (H_{\mathbb{Z}})_{s_0} = H^n(X_{s_0}, \mathbb{Z})$, $H^{p,q} = (H^{p,q})_{s_0}$, $Q = Q_{s_0}$. Because f is a locally trivial fibration (in the topological and C^{∞} categories), there exists a family diffeomorphisms $g_s : X_{s_0} \longmapsto X_s$. The induced homomorphisms g_s^* allow to identify the lattice $(H_{\mathbb{Z}})_s$ with $H_{\mathbb{Z}}$ and the form Q_s with Q. In fact, this identification is unique modulo the monodromy, i.e., the action of the fundamental group $\pi_1(S, s_0)$.
However the spaces $H_s^{p,q} = g_s^*(H^{p,q})_s$ can differ from $H^{p,q}$ (because the fibration should not be an analytic bundle). We obtain a family of polarized Hodge structures, i.e., a family of elements in D modulo the action of $\pi_1(S, s_0)$. Denote by Γ the image of $\pi_1(S, s_0)$ in $G_{\mathbb{Z}}$. The resulting map

$$\Phi_{\Gamma} : S \longmapsto \Gamma \setminus D$$

is the **period mapping of the family** X_s.
In applications one uses $S = \mathbb{D}^*$ or $S = \mathcal{M}_Y$, where $\mathcal{M}_Y$ is the space of moduli of analytic varieties topologically equivalent to Y.
P. Griffiths [**Gri1**] proved that Φ_{Γ} is locally liftable to D. It means that there are neighborhoods U (of points in S) and local maps $\Phi = \Phi_U : U \longmapsto D$ such that $\Phi_{\Gamma}|_U$ is the composition of Φ and of the quotient map.

We have the locally trivial vector bundles: $\mathcal{H} \longmapsto S$ and $\mathcal{F}^p \longmapsto S$, with the fibers $H(X_s, \mathbb{C})$ and $F^pH(X_s, \mathbb{C})$ respectively, where $F^\bullet$ is the Hodge filtration. The cohomological bundle is equipped with the Gauss–Manin connection; it is the holomorphic bundle.

7.52. The period map and the Gauss–Manin connection. (a) Assume that we have a family $f : X \longmapsto S$ of algebraic varieties X_s, as above, and let $\Phi : U \longmapsto D$ be a local lift of the period mapping. We want to differentiate Φ with respect to vectors $v \in T_{\mathrm{hol},s_0}S = T_{s_0}S$ (vectors of type $(1,0)$). Because the lattice $H_\mathbb{Z}$ does not depend on s, such derivative is in fact the derivative along the Gauss–Manin connection, $\partial\Phi/\partial v = \nabla_v\Phi$. Let us firstly differentiate the classes $[\eta] \in H^n(X_{t_0}, \mathbb{C})$.

(b) **Griffiths transversality Theorem.** ([**Gri1**]) *If* $[\eta] \in F^p$ *and* $v \in T_{s_0}S$, *then* $\nabla_v[\eta] \in F^{p-1}$.

The property $\nabla : \mathcal{O}_S(\mathcal{F}^p) \longmapsto \Omega^1_S(\mathcal{F}^{p-1})$ is called the *transversality condition* (or the *infinitesimal period relations*). It says that, although the subbundles $\mathcal{F}^p$ are not horizontal with respect to the Gauss–Manin connection, this non-horizontality is relatively small.

(c) **Corollary.** *The subbundles* $\mathcal{F}^p$ *are holomorphic and the period map* Φ *is also holomorphic.*

Proof. $\nabla_v\Phi$ is a tangent vector to the complexification of D, treated as C^∞ real manifold. The affine part of the Grassmann variety $G(H, f^p)$, of f^p-dimensional subspaces F^p of H, consists of the graphs of linear maps from $F^p_{s_0} = F^p$ to the quotient space H/F^p. Therefore the (holomorphic) tangent space to this Grassmann variety at the point $d = \Phi(s_0)$ is identified with the space $L(F^p, H/F^p)$ of linear maps. We have $\nabla_v\Phi(s_0) \in L(F^p, H/F^p)_\mathbb{C}$. The transversality theorem says that $\nabla_v\Phi(s_0) \in L(F^p, F^{p-1}/F^p)_\mathbb{C}$ in fact.
The derivative of a section $[\eta]$ along the conjugate vector $\bar{v} \in (T_{s_0}S)_\mathbb{C}$ is equal $\partial[\eta]/\partial\bar{v} = \overline{\partial\overline{[\eta]}/\partial v}$, where $\overline{[\eta]} \in F^{n-p}$ and $\partial\overline{[\eta]}/\partial v \in F^{n-p-1}$. Hence $\partial[\eta]/\partial\bar{v} \in F^{p+1}$. This shows that $\partial\Phi/\partial\bar{v} \in L(F^p, F^{p+1}/F^p) = 0$ (as $F^{p+1} \subset F^p$).
(The analyticity of the bundles $\mathrm{Gr}^p_\mathcal{F}$ follows also from their definition as the Leray sheaves $R^q f_*\Omega^p_{X/S}$.) □

(d) *Proof of Griffiths transversality Theorem.* Let us choose the representative η as a harmonic form on X_{s_0}, a sum of components of the type $(n,0), (n-1,1), \ldots, (p, n-p)$. We extend η to a form $\tilde{\eta}$ in a neighborhood of X_{s_0} in X with components of the same type. We use the C^∞ trivialization $f^{-1}(U) \simeq U \times X_{s_0}$. The (Gelfand–Leray) form

$$i_v \mathrm{d}\tilde{\eta},$$

where i_v is the internal product, defines the class of $\nabla_v[\eta]$. Because $\mathrm{d}\tilde{\eta}$ is of the type $(n+1,0) + \ldots + (p, n-p+1)$ and v is of the type $(1,0)$, the form $i_v\mathrm{d}\tilde{\eta}$ is of the type $(n,0) + \ldots + (p-1, n-p+1)$, i.e., it lies in F^{p-1}. □

(e) **Definition.** The **horizontal subspace** $T_{h,d}(D)$ of the (holomorphic) tangent space T_dD is defined as

$$L(F^p, F^{p-1}/F^p) \subset L(F^p/H/F^p),\ p = 1, 2, \ldots$$

The **horizontal subbundle** $T_h(D)$ is the subbundle of TD with the fibers $T_{h,d}(D)$. Griffiths Transversality Theorem says that Φ_* takes values in the horizontal subbundle.

7.53. Kodaira–Spencer mapping. Again we have the family $f : X \longmapsto S$, deformation of $Y = X_{s_0}$. The **Kodaira–Spencer map** is a linear map

$$\rho : T_{s_0}S \longmapsto H^1(Y, TY),$$

defined by means of the connecting homomorphism $H^0(Y, NY) \longmapsto H^1(Y, TY)$ associated with the short exact sequence $0 \longmapsto TY \longmapsto TX|_Y \longmapsto NY \longmapsto 0$. Here NY is the normal bundle to Y in X and one identifies the vectors $\xi \in T_{s_0}S$ with global sections $\sigma(y)$ of the normal bundle; $\sigma(y) = [w]$ with any $w \in T_yX$ such that $f_*w = \xi$.

One can describe the Kodaira–Spencer class $\rho(\partial/\partial s)$ more explicitly. Let $\mathcal{U} = (U_\alpha)$ be a covering of a neighborhood of Y in X by sets U_α, equipped with holomorphic coordinate systems $(x_{\alpha,1}, \ldots, x_{\alpha,d}, s)$, $d = \dim Y$. The vector $\partial/\partial s$ from $T_{s_0}S$ has representations $\boldsymbol{V}_\alpha$ in the coordinate system in U_α. $\boldsymbol{V}_\alpha$ is a vector field in U_α which allows us to construct a holomorphic diffeomorphism between $X_{s_0} \cap U_\alpha$ and $X_s \cap U_\alpha$. If we had a global vector field $\boldsymbol{V}$ defined in a neighborhood of Y (projecting itself to $\partial/\partial s$), then the phase flow maps $g_{\boldsymbol{V}}^{s-s_0} : X_{s_0} \longmapsto X_s$ would define a local analytic trivialization of the bundle f. The obstacles to prolongation of $\boldsymbol{V}_\alpha$ are the differences $\boldsymbol{T}_{\alpha\beta} = \boldsymbol{V}_\alpha - \boldsymbol{V}_\beta$ in $U_\alpha \cap U_\beta$. These differences satisfy a 1-cocycle condition and take values in TY (the s-components are the same). The cohomology class of the cocycle $(\boldsymbol{T}_{\alpha\beta}) \in C^1(\mathcal{U}, TY)$ is the value of $\rho(\partial/\partial s)$.

We see that the Kodaira–Spencer class constitutes the first obstruction to the holomorphic triviality of the bundle $f : X \longmapsto S$. The elements of $H^1(Y, TY)$ are called the *infinitesimal deformations of* Y.

It turns out that the component $\Phi_*^p : T_{s_0}S \longmapsto L(\mathrm{Gr}_F^p, \mathrm{Gr}_F^{p-1}) = L(H^{p,q}, H^{p-1,q+1})$ of the tangent Φ_* to the period map is equal to the composition

$$T_{s_0}S \overset{\rho}{\longmapsto} H^1(Y, TY) \longmapsto L(H^q(\Omega^p), H^{q+1}(\Omega^{p-1}),$$

where the first map is the Kodaira–Spencer map and the second map is the internal product in cohomologies.

The space $H^1(Y, TY)$ is a candidate to the space of moduli of analytic varieties, which have the same topological invariants as Y. Here by a moduli space $\mathcal{M}_Y$ we mean the following.

Either one associates to a point from $\mathcal{M}_Y$ a variety (like in the case of the moduli space $\mathcal{M}_g$ of curves) or one fixes the manifold Y, treated as a C^∞ manifold,

and varies the complex structure on Y. The second approach is better, because the variation of various invariants (of the complex structure) can be expressed in terms of objects (e.g., cohomologies) living on Y. One looks for deformations, which are versal; it means that any complex structure on Y is among those from the deformation (like versal deformation of a germ of holomorphic function).
Not always $H^1(Y, TY)$ is a base of the versal deformation. Sometimes an infinitesimal deformation, which forms the first term of the Taylor series of a deformation, cannot be completed to the whole Taylor series. The obstacles lie in the group $H^2(Y, TY)$. In [**KNS**] it is proved that, if $H^2(Y, TY) = 0$, then the Kodaira–Spencer mapping is a isomorphism between the base space of versal deformation (which exists) and $H^1(X, TX)$.
The versal deformation was constructed by Kuranishi [**Kur**]. It turns out that there exists a mapping $\gamma : H^1(Y, TY) \longmapsto H^2(Y, TY)$ such that $\mathcal{M}_Y = \gamma^{-1}(0)$.
Another approach to the problem of moduli of algebraic varieties runs through definition of quotients of actions of algebraic groups and was developed by D. Mumford [**Mum1**].

7.54. Mirror symmetry. The deformations of complex structures play crucial role in the mirror symmetry conjecture between Calabi–Yau varieties (see [**Dub**], [**Wit**], [**CoKa**] and [**Voi**]).
A Kähler manifold X is called *Calabi–Yau* iff its *canonical bundle* $K_X = \Omega_X^{\dim X}$ is trivial. Calabi–Yau manifolds with the same Hodge numbers as X form a family with the moduli space $\mathcal{M}_X$. Examples of the Calabi–Yau manifolds constitute the K-3 surfaces (i.e., 2-dimensional analytic varieties with $H^1(X, \mathbb{Z}) = 0$ and trivial canonical bundle $K_X = \Omega^2 \simeq \mathcal{O}_X$).
The *mirror symmetry* relies on the 'existence' of a family X_s', $s \in \mathcal{M}_{X'}$, of Calabi–Yau varieties such that the moduli space $\mathcal{M}_{X'}$ is parametrized by a subset $K(X) \subset H^{1,1}(X)$ (Kähler cone) and also $\mathcal{M}_X$ is parametrized by $K(X')$.
Existence of pairs (X, X') of families of (complex) 3-dimensional Calabi–Yau varieties was discovered by physicists [**COGP**] and it lead to fantastic mathematical predictions. Namely, if $\Omega \in \Gamma(X, \Omega_X^3)$ is a global holomorphic 3-form, then one can define the following Yukawa potential on $(T\mathcal{M}_X)^{\otimes 3} = (H^1(X, T^{1,0})^{\otimes 3}$:

$$(\xi_1, \xi_2, \xi_3) \longmapsto \int_X \Omega \wedge \nabla_{\xi_1} \nabla_{\xi_2} \nabla_{\xi_3} \Omega,$$

where ∇ is the Gauss–Manin connection. By some arguments, taken from Supersymmetric String Theory, it turns out that the Yukawa potential should 'coincide' with the following Gromow–Witten potential on $\left(H^2(X', \mathbb{Z})\right)^{\otimes 3}$:

$$(\eta_1, \eta_2, \eta_3) \longmapsto \sum_\beta \mathrm{e}^{-\langle \beta, \omega \rangle} N(\alpha_1, \alpha_2, \alpha_3; \beta).$$

Here $\omega \in H^{1,1}(X')$ is a symplectic form (imaginary part of the Kähler form), α_j are cycled dual to η_j, the summation runs over homology classes $\beta \in H^2(X', \mathbb{Z})$

and $N(\alpha_1, \alpha_2, \alpha_3; \beta)$ is the number of holomorphic curves $\phi : \mathbb{CP}^1 \longmapsto X'$ with $\phi_* [\mathbb{CP}^1] = \beta$ and incident to the cycles α_j (the Gromow–Witten invariants).
In [**COGP**] the case when X' is a quintic hypersurface in $\mathbb{CP}^4$ and

$$X = \left\{x_1^5 + \ldots + x_5^5 + sx_1 \ldots x_5 = 0\right\} / (\mathbb{Z}_5)^3$$

is considered. Here $H^2(X')$ and $\mathcal{M}_X$ are 1-dimensional; so, for fixed generators $\xi = \xi_{1,2,3}$ and $\eta = \eta_{1,2,3}$. The above potentials depend essentially on one variable. The mirror symmetry predicted equality of the Yukawa and Gromow–Witten potentials as functions of one variable. P. Candelas, X. de la Ossa, P. Green and L. Parkes calculated the Yukawa potential (it is rather standard) and predicted the number of rational curves in a quintic. The rigorous proof of all this was eventually found by A. Givental [**Giv2**].
For more details on this subject we refer the reader to [**CoKa**] and [**Voi**].

7.55. Torelli theorems. By the notion of Torelli theorem we have in mind certain analogues of the Torelli theorem for curves (Theorem 7.48). These are results about embeddings of the moduli space $\mathcal{M}_Y$ into the period matrix space D. If $\Phi : \mathcal{M}_Y \longmapsto D$ is an embedding, then we say that the *global Torelli theorem* holds. If $\Phi_*(s)$ is embedding (for any $s \in \mathcal{M}_Y$), then we say that the *infinitesimal Torelli theorem holds.*
The global Torelli theorems were proved for: curves, cubics in $\mathbb{CP}^4$ and Calabi–Yau manifolds.
The infinitesimal Torelli theorem holds for hypersurfaces in $\mathbb{CP}^n$.
For more informations we refer the reader to [**KK**] and [**Voi**].

7.56. Hyperbolic geometry methods. (a) Consider a degeneration $f : X^* \longmapsto \mathbb{D}^*$ of varieties, parametrized by the punctured disc and associated with it the period mapping $\Phi_\Gamma : \mathbb{D}^* \longmapsto \Gamma\backslash D$. Here Γ is the cyclic group generated by the monodromy automorphism.
The space D is the homogeneous space $G_\mathbb{R}/K$. It is equipped with invariant metric $\mathrm{d}s_D^2$, induced from the Cartan–Killing form on the Lie algebra $\mathfrak{g} = T_e G_\mathbb{R}$. (The Cartan–Killing form $\langle\cdot,\cdot\rangle$ satisfies $\langle[x,y],z\rangle + \langle y,[x,z]\rangle = 0$, defines invariant metric on the group and is unique up to normalization for a simple Lie algebra, see [**Ser1**]).

(b) It turns out that (see [**Gri1**], [**Gri2**]):

The period mapping is negatively curved. It means that the sectional curvature in the direction $\Phi_* T_t \mathbb{D}^*$ is negative; after suitable normalization we have

$$K_D(v, \bar{v}) \leq -|v|_D^2, \quad v \in \Phi_* T_t \mathbb{D}^*.$$

In other words, the horizontal subbundle $T_h(D)$ (see 7.52(e)) is negatively curved. On the other hand, the punctured disc $\mathbb{D}^*$ also admits negatively curved metric, the metric $|\mathrm{d}_h t|^2 = |\mathrm{d}t|^2/(|t| \ln |t|)^2$ (induced from the hyperbolic metric $|\mathrm{d}z|^2/(\mathrm{Im} z)^2$ on the universal covering $\mathbb{H}$).

(c) There is a generalization of the classical Schwarz theorem, which says that if $f : (\mathbb{D}, 0) \longmapsto (\mathbb{D}, 0)$ is holomorphic, then $|f(t)| \le |t|$. This generalization says that:

The period map is hyperbolic distance decreasing, $d(\Phi_\Gamma(t), \Phi_\Gamma(t')) \le d_H(t, t')$.

We cannot explain here the above two results. We refer the reader to the works of Griffiths [**Gri1**] and [**Gri2**]. Below we have their application to the monodromy operator.

(d) **Borel's proof of the first part of monodromy theorem.** We want to show that the eigenvalues of the monodromy operator M are roots of unity.
We apply the covering map $\widetilde{\Phi} : \mathbb{H} \longmapsto D = G_\mathbb{R}/K$. We have $\widetilde{\Phi}(z+1) = M\widetilde{\Phi}(z), \ z \in \mathbb{H}$.
Consider the points $ni, ni+1 \in \mathbb{H}, \ \mathrm{i} = \sqrt{-1}, \ n = 1, 2, \ldots$. The elements $\widetilde{\Phi}(ni) \in D$ are cosets $g_nK, \ g_n \in G_\mathbb{R}$, and $\widetilde{\Phi}(ni + 1) = Mg_nK$. The hyperbolic distance between ni and $ni+1$ tends to zero, $d_H(ni, ni+1) = 1/n$. By the point (c) above, $d(g_nK, Mg_nK) \le 1/n$, which means that $d(g_n^{-1}Mg_nK, K) \to 0$. Thus the elements $g_n^{-1}Mg_n$, from the conjugacy class of M, have accumulation point in the subgroup K.
The group K is compact. If M would have an eigenvalue λ_i with module different from 1, then the groups generated by $g_n^{-1}Mg_n$ would be non-compact. This shows that all $|\lambda_i| = 1$.
Because λ_i are algebraic numbers, they are roots of unity. □

7.57. The Baily–Borel compactification and limit of period mapping. Let $f : X \longmapsto \mathbb{D}$ be a degeneration of a family of algebraic varieties. It means that $X_t, \ t \neq 0$, are compact and smooth and X_0 is singular. We have the period mapping $\Phi_\Gamma : \mathbb{D}^* \longmapsto \Gamma \setminus D$.
The punctured disc $\mathbb{D}^*$ is open. Open is also the space $\Gamma \setminus D$, because of openness of the second Hodge–Riemann relation.
If the monodromy operator were periodic, i.e., $M = M_s$ and $N = \log M_u = 0$, then after applying the semi-stable reduction, one would obtain a situation with holomorphic single valued map $\mathbb{D}^* \longmapsto D$. After showing that it is bounded, one would obtain an analytic prolongation of Φ_Γ to the full disc with values in $\Gamma \setminus D$ (see [**Gri2**]).
In general case one cannot obtain prolongation with values in $\Gamma \setminus D$. However, one can try to obtain a prolongation in the form $\overline{\Phi}_\Gamma : \mathbb{D} \longmapsto \{\Gamma \setminus D\}^*$, where $\{\Gamma \setminus D\}^*$ is some (partial) compactification of the space $\Gamma \setminus D$. A natural candidate for $\{\Gamma \setminus D\}^*$ is the so called Baily–Borel compactification (see [**BaiB**]), which can be described as follows.
The boundary $\partial D = \overline{D} \setminus D \subset \check{D}$ is a union of disjoint subsets $F_1, \ldots, F_r$. They are complex submanifolds in $\check{D}$. The action of $G_\mathbb{R}$ prolongs itself to an action on ∂D, but the varieties F_j can be permuted. Denote $N(F_j) = \{g : g(F_j) = F_j\}$, the *normalizer of* F_j, and $Z(F_j) = \{g : g|_{F_j} = id\}$, the *centralizer of* F_j. We have $F_j \simeq G_{F_j}/K_{F_j}$, where $G_{F_j} = N(F_j)/Z(F_j)$ and K_{F_j} are their compact subgroups.

Assume also that the discrete groups $\Gamma_j = \Gamma \cap N(F_j)$ act properly discontinuously on F_j; (i.e., elements from $\Gamma_j \setminus Z(F_j)$ have no fixed points in F_j).
The **Baily–Borel compactification** (see [**BaiB**]) is equal to

$$\{\Gamma \setminus D\}^* = (\Gamma \setminus D) \cup \bigcup_j (\Gamma_j \setminus F_j,$$

where the induced topologies on $\Gamma \setminus D$ and on $\Gamma_j \setminus F_j$ are the natural topologies and $\Gamma \setminus D$ is an open subset in $\{\Gamma \setminus D\}^*$.
The following result belongs to A. Borel (accordingly to Griffiths [**Gri2**]).

Theorem. *The period mapping can be extended to a map* $\overline{\Phi}_\Gamma : D \longmapsto \{\Gamma \setminus D\}^*$ *in the cases when the varieties* X_s *are: curves, cubic threefolds and K-3 surfaces.*

7.58. Example (Extension of the period mapping for curves). This result is due to D. Mumford (unpublished). We follow the exposition in [**Gri2**].

(a) We have $F : X \longmapsto \mathbb{D}$ with fibers X_t, $t \neq 0$, which are curves of genus g. Without loss of generality we can assume that the monodromy operator is unipotent, $(M - I)^2 = 0$.

(b) We begin with a proper choice of the basis of rational cycles in $H_1(X_t, \mathbb{Q})$. By the monodromy assumption there exists a basis $x_1, \ldots, x_m$, $y_1, \ldots, y_m$, $z_1, \ldots, z_{2g-2m}$ such that

$$Mx_j = x_j + y_j, \quad My_j = y_j, \quad Mz_j = z_j. \tag{7.8}$$

Because the intersection form is monodromy invariant, we have $Q(x_i, y_j) = Q(Mx_i, My_j) = Q(x_i+y_i, y_j)$ and hence $Q(y_i, y_j)=0$. Similarly we get $Q(y_j, z_k) = 0$. Next, from $Q(x_i, x_j) = Q(x_i+y_i, x_j+y_j)$ it follows that $Q(x_i, y_j)+Q(y_i, x_j) = 0$. Therefore,

$$Q = \begin{pmatrix} 0 & A & 0 \\ -A^\top & B & C \\ 0 & -C^\top & D \end{pmatrix}, \quad A = A^\top,$$

in the basis (y_j, x_i, z_k). By means of the change $x_i' = x_i + (1/2)\sum(A^{-1}B)_{ji}y_j$, we reduce B to 0. Similarly we reduce C to 0. Then $D = -D^\top$.
The further substitution $x' = Ex$, $y' = Fy$ gives a similar matrix Q, with $B = 0$, $C = 0$ and $A' = EAF^\top$. Thus one can transform A to I_m, using symmetric and commuting matrices E and F with rational coefficients. Because Q, restricted to $\{z_1, \ldots, z_{2g-2m}\}$ (the space spanned by z_k), is symplectic, we can assume that $D = \begin{pmatrix} 0 & I \\ -I & 0 \end{pmatrix}$
So, in the basis $y_1, \ldots, y_m, z_1, \ldots, z_{g-m}, x_1, \ldots, x_m, z_{g-m+1}, \ldots, z_{2g-2m}$ we obtain the standard form for Q.

However, the formulas (7.8) for the monodromy do not hold now, because we have replaced x_i and y_j. In the new basis $\{y_j, z_k, x_i, z_{g-m+l}\}$ we have

$$M = \begin{pmatrix} I_m & 0 & \Lambda & 0 \\ 0 & I_{g-m} & 0 & 0 \\ 0 & 0 & I_m & 0 \\ 0 & 0 & 0 & I_{g-m} \end{pmatrix}, \quad \Lambda = EF^{-1} = \Lambda^\top. \tag{7.9}$$

(c) Let us pass to the cohomologies and the period mapping. We choose a basis $\psi_1, \ldots, \psi_{2g}$ of $H^1(X_t, \mathbb{C})$, depending holomorphically on $t \in \mathbb{D}^*$ and such that the integrals $\int_{\delta_i} \psi_j$ along horizontal cycles (e.g., x_i, y_j, z_k) have the form

$$a(t) + b(t) \cdot (2\pi \mathrm{i})^{-1} \log t,$$

with holomorphic $a(t)$ and $b(t)$ near $t = 0$. The existence of such basis follows from the regularity of the Gauss–Manin connection. In Chapter 8 it is proved that a germ of regular linear non-autonomous system with unipotent monodromy is analytically reduced to the system $t\dot{u} = (C_0 + O(t))u$, C_0 nilpotent. The solutions of such system have just the above type singularities.

The period matrix, i.e., the matrix of integrals of holomorphic forms along the cycles from the above basis, takes the form $\Omega(t) = \Omega_0(t) + (2\pi\mathrm{i})^{-1}\widetilde{\Omega}_0(t) \ln t = \Omega_0 + (2\pi\mathrm{i})^{-1} \ln t \cdot \Omega(M - I)$, where Ω_0 and $\widetilde{\Omega}_0$ are holomorphic. (This follows from the monodromy action $\Omega \longmapsto \Omega M = \Omega + \widetilde{\Omega}_0$.) Writing $\Omega = (\Omega_1|\Omega_2)$ (Ω_j – $g \times g$ matrices) one finds from Eq. (7.9) that Ω_1 is holomorphic and single-valued. The Hodge–Riemann relations imply $\det \Omega_1(t) \neq 0$ for $t \neq 0$. Application of Ω_1^{-1} gives the standard form $\Omega = (I_g|Z(t))$, where

$$Z(t) = Z_0(t) + \frac{\ln t}{2\pi\mathrm{i}} \begin{pmatrix} \Lambda & 0 \\ 0 & 0 \end{pmatrix},$$

with meromorphic Z_0 and Λ the same as above (the only nonzero part in $M - I$). Next, we know that $Z(t)$ takes values in the Siegel upper half-plane $\mathbb{H}_g$. If $Z_0(t)$ would have a pole at $t = 0$, then the image of $Z(t)$ could not be restricted to $\mathbb{H}_g$, which is biholomorphically equivalent to a bounded domain (like $\mathbb{H} \simeq \mathbb{D}$). Thus Z_0 is holomorphic. But then the dominating term in $Z(t)$, $t \to 0$, is $\Lambda \frac{\ln t}{2\pi\mathrm{i}}$. Because $\mathrm{Im} Z > 0$, we find that $\Lambda > 0$.

(d) Recall the weight filtration of the limit mixed Hodge structure associated with the degeneration f. It is the monodromy weight filtration with central index 1 and is defined by $W_0 = \mathrm{Im}\,(M - I)$, $W_1 = \ker(M - I)$, $W_2 = H^1(X_t)$.

We choose a rational basis $\eta_1, \ldots, \eta_g, \mu_1, \ldots, \mu_g$ of $H^1(X_t, \mathbb{Q})$ such that the period matrix is equal $(I|Z(t))$, the exterior product form Q has the standard form and the monodromy acts as follows:

$$
\begin{aligned}
M\eta_i &= \eta_i + \sum_{j=1}^{m} \Lambda_{ij}\mu_j, \quad i = 1, \dots, m, \\
M\eta_j &= \eta_j, \quad j > m, \\
M\mu_i &= \mu_i.
\end{aligned}
$$

Then we have

$$
\begin{aligned}
\mathrm{Gr}_2^W &= \{\eta_1, \dots, \eta_m\}, \\
\mathrm{Gr}_1^W &= \{\eta_{m+1}, \dots, \eta_g; \mu_{m+1}, \dots, \mu_g\}, \\
W_0 &= \{\mu_1, \dots, \mu_m\}.
\end{aligned}
$$

(e) The Hodge filtration $F_t^0 \supset F_t^1 \supset \{0\}$ defines subspaces $F_t^1 = H^0(X_t, \Omega^1)$ in $H^1(X_t)$. With fixed basis we obtain a family of g-dimensional subspaces $S_t \subset \mathbb{C}^{2g}$. Each such subspace $S = S_t$ is an element of the Grassmann variety $G(2g, g)$ and is isotropic with respect to the symplectic form Q. The isotropic subspaces form the subvariety $\check{D} \subset G(2g, g)$. We know also that if $t \neq 0$, then the form $\mathrm{i}Q(S, \overline{S}) > 0$, i.e., $S \in D$.
The basis of the space S_t constitutes of the forms $\omega_j = \eta_j + \sum_{k=1}^{g} (Z(t))_{jk}\mu_k$.
We pass to the limit $t \to 0$ with S_t. Thus we normalize the above basis of S_t as follows: $\psi_j(t) = \sum_{k=1}^{m} \Lambda_{jk}\mu_k + O(\ln^{-1} t)$, $j = 1, \dots, m$; $\psi_j(t) = \eta_j + \sum_{k=1}^{g} (Z_0(t))_{jk}\mu_k$, $j = m+1, \dots, g$. Here the vectors η_i, μ_k are fixed and the matrix $Z_0(t)$ is holomorphic.
We see that the limit $S_0 = \lim S_t$ is a subspace containing W_0.

(f) The limit S_0 turns out to be an element from the boundary $\overline{D} \setminus D$. The boundary $\overline{D} \setminus D \subset \check{D}$ is divided into components as follows. Each component F_j is characterized by the dimension of the kernel of the form $\mathrm{i}Q(S, \overline{S})$.
If W_0 is a rational (i.e., defined over $\mathbb{Q}$) isotropic subspace (i.e., $Q(W_0, W_0) = 0$) of $(\mathbb{C}^{2g}, Q)$ of dimension m, then we put $F(W_0)$ as the set of all $S \in \check{D}$ with $W_0 \subset S$ and such that $\mathrm{i}Q(S, \overline{S})$ has rank $g - m$. Equivalent conditions state that $W_0 = S \cap \overline{S}$, or that $(W_0)^{\perp} = S + \overline{S}$.
We have:

$S_0 \in F(W_0)$ in the above limit.

Indeed, we can assume that

$$
\begin{aligned}
S_0 &= \{\mu_1, \dots, \mu_m, \eta_{m+1} + \sum_{k=m+1}^{g} z_{m+1,k}(0)\mu_k, \dots, \eta_g + \sum_{k=m+1}^{g} z_{g,k}(0)\mu_k\}, \\
\overline{S}_0 &= \{\mu_1, \dots, \mu_m, \eta_{m+1} + \sum_{k=m+1}^{g} \overline{z}_{m+1,k}(0)\mu_k, \dots, \eta_g + \sum_{k=m+1}^{g} \overline{z}_{g,k}(0)\mu_k\}.
\end{aligned}
$$

Moreover, the matrix $\operatorname{Im}(z_{jk}(0))_{j,k=m+1}^{g}$ is positive definite. This means that $S_0 \cap \overline{S}_0 = \{\mu_1, \ldots, \mu_m\} = W_0$.

(h) Each space $F(W_0)$ is isomorphic to the Siegel upper half-plane $\mathbb{H}_{g-m}$: $F(W_0) \ni S \longmapsto S/W_0 \subset (W_0)^{\perp}/W_0$.
Moreover, any two m-dimensional isotropic rational subspaces of $(\mathbb{C}^{2g}, Q)$, related by a transformation from the modular group $\Gamma_g \simeq Sp\,(2g, \mathbb{Z})$, are treated equivalent. This means that we have set theoretically

$$\{\Gamma \setminus D\}^* = (\Gamma_g \setminus \mathbb{H}_g) \cup (\Gamma_{g-1} \setminus \mathbb{H}_{g-1}) \cup \ldots \cup (\Gamma_0 \setminus \mathbb{H}_0),$$

where Γ_p is the modular group acting on $\mathbb{H}_p$.
This is the *Borel–Baily compactification* of $\Gamma \setminus \mathbb{H}_g$. In this special case it is also called the *Satake compactification* (see [**Gri2**]).

Chapter 8

Linear Differential Systems

§1 Introduction

The subject of investigations in this chapter are the non-autonomous linear differential systems

$$\dot{z} = A(t)z, \quad z \in \mathbb{C}^n, \tag{8.1}$$

and the linear higher order differential equations

$$x^{(n)} + a_1(t)x^{(n-1)} + \ldots + a_n(t)x = 0, \;\; x \in \mathbb{C}. \tag{8.2}$$

In the above two equations the 'time' t usually takes value in the complex plane $\mathbb{C}$. Sometimes $t \in S$, where S is some Riemann surface. In the latter case Eq. (8.1) is treated as an equation for horizontal sections of some holomorphic vector bundle with respect to certain connection in it. Often we will consider Eqs. (8.1)–(8.2) locally; then $t \in (\mathbb{C}, 0)$.
The entries of the matrix $A(t)$ and the coefficients $a_i(t)$ are meromorphic functions. Equations of this type are met very often, as the following examples show.

8.1. Examples.

(a) The Gauss–Manin connection on the cohomological Milnor bundle. Recall that, if $\Delta(t)$ is a family of vanishing cycles (in fibres of the Milnor fibration associated with isolated critical point of a holomorphic function) and $\omega_1, \ldots, \omega_\mu$ define a trivialization (of the cohomological Milnor bundle), then the vector function $J_i = (\int_{\Delta(t)} \omega_1/\mathrm{d}f, \ldots, \int_{\Delta(t)} \omega_\mu/\mathrm{d}f)^\top$ satisfies the Picard–Fuchs differential system (see Theorem 5.29)

$$\dot{J} = A(t)J.$$

(b) The particular case of the Picard–Fuchs equations for the elliptic integrals are the following (see Lemmas 5.22 and 5.23): $2(9t^2-4)\dot{I}_0 = 45tI_0 - 42I_1$, $\;2(9t^2-4)\dot{I}_1 = -10I_0 + 63tI_1$, or $4(9t^2-4)\ddot{I}_0 = -15I_0$, $\;4(9t^2-4)\ddot{I}_1 = 21I_1$.

(c) The *Gauss hypergeometric equation*

$$t(t-1)\ddot{x} + [(\alpha+\beta+1)t - \gamma]\dot{x} + \alpha\beta x = 0.$$

We shall study this equation and its generalizations later. Most of the equations written below are obtained from the hypergeometric equation by application of some limit process.

H. Żołądek, *The Monodromy Group*, Monografie Matematyczne 67,
https://doi.org/10.1007/978-3-031-91270-2_8

(d) The *Bessel equation*

$$t^2\ddot{x} + t\dot{x} + (t^2 - \nu^2)x = 0.$$

(e) The *Weber equation*

$$\ddot{x} + (a^2 - t^2)x = 0.$$

(f) The *Legendre equation*

$$(t^2 - 1)\ddot{x} + 2t\dot{x} - \nu(\nu + 1)x = 0.$$

(g) The *confluent hypergeometric equation*

$$t\ddot{x} + (c - t)\dot{x} - ax = 0.$$

(h) The *Hermite equation*

$$\ddot{x} - 2t\dot{x} + 2nx = 0.$$

(i) The *Laguerre equation*

$$t\ddot{x} + (\alpha + 1 - t)\dot{x} + nx = 0.$$

(j) The *Gegenbauer equation*

$$(1 - t^2)\ddot{x} + (2\lambda + 1)t\dot{x} + n(n + 2\lambda)x = 0.$$

(k) The *Jacobi equation*

$$(1 - t^2)\ddot{x} + [\beta - \alpha - (\alpha + \beta + 2)t]\dot{x} + n(n + \alpha + \beta + 1)x = 0.$$

(l) The *Mathieu equation* $\mathrm{d}^2u/\mathrm{d}z^2 + (1 + a\cos(2z))u = 0$, after the substitution $t = \sin^2 z$, $x(t) = u(z)$, takes the form

$$4t(1 - t)\ddot{x} + 2(1 - 2t)\dot{x} + (1 + a - 2at)x = 0.$$

(m) The *Airy equation*

$$\ddot{x} - tx = 0.$$

(n) The algebraic function $x = \sqrt[n]{t}$ satisfies the equation

$$nt\dot{x} = x.$$

(o) The algebraic function $x(t) = \sqrt{t} + \sqrt[3]{t-1}$ also satisfies the linear differential equation

$$(t-3)\frac{\mathrm{d}}{\mathrm{d}t}(6t(t-1)\dot{x}) = 3(3t-1)[3(t-1)\dot{x} - x] + 2(4t-3)[2t\dot{x} - x].$$

(Hint. From the formula for x and from $6t(t-1)\dot{x} = 3(t-1)\sqrt{t} + 2t\sqrt[3]{t-1}$ express $\sqrt{t}$ and $\sqrt[3]{t-1}$ by means of x and $\dot{x}$).

It turns out that any algebraic function is a solution of some linear differential equation. This follows from the Riemann theorem about multivalued holomorphic functions with regular singularities (see Theorem 8.35 below).

8.2. Definition. A point t_j, at which the matrix $A(t)$ in Eq. (8.1), or one of the coefficients $a_i(t)$ in Eq. (8.2), has a pole is the **singular** point for the equation. In the case of local situation we will assume that $t_j = 0$. The first rough classification of singular points is the following.
The singular point $t = 0$ of Eq. (8.1) or of Eq. (8.2) is called **regular** iff any its solution $\phi(t)$ has at most polynomial growth in any sector with vertex at $t = 0$. It means that

$$|\phi(t)| < C_N/|t|^N, \quad |t| \to 0, \quad \alpha < \arg t < \beta.$$

Otherwise the singular point is called **irregular**.

Example. The Picard–Fuchs equations have regular singularities at their singular points (which are the critical values of the corresponding holomorphic function, see Theorem 5.40).

8.3. Definition. Let t_0 be a non-singular point of Eq. (8.1), or of Eq. (8.2), (defined near the singularity $t = 0$). Let $\mathcal{V}$ be the space of its solutions defined near t_0. As t turns around the point 0, with the beginning and end at t_0, the solutions are prolonged analytically and define the **monodromy operator** $\mathcal{M} : \mathcal{V} \longmapsto \mathcal{V}$: if $\phi(t)$, t near t_0, is a solution, then $\mathcal{M}\phi(t) = \phi(\mathrm{e}^{2\pi\mathrm{i}}t)$.
Let $\phi_1(t), \ldots, \phi_n(t)$ form a basis of $\mathcal{V}$. We define the *fundamental matrix* $\mathcal{F} = (\phi_1, \ldots, \phi_n)$ in the case (8.1) and

$$\mathcal{F} = \begin{pmatrix} \phi_1 & \cdots & \phi_n \\ \cdots & \cdots & \cdots \\ \phi_1^{(n-1)} & \cdots & \phi_n^{(n-1)} \end{pmatrix}$$

in the case (8.2). Then the monodromy operator, written in this basis, is the monodromy matrix M satisfying $\mathcal{F}(t)M = \mathcal{F}(\mathrm{e}^{2\pi\mathrm{i}}t)$.
Here the matrix M depends on the choice of $\mathcal{F}$ but not on the point t_0. Note also that the matrix M acts in the same way onto any row of the fundamental matrix. If system (8.1), or equation (8.2), is defined on a compact Riemann surface S with singular points $t_1, \ldots, t_m$, then any loop $\gamma \in \pi_1(S\setminus\{t_1, \ldots, t_m\}, t_0)$ defines (in the same way as above) the corresponding monodromy operator $\mathcal{M}_\gamma \in \operatorname{Aut}\mathcal{V}$. These

operators generate a subgroup in $\operatorname{Aut}\mathcal{V}$ called the **monodromy group**. The correspondence $\gamma \longmapsto \mathcal{M}_\gamma$ satisfies $\mathcal{M}_{\gamma\delta} = \mathcal{M}_\gamma \mathcal{M}_\delta$, i.e., it is an anti-representation. The operators $\mathcal{M}_\gamma$ define also a certain vector bundle on $S\setminus\{t_1, \dots, t_m\}$, with locally constant transition operators (equal to $\mathcal{M}_\gamma$). It is the **local system** associated with the linear equation.

§2 Regular Singularities

We begin the study of singularities of meromorphic differential equations (8.2) with one-dimensional phase space, $n = 1$. It means that we have the equation

$$t^r \dot{z} = a(t)z,$$

where $a(t)$ is an analytic function near 0.
If $r = 0$, then the solutions are analytic (by the theorem on existence and uniqueness of solutions in the analytic version).
If $r > 0$, then we assume that the coefficient $a_0 \neq 0$ in the expansion $a(t) = a_0 + a_1 t + \dots$. Of course, the general solution is written in the form

$$z(t) = C \cdot \exp\left[\int^t a(s)s^{-r}\mathrm{d}s\right].$$

If $r = 1$, then we get $z(t) = Ct^{a_0} \times$ (analytic function). This means regularity of the singular point.
If $r > 1$, then $z(t) = C\mathrm{e}^{P(1/t)}t^b \times$ (analytic function), where

$$P(\lambda) = \frac{a_0}{1-r}\lambda^{r-1} + \frac{a_1}{2-r}\lambda^{r-2} + \dots - a_{r-2}\lambda,$$

$b = a_{r-1}$. Here the singularity is irregular, because for $\operatorname{Re}\left(a_0 t^{1-r}\right) < 0$ the solution diverge faster than any power of t. We have proved the following result.

8.4. Proposition. *If $n = 1$, then the singular point of Eq. (8.2) is regular iff the function $A(t)$ has pole of order exactly 1.*
In general, Eq. (8.2) with $n = 1$ is analytically equivalent to the equation $t^r \dot{z} = [\sum_0^{r-1} a_j t^j]z$.

Consider now the case of n-dimensional system with the following singularity

$$t\dot{z} = Az,$$

where A is a constant matrix. The fundamental matrix of solutions of this system is

$$t^A = \mathrm{e}^{A\ln t},$$

which grows at most polynomially (regularity). This result is generalized in the following way.

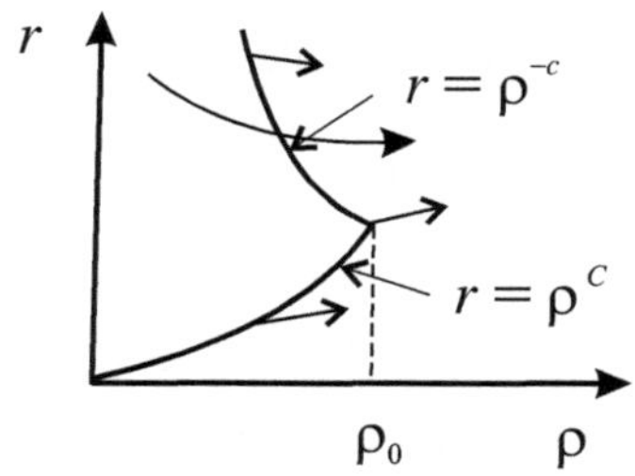

Figure 8.1: Proof of Prop. 8.5

8.5. Proposition. *If $A(t) = A_0/t + A_1(t)$, where $A_0 = \text{const}$ and $A_1(t)$ is a holomorphic matrix-valued function, then the point $t = 0$ is regular.*

Proof. We put $t = \rho e^{i\theta}$, where $\theta = \text{const}$ and ρ tends to 0. Let $\phi(\rho) = \arg z(t)$ and $r(\rho) = |\phi(\rho)|$.
From the identity $d\phi/d\rho = e^{i\theta}\dot{z} = (\widetilde{A}(t)/\rho)\phi$ (where $\widetilde{A}$ is holomorphic), we get the inequality

$$-\frac{Cr}{\rho} < \frac{dr}{d\rho} < \frac{Cr}{\rho}$$

for $0 < \rho < \rho_0$ with some constant C and ρ_0.
From this and Fig. 8.1 we see that the graphic of the function $r(\rho)$ must lie in the distinguished domain. Thus $r < \rho^{-C}$ near $\rho = 0$. □

8.6. Definition. System (8.1), i.e., $\dot{z} = A(t)z$, has singularity of the **Fuchs type** at $t = 0$ iff $A(t)$ has simple pole at 0. Eq. (8.2), i.e., $x^{(n)} + \ldots + a_n(t)x = 0$, has singularity of the **Fuchs type** at $t = 0$ iff all the functions $t^j a_j(t)$ are holomorphic near 0.

Proposition 8.5 says that, if system (8.1) has singularity of the Fuchs type, then this singularity is regular. The following example shows that the converse is not true.

8.7. Example (Euler equation).

$$t^n x^{(n)} + b_1 t^{n-1} x^{(n-1)} + \ldots + b_n x = 0,$$

where b_i are constants, by means of the change $t = e^u$, $\frac{d}{dt} = e^{-u}\frac{d}{du}$, $\frac{d^2}{dt^2} = e^{-2u}(\frac{d^2}{du^2} - \frac{d}{du}), \ldots$ is reduced to the equation

$$\frac{d^n x}{du^n} + c_1 \frac{d^{n-1}x}{du^{n-1}} + \ldots + c_0 x = 0$$

with constant coefficients. Its general solution has the form $x = \sum d_{\alpha,k} e^{\alpha u} u^k = \sum d_{\alpha,k} t^\alpha (\ln t)^k$. Thus the point $t = 0$ is regular.

On the other hand, if we rewrite the Euler equation in the form of the linear system

$$\dot{z} = \begin{pmatrix} 0 & 1 & 0 & \dots & 0 \\ 0 & 0 & 1 & \dots & 0 \\ \dots & \dots & \dots & \dots & \dots \\ -b_n/t^n & -b_{n-1}/t^{n-1} & \dots & \dots & -b_1/t \end{pmatrix} z$$

(where $z_1 = x, z_2 = \dot{x}, \dots$), then the corresponding matrix $A(t)$ has non-simple pole.

In the case of linear equation of nth order the situation is much more clear.

8.8. Theorem. *The singular point $t = 0$ for Eq.* (8.2) *is regular iff it is of the Fuchs type.*

Proof. (We follow the proof of Theorem 5.2 in [**CL**]) 1. Assume that the Fuchs conditions, i.e., all $t^j a_j(t)$ are holomorphic, hold. Following the latter example we perform a transformation reducing Eq. (8.2) to system (8.1) with the Fuchs type singularity. Then Proposition 8.4 will give us the regularity property.
The change is following

$$z_1 = x, \quad z_2 = (t\mathrm{d}/\mathrm{d}t)x, \quad z_3 = (t\mathrm{d}/\mathrm{d}t)^2 x, \dots, \quad z_n = (t\mathrm{d}/\mathrm{d}t)^{n-1} x.$$

It gives $\dot{z}_1 = (1/t)z_2$, $\dot{z}_2 = (1/t)z_3, \dots, \dot{z}_{n-1} = (1/t)z_n$ and $z_j = t^j x^{(j)} +$ combination of $t^k x^{(k)}$, $k < j$. Thus, $t^j x^{(j)}$ can be expressed as linear combination of z_k, $k \leq j$.
The last equation from the promised system is $\dot{z}_n = (t^{n-1}x^{(n-1)}) +$ (combination of $\dot{z}_i$, $i < n$). The term $\frac{\mathrm{d}}{\mathrm{d}t}(t^{n-1}x^{(n-1)})$ contains $(n-1)z_n/t$ and $t^{n-1}x^{(n)} = -t^{n-1}\sum a_j x^{(n-j)} = -t^{-1}\sum(a_j t^j)(x^{(n-j)}t^{n-j})$. The functions $a_j t^j$ are holomorphic, by the assumption, and $x^{(n-j)}t^{n-j}$ are expressed by means of z_i's. All this shows that $t\dot{z}_n$ is analytic in t and z.

2. Assume that $t = 0$ is regular point. We want to show that the functions

$$b_j(t) = a_j t^j$$

are analytic.
We will use the monodromy operator. If $\mathcal{F}$ is the fundamental matrix, $\mathcal{F} \longmapsto \mathcal{F}M$ is the monodromy operation and $G(t) = t^{-\ln M/2\pi\mathrm{i}}$, then the matrix-function $\mathcal{F}G$ is univalent. By regularity of $\mathcal{F}$ it is meromorphic. Thus $\mathcal{F} =$ (merom. matrix) $\times$ $t^{\ln M/2\pi\mathrm{i}}$.
The matrix M has always some (left) eigenvector $v = (v_1, \dots, v_n)$, $vM = \alpha v$. Using it, we can find a solution $\phi(t) = \sum v_i \phi_i$ of the form $t^\lambda \times$ (analytic function); here $2\pi\mathrm{i}\lambda = \alpha -$ (integer). The importance of this solution is that it does not contain logarithms in its expansion.
Other solutions are searched for in the form $x = \phi \cdot y$. For $z = \dot{y}$ we obtain an equation of order $n - 1$.

Indeed, if $D = (\mathrm{d}/\mathrm{d}t)^n + \sum a_j (\mathrm{d}/\mathrm{d}t)^{n-j}$, then $D(\phi y) = D(\phi)\cdot y +$ (expres. dep. on $\dot{y}$). This equation is of the form

$$\sum d_{n-1-j}(t) z^{(j)} = 0,$$

where

$$d_j = a_j + \mathrm{const}\cdot a_{j-1}\cdot \dot{\phi}/\phi + \mathrm{const}\cdot a_{j-2}\cdot \ddot{\phi}/\phi + \ldots, \quad d_0 = a_0 = 1$$

and $\phi^{(l)}/\phi \sim \mathrm{const}\cdot t^{-l}$
Next, we use induction with respect to the order n. In the case $n = 1$ the statement was proved in Proposition 8.4; the regularity implies the Fuchs property.
Assume that we know that any regular equation of order $n-1$ satisfies the Fuchs conditions and consider our system.
Let ψ_i be the system of independent solutions of the equation for z. Then $\psi_i = \frac{\mathrm{d}}{\mathrm{d}t}(\phi_i/\phi)$, where ϕ_i are independent solutions of the equation for x. Because all ϕ_i are regular (by assumption) and $\phi \sim t^\lambda$, then also $\psi_i(t)$ behave regularly. Thus the equation for z has regular singularity at $t = 0$. By the induction assumption the coefficients d_j have good behavior, $d_j t^j$ are analytic. Because of the above relation between a_j's and d_k's, we find that also $a_j t^j$ are holomorphic. □

8.9. Definition. If $t = 0$ is regular for Eq. (8.2) and $a_i(t) = \mu_i t^{-i} + \ldots$, then the equation

$$P(\lambda) = \lambda(\lambda-1)\ldots(\lambda-n+1) + \mu_1\lambda\ldots(\lambda-n+2) + \ldots + \mu_n = 0$$

is called the *defining equation* for Eq. (8.2). In particular, its roots and multiplicities allow to determine the first terms of the asymptotic of solutions of the differential equation.

8.10. Example (Elliptic integrals revisited). (See [**BE**, vol. 3], and [**Gol**]).
(i) The complete elliptic integrals (those which are given in tables and are used by engineers) are defined by the formulas

$$K(k) = \int_0^1 \frac{\mathrm{d}s}{\sqrt{(1-s^2)(1-ks^2)}},$$

$$\mathrm{i}K'(k) = \int_1^{1/k} \frac{\mathrm{d}s}{\sqrt{(1-s^2)(1-ks^2)}},$$

$$E = \int_0^1 \sqrt{\frac{1-k^2s^2}{1-s^2}}\mathrm{d}s,$$

$$\mathrm{i}E' = \int_1^{1/k} \sqrt{\frac{1-k^2s^2}{1-s^2}}\mathrm{d}s.$$

As one can easily see, they define the periods of the incomplete elliptic integrals

$$F(k,\phi)=\int_0^{\phi}\left(1-k^2\sin^2 t\right)^{-1/2}\mathrm{d}t,$$
$$E(k,\phi)=\int_0^{\phi}\left(1-k^2\sin^2 t\right)^{1/2}\mathrm{d}t.$$

One should apply the substitution $s=\sin t$. The periods are $4K$ and $2\mathrm{i}K'$ (or $4E$ and $2\mathrm{i}E'$ respectively). They are integrals along two cycles in the Riemann surface $y^2=(1-x^2)(1-k^2x^2)$.

(ii) One has the formulas $K=F(k,\pi/2)$, $K'=F(k',\pi/2)$, $E=E(k,\pi/2)$, $E'=E(k',\pi/2)$, where $k'=\sqrt{1-k^2}$. (One should apply the changes $1-k^2s^2=u^2$ and $u=k'v$.)

(iii) These integrals can be expressed by means of the elliptic integrals along the curve $y^2=P_3(z)$, which have been investigated in Chapter 5. Namely, the substitution $s=1-1/z$ gives $(1-s^2)(1-k^2s^2)=P_3(z)/z^4$ and the integral K becomes $\int(P_3(z))^{-1/2}\mathrm{d}z$, with whom we are acquainted.

(iv) The functions K,K', as functions of $t=k^2$, satisfy the differential equation

$$t(1-t)\ddot{w}+(1-2t)\dot{w}-w/4=0. \tag{8.3}$$

This equation, after the change $t=(1-\tau)/2$, becomes the Legendre equation

$$(\tau^2-1)\frac{\mathrm{d}^2w}{\mathrm{d}\tau^2}+2\tau\frac{\mathrm{d}w}{\mathrm{d}\tau}-\nu(\nu+1)w=0$$

for $\nu=-1/2$.
To obtain Eq. (8.3), one takes the expressions

$$K=\int_0^1\left((1-s^2)(1-ts^2)\right)^{-1/2}\mathrm{d}s,\quad \dot{K}=(1/2)\int s^2\left(1-s^2\right)^{-1/2}\left(1-ks^2\right)^{-3/2},$$
$$\ddot{K}=(3/4)\int s^4\left(1-s^2\right)^{-1/2}\left(1-ts^2\right)^{-5/2},$$

and substitutes them into the left-hand-side of the differential equation. One arrives to an integral of the expression $[tu^2+2(1-t)u-1]\cdot(u(1-u))^{-1/2}\cdot(1-tu)^{-5/2}$, where $u=s^2$. The latter is the complete differential of the function $u^{1/2}(1-u)^{1/2}(1-tu)^{-3/2}$ and gives zero integral.

(v) Here is a good place to explain to the reader the origins of the name Gauss–Manin connection. Yu. Manin in [**Man1**] considered the family of elliptic curves $y^2=x(x-1)(x-t)$; he treated this family as an elliptic curve with the basic field $\mathbb{C}(t)$. He studied integrals of the holomorphic form $\omega=\mathrm{d}x/y$ along integer cycles in this curve. These integrals are integer combinations of two basic periods (integrals along the generators of the first homology group) and satisfy

the Legendre differential equation $4t(t-1)\ddot{I} - 4(1-2t)\dot{I} + I = 0$ (the same as Eq. (8.3)). Because this equation is a special case of the Gauss hypergeometric equation, Manin called it the Gauss equation. The form ω satisfies the equation $[4t(t-1)(\mathrm{d}^2/\mathrm{d}t^2) - 4(1-2t)(\mathrm{d}/\mathrm{d}t) + 1]\omega = \mathrm{d}(-2y/(x-2)^2)$.
In the same paper Manin generalized this situation,. He considered integrals of a holomorphic 1-form ω along integer cycles in a family C_t of algebraic curve of genus g, and called the obtained differential equations $P\omega = \mathrm{d}f$ and $PI = 0$ the Picard–Fuchs equations; here $P = P(t, \mathrm{d}/\mathrm{d}t)$ is a linear differential operator, f is a function on C_t and I is an integral. Next, he introduced the Picard–Fuchs equations in the case of a family of multidimensional varieties depending algebraically on a parameter; the latter was formulated in algebraic terms of functional fields and their extensions. Since then the operator P is called the Gauss–Manin operator, the corresponding connection in the cohomological bundle is called the Gauss–Manin connection and the equation $PI = 0$ is called the Picard–Fuchs equation (see also [**Dav**]).

(vi) Eq. (8.3) has three singular points $t = 0, 1, \infty$; (we treat it as an equation in $\overline{\mathbb{C}}$).
The point $t = 0$ is regular, because we have $\ddot{w} + (t^{-1} + \ldots)\dot{w} + (-t^{-1}/4 + \ldots)w = 0$. Its defining equation is $\lambda^2 = 0$, which implies that one of its solutions is analytic (it is K) and the other solution contains logarithm (it is $\mathrm{i}K'$).
Similarly, the point $t = 1$ is regular with the defining equation $\lambda^2 = 0$.
After some transformations, one checks that the point $t = \infty$ is also regular with the defining equation $(\lambda - 1/2)^2 = 0$

(vii) The elliptic functions appear in the Newton's equation $\ddot{x} = -V'(x)$, where the potential $V = x^4 + ax^3 + bx^2 + cx$ has two local minima. If $\dot{x}^2/2 + V(x) = E$ is the equation of a trajectory oscillating around one such minimum, then the period of the oscillations is given by the elliptic integral $T = \int \mathrm{d}x/\sqrt{2E - V(x)}$, which (after some changes) can be expressed by means of K'. The reader can show that the periods of oscillations around both minima of the potential, with the same total energy E, are equal.

(viii) The Weierstrass function $\wp(t)$ is the inverse of the elliptic integral $t = \int_\infty^x \mathrm{d}s/\sqrt{4s^3 - g_2 s - g_3}$. It satisfies the equation $(\dot{\wp})^2 = 4\wp^3 - g_2\wp - g_3$ (which is obvious). It represents a meromorphic function of t and is two-periodic. Namely, when one adds to t one of the two periods $2\omega = \int_\gamma \mathrm{d}s/y$, $2\omega' = \int_\delta \mathrm{d}s/y$, $y = \sqrt{4s^2 - g_2 s - g_3}$, represented as the complete elliptic integrals along two basic cycles γ, δ, then the value of $\wp(t)$ does not change. Usually one chooses the periods in such a way that $\mathrm{Im}\omega'/\omega > 0$. Another representation of the Weierstrass function is

$$\wp(t) = \frac{1}{t^2} + \sum_w{}' \left(\frac{1}{(z-w)^2} - \frac{1}{w^2} \right),$$

where the sum $\sum'$ runs over the incomplete lattice $\{w = 2m\omega + 2n\omega' : (m,n) \neq (0,0)\}$. In this case $g_2 = 60 \sum' w^{-4}$, $g_3 = 140 \sum' w^{-6}$.

8.11. Definition. Two local differential systems $\dot{z}_1 = A_1(t)z_1$ and $\dot{z}_2 = A_2(t)z_2$ are called **holomorphically equivalent** (respectively, **meromorphically equivalent** or **formally equivalent** or **formally meromorphically equivalent**) near the singular point $t = 0$ iff there is a holomorphic matrix $H(t)$ (respectively, meromorphic matrix $H(t)$, or the formal series $H(t) \sim \sum_{j=0}^{\infty} H_j t^j$ or $\sum_{j=-m}^{\infty} H_j t^j$) such that, after the change $z_2 = H(t)z_1$, the first system is transformed to the other system. In the formal cases this holds at the level of formal expansions of these systems in powers of t.

Remark. If the equation $\dot{z} = A(t)z$ is treated as the equation for horizontal section with respect to the connection $\mathrm{d}/\mathrm{d}t - A(t)$, then the application of the change $z \longmapsto H(t)z$ means application of the gauge transformation

$$A \longmapsto B = \dot{H}H^{-1} + HAH^{-1}.$$

8.12. Theorem. *The point $t = 0$ is regular singular point for the system $\dot{z} = A(t)z$ if an only if this system is meromorphically equivalent to the system $\dot{z} = \frac{C}{t}z$, where $M = \mathrm{e}^{2\pi \mathrm{i} C}$ is the monodromy transformation.*

Proof. The proof that regularity implies meromorphic reducibility repeats the analogous proof of Theorem 5.40. If $\mathcal{F}(t)$ is the fundamental matrix of the system with the matrix A, then the conjugating transformation is equal to $H(t) = \mathcal{F}(t)t^{-C}$. It is univalent and meromorphic (by regularity).
The reverse implication is obvious. □

In order to develop the theory of formal classification of meromorphic systems, we need to say something about the *Poincaré–Dulac theory of normal forms.* This theory will be used in the further parts of the book.
Consider a germ of vector field in $(\mathbb{C}^n, 0)$

$$\dot{x} = Ax + \ldots,$$

where A is a constant matrix of linearization of the vector field at 0. Let $\lambda_1, \ldots, \lambda_n$ be the eigenvalues of A.

8.13. Definition. We say that the eigenvalues satisfy the **resonant relation of the type** $(i; \boldsymbol{k})$ iff

$$\lambda_i = \lambda_1 k_1 + \ldots + \lambda_n k_n,$$

where $\boldsymbol{k} = (k_1, \ldots, k_n)$ and k_j are non-negative integers.

8.14. Poincaré–Dulac normal form theorem. *There exists a formal change $y = x + \ldots$ such that*

$$\dot{y}_i = (Ay)_i + \sum a_{i,\boldsymbol{k}} y^{\boldsymbol{k}}, \quad i = 1, \ldots, n,$$

where the sum runs over the multi-indices $\boldsymbol{k}$ such that the eigenvalues satisfy the resonant relation of the type $(i; \boldsymbol{k})$.

Proof. We apply a series of changes of the type $x \longmapsto x' = x + \phi(x)$, where ϕ is homogeneous transformation of degree m,

$$(\phi)_i = \sum_{|\boldsymbol{k}|=m} b_{i,\boldsymbol{k}} x^{\boldsymbol{k}}.$$

The inverse map has the form $x' \longmapsto x = x' - \phi(x') + \ldots$. We strive to cancel all possible terms in the vector field of homogeneous degree m.
We have $\dot{x}'_i =$ (old part of degree $< m$) + (old part of degree m) + $(L\phi)_i(x') + \ldots$, where

$$(L\phi)_i = \partial\phi_i(x)/\partial x \cdot Ax' - (A\phi)_i$$

is called the *homological operator.*
If the matrix A is diagonal, then L is also diagonal:

$$Lx^{\boldsymbol{k}} e_i = \left(\sum k_j\lambda_j - \lambda_i\right) \cdot x^{\boldsymbol{k}} e_i,$$

where e_i form the standard basis in $\mathbb{C}^n$. Because the eigenvalues of L vanish only at the resonant terms, then all the non-resonant terms in the vector field of degree m can be cancelled in this way.
If A is upper-triangular, then, taking the suitable ordering of the basis $x^{\boldsymbol{k}} e_i$,we get the triangular form of the operator L

$$Lx^{\boldsymbol{k}} e_i = \Lambda_{i,\boldsymbol{k}} x^{\boldsymbol{k}} e_i + \sum_{j<i,\boldsymbol{l}} \text{const} \cdot x^{\boldsymbol{l}} e_j. \qquad \square$$

8.15. Examples for $n = 2$. The resonance means

$$\text{(i) } \lambda_1 = k_1\lambda_1 + k_2\lambda_2 \quad \text{and/or} \quad \text{(ii) } \lambda_2 = k_1\lambda_1 + k_2\lambda_2.$$

Next, we divide the problem into several subcases.

(a) If $\lambda_1 = \lambda_2 = 0$, then all terms in the expansion of the vector field are resonant and nothing can be reduced (by means of Theorem 8.14).

(b) If $\lambda_1 = 0 \neq \lambda_2$, then in (i) we must have $k_2 = 0$ and in (ii) we must have $k_2 = 1$. Thus the Poincaré–Dulac formal normal form is

$$\dot{x} = x^2 f(x), \quad \dot{y} = \lambda_2 y(1 + g(x)).$$

(c) If $\lambda_2/\lambda_1 \geq 1$ is a rational number, then (i) cannot hold and (ii) can hold with $k_1 = \lambda_2/\lambda_1$, $k_2 = 0$, provided that $\lambda_2/\lambda_1 = m$ is integer. The normal form is either the linear system or

$$\dot{x} = \lambda_1 x, \quad \dot{y} = m\lambda_1 y + ax^m.$$

(d) If $\lambda_2/\lambda_1 = -p/q$ is rational, then the normal form is

$$\dot{x} = \lambda_1 x(1 + f(x^q y^p)), \quad \dot{y} = \lambda_2 y(1 + g(x^q y^p)).$$

(e) If λ_2/λ_1 is irrational, then the normal form is linear.

We apply the Poincaré–Dulac theorem to the non-autonomous system

$$t^r \dot{z} = A(t)z, \quad A(0) \neq 0, \tag{8.4}$$

with an analytic matrix $A(t)$. This system is associated with the autonomous system in the extended phase space

$$z' = A(t)z, \quad t' = t^r, \tag{8.5}$$

where $' = \mathrm{d}/\mathrm{d}\tau$. (The graphs of solutions of Eq. (8.4) form the phase curves of Eq. (8.5)). We apply Theorem 8.14 to the second system. Assume also that $\lambda_1, \ldots, \lambda_n$ are the eigenvalues of the matrix $A(0)$.
If $r=1$, i.e., in the Fuchs case, then the eigenvalues of system (8.5) are $\lambda_1, \ldots, \lambda_n, 1$. If $r > 1$, then the eigenvalues of system (8.5) are $\lambda_1, \ldots, \lambda_n, 0$.

8.16. Definition. The system $\lambda_1, \ldots, \lambda_n$ of eigenvalues of $A(0)$ is **resonant** if:

(a) $\lambda_i - \lambda_j \in \mathbb{Z}$ for some $i \neq j$ in the case $r = 1$;

(b) $\lambda_i - \lambda_j = 0$ for some $i \neq j$ in the case $r > 1$.

The reader can notice that these resonant relations imply the resonant relations between the eigenvalues of system (8.5). They are of the type $(i; 0, \ldots, 0, \overset{j}{1}, 0, \ldots, 0, l)$, $i, j = 1, \ldots, n$, because in the linear changes we use only the ϕ's of the form $t^l z_j e_i$. If the Fuchsian singular point is non-resonant, then the Poincaré–Dulac theorem says that system (8.4) is formally equivalent to the system $t\dot{z} = A(0)z$.
On the other hand, Theorem 8.12 says about meromorphic equivalence. But the formal equivalence means that, in the meromorphic matrix $H(t) = H_{-d}t^{-d} + \ldots$, all the terms with negative power of t vanish. This gives the following result.

8.17. Theorem (normal form for Fuchsian singularity). *The system $t\dot{z} = A(t)z$ with non-resonant Fuchsian singular point is analytically equivalent to $t\dot{z} = A(0)z$.*

8.18. Remark. If a Fuchsian singular point is resonant, then it is analytically equivalent to the system

$$t\dot{z}_i = \lambda_i z_i + \mathrm{const} \cdot t^k z_j, \quad i = 1, \ldots, n.$$

After reordering the variables z_i, we obtain a system, where $A(t)$ is of the form (constant diagonal) + (strictly triangular). Such system can be integrated (see also [**CL**]).

8.18. Regularity of the Gauss–Manin connection. Consider a degeneration of algebraic manifolds $f : X \longmapsto \mathbb{D}$ such that $X_t = f^{-1}(t)$, $t \neq 0$, are smooth compact and X_0 is a divisor with normal intersections (like in Chapter 7). For given n we have the cohomological bundle $\mathcal{H}^n$ over $\mathbb{D}^*$ with the fiber $H^n(X_t, \mathbb{C})$ and the Gauss–Manin connection; such that the integer cocycles form horizontal section of the cohomological bundle. The Gauss–Manin connection is regular iff the corresponding Picard–Fuchs differential system (for integrals along horizontal families of cycles $\Delta(t)$) is regular. The regularity of the Gauss–Manin connection can be proved in three ways.

The first method relies on description of the integer cycles $\Delta_j(t)$ in $H_n(X_t, \mathbb{Z})$ by means of real semi-algebraic subsets of X_t depending in regular way on t. Such description is done in the book of S. Lefschetz [**Lef**]. One can use the geometrical sections $s[\omega_i] = \langle \cdot, \omega_i/\mathrm{d}f \rangle$ as basis in $H^n(X_t)$ (ω_i – holomorphic $(n+1)$-forms in X). Thus the integrals $I_{ij}(t) = \langle \delta_i(t), s[\omega_j] \rangle$ have regular singularities.

The second method uses the regularity criterion following from Theorem 8.12: a system $\dot{y} = A(t)y$, $y \in \mathbb{C}^m$, has regular singular point iff there is a meromorphic change $z = B(t)y$ such that $t\dot{z} = C(t)z$ with holomorphic $C(t)$. This means that one should find a system of vector-valued functions $v_j(t)$ satisfying $t\nabla_{\partial/\partial t} v_j \subset \mathrm{span}\{v_1, v_2, \ldots\}$ over $\mathcal{O}_0(t)$. In [**Brie**] and [**Kul**] it is said that there is a lattice in the $\mathcal{O}_0(t)$-module of germs of holomorphic vector-functions $y(t)$, generated by v_i and invariant with respect to the operator $t\nabla_{\partial/\partial t}$. This lattice is called the *Brieskorn lattice*.

The Gauss–Manin connection is defined also as the connecting homomorphism, associated with the short exact sequence $0 \longmapsto \Omega^1_{\mathbb{D}^*} \otimes \Omega^{\bullet-1}_{X^*/\mathbb{D}^*} \longmapsto \Omega^{\bullet}_{X^*} \longmapsto \Omega^{\bullet}_{X^*/\mathbb{D}^*} \longmapsto 0$ of sheaf complexes over the punctured disc (see the point 7.37). This exact sequence is completed to the exact sequence of complexes of holomorphic sheaves with logarithmic singularities, i.e., $\Omega^{\bullet}_{X^*}(\log)$ instead of $\Omega^{\bullet}_{X^*}$ and $\Omega^1_{\mathbb{D}}(\log 0)$ instead of $\Omega^1_{\mathbb{D}^*}$. The corresponding cohomological bundle is prolonged to a bundle (denoted also by $\mathcal{H}^n$) on the whole disc $\mathbb{D}$. This bundle is trivial $\mathcal{H}^m \simeq \mathbb{D} \times \mathbb{C}^M = \{(t, y)\}$ and the condition $\nabla_{\partial/\partial t} y = 0$ is the corresponding Picard–Fuchs equation (for components in certain basis).

However, in our case the sheaf $\mathcal{O}(\mathcal{H}^n) = R^n f_* \Omega^{\bullet}_{X/\mathbb{D}}(\log)$ (the Leray hypercohomological sheaf) and the connection ∇ already satisfy the property $\nabla : \mathcal{O}(\mathcal{H}^n) \longmapsto \Omega^1_{\mathbb{D}}(\log 0) \otimes \mathcal{O}(\mathcal{H}^n)$. This means that $\nabla_{\partial/\partial t}$ has only first order pole, when acting upon sections of $\mathcal{H}^n$. Any $\mathcal{O}_0(t)$-basis of the module $\mathcal{O}(\mathcal{H}^n)$ satisfies the regularity criterion following from Theorem 8.12.

We see that the essential point in this proof lies in the proper extension of the cohomological bundle to the whole disc. Similar arguments work also in the case of Gauss–Manin connection in vanishing cohomologies.

Interesting proof of regularity of the Gauss–Manin connection , using its 'nilpotency in characteristic p', was given by N. Katz (see [**Kat2**] and [**Gri2**]).

§3 Irregular Singularities

8.20. Theorem (formal classification in non-resonant case). *If the* ***rank*** $r > 1$ *and* $A(0)$ *is non-resonant, then the system* $t^r\dot{z} = A(t)z$ *is formally equivalent to the system* $t^r\dot{w} = B(t)w$*, where the matrix* $B(t)$ *is diagonal,* $B = \mathrm{diag}(b_1(t), \ldots, b_n(t)$ *and* $b_i(t)$ *are polynomials of degree at most* $r - 1$*.*

Proof. By the Poincaré–Dulac theorem all the terms $t^k z_j e_i$, $i \neq j$, can be cancelled (are non-resonant). There remain only the terms $t^k z_i e_i$, which are diagonal. Next one applies Proposition 8.4. □

If we write $B(t)t^{-r} = D_r t^{-r} + D_{r-1}t^{-r+1} + \ldots + D_2 t^{-2} + Et^{-1}$, where D_j and E are diagonal matrices, then the formal fundamental matrix of the system from Theorem 8.20 takes the form

$$\mathcal{F}(t) = \widehat{G}(t)t^E \exp\left[\sum D_j/(j+1)t^{j+1}\right],$$

where $\widehat{G}$ is a formal power series.

In the resonant case there is also diagonal formal normal form, but the gauge transformation generally cannot be expanded into formal power series in t. Instead, one should change the local ring $\mathbb{C}[[t]]$ (of formal power series in t) by the ring $\mathbb{C}[[t^{1/b}]]$, where b is some positive integer. This was firstly observed by E. Fabry in [**Fab**].

8.21. Example of Fabry. ([**Fab**], [**Vara**]) The system

$$\dot{z}_1 = (a/t)z_3, \quad \dot{z}_2 = (1/t^2)z_1, \quad \dot{z}_3 = (1/t^2)z_2$$

(with $a \neq 0$) is of the form $t^2\dot{z} = B(t)z$, where the matrix $B(0)$ is nilpotent. Let us try the substitution $y_i = t^{a_i}z_i$. Simple calculations give

$$\begin{aligned}\dot{y}_1 &= at^{a_1-a_3-1}y_3 + (a_1/t)y_1,\\ \dot{y}_2 &= t^{a_2-a_1-2}y_1 + (a_2/t)y_2,\\ \dot{y}_3 &= t^{a_3-a_2-2}y_2 + (a_3/t)y_3.\end{aligned}$$

If we put $a_1 = -\frac{1}{3}$, $a_2 = 0$, $a_3 = \frac{1}{3}$, then the first terms in the right-hand-sides of the above equations have the same power of t; namely $-\frac{5}{3}$. Thus we get $\dot{y} = (t^{-5/3}D + t^{-1}E)y$, where

$$D = \begin{pmatrix} 0 & 0 & a \\ 1 & 0 & 0 \\ 0 & 1 & 0 \end{pmatrix}, \quad E = \mathrm{diag}\,(-1/3, 0, 1/3).$$

We see that the leading matrix D has three distinct eigenvalues, the cubic roots of $-a$. Transforming it to the diagonal form and applying the series of transformations from the proof of Theorem 8.20 (with power series in $\zeta = t^{1/3}$), we can diagonalize also the remaining part of the system for y.

The general result concerning normal forms in the case of resonant irregular singular point belongs to M. Hukuhara [**Huk**], H. Turrittin [**Tur**] and A. Levelt [**Lev**] and is formulated in the next theorem.

8.22. Theorem (Formal normalization). *If $t = 0$ is irregular singular point of the system $\dot{z} = A(t)z$, then there exist:*

- *a positive integer b,*
- *rational numbers (with the common denominator b) $r_1 < r_2 < \ldots < r_m < -1$,*
- *diagonal matrices $D_1, \ldots, D_m, E$,*
- *a change $y = H(t^{1/b})$ in the class of formal series in $t^{1/b}$*

such that H transforms the system for z to the system

$$\dot{y} = (D_1 t^{r_1} + \ldots + D_m t^{r_m} + E t^{-1})y. \tag{8.6}$$

Two forms (8.6) with D_j, E and D'_j, E' are formally equivalent (over $C[[t^{1/b}]]$) iff the matrices D_j and D'_j and E and E' are mutually spontaneously conjugated by means of some matrix from $GL(n, C)$.
In the variable $s = t^{1/b}$ system (8.6) takes the form

$$\mathrm{d}y/\mathrm{d}s = b(D_1 s^{-R_1} + \ldots + D_m s^{-R_m} + E s^{-1})y$$

for some integers $R_1 > R_2 > \ldots > R_m > 1$.

Remark. The numbers r_j are called the *canonical levels*, the principal level r_1 is called the *Katz invariant* and the smallest integer b is called the *ramification index*. Katz in [**Kat2**] gave an algebraic interpretation of the canonical level r_1 (see also [**Del2**]).

Problem. Calculate the Katz invariant in the case of the equation $P(t, t\frac{\mathrm{d}}{\mathrm{d}t})x = 0$ in terms of the Newton's diagram of the function $P(x, y)$.

Proof of Theorem 8.22. We present a sketch of this proof based on the paper [**Sib1**] of Y. Sibuya.
Assume that we have $t^r \dot{z} = A(t)z$.

1. Repeating the proof of Theorem 8.20 we show that division of the matrix $A(0)$ into diagonal blocks corresponding to different eigenvalues λ_i can be prolonged to formal splitting of the system into several independent systems, which are characterized by having only one eigenvalue.

2. Let $\lambda_1 = \ldots = \lambda_n = \lambda$. The change $z = \mathrm{e}^{\lambda/(1-r)t^{r-1}} y$ leads to the system $t^r \dot{y} = B(t)y$, where

$$B(0) = \begin{pmatrix} 0 & d_1 & \ldots & 0 \\ 0 & 0 & \ldots & 0 \\ \ldots & \ldots & \ldots & \ldots \\ 0 & 0 & \ldots & d_{n-1} \\ 0 & 0 & \ldots & 0 \end{pmatrix},$$

where $d_i = 0, 1$.

3. Assume that all $d_j = 1$. We introduce the variables $u_1 = y_1,\ u_2 = t^r \dot{u}_1, \dots, u_n = t^r \dot{u}_{n-1}$. Then we obtain the system $t^r \dot{u} = C(t)u$ where

$$C(t) = \begin{pmatrix} 0 & 1 & \dots & 0 \\ 0 & 0 & \dots & 0 \\ \dots & \dots & \dots & \dots \\ 0 & 0 & \dots & 1 \\ c_1(t) & c_2(t) & \dots & c_n(t) \end{pmatrix},$$

where each $c_j(t) = O(t)$

Now the substitution $u_j = t^{\alpha_j} v_j$, with suitable rational exponents α_j leads to the equation $\dot{v} = (t^{-\alpha} D + \dots)v$, where the matrix D has the form like the matrix $C(0)$, but with at least one constant $c_j(0) \neq 0$. This means that C has at least one nonzero eigenvalue and we can apply the induction with respect to the dimension of the eigenspace.

4. If some of the d_i's in the matrix $B(0)$ are equal to zero, then, using the change $y \longmapsto u$ (analogous as in the point 3), we reduce the matrix $B(t)$ to the matrix

$$\begin{pmatrix} C_1 & E_{12} & \dots \\ E_{21} & C_2 & \dots \\ \dots & \dots & \dots \end{pmatrix},$$

where $C_j = C_j(t)$ are of the form of the matrix $C(t)$ from the point 3 and

$$E_{ij} = \begin{pmatrix} & 0 & \\ e_1(t) & \dots & e_s(t) \end{pmatrix}.$$

One can show that, after applying a suitable change $u \longmapsto v$ of the same form from 3, we obtain $\dot{v} = (t^{-\alpha} D + \dots)v$, where either:

(i) the matrix D has nonzero eigenvalue (if some $b_j(0)$ becomes $\neq 0$ or some $e_l(0) \neq 0$ in some $E_{ij},\ i > j$), or

(ii) the matrix D is nilpotent but its Jordan normal form contains more nonzero d_j's than in the matrix $B(0)$ from the point 2 (here some $e_l(0) \neq 0$ in a $E_{i,j},\ i < j$)

Next we apply the induction with respect to the dimension of the eigenspace and to the number of nonzero d_j's. (See also [**Vara**]). □

Remark. From the proof of Theorem 8.22 it follows that the case when the matrix D_1 (from its thesis) has a pair of coinciding eigenvalues occurs when some row is identically equal zero in the block-matrix from the point 4. This means that the system is analytically separated into two independent subsystems. Such phenomenon has infinite codimension.

So, in what follows we assume that the eigenvalues of D_1 are different.

Having solved the problem of formal classification of meromorphic systems with irregular singularity, we pass to the analytic classification. It turns out that formal solutions of such systems (or higher order equations) can be divergent. The series from the formal normal forms are only asymptotic in the sense of Definition 5.51.

8.23. Example of Euler. The series $\sum k!t^k$ formally satisfies the equation $t^2\ddot{x} + (3t-1)\dot{x} + x = 0$, but is divergent.

As we shall see, the equations with irregular singular point can be reduced to their formal normal form (from Theorem 8.20 or Theorem 8.22) by means of analytic gauge transformation only in some sector in the complex plane of variable t. However, two such gauge transformations (defined in adjacent sectors) are different in the intersections of sectors. This phenomenon is called the *Stokes phenomenon.* Stokes discovered it in studying the Airy equation and the Bessel equation in [**Sto1**], [**Sto2**] and [**Sto3**] (see also [**Hea**]). A rigorous treatment of the Stokes phenomenon allows to construct certain functional invariants of analytic classification of such systems.

Assume that the system $t^r\dot{z} = A(t)z$ has irregular singular point. In the case, as in Theorem 8.22, we apply the change of time $t \longmapsto t^{1/b}$; and the new time we again denote by t. We apply a polynomial change $z \longmapsto \sum_{j=0}^{N} H_j(t)z$, which reduces the polar part of the matrix $A(t)/t^r$ to the diagonal form. Thus we get a system $t^r\dot{z} = A(t)z$, where $A(t) = B(t) + O(t^r)$ (we do not change notation of A) and whose formal normal form is

$$t^r\dot{y} = B(t)y \ (B(t) \text{ diagonal and polynomial}).$$

Assume also that the eigenvalues of $A(0)$ satisfy the condition (see the remark above)

$$\lambda_i \neq \lambda_j, \quad i \neq j.$$

8.24. Definition. The rays in the complex t-plane defined by

$$\mathrm{Re}[(\lambda_i - \lambda_j)t^{1-r}] = 0, \quad i \neq j,$$

are called the **rays of division corresponding to the pair**(λ_i, λ_j).

8.25. Sectorial normalization theorem. *Let $S = \{|t| < \epsilon,\ \alpha < \arg t < \beta\}$ be a sector in the t-plane not containing two rays of division (corresponding to any pair (λ_i, λ_j)). Then there exists a matrix function $H(t) = I + O(t)$ analytic in S and transforming the equation $t^r\dot{z} = A(t)z$ to the formal normal form system $t^t\dot{y} = C(t)y$.*

Proof. We follow the book of W. Wasow [**Was**]. The essential fact, which has to be proved, is the following. □

Proposition. *Let the eigenvalues of $A(0)$ be divided into two groups $\lambda_1, \ldots, \lambda_p)$ and $(\lambda_{p+1}, \ldots, \lambda_n)$ such that $\lambda_i \neq \lambda_j$ for $i \leq p < j$. Then there is a matrix $H(t)$*

holomorphic in S, which transforms the system to an analogous system with the matrix $B(t)$ of the block-diagonal form

$$\begin{pmatrix} B^{11}(t) & 0 \\ 0 & B^{22}(t) \end{pmatrix}.$$

Moreover, we can assume that the sector S is symmetric with respect to the real positive semi-axis and with the magnitude $< \pi/(r-1)$.

Proof. If $z = H(t)y$, then y satisfies the equation $t^r H\dot{y} = (AH - t^r\dot{H})y$ and we have

$$HB = AH - t^r\dot{H}.$$

So, we assume that $A(0)$ is in the block-diagonal form $A(0) = \operatorname{diag}(A^{11}(0), A^{22}(0))$, and we seek $H(t)$ in the form

$$\begin{pmatrix} I & H^{12}(t) \\ H^{21}(t) & I \end{pmatrix}.$$

In what follows we will deal only with H^{12}. The analysis for H^{21} is quite analogous. The matrix $X(t) = H^{12}(t)$ satisfies the following nonlinear differential equation

$$t^r\dot{X} = A^{11}X - XA^{22} + A^{12} - XA^{21}X, \tag{8.7}$$

where $A^{ij}(t)$ are the corresponding block-matrices.
The latter equation is a particular case of the nonlinear equation

$$t^r\dot{u} = \Lambda u + p(t,u), \tag{8.8}$$

where Λ has nonzero eigenvalues Λ_k (equal to $\lambda_i - \lambda_j$ for (8.7) and the term p is nonlinear.

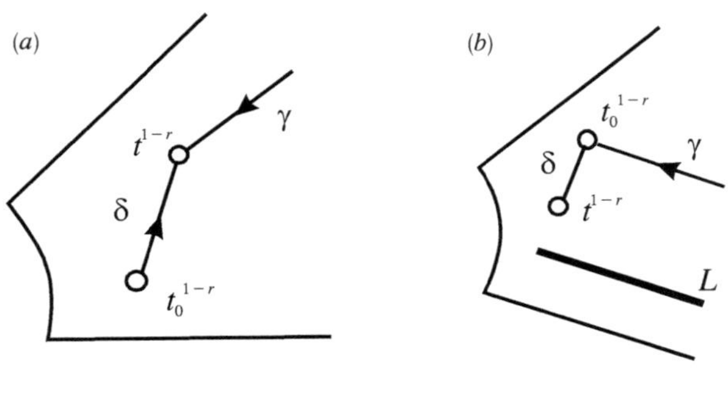

Figure 8.2: Integration paths in the cases without rays of division (a) and with one ray L (b)

Eq. (8.8) is equivalent to the integral equation

$$u(t) = V(t)D + \int_{t_0}^{t} V(t)V^{-1}(s)s^{-r}p(s, u(s))\mathrm{d}s, \tag{8.9}$$

where $V(t) = \mathrm{e}^{\Lambda/(r-1)t^{r-1}}$ and D is some constant vector.
In fact, the constant vector D and the limit t_0 of integration are in some way connected one with another. One can replace the integration in Eq. (8.9) by integration along some contour $\Gamma(t)$ in the complex s-plane in such a way that it ends at t. In this case the constant vector can be put equal to zero. Moreover, for each component of Eq. (8.9) we can choose the integration path independently.
The main trick of the proof is to choose the path $\Gamma(t)$ in such a way that the expression $V(t)V^{-1}(s)$ is not too big along the path of integration; then we will be able to apply the principle of contracting maps. In particular, we want that the expression

$$\exp\left[\operatorname{Re}\left(\Lambda_k(t^{1-r} - s^{1-r})/(1-r)\right)\right]$$

associated with the kth component, is not big.
We fix for a while the index k and some point $t_0 \in \mathbb{R} \cap S$. We shall consider only such t's that $|t| << t_0$. Moreover, it is useful to pass to the chart $\zeta = s^{1-r}$; then the sector S is replaced by a sector with vertex at infinity.
Assume that S does not contain any ray of division corresponding to Λ_k. It means that $\operatorname{Re}\Lambda_k\zeta/(1-r)$ has definite sign in the whole S. Depending on this we choose either:

- the straight semi-line γ from $\zeta = \infty$ to $\zeta = t^{1-r}$ with constant $\arg\zeta$, if the sign is positive, or
- the interval δ joining t_0^{1-r} with t^{1-r} otherwise (see Fig. 8.2(a)).

In both cases we have $\operatorname{Re}\Lambda_k(t^{1-r} - \zeta)/(1-r) < 0$.
If S contains a ray of division L (only one by the assumption) then the choice of the path δ remains unchanged but the path γ passes along a straight half-line parallel to L (from infinity to t^{1-r}, see Fig. 8.2(b)). Here we have $\operatorname{Re}\Lambda_k(t^{1-r}-\zeta)/(1-r) = 0$ along γ.
The further proof uses the standard analytic methods and we describe it only shortly, without detailed estimates.
We solve the fixed point equation

$$u = T(u),$$

where $T(u)$ is the nonlinear integral operator $\int V(t)V^{-1}(s)s^{-r}p(s, u(s))\mathrm{d}s$ and the integration runs along the paths described above.
In order to apply the method of contracting maps, we must define some Banach space of holomorphic functions u and get an estimate for $||T(u_1) - T(u_2)||$ (i.e., the Lipschitz continuity of $p(t, u)$ with respect to u).

The class of u's consists of those which satisfy the estimate $|u(t)| < c|t|^m$ with certain fixed m. This m is defined as the order of $p(t,0) = \text{const} \cdot t^m + \ldots$. One shows that, if u satisfies this estimate, then $|T(u)(t)| < Kc|t|^m$ with a constant K not depending on u (see Lemma 14.2 in [**Was**]).
The Lipschitz estimate $|p(t,u_1) - p(t,u_2)| < \mu|u_1 - u_2|$, with μ arbitrarily small, follows from the fact that p contains only terms of the form $O(t^m)u^0$, $O(t)u$ and $O(|u|^2)$.
Next, one chooses the sequence $u_0 = 0$, $u_1 = T(u_0)$, $u_2 = T(u_1) \ldots$ of successive approximations. We get $|u_1| < c|t|^m$ and $|u_{n+1} - u_n| < \mu \cdot K \cdot |u_n - u_{n-1}|$. If $\mu \cdot K < 1$, then this series converges to a function, holomorphic in some sector S_0 of small radius.
Another proof of Theorem 8.25 uses the Gevrey expansions and multi-summability (see Section 4 in Chapter 9 below) and is given in the paper of B. Braaksma [**Bra**]. □

8.26. Definition (Stokes operators). Let S_1 and S_2 be two adjacent sectors (i.e., with non-empty intersection) satisfying the assumptions of Theorem 8.25. Let H_{S_1} and H_{S_2} be the corresponding operators guaranteed by this theorem. From Theorem 8.20 it follows that these two matrix-functions have the same Taylor expansions. Because these Taylor expansions are only asymptotic, then H_{S_i} are generally different in the intersection of the two sectors.
One defines the matrix function

$$G_S(t) = H_{S_1} H_{S_2}^{-1}, \quad t \in S = S_1 \cap S_2.$$

It has the property $G_S(t) = I + O(t^N)$ as $t \to 0$ for any N. Moreover, it preserves the normalized system, i.e., $t^r \dot{y} = B(t)y$ with diagonal $B(t)$.
Let $\mathcal{L}_S$ be the space of solutions of the normalized system. The gauge transformation induced by G_S transforms solutions from $\mathcal{L}_S$ to solutions from the same space. Therefore it defines certain automorphism $\mathcal{C}_S : \mathcal{L}_S \longmapsto \mathcal{L}_S$, which is called the **Stokes operator**.

8.27. Properties of the Stokes operators. We choose a basis of $\mathcal{L}_S$ in the form

$$\phi_j = a_j(t) e_j, \quad a_j(t) = \exp\left[\lambda_j t^{1-r}/(1-r) + \ldots\right],$$

where (e_j) is the standard basis of $\mathbb{C}^n$. The Stokes operator expressed in this basis is a constant matrix $C_S = (c_{ij})$ and is called the **Stokes matrix**.
If $\psi_i = G_S(t)\phi_i(t) = C_S \phi_i = \sum_j c_{ij}\phi_j$, then the matrix G_S acts as follows: $G_S e_i = \psi_i / a_i(t)$. We have

$$(G_S)_{ij} = a_i^{-1}(t) \cdot c_{ij} \cdot a_j(t) = c_{ij} \cdot \exp\left[(\lambda_j - \lambda_i) t^{1-r}/(1-r) + \ldots\right].$$

Because $G_S \sim I$, then $c_{ii} = 1$ and if $\operatorname{Re}(\lambda_i - \lambda_j)t^{1-r} \to -\infty$, then $c_{ij} = 0$.
After suitable ordering of the basis (e_i), we obtain the property that:

The Stokes operators C_S are unipotent.

8.28. Definition (Stokes sheaf). The system of Stokes operators can be naturally described using the cohomological language. Take the circle $\mathbb{S}^1$, which we treat as the circle $\{r = 0\}$ in $\mathbb{R}_+ \times \mathbb{S}^1 = \{(r, \theta)\}$, where r, θ are the polar coordinates of the $t = re^{i\theta} \in \mathbb{C}$ (the boundary circle in the polar blowing-up).
The **Stokes sheaf** St is defined by the presheaf of groups $St(U)$, where $U \subset \mathbb{S}^1$ are open connected arcs. With each such U one associates a germ of sector $S = S_U$ in $(\mathbb{C}, 0)$ with base at U (see Fig. 8.3).

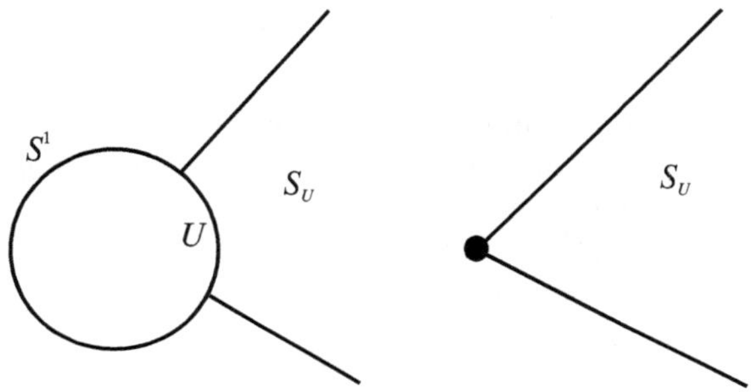

Figure 8.3: Domains U of the Stokes presheaf in the circle correspond to cones S_U

$St(U)$ consists of matrix functions $S \ni t \longmapsto G_S(t)$ satisfying the following properties:

(i) $G_S(t) \sim I$,

(ii) $G_S(t)$ preserve the normalized system.

Note that the Stokes sheaf is associated with some normalized diagonal system. There are many such functions G_S; for example, if $\dot{x}_1 = \dot{a}_1(t)x_1$, $\dot{x}_2 = \dot{a}_2(t)x_2$, then the changes are of the type $(x_1, x_2) \longmapsto (x_1, x_2 + c\exp(a_2(t) - a_1(t))x_1))$.

8.29. Remark. St is not a standard sheaf, with which we are acquainted. The groups $St(U)$ are (generally) non-abelian groups and the Čech coboundary operator, cocycles and cohomologies must take another meaning. Here we recall the definition of the first cohomology group of sheaves $\mathcal{F}$ of non-abelian groups (see Example 2 in 3.27).
The cochains are the same as in the abelian theory. In particular, a 0-cochains associated with a covering $\mathcal{U} = \{U\}$ is a system $(G_U)_{U \in \mathcal{U}}$ with $G_U \in \mathcal{F}(U)$. The 1-cochains are the systems (G_{UV}) with $G_{UV} \in \mathcal{F}(U \cap V)$ such that $G_{VU} = G_{UV}^{-1}$. This 1-cochain is a cocycle iff $G_{UV}G_{VW}G_{WU} = e$. Two 1-cochains (G_{UV}) and (G'_{UV}) are equivalent iff there is 0-cochain (K_U) such that

$$G'_{UV} = K_U G_{UV} K_V^{-1}.$$

The first cohomology group $H^1(\mathcal{F}, \mathcal{U})$ is defined as the set of classes of 1-cocycles with respect to the this equivalence.

Next, one takes the direct limit with respect to the coverings.

Now we pass to the description of moduli of analytic classification of linear systems with irregular singular point. We fix the normalized system $\mathcal{S}_0 : t^r\dot{z} = B(t)z$, and consider the space $\mathcal{M} = \mathcal{M}_B$ of systems $\mathcal{S} : t^r\dot{z} = A(t)$, which are formally equivalent to $\mathcal{S}_0$. On the space $\mathcal{M}$ we introduce the following equivalence relation: $\mathcal{S} \sim \mathcal{S}'$ iff there is a matrix function $H(t)$ holomorphic in a (whole) neighborhood of $t = 0$ and realizing the equivalence between $\mathcal{S}$ and $\mathcal{S}'$.
The following, fundamental in this theory, result was proved by B. Malgrange [**Mal3**] and Y. Sibuya [**Sib2**]..

8.30. Theorem of Malgrange–Sibuya (Analytic classification). *The space of equivalence classes defined above coincides with the first cohomology group of the circle with coefficients in the Stokes sheaf.*
In other words, the group $H^1(\mathbb{S}^1, St)$ parameterizes the orbits of the action of the group of analytic equivalences on the space of germs of meromorphic linear systems with fixed formal normal form.
This property holds in the non-resonant case as well as in the resonant case with the normal form the thesis of Theorem 8.22.

Proof. Let $\mathcal{M}$ be the above space of the equivalence classes. We define a natural map from $\mathcal{M}$ to $H^1(\mathbb{S}^1, St)$.
Take a system $\mathcal{S} \in \mathcal{M}$. By the sectorial normalization theorem, there exists a covering of a neighborhood of the point $t = 0$ by sectors S_i such that $\mathcal{S}$ is transformed to $\mathcal{S}_0$ by means of holomorphic maps $H_i(t)$ in S_i. Let U_i be the bases of the sectors S_i. Then the family of operators $H_iH_j^{-1}$ defines a Čech cocycle associated with the covering $\{U_i\}$.
If two systems $\mathcal{S}$ (defining a cocycle $H_iH_j^{-1}$) and $\mathcal{S}'$ (defining a cocycle $H_i'(H_i')^{-1}$) are equivalent by means of a holomorphic matrix $H(t)$, then $H_i' = H_iH$ (by the uniqueness in Theorem 8.25) and the cocycles define the same cohomology class. This gives the map $\Phi : \mathcal{M} \longmapsto H^1(\mathbb{S}^1, St)$.
The injectivity of the map Φ is simple. If $H_i'(H_j')^{-1} = G_i^{-1}H_iH_j^{-1}G_j$, then $H_i^{-1}G_iH_i' = H_j^{-1}G_jH_j'$. The latter matrix is thus univalent and defines the holomorphic conjugation between the systems $\mathcal{S}'$ and $\mathcal{S}$.
The surjectivity of the map Φ is the most difficult part of the proof. Here we follow [**BV**], but we omit the details. Probably the first proof of this surjectivity was given by G. Birkhoff [**Bir1**].
Let $(G_{ij}) = (G_{U_iU_j})$ be a cocycle with values in the Stokes sheaf. If we could find a system $H_i = I + O(t)$ of holomorphic matrices in sectors S_i (with bases U_i) and such that $G_{ij} = H_j^{-1}H_i$, then this would give us a corresponding meromorphic linear equation $\dot{z} = Az$.
Indeed, let $\mathcal{F}_0$ be a fundamental matrix for the normalized system $\dot{z} = B_0z$, $B_0 = t^{-r}B(t)$. The matrices $\mathcal{F}_i = H_i\mathcal{F}_0$ define the fundamental systems with matrices $A_i = \dot{\mathcal{F}}_i\mathcal{F}_i^{-1}$ in the sectors S_i. We have $A_i = \dot{H}_iH_i^{-1} + H_iB_0H_i^{-1}$, $B_0 = \dot{\mathcal{F}}_0\mathcal{F}_0^{-1}$. Calculations, which use the fact that G_{ij} preserve B_0 (as gauge transformations)

show that $A_i = A_j$ at the intersections of sectors. Thus A_j's define one univalent matrix $A(t)$. Because $H_i = I + O(t)$, the matrix $A(t)$ has the same order of pole as B_0, $A(t) = t^{-r}B(t)$ with analytic B.

To prove the existence of such H_j's we use the apparatus of the sheaf theory. It turns out that the fact that the Stokes sheaf (and the sheaf $\mathcal{E}$ defined below) is not abelian is not a serious obstacle to apply the results from the abelian theory to our purposes.

Firstly we solve the system of equations $G_{ij} = G_j^{-1}G_i$ in the class of smooth (C^∞) matrix-valued functions $G_i = I + O(t)$ in sectors S_i. To do this one introduces the sheaf $\mathcal{E}$ (on $\mathbb{S}^1$) of germs of smooth matrix-functions in sectors, $\mathcal{E}(U) = C^\infty(S_U, GL(n))$. It is a flabby sheaf, admitting the partition of unity, and has trivial higher cohomology groups, $H^i(S^1, \mathcal{E}) = 0,\ i > 0$. In the abelian case it is Lemma 3.27, in the non-abelian case it is proved in [**BV**].

Assume that we have such G_i's. Because $\bar{\partial}G_{ij} = 0$, we have $\bar{\partial}G_i \cdot G_i^{-1} = \bar{\partial}G_j \cdot G_j^{-1}$. The latter expression defines a, univalent in a punctured neighborhood of $t = 0$, smooth function F.

Now, applying a variant of the Poincaré $\bar{\partial}$-lemma, we find a matrix function $G = I + O(t)$ such that $\bar{\partial}G \cdot G^{-1} = F$.

The functions $H_i = G^{-1}G_i$ are holomorphic and satisfy the relation $H_j^{-1}H_i = G_{ij}$. □

Remark. The Stokes cochain can be interpreted as a cocycle of Stokes operators. If $\mathcal{F}_0$ is the fundamental matrix of the normalized system, then

$$C_{ij} = \mathcal{F}_0 G_{ij} \mathcal{F}_0^{-1}$$

is a cocycle of constant matrices. Two Stokes cocycles (C_{ij}) and (C'_{ij}) are called equivalent iff there exists a system (C_i) of constant matrices such that $C'_{ij} = C_i^{-1}C_{ij}C_j$.

If one calculates the cohomology group of the Stokes sheaf using C_{ij}, then we see that it is finite-dimensional.

Moreover, for generic system with irregular singular point the formal normalizing transformation is not convergent.

Theorem 8.30, together with Theorems 8.20 and 8.22, gives a complete solution of the problem of analytic classification of systems with irregular singularity. The formal normal form is determined in analytic way (using finite number of polynomial transformations). This formal normal form gives the moduli space. A separate problem constitutes the task of computations of the Malgrange–Sibuya moduli, i.e., how to calculate the elements from $H^1(\mathbb{S}^1, St)$ from data of the initial analytic system. In the below examples we show how it is done in special cases.

8.31. Examples. (a) The first example comes from the work [**Zo7**].

(i) Consider the function

$$F(t) = \int_{0+}^{t} s^a \mathrm{e}^{-1/s^k} \varphi(s) \mathrm{d}s = \int f(s) \mathrm{d}s.$$

We ask when this function is of the local Darbouxian type

$$F(t) = t^{a+k+1}e^{-1/t^k}\psi(t), \tag{8.10}$$

with analytic ψ.

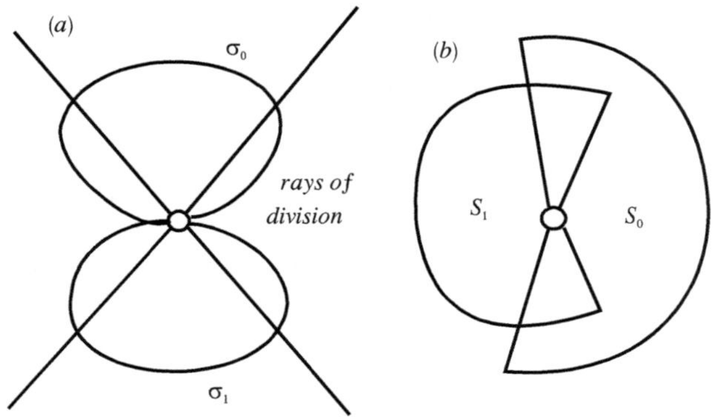

Figure 8.4: (a) rays of division and loops; (b) Stokes covering

To get the answer, we introduce the loops σ_j as in the Fig 8.4(a). The punctured neighborhood of $t = 0$ is divided into sectors of fall and of jump of e^{-1/s^k}, as $|t| \to 0$ and $\arg t = \text{const}$. Each path starts and ends at successive sectors of fall (at $t = 0$) and 'surrounds' one sector of jump.
We define the quantities

$$A_j = \int_{\sigma_j} f(s)\mathrm{d}s.$$

We have the following property:

The function F has the form (8.10) *iff all $A_j = 0$.*

The values of $F(t)$ in neighboring sectors (from Fig. 8.4(b)) differ by the constants A_j.
(ii) Consider the linear system

$$t^{k+1}\dot{z} = \begin{pmatrix} -k & t^{k+1}\varphi(t) \\ 0 & at^k \end{pmatrix} z. \tag{8.11}$$

Its general solution is $z_1(t) = De^{1/t^k} + Ce^{1/t^k}\int^t f(s)$, $z_2(t) = Ct^a$. Because the singular part of the matrix $A(t)$ is in the diagonal form $A_0(t) = \operatorname{diag}(-kt^{-k-1}, at^{-1})$, then such is also the normalized system.
We choose the system of fundamental solutions of the normalized system in the form $w_1 = e^{1/t^k}$, $w_2 = t^a$. Therefore $z_1 = w_1 + \eta(t)w_2$, $z_2 = w_2$, where $\eta(t) = t^{-a}e^{1/t^k}F(t)$.

In different sectors S_j (see Fig. 8.4(b)) we have different branches of the function F (see the point (i)) and different branches η_j of η. We have $\eta_{j+1} = \eta_j + A_j t^{-a} \mathrm{e}^{1/t^k}$. Therefore the normalizing matrices take the form $H_j = \begin{pmatrix} 1 & \eta_j \\ 0 & 1 \end{pmatrix}$ and the Stokes cocycle $C_{j,j+1}$ is given by the matrices $\begin{pmatrix} 1 & A_j \\ 0 & 1 \end{pmatrix}$.

We have $H^1(\mathbb{S}^1, St) \simeq \mathbb{C}^k$ and the Stokes cocycle associated with system (8.11) is equal to $G_{j,j+1} = \begin{pmatrix} 1 & A_j t^{-a} \exp(1/t^k) \\ 0 & 1 \end{pmatrix}$.

(b) **The Bessel equation.** We follow [**BV**] and [**Hea**]. Recall the Bessel equation $t^2\ddot{x} + t\dot{x} + (t^2 - \nu^2)x = 0$.

We see that the point $t = 0$ is regular (see Definition 8.6 and Theorem 8.7). The point $t = \infty$ turns out irregular.

We introduce the notations $\tau = 1/t$, $u(\tau) = (x(1/\tau), \dot{x}(1/\tau))^\top$, and below by the dot we will denote the derivative with respect to τ. Then we get the system $\tau^2\dot{u} = A(\tau)u$, where

$$A = \begin{pmatrix} 0 & -1 \\ 1 & 0 \end{pmatrix} + \begin{pmatrix} 0 & 0 \\ 0 & 1 \end{pmatrix}\tau + \begin{pmatrix} 0 & 0 \\ -\nu^2 & 0 \end{pmatrix}\tau^2.$$

In order to get the formal normal form, we have to transform only the first two terms of $B(\tau)$ to the diagonal form. The matrix A_0 has distinct eigenvalues $\pm\mathrm{i}$, i.e., we have the non-resonant case. The transformation matrix is chosen in the form $H(t) \sim H_0(I + H_1\tau + \ldots)$, where only the first two terms are important. We have $H_0 = \begin{pmatrix} -1 & -\mathrm{i} \\ -\mathrm{i} & -1 \end{pmatrix}$ and $H_0 A_0 H_0^{-1} = B_0 = \mathrm{diag}(\mathrm{i}, -\mathrm{i})$. Next, $B_1 = H_0(A_1 + [H_1, A_0])H_0^{-1} = H_0 A_1 H_0^{-1} + [\widetilde{H}_1, B_0]$, $\widetilde{H}_1 = H_0 H_1 H_0^{-1}$. The commutator in the latter expression acts on the non-diagonal entries (because B_0 is diagonal) and, choosing suitable $\widetilde{H}_1$, we can obtain B_1 in the diagonal form. Because $H_0 A_1 H_0^{-1} = \frac{1}{2}\begin{pmatrix} 1 & * \\ * & 1 \end{pmatrix}$, then $B_1 = \frac{1}{2}I$.

Thus the basis of solutions of the normalized system is $u_{1,2} = \tau^{1/2}\mathrm{e}^{\pm\mathrm{i}/\tau}e_{1,2} = t^{-1/2}\mathrm{e}^{\pm\mathrm{i}t}e_{1,2}$.

Recall that the Stokes sheaf consists of such matrix functions $G(t)$ that $G \sim I$ and $GBG^{-1} + \tau^2\dot{G}G^{-1} = B$ (preservation of the normalized system). In view of the fact that $2B_1 = I$, this is equivalent to the following condition. Let $\chi(\tau) = \mathrm{diag}(\mathrm{e}^{-\mathrm{i}/\tau}, \mathrm{e}^{\mathrm{i}/\tau})$. Then $\frac{\mathrm{d}}{\mathrm{d}t}(\chi^{-1}G\chi) = 0$.

Therefore the map $G \longmapsto G = \chi^{-1}G\chi$ defines an isomorphism between the sheaf St and some sheaf of subgroups of $GL(2, \mathbb{C})$. The first condition for the Stokes sheaf $\chi F \chi^{-1} \sim I$ implies that the matrix elements of F fulfill the conditions $F_{11} = F_{22} = 1$, $F_{21}\mathrm{e}^{-2\mathrm{i}/\tau} \sim 0$, $F_{12}\mathrm{e}^{2\mathrm{i}/\tau} \sim 0$. Therefore, depending in which part of the τ-plane we are, the matrix F takes one of the two forms

$$\begin{pmatrix} 1 & f_+ \\ 0 & 1 \end{pmatrix}, \ \mathrm{Im}(t) > 0; \quad \begin{pmatrix} 1 & 0 \\ f_- & 1 \end{pmatrix}, \ \mathrm{Im}(t) < 0. \tag{8.12}$$

The parameters $f_\pm$ parameterize the space $H^1(\mathbb{S}^1, St)$ and form the moduli of analytic classification of equations which are formally equivalent to the Bessel equation.

In the above description of the moduli space we have followed the book of D. Babbitt and V. Varadarajan [**BV**]. Unfortunately, the authors of [**BV**] do not give any hint how to compute the moduli $f_\pm$ for the most interesting case of the Bessel equation.

These constants were in fact computed by Stokes himself in [**Sto2**]. We present these calculations following the book of J. Heading [**Hea**].

One uses the basis of solutions of the Bessel equation consisting of the *Bessel functions*

$$J_{\pm\nu}(t) = \sum_{j=0}^{\infty} \frac{(-1)^j (t/2)^{2j\pm\nu}}{j!\Gamma(\pm\nu + j + 1)} = t^{\pm\nu} P_\pm(t^2). \tag{8.13}$$

Here the function $P_\pm$ are entire. Assume that a solution $AJ_\nu(t) + BJ_{-\nu}(t)$ has the asymptotic expansion $\sim t^{-1/2}\mathrm{e}^{\mathrm{i}t}$ for $\arg t = 0$ and $t \to \infty$. After analytic prolongation of this solution to the rays $\arg t = \pi$ and $\arg t = 2\pi$, with use of the formulas (8.12) and (8.13), we obtain

$$\begin{aligned} A\mathrm{e}^{\mathrm{i}\pi\nu} J_\nu(t) + B\mathrm{e}^{-\mathrm{i}\pi\nu} J_{-\nu}(t) &\sim \mathrm{e}^{-\mathrm{i}\pi/2} t^{-1/2} \mathrm{e}^{-\mathrm{i}t}, \\ A\mathrm{e}^{2\mathrm{i}\pi\nu} J_\nu(t) + B\mathrm{e}^{-2\mathrm{i}\pi\nu} J_{-\nu}(t) &\sim \mathrm{e}^{-\mathrm{i}\pi} t^{-1/2} \mathrm{e}^{-\mathrm{i}t} + f_- \mathrm{e}^{-\mathrm{i}\pi} t^{-1/2} \mathrm{e}^{-\mathrm{i}t}, \end{aligned}$$

as $t \to +\infty$. Together with $AJ_\nu(t) + BJ_{-\nu}(t) \sim t^{-1/2}\mathrm{e}^{\mathrm{i}t}$ this gives the value

$$f_- = 2\mathrm{i}\cos(\pi\nu).$$

Similar calculations give $f_+ = -2\mathrm{i}\cos(\pi\nu)$.

We encourage the reader to find the moduli space and the Stokes operators in the cases of the Airy equation and the Weber equation.

Remark. We see that the Stokes constants are nontrivial for some classical equations. The problem of explanation of the non-uniqueness of representation of solutions in a given (asymptotic) basis constituted a great challenge and mystery for the nineteenth century mathematicians.

In March 19, 1857 Stokes was writing to his future wife: "... I have been doing what I guess you won't let me when we are married, sitting up till 3 o'clock in the morning fighting against a mathematical difficulty. Some years ago I attacked an integral of Airy's, and after a severe trial reduced it to a readily calculable form. But there was one difficulty about it which, though I tried till I almost made myself ill, I could not get over, and at least I had to give it up and profess myself unable to master it. I took it up again a few days ago and after two or three days' fight, the last of which I sat up till 3, I at least mastered it. ..." (The citation comes from [**Hea**] and [**StoH**]).

8.32. Note about deformations and their moduli. 1. In the theory of singularities of holomorphic functions (described in Chapter 2), not only the invariants of functions (under holomorphic changes of coordinated), but also invariants of deformations of such singularities are described. In fact, there are only finitely many parameters (moduli) of such classification. So, one can ask about classification of singularities of linear diffeomorphic equations with meromorphic coefficients.
In the case of regular singularities the situation is practically the same as in the case of functions; however, the corresponding theory is not presented in monographs, also in this book.
But the case of irregular singularities is more tricky. Recall that there the number of moduli of the classification is finite: the formal normal form and the Stokes matrices (modulo simultaneous conjugation by a fixed constant matrix).
A. Glyutsuk [**Gl1**] was the first who started to study this subject, under some restriction. Next, a corresponding theory was developed by C. Rousseau with collaborators, see [**LR1, LR2**] (for example). It must be admitted that the results are not conclusive; the question of realization of corresponding functional moduli is still not answered.
Let me present the subject following my seminal paper [**Zo14**]. When the initial singularity has rank $r = 1$ then a natural deformation looks as follows:

$$\begin{aligned}\left(t^2 - \mu\right) \dot{x} &= B(t,\mu)x, \\ B &= \operatorname{diag}(\lambda_1(\mu), \ldots \lambda_n(\mu)) + O(t^2 - \mu)\end{aligned}$$

It has two singular points $t_{1,2} = \pm\sqrt{\mu}$. The corresponding $(n+1)$-dimensional (nonlinear) system $\frac{\mathrm{d}t}{\mathrm{d}\tau} = t^2 - \mu$, $\frac{\mathrm{d}x}{\mathrm{d}\tau} = Bx$ (with new 'time'), has eigenvalues $\lambda_0(t_{1,2}) = \pm 2\sqrt{\mu}$ and the eigenvalues $\lambda_j(\mu) = \lambda_j + \ldots$ of $B(t_{1,2},\mu)$.
It turns out that one can generalize the proof of Theorem 8.25 (about sectorial normalization) to the case of deformed family. It was done firstly by A. Glyutsuk (under an assumption of non-existence of resonances for the singular points $t_{1,2}$, see Remark 2(a) below) and by C. Rousseau with collaborators in general case. Let us try such generalization in the case $k = 1$ and $n = 2$.
For the entry $X(t) = X(t;\mu) = H^{12}$ (of a matrix H realizing equivalence with corresponding formal normal form) we get the following generalization of Eq. (8.9):

$$X(t) = \int_{\Delta(t)} V(t)V(s)^{-1} p(s, X(s))\mathrm{d}s,$$

where

$$V(t) = \left(t - \sqrt{\mu}\right)^{\Lambda/2\sqrt{\mu}+\beta/2} \left(t + \sqrt{\mu}\right)^{-\Lambda/2\sqrt{\mu}+\beta/2},$$

$\Lambda = \Lambda(\mu) = \lambda_1 - \lambda_2 + \ldots$, $\beta = \beta(\mu) \in \mathbb{C}$ and $\Lambda(0) > 0$ (we assume this). As in the proof of Theorem 8.25, the choice of the path $\Delta(t)$ is essential.

We assume that the parameter varies in

$$D = \{-\pi \leq \arg(\mu) \leq \pi\};$$

so, the positions of the singular points vary in the angles $\{-\pi/2 \leq \arg(t_1) \leq \pi/2\}$ and $\{\pi/2 \leq \arg(t_2) \leq 3\pi/2\}$ and, when μ makes a full turn, then they exchange their positions.
For $\mu = 0$ we have sectors S_u (up) and S_d (down) in the $t-$plane, where the solutions are realized. Now we have families of domains, $S_u(\mu)$ and $S_d(\mu)$, in the $t-$plane with the singular points at their boundaries and intersecting along three domains $S^r(\mu)$ (right), $S^l(\mu)$ (left) and $S^c(\mu)$ (central).
If $t \in S_u(\mu)$ then the path $\Delta(t) \subset S_u(\mu)$ should start at $s = t_2$ and end at $s = t$; namely, in this case the factor $\left(s+\sqrt{\mu}\right)^{\Lambda/2\sqrt{\mu}} = \left|s+\sqrt{\mu}\right|^{\mathrm{Re}\left(\Lambda/2\sqrt{\mu}\right)} \times$(bounded function) in $V^{-1}(s)$ remains bounded. Analogously, if $t \in S_d(\mu)$, then $\Delta(t) \subset S_d(\mu)$ joins $s = t_2$ with $s = t$. One can check that the integral operator on the right-hand side of the above integral equation is contracting and depends regularly on μ.
Therefore one can define the unfolded matrices $H_{u,d}(t;\mu)$, which realize gauge equivalences of the deformed system with suitable diagonal model family, and the matrices $G_r = H_u H_d^{-1}$ (in $S^r(\mu)$) and $G_l = H_d H_u^{-1}$ (in $S^l(\mu)$). The latter define so-called unfolded Stokes matrices $C_r(\mu)$ and $C_l(\mu)$ which tend to the genuine Stokes matrices as $\mu \to 0$ in the 'interior' of the domain M. One can relate the latter matrices with corresponding monodromy matrices $M_r(\mu)$ and $M_u(\mu)$ associated with loops around the singular points t_1 and t_2.

2. **Remarks.** (a) One could ask about the influence of possible resonances on this picture. For $\mu = 0$ there are no resonances ($\lambda_i \neq \lambda_j$), but in the perturbed system there is a sequence of values of the parameter for which the system at some of the singular points cannot be diagonalizable. Indeed, the term $(t-t_1)^N y_i$ in the equation for $\dot{y}_j$ is resonant when $\lambda_0(t_1) = \left(\lambda_j(\mu) - \lambda_i(\mu)\right)/N$, $\lambda_0 = \sqrt{\mu}$. We see that there is a sequence $\mu_N^{(i,j)} \approx -4\left(\lambda_i - \lambda_j\right)^2/N^2$ corresponding to such resonances. The limit direction of such sequence lies in the interior of the domain D of parameters. Anyway, these resonances do not spoil our construction, only the corresponding monodromy matrices cannot be diagonalizable (like general Stokes matrices).

(b) In the works of Glyutsuk and Rousseau with collaborators another approach to the sectorial normalization is developed. Firstly, instead of a linear system in $\mathbb{C}^n$, a corresponding Riccati type system in the projectivization $\mathbb{CP}^{n-1}$ of $\mathbb{C}^n$ is considered. After some preparation a corresponding vector field in $(\mathbb{C},0) \times \mathbb{CP}^{n-1}$ has singular points $p_{j,l} = \left(t_j, [e_l]\right)$, $j = 1, 2$, $l = 1, \ldots, n$.
Glyutsuk associates so-called Levelt solutions near t_j, of the form $(t-t_j)^{\alpha_l} \times$ (holomorphic function), with special separatrices of the corresponding singular points. For this he requires that the singular points are non-resonant (no logarithms in the basic solutions).

Rousseau uses a version of invariant manifold theorem (Hadamard–Perron theorem for holomorphic flows from [IIY3] (see Remark 9.49 in the next chapter) in $(\mathbb{C},0 \times \mathbb{CP}^{n-1}$ to find such separatrices. For this she normalizes (rotates) the $t-$coordinate in a way that $\mathrm{Re}(\lambda_1) < \ldots < \mathrm{Re}\,(\lambda_n)$ and the sectors S_u and S_d become fixed independently on the dimension n. Corresponding separatrices are intersections of suitable invariant manifolds of $p_{1,l}$ and $p_{2,l}$ and join these singular points (above suitable sector) and converge to some special separatrices of the degenerate singular point $(0,[e_l])$ as $\mu \to 0$. These separatrices correspond to entries in the sectorial normalization matrices.

But there is a special feature of the above construction of the unfolded Stokes matrices. Recall that, after turning of μ from the negative semi-axis to itself, the singular points exchange their positions. This results in the change of the integration path: $\Delta(t) \longmapsto \Delta(t) + \Delta(t_1,t_2)$, adjoining a path between the singular points. This implies a corresponding change of the Stokes matrices (which vanishes at $\mu = 0$). In [**LR1**, **LR2**] this phenomenon is called the 'auto-intersection phenomenon'. As the below example shows, the result depends on the way the paths approach the singular points.

3. **Example.** Consider the system

$$t^2\dot{x}_1 = -x_1 + t^2\varphi(t)x_2, \quad t^2\dot{x}_2 = \alpha t x_2,$$

for some holomorphic germ φ, with the solutions $x_1(t) = C_1\exp(1/t) + C_2\exp(1/t) \int^t f(s)\mathrm{d}s$, $x_2(t) = C_2 t^\alpha$, where $f(s) = s^\alpha \exp(-1/s)\varphi(s)$ (compare with Example 8.31(a)). Because the singular part of the matrix $B(t)$ is diagonal, the same is the normalized system $t^2\dot{w}_1 = -kw_1$, $t\dot{w}_2 = \alpha w_2$; the basis of solutions is $w_1 = \exp(1/t)$, $w_2 = t^\alpha$. In the sectors $\widetilde{S}_j = S_{u,d}$, $j = 1,2$, (like in the example) we have functions $\chi_j = t^{-\alpha}\exp(1/t)\int_0^t f(s)\mathrm{d}s$, where the integration path begins at $s=0$ along the ray $\{\arg(s) = 0\}$ and ends at t. Therefore $x_1 = w_1 + \chi_j w_2$, $x_2 = w_2$. We see that the normalization matrices take the form $H_j(t) = \begin{pmatrix} 1 & \chi_j(t) \\ 0 & 1 \end{pmatrix}$ and the only nontrivial Stokes matrix is $C_1 = \begin{pmatrix} 1 & a_1 \\ 0 & 1 \end{pmatrix}$, where $a_1 = \int_{\sigma_1} f(s)\mathrm{d}s$ and the loop σ_1 is like in Example 8.31(a).

Consider now the deformed system

$$\left(t^2-\mu\right)\dot{x}_1 = -x_1 + \left(t^2-\mu\right)\varphi(t)x_2, \quad \left(t^2-\mu\right)\dot{x}_2 = \alpha t x_2,$$

with the solutions $x_1 = C_1F_\mu(t) + C_2F_\mu(t)\int^t f_\mu(s)\mathrm{d}s$, $x_2 = C_2\left(t^2-\mu\right)^{\alpha/2}$, where $f_\mu(s) = F_\mu^{-1}(s)\left(s^2-\mu\right)^{\alpha/2}\varphi(s)$ and

$$F_\mu(t) = \left(\frac{t+\sqrt{\mu}}{t-\sqrt{\mu}}\right)^{1/2\sqrt{\mu}}.$$

One finds that, for $|\arg(\mu)| < \pi$, the unique unfolded Stokes matrix is $C_1(\mu) = \begin{pmatrix} 1 & a_1(\mu) \\ 0 & 1 \end{pmatrix}$ with $a_1(\mu) = \int_{\sigma_1(\mu)} f_\mu(s)\mathrm{d}s$, where $\sigma_1(\mu)$ is the loop with the base point $s = \sqrt{\mu}$ which surrounds the point $s = -\sqrt{\mu}$ in the positive direction.
The two limits $\lim_{\mu\to\mu_0\pm \mathrm{i}0} a_1(\mu)$, for $\mu_0 < 0$, are defined by integrals along two different loops (one around $-\mathrm{i}\sqrt{\mu_0}$ and another around $\mathrm{i}\sqrt{\mu_0}$). The angles the loops enter at their base points influence the limits. We see that the difference between the two limits is weakly controllable.

4. In [**LR1**] it was proved that the data $\{\mathrm{diag}(\lambda_1, \dots \lambda_n), C_l, C_r\}$, satisfying suitable 'auto-intersection relation' and modulo simultaneous conjugation, constitute the complete system of invariants of families as above. In [**LR2**] the question of realization of these invariants is discussed.

8.33. Note about the algebraic approach. Above we have presented the theory of local meromorphic linear equations in analytic and geometrical terms. Some specialists, usually from algebraic geometry, reformulated this theory in purely algebraic terms. This theory is now called the theory of $\mathcal{D}$-moduli.
In particular, the books of P. Deligne [**Del2**] and of Babbitt and Varadarajan [**BV**], or the articles [**Man2**], [**Vara**], [**Ber**], are written in the algebraic language. We present here one result of N. Katz obtained in the algebraic way.
Let $\mathcal{O}$ be the local ring of germs of holomorphic functions in $(\mathbb{C}, 0)$ and let $\mathcal{K}$ be its field of quotients, i.e., the field of germs of meromorphic functions.
Let V be a finite-dimensional space over $\mathcal{K}$, $V = \mathcal{K}^n$; it corresponds to the space of germs of meromorphic vector-valued functions.
The linear non-autonomous differential system takes the form $\nabla_{\mathrm{d}/\mathrm{d}t} Y = 0$, where $\nabla_{\mathrm{d}/\mathrm{d}t} = \frac{\mathrm{d}}{\mathrm{d}t} - A(t)$ is a connection in $V = \mathcal{K}^n$; here A is a matrix with coefficients in $\mathcal{K}$.
Next one fixes certain lattice $L_0 \subset V$ which is a free $\mathcal{O}$-module of rank n. In our situation it can be $\mathcal{O}^n$, the space of germs of holomorphic vector functions.
The ring $\mathcal{O}$ is a ring of discrete valuation. The valuation $v : \mathcal{O} \longmapsto \mathbb{Z}$ is defines as $v(g) = r$, if $g(t) = t^r h(t)$, $h(0) \neq 0$. If $L \subset V$ is any lattice, then one defines $v(L) = \max\{\nu : L \subset t_0^\nu L_0\}$.
Let $L^i_{t\mathrm{d}/\mathrm{d}t} = L + \nabla_{t\mathrm{d}/\mathrm{d}t} L + \dots + \nabla^i_{t\mathrm{d}/\mathrm{d}t} L$. Katz [**Kat1**], [**Del2**] has proved the following result.

There exists rational negative r such that for any lattice L the series of numbers $\left| -ri - v(L^i_{t\mathrm{d}/\mathrm{d}t}) \right|$, $i = 1, 2, \dots$, is bounded.

The number r is called the Katz rank. The principal level (or the Katz invariant) r_1 from Theorem 8.22 is expressed by means of the Katz rank. The case $r = 0$ corresponds to the case of regular connection. Therefore, the irregularity can be detected asymptotically; in that case the application of the connection $\nabla_{t\mathrm{d}/\mathrm{d}t}$ many times to holomorphic vector-functions gives meromorphic functions with linearly growing order of the pole.

One says that a module V is *Fuchsian* iff there exists a lattice L such that $\nabla_{\mathrm{td/dt}} L \subset L$ (see [**Man2**]). In our (analytic) case it means that $t\dot{z} = D(t)z$ (D – holomorphic), after some change of coordinates (depending meromorphically on t). Thus the definition of the 'module Fuchsiani' is an algebraic reformulation of Theorem 8.12.
All this is generalized to the situations, where the objects $\mathcal{O}$, V, ∇, L, L_0 are replaced by abstract algebraic objects. We will not develop this subject.

§4 Global Theory of Linear Equations

In this section we consider the differential systems

$$\dot{z} = A(t)z$$

and the differential equations

$$x^{(n)} + a_1(t)x^{(n-1)} + \ldots + a_n(t)x = 0,$$

where the 'time' t belongs to the Riemann sphere $\mathbb{CP}^1$ and $A(t)$, $a_j(t)$ are meromorphic.

Remark. The above situation has natural generalization to the case when the 'time' t takes values in a general Riemann surface S. In that case the operator $\mathrm{d}/\mathrm{d}t - A(t)$ is treated as a connection in certain vector bundle over S. In different charts we have different operators $A(t)$ and in intersections of such two charts the corresponding operators are related by means of gauge transformations. The space of horizontal sections of this connection defines a local system on $S\setminus$(singularities). This leads to some general theory, developed in [**Kat3**] (for example).
We shall not study this general subject.

8.34. Lemma.

(a) *The equation* $\dot{z} = A(t)z$ *has singular points* $t_1, \ldots, t_m$ *of only Fuchs type iff*

$$A(t) = \sum \frac{A_j}{t - t_j}.$$

(b) *The equation* $x^{(n)} + b_1(t)x^{(n-1)} + \ldots + b_n(t)x = 0$ *has singular points* $t_1, \ldots, t_m$ *of only Fuchs type iff*

$$b_j(t) = \frac{P_j(t)}{Q^j(t)},$$

where $Q(t) = (t - t_1)\ldots(t - t_m)$ *and* P_j *are polynomials of degree* $\leq (m-1)j$.

Proof. This proof includes the definition of the Fuchs singularity at the point $t = \infty$. One has to introduce the change $\tau = 1/t$ and the transformed system

$\mathrm{d}z/\mathrm{d}\tau = -\tau^{-2}A(1/\tau)z$ should have the Fuchs type singularity at $\tau = 0$. Thus $\mathrm{d}z/\mathrm{d}\tau = (A_\infty/\tau + \ldots)z$, which means that $A(t) \sim -A_\infty/t$ as $t \to \infty$.
If A_j are the residues of $A(t)$ at the points t_j, then the above shows that $A(t) = \sum A_j/(t-t_j)$ and

$$A_\infty = -\sum A_j.$$

The case of differential equation is treated analogously. □

8.35. Definition. The systems and equations satisfying the conditions of Lemma 8.33 are called **systems of the Fuchs class** and **equations of the Fuchs class.**

Consider a meromorphic system, or a meromorphic equation in $\mathbb{CP}^1$, with singular points $t_1, \ldots, t_m, \infty$. (Usually, the point at infinity will be singular; if not, then we underline it). Thus we have an analytic equation in $\Omega = \mathbb{C}\setminus\{t_1, \ldots, t_m\}$. By the theorem on analytic dependence of solutions of differential equations on initial conditions, the solutions $z = \phi(t)$ (or $x = \phi(t)$) are locally analytic functions in Ω. They can be prolonged to multivalued functions in Ω, or to single valued functions $\phi(\hat{t})$ in the universal covering $\widehat{\Omega}$ of Ω. Here $\hat{t} \in \widehat{\Omega}$ is projected to $t \in \Omega$.
If $t_0 \in \Omega$ and we choose some basis of the space of solutions of the equation in a neighborhood of t_0, then we can define the monodromy group of the equation as a subgroup of $GL(n, \mathbb{C})$ (see Definition 8.3 above). Let γ_j be simple loops in Ω starting at t_0 and surrounding just one point t_j (in the counterclockwise direction). In the basis of solutions the monodromy operators associated to γ_j are matrices $M_j \in GL(n, \mathbb{C})$. They generate the monodromy group, a subgroup Mon of $GL(n, \mathbb{C})$.
Thus the differential equation gives rise to monodromy groups of a multivalued functions, solutions of this equation. It turns out that this situation can be reversed in some sense. The corresponding result belongs to B. Riemann [**Rie**] (see also [**AI**]).
Recall that a multivalued function ϕ in Ω is regular iff it has at most polynomial growth in sectors near singular points.

8.36. Theorem of Riemann (Multivalued functions). ([**Rie**]) *Let $\phi_1(t), \ldots, \phi_n(t)$ be a system of multivalued and regular holomorphic functions on Ω such that its Wronskian* $\det(\phi_i^{(j)}) \neq 0$ *and such that the prolongations of ϕ_j's along the loops γ_j define automorphisms of the space of functions spanned by ϕ_k's. Then there exists an nth order differential equation of the Fuchs type such that the system $\phi_1, \ldots, \phi_n$ of functions is its fundamental system.*

Proof. For each $t \in \Omega$ consider $n+1$ vectors $\phi^{(0)}(t), \phi^{(1)}(t), \ldots, \phi^{(n)}(t)$, where $\phi^{(j)}$ has components $\mathrm{d}^j\phi_i/\mathrm{d}t^j$, $j = 0, \ldots, n$. They are linearly dependent. By assumption the vectors $\phi^{(0)}(t), \ldots, \phi^{(n-1)}(t)$, $t \in \Omega$, are linearly independent. So, $\phi^{(n)} = \sum b_j(t)\phi^{(j)}$ and this representation is unique. Because the vector functions $\phi^{(j)}(t)$ have the same monodromies, the functions $b_j(t)$ are single-valued. Because b_j have representations as quotients of determinants (the Cramer's formula) of matrices with entries $\phi^{(k)}(t)$ and ϕ_i are regular, also b_j are regular and hence meromorphic. □

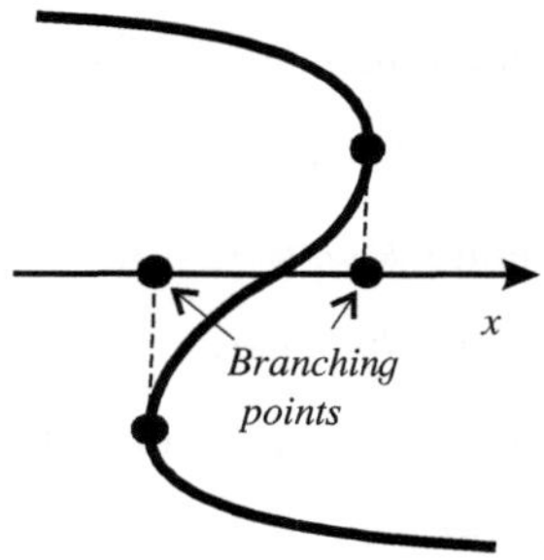

Figure 8.5: Ramified covering associated with a multi-valued function

Corollaries.

(a) *Algebraic functions satisfy differential equations of the Fuchs class.*

(b) *Abelian integrals satisfy differential equations of the Fuchs type.*

Proof. (a) Recall that algebraic functions $y(x)$ are solutions of equations of the type $F(x,y) = 0$, where F is a polynomial. This equation has several local branches of solutions $y = \psi_1, \ldots, \psi_p$. They are multivalued with branching at the finite set $F = F'_y = 0$ (see Fig. 8.5). It is possible that the branches are linearly dependent, e.g., for the equation $y^m - x = 0$. We choose the maximal system $\phi_1, \ldots, \phi_n$ of independent branches. We restrict them to the set $\Omega = \mathbb{C} \setminus (\text{ branching points }) \setminus \{W = 0\}$, where W is the Wronskian. It is clear that the assumptions of the theorem of Riemann (i.e., the regularity, $W \neq 0$ and the monodromy) are satisfied.
(b) It was proved in Theorem 5.29(a). □

§5 Riemann–Hilbert Problem

The fundamental problem in the theory of meromorphic differential equations on the Riemann sphere is the Riemann–Hilbert problem. It was formulated explicitly by D. Hilbert, as the 21th of his famous list of problems [**Hil**]. Later, people working upon it (H. Röhrl and others) realized that the questions of the same kind were earlier investigated by B. Riemann (see [**Rie**]) and began to call it the Riemann–Hilbert problem.
It is striking that this problem has turned out very productive in mathematics, especially in algebraic geometry. Such tools like Grothendieck's étale cohomology theory, Deligne's mixed Hodge structures and his proof of the Weil conjectures, $\mathcal{D}$-moduli and perverse sheaves owe their existence to investigations on this problem. The formulation of the Riemann–Hilbert problem, which we present below, belongs to Yu. Ilyashenko and was firstly formulated in the survey article [**AI**]. We follow this survey.

8.37. The Riemann–Hilbert problem. *Find:*

(A) *an equation* $x^{(n)} + \ldots = 0$ *of the Fuchs class,*

(B) *a system* $\dot{z} = Az$ *with regular singularities,*

(C) *a system* $\dot{z} = Az$ *from the Fuchs class,*

such that its singular points and monodromy operators are given.

Remarks. 1. The original Hilbert's formulation of his problem is the following. "Prove that there always exists a Fuchsian system with given singularities and a given monodromy" . It caused some misunderstandings among mathematicians studying it. For example, H. Röhrl [**Roh**] and J. Plemelj [**Ple**] proved the version B of this problem and many authors treated the Riemann–Hilbert problem as finished. However, the version C of the problem, which can be formulated as to find the residue matrices A_j (of the Fuchs system) as functions of the positions of singular points t_j and monodromy matrices M_j, may have no solutions. A. Bolibruch [**AB**] constructed such an example. (Below we present the proof of the Plemelj–Röhrl theorem and the Bolibruch's example).

2. In view of Theorem 8.7 above, the solution of the problem B implies the solution of the problem A. Indeed, having the regular differential system, we can easily associate with it a differential equation of higher order; (for example, for the first component $z_1(t)$). The regularity of this equation is equivalent to the fact that it belongs to the Fuchs class (Theorem 8.7).

8.38. Theorem of Plemelj and Röhrl. *The problem B always has a solution.*

Below we present the Röhrl's proof [**Roh**] of this result. The Plemelj's proof of this theorem is given in [**Ple**] and is different. It relies on a theory of integral operators developed by him especially for this purpose.

Proof. 1. We use the following covering of the Riemann sphere $\mathbb{CP}^1$. The point ∞ is covered by the disc

$$K_\infty = \{|t| > r\} \cup \{\infty\}$$

and the disc

$$K_0 = \{|t| < R\}, \quad r < R,$$

is covered by domains U_i, each homeomorphic to a disc, each containing exactly one singular point t_i and with only non-empty intersections $U_i \cap U_{i+1}$, which are simply connected. Moreover, the sets $U_1 \cup U_2 \cup \ldots U_j$ are also simply connected (see Fig. 8.6). The basic point t_0 and the loops $\gamma_i \subset U_i$ are also presented in this picture.

2. Firstly we solve the Riemann–Hilbert problem for the disc K_0. Instead of looking for the matrix function $A(t)$ (in the equation $\dot{z} = Az$), we try to find a fundamental matrix $\mathcal{F}(t)$.

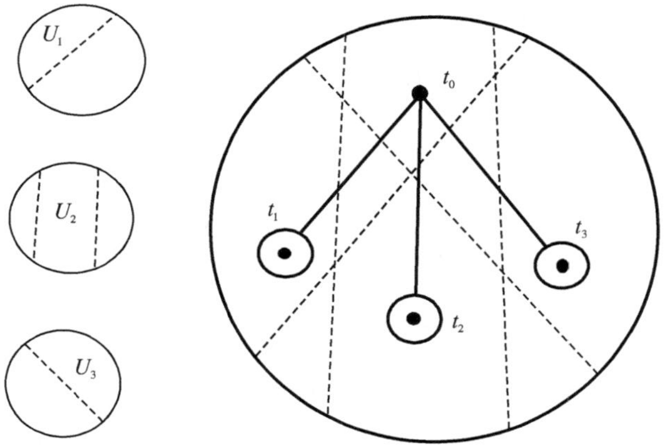

Figure 8.6: Covering of the disc and distinguished loops

In fact, this matrix is a multivalued function; so, we will write $\mathcal{F}(\hat{t})$. Here $\hat{t}$ belongs to the universal covering of $K_0 \diagdown \{t_1, \ldots, t_m\}$. When some functions are univalent, then we use the usual argument t.
We look for $\mathcal{F}$ such that, above each $U_j \setminus t_j$, we have

$$\mathcal{F}(\hat{t}) = \Phi_j(t)(\hat{t} - t_j)^{C_j}, \quad C_j = \ln M_j / 2\pi\mathrm{i}, \tag{8.14}$$

where Φ_j are holomorphic in U_j matrix functions.
Thus, in the intersection $U_j \cap U_{j+1}$ we have the equality

$$\Phi_j^{-1}(t)\Phi_{j+1}(t) = (\hat{t} - t_j)^{C_j}(\hat{t} - t_{j+1})^{-C_{j+1}} =: F_{j,j+1}.$$

On the right-hand-side of this equality stays the known function $F_{j,j+1}$ (which now can be treated as function of t). The collection $\{F_{i,i+1}\}$ can be treated as a Čech 1-cocycle associated with the covering of K_0 by U_j's (with values in the sheaf of invertible matrix functions). The existence of the above Φ_j's means that this cocycle represents zero cohomology class. In [**For**], [**GrRe**] and [**GuRo**] it is shown that the first cohomology group of any open Riemann surface (or Stein manifold) with coefficients in this sheaf is zero. Because the open disc K_0 is such a manifold, then the system of equations (8.14) has a solution.
The proof of vanishing of the cohomology groups uses the Lemma B of Cartan. Therefore, we present this lemma (with the proof) and, applying it to our situation, we give a direct solution of the Riemann–Hilbert problem for a disc.

8.39. Cartan lemma B. *Let U_1, U_2 be simply connected domains in $\mathbb{C}$ with piecewise smooth boundaries and such that $U_1 \cap U_2$ is simply connected. Let $F \in$* Hol $(\overline{U_1 \cap U_2}, GL(n, \mathbb{C}))$, *i.e., F is holomorphic in a neighborhood of the closure of $U_1 \cap U_2$.*

Then there exist $F_{1,2} \in \mathrm{Hol}\,(\overline{U_{1,2}}, GL(n,\mathbb{C}))$ *such that*

$$F(t) = F_1(t)F_2(t)$$

in $U_1 \cap U_2$.

We apply Lemma 8.39 to the domains U_1 and U_2. We find $\Phi_1 = F_1^{-1}$ and $\Phi_2 = F_2$ satisfying relation (8.14). Thus we have defined a fundamental system $\mathcal{F}_2$ in $U_1 \cup U_2$; more precisely, in the suitable universal covering.
In U_3 we have the fundamental system $(\hat{t} - t_3)^{C_3}$. We glue together these two fundamental systems to a one system in $U_1 \cup U_2 \cup U_3$ by means of the formula

$$\Psi_2(t)\mathcal{F}_2(\hat{t}) = \Phi_3(t)(\hat{t} - t_3)^{C_3},$$

where Ψ_2 and Φ_3 are holomorphic in $U_1 \cup U_2$ and in U_3 respectively (here we again use Cartan Lemma B).
Repeating this procedure, we obtain a fundamental matrix defined in the whole K_0 and such that near each singularity we have $\mathcal{F} = \Phi_j(\hat{t} - t_j)^{C_j}$. Because $\dot{\mathcal{F}} = A\mathcal{F}$, we get the following formula for the matrix $A(t)$

$$A(t) = \dot{\mathcal{F}}\mathcal{F}^{-1} = \dot{\Phi}_j\Phi_j^{-1} + \Phi_j C_j \Phi_j^{-1}/(t - t_j).$$

We see that it has only first order pole at t_j. From this proof we get the following corollary.

8.40. Corollary. *The Riemann–Hilbert problem in version C has positive solution in a disc.*

3. *The case of the Riemann sphere.* Let $\mathcal{F}_0$ be a fundamental matrix in K_0 with only Fuchsian singularities; its existence was proved above. In K_∞ we have the fundamental matrix $\mathcal{F}_\infty = t^{-C_\infty}$. If one could find matrix functions $\Phi_\pm$ holomorphic in $K_\pm$ and such that $\Phi_0\mathcal{F}_0 = \Phi_\infty\mathcal{F}_\infty$ in $K_0 \cap K_\infty$, then one would have a solution of our problem with all singularities of the Fuchsian type. That property is equivalent to the solution of the equation $F_0F_\infty = F$ for given holomorphic F in $K_0 \cap K_\infty$.
The latter equation usually does not have solution. Instead, we have the following result belonging to G. Birkhoff [**Bir1**] and A. Grothendieck [**Gro1**]. (In algebraic version it first appeared in the works of R. Dedekind and W. Weber [**DW**], see also [**OSS**].)

8.41. Theorem of Grothendieck and Birkhoff. *For any holomorphic function* F *in* $K_0 \cap K_\infty$ *with values in* $GL(n,\mathbb{C})$, *there exist holomorphic matrix-valued functions* $F_{0,\infty} : K_{0,\infty} \longmapsto GL(n,\mathbb{C})$ *and a unique (modulo permutation of entries) diagonal integer matrix* $J = \mathrm{diag}\,(j_1, \ldots, j_n)$ *such that*

$$F = F_0 t^J F_\infty$$

in $K_0 \cap K_\infty$.

Having this result we can finish the proof of the Röhrl–Plemelj theorem. The matrix-function

$$\mathcal{F} = F_0^{-1}\mathcal{F}_0 = t^J F_\infty \mathcal{F}_\infty$$

defines a regular fundamental matrix in $\widehat{\Omega}$. It satisfies the differential equation $\dot{\mathcal{F}} = A\mathcal{F}$, where $A(t)$ is a meromorphic matrix function. A has simple poles in t_j (in K_0) and pole in ∞ (maybe non-simple). It is because, near ∞, we have $A = J/t + t^J \dot{F}_\infty F_\infty^{-1} t^{-J} - t^J F_\infty C_\infty F_\infty^{-1} t^{-J}$ and the action Ad_{t^J} may lead to a non-simple pole at infinity.
Theorem 8.38 is complete modulo the proofs of Lemma 8.39 and Theorem 8.41. □

8.42. Proof of the lemma B of Cartan. We will follow the proof from the book of R. Gunning and H. Rossi [**GuR**]. Another proof is given in [**GrR**].

1. Firstly we consider the abelian case.

Lemma. *Let U_1 and U_2 be as in assumptions of Theorem 8.38, simply connected with simply connected intersection. Let $f \in \mathrm{Hol}\,(\overline{U_1 \cap U_2}, \mathbb{C}^k)$. Then there exist two functions $f_{1,2} \in \mathrm{Hol}\,(\overline{U_{1,2}}, \mathbb{C}^k)$ such that*

$$f = f_1 + f_2$$

(in the intersection) and we have an estimate

$$||f_{1,2}|| < K\,||f||,$$

where the norms are the supremum norms in the suitable domains and the constant K depends only on the domains.

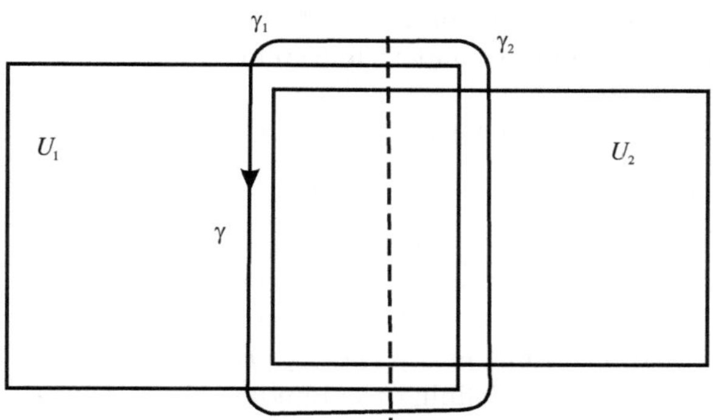

Figure 8.7: Domains and paths used in the proof of the lemma

Proof. Let V be a neighborhood of $\overline{U_1 \cap U_2}$, where f is defined. Let γ be a loop in $V \setminus \overline{U_1 \cap U_2}$. Then we have $f(z) = \frac{1}{2\pi \mathrm{i}} \int_\gamma f(\zeta)\mathrm{d}\zeta/(\zeta - z)$ for $z \in U_1 \cap U_2$.

We divide γ into two parts γ_1 and γ_2 (as in Fig. 8.7). Thus γ_1 does not pass through $U_2 \setminus U_1$ and γ_2 has the analogous property.
Then the functions

$$f_{1,2}(z) = \frac{1}{2\pi \mathrm{i}} \int_{\gamma_{1,2}} \frac{f(\zeta)\mathrm{d}\zeta}{\zeta - z}$$

satisfy the thesis of the lemma. □

2. Consider now the case when F is a matrix function in a domain larger than $U_1 \cap U_2$ and is close to the identity matrix, $F(t) = I + F_1(t)$.
Using the arguments from the abelian case we can represent F_1 as the sum G_1+H_1, where G_1 is holomorphic in $\overline{U_1}$ and H_1 is holomorphic in $\overline{U_2}$. Moreover, $||G_1|| \sim O(||F_1||)$, $||H_1| \sim O(||F_1||)$.
We can write

$$I + F_1 = (I + G_1)(I + F_2)(I + H_1),$$

where one has the estimate $||F_2|| \sim O(||F_1||^2)$.
Now we repeat this procedure with F_2. We represent it as $G_2 + H_2$, where G_2 and H_2 are of the second order with respect to $||F_1||$. One obtains F_3 of order $O(||F_1||^4)$, and so on. Finally, we get

$$F = (I + G_1)(I + G_2)\ldots\ldots(I + H_2)(I + H_1),$$

where the two infinite products converge uniformly in the domains $\overline{U_i}$.

3. The general case is reduced to the Case 2 by a polynomial approximation $K(t)$ of $F(t)$. If $||K - F||$ is small, then $F_1 = K^{-1}(F - K)$ is also small and we have

$$F = K(I + F_1). \qquad \square$$

8.43. A few words about vector bundles. Before presenting the proof of the Grothendieck–Birkhoff Theorem, we make some comments about algebro-geometrical interpretation of this result.
In the language of holomorphic vector bundles this theorem states that:

Every holomorphic vector bundle over $\mathbb{CP}^1$ is isomorphic to a Whitney sum of line bundles.

The above needs some explanations. A vector bundle $E \longmapsto X$ is *holomorphic* iff its transition maps $(U_i \cap U_j) \times \mathbb{C}^n \ni (x, z) \longmapsto (x, h_{ij}(x)z)$ are holomorphic. E is a *Whitney sum* (or direct sum) of E_1 and E_2, $E = E_1 \oplus E_2$, iff $h_{ij} = f_{ij} \oplus g_{ij}$, where f_{ij}, g_{ij} are the cocycles defining the bundles E_1, E_2. A vector bundle is the *line bundle* iff its fiber is one-dimensional, $n = 1$.
If $X = \mathbb{CP}^1$, then it has the covering by two open discs K_0, K_∞, where the restrictions of the bundles to each of them is trivial. (It is because the corresponding first cohomology group vanishes, which follows from the Lemma B of Cartan.) In algebraic geometry the roles of $K_{0,\infty}$ play the Zariski open subsets $U_1 = \mathbb{C}$ $(= \mathbb{CP}^1 \setminus \infty)$ and $U_2 = \mathbb{C}^* \cup \infty$ with $U_1 \cap U_2 = \mathbb{C}^* = \mathbb{C} \setminus 0$. If the bundle is

a line bundle, then the cocycle h_{12} is a scalar nonzero function on $\mathbb{C}^*$. We can multiply it by a nonzero (in $\mathbb{C}^*$) functions which are analytic in U_1 or in U_2, i.e., we can add to it a coboundary. The only invariant of such equivalence relation is the topological degree of $h_{12} : \mathbb{C}^* \longmapsto \mathbb{C}^*$. This means that:

Any line bundle over $\mathbb{CP}^1$ *is isomorphic to* $\mathcal{O}(j)$ *with the cocycle* $h_{12}(t) = t^j$.

The bundles $\mathcal{O}(j)$ are usually interpreted in terms of sheaves. The sheaf of local sections of such bundle is also denoted by $\mathcal{O}(j)$. In particular, $\mathcal{O}(0)$ is the same as the sheaf $\mathcal{O}_X$ of germs of holomorphic functions on $X = \mathbb{CP}^1$.
If t_0 is some distinguished point (*divisor*), e.g., $t_0 = 0$, then we can consider the sheaf $\mathcal{J}(t_0)$ of ideals consisting of germs vanishing at t_0. Because in the affine part $U_1 = \mathbb{C}$, $\mathcal{J}(0)$ is generated by $s_1(t) = t$ and in U_2 by $s_2(t) = 1$, then the transition map $h_{12} = t^{-1}$, $(s_1 \circ h_{12} = s_2)$, and this sheaf is isomorphic to $\mathcal{O}(-1)$.
If we take the bundle associated with the divisor $D = t_0$ with the transition map $h_{12} = f_1/f_2$, where $f_i = 0$ are the equations for the divisor in U_i, then we obtain the bundle $\mathcal{O}(1)$. If the divisor is more general $D = \sum n_i t_i$, $n_i \in \mathbb{Z}$, $t_i \in \mathbb{CP}^1$, then the bundle $\mathcal{O}(D)$ associated with it has the transition maps $\prod(f_{i,1}/f_{i/2})^{n_i}$ and the bundle is isomorphic to $\mathcal{O}(\sum n_i)$. Here $f_{i,j} = 0$ are the equations for the divisors t_i in U_j.
The *canonical bundle* $K_X = \Omega_X$ of holomorphic 1-forms on $X = \mathbb{CP}^1$ is isomorphic to $\mathcal{O}(-2)$.
The bundles $\mathcal{O}(j)$ have natural generalizations for higher dimensional projective spaces $\mathbb{CP}^d$. We have the natural covering of the projective space, with the homogeneous coordinates $(t_0 : t_1 : \ldots : t_d)$, by the sets $U_j = \{t_j \neq 0\}$ and parameterized by $x_0^{(j)} = t_0/t_j, \ldots x_n^{(j)} = t_n/t_j$. The cocycle functions defining $\mathcal{O}(j)$ are equal to $h_{kl} = (x_k/x_l)^j$. Here, as in the one-dimensional space, the sheaf of ideals of germs of functions vanishing at a hyperplane is equal to $\mathcal{O}(-1)$ and the sheaf associated with the hyperplane divisor is equal to $\mathcal{O}(1)$. The space of global section of the bundle $\mathcal{O}(j)$, $j \geq 0$, is isomorphic to the space of homogeneous polynomials of $t_0, \ldots, t_n$ of degree j. Every line bundle on $\mathbb{CP}^d$ is isomorphic to one of $\mathcal{O}(j)$. The *canonical bundle* $K_X = \Omega_X^d$ on $X = \mathbb{CP}^d$ is isomorphic to $\mathcal{O}(-d-1)$.
The theory of bundles of higher rank, i.e., $n > 1$, on projective spaces is an advanced and difficult theory. We refer the reader to the book [**OSS**].

8.44. Proof of the Grothendieck–Birkhoff theorem. 1. Recall that having a holomorphic function $F : K_0 \cap K_\infty \longmapsto GL(n, \mathbb{C})$, we have to find matrix-valued holomorphic functions $F_\pm$ on $K_\pm$ and a diagonal integer matrix J such that $F = F_0 t^J F_\infty$. Here K_0 is a disc containing $t = 0$ and K_∞ contains ∞.
If we treat F as a cocycle defining a holomorphic bundle in $\mathbb{CP}^1$, then the statement of this theorem says that this cocycle is equivalent to the diagonal cocycle $t^J = t^{j_1} \oplus \ldots \oplus t^{j_n}$. This means that the bundle is isomorphic to $\mathcal{O}(j_1) \oplus \ldots \oplus \mathcal{O}(j_n)$.
Let us pass to the proof, which belongs to Birkhoff and which we have taken from the book of D. Anosov and A. Bolibruch [**AB**]. We divide it into some steps.

2. *The case $F = I + F_1$, where F_1 is small.* We define the operators

$$\Pi_\pm : \mathrm{Hol}\,(K_0 \cap K_\infty, GL) \longmapsto \mathrm{Hol}\,(K_\pm, GL),$$

where $GL = GL(n, \mathbb{C})$, as follows. If $G(t)$ has the Laurent expansion $G = \sum_{-\infty}^{\infty} G_j t^j$, then

$$\Pi_0 G = \sum_{0}^{\infty} G_j t^j, \quad \Pi_\infty G = \sum_{-\infty}^{-1} G_j t^j.$$

We seek the solution in the form

$$F_0 = F(I + X), \quad J = I, \quad F_\infty = (I + X)^{-1},$$

with X small and holomorphic in K_∞. The condition $F_0 \in \mathrm{Hol}\,(K_0)$ means that $\Pi_\infty(F(I + X)) = 0$. It leads to

$$TX = -\Pi_\infty F_1, \tag{8.15}$$

where T is the linear operator (in the space of matrix functions) defined as $TX = X + \Pi_\infty(F_1 X)$. Because F_1 is small, the operator T is close to the identity. Thus T is invertible and Eq. (8.15) has a solution.

3. *Reduction to the case with rational F and such that its only pole in K_0 is $t = 0$.* Using a sufficiently long finite Laurent expansion of F, we approximate F by a matrix $G(t)$ with only poles at 0 and ∞. Thus FG^{-1} is close to the identity and, using the previous point, we get the representation $\Phi_0\Phi_\infty = FG^{-1}$, where $\Phi_{0,\infty}$ are holomorphic in $K_{0,\infty}$. Thus we have

$$F = \Phi_0(\Phi_\infty G). \tag{8.16}$$

Now we apply the same to $\Phi_\infty G$. We approximate it by a rational H (with poles at $0, \infty$) and get the representation $H^{-1}\Phi_\infty G = \Psi_0\Psi_\infty$ with $\Psi_{0,\infty}$ analytic in $K_\pm$. This gives

$$\Psi_0 = H^{-1}\Phi_\infty G \Psi_\infty^{-1}.$$

On the left-hand-side of the latter formula we have a function holomorphic in K_0 and on the right we have a function meromorphic in K_∞. Thus Ψ_0 is rational. Now, from Eq. (8.16) we get the formula for F

$$F = \Phi_0(H\Psi_0)\Psi_\infty,$$

where $F_1 = H\Psi_0$ is rational with the only possible pole in K_0 at $t = 0$. We see that the problem is to represent F_1 in the needed form.

4. *Determination of J.* Recall our equation $FF_0^{-1} = t^J F_\infty$, where F is rational with the only pole in K_0 at $t = 0$. Denote the columns of the matrix F_0^{-1} by $C_k(t)$ and of F_∞ by $D_k(t)$. Therefore, we have to solve the series of vector equations $FC_k = t^{j_k} D_k$.

Consider the general equation

$$F(t)C(t) = t^j D(t), \tag{8.17}$$

with the boundary condition

$$C(\infty) = y. \tag{8.18}$$

Here $C(t)$ is a rational vector-valued function holomorphic in K_0 and $D(t)$ is a rational vector function holomorphic in K_∞.

Definition. A triple (j, C, D) is called an *admissible triple for* $y \in \mathbb{C}^n \setminus 0$ iff Eqs. (8.17)–(8.18) hold.

Below we present some properties of the admissible triples.

(a) *The admissible triples exist for any* y.

Indeed, let

$$F(t) = t^{-\beta} q^{-1}(t) P(t), \ \deg P = \alpha,$$

where $q(t)$ is a scalar polynomial with zeroes outside K_0 and $P(t)$ is a polynomial matrix. Then the triple $(-\beta, y, q^{-1}Py)$ is admissible for y.

(b) *We have* $j \leq \alpha - \beta$ *for any admissible triple;* (α, β are above).

It follows from the equation

$$P(t)C(t) = t^{\beta+j} q(t) D(t),$$

where $P(t) \sim \text{const} \cdot t^\alpha$ and $C(t) \to y$ as $t \to \infty$.

Definition. An admissible triple (j, C, D) is called *maximal* iff the integer j is maximal possible.

(c) *If the triple* (j, C, D) *is maximal, then* $C \neq 0$ *in* K_0 *and* $D \neq 0$ *in* K_∞.

If $C(a) = 0$ (or $D(a) = 0$) for $a \in K_0 \cap K_\infty$, then $D(a) = 0$ (respectively $C(a) = 0$). Thus $C(t) = (t-a)C_1(t)$, $D(t) = (t-a)D_1(t)$, and, replacing C by $\frac{t}{t-a}C$ and D by $\frac{D}{t-a}$, we get

$$F \cdot \frac{t}{t-a} \cdot C = t^{j+1} \cdot \frac{D}{t-a}, \tag{8.19}$$

which means that $(j+1, \frac{tC}{t-a}, \frac{D}{t-a})$ is admissible for y.
The formula (8.19) remains true when $D(a) = 0$, $a \in K_\infty \setminus K_0$, and when $C(a) = 0$, $a \in K_0 \setminus K_\infty$.

Definition. The function $y \longmapsto v(y) = \max\{j : (j, C, D) \text{ is admissible for } y\}$ (with $v(0) = \infty$) defines a valuation on $\mathbb{C}^n$, which we call the **Levelt valuation.** It has the usual properties of valuation:

$v(\lambda y) = v(y)$, $\lambda \neq 0$ (obvious) and

$v(y_1 + y_2) \geq \min(v(y_1), v(y_2))$.

Indeed, if (j_1, C_1, D_1) and (j_2, C_2, D_2) are admissible for y_1 and y_2 and $j_1 \leq j_2$, then $(j_1, C_1 + C_2, D_1 + t^{j_2 - j_1} D_2)$ is admissible for $y_1 + y_2$.
The valuation $v(\cdot)$ defines the filtration

$$0 = E_0 \subset E_1 \subset \ldots \subset E_h = \mathbb{C}^n,$$

such that $v|_{E_i \setminus E_{i-1}} = const.$
The integers j_k in the matrix $\operatorname{diag}(j_1, \ldots, j_n)$, $j_1 \geq j_2 \cdots \geq j_n$, are the values of the valuation function $v(\cdot)$, where the ith value is taken $\dim E_i - \dim E_{i-1}$ times.

5. *Construction of the matrices* $F_{0,\infty}$. We choose a basis $y_1, \ldots, y_n$ in $\mathbb{C}^n$ associated with the above filtration. It means that each y_k has a maximal triple (j_k, C_k, D_k). We define

$$\begin{aligned} F_0^{-1} &= (C_1, \ldots, C_n), \\ F_\infty &= (D_1, \ldots, D_n). \end{aligned}$$

It remains only to show that these matrices are invertible.
Assume that $\det F_+^{-1}(a) = 0$ (the proof in the case $\det F_\infty(a) = 0$ is analogous). Then there exists a nontrivial combination $C(t) = \sum \lambda_i C_i(t)$ which vanishes at $t = a$. Let $y = \sum \lambda_i y_i$ and let $y \in E_k \setminus E_{k-1}$. The valuation $v(y)$ is equal to $v|_{E_k \setminus E_{k-1}} = \min\{j_i : \lambda_i \neq 0\}$. Let $D(t) = \sum \lambda_i t^{j_i - v(y)} D_i(t)$. It is clear that the triple $(v(y), C, D)$ is a maximal admissible triple for $y \neq 0$.
By the property (c) of maximal triples (see above), we have $C(t) \neq 0$ in K_0. This gives the contradiction.
From the above analysis it follows that the integers $j_1, \ldots, j_n$ are defined uniquely. The proof of Theorem 8.41 is complete. □

Now we pass to the version C of the Riemann–Hilbert problem. One asks whether it is possible to improve the result of Röhrl and Plemelj and to get a Fuchsian system with prescribed singularities and monodromy. As we shall see later, it is not possible in general. However, when one adds some restriction (on dimension of the system or on the monodromy), then one gets the positive result in the problem C. One of them is presented in the next theorem belonging to J. Plemelj [**Ple**] (see also [**AI**]).

8.45 Theorem. *If one of the monodromy matrices M_j is diagonalizable, then the problem C has positive solution.*

Proof. We can assume that diagonalizable is the matrix $M_\infty = (M_m \ldots M_1)^{-1}$, associated with the loop around ∞. We assume that M_∞ and $C_\infty = \frac{1}{2\pi i} \ln M_\infty$ are diagonal.
From the proof of Plemelj–Röhrl theorem we can find fundamental matrix $\mathcal{F}$ such that $\mathcal{F} = \Phi_0 \mathcal{F}_0 = \Phi_\infty t^{-C_\infty}$, where Φ_0 is analytic in the disc K_0, $\mathcal{F}_0$ is fundamental matrix in K_0 with Fuchsian singularities, and Φ_∞ is *meromorphic* in the disc K_∞ with the pole at ∞.

8.46. Sauvage lemma. *There exist matrix-valued functions* $P(t), \Psi(t), D$ *such that* $P(t)$ *is polynomial with* $\det P = 1$, $\Psi(t)$ *is analytic and invertible in* K_∞, $D =$ const *is integer and diagonal, and the identity*

$$\Phi_\infty = P(t) \cdot \Psi(t) \cdot t^D$$

holds.

Using this lemma we easily finish the proof of Theorem 8.45. We put $\mathcal{G} = P^{-1}\mathcal{F}$. It is a fundamental matrix with Fuchsian singularities in K_0 (because P^{-1} is analytic and invertible there). In K_∞ we have $\mathcal{G} = \Psi t^{D-C_\infty}$ (here we use the fact that D and C_∞ commute). This matrix also gives Fuchsian singularity (with the residuum $\Psi(\infty)(C_\infty - D)\Psi^{-1}(\infty)$). □

Remark. The commutativity property of D and C_∞ is essential in the proof. For example, if $\mathcal{X} = t^D t^C$, $C = \begin{pmatrix} 0 & 1 \\ 0 & 0 \end{pmatrix}$, $D = \begin{pmatrix} 0 & 0 \\ 0 & 1 \end{pmatrix}$, then $\dot{\mathcal{X}}\mathcal{X}^{-1}$ has a second order pole.

Proof of the Sauvage lemma. 1. We borrowed the name of this lemma from [**AI**], where the reference to the book of P. Hartman [**HaP**] is given. In fact, in [**HaP**] another result is proved. It is shown that any holomorphic matrix-function $F(t)$ in the disc K_∞ has a representation in the form $F = P(t)t^D G(t)$, where $P(t)$ is polynomial with $\det P = 1$, D is integer diagonal matrix and $G(t)$ is invertible. The proof of Lemma 8.45 is much more complicated than the proof from [**HaP**]. In the book of Plemelj [**Ple**] the proof of Theorem 8.44 is based on other properties of fundamental systems than the ones used in our above proof.
We shall not present the complete proof of the Sauvage lemma. We limit ourselves to the case $n = 2$.

2. Firstly, by applying the change $t \longmapsto 1/t$, we can assume that we are in the disc K_0 around 0. Thus, given a meromorphic F with pole at 0, we look for the representation $F = P(1/t)\Psi(t)t^D$ with polynomial P, $\det P = 1$, analytic Ψ, $\det \Psi \neq 0$, and diagonal integer D. Of course, it is enough to obtain this representation with $\det P(1/t) = \text{const} \neq 0$.

3. The action of a diagonal matrix $\operatorname{diag}(\lambda_1, \ldots, \lambda_n)$ on a given matrix from the right means multiplying its columns by the numbers λ_i.
Note also that diagonal matrices commute and the automorphism $Z \longmapsto QZQ^{-1}$, with Q a matrix of permutation of coordinates, sends diagonal matrices to diagonal matrices with permuted columns. Therefore we apply from the right: the diagonal matrices t^A, constant diagonal matrices and permutations of columns.
An action of some matrix onto a given matrix A from the left means some action on the rows of A. We apply the following three types of action onto the rows of meromorphic matrices.

(i) Transpositions of rows. (ii) Multiplication of rows by constants; it is the action of a diagonal matrix. (iii) Adding to a row another row multiplied by a polynomial of $1/t$;(this means action of a triangular matrix with 1 at the diagonal and a polynomial above or below the diagonal.

We see that in all the cases we multiply from the left by a matrix with constant determinant.

4. Let $F = (t^{a_1} f_1, t^{a_2} f_2) = (f_1, f_2) t^A$, where f_i are holomorphic columns, $f_i(0) \neq 0$, and A is diagonal. If $\det(f_1, f_2) \neq 0$, then we have the result.

5. Assume that this determinant vanishes. Using eventual transposition of rows and multiplications of rows and columns by constant numbers, we can assume that the components of the first row start from 1. After extracting a multiple of the first row from the second row and eventual permutation of columns, we arrive to the matrix $\begin{pmatrix} 1+\dots & 1+\dots \\ at^\alpha+\dots & bt^\beta+\dots \end{pmatrix}$, $ab \neq 0$, $\alpha \geq \beta$. We multiply the second column by $t^{-\beta}$ and, after little modification, we obtain the matrix $\begin{pmatrix} 1 & ct^{-\beta}+\dots \\ at^\alpha+\dots & 1 \end{pmatrix}$.

6. Now we multiply this matrix from the left by an upper-triangular polynomial matrix $\begin{pmatrix} 1 & p(1/t) \\ 0 & 1 \end{pmatrix}$ of the type (iii). The polynomial $p(t)$ is chosen in such a way that, after multiplication, the upper-right corner of the resulting matrix becomes analytic.

7. If $\alpha > \beta$, or $\alpha = \beta$ but $ac \neq 1$, then we obtain the matrix $\begin{pmatrix} d+\dots & O(1) \\ at^\alpha+\dots & 1 \end{pmatrix}$, $d \neq 0$, which is invertible and the lemma is proved.

8. If $\alpha = \beta$ and $ac = 1$, then we get the matrix $\begin{pmatrix} dt^\gamma+\dots & O(1) \\ at^\alpha+\dots & 1 \end{pmatrix} = F_1 \cdot \begin{pmatrix} t^\delta & 0 \\ 0 & 1 \end{pmatrix}$, where $\det F_1$ has lower order than the determinant of the matrix from 4.

We apply the above (5., 6. and 7.) to the matrix F_1. After finite number of such steps we obtain a matrix which is holomorphic and invertible. □

§6 Bolibruch's Example

The negative answer to the Riemann–Hilbert problem in version C (i.e., the existence of a Fuchsian system with given monodromy) is given in the below result of A. Bolibruch from [**AB**]. The proof of the Bolibruch's theorem is relatively long. But the reader can learn from it a lot of methods and tricks. For this reason we included it into the book.

8.47. Theorem (The counter-example of Bolibruch). *Consider the system $\dot{z} = A(t)z$ with*

$$A = \frac{1}{t^2}\begin{pmatrix} 0 & 1 & 0 \\ 0 & t & 0 \\ 0 & 0 & -t \end{pmatrix} + \frac{1}{6(t+1)}\begin{pmatrix} 0 & 6 & 0 \\ 0 & -1 & 1 \\ 0 & -1 & 1 \end{pmatrix}$$
$$+ \frac{1}{2(t-1)}\begin{pmatrix} 0 & 0 & 2 \\ 0 & -1 & -1 \\ 0 & 1 & 1 \end{pmatrix} + \frac{1}{3(t-1/2)}\begin{pmatrix} 0 & -3 & -3 \\ 0 & -1 & 1 \\ 0 & -1 & 1 \end{pmatrix}.$$

This system has only regular singular points and there does not exists any system from the Fuchs class, whose singular points and monodromy group are the same as for the above system.

Proof. 1. Firstly we investigate the singular points of the system $\dot{z} = Az$. Of course, the points $t_1 = 0$, $t_2 = 1$, $t_3 = -1$, $t_4 = \frac{1}{2}$ are singular. As $t \to \infty$ we have $A(t) \sim \frac{1}{t}\sum \operatorname{Res}_{t_i} A(t)$ and, because $\sum \operatorname{Res}_{t_i} A(t) = 0$, the point $t = \infty$ is not singular.
$A(t)$ has simple poles at $t_{2,3,4}$; so these points are Fuchsian and hence regular. At t_1, $A(t)$ has the second order pole. The regularity of t_1 is obtained in the following way. Let $z = \binom{x}{y}$, where $x \in \mathbb{C}$, $y = (y_1, y_2)^\top \in \mathbb{C}^2$. The system for z is equivalent to the system

$$\dot{x} = a(t)y_1 + b(t)y_2,$$
$$\dot{y} = B(t)y,$$

where

$$B(t) = \sum B_i/(t - t_i).$$

Because the system for y is Fuchsian, then its solutions $y_j(t)$ are regular, of polynomial growth at singular points (in particular, at $t = 0$). Because $a(t)$ and $b(t)$ are rational, then integrating the equation for x we obtain the regularity of the solutions $x(t)$.

2. Note that the matrices $B_j = \operatorname{Res}_{t_j} B(t)$, $j = 2, 3, 4$, are nilpotent. Indeed, $\operatorname{tr} B_j = \det B_j = 0$ (so the eigenvalues are equal to zero) but $B_j \neq 0$. This means that we have resonances at t_j, $\lambda_1 - \lambda_2 \in \mathbb{Z}$ (see Definition 8.16). Using Theorem 8.17 and Remark 8.18, we obtain that the normal forms of the system for y at $t_{2,3,4}$ are given by triangular matrices: $\dot{\tilde{y}}_1 = (e_i/(t - t_i))\tilde{y}_2$, $\dot{\tilde{y}}_2 = 0$. This leads to triangular monodromy matrices.
At the point t_1 we have the eigenvalues $\lambda_1 = 1$, $\lambda_2 = -1$. So here we have also resonance but the normal form is slightly different: $\dot{\tilde{y}}_1 = (1/t)\tilde{y}_1 + e_1 t\tilde{y}_2$, $\dot{\tilde{y}}_2 = (-1/t)\tilde{y}_2$. This also gives triangular monodromy matrix.
Because the system for z is block-triangular, all its monodromy matrices are equivalent to triangular matrices. Thus the assumptions of Theorem 8.44 are not satisfied and there is a hope that we will obtain a good counter-example.

We will prove the above facts about the normal forms and local monodromies later.

3. Here we develop a theory which describes the local behavior of fundamental systems near regular singular points. In particular, we shall construct certain invariants which allow us to distinguish fundamental matrices of local Fuchsian systems.
If a fundamental matrix $\mathcal{F}(\hat{t})$ is regular near the singular point $t=0$, then it has the form $\mathcal{F}(\hat{t})=\mathcal{G}(t)\hat{t}^{C}$, where $\mathcal{G}$ is meromorphic (univalent), $C=\frac{\ln M}{2\pi \mathrm{i}}$ and M is the monodromy matrix (in the basis given by the columns of $\mathcal{F}$). It follows from the proofs of Theorems 8.12 and 8.17.
The logarithm of the monodromy matrix is not chosen uniquely. In order to ensure this uniqueness we assume the following

$$0 \leq \mathrm{Re}\mu_j < 1,$$

where μ_j are the eigenvalues of the matrix C.
It turns out that the above formula for $\mathcal{F}$ can be canonically improved. We shall show the representation

$$\mathcal{F}(\hat{t}) = U(t)t^V\hat{t}^C, \tag{8.20}$$

where $V=\mathrm{diag}\,(v_1,\ldots,v_n)$ is integer matrix such that $v_1\geq v_2\geq \ldots \geq v_n$ and $U(t)$ is holomorphic. The integers v_j will be defined canonically as values of certain 'valuation' in the space of solutions of the system $\dot{y}=By,\ B=\dot{\mathcal{F}}\mathcal{F}^{-1}$.

4. **Definition (Levelt valuation).** If $y(t)\sim Ct^{\alpha}(\ln t)^k$ in a sector with vertex at $t=0$, then we put

$$v(y) = [\alpha],$$

i.e., the integer part of the exponent α, and call it the **Levelt valuation** on $y(t)$. For example, $v(\sqrt{t})=0,\ v(t\ln t)=1$.
If $\mathcal{X}$ is the space of solutions of $\dot{y}=By$, then the function $v:\mathcal{X}\longmapsto\mathbb{Z}$ has the usual properties of valuation:

(i) $v(0)=\infty$ (by definition);

(ii) $v(\lambda y)=v(y),\ \lambda\in\mathbb{C}\setminus 0$;

(iii) $v(y_1+y_2)\geq \min(v(y_1),v(y_2))$, with the equality if $v(y_1)\neq v(y_2)$.

The valuation defines a natural filtration of $\mathcal{X}$:

$$0=\mathcal{X}_0\subset\mathcal{X}_1\subset\ldots\subset\mathcal{X}_h=\mathcal{X},$$

such that $v|_{\mathcal{X}_j\setminus\mathcal{X}_{j-1}}=\mathrm{const}$.
Moreover, the monodromy operator M preserves the above filtration, $M\mathcal{X}_j\subset\mathcal{X}_j$. This means that it has a block-triangular form. At the diagonal stay the operators Gr_jM (of the gradation of M), acting on the graded space $\mathrm{Gr}\,\mathcal{X}=\bigoplus\mathcal{X}_j/\mathcal{X}_{j-1}$.

By choosing Jordan bases in $\mathcal{X}_j/\mathcal{X}_{j-1}$ the operators $\mathrm{Gr}_j M$ can be represented as triangular matrices. Using this, we can choose a basis $y_1, \ldots, y_n$ in $\mathcal{X}$ such that the sequence $v_j = v(y_j)$ is decreasing and the matrix of monodromy operator (written in this basis) is triangular. Such basis is called the *Levelt's basis.*
One can see that, if we take the fundamental matrix $\mathcal{F}$ as composed of the vectors $y_j(t)$, then it has representation (8.20).

5. **Lemma**. *The singular point $t = 0$ of the regular system $\dot{y} = By$ is Fuchsian iff the matrix $U(t)$ from Eq.* (8.20) *is invertible.*

Proof. (a) If $U(0)$ is invertible, then we have

$$B = \dot{\mathcal{F}}\mathcal{F}^{-1} = \dot{U}U^{-1} + \tfrac{1}{t}UVU^{-1} + \tfrac{1}{t}Ut^V Ct^{-V}U^{-1} = \tfrac{1}{t}[ULU^{-1} + O(t)],$$

where

$$L = V + t^V Ct^{-V}.$$

Note that the matrix $\widetilde{C} = t^V Ct^{-V}$ is analytic. Its ij^{th} entry is equal to $c_{ij}t^{v_i - v_j}$, where the entries c_{ij} of C vanish for $i > j$. So, the block-diagonal submatrices of $\widetilde{C}$ remain the same as in C and the off-diagonal submatrices acquire positive powers of t; because $v_i - v_j \geq 0$ for $i < j$.
Now the Fuchsian property is obvious. Moreover, the residuum of $B(t)$ at 0 is $B_0 = U(0)L(0)U(0)^{-1}$, $L(0) = V + (\text{block-diagonal part of } C)$.

(b) Assume that U is not invertible. In particular, $\ker U(0) \neq 0$. Suppose also that the system is Fuchsian, $\dot{y} = (B_0/t + \ldots)y$. We will get a contradiction.
The formula $B_0 = U(0)L(0)U(0)^{-1}$ from the previous point cannot hold any longer. However, its analogue

$$B_0 U(0) = U(0)L(0)$$

should be true.
This means that $L(0)(\ker U(0)) \subset \ker U(0)$; if $y \in \ker U(0)$ and $z = L(0)y$, then $U(0)z = B_0 U(0)y = 0$. This leads to the inclusion

$$\hat{t}^{L(0)} \ker U(0) \subset \ker U(0).$$

Let $y_0 \in \ker U(0) \setminus 0$ and let $y(\hat{t}) = \mathcal{F}(\hat{t})y_0$. Let $v(y(t)) = v_m$. We shall show that $v(y)$ should be greater than v_m, which will give us the needed contradiction.
We have

$$\begin{aligned} y(\hat{t}) &= U(t)t^V \hat{t}^C y_0, \\ &= U(0)\hat{t}^{L(0)}y_0 + (U(t) - U(0))\hat{t}^{L(0)}y_0 + U(t)(t^V\hat{t}^C - \hat{t}^{L(0)})y_0. \end{aligned}$$

The first term in the second row of the above formula is equal to zero (see above). The second term is of the form $O(t)\hat{t}^{L(0)}y_0$, where $L(0)$ is equal to V (with $Vy_0 = v_m y_0$) plus the block-diagonal part of C (with positive real parts of the

eigenvalues). Thus $|\hat{t}^{L(0)}y_0| \sim |t|^{v_m+\epsilon}$, $\epsilon \geq 0$, and the valuation of the second term is greater than v_m.
Finally, the matrix $t^V\hat{t}^C - \hat{t}^{L(0)}$ contains only off-block-diagonal terms which are of higher order than the block-diagonal part of $t^V\hat{t}^C$. This implies that also the third term is of order $|t|^{v_m+1+\epsilon}$. □

6. **Lemma.** *Consider a global system $\dot{z} = A(t)z$, $t \in \mathbb{CP}^1$, with regular singular points $t_1, \ldots, t_m$. Let $L_j = V_j + t^{V_j}C_j t^{-V_j}$ be the matrices associated with the corresponding local systems and constructed in the previous point.*
Then

$$\sum \operatorname{tr} L(t_j) \leq 0$$

and the equality holds for systems from the Fuchs class.
In particular, if the matrices C_j are nilpotent and the system is Fuchsian, then $\sum \operatorname{tr} V_j = \sum_{i,j} v_{i,t_j} = 0$ (where v_{i,t_j} are the Levelt's valuations, i.e., the eigenvalues of V_j).

Proof. Consider the 1-form $\operatorname{tr} A(t)\mathrm{d}t = \mathrm{d}(\ln \det \mathcal{F})$, which is meromorphic in the Riemann sphere. We have $\sum \operatorname{Res}_{t_j}(\operatorname{tr} A\mathrm{d}t) = 0$.
On the other hand,

$$\operatorname{Res}_{t_j} \mathrm{d}(\ln \det \mathcal{F}) = \operatorname{tr} L(t_j) + \text{order of zero of } \det U_j(t)$$

for $\mathcal{F} = U_j(t)(t-t_j)^{V_j}(\hat{t}-t_j)^{C_j}$, $L(t_j) = V_j + \operatorname{diag}\widetilde{C}_j$. (Here $\operatorname{diag}\widetilde{C}_j$ is the block-diagonal part of C_j).
The result follows from the fact that U_j are holomorphic. □

7. Now we begin to apply the tools and results obtained in the previous two points to the Bolibruch's system. In this point we consider the 2-dimensional Fuchsian system $\dot{y} = By$ from 2, i.e., the subsystem of the Bolibruch system.
Recall that, if, at the singular point $t = 0$ of a system $\dot{y} = (C/t + \ldots)y$, there is a resonant relation $\lambda_i = \lambda_j + k$ with non-negative integer k, then the term $t^{k-1}y_j$ cannot be removed from the ith equation (see Definition 8.16).
Applying this fact to the singular points $t_{2,3,4}$ (which are resonant $\lambda_1 = \lambda_2 + 0$), we can assume that the system is locally equivalent to systems $\dot{\tilde{y}} = (C_j/(t-t_j))\tilde{y}$ with $C_j = \begin{pmatrix} 0 & e_j \\ 0 & 0 \end{pmatrix}$, $e_j \neq 0$. The representations (8.20), for the fundamental matrix, take the forms $\mathcal{F}(\hat{t}) = U_j(t)(\hat{t}-t_j)^{C_j}$ with invertible U_j. We have $V_j = 0$, $L(t_j) = C_j$.
Note that $(\hat{t}-t_j)^{C_j} = \begin{pmatrix} 1 & e_j \ln(\hat{t}-t_j) \\ 0 & 1 \end{pmatrix}$, with the monodromy maps $M_j = \begin{pmatrix} 1 & 2\pi \mathrm{i} e_j \\ 0 & 1 \end{pmatrix}$.
At the point $t = t_1 = 0$, we have $B_0 = \operatorname{diag}(1, -1)$ and the normal form is the following: $\dot{\tilde{y}}_1 = (1/t)\tilde{y}_1 + e_1 t\tilde{y}_2$, $\dot{\tilde{y}}_2 = (-1/t)\tilde{y}_2$.

We must calculate e_1. For this reason we calculate the first terms of the expansion of the matrix $B(t)$ near 0. We have

$$B = \frac{1}{t}\left[(1+\ldots)\begin{pmatrix} 1 & 0 \\ 0 & -1 \end{pmatrix} - (t^2+\ldots)\begin{pmatrix} 0 & 1 \\ -1 & 0 \end{pmatrix}\right].$$

The diagonal terms are reduced to their normal form using diagonal changes of y. The off-diagonal terms are reduced using off-diagonal changes, like $(y_1, y_2) \longmapsto (y_1, y_2 + ct^2 y_2)$ (and higher order changes). It is clear that, during such changes, the first term in the upper-right corner of $B(t)$ remains unchanged. Therefore

$$e_1 = -1.$$

In this case we have $U_1(t) = (1+\ldots)I + O(t^2)$, $V = \operatorname{diag}(1,-1)$, $C_1 = \begin{pmatrix} 0 & -1 \\ 0 & 0 \end{pmatrix}$. Thus

$$\begin{aligned} \mathcal{F} &= \begin{pmatrix} 1+\ldots & O(t^2) \\ O(t^2) & 1+\ldots \end{pmatrix}\begin{pmatrix} t & 0 \\ 0 & t^{-1} \end{pmatrix}\begin{pmatrix} 1 & -\ln\hat{t} \\ 0 & 1 \end{pmatrix} \\ &= \begin{pmatrix} t+\ldots & -t\ln\hat{t}+\ldots \\ O(t^3) & t^{-1}+\ldots \end{pmatrix}. \end{aligned}$$

8. **Definition.** Introduce the following invariant of a 2-dimensional Fuchsian system $\dot{y} = K(t)y$, called the *Fuchsian weight of* K:

$$\gamma_K = \sum_j (v_{1,t_j} - v_{2,t_j}),$$

where $v_{1,t_j} \geq v_{2,t_j}$ are the values of the Levelt's valuations at the singular points t_j. We will also denote them by $v_{i,t_j}(K)$, in order to underline their dependence on K.

Of course, we have $\gamma_B = 2$ for the 2-dimensional subsystem of the Bolibruch system.

Let Mon $=$ Mon (K) be the monodromy group associated with a 2-dimensional Fuchsian system $\dot{y} = Ky$. The *Fuchsian weight of the monodromy group* Mon is the number

$$\gamma_{\mathrm{Mon}} = \min\{\gamma_D : \mathrm{Mon}\,(D) = \mathrm{Mon}\}.$$

9. **Lemma.** *We have* $\gamma_{\mathrm{Mon}(B)} = 2$ *for the matrix* $B(t)$ *from the point* 1.

In 12. below we will show that the Fuchsian weight of certain 2-dimensional Fuchsian system associated with any 3-dimensional Fuchsian system must be equal to zero. Therefore $\gamma_{\mathrm{Mon}(B)}$ is the invariant deciding in the Bolibruch counter-example.

10. *Proof of Lemma* 9. Suppose that there exists a system $\dot{y} = Dy$ with $\gamma_D < 2$ and Mon $(D) =$ Mon (B), where B is the matrix of the 2-dimensional subsystem of the Bolibruch system. So, the monodromy matrices associated with D are unipotent and the corresponding matrices C_j in Eq. (8.20) are nilpotent.

We have $\gamma_D = \sum_j v_{1,t_j}(D) - \sum v_{2,t_j}(D)$. But $\sum v_{1,t_j}(D) = -\sum v_{2,t_j}(D)$, by Lemma 6 applied to the Fuchsian system defined by the matrix D with nilpotent C_j. This means that γ_D is an even number and, because $v_{1,t_j} \geq v_{2,t_j}$, we have

$$\gamma_D = 0, \ v_{1,t_j} = v_{2,t_j}.$$

Denote the latter integers v_{1,t_j} by m_j and apply the change $y = (\prod(t-t_j)^{m_j})u$. If y satisfies the Fuchsian equation defined by the matrix D, then u satisfies the Fuchsian equation

$$\dot{u} = F(t)u, \quad F = D(t) - \sum \frac{m_i}{t-t_i}.$$

Moreover, we have $v_{i,t_j}(F) = 0$.
Let $\mathcal{G}$ be a fundamental matrix associated with the equation $\dot{u} = Fu$. Near the points t_i we have the representations

$$\mathcal{G}(\hat{t}) = W_i(t)(\hat{t} - t_i)^{C_i}, \quad \det W_i \neq 0,$$

with analytic W_i. Let $\mathcal{F}$ be the fundamental matrix for $\dot{y} = By$ with the local representations

$$\mathcal{F} = U_i(\hat{t} - t_i)^{C_i}, \ i = 2,3,4; \quad \mathcal{F} = U_1 t^{V_1} \hat{t}^{C_1},$$

and with

$$U_1 = \begin{pmatrix} 1+\ldots & O(t^2) \\ O(t^2) & 1+\ldots \end{pmatrix}.$$

Because $\mathcal{F}$ and $\mathcal{G}$ have the same monodromy, the function $\mathcal{F}\mathcal{G}^{-1}$ is univalent and then meromorphic (regularity). Near t_i, $i = 2,3,4$, the matrices $\mathcal{F}\mathcal{G}^{-1} = U_i W_i^{-1}$ are analytic and invertible. Near t_1 we have

$$\mathcal{F}\mathcal{G}^{-1} = U_1 \cdot \operatorname{diag}(t, t^{-1}) \cdot W_1^{-1} = \begin{pmatrix} O(t) & O(t) \\ * & * \end{pmatrix}.$$

We see that the first row of the matrix $\mathcal{F}\mathcal{G}^{-1}$ is holomorphic in the Riemann sphere. So, it is constant and, because it vanishes at $t = 0$, it should be identically zero. But then the matrix $\mathcal{F}\mathcal{G}^{-1}$ could not be invertible outside singularities.
This contradiction proves Lemma 9. □

11. Here we study the 3-dimensional Bolibruch system. In the points $t_{2,3,4}$ the residua of the matrix $A(t)$ are of the form

$$A_j = \begin{pmatrix} 0 & a\ b \\ 0 & B_j \end{pmatrix}$$

with nilpotent 2×2 matrices B_j. Thus A_j are nilpotent and the 3-dimensional monodromy maps are unipotent, $M_j = \begin{pmatrix} 1 & * \\ 0 & N_j \end{pmatrix}$, where N_j are the monodromy matrices of 2-dimensional subsystem.

Consider the point t_1. We shall choose a Levelt's basis in the form $\phi_1(t) = (1,0,0)^\top$, $\phi_{2,3}(t) = \binom{x_{2,3}}{\psi_{2,3}}$, where $(\psi_2, \psi_3) = \begin{pmatrix} t + \dots & -t\ln\hat{t} + \dots \\ O(t^3) & t^{-1} + \dots \end{pmatrix}$ is the fundamental system from the end of 7. Of course, ϕ_1 is a solution with the valuation $v(\phi_1) = 0$. Note also that ϕ_1 is an invariant vector for any operator from the monodromy group generated by the Bolibruch's system.

In order to find the first components of $\phi_{2,3}(t)$ we integrate the equation $\dot{x} = a(t)y_2 + b(t)y_3$ with $a(t) = t^{-2} + \dots$, $b(t) = 1 + \dots$ (see the point 1). This gives $x_2(\hat{t}) = \ln\hat{t} + \dots$, $x_3(\hat{t}) = -\ln^2\hat{t}/2 + \dots$. Hence $v(\phi_2) = 0$, $v(\phi_3) = -1$.

The Levelt's filtration of the space of $\mathcal{X}$ of solutions consists of the 2-dimensional subspace $\mathcal{X}_1$ (with valuation 0) and of $\mathcal{X}$ (with valuation -1).

The monodromy matrix M_1 and its 'logarithm' C_1 in the above basis take the forms

$$M_1 = \begin{pmatrix} 1 & 2\pi \mathrm{i} & * \\ 0 & 1 & -2\pi \mathrm{i} \\ 0 & 0 & 1 \end{pmatrix}, \quad C_1 = \begin{pmatrix} 0 & 1 & * \\ 0 & 0 & -1 \\ 0 & 0 & 0 \end{pmatrix}.$$

12. Let $\dot{w} = E(t)w$ be a supposed Fuchsian system with the same singularities and monodromy as the Bolibruch's system $\dot{z} = Az$. This means that there is an isomorphism Θ between fundamental systems of both systems. Thus, if $\chi_j = \Theta\phi_j$, then $\mathcal{E}(t) = (\chi_1, \chi_2, \chi_3)$ is fundamental system for the equation for w. (Here ϕ_j are the solutions from the previous point).

Consider $\chi_1 = \Theta\phi_1$. Because ϕ_1 is the unique vector fixed by any monodromy operator, χ_1 also has this property. This means that $\chi_1(t)$ is a univalent function and therefore rational (regularity).

Moreover, χ_1 is the first vector in the Levelt's basis; because of triangularity of M_j. This implies that the valuations $v_{t_j}(\chi_1)$ at the singular points t_j take the maximal values, which we denote by v_{1,t_j}.

We have the inequality $\sum v_{t_j}(\chi_1) \leq 0$. (For each component the sum of orders is zero, but $v_{t_j}(\chi_1)$ is the minimum of orders of the components).

By Lemma 6

$$\sum_j \sum_i v_{t_j}(\chi_i) = 0,$$

because we have the Fuchsian case. But the above double sum is a sum of three partial sums $s_1 = \sum_j v_{1,t_j}$, $s_2 = \sum_j v_{2,t_j}$, $s_3 = \sum_j v_{3,t_j}$. Because $s_1 \geq s_2 \geq s_3$ all three sums are equal to zero, $s_i = 0$ and, moreover, $v_{1,t_j} = v_{2,t_j} = v_{3,t_j}$, $j = 1,2,3,4$. We denote the latter numbers by n_j.

Let $u = \left(\prod_j (t-t_j)^{n_j}\right) w = f(t)w$. It is easy to see that u satisfies a Fuchsian equation $\dot{u} = Fu$. In the latter system we have all valuations equal to zero.

The vector-function $\eta_1 = f\chi_1$ is holomorphic in the whole Riemann sphere. So, it is a constant vector and, after some transformation, we can assume that $\eta_1 = (1,0,0)^\top$. This means that $\dot{u} = Fu = \begin{pmatrix} 0 & * \\ 0 & G \end{pmatrix} u$. Here G is 2×2 matrix defining a 2-dimensional subsystem.

13. **Lemma.** *We have* $\gamma_G = 0$.

Proof. Recall that γ_G is the Fuchsian weight of the 2-dimensional system defined by $G(t)$ and is equal to $\sum_j [v_{1,t_j}(G) - v_{2,t_j}(G)]$. We proved that $v_{i,t_j}(F) = 0$. Because $v_{1,t_j}(F) = v_{t_j}(u_1)$, then $v_{2,t_j}(F)$ and $v_{3,t_j}(F)$ are the Levelt valuations of the system defined by G. Hence $\gamma_G = 0$. □

14. Lemma 13 provides the contradiction with Lemma 8. This completes the proof of the Bolibruch theorem. □

8.48. Another approach. Yu. Ilyashenko and S. Yakovenko in [**IlY3**] present somewhat alternative and visibly simpler approach to the Bolibruch type examples.

1. **New example.** Consider the matrices, denoted M_1, M_2 and M_3 respectively:

$$\begin{pmatrix} 1 & 1 & 0 & 0 \\ 0 & 1 & 1 & 0 \\ 0 & 0 & 1 & 1 \\ 0 & 0 & 0 & 1 \end{pmatrix}, \quad \begin{pmatrix} 3 & 1 & 1 & -1 \\ -4 & -1 & 1 & 2 \\ 0 & 0 & 3 & 1 \\ 0 & 0 & -4 & -1 \end{pmatrix}, \quad \begin{pmatrix} -1 & 0 & 2 & -1 \\ 4 & -1 & 0 & 1 \\ 0 & 0 & -1 & 0 \\ 0 & 0 & 4 & -1 \end{pmatrix}.$$

It is easy that they satisfy the following properties (in [**IlY3**] is said that this triple is of *class B*):

(i) The Jordan normal form for each of them consists of one 4-dimensional cell with the eigenvalues 1, 1, -1 respectively.

(ii) Each of them has the same invariant plane $\{x_3 = x_4 = 0\}$.

(iii) Their product is the identity, $M_1 M_2 M_3 = I$.

But the most important their property is the following.

Theorem of Ilyashenko–Yakovenko. [**IlY3**, Theorem 16.33] *These matrices cannot be realized as a monodromy data for a Fuchsian system.*

In the sequel points we sketch two proofs of this result.

2. *The analytic proof.* Suppose that such Fuchsian system $\dot{x} = \left(\sum (t-a_j)^{-1} A_j\right) x$, $\sum A_j = 0$, exists. We claim that

Proposition 1. *Each residue matrix A_j has only one eigenvalue λ_j of multiplicity* 4.

This implies that each monodromy matrix M_j has eigenvalue $\nu_j = \mathrm{e}^{2\pi\mathrm{i}\lambda_j}$ of multiplicity 4. Because $0 = \sum \mathrm{tr} A_j = 4 \sum \lambda_j$, we have $\prod \nu_j = 1$. The above matrices do not satisfy this and we have a contradictions. □

Proof of the proposition. Assume that the eigenvalues $\lambda_{i.j}$ of the residuum matrices A_j are ordered, $\lambda_{i,j} - \lambda_{i+1,j} \in \mathbb{Z}_+$. The fundamental matrices of the normalized systems near a_j, like in Remark 8.18, take the form

$$\mathcal{F}_j = (t-a_j)^{\Lambda_j} (t-a_j)^{N_j},$$

where Λ_j is diagonal and N_j is nilpotent upper triangular (thus $(t-a_j)^{N_j}$ is a polynomial in $\ln(t-a_j)$).
Let $\varphi_1(t)$ and $\varphi_2(t)$ be two solutions, which span a plane preserved by the monodromy matrices. Assume that at a base point $t=a$ we have $\varphi_1(a)=e_1$ and $\varphi_2(a)=e_2$. Let $\mathcal{F}(t)$ be the global fundamental matrix such that $\mathcal{F}(a)=I$.
Near each point a_j we have a gauge transformation $y_j=H_j(t)x$ bringing our system to the Levelt type normal form. It turns out that:

$H_j\varphi_1$ and $H_j\varphi_2$ belong to the plane spanned by the first two columns $(\mathcal{V}_j|0)^\intercal$ of the matrix $\mathcal{F}_j$; here $\mathcal{V}_j$ are 2×2 matrices.

This follows from the analysis of the 4×2 matrices $(\mathcal{V}_j|0)^\intercal$; a linear operator equivalent to one Jordan cell has unique invariant subspace of any dimension.
We have $(\varphi_1,\varphi_2)=H_j^{-1}(\mathcal{V}_j|0)^\intercal$. Let $\mathcal{V}(t)$ be the upper-left 2×2 minor of $\mathcal{F}(t)$. We have $\mathcal{V}(t)=G_j(t)\mathcal{V}_j$ with $\det G_j\neq 0$. The monodromy matrices for $\mathcal{V}_j$ are of the form

$$\begin{pmatrix} \nu_j & \alpha_j \\ 0 & \nu_j \end{pmatrix}.$$

Therefore, the matrix $\mathcal{V}(t)$ is a fundamental matrix for a regular system and satisfies $\dot{\mathcal{V}}=B(t)\mathcal{V}$, with meromorphic B in $\mathbb{CP}^1$. Let $W(t)=\det\mathcal{V}(t)$ be the corresponding Wronskian; by the Liouville theorem we have $\frac{\mathrm{d}}{\mathrm{d}t}\ln W=\mathrm{tr}B$.
But, near a_j, we have $W=(t-a_j)^{\lambda_{1,j}+\lambda_{2,j}}\det G_j$ and hence

$$\mathrm{Res}_{a_j}\frac{\mathrm{d}}{\mathrm{d}t}\ln W\geq\lambda_{1,j}+\lambda_{2,j}\geq\frac{1}{2}\mathrm{tr}A_j$$

(by the ordering of eigenvalues). Hence $\sum\mathrm{Res}_{a_j}\mathrm{tr}B\geq\frac{1}{2}\sum\mathrm{tr}A_j=0$. On the other hand, $\sum\mathrm{Res}_{a_j}\mathrm{tr}B=0$.
This implies that all above inequalities are, in fact, equalities and hence all eigenvalues of A_j are equal. □

3. *The geometric proof.* Recall that the Fuchs type equation $\dot{x}=A(t)x$ can be interpreted as the horizontality of a section $t\longmapsto x(t)$ of a vector bundle L over $\mathbb{CP}^1$ with respect to the connection $\triangledown=\mathrm{d}-A\mathrm{d}t$. The residuum of this connection at a point $t=a$ equals $\mathrm{Res}_a\triangledown=\mathrm{Res}_aA$.

Assume a local situation, i.e., a connection on a bundle L of rank n above a neighborhood of $t=0$. As usual, one defines the local monodromy M.
Suppose that the Jordan form of M contains only one cell. Then, for each $0<k<n$, there exists a unique subbundle L_k of rank k, invariant for $\triangledown$, and the restriction $\triangledown_k$ of $\triangledown$ to L_k satisfies

$$\frac{1}{k}\mathrm{tr}\ \mathrm{Res}_0\triangledown_k\geq\frac{1}{n}\mathrm{tr}\ \mathrm{Res}_0\triangledown\,.$$

In the proof one uses the fact that the Levelt type normal form of $A(t)$ is upper-triangular with ordered eigenvalues $\lambda_1\geq\ldots\geq\lambda_n$. We have $L_k=\{x_{k+1}=\ldots=x_n=0\}$ and $\mathrm{Res}_0\triangledown_k$ has eigenvalues $\lambda_1,\ldots,\lambda_k$.
Suppose that a bundle L over $\mathbb{CP}^1$ has a so-called *Bolibruch connection.* Thus:

(a) all singular points are Fuchsian;

(b) all local monodromies have only one Jordan cell;

(c) it has an invariant subbundle $0 \neq K \neq L$.

Proposition 2. *For a Bolibruch connection the spectra of all residua matrices are singletons and the invariant subbundle K is trivial.*

Proof. It is based on the inequalities

$$\frac{1}{k}\deg K = \frac{1}{k}\sum \mathrm{Res}_a \mathrm{tr} \bigtriangledown |_K \geq \frac{1}{n}\sum \mathrm{Res}_a \mathrm{tr} \bigtriangledown .$$

Since L is trivial, then $\deg L = 0$; so, $\deg K \geq 0$.
But any subbundle of a trivial bundle has non-positive degree. For more details see [**IlY3**, Section 18]. □

The further proof runs more or less like in the analytic proof. □

Before ending this subsection we present some results about positive solutions of the Riemann–Hilbert problem (version C). We do not present the proofs.

8.49. Theorem of Lappo-Danilevskij. ([**Lap**]) *If the monodromy operators M_i are sufficiently close to identity, then the answer to the problem C is positive.*

In his proof in [**Lap**] P. Lappo-Danilevskij finds the expressions of the monodromy matrices M_j as series expansions of the residue matrices A_i (for fixed loops γ_i surrounding the singular points). If A_i are small, then the series are convergent and can be reversed, A_j are given by series in $I - M_i$.

8.50. Theorem of Dekkers. ([**Dek**]) *If $n = 2$ (dimension of the system), then the answer to the problem C is positive.*

The idea of the proof is following. If all monodromy operators M_j are non-diagonalizable, then they turn out to be simultaneously transformed to upper-triangular Jordan cells. Such cells commute one with another. Then one can choose the matrices A_j also in such upper-triangular forms.

8.51. Theorem of Kostov and Bolibruch. ([**Kos**], [**AB**]) *If the monodromy group is irreducible, i.e., without proper invariant subspaces, then the answer to problem C is positive.*

§7 Isomonodromic Deformations

As in other theories, also in the theory of Fuchsian systems it is natural to consider families of Fuchsian systems and to investigate the variation of their invariants. One of such invariants is the monodromy group. One can ask which deformations are isomonodromic, i.e., when the monodromy group of a family of Fuchsian systems remains constant. (This notion will be defined.)

The natural parameters of Fuchs systems are the positions of their singular points. When we assume that the matrices-residues at these singular points also depend on these positions, then we obtain a simplest natural family. It can be written in the form

$$\dot{z} = \left(\sum_{i=1}^{m} \frac{A_i(a)}{t - a_i} \right) z, \quad z \in \mathbb{C}^n, \tag{8.21}$$

where $A_i(a) = A_i$ depend on $a = (a_1, \dots, a_m)$. We assume also that the point $t = \infty$ is singular with the residuum $A_\infty = -\sum A_i$.
We shall study this system under the following fundamental assumption.

8.52. Assumption. The local systems near the singular points a_i are non-resonant. This means that the eigenvalues of any of the matrices $A_1(a_k), \dots, A_\infty$ do not satisfy any resonant relation of the type $\lambda_i - \lambda_j \in \mathbb{Z}$.
(By Theorem 8.17 this means that the local systems are analytically equivalent to $\dot{z} = (A_i/(t - a_i))z$).

Therefore, all the matrices A_j are diagonalizable and we can find invertible matrices $G_j = G_j(a)$ (depending on a) such that the matrices

$$A_j^0 = G_j^{-1} A_j G_j$$

are diagonal. Because we consider the case with constant monodromies, then the matrices A_j^0 do not depend on a.
We choose the fundamental matrix $\mathcal{F}(\hat{t})$ such that

$$\mathcal{F} = \Phi_\infty(t)\hat{t}^{-A_\infty^0}, \quad \Phi_\infty(\infty) = I,$$

in a neighborhood of $t = \infty$ (here Φ_∞ is holomorphic and invertible near infinity). It is clear that this condition defines the fundamental matrix uniquely.
By the theory developed in the previous sections, the behavior of $\mathcal{F}$ near other singular points is following:

$$\mathcal{F} = G_j \Phi_j(t)(\hat{t} - a_j)^{A_j^0} H_j, \quad \Phi_j(a_j) = I,$$

where $\Phi_j(t)$ are holomorphic and invertible and $H_j = H_j(a)$ are constant invertible matrices. (One can check that $A(t,a) = \dot{\mathcal{F}}\mathcal{F}^{-1} = G_j A_j^0 G^{-1}/(t - a_j) + \dots$).
The monodromy matrices are now well defined and are of the form

$$M_j = H_j^{-1} \mathrm{e}^{2\pi \mathrm{i} A_j^0} H_j.$$

8.53. Definition. Deformation (8.21) is called **isomonodromic** iff the matrices M_j do not depend on a. Equivalently: iff the matrices H_j can be chosen not depending on a.

Remark. We have here two kinds of data associated with Eq. (8.21): the *singularity data*

$$\{a_j,\ A_j^0,\ G_j(a)\}$$

and the *monodromy data*

$$\{a_j,\ A_j^0,\ H_j(a),\quad j=1,\dots,\infty\},$$

where $G_\infty = I,\ H_\infty = I$, and the relations $A_1 + \ldots + A_\infty = 0,\ M_1 \ldots M_\infty = I$, hold.
The following theorem of Schlesinger [**Sch2**] answers the question of conditions needed to ensure the isomonodromicity of deformation (8.21).

8.54. Theorem of Schlesinger (Isomonodromic deformations). *Let the assumption* 8.52 *hold. The deformation* (8.21) *is isomonodromic iff one of the two equivalent conditions holds:*

(a) *The fundamental matrix* $\mathcal{F}(t,a)$ *constructed above satisfies the system of equations*
$$\frac{\partial \mathcal{F}}{\partial a_j} = \frac{-A_j}{t-a_j}\mathcal{F}, \quad j=1,\dots,m.$$

(b) *The matrices* A_j *satisfy the system of equations, called the* ***Schlesinger equations***,
$$\frac{\partial A_i}{\partial a_j} = \frac{[A_j, A_i]}{a_j - a_i}, \quad i \neq j,$$
$$\frac{\partial A_j}{\partial a_j} = -\sum_{i\neq j} \frac{[A_j, A_i]}{a_j - a_i}.$$

8.55. Remark (Connections and curvatures). Before proving this theorem, we must comment its formulation. Condition (a) is a condition formulated in terms of the fundamental matrix, whereas condition (b) concerns only the matrices A_j. The explanation of their equivalence goes through connections and their curvatures. Introduce the operators of external derivative

$$\mathrm{d}' = \sum \mathrm{d}a_j \frac{\partial}{\partial a_j}, \quad \mathrm{d} = \mathrm{d}t\frac{\partial}{\partial t} + \mathrm{d}',$$

and the 1-forms with values in the space of matrices

$$\Omega' = -\sum \frac{A_j(a)}{t-a_j}\mathrm{d}a_j, \quad \Omega = \sum A_j \mathrm{d}\ln(t-a_j).$$

Then the condition (a) of the Schlesinger theorem says that $\mathrm{d}'\mathcal{F} = \Omega'\mathcal{F}$. This equation together with the equation $\dot{\mathcal{F}} = A(t,a)\mathcal{F}$ can be written in the form

$$(\mathrm{d} - \Omega)\mathcal{F} = 0.$$

Consider the trivial (principal) fibre bundle $E : U \times gl(n,\mathbb{C}) \longmapsto U$, where $U \subset \mathbb{CP}^1 \times \mathbb{C}^m$ is an open set with variables t and a_j. Introduce the operator $D = \mathrm{d} - \Omega : \mathcal{O}_U(E) \longmapsto \mathcal{O}_U(E) \otimes \Omega_U^1 = \Omega_U^1(E)$ on the sheaf of germs of section of E. It defines a connection in E. The equation $D\mathcal{F} = 0$ means that $\mathcal{F}$ is a horizontal section of E with respect to this connection.
Not every connection admits horizontal sections. (Usually we can only prolong sections along curves in the base, but prolongation along a closed curve may lead to a point (in the initial fiber) which is different than the initial point). The obstruction to existence of horizontal sections (above open subsets of the base) is the curvature defined as follows.
We can extend the connection operator D to an operator on $\Omega_U^1(E)$, with values in $\Omega_U^2(E)$. Then we have $D \circ D = \mathrm{d} \circ \mathrm{d} - \mathrm{d} \circ \Omega - \Omega \circ \mathrm{d} + \Omega \wedge \Omega = -(\mathrm{d}\Omega - \Omega \wedge \Omega)$ (because $\mathrm{d} \circ \Omega = \Omega \circ \mathrm{d} - (\mathrm{d}\Omega) \wedge (\cdot)$). The 2-form (with values in $gl(n,\mathbb{C})$)

$$R = \mathrm{d}\Omega - \Omega \wedge \Omega$$

is the *curvature tensor* of the connection.

Lemma. *The Schlesinger equations are equivalent to the condition* $R = 0$.

Proof. We should show that:

(i) if the $\mathrm{d}t \wedge \mathrm{d}a_i$-components of R vanish, then the Schlesinger equations hold;

(ii) if the Schlesinger equations hold, then $R = 0$.

The $\mathrm{d}t \wedge \mathrm{d}a_i$-component of $\mathrm{d}\Omega$ is equal to $-\sum_j \frac{\partial A_j/\partial a_i}{t-a_j}$. The analogous component of $\Omega \wedge \Omega$ is equal to $\left(\sum \frac{A_j}{t-a_j}\right) \cdot \left(\frac{-A_i}{t-a_i}\right) + \left(\frac{A_i}{t-a_i}\right) \cdot \left(\sum \frac{A_j}{t-a_j}\right)$; its residuum at the point $t = a_j$ is equal to $[A_j, A_i]/(a_j - a_i)$ for $j \neq i$ and $-\sum_{k\neq i}[A_k, A_i]/(a_k - a_i)$ for $j = i$. This gives the Schlesinger equations.
We have just proved that the Schlesinger equations are equivalent to vanishing of the $\mathrm{d}t \wedge \mathrm{d}a_j$-components of R. So, it remains to show that in this case the $\mathrm{d}a_i \wedge \mathrm{d}a_j$-components of R vanish too.
The $\mathrm{d}a_i \wedge \mathrm{d}a_j$-component of $\mathrm{d}\Omega$ is equal to $(\partial A_i/\partial a_j)/(t-a_i) - (\partial A_j/\partial a_i)/(t-a_j)$ which, in view of the Schlesinger equations, is equal to $[A_i, A_j]/(t - a_i)(t - a_j)$. The same is the $\mathrm{d}a_i \wedge \mathrm{d}a_j$-component of $\Omega \wedge \Omega$. □

Proof of Theorem 8.54. *Equivalence of the conditions* (a) *and* (b):
If (a) holds, then the fibration E has a section $\mathcal{F}$ horizontal with respect to the connection D. So, the curvature of the connection vanishes and the Schlesinger equations hold.
If (b) holds, then the curvature of the connection $\mathrm{d} - \Omega$ vanishes and the distribution defined by it is integrable. This means that there is a horizontal section $\mathcal{F}$ of the connection.

Isomonodromicity $\Rightarrow$ *the condition* (a): We have to show that the matrix function $\mathrm{d}'\mathcal{F}\mathcal{F}^{-1}$ is equal to the meromorphic form Ω'.

Firstly, from the isomonodromicity, it follows that the operation of prolongation of $\mathcal{F}(\hat{t}, a)$ along a closed loop (in the t-plane) commutes with the operation of external derivation d' (with respect to a). Thus the prolongation along loops around a_i leads to the transformations $\mathcal{F} \longmapsto \mathcal{F}M_i$, $\mathrm{d}'\mathcal{F} \longmapsto \mathrm{d}'\mathcal{F}M_i$. This implies that the form $\mathrm{d}'\mathcal{F}\mathcal{F}^{-1}$ is univalent and hence meromorphic.
In order to find the formula for Ω', we investigate its behavior near the singular points a_i and ∞. We shall look only for the polar parts of Ω' at these points. Using the asymptotic formulas for $\mathcal{F}$ (before Definition 8.53), it is easy to check that, near a_i: $\frac{\partial}{\partial a_j}\mathcal{F}\mathcal{F}^{-1} = -A_j/(t-a_i) + \ldots$, if $j = i$, and that $\frac{\partial}{\partial a_j}\mathcal{F}\mathcal{F}^{-1}$ is analytic, if $j \neq i$. Moreover, $\mathrm{d}'\mathcal{F}\mathcal{F}^{-1} = O(1/t)$, as $t \to \infty$. This implies that Ω' has only simple poles with definite residues and the formula $\Omega' = -\sum \mathrm{d}a_i A_i/(t-a_i)$ follows.

Univalency of $\Omega \Rightarrow$ *isomonodromicity*: We have to show that $\mathrm{d}'M_j = 0$.
Univalency of Ω implies $\mathcal{F} \longmapsto \mathcal{F}M_j$, $\mathrm{d}'\mathcal{F} = \Omega'\mathcal{F} \longmapsto \Omega'\mathcal{F} \cdot M_j = \mathrm{d}'\mathcal{F} \cdot M_j$. But $\mathrm{d}'\mathcal{F}$ is transformed to $\mathrm{d}'(\mathcal{F}M_j) = \mathrm{d}'\mathcal{F} \cdot M_j + \mathcal{F} \cdot \mathrm{d}'M_j$. So, $\mathrm{d}'M_j = 0$. □

The problem of isomonodromic deformations is trivial when the number of singular points is two or three. In both cases we can move these singular points to fixed ones, using a Möbius transformation. So, we can assume that these points are either 0 and ∞, or $0, 1$ and ∞, and there are no parameters of deformation. Also the case, when the dimension n of the Fuchsian system is one, is elementary.
The first nontrivial case is when $m = 3$, $n = 2$.

8.56. Isomonodromic deformation in the case $m = 3$, $n = 2$ **and Painlevé 6.** It turns out that this problem leads to the **equation of Painlevé 6**

$$y'' = \frac{1}{2}\left(\frac{1}{y} + \frac{1}{y-1} + \frac{1}{y-x}\right)(y')^2 - \left(\frac{1}{x} + \frac{1}{x-1} + \frac{1}{y-x}\right)y' \\ + \frac{y(y-1)(y-x)}{x^2(x-1)^2}\left[\alpha + \beta\frac{x}{y^2} + \gamma\frac{x-1}{(y-1)^2} + \delta\frac{x(x-1)}{(y-x)^2}\right].$$

Let us pass to the formulation of the problem. We have the system

$$\dot{z} = \left[\frac{A_0(x)}{t} + \frac{A_1(x)}{t-1} + \frac{A_2(x)}{t-x}\right] z$$

and we must write down the equations for $A_i(x)$ which imply the isomonodromicity.
Firstly, using the change $z \longmapsto w = \left(\prod (t-a_i)^{\mu_i}\right) z$, we can reduce the problem to the analogous problem with traceless matrices A_i.
Assume that the numbers

$$\pm\theta_0/2,\ \pm\theta_1/2,\ \pm\theta_2/2,\ \pm\theta_\infty/2$$

are the eigenvalues of A_0, A_1, A_2, A_∞, where $A_0 + A_1 + A_2 + A_\infty = 0$ and A_∞ is diagonal. The numbers θ_i are constant parameters of the problem.

So, we have to determine the matrices A_0 and A_1. We choose them in the form

$$A_0 = \frac{1}{2}\begin{pmatrix} z_0(x) & u_0(x) \\ v_0(x) & -z_0(x) \end{pmatrix}, \quad A_1 = \frac{1}{2}\begin{pmatrix} z_1(x) & u_1(x) \\ v_1(x) & -z_1(x) \end{pmatrix}.$$

Thus

$$A_2 = \frac{1}{2}\begin{pmatrix} -(\theta_\infty + z_0 + z_1) & -u_0 - u_1 \\ -v_0 - v_1 & \theta_\infty + z_0 + z_1 \end{pmatrix}.$$

The six functions $z_{0,1}(x), u_{0,1}(x), v_{0,1}(x)$ are dependent. The conditions for eigenvalues imply that

$$z_0^2 + u_0 v_0 = \theta_0^2, \quad z_1^2 + u_1 v_1 = \theta_1^2, \tag{8.22}$$

$$(\theta_\infty + z_0 + z_1)^2 + (u_0 + u_1)(v_0 + v_1) = \theta_2^2. \tag{8.23}$$

Therefore there are only three independent functions. We choose them as

$$z_1(x),\ z_2(x), \quad y = y(x) = \frac{x u_0}{x u_0 + (x-1) u_1}.$$

We put also $u = u_0/u_1$. We have

$$u = -\frac{y \cdot (x-1)}{x \cdot (y-1)}. \tag{8.24}$$

Let us pass to differentiations, where the $'$ will denote the differentiation with respect to x. The Schlesinger equations are $A_0' = [A_2, A_0]/x,\ A_1' = [A_2, A_1]/(x-1)$. The calculation of the upper-left and upper-right entries gives the four equations:

$$\begin{aligned} &z_0' = \Phi/2x, \quad z_1' = -\Phi/2(x-1), \\ &\Phi = u_0 v_1 - u_1 v_0 = u(\theta_1^2 - z_1^2) - u^{-1}(\theta_0^2 - z_0^2), \end{aligned} \tag{8.25}$$

and

$$\begin{aligned} &u_0' = -(\theta_\infty/x) u_0 + \Psi/x, \quad u_1' = -(\theta_\infty/(x-1)) u_1 - \Psi/(x-1), \\ &\Psi = u_1(z_0 - u z_1). \end{aligned} \tag{8.26}$$

Above we have expressed v_i as $(\theta_i^2 - z_i^2)/u_i$ from Eq. (8.22).

We use Eq. (8.26) to calculate the derivative of $u = u_0/u_1$ which, after applying Eq. (8.24), gives

$$u' = \frac{-1}{x(y-1)}\left[\theta_\infty \frac{y}{x} + (z_0 - u z_1)\right]. \tag{8.27}$$

The calculation of the same derivative from Eq. (8.24) gives

$$u' = \frac{1}{x(y-1)}\left[\frac{x-1}{y-1} y' - \frac{y}{x}\right].$$

This, together with (8.27), gives

$$z_0 - uz_1 = (1-\theta_\infty)\frac{y}{x} - \frac{x-1}{y-1}y'. \tag{8.28}$$

Eq. (8.23) can be rewritten in the form

$$2\theta_\infty(z_0+z_1) = -\theta_\infty^2 - (1+u^{-1})\theta_0^2 - (1+u)\theta_1^2 + \theta_2^2 + u^{-1}(z_0-uz_1)^2. \tag{8.29}$$

This and Eq. (8.28) allow to express z_0 and z_1 in terms of u, θ_i and $(z_0-uz_1)^2$. Namely, we find

$$\begin{aligned}\theta_\infty z_1 = \theta_\infty \frac{x(x-1)}{y-x}y' - \left[\theta_\infty^2 - 2\theta_\infty(1-\theta_\infty)\frac{y}{x}\right]\frac{x(y-1)}{2(y-x)} \\ -\frac{\theta_0^2}{2u} - \frac{\theta_1^2}{2} + \frac{\theta_2^2 x(y-1)}{2(y-x)} - \frac{x^2(y-1)^2}{y(x-1)(y-x)}(z_0-uz_1)^2.\end{aligned} \tag{8.30}$$

Of course, $z_0 - uz_1$ can be expressed by the right-hand side of Eq. (8.28), but we do not do it now.

Finally, we calculate y'' from Eq. (8.28). This gives

$$\begin{aligned}y'' = \frac{(y')^2}{y-1} + \left[(1-\theta_\infty)\frac{y-x}{x(x-1)} - \frac{1}{x-1}\right]y' \\ -(1-\theta_\infty)\frac{y(y-1)}{x^2(x-1)} - \frac{y-1}{x-1}(z_0-uz_1)'.\end{aligned} \tag{8.31}$$

But, from Eq. (8.25), we find that

$$(z_0-uz_1)' = \frac{1}{x(y-1)}\left[\frac{\theta_0^2}{2u} - \frac{\theta_1^2 u}{2} - \frac{(z_0-uz_1)^2}{2u} - \frac{y}{x}\theta_\infty z_1\right]. \tag{8.32}$$

Substituting (8.30) to (8.32), next (8.32) to (8.31) and then using Eq. (8.28) (for $(z_0-uz_1)^2$), after exhaustive calculations, one obtains the equation Painlevé 6. (If the reader wants to repeat the computations himself, then we have one advise for him: calculate separately the expressions before terms like $(y')^2$, y', θ_2^2, etc.) The coefficients of the Painlevé 6 take the forms

$$\alpha = (1-\theta_\infty)^2/2, \quad \beta = -\theta_0^2/2, \quad \gamma = \theta_1^2/2, \quad \delta = (1-\theta_2^2)/2.$$

Having a solution $y(x)$ of the Painlevé 6, we can find $z_i = z_i(x)$ and $u(x) = u_0/u_1$. The differential equation for u_0 gives $u_0'/u_0 = (-\theta_\infty + u^{-1}z_0 - z_1)/x$, which allows to determine $u_0(x)$ (and than also $u_1(x)$). The functions $v_{0,1}(x)$ are calculated from Eq. (8.22).

We have proved the following result.

8.57. Theorem. *The problem of solution of the Schlesinger equations for isomonodromic deformation of the 2-dimensional Fuchsian system* $\dot{z} = A(t)z$, $A = \sum_{i=0}^{3} \frac{A_i(a)}{t-a_i}$, *is equivalent to the problem of integration of the equation of Painlevé 6 and usual quadratures.*

Remark. The above theorem is taken from the lecture notes of G. Mahoux [**Mah**]. However the first author who associated the Painlevé 6 with isomonodromic deformations was R. Fuchs [**Fuch**]. He considered the second order differential equations of the Fuchs type with four singular points $t = 0,\ 1,\ x,\ \infty$, where the coefficients are functions of t, x (rational in t). The condition of isomonodromicity also leads to the Painlevé 6.

8.58. Movable and non-movable critical points. The origins of the Painlevé equations. Let

$$F(x, y, y', \ldots, y^{(n)}) = 0$$

be an analytic differential equation defined by means of a polynomial F. Here the argument x and the function y take complex values. y can take values in the complex projective plane $\mathbb{CP}^1$.

If $\phi(x)$ is a solution of this equation, then it is usually a multivalued holomorphic function with some branching points, poles, essential singularities, etc. A point x_0 is called **critical** iff it is neither regular nor a pole of the solution ϕ; after turning the argument around x_0 we arrive at another branch of ϕ. The critical point is called **movable** iff its position changes with the change of the solution. Otherwise the critical point is called **non-movable**.

Examples. (a) The equation $y' = y^2$ has the solutions $y(x) = C/(1 - Cx)$. The solutions have poles which are not critical points in the sense of the above definition.

(b) The equation $2y' = y^3$ has the solutions $y(x) = (c - x)^{-1/2}$. The points $x = c$ are movable critical points.

(c) The linear equation $a(x)y' = b(x)y$ has only non-movable critical points.

The situation with equation of first order $F(x, y, y') = 0$, e.g., $y' = P(x, y)/Q(x, y)$, is more or less clear (see [**Gol**] and [**AI**]). The only critical points of its solutions are algebraic branching points, like $y = (x - x_0)^{p/q} + \ldots$ (Painlevé). The equation $y' = P/Q$ has no movable critical points iff it is the **Riccati equation**

$$y' = a_0(x)y^2 + a_1(x)y + a_2(x).$$

(The proof relies upon the fact that any holomorphic vector field on $\mathbb{CP}^1$ is quadratic, the tangent bundle is equal to $\mathcal{O}(2)$).

P. Painlevé and B. Gambier classified the equations

$$y'' = R(x, y, y')$$

with rational R without movable critical points. They treated two such equations equivalent iff one is obtained from the other by means of changes of the variable x and Möbius type transformations in each fiber $x = \text{const}$.
Firstly Painlevé found 14 such equations (omitting Painlevé 6) and later Gambier finished this classification. There is a list of 50 equations (see [**Gol**]). 44 of them are either integrable in quadratures or are reduced to equations of the type $F(x, y, y') = 0$. The remaining six are the **Painlevé equations**. They are the following:

$$\begin{aligned}
P_1 &: \ y'' = 6y^2 + x, \\
P_2 &: \ y'' = 2y^3 + xy + \alpha, \\
P_3 &: \ y'' = (y')^2/y - y'/x + \left(\alpha y^2 + \beta\right)/x + \gamma y^3 + \delta/y, \\
P_4 &: \ y'' = (y')^2/2y + \frac{3}{2}y^3 + 4xy^2 - 2(x^2 - \alpha)y + \beta/y, \\
P_5 &: \ y'' = \left(\frac{1}{2y} + \frac{1}{y-1}\right)(y')^2 - \frac{y'}{t} + \frac{(y-1)^2}{t^2}\left(\alpha y + \frac{\beta}{y}\right) + \gamma\frac{y}{t} + \delta\frac{y(y+1)}{y-1}
\end{aligned}$$

and P_6.
The solutions of the equations of Painlevé are called the *Painlevé transcendents.* They form new class of special functions, different form elliptic functions, hypergeometric functions, Bessel functions, etc. Their investigation is not finished (see [**GL**]). They appear in many branches of mathematics and physics. For example, if $y(x)$ is the second transcendent of Painlevé, then the function $u(t, x) = t^{-2/3}[y'(t)+y^2(t)]$ is a solution of the *Korteweg–de Vries equation* $u_t = 6uu_x - u_{xxx}$.
Recently, K. Okamoto [**Oka**] proved that with the Painlevé equations an interesting group action can be associated. For example, on the 4-dimensional space of Painlevé 6 (parametrized by $\alpha, \beta, \gamma, \delta$) the affine Weyl group with the root system $\mathbf{D}_4$ acts (see 4.28 above and 10.31(c) below). This is infinite group generated by reflections with respect to mirrors. If the parameter belongs to such mirror, then the corresponding Painlevé 6 is expressed by means of hypergeometric functions. Very nice explanation of these symmetries was given recently by Yu. Manin in [**Man3**], where he used Fuchs' formulas (from [**Fuch**]), which connect the Painlevé 6 with elliptic integrals and he related the Painlevé 6 with the Gromov–Witten invariants. Manin has rewritten the Painlevé 6 in a form which is much easier to remember:

$$\frac{\mathrm{d}^2 z}{\mathrm{d}\tau^2} = \left(\frac{1}{2\pi \mathrm{i}}\right)^2 \sum_{j=0}^{3} \alpha_j \mathcal{P}'_z(z + T_j/2, \tau),$$

where $\mathcal{P}(z, \tau)$ is the Weierstrass function associated with the elliptic curve $\mathbb{C}/(\mathbb{Z} + \tau\mathbb{Z})$ (see 8.10.(viii)), $T_0 = 0$, $T_1 = 1$, $T_2 = \tau$ and $T_3 = 1 + \tau$ are periods and the parameters α_j are defined by $(\alpha_0, \ldots, \alpha_3) = (\alpha, -\beta, \gamma, \frac{1}{2} - \delta)$.
In an impressive paper [**DuMa**] B. Dubrovin and M. Mazzocco found algebraic solutions to P_6; We describe it in 12.20.

8.59. Other Painlevé equations and isomonodromic deformations. The natural question is: why we obtain just Painlevé 6 as an equation for isomonodromic deformation in the simplest nontrivial case? Do the equations P_1-P_5 play some role in the theory of isomonodromic deformations?
It turns out that these questions have positive answers. They imply isomonodromicity of deformations with irregular singular points. We explain shortly what does it mean. We begin with the Painlevé 5.

Lemma. *The limit transition*

$$(t, y, \alpha, \beta, \gamma, \delta) \longmapsto (1 + \epsilon t, y, \alpha, \beta, \gamma/\epsilon - \delta/\epsilon^2, \delta/\epsilon^2),$$

as $\epsilon \to 0$ *in the equation* P_6*, gives the equation* P_5*.*

Remark. It turns out that there are more limit relations between the Painlevé equations; P_4 is a limit of P_5, P_3 is a limit of P_4, etc.

When one applies the limit procedure from the above lemma to the Fuchsian system from 8.54, then he gets the system

$$\dot{z} = \left(\frac{A_0}{t} - \frac{A_0 + A_\infty}{t-1} - \frac{xA_1}{(t-1)^2} \right) z.$$

The points $t = 1$ and $t = x$ tend to $t = 1$ as $\epsilon \to 0$ and the point $t = 1$ becomes non-Fuchsian. It is irregular, because the matrix xA_1 standing before $(t-1)^{-2}$ has generally non-resonant eigenvalues. Then the system can be locally formally transformed to a diagonal system (or to two independent equations) with second order pole at $t = 1$. (Recall that resonances correspond to $\lambda_i = \lambda_j$).
Therefore one should extend the Schlesinger theory of isomonodromic deformations to the case of systems with irregular singularities. Such theory was initiated by H. Flaschka and A. Newell in [**FN**], where they obtained the equation Painlevé 2.
The general theory was constructed by M. Jimbo, T. Miwa and K. Ueno [**JMU**] (see also [**Mah**]). There the role of parameters of deformation play the poles a_j and certain coefficients in the formal normal forms near irregular singularities. If this normal form is equal to

$$\dot{\tilde{z}} = \sum \left(\frac{T_{j,r_j}}{(t-a_j)^r} + \ldots + \frac{T_{j,2}}{(t-a_j)^2} + \frac{A_j^0}{t-a_j} \right) \tilde{z},$$

with diagonal $T_{j,i}$, A_j^0, then the additional parameters are the diagonal elements of the matrices $T_{j,i}$.
Into the set of monodromy matrices M_j, associated with a fixed fundamental matrix, one adds the set of Stokes operators $S_{j,i}$, corresponding to a suitable coverings of the punctured neighborhoods of the irregular singularities a_j by sectors

(see Definition 8.26). Like the monodromy operators, the operators $S_{j,i}$ are associated with a fixed basis of the space of solutions (fundamental matrix); we do not give the precise definitions.

The deformation is called isomonodromic iff M_j = const and $S_{j,i}$ = const as functions of a_j and $T_{j,i}$. The analogue of the Schlesinger theorem, proved by Jimbo, Miwa and Ueno, states that the deformation is isomonodromic iff the fundamental matrix $\mathcal{F}(t,a,T)$ satisfies certain system of linear partial differential equations (like in the condition (a) of Theorem 8.54). Also analogues of the Schlesinger equations hold (see [**Mah**]).

All the equations of Painlevé play roles of isomonodromic deformation equations for suitable linear meromorphic systems with irregular singularities.

As an example, we present the system leading to the Painlevé 1 $y'' = 6y^2 + x$:

$$\dot{z} = \left[\begin{pmatrix} -v & y^2 + x/2 \\ -4y & v \end{pmatrix} + t\begin{pmatrix} 0 & y \\ 4 & 0 \end{pmatrix} + t^2\begin{pmatrix} 0 & 1 \\ 0 & 0 \end{pmatrix}\right] z.$$

It turns out that the property of absence of movable critical points in the equations of isomonodromicity of deformations, which we observed in the above examples, is a general rule.

8.60. Theorem of Malgrange and Miwa. ([**Mal4**], [**Miw**]) *Solutions of the equations for isomonodromic deformation (in the Fuchsian and irregular cases) do not have movable critical points.*

§8 Relation with Quantum Field Theory

There exists an interesting connection between monodromy theory of Fuchsian linear meromorphic differential systems and holonomic quantum fields. This theory was developed by M. Sato, M. Jimbo and T. Miwa in the series of papers [**SMJ1**] and [**SMJ2**].

The theory of holonomic quantum fields is rather technical and does not concentrate only on applications to differential equations. Its constructions deal with fermionic fields with values in Clifford algebra and has applications in such areas like the 2-dimensional Ising model.

Unfortunately, the constructions are highly technical and it is difficult to follow the proofs. In particular, the authors give explicit formulas for the solution of the Riemann–Hilbert problem, but without stating explicit assumptions. More precisely, they give a formula for the (regular) fundamental matrix $\mathcal{F}(\hat{t})$; this gives a solution of the problem B. They claim that they obtain a solution of the problem C by making references to the Schlesinger equations. Probably the assumption, that the monodromy matrices M_j are close to identity, is needed.

Below we present only the general ideas of the construction of the mentioned solution without pretending to full explanation and without proofs. We begin with some basic notions of the quantum field theory.

8.61. The origins of the quantum field theory. Quantum field theory was created, because the quantum mechanics has turned out not compatible with the relativity theory. In quantum mechanics one can determine the position of a particle by means of the square of the absolute value of the wave function. It is the density of the probability of finding the particle in a fragment of the configuration space. In this approach one loses the information about the momentum of the particle. It turned out that, when the velocities are limited (as in the case of relativistic particles), one cannot determine even the position of the particle. The wave function of an individual particle does not have sense. The only outcome of this trap is to allow particles to appear and disappear in reactions with other particles. This leads to a construction called the *second quantization.*

8.63. Definition (bosonic and fermionic Fock spaces and operators of creation and annihilation). Let H be a Hilbert space. Define the spaces

$$F_s(H) = \mathbb{C} \oplus H \oplus P_s(H \otimes H) \oplus \dots$$
$$F_a(H) = \mathbb{C} \oplus H \oplus P_a(H \otimes H) \oplus \dots,$$

where P_s and P_a are the symmetrization and anti-symmetrization operators acting on $H^{\otimes n} = H \otimes \dots \otimes H$ (n times). For example, $P_a(f_1 \otimes \dots \otimes f_n) = \frac{1}{n!}\sum_\sigma \epsilon(\sigma) f_{\sigma(1)} \otimes \dots \otimes f_{\sigma(n)}$, where the summation runs over permutations of the set of indices and $\epsilon(\sigma) = \pm 1$ is the sign of the permutation. (Above $F_\#$, $\# = s, a$, are understood as closures of the infinite direct sums).
$F_s(H)$ is called the **Bose–Einstein Fock space** and $F_a(H)$ is the **Fermi–Dirac Fock space**. We denote by $F_n = F_{\#,n}$ the nth component of $F_\#$.
If $f \in H$, then we define the **creation operator** $a^*(f)$ as

$$a^*(f)g = \sqrt{n+1} P_\#(f \otimes g), \ g \in F_{\#,n}.$$

The **annihilation operator** is equal to $a(f) = (a^*(f))^*$; it can be written in the form

$$a(f) g_1 \otimes \dots \otimes g_n = \sqrt{n}(f, g_n) g_1 \otimes \dots \otimes g_{n-1}.$$

The generator of the component $\mathbb{C}$ of unit norm is called the **vacuum vector** and is denoted by Ω.

8.64. Example. If $H = L_2(\mathbb{R}^n)$ is the Hilbert phase space of a quantum particle, then $F_{\#,n}$ consist of functions $f(x_1, \dots, x_n)$ (on $\mathbb{R}^{mn}$) which are symmetric (or anti-symmetric) with respect to permutations of arguments. The vector Ω represents a function on the 0-dimensional space $(\mathbb{R}^n)^0$ (a point). The function $f_1(x_1) \dots f_n(x_n)$, $||f_j|| = 1$, can be interpreted as a wave function of n independent particles, each in the state defined by f_i. The creation and annihilation operators can be interpreted as generalized functions (distributions) with values in the space of operators on $F_\#$:

$$a(f) = \int a(x) f(x) \mathrm{d}x, \quad a^*(f) = \int a^*(x) f(x) \mathrm{d}x,$$

where f is a test function (smooth with fast decaying at infinity). The latter operators are called the *field operators*.

8.65. Proposition.

(a) *In the Bose–Einstein case the following commutation relations hold:*

$$[a(f), a(g)] = [a^*(f), a^*(g)] = 0, \quad [a(f), a^*(g)] = (f, g)I.$$

(b) *In the Fermi–Dirac case the following anticommutation relations hold:*

$$\{a(f), a(g)\} = \{a^*(f), a^*(g)\} = 0, \quad \{a(f), a^*(g)\} = (f, g)I,$$

where $\{A, B\} = AB + BA$ *is the anticommutator.*

(c) *In both cases* $a(f)\Omega = 0$.

From Proposition 8.65 it follows that the vectors of the form $a^*(f_n)\dots a^*(f_1)\Omega$ form a dense subset of $F_{\#}$.
In the Bose–Einstein case there can be many particles in one state, e.g., the state $a^*(f)a^*(f)\dots a^*(f)\Omega$ corresponds to the situation when several particles are in the same state defined by the wave function f.
In the Fermi–Dirac case this is impossible because $a(f)a(f) = 0$ and, if a vector $\theta \in F_a$ contains a particle in the state f, then $a(f)\theta$ does not contain such particles. Thus θ can contain at most one particle in the state f. This property constitutes the known *Pauli exclusion principle.*
In terms of the field operators $a(x)$, $a^*(x)$ the commutation and anti-commutation relations mean that

$$[a(x), a(y)] = [a^*(x), a^*(y)] = 0, \quad [a(x), a^*(y)] = \delta(x - y)$$

and

$$\{a(x), a(y)\} = \{a^*(x), a^*(y)\} = 0, \quad \{a(x), a^*(y)\} = \delta(x - y).$$

Usually in theory of elementary particles the particles have spin and their wave functions are vector-valued. For example, the field operators associated with electrons have four components (spinors). Associated with them are the field operators for the positrons (also four components).
The authors in [**SMJ1**] consider the field operators taking values in Clifford algebra.

8.66. Definition (Clifford algebra, Clifford group and normal ordering). Let W be a complex vector space equipped with a scalar product $\langle \cdot, \cdot \rangle$. The algebra $A(W)$, generated by elements of W with the relations

$$ww' + w'w = \langle w, w' \rangle \in \mathbb{C}, \tag{8.33}$$

for $w, w' \in W$, is called the **Clifford algebra**.

The group

$$G(W) = \{g \in A(W) : \exists g^{-1} \text{ and } gwg^{-1} \in W \text{ for } w \in W\}$$

is called the **Clifford group**. If $w_1, \dots, w_N$ is a fixed basis of W, then we have

$$gw_ig^{-1} = \sum_j t_{ij}w_j,$$

where $T = (t_{ij})$ is an orthogonal matrix. This mapping induces the exact sequence of Lie groups $1 \longmapsto \mathbb{S}^1 \longmapsto G(W) \longmapsto O(W) \longmapsto 1$, where $O(W)$ is the orthogonal group.
In [**SMJ1**] the authors consider the situation when the space W is expanded into the direct sum $V^* \oplus V$ such that

$$\langle v_1^*, v_2^* \rangle = \langle v_1, v_2 \rangle = 0, \;\; v_{1,2}^* \in V^*, \; v_{1,2} \in V.$$

In analogy with the field operators, the elements of V^* can be called the creation operators and the elements of V are the annihilation operators. $1 \in A(w)$ can be treated as the vacuum. The subspaces V^*, V are the maximal subspaces isotropic with respect to the scalar product. Such subspaces are sometimes called *holonomic*. Here lies the origin of the notion 'holonomic quantum fields'; these are the fields taking values in a Clifford algebra with holonomic decomposition.
The Clifford algebra, treated as a vector space, is isomorphic to $\bigwedge W$, the exterior algebra over W. The latter consists of vectors $v_1^* \wedge \ldots \wedge v_r^* \wedge v_1 \wedge \ldots \wedge v_s$ and has the gradation $\bigwedge W = \mathbb{C} \oplus W \oplus \bigwedge^2 W \ldots$. The isomorphism between the two algebras, which is called the *norm mapping* and is denoted by Nr, is constructed as follows. In any monomial we move the creation operators to the left and the annihilation operators to the right, using the relations (8.33). For example, $\mathrm{Nr}\, 1 = 1$, $\mathrm{Nr}\, vv^* = \langle v, v^* \rangle - v^* \wedge v$, $\mathrm{Nr}\, w(v + v^*)^2 = \langle v, v^* \rangle w$.
One defines the so-called *normal ordering mapping* $: \cdot :$, from $\bigwedge W$ to $A(W)$, as the inverse to Nr. Thus, $: 1 := 1$, $: v^* \wedge v := v^*v$, etc.
One denotes by $\langle a \rangle$ the component of 0th degree of $\mathrm{Nr}\, a$ in $\bigwedge W$. It is a kind of vacuum average. For example, $\langle (v^* + v)^2 \rangle = \langle v, v^* \rangle$, $\langle v \rangle = 0$.

8.67. Remark. If we replace the field $\mathbb{C}$ of coefficients by $\mathbb{R}$, then the group $SO(n, \mathbb{R})$ has two-fold 'spinor representation' in $G(\mathbb{R}^n)$ (where $\mathbb{R}^n$ has the standard scalar product). Its image becomes the *spinor group* $Spin(n, \mathbb{R})$ which forms the two-fold covering of the special orthogonal group $SO(n, \mathbb{R})$. In the case $n = 3$ the Clifford algebra is generated by the *Pauli matrices*

$$\sigma_1 = \begin{pmatrix} 0 & 1 \\ 1 & 0 \end{pmatrix}, \quad \sigma_2 = \begin{pmatrix} 0 & -\mathrm{i} \\ \mathrm{i} & 0 \end{pmatrix}, \quad \sigma_3 = \begin{pmatrix} 1 & 0 \\ 0 & -1 \end{pmatrix}$$

It is the algebra $M(2, \mathbb{C})$ of 2×2 matrices. There is a map $(x_1, x_2, x_3) \longmapsto x_1\sigma_1 + x_2\sigma_2 + x_3\sigma_3$, which has the property that any rotation T from $SO(3)$ induces

an automorphism of $M(2,\mathbb{C})$. Any such automorphism is internal, $B \longmapsto gBg^{-1}$, where $g = g_T$ can be chosen unitary and is defined modulo const $\cdot\, I$. We have $SO(3,\mathbb{R}) \approx \mathbb{RP}^3$ and $Spin(3) = SU(2) \approx \mathbb{S}^3$ is its universal covering (see [**DNF**]). We have also $SO(4,\mathbb{R}) \approx \mathbb{RP}^3 \times \mathbb{S}^3$ and $Spin(4) = SU(2) \times SU(2)$.
If W is the (real) Minkowski space and $SO(W) = SO(3,1)$ is the proper Lorentz group, then the corresponding Clifford algebra is generated by the Dirac matrices $\gamma^0 = \begin{pmatrix} I & 0 \\ 0 & -I \end{pmatrix}$, $\gamma_j = \begin{pmatrix} 0 & \sigma_j \\ -\sigma_j & 0 \end{pmatrix}$ and there is a 'spinor representation' of $SO(3,1)$ in the Clifford group $G(W)$. The corresponding two-fold covering of $SO(3,1)$ is equal to $SL(2,\mathbb{C})$. It is realized as certain sub-representation of the spinor representation (see [**DNF**]).
The Clifford algebra $A(\mathbb{R}^{2n})$ and the corresponding Clifford group is used in the Onsager's solution of the two-dimensional Ising model (see [**Hua**]). The transfer-matrix of this model is an operator from the Clifford algebra and, in order to find its eigenvalues, one applies the automorphisms g_T induced by orthogonal transformations of $\mathbb{R}^{2n}$. In this way the free energy of the Ising model is calculated. In [**SMJ1**] this method is developed further and the correlation functions of the Ising model are calculated.

8.68. Theorem (A formula for g_T). *Let $w_1, \ldots, w_N$ be some fixed basis of $W = V^* \oplus V$. Let K be the matrix with the entries*

$$\langle w_i \cdot w_j \rangle.$$

Assume that we have an orthogonal matrix $T \in O(W)$. Define a matrix $R = (r_{ij})$ as

$$(T - I)(K + K^{\top} T)^{-1}$$

and an element $\rho \in \bigwedge^2 W$ as

$$\rho = \sum_{i,j} r_{ij} w_i \wedge w_j.$$

Then the transformation T is induced by the element

$$g =: \mathrm{e}^{\rho/2} :$$

from $G(W)$. In other words, $gwg^{-1} = Tw$ (or $g = g_T$).

We will not give the proof of this theorem. We refer the reader to [**SMJ2**]. Now we pass to the promised solution of the Riemann–Hilbert problem.

8.69. Reformulation of the Riemann–Hilbert problem. The authors of [**SMJ1**] make the restriction that the singular points $t_1, \ldots t_m$ lie all on the real axis (in the complex t-plane). Assume that $t_1 < t_2 < \ldots < t_m$ (see Fig. 8.8). With them certain monodromy matrices M_j are associated. The task is to find a fundamental matrix $\mathcal{F}(t)$ which is regular and undergoes the transformations $\mathcal{F}(t) \longmapsto \mathcal{F}(t) M_j$ as t varies along closed loops surrounding just one point t_j.

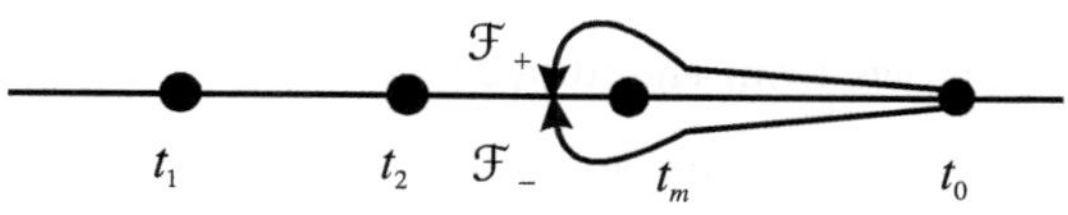

Figure 8.8: Jumps of the fundamental matrix $\mathcal{F}$ at the real axis

(In order to have Fuchsian singularities, one should require that $\dot{\mathcal{F}}\mathcal{F}^{-1}$ has only first order poles. In [**SMJ1**] and [**SMJ2**] this issue is not discussed. Therefore, only the version B of the Riemann–Hilbert problem is solved here.)
The point at infinity is supposed to be non-singular. Thus one has to find a multivalued matrix $\mathcal{F}(t)$, which has definite behavior at the upper and lower cuts of the t-plane along the real axis. We normalize it by putting $\mathcal{F}(t_0) = I$ for some real point $t_0 > t_m$.
If we denote by $\mathcal{F}_\pm(x) = \mathcal{F}(x + \mathrm{i}0^\pm)$, $x \in \mathbb{R}$, then we have $\mathcal{F}_\infty(x) = \mathcal{F}_0(x)$ for $x > t_m$, $\mathcal{F}_+ = \mathcal{F}_\infty M_m$ for $t_{m-1} < x < t_m$, etc.
Define $M_j(x) = I$ for $x > t_j$ and $M_j(x) = M_j$ for $x < t_j$, and

$$M(x) = M_1(x) \ldots M_m(x).$$

Then we should have

$$\mathcal{F}_+(x) = \mathcal{F}_\infty(x) M(x), \quad x \in \mathbb{R}.$$

8.70. Reformulation in the language of field theory. Let $\psi^j(x)$, $\psi^{j*}(x)$ be fermionic fields taking values in the Clifford algebra generated by $W = \mathbb{R}^n \oplus \mathbb{R}^n$ with the scalar product defined by the matrix $\begin{pmatrix} 0 & I \\ I & 0 \end{pmatrix}$. Here n is the dimension of the fundamental matrix $\mathcal{F}$. Thus we have the relations $\{\psi^i(x), \psi^j(y)\} = \{\psi^{*i}(x), \psi^{*j}(y)\} = 0$, $\{\psi^i(x), \psi^{*j}(y)\} = \delta_{ij}\delta(x - y)$.
Assume that there is a field operator $\phi(x)$ satisfying the following relations:

$$\phi(x)\psi^j(x) = \sum_i \psi^i(x)\phi(x)m_{ij}(x),$$

$$\phi(x)\psi^{*j}(x) = \sum_i \psi^{*i}(x)\phi(x)m^*_{ij}(x),$$

where $(m_{ij}) = M(x)$ and $(m^*_{ij}) = (M(x)^\top)^{-1}$. These relations can be written shortly as $\phi\vec{\psi} = \vec{\psi}\phi M$ and $\phi\vec{\psi}^* = \vec{\psi}^*\phi M^*$, where $\vec{\psi}$ is the column vector with components $\psi^i(x)$.
We define the matrices

$$(\mathcal{F}_-(x))_{ij} = -2\pi\mathrm{i}\langle\psi^{*i}(x_0)\phi(x)\psi^j(x)\rangle / \langle\phi(x)\rangle,$$

$$(\mathcal{F}_+(x))_{ij} = -2\pi\mathrm{i}\langle\psi^{*i}(x_0)\psi^j(x)\phi(x)\rangle / \langle\phi(x)\rangle.$$

The following result follows from the above commutation relations.

Proposition. *The functions $\mathcal{F}_\pm$ can be prolonged to the upper and lower complex half-planes, respectively, and they satisfy the relation $\mathcal{F}_+ = \mathcal{F}_- M$.*

8.71. The solution. Thus the problem is reduced to the problem of solving the equations

$$\phi\vec{\psi}\phi^{-1} = T_1\vec{\psi}, \quad \phi\vec{\psi}^*\phi^{-1} = T_2\vec{\psi}^* \tag{8.34}$$

with respect to the field operator $\phi(x)$. Here $T_1 = T_1(x)$ and $T_2 = T_2(x)$ are matrices such that the block-diagonal matrix $T = \mathrm{diag}(T_1, T_2)$ is orthogonal. Note also that the argument x is fixed in Eqs. (8.34).

However, this is the problem of inversion the map $g \longmapsto g_T$ from the Clifford group to the orthogonal group. Its solution is given in Theorem 8.65. It remains only to apply it to our situation. We present the answer without proof.

Let L be a fixed matrix (it will be one of $L_j = \ln M_j$) and let $a \in \mathbb{R}$ (it will be one of t_j). Define the field operator

$$\phi(a, L) =: \mathrm{e}^{\rho(a,L)/2} :,$$

where

$$\frac{1}{2}\rho(a, L) = \frac{-1}{\pi}\int_{-\infty}^{a}\int_{-\infty}^{a} \vec{\psi}(x) \wedge R^\top(x-a, y-a)\vec{\psi}^*(y)\mathrm{d}x\mathrm{d}y,$$

$$R(x, y) = \sin(\pi L)\left(\frac{\exp(\mathrm{i}\pi L)}{x - y + \mathrm{i}0} - \frac{\exp(-\mathrm{i}\pi L)}{x - y - \mathrm{i}0}\right).$$

Theorem. *If there is only one singular point t_1 with the monodromy matrix M_1, then the operator $\phi(t_1, L_1)$ satisfies Eqs.* (8.34).

In the general case the solution to Eqs. (8.34) *is given by the field operator*

$$\phi = \frac{\phi(t_1, L_1)\ldots\phi(t_m, L_m)}{\langle\phi(t_1, L_1)\ldots\phi(t_m, L_m)\rangle}.$$

8.72. Remark. The notion of normal ordering is used also in the bosonic field theory. In particular, the canonical commutation relations (from Definition 8.61) have representation in such large space as $L_2(\mathcal{S}'(\mathbb{R}^m), \mu)$, where $\mathcal{S}'$ is the space of tempered distributions on $\mathbb{R}^m$ and μ is a certain Gaussian measure. The field operators have natural interpretations in terms of multiplication operators by Gaussian random variables and variational derivatives. In particular, the normal ordering of the product of a Gaussian variable f, $: f^n :$, is the value of the nth Hermite polynomial on f.

The reader interested in these topics is referred to [**GJ**].

8.73. Frobenius manifolds in conformal field theory and isomonodromic deformations. B. Dubrovin in [**Dub**] introduced the following notion of Frobenius manifold.

A **Frobenius algebra** is a commutative algebra A over $\mathbb{C}$ with unity e and equipped with a $\mathbb{C}$-bilinear symmetric non-degenerate inner product $\langle\cdot,\cdot\rangle : A \times A \longmapsto \mathbb{C}$, being invariant in the sense that

$$\langle ab, c\rangle = \langle a, bc\rangle.$$

A manifold M is called **Frobenius manifold** iff each its tangent plane T_tM is equipped with a structure of Frobenius algebra and (additionally):

(i) the invariant inner product defines a flat metric (i.e., with zero curvature) and with the Levi-Civita connection ∇;

(ii) $\nabla e \equiv 0$;

(iii) the 4-tensor $\nabla_z\langle uv, w\rangle$ is symmetric;

(iv) there is an Euler vector field $\boldsymbol{E}$ such that $\nabla(\nabla \boldsymbol{E}) = 0$ and whose flow acts on T_tM's by rescaling these algebras.

An example of Frobenius manifold appears, when we have a function $F(t)$, $t = (t^1, \ldots, t^n)$ such that its third derivatives $c_{\alpha\beta\gamma} = \partial^3 F/\partial t^\alpha \partial t^\beta \partial t^\gamma$ obey the following relations:

(a) The matrix $\eta_{\alpha\beta} = c_{1\alpha\beta}$ is non-degenerate with inverse $\eta^{\alpha\beta}$; we use η to rise and lower the indices.

(b) The constants $c_{\alpha\beta}^{\gamma} = \sum_\epsilon \eta^{\gamma\epsilon} c_{\epsilon\alpha\beta}$ define an algebra $A_t \simeq T_t\mathbb{C}^n = \text{span}\,[e_1, \ldots, e_n]$, $e_\alpha = \partial/\partial t^\alpha$, by the formula $e_\alpha \cdot e_\beta = \sum_\gamma c_{\alpha\beta}^{\gamma} e_\gamma$; A_t is associative iff the so-called *WDVV equations* (E. Witten, R. Dijkgraaf, E. Verlinde, H. Verlinde) hold.

(c) F is quasi-homogeneous.

Then the unit of A_t is $e = e_1$, $\langle e_\alpha, e_\beta\rangle = \eta_{\alpha\beta}$, and the Euler vector field is $\boldsymbol{E} = \sum d_\alpha \partial/\partial t_\alpha$, the field associated with the quasi-homogeneous action of $\mathbb{C}^*$.

The natural situation, when a function F satisfies the WDVV equations, is the case when F is the generating function for correlations in the conformal field theory. Such theory arises in the case when the 'fields' are holomorphic maps $\psi : \mathbb{CP}^1 \longmapsto X$, where X is a Calabi–Yau manifold and the role of the space of configurations is played by the moduli space of such maps. The generating function F for correlations turns out to be a function on $M = \bigoplus H^{2i}(X)$ and its coefficients are important invariants of enumerative geometry of X, called the Gromov–Witten invariants.

Another example of Frobenius manifold is the space of isomonodromic deformations of linear differential equation of the form

$$\mathrm{d}z/\mathrm{d}t = (U + V/t)z,$$

where U and V are constant complex matrices; $U = \operatorname{diag}(u_1, \ldots, u_n)$, $u_i \neq u_j$ is diagonal, and V is skew symmetric.
The isomonodromicity condition implies that $V = V(u)$. Thus the space M is parametrized by u. The Frobenius structure is given by: $\partial/\partial u_i \cdot \partial/\partial u_j = \delta_{ij} \cdot \partial/\partial u_j$, $\langle \cdot, \cdot \rangle = \sum \psi^2 \mathrm{d}u_i^2$, where $(\psi_1, \ldots, \psi_n)^\top$ is an eigenvector of $V = V(u)$, $e = \sum \partial/\partial u_i$, $\boldsymbol{E} = \sum u_i \partial/\partial u_i$.
The latter Frobenius manifold admits an action of the monodromy group, i.e., the representation of $\pi_1(M, t_0)$, $M = \mathbb{C}^n \setminus \{u_i = u_j,\ i \neq i\}$, in $T_{t_0}M$. There appear Stokes operators and groups generated by reflections. We cannot describe all this here and we refer the reader to the Dubrovin's article [**Dub**].

Chapter 9

Holomorphic Foliations. Local Theory

The previous chapter was devoted to linear analytic differential equations. In this chapter we develop the theory of analytic nonlinear differential equations, i.e., holomorphic vector fields.

Therefore, the time is treated as complex and the phase curves become Riemann surfaces. The phase portrait of a holomorphic vector field defines a foliation into phase curves. This foliation may have singularities, e.g., at the singular points of vector fields. Also, in applications, one often deals with algebraic manifolds and the foliations are defined by means of polynomial vector fields (in affine charts). Such foliations are called holomorphic foliations. Our analysis will be principally restricted to the case of holomorphic foliations in $\mathbb{CP}^2$.

We study holomorphic foliations in $\mathbb{CP}^2$ from the local and global point of view; in this chapter we shall concentrate on the local theory.

The local analysis includes the Bendixson–Seidenberg–Dumortier–van den Essen theorem on resolution of singular point of a holomorphic planar vector field. We present the proof of this theorem. It is an analogue of the Hironaka resolution theorem 4.56, but it is restricted to the dimension 2. After resolution one obtains only elementary singularities.

An elementary singular point has a separatrix, a leaf diffeomorphic to the punctured disc $\mathbb{D}^*$. A loop in $\mathbb{D}^*$ defines a holonomy map, which constitutes a nonlinear analogue of the monodromy map from the previous chapter.

The analysis of geometry of leaves of a holomorphic foliation near elementary singular point leads to the theory of analytic orbital normal forms (for the holonomy diffeomorphisms of $(\mathbb{C}, 0)$ and for germs of vector fields in $(\mathbb{C}^2, 0)$). Due to investigations of French (J. Martinet, J.-P. Ramis, J. Ecalle, B. Malgrange, J.-C. Yoccoz, R. Perez-Marco), Russian (Yu. Ilyashenko, A. Briuno, S. Voronin, P. Elizarov A. Shcherbakov) and Brazilian (C. Camacho, A. Lins-Neto, P. Sad) mathematicians this theory is now practically completed. In the resonant cases the analytic classification is described by the Ecalle–Voronin functional moduli (for diffeomorphisms) and by the Martinet–Ramis functional moduli for saddle-nodes and in the non-resonant case there is the so-called Briuno condition (with its dynamical interpretation by Yoccoz). We present this theory with details.

In this new version I allow myself to present a quite novel approach to normal forms of vector fields developed by myself with E. Stróżyna. In particular, we have found a relation with the Invariant Theory.

We begin with an auxiliary section containing main notions of foliations and almost complex structures.

H. Żołądek, *The Monodromy Group*, Monografie Matematyczne 67,
https://doi.org/10.1007/978-3-031-91270-2_9

§1 Foliations and Complex Structures

Recall the definition of a foliation of codimension k in an n-dimensional manifold M. Here $\mathbb{K} = \mathbb{R}$, if the manifold is real, and $\mathbb{K} = \mathbb{C}$, if it is complex.

9.1. Definition. Such a **foliation** $\mathcal{F}$ is defined by means of an atlas $\phi_\alpha : U_\alpha \longmapsto \mathbb{K}^{n-k} \times \mathbb{K}^k$ such that the transition maps $\phi_{\alpha,\beta}$ have the form

$$(x, y) \longmapsto (f(x, y), g(y)).$$

The **leaves** of the foliation $\mathcal{F}$ are locally defined as $\phi_\alpha^{-1}(\{y = \text{const}\})$.
If $\mathbb{K} = \mathbb{C}$, M is an analytic manifold and the transition maps are holomorphic, then $\mathcal{F}$ is a holomorphic foliation without singularities.
Another definition of a foliation involves the notion of **distribution in the tangent bundle.** It is a system $V_x \subset T_x M$ of codimension k subspaces of the tangent space. Such distribution is called *integrable* iff it arises from a certain foliation, i.e., $V_x = T_x L$, where L is a leaf of the foliation passing through x.
The subspaces V_x can be defined as kernels of 1-forms $V_x = \{\omega_1(x) = \ldots = \omega_k(x) = 0\}$. The Frobenius theorem (see [**KN**] and below) says that the foliation is integrable iff $\mathrm{d}\omega_j$ vanishes at V_x. In other words, $\mathrm{d}\omega_j = \sum \alpha_{ij} \wedge \omega_i$.
When the situation is holomorphic and ω_j are holomorphic forms, then it is natural that $\omega_j(x)$ can vanish at some points and V_x can change its dimension.
We say that a foliation $\mathcal{F}$, which is locally defined by means of zeroes of holomorphic 1-forms, is **holomorphic foliation** iff its set of singular points has complex codimension ≥ 2.

9.2. Theorem of Frobenius. *A distribution $\{V_x\} \subset TM$ is integrable iff it is involutive.*
The latter means that, for any two of vector fields $\boldsymbol{X}, \boldsymbol{Y}$ lying in this distribution, $\boldsymbol{X}(x) \in V_x$, $\boldsymbol{Y}(x) \in V_x$, their commutator $[\boldsymbol{X}, \boldsymbol{Y}]$ also lies in the distribution. Equivalent condition: for any 1-form ω vanishing on the distribution, $V_x \subset \ker \omega(x)$, its external derivative also vanishes on the distribution.

Proof. The equivalence of the two definitions of the involutivity is a consequence of the E. Cartan's formula

$$2\mathrm{d}\omega(\boldsymbol{X}, \boldsymbol{Y}) = \boldsymbol{X}(\langle \omega, \boldsymbol{Y} \rangle) - \boldsymbol{Y}(\langle \omega, \boldsymbol{X} \rangle) - \omega([\boldsymbol{X}, \boldsymbol{Y}]).$$

If the distribution is integrable and the vector fields X, Y are as in the statement of Theorem 9.2, then $\boldsymbol{X}, \boldsymbol{Y}$ are tangent to a leaf of the corresponding foliation. Such is also the commutator $[\boldsymbol{X}, \boldsymbol{Y}]$, which shows the involutivity.
If the foliation is involutive, then, near $x_0 \in M$, one defines a system $\boldsymbol{X}_1, \ldots, \boldsymbol{X}_{n-k}$ of vector fields such that $(\boldsymbol{X}_i(x))$ define a basis of V_x. Denote g_i^t the phase flows generated by $\boldsymbol{X}_i$. The map

$$(t_1, \ldots, t_{n-k}) \longmapsto g_1^{t_1} \circ \ldots \circ g_{n-k}^{t_{n-k}}(x_0)$$

defines a leaf of the needed foliation passing through x_0. The involutivity of the distribution ensures that this definition of leaf does not depend on choice of the initial point x_0. □

9.3. Example (Holomorphic foliation in $\mathbb{CP}^2$ defined by a polynomial vector field). Any vector field, which is holomorphic in the whole $\mathbb{CP}^2$, is a vector field from the Lie algebra of the Lie group of projective transformations.
(Recall that any vector field holomorphic in the whole Riemann sphere $\mathbb{CP}^1$ is quadratic. The vector field $z\frac{\partial}{\partial z}$ generates the group of hyperbolic transformations $e^t z$, the field $(z^2+1)\frac{\partial}{\partial z}$ generates the elliptic transformations $(z\cos t + \sin t)/(-z\sin t + \cos t)$ and the field $\frac{\partial}{\partial z}$ generates translations $z+t$ (parabolic transformation) These three 1-parameter groups generate the group $PSL(2,\mathbb{C})$. An analogue holds in the two dimensional case.)
If $\boldsymbol{V}(x,y) = P(x,y)\frac{\partial}{\partial x} + Q(x,y)\frac{\partial}{\partial y}$ is a polynomial vector field on $\mathbb{C}^2$ of degree n, then it defines a field of directions in $\mathbb{CP}^2$. In the chart $(z,u) = (1/x, y/x)$ the field $\boldsymbol{V}$ takes the form $-z^2P(1/z,u/z)\frac{\partial}{\partial z} + z[Q(1/z,u/z) - uP(1/z,u/z)]\frac{\partial}{\partial u}$ and has a pole along the line $z=0$ (provided $n>2$). After multiplying it by a suitable power of z, we get a holomorphic vector field in the affine (z,u)-plane. This power is either $n-2$ or $n-1$ depending on whether the leading terms of Q and uP in $Q-uP$ cancel themselves or not. However, multiplying a vector field by a function does not change its phase portrait; only the velocities along phase curves undergo changes.
One says about the *field of directions*, which is a well defined object. This field of directions defines a holomorphic foliation with singularities. The leaves of this foliation are either singular points of the field of directions or Riemann surfaces (complex phase curves of $\boldsymbol{V}$).
The fields of directions (or the holomorphic foliations) are well defined objects in algebraic compact projective surfaces $S \subset \mathbb{CP}^N$. In affine charts they are given by means of polynomial vector fields tangent to the surface.
The above statement can be reversed. If a field of directions in $\mathbb{CP}^2$ is locally defined by means of local holomorphic vector fields outside a subset of (complex) codimension 2, then this vector field must be polynomial in the affine charts.
With the field of directions, defined locally by means of the vector field $P\frac{\partial}{\partial x}+Q\frac{\partial}{\partial y}$, one associates the holomorphic 1-form

$$Q\mathrm{d}x - P\mathrm{d}y.$$

Its kernels are exactly the directions defined by the field of directions.
One can also give description of the above picture in terms of the homogeneous coordinates $(x:y:z)$ in $\mathbb{CP}^2$. With the homogeneous vector field $X\frac{\partial}{\partial x}+Y\frac{\partial}{\partial y}+Z\frac{\partial}{\partial z}$ one associates the homogeneous 1-form $P\mathrm{d}x + Q\mathrm{d}y + R\mathrm{d}z$, where

$$\begin{pmatrix} P \\ Q \\ R \end{pmatrix} = \begin{pmatrix} X \\ Y \\ Z \end{pmatrix} \wedge \begin{pmatrix} x \\ y \\ z \end{pmatrix} = \begin{pmatrix} Yz - Zy \\ Zx - Xz \\ Xy - Yx \end{pmatrix}$$

(see [**Jou**]).

We pass to the presentation of some facts from the complex geometry, which will be needed in the sequel.

9.4. Definition. Consider the following situation. Let M^n be a n-dimensional complex manifold. Let us treat it as a $2n$-dimensional real manifold ${}^{\mathbb{R}}M^{2n}$. Assume now that the complex atlas on M is replaced by a C^1-smooth atlas on ${}^{\mathbb{R}}M^{2n}$. Analogous situation occurs, when there is a C^1-diffeomorphism ϕ from a complex manifold ${}^{\mathbb{C}}N^n$ to a real manifold ${}^{\mathbb{R}}M^{2n}$. It turns out that the complex structure from M is not completely lost on ${}^{\mathbb{R}}M^{2n}$. There remains its trace called the almost complex structure. Also such an almost complex structure appears on M (in the second example).

An **almost complex structure** on a real manifold M is a field $J=\{J_x\}$ of endomorphisms of the tangent spaces T_xM satisfying the identity

$$J_x^2+I=0.$$

In other words, it is a field of splittings of the complexifications of tangent spaces into eigenspaces of J_x with eigenvalues $\mathrm{i}=\sqrt{-1}$ and $-\mathrm{i}$

$${}^{\mathbb{C}}T_xM=T_xM\otimes\mathbb{C}=T_x^{1,0}\oplus T_x^{0,1}$$

of equal (complex) dimension n.

9.5. Example. The standard complex structure on a complex manifold N, with local complex coordinates $z_j=x_j+\mathrm{i}y_j$, induces the almost complex structure by the formulas $J\partial_{x_j}=\partial_{y_j}$, $J\partial_{y_j}=-\partial_{x_j}$.

9.6. Definition. An almost complex structure J is called **integrable** iff locally it is an image of complex structure by means of a C^1-mapping. It means that in neighborhood of any point on M there exists a local system of C^1-coordinates x_j,y_j such that the formulas from Example 9.5 hold.

If M is a real analytic $2n$-dimensional manifold and ${}^{\mathbb{C}}M$ is its local complexification (a complex $2n$-dimensional manifold containing M as its real part), then the integrability of the almost complex structure means the simultaneous integrability of the two distributions $T^{1,0}=\bigcup T_x^{1,0}({}^{\mathbb{C}}M)$ and $T^{0,1}$. The local leaves of the corresponding foliations are $\{x_1+\mathrm{i}y_1=c_1,\dots,x_n+\mathrm{i}y_n=c_n\}$ and $\{x_1-\mathrm{i}y_1=d_1,\dots,x_n-\mathrm{i}y_n=d_n\}$, $x_j,y_j,c_j,d_j\in\mathbb{C}$.

The splitting of the complexifications of the tangent spaces induces analogous splitting of the complexifications of the cotangent spaces into forms of the type $(1,0)$ and of the type $(0,1)$,

$${}^{\mathbb{C}}T^*M={}^{\mathbb{C}}\mathcal{E}^1=\mathcal{E}^{1,0}\oplus\mathcal{E}^{0,1};$$

ω is of the type $(1,0)$ iff it vanishes at $T^{0,1}$. This induces splitting of all differential k-forms (with complex coefficients) into forms of the type (p,q).

As we know from the Frobenius theorem 9.2, the distributions $T^{1,0}$ and $T^{0,1}$ (of vectors of the type $(1,0)$ and $(0,1)$, respectively) are integrable iff they are involutive. This means that the commutator of any two vector fields from $T^{1,0}$ (respectively from $T^{0,1}$) is a vector field from $T^{1,0}$ (respectively from $T^{0,1}$). This condition can be written in the following compact form.

9.7. Definition. The **torsion** (or the **Nijenhuis tensor**) of the almost complex structure J is the following bilinear form on the space of vector fields on M :

$$N(\boldsymbol{X},\boldsymbol{Y}) = 2\{[J\boldsymbol{X},J\boldsymbol{Y}] - [\boldsymbol{X},\boldsymbol{Y}] - J[\boldsymbol{X},J\boldsymbol{Y}] - J[J\boldsymbol{X},\boldsymbol{Y}]\}.$$

This tensor, written in the language of differential forms, is

$$\omega \longmapsto \mathrm{d}\omega \circ (J \wedge J) - \mathrm{d}(\omega) - \mathrm{d}(\omega \circ J) \circ (J \wedge id + id \wedge J).$$

Vanishing of the torsion is equivalent to the fact that the derivative of a form ω of the type $(1,0)$ does not have component of the type $(0,2)$.

9.8. Theorem of Newlander and Nirenberg. ([NN]) *An almost complex structure is integrable iff its torsion vanishes.*

In the case of real analytic manifold M this result follows from the above remarks and the Frobenius theorem. Newlander and Nirenberg proved it in the finitely smooth case.

Now we pass to the 1-dimensional case. Here the problem of integrability of almost complex structures is treated differently. Assume that we are in a 2-dimensional plane (local chart), which we identify with the complex plane $\mathbb{C}$. Here the splitting of ${}^{\mathbb{C}}T^*\mathbb{C}$ can be defined by means of one 1-form

$$\omega = \mathrm{d}z + \mu \mathrm{d}\bar{z},$$

where $\mu = \mu(z,\bar{z})$ is a function. We have $\mathcal{E}^{1,0} = \mathbb{C}\omega$ and the other space $\mathcal{E}^{0,1}$ is its conjugate.
The function μ is not defined invariantly. Note that, when we replace the complex variable z by $w = \phi(z)$ (ϕ holomorphic), then the analogous 1-form $\tilde{\omega} = \mathrm{d}w + \tilde{\mu}\mathrm{d}\bar{w}$ takes the form

$$\phi'_z \mathrm{d}z + \tilde{\mu}\overline{\phi'_z}\mathrm{d}\bar{z} = \phi'_z[\mathrm{d}z + \tilde{\mu}\cdot(\overline{\phi'_z}/\phi'_z)\cdot \mathrm{d}\bar{z}]. \qquad (9.1)$$

This rule of transformation of μ suggests the following definition.

9.9. Definition. The quantity

$$\mu \cdot (\mathrm{d}\bar{z}/\mathrm{d}z)$$

is called the **Beltrami differential** defining the almost complex structure. In particular, $|\mu|$ is a well defined function.

The integrability of the almost complex structure, defined by the Beltrami differential, means existence of a differentiable complex function $\phi = \phi(z,\bar{z})$, which is a homeomorphism, and such that the almost complex structure is defined by $\phi^* \mathrm{d}u = \mathrm{d}\phi = \phi_z \mathrm{d}z + \phi_{\bar{z}} \mathrm{d}\bar{z}$ (where $\phi_z = \frac{\partial \phi}{\partial z}$),i.e., the following **Beltrami equation**

$$\phi_{\bar{z}} = \mu \phi_z$$

holds. The Beltrami equation is a partial differential equation and is solved in a suitable Sobolev space.
(Note that the torsion of any almost complex structure in $\mathbb{C}$ vanishes. It is because $\bigwedge^2 \mathcal{E}^1 \otimes \mathbb{C} = \mathcal{E}^{1,0} \wedge \mathcal{E}^{0,1}$ is one dimensional and contains only 2-forms of the type $(1,1)$. Therefore here we have different kind of problems than in the multi-dimensional case. In particular, one condition of integrability would be the preservation of the orientation by the map ϕ; it is equivalent to the fact that $|\mu| < 1$.)
The next theorem has many authors (L. Ahlfors, B. Bojarski, L. Bers, A. Newlander, L. Nirenberg).

9.10. Theorem. ([**Ahl**]) *If μ is continuous and satisfies the estimate*

$$|\mu(x)| \leq k < 1,$$

then the Beltrami equation has a solution and the almost complex structure defined by $\mu \mathrm{d}\bar{z}/\mathrm{d}z$ is integrable.

9.11. Remarks.

(i) In the proof of this theorem one uses the fact that, if $||\mu||_\infty = k < 1$, then the operator $\frac{\partial}{\partial \bar{z}} - \mu \frac{\partial}{\partial z}$ is invertible in a suitable Sobolev space. Here it is enough to assume that μ is only essentially bounded, $\mu \in L^\infty$, and then ϕ is a homeomorphism. This proof can be found in [**Ahl**].

The ϕ's satisfying the Beltrami equation are called the **quasi-conformal maps** and $\mu = \mu_\phi$ is called the *modulus* of ϕ.

(ii) If g is a conformal (i.e., analytic) mapping, then $\phi \circ g$ is quasi-conformal with the modulus $\mu_{\phi \circ g} = \mu_\phi \circ g \cdot (\overline{g'}/g')$ (see (9.1)).

(iii) If g is conformal, then $g \circ \phi$ is quasi-conformal with $\mu_{g \circ \phi} = \mu_\phi$. This means that the Beltrami equation has many solutions.

(iv) If $||\mu_\phi||_\infty = k < 1$, then the map ϕ transforms infinitesimally small sphere around any z_0 to an infinitesimally small ellipse with the ratio of lengths of the semi-axes (longer to shorter) $K(z_0)$. We have $K = \sup K(z_0) = \frac{1+k}{1-k}$.

(v) If $\mu_\phi = \mu_\psi$, then the map $\phi \circ \psi^{-1}$ is conformal. (ψ^{-1} maps circles to ellipses, which are sent again to circles by ϕ.)

(vi) If $||\mu_\phi||_\infty = k < 1$, then the map ϕ is Hölder continuous with the exponent $\frac{1-k}{1+k}$ (A. Mori). The proof can be found in [**Ahl**].

9.12. Example. Let $f : \mathbb{C}^* \longmapsto S$ be a smooth map to a Riemann surface S. Define $\mu = f'_{\bar{z}}/f'_z$.
We say that f is *regular* iff $\mu(z)$ can be continuously prolonged to $\overline{\mathbb{C}}$ with $\mu(0) = \mu(\infty) = 0$.

9.13. Proposition. *If f is regular, then there is a conformal mapping $g : \mathbb{C}^* \longmapsto S$. This means that S has conformal type of $\mathbb{C}^*$; not of an annulus or of a punctured disc.*

Proof. We prolong μ to $\overline{\mathbb{C}}$ and integrate the corresponding Beltrami equation in the Riemann sphere. Using an eventual composition with a Möbius automorphism, we can assume that the solution $u : \overline{\mathbb{C}} \longmapsto \overline{\mathbb{C}}$ is a homeomorphism such that $u(0) = 0$, $u(\infty) = \infty$.
We have $\mu_u = \mu_f$ in $\mathbb{C}^*$ and the map $u \circ f^{-1} : S \longmapsto \mathbb{C}^*$ is conformal (see 9.11(v)). □

9.14. Moduli of quadrangles and annuli. If $Q \subset \mathbb{C}$ is a (curved) quadrangle, considered together with a pair of arcs on its boundary (b-arcs), then it can be transformed conformally onto a rectangle (with sides a, b) such that the b-arcs are sent to the vertical sides. The ratio of the lengths of sides is called the **modulus** of Q (see [**Ahl**])

$$m(Q) = a/b.$$

$m(Q)$ is an invariant of conformal transformations; by applying the Schwarz lemma, one easily shows that two rectangles are conformally equivalent iff their moduli are the same.
The condition of quasi-conformality (for a map ϕ) says that $m(Q') < Km(Q)$ for any quadrangle Q transformed to a quadrangle Q'. Here $K = \frac{1+k}{1-k}$ is the constant from the point (iv) of 9.11. It means that K measures the distortion of small squares.
There is another definition of the modulus of quadrangle, which does not make use of conformal mapping to a rectangle (see [**Ahl**]):

$$m = \sup_{\rho} L^2(\rho)/A(\rho), \tag{9.2}$$

where ρ are square integrable non-negative functions, $A(\rho) = \int\int_Q \rho^2 \mathrm{d}x\mathrm{d}y$ and $L(\rho) = \inf \int_\gamma \rho|\mathrm{d}z|$ and γ are paths joining the b-arcs.

If $P \subset \mathbb{C}$ is a domain, which can be conformally transformed to an annulus $\{r < |z| < R\}$, then we define its **modulus** as

$$m(P) = \ln(R/r).$$

Note that the map $\ln z$ transforms the standard annulus, cut along a radius, into the rectangle with the sides $\ln R - \ln r$ and 2π. Thus, up to a constant, the modulus of a ring is the same as the modulus of corresponding quadrangle.
The formula (9.2) holds also in the case when the quadrangle is replaced by the ring.

§2 Resolution for Vector Fields

9.15. Definition. A singular point of a germ of holomorphic planar vector field is called **elementary** (or reduced) iff at least one of its eigenvalues is nonzero.
Thus we can assume that we have

$$\dot{x} = \lambda_1 x + \ldots, \quad \dot{y} = \lambda_2 y + \ldots, \qquad \lambda_1 \neq 0.$$

Important invariant of the singular point is the **ratio of eigenvalues**

$$\lambda = \lambda_2/\lambda_1.$$

If $\lambda \in \mathbb{C} \setminus \mathbb{R}$, then we say that the singular point is a **focus**; if $\lambda > 0$, then it is a **node**; if $\lambda < 0$, then it is a **saddle**; if $\lambda = 0$, then it is **saddle-node**.
The singular point is called **resonant** iff λ is a rational number. The resonant node is called **dicritical** iff it is formally equivalent to its linear part.
(Recall that, by Poincaré–Dulac Theorem 8.14 and Example 8.15, the resonant node can be formally reduced to the form $\dot{x} = \lambda_1 x,\ \dot{y} = \lambda_2 y + a x^{\lambda}$, if $\lambda = p/q \geq 1$. In Remark 9.25 below it is proved that this change can be made analytic. The dicriticality means that $a = 0$. In that case a whole family of invariant analytic curves $x^q = \mathrm{const} \cdot y^p$ pass through the singular point.)

The definition of resolution of singularity of a hypersurface in $\mathbb{C}^n$, or of a function, is given in the point 4.55 (in Chapter 4). Below we repeat it in the 2-dimensional case and in applications to vector fields.

9.16. Definition. The **elementary blowing-up map** (or the **σ-process**) is the map

$$\pi : (M, E) \longmapsto (\mathbb{C}^2, 0),$$

where M is a 2-dimensional complex manifold equal (as a set) to $(\mathbb{C}^2 \setminus 0) \cup E$, where $E \simeq \mathbb{CP}^1$ is the set of lines passing through the origin. E is called the **exceptional divisor**. If $(u_1 : u_2)$ are the homogeneous coordinates in $\mathbb{CP}^1$, then the blowing-up map is locally given by the formulas

$$(x, u) \longmapsto (x, xu), \quad u = u_2/u_1, \quad (u_1 \neq 0),$$

$$(y, z) \longmapsto (yv, y), \quad v = u_1/u_2, \quad (u_2 \neq 0).$$

If $\boldsymbol{V} : (\mathbb{C}^2, 0) \longmapsto (\mathbb{C}^2, 0)$ is a germ of holomorphic vector field, then after the elementary blowing-up, we obtain usually not a vector field in M but rather a field of directions $\tilde{\boldsymbol{V}}$ near the distinguished divisor E. In the local (x, u)-chart (or (y, v)-chart) in M this field of directions is defined by means of an analytic vector field.
Let $\boldsymbol{V} : (\mathbb{C}^2, 0) \longmapsto (\mathbb{C}^2, 0)$ be a germ of holomorphic vector field. We say that it admits a **good resolution** iff after a *finite* number of elementary blowing-ups we obtain a field of directions in a 2-dimensional complex manifold with only elementary and no-dicritical singular points.

9.17. Examples. (a) *A general homogeneous vector field of degree n*

$$\dot{x} = P_n(x,y), \quad \dot{y} = Q_n(x,y)$$

has $n+1$ invariant lines.

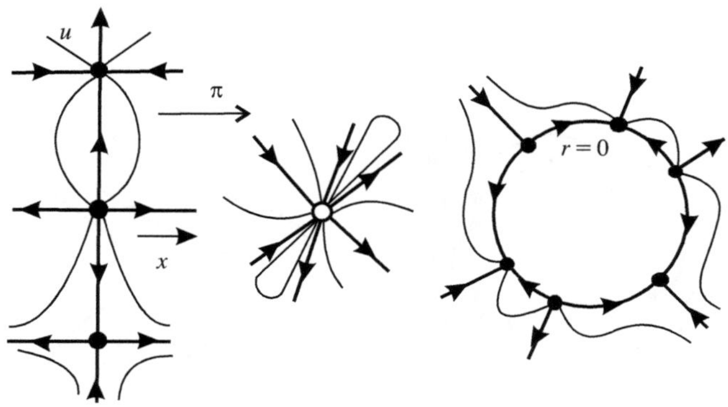

Figure 9.1: Example of the resolution using one blow-up

Indeed, in the blowing-up coordinates $x, u = y/x$, we get the system $\dot{x} = x^n P_n(1,u)$, $\dot{u} = x^{n-1}[Q_n(1,u) - uP_n(1,u)] = x^{n-1}H_{n-1}(u)$, where the polynomial $H_{n+1}(u)$ in the square brackets is (generally) nonzero. In order to obtain a field of directions, one has to divide the latter system by x^{n-1}. The zeroes u_i of $H_{n+1}(u)$ define the invariant lines $y = u_i x$.

In order to draw real pictures, one uses the **polar blowing-up**, which relies on rewriting the vector field in the polar coordinates $x = r\cos\theta$, $y = r\sin\theta$ and dividing it by a suitable power of r. In the above example we get

$$\dot{r} = r[\cos\theta P_n(\cos\theta, \sin\theta) + \sin\theta Q_n(\cos\theta, \sin\theta)],$$
$$\dot{\theta} = \cos\theta Q_n(\cos\theta, \sin\theta) - \sin\theta P_n(\cos\theta, \sin\theta).$$

The corresponding pictures in the case $n = 2$ are given in Fig. 9.1.

If the zeroes u_i are isolated, then we can perturb this vector field by adding terms of degree $> n$ and the above blowing-up gives the resolution of the perturbed vector field; the singular points $(u_i, 0)$ remain elementary.

If $P_n = xR_{n-1}, Q_n = yR_{n-1}$, then $H_{n+1} \equiv 0$ and the foliation in the (x,u)-coordinates is the flow-box foliation. If additionally we consider perturbation, with terms of higher degree of general type, then in the resolution we must divide the blown-up vector field by x^n. The blowing-up defines the resolution of the singularity, but the exceptional divisor $x = 0$ is not invariant for the corresponding foliation. In such a case we say that the divisor E is **dicritical**.

(b) Consider the *Bogdanov–Takens singularity*

$$\dot{x} = y, \quad \dot{y} = ax^2 + bxy + cy^2 + \dots,$$

where we assume that $a \neq 0$. This system can be treated as a small perturbation of the Hamiltonian system $\boldsymbol{X}_H$, with the Hamilton function $H = \frac{1}{2}y^2 - \frac{a}{3}x^3$ with the cusp singularity. Indeed, in the quasi-homogeneous grading with the exponents $\alpha_1 = \deg(x) = 2$, $\alpha_2 = \deg(y) = 3$, the Hamiltonian part has degree 1 and the remaining terms have greater degrees.

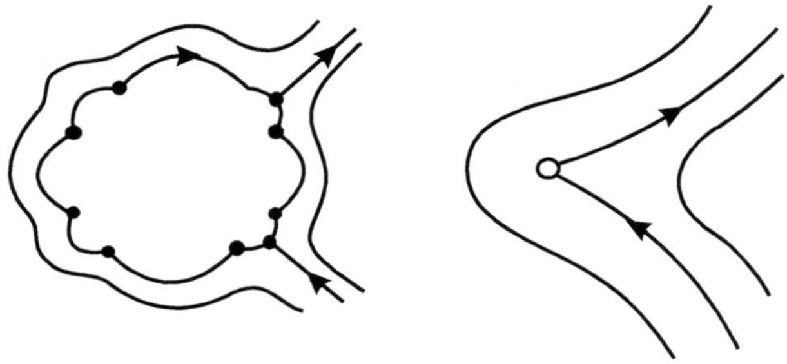

Figure 9.2: Resolution of the cusp singularity

The blowing-up of the cusp curve is presented at Fig. 4.32 in Chapter 4. It is easy to see that the same elementary blowing-up maps can be applied to the Bogdanov–Takens singularity. The resolved field of directions has the same singularities as $\boldsymbol{X}_H$, which are also hyperbolic saddles. The curve Γ is an invariant curve of the vector field (separatrix) and has the cusp singularity. The real picture with polar blowing-ups is given in Fig. 9.2.
We shall return to this singularity in the next chapter.

9.18. Theorem of Bendixson–Seidenberg–Dumortier–van den Essen. *Any germ of analytic planar vector field with isolated singularity admits a good resolution.*

Remark. I. Bendixson [**Ben**] announced this theorem under the assumption that the vector field is real and analytic. In the thesis (of his theorem) there was only the statement that the resolved field of directions has only elementary singularities (they can be dicritical).
F. Dumortier [**Dum**] also considered the real case with only elementary singularities at the end. Moreover, he assumed that the initial vector field is only smooth. The assumption that the singular point is isolated is replaced by the following *Łojasiewicz condition*:

$$|V(x)| > C|x|^{\alpha}.$$

Our version of the desingularization theorem was firstly formulated by A. Seidenberg [**Sei**]. However its final proof, which is the just proof, was given by A. van den Essen [**Ess**] and can be found in the paper of J. Mattei and R. Moussu [**MM**].

We note that there is an analogous theorem (about desingularization) for germs of holomorphic foliations of codimension 1 in $(\mathbb{C}^3, 0)$ (see [**Can**]). But the analogous result about the codimension 1 foliations is not known in dimensions > 3. Also there is no desingularization theorem for holomorphic vector fields in dimensions > 2.

Theorem 9.18 has important consequences. One of them is the proof of the Hironaka's resolution theorem 4.56 in the 2-dimensional case.

9.19. Corollary. *Let $f : (\mathbb{C}^2, 0) \longmapsto (\mathbb{C}, 0)$ be a germ of holomorphic function with isolated singularity at 0. Then the resolution of the Hamiltonian vector field $\boldsymbol{X}_f$ defines also the resolution of the germ f and of the hypersurface $f^{-1}(0)$.*

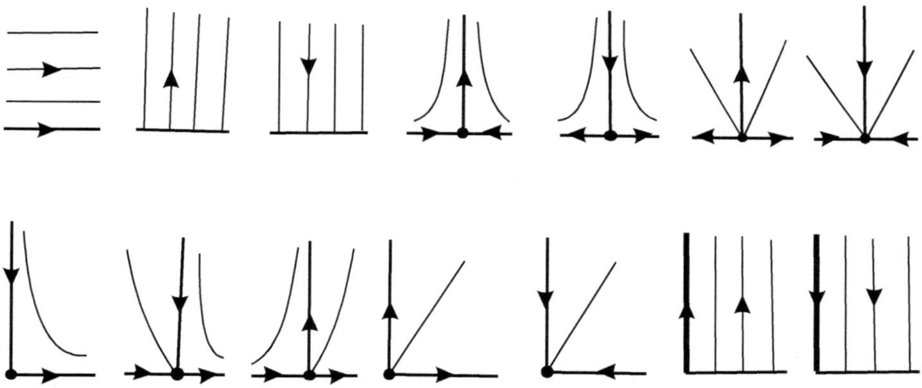

Figure 9.3: Standard phase portraits

9.20. Corollary. *The resolution of singular point of a real vector field allows to determine its topological type modulo the solution of the center-focus problem. This means that, after polar blowing-ups, the phase portrait near the distinguished divisors can be composed from the finite collection of standard portraits (presented at Fig. 9.3). If there is a characteristic trajectory of the resolved field (a separatrix tending to a point in the distinguished divisors), then the topological type of the vector field is completely determined; if there is no characteristic trajectory, then we say that the singular point is monodromic and the topological type is determined by the stability type of the singularity.*

Proof of Theorem 9.18. (We follow the article [**MM**].)

1. *Resolution of dicritical singular points.* If we have a $p : q$ resonant node

$$\dot{x} = px + \dots, \quad \dot{y} = qy + \dots, \qquad p < q,$$

then the resolved vector field has two singular points in the divisor $E = \{(u_1 : u_2)\}$: the $p : (q-p)$ resonant node $(1 : 0)$ (i.e., $\dot{x} \approx px$, $\dot{u} \approx (q-p)u$) and $p : (p-q)$

resonant saddle $(0:1)$. If the node $x=y=0$ is dicritical, then the node $(1:0)$ is also dicritical.
Repeating this several times, we arrive to a $1:1$ resonant node which, after elementary blowing-up, is transformed to the flow-box (see Example 9.17(a)).

2. *Two invariants: index and rank.* We shall work with the vector fields

$$\boldsymbol{V} = b(x,y)\partial_x - a(x,y)\partial_y$$

and with the equivalent to them 1-forms

$$\omega = a\mathrm{d}x + b\mathrm{d}y.$$

The **index** of the vector field (or of the 1-from) $i_0(\boldsymbol{V}) = i_0(\omega) = i(a,b;0)$ is the same as defined in Section 2. Here we use the definition 2.25(d). We recall it.
Let $b(x,y)=0$ be an irreducible (analytic) curve passing through the origin. It can be locally parametrized by an analytic map $(\mathbb{C},0)\ni t\longmapsto\gamma(t)\in(\mathbb{C}^2,0)$ such that $b\circ\gamma\equiv 0$. There is the so called *primitive parametrization* such that any other parametrization of the same curve is induced from the primitive parametrization by a change of parameters. (Example: the primitive parametrization of the cusp $y^2=x^3$ is $x=t^2, y=t^3$.) Then we have $a\circ\gamma(t)=\alpha t^d+\ldots$, $\alpha\neq 0$. We get $i(a,b;0)=d$.
If the function b has the representation $b=b_1^{l_1}\ldots b_r^{l_r}$ with irreducible b_j's, then $i(a,b;0)=\sum l_j\cdot i(a,b_j;0)$.
Note that $i_0(\boldsymbol{V})=1$, if 0 is a non-degenerate elementary singular point, i.e., not a saddle-node.
The **rank** of the vector field (or of the 1-form) $\nu(\boldsymbol{V})=\nu(\omega)=\nu_0(\omega)$ equals to the order of the first nonzero term in the Taylor expansion of $\boldsymbol{V}$ at 0. Analogously one defines the rank of a function.

3. **Lemma**. *Let $\pi:(x,u)\longmapsto(x,xu)$ be the blowing-up map with the distinguished divisor $E=\pi^{-1}(0)$. We define the functions $\tilde{a},\tilde{b}$ by the formulas*

$$a\circ\pi = x^{\nu(a)}\tilde{a},\ \ b\circ\pi = x^{\nu(b)}\tilde{b}.$$

Then we have

$$i(a,b;0) = \nu(a)\nu(b) + \sum_{c\in E} i(\tilde{a},\tilde{b};c).$$

Proof. When the polynomial $\tilde{b}(0,u)$ has only simple zeroes $c\in E$ (there is $\nu(b)$ of them), then each such zero gives the contribution $\nu(a)+i(\tilde{a},\tilde{b};c)$ to the index $i(a,b;0)$. The general case is a consequence of the additivity of the index. □

4. **Proposition**. *There exists a sequence of elementary blowing-ups such that at the end we obtain only singularities with rank ≤ 1.*

Proof. Let $\omega_\nu = a_\nu\mathrm{d}x + b_\nu\mathrm{d}y$ be the homogeneous part of ω of order $\nu=\nu(\omega)=\min(\nu(a),\nu(b))$. We put

$$P_{\nu+1} = xa_\nu + yb_\nu.$$

Then we have

$$\begin{aligned}\pi^*\omega &= a(x,ux)\mathrm{d}x + b(x,ux)\mathrm{d}(ux)\\ &= (a+ub)\mathrm{d}x + xb\mathrm{d}u\\ &= x^\nu\{[P_{\nu+1}(1,u)+\ldots]\mathrm{d}x + x[b_\nu(1,u)+\ldots]\mathrm{d}u\}.\end{aligned}$$

Here the case with $P_{\nu+1}\equiv 0$ is dicritical.
For any $c\in E$ we define the germ $\tilde{\omega}=\tilde{\omega}_c$ as the germ of $\pi^*\omega$ at c, divided by the greatest common multiplier of the germs its components. Thus, if $\pi^*\omega = a_1\mathrm{d}x + b_1\mathrm{d}u$, then $\tilde{\omega} = \pi^*\omega/\gcd(a_1,b_1)$. The Pfaff equation $\tilde{\omega}=0$ defines the blown-up foliation near c.
Proposition 4 follows from the following result. □

5. **Lemma.** *If* $\nu>1$*, then for any* $c\in E$ *we have*

$$i_c(\tilde{\omega}) < i_0(\omega).$$

Proof. Consider firstly the dicritical case, i.e., $P_{\nu+1}\equiv 0$. We shall show prove that

$$i_0(\omega) = \nu^2+\nu-1+\sum i_c(\tilde{\omega}).$$

We can choose a local system of coordinates $(x,y)\longmapsto(x+\ldots,y+\ldots)$ such that the following conditions hold:

(i) $P(x,y)=xa+yb$ has rank exactly $\nu+2$;

(ii) $b(0,y)=y^{\nu+1}$ (then $P(0,y)=y^{\nu+2}$).

We have

$$i(P,b;0) = i(x,b;0)+i(a,b;0),$$

where (by Lemma 3)

$$i(P,b;0) = \nu(\nu+2)+\sum i(\tilde{P},\tilde{b};c),$$

and $i(x,b;0)=\nu+1$ (by (ii)). Now one must only notice that $\tilde{\omega}=\tilde{P}\mathrm{d}x+\tilde{b}\mathrm{d}u$ and thus $i(\tilde{P},\tilde{b};c)=i_c(\tilde{\omega})$.

Consider now the non-dicritical case $P_{\nu+1}\not\equiv 0$. We shall prove that

$$i_0(\omega) = \nu^2-\nu-1+\sum i_c(\tilde{\omega}).$$

We have (by Lemma 3)

$$i(a,b;0) = \nu^2+\sum i(\tilde{a},\tilde{b};c). \tag{9.3}$$

Next,

$$i_c(\tilde{\omega}) = i(\tilde{a}+u\tilde{b}, x\tilde{b};c) = i(\tilde{a}+u\tilde{b},x;c)+i(\tilde{a},\tilde{b};c),$$

where $i(\tilde{a}+u\tilde{b},x;c)$ is the multiplicity of c as zero of the polynomial $P_{\nu+1}$; the sum of these multiplicities is $\nu+1$.
Summing up the latter equality over c's and taking into account (9.3), we get the result. □

6. *The case* $\nu=1$. The only case with non-elementary singular point of rank one is the case with nilpotent linear part $\dot{x}=y+\ldots,\dot{y}=\ldots$. The corresponding 1-form is

$$\omega=[y+A_1(x,y)]\mathrm{d}y+B_1(x,y)\mathrm{d}x,$$

with $\nu(A_1),\nu(B_1)\geq 2$.
After blowing-up in the x-direction, i.e., with $y=xu$, we get $\tilde{\omega}=x(u+xA_2)\mathrm{d}u+(u^2+xB_2)\mathrm{d}x$, where A_2, B_2 are analytic functions (without restrictions onto ranks). The latter form can be included in the series of forms

$$\eta=x(y+xA)\mathrm{d}y+(ny^2+xB)\mathrm{d}x, \tag{9.4}$$

where n is a natural number. In what follows we deal with (9.4).
Note also that $\nu(\eta)=2$.

7. *The case* $B(0)=b_0\neq 0$. Here we blow-up η in the y-direction by putting $x=vy$. We obtain

$$\tilde{\eta}=v[(n+1)y+b_0v+A']\mathrm{d}y+y[ny+b_0v+B']\mathrm{d}v.$$

This form has good quadratic part. Note that the polynomial

$$P_3(\tilde{\eta})=vy[(2n+1)y+2b_0v]$$

has distinct factors. After the next blowing up we obtain three elementary singular points (see Example 9.17(a)).

8. *The case* $B(0)=0$. From Eq. (9.4) we get

$$P_3(\eta)=x[(n+1)y^2+\alpha xy+\beta y^2]=(n+1)x(y-c_1x)(y-c_2x).$$

If $c_1\neq c_2$, then the next blowing-up solves the problem (like in 7).
Let $c_1=c_2$. Eventually applying the change $y\longmapsto y-cx$, we can assume that $c_1=c_2=0$, i.e., $P_3(\eta)=(n+1)xy^2$. The singular point $(0:1)\in E$, corresponding to the line $x=0$ in $P_3=0$, is elementary; the point $(1:0)$ is non-elementary.
Then the calculations of the blowing up with $y=ux$ give

$$\tilde{\eta}=x(u+A')\mathrm{d}u+[(n+1)u^2+xB']\mathrm{d}x,$$

where $A'=A(x,ux)$, $B'=B(x,ux)/x+uA(x,ux)$.
Here, if $A(0)\neq 0$, then the point $x=u=0$ is elementary.

9. *Let* $A(0)=0$. Then $A'=xA''$ and $\tilde{\eta}$ has the same form as η in Eq. (9.4), but with n replaced by $n+1$. So, we can repeat the analysis from 7 and 8. We have two possibilities: either, after finitely many steps, we obtain elementary singular point, or infinitely many times we obtain the form (9.4).
However, the second possibility contradicts Proposition 4, because we would get infinitely many times $\nu>1$. This contradiction proves Theorem 9.18. □

§3 One-dimensional Analytic Diffeomorphisms

After the resolution theorem our next task is to describe the analytic classifications of planar holomorphic foliations near elementary singular points. As we have said, this theory is practically completed. However, before classifying the two-dimensional vector fields we shall present the classification of analogous one-dimensional objects.

These objects are the germs of analytic diffeomorphisms $f : (\mathbb{C}, 0) \longmapsto (\mathbb{C}, 0)$:

$$z \longmapsto f(z) = \mu z + a_2 z^2 + \ldots$$

9.21. Definition. Two germs f and $\tilde{f}$ of conformal diffeomorphisms of $(\mathbb{C}, 0)$ are **analytically equivalent** (respectively **formally equivalent, topologically equivalent)** iff there is a germ $h(z)$ of analytic diffeomorphism (respectively a formal power series $h(z) \sim h_1 z + h_2 z^2 + \ldots$, a homeomorphism $h(z)$) conjugating f with $\tilde{f}$. It means that

$$\tilde{f} = h^{-1} \circ f \circ h.$$

Analogous definition takes place in the case of multi-dimensional diffeomorphisms of $(\mathbb{C}^n, 0)$.

The diffeomorphism h transforms the orbits of $\tilde{f}$, i.e., the sets $\{\ldots, (\tilde{f})^{-1}(z), z, \tilde{f}(z), (\tilde{f})^2(z) = \tilde{f} \circ \tilde{f}(z), \ldots\}$, to the orbits of f. Thus f and $\tilde{f}$ have diffeomorphic sets of fixed points and of periodic points of given period.

9.22. Definition. Let $f(x) = Bx + \ldots$ be a local diffeomorphism of $(\mathbb{C}^n, 0)$ and let $\mu_1, \ldots, \mu_n$ be the eigenvalues of the matrix B. The system $(\mu_1, \ldots, \mu_n)$ satisfies the **resonant relation for diffeomorphism of the type** $(i; k)$, $i = 1, \ldots, n$, $k = (k_1, \ldots, k_n) \in \mathbb{Z}^n$, $k_j \geq 0$, iff

$$\mu_i = \mu_1^{k_1} \ldots \mu_n^{k_n}.$$

9.23. The Poincaré–Dulac theorem for diffeomorphisms. *Let $f(x) = Bx + \ldots$ be a local diffeomorphism of $(\mathbb{C}^n, 0)$.*

There is a formal change $x \longmapsto y = x + \ldots$, which formally conjugates f with the map

$$y \longmapsto By + \sum a_{i;k} y^k e_i,$$

where the sum runs over the set of resonant relations for diffeomorphism satisfied by the system of eigenvalues of B.

This theorem is proved in the same way as Theorem 8.14. Here the map f can be treated as the time one phase flow map associated with the vector field $\dot{x} = Ax + \ldots$, with $B = \mathrm{e}^A$.

Applying Theorem 9.23 to the one-dimensional case, we find the following result.

9.24. Theorem.

(a) *If μ is not a root of unity, then the map f is formally linearizable, i.e., it is formally equivalent to the linear map μz.*

(b) *If $|\mu| \neq 1$, then f is analytically linearizable.*

Proof. Only the statement (b) needs some comments. The reduction to the linear form is obtained by means of infinite composition of conjugating maps of the form $z \longmapsto z + g_k z^k$ applied to the maps $\mu z + a_k z^k + \ldots$. The homological equation gives $g_k = a_k/(\mu - \mu^k)$, where $|\mu - \mu^k|$ are bounded from zero by a positive constant. If the coefficients a_k grow at most exponentially, then the coefficients g_k behave in the same way. From this it is easy to obtain a rigorous proof. □

9.25. Remark. Point (b) of Theorem 9.24 has the following analogue in the case of vector fields.

We say that a germ of vector field $\boldsymbol{V} : \dot{x} = Ax + \ldots$ is in the *Poincaré domain* iff the convex hull (in $\mathbb{C}$) of the set of its eigenvalues $\lambda_1, \ldots, \lambda_n$ is separated from 0. Then also the phase flow diffeomorphism $g^1_{\boldsymbol{V}}(x) = \mathrm{e}^A x + \ldots$ is in the Poincaré domain.

If $\boldsymbol{V}$ is in the Poincaré domain, then there is only finitely many resonant relations between the eigenvalues. Thus the homological operator L_A, from the proof of Theorem 9.14, has few eigenvalues $\lambda_i - \sum_j k_j\lambda_j$ equal to zero and all its other eigenvalues are estimated as follows: $|\lambda_i - \sum_j k_j\lambda_j| > \delta \cdot |k|$, where $\delta > 0$ is a constant not depending on $(i; k)$. If the coefficients of the Taylor expansion of $\boldsymbol{V}$ grow at most exponentially with $|k|$, then the coefficients of the Taylor expansion of the conjugating diffeomorphism also grow at most exponentially. This gives the following statement (see also [**Arn5**]):

If an analytic vector field $\dot{x} = Ax + \ldots$ is in the Poincaré domain, then the change $y = x + \ldots$, reducing it to the Poincaré–Dulac normal form, is convergent.

The problem of analytic classification of diffeomorphisms with $|\mu| = 1$, but $\mu^N \neq 1$ for any N, will be considered later (see §7 below).

Now we focus our attention on *germs of analytic diffeomorphisms tangent to identity*

$$f(z) = z + az^{p+1} + \ldots, \quad a \neq 0.$$

Note that, after suitable normalization, we can assume that $a = 1$. Following Ilyashenko in [**Il5**], we denote the space of such germs by $\mathcal{A}_p$.

9.26. Theorem (Formal classification). *Any germ from $\mathcal{A}_p$ is formally equivalent to the germ*

$$f_{p,\lambda} = g^1_{\boldsymbol{W}},$$

where $g^t_{\boldsymbol{W}}$ is the phase flow map after time t of the vector field

$$\boldsymbol{W} = \boldsymbol{W}_{p,\lambda} = \frac{z^{p+1}}{1 + \lambda z^p}\frac{\partial}{\partial z}$$

and $\lambda \in \mathbb{C}$ is the invariant of the formal classification.

Proof. We cancel the suitable terms in f by means of the conjugating maps $z \longmapsto z + b_l z^l$. Simple calculations show that the leading contribution from this change is the term

$$(l - p - 1) b_l z^{l+p}.$$

If $l \neq p+1$, then the corresponding power of z in f can be cancelled. There remains only the power z^{2p+1}.
Thus the formal normal form is $z + z^{p+1} + \nu z^{2p+1}$. Expanding $f_{p,\lambda}$ into powers of z, we see that the latter form is formally equivalent to $f_{p,\lambda}$ for $\lambda = p + 1 - \nu$. □

The space of germs f which are formally equivalent to $f_{p,\lambda}$ is denoted by $\mathcal{A}_{p,\lambda}$.

9.27. Theorem (Topological classification). ([**CS1**], [**Shc1**]) *Any germ from $\mathcal{A}_p$ is topologically equivalent to $g = f_{p,0} = z(1 - z^p)^{-1/p}$.*

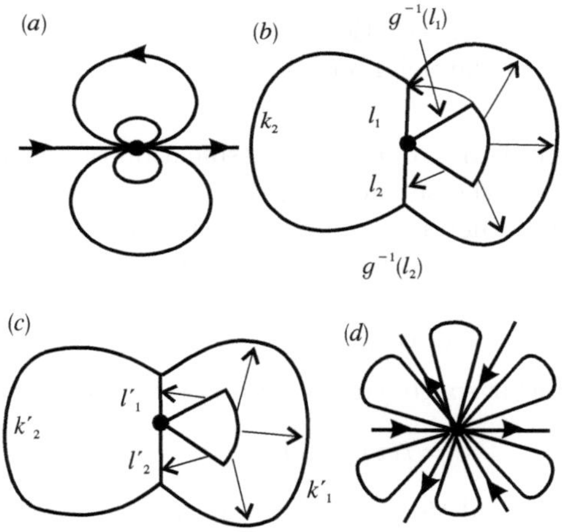

Figure 9.4: Singular vector fields of codimension 1 (a) and 3 (d) and construction of conjugation ((b) and (c))

Proof. We present the proof in the case $p = 1$. Assume that $f(z) = z + z^2 + \ldots$; it is close to $g(z) = \frac{z}{1-z} = z + z^2 + z^3 + \ldots$.
The phase portrait of the vector field $\dot{z} = z^2$ (with real time) is presented at Fig. 9.4(a).
Consider the partition of the punctured neighborhood of the origin by sectors bounded by the curves l_1, l_2 and k_1, k_2, as in Fig. 9.4(b). Here l_1, l_2, k_1, k_2 are transversal to the integral curves of the field $z^2 \frac{\partial}{\partial z}$, ($l_1, l_2$ are straight segments).
The same partition is associated with the map $f \in \mathcal{A}_2$, where we denote the curves by primes, $l'_1 = l_1$, etc. We need that the images and preimages of l'_1, l'_2 under f lie

completely inside the corresponding sectors. Also $f(k_1')$ does not intersect k_1' and $f^{-1}(k_2')$ does not intersect k_2'. This construction is possible due to the fact that f and g are close one to the other.
We construct the conjugating homeomorphism using the above two partitions. Firstly, one defines the homeomorphisms between l_1 and l_1', between l_2 and l_2' and between k_1 and k_1'; it is equal to identity. Then the conjugation condition $h = f^{-1}hg$ implies that we have also a homeomorphism between the curves $l_3 = g^{-1}(l_1)$, $l_4 = g^{-1}(l_2)$, $k_3 = g^{-1}(k_1)$ and the corresponding curves l_3', l_4', k_3'. We prolong this homeomorphism (arbitrarily) to a homeomorphism between the thin domain, bounded by the six curves $l_1, l_2, k_1, l_3, l_4, k_3$, and the analogous domain in Fig. 9.4(c).
This homeomorphism is extended to a homeomorphism of the sector bounded by l_1, l_2, k_1, (here we use the conjugation condition).
After defining the identity homeomorphism between k_2 and k_2', we extend it to the sector bounded by l_1, l_2, k_2 in the same way as before.
In the case of general $p > 2$ the phase portrait of $z^{p+1}\frac{\partial}{\partial z}$ contains $2p$ elliptic sectors (see Fig. 9.4(d)) and the construction of the conjugating homeomorphism needs more refined partition of the neighborhood of $z = 0$. However, the proof remains the same. □

The functional moduli of the analytical classification of germs of conformal mappings are connected with the structure of the space of orbits of the action of such a mapping on a neighborhood of the fixed point. It means that we look at the space U^*/f, where $U^* = U \setminus 0$ is a punctured neighborhood of $z = 0$ and points z and $f(z)$ are treated equivalent. (The space U^*/f is Hausdorff, but the analogous space U/f space is not Hausdorff.)

The structure of the orbit space can be better seen in the t-chart, where

$$t = t(z) = -1/pz^p + \lambda \ln z$$

is the time in the vector field $\boldsymbol{W}_{p,\lambda}$. Because the solution of the equation $\dot{z} = \boldsymbol{W}_{p,\lambda}$ is $t = t(z) + \text{const}$, then the map $f_{p,\lambda} = g^1_{\boldsymbol{W}}$ takes the simple form

$$t \longmapsto t + 1.$$

Assume for a while that $\lambda = 0$ (then $t(z)$ is single-valued) and that $p = 1$. Then the neighborhood $U^* = \{|z| < r\}$ replaced by $W = \{|t| > R\}$. The set W is divided into the vertical strips of width 1. Identifying two sides of such strip, we get a cylinder which can be identified with the doubly punctured sphere $\mathbb{C}^*$: $-\mathrm{i}\infty \longmapsto 0$, $+\mathrm{i}\infty \longmapsto \infty$. The space $W/(id + 1)$ consists of two copies of $\overline{\mathbb{C}}$ with identifications near 0 and ∞, as in Fig. 9.5. The vertical strips from the t-chart correspond to the crescents in the z-chart.
In the general case (λ and p arbitrary and f close to $f_{p,\lambda}$) the chart $t(z)$ is a well defined diffeomorphism in sectors

$$S_j = \{(j/p)\pi - \alpha < \arg z < (j/p)\pi + \alpha\},$$

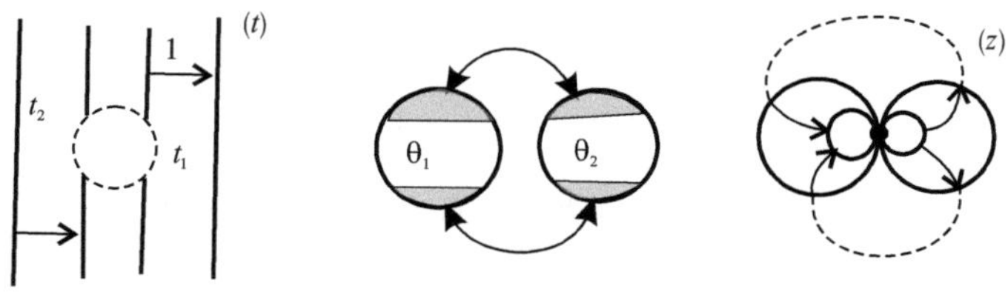

Figure 9.5: Actions of diffeomorphisms in the t-plane and the z-plane and gluing diffeomorphisms in the θ coordinates

$j = 1, \dots, 2p$, where $\alpha > 0$ is small.

The function $t(z)$ sends S_j to $\widetilde{S}_j$. In $\widetilde{S}_j$ we denote $t_j = t(z)|_{S_j}$ the local charts, where the branches of $\ln z$ are such that they are real in $S_1 \cap \mathbb{R}$, $t_j = t_{j+1}$ in the intersections $S_j \cap S_{j+1}$ for $j < 2p$, and $t_{2p} = t_1 + 2\pi \mathrm{i}\lambda$.
The maps $f_{p,\lambda}$ and f are replaced by the mappings

$$\begin{aligned} \tilde{f}_{p,\lambda} : \quad t_j &\longmapsto \quad t_j + 1, \\ \tilde{f} : \quad t_j &\longmapsto \quad t_j + 1 + R_j(t_j), \end{aligned}$$

in $\widetilde{S}_j$. Here we can assume that

$$R_j(t) = O(t^{-M})$$

for some large M. The vertical strips from Fig. 9.5 and orbit spaces will be only slightly perturbed. The vertical lines are replaced by the iterations of one line, $f^n(l_0)$ where, due to the fact that $R(t)$ is small, there are no intersections between them.
Therefore, U^*/f is topologically equivalent to $2p$ copies of $\overline{\mathbb{C}}$ with suitable identifications of the neighborhoods of 0 and ∞. More precisely, we have the spheres $\overline{\mathbb{C}} \times \{j\}$, $j = 1, \dots, 2p$, and identifying maps of the form:

$$\begin{aligned} \phi_j : \quad (\overline{\mathbb{C}}, 0) \times \{j\} \quad &\longmapsto \quad (\overline{\mathbb{C}}, 0) \times \{j+1\}, \quad j \text{ even}, \\ \phi_j : \quad (\overline{\mathbb{C}}, \infty) \times \{j\} \quad &\longmapsto \quad (\overline{\mathbb{C}}, \infty) \times \{j+1\}, \quad j \text{ odd}. \end{aligned}$$

The functional invariants of the classification of germs of maps are the invariants of the holomorphic structure on the space U^*/f.

9.28. Proposition. *The quotient spaces $\widetilde{S}_j/\tilde{f}$ are conformally equivalent to $\mathbb{C}^*$.*

Proof. This quotient space, which we denote by T, is obtained from the (non-straight) strip between the lines $l_0 = \{\mathrm{Re}(t) = c\}$ and $\tilde{f}(l_0)$ by identifying its sides. Here $\tilde{f}(c + \mathrm{i}\tau) = c + \mathrm{i}\tau + 1 + R_j(c + \mathrm{i}\tau)$, where $R(t)$ is small.
We treat $\mathbb{C}^*$ as the straight strip $c \le \mathrm{Re}(t) \le c + 1$ with identified sides. It is parametrized by $\theta \in [0, 1]$ and $\tau \in \mathbb{R}$: $t = c + \theta + \mathrm{i}\tau$.

We construct a map from $\mathbb{C}^*$ to T

$$F(c+\theta+\mathrm{i}\tau)=c+\theta+\mathrm{i}\tau+\theta R_j(t).$$

Because $F'_t = 1+R(t)/2+\theta R'_j(t)$, $F'_{\bar{t}} = R_j(t)/2$, the map F is quasi-conformal and satisfies the condition of regularity from Example 9.12. The modulus of quasi-conformality $\mu_F = F'_{\bar{t}}/F'_t$ is small and tends to zero at the ends of the strip $\{c<\mathrm{Re}(t)<c+1\}$; they correspond to 0 and ∞ in $\overline{\mathbb{C}}$.
Applying the Proposition 9.13 form the previous section, we get the assertion of Proposition 9.28. □

In the spheres one uses conformal coordinates ϑ_j. (If $f=f_{p,\lambda}$, then one can take $\vartheta_j = \mathrm{e}^{2\pi \mathrm{i} t_j}$.)
ϑ_j are defined uniquely modulo multiplication by a constant. If we have chosen ϑ_1, then ϑ_2 can be taken in such a way that $\vartheta_2=\phi_1(\vartheta_1)=\vartheta_1+o(\vartheta_1)$ as $\vartheta_1\to\infty$, $\vartheta_3=\phi_2(\vartheta_2)=\vartheta_2+o(\vartheta_2)$ is such that $\phi'_2(0)=1$, etc. However the charts ϑ_{2p} and ϑ_1 near ∞ cannot be correlated in this way. We have $\vartheta_{2p}=\mathrm{e}^{2\pi\mathrm{i}t_{2p}}=\mathrm{e}^{2\pi\mathrm{i}t_1-4\pi^2\lambda}=\nu\vartheta_1$. Thus $\phi_{2p}(\vartheta_{2p})=\nu\vartheta_{2p}+o(\vartheta_{2p})$, $\nu=\exp(-4\pi^2\lambda)$.
Changing the chart $\vartheta_1\longmapsto C\vartheta_1$ induces the same changes in the spheres $\overline{\mathbb{C}}_j$ and the maps ϕ_j become conjugated by means of C. The above suggests the following definition.

9.29. Definition (Moduli space). $\mathcal{M}^*_{p,\lambda}$). Consider the space of collections $\phi=(\phi_1,\ldots,\phi_{2p})$ of germs of maps $\phi_j:(\overline{\mathbb{C}},0(\infty))\times\{j\}\longmapsto(\overline{\mathbb{C}},0(\infty))\times\{j+1\}$ with identity linear parts for $j<2p$ and $\phi'_{2p}(\infty)=\nu$. Two such collections ϕ,ϕ' are called *equivalent* $\phi\sim\phi'$ iff there exists $C\in\mathbb{C}^*$ such that

$$\phi_i\circ C = C\phi'_i.$$

The set of equivalence classes of such ϕ's is denoted by $\mathcal{M}^*_{p,\lambda}$.

If $f\in\mathcal{A}_{p,\lambda}$, then the construction (with the strips in $\widetilde{S}_j$'s and identifications of U^*/f with a collection of glued together Riemann spheres) defines a collection ϕ. Thus, we have a mapping from $\mathcal{A}_{p,\lambda}$ to $\mathcal{M}^*_{p,\lambda}$. It is the first statements of the following theorem.

9.30. Ecalle–Voronin classification theorem. *There is a map $f\longmapsto\mu^*_f$ from $\mathcal{A}_{p,\lambda}$ to $\mathcal{M}^*_{p,\lambda}$ with the following properties:*

(a) *If f and $\tilde{f}$ are holomorphically equivalent, then $\mu^*_f=\mu^*_{\tilde{f}}$ (invariance).*

(b) *If $\mu^*_f=\mu^*_{\tilde{f}}$, then f and $\tilde{f}$ are holomorphically equivalent (equimodality).*

(c) *For any class $[\phi]\in\mathcal{M}^*_{p,\lambda}$ there exists $f\in\mathcal{A}_{p,\lambda}$ such that $[\phi]=\mu^*_f$ (realization).*

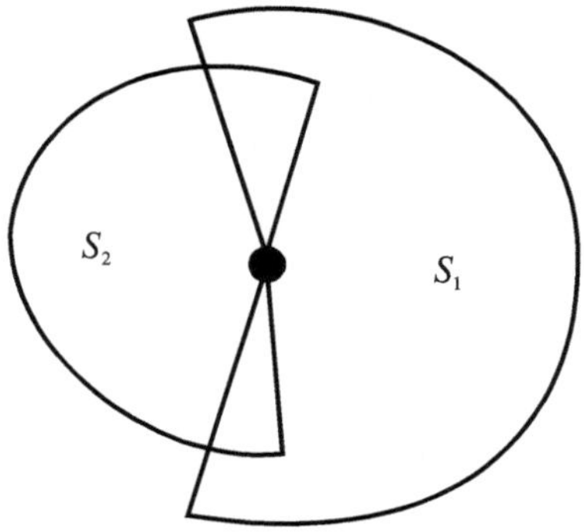

Figure 9.6: Sectors for normalizations in Th. 9.31

Remark. J. Ecalle [**Ec1**] proved a variant of this result in 1975. It was a starting point of his theory of resurgent functions [**Ec2**]. That theory relies on exploiting the Borel transform. It has applications in other fields (linear meromorphic systems with irregular singularities, the problem of limit cycles, relaxational oscillations, quantum field theory). It stimulated development of the Gevrey analysis and summability theories.
S. Voronin's proof appeared in 1981 [**Vor**] and it relies on application of analytic methods (almost complex structures and quasi-conformal mappings). Below we present the Voronin's proof (following the Ilyashenko's paper [**Il5**]). In the next section we present the main ideas lying behind the Ecalle's approach.
However, it must be said that the Ecalle–Voronin moduli were already described by G. Birkhoff in 1939 in [**Bir2**] and the sectorial normalization (i.e., conjugation with translation) was constructed about century ago by L. Leau [**Lea**]. Thus the theory is not as new and as extraordinary as it was publicized. Only some old works fell into oblivion.
Among other mathematicians who contributed into this field we mention N. Abel, P. Fatou, G. Szekers, P. Erdös, E. Jabotinsky, J. Hadamard, J. Baker, T. Kimura, J. Rey, L. Liverpool, and M. Kuczma [**Ku**]. Very interesting review of the history of the problem of analytic classification is presented in the lectures of F. Loray [**Lor2**].

The essential part of the proof of the Ecalle–Voronin theorem is the fact that the dynamics defined by the map f and by the model map are conjugated holomorphically in sectors (see Fig. 9.6).

9.31. Sectorial normalization theorem. *In every sector S_j there is an unique analytic diffeomorphism of the form*

$$H_j(z) = z + h_j(z), \quad h_j = o(z^{p+1}),$$

conjugating f with $f_{p,\lambda}$.

Proof. It is enough to construct the conjugating diffeomorphisms in the t-charts, i.e., $\widetilde{H}_j(t) = t + \tilde{h}_j(t)$, $t \in \widetilde{S}_j$. Then $H_i = t^{-1} \circ \widetilde{H}_j \circ t$.
The conjugation condition $\widetilde{H}_j \circ \tilde{f} = \tilde{f}_{p,\lambda} \circ \widetilde{H}_j$, i.e., $t+1+R_j+\tilde{h}_j \circ \tilde{f} = t+\tilde{h}_j+1$, means that

$$\tilde{h}_j = \tilde{h}_j \circ \tilde{f} + R_j.$$

Iterating this equation (with respect to $\tilde{h}_j$) infinitely many times, we find the solution

$$\tilde{h}_j(t) = \sum_0^\infty R_j \circ \tilde{f}^n(t).$$

The condition $|R_j| < |t|^{-M}$ and the property $\tilde{f}^n(t) \sim O(n)$ ensures that this series is convergent and represents an analytic function of t. □

Proof of Theorem 9.30. 1. *The normalizing cochain and its coboundary.* Take the system H_j of germs diffeomorphisms from Theorem 9.30. It can be interpreted as certain Čech cochain. Namely, we consider the (non-Abelian) sheaf on $\mathbb{S}^1$ of germs of analytic diffeomorphisms strongly tangent to identity. If U is an open arc in $\mathbb{S}^1$, then the group of sections of this sheaf at U consists of diffeomorphism $H_{S(U)} = id + h_{S(U)}$, $h_{S(U)}(z) = O(z^{p+1})$ defines on a sector $S(U)$ with base at U (in the polar coordinates $S(U) = \{z = r\mathrm{e}^{\mathrm{i}\theta},\ \theta \in U\}$, see [**MR1**]).
The system

$$H = (H_1, \dots, H_{2p})$$

defines the Čech cochain associated with the covering of $\mathbb{S}^1$ induced by the sectors S_i.
We consider the coboundary of this cochain, i.e., $\delta H = (\Phi_1, \dots, \Phi_{2p})$,

$$\Phi_j = H_{j+1} \circ H_j^{-1}$$

in $S_j \cap S_{j+1}$. (The system (Φ_j) will correspond to the functional modulus $\phi \in \mathcal{M}^*_{p,\lambda}$ from definition 9.29).

2. **Remark.** *If we skip the assumption $h_j = o(z^{p+1})$ in the sectorial normalization theorem, then the normalizing maps H_j cease to be unique. We can replace H_j by $g^\theta_{\boldsymbol{W}} \circ H_j$ for some $\theta \in \mathbb{C}$. (Here $g^\theta_{\boldsymbol{W}}$ is the phase flow generated by the vector field $\boldsymbol{W} = [z^{p+1}/(1+\lambda z^p)]\partial_z$).*

In the t-charts this non-uniqueness means that the coordinates t_j are defined uniquely up to a translation; note that $g^\theta_{\boldsymbol{W}} = id + \theta$. It is the whole non-uniqueness in the definition of $\widetilde{H}_j$.
Indeed, if $\widetilde{G}_j$ is another normalizing diffeomorphism, then the map $\widetilde{G}_j \circ \widetilde{H}_j^{-1}$ is defined on a right half-plane $\mathrm{Re}(t) > c$, commutes with translation $id+1$ and tends to $id+$const at infinity. Therefore, it can be prolonged to the whole complex t-plane with the same properties. But any such map is a translation (apply the Liouville theorem to $\widetilde{G}_j \circ \widetilde{H}_j^{-1} - id$). □

3. **Lemma.**

(a) *The maps* Φ_j *commute with the standard map* $f_{p,\lambda}$.

(b) *They are exponentially close to identity*

$$|\Phi_j(z) - z| < \mathrm{e}^{-c/|z|^p};$$

equivalently, $H_j - H_{j+1}$ *are exponentially small.*

(c) *They are defined uniquely up to conjugation with the map* $g^{\theta}_{\mathbf{W}}$ *for some* $\theta \in \mathbb{C}$.

Proof. (a) The first statement follows from the identity $H_j \circ f = f_{p,\lambda} \circ H_j$.

(b) The second property follows from the first one. Indeed, in the t-chart the commutativity of the induced map

$$\widetilde{\Phi}_j(t) = t + \tilde{\phi}_j(t)$$

with the standard map $t \longmapsto t+1$ means that the functions $\tilde{\phi}_j(t)$ are periodic, $\tilde{\phi}_j(t+1) = \tilde{\phi}_j(t)$. Thus $\tilde{\phi}_j$ are expanded into the Fourier series $\sum c_{jl}\mathrm{e}^{2\pi \mathrm{i} lt}$. The function Φ_j is analytic in $S_j \cap S_{j+1}$, which corresponds to the domain $\mathrm{Im}(t) > t_0 > 0$ in the $t = t_j$-chart in $\widetilde{S}_j$, if j is odd, and to the domain $\mathrm{Im}(t) < -t_0$, if j is even. Moreover, $\widetilde{\Phi}_j(t) = t + o(t)$ as $t \to \infty$ (because $\widetilde{H}_j$ have this property). This implies that

$$\begin{aligned}
\widetilde{\Phi}_j(t_j) &= t_j + \sum_{l>0} c_{jl}\mathrm{e}^{2\pi \mathrm{i} lt_j}, \quad j \text{ odd},\\
\widetilde{\Phi}_j(t_j) &= t_j + \sum_{l<0} c_{jl}\mathrm{e}^{2\pi \mathrm{i} lt_j}, \quad j \text{ even},\\
\widetilde{\Phi}_{2p}(t_{2p}) &= -2\pi \mathrm{i}\lambda + t_{2p} + \sum_{l<0} c_{jl}\mathrm{e}^{2\pi \mathrm{i} lt_j}.
\end{aligned} \tag{9.5}$$

The constant in the latter formula follows from the difference between the local charts $t_{2p} = t_1 + 2\pi \mathrm{i}\lambda$ and t_1.

Now we have $\Phi_j = t_{j+1}^{-1} \circ \widetilde{\Phi}_j \circ t_j$ with $t_j = -1/(pz^p) + \lambda \ln z$. If j is odd, then we see that $\Phi_j(z) = z + O(|c_{j1}z^{\alpha}\mathrm{e}^{-2\pi \mathrm{i}/(pz^p)}|)$. Analogous estimates hold in the case of even j.

(c) This point follows from Remark 2. □

4. **Remark.** The coboundary 1-cochain δH takes values in a sheaf analogous to the Stokes sheaf defined in Definition 8.28.

5. **Definition (space $M^{+}_{p,\lambda}$ of functional moduli).** Consider the space of systems $\widetilde{\Phi} = (\widetilde{\Phi}_1, \ldots, \widetilde{\Phi}_{2p})$ of the form (9.5). Two systems are called equivalent, $\widetilde{\Phi} \sim \widetilde{\Phi}'$ iff $\widetilde{\Phi}'_j = (id - \theta) \circ \widetilde{\Phi}_j \circ (id + \theta))$. The space of equivalence classes of the systems $\widetilde{\Phi}$ is denoted by $\mathcal{M}^{+}_{p,\lambda}$.

The construction of the functional coboundary $\widetilde{\Phi}$ defines a map

$$f \longmapsto \mu_f^+ \in \mathcal{M}_{p,\lambda}^+.$$

6. **Lemma**. *The moduli space* $\mathcal{M}_{p,\lambda}^+$ *is diffeomorphic with the moduli space* $\mathcal{M}_{p,\lambda}^*$ *from Definition* 9.29.

Proof. Recall that the moduli space $\mathcal{M}_{p,\lambda}^*$ is the space of conformal structures on the orbit space U^*/f. Thus we should associate with a system $\widetilde{\Phi}$ a system of conformal charts on the spaces S_j/f (or $\widetilde{S}_j/\tilde{f}$) and a system of gluing maps between them.

We introduce new charts $\tau_j := \widetilde{H}_j(t_j) = t_j + \dots$ in $\widetilde{S}_j$ having the property that they conjugate f with the translation $id+1$. They are defined by the formulas

$$\tau_j = t_j \circ H_j.$$

One easily checks that $\tau_j \circ f \circ \tau_j^{-1} = id + 1$. It is also clear that the space $\widetilde{S}_j/\tilde{f}$ is conformally equivalent to the τ_j-plane divided by the action of the translation $id+1$.

The identification of $\widetilde{S}_j/\tilde{f}$ with $\mathbb{C}^* = \mathbb{C}^* \times \{j\}$ is given by the formula

$$\tau_j \longmapsto \vartheta_j = \mathrm{e}^{2\pi\mathrm{i}\tau_j}.$$

The gluing maps $\vartheta_j \longmapsto \vartheta_{j+1} = \phi_j(\vartheta_j)$ (see Definition 9.29) arise from the differences between τ_j and τ_{j+1} in $\widetilde{S}_j \cap \widetilde{S}_{j+1}$. We have $\phi_j = [\exp 2\pi\mathrm{i}(\cdot)] \circ (\tau_{j+1} \circ \tau_j^{-1}) \circ [(1/2\pi\mathrm{i})\ln(\cdot)]$. Therefore we should find an expression for $\tau_{j+1} \circ \tau_j^{-1}$. We have

$$\tau_{j+1} \circ \tau_j^{-1} = t_{j+1} \circ H_{j+1} \circ H_j^{-1} \circ t_j^{-1} = \widetilde{\Phi}_j.$$

If $\widetilde{\Phi}_j$ are defined by formulas (9.5), then $\phi_j(\vartheta_j) = \exp\{2\pi\mathrm{i}[\tau_j + \sum c_{jl}\vartheta_j^l]\}$, or

$$\phi_j = \vartheta_j + \sum d_{j,l}\vartheta_j^l,$$
$$\phi_{2p} = \nu\vartheta_{2p} + \sum d_{jl}\vartheta_{2p}^l.$$

(Above the sums run either over positive or over negative l's).

Like t_j, the variables τ_j are defined uniquely modulo translation by a constant θ. This means that the charts ϑ_j in $\mathbb{C}^* \times \{j\}$ are defined uniquely modulo multiplication by the constant $C = \mathrm{e}^{2\pi\mathrm{i}\theta}$. □

7. *Invariance of the class* μ_f^+. Let f, g be two analytically equivalent germs from the class $\mathcal{A}_{p,\lambda}$, h being the conjugating holomorphism: $h \circ f = g \circ h$. Let H, G be the cochains normalizing f, g respectively. Then the cochain $G \circ h$ normalizes f. From Remark 2 we find that $g \circ h = g_w^\theta \circ H$, (as H and $G \circ h$ conjugate the same f with $f_{p\lambda}$). So,

$$G_{j+1} \circ G_j^{-1} = g_w^\theta \circ (H_{j+1} \circ H_j^{-1}) \circ g_w^{-\theta}.$$

This means that, if $\widetilde{\Phi}$, $\widetilde{\Psi}$ are the coboundaries associated with f, g (expressed in the t-charts), then they are equivalent by means of the map $id + \theta$ (in the sense of Definition 6).

8. *Equimodality.* Let $\widetilde{\Phi}$, $\widetilde{\Psi}$ be two collections associated with maps f, g respectively. Assume that they are equivalent in the sense of Definition 6. Therefore, choosing another normalizing cochain, we can obtain the coincidence of the cochains $\widetilde{\Phi}$ and $\widetilde{\Psi}$.
But the identity $G_{j+1} \circ G_j^{-1} = H_{j+1} \circ H_j^{-1}$ is equivalent to

$$H_j^{-1} \circ G_j = H_{j+1}^{-1} \circ G_{j+1}.$$

This means that the maps $H_j^{-1} \circ G_j$ are prolonged to a punctured neighborhood of $z = 0$ as a single valued mapping. By the Riemann removability of singularities theorem we prolong it to an analytic map h in a whole neighborhood of 0.
Moreover, because G_j conjugates g with $f_{p,\lambda}$ and H_j^{-1} conjugates $f_{p,\lambda}$ with f, then h conjugates g with f.

9. *Realization.* Let $\mu^+ \in \mathcal{M}_{p,\lambda}^+$ be an element of the moduli space. We have to show that $\mu^+ = \mu_f^+$ for some map f.
We choose a representative $\widetilde{\Phi}$ of μ^+ and associated with it the 1-cocycle $\Phi = (\Phi_j)$ of germs of holomorphic maps in the sectors $S_j \cap S_{j+1}$, strongly tangent to the identity.
Consider the disjoint collection of the sectors, i.e., $S_j \times \{j\}$. Denote by z_j the charts induced by (z, j). We glue these sectors using the mappings $\Phi_j : \ z_{j+1} \sim \Phi_j(z_j)$, as in Fig. 9.7. We obtain certain topological space $\mathcal{S}$ homeomorphic to a punctured disc. This space has a conformal structure arising from the coordinates z_j. The space $\mathcal{S}$ (and its closure $\overline{\mathcal{S}}$) admits a homeomorphism f_0 induced from $f_{p,\lambda}$ at each $S_j \times \{j\}$. f_0 is analytic with respect to the conformal structure. Our task is to show that the conformal structure of $\overline{\mathcal{S}}$ coincides with the conformal structure of the punctured disc and that f_0 belongs to $\mathcal{A}_{p,\lambda}$.
Let $\chi_j(z)$ be an infinitely smooth partition of unity, associated with the covering S_j. We need that the functions χ_j and their gradients have at worst power type singularities as $z \to 0$.
We construct the map

$$H_0 = \sum \chi_j(z_j) z_j$$

from $\mathcal{S}$ to a punctured disc $\mathbb{D}^* \subset \mathbb{C}$. The mapping H_0 is quasi-conformal (see §1). Indeed, in $S_j \cap S_{j+1}$ we have $H_0 = \chi_j z_j + \chi_{j+1} z_{j+1} = z_j + (\Phi_j(z_j) - z_j)\chi_{j+1}$, which implies

$$\partial H_0/\partial z_j = 1 + o(1), \ \partial H_0/\partial \bar{z}_j = (\Phi_j - z_j)\partial \chi_j/\partial \bar{z}_j = o(1).$$

This means that the almost complex structure, induced on $\mathbb{D}^*$ by H_0, is defined by the Beltrami differential $\mu(\mathrm{d}\bar{z}/\mathrm{d}z)$, where $\mu = \mu(H_0^{-1})$ is exponentially small together with its derivatives.

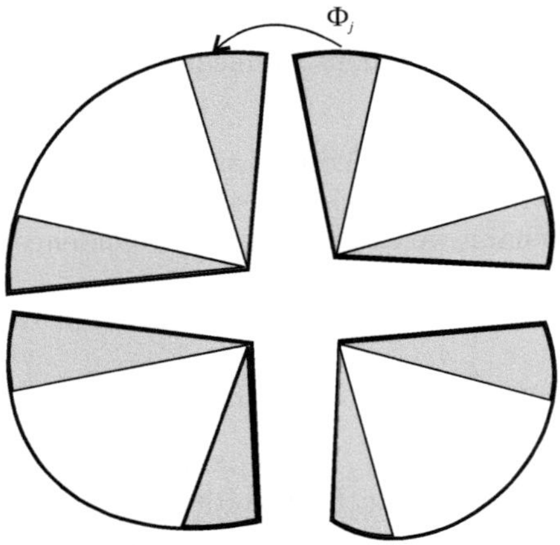

Figure 9.7: Domains which are glued

Let us prolong μ to the whole disc $\mathbb{D}$. By Theorem 9.10 the almost complex structure on $\mathbb{D}$ is integrable and there exists a diffeomorphic mapping $G : \mathbb{D} \longmapsto \mathbb{C}$ such that $\mu_G = \mu$. This means that the that composition $H = G \circ H_0 : \mathcal{S} \longmapsto \mathbb{C}$ is conformal and $\mathcal{S}$ is conformally equivalent to $\mathbb{D}^*$.
The mapping f_0 in $\mathcal{S}$, induced from $f_{p,\lambda}$, is prolonged to a mapping of $\mathbb{D}$. We denote it by f. Of course, it is analytic.
Because the functions μ, G, H are smooth and f is conjugated by means of H with $f_{p,\lambda}$ at the sectors S_j, then their linear parts are also conjugated. So, $f'(0) = 1$ and $f \in \mathcal{A}_l$ for some l.
But the index l is a topological invariant (see Theorem 9.27), it is the number of 'elliptic sectors' (divided by 2). So $l = p$.
Finally, the formal classification invariant λ is included into the functional modulus $\widetilde{\Phi}_{2p} = -2\pi \mathrm{i}\lambda + t_{2p} + \ldots$.
Note that the maps $H_j = z_j \circ H^{-1}$ form the normalizing maps for f (the normalizing cochain). Its coboundary maps are $z_{j+1} \circ {z_j}^{-1} = \Phi_j$. This means that the collection $\widetilde{\Phi}$ is the functional modulus for f.

Theorem 9.30 is complete. □

Using Ecalle–Voronin Theorem we can obtain other results about analytic classification of germs of 1-dimensional diffeomorphisms. These results say about extraction of roots from such maps, about their embeddings into flows and about analytic classification of germs with resonant eigenvalue.

9.32. Theorem (Extraction of roots and embedding). *Let a germ $f \in \mathcal{A}_{p,\lambda}$ has the Ecalle–Voronin modulus $\mu_f^* = [(\phi_1, \ldots, \phi_{2p})] \in \mathcal{M}_{p,\lambda}^*$.*

(a) *f admits extraction of the n-the root tangent to identity,*

$$\exists g: \ g^n = f, \ \ g'(0) = 1,$$

iff the modulus μ_f^* commutes with the multiplication by $\mathrm{e}^{2\pi \mathrm{i}/n}$; i.e., the gluing maps take the form $\phi_j(\vartheta) = \vartheta\psi_j(\vartheta^n)$.

(b) *f can be imbedded into certain flow g_v^t (1-parameter family of diffeomorphisms generated by a vector field v) iff f has trivial Ecalle–Voronin modulus, $\mu_f^* = (id, \ldots, id, \nu)$; i.e., f is analytically equivalent to $f_{p,\lambda} = g_w^1$.*

Proof. (a) If $f = g^n$, then g commutes with f and maps orbits of f to orbits of g. The latter induces the mapping $\pi_* g : U^*/f \longmapsto U^*/f$. Recall that U^*/f is a collection of doubly punctured Riemann spheres $\mathbb{C}^* \times \{j\}$, glued into a chain by means of the diffeomorphisms ϕ_j.
The map $\pi_* g$, restricted to each sphere, is holomorphic and its nth iterate is identity. Thus $\pi_* g(\vartheta_j) = \mathrm{e}^{2\pi \mathrm{i}/n}\vartheta_j$. This action of $\pi_* g$ is compatible with the gluing maps ϕ_j, which means the commutativity of ϕ_j with $\mathrm{e}^{2\pi \mathrm{i}/n}$.
(b) If f is embeddable into flow, then the gluing maps commute with multiplication by any root of unity. So, they commute with multiplication by any number ζ with absolute value 1: if $\phi_j(z) = \sum a_k z^k$, then $\zeta \sum a_k z^k = \sum a_k z^k \zeta^k$. Only linear maps satisfy this condition. This implies the triviality of μ_f^*.
On the other hand, any map holomorphically equivalent to the $f_{p,\lambda}$ is embeddable. □

Now we consider conformal maps of the form

$$f(z) = \mathrm{e}^{2\pi \mathrm{i} m/n} z + \ldots,$$

i.e., $f'(0)$ is a primitive root of unity of order n.

9.33. Proposition. *Any germ of this form is formally equivalent to*

$$\mathrm{e}^{2\pi \mathrm{i} m/n} g_{\boldsymbol{W}}^1, \quad \boldsymbol{W} = \frac{z^{nk+1}}{1+\lambda z^{nk}} \frac{\partial}{\partial z}.$$

Proof. The Poincaré–Dulac theorem for diffeomorphisms allows to reduce f to the form $\mathrm{e}^{2\pi \mathrm{i} m/n} z(1 + \sum a_l z^{nl})$. Let k be the first index l with nonzero a_l. We can assume that the coefficient $a_k = 1$.
Next, repeating the proof of Theorem 9.26, we successively cancel the terms $a_l z^{nl+1}$ using the changes $z \longmapsto z + b_l z^{n(l-k)}$. It is possible to do it in all cases but $l = 2k$. □

One denotes by $\mathcal{A}_{m,n,k,\lambda}$ the space of germs with the formal normal form having the indicated indices.
Let us pass to analytic classification of the germs from $\mathcal{A}_{m,n,k,\lambda}$. Notice firstly that f^n is tangent to identity, it belongs to $\mathcal{A}_{p,\beta}$, $p = nk$, with some β (defined by λ).

Consider the orbit space U^*/f^n, consisting of $2nk$ Riemann spheres $\mathbb{C}^* \times \{j\}$. The map f transforms the orbits of f^n to orbits of f^n and induces the map $\pi_* f : U^*/f^n \longmapsto U^*/f^n$. We describe the action of this map on the Riemann spheres.
Note that the sphere $\mathbb{C}^* \times \{j\}$ is the quotient of the sector S_j with bisectrix $\arg z = \pi(j-1)/(2kn)$ and the transformation f is approximately the rotation by the angle $2\pi m/n$. Therefore, $f(S_j) \approx S_{j+2km}$ and $\pi_* f : \mathbb{C}^* \times \{j\} \longmapsto \mathbb{C}^* \times \{j + 2km\}$. Moreover, $\pi_* f$ is a holomorphic diffeomorphism and takes the form $\pi_* f(\vartheta_j) = C_j \vartheta_{j+2km}$. It means also that $\vartheta_{j+2kn} \circ f = C_j \vartheta_j$. One can see that $C_j = \mathrm{e}^{2\pi \mathrm{i}/n}$. Indeed, because $\phi_j = \vartheta_{l+1} \circ \vartheta_l^{-1}$ are close to identity (or to multiplication by the constant ν for $l = 2p = 2kn$), then $\vartheta_{j+2kn+1} \circ \vartheta_{j+2kn}^{-1} = \vartheta_{j+2kn+1} \circ f \circ f^{-1} \circ \vartheta_{j+2kn}^{-1} = C_{j+1} \vartheta_{j+1} \circ \vartheta_j^{-1} \circ C_j^{-1} = C_{j+1} \phi_j \circ C_j^{-1} = id + \ldots$. So, $C_j = C_{j+1} = \ldots = C$. Because $\pi_* f^n = id$, $C^n = 1$.
From the above it follows that only $2k$ first gluing maps are needed to determine the functional modulus $\mu^*_{f^n}$. Indeed, we have $\phi_{j+2kn} = \mathrm{e}^{2\pi \mathrm{i}/n} \phi_j \circ \mathrm{e}^{-2\pi \mathrm{i}/n}$. For the last gluing map we get $\phi_{2kn} = \nu \mathrm{e}^{2\pi \mathrm{i}/n} \phi_{2(k-1)n} \circ \mathrm{e}^{-2\pi \mathrm{i}/n}$, where $\nu = \phi'_{2kn}(\infty)$ is the constant expressed by the formal modulus λ.
All this leads to the following result.

9.34. Theorem (Analytic classification of germs of resonant one dimensional diffeomorphisms). *The space $\mathcal{A}_{m,n,k,\lambda}$ of germs of conformal diffeomorphism modulo analytic equivalence is in one-to-one correspondence with the space $\mathcal{M}^*_{m,n,k,\lambda}$ of systems*

$$\phi = (\phi_1, \ldots, \phi_{2k})$$

(of germs of analytic diffeomorphisms $\phi_j(\overline{\mathbb{C}}, 0) \longmapsto (\overline{\mathbb{C}}, 0)$ (j odd) or $\phi_j(\overline{\mathbb{C}}, \infty) \longmapsto (\overline{\mathbb{C}}, \infty)$ (j even) with identity linear parts) modulo the equivalence $\phi \sim D\phi \circ D^{-1}$.

9.35. Note on unfoldings. Recall that in the point 8.32 we presented efforts aimed at generalization of the Malgrange–Sibuya moduli of irregular singularities of linear meromorphic systems to unfolding of such systems.
Here we present analogous activity associated with unfoldings of 1−dimensional local diffeomorphisms, due to A. Glyutsuk [**Gl2**] and C. Rousseau [**MRR**] with collaborators. Below I use a my analysis of this case from the seminal paper [**Zo14**].

Let us consider a deformation of a resonant singularity. For simplicity assume that it is tangent to identity and of codimension 1. Thus we have $f_\mu(z) = f_0(z) + \mu f_1(z, \mu) = a_0(\mu) + a_1(\mu) z + \ldots$, where $f_0(z) = z + z^2 + bz^3 + \ldots$ and $\mu \in (\mathbb{C}, 0)$ is a parameter. A natural non-degeneracy assumption is following:

$$a_0'(0) \neq 0.$$

So, after some change of the parameter (induction) and of the coordinate variable z depending on μ (equivalence), we get

$$f_\mu(z) = -\mu + z + z^2 + c\mu z + d\mu^2 + \ldots.$$

We have the curve $\{f_\mu(z) = z\} = \{\mu = z^2 + \ldots\}$ of fixed points: $z_{1,2} = \pm\sqrt{\mu} + \ldots$. Moreover,

$$\lambda_{1,2} = f_\mu'(z_{1,2}) = 1 + 2z_{1,2} + c\mu + \ldots = 1 \pm 2\sqrt{\mu} + \ldots.$$

If the parameter μ is real positive, then the fixed points $z_{1,2}(\mu)$ are hyperbolic, z_1 is an attractor and z_2 is a repeller. One can construct (ring like) fundamental domains $D_1(f_\mu)$ around z_1 and $D_2(f_\mu)$ around z_2, they tend to the crescents $D_{1,2}(f_0)$ as $\mu \to 0$. Also $f^N(D_1(f_\mu))$ intersect $D_2(f_\mu)$ along suitable segment. So, one can construct two diffeomorphisms $h_1^{f_\mu}$ and $h_2^{f_\mu}$ in neighborhoods of the fixed points, which conjugate f_μ with maps from some model family, e.g. the time one flow map $g^1_{\boldsymbol{V}_\mu}$ for the vector field

$$\boldsymbol{V}_\mu(z) = \frac{z^2 - \mu}{1 + az}\frac{\partial}{\partial z}.$$

(One should make some preparations in order that the multipliers of the fixed points for the both families coincide; this can be achieved by additional change of the parameter and of the coordinate.) Therefore a version of functional moduli $\Phi_1^{f_\mu} = h_2^{f_\mu} \circ \left(h_1^{f_\mu}\right)^{-1}$ and $\Phi_2^{f_\mu} = h_1^{f_\mu} \circ \left(h_2^{f_\mu}\right)^{-1}$ for deformations seems possible. More precisely, it is possible when the points $z_{1,2}$ are hyperbolic; it takes place when μ is outside a small sector with the bisectrix along the ray $\{\mu < 0\}$. Moreover, since the model map depends regularly on μ, one has defined the limits $\lim_{\mu\to 0}\Phi_1^{f_\mu}$ and $\lim_{\mu\to 0}\Phi_2^{f_\mu}$ in suitable sectors $S_{1,2}$ and $S_{2,1}$. I refer to the paper [**Gl2**] by A. Glyutsuk for more details.

But one cannot avoid the possibilities, when one of the multipliers lies in the unit circle. For definiteness let $\lambda_2 = e^{2\pi i\alpha}$, $\alpha \in (\mathbb{R}, 0)$; this holds along the curve

$$\Gamma = \{\mu = \mu(\alpha) = -\pi^2\alpha^2 + \ldots\}$$

in the μ-plane.
If $\alpha = m/n$ is rational (and n is very large) then we have the Ecalle–Voronin moduli associated with very thin sectors. I do not see how to relate the fundamental domains associated with $f_{\mu(m/n)}$ with the fundamental domains for f_0.
If α is irrational and weakly approximated by rationals then $f_{\mu(\alpha)}$ has first integral (a function invariant for the map) and orbits of points near the fixed point are dense in the level curves of the first integral; anyway, there are no sensible fundamental domains. If α fails to obey the Briuno condition then it is impossible to predict its dynamics near z_2. See Section 7 for details.
Note also that the vector field $\boldsymbol{V}_\mu$ is holomorphic and hence it is linearizable near each nondegenerate singular point. Moreover, the radius of the linearization disc depends only on the distance between the equilibrium points. In particular, for $\mu \in \Gamma$, it has a real analytic first integral near the elliptic point z_1 (or z_2). So the map $g^1_{\boldsymbol{V}_\mu}$ is in some sense regular, contrary to the map f_μ.

One can try to construct families of fundamental domains $D_j(f_\mu)$ in a way that near the elliptic point $z_1(\mu)$ (or $z_2(\mu)$), $\mu \in \Gamma$, they become approximate sectors of the limit angle $2\pi/|\alpha|$. This idea was firstly used in [**Shi**] by M. Shishikura. He noted that the corresponding transition maps can be defined also in the case when the points $z_{1,2}(\mu)$ are non-hyperbolic, i.e., elliptic. His construction uses quasi-conformal tools and is definitely not dynamical. He was interested in the quadratic family $z \longmapsto z^2 + c$, but his paper was somehow forgotten in the holomorphic dynamics. In fact, only for rational α's there is a chance that they become genuine fundamental domains. But then one cannot expect that the solution to an equation for conjugation, defined by iterations of f_μ^{-1} (or of f_μ) would converge to a holomorphic map.
(Here I mean the equation $\tilde{h}_j = \tilde{h}_j \circ \tilde{f} + R_j$, with respect to $\tilde{h}_j$, and its iterations infinitely many times, giving the solution $\tilde{h}_j(t) = \sum_0^\infty R_j \circ \tilde{f}^n(t)$ form the proof of Theorem 9.31.)
Nevertheless, P. Mardesič, R. Roussarie and C. Rousseau in [**MRR**] explored this idea. They somehow combined the Glyutsuk and Shishikura construction and defined some moduli. In part, the corresponding moduli space is of the form:

{germs of diffeomorphisms (return maps) near elliptic points $z_{1,2}$}/{germs of conjugating diffeomorphisms}

But such moduli are rather useless from the dynamical point of view. These problems are well known in the holomorphic dynamics.

Example (Quadratic family). A. Douady and J. Hubbard [**DH**] initiated the study of the following quadratic map

$$g_c(z) = z^2 + c$$

in the complex plane. It has two fixed points $z_{1,2} = \frac{1}{2}\left(1 \pm \sqrt{1-4c}\right)$ with the multipliers $\lambda_{1,2} = 2z_{1,2}$. The curve of parameters with non-hyperbolic fixed point is the cardioid $\Gamma_{0/1} = \left\{c = c(\alpha) = \mathrm{e}^{2\pi \mathrm{i}\alpha}\left(2 - \mathrm{e}^{2\pi \mathrm{i}\alpha}\right)/4 : \alpha \in [-1/2, 1/2]\right\}$ (which bounds a domain $\Delta_{0/1}$). This cardioid has cuspidal singularity at $c(0) = 1/4$ and $g_{1/4}$ has parabolic singularity at $z_1 = z_2 = 1/2$.
The point $c(1/2) = -3/4$ corresponds to to $g_{-3/4}$ with fixed point $z_2 = -1/2$ with the multiplier $\lambda = -1$ (a so-called flip). There exists a bulb $\Delta_{1/2}$ tangent to $\Delta_{0/1}$ at $c = -3/4$ from the left, such that, if $c \in \Delta_{1/2}$, then g_c has a hyperbolic periodic orbit of period 2 (besides the fixed points).
The boundary $\Gamma_{1/2}$ of $\Delta_{1/2}$ consists of those parameters that this periodic orbit ceases to be hyperbolic. Generally, for any rational $\alpha = p/q$ there is a bulb $\Delta_{p/q}$ tangent to $\Delta_{0/1}$ at $c(p/q)$ such that g_c, $c \in \Delta_{p/q}$, has a hyperbolic periodic orbit of period q. Also the periodic points of g_c with $c \in \partial\Delta_{p/q}$ bifurcate generating periodic points of second generation, etc.
The union of the domains $\Delta_{p/q}$ lies in so-called Mandelbrot set, which consists of those parameters c for which the orbits of the critical point $z = c$ is bounded.

The Mandelbrot set is very complicated and its structure is still investigated, as well as the dynamics of the maps g_c.
When looking for deformations of the map $g_{c(0/1)} = g_{1/4}$ one should take into account the long-periodic orbits of the map g_c, $c \in \Delta_{p/q}$, with small $|p/q|$ and large q. It is impossible to organize fundamental domains for g_c in a continuous and controllable way.

It seems that there is no good moduli for this classification problem.
Of course, problems of the same sort are encountered when ufolding singularities of higher codimension and/or order of resonance.
Finally, we note that C. Rousseau (with other collaborators) explored the above construction to define some moduli in other situations, e.g., in unfoldings of saddle-nodes.

§4 Ecalle's Approach

J. Ecalle proved an analogue of Ecalle–Voronin Theorem in 1975 in [**Ec1**]. On this occasion he created a new branch of complex analysis, the theory of *resurgent functions* (see [**Ec2**]). Unfortunately, for long time his results were unknown to wider mathematical community; probably due to unconventional language and notions. Later French mathematicians (B. Malgrange, F. Pham, J.-C. Tougeron, J.-P. Ramis, J. Martinet and others) detected the leading ideas from the Ecalle's theory and presented them in the standard mathematical language. They also developed the Ecalle's ideas further.
Below we present the Ecalle's approach based on the article of Malgrange [**Mal5**].

9.36. The Borel transform of the formal series conjugating a map from $\mathcal{A}_2$ to its formal normal form. We consider germs of the form $f(z) = z+z^2+z^3+O(z^4)$, i.e., $f = g^1_{z^2} + O(z^3) \in \mathcal{A}_{2,0}$. It is more convenient to work in the variable $t = -1/z$, where we have

$$t \longmapsto g(t) = t + 1 + a(t), \quad a(t) = \sum_{n\geq 2} a_n t^{-n}.$$

Theorem 9.26 asserts that there exists a formal power series

$$h(t) = t + b(t) = t + \sum_{n\geq 1} b_n t^{-n},$$

which formally conjugates g with $g_0 = id + 1$,

$$h \circ g \simeq g_0 \circ h.$$

We know that (usually) the series $b(t)$ diverges. However this divergence can be controlled. It turns out that the coefficients b_n grow no faster than $n!$.

One associates with b its **Borel transform** $\mathcal{B}b(\xi) = \sum \frac{b_n}{(n-1)!}\xi^{n-1}$, which has good properties. It is analytic near $\xi = 0$ and prolongs to a multi-valued holomorphic function in $(\mathbb{C} \setminus 2\pi i\mathbb{Z}) \cup \{0\}$, with exponential growth at infinity.
The 'inverse' to the Borel transform is the **Laplace transform**: if $\Phi(\xi)$ is defined in $(\mathbb{C}\setminus 2\pi i\mathbb{Z}) \cup \{0\}$ and grows at most exponentially at infinity $|\Phi(\xi)| < e^{R|\xi|}$, then the Laplace transform of Φ equals $\phi(t) = \mathcal{L}\Phi(t) = \int_0^\infty \Phi(\xi)e^{-t\xi}d\xi$.

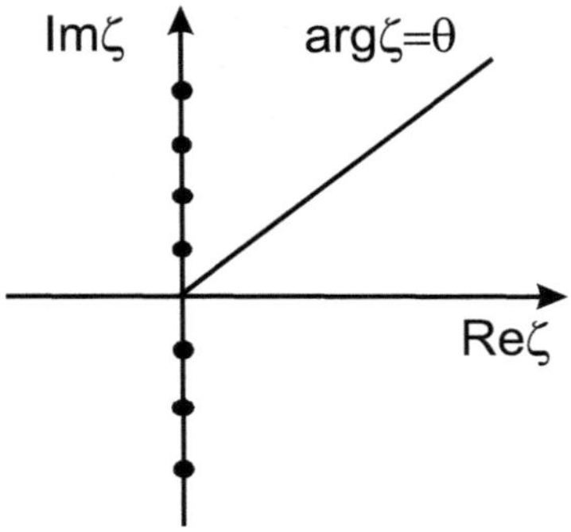

Figure 9.8: Singular points and a ray in the ζ plane

Note that this integral is convergent for $\operatorname{Re} t > R$ and represents there an analytic function $\phi(t)$. In the case $\Phi = \mathcal{B}b$ the function ϕ can be prolonged analytically to a larger domain. Namely, we can vary the path of integration in $\mathcal{L}\Phi$: instead of the ray $\arg\xi = 0$ we take the rays $\arg\xi = \theta$, $\theta \in (-\pi/2+\epsilon, \pi/2-\epsilon)$ (see Fig. 9.8). The result is a prolongation of ϕ to the sector $\widetilde{S}_1 = \{|t| > R, \ \arg t \in (-\pi + \epsilon, \pi - \epsilon)\}$; it is the same as the sector $\widetilde{S_1}$ from the previous section. Denote the function obtained by b_1.
If we apply the integration along the rays $\arg t = \theta$, $\theta \in (\pi/2+\epsilon, 3\pi/2-\epsilon)$, to the formula for the Laplace transform, then we obtain an analytic function b_2 in the analogous sector $\widetilde{S}_2$.

It turns out that:

(i) $b_{1,2}(t)$ have the same asymptotic power series expansions as the series $b(t)$ and the maps $h_{1,2} = t + b_{1,2}$ represent analytic diffeomorphisms from the Sectorial Normalization Theorem.

(ii) The Ecalle–Voronin moduli can be expressed by means of the singularities of the function $\mathcal{B}b$ at its singular points $\xi = \pm 2\pi i, \pm 4\pi i, \ldots$.

This approach is in the spirit of Gevrey expansions and summability. We shall return to it later, but first we will get acquainted with the strictly Ecalle's approach.

9.37. Definition (Space of resurgent functions). Let $\mathbb{D} = \mathbb{D}(r) = \{|\xi| < r\}$ be a disc in the complex ξ-plane (of relatively small radius) and let $\mathbb{D}^* = \mathbb{D} \setminus 0$. Let $\xi_0 > 0$ be some base point of $\mathbb{D}^*$.

We denote $\mathcal{O} = \mathcal{O}(r)$ the space of holomorphic functions on $\mathbb{D}$. Let $\widetilde{\mathcal{O}} = \widetilde{\mathcal{O}}(r)$ be the space of multivalued holomorphic functions on $\mathbb{D}^*$; they are obtained by prolongation of analytic germs at ξ_0 along paths in $\mathbb{D}^*$ (they are holomorphic in the universal covering $\widehat{\mathbb{D}^*}$). Finally we put

$$\mathcal{C} = \mathcal{C}(r) = \widetilde{\mathcal{O}}(r)/\mathcal{O}(r).$$

There is a natural *monodromy operator* $M : \widetilde{\mathcal{O}} \longmapsto \widetilde{\mathcal{O}}$ which associates to a germ of a function at ξ_0 its prolongation along the loop in $\mathbb{D}^*$ surrounding 0. The map $M - id$ defines the *variation operator* var : $\mathcal{C} \longmapsto \widetilde{\mathcal{O}}$.
If $\Omega \subset \mathbb{C}$ is a discrete set with one-point intersection with $\mathbb{D}$, $\Omega \cap \mathbb{D} = 0$, then one defines the **space of resurgent functions** as

$$\mathcal{C}(\Omega) = \{F \in \mathcal{C} : \ \mathrm{var}(F) \ \text{ is } \ \text{holomorphic in } \widehat{\mathbb{C} \setminus \Omega}\}.$$

Equivalently: $F \in \mathcal{C}(\Omega)$ iff there exists $\widetilde{F}$ holomorphic in $\mathbb{C} \setminus \Omega$ and such that $\widetilde{F}|_D = F +$ (holom. function).

9.38. Convolution and its properties. For two elements F, G of $\mathcal{C}$, with representatives $\widetilde{F}, \widetilde{G} \in \widetilde{\mathcal{O}}(r)$, we define the function

$$H(\xi) = \int_\gamma \widetilde{F}(\xi - \zeta)\widetilde{G}(\zeta)\mathrm{d}\zeta, \quad \zeta \in \mathbb{C} \diagdown \mathbb{R},$$

where the integral runs along the loop $\gamma = \gamma_{\xi_0,\epsilon}$, $0 < \epsilon << \xi_0$, starting at $\xi_0 > 0$, and running firstly along upper ridge of the cut along the interval $[0, r]$, next making the counterclockwise turn along the circle $|\zeta| = \epsilon$ and then going back along the lower ridge of the cut along $[0, r]$. H prolongs itself to a holomorphic multivalued function, $H \in \widetilde{\mathcal{O}}$, and its class in $\mathcal{C}$ does not depend on the representatives of F and of G and on the point ξ_0. We denote this class by

$$F * G = [H],$$

and call it the **convolution** of F and G.
(Up to a constant this convolution is the same as the convolution of two distributions defined by functions $f(x)$ and $g(x)$ with support in $\mathbb{R}_+$: $f * g(x) = \int_0^x f(x - y)g(y)\mathrm{d}y$. It turns out that the functions $f(x)$, defined for $x > 0$ and locally integrable, can be included into $\mathcal{C}$ in the following way: let $F(\xi) = \frac{1}{2\pi \mathrm{i}} \int_0^a f(x)\mathrm{d}x/(x-\xi)$ define a holomorphic function outside real axis, then one defines $\mathrm{Pf}(f) = [F]$. For example, $\mathrm{Pf}(x^\alpha) = [\xi^\alpha/(\mathrm{e}^{2\pi \mathrm{i}\alpha} - 1)]$.)
The convolution operator has the following properties:

(i) associativity and commutativity;

(ii) $\delta = \left[\frac{-1}{2\pi \mathrm{i}\xi}\right]$ is the convolution unity (it follows from the Cauchy formula);

(iii) $\frac{\mathrm{d}}{\mathrm{d}\xi}(F * G) = \frac{\mathrm{d}F}{\mathrm{d}\xi} * G$;

(iv) $\xi(F * G) = (\xi F) * G + F * (\xi G)$.

The latter formula means that the multiplication operator by $-\xi$ is a derivation of the convolution algebra. It is denoted by

$$\partial = -\xi.$$

It turns out that $\mathcal{C}(\Omega)$ is a convolution subalgebra of $\mathcal{C}$ (see [**Mal5**]).

9.39. The Laplace transform and its properties. If $F \in \mathcal{C}$ with a representative $\widetilde{F} \in \widetilde{\mathcal{O}}$, then we define its **Laplace transform** as

$$\mathcal{L}F(t) = \int_{\gamma} \widetilde{F}(\xi)\mathrm{e}^{-t\xi}\mathrm{d}\xi,$$

where $\gamma = \gamma_{\xi_0,\epsilon}$ is the same loop as in the definition of convolution (in the previous point). One shows that the Laplace transform does not depend on the representative $\widetilde{F}$ and on ξ_0.
(Usually, this definition is applied to functions of the form $\xi^{\alpha}(\ln\xi)^p G(\xi)$ with analytic G. This class of functions is sometimes called the *Nilsson class*.)
For example, $\mathcal{L}([\xi^{\alpha}]) = (\mathrm{e}^{2\pi\mathrm{i}\alpha} - 1)\Gamma(\alpha+1)/t^{\alpha+1}$ and

$$\mathcal{L}\delta^{(n)} = \mathcal{L}\left(\mathrm{d}^n\delta/\mathrm{d}\xi^n\right) = t^n.$$

We see that, up to some normalization, this definition of Laplace transform agrees with the definition 9.36.
The inverse to the Laplace transform is called the **Borel transform,** $\mathcal{B} = \mathcal{L}^{-1}$.
The Laplace transform has the following properties:

(i) $\mathcal{L}(F * G) = \mathcal{L}F \cdot \mathcal{L}G$,

(ii) $\mathcal{L}(\partial F) = \frac{\mathrm{d}}{\mathrm{d}t}(\mathcal{L}F)$,

(iii) $\mathcal{L}(\frac{\mathrm{d}F}{\mathrm{d}\xi}) = t(\mathcal{L}F)$.

9.40. The Ecalle composition. Let $F \in \mathcal{C}(2\pi\mathrm{i}\mathbb{Z})$ and $H = \delta' + E,\ E \in \mathcal{C}(2\pi\mathrm{i}\mathbb{Z})$. Then one defines the series

$$F \otimes_{\mathrm{e}} H = \sum_{n\geq 0} \frac{1}{n!}(\partial F) * E^{*n}$$

It turns out that, under assumption of local integrability of F and E, this series is convergent and represents an element of $\mathcal{C}(2\pi\mathrm{i}\mathbb{Z})$, which we call the *Ecalle's composition of F and H*.
The Ecalle's composition is the Borel transform of the usual composition of series $f = \mathcal{L}F \sim \sum f_k t^{-k}$ and $h = \mathcal{L}H \sim t + \sum b_k t^{-k} = t + b(t)$. It follows from the Taylor expansion $f \circ (t+b) = \sum_n (\mathrm{d}^n f/\mathrm{d}t^n) b(t)^n$.

9.41. Application to conformal germs tangent to identity. Recall that in the point 9.36 we had two maps $g(t) = t + 1 + a(t)$, $g_0(t) = t + 1$, which are formally conjugated by the series $h \sim t + b(t)$. We shall show the following.

Proposition. *The Borel transform of $b(t)$ belongs to the resurgent space $\mathcal{C}(2\pi\mathrm{i}\mathbb{Z})$ and the Borel transform of h belongs to $\delta' + \mathcal{C}(2\pi\mathrm{i}\mathbb{Z})$. It means that $\mathcal{B}h$ is analytic in $\mathbb{C} \setminus (2\pi\mathrm{i}\mathbb{Z} \setminus 0)$.*

Proof. The conjugation condition $h \circ g = g_0 \circ h$ means that $t+1+a(t)+b(t+1+a) = t + b(t) + 1$ or, after replacing t by $t - 1$,

$$b(t-1) - b(t + \tilde{a}(t)) = \tilde{a}(t). \tag{9.6}$$

It is an equation for b.
We introduce two linear operators $K\psi(t) = \psi(t+\tilde{a}) - \psi(t)$, $L\psi(t) = \psi(t-1) - \psi(t)$. Then Eq. (9.6) takes the operator form $[L - K]b = \tilde{a}$. Note that the operator L is dominating here; if $\psi = t^{-n}$, then $L\psi \sim nt^{-n-1}$ and $K\psi \sim O(t^{-n-3})$. Thus the solution to Eq. (9.6) takes the form

$$b = \sum [L^{-1}K]^n L^{-1}\tilde{a}, \tag{9.7}$$

where the series is well defined.
Now we rewrite the formula (9.7) in the Borel chart. There the operator $I + L$ is replaced by the multiplication operator by the function e^{ξ}: $\psi(t-1) = \sum(-\mathrm{d}/\mathrm{d}t)^n \psi/n!$, where $-\mathrm{d}/\mathrm{d}t$ is replaced by ξ (see the formula (ii) from 9.38). The operator $I + K$ is calculated using the formula from 9.40: $\mathcal{B}(I+K)\mathcal{B}^{-1}\Psi = \Psi \otimes_{\mathrm{e}} A = \sum((\xi)^m \Psi) * A^{*m}$, $A = \mathcal{B}\tilde{a}$.
Because $(\mathrm{e}^{\xi} - 1)^{-1}$ has poles in $2\pi\mathrm{i}\mathbb{Z}$ and $A(\xi)$ is integer function, the Borel transform of b, i.e., $B = \mathcal{B}b$, is meromorphic in $\mathbb{C} \setminus 2\pi\mathrm{i}\mathbb{Z}$.
One can also show that $|B(\xi)| < \mathrm{const} \cdot \mathrm{e}^{\mathrm{const}\,|\xi|}$. □

9.42. The alien derivations and the Ecalle's invariants of holomorphic classification of germs tangent to identity. J. Ecalle introduced the following operators acting on the space of resurgent functions $\mathcal{C}(2\pi\mathrm{i}\mathbb{Z})$.
Denote by $\Gamma_{2\pi\mathrm{i}n}$ the set of 2^{n-1} paths in $\mathbb{C} \setminus 2\pi\mathrm{i}\mathbb{Z}$, with the beginning at ξ_0 and the end at $\xi_0 + 2\pi\mathrm{i}n$; the paths differ in the way they omit the points $2\pi\mathrm{i}j$, $j = 1, \ldots n$. If $\gamma \in \Gamma_{2\pi\mathrm{i}n}$ and $f \in \mathcal{C}(2\pi\mathrm{i}\mathbb{Z})$, then

$$\Delta_\gamma f$$

is the germ of prolongation of f along γ with argument translated by $-2\pi\mathrm{i}n$, i.e., to a neighborhood of ξ_0. If $n > 0$, then such path γ can avoid the singularities $2\pi\mathrm{i}j$, $0 < j < n$, form the right or from the left. Let $\gamma_{2\pi\mathrm{i}n,+}$ be the path running on the right. We define the formal operator

$$\Delta^+[[\lambda]] = id + \sum_{n \geq 1} \Delta_{\gamma_{2\pi\mathrm{i}n,+}} \lambda^n.$$

It turns out that $\Delta^+[[\lambda]]$ is a homomorphism of the algebras $\mathcal{C}(2\pi\mathrm{i}\mathbb{Z})$ and $\mathcal{C}(2\pi\mathrm{i}\mathbb{Z})[[\lambda]]$ (with natural extension of the convolution).
Moreover, if one defines the operators $\Delta_{2\pi in}$ by the formula

$$\ln(\Delta^+[[\lambda]]) = \sum \Delta_{2\pi in}\lambda^n,$$

then:

(i) $\Delta_{2\pi in}$ *are derivations of* $\mathcal{C}(2\pi\mathrm{i}\mathbb{Z})$ *(called the* ***alien derivations*** *by Ecalle), and*

(ii) $\Delta_{2\pi\mathrm{i}n}(F \otimes_{\mathrm{e}} H) = (\partial F \otimes_{\mathrm{e}} H) * \Delta_{2\pi\mathrm{i}n} H + (\sum(-nH)^{*m}) * ((\Delta_{2\pi\mathrm{i}n} F) \otimes_{\mathrm{e}} H).$

Analogously one defines the alien derivations for $n < 0$. (The alien derivations can be expressed by means of the maps Δ_γ, $\gamma \in \Gamma_{2\pi in}$ (see [**Mal5**])).

Applying the alien derivations to the both sides of the equation $G \otimes_{\mathrm{e}} H = H \otimes_{\mathrm{e}} G_0$ ($G = \delta' + \delta + A$, $G_0 = \delta' + \delta$ – Borel transforms of g, g_0), one obtains the equation $(\partial G \otimes_{\mathrm{e}} H) * \Delta_\omega H = \Delta_\omega H \otimes_{\mathrm{e}} G_0$. This implies that the function $\psi = \psi(t) = \mathcal{L}(\Delta_\omega H$, $\omega = 2\pi in$, satisfies the linear formal functional equation $(g' \circ h)\psi = \psi \circ (id + 1)$, the same as the derivative $h'(t)$. By uniqueness, there exists a constant $A_\omega \in \mathbb{C}$ such that $\psi = A_\omega h'$, or

$$\Delta_\omega H = A_\omega \partial H;$$

(Ecalle calls the latter equation the *bridge equation*).
Namely, these constants A_ω, $\omega \in \Omega \setminus 0$, $\Omega = 2\pi\mathrm{i}\mathbb{Z}$, form the complete infinite set of invariants of analytic conjugation of germs $f = z + z^2 + z^3 + \ldots$ from $\mathcal{A}_{2,0}$:

Two germs $f, f' \in \mathcal{A}_{2,0}$ *are analytically equivalent iff* $A_\omega = A'_\omega$, $\omega \in \Omega$. *Moreover, the coefficients* A_ω *can be expressed in terms of the functional coboundary operators:* $\widetilde{H}_2\widetilde{H}_1^{-1}(t) = t + \sum c_{1l}\mathrm{e}^{2\pi\mathrm{i}lt}$, $\mathrm{Im}t >> 0$ *and the analogous operator for* $\mathrm{Im}t << 0$ *(see the proof of Theorem* 9.31*).*

In [**Mal5**] there is a formula expressing the coefficients c_{jl} in terms of A_ω.
The Ecalle's theory has a natural generalization to a theory, which can be applied to the germs from the classes $\mathcal{A}_p$ for any p. We present it in the next point.

9.43. The Gevrey series and expansions. Let $h(t)$ be a germ of function holomorphic in a sector S with vertex at ∞, $h \in \mathcal{O}(S)$, with the asymptotic expansion $h \sim \sum a_n t^{-n}$. We say that h is of **Gevrey type of order** s iff

$$\left| h(t) - \sum_0^{k-1} a_j t^{-j} \right| \cdot |t|^k < C(k!)^{s-1} A^k,$$

where the constants C, A depend on f and on the sector S.
A formal series $\hat{h} = \sum a_n t^{-n} \in \mathbb{C}[[1/t]]$ is of **Gevrey type of order** s iff

$$|a_n| < C(n!)^{s-1} A^n.$$

The following result is a generalization of the Borel–Ritt theorem (see Theorem 5.52 in Chapter 5).

Theorem of Borel, Ritt and Gevrey. ([**MR1**]) *Let a series $\hat{h} \in \mathbb{C}[[1/t]]$ be of Gevrey type of order $s > 1$ and let S be a sector with angle $< (s-1)\pi$. Then there exists a function $h \in \mathcal{O}(S)$ of Gevrey type of order s and such that its Taylor expansion at infinity coincides with $\hat{h}$.*

Proof. One can reduce the problem to the case when S is a sector with bisectrix along the positive real axis. Then h is given by the explicit formula

$$h(t) = pt^p \int_0^r \Phi(\xi) \exp\left(-\xi^p t^p\right) \xi^{p-1} \mathrm{d}\xi,$$

where $p = \frac{1}{s-1}$ and $\Phi(\xi) = \sum \frac{a_n}{\Gamma(1+n/p)} \xi^n$ is analytic function near $\xi = 0$ with radius of convergence $> r$.
Note that, when $s = 2$, the function $\Phi(\xi)$ equals to the derivative of the Borel transform of $\hat{h}$. Also the integral defining h reminds the Laplace transform. □

As in the case $p = 1$, the functions h from this theorem depend on the sector and the representatives corresponding to different sectors do not need to coincide.
The series $\hat{h}$ (of Gevrey type of order $s = 1 + 1/p$) is p-**summable in the direction** θ iff $\hat{h}$ forms asymptotic expansion of some function $h_S \in \mathcal{O}(S)$, where S is a sector with bisectrix $\arg t = \theta$ and with angle $> \pi/p = \pi/(s-1)$. Note that this inequality is reverse to the inequality from the Borel–Ritt–Gevrey theorem.
The series $\hat{h}$ is p-**summable** iff it is summable for all but finite number of directions θ.

Example. The formal series $h(t)$, conjugating the diffeomorphisms g and $g_0 = t+1$ in 9.36, is 1-summable. It is summable in all directions but $\theta = \pm\pi/2$.

If $\hat{h}$ is p-summable, then we can associate with it certain functional cochain. Namely, we take a covering of a neighborhood of infinity by sectors S_j with angles $> \pi/p$ such that in S_j there is a function h_j (of Gevrey class of order s) with $\hat{h}$ as the Taylor expansion. The system (h_j) is a functional Čech cochain associated with the covering (S_j). The coboundary of this cochain is the cocycle $h_{j+1} \circ h_j^{-1}$ with values in a sheaf resembling the Stokes sheaf.
All this is applied to holomorphic classification of germs of analytic diffeomorphisms (see the next point).

9.44. Germs of the type $z + z^{p+1} + \ldots$. Let $f \in \mathcal{A}_{p,\lambda} = f_{p,\lambda} + \ldots$, $f_{p,\lambda} = g^1_{z^p/(1+\lambda z^p)}$ be two germs of holomorphic diffeomorphisms (from Section 3). In the t-chart, $t = -1/(pz^p) + \lambda \ln z$, we have the maps $g = t + 1 + \ldots$ and $g_0 = t + 1$.
There is a formal power series $\widehat{h} = t + \sum b_j t^{-j}$ conjugating g with g_0.

Theorem. ([**MR1**]) *The series $\hat{h}$ is of Gevrey type of order $s = 1 + 1/p$ and is p-summable. Associated with it the functional cocycle $h_{j+1} \circ h_j^{-1}$ is the Ecalle–Voronin modulus of the analytic classification of germs f.*

§5 Martinet–Ramis Moduli

The saddle-node singularity of an analytic planar vector field is such that one of its eigenvalues vanishes and the other eigenvalue is nonzero. First we present the formal classification of saddle-nodes and next we describe the moduli of their analytic classification.

9.45. Definition. Two germs of vector fields $\boldsymbol{V}$ and $\boldsymbol{W}$ (in $\mathbb{R}^n$, $\mathbb{C}^n$, smooth or analytic) are called **(topologically, formally, analytically) orbitally equivalent** iff there exists a (C^k, formal, analytic) diffeomorphism H transforming the phase curves of $\boldsymbol{V}$ to phase curves of $\boldsymbol{W}$. It means that the corresponding foliations, defined by the phase portraits of $\boldsymbol{V}$ and of $\boldsymbol{W}$, are transformed one to the other by means of H. In the smooth and analytic cases we have

$$H_*\boldsymbol{V} = F \cdot \boldsymbol{W},$$

where $F \neq 0$ is some function (of definite class).

9.46. Theorem (Formal normal form of saddle-nodes). *Any complex analytic saddle-node is formally orbitally equivalent to*

$$\dot{x} = x^{p+1}, \quad \dot{y} = -y(1+\lambda x^p).$$

Proof. The Poincaré–Dulac theorem 9.14 and Example 9.15(b) give the following (non-orbital) normal form $\dot{x} = a_{p+1}x^{p+1} + a_{p+2}x^{p+2} + \ldots$, $\dot{y} = \lambda_2 y(1 + b_1 x + b_2 x^2 + \ldots)$, where $a_{p+1} \neq 0$; if all $a_j = 0$, then the singular point would have infinity multiplicity (non-isolated in the complex analytic case). By linear changes of variables and time we can assume that $a_{p+1} = 1$, $\lambda_2 = -1$.
Now we divide the field by $1 + b_1 x + \ldots$, which gives $\dot{y} = -y$ and

$$\dot{x} = x^{p+1}(1 + c_1 x + \ldots). \tag{9.8}$$

Therefore the problem is reduced to formal *non-orbital* classification of 1-dimensional saddle-nodes (9.8).
The latter problem is solved in the same way as in the proof of Theorem 9.26 (about formal classification of diffeomorphisms $z + z^{p+1} + \ldots$). Using the changes $x \longmapsto x + d_l x^l$, we obtain the (leading) term $(l-p-1)d_l x^{l+p}\frac{\partial}{\partial x}$ in the transformed vector field. If $l \neq p+1$, then the corresponding monomial vector field can be cancelled. There remains only the field $\boldsymbol{W}_{p,\nu} = x^{p+1}(1+\nu x^p)\frac{\partial}{\partial x}$, where ν is the formal invariant.
Of course, one can fix also other terms in $\frac{\partial}{\partial x}$, like $x^{p+1}(1 - \lambda x^p + (\lambda x^p)^2 - \ldots) = x^{p+1}(1+\lambda x^p)^{-1}$, $\lambda = -\nu$. Multiplying this vector field by $1 + \lambda x^p$ gives the vector field from the thesis of Theorem 9.46. □

The integer p is called the **codimension** of saddle-node and $\lambda \in \mathbb{C}$ is the **modulus of formal classification**.The space of germs with the formal normal form as in Theorem 9.45 will be denoted by $\mathcal{E}_{p,\lambda}$.

9.47. Problem. Show that, if (9.8) is analytic, then the change reducing it to $\dot{x} = x^{p+1} + \nu x^{2p+1}$ is analytic.
This means that in one-dimensional case the (non-orbital) formal and analytic classifications of saddle-nodes coincide.
Hint: under the assumption that $(2p+1)^{\text{th}}$ jet of Eq. (9.8) is like in the normal form, solve a functional equation for the change $x \longmapsto x + x^{2p+2}\phi(x)$ in some space of analytic functions (of ϕ's).

As we shall see, the normal form from Theorem 9.46 is not analytic. However some of its statements have analytic equivalents. Note that it has two formal invariant curves: the **strong manifold** $x = 0$ and the formal **center manifold** $y = 0$. It turns out that the strong manifold is always analytic (see the next result), but usually the center manifold is not analytic. The next theorem, which concerns the cases of saddle-node and of saddle, was firstly proved by C. Briot and J. Bouquet [**BB**].

9.48. The analytic Hadamard–Perron theorem. *Consider an analytic planar system*

$$\dot{x} = \lambda_1 x + \ldots, \qquad \dot{y} = \lambda_2 y + \ldots,$$

such that $\lambda_2 \neq 0$ and $\lambda_1/\lambda_2 \leq 0$. Then there exists an invariant analytic curve tangent to $x = 0$ at the origin.

Proof. In the case of a saddle-node we can assume that $\frac{\mathrm{d}x}{\mathrm{d}y} = \frac{f(x,y)}{\lambda_2 y + g(x,y)}$, with $f, g \in \mathfrak{m}^2$ (square of the maximal ideal) and $f(0,y) = O(y^3)$. In the case of a saddle we can assume that $\frac{\mathrm{d}x}{\mathrm{d}y} = \frac{\lambda_1 x + f(x,y)}{\lambda_2 y + g(x,y)}$, with $f, g \in \mathfrak{m}^3$ (because the resonant terms in the Poincaré–Dulac normal form start from cubic monomials).
We put the equation of the invariant curve in the form $x = y^2\phi(y)$. In the case of saddle-node we get the equation $y^2\phi' + 2y\phi = f(y^2\phi, y)/(\lambda_2 y + g(y^2\phi, y))$, which (after division by y^2) takes the form $\phi' + 2\phi/y = F(y, \phi)$ (F – analytic) and can be written in the integral form

$$\phi(y) = y^{-2} \int_0^y s^2 F(s, \phi(s)) \mathrm{d}s.$$

The latter equation is solved using the contraction principle in some ball in the Banach space of analytic functions ϕ defined in a disc $|y| < \epsilon$ with the sup-norm.
In the saddle case we arrive at the equation $\phi' = (\lambda - 2)\phi/y + F(y, \phi)$, $\lambda = \lambda_1/\lambda_2 < 0$, or at the integral equation $\phi = y^{\lambda-2} \int_0^y s^{2-\lambda} F(s, \phi(s)) \mathrm{d}s$. □

9.49. Remarks. (a) In [**IlY3**, Theorem 7.4] Yu. Ilyashenko and S. Yakovenko generalized the Briot–Bouquet theorem as follows.
Assume that we have an analytic vector field $\dot{x} = Ax + \ldots$ in a neighborhood of $x = 0$ in $\mathbb{C}^n$. Assume also that the spectrum of the matrix A is divided into two groups separated by a line through the origin. Then there exist two invariant analytic submanifolds tangent to the corresponding invariant subspaces of A.

(b) The reader can observe that the above proof can fail when the ratio of eigenvalues λ is a positive integer (the resonant node); one can get the integral $\int \mathrm{d}s/s$.

One can see that the phase curves are of the form $x = Cy^\lambda + \sigma y^\lambda \ln y$, and are not analytic, if $\sigma \neq 0$. However in this case the normal form is polynomial $\dot{x} = \lambda x + \sigma y^\lambda$, $\dot{y} = y$ and is analytic; we are in the Poincaré domain and we can use Remark 9.25.
Also the proof of Theorem 9.47 does not work in the case of center manifold of a saddle-node. Already L. Euler noticed that the center manifold can be non-analytic.

(c) **Euler's example.** ([**Eu1**]) The equation

$$\mathrm{d}y/\mathrm{d}x = (y - x)/x^2$$

has the following unique formal solution $y = \sum_1^\infty (n-1)! x^n$ passing through 0.

9.50. A natural approach to the non-analyticity proofs. This approach is somehow inspired by the Euler's example. It was developed in a recent paper [**SZ4**] by the author with E. Stróżyna.

(a) **Homological operators.** We consider vector fields of the form

$$\boldsymbol{V} = \boldsymbol{V}_0 + \ldots,$$

where $\boldsymbol{V}_0 = \lambda_1 x \frac{\partial}{\partial x} + \lambda_2 \frac{\partial}{\partial y}$, $\lambda_1 \neq 0$, i.e., with elementary singularity at $x = y = 0$. In the reduction of higher order terms, via the Poincaré–Dulac theorem 8.14, one uses the **homological operator**

$$\boldsymbol{Z} \longmapsto \mathrm{ad}_{\boldsymbol{V}} \boldsymbol{Z} = [\boldsymbol{V}, \boldsymbol{Z}],$$

i.e., the commutator. ($\mathrm{ad}_{\boldsymbol{Z}} \boldsymbol{V} = -\mathrm{ad}_{\boldsymbol{V}} \boldsymbol{Z}$ is the linearization of the derivative $(\mathrm{Ad}_{\exp \boldsymbol{Z}})_*$, $\exp \boldsymbol{Z} = g^1_{\boldsymbol{Z}}$, of the adjoint action $g \longmapsto \exp \boldsymbol{Z} \circ g \circ \exp(-\boldsymbol{Z})$ in the group Diff $(\mathbb{C}, 0)$.) We split the 2-dimensional homological operator $\mathrm{ad}_{\boldsymbol{V}}$ into simpler 1-dimensional homological operators.
Let

$$\mathcal{F}_d = \{f \in \mathbb{C}[x,y] : \deg f = d\},$$
$$\mathcal{Z}_d = \{\boldsymbol{Z} \in \mathcal{Z} : \deg \boldsymbol{Z} = d\},$$

denote spaces of quasi-homogeneous polynomials and of quasi-homogeneous polynomial vector fields of degree d (with the agreement that $\deg \frac{\partial}{\partial x} = \deg \frac{\partial}{\partial y} = -1$). Analogously we define the spaces $\mathcal{F}$, $\mathcal{F}_{\geq d}$, $\mathcal{Z}$ and $\mathcal{Z}_{\geq d}$ of formal power series and of formal vector fields respectively.
Introduce the operators

$$A(\boldsymbol{V}) f = f \cdot \boldsymbol{V},$$
$$B(\boldsymbol{V}) \boldsymbol{Z} = \boldsymbol{V} \wedge \boldsymbol{Z} / \partial_x \wedge \partial_y,$$
$$C(\boldsymbol{V}) f = \boldsymbol{V}(f) = \partial f / \partial \boldsymbol{V},$$
$$D(\boldsymbol{V}) f = \boldsymbol{V}(f) - \mathrm{div}(\boldsymbol{V}) f.$$

In the right-hand side of the first of these equations we have vector fields tangent to $\boldsymbol{V}$ and in the second we encounter so-called bi-vector fields, which measure the part of the vector field $\boldsymbol{Z}$ transversal to $\boldsymbol{V}$.
Consider the following diagram, whose rows form the so-called *Koszul complexes*:

$$\begin{array}{ccccccccc} 0 & \longrightarrow & \mathcal{F} & \overset{A(\boldsymbol{V})}{\longrightarrow} & \mathcal{Z} & \overset{B(\boldsymbol{V})}{\longrightarrow} & \mathcal{F} & \longrightarrow & 0 \\ & & \downarrow C(\boldsymbol{V}) & & \downarrow \mathrm{ad}_{\boldsymbol{V}} & & \downarrow D(\boldsymbol{V}) & & \\ 0 & \longrightarrow & \mathcal{F} & \overset{A(\boldsymbol{V})}{\longrightarrow} & \mathcal{Z} & \overset{B(\boldsymbol{V})}{\longrightarrow} & \mathcal{F} & \longrightarrow & 0 \end{array}$$

The operators $C(\boldsymbol{V})$ and $D(\boldsymbol{V})$ are also called the **homological operators.**
We note the following facts:

(i) *The above diagram is commutative.*

(ii) *If $\lambda = \lambda_2/\lambda_1 \neq 0$, i.e., when $x = y = 0$ is an isolated singularity of V_0, then the Koszul complexes are exact.*

Therefore, in the case $\lambda \neq 0$, we have a 'splitting' $\mathcal{Z} \simeq A(\boldsymbol{V})\mathcal{F} + B(\boldsymbol{V})^{-1}\mathcal{F}$, where by $B(\mathbf{V})^{-1}$ we mean an operator to some subspace complementary to $\mathrm{Im}A(\boldsymbol{V})$ (a right inverse to $B(\boldsymbol{V})$). Moreover, the operator $\mathrm{ad}_{\boldsymbol{V}}$ 'splits' as $C(\boldsymbol{V}) + D(\boldsymbol{V})$ and hence $\mathrm{Im}\,\mathrm{ad}_{\boldsymbol{V}} \simeq A(\boldsymbol{V})\mathrm{Im}C(\boldsymbol{V}) + B(\boldsymbol{V})^{-1}\mathrm{Im}D(\boldsymbol{V})$ and $\mathrm{coker}\mathrm{ad}_{\boldsymbol{V}} \simeq A(\boldsymbol{V})\mathcal{N}(C) + B(\boldsymbol{V})^{-1}\mathcal{N}(D)$, where $\mathcal{N}(C) \simeq \mathrm{coker}C(\boldsymbol{V})$ and $\mathcal{N}(D) \simeq \mathrm{coker}D(\boldsymbol{V})$.
If $\lambda = 0$, then

$$\boldsymbol{V}_0 = G\boldsymbol{X},$$

where $G = \lambda_1 x$ and $\boldsymbol{X} = \frac{\partial}{\partial x}$ and, instead of the operators $C(\boldsymbol{V}_0)$, we rather use the operator

$$C'(\boldsymbol{V}_0)f = G\boldsymbol{X}(f) - \boldsymbol{X}(G)f.$$

Note that $\ker B(\boldsymbol{V}_0) = \ker B(\boldsymbol{X}) = \mathrm{Im}C(\boldsymbol{X})$ and we have different splitting: $\mathcal{Z} \simeq \mathrm{Im}A(\boldsymbol{X}) + B(\boldsymbol{V}_0)^{-1}\mathcal{F}$. Note also the formula $\mathrm{ad}_{\boldsymbol{V}_0} f\boldsymbol{X} = [G\boldsymbol{X}, f\boldsymbol{X}] = C'(\boldsymbol{V}_0)f \cdot \boldsymbol{X}$.
It is quite obvious that:

$\ker C(\boldsymbol{V})$ consists of the *first integrals* F of $\boldsymbol{V}$, and

$\ker D(\boldsymbol{V})$ consists of the *inverse integrating multipliers* M for $\boldsymbol{V}$, i.e., $\mathrm{div} M^{-1}\boldsymbol{V} = 0$.

In the below table we present the first integrals and inverse integrating multipliers for the vector field $\boldsymbol{V}_0$:

Type I	$\ker C = \ker D = 0$;
Type II	$\ker C = 0$, $\ker D = \mathbb{C}x^{k+1}$;
Type III	$\ker C = \mathbb{C}[[F]]$, $\ker D = xy\mathbb{C}[[F]]$;
$\lambda = 0$	$\ker C' = \mathbb{C}[[y]]$.

Here the *Type I* denotes either non-resonant case, $\lambda = \lambda_2/\lambda_1 \notin \mathbb{Q}$, or resonant node $\lambda = \frac{k}{l}$ with $\gcd(k,l) = 1$ and $k, l > 1$. The *Type II* denotes $1 : k$ resonant node, i.e., $\lambda = k \in \mathbb{N}$. The *Type III* denotes $k : -l$ resonant saddles, $\lambda = -\frac{k}{l} \in \mathbb{Q}_{<0}$ with the first integral $F = x^k y^l$. The last case (also in the *Type III*) is the saddle-node. We note also the following identity

$$B(\boldsymbol{V}_0)\,\boldsymbol{E} = (\lambda_1 - \lambda_2)\,xy,$$

where $\boldsymbol{E} = x\frac{\partial}{\partial x} + y\frac{\partial}{\partial y}$ is the standard Euler vector field.
The homological operator $\mathrm{ad}_{\boldsymbol{V}}$ is a perturbation of $\mathrm{ad}_{\boldsymbol{V}_0}$; so, we consider firstly the three homological operators associated with $\boldsymbol{V}_0$. After restriction to the spaces $\mathcal{F}_d$ of degree d homogeneous polynomials, they become endomorphisms of these spaces. We denote by $C_d(\boldsymbol{V}_0)$ and $D_d(\boldsymbol{V}_0)$ the corresponding 1-dimensional operators. In the monomial basis they become diagonal:

$$\begin{aligned} C_d(\boldsymbol{V}_0)\,x^i y^j &= (\lambda_1 i + \lambda_2 j)\,x^i y^j, \\ D_d(\boldsymbol{V}_0)\,x^i y^j &= (\lambda_1(i-1) + \lambda_2(j-1))\,x^i y^j. \end{aligned}$$

(b) **The norm.** We will use the norm

$$\|f\| = \|f\|_\rho = \sum |a_{i,j}|\,\rho^{i+j}.$$

Note that the existence of bounded inverses to the homological operators $C(\boldsymbol{V})$ and $D(\boldsymbol{V})$ (i.e., assuming that they are invertible) implies the analyticity of the reduction to the linear normal form; see [**SZ4**] for more details.

(c) **The focus case.** Recall that in this case, $\lambda = \lambda_2/\lambda_2 \notin \mathbb{R}$, the normal form is linear, because the operators $C_d(\boldsymbol{V}_0)$ and $D_d(\boldsymbol{V}_0)$ are isomorphisms. We claim that this normal form is analytic, i.e., the corresponding formal change is analytic. So, we can assume $\boldsymbol{V} = \boldsymbol{V}_0 + \boldsymbol{W}$, where $\boldsymbol{W} = O\left(|(x,y)|^D\right)$ is of high order. To prove the analyticity of the reduction to the normal form it is enough to show that the operators $C(\boldsymbol{V})$ and $D(\boldsymbol{V})$ are invertible and that their inverse are bounded (see [**SZ4**] for details).
We have $C(\boldsymbol{V}) = C(\boldsymbol{V}_0)(I-K)$, $K = -C(\boldsymbol{V}_0)^{-1}C(\boldsymbol{W})$, and hence

$$C(\boldsymbol{V})^{-1} = \left(\sum K^n\right) C(\boldsymbol{V}_0)^{-1}.$$

But, on $\mathcal{F}_d$, we have $C(\boldsymbol{W}) : \mathcal{F}_d \longmapsto \mathcal{F}_{\geq d+D}$ with $\|C(\boldsymbol{W})\| \leq C_1 d\rho^D$, for some constant C_1 and with $\rho < 1$ (the factor d comes from the differentiation). Next, $\left\|C_k(\boldsymbol{V}_0)^{-1}\right\| \leq C_2 k^{-1}$ for a constant C_2; indeed,

$$|\lambda_1 i + \lambda_2 j| = |\lambda_1| \cdot |i + \lambda j| \geq \frac{1}{2}|\lambda_1| \cdot \{|i + \mathrm{Re}\lambda \cdot j| + |\mathrm{Im}\lambda| \cdot j\} \geq C_3(i+j).$$

It follows that

$$\|K_d\| \leq C_4 \frac{d}{d+D}\rho^D \leq \nu < 1$$

and the von Neumann series $\sum K^n$ is absolutely convergent. Similar estimates hold for $D(\boldsymbol{V})$.

(d) **The node case.** Recall that here ($\lambda \geq 1$) the normal form is polynomial: $\boldsymbol{V}_0 + \boldsymbol{V}_1$, $\boldsymbol{V}_1 = ax^\lambda \frac{\partial}{\partial y}$, where $a = 0$, if λ is not integer, and a can be assumed very small (by rescaling of the coordinates) in the case of $k:1$ resonance, $k = \lambda$. Compare also Remark 9.25 above.
Therefore, in the iteration step, we deal with the vector field $\boldsymbol{V} = \boldsymbol{V}_0 + \boldsymbol{V}_1 + \boldsymbol{W}$, where $\boldsymbol{W} = O\left(|(x,y)|^D\right)$ for a large D.
The operators $C_d(\boldsymbol{V}_0)$ and $D_d(\boldsymbol{V}_0)$ are invertible, with the norms of their inverses of order $O(1/d)$, while the operators $C_d(\boldsymbol{V}_1)$ and $D_d(\boldsymbol{V}_1)$ are their small perturbations.
The further analysis is like in the previous case.

(e) **The nonresonant saddle case.** Here we introduce an important tool for the non-analyticity proof.
Recall that here $\lambda \in \mathbb{R}_- \setminus \mathbb{Q}$. By an analytic version of the Hadamard–Perron theorem 9.48 there exist two analytic separatrices, which can be assumed equal $\{x=0\}$ and $\{y=0\}$. Thus we can assume that the perturbation part of $\boldsymbol{V} = \boldsymbol{V}_0 + \boldsymbol{W}$ equals

$$\boldsymbol{W} = g_1 \boldsymbol{V}_0 + g_2 \boldsymbol{E}.$$

where $\boldsymbol{E}$ is the standard Euler vector field. Also the normalizing series $G = \exp(\boldsymbol{Z})$ is such that

$$\boldsymbol{Z} = f_1 \boldsymbol{V}_0 + f_2 \boldsymbol{E}.$$

In terms of the homological operators we have

$$g_1 \boldsymbol{V}_0 = A(\boldsymbol{V}_0) g_1, \quad B(\boldsymbol{V}_0)\boldsymbol{W} = (\lambda_1 - \lambda_2)\, xy g_2,$$
$$f_1 = C(\boldsymbol{V}_0)^{-1} g_1, \quad f_2 = (xy)^{-1} D(\boldsymbol{V}_0)^{-1} xy g_2.$$

Unfortunately, it is hard to prove the divergence in the case of a concrete perturbation. Yu. Ilyashenko (see [**IlY4**]) proposed to consider the following 1-parameter family of perturbed vector fields

$$\boldsymbol{V}_\zeta = \boldsymbol{V}_0 + \zeta \boldsymbol{W}, \quad \zeta \in \mathbb{C}.$$

It turns out that: either

(1) *for all $\zeta \in \mathbb{C}$ the normalizing series converges in some domain D_ρ, $\rho=\rho(\zeta) > 0$, or*

(2) *this series diverges for all ζ except a set $K_{\boldsymbol{W}}$ of capacity zero.*

To exclude the alternative (1) it is enough to assume that:

$$\sum_{i,j} \frac{x^i y^j}{\lambda_1 i + \lambda_2 j} \text{ is divergent for any } (x,y) \neq (0,0),$$

and

the moduli of all coefficients of g_1 and g_2 are bounded from below by some geometric series.

Indeed, assuming that the series G converges for a positive capacity set of ζ's, it should converge for all values of ζ. In particular, G is holomorphic in some polydisc $\{|x| < \rho_1, |y| < \rho_1, |\zeta| < \rho_2\}$. Then the derivative $\frac{\partial}{\partial\zeta}|_{\zeta=0}G = \boldsymbol{Z} = f_1\boldsymbol{V}_0 + f_2\boldsymbol{E}$ satisfies $\mathrm{ad}_{\boldsymbol{V}_0}\boldsymbol{Z} = -\boldsymbol{W}$ and is defined by the series $C(\boldsymbol{V}_0)^{-1}g_1$ and $D(\boldsymbol{V}_0)^{-1}xyg_2$. The first of the above conditions is a certain assumption on the number λ (it is quickly approximated by rationals) and holds for a dense set (but of zero Lebesque measure) of λ's. The second condition is generic.

(f) **The resonant saddle case.** Here $\lambda = -\frac{k}{l}$, $k, l \in \mathbb{Z}_{>0}$, $\gcd(k,l) = 1$.
Recall that $\ker C(\boldsymbol{V}_0) = \mathbb{C}[[F]]$ and $\ker D(\boldsymbol{V}_0) = xy\mathbb{C}[[F]]$, $F = x^k y^l$. So, in $\mathcal{F}_d$, we have

$$\mathrm{coker} C_d(\boldsymbol{V}_0) \simeq \mathcal{N}(C_d) = \mathbb{C}\cdot F^m, \text{ if } d = mr_0,$$

and $= 0$ otherwise, and

$$\mathrm{coker} D_d(\boldsymbol{V}_0) \simeq \mathcal{N}(D_d) = \mathbb{C}\cdot xyF^m, \text{ if } d = mr_0 + 2,$$

and $= 0$ otherwise, where

$$r_0 = k + l.$$

We have $xyF^m = (\lambda_1 - \lambda_2)^{-1} B(\boldsymbol{V}_0) F^m\boldsymbol{E}$ (see (a)). It follows that the normal form is

$$\Psi(F)\{\boldsymbol{V}_0 + \Phi(F)\boldsymbol{E}\},$$

$\Psi = 1 + F^p\psi(F)$, $\Phi = F^n\varphi(F)$ (this can be made more precise).
It is known that this normal form is divergent in general and that the obstacles to the divergence lie in so-called nonlinear Stokes phenomena (see below). Here we present a simpler argument.
Assume $\Phi = aF^n + \ldots \not\equiv 0$. Let

$$\boldsymbol{V}_1 = aF^n\boldsymbol{E}.$$

Introduce the projection operators: P onto the space complementary to $\mathcal{N}(C) = \mathbb{C}[[F]]$ and Q onto the space complementary to $\mathcal{N}D = xy\mathbb{C}[[F]]$, where we use the monomial basis. Then the operator $C(\boldsymbol{V}_0)$ restricted to $\mathrm{Im}P$ (respectively, $D(\boldsymbol{V}_0)$ restricted to $\mathrm{Im}Q$) is invertible. Denote these restrictions by

$$C^{\mathrm{res}}(\boldsymbol{V}_0) \text{ and } D^{\mathrm{res}}(\boldsymbol{V}_0).$$

We have the following analogues of the operator K from (c):

$$K = -C^{\mathrm{res}}(\boldsymbol{V}_0)^{-1} PC(\boldsymbol{V}_1), \quad L = -D^{\mathrm{res}}(\boldsymbol{V}_0)^{-1} QD(\boldsymbol{V}_1).$$

We claim that they behave badly under iterations.
Indeed, take the function

$$f = xF^m.$$

Then $\boldsymbol{E}f = (mr_0+1)\,f$, $C(\boldsymbol{V}_1)\,f = aF^n \cdot (mr_1+1)\cdot xF^m$ and $Kf = -\lambda_1^{-1} \cdot a\,(mr_0+1)\cdot xF^{m+n}$. Generally,

$$K^N f = a(-\lambda_1)^{-N}\,(mr_0+1)\cdots((m+Nn-n)\,r_0+1)\,xF^{m+Nn}.$$

Thus the von Neumann series is divergent. We see that, like in the Euler's example, the perturbation $\boldsymbol{V}_1$ is not small with respect to leading field $\boldsymbol{V}_0$.
For the operator L we use the function $g = x^2yF^m$, instead of f.

To extract from the latter divergence the generic non-analyticity of the normal form reduction we use the method from the previous point. Consider a perturbed vector field of the form $\boldsymbol{V}_0 + \boldsymbol{V}_1 + \boldsymbol{W}$ such that $\boldsymbol{V}_1 = aF^n\boldsymbol{E}$ and the analytic perturbation $\boldsymbol{W}$, of order $> \mathrm{ord}\boldsymbol{V}_1 = nr_0$, is such that

$$\boldsymbol{W} = f\,(\boldsymbol{V}_0+\boldsymbol{V}_1) + \boldsymbol{W}_1,\ \ B\,(\boldsymbol{V}_0+\boldsymbol{V}_1)\,\boldsymbol{W}_1 = g,$$

where $f = Pf$ and $g = Qg$ are such that the series$\{K^N f\}$ and $\{L^N g\}$ diverge.
Next, we take the family

$$\boldsymbol{V}_\zeta = \boldsymbol{V}_0 + \boldsymbol{V}_1 + \zeta\boldsymbol{W},\quad \zeta\in\mathbb{C},$$

(analogous to the family from the previous point) and reduce it to the normal form.
The reduction process is inductive; one applies the homological operators (i.e., their restricted inverses) associated with polynomial parts of the normal form of higher and higher degree. It follows that the homogeneous parts of the reduction transformations depend polynomially on the parameter ζ, with good estimate for the degree.
Therefore, we have the 'total convergence or general divergence dichotomy'. But the linear in ζ part of the reduction is divergent by our choice of $\boldsymbol{W}$ (i.e., of f and g) and we remain with the second choice.

(g) **The saddle-node case.** Let $\lambda = 0$, i.e., $\boldsymbol{V}_0 = \lambda_1 x\frac{\partial}{\partial x}$. Then we have that $B\,(\boldsymbol{V}_0)\,(z_1\frac{\partial}{\partial x} + z_2\frac{\partial}{\partial y}) = \lambda_1 xz_2$ and hence $\ker B\,(\boldsymbol{V}_0) = \{f\frac{\partial}{\partial x}\}$. Next, $[\boldsymbol{V}_0, f\frac{\partial}{\partial x}] = \lambda_1\,(xf'_x - f)\,\frac{\partial}{\partial x} = C'\,(\boldsymbol{V}_0)\,f$, $D\,(\boldsymbol{V}_0)\,g = \lambda_1\,(xg'_x - g)$ and $B\,(\boldsymbol{V}_0)\,\boldsymbol{E} = \lambda_1 xy$. Therefore, we have the splitting $\mathcal{Z} \simeq \mathrm{Im}A\left(\frac{\partial}{\partial x}\right) + B^{-1}\mathcal{F}$ and we should replace the operator $C\,(\boldsymbol{V}_0)$ with $C'\,(\boldsymbol{V}_0)$ (see also the point (a) above). It follows that $\ker C'\,(\boldsymbol{V}_0) \simeq \mathrm{coker}C'\,(\boldsymbol{V}_0) \simeq x\mathbb{C}\,[[y]]$ and $\ker D\,(\boldsymbol{V}_0) \simeq \mathrm{coker}D\,(\boldsymbol{V}_0) \simeq x\mathbb{C}\,[[y]]$. Therefore, the normal form is like in in the previous case with $F = y$.
The further analysis is like above.

(h) **Final remarks.** Recall that, in the cases of resonant node and saddle-node ($-0:1$ resonant node), the homological operators $C(\boldsymbol{V}_0)$ and $D(\boldsymbol{V}_0)$ are not

invertible. This implies a nontrivial normal form; in [**SZ4**] it is called the *first level normal form.*
But these operators have nontrivial kernels and these kernels can be used in reduction of some terms from the first level normal form. For this one defines suitable second level homological operators and study their kernels and images (I skip the details).
The result of that analysis is that the second level normal form is $\Psi(F)\{\boldsymbol{V}_0 + \Phi(F)\boldsymbol{E}\}$ with $\Phi = aF^n + bF^{2n}$ and Ψ a polynomial of degree $\leq n$; it is the final normal form. Here $\boldsymbol{V}_0 + \Phi(F)\boldsymbol{E}$ is the so-called *orbital normal form.*
Finally, note that, if the orbital normal form is trivial, i.e., $\Phi \equiv 0$, then the normal form is analytic.

Continuation of the analysis of the saddle-node singularity. By Theorem 9.48 we can assume that the vector field has the form (in some analytic coordinates) $\dot{x} = xf(x,y)$, $\dot{y} = -y + g(x,y)$. This preliminary form is not satisfactory to our purposes. We want to get the factor x^{p+1} in the first component. This is done using series of the analytic changes $x \longmapsto x + x^j\psi_j(y)$, $j = 1,\ldots,p$: if $\dot{x} = x^j[\phi_j(y)+\ldots]$, then we get the equation $(-y+\ldots)\psi_j' + \phi_j(y) = 0$, solvable provided $\phi_j(0) = 0$. Thus we can put $\dot{x} = x^{p+1}$. Moreover, cancelling some initial terms from $\dot{y}$ and applying the result from Problem 9.47, we can assume that g has good properties. More precisely, we have the following.

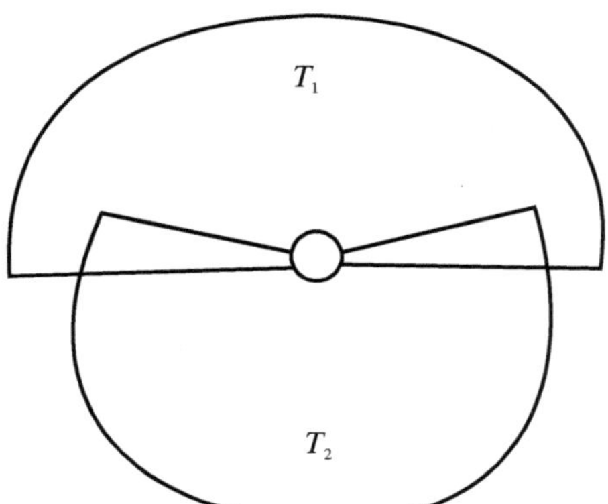

Figure 9.9: Extended sectors

9.51. Lemma. *The initial analytic form is the following*

$$\dot{x} = x^{p+1}, \quad \dot{y} = -y(1+\lambda x^p) + g_0(x) + y^2 g_2(x,y),$$

where $g_0, y^2 g_2 \in \mathfrak{m}^{p+2}$.

We define the extended sectors

$$T_j = \left\{ |x| < \epsilon, |y| < \epsilon, \ \frac{(2j-1)\pi}{2p} - \alpha < \arg x < \frac{(2j-1)\pi}{2p} + \alpha \right\},$$

where α is some angle between $\pi/2p$ and π/p (see Fig. 9.9). Note that the projections of these sectors onto the x-plane are different than the sectors S_j (from Section 3); they have the same angle but different bisectrices.

9.52. Theorem of Hukuhara, Kimura and Matuda (Sectorial normalization). ([**HKM**]) *In each extended sector T_j there exists an unique analytic diffeomorphism*

$$H_j(x,y) = (x, y + h_j(x,y)), \ h_j = O(x^{p+1}),$$

conjugating the system from Lemma 9.51 *with its formal normal form from Theorem* 9.45.

We will present the main ideas in the proof of this theorem later. Now we pass to the description of the Martinet–Ramis moduli of analytical orbital equivalence of germs from $\mathcal{E}_{p,\lambda}$.
The reader can guess that these moduli are constructed using the differences between the normalizing diffeomorphisms,

$$\Phi_j = H_{j+1} \circ H_j^{-1} \quad \text{in} \quad T_j \cap T_{j+1}.$$

However, the description of the Φ_j's can be simplified.
Notice that the diffeomorphisms Φ_j keep the first coordinate fixed and preserve the formal normal form $\omega_0 = x^{p+1}\mathrm{d}y + y(1+\lambda x^p)\mathrm{d}x = 0$ (of the corresponding Pfaff equation). The latter fact gives series of restrictions on possible forms of Φ_j's. Let us look at the behavior of the leaves of the foliation defined by $\omega_0 = 0$ in the sectors $T_j \cap T_{j+1}$. These leaves are given by

$$y = C\mathrm{e}^{-t(x)}, \quad t(x) = t_j(x) = -1/(px^p) + \lambda \ln x, \quad C \in \mathbb{C},$$

and the first integral takes the form

$$u(x,y) = u_j(x,y) = y\mathrm{e}^{t(x)} = y\mathrm{e}^{t_j(x)}.$$

(Because of $\ln x$ in the definition of $t(x)$, we index the 'times' and first integrals in different sectors. Thus $t_{2p} = t_1 + 2\pi\mathrm{i}\lambda$, $u_{2p} = \mathrm{e}^{2\pi\mathrm{i}\lambda}u_1$.)
The neighborhood of $x = y = 0$ is divided into sectors of two types:

- the sectors of *jump*, where $\mathrm{Re}\, x^{-p} \to \infty$,
- the sectors of *fall*, where $\mathrm{Re}\, x^{-p} \to -\infty$.

In the sectors of jump only one leaf approaches the singularity, it is the *local center manifold*. Other leaves diverge as $x \to 0$.

In the sectors of fall all the leaves are indistinguishable. All approach the singularity.
We have presented this schematic behavior of the complex phase portrait in Fig. 9.10. We see also that the extended sectors $T_{j+1} \cap T_j$ lie wholly either in the sectors of jump or in the sectors of fall.

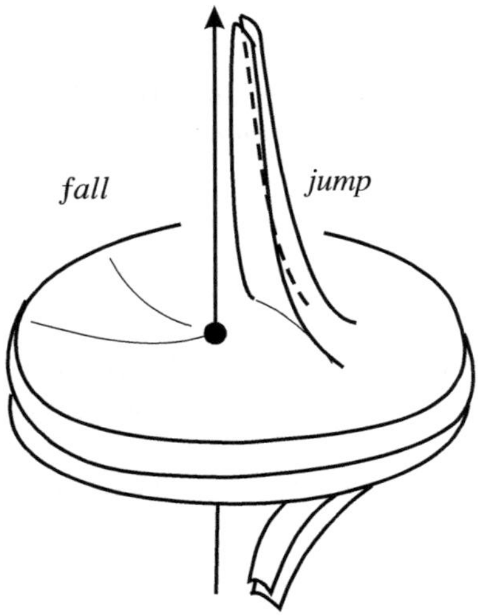

Figure 9.10: Sectors of fall and jump

9.53. Proposition. *The transition diffeomorphisms Φ_j take the forms*

$$\Phi_j(x, y) = (x, \phi_j(u_j)\mathrm{e}^{-t_{j+1}}),$$

where

$$\phi_j(u) = a_j + u, \quad a_j \in \mathbb{C},$$

if $T_{j+1} \cap T_j$ lies in a sector of fall, and

$$\phi_j(u) = u + \alpha_{j,2}u^2 + \ldots,$$

if $T_{j+1} \cap T_j$ lies in a sector of jump and $j \neq 2p$, and

$$\phi_{2p}(u) = \mathrm{e}^{-2\pi\mathrm{i}\lambda}u + \alpha_{2p,2}u^2 + \ldots$$

Moreover, Φ_j tend to id exponentially fast as $x \to 0$.

Proof. The first statement follows from the fact that Φ_j preserve the formal normal form. So, if u_{j+1} is the first integral in T_{j+1}, then the function $u_{j+1} \circ \Phi_j$ is the first integral in $T_j \cap T_{j+1}$. But also u_j is a first integral. Thus the first first integral is a function of the second first integral, $u_{j+1} \circ \Phi_j = \phi_j \circ u_j$.

In the coordinates (x,u) the map Φ_j is equal to $(x,\phi_j(u))$. Passing to the (x,y) coordinates we get the formula for Φ_j.
The proof of the next statements relies on the fact that $\Phi_j - id = O(x^{p+1})$.
Let $T_j \cap T_{j+1}$ be a sector of fall. We have $t_j = t_{j+1}$ (because $j \neq 2p$) and $t(x) \to +\infty$. Let the Taylor expansion of ϕ_j be $\sum_{n=0}^{\infty} \alpha_{j,n} u^n$. Then

$$\Phi_j = \left(x, \alpha_{j,0}\mathrm{e}^{-t(x)} + \alpha_{j,1} y + \sum \alpha_{j,n} y^n \mathrm{e}^{(n-1)t(x)}\right).$$

The condition $\Phi_j \to id$ implies that $\alpha_{j,1} = 1$ and $\alpha_{j,2} = \alpha_{j,3} = \ldots = 0$. This means that ϕ_j is a translation. Note also that in this case $\Phi_j - id = (0, O(\mathrm{e}^{-t(x)}))$ is exponentially small.
Let $T_j \cap T_{j+1}$ be a sector of jump and $j \neq 2p$. Then, acting exactly in the same manner as in the previous case, we get $\alpha_{j,0} = 0, \alpha_{j,1} = 1$. It means that ϕ_j is a germ of diffeomorphism of $(\mathbb{C},0)$ with $\phi'(0) = 1$. Also in this case $\Phi_j - id$ is exponentially small.
Let $j = 2p$. We are in the sector of jump with $\mathrm{e}^{t_{2p}} = \mathrm{e}^{2\pi\mathrm{i}\lambda}\mathrm{e}^{t_1}$. We have $\Phi_{2p} = (x, \phi_{2p}(u_{2p})\mathrm{e}^{-t_1}) = (x, \mathrm{e}^{2\pi\mathrm{i}\lambda}\phi_{2p}(u_{2p})\mathrm{e}^{-t_{2p}})$. Analogously as in the previous case, we get the formula for ϕ_j and the exponential closeness to the identity. □

The fact that the diffeomorphisms Φ_j have different forms can be explained geometrically. In the sector of jump the values of the first integral are bounded, $u \in (\mathbb{C},0)$. In the sectors of fall the first integral takes values in the Riemann sphere $\overline{\mathbb{C}}$ and $u = \infty$ along the strong manifold $x = 0$; the corresponding map ϕ_j is an automorphism of this sphere satisfying $\phi_j(\infty) = \infty$.
The collection of ϕ_j's satisfying the properties of Proposition 9.53 do not yet form a complete set of invariants. The reader can notice that the coordinates in the normalized Pfaff equation $\omega_0 = 0$ are not unique. Its non-uniqueness lies in the possibility to multiplication of y by a constant. This gives the non-uniqueness of the choice of the local first integral: $u_j \longmapsto Cu_j$, where C is the same in each sector.

9.54. Definition. Two collections $\phi = (\phi_1, \ldots, \phi_{2p})$ and $\tilde{\phi} = (\tilde{\phi}_1, \ldots, \tilde{\phi}_{2p})$, satisfying the conditions from Proposition 9.52, are called *equivalent* iff they are conjugated by a linear transformation: $\phi_j \circ C = C\tilde{\phi}_j$.
The equivalence class $[\phi]$ of such a collection is called the **Martinet–Ramis modulus**. The space of Martinet–Ramis moduli is denoted by $\mathcal{N}_{p,\lambda}$.
The above construction of the Martinet–Ramis modulus, associated to saddle-node, defines a map from $\mathcal{E}_{p,\lambda} \longmapsto \mathcal{N}_{p,\lambda}$

$$V \longmapsto \epsilon_V.$$

9.55. Martinet–Ramis classification theorem for saddle-nodes). *The map ϵ_V has the following properties:*

(a) *It sends orbitally analytically equivalent germs of vector fields to the same point (invariance);*

(b) *If* $\epsilon_{\boldsymbol{V}} = \epsilon_{\boldsymbol{V}'}$, *then* $\boldsymbol{V}$ *and* $\boldsymbol{V}'$ *are orbitally analytically equivalent (equimodality);*

(c) *For any* $\epsilon \in \mathcal{N}_{p,\lambda}$ *there exists* $\boldsymbol{V} \in \mathcal{E}_{p,\lambda}$ *such that* $\epsilon = \epsilon_{\boldsymbol{V}}$ *(realization).*

Proof. The points (a), (b) and (d) are proved in the same way as the analogous points of the Ecalle–Voronin theorem 9.30. Only the realization property needs separate arguments.

Assume that we have an element from $\mathcal{N}_{p,\lambda}$, which is an equivalence class with some representative $\phi = (\phi_j)$. These ϕ_j define the diffeomorphisms Φ_j in the extended sectors $T_j \cap T_{j+1}$ (see Proposition 9.53).

Take disjoint union of the extended sectors, $\bigsqcup T_j \times \{j\}$, and glue them using the Φ_j as the gluing maps. We obtain certain (real) 4-dimensional manifold $\mathcal{S}$. In each $T_j \times \{j\}$ we have the 'coordinates' $z_j = (x, y_j)$ and $z_{j+1} = \Phi_j(z_j)$ at the glued part.

Note also that $\mathcal{S}$ is equipped with a foliation $\mathcal{F}_0$ such that, in each T_j, it is the foliation of the vector field in the formal normal form. We have to define certain complex structure on $\mathcal{S}$ such that the foliation $\mathcal{F}_0$ arises from a holomorphic vector field with saddle-node singularity.

Firstly one defines the map $H_0 : \mathcal{S} \longmapsto W \subset \mathbb{C}^2 \setminus \{x = 0\}$ as

$$H_0 = \sum \chi_j z_j,$$

where $\{\chi_j\}$ is a partition of unity of $\mathcal{S}$ associated with the covering (T_j) with regular properties as $x \to 0$.

H_0 defines an almost complex structure on W, the image of the complex structures on T_j. By definition, this almost complex structure is integrable (see Definition 9.6). By Theorem of Newlander–Nirenberg 9.8, the torsion of this almost complex structure vanishes identically.

It turns out that the almost complex structure from W can be prolonged smoothly to an almost complex structure on a full neighborhood of 0 in $\mathbb{C}^2$. Indeed, the almost complex structure is defined by $(1,0)$-forms $\mathrm{d}x$ and $\mathrm{d}y$, where $\mathrm{d}x$ is good (with zero derivative). Since $y = \sum \chi_j y_j$, in $T_j \cap T_{j+1}$ we have $y = \chi_j y_j + \chi_{j+1} y_{j+1} = y_j + h_j(x, y_j)\chi_j$ (see the sectorial normalization theorem 9.52). Thus $\mathrm{d}y = \mathrm{d}y_j + o(1)$, where the term $o(1)$ is exponentially small as $x \to 0$.

The Nijenhuis tensor (torsion) of the latter almost complex structure is also equal to zero. Applying again the Newlander–Nirenberg theorem, we show that this almost complex structure is integrable. It means that it is equal to the transportation of the standard complex structure by means of some differentiable map $G : (\mathbb{C}^2, 0) \longmapsto (\mathbb{C}^2, 0)$

The composition $H = G^{-1} \circ H_0 : \mathcal{S} \longmapsto \mathbb{C}^2$ is holomorphic on $\mathcal{S}$ and prolongs to the closure of $\mathcal{S}$. It transforms the foliation $\mathcal{F}_0$ to a foliation $\mathcal{F}$ in $(\mathbb{C}^2, 0)$.

The foliation $\mathcal{F}$ is a saddle-node foliation with the Martinet–Ramis invariant equal to $[\phi]$. □

Theorem of Martinet and Ramis allows to solve the problem of analyticity of the center manifold and to explain the reason of non-analyticity of the Euler's solution. Note that in each sector of jump the local center manifold is defined uniquely, using topological criteria. It is the only leaf of the holomorphic foliation containing the singular point in its closure. This local center manifold prolongs itself (uniquely) to the adjacent sectors of fall. The gluing map Φ_j in the sectors of jump leave this local center manifold invariant, $\phi_j(0) = 0$.
The problem is whether the local center manifolds from different sectors of jump lie in one leaf of the foliation. It means, whether the gluing maps in the sectors of fall preserve the two prolongations of the local center manifolds (from the adjacent sectors of jump). This condition reads as $\phi_j(0) = 0$, where $\phi_j(u) = u + a_j$.
We have proved the following result.

9.56. Corollary. *A saddle-node singularity has analytic center manifold iff all $a_j = 0$ in its Martinet–Ramis modulus. In other words this modulus has the form $(id, \phi_2, id, \phi_4, \ldots, \phi_{2p})$.*

Problem. Unlike the Ecalle–Voronin moduli, the Martinet–Ramis moduli can be sometimes calculated explicitly. Show that the Martinet–Ramis modulus of the singularity

$$\dot{x} = -kx^{k+1}, \quad \dot{y} = y(1 + akx^k) + kx^{k+1}\varphi(x)$$

(with an analytic function φ) is the following

$$(id + A_1, id, id + A_2, id, ..., id + A_k, \mathrm{e}^{-2\pi \mathrm{i} a} id),$$

where the constants $A_j = \int_{\sigma_j} s^a \mathrm{e}^{-1/s^k}\varphi(s)\mathrm{d}s$ are the same as in Example 8.31(a) in Chapter 8 (see also [**Zo7**]).
Especially successful in calculations of the Martinet–Ramis moduli is P. Elizarov. In particular, he succeed to associate the Martinet–Ramis moduli with some special perturbations of the formal normal form. He localized the Newton supports of this perturbations; see [**Eli2**].
Also L. Teyssier [**Tey**] has calculated explicitly some Martinet–Ramis moduli.

There is a certain relation between the Martinet–Ramis moduli and the Ecalle–Voronin moduli. This connection goes through the monodromy transformation associated with the smooth analytic separatrix, i.e., the strong manifold.

9.57. Definition (Monodromy transformation for holomorphic foliation). Let $\mathcal{F}$ be a holomorphic foliation in a 2-dimensional complex manifold, L be a non-singular leaf of $\mathcal{F}$ and $\gamma \subset L$ be a closed loop with beginning at a point x_0. Take a small holomorphic disc D transversal to L at x_0 and parameterized by $z \in (\mathbb{C}, 0)$, $z(x_0) = 0$. We can cover the loop γ by open sets U_i, where the foliation is the flow-box foliation: $\mathcal{F}|_{U_i} = \{\mathrm{d}y = 0\}$.
Let $z \in D$. Take a leaf $L(z)$ passing through z. In each U_i we can lift the path $\gamma \cap U_i$ to the leaf $L(z) \cap U_i$. After returning of γ to the point x_0, the lifted curve intersects the disc D again. However, the intersection point can differ from z (see

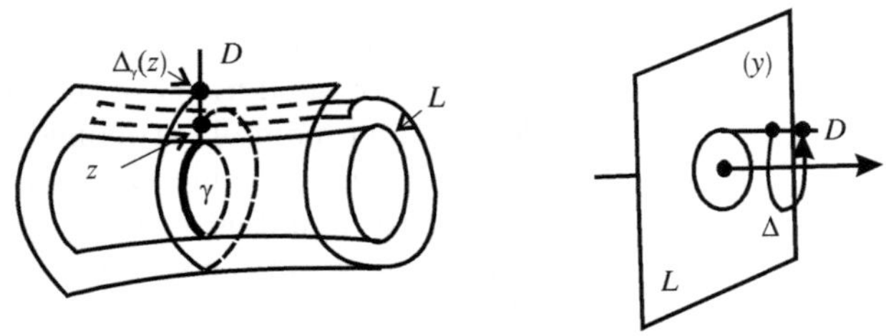

Figure 9.11: Monodromy transformations

Fig. 9.11). We denote it by $\Delta_\gamma(z)$. We define the **monodromy transformation** (or the **holonomy transformation)** associated with the loop γ as the map

$$z \longmapsto \Delta_\gamma(z).$$

Δ_γ is a local holomorphic diffeomorphism (by analytic dependence of solutions on initial conditions). It does not depend on the homotopy class of the loop γ (for fixed D); it is consequence of the theorem about monodromy 1.2 (in Chapter 1). The change of the section D and of the initial point x_0 results in internal conjugations of Δ_γ in the group Diff $(\mathbb{C}, 0)$ of germs of holomorphic diffeomorphisms of $(\mathbb{C}, 0)$. Of course, one can generalize this definition to more general foliations. We do not do it here.

The strong invariant manifold of a saddle-node singularity $\boldsymbol{V}$ consists of two leaves of the holomorphic foliation defined by $\boldsymbol{V}$: the singular point and the leaf L diffeomorphic to a punctured disc. Take a loop $\gamma \subset L$ generating its fundamental group. Using Definition 9.57, we associate with γ the monodromy transformation $\Delta = \Delta_\gamma$.

Recall that $\mathcal{A}_{p,\alpha}$ is the space of germs from Diff$(\mathbb{C}, 0)$ formally equivalent to the time 1 flow map generated by the vector field $\boldsymbol{W} = [z^{p+1}/(1 + \alpha z^p)]\partial_z$ (see Theorem 9.26).

9.58. Theorem.

(a) *If* $\boldsymbol{V} \in \mathcal{E}_{p,\lambda}$, *then* $\Delta \in \mathcal{A}_{p,\alpha}$, $\alpha = 2\pi \mathrm{i}\lambda$.

(b) *The Martinet–Ramis modulus of* $\boldsymbol{V}$ *coincides with the Ecalle–Voronin modulus of* Δ. *It means that two vector fields are orbitally analytically equivalent iff their monodromy maps are analytically conjugate. In particular, we have an embedding of the set of equivalence classes of vector fields* $\mathcal{E}_{p,\lambda}/\text{anal}$ *into the set of equivalence classes of diffeomorphisms* $\mathcal{A}_{p,\alpha}/\text{anal}$ *(i.e., embedding of the moduli space* $\mathcal{N}_{p,\lambda}$ *into the moduli space* $\mathcal{M}^*_{p,\alpha}$*).*

Proof. The first statement can be checked for the vector field in its formal normal form. Then we have $L = \{x = 0 \neq y\}$, $\gamma = \{(0, \mathrm{e}^{\mathrm{i}t})\}$, and the foliation is given by

$\frac{dx}{dt} = \frac{dx}{d\ln y} = x^{p+1}/y(1+\lambda x^p) = \boldsymbol{W}_{p,\lambda}$. The monodromy map Δ is given by the time $2\pi i$ flow map $\Delta = \exp[2\pi i \boldsymbol{W}_{p\,\lambda}] = \exp[\boldsymbol{W}_{p,\alpha}]$.
To prove the second statement, let us keep Δ formally equivalent to $g_{\boldsymbol{W}}^{2\pi i}$. Then the sectors S_j (from the sectorial normalization theorem 9.31) are turned by the angle $\pi/2p$; we denote them still by S_j. The extended sectors T_j are of the form $S_j \times \{|y| < \epsilon\}$. We fix also the disc $D = \{y = y_0\}$ transversal to the axis $\{x = 0\}$. The extended sectors contain complete lifts (to the leaves of the foliation) of the loop in the strong manifold appearing in the definition of the holonomy map.
The normalization diffeomorphisms $H_j(x, y)$ in T_j (from Theorem 9.51) define altogether the normalization diffeomorphisms for the monodromy map Δ in $S_j \times \{y_0\} \subset D$. They define the local parametrizations of S_j by means of the first integral $u_j(x) = u \circ H_j|_{S_j \times \{y_0\}}$. In the charts u_j the map Δ is equal to the formal normal form.
The transition diffeomorphisms $H_{j+1} \circ H_j^{-1}$ are expressed by means of the one-dimensional transition maps $u_{j+1} \circ u_j^{-1}$. The latter define the Martinet–Ramis moduli as well as the Ecalle–Voronin moduli. □

The image of the embedding from the point (b) of the last theorem forms a proper subset of the set $\mathcal{M}^*{}_{p,\lambda}$. It is because the diffeomorphisms ϕ_j are subject to the restrictions from Proposition 9.52, $\phi_j = u + a_j : (\overline{\mathbb{C}}, \infty) \longmapsto (\overline{\mathbb{C}}, \infty)$ (j odd).
It turns out that the correspondence $\boldsymbol{V} \longmapsto \Delta$ is not one-to-one even at the level of topological equivalences; i.e., the sets $\mathcal{E}_{p,\lambda}/\mathrm{top}$ and $\mathcal{A}_{p,\lambda}/\mathrm{top}$ are really different. (At the level of formal equivalences it is one-to-one).
As the first example we consider the case with the Martinet–Ramis modulus equal to $(id + a_1, id, id + a_2, id, \ldots, id)$ with some $a_j \neq 0$. The monodromy map Δ is topologically equivalent to its formal normal form (see Theorem 9.27), but the vector field is not equivalent to its formal normal form, because it does not have analytic center manifold.
Another example concerns the case with $\lambda = 0$ and with all $a_j = 0$, i.e., with analytic center manifold. We consider the monodromy map Δ_c associated with the leaf L_c containing the center manifold. The topology of the foliation determines the topological type of Δ_c, which is determined by the order of tangency of Δ_c to identity (see Theorem 9.27). So, the class $[\Delta]_{\mathrm{top}}$ is fixed, but there can be infinitely many classes $[\Delta_c]_{\mathrm{top}}$.
The problem of topological classification of saddle-nodes was solved (almost completely) by P. Elizarov.

9.59. Theorem of Elizarov. ([**Eli1**])

(a) *If there is no analytic center manifold, then the class of the sequence of* 0*'s and* 1*'s*

$$(\mathrm{sign}\,|a_1|, \mathrm{sign}\,|a_3|, \ldots, \mathrm{sign}\,|a_{2p-1}|),$$

modulo cyclic permutations, constitutes an invariant of the topological classification.

(b) *If there is analytic center manifold, then we have the following possibilities:*

(i) *If $\lambda \in \mathbb{C}\backslash\mathbb{R}$ or $\lambda \in \mathbb{R}\backslash\mathbb{Q}$ is a Diophantine number (weakly approximated by rationals), then the topological type of the foliation is determined by the topological type of the map Δ_c. (The invariants are* sign $(\mathrm{Im}\lambda)$, *and λ for $\lambda \in \mathbb{R}$).*

(ii) *If $\lambda = m/n$ is rational, then the vector field is topologically equivalent either to the formal normal form or to the system*

$$\dot{x} = x^{p+1}, \quad \dot{y} = y(1 + \lambda x^p) + x^l y^{kn+1},$$

with the conditions $p + 1 < l \leq 2p + 1$, $2p + 1 - l = km \pmod p$ (i.e., indexed by the discrete invariant k).

9.60. Proof of the Hukuhara–Kimura–Matuda theorem. (We follow the book [**HKM**]). We shall present the proof only in the case $p = 1$. Therefore we assume that $x^2\mathrm{d}y/\mathrm{d}x = g_0(x) + y(1 + \lambda x) + y^2 g_2(x, y)$.
Let $T_1 = \{-\pi/2 + \alpha < \arg x < 3\pi/2 - \alpha,\ |x| < \epsilon\}$ be a sector in the x-plane (see Fig. 9.12(a)). We seek a change $(x, y) \longmapsto (x, z)$ such that $x^2\mathrm{d}z/\mathrm{d}x = z(1 + \lambda z^2)$ in T_1.
The proof is divided into two parts: the existence of the analytic central manifold in T_1 and the analyticity of $z(x, y)$.

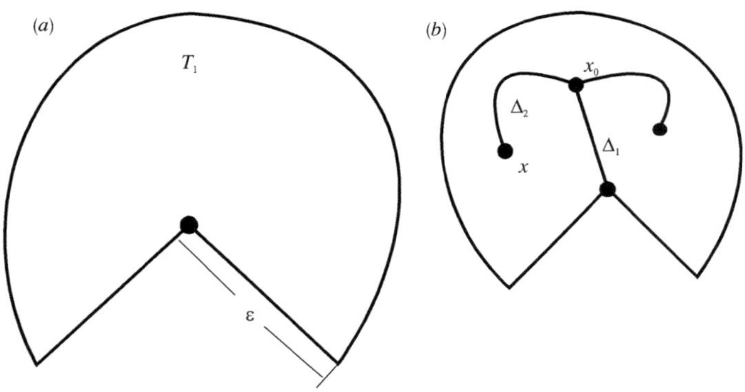

Figure 9.12: Sectors and integration paths used in the proof

1. *Existence of the center manifold.* We seek it in the form of a graphic

$$y = v(x)\mathrm{e}^{-1/x}, \quad x \in T_1,$$

where $v(x)\mathrm{e}^{-1/x} = O(x)$. We obtain the following equation onto v :

$$x^2 v' = G(x, v), \tag{9.9}$$

where $G = g_0 e^{1/x} + \lambda x v + v^2 e^{-1/x} g_2(x, v e^{1/x})$. We seek the solution to Eq. (9.9) in the class of functions satisfying the estimate

$$|v| < K|x|e^{\mathrm{Re}(1/x)}, \tag{9.10}$$

for some constant K.
The right-hand-side of Eq. (9.9) satisfies the estimates: $|G(x, v_1) - G(x, v_2)| < \delta|v_1 - v_2|$ and $|G| < (K_0 + \delta K)|x|e^{\mathrm{Re}(1/x)}$ for some small δ and fixed K_0. The first (Lipschitz) condition will guarantee the uniqueness of the solution.
Eq. (9.9) is replaced by the integral equation

$$v(x) = \int_0^x G(s, v(s)) s^{-2} \mathrm{d}s,$$

which is a fixed point equation $v = \mathcal{T}(v)$ for a nonlinear operator.
Usually one proves the existence of fixed point of $\mathcal{T}$ using the contraction principle. Here we can use some stronger result, the *Leray–Schauder–Tikhonov fixed point theorem* (see [**RS**]):

Any continuous map of a compact and convex subset of a locally convex space into itself contains a fixed point.

In our case the set of v's satisfying (9.10) is convex and compact in the topology of almost uniform convergence; (it is a normal family). It remains to show that the operator $\mathcal{T}$ maps (9.10) into itself.
The latter statement is proved by appropriate choosing the path Δ of integration in the formula defining $\mathcal{T}$. This path is presented at Fig. 9.12(b).
Let $s = re^{i\theta}$. We have the straight segment Δ_1, joining 0 with $x_0 = r_0 e^{i\pi/2 + 2\gamma}$, where $\gamma > 0$ is a small constant and r_0 will be defined later. In the second part Δ_2, we have we have either $r(\theta) = r_0\sqrt{\sin(\theta - \pi/2 - \gamma)}$ or $r_0\sqrt{\sin(\pi/2 + 3\gamma - \theta)}$, depending on whether $\arg x > \pi/2 + 2\gamma$ or not. If the angle α (defining the sector T_1, see above) is greater than 4γ, then $r_0 < r(\theta) < Mr_0$ (where $M = M(\gamma)$). Now we chose r_0 such that $Mr_0 < \epsilon$.
One can show that for large K, the quantity $|Gs^{-2}|$ is bounded by the absolute values of the derivatives of $K|s|e^{\mathrm{Re}(1/s)}$ along $\Delta_{1,2}$ (see [**HKM**]). This gives $|\mathcal{T}(v)| = |\int Gs^{-2}| \leq K|x|e^{\mathrm{Re}(1/x)}$.

2. Now we can assume that $x^2 \mathrm{d}y/\mathrm{d}x = y(1 + \lambda x + y g_2)$ in S. We have to find the change $y \longmapsto z$ giving the normal form $x^2 z' = z(1 + \lambda x)$.
We represent y as a function of z, $y = z + \phi(x, z)$. The ϕ satisfies the following partial differential equation:

$$x^2 \phi_x + (1 + \lambda x) z \phi_z = F(x, z, \phi).$$

We solve it using the method of characteristics: $\phi(x, z) = \psi(t)$, where $\dot{x} = x^2$, $\dot{z} = z(1 + \lambda x)$, $\dot{\psi} = F(x, z, \psi)$. Here t can be replaced by x which gives $x^2 \psi'_x = F(x, z, \psi)$, $z = Cx^{\lambda} e^{-1/x}$. This is a family of equations depending on C.

Now we proceed as in the previous point. We put $\psi = w\mathrm{e}^{-1/x}$, where $w = w(x, C)$ satisfies the equation

$$x^2 w' = G(x, z, w), \; z = Cx^\lambda \mathrm{e}^{-1/x},$$

analogous to Eq. (9.9). We replace it by the integral equation

$$w(x, C) = \int_0^x G(s, Cs^\lambda \mathrm{e}^{-1/s}, w(s, C)) s^{-2} \mathrm{d}s,$$

which we solve in the space of v's satisfying the estimate $|v| < K|z|^2$.
Using the same estimates as in the previous point (with the same choice of path of integration), one shows that this integral operator satisfies the assumption of the Leray–Schauder–Tikhonov theorem.
We refer the reader to [**HKM**] for more details. □

9.61. Remark. There is an alternative proof of the theorem about sectorial normalization. It is based on the theory of Gevrey expansions. It turns out that:

(i) *The series $\phi(x)$, appearing in the definition of the center manifold $y \sim \phi(x)$, admits Gevrey expansion of order $s = 1 + 1/p$ and is p-summable.*

(ii) *The series $h(x, y) = \sum h_n(y) x^n$, appearing in the formal change $H(x, y) = (x, y + h(x, y))$ and reducing the saddle node to its normal form, is of Gevrey type of order s and is p-summable.*

The proof of this is sketched in [**MR1**] for $p = 1$. As in the case of Ecalle–Voronin moduli, this proof relies on application of the Borel transform to ϕ and to $h(x, y)$, solving the corresponding equations in the Laplace image and then returning to the phase space by means of the Laplace transform.

§6 Normal Forms for Resonant Saddles

Here we finish the description of analytic normal forms for germs of planar vector fields with elementary singular points. In the previous section we classified the saddle-nodes.
The singular points from the *Poincaré domain,* i.e., when the convex hull of the eigenvalues (in $\mathbb{C}$) does not contain the origin (which means that the ratio of eigenvalues is not a negative number), have analytic Poincaré–Dulac normal form. This means that the normalizing transformation can be chosen analytic. The proof of this fact is given in [**A5**] (see also Remark 9.25).
Therefore, it remains to investigate the saddles, i.e., singularities with negative ratio of eigenvalues $-\lambda < 0$. The saddles are of two kinds: **resonant** (i.e., with rational $\lambda = m/n$) and non-resonant. The nonresonant saddles are considered in the next section.
The description of moduli of resonant saddles relies on the classification of germs of resonant 1-dimensional diffeomorphisms and was finished firstly by J. Martinet and

J.-P. Ramis [**MR2**]; although Yu. Ilyashenko, P. Elizarov and S. Voronin arrived independently to the same results (see [**Il5**]). We describe these results in this section.

The problem of analytic conjugation of non-resonant saddles is analogous to the problem of analytic linearization of germs of non-resonant 1-dimensional diffeomorphisms and was solved by J.-C. Yoccoz[**Yoc**]. Those results will be described in the next section.

9.62. The monodromy transformation. The connection between saddles and one-dimensional maps lies in the holonomy transformation associated with any of its separatrices. (Recall that, by the analytic Hadamard–Perron theorem 9.47, the saddle has both separatrices analytic). We can assume that

$$\dot{x} = \lambda x(1 + \ldots), \quad \dot{y} = -y(1 + \ldots).$$

Consider the monodromy transformation $\Delta(x)$ corresponding to a loop in $x = 0$. One can easily check that

$$\Delta(x) = \mathrm{e}^{-2\pi \mathrm{i}\lambda} x + \ldots$$

The next result was firstly proved by J. Mattei and R. Moussu [**MM**].

9.63. Theorem of Mattei and Moussu. *Two germs with saddle singularity with the same ratio $-\lambda$ are analytically orbitally equivalent iff the corresponding germs of monodromy maps are analytically equivalent.*

Proof. If two germs of vector fields are orbitally equivalent, then restriction of the conjugating map H to a disc D transversal to $\{x = 0\}$ defines a conjugation of monodromies (in D and in $H(D)$).

Let the monodromies be conjugated. Consider two copies of $(\mathbb{C}^2, 0)$ with foliations $\mathcal{F}$ and $\mathcal{F}'$ (by the phase curves L, L' of the two vector fields $\boldsymbol{V}$ and $\boldsymbol{V}'$). In each copy we consider the fibration $\pi : (x, y) \longmapsto y$. Their fibers outside $y = 0$ are transversal to the leaves of the foliations $\mathcal{F}$ and $\mathcal{F}'$. We denote by $\mathcal{F}_0$ the foliation by the fibers $y = \mathrm{const}$.

Take two discs $D = \{y = \epsilon, |x| < \epsilon\}$ and $D' = \{y = \epsilon, |x| < 2\epsilon\}$. Let $h : D \longmapsto D'$ be the map conjugating the monodromies.

We extend the map h to a map $H_0 : A \longmapsto A'$, where $A = \{|x| < \epsilon, \ \epsilon/2 < |y| < \epsilon\}$, $A' = \{|x| < 2\epsilon, \ \epsilon/2 < |y| < \epsilon\}$ are ring-like domains. H_0 preserves the foliation $\mathcal{F}_0$, i.e., $H_0(x, y) = (F(x, y), y)$ and sends leaves L of the foliation $\mathcal{F}$ to leaves L' of the foliation $\mathcal{F}'$. Because of the transversality of the fibers $y = \mathrm{const}$ to the leaves L, L' in the domains A and A', the prolongation H_0 is locally analytic and unique.

H_0 is single-valued in the whole domain A. Indeed, the change of H_0, as the argument turns around the axis $y = 0$, is equal to the composition of the monodromy map (defined by $\mathcal{F}$) and of H_0. Because $h = H_0|_D$ conjugates the monodromies, H_0 is univalent.

Therefore we have the Laurent expansion

$$H_0(x,y) = \sum_{j=-\infty}^{\infty} h_j(x) y^j, \tag{9.11}$$

with analytic coefficients $h_j(x) = \frac{1}{2\pi \mathrm{i}} \int_{|y|=\epsilon} H_0(x,y) y^{-j-1} \mathrm{d}y$. If we knew that the sum in Eq. (9.11) contains only positive powers of y, then we would obtain the prolongation of H_0 to the whole neighborhood of 0. We could use the Cauchy formula $H(x,y) = (2\pi \mathrm{i})^{-1} \int_{|\zeta|=\epsilon} H_0(x,\zeta)(\zeta - y)^{-1} \mathrm{d}\zeta$.

To show that the sum in Eq. (9.11) runs over positive powers of y, we use the assumption $\lambda = \lambda'$ (equality of the ratios of eigenvalues). Contrary to [**MM**] and [**MR1**], we apply topological arguments (taken from [**Str**]). It is enough to show that any circle $C_r = \{x = r,\ y = \epsilon \mathrm{e}^{\mathrm{i}\theta},\ 0 \le \theta \le 2\pi\}$ (with the linking number with the axis $y = 0$ equal to 1) is sent to a circle with the linking number with the axis $y = 0$ also equal to 1. But the assumption $\lambda = \lambda'$ means that the vector fields $\boldsymbol{V}$ and $\boldsymbol{V}'$ are close near the origin. A point $x = r,\ y = \epsilon \mathrm{e}^{\mathrm{i}\theta}$ from C_r belongs to a leaf L starting at the point $x = x_0(\theta), y = \epsilon$ in D, where $x_0(\theta) \approx r \mathrm{e}^{\mathrm{i}\lambda\theta}$ (because $\mathrm{d}x/\mathrm{d}y \approx -\lambda x/y$ and $x \approx \mathrm{const} \cdot y^{-\lambda}$). We have

$$H_0(r, \epsilon \mathrm{e}^{\mathrm{i}\theta}) \approx (h(x_0) r^{-\mathrm{i}\lambda\theta}, \epsilon \mathrm{e}^{\mathrm{i}\theta}) \approx (ar, \epsilon \mathrm{e}^{\mathrm{i}\theta}),$$

provided $h(x) \approx ax$. This shows that $H_0(C_r)$ is close to C_{ar}. □

Consider firstly the case when $\lambda = -m/n$, $\gcd(m,n) = 1$, is a rational number. Then the monodromy map $\Delta = \mathrm{e}^{2\pi \mathrm{i} m/n} x + \ldots$ is a resonant one-dimensional map considered in Section 3. Its formal normal form is of one of the following types $\mathrm{e}^{2\pi \mathrm{i} m/n} g^1_{\boldsymbol{W}}$, $\boldsymbol{W} = [z^{nk+1}/(1+\lambda z^{nk})] \frac{\partial}{\partial z}$ (see Proposition 9.33). The class of such maps was denoted by $\mathcal{A}_{m,n,k,\lambda}$.

The above form for diffeomorphism corresponds to the following formal orbital normal form for vector fields

$$\dot{x} = x, \quad \dot{y} = y(-m/n + u^k(1+\alpha u^k)^{-1}), \quad u = x^m y^n,$$

where $\alpha = 2\pi \mathrm{i} n\lambda$; compare also the points 9.50(f)–(h). (The proof relies on passing to the variables x, u in the Poincaré–Dulac normal form and further applying the reduction as in the case of saddle-node.) We denote the class of analytic saddles with such formal orbital normal form by $\mathcal{B}_{m,n,k,\lambda}$.

9.64. Theorem (Analytic classification of resonant saddles).([**MR2**]) *The space of classes of analytical orbital equivalence of vector fields from the class $\mathcal{B}_{m,n,k,\lambda}$ is the same as the space of classes of analytical equivalence of diffeomorphisms from the class $\mathcal{A}_{m,n,k,\lambda}$ described in Theorem* 9.34.

Remarks about the proof. In view of Theorem 9.62, the proof of this theorem is reduced to demonstration of the surjectivity of the map $\boldsymbol{V} \longmapsto \Delta$ from $\mathcal{B}_{m,n,k,\lambda}$ to $\mathcal{A}_{m,n,k,\lambda}$. One has to show that any resonant map arises from some resonant saddle as its holonomy. The construction of $\boldsymbol{V}$ is analogous to the construction of the realization parts of the Ecalle–Voronin and Martinet–Ramis classification theorems. We shall not do it here and refer the reader to [**I15**]. □

9.65. Bogdanov–Takens singularity. Here we present results of the author with E. Stróżyna [**SZ3, SZ4**] about normal forms for the nilpotent singularity in $(\mathbb{C}^2, 0)$ and its analytic properties.

For long time the problem of 'unique normal form' was not solved. Visibly the reason lied in correct splitting of the homological operator associated with the leading part of the vector field $\boldsymbol{V}_0$ (see 9.50). Usually, the authors tried to divide the vector fields into Hamiltonian (divergenceless) and to parallel to a corresponding quasi-homogeneous Euler vector field. But this is associated with heavy computations in suitable bases.

We use another splitting: to vector fields parallel to $\boldsymbol{V}_0$ and transversal to $\boldsymbol{V}_0$. We follow the point 9.50 above.

(a) **Takens normal form.** In the case of the **Bogdanov–Takens singularity** (B–T singularity), i.e., with nonzero nilpotent linear part, the classical **Takens prenormal form** [**Tak**]

$$(y + a(x))\,\frac{\partial}{\partial x} + b(x)\frac{\partial}{\partial y}$$

is analytic (see [**SZ1**, **Lor**]). Using the Takens result we can assume that we deal with vector fields of the form $\boldsymbol{V}_0 + \boldsymbol{W}$, where

$$\boldsymbol{V}_0 = (y + (\lambda+1)\,x^r)\,\frac{\partial}{\partial x} - \lambda r x^{2r-1}\frac{\partial}{\partial y} \tag{9.12}$$

is a quasi-homogeneous vector field with respect to the grading $\deg_H$ such that

$$\deg_H x = \deg x = 1, \quad \deg_H y = r,$$

and $\boldsymbol{W}$ contains higher degree terms, where $r \in \frac{1}{2}\mathbb{Z}$. In particular, $\deg V_0 = r - 1$. If $r \notin \mathbb{Z}$ (is a half-integer) then we have so-called *generalized cusp singularity* and here $\lambda = -1$.

The relation with elementary singularities follows from the fact that, after putting $z = x^r$ and dividing by rx^{r-1}, one arrives at the linear vector field

$$(y + (\lambda+1)\,z)\,\frac{\partial}{\partial z} - \lambda z\frac{\partial}{\partial y},$$

such that λ is the ratio of its eigenvalues.

In [**SZ3**] the B–T singularities are divided into three types. *Type I* includes the cases with $\lambda \notin \mathbb{Q}$ (nonresonant) and $\lambda = \frac{k}{l} \in \mathbb{Q}_{>0}$, $k, l > 1$, $\gcd(k,l) = 1$ (analogues of the $k : l$ resonant nodes). *Type II* includes analogues of resonant nodes with $l = 1$ ($\lambda = k > 0$) and *Type III* includes the cases corresponding to the $k : -l$ resonant saddles ($\lambda = -\frac{k}{l} \in \mathbb{Q}_{\leq 0}$, $\gcd(k,l) = 1$, including $\lambda = 0 = -\frac{0}{1}$). It is analogous to the division from 9.50.

(b) **Homological operators.** We consider vector fields of the form

$$\boldsymbol{V} = \boldsymbol{V}_0 + \dots,$$

where $\boldsymbol{V}_0$ is the vector field (9.12).
Let, as before,

$$\mathcal{F}_d = \{f \in \mathbb{C}[x,y] : \deg f = d\}, \quad \mathcal{Z}_d = \{\boldsymbol{Z} \in \mathcal{Z} : \deg \boldsymbol{Z} = d\},$$

denote spaces of quasi-homogeneous polynomials and of quasi-homogeneous polynomial vector fields of degree d. Analogously we define the spaces $\mathcal{F} = \mathcal{F}_{\geq d}$ and $\mathcal{Z} = \mathcal{Z}_{\geq d}$ of formal power series and of formal vector fields respectively.
Recall the operators

$$A(\boldsymbol{V})f = f \cdot \boldsymbol{V}, \qquad B(\boldsymbol{V})\boldsymbol{Z} = \boldsymbol{V} \wedge \boldsymbol{Z}/\partial_x \wedge \partial_y, \tag{9.13}$$

$$C(\boldsymbol{V})f = \boldsymbol{V}(f), \qquad D(\boldsymbol{V})f = \boldsymbol{V}(f) - \operatorname{div}(\boldsymbol{V})f. \tag{9.14}$$

Again we have the commutative diagram

$$\begin{array}{ccccccccc} 0 & \longrightarrow & \mathcal{F} & \overset{A(\boldsymbol{V})}{\longrightarrow} & \mathcal{Z} & \overset{B(\boldsymbol{V})}{\longrightarrow} & \mathcal{F} & \longrightarrow & 0 \\ & & \downarrow C(\boldsymbol{V}) & & \downarrow \operatorname{ad}_{\boldsymbol{V}} & & \downarrow D(\boldsymbol{V}) & & \\ 0 & \longrightarrow & \mathcal{F} & \overset{A(\boldsymbol{V})}{\longrightarrow} & \mathcal{Z} & \overset{B(\boldsymbol{V})}{\longrightarrow} & \mathcal{F} & \longrightarrow & 0 \end{array} \tag{9.15}$$

The operators $C(\boldsymbol{V})$, $\operatorname{ad}_{\boldsymbol{V}}$ and $D(\boldsymbol{V})$ are called the **homological operators**.
We note the following facts:

(i) *The above diagram is commutative.*

(ii) *If $\lambda = \lambda_2/\lambda_1 \neq 0$, i.e., when $x = y = 0$ is an isolated singularity of V_0, then the Koszul complexes are exact.*

Recall that:

$\ker C(\boldsymbol{V})$ consists of the *first integrals* (FIs) F of $\boldsymbol{V}$ and

$\ker D(\boldsymbol{V})$ consists of the *inverse integrating multipliers* (IIMs) M for $\boldsymbol{V}$, i.e., $\operatorname{div} M^{-1}\boldsymbol{V} = 0$.

In the below table we present the FIs and IIMs for the vector field $\boldsymbol{V}_0$ in the elementary case (left) and in the B–T case (right):

Type I	$\ker C = \ker D = 0$
Type II	$\ker C = 0,\ \ker D = \mathbb{C}\,(y + x^r)^{k+1}$,
Type III	$\ker C = \mathbb{C}\,[[F]],\ \ker D = F_1F_2\mathbb{C}\,[[F]]$,

where $F_1 = y + x^r$, $F_2 = y + \lambda x^r$ and $F = F_1^k F_2^l$. In the case of half-integer $r = \frac{1}{2}r_0$ we have $\lambda = -1$ and $F = y^2 - x^{r_0}$. Below we do not consider here the *generalized saddle-node case* $\lambda = 0$ and refer the reader to [**SZ3**, **SZ4**].

Indeed, in the blow-up coordinates $x, u = y/x^r$, we have the equation

$$\frac{\mathrm{d}x}{\mathrm{d}u} = \frac{1}{r(\lambda-1)}\left(\frac{1}{u+\lambda} - \frac{\lambda}{u+1}\right)x,$$

with the first integral

$$H = x^{r(1-\lambda)}\frac{u+\lambda}{(u+1)^{\lambda}} = \frac{y+\lambda x^r}{(y+x^r)^{\lambda}} = \frac{F_2}{F_1^{\lambda}}, \quad \lambda \neq 1. \tag{9.16}$$

(When $\lambda = -1$ the first integral contains an exponent.)
A quasi-homogeneous FI is a power of H; it can be polynomial only for *Type III*, where $H^l = F = F_1^k F_2^l$.
A polynomial IIM is related with a rational first integral and occurs for the *Types II* and *III*; we have $H = F_1 F_1^{-k}$ and $\boldsymbol{X}_H = M^{-1}\boldsymbol{V}_0$, $M = F_1^{k+1}$ for the *Type II*, and $H^{-ml} = F^{-m} = F_1^{-km}F_2^{-lm}$ and $\boldsymbol{X}_H = M_m^{-1}\boldsymbol{V}_0$, $M_m = F_1F_2F^m$, for the *Type III* (see also [**SZ3**]).
We note also the following identity

$$B(\boldsymbol{V}_0)\,\boldsymbol{E} = rF_1F_2, \tag{9.17}$$

where $\boldsymbol{E} = x\frac{\partial}{\partial x} + ry\frac{\partial}{\partial y}$ is the corresponding quasi-homogeneous Euler vector field.

(c) **Remark (Images of homological operators).** In [**SZ3**] the images of the above homological operators are described explicitly in terms of **periods** of certain Schwarz–Christoffel functions $\int_{\infty}^{u}\tilde{g}(v)(v+1)^{-\alpha-1}(v+\lambda)^{-\beta-1}\mathrm{d}v$. It was next used in the derivation of the normal forms. Those periods are principal values of some definite integrals, like

$$\Omega_{C,D}(g) = \mathrm{P.V.}\int_{-1}^{-\lambda}\frac{\tilde{g}(u)\mathrm{d}u}{(u+1)^{\alpha+1}(u+\lambda)^{\beta+1}} \tag{9.18}$$

where $\tilde{g}(u)$ is a polynomial, associated with the right-hand side of a homological equation $C(\boldsymbol{V}_0)f = g$ or $D(\boldsymbol{V}_0)f = g$, and α, β depend on the degree d on r on λ and on the case (C or D). We have

$$\operatorname{Im} C = \{g : \Omega_C(g) = 0\} \quad \text{and} \quad \operatorname{Im} D = \{g : \Omega_D(g) = 0\};$$

see also [**SZ3**]. This is used in the derivation of the normal forms.

(d) **First level reduction.** It is the reduction using the homological operators $C(\boldsymbol{V}_0)$ and $D(\boldsymbol{V}_0)$.
If $\mathcal{F}_d$ are subspace of $\mathcal{F}$ consisting of polynomials of quasi-homogeneous degree d, then the restrictions of the above operators to these subspaces $C_d(\boldsymbol{V}_0)$ and $D_d(\boldsymbol{V}_0)$ act as follows:

$$C_d(\boldsymbol{V}_0), D_d(\boldsymbol{V}_0) : \mathcal{F}_d \longmapsto \mathcal{F}_{d+r-1}.$$

Consider the *Type I*. Recall that the homological operators have trivial kernels. For $d = d_0 + rd_1$, $d_0 = 0, \ldots, r-1$, we have $\mathcal{F}_d = \mathrm{span}\{x^d, x^{d-r}y, \ldots, x^{d_0}y^{d_1}\}$. Therefore, these operators act between:

(i) $\mathbb{C}^{d_1+1}$ and $\mathbb{C}^{d_1+2}$ if $d_0 \neq 0$, and

(ii) $\mathbb{C}^{d_1+1}$ and $\mathbb{C}^{d_1+1}$ otherwise.

In the case (i) the complementary to ImC_d (or ImD_d) subspace is 1−dimensional, in [**SZ3**] it was chosen as $\mathcal{N}(C_d) = \mathbb{C} \cdot x^{d+r-1}$; in the case (ii) these operators are isomorphisms. Using Eq. (9.17) one gets the following.

Theorem (Normal form for Type I). *The complete normal form for the Type I is*

$$\Psi \{\boldsymbol{V}_0 + \Phi \boldsymbol{E}\},$$

where

$$\Phi = \varphi_0(y) + x\varphi_1(y) + \ldots + x^{r-2}\varphi_{r-2}(y), \;\; \Psi = \psi_0(y) + x\psi_1(y) + \ldots + x^{r-2}\psi_{r-2}(y),$$

and φ_j and ψ_j are formal power series.

Recall that the vector field $\boldsymbol{V}_0 + \Phi\boldsymbol{E}$ is the **orbital normal form** and Ψ is the **orbital factor**.

Consider the *Type II*. Recall that $\lambda = k$, $\ker C(\boldsymbol{V}_0) = 0$ but $\ker D(\boldsymbol{V}_0) = \mathbb{C} \cdot rF_1^{k+1}$, $F_1 = y + x^r$.
It follows that, contrary to the previous case, one additional term in the first level orbital normal form remains.
However, the corresponding IIM $M = rF_1^{k+1}$ from the kernel of $D(\boldsymbol{V}_0)$ can be used in killing some other term from this form. It is of the form

$$M = B(\boldsymbol{V}_0)\boldsymbol{T}, \;\; \boldsymbol{T} = F_2 \{kx\partial_x + r(y + (1-k)x^r)\partial_y\}, \tag{9.19}$$

$F_2 = y + \lambda x^r = y + kx^r$. Next, we have

$$\mathrm{ad}_{\boldsymbol{V}_0}\boldsymbol{T} = (k - r)F_1^{k-1}\boldsymbol{V}_0; \tag{9.20}$$

hence, a corresponding term can be removed from the orbital factor, provided $r \neq k$. The latter reduction uses only $\boldsymbol{V}_0$ and lies in the first level.
But the first level normal form is not like in the Type I. It is either of the form

$$\Psi(x)\{\boldsymbol{V}_0 + \Phi(x)\partial/\partial y\},$$

where $\Phi(x) = \sum_{j \in \mathcal{I}(\Phi)} a_j x^j$ and $\Psi(x) = \sum_{j \in \mathcal{I}(\Psi)} b_j x^j$are formal series with indicated supports $\mathcal{I}(\Phi)$ and $\mathcal{I}(\Psi)$ of coefficients (we do not present here the rather complicated formulas). If $\Phi \equiv 0$, then the orbital normal form is trivial.
The reason for the difference between the Type I lies in the fact that in the Type I there are always two analytic separatrices $F_1 + \ldots = 0$ and $F_2 + \ldots = 0$, while for the Type II generically there is only one separatrix $F_1 + \ldots = 0$. After resolution of the singularity, at the final exceptional divisor, one finds a $r(k-1):1$ resonant node with unique separatrix along the exceptional divisor.

Consider the *Type III*. The general normal form (of all levels) take the form

$$\Psi(x)\left\{\boldsymbol{V}_0+\Phi(x)\boldsymbol{E}\right\},$$

with concrete supports $\mathcal{I}(\Phi)$ and $\mathcal{I}(\Psi)$ of coefficients.
Recall that $\lambda=-\frac{k}{l}$, $\ker C(\boldsymbol{V}_0)=\mathbb{C}\left[[F]\right]$ and $\ker D(\boldsymbol{V}_0)=F_1F_2\mathbb{C}\left[[F]\right]$, $F=F_1^kF_2^l$. It follows that many terms from the series $\Phi(x)$ in the orbital normal form and from the orbital factor Ψ remain; we do not indicate them. We cannot use the FIs form the kernel of $C(\boldsymbol{V}_0)$ in the first level reduction, but we can use corresponding IIMs $M_m=rF_1F_2F^m$, $m=1,2,\dots$, form $\ker D(\boldsymbol{V}_0)$. We have

$$M_m=B(\boldsymbol{V}_0)\boldsymbol{T}_m,\ \boldsymbol{T}_m=F^m\boldsymbol{E}$$

and

$$\mathrm{ad}_{\boldsymbol{V}_0}\boldsymbol{T}_m=(1-r)F^m\boldsymbol{V}_0.$$

So, some series of coefficients b_j can be removed from the orbital factor Ψ (but we do not indicate them here); moreover, using only $\boldsymbol{V}_0$.

(e) **Second level reduction.** Here one considers the homological operators associated with vector fields of the form

$$\boldsymbol{V}_0+\boldsymbol{V}_1,$$

where $\boldsymbol{V}_1$ is some quasi-homogeneous vector field.
If the first level orbital normal form is nontrivial, then $\boldsymbol{V}_1$ is the lowest degree term from the orbital normal form: $\boldsymbol{V}_1=a_0x^p\frac{\partial}{\partial y}$ (for the Type II) or $\boldsymbol{V}_1=a_0x^p\boldsymbol{E}$ (for the Type III). If the first level orbital normal form is $\boldsymbol{V}_0$ (trivial), then $\boldsymbol{V}_1=b_0x^q\boldsymbol{V}_0$ is the lowest degree term parallel to $\boldsymbol{V}_0$.
The homological operators $C(\boldsymbol{V}_0+\boldsymbol{V}_1)$ and $D(\boldsymbol{V}_0+\boldsymbol{V}_1)$ associated with the above vector field act on spaces of functions of the form

$$\mu F+f,\quad \text{or}\quad \nu M+f,$$

$\mu,\nu\in\mathbb{C}$ and $f\in\mathcal{F}_d$. Here F and M are FI and IIM for $\boldsymbol{V}_0$ and the degree d is suitably chosen. More precisely, we are interested in the quasi-homogeneous functions

$$\mu C(\boldsymbol{V}_1)F+C(\boldsymbol{V}_0)f,$$
$$\nu D(\boldsymbol{V}_1)M+D(\boldsymbol{V}_0)f.$$

The question is: does $C(\boldsymbol{V}_1)F$ (respectively, $D(\boldsymbol{V}_1)M$) belong to $\mathrm{Im}C(\boldsymbol{V}_0)$ (respectively, to $\mathrm{Im}D(\boldsymbol{V}_0)$)? For the answer one uses the periods (9.18) from the Remark (c).
But also, when the first level orbital normal form is trivial (equal $\boldsymbol{V}_0$) and some IIM $M\in\ker D(\boldsymbol{V}_0)$ was not used in killing a term from the orbital factor using

$\boldsymbol{V}_0$ (first level), then this M can be used in killing another term from the orbital factor using $\boldsymbol{V}_1 = b_0 x^q \boldsymbol{V}_0$.

Consider the *Type II*. Recall that we have only one nontrivial kernel, which is $\ker D_{r(k+1)}(\boldsymbol{V}_0) = \mathbb{C} \cdot M$, $M = rF_1^{k+1}$.
If $r \neq k$, then this M was used in the first level reduction for killing some term from the orbital factor. Indeed, we have $M = B(\boldsymbol{V}_0)\boldsymbol{T}$, $\boldsymbol{T} = F_2\{kx\partial_x + r(y + (1-k)x^r)\partial_y\}$ and $\mathrm{ad}_{\boldsymbol{V}_0}\boldsymbol{T} = (k-r)F_1^{k-1}\boldsymbol{V}_0 \neq 0$ (see Eqs. (9.19)–(9.20)).
If $r = k$ and the orbital normal form is nontrivial, $\boldsymbol{V}_1 = ax^p\frac{\partial}{\partial y}$, then one checks that

$$D(\boldsymbol{V}_1)M = ar(k+1)x^p F_1^k \notin \mathrm{Im} D(\boldsymbol{V}_0)$$

(using Lemma (c)), and an additional term from Φ can be killed.
If $r = k$, the orbital normal form is $\boldsymbol{V}_0$ and $\boldsymbol{V}_1 = bx^q\boldsymbol{V}_0 \neq 0$, then one uses the identity

$$[x^q\boldsymbol{V}_0, \boldsymbol{T}] = -kqx^qF_1^{k-1}\boldsymbol{V}_0.$$

Here the coefficient function $x^qF_1^{k-1}$ does not lie in $\mathrm{Im} C(\boldsymbol{V}_0)$ and hence one term from Ψ can be killed.
The corresponding normal forms are complete, but the supports $\mathcal{I}(\Phi)$ and $\mathcal{I}(\Psi)$ are rather complicated. Instead, we present them in another way; that way was used in [**SZ4**] in the proofs of analyticity of some normal forms (see the point (h) below).
In place of y we put z defined by the formula

$$y = \varepsilon z - x^r, \tag{9.21}$$

where $\varepsilon \neq 0$ is a small constant and $r \in \mathbb{Z}$. The principal vector field $\boldsymbol{V}_0$ becomes

$$\boldsymbol{U}_H = x^{r-1}\left(\lambda x\frac{\partial}{\partial x} + rz\frac{\partial}{\partial z}\right) + \varepsilon x\frac{\partial}{\partial x} = x^{r-1}\boldsymbol{U}_0 + \varepsilon\boldsymbol{U}_1. \tag{9.22}$$

Theorem (Normal form for Type II). *The complete normal form for the Type II is either*

(i)

$$\Psi(x,z)\left\{\boldsymbol{U}_H + ax^{r-1}z^k\frac{\partial}{\partial z} + \Phi(x,z)\boldsymbol{E}\right\},$$

with $\Phi = \varphi_0(z) + x\varphi_1(z) + \ldots + x^{r-2}\varphi_{r-2}(z) = ax^pz^n + \ldots$ *and* $\Psi = \psi_0(z) + x\psi_1(z) + \ldots + x^{r-2}\psi_{r-2}(z) = bx^qz^s + \ldots$, *where* $\varphi_p^{(n+k)}(0) = 0$*,or*

(ii)

$$\Psi(x,z)\boldsymbol{U}_H$$

with Ψ *as above, but* $\psi_0^{(k)}(0) = 0$ *if* $k \neq r$*, and* $\psi_q^{(s+k)}(0) = 0$ *if* $k = r$*.*

Consider the *generalized cusp case* (*Type III*). Recall that we have $r = \frac{1}{2}r_0 \in \frac{1}{2}\mathbb{Z}\diagdown\mathbb{Z}$ and $\lambda = -1$. The principal vector field is Hamiltonian, $\boldsymbol{V}_0 = \frac{1}{2}\boldsymbol{X}_F$, $F = y^2 - x^{r_0}$. Since $\operatorname{div}\boldsymbol{V}_0 = 0$, our homological operators coincide: $C(\boldsymbol{V}_0) = D(\boldsymbol{V}_0) = \partial/\partial\boldsymbol{V}_0$.
One can check that the *first level normal form* is: either $\Psi(x,F)\{\boldsymbol{V}_0 + \Phi(x,F)\boldsymbol{E}\}$, with $\Phi = \varphi_0(F) + \ldots + x^{r_0-2}\varphi_{r_0-2}(F) = ax^pF^n + \ldots$ and $\Psi = \psi_0(F) + \ldots + x^{r_0-2}\psi_{r_0-2}(F)$, or $\Psi(x,F)V_0$,with $\Psi = 1 + x\psi_1(F) + \ldots x^{r_0-2}\psi_{r_0-2}(F)$.
If the orbital normal form is non-trivial, $\Phi = ax^pF^n + \ldots \neq 0$, then one uses the IIMs $M = F^m$, $m = 1, 2, \ldots$, from $\ker D(\boldsymbol{V}_0)$ and FIs F^k, $k = 1, 2, \ldots$, from $\ker C(\boldsymbol{V}_0)$;one finds that the polynomials $D(\boldsymbol{V}_1)F^m$ and $C(\boldsymbol{V}_1)F^k$ do not belong to the images of the operators $D(\boldsymbol{V}_0)$ and $C(\boldsymbol{V}_0)$.
If the orbital normal form is trivial, $\Phi \equiv 0$, then the kernel of $D(\boldsymbol{V}_0)$ was used in the first level reduction and the use of the kernel of $C(\boldsymbol{V}_0)$ leads to nothing.

Theorem (Normal form for generalized cusp). *The complete normal form for generalized cusp is either*

(i)

$$\Psi(x,F)\{\boldsymbol{V}_H + \Phi(x,F)\boldsymbol{E}_H\} \tag{9.23}$$

with $\Phi = ax^pF^n + \sum_{p\neq i<r_0} x^i\varphi_0(F)$ *and* $\Psi = \sum_{p\neq i<r_0} x^i\psi_i(F)$,*or*

(ii)

$$\Psi(x,F)\boldsymbol{V}_H$$

with $\Psi = 1 + \sum_{0<i<r_0} x^i\psi_i(F)$.

Corollary (Normal form for cusp). *In the case of the cusp singularity,* $r = \frac{3}{2}$, *we have the form* (9.23) *with either*
$\Phi = aF^n + x\varphi_1(F)$ *and* $\Psi = x\psi_1(F)$, $\varphi_1 = O\left(F_0^n\right)$, *or*
$\Phi = \varphi_0(F) + axF^n$, $\Psi = \psi_0(F)$, $\varphi_0 = O\left(F^{n+1}\right)$, *or*
$\Phi \equiv 0$, $\Psi = 1 + x\psi_1(F)$.

We skip analysis of the second level normal form for other singularities of the Type III, i.e., the *generalized* $k:-l$ *resonant saddles.*

(f) **Third level reduction**. This concerns only the cases of generalized $k:-l$ resonant saddles, which was omitted in the second level analysis. Here we concentrate only on the third level homological operator.
This is the situation when the second level orbital normal form is

$$\boldsymbol{V}_0 + \boldsymbol{V}_1 + \boldsymbol{V}_2 + \ldots,$$

where $\boldsymbol{V}_1 = ax^{r-1}F^n\boldsymbol{E}$, $F = F_1^kF_2^l$, and $\boldsymbol{V}_2 = \Phi_q\boldsymbol{E}$ with quasi-homogeneous polynomial Φ_q.
For $M_n = F_1F_2F^n$ from $\ker D(\boldsymbol{V}_0)$ we have

$$D(\boldsymbol{V}_1)M_n = 0,$$

which means that M_n is an IIM for $\boldsymbol{V}_0 + \boldsymbol{V}_1$. This M_n was not be used in the second level reduction of the orbital factor, because there are higher degree terms in the orbital normal form; in fact, it could used in the second level only in the case the orbital normal form is $\boldsymbol{V}_0 + ax^{r-1}F^n\boldsymbol{E}$.
So, it is used in the reduction, via $D(\boldsymbol{V}_2)M_n$, of a terms in the orbital normal form, provided $\boldsymbol{V}_2 \neq 0$.
For details and complete normal form I refer the reader to [**SZ3**, **SZ4**].

(g) **Generalizations.** We begin with a vector field $\boldsymbol{V}_0$ which is quasi-homogeneous with respect to a grading like $\deg x = \deg_H x = s$, $\deg y = r$, $\gcd(r,s) = 1$.If $\boldsymbol{E} = sx\frac{\partial}{\partial x} + ry\frac{\partial}{\partial y}$ is the corresponding quasi-homogeneous Euler vector field, then we have $[\boldsymbol{E}, \boldsymbol{V}_0] = n\boldsymbol{V}_0, \quad n = \deg \boldsymbol{V}_0$.
Introduce quasi-homogeneous coordinates as follows:

$$x = v^s u^p, \quad y = v^r u^q, \quad sq - rp = 1;$$

thus $u = y^s/x^r$. The equation for phase curves of $\boldsymbol{V}_0$ in the (u,v) coordinates takes the form

$$\frac{\mathrm{d}v}{\mathrm{d}u} = \frac{P(u)}{(u-u_1)\ldots(u-u_k)}v$$

(P – polynomial) and we have a FI in the Darbouxian form:

$$H(x,y) = v \cdot (u-u_1)^{\kappa_1} \ldots (u-u_k)^{\kappa_k} = x^{\kappa_0}(y^s - u_1 x^r)^{\kappa_1} \ldots (y^s - u_k x^r)^{\kappa_k} y^{\kappa_\infty}; \tag{9.24}$$

an analogue of Eq. (9.16). If $\boldsymbol{V}_0$ depends on some parameters, then the exponents κ_j vary and for some values of the parameters an exponential factor may appear.
Our problem is to reduce a vector field of the form

$$\boldsymbol{V} = \boldsymbol{V}_0 + \boldsymbol{W}, \quad \deg \boldsymbol{W} > n,$$

to a 'normal form'. Like in the point (b) we have the diagram (9.15) with the Koszul complexes and the operators $A(\boldsymbol{V}), \ldots, D(\boldsymbol{V})$ defined in Eqs. (9.14). The Koszul complexes are exact when the origin is an isolated singular point of $\boldsymbol{V}_0$.
Like in the point (d) the first level analysis deals with the case $\boldsymbol{V} = \boldsymbol{V}_0$. Here $C_d(\boldsymbol{V}_0), D_d(\boldsymbol{V}_0) : \mathcal{F}_d \longmapsto \mathcal{F}_{d+n}$, and the dimensions of the spaces $\mathcal{F}_d$ and $\mathcal{F}_{d+n}$ should be thoroughly calculated.
Recall that $\ker C_d(\boldsymbol{V}_0)$ consists of quasi-homogeneous polynomial FIs. It means that the exponents κ_j in Eq. (9.24) are all rational and of the same sign (say, all $\kappa_j \geq 0$). Then the polynomial FIs are some powers of some principal FI $F_0(x,y)$, $F_m = F_0^m$, $m = 1, 2, \ldots$.
If $\kappa_j \geq 0$ and $\kappa_j \in \mathbb{Q}$, then we have a series of polynomial quasi-homogeneous IIMs $M_m = F_m \prod (y^s - u_1 x^r) \ldots (y^s - u_k x^r) \in \ker D_d(\boldsymbol{V}_0)$. Polynomial IIMs exist also in some cases when polynomial FIs do not exist; for example, when F^l is a product of powers of simple factors but all negative exponents equal -1.

When calculating the cokernels of $C_d(\boldsymbol{V}_0)$ and of $D_d(\boldsymbol{V}_0)$, one arrives to Schwarz–Christoffel functions with *periods*

$$\mathrm{P.V.}\int_{u_i}^{u_j}(\tau-u_1)^{\delta_1-1}\ldots(\tau-u_k)^{\delta_k-1}\tilde{g}(\tau)\mathrm{d}\tau.$$

The latter are known as hypergeometric integrals, like in Chapter 12 below.
This allows to complete the first level analysis, which should be sufficient when the exponents κ_j are generic.
The higher level analysis is necessary when $\ker D_d(\boldsymbol{V}_0)\neq 0$ and/or $\ker C_d(\boldsymbol{V}_0)\neq 0$. Of course, the details depend on the case.

In [**Str2**] E. Stróżyna studied the case of homogeneous quadratic principal part,

$$\boldsymbol{V}_0=\left(ax^2+bxy+cy^2\right)\partial_x+(dx^2+exy+fy^2)\partial_y.$$

There there are 12 types, depending on the form of the principal FI (9.25), and there are 28 different complete normal forms. The highest level of reduction is 4. At some step a 2×2 matrix of periods of Schwarz–Christoffel functions is used.

(h) **Analyticity results.** The next two theorems are B–T analogues of corresponding statements from the points (c) and (d) of 9.50 about elementary singularities.

Theorem (Generalized focus). *If $\lambda\in\mathbb{C}\diagdown\mathbb{R}$, then the above normal for the Type I is analytic.*

Theorem (Generalized node). *If $\lambda>0$, the normal form, like for the Type II, is analytic.*

In the proofs, given in [**SZ4**], one works with the variables x,z defined in Eq. (9.21); so, in the formulas from the theorem about normal form for the Type I one replaces y with z. One uses the norm

$$\|f\|=\|f\|_\rho=\sum|a_{i,j}|\,\rho^{i+rj}$$

for a series $f=\sum a_{i,j}x^iz^j$.
Recall that the principal vector field $\boldsymbol{V}_0$ takes the form $\boldsymbol{U}_H=x^{r-1}\boldsymbol{U}_0+\varepsilon\boldsymbol{U}_1$ (see Eq. (9.22)). Here $x^{r-1}\boldsymbol{U}_0$ is diagonal like and $\varepsilon\boldsymbol{U}_1$ is its small perturbation.
We have

$$\begin{aligned}C_d\left(x^{r-1}\boldsymbol{U}_0\right)x^iz^j&=(\lambda i+rj)x^{r-1}\cdot x^iz^j,\\ D_d(x^{r-1}\boldsymbol{U}_0)x^iz^j&=(\lambda(i-r)+r(j-1))x^{r-1}\cdot x^iz^j.\end{aligned}$$

Next, one introduces suitable projection operators onto spaces complementary to the cokernels and kernels of these homological operators. Composing them suitably with C_d and D_d, one obtains restricted homological operators $C_d^{\mathrm{res}}(U_H)$

and $D_d^{\text{res}}(U_H)$ which are invertible, provided $|\varepsilon|$ and $|\rho|$ are sufficiently small, with the norms of the inverses bounded by

$$\text{const}/d\rho^{r-1}.$$

The further proof runs like in 9.50(c)–(d).

The next statement belongs to the folklore and follows from the resolution of singularity of a formal function.

Theorem (Trivial orbital normal form for Type III). *If the orbital normal form for the Type III is trivial, then the complete normal form* $\Psi \boldsymbol{V}_0$ *is analytic.*

(i) **Non-analyticity results.** The next results are B–T analogues of the corresponding statements from the points (e)–(g) of 9.50.

Theorem (Nonresonant generalized saddle). *In the case* $\lambda \in \mathbb{R}_{<0} \setminus \mathbb{Q}$, *for* λ *from a dense set a a generic perturbation* $\boldsymbol{V}_0 + \boldsymbol{W}$ *of* $\boldsymbol{V}_0$ *the above normal form for the Type I is non-analytic.*

In the proof one considers the 1-parameter family

$$\boldsymbol{V}_0 + \zeta \boldsymbol{W},$$

and applies the Ilyashenko's 'total convergence or general divergence dichotomy' (like in 9.50(e)). We skip the details.

Theorem (Generalized resonant saddle). *In the case of* $k : -l$ *resonance and non-trivial orbital normal form the complete normal form is non-analytic in general.*

Proof. We sketch the proof only for the generalized cusp and refer to [**SZ4**] for other cases. (Another, analytic proof in the case of cusp, $r = \frac{3}{2}$, was given by M. Canalis–Durand and R. Shäfke in [**C-DS**].) In fact, we demonstrate the non-analyticity of the first level reduction.
Recall that $C(\boldsymbol{V}_0) = D(\boldsymbol{V}_0) = \partial/\partial \boldsymbol{V}_0$; we will deal mainly with $C(\boldsymbol{V}_0)$.
We introduce new bases in the spaces $\mathcal{F}_d$. If $d = d_0 + d_2 r_0$, $d_0 = 0, \ldots, r_0 - 1$, then

$$\mathcal{F}_d = \text{span}\left\{x^{d_0} F^{d_2}, x^{d_0 + r_0} F^{d_2 - 1}, \ldots, x^d\right\} \simeq \mathbb{C}^{d_2+1},$$

and, if $d = d_0 + r + d_2 r_0$, then

$$\mathcal{F}_d = \text{span}\left\{x^{d_0} y F^{d_2}, \ldots, x^{d_0 + d_2 r_0} y\right\} \simeq \mathbb{C}^{d_2+1}.$$

Since the homological operators increase the degree by $r-1$ and since $C(\boldsymbol{V}_0) F = 0$, we have the following possibilities for the action of $C_d(\boldsymbol{V}_0)$:

$$x^i F^m \longmapsto i x^{i-1} y F^m, \ i \neq 0 \bmod r_0, \tag{9.25}$$
$$x^i F^m \longmapsto i x^{i-1} y F^m, \ i = 0 \bmod r_0, \tag{9.26}$$
$$x^i y F^m \longmapsto i x^{i-1} F^{m+1} + (i+r) x^{i+r_0-1} F^m, \ i \neq 0 \bmod r_0, \tag{9.27}$$
$$x^i y F^m \longmapsto i x^{i-1} F^{m+1} + (i+r) x^{i+r_0-1} F^m, \ i = 0 \bmod r_0. \tag{9.28}$$

The operator (9.25) is an isomorphism and is represented by a diagonal matrix. The operator (9.26) acts between spaces of dimensions d_2+1 and d_2, has kernel $\mathbb{C}\cdot F^{d_2}$; after factorization to $\mathcal{F}_d/\ker C_d$ it is also represented by a diagonal matrix. The operator (9.27) acts between $\mathbb{C}^{d_2+1}$ and $\mathbb{C}^{d_2+2}$, is injective and its cokernel will be identified with $\mathcal{N}(C_d)=\mathbb{C}\cdot x^{d_0-1}F^{d_2+1}$. After suitable projection, i.e., after taking PC_d, it is represented by the upper triangular matrix

$$\mathcal{A}=\begin{bmatrix} d_0+r & d_0+r_0 & 0 & \cdots & 0 \\ 0 & d_0+3r & d_0+2r_0 & \cdots & 0 \\ 0 & 0 & d_0+5r & \cdots & 0 \\ \cdots & \cdots & \cdots & \cdots & \cdots \\ 0 & 0 & 0 & 0 & d_0+(2d_2+1)\,r \end{bmatrix}. \tag{9.29}$$

The operator (9.28) acts between spaces of the same dimension d_2+1, is an isomorphism and is represented by the upper triangular matrix

$$\mathcal{A}=\begin{bmatrix} r & r_0 & 0 & \cdots & 0 \\ 0 & 3r & 2r_0 & \cdots & 0 \\ 0 & 0 & 5r & \cdots & 0 \\ \cdots & \cdots & \cdots & \cdots & \cdots \\ 0 & 0 & 0 & \ldots & (2d_2+1)\,r \end{bmatrix}. \tag{9.30}$$

Denote by P the projection operator onto the space complementary to the first level normal form and defined by our basis, i.e., x^iF^m and x^iyF^m. Denote also by R the projection operator onto the space complementary to $\ker C(\boldsymbol{V}_0)$ and also defined via our basis. The operators

$$C_d^{\mathrm{res}}(\boldsymbol{V}_0):=PC_d(\boldsymbol{V}_0)\,R:\mathrm{Im}R\longmapsto\mathrm{Im}P$$

are invertible and represented either by diagonal matrices (via Eqs. (9.25)–(9.26)) or by the matrices (9.29) or (9.30); recall that the matrix (9.29) corresponds to the composition $PC_d(\boldsymbol{V}_0)$.

Let us rewrite the matrix $\mathcal{A}$, form Eqs. (9.29)–(9.30), as

$$\mathcal{A}=S(I+N)$$

where the matrix $S=\operatorname{diag}(\sigma_1,\ldots,\sigma_k)$ is diagonal and N is a nilpotent matrix with the only nonzero entries $N_{i,i+1}=\nu_{i,i+1}$, $i=1,\ldots,k-1$. Note that $\sigma_j=\sigma_0+j\Delta\sigma$, i.e., grow linearly with j, and that $\nu_{i,i+1}>1$.

The following statements do not require proofs.

Lemma 1. *For $\mathcal{A}$ as above we have:*

(i)
$$\mathcal{A}^{-1}=\left(I-N+N^2-\ldots\pm N^{k-1}\right)S^{-1}$$

where N^j have the only nonzero entries

$$\left(N^j\right)_{i,i+j}=\nu_{i,i+1}\nu_{i+1,i+1}\cdots\nu_{i+j-1,i+j}>1;$$

(ii) *if a vector* $f = (x_1, \ldots, x_k)$ *has entries with alternating signs,* $x_j x_{j+1} \leq 0$, *then also the vector* $\mathcal{A}^{-1} f$ *has entries with alternating signs;*

(iii) *for such a vector we have the inequality*

$$\left\|\mathcal{A}^{-1} f\right\| > C \|f\|,$$

where $\|f\| = \sum |x_j|$.

Lemma 2. *If* $C_d^{\mathrm{res}}(\boldsymbol{V}_0)$ *is represented by a diagonal matrix, then its diagonal entries are positive and grow linearly. Moreover,*

$$\left\|C_d^{\mathrm{res}}(\boldsymbol{V}_0)^{-1} f\right\| \geq (C/q)\|f\|,$$

where $q = q(f)$ *is minimal such that* $f = F^{[(d-q)/r_0]} \tilde{f}$ *and the constant* C *does not depend on* d *and* q.

Let

$$\boldsymbol{V}_1 = a x^p F^n \boldsymbol{E}$$

be the lowest order term from the orbital normal form. Define the operator

$$K = -C^{\mathrm{res}}(\boldsymbol{V}_0)^{-1} PC(\boldsymbol{V}_1).$$

We claim that the iterations K^N of this operator grow super-exponentially with N. Due to the Ilyashenko type argument, this would be sufficient.

Recall that $p = 0, 1, \ldots, r_0 - 2$ and $\deg \boldsymbol{V}_1 = p + n r_0 > \deg \boldsymbol{V}_0 = r - 1$. We can distinguish two cases:

Case 1: $p > r - 1$ and **Case 2:** $p < r - 1$.

We consider only the **Case 1** and refer to [**SZ4**] for the second one. We take

$$f = f_0 = x^q \in \mathcal{F}_{\delta_0}, \quad \delta_0 = q = q_0$$

for some sufficiently large q, e.g., $p + q > r_0 - 1$.
We have

$$C(\boldsymbol{V}_1) f = a q x^{p+q} F^n \in \mathcal{F}_{p+q+n r_0}.$$

The latter polynomial is represented by a vector of length $\sim n + (p+q)/r_0$ with one entry equal to aq and other equal to 0. Here the operator $C_{\delta_1}^{\mathrm{res}}(\boldsymbol{V}_0)^{-1}$, $\delta_1 = p + q + n r_0 - r + 1$, is represented by a triangular matrix and from Lemma 1 it follows that the polynomial $f_1 = K f \in \mathcal{F}_{\delta_1}$ is represented by a vector with nonzero entries (of alternating signs) at the initial places, from 1 to $\sim (p+q)/r_0 := q_1/r_0$. Moreover, we have

$$\|K f\| \geq C_1 \delta_0 \|f\|, \quad C_1 = C|a|.$$

Next, $C(\boldsymbol{V}_1)Kf = a\delta_1 x^p F^n \cdot Kf$ and $C_d^{\text{res}}(\boldsymbol{V}_0)^{-1}$ is diagonal with entries growing linearly. We find that $f_2 = K^2 f \in \mathcal{F}_{\delta_2}$, $\delta_2 = q + 2(p + nr_0 - r + 1)$, is represented by a vector of length $\sim \delta_2/r_0$ and has nonzero entries (with alternating signs) at places up to $\sim q_2/r_0$ $q_2 = q + 2(p - r + 1)$. By Lemma 2 we have

$$\left\|K^2 f\right\| \geq \frac{C_1}{q_2}\delta_1 \|Kf\| \geq \frac{\delta_1\delta_0}{q_2} C_1^2 \|f\|.$$

It is clear how the further iterations of K, i.e., products of K_d's (with growing d's), act on f.
Half of them are represented by triangular type matrices, and we get factors which grow linearly with the iteration; the large factors are due to $\boldsymbol{E}$ in $C(\boldsymbol{V}_1)$), whereas the action of $C_d^{\text{res}}(\boldsymbol{V}_0)^{-1}$ does not decrease the norm essentially.
Another half of the K_d's are represented by diagonal type matrices and do not decrease the norm essentially, $\|K_d f\| / \|f\| \geq C_1$.
This implies that the sequence $\left\{\left\|K^N f\right\| / \|f\|\right\}_{N=1,2,\ldots}$ grows super exponentially. In particular, we have $\left\|K^3 f\right\| \geq \frac{\delta_2\delta_1\delta_0}{q_2} C_1^3 \|f\|$ and $\left\|K^4 f\right\| \geq \frac{\delta_3\delta_2\delta_1\delta_0}{q_4 q_2} C_1^4 \|f\|$. Here δ_j and q_j grow linearly with j:

$$\delta_{j+1} = \delta_j + p + nr_0 - r + 1, \quad q_{j+1} = q_j + p - r + 1. \qquad \square$$

9.66. Multidimensional Takens normal form and the Invariant Theory. A natural question is about normal forms for equilibrium points of higher dimensional vector fields. This problem is generally difficult, because its progress is associated with some classical unsolved problems.
For example, if we have a 3-dimensional germ of the form $\boldsymbol{V}_0 + \ldots$, where $\boldsymbol{V}_0$ is homogeneous of degree $n-1$, then the resolution of the singularity leads to a holomorphic foliation $\mathcal{F}$ on the exceptional divisor $E \simeq \mathbb{CP}^2$. This foliation can be highly complicated (see the Section 10.2 below). In the real case we encounter the 16th Hilbert problem (see 6.11 above).

(a) **Generalized Takens normal form.** Another situation concerns vector fields in $(\mathbb{C}^{n+1}, 0)$ of the form $\boldsymbol{Y} + \ldots$, where

$$\boldsymbol{Y} = nx_1\partial/\partial x_0 + (n-1)x_2\partial/\partial x_1 + \ldots + x_n\partial/\partial x_{n-1}$$

is a linear vector field defined by a nilpotent upper-triangular matrix (a nilpotent vector field or a locally nilpotent derivation).
There is a representation theory aspect of the vector fields $\boldsymbol{Y} + \ldots$. We introduce another locally nilpotent derivation

$$\boldsymbol{X} = x_0\partial/\partial x_1 + 2x_1\partial/\partial x_2 + \ldots + nx_{n-1}\partial/\partial x_n;$$

we call it the **Weitzenböck vector field** (or the *Weitzenböck derivation*). The vector fields $\boldsymbol{X}$, $\boldsymbol{Y}$ and

$$\boldsymbol{H} = [\boldsymbol{X}, \boldsymbol{Y}] = nx_0\partial/\partial x_0 + (n-2)x_1\partial/\partial x_1 + \ldots + (-n)x_n\partial/\partial x_n$$

generate a representation of the Lie algebra $\mathrm{sl}(2,\mathbb{C}) = \mathrm{sl}(2)$ in $\mathbb{C}^{n+1}$; we have $[\boldsymbol{H},\boldsymbol{X}] = 2\boldsymbol{X}$ and $[\boldsymbol{H},\boldsymbol{Y}] = -2\boldsymbol{Y}$. The vector field $\boldsymbol{H}$ is a **quasi-homogeneous Euler vector field**; it defines a quasi-homogeneous grading in $\mathbb{C}[x]$, with the **weights**

$$\mathrm{weight}(x_0) = n, \quad \mathrm{weight}(x_1) = n-2, \ldots, \mathrm{weight}(x_n) = -n,$$

thus $\boldsymbol{H}F = wF$ for a quasi-homogeneous polynomial F of weight $w = \mathrm{weight}F$.

Theorem (Generalized Takens normal form). *Using the homological operators* $\mathrm{ad}_{\boldsymbol{Y}}$ *one can reduce the germ* $\boldsymbol{Y} + \ldots$ *to*

$$\boldsymbol{Z}_{\text{Takens}} = \boldsymbol{Y} + a_0(x)\partial/\partial x_0 + \ldots + a_n(x)\partial/\partial x_n,$$

where $a_j(x)$ *are formal power series which are first integrals for* $\boldsymbol{X}$; *moreover, with only positive weights for* $j = 0, \ldots, n-1$ *(but not for* $j = n$*).*

For the proof we refer the reader to [**SZ5**]).

(b) **Rings of invariants of linear vector fields.** To get more precise understanding of the vector field Z_{Takens} one needs more informations about the first integrals for the Weitzenböck vector field, i.e., about the ring

$$\mathbb{C}[x]^{\boldsymbol{X}} = \{F \in \mathbb{C}[x] : \boldsymbol{X}F = 0\} = \ker \boldsymbol{X}$$

of polynomial **first integrals** of the Weitzenböck vector field (or the subring of $\mathbb{C}[x]$ of **invariants** of the Weitzenböck derivation). But it is a classical and still unsolved problem.

The problem of invariants of groups actions has long history, see the 14th Hilbert problem. In the linear case, i.e., of linear representations of groups in vector spaces, we have two works by L. Maurer in 1903 and of R. Weitzenböck in 1932. It was claimed that for any linear derivation in $\mathbb{C}[x]$, i.e., defined by a linear vector field, the ring of its first integrals is finitely generated. The Maurer and Weitzenböck proofs are incorrect and the just proof was given only in 1962 by C. Seshadri [**Ses**]; moreover, only in the nilpotent case. In [**SZ5**] a constructive proof was found, moreover, in general case; this proof is presented in the point (d) below. Let us state this result.

Theorem of Maurer and Weitzenböck. *For a linear differential system* $\dot{x} = Ax$, $x \in \mathbb{C}^m$, *the ring of polynomial first integrals* $\mathbb{C}[x]^A$ *has finite number of generators.*

(c) **The Weitzenböck vector field and symmetric polynomials.** Let

$$\mathbb{V} \simeq \mathbb{C}^2$$

be the standard representation of the Lie algebra $\mathrm{sl}(2)$ with $\boldsymbol{X} = x_0\partial/\partial x_1$, $\boldsymbol{Y} = x_1\partial/\partial x_0$ and $\boldsymbol{H} = x_0\partial/\partial x_0 - x_1\partial/\partial x_1$ (corresponding to lower-triangular, upper-triangular and diagonal matrices).

The above representation of $\mathrm{sl}(2)$ in $\mathbb{C}^{n+1}$ is irreducible; it is called the **representation with highest weight** n and x_0 is the **highest weight vector**. In fact, it is isomorphic to a symmetric tensor power of the standard 2−dimensional representation; we have

$$\mathrm{Sym}^n\mathbb{V} \simeq \mathbb{C}^{n+1}.$$

It is known that any finite dimensional irreducible representation of sl(2) is of such type, a representation with a highest weight (see [**SZ5**]).
Note that the representation with the highest weight 0 is 1-dimensional. The corresponding highest weight vector is sl(2)−invariant
The representation $\mathrm{Sym}^n\mathbb{V}$ of sl(2) implies representations of the same Lie algebras in the spaces $\mathrm{Sym}^d(\mathrm{Sym}^n\mathbb{V}) \simeq \mathbb{C}_d[x]$ of homogeneous polynomials of degree d in $x \in \mathbb{C}^{n+1}$. Any of them is split into irreducible representations; each a highest weight representation with a highest weight vector, a homogeneous polynomial annihilated by $\boldsymbol{X}$, i.e., a degree d homogeneous first integral of the vector field $\boldsymbol{X}$. Therefore, our problem becomes the problem of effective description of the highest weight vectors in the irreducible summands of the representation $\mathrm{Sym}^d(\mathrm{Sym}^n\mathbb{V})$. In particular, such highest weight vectors of weight 0 correspond to sl(2)−invariant polynomials.
Here we explore the geometric isomorphism

$$\mathbb{P}(\mathrm{Sym}^n\mathbb{V}) \simeq \mathrm{Sym}^n(\mathbb{P}\mathbb{V}),$$

where $\mathbb{P}$ denotes the projectivization operation and the nth symmetric power of a variety M is $\mathrm{Sym}^n M = M \times \ldots \times M/S(n)$ (the quotient of the n−fold Cartesian product of M by the action of the symmetric group $S(n)$). Recall that $\mathbb{V} = \mathbb{C}^2$ and $\mathrm{Sym}^n\mathbb{V} = \mathbb{C}^{n+1}$; thus $\mathbb{P}(\mathrm{Sym}^n\mathbb{V}) = \left(\mathbb{C}^{n+1}\diagdown 0\right)/\mathbb{C}^* = \mathbb{CP}^n$, $\mathbb{PV} = (\mathbb{V}\diagdown 0)/\mathbb{C}^* = \mathbb{CP}^1$ and $\mathrm{Sym}^n(\mathbb{PV}) = (\mathbb{V}\diagdown 0)^n/(\mathbb{C}^*)^n \times S(n)$.
It is expressed in the factorization formula for a degree n homogeneous polynomial,

$$\prod_{j=1}^{n}(u_j\lambda + v_j\mu) = \sum_{k=0}^{n} x_k \frac{\lambda^{n-k}}{(n-k)!}\frac{\mu^k}{k!},$$

which defines the **elementary expansion map**

$$\mathcal{E}_{S(n)} : \mathbb{V}^n \longmapsto \mathrm{Sym}^n\mathbb{V} = \mathbb{C}^{n+1}.$$

For example, if $n = 2$ then we have $x_0 = 2u_1u_2$, $x_1 = u_1v_2 + u_2v_1$ and $x_2 = 2v_1v_2$.

(Recall the generating function for the **elementary symmetric polynomials** $e_k = e_k(t_1, \ldots, t_n)$, $E(z) = \prod(1 + t_jz) = \sum_{k\geq 0} e_k z^k$. Therefore, if we introduce the *local times*

$$t_j = v_j/u_j \tag{9.31}$$

and denote $z = \mu/\lambda$, then we find that $x_k = k!\,(n-k)! \cdot u_1 \cdots u_n \cdot e_k(t_1, \ldots, t_n)$.)
The group $SL(2)$ acts on the variables λ, μ as follows:

$$g_{\mathbf{X}}^{\tau} : (\lambda, \mu) \longmapsto (\lambda + \tau\mu, \mu), \quad g_{\mathbf{Y}}^{\tau} : (\lambda, \mu) \longmapsto (\lambda, \mu + \tau\lambda),$$

and induces an action on the space of coefficients. The corresponding vector fields

$$\boldsymbol{X} = \sum u_j \frac{\partial}{\partial v_j}, \quad \boldsymbol{Y} = \sum v_j \frac{\partial}{\partial u_j}$$

act on the coefficients u_j, v_k. The corresponding vector fields, acting in the space of $x-$variables, are $\boldsymbol{X}$ and $\boldsymbol{Y}$ given above; we denote the vector fields acting in different spaces by the same letters (it should not lead to a confusion).
So, the variables in $\mathbb{V}^n$ have the weights

$$\text{weight}(u_j) = 1, \quad \text{weight}(v_j) = -1.$$

We have the commuting diagram

$$\begin{array}{ccc} \mathbb{V}^n & \stackrel{\mathcal{E}_{S(n)}}{\longrightarrow} & \text{Sym}^n\mathbb{V} \\ \downarrow g^t_{\boldsymbol{X}} & & \downarrow g^t_{\boldsymbol{X}} \\ \mathbb{V}^n & \stackrel{\mathcal{E}_{S(n)}}{\longrightarrow} & \text{Sym}^n\mathbb{V} \end{array}$$

Analogous commuting diagrams are like this, but with the vertical arrows being $g^t_{\boldsymbol{Y}}$ and $g^t_{\boldsymbol{H}}$.
We have the induced homomorphism

$$\mathcal{E}^*_{S(n)} : \mathbb{C}\left[\mathbb{C}^{n+1}\right] \longmapsto \mathbb{C}\left[\mathbb{V}^n\right]. \tag{9.32}$$

We are interested in its image.

Lemma 1.

(i) *The homomorphism $\mathcal{E}^*_{S(n)}$ is injective (as $\mathcal{E}_{S(n)}$ is surjective).*

(ii) *Its image is generated by 'degree d' homogeneous polynomials of the form*

$$F(u, v) = f(t_1, \ldots, t_n)\left(u_1 \cdots u_n\right)^d, \tag{9.33}$$

where f is a symmetric polynomial with $\deg_{t_j} f \le d$ (where $\deg_{t_j}$ is the degree in the variable t_j).

Proof. Indeed, any polynomial in $\mathbb{V}^n$ is a combination of polynomials which are homogeneous with respect to each pair u_j, v_j; the latter can be written as

$$F_1(u_1, v_1) \cdots F_n(u_n, v_n) = f_1(t_1) \cdots f_n(t_n) u_1^{d_1} \cdots u_n^{d_n}, \tag{9.34}$$

where f_j are polynomials of degree $\le d_j$ in the local times. If a combination of such polynomials is $S(n)-$invariant, then $d_1 = \ldots = d_n = d$ and the sum of coefficients must be a symmetric function of t_j. Note also that $\left(u_1 \cdots u_n\right)^d = \left(x_0/n!\right)^d$ which justifies the notion of degree. □

The following statement is the same as Proposition 2 in the point (d.1) below and is proved there.

Proposition 1. *If a polynomial F of the above form from Lemma 1 is a first integral for $\boldsymbol{X}$, then the polynomial f depends only on the differences $t_i - t_j$ between the local times.*

We want to give a more precise description of the form the polynomials f can take. Let

$$s_j = (n-1)\, t_j - \sum_{i \neq j} t_i = (t_j - t_1) + \ldots + (t_j - t_n)\,, \quad j = 1, \ldots, n.$$

Since weight (t_j) = weight (v_j) − weight $(u_j) = (-1) - 1 = -2$, also weight $(s_j) = -2$. Recall also that weight $(x_0) = n \cdot$ weight $(u_j) = n$.

Lemma 2. *The subring of $\mathbb{C}\,[t_1, \ldots, t_n]$ consisting of symmetric polynomials which depend only on differences between the local times is the ring of symmetric polynomials in the quantities s_j); in fact, it is a polynomial ring generated by the basic symmetric polynomials*

$$p_2(s) = s_1^2 + \ldots + s_n^2, \ldots, p_n(s) = s_1^n + \ldots + s_n^n,$$

with weight $(p_k) = -2k$.

Proof. Indeed, any such polynomial $f(t_1, \ldots, t_n)$ is a function of s_j's and, moreover, $S(n)$−symmetric. The variables s_j satisfy the relation

$$p_1(s) = s_1 + \ldots + s_n = 0.$$

It is known that the ring of symmetric polynomials in $s_1, \ldots, s_n$ is the polynomial ring generated by $p_1, \ldots, p_n$ Restrictions of the latter to the above hyperplane give the generators from Lemma 2. □

Therefore we have the following *principal first integrals for $\boldsymbol{X}$* : x_0 (of degree 1 and weight n) and

$$E_k =_n E_k =_n E_{k,w} = p_k(s) x_0^k \tag{9.35}$$

of degree k and of weight $w = (n-2)k$. One can show that $\mathbb{C}\left[\mathbb{C}^{n+1}\right]^{\boldsymbol{X}} = \mathbb{C}\left[\mathbb{C}^{n+1}\right] \cap \mathbb{C}\left[E_2, \ldots, E_n, x_0, x_0^{-1}\right].$

But we can define another set of first integrals for $\boldsymbol{X}$, which also deserve the name principal. They are obtained from the integration constants (initial conditions) of the corresponding differential system

$$\dot{x}_0 = 0, \quad \dot{x}_1 = x_0, \ldots, \quad \dot{x}_n = n x_{n-1}.$$

Assuming that $x_0(0) = x_0 \neq 0$ we can choose the time such that $x_1(0) = 0$, i.e., $x_1(t) = x_0 t$, and we have the local time

$$t = x_1 / x_0.$$

Then, assuming the initial conditions $x_j(0) = C_j$, we find the solutions

$$\begin{aligned}
x_2(t) &= C_2 + 2\int_0^t x_1(\tau)d\tau = C_2 + x_0 t^2 = C_2 + x_1^2/x_0, \\
x_3(t) &= C_3 + 3\int x_2(\tau)\mathrm{d}\tau = C_3 + 3C_2x_1/x_0 + x_1^3/x_0^2, \\
&\cdots\cdots\cdots\cdots\cdots\cdots\cdots\cdots\cdots\cdots\cdots\cdots \\
x_n(t) &= C_n + n\int x_{n-1}(\tau)\mathrm{d}\tau.
\end{aligned}$$

From these equations we express the constants C_j as rational functions of $x_j = x_j(t)$ with pole at $x_0 = 0$. Namely, we get

$$C_j = F_j(x)/x_0^{j-1}, \quad j = 2, 3, \ldots, \tag{9.36}$$

where $F_j(x) =_n F_j =_n F_{j,w}$, $w = (n-2)j$, are 'universal' homogeneous polynomials depending only on $x_0, \ldots, x_j$ and not depending on n. We have $\deg(F_j) = j$ and $\mathrm{weight}(F_j) = (n-2)\,j$. They also deserve the name *standard first integrals*. We have

$$\begin{aligned}
F_2 &= x_0x_2 - x_1^2, \\
F_3 &= x_0^2x_3 - 3x_0x_1x_2 + 2x_1^3, \\
F_4 &= x_0^3x_4 - 4x_0^2x_1x_3 + 6x_0x_1^2x_2 - 3x_1^4.
\end{aligned}$$

It follows that the polynomials F_j are expressed via E_j's and vice versa. For example, we have

$$\begin{aligned}
{}_2E_2 &= \left(s_1^2 + s_2^2\right)x_0^2 = 2\left(t_1 - t_2\right)^2\left(2u_1u_2\right)^2 = 8\Delta_{12}^2, \\
{}_2F_2 &= \left(2u_2u_2\right)\left(2v_1v_2\right) - \left(u_1v_2 + u_2v_1\right)^2 = -\Delta_{12}^2,
\end{aligned}$$

where $\Delta_{ij} = u_iv_j - u_jv_i$. Next,

$$\begin{aligned}
{}_3E_2 &= \left(s_1^2 + s_2^2 + s_3^2\right)x_0^2 = 6\left(t_1^2 + t_2^2 + t_3^2 - t_1t_2 - t_2t_3 - t_3t_1\right)x_0^2 \\
&= 6\left\{\left(t_1 + t_2 + t_3\right)^2 - 3\left(t_1t_2 + t_2t_3 + t_3t_1\right)\right\}\left(6u_1u_2u_3\right)^2 \\
&= 216\left\{(x_1/2)^2 - 3(x_2/2)(x_0/6)\right\} = -54 \cdot {}_3F_2.
\end{aligned}$$

Below we present the generators of the rings $\mathbb{C}\left[\mathbb{C}^{n+1}\right]^{\boldsymbol{X}} = \mathbb{C}\left[\mathrm{Sym}^n\mathbb{V}\right]^{\boldsymbol{X}}$ and of $\mathbb{C}\left[\mathbb{C}^{n+1}\right]^{\mathrm{sl}(2)}$ and relations between them for small n's. These cases are pretty easy.

The higher dimensional cases are not that simple. In particular, $\mathbb{C}\left[\mathbb{C}^{10+1}\right]^{\mathrm{sl}(2)}$ is generated by 106 concrete polynomials expressed via so-called iterated transvectants. The ring $\mathbb{C}\left[\mathbb{C}^{n+1}\right]^{\mathrm{sl}(2)}$ is not polynomial for $n \geq 5$, there are relations between generators. I have learned that generators of the rings $\mathbb{C}\left[\mathbb{C}^{n+1}\right]^{\mathrm{sl}(2)}$ are known for $n \le 12$ and of the rings $\mathbb{C}\left[\mathbb{C}^{n+1}\right]^{\boldsymbol{X}}$ for $n \le 8$.

The case $n = 2$. For $n = 2$ the ring $\mathbb{C}\left[\mathbb{C}^{2+1}\right]^{\boldsymbol{X}}$ is polynomial,

$$\mathbb{C}\left[\mathbb{C}^{2+1}\right]^{\boldsymbol{X}} = \mathbb{C}\left[x_0, F_2\right],$$

where $F_2 = F_{2,0} = \mathrm{const} \cdot \Delta_{12}^2$, $\Delta_{12} = u_1 v_2 - u_2 v_1$. Therefore, its degree d homogeneous parts are

$$\mathbb{C}_d\left[\mathbb{C}^{2+1}\right]^{\boldsymbol{X}} = \mathbb{C}F_2^e + \mathbb{C}x_0^2 F_2^{e-1} + \ldots + \mathbb{C}x_0^{2e}, \quad d = 2e,$$
$$\mathbb{C}_d\left[\mathbb{C}^{2+1}\right]^{\boldsymbol{X}} = \mathbb{C}x_0 F_2^{e-1} + \ldots + \mathbb{C}x_0^{2e-1}, \quad d = 2e-1.$$

Also we have $\mathbb{C}\left[\mathbb{C}^{2+1}\right]^{\mathrm{sl}(2)} = \mathbb{C}\left[F_2\right].$

The case $n = 3$. Here we have the standard integrals $x_0 = x_{0,3}$, $F_2 = F_{2,2}$ and $F_3 = F_{3,3}$ (of weights 3, 2 and 3 respectively). Note that $F_2 = -x_1^2 + O(x_0)$ and $F_3 = 2x_1^3 + O(x_0)$, i.e., they are the first integral for the restriction of $\boldsymbol{X}$ to the invariant hypersurface $x_0 = 0$. Therefore,

$$F_3^2 + 4F_2^3 = x_0^2 G_4,$$

where

$$G_4 = G_{4,0} = 4x_1^3 x_3 - 3x_1^2 x_2^2 + 4x_0 x_2^3 + x_0^2 x_3^2 - 6x_0 x_1 x_2 x_3$$

is a quartic polynomial of weight 0 (hence G_4 is an $\mathrm{sl}(2)$−invariant).

It follows that

$$\mathbb{C}\left[\mathbb{C}^{3+1}\right]^{\boldsymbol{X}} = \mathbb{C}\left[x_0, F_2, F_3, G_4\right], \quad \mathbb{C}\left[\mathbb{C}^{3+1}\right]^{\mathrm{sl}(2)} = \mathbb{C}\left[G_4\right],$$

with the only one relation. There cannot be other relations (and other generators), because the $\boldsymbol{X}$−invariants have weight ≥ 0.
In terms of the u_i, v_i variables we have

$$G_{4,0} = \mathrm{const} \cdot \Delta_{12}^2 \Delta_{23}^2 \Delta_{31}^2.$$

Indeed, the right-hand side is manifestly $\boldsymbol{X}$−invariant, of weight 0, and it has degree $3 \cdot 2 \cdot 2 = 12$ with respect to u_i, v_j and hence is of degree 4 with respect to x_k (which are of degree $n = 3$).

The case $n = 4$. Now we have the standard integrals $x_0 = x_{0,4}$, $F_{2,4}$, $F_{3,6}$ and $F_{4,8}$. Also G_4, previously $= G_{4,0}$, should be replaced with $G_{4,4}$. Moreover, we have the new relation

$$F_4 + 3F_2^2 = x_0^2 G_2,$$

where

$$G_2 = G_{2,0} = 3x_2^2 - 4x_1 x_3 + x_0 x_4$$

has weight 0, and another new relation

$$G_4 - F_2 G_2 = x_0 G_3,$$

where $G_3 = G_{3,0}$ is also $\mathrm{sl}(2)-$ invariant. It follows that

$$\mathbb{C}\left[\mathbb{C}^{4+1}\right]^{\boldsymbol{X}} = \mathbb{C}\left[x_0, F_2, F_3, G_2, G_3\right], \quad \mathbb{C}\left[\mathbb{C}^{4+1}\right]^{\mathrm{sl}(2)} = \mathbb{C}\left[G_2, G_3\right]$$

and there is only one relation

$$F_3^2 + 4F_2^3 = x_0^2 F_2 G_2 + x_0^3 G_3.$$

(d) **Proof of the Maurer–Weitzenböck theorem.**

1. *Conjecture of Nowicki.* Consider the case when a nilpotent matrix A is composed of 2-dimensional Jordan cells. In suitable variables we have the system

$$\dot{u}_1 = 0, \quad \dot{v}_1 = u_1, \ldots, \quad \dot{u}_s = 0, \quad \dot{v}_s = u_s.$$

It is easily solved: $u_j(t) = u_j \equiv \mathrm{const}$, $v_j(t) = C_j + u_j t$, $j = 1, \ldots, s$, $C_j = \mathrm{const}$. Of course, the coordinates u_j are first integrals and the quadratic polynomials

$$\Delta_{jk} = u_j v_k - u_k v_j$$

are also first integrals. Let us introduce the *local times* $t_j = v_j/u_j$; in each domain $\{u_j \neq 0\}$ the time t can be replaced with the local time t_j. We have

$$\Delta_{jk} = u_j u_k \left(t_k - t_j\right),$$

which means that the differences between the local times are constants of the evolution.

The next proposition was conjectured by A. Nowicki. It has now several proofs: by S. Khoury (in his dissertation), by J. Kuroda and by V. Drensky with L. Makar-Limanov. In [**SZ4**] a geometric approach to this conjecture was proposed.

Proposition 2. *The ring of polynomial invariants* $\mathbb{C}[u,v]^A$ *of the above system equals*

$$\mathbb{C}\left[\{u_i\}, \{\Delta_{jk}\}\right],$$

where the generators are subject to the relations

$$\begin{gathered} \Delta_{ij} + \Delta_{ji} = 0, \\ \Gamma_{ijk} := u_i \Delta_{jk} + u_j \Delta_{ki} + u_k \Delta_{ij} = 0. \end{gathered} \tag{9.37}$$

Proof. Let us replace the time with the local time $t_1 = v_1/u_1$ in the domain $\{u_1 \neq 0\}$. Thus we put $C_1 = 0$ and we get

$$C_j = v_j - u_j t_1 = \Delta_{1j}/u_1 = u_j\left(t_j - t_1\right) \quad (j \neq 1).$$

Therefore the space of orbits of the action of the group $\mathbb{C}$, i.e., the linear phase flow $\{\exp\tau A\}_{\tau\in\mathbb{C}}$ on the invariant set $\{u_1\neq 0\}$, is parameterized by $u_1,\ldots,u_s$, $C_2,\ldots,C_s$. Any invariant polynomial $F(u,v)$ is a polynomial in these quantities, $F(u,v) = f(u,C_2,\ldots,C_s)$; the polynomial dependence on C_j follows from the polynomial dependence on v_j.
On the other hand, in the domain $\{u_1u_2\neq 0\}$ we analogously find that $F(u,v) = g(u,D_1,D_3,\ldots,D_s)$,

$$D_j = \Delta_{2j}/u_2 = u_j\,(t_j - t_2)\,,$$

for a polynomial g. We have $D_1 = u_1\,(t_1-t_2) = -u_1(C_2/u_2)$ and $D_j = u_j\,(t_j-t_1) - u_j(t_2-t_1) = C_j - u_j(C_2/u_2)$ for $j\geq 3$. Therefore, we can write

$$g = g\,(u, 0 + u_1 z, C_3 + u_3 z, \ldots, C_s + u_s z)\,,$$

$z = -C_2/u_2$, where the $0+u_1z$ in the place of D_1 is explained by the above choice $C_1 = 0$. The terms with (powers of) z in g must cancel themselves, because F does not have a pole at $u_2 = 0$.
But this means exactly the invariance of g under the action of $\mathbb{C} : D\longmapsto D + zu$, $u = (u_1,u_3,\ldots,u_s)$. The latter action corresponds to the action $v\longmapsto v+\tau u$, $u,v\in\mathbb{C}^{s-1}$, i.e., with reduced number of cells.
Using the induction with respect to s, we conclude that g depends on u and on $u_jD_k - u_kD_j$. But, it is easy to check that

$$\begin{aligned} u_1D_k - u_kD_1 &= u_1C_k = \Delta_{1k} \quad (k\geq 3),\\ u_jD_k - u_kD_j &= -\Delta_{jk} \quad (j,k\geq 3). \end{aligned}$$

The first relations from Eqs. (9.37) are obvious, and the source of the other relations is the identity $(t_i - t_j) + (t_j - t_k) + (t_k - t_i) = 0$.

2. *Reynolds operators.* Recall that a ring $\mathbf{R}$ is *Noetherian* if any increasing sequence of its ideals is stabilizing. Another, equivalent, definition states that a ring is Noetherian if any its ideal has a finite basis.
We will use the second definition to the rings which are subrings of polynomial rings $\mathbb{C}\,[x]$, i.e., graded algebras.

Lemma 3. *If a graded subalgebra* $\mathbf{R}\subset\mathbb{C}\,[x]$ *is Noetherian, then* $\mathbf{R}$ *is finitely generated as a* $\mathbf{C}$*-algebra.*

Proof. Consider the maximal ideal $I = \{F\in\mathbf{R} : F(0) = 0\}$. It has a finite basis $F_1,\ldots,F_k$. Then any element of $\mathbf{R}$ equals $G = G\,(0) + G_1F_1 + \ldots + G_kF_k$ with $G_j\in\mathbf{R}$. We apply the same representation to each G_j, etc. □

Let $\mathbf{R}$ be a ring (a commutative algebra over the field $\mathbb{C}$) and $\mathbf{S}\subset\mathbf{R}$ be a subring. A map

$$R : \mathbf{R}\longmapsto\mathbf{S}$$

is a *Reynolds operator* if it

– is a linear projection (i.e., is surjective and satisfies $R(F) = F$ for $F \in \mathbf{S}$),

– satisfies $R(FG) = FR(G)$ for $F \in \mathbf{S}$ and $G \in \mathbf{R}$.

The importance of the above notion lies in the following classical result.

Proposition 3. *If $R : \mathbf{R} \longmapsto \mathbf{S}$ is a Reynolds operator and the ring $\mathbf{R}$ is Noetherian, then the subring $\mathbf{S}$ is also Noetherian.*

Proof. Assume I an ideal of $\mathbf{S}$. Of course, $I\mathbf{R}$ is an ideal of $\mathbf{R}$. Since $\mathbf{R}$ is Noetherian, the ideal $I\mathbf{R}$ has finite basis. But each element of this basis is a finite combination $\sum E_i H_i$ with $E_i \in I$ and $H_i \in \mathbf{R}$. It follows that there exist $F_1, \ldots, F_k \in I$ and $G_1, \ldots, G_k \in \mathbf{R}$ such that $I\mathbf{R} = F_1 G_1 \mathbf{R} + \ldots + F_k G_k \mathbf{R} \subset F_1 \mathbf{R} + \ldots + F_k \mathbf{R} \subset I\mathbf{R}$. We will show that I, as an ideal in $\mathbf{S}$, is generated by $F_1, \ldots, F_k$.
Indeed, if $F \in I \subset I\mathbf{R}$ then $F = F_1 D_1 + \ldots + F_k D_k$ for some $D_j \in \mathbf{R}$. Applying the operator R to this identity we find that $F = R(F) = F_1 R(D_1) + \ldots + F_k R(D_k)$ belongs to $(F_1, \ldots, F_k) \subset \mathbf{S}$. Thus $I = (F_1, \ldots, F_k)$. □

Example 1. The map $\mathcal{E}_{S(n)} : \mathbb{V}^n \longmapsto \mathbb{C}^{n+1}$ from the point (c) induces the ring homomorphism

$$\mathcal{E}^*_{S(n)} : \mathbb{C}[x] \longmapsto \mathbb{C}[\mathbb{V}^n]$$

(see Eq. (9.32)). It turns out that the image of this map, i.e., the graded ring of polynomials (9.33), is related with some Reynolds operators.
The first operator $R_{\mathrm{hom}} : \mathbb{C}[\mathbb{V}^n] \longmapsto \mathrm{Im} R_{\mathrm{hom}}$ sends a polynomial $f_1(t_1) \ldots f_n(t_n)$ $u_1^{d_1} \cdots u_n^{d_n}$ (see Eq. (9.34)) to

$$f_1(t_1) \ldots f_n(t_n) (u_1 \cdots u_n)^d \tag{9.38}$$

where $d = \max(d_1, \ldots, d_n)$. Thus R_{hom} is a sort of homogenization and its image is spanned by polynomials of the form $f(t_1, \ldots, t_n)(u_1 \cdots u_n)^d$, where $\deg_{t_j} f \leq d$ (but f may be not symmetric).
The second operator $R_{S(n)} : \mathrm{Im} R_{\mathrm{hom}} \longmapsto \mathcal{E}^*_{S(n)} \mathbb{C}[x]$ is the standard averaging over the symmetric group,

$$R_{S(n)}(F) = \frac{1}{n!} \sum_{\gamma \in S(n)} \gamma^* F; \tag{9.39}$$

here the elements γ permute the variables t_j in $f(t_1, \ldots, t_n)$,

$$\gamma^* F = f\left(t_{\gamma(1)}, \ldots, t_{\gamma(n)}\right) (u_1 \cdots u_n)^d .$$

It is easy to see that the maps (9.38) and (9.39) are Reynolds operators.

3. *Proof of the Maurer–Weitzenböck theorem in the nilpotent case.* We present the detailed proof in the case when the matrix A is nilpotent and consists of one Jordan cell, i.e., when the corresponding vector field is the Weitzenböck field $\boldsymbol{X}$. We are interested in the rings of invariants of the vector field $\boldsymbol{X}$. Denote the corresponding rings as

$$\mathbf{S} = \mathcal{E}^*_{\Gamma} \mathbb{C}\left[\mathbb{C}^{n+1}\right]^{\boldsymbol{X}}, \quad \mathbf{R} = \mathbb{C}[\mathbb{V}^n]^{\boldsymbol{X}} .$$

By Proposition 2 (Nowicki conjecture) we have $\mathbf{R} = \mathbb{C}[\{u_i\}, \{\Delta_{jk}\}]$, $\Delta_{jk} = u_j v_k - u_k v_j$, with the relations (9.37); note that $u_{j,k}$ and $v_{j,k}$ do not need to come from one and the same space $\mathbb{V}^n$. So, the ring $\mathbf{R}$ is Noetherian, because it is a ring of regular functions on a finite dimensional affine algebraic variety spec$\mathbf{R}$.
We want to apply the Reynolds operators R_{hom} and $R_{s(n)}$, defined in Eqs. (9.38)–(9.39), to the ring $\mathbf{R}$. Because the homogenization operator R_{hom} acts by multiplication of monomials by powers of the 'constants' u_i and the symmetrization commutes with the action of the flow $\{g^t_{\mathbf{X}}\}$, they act in the ring $\mathbf{R}$.
But we also have to ensure that the left-hand sides of relation equations (9.37) are sent to 0. Also to 0 should be sent those sides multiplied by polynomials.
This is obvious in the case of the first of these relations (anti-symmetry). In the case of the relations $\Gamma_{ijk} = u_i\Delta_{jk} + u_j\Delta_{ki} + u_k\Delta_{ij} = 0$ the operator R_{hom} acts correctly, because each summand is homogeneous (of degree 1) with respect to each pair (u_i, v_i) , (u_j, v_j) and (u_k, v_k) .
Any permutation from the group $S(n)$ acts on the homogeneous polynomial Γ_{ijk} by a permutation of the indices, it sends Γ_{ijk} to another polynomial $\Gamma_{i'j'k'}$. Therefore, the ideal of relations (9.37) is preserved by the operator $R_{S(n)}$.
Now Proposition 3 implies that the ring $\mathbf{S}$, which is isomorphic with the ring $\mathbb{C}[\mathbb{C}^m]^{\boldsymbol{X}}$ of polynomial first integrals of the nilpotent vector field $\boldsymbol{X}$, is finitely generated. □

The case of several Jordan cells is treated as follows. Assume that the nilpotent matrix A defining a nilpotent vector field, which we still denote $\boldsymbol{X}$, consists of Jordan cells of dimensions $n_1 + 1, \ldots, n_s + 1$. Thus the dimension of the vector space is $m = n_1 + \ldots + n_s + s$. Here some n_j's can be zero.
We consider the space $\mathbb{W} = \mathbb{V}^{n_1} \times \ldots \times \mathbb{V}^{n_s} \simeq \mathbb{V}^n$, $n = n_1 + \ldots + n_s$, $\mathbb{V} = \mathbb{C}^2$, with the map

$$\mathcal{E}_\Gamma : \mathbb{W} \longmapsto \mathbb{C}^m \simeq \mathbb{C}^{n_1+1} \times \ldots \times \mathbb{C}^{n_s+1},$$

induced from the elementary expansion maps $\mathcal{E}_{S(n_j)} : \mathbb{V}^{n_j} \longmapsto \mathbb{C}^{n_j+1}$. This map is surjective and is invariant with respect to the group $\Gamma = S(n_1) \times \ldots \times S(n_s)$ acting naturally on the space $\mathbb{W}$.
We find that the induced homomorphism

$$\mathcal{E}^*_\Gamma : \mathbb{C}[\mathbb{C}^m] \longmapsto \mathbb{C}[\mathbb{W}]$$

is injective and its image $\mathcal{E}^*_\Gamma \mathbb{C}[\mathbb{C}^m]$ is spanned by polynomials generalizing functions (9.38), they are of the form

$$f_1(t_1, \ldots, t_{n_1}) (u_1 \cdots u_{n_1})^{d_1} \ldots f_s(t_{n-n_s+1}, \ldots, t_n) (u_{n-n_s+1} \cdots u_n)^{d_s}$$

where f_j are symmetric polynomials with $\deg_{t_i} f_j \leq d_j$.
Next, one defines Reynolds operators R_{hom} and R_Γ which generalize operators (9.38)–(9.39) and whose composition projects $\mathbb{C}[\mathbb{W}]$ to $\mathcal{E}^*_\Gamma \mathbb{C}[\mathbb{C}^m]$.
The further proof is as in the case of Weitzenböck derivation.

4. *The case of diagonal A*. Consider firstly the semi-simple case, $A = \text{diag}(\lambda_1, \ldots, \lambda_n)$. Here we consider the monomials $x^\alpha = x_1^{\alpha_1} \ldots x_m^{\alpha_m}$. We have $\frac{\mathrm{d}}{\mathrm{d}t} x^\alpha = (\alpha, \lambda) x^\alpha$, $x^\alpha(t) = \mathrm{e}^{(\alpha,\lambda)t} x^\alpha(0)$, where $x^\alpha(t)$ denotes the evolution of the quantity x^α with time and $(\alpha, \lambda) = \alpha_1\lambda_1 + \ldots + \alpha_m\lambda_m$. Therefore the ring of polynomial first integrals is

$$\mathbb{C}[x]^A = \text{span}\{x^\alpha : (\alpha, \lambda) = 0\},$$

and can be identified with the sub-semigroup (monoid) of $(\mathbb{Z}_{\geq 0})^m : \mathbf{M}_A = \{\alpha \in (\mathbb{Z}_{\geq 0})^m : (\alpha, \lambda) = 0\}$.
This monoid lies in the sublattice $\mathbf{L}_A = \{\alpha \in \mathbb{Z}^m : (\alpha, \lambda) = 0\}$ of the integer lattice $\mathbb{Z}^m$. In fact, the monoid $\mathbf{M}_A$ is the intersection of $\mathbf{L}_A$ with the closed convex cone $(\mathbb{R}_{\geq 0})^m$.
The sublattice $\mathbf{L}_A$ is finitely generated, $\mathbf{L}_A = \alpha^{(1)}\mathbb{Z} + \ldots + \alpha^{(r)}\mathbb{Z} \simeq \mathbb{Z}^r = \{\beta = (\beta_1, \ldots, \beta_r) : \beta_j \in \mathbb{Z}\}$ where $\alpha^{(j)}$ are the generators. With this identification, the monoid $\mathbf{M}_A$ becomes the intersection

$$\mathbf{M}^C = \mathbb{Z}^r \cap C,$$

where $C \subset \mathbb{R}^r$ is a closed strictly convex cone; the intersection of C with a suitable hyperplane is a bounded convex polyhedron.
Important property of the cone C is its 'rationality'. It is defined by a finite number of inequalities $L_j(\beta) \geq 0$, where L_j are linear functions with rational coefficients.

Proposition 4. *Any 'rational' monoid of the above form is finitely generated, i.e., there exists a finite collection* $\left\{\beta^{(1)}, \ldots, \beta^{(s)}\right\} \subset \mathbf{M}^C \setminus 0$ *such that any* $\beta \in \mathbf{M}^C$ *is of the form* $k_1\beta^{(1)} + \ldots + k_s\beta^{(s)}$, $k_j \in \mathbb{Z}_{\geq 0}$.

Proof. It is enough to find a system $\beta^{(1)}, \ldots, \beta^{(s)}$ such that $\mathbf{M}^C \setminus 0 = \left(\beta^{(1)} + \mathbf{M}^C\right) \cup \ldots \cup \left(\beta^{(s)} + \mathbf{M}^C\right)$.
The boundary of C is a union of convex cones of dimensions between 1 and $r-1$. Each $1-$dimensional cone, a half-line $C_j \subset C$, contains elements of $\mathbf{M}^C$; this follows from the rationality of the cone C. Take generators $\beta^{(1)}, \ldots, \beta^{(t)}$ of the corresponding $1-$dimensional sub-monoids $\mathbf{M}^{C_j} = \mathbb{Z}^r \cap C_j$.
The cones $\beta^{(1)} + C, \ldots, \beta^{(t)} + C$ almost cover the cone C, the complement $C \setminus \bigcup \left(\beta^{(j)} + C\right)$ is of finite volume. So, the set $\left(\beta^{(1)} + \mathbf{M}^C\right) \cup \ldots \cup \left(\beta^{(t)} + \mathbf{M}^C\right)$ is $\mathbf{M}^C \setminus 0$ deprived of finite number of points $\beta^{(t+1)}, \ldots, \beta^{(s)}$. □

We underline that the generators $\alpha^{(1)}, \ldots, \alpha^{(s)}$ of the monoid $\mathbf{M}_A$ are subject to some relations. Therefore, also the generators $z_1 = x^{\alpha^{(1)}}, \ldots, z_s = x^{\alpha^{(s)}}$ of the ring $\mathbb{C}[x]^A$ of polynomial first integrals obey some relations. The ideal of these relations is generated by polynomials of the form of the form $z^{\gamma^{(j)}} - z^{\delta^{(j)}} = 0$, $j = 1, \ldots, q$, where the multiindices $\gamma^{(j)}, \delta^{(j)} \in (\mathbb{Z}_{\geq 0})^s$ have disjoint 'supports' (if $\gamma_l^{(j)} \neq 0$ then $\delta_l^{(j)} = 0$ and vice versa).

Recall that the existence of relations between the generators $z_i \in \mathbb{C}[x]$ means that the spectrum $Z = \mathrm{spec}\mathbb{C}[x]^A$ of the ring of first integrals, the so-called algebraic quotient of the action of the group $\{\exp At\}_{t\in\mathbb{C}}$ on $\mathbb{C}^m$, is an algebraic subvariety of $\mathbb{C}^s$. Namely, we consider the ideal of $\mathbb{C}[z]$ generated by the relations, $I = \left(z^{\gamma^{(1)}} - z^{\delta^{(1)}}, \ldots, z^{\gamma^{(q)}} - z^{\delta^{(q)}}\right)$. Then

$$Z = V(I) \subset \mathbb{C}^s$$

is the zero locus of the ideal I and $\mathbb{C}[x]^A$ is the ring $\mathbb{C}[Z] = \mathbb{C}[z]/I$ of regular functions on the affine algebraic variety Z.

5. *Reduction to the nilpotent case.* Assume that the matrix $A = S + N$, $S = \mathrm{diag}(\lambda_1, \ldots, \lambda_n)$, in the Jordan form, has nontrivial Jordan cells. Recall that the variables evolve as follows: $x_j(t) = e^{\lambda_j t} p_i(t)$,where p_j are polynomials.
Again we consider the monoid $\mathbf{M}_A$ defined above. But now the monomials x^α, $\alpha \in \mathbf{M}_A$, generate the ring of regular functions which evolve with time at most polynomially. Of course, it contains the ring of invariants.
It turns out that one can order generating monomials x^α in a way that the corresponding differential equations become linear in these generators and defined by a lower triangular matrix with zeroes on the diagonal.

Example 2. For the system

$$\dot{x}_1 = \lambda x_1, \quad \dot{x}_2 = \lambda x_2 + x_1, \quad \dot{x}_3 = -\lambda x_3, \quad \dot{x}_4 = -\lambda x_4 + x_3,$$

i.e., the vector field $\boldsymbol{U} = \lambda x_1 \frac{\partial}{\partial x_1} + (\lambda x_2 + x_1) \frac{\partial}{\partial x_2} - \lambda x_3 \frac{\partial}{\partial x_3} + (-\lambda x_4 + x_3) \frac{\partial}{\partial x_4}$, define the quantities $z_1 = x_1 x_3$, $z_2 = x_1 x_4$, $z_3 = x_2 x_3$, $z_4 = x_2 x_4$. This leads to the system

$$\dot{z}_1 = 0, \quad \dot{z}_2 = z_1, \quad \dot{z}_3 = z_1, \quad dot z_4 = z_2 + z_3,$$

i.e., the vector field $\boldsymbol{X} = z_1 \partial/\partial z_2 + z_1 \partial/\partial z_3 + (z_2 + z_3)\partial/\partial z_4$. Here we have the first integrals: $z_2 - z_3$, z_4, $F = (z_2 + z_3)^2 - 4 z_1 z_4$ and one relation $f(z) = z_1 z_4 - z_2 z_3 = 0$; so, $F = (z_2 - z_3)^2$. One also directly checks that $\boldsymbol{X} f = 0$, i.e., the relation is defined by the $\boldsymbol{X}$-invariant polynomial f.

In this way we reduce the problem of invariants for a general matrix A to the analogous problem for a nilpotent matrix, but with possibly enlarged dimension. We denote the corresponding vector field by $\boldsymbol{X}$.
Of course, in general, our new variables (like z_j from Example 2) are subject to some relations (like in the above example). But one can say more.
If the generators of the ring of relations are $f_i(z) = z^{\gamma^{(i)}} - z^{\delta^{(i)}}$ (compare the previous case), then we have

$$\boldsymbol{X} f_i = 0, \tag{9.40}$$

i.e., f_i are $\boldsymbol{X}$-invariants.
Recall that the equations $f_i(z) = 0$ define an affine algebraic variety Z, a subvariety of $\mathbb{C}^s$, which is an image of $\mathbb{C}^n$ under a polynomial mapping φ. The vector field

$\boldsymbol{X}$ is a push forward of a linear vector field $\boldsymbol{U}$ associated with the system $\dot{x} = Ax$. The properties $\boldsymbol{X} f_i = 0$ mean that the vector field $\boldsymbol{X}$ is tangent to Z. So, we are interested in the ring $\mathbb{C}[Z]^{\boldsymbol{X}}$ of invariants of the restriction of the nilpotent vector field $\boldsymbol{X}$ to Z.
But it is not difficult to show, using Eqs. (9.40), that any $\boldsymbol{X}$–invariant regular function on Z comes from an $\boldsymbol{X}$–invariant polynomial in $\mathbb{C}^s$. Since the ring $\mathbb{C}[\mathbb{C}^s]^{\boldsymbol{X}}$ is finitely generated, also the ring $\mathbb{C}[Z]^{\boldsymbol{X}}$ is finitely generated. This completes the proof of the Maurer–Weitzenböck theorem. □

§7 Theorems of Briuno and Yoccoz

Now we pass to the non-resonant saddles. Their one-dimensional equivalents are the germs of non-resonant analytic diffeomorphisms f of $(\mathbb{C}, 0)$ of the form

$$z \longmapsto \mathrm{e}^{2\pi \mathrm{i}\alpha} z + O(z^2),$$

where $\alpha \notin \mathbb{Q}$, $0 < \alpha < 1$. The Poincaré–Dulac theorem for diffeomorphisms says that any such f is formally linearizable (see 9.23 and 9.24); there exists a formal power series $h(z) \sim z + h_2 z^2 + \dots$ such that $h^{-1} \circ f \circ h \sim \lambda z$, $\lambda = \mathrm{e}^{2\pi \mathrm{i}\alpha}$. The problem is:

Under which assumptions f is analytically linearizable, i.e., when the series defining h is convergent?

The positive or negative answer to this question depends on how quickly the number α can be approximated by rational numbers. If α is slowly approximated by rationals, then any such diffeomorphism is analytically linearizable. If α is quickly approximated by rationals, then there should exist diffeomorphisms f for which the series h is divergent. Let us make these results precise.

9.67. Theorem of Siegel. ([**Sie**]) *If there exist two constants C, μ such that, for any integer p, q,*

$$|\alpha - p/q| > C|q|^{-\mu}, \tag{9.41}$$

then any $f = \mathrm{e}^{2\pi \mathrm{i}\alpha} z + \dots$ is analytically linearizable.

The short proof of this result, based on the Newton method, can be found in [**Arn5**].

Remark. If $\mu > 2$, then the set of those α, for which there exists C such that the inequalities (9.41) hold (for any p, q), is a set of full Lebesque measure. (The reader can prove this by himself.)

9.68. Theorem of Briuno. ([**Briu**]) *Let q_n be the numerators of the nth reducts in expansion of α into continued fractions. If the following* **Briuno condition**

$$\sum \frac{\ln q_{n+1}}{q_n} < \infty \tag{9.42}$$

holds, then any analytic diffeomorphism f is analytically linearizable.

The proof of Briuno's theorem is an example of a complete exploration of the Newton's method. We shall present its main idea below.
Already A. Briuno and H. Cremer had noticed that the Briuno condition is close to an optimal condition for convergence.

9.69. Theorem. ([**Briu**], [**Cre**]) *If*

$$\limsup \frac{\ln q_{n+1}}{q_n} = \infty,$$

then there exists an analytic diffeomorphism f which is not analytically linearizable.

The proofs of this result are based on estimations. V. Arnold has commented them in the following way (see [**Arn5**]):
"The proofs of divergence (Poincaré, Siegel, Briuno), preceding the Pyartli's work, are based on calculation of the growth of the coefficients and do not explain the reason of divergence in the same sense as the calculation of the coefficients of the series $\arctan z$ shows its divergence for $|z| > 1$, but does not reveal the reason – the singularities at $z = \pm\mathrm{i}$" .
These words were written about the proofs of divergence of normalizing series for non-resonant singular points of vector fields. In the work of A. Pyartli [**Pya**] finite parameter perturbations of such vector fields are considered. When we pass through resonance, an invariant analytic manifold is born. If the resonances approach quickly the unperturbed non-resonant case, then these invariant manifolds cannot be destroyed and form obstructions to the linearization.
Of course, the Arnold's words are relevant also in the existing proofs of convergence.
The main achievement of J.-C. Yoccoz was to give a geometrical proof of the theorem of Briuno and to show that Briuno condition (9.42) is also necessary for convergence in the following sense.

9.70. Theorem of Yoccoz. ([**Yoc**]) *If the irrational α does not satisfy the condition* (9.42), *then there exists an analytic diffeomorphism $f = \mathrm{e}^{2\pi\mathrm{i}\alpha}z + \ldots$ which is not analytically linearizable.*

Before passing to the proofs we say few words about expansions of real numbers into continued fractions and present conditions equivalent to (9.42).

9.71. The continued fractions. In books on the theory of numbers the continued fractions are introduced using a lot of formulas. We prefer the geometrical definition such as in [**Arn5**] (but we cannot avoid formulas).

Let $0 < \alpha < 1$ be an irrational. Consider the real plane with coordinates x, y (see Fig. 9.13). We draw the line $y = \alpha x$. We distinguish the points with integer coordinates (q, p) in the first quadrant. They all, except $(0, 0)$, do not belong to

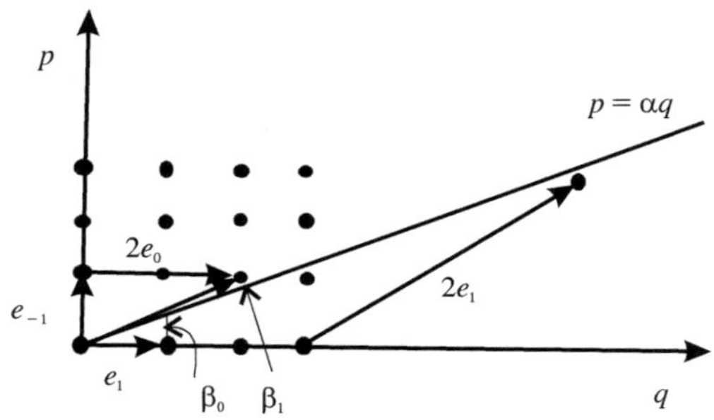

Figure 9.13: Geometric construction of the continued fraction

our line. We consider the convex hulls of the sets of integer points of the quadrant, which lie at one side of our line (below and above respectively). (One can imagine the integer points as nails and our line as a thread with fixed one end at infinity; next one stretches the thread in the down direction and in the up direction.) The vertices (q, p) of the broken lines just constructed define the best approximations of the number α by rationals p/q. It turns out that $|\alpha - p/q| < 1/q^2$.
We describe the construction of our broken lines in another way. Denote the basic vectors $e_{-1} = (0, 1)$, $e_0 = (1, 0)$. They lie in different sides of the line $y = \alpha x$. We shall construct the sequence of vectors $e_1, e_2, \ldots$ as follows. Let e_{n-1} and e_n be already constructed and lie in different sides of the line. We shall add the vector e_n to e_{n-1} as many times a_n as it is possible with the condition that the sum lies in the same side of the line as e_{n-1}.
In this way we obtain the sequence of natural numbers a_n and the sequence of integer vectors

$$e_1 = e_{-1} + a_1 e_0, \ldots, \quad e_{n+1} = e_{n-1} + a_{n+1} e_n. \tag{9.43}$$

(Note that $a_1 = [1/\alpha]$, the integer part of $1/\alpha$). The endpoints of the vectors $e_n = (q_n, p_n)$ are the vertices of the above two convex hulls.
We denote by β_n the 'distance' of the point e_n to the line $y = \alpha x$,

$$\beta_n = (-1)^n (q_n \alpha - p_n). \tag{9.44}$$

From Fig. 9.13 it is seen that $\beta_{-1} = 1$, $\beta_0 = \alpha$, the fractional part of α, and

$$\beta_{n+1} = \beta_{n-1} - a_{n+1} \beta_n, \tag{9.45}$$

which implies that the ratios p_n/q_n form the best approximations to α, in the sense that $|\alpha - p_n/q_n| < |\alpha - p/q|$ for any p and any $q < q_n$. In fact, we have the stronger statement $\beta_n < |q\alpha - p| < \beta_{n-1}$ for $q_{n-1} < q < q_n$.
We define also the numbers

$$\alpha_n = \beta_n / \beta_{n-1}. \tag{9.46}$$

Thus $\alpha_0 = \alpha$. They lie between 0 and 1.
Dividing Eq. (9.45) by β_n, we obtain

$$\alpha_{n-1}^{-1} = a_n + \alpha_n. \tag{9.47}$$

This means that $\alpha_0 = \{\alpha\}, \ldots, \alpha_{n+1} = \{\alpha_n^{-1}\}$, where $\{\cdot\}$ denotes the fractional part. Moreover, iterating Eq. (9.47) we recognize the continued fraction

$$\alpha = a_0 + \alpha_0 = a_0 + \frac{1}{a_1 + \alpha_1} = a_0 + \frac{1}{a_1 + \frac{1}{a_2+\alpha_2}} = \ldots$$

We have also other properties of the introduced quantities.

Lemma.

(a) *The (oriented) area of the parallelogram with sides e_n, e_{n-1} is equal to $(-1)^n$.*

(b) *We have*

$$q_{n+1}\beta_n + q_n\beta_{n+1} = 1. \tag{9.48}$$

Proof. (a) For the initial parallelogram (e_0, e_{-1}) it is obvious. In the general case this area equals to

$$e_{n+1} \times e_n = q_{n+1}p_n - q_n p_{n+1}$$

and is proved by induction, using Eqs. (9.43).
(b) We have $q_0 = 0,\ q_1 = 1,\ \beta_0 = 1$, which gives the formula (9.48) for $n = 1$. The general case is proved by induction, using Eqs. (9.45) and $q_{n+1} = q_{n-1} + a_n q_n$. □

Corollary. *We have $|\alpha - p_n/q_n| < 1/q_n^2$.*

This property follows from the point (a) of Lemma; because $p_n/q_n - p_{n+1}/q_{n+1} = (-1)^n/(q_n q_{n+1})$ and α lies between these two rationals.
(b) of Lemma (i.e., Eq. (9.48)) is equivalent to

$$\beta_n(q_{n+1} + q_n\alpha_{n+1}) = 1. \tag{9.49}$$

9.72. Proposition. *The following conditions are equivalent:*

(i) $\sum q_n^{-1} \ln q_{n+1} < \infty$;

(ii) $\Phi(\alpha) := -\sum \beta_n \ln \alpha_{n+1} < \infty$;

(iii) $-\sum 2^{-k} \ln \omega_k < \infty$, *where* $\omega_k = \min_{q<2^k} |q\alpha - p|$.

Proof. Because $q_n < q_{n+1},\ a_n \geq 1$, from Eqs. (9.43) we get $2q_n < q_{n+2}$. Similarly we show that $\beta_{n+1} < \beta_n,\ \beta_{n+2} < \beta_n/2$. This means that the series q_n grows exponentially and the series β_n decreases exponentially. The latter implies that the condition $\Phi(\alpha) < \infty$ is equivalent to the condition $-\sum \beta_n \ln \beta_{n+1} < \infty$.
We estimate the expression

$$\left| \sum q_n^{-1} \ln q_{n+1} + \sum \beta_{n-1} \ln \beta_n \right|. \tag{9.50}$$

Using the formula $q_{n+1}^{-1} = \beta_n + (q_n/q_{n+1})\beta_{n+1} = \beta_n(1 + (q_n/q_{n+1})\alpha_n)$ (following from Eq. (9.49)), we have $|\ln \beta_n| - \ln 2 < \ln q_{n+1} < |\ln \beta_n|$. Thus the expression (9.50) is estimated by

$$\ln 2 \sum q_n^{-1} + \sum (q_{n-1}/q_n)\beta_n |\ln \beta_n| < \infty.$$

This gives the equivalence of the conditions (i) and (ii).

We see also that (i) is equivalent to convergence of the series $\sum q_n^{-1} \ln \beta_n$. We shall show the equivalence of the latter condition with (iii).

It follows from the estimates

$$\sum_{2^{k-1} \le q_n < 2^k} q_n^{-1} \ln \beta_n < 2 \cdot (2^{k-1})^{-1} \cdot \ln \omega_k$$

(because the sum contains at most two terms) and

$$\sum_{q_n \le 2^k < q_{n+1}} 2^{-k} \ln \omega_k \le \ln \beta_n \sum_{q_n \le 2^k} 2^{-k} < \text{ const } \cdot q_n^{-1} \cdot \ln \beta_n.$$

Remark. In his formulation of Theorem 9.68 in [**Briu**] A. Briuno writes down the assumption in the form $\limsup 2^{-k} \ln \omega_k = \infty$ and H. Cremer in [**Cre**] formulates it in the form $\sup q_n^{-1} \ln q_{n+1} = \infty$.

9.73. Briuno's proof of the Briuno theorem. Here we present the main ideas lying behind the proof of Theorem 9.68.

Let $f(z) = \lambda z + a_2 z^2 + \ldots$, $\lambda = \mathrm{e}^{2\pi \mathrm{i}\alpha}$. Instead of applying the infinite series of changes $z \longmapsto z + b_j z^j$, we shall apply an infinite series of changes of the form

$$h_k = z + b_m z^m + b_{m+1} z^{m+1} + \ldots + b_{2m-2} z^{2m-1}, \quad m = 2^k + 1.$$

It means that we group the terms in series of lengths about 2^k.

If $f_k = h_{k-1}^{-1} \circ \ldots \circ h_1^{-1} \circ h_0^{-1} \circ f \circ h_0 \circ \ldots \circ h_{k-1} = \lambda z + a_m z^m + \ldots$, $m = 2^k + 1$, was obtained in the previous steps, then $f_{k+1} = h_k^{-1} \circ f_k \circ h_k = \lambda z + O(z^{2m-1})$. Moreover,

$$b_j = a_j / (\lambda^j - \lambda),$$

where $|\lambda^j - \lambda| > \text{const} \cdot \inf\{|q\alpha - p| : q = j - 1\} > \text{const} \cdot \omega_{k+1}$.

Let the series defining f_k be convergent in a disc with radius ρ_k, i.e., $|a_j| < \text{const} \cdot \rho_k^{-j}$. Then $|b_j| < \text{const} \cdot (\rho_k^j \omega_{k+1})^{-1}$ for $2^k < j \le 2^{k+1}$. Taking the jth root of $|b_j|$'s and neglecting the higher terms, we obtain the following rough rule of decreasing of the radii of convergence

$$\rho_k \sim \rho_{k-1} \cdot \omega_k^{1/2^k} = \rho_{k-1} \cdot \exp[2^{-k} \ln \omega_k].$$

If the Briuno condition holds, then the product of the diminishing factors is convergent to a positive constant.

The rigorous proof needs more estimates and is done using the Newton's method. □

9.74. Yoccoz' proof of the Briuno theorem. 1. After rescaling the variable z we can assume that we are dealing with functions which are analytic diffeomorphisms of the unit disc $\mathbb{D} = \{|z| < 1\}$ into $\mathbb{C}$ with the fixed first term $e^{2\pi i\alpha}z$. We denote the space of such functions by $S(\alpha)$. Of course, we can (and shall) assume that $0 < \alpha < 1$.
The space $S(\alpha)$ has nice properties based on the following **Theorem of Koebé**:

If $f : (\mathbb{D}, 0) \longmapsto (\mathbb{C}, 0)$ *is an univalent holomorphic function with* $|f'(0)| = 1$, *then*

$$|f(z)| < |z|(1 - |z|)^{-2}, \quad |f'(z)| < (1 + |z|)(1 - |z|)^{-3}.$$

(The proof of this theorem can be found in [**PoSz**]. One passes to the chart near infinity. There the univalency gives restriction on the growth of the coefficients of the Laurent expansion. The estimates for coefficients give the estimates for the function and its derivative. Note here that, according to the Bieberbach conjecture (proved by L. de Branges), one has the estimates $|a_n| < n$ for the coefficients of the Taylor expansion of f.)
The important corollary of the Koebé estimates is the following.

2. **Lemma.** *The spaces* $S(\alpha)$ *are compact in the topology of almost uniform convergence. In other words, the family* $S(\alpha)$ *is normal.*

The next lemma gives a nice criterion of analytic linearizability.

3. **Lemma.** f *is linearizable iff it is stable, i.e., iff for any neighborhood* U *of* 0 *there is a neighborhood* V *such that* $f^n(V) \subset U$ *for all* $n \geq 0$.

Proof. If f is linearizable, then it is stable because the linear map is stable.
Assume that f is stable and let V be such that $f^n(V) \subset U = \mathbb{D}$, $n \geq 0$. Then the family $f^n|_V$ is normal. Define $h_n = \frac{1}{n}\sum_{i=0}^{n-1} \lambda^{-i} f^i$. It is also normal family. We have $h_n \circ f = \lambda h_n + \frac{1}{n}\lambda(\lambda^{-n} f^n - 1)$.
Now it is enough to choose a convergent subsequence from $\{h_n\}$, and the limit conjugates f with λz. □

The authorship of the next lemma belongs to A. Douady and E. Ghys.

4. **Lemma.** *The set of those* α, *for which any* $f \in S(\alpha)$ *is linearizable, is invariant with respect to the action of the group* $SL(2, \mathbb{Z}) : \alpha \longmapsto \frac{a\alpha+b}{c\alpha+d}$.

Proof. The group $SL(2, \mathbb{Z})$ is generated by transformations of the form $\alpha \longmapsto \alpha + a$ and $\alpha \longmapsto -1/\alpha$. The invariance with respect to the first transformations is obvious.
Let $0 < \alpha < 1$. We shall associate with any $f \in S(\alpha)$ a map $g \in S(\beta)$, $\beta = -1/\alpha$. This construction is presented in Fig. 9.14(a). If l is a ray and $f(l)$ is its image, then we add a curve l' closing-up a triangle-like region Δ. We glue l with $f(l)$ by

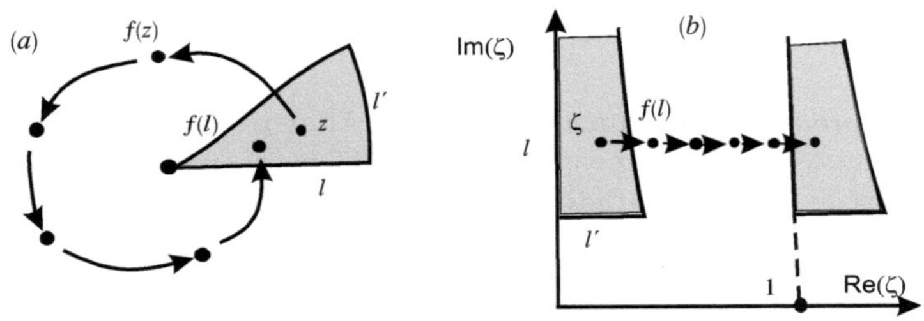

Figure 9.14: The return map in the z-plane and in the ζ-plane

means of f and obtain a surface, which admits a holomorphic structure of the disc $\mathbb{D}$ (see Section 1 above). This equivalence is of the form $z \longmapsto z_1 = z^{1/\alpha} + \dots$. Denote by g the map of first return to Δ. It turns out to be holomorphic with respect to the complex structure on $\Delta/\sim$. Expressed in the chart $z_1 = \rho e^{i\theta}$ on $\Delta/\sim$, it takes the form

$$e^{2\pi i(j\alpha-1)/\alpha}\rho + \dots = e^{2\pi i\beta} z_1 + \dots,$$

where $j = j(z_1)$ is the moment of the first return. □

It is useful to pass to the chart, where the punctured disc $\mathbb{D}^*$ is replaced by its infinite covering; the upper half-plane $\mathbb{H} = \{\mathrm{Im}\zeta > 0\}$ with the covering map $\zeta \longmapsto z = e^{2\pi i\zeta}$.
The lift of the map f to $\mathbb{H}$ has the form

$$F(\zeta) = \zeta + \alpha + \sum a_n e^{2\pi i n\zeta},$$

i.e., it commutes with the translation $id + 1$. Using the compactness of the space $S(\alpha)$ (or estimates for the coefficients a_n), it is possible to get the following uniform bound for the difference between F and the pure translation.

5. **Lemma.** *There exists a constant C, not depending on α and F, such that the inequality*

$$|F(\zeta) - \zeta - \alpha| < \alpha/4$$

holds for

$$\mathrm{Im}\zeta > t(\alpha) := (1/2\pi) \cdot \ln \alpha^{-1} + C.$$

The construction from the proof of Lemma 4 can be implemented for maps on $\mathbb{H}$. It is presented at Fig. 9.14(b). The line $L = \{\zeta = it,\ t > t(\alpha)\}$ is identified with $F(L)$. The domain $\widetilde{\Delta}$ between L, $F(L)$ and the straight segment (joining $it(\alpha)$ with $F(it(\alpha))$ is replaced by the domain $\{0 \le \mathrm{Re}\zeta_1 \le 1, \mathrm{Im}\zeta_1 \ge 0\} \subset \mathbb{H}$: $\zeta_1 \approx \alpha^{-1}(\zeta - it(\alpha))$. The map $G \in S(\beta)$, $\beta = -1/\alpha$, is induced by the return map to $\widetilde{\Delta}$ with the chart ζ_1.

Because we use the expansions of α into continuous fractions, we prefer to get some map from $S(1/\alpha)$. This is done by conjugation of the map G with the map $\zeta = s + it \longmapsto -s + it$, reversing the orientation. The new map will be denoted also by G.
Therefore, we have the transformation from $S(\alpha)$ to $S(1/\alpha)$: $F \longmapsto G = G(F)$.
Note that, if F is unstable, i.e., iterations of some points escape from the upper half-plane, then G is also unstable and the escape is quicker that for F.

6. **Lemma.** *If a point ζ with $\mathrm{Im}\zeta > t(\alpha)$ escapes from $\mathbb{H}$ after m iterations of F, i.e., $F^m(\zeta) \notin \mathbb{H}$ and $F^i(\zeta) \in \mathbb{H}$, $i < m$, then there exists a point ζ_1 with $\mathrm{Im}\zeta_1 > \alpha^{-1}[\mathrm{Im}\zeta - t(\alpha) - C_1]$ which escapes from $\mathbb{H}$ after $m_1 < m$ iterations of G. Here C_1 is an universal constant such that G is well defined for $\mathrm{Im}\zeta > t(\alpha) + C_1$.*

7. Let us start to iterate the property of Lemma 6. Let the formulas

$$\alpha_0 = \alpha, \ \alpha_n^{-1} = a_{n+1} + \alpha_{n+1}$$

define the sequence of natural numbers a_n and $\alpha_n \in (0, 1)$. We put $F_0 = F$, $F_1 = G(F_0) - a_1 \in S(\alpha_1), \ldots, F_{n+1} = G(F_n) - a_n$
If the point ζ_1 (from Lemma 6) satisfies the inequality $\mathrm{Im}\zeta_1 > t(\alpha_1)$, which is equivalent to

$$\mathrm{Im}\zeta > t(\alpha_0) + \alpha_0 t(\alpha_1) + C_1, \tag{9.51}$$

then $F_1^{m_1}(\zeta_1) \notin \mathbb{H}$, $F_2^{m_2}(\zeta_2) \notin \mathbb{H}$ for some ζ_1, ζ_2 and $0 < m_2 < m_1 < m$.
At the nth repeating of this procedure, the inequality (9.51) should be replaced by $\mathrm{Im}\zeta > t(\alpha_0) + \alpha_0 t(\alpha_1) + \alpha_0\alpha_1 t(\alpha_2) + \ldots + \alpha_0 \ldots \alpha_n t(\alpha_{n+1}) + C_1$. Because $t(\alpha) = -\ln \alpha/(2\pi) + C$, $\beta_n = \alpha_0 \ldots \alpha_n$ (see Eq. (9.46) above), the latter finite series equals (up to constants) to the partial sum defining the series $\Phi(\alpha) = -\sum \beta_n \ln \alpha_{n+1}$ (defined in Proposition 9.72). By the Briuno condition this series is convergent.
We claim that the region $\mathrm{Im}\zeta > \sum \beta_n t(\alpha_{n+1}) + C_1$ is the region of stability of the map F. Assuming contrary, it should contain a point ζ_0 escaping from $\mathbb{H}$ after n_0 iterations of $F = F_0$. This would imply existence of infinite sequence of points ζ_j escaping after n_j iterations of F_j. Moreover one should have infinite sequence of inequalities

$$0 < \ldots < n_2 < n_1 < n_0.$$

Of course, it is not possible. □

9.75. Remark. The above proof allows to estimate the radius $R(\alpha)$ of convergence of any map from $S(\alpha)$. Namely we have $|\Phi(\alpha) + \ln R(\alpha)| < \mathrm{const.}$

9.76. Proof of the theorem of Yoccoz. 1. The idea of the proof of Theorem 9.70 is to construct a map from $S(\alpha)$ with a sequence of periodic orbits approaching the fixed point 0. By reversing and modifying the construction of Douady and Ghys, one successively inserts fixed points near 0. In the next modification this fixed point becomes periodic and a new fixed point is inserted.
The reversing of the construction $F \longmapsto G(F)$, i.e., the transformation $G \longmapsto F(G)$, is presented in Fig. 9.15.

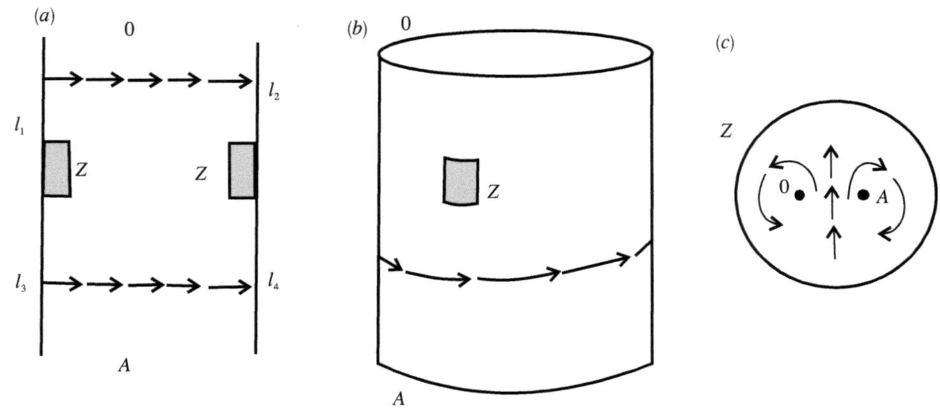

Figure 9.15: Construction of a new map

Let $G \in S(\beta)$, $0 < \beta < 1$ and let a be a positive integer. We put $\alpha = (\beta + a)^{-1}$ and let $\widetilde{G} = G + a$.
We have the half-line $l_1 = \{\zeta = \mathrm{i}t,\ t > t_0\}$, its image $l_2 = \widetilde{G}(l_1)$, the half-line $l_3 = \{\mathrm{i}t,\ t < t_0\}$ and the half-line $l_4 = l_3 + [\widetilde{G}(\mathrm{i}t_0) - \mathrm{i}t_0]$. These curves bound a domain Λ, close to a vertical strip. Identifying the sides of this non-straight strip, we would obtain a surface $\Lambda/\sim$ with a holomorphic structure of $\mathbb{C}^*$. Because we want to have a disc, we delete form $\Lambda/\sim$ a disc Z around $\mathrm{i}t_0$ of size $O(1)$, as in Fig. 9.15(a) and 9.15(b). Thus $(\Lambda/\sim) \setminus Z$ is the same as a doubly punctured disc $\mathbb{D} \setminus \{0, A\}$: $\mathrm{i}\infty$ is sent to 0 and $-\mathrm{i}\infty$ is sent to a point A. The set Z is sent to the outside of $\mathbb{D}$. (Of course, $(\Lambda/\sim) \setminus Z$ can be next covered by $\mathbb{H}$).
Let is define the dynamics on $(\Lambda/\sim) \setminus Z$, as induced by the translation $id + 1$. Because G commutes with $id + 1$ and l_4 is a translation of l_3, we obtain a well defined map of the quotient $(\Lambda/\sim) \setminus Z$. We denote it by $F = F(G)$. It is easy to see that $F \in S(\alpha)$ and that F has two fixed points in $\mathbb{D}$: 0 and A (see Fig. 9.15(c)). Therefore we have inserted a new fixed point.
Before exploring this construction, we have to perform some estimates. We want to insert the fixed point A as close to 0 as possible. However, there are some restrictions implied by the geometry of the closure of $(\Lambda/\sim) \setminus Z$. When we delete from it a disc containing 0 and A, then we obtain a domain Λ_1 in form of an annulus.
There is a conformal invariant of annulus, the modulus equal to $m = \ln(R/r)$ (see 9.14 above).
The modulus of the ring Λ_1 is of order $-\ln|A|$. On the other hand, this modulus can be estimated from Fig. 9.15(a) and 9.15(b). The domain Z is of order $O(1)$ and the domain Λ is of order $O(a) = O(\alpha^{-1})$; it is the width of Λ. This implies that the modulus of Λ_1 is of order $\ln \alpha$. The rigorous proof uses the estimate from

the point 9.14: $m \geq L^2/S$, where $S = \int\int_{\Lambda - \Lambda_1} \rho$, $L = \inf \int_\gamma \rho$, where $\rho(\zeta) > 0$ is some function and γ's are paths joining the two components of the boundary of Λ_1. Yoccoz [**Yoc**] uses $\rho(\zeta) = 1/|\zeta - \zeta_j|$ (in the ζ-plane).
This allows to get the estimate $|A| < \text{const} \cdot \alpha$. This estimate, expressed in terms of coordinates at the upper half-plane, says that we have the fixed point ζ_0 of F and satisfying

$$\text{Im}\zeta_0 > t(\alpha) - C_2,$$

where $t(\beta) = -\frac{\ln \beta}{2\pi} + C$ is the same as in the previous proof and C_2 is some universal constant.
Notice also that, if G had some periodic orbit, then F also has a periodic orbit.

2. **Lemma.** *Let $0 < \beta < 1$, a be a positive integer, $\alpha = (\beta + a)^{-1}$ and $G \in S(\beta)$. There exists $F \in S(\alpha)$ such that:*

(i) *F has fixed point ζ_0 with $\text{Im}\zeta_0 > t(\alpha) - C_2$.*

(ii) *If G has a periodic orbit with the rotation number p/q (i.e., $G^q(\zeta) = \zeta + p$), then F has a periodic orbit with the rotation number $(a + p/q)^{-1}$ and such that*

$$\min \text{Im}F^j(\zeta') > \alpha \cdot \min \text{Im}G^i(\zeta) + t(\alpha) - \text{ const.}$$

3. Let α_n and a_n be the sequences defining the continued fraction expansion. For any natural m we construct series of maps $F_{m,m+1}, F_{m,m}, \ldots, F_{m,0}$ as follows.
$F_{m,m+1} = \zeta + \alpha_{m+1} \in S(\alpha_{m+1})$, $F_{m,m} = F(F_{m,m+1})$ (with $a = a_{m+1}$) and other maps are constructed in the same way: $F_{m,l} = F(F_{m,l+1})$. Thus $F_{m,0} \in S(\alpha)$.
Moreover, each $F_{m,0}$ has periodic orbits with rotation numbers equal to the reduct p_n/q_n of the continuous fraction. The imaginary parts of such orbit are estimated from below by $\sim \sum_{i=1}^n \beta_{i-1}|\ln \alpha_i|$. Because the Briuno condition fails, the latter series tend to infinity.
Now we choose a convergent subsequence from the sequence $F_{m,0}$. Its limit is just the F we are looking for. It has infinite series of periodic orbits with arbitrary large imaginary part. These periodic orbits form obstacle to the linearizability of F. □

We shall apply the theorems of Briuno and of Yoccoz to germs of analytic planar vector fields with singular points of saddle type without resonance

$$\dot{x} = \alpha x(1 + \ldots), \quad \dot{y} = -y(1 + \ldots), \tag{9.52}$$

where α is irrational.
As in the resonant case (see Section 6) one associates with the analytic separatrix $x = 0$ the monodromy map $\Delta : (\mathbb{C}, 0) \longmapsto (\mathbb{C}, 0)$, $\Delta(x) = e^{-2\pi i \alpha} x + \ldots$. Theorems 9.62 and 9.68 imply the following.

9.77. Theorem. *If α satisfies the Briuno condition, than any germ (9.52) is analytically orbitally linearizable.*

If α does not satisfy the Briuno condition, then we cannot simply apply the Theorem of Yoccoz, Theorem 9.70. We should have the realization theorem; any germ of non-resonant one-dimensional map is realized as a holonomy map of certain holomorphic foliation with saddle singularity. Such theorem was proved by R. Perez-Marco and J.-C. Yoccoz [**P-MY**] (see also [**EISV**]).

9.78. Theorem.. *Let α be irrational. The correspondence:*

$$\textit{germ of vector field } (9.52) \longmapsto \textit{the germ } \Delta \textit{ of its monodromy}$$

is surjective. This implies that, if α does not satisfy the Briuno condition (9.42), *then there exists a germ* (9.52) *of vector field which is not analytically linearizable.*

Like in the case of resonant nodes, we do not present the proof of this result. It relies on gluing together certain domains with standard foliations and application of the theory of almost complex structures to get a holomorphic foliation.

Chapter 10

Holomorphic Foliations. Global Aspects

§1 Algebraic Leaves

We study holomorphic foliations in $\mathbb{CP}^2$ (and in other general projective algebraic surfaces). Any such foliation $\mathcal{F}$ in $\mathbb{CP}^2$ is defined as a phase portrait of a polynomial vector field in each affine chart

$$\boldsymbol{V} = P\partial/\partial x + Q\partial/\partial y,$$

or of a Pfaff equation $\omega = 0$,

$$\omega = Q\mathrm{d}x - P\mathrm{d}y$$

$(P(x, y), Q(x, y)$ – polynomials). The leaves (i.e., the phase curves with complex 'time') are either Riemann surfaces or singular points.

10.1. Definition. The **degree** of the foliation $\mathcal{F}$ in $\mathbb{CP}^2$ is the number of tangency points of the leaves of the foliations with a typical projective line. It is denoted by $\deg \mathcal{F}$.
The equivalent definition relies on the following alternative:

(i) either $\deg \boldsymbol{V} = n$ and the highest degree homogeneous part of $\boldsymbol{V}$ is not of the form $K\boldsymbol{E}$, where

$$\boldsymbol{E} = x\partial/\partial x + y\partial/y$$

is the Euler vector field, and K is a homogeneous polynomial of degree $n-1$;

(ii) or $\boldsymbol{V} = K\boldsymbol{E} + (\text{lower order terms})$.

In the case (i) we have $\deg \mathcal{F} = n$ and in the case (ii) we have $\deg \mathcal{F} = n - 1$.
The space of foliations of given degree can be identified with some Zariski open subset of a projective space $\mathbb{CP}^N$; we take the space of polynomial vector fields of given form and identify two vector fields which differ by multiplication by a constant.

Problem. Prove equivalence of the two definitions.

10.2. Proposition. *Any holomorphic foliation of degree n has $n^2 + n + 1$ singular points (multiplicity counting).*

H. Żołądek, *The Monodromy Group*, Monografie Matematyczne 67,
https://doi.org/10.1007/978-3-031-91270-2_10

Proof. It is enough to prove this proposition separately in the cases (i) and (ii) with generic P, Q.
In the case (i) there are n^2 finite critical points. At the line at infinity, denoted by

$$L_\infty = \{(x : y : 0)\} \simeq \mathbb{CP}^1,$$

there are $n+1$ critical points; in the projective coordinates $z = 1/x, y = y/x$ near $L_\infty = \{z = 0\}$ the foliation is given by

$$\dot{z} = z\widetilde{P}, \quad \dot{u} = \widetilde{Q}_n - u\widetilde{P}_n + z(\ldots),$$

where the polynomial $\widetilde{Q}_n - u\widetilde{P}_n$ is of degree $n+1$. (Here $\widetilde{P} = z^n P(1/z, u/z) = \widetilde{P}_n + z(\ldots)$).
In the case (ii) the intersection number (in $\mathbb{CP}^2$) of the algebraic curves, defined (locally) by $P = 0$ and by $Q = 0$, is equal to n^2 (by the Bezout theorem). However, because $\mathcal{F}$ is non-singular at L_∞, we must delete from this intersection number the intersections at infinity. These intersections are given by

$$\widetilde{P}_n = \widetilde{Q}_n = 0,$$

where $\widetilde{P}_n = \widetilde{K}$, $\widetilde{Q}_n = u\widetilde{K}$. Thus, we have $n^2 - (n-1) = (\deg \mathcal{F})^2 + \deg \mathcal{F} + 1$ singular points. □

10.3. Definition. An algebraic curve S is called **invariant** for the foliation $\mathcal{F}$ iff it forms an union of leaves of $\mathcal{F}$.
A (complex) 1-dimensional leaf L is called **algebraic** iff its topological closure forms an invariant algebraic curve.
If the invariant irreducible curve is given (in an affine chart) as zero of a polynomial $S = \{f = 0\}$ (f – irreducible), then we have

$$\dot{f} = g \cdot f,$$

where the dot means the derivative with respect to the vector field, $f'_x P + f'_y Q$, and g is a polynomial of degree $n-1$.
(Note that when two curves $f_{1,2} = 0$ are invariant, then $\frac{\mathrm{d}}{\mathrm{d}t}(f_1 f_2) = (g_1 + g_2) f_1 f_2$, and $f_1 f_2 = 0$ is also invariant. One can take also powers f^j and extend the definition of invariant algebraic curves to non-reduced curves (with nilpotent elements in their local rings.)

10.4. Examples.

(a) The vector fields with invariant line are of the form $\dot{x} = P$, $\dot{y} = yQ_1$. The corresponding Pfaff forms are $yQ_1\mathrm{d}x - P\mathrm{d}y$.

(b) The line at infinity is invariant in the case (i) and is not invariant in the case (ii) (see Definition 10.1).

(c) The level curves $H(x, y) = \text{const}$ are invariant for the Hamiltonian vector field $\boldsymbol{X}_H = H_y\partial_x - H_x\partial_y$ (or for the Pfaff equation $\mathrm{d}H = 0$).

These curves are also invariant for $M^{-1}\boldsymbol{X}_H$ (and $M^{-1}\mathrm{d}H = 0$), where M is the *integrating multiplier* and H is the *first integral*.

(d) The Pfaff equation $g\mathrm{d}f + f\alpha = 0$, with a polynomial g and a polynomial 1-form α, has invariant curve $f = 0$.

10.5. Theorem of Darboux. ([**Dar**]) *If a polynomial vector field of degree n has $r \geq n(n+1)/2$ different affine invariant algebraic curves $f_i = 0$, then it has a first integral of the form*

$$f_1^{a_1} \dots f_r^{a_r}, \tag{10.1}$$

where $a_j \in \mathbb{C}$.

Proof. We have $\dot{f}_i = g_i f_i$, where g_i belong to the $n(n+1)/2$-dimensional linear space of polynomials of degree $\leq n - 1$. Thus, they are dependent, $\sum a_i g_i = 0$. But this means that $\sum a_i(\dot{f}_i/f_i) = \frac{\mathrm{d}}{\mathrm{d}t}(\ln \prod f_i^{a_i}) = 0$. □

10.6. Definition. The functions of the form (10.1) are called the **Darbouxian functions**.

The functions of the form

$$\mathrm{e}^g \prod f_i^{a_i},$$

with rational g and polynomial f_i, are called the **generalized Darbouxian functions**. They can be obtained as limits of Darbouxian functions.

If H is a Darbouxian function, then the vector field

$$\boldsymbol{V}_H = M^{-1}\boldsymbol{X}_H = \sum_i a_i X_{f_i} \prod_{j \neq i} f_j,$$

where $M = \prod f_i^{a_i - 1}$, has first integral H and invariant algebraic curves $f_i = 0$. If H is a generalized Darbouxian function, then also $\boldsymbol{X}_H$ is of the form M^{-1} times a polynomial vector field, with H as first integral and the curves $f_i = 0$ as invariant algebraic curves. Also the polar curves of g (i.e., $g = \infty$) are invariant.

One can generalize the Darboux theorem to smaller number of curves, but of rather general type. In order to do it, we begin with demonstration that Example 10.4(d) is rather typical.

10.7. Proposition.

(a) *If the affine algebraic curve $f = 0$ is smooth and invariant for the Pfaff equation $\omega = 0$, then there exist a polynomial g and a polynomial 1-form α such that*

$$\omega = g\mathrm{d}f + f\alpha.$$

(b) *If the curve* $f=0$ *is not smooth, then we have*

$$h\omega = g\mathrm{d}f + f\alpha$$

for some polynomial h.

(c) *If* $f=0$ *has only double points (i.e., transversal self-intersections) as singularities, then we can choose the function* h *in (b) as any polynomial vanishing at the double points.*

Proof. (a) The 1-forms $\omega|_{f=0}$ and $\mathrm{d}f|_{f=0}$ have the same kernels and hence are proportional. Moreover, by the smoothness assumption, $\mathrm{d}f|_{f=0}$ is nonzero. Thus $(\omega/\mathrm{d}f)|_{f=0}$ is a regular function on $f=0$ with polynomial growth at infinity. It is a polynomial function on an affine algebraic curve. The ring of such functions is the quotient $\mathbb{C}[x,y]/(f)$; note that due to the smoothness the curve $f=0$ is irreducible. This means that $(\omega/\mathrm{d}f)|_{f=0}$ is a restriction of some polynomial g to the curve. The 1-form $\omega - g\mathrm{d}f$ (defined in $\mathbb{C}^2$) vanishes at $f=0$ and must be of the form $f\alpha$.

(c) If $f=0$ has a double singular point p, with two local branches, then the function $(\omega/\mathrm{d}f)|_{f=0}$ is regular at any of the branches separately. However, their values at p may disagree. In order to avoid this difficulty, one multiplies ω by a function vanishing at p.

(b) We choose the affine coordinates such that $f_y' \not\equiv 0$. Let $\omega = Q\mathrm{d}x - P\mathrm{d}y$. From the formula $\mathrm{d}f = f_x'\mathrm{d}x + f_y'\mathrm{d}y$ we calculate $\mathrm{d}y = (\mathrm{d}f - f_x'\mathrm{d}x)/f_y'$ and substitute it to ω. We find $f_y'\omega = -P\mathrm{d}f + (Pf_x' + Qf_y')\mathrm{d}x$, where $Pf_x' + Qf_y' = \dot{f} = kf$ (the invariance of $f=0$). This gives the formula from the point (b) with $h = f_y'$, $g = -P$, $\alpha = k\mathrm{d}x$. □

The above proposition has generalization to the case of many invariant affine algebraic curves $f_1 = 0, \ldots, f_r = 0$ of degrees $k_1, \ldots, k_r$. We put here the following restrictions onto their relative positions:

(i) they are smooth (in $\mathbb{CP}^2$), or

(ii) ¢(i)′] they have only double points as singularities;

(ii) they intersect each other transversally in $\mathbb{C}^2$ and there are no triple intersections;

(iii) they intersect the line at infinity transversally and at different points.

10.8. Proposition.

(a) *Under the assumptions* (i) *and* (ii) *we have*

$$h\omega = \alpha \prod f_i + \sum_i g_i \left(\prod_{j\neq i} f_j\right) \mathrm{d}f_i, \tag{10.2}$$

where $h \equiv 1$, α *is a polynomial* 1*-form and* g_i *are polynomials.*

(b) *Under assumption* (i)′ *and* (ii) *representation* (10.2) *holds for any polynomial* h *vanishing at the double singular points.*

(c) *Let* $n = \deg h\omega$ *and* $k = \sum k_i$. *If, additionally, condition* (iii) *holds, then* α *and* g_i *in Eq.* (10.2) *can be chosen such that*

$$\deg \alpha \leq n - k, \ \deg g_i \leq n - k + 1.$$

In particular, if $n = k - 1$, *then there is a Darbouxian first integral and, if* $n < k - 1$, *then* $\omega \equiv 0$.

Proof. (a) We repeat the proof of the point (a) of Proposition 10.7 and we obtain a holomorphic function $(\omega/\mathrm{d}_i f)|_{f_i=0}$ on any curve $f_i = 0$. The intersection points $f_i = f_j = 0$ are singular points of the foliation. ω vanishes there and also $(\omega/\mathrm{d}f_i)|_{f_i=0}$ vanishes there. Thus $(\omega/\mathrm{d}f)|_{f_i=0} = g_i(\prod_{j\neq i} f_j)|_{f_i=0}$, where g_i is holomorphic on the curve $f_i = 0$. As before, we extend g_i to a polynomial in $\mathbb{C}^2$ (denoted still by g_i). The 1-form $\omega - \sum_i g_i \left(\prod_{j\neq i} f_j\right) \mathrm{d}f_i$ vanishes at each $f_j = 0$ and is divisible by the product of f_j's.

(b) As in the proof of the point (c) of Proposition 10.7, we show that $(h\omega/\mathrm{d}f_i)|_{f_i=0}$ are holomorphic at $f_i = 0$. The further proof is as in the point (a).

(c) Of course, the representation (10.2) is not unique; one can replace g_i by $g_i - f_i e_i$ with simultaneous replacing of α by $\alpha + e_i \mathrm{d}f_i$.

The transversality condition at infinity means that the highest order homogeneous parts f_{i,k_i} of the polynomials f_i have only simple linear factors $\alpha_j x - \beta_j y$ (corresponding to the points $[\beta_j : \alpha_j : 0] \in L_\infty$). Moreover, the sets of intersection points for different curves do not intersect each other. This implies the following: Along any branch of $f_i = 0$, going to infinity, we have

$$\mathrm{d}f_i|_{f_i=0} \sim |x|^{k_i - 1}, \quad f_j|_{f_i=0} \sim |x|^{k_j};$$

here x is a point on $\{f_i = 0\} \subset \mathbb{C}^2$ tending to infinity.

We have $(h\omega)|_{f_i=0} = g_i \left(\prod_{j\neq i} f_j\right) \mathrm{d}f_i|_{f_i=0}$. This implies $g_i|_{f_i=0} = O(|x|^{n-k+1})$. If $\deg g_j > n - k + 1$, then its highest order homogeneous part is divisible by the part f_{i,k_i} of f_i. We can represent g_i as $g' f_i + g''$ with $\deg g'' < \deg g$. Next, we apply the same to g'', etc.; until we obtain the degree $\leq n - k + 1$.

If $\deg g_i \leq n - k + 1$, then the form $\omega - \sum g_i \left(\prod_{j\neq i} f_j\right) \mathrm{d}f_i$ is of degree $\leq n$. This means that α (from Eq. (10.2)) has degree $\leq n - k$.

If $n = k + 1$, then $\alpha = 0$ and g_i are constants. The function $\prod f_i^{g_i}$ is the first integral. If $n < k + 1$, then $\alpha = 0, \ g_i = 0$. □

Now we are prepared to prove the result of Jouanolou about absence of algebraic invariant curves for certain foliations in $\mathbb{CP}^2$. Namely, J. Jouanolou considered the following system (written in the homogeneous coordinates):

$$\dot{x}_1 = x_2^n, \quad \dot{x}_2 = x_1^n, \quad \dot{x}_3 = x_2^n.$$

In the affine coordinates $x = x_1/x_3, y = x_2/x_3$ we have

$$\dot{x} = 1 - xy^n, \quad \dot{y} = x^n - y^{n+1}. \tag{10.3}$$

We see that the degree of the corresponding foliation is equal to n.
Note that, if $n = 1$, then the line $x_1 + x_2 + x_3 = 0$ is invariant.
(Generally, any foliation of degree 1 has invariant algebraic curve. Indeed, with a singular point at $x = (x_1, x_2) = (0,0)$, we have $\dot{x} = Ax + l(x)x$, where A is a matrix and $l(x)$ is a linear functional. Transforming A to the Jordan form, we obtain an invariant line.)

10.9. Theorem of Jouanolou. ([**Jou**]) *If $n \geq 2$, then vector field* (10.3) *does not have algebraic invariant curves in* $\mathbb{CP}^2$.

Proof. (We follow the work [**Zo5**].)

1. **Lemma.** *The vector field* (10.3) *admits actions of the three groups* $\mathbb{Z}_2$, $\mathbb{Z}_3$ *and* $\mathbb{Z}_{n^2+n+1}$.

Proof. The generator of $\mathbb{Z}_2$ is the usual conjugation $\rho : (x, y) \longmapsto (\bar{x}, \bar{y})$. The generator of $\mathbb{Z}_3$ is obtained from the cyclic permutation (in reverse order) of the homogeneous coordinates and reads as $(x, y) \longmapsto (y/x, 1/x)$; After this transformation we get a rational vector field and we have to multiply it by some polynomial, but such operation does not change the phase portrait. The generator of $\mathbb{Z}_{n^2+n+1}$ is given by

$$\sigma : (x, y) \longmapsto (\zeta^{-n}x, \zeta y), \quad \zeta = \mathrm{e}^{2\pi \mathrm{i}/(n^2+n+1)}.$$

2. **Lemma.**

(a) *The vector field* (10.3) *has $n^2 + n + 1$ singular points $p_0 = (1, 1)$, $p_j = \sigma^j p_0$, $j = 1, \ldots, n^2 + n$.*

(b) *The eigenvalues of the linear parts of the vector field* (10.3) *at the points p_j are equal to $\lambda_{1,2} = \frac{1}{2}(-n - 2 \pm \mathrm{i}n\sqrt{3})\zeta^{nj}$, $\mathrm{i} = \sqrt{-1}$.*

(c) *For any p_j there are only two locally analytic curves tangent to vector field* (10.3) *and passing through p_j; they intersect one another transversally.*

(d) *The line at infinity l_∞ is not invariant for the vector field* (10.3). *The trajectories of* (10.3) *cross L_∞ transversally, except the trajectory γ going through the point $(1 : 0 : 0)$. The γ has its real part $\gamma^{\mathbb{R}}$, part of which lies in the first quadrant $\Delta = \{x, y, z > 0\}$.*

(e) *The only singular point of the vector field* (10.3) *in $\mathbb{R}^2$, p_0, is a stable focus and all trajectories starting at the first quadrant tend to p_0. None of these trajectories is algebraic. If n is even, then none of the real 1-dimensional trajectories is algebraic.*

Proof. (a), (b) and (c) are standard.

(d) In the (x, z)-chart we have $\dot{z}|_{L_\infty} = 1 > 0$. In the (y, z)-chart we have $\dot{z}|_{L_\infty} = y^n$ and, near the point $z = y = 0$, we have $\frac{dz}{dy} = y^n + \dots$. So the equation of the trajectory γ is $z = \frac{1}{n+1}y^{n+1} + \dots$ and $\gamma^{\mathbb{R}} \cap \{y > 0\}$ is in Δ.

(e) The type of p_0 follows from the point (b). To prove that p_0 is contracting all points of Δ, we notice that at the boundary of Δ the vector field (10.3) is directed into interior of Δ. Next, the divergence of this vector field is equal to $-(n+2)y^n < 0$ and, by the Dulac criterion (Theorem 6.8), Δ does not contain limit cycles and the point p_0 is contracting any point from Δ. Therefore, the trajectories in Δ are spirals and cannot be algebraic.

If n is even, then all this holds for any trajectory starting at $\mathbb{RP}^2 \setminus [1:1:1]$. □

3. **Corollary.** *Any irreducible invariant algebraic curve* $f_i = 0$ *of the vector field* (10.3) *satisfies the properties* (i)′, (ii) *and* (iii) *preceding Proposition* 10.8.

4. Suppose that S_0 is an invariant irreducible algebraic curve of the vector field (10.3). Of course, the curves $\sigma^i(S_0) = \sigma^i S_0$ and $\rho\sigma^i S_0$ are also invariant. (It may happen that $\sigma^i S_0 = S_0$ or $\rho\sigma^i S_0 = \sigma^i S_0$).

Let us estimate the total number of intersections of these curves with L_∞. If $a_0 \in S_0 \cap L_\infty$, then we get the points $\sigma^i a_0$ and $\rho\sigma^i a_0$, $i = 1, \dots, n^2 + n$.

We have $a_0 = [x_0 : y_0 : 0]$, $x_0 y_0 \neq 0$, because otherwise a_0 would be one of the corners of Δ and its phase curve would be a spiral. Thus $a_0 = [1 : y_0 : 0]$ and $\sigma a_0 = [1 : \zeta^{n+1} y_0 : 0]$.

Next, it is possible that the set $\{\rho\sigma^i a_0\}_{i=0,1,\dots}$ intersects the set $\{\sigma^i a_0\}$. But then it includes a real point, which cannot occur for n even, (see Lemma 2(e)).

Let $S_0, S_1, \dots$ be all irreducible invariant algebraic curves for the vector field (10.3), $\bigcup \sigma^i S_0 \cup \bigcup \rho\sigma^i S_0 \subset \bigcup S_i$. The sum of their degrees is equal to the number of their intersections with the line at infinity.

Therefore $k = \sum \deg S_i \geq n^2 + n + 1$, for n odd, and $k \geq 2n^2 + 2n + 2$, for n even.

By Corollary 4 the assumptions of Proposition 10.8 hold. We apply it with $h = (1 - xy^n)$ (the first components of vector field (10.3)) and with $\omega = (x^n - y^{n+1})dx - (1 - xy^n)dy$. We get $\deg h\omega = 2n + 2$. It is smaller than $k - 1$ for $n \geq 2$. Therefore $\omega \equiv 0$ (see the point (c) of Proposition 10.8).

This contradiction completes the proof of Theorem 10.9. □

Before formulating and proving the next result (genericity of foliations without algebraic leaves) we present some information about complex line bundles, their Chern classes and divisors on complex algebraic manifolds. These are the standard tools of studying algebraic manifolds. We shall treat them shortly; for more details we refer the reader to the book of P. Griffiths and J. Harris [**GH**].

10.10. Line and vector bundles, Chern classes and divisors.

1. *The first Chern class of a line bundle.* A **line bundle** L on a complex manifold M is a complex holomorphic vector bundle with 1-dimensional fiber. Locally (over $U_\alpha \subset M$) it is isomorphic to $U_\alpha \times \mathbb{C}$ with the transition diffeomorphisms of the

form $(x,\zeta)\longmapsto(x,\xi_{\alpha,\beta}(x)\zeta)$ over $U_\alpha\cap U_\beta$. Here $\xi_{\alpha,\beta}(x)$ are nonzero holomorphic functions, sections of the sheaf $\mathcal{O}^*$. The line bundle L has its inverse (or dual) $L^{-1}=L^\vee=L^*$ with the transition functions $\xi_{\alpha,\beta}^{-1}$. The line bundle $L\otimes L'$ has transition functions $\xi_{\alpha,\beta}\cdot\xi'_{\alpha,\beta}$.

The system $\xi=(\xi_{\alpha,\beta})$ defines a 1-cochain with values in $\mathcal{O}^*$. It is a cocycle, $\xi_{\alpha,\beta}\xi_{\beta,\gamma}\xi_{\gamma,\alpha}=1$ and defines an element of the first Čech cohomology group $[\xi]\in H^1(M,\mathcal{O}^*)$. Note also that, if ξ is a coboundary, i.e., $\xi_{\alpha,\beta}=\zeta_\alpha/\zeta_\beta$, then the mappings $(x,z)\longmapsto(x,\zeta_\alpha(x)z)$ in $U_\alpha\times\mathbb{C}$ define diffeomorphism of L with trivial bundle $M\times\mathbb{C}$. Thus the space of line bundles modulo isomorphism is the same as the above cohomology group.

We have the exact sequence of sheaves

$$0\longmapsto\mathbb{Z}\longmapsto\mathcal{O}\overset{\exp}{\longmapsto}\mathcal{O}^*\longmapsto 0,$$

where $\exp=\exp(2\pi\mathrm{i}\cdot)$ and $\mathbb{Z}$ is the locally constant sheaf. By Lemma 3.29 (in Chapter 3), it induces the long exact sequence of cohomology groups. In particular, we get the homomorphism

$$H^1(M,\mathcal{O})\overset{\delta}{\longmapsto}H^2(M,\mathbb{Z}).$$

By definition, the **first Chern class of the bundle** L is $c_1(L)=\delta[\xi]$.

We describe this class in terms of the transition cocycle $\xi_{\alpha,\beta}$. This cocycle is an image of some cocycle with values in $\mathcal{O}$, i.e., $h_{\alpha,\beta}=(2\pi\mathrm{i})^{-1}\ln\xi_{\alpha,\beta}$. We take its Čech coboundary $k_{\alpha,\beta,\gamma}=h_{\alpha,\beta}+h_{\beta,\gamma}+h_{\gamma,\alpha}$. We have $\exp k=0$ (it equals to $\delta^1\circ\exp h=\delta^1\xi=0$), which means that $k_{\alpha,\beta,\gamma}$ takes integer values and defines a 2-cocycle with values in $\mathbb{Z}$. We have $c_1(L)=[k_{\alpha,\beta,\gamma}]$.

2. *Expression of c_1 in terms of the connection and the higher Chern classes for vector bundles.*

By the de Rham Theorem 3.24, the group $H^2(M,\mathbb{Z})$ is included into the group $H^2_{\mathrm{dR}}(M,\mathbb{C})$ of de Rham cohomologies. It turns out that the 2-form representing $c_1(L)$ is the curvature form of any holomorphic connection of the bundle.

A connection in L says how to lift vectors from T_xM to $T_{(x,\zeta)}L$ and is locally (over U_α) given by the formula $v\longmapsto(v,\langle\theta,v\rangle\zeta)$, where $\theta=\theta_\alpha=\sum\theta^i\mathrm{d}x_i$ (without $\mathrm{d}\bar{x}_i$'s) is a smooth 1-form called the *connection form.* Such connection is *compatible with the complex structure of M.* The connection says how to differentiate local sections $s(x)$ of L: $\nabla s=\mathrm{d}s-\theta s$. This differentiation extends to differentiations of k-forms on M with values in L. If L is equipped with a Hermitian metric (we shall assume this), then the connection is *compatible with the metric* iff $\mathrm{d}(s,s')=(\nabla s,s')+(s,\nabla s')$. The operator $-\nabla^2=\mathrm{d}\theta-\theta\wedge\theta$ is the 2-form of the curvature Θ of the connection (see the remark in Definition 5.34). Because θ is of the type $(1,0)$, the compatibility with the metric implies that Θ is a Hermitian matrix of $(1,1)$-forms. In one dimension $\Theta=\mathrm{d}\theta$ is a $(1,1)$-form with real coefficients (its $(0,2)$-part is zero and $\Theta+\overline{\Theta}^\top=0$).

In the case of vector bundles (many dimensional fiber), the 1-form θ takes values in the Lie algebra $gl(m,\mathbb{C})$ (or in the Lie algebra of the structural group of the bundle). In such case Θ is a $(1,1)$-form with values in this Lie algebra.
The connection forms, written in different trivializations, are related by the gauge transformations: $\theta_\alpha = \xi_{\alpha,\beta}\theta_\beta\xi_{\alpha,\beta}^{-1} + \mathrm{d}\xi_{\alpha,\beta}\cdot\xi_{\alpha,\beta}^{-1}$ (or $\theta_\alpha = \theta_\beta + \mathrm{d}\ln\xi_{\alpha,\beta}$ after identifications). This implies that the differences between connection forms in different trivializations are equal to $\theta_\alpha - \theta_\beta = \mathrm{d}\ln\xi_{\alpha,\beta} = h_{\alpha,\beta}$ and that Θ is well defined, $\mathrm{d}\theta_\alpha = \mathrm{d}\theta_\beta$.
To show that $c_1(L) = \Theta$, we must repeat the proof of the de Rham theorem. We have two exact sequences of sheaves $0 \longmapsto \mathcal{Z}^1 \longmapsto \mathcal{E}^1 \longmapsto \mathcal{Z}^2 \longmapsto 0$ and $0 \longmapsto \mathbb{R} \longmapsto \mathcal{E}^0 \longmapsto \mathcal{Z}^1 \longmapsto 0$, where $\mathcal{E}^j$ are sheaves of smooth forms (with zero positive cohomology groups) and $\mathcal{Z}^j$ are closed forms. The long exact sequences give the isomorphisms $H^0(\mathcal{Z}^2)/\mathrm{d}H^0(\mathcal{E}^1) \overset{\delta_1}{\longmapsto} H^1(\mathcal{Z}^1)$ and $H^1(\mathcal{Z}^1) \overset{\delta_2}{\longmapsto} H^2(\mathbb{R})$. One checks that $\delta_1(\Theta) = (\theta_\beta - \theta_\alpha) = -(h_{\alpha,\beta})$ and $\delta_2(h) = (k_{\alpha,\beta,\gamma})$ (see the previous point). Thus we get

$$c_1(L) = (\mathrm{i}/2\pi)[\Theta].$$

In the case of many dimensional complex vector bundle one defines the higher *Chern classes* c_j as the de Rham classes of the (j,j)-forms $\mathrm{tr}\bigwedge^j(\mathrm{i}\Theta/2\pi)$. It turns out that these forms are closed and their cohomology classes do not depend on the choice of the connection.

3. *Line bundles associated with divisors.* If $V \subset M$ is a codimension 1 algebraic irreducible subvariety of an algebraic variety M, then we can associate with it certain line bundle $[V]$ as follows. If locally V is given by the equations $f_\alpha = 0$, then we define the cocycle $\xi_{\alpha,\beta} = f_\alpha/f_\beta$ with values in $\mathcal{O}^*$. The cocycle (ξ) defines the line bundle $L = [V]$.
The **divisor** in M is a finite formal sum $D = \sum n_j V_j$ of hypersurfaces with integer coefficients. If $f_{j,\alpha} = 0$ are the local equations of the hypersurfaces V_j, then the cocycle $\xi_{\alpha,\beta} = \prod (f_{j,\alpha}/f_{j,\beta})^{n_j}$ defines the linear bundle $[D]$.
If $f \in \mathbb{C}(M)$ is a meromorphic function on M, then we associate with it the divisor $(f) = \sum_V \mathrm{ord}_V f \cdot V$ of zeroes and poles; here $\mathrm{ord}_V f$ is the order of zero of f at V minus the order of pole of f at V. The divisor (f) is called the *principal divisor*. Two divisors, whose difference is a principal divisor, are called *linearly equivalent*. It is easy to see that the line bundle associated with a principal divisor is trivial. Thus, we have a homomorphism from the group of classes of divisors $\mathrm{Div}\,(M)$ to the group (with respect to $\otimes$) of line bundles, called also the Picard group $\mathrm{Pic}\,(M)$. There is a reverse homomorphism defined as follows. Let $s(x)$ be a meromorphic section of a line bundle L. Then we associate with L the divisor (s) of zeroes and poles of the section s. We have $[(s)] = L$.

4. **Theorem**. *The Chern class of the line bundle associated with a hypersurface* $[V]$ *is the coorientation class of* V.

We prove this theorem only in the case when M is 1-dimensional. Then $V = p$ is a point and the thesis of Theorem 4 means that $\frac{\mathrm{i}}{2\pi}\int_M \Theta = 1$.
Note that we can choose the covering consisting of two open sets $U_1 = \{|x| < 1\}$ and $U_2 = M\diagdown\{|x| \leq 1/2\}$, where x is a local coordinate near p. Then $f_1 = x,\ f_2 = 1,\ \xi_{1,2} = x$, and the (local) sections of the bundle $L = [p]$ can be identified with germs of holomorphic functions vanishing at p. (Note that any global section of L is identically zero, $H^0(M, \mathcal{O}(L)) = 0$).
Let $e = e(x)$ be a local nonzero section of a line bundle L (defining local trivialization) and let θ be the connection form expressed in the basis defined by e, i.e., $\nabla(\lambda \cdot e) = (\mathrm{d}\lambda + \theta\lambda) \cdot e$. Then we have $\theta = \partial \ln |e|^2$ and

$$\Theta = \bar{\partial}\partial \ln |e|^2 = 2\pi\mathrm{i}\ \mathrm{dd}^c \ln |e|^2. \tag{10.4}$$

It is proved by developing the formula $\mathrm{d}|\lambda e|^2 = (\nabla(\lambda e), \lambda e) + (\lambda e, \nabla(\lambda e)$ for any section $s = \lambda e(x)$. Above, $\partial = \mathrm{d}z\partial_z,\ \bar{\partial} = \mathrm{d}\bar{z}\partial_{\bar{z}},\ \mathrm{d} = \partial + \bar{\partial},\ \mathrm{d}^c = \frac{\mathrm{i}}{4\pi}(\bar{\partial} - \partial)$. (Moreover, formula (10.4) holds also in many dimensional case.) Note that $e(x) \not\equiv 0$ can be chosen arbitrarily.
We choose $e(x) = x$ in U_1 and $e(x) = 1$ in U_2. We have

$$\frac{\mathrm{i}}{2\pi}\int_M \Theta = -\lim_{\epsilon\to 0}\int_{M\setminus\{|x|<\epsilon\}} \mathrm{dd}^c \ln |e|^2 = \lim \int_{|x|=\epsilon} \mathrm{d}^c \ln |x|^2 = 1.$$

The proof in general case is not much more complicated (see [**GH**]).

5. *Chern classes of line bundles over curves.* If L is a line bundle over Riemann surface M, then $c_1(L)$ can be treated as an integer number, the value of the Chern class on the fundamental cycle of M (or as the integral of the curvature form over M). In this case $c_1(L)$ is the number of zeroes minus the number of poles (with multiplicities) of any meromorphic section of the bundle.
Here $c_1(L)$ is called also the *Euler class* of the bundle or the *degree* of the bundle. One can see that $c_1(L)$ is equal to the self-intersection index of the zero section $M \longmapsto L,\ x \longmapsto (x, 0)$. Indeed, the self-intersection index $\langle M, M\rangle_L$ is the index of intersection of M with a typical small continuous section $t(M)$ of L. One can construct such section from a meromorphic section $s(x)$.
We put $t(x) = s(x)$ outside small discs around poles of s. In any disc around a pole we have $s(x) = x^{-m}$; we replace it by $t(x) = \mathrm{const} \cdot \bar{x}^m$. $t(M)$ intersects M in the zeroes of s, with the index equal to the multiplicity of the zero, and in the poles of s, with the index equal to minus the multiplicity of the pole.
This topological definition of the Chern class allows to extend it to topological fibrations, whose fibers are either real planes or circles and with transition maps taking values in the group $SO(2)$.

6. *The Chern class of a normal bundle of a curve in a surface.* Let $V \subset M$ be a smooth algebraic hypersurface in algebraic manifold M. The normal bundle $N_V = T_V M/T_V V$ on V is a line bundle. It turns out that

$$N_V = [V]|_V$$

(the *adjunction formula*), i.e., the bundle (on M) defined by the divisor $V \subset M$ and restricted to V.

We shall show that $N_V^\vee \otimes [V]$ is a trivial bundle, which is equivalent to the adjunction formula. Let V be locally defined by $f_\alpha = 0$. The forms $\mathrm{d}f_\alpha$ define local sections of N_V^*. Moreover, $\mathrm{d}f_\alpha = \mathrm{d}(\xi_{\alpha,\beta} f_\beta) = \xi_{\alpha,\beta}\mathrm{d}f_\beta$ on V. Thus $\mathrm{d}f_\alpha$ are transformed in the same way as the sections of the bundle $[V]$, defined by $\xi_{\alpha,\beta} = f_\alpha/f_\beta$. This means that they should be treated as sections of $[V]|_V$ with coefficients in N_V, $\mathrm{d}f_\alpha = \mathrm{d}f_\alpha \otimes 1$. In this way we get a global section s of $N_V^\vee \otimes [V]$. This section nowhere vanishes. The latter fact allows to construct trivialization of $N_V^\vee \otimes [V]$: $V \times \mathbb{C} \ni (x,z) \longmapsto zs(x)$.

Let now V be a curve S in an algebraic surface M. By 5., $c_1(N_V)$ is the self-intersection index of S in N_S, as its zero section; (it follows also from the adjunction formula). This is the same as the index of self-intersection of S in M

$$c_1(N_S) = \langle S, S\rangle_M.$$

7. *Examples.* If $M = \mathbb{CP}^n$ and $H \simeq \mathbb{CP}^{n-1}$ is its hypersurface, then the corresponding line bundle $[H]$ is $\mathcal{O}(1)$ (see also the 8.42). For example, let $H = \{x_1 + x_2 + x_3 = 0\}$ in $\mathbb{CP}^2 = \{(x_1 : x_2 : x_3)\}$. We take the covering of M by: $U_3 = \{x_3 \neq 0\}$ (with the coordinates $x = x_1/x_3, y = x_2/x_3$), $U_2 = \{x_2 \neq 0\}$ (with $v = x_1/x_2, w = x_3/x_2$) and $U_1 = \{x_1 \neq 0\}$ (with $u = x_2/x_1, z = x_3/x_1$). Then the line H is given by the equations $f_j = 0$, where $f_3 = x + y + 1$, $f_2 = v + 1 + w$, $f_1 = 1 + u + z$. The line bundle $[H]$ is given by the cocycle $\xi_{ij} = f_i/f_j = x_j/x_i$.

All linear homogeneous functions $l(x_1, x_2, x_3)$ can be treated as global sections of $\mathcal{O}(1)$. Indeed, putting $l_i = l(x_1/x_i, x_2/x_i, x_3/x_i)$ as local sections of $\mathcal{O}(1)$ in U_i, we find that $l_i = \xi_{ij} l_j$.

If V is a hypersurface of degree d, then $[V] = \mathcal{O}(d) = \mathcal{O}(1)^{\otimes d}$ with the transition cocycle $\xi_{ij} = (x_j/x_i)^d$ and with the space of global section isomorphic to the space of homogeneous polynomials of $n+1$ variables of degree d.

If H is as above, then the space tangent to H above U_3 is isomorphic to the space of pairs $((x,y),(\eta,\rho)) \in \mathbb{C}^2 \times \mathbb{C}^2$, such that $f_3(x,y) = 0$ and $\langle \mathrm{d}f_3, (\eta,\rho)\rangle = 0$, i.e., $\eta + \rho = 0$. The normal bundle N_H (with the projection $N_H \longmapsto H$) can be identified with the space of points $((x,y), \theta\nabla f_3) = ((x,y),(\theta,\theta))$ (with the projection $(\eta,\rho) \longmapsto \frac{1}{2}(\eta+\rho)(1,1)$). Above U_1 we have also $N_H = \{((u,z), \lambda(1,1))\}$ with the same projection.

The transition function $(x,y) \longmapsto (u,z) = \Phi_{31} = (y/x, 1/x)$ has its derivative $D\Phi_{31} = \begin{pmatrix} -uz & z \\ -z^2 & 0 \end{pmatrix}$ as its transition matrix for the tangent bundle. Because $u = -z-1$ at H, one sees that the action of $D\Phi_{31}$ on N_H is the following $\theta(1,1) \longmapsto z\theta(1,1) \pmod{TH}$; the vector $\theta(1,1)$ above U_1 is identified with $-z^{-1}\theta(1,1)$. Thus, the transition map is of the form $\xi_{31} = z^{-1} = x_1/x_3$, the same as for $[H]$. This explains the adjunction formula.

The bundle $K_M = \Omega_M^n = \bigwedge^n T^*_{\mathrm{hol}}$, $n = \dim M$, is called the *canonical bundle*. We have $K = \mathcal{O}(-n-1)$ for $M = \mathbb{CP}^n$. In the case $n = 2$ we have $\mathrm{d}u \wedge \mathrm{d}z = z^3 \mathrm{d}x \wedge \mathrm{d}y$ in $U_1 \cap U_3$ and the transition map for K is $\xi_{ij} = (x_j/x_i)^{-3}$

We look also a little bit at the tangent bundle to $\mathbb{CP}^2$, $T_{\text{hol}}\mathbb{CP}^2$. Its transition operators are the matrices $D\Phi_{ij}$ defined above. This bundle has global sections. For example, the constant vector field $a\frac{\partial}{\partial x} + b\frac{\partial}{\partial y}$ in U_3 is identified with $z(b - au)\frac{\partial}{\partial u} - az^2\frac{\partial}{\partial z}$ in $U_1 \cap U_3$. The latter vector field is divisible by z.
If we want to represent foliations of $\mathbb{CP}^2$ as sections of certain holomorphic vector bundle over $\mathbb{CP}^2$, then the bundle $T = T_{\text{hol}}$ is not a good choice. However, when we take the tensor product $T \otimes \mathcal{O}^{-1}$, then the tensor field $(a\frac{\partial}{\partial x} + b\frac{\partial}{\partial y}) \otimes 1$ in U_3, $(b - au)\frac{\partial}{\partial u} - az\frac{\partial}{\partial z}) \otimes 1$ in U_1 (and analogously in U_2) represents a global section of the bundle $T \otimes \mathcal{O}(-1)$. On the other hand, this system of vector fields represents a holomorphic foliation of degree 0. One can see that, instead of the constant initial vector field, we could use the general vector field giving foliation of degree 0: $(a + cx)\frac{\partial}{\partial x} + (b + cy)\frac{\partial}{\partial y}$, transformed to $(b - au)\frac{\partial}{\partial u} - (c + az)\frac{\partial}{\partial z}$.
Generally, a foliation of degree d is a global holomorphic section of the vector bundle

$$T_{\text{hol}}\mathbb{CP}^2 \otimes \mathcal{O}(d-1).$$

Now we pass to holomorphic foliations on algebraic surfaces and their invariant algebraic curves.
Let $\mathcal{F}$ be a holomorphic foliation defined in a smooth algebraic surface M by a Pfaff equation $\omega = 0$. Assume that $S \subset M$ is an algebraic curve invariant with respect to $\mathcal{F}$.

10.10. Definition. Let $p \in S$ be a singular point of the foliation; p can be singular point of the curve S or not. In any case, S can be decomposed near S into certain irreducible local components B_j, $j = 1, \ldots, s$, each homeomorphic to a disc; there is a parameterization $\mathbb{D} \longmapsto B_j$ (homeomorphism analytic outside 0).
Let $B_j = \{f_j = 0\}$. By Proposition 10.7(b) we have $h\omega = g\mathrm{d}f_j + f_j\alpha$, where g, h are analytic functions and α is an analytic 1-form.
We define the **Camacho–Sad index of the branch B_j with respect to $\mathcal{F}$** as

$$i(B_j, \mathcal{F}) = \frac{-1}{2\pi \mathrm{i}} \int_\gamma \frac{\alpha}{g},$$

where γ is a positively oriented loop in B_j, generating the fundamental group of $B_j \setminus p$. One can see that this definition does not depend on the above representation of the Pfaff form, i.e., on the change $g \longmapsto g - kf_j$, $\alpha \longmapsto \alpha + k\mathrm{d}f_j$.
We define the *Camacho–Sad index of the curve S with respect to $\mathcal{F}$* as

$$C(S, \mathcal{F}) = \sum_p \sum_{B_j} i(B_j, \mathcal{F})$$

(the sum over singular points of $\mathcal{F}$ and local branches of S).

10.12. Examples. If S is smooth near p, i.e., $S = B = \{y = 0\}$, then we have $\omega = y(Q_0(x) + \ldots)\mathrm{d}x - (P_0(x) + \ldots)\mathrm{d}y$, or $h \equiv 1$, $g = -P_0 + O(y)$ and$\alpha = (Q_0 + O(y))\mathrm{d}x$. Thus, $i(B, \mathcal{F}) = \mathrm{Res}_p(Q_0/P_0)$, the residuum at p.

Assume that $p=(0,0)$ and that $P_0=\lambda_1 x+\ldots$, $Q_0=\lambda_2+\ldots$. $\lambda_{1,2}$ are the eigenvalues of the linearization of the corresponding vector field at the critical point. In this case we have

$$i(B,\mathcal{F})=\lambda_2/\lambda_1.$$

Therefore, for smooth S we have $C(S,\mathcal{F})=\sum_p(\lambda_2/\lambda_1)(p)$.

10.13. Theorem.

(a) (C. Camacho, P. Sad [**CS2**]) *If S is smooth, then*

$$C(S,\mathcal{F})=c_1(N_S).$$

(b) (A. Lins-Neto [**L-N1**]) *If S is arbitrary and $M=\mathbb{CP}^2$, then*

$$C(S,\mathcal{F})=3\deg S-\chi(S)+\sum_p\sum_{B_j}\mu(B_j),$$

where $\chi(S)=2-2g(S)$ is the intrinsic Euler characteristic of S and $\mu(B_j)$ is the Milnor number of the germ of hypersurface B_j. In particular, $C(S,F)$ is always positive.

10.14. Remark. The **intrinsic Euler characteristic** and the **intrinsic genus** $g(S)$ of an irreducible algebraic curve can be defined as the Euler characteristic and the genus of its smooth normalization. In the process of desigularization we take only the proper preimages of the blown-up curve (we delete the pasted divisors) and we blow-up also double points. For example, the generic elliptic curve $y^2=P_3(x)$ has genus 1 but the singular elliptic curve $y^2=x^2(x-1)$ is parameterized by the rational curve $\mathbb{CP}^1$, $t\longmapsto(1+t^2,t(1+t^2)^2)$, and has genus 0.
If $S\subset\mathbb{CP}^2$ is smooth of degree d, then we have the formula for genus $g(S)=\frac{1}{2}(d-1)(d-2)$. Putting it to the Lins-Neto formula, we get $C(S,\mathcal{F})=d^2$, which is equal to the index of self-intersection of S (see above).

Proof of Theorem 10.13. (a) Locally, in a chart U_α (which we assume contains all critical points) in M, the foliation is given in the form of the equation

$$\frac{\mathrm{d}y}{\mathrm{d}x}=\frac{P(x,y)}{Q(x,y)}=y\frac{P_0(x)}{Q_0(x)}+\ldots,$$

where $S=\{y=0\}$. Here P_0/Q_0 depends on the chart, i.e., on α. The linear part of this equation is the equation for sections $y(x)=y_\alpha(x)$ of the normal bundle to S. We have $y_\alpha(x)=C\exp\left(\int^x\eta_\alpha\right)$, where $\eta_\alpha=(Q_0/P_0)\mathrm{d}x$. Thus, $y_\alpha=\xi_{\alpha,\beta}y_\beta$, with $\xi_{\alpha,\beta}=\exp\left(\int\eta_\alpha-\eta_\beta\right)$. Therefore

$$\eta_\alpha-\eta_\beta=\mathrm{d}\ln\xi_{\alpha,\beta}.$$

This formula allows to calculate the 2-form representing the Chern class of the normal bundle to S. Recall that, by 10.10.2, we should represent $\mathrm{d}\ln\xi_{\alpha,\beta}$ as $\theta_\alpha-\theta_\beta$, where θ_α are smooth $(1,0)$-forms; next, we take $\Theta = \mathrm{d}\theta_\alpha$ and the Chern form is $\mathrm{i}\Theta/2\pi$.

We have $\eta_\alpha = \sum a_i \mathrm{d}\ln(x-x_i) + \mathrm{d}\psi + \tau_\alpha$, where $a_i = \mathrm{Res}_{x_i}(P_0/Q_0)$, x_i and the meromorphic function ψ are the same as for other η_β (the same polar parts) and τ_α is a smooth $(1,0)$-form. We can choose $\theta_\alpha = \tau_\alpha$. Thus, $\Theta = \mathrm{d}\tau_\alpha = \mathrm{d}\eta_\alpha$.

We have

$$\int_M \Theta = \lim_{\epsilon\to 0} \int_{M\setminus \cup B(x_i,\epsilon)} \Theta = -\lim \sum \int_{\partial B(x_i,\epsilon)} \eta_\alpha,$$

where $B(x_i,\epsilon)$ are ϵ-balls with centers at x_i. The integrals of $\mathrm{d}\psi$ are zero, the integrals of τ_α's tend to zero and the integrals of $a_i \mathrm{d}\ln(x-x_i)$ give $2\pi\mathrm{i}a_i$.

This gives the point (a), i.e., the theorem of Camacho and Sad.

(b) We do not present all the details, only the general outline of the proof.

The idea is to resolve all the singularities of $S \subset \mathbb{CP}^2$. Then we obtain some smooth curve $\widetilde{S}$ in an algebraic variety M, the latter equipped with a foliation $\widetilde{\mathcal{F}}$. We apply the Camacho–Sad theorem to $\widetilde{S}$, which gives the index of the normalized curve. This index is an integer number. We shall show that, during the resolution process, the index of the curve changes by integers and does not depend on the foliation (with the only condition that the curve is invariant). Next, we calculate the index of the initial curve with respect to some special foliation defined by a Hamiltonian vector field.

Lemma. *Let $(x,u) \overset{\pi}{\longmapsto} (x,y) = (x,ux)$ be the blowing-up of the singular point of a local irreducible component $B = \{f=0\}$ of S, let $\widetilde{B} = \overline{\pi^{-1}(B\setminus 0)}$ be the proper preimage of B and let $\widetilde{\mathcal{F}} = \pi^*\mathcal{F}$ be the pull-back of the foliation $\mathcal{F}$ defined by $\omega = 0$ (tangent to B). Then*

$$i(B,\mathcal{F}) = i(\widetilde{B},\widetilde{\mathcal{F}}) + m,$$

where m is a positive integer.

Proof. We have $\pi^* f = x^l \cdot \tilde{f}(x,u)$. Let $h\omega = g\mathrm{d}f + f\alpha$ be the decomposition from Definition 10.11. Then we have the analogous decomposition for $\pi^*\omega$: $\pi^*(h\omega) = \pi^*h \cdot \pi^*\omega = (\pi^*g\cdot x^l)\mathrm{d}\tilde{f} + [l\pi^*g\cdot x^{l-1}\mathrm{d}x + \pi^*\alpha\cdot x^l]\tilde{f} = g_1 \mathrm{d}\tilde{f} + \tilde{f}\alpha_1$. We see that $i(B,\mathcal{F}) = i(\widetilde{B},\widetilde{\mathcal{F}}) + \frac{l}{2\pi\mathrm{i}}\int \mathrm{d}x/x$, where the integral runs along a loop in $\widetilde{B}\setminus 0$. If B projects regularly onto the x-plane, then the integral is $2\pi\mathrm{i}$; otherwise, it is equal to the ramification index of this projection times $2\pi\mathrm{i}$. □

Proposition 1. *Let the curve $S = \{f=0\}$ be transversal to the line at infinity and let $\mathcal{F}_0$ be the Hamiltonian foliation, defined by the polynomial Pfaff form $\omega = \mathrm{d}f$ (in the affine part). Then the index of S with respect to $\mathcal{F}_0$ is equal to the number from thesis of Theorem* 10.13(b).

We shall not prove this result in its full generality; the reader can find it in the A. Lins-Neto's paper [**L-N1**]. We prove the weaker version of this proposition.

Proposition 2. *If the curve S of degree d is transversal to L_∞ and has only δ finite double points as its singularities, then:*

(i) $C(S, \mathcal{F}_0) = d^2 - 2\delta$ *and*

(ii) $\chi(S) = 3d - d^2 + 2\delta$.

In this case Theorem 10.13(b) follows because $\mu(B_j) = 0$. □

Proof of Proposition 2. (i) The singular points of the foliation $\mathcal{F}_0$ are finite and infinite.
The finite singular points in S are the critical points of the function f at $f = 0$; they are the double points of the curve S. Each such singularity is non-degenerate with two local smooth branches of S. The ratios of eigenvalues of the Hamiltonian vector field are equal to -1 (both). This means that the contribution to $C(S, \mathcal{F}_0)$ from finite singularities is equal to -2δ.
Any infinite singularity of the vector field $\boldsymbol{X}_f$ is a $d:1$ resonant node, where the greater eigenvalue is in the direction of the line at infinity. Indeed, if $[1:0:0] \in S \cap L_\infty$ is such point, then $f(x,y) = y\prod_1^{d-1}(x - v_i y) +$ (lower order terms). In the projective coordinates we get $f = z^{-d}(u + \ldots) =$ const for the phase curves, like in the resonant node.
This means that each infinite singularity gives the contribution d to the index and all infinite singularities give d^2.
Summing-up these two numbers, we obtain the result.

(ii) We use the Poincaré–Hopf theorem (Theorem 1.14 in Chapter 1) saying that $\chi(\widetilde{S})$ is equal to the sum of indices of any vector field on $\widetilde{S}$. Here $\widetilde{S}$ is a normalization of S (the double points are unglued) and is treated as a real surface.
As a candidate for the vector field we choose

$$\boldsymbol{X} = \rho \boldsymbol{X}_f,$$

where $\rho : S \longmapsto \mathbb{R}$ is a smooth function, positive on the finite part of S and having sufficiently high order zero at $S \cap L_\infty$. This vector field is defined on the singular curve S, but it is lifted without obstacles to $\widetilde{S}$.
The index of X at a local branch of a finite singularity is equal to 1. Indeed, if the branch is parameterized by u and $\frac{\mathrm{d}x}{\mathrm{d}u} \neq 0$, then $\dot{u} = (\frac{\mathrm{d}x}{\mathrm{d}u})^{-1}\dot{x} = (\frac{\mathrm{d}x}{\mathrm{d}u})^{-1}\rho f_y' = Au + \ldots$, with the complex number $A \neq 0$. This means that the contribution to $\chi(S)$ from finite singularities is 2δ.
Near a point at infinity, e.g., $[1:0:0]$, the local parameter is $z = 1/x$. We have $\dot{z} = -z^2\dot{x} = -\rho z^2 f_y' = -\rho z^{3-d}(1 + \ldots) = -\rho|z|^{6-2d}(1 + \ldots)\bar{z}^{d-3}$. Here ρ is so flat that the non-negative factor $\rho|z|^{6-2d}$ is smooth and does not affect the index. Thus the index of an individual singular point is $3 - d$ and the contribution to $\chi(S)$ from all infinite singularities of $\boldsymbol{X}$ equals to $3d - d^2$. □

10.15. Remark. If S had a cusp type singularity, i.e., $y^2 - x^3 + \ldots = 0$ (parameterized by $\tau:\ x = \tau^2 + \ldots,\ y = \tau^3 + \ldots$), then the index of $\boldsymbol{X}_f$ at such singularity would be equal to 2: $\dot{\tau} \approx (2\tau)^{-1}(2y + \ldots) = \tau^2 + \ldots$.

This gives the **first Plücker formula** for planar projective curves with only singularities of double point and cusp types

$$g(S)=\frac{1}{2}(d-1)(d-2)-\delta-\kappa,$$

where κ is the number of cusps.

10.16. Examples. (a) Let $\dot{x}=P_0(x)+O(y),\ \dot{y}=yQ_0(x)+O(y^2)$ be a polynomial vector field in $\mathbb{C}^2$ of degree n with the invariant line $S:y=0$. We shall compactify $\mathbb{C}^2$ in two different ways: to $\mathbb{CP}^1\times\mathbb{CP}^1$ and to $\mathbb{CP}^2$ (with prolongation of the holomorphic foliations to $\mathcal{F}_0$ and to $\mathcal{F}_1$ respectively).
In $\mathbb{CP}^1\times\mathbb{CP}^1$ we choose the local coordinates near $(x,y)=(\infty,0)$ as $z=1/x,y$. The index of the line S with respect to $\mathcal{F}_0$ is the sum of finite residua of $Q_0(x)/P_0(x)$ plus the index of the singularity at infinity. We have $\dot{z}=-z^{2-n}(\widetilde{P}_0(z)+O(y))$, $\dot{y}=z^{-n}(\widetilde{Q}(z)+O(y))$ and the index at $z=y=0$ is equal to the residuum of Q_0/P_0 at $x=\infty$. The sum of residua of a meromorphic function on $\mathbb{CP}^1$ is zero. Thus $C(S,\mathcal{F}_0)=0$. Because S can be translated in the y-direction, also its self-intersection index is 0.
In $\mathbb{CP}^2$, near infinity, we have the coordinates $z=1/x$, $u=y/x$, and the equation $\frac{\mathrm{d}u}{\mathrm{d}z}=(-z^{-2}\widetilde{Q}_0/\widetilde{P}_0+z^{-1})u+\dots$. We see that $C(S,\mathcal{F}_1)=1$. The same is the self-intersection index.

(b) Let $\dot{x}=P_d(x,y)+\dots,\ \dot{y}=Q_d(x,y)+\dots$ be a holomorphic system with first terms of degree d. In the blow-up coordinates $u=y/x,x$ we have $\frac{\mathrm{d}x}{\mathrm{d}u}=R(u)x+\dots$. Here $R=P_d(1,u)/(Q_d(1,u)-uP_d(1,u))$ can be written as $\sum_1^{d+1}a_i/(u-u_i)$ with $\sum a_i=-1$ (because $R\sim-1/u$ as $u\to\infty$). Thus, the Camacho–Sad index of the exceptional divisor $E=\mathbb{CP}^1=\{x=0\}$ is equal to $\sum a_i=-1$. Also $E\cdot E=-1$.

10.17. Theorem of Lins-Neto. ([**L-N1**]) *The space of foliations of degree $n>1$ of $\mathbb{CP}^2$ without algebraic leaves constitutes an open and dense subset of the space of all foliations of degree n.*

(The topology of the space of foliations is induced from the usual topology of projective space of suitable vector fields, see Definition 10.1.)

Proof. 1. Let $\mathcal{F}_n$ be the space of all foliations of degree n. Let $\mathcal{AN}_n$ consist of foliations with singular points which are elementary and not of the node type (anti-nodes). It means that the ratio of eigenvalues of the vector field is not a non-negative number, $\lambda_2/\lambda_1\not\geq 0$. Using the Thom transversality theorem, one easily shows that:

$\mathcal{AN}_n$ is open and dense in $\mathcal{F}_n$.

2. From the local theory, developed in the previous section, we get that, if p is a singular point p of a foliation $\mathcal{F}$ from $\mathcal{AN}_n$, then it has only two analytic

separatrices, i.e., local analytic invariant curves passing through p. We denote these separatrices by $B(p,1)$ and $B(p,2)$ and we put

$$i(p,1) = i(B(p,1),\mathcal{F}) = \lambda_2/\lambda_1, \quad i(p,2) = i(B(p,1),\mathcal{F}) = \lambda_1/\lambda_2.$$

Denote by Z the set of local separatrices $B(p,j)$ (i.e., the pairs (p,j)).
Thus, we have $2(n^2+n+1)$ indices $i(p,j)$. If there is an invariant algebraic curve S, then we have the subset $Z(S)$ of separatrices belonging to S. The index of S is the sum of indices in $Z(S)$ and should be integer. One would like to find foliations such that for any $Z_1 \subset Z$ the sum of $i(p,j)$'s over Z_1 is not an integer. That would imply that there are no a priori relations between $i(p,j)$'s. However one such relation exists.

3. **Proposition.** *We have*

$$\sum_{p,j} i(p,j) = -n^2 + 2n + 2.$$

Proof. First we check this formula for the Jouanolou's example. We have $i(p,1) + i(p,2) + 2 = (\mathrm{tr}D\boldsymbol{V})^2/\det D\boldsymbol{V}$. By Lemma 2 from the proof of the Jouanolou theorem, we have $\mathrm{tr}D\boldsymbol{V}(p_0) = -(n+2)$, $\det D\boldsymbol{V}(p_0) = n^2+n+1$, and $i(p_j,1) + i(p_j,2) = (n+2)^2/(n^2+n+1)$ for all singular points $p_j = \sigma^j(p_0)$. Thus $\sum_{p,j} i(p,j) = (n^2+n+1)[(n+2)^2/(n^2+n+1) - 2] = -n^2+2n+2$.
The proof of this formula for general foliation needs application of the Baum–Bott formula, which allows to express characteristic classes of certain holomorphic vector bundles by means of local invariants of its global sections. It is an analogue of the formula, expressing the Chern class of a line bundle on a Riemann surface by the degree of divisor of zeroes and poles of any of its meromorphic sections.
The Baum–Bott formula is applied to bundles of the form $E = L \otimes T_{\mathrm{hol}}M$, where L is a line bundle over a m-dimensional manifold M. Let $P(A)$ be a polynomial function on $GL(m,\mathbb{C})$ of degree m, which is invariant with respect to the changes $A \longmapsto CAC^{-1}$; for example, $\det A$. Let also Θ be the curvature tensor of some connection on E, compatible with the holomorphic structure.

Theorem of Baum and Bott. ([**BauB**]) *For any generic global holomorphic section* $\boldsymbol{X}$ *of the bundle* E *we have*

$$\sum_{\boldsymbol{X}(p)=0} P(D\boldsymbol{X}(p))/\det D\boldsymbol{X}(p) = \int_M P(\mathrm{i}\Theta/2\pi).$$

The proof of this formula can be found in the book of Griffiths and Harris [**GH**]. We apply this formula to the bundle $T_{\mathrm{hol}}\mathbb{CP}^2 \otimes \mathcal{O}(n-1)$, whose sections represent holomorphic foliations of degree n (see the examples in the point 10.10). The Baum–Bott formula says that the sum $\sum i(p,j)$ does not depend on the foliation. □

4. It turns out that, besides the Baum–Bott formula, there are no other relations between the quantities $i(p,j)$.

The set of foliations from $\mathcal{AN}_n$, such that for any proper subset Z_1 of Z the sum $I(Z_1)=\sum_{Z_1} i(p,j)$ is not a positive integer, is open and dense in $\mathcal{F}_n$.

This property follows from the transversality arguments, provided that we know that the mentioned set is non-empty.
Consider the Jouanolou's example. We have either $i(p,j)=\lambda$ or $i(p,j)=\bar{\lambda}$, where $\lambda=\lambda_2/\lambda_1\in\mathbb{C}\setminus\mathbb{R}$. The considered sum equals to $k\lambda+l\bar{\lambda}$. If it is integer, then $k=l$ and the sum $I(Z_1)$ equals to $(k/(n^2+n+1))\cdot(-n^2+2n+2)$. If $n\geq 3$, then this number is negative. If $n=2$, then $n^2+n+1=7$, $k<7$, and $I(Z_1)=2k/7$ cannot be integer.

5. Let $\mathcal{F}$ be a foliation satisfying the condition from the previous point of the proof.
Assume that some invariant algebraic curve S of degree d exists. By 4., $Z(S)$ cannot be a proper subset of Z and, by Proposition 3, we have $C(S,\mathcal{F})=I(Z)=-n^2+2n+2$. If $n>2$, than this number becomes negative, which contradicts the positivity of $C(S,\mathcal{F})$.
If $n=2$, then $C(S,\mathcal{F})=d^2-2\delta=2$ (see Proposition 2 in the proof of Theorem 10.13(b)). Because the number of double points of S is the same as the number of all singular points of $\mathcal{F}$, i.e., $\delta=7$, then we obtain $d=4$. To complete the proof it is enough to delete from the set of foliations, defined in 4., foliations of degree 2 with invariant algebraic curves of degree 4. □

Remarks. 1. Note that, by the way, we obtained another proof of the Jouanolou's theorem for $n>2$. To get the Jouanolou's theorem for $n=2$, it is enough to check that the Jouanolou foliation does not have invariant curves of degree 4; it can be done by direct calculations.

2. Some works were devoted to the multi-dimensional Jouanolou system

$$\dot{x}_0=x_n^s,\quad \dot{x}_1=x_0^s,\ldots,\quad \dot{x}_n=x_0^s.$$

It admits the symmetry groups: $\mathbb{Z}/2\mathbb{Z}$, $\mathbb{Z}/(n+1)\mathbb{Z}$ and $\mathbb{Z}/N\mathbb{Z}$ for $N=(s^{n+1}-1)/(s-1)$. Here one looks for invariant algebraic surfaces, especially curves and hypersurfaces.
If n is odd, then there are $(s^{n+1}-1)/(s^2-1)$ invariant lines, obtained from $l_0: x_0=x_2=\ldots=x_{n-1}, x_1=x_3=\ldots=x_n$ by application of symmetries from $\mathbb{Z}/N\mathbb{Z}$; they are the only invariant algebraic curves. But when one adds a perturbation $-\mu x_n^s\partial_{x_n}$, $0<|\mu|<<1$, then there are no invariant algebraic curves at all (see [**So**] and [**L-NS**]). In [**L-NS**] openness and the density of the set of 1-dimensional foliations without algebraic leaves is proven.
In [**Zo12**] it is proved that the multi-dimensional Jouanolou system does not have any invariant algebraic hypersurface. Partial results in this direction were earlier obtained in [**So**] (for $n=3$) and in [**MMNS**] (for prime $n+1$).

Theorem 10.13(a) was used by C. Camacho and P. Sad in [**CS2**] to obtain the following result, whose proof will not be included here. It relies on controlling of the indices of exceptional divisors appearing in the process of desingularization of vector field, with the use of the property that any such divisor has self-intersection index $-k$, where k is the depth of desingularization of its singularities.

10.18. Theorem of Camacho and Sad. ([**CS2**]) *A singular point of a germ of planar holomorphic vector field has an analytic separatrix.*

§2 Monodromy of the Leaf at Infinity

10.19. Introduction. If L is a leaf of a holomorphic foliation $\mathcal{F}$ in a complex surface M, then one can associate with it its **monodromy group** (or the **holonomy group**) as the group of germs of holomorphic diffeomorphisms of a holomorphic disc D (transversal to L) at a point p and generated by the monodromy maps Δ_γ, $\gamma \in \pi_1(L,p)$ (see Definition 9.56)
It is a subgroup of the group $\mathrm{Diff}(\mathbb{C},0)$ of germs of conformal diffeomorphisms of $(\mathbb{C},0)$. Changing the disc D, we replace the monodromy group by a group internally conjugated in $\mathrm{Diff}(\mathbb{C},0)$.
Two examples are important:

(i) when $M = \mathbb{CP}^2$, with the line at infinity L_∞ invariant, and $L = L_\infty^* = L_\infty \setminus$(singular points);

(ii) when M is a resolution of a singularity of a foliation in $(\mathbb{C}^2,0)$, with some non-dicritical divisor E, and $L = E^* = E \setminus$ (singular points).

In the cases when L is a projective line deprived of singular points, its holonomy group is also called the *projective holonomy group*.
Most results concern these two situations. Foliations with other algebraic leaves and their monodromies are less investigated.
The analysis of the monodromy group in the case (i) gives some global results about qualitative properties of the foliation: density of leaves in $\mathbb{CP}^2$ and infinite number of limit cycles for generic foliation (see Theorems 10.22 and 10.24 below).
The problem of studying the projective monodromy groups associated with divisors of resolution of a germ of holomorphic foliations in $(\mathbb{C}^2,0)$ was stated by R. Thom. Another natural problem is the problem of realization of a given holonomy group (of an algebraic curve) by a foliation.

10.20. Problems of R. Thom.

(a) *Assume that the singular point has only finitely many separatrices (e.g., analytic invariant hypersurfaces through singularity) and all its other leaves are analytic. Does there exist a local analytic first integral?*

(b) *Assume that all leaves are analytic and their closures contain singularity. Does there exist a meromorphic first integral?*

(c) *Assume that there are only finitely many separatrices and all other leaves contain singularity in their accumulations. Do the separatrices and their holonomies describe the analytic type of the foliation?*

(d) *Describe the complete set of invariants of holomorphic orbital classification of such planar germs.*

(Recall that a subset $Z \subset M$ is *analytic* iff near any point $x \in M$ it is given as a zero set of a system of functions analytic near x. Thus a spiral near a planar focus is not an analytic curve.)
The problem (a) was solved positively by J. Mattei and R. Moussu in [**MM**]. The problem (b) has negative answer: M. Suzuki [**Suz**] has found the example $\dot{x} = x(y - x + 2y^2)$, $\dot{y} = y(y - x + y^2)$ with dicritical singularity at 0 but with non-rational first integral $(x/y)\mathrm{e}^{y(y+1)/x}$ (see also [**CeMa**]). The question (c) has negative answer (see [**CeMo**] and Examples 10.34.3 below).

10.21. Nonlinear Riemann–Hilbert problem. *Given a Riemann sphere $\overline{\mathbb{C}}$, with finite number of points $t_1, \ldots, t_m$ deleted, and a couple of germs of holomorphic maps $\Delta_i : (\mathbb{C}^n, 0) \longmapsto (\mathbb{C}^n, 0)$, with the restriction $\Delta_m \circ \ldots \circ \Delta_1 = id$, find:*

(A) *a foliation in a neighborhood of $\overline{\mathbb{C}} \times (\mathbb{C}^n, 0)$ into analytic curves such that the monodromy maps of the leaf $\overline{\mathbb{C}} \setminus \{t_1, \ldots, t_m\}$ are equal to Δ_i;*

(B) *a nonlinear Fuchsian foliation with this property.*

Here, by the *nonlinear holomorphic Fuchsian foliation* we mean a foliation, defined locally by means of a vector fields with the following singularities

$$\dot{t} = t - t_i, \quad \dot{x} = v(t, x),$$

where v is holomorphic near $t = t_i$, $x = 0$.
The problem A was stated in the article [**EISV**] by P. Elizarov, Yu. Ilyashenko, A. Shcherbakov and S. Voronin and the problem B was stated in the article [**Il6**] by Ilyashenko.
The problem B not always has a solution. Below we present the suitable example from [**Il6**]:
Assume that $n = 1$ and that the germs $\Delta_j = \mathrm{e}^{2\pi \mathrm{i} \alpha_j} x + \ldots$, $\alpha_j \in \mathbb{R} \setminus \mathbb{Q}$, are analytically non-linearizable. Then the nonlinear Riemann–Hilbert problem B with these data does not have any solution.
Indeed, near the singular points $(t_j, 0)$ we have $\frac{\mathrm{d}x}{\mathrm{d}t} = (\lambda_j x + \ldots)/(t - t_j)$, where the Camacho–Sad indices λ_j satisfy the relations: $\mathrm{e}^{2\pi \mathrm{i} \lambda_j} = \Delta_j'(0)$ and

$$\sum \lambda_j = 0.$$

Thus there should exist positive λ_j's. The corresponding singular points of the foliation are nodes and are analytically linearizable (in the Poincaré domain). So, the corresponding monodromy maps should be analytically linearizable too.

In the case $n = 1$, Ilyashenko gave necessary and sufficient conditions to solve the problem B:

Either (i) *one of* Δ_j *is linearizable, or* (ii) $\Delta_1, \ldots, \Delta_k$, $k < m$, *are the monodromy maps of resonant non-linearizable nodes and* $k + \sum_1^k \lambda_j \geq 0$ *(where* $\mathrm{Re}\lambda_j \in [-1, 0)$*).*

In the case $n > 1$, the solution of the problem B is given in [**Il6**], under the following assumption:

Δ_1 *is analytically linearizable and the linear Riemann–Hilbert problem with the data* $\Delta_j'(0)$ *has a solution in the class of linear foliations of Fuchs type.*

Before Ilyashenko's work, A. Lins-Neto [**L-N2**] gave a construction of a holomorphic foliation near an exceptional divisor $E \subset M$ (where M is a surface and $E \simeq \mathbb{CP}^1$, $\langle E, E \rangle = -1$) using the following data:

(i) *the classes* $[F_j]$, $j = 1, \ldots, m$, *of local analytic equivalence of foliations near singularities* $t_j \in E$ *(with the sum of indices equal to* -1*);*

(ii) *the holonomy maps* Δ_j *(compatible with* $[F_j]$*).*

In this way he constructed a foliation with non-elementary singular point, which needs only one blowing-up. In fact he proves more general theorem, about construction of foliation with more general resolution process (with many divisors E_j with transversal intersections and with local models of foliations near the corner points).

Later J. Mattei and E. Salem [**MaSa**] proved that the versal deformation of a foliation, with fixed the Lins-Neto's data (during the deformation), is finite dimensional. They calculated also the number of parameters of the versal deformation.

Now we pass to presentation and proofs of the most interesting (in the author's opinion) results.

10.22. Theorem of Hudai-Verenov. ([**HV**]) *A generic polynomial planar vector field of degree* $n \geq 2$, *with the line at infinity invariant, has all its non-singular solutions dense in* $\mathbb{C}^2$.

Proof. We follow [**Il2**]. 1. Denote by $\mathcal{V}_n$ the space of polynomial vector fields of degree n with the line at infinity invariant. It means that the highest degree homogeneous part is not proportional to the Euler vector field.
We choose a subset $\mathcal{V}'$ of $\mathcal{V}_n$ such that each $\boldsymbol{V} \in \mathcal{V}'$ satisfies the following conditions:

(i) There exist $n + 1$ non-degenerate singular points p_i at L_∞ with ratios of eigenvalues $\lambda_i \in \mathbb{C} \setminus \mathbb{R}$. Thus, the monodromy maps Δ_i (associated with simple loops around p_i) take the forms $z \longmapsto \nu_i z + \ldots$, $|\nu_i| \neq 1$.

(ii) The ν_i's generate a dense subgroup in $\mathbb{C}^*$.

(iii) There are no algebraic invariant curves besides L_∞.

2. **Lemma.** *The conditions (i), (ii) and (iii) hold for general* $\boldsymbol{V} \in \mathcal{V}_n$.

Proof. If $\dot{x} = P_n(x,y) + \ldots$, $\dot{y} = Q_n(x,y) + \ldots$, where P_n, Q_n are homogeneous of degree n, then in the projective coordinates $z = 1/x, u = y/x$ we get

$$\mathrm{d}z/\mathrm{d}u = A_1(u)z + \ldots,$$

where $A_1 = P_n(1,u)/\left(uP_n(1,u) - Q_n(1,u)\right)$. It is clear that for general P_n, Q_n, the polynomial $R(u) = uP_n(1,u) - Q_n(1,u)$ has $n+1$ isolated zeroes $u_1, \ldots, u_{n+1}$ and

$$A_1 = \sum \lambda_i/(u - u_i),$$

with the restriction $\sum \lambda_i = 1$ (see the Theorem of Camacho–Sad). The $2n+1$ numbers u_i, λ_j, $i = 1, \ldots, n+1$, $j = 1, \ldots, n$, can vary independently with the variation of $2n+2$ coefficients in P_n, Q_n. Because $\nu_j = \mathrm{e}^{2\pi \mathrm{i} \lambda_j}$, the condition (i) is generic.

The density in $\mathbb{C}^*$ of the multiplicative group generated by $\nu_1, \ldots, \nu_n$ is equivalent to the density in $\mathbb{C}$ of the additive group generated by $\lambda_1, \ldots, \lambda_n$ and by 1. If $n \geq 2$ and $\lambda_1, \ldots, \lambda_n$ are generic, then the density property holds.

If $\mathrm{Re}\lambda_i \neq 0$, then the singular points p_i are complex foci. Thus they are analytically linearizable and have only two analytic separatrices. One of the separatrices is the line at infinity; the other is transversal to L_∞.

If the vector field has an invariant algebraic curve S, then it can meet L_∞ only along the separatrices of the singular points. It should intersect L_∞ transversally and its degree should be bounded by $n+1$. So, if we delete from $\mathcal{V}_n$ the proper algebraic subset of fields with invariant algebraic curves of degree $\leq n+1$, then also the condition (iii) will be satisfied. □

Below we fix the chart ζ linearizing Δ_1, $\Delta_1(\zeta) = \nu_1 \zeta$, and we assume that $|\nu_1| < 1$. Denote by G the group generated by Δ_i (in the chart ζ).

3. **Lemma.** *For any* $\nu \in \mathbb{C}^*$, *there exists a sequence* $g_m \in G$ *such that* $g_m \longmapsto \nu\zeta$. *This means that the orbits of* G *are dense in a neighborhood of* $\zeta = 0$.

Proof. Because the closure of the set of exponents $f'(0)$ for $f \in G$ is dense in $\mathbb{C}^*$, it is enough to prove the first point of the lemma for ν such that some $f = \nu\zeta + \sum a_j \zeta^j \in G$. We put $g_m = \Delta_1^{-m} \circ f \circ \Delta_1^m = \nu\zeta + \sum a_j \nu_1^{m(j-1)} \zeta^j \longmapsto \nu\zeta$.

Note that the domains of definition of the maps g_m remain fixed; a point from a fixed disc D is firstly sent by means of Δ_1^m to a point very close to 0, where f is almost linear, and then is sent away from 0.

Let $\zeta_1, \zeta_2 \in D$ and $\nu = \zeta_2/\zeta_1$. Then we have $g_m(\zeta_1) \longmapsto \nu\zeta_1 = \zeta_2$. This means that ζ_2 is an accumulation point of the orbit of ζ_1. □

Lemma 3 implies that the leaves of the foliation defined by $\boldsymbol{V}$, which pass near the infinite leaf L_∞^*, are dense in this neighborhood. Indeed, the element $g_m(\zeta_1)$

represents a very long paths in a leaf $L(\zeta_1)$ (passing through $\zeta_1 \in D$) and the leaf $L(\zeta_2)$ lies in the closure of $L(\zeta_1)$.

4. **Lemma on extension.** *Any* 1*-dimensional phase curve* L *of a polynomial vector field can be extended to any neighborhood of* L_∞.

Proof. We represent this phase curve as a graph of a multivalued function $y = y(x)$. The function $y(x)$ is defined outside the projections of the finite singular points onto the x-plane and has ramification points, where L meets the curve $P = 0$ (i.e., $\dot{x} = 0$). There are at most countably many ramification points.
Take a line l in the x-plane, avoiding the above singularities. We prolong the function $y(x)$ along this line. Either the solution escapes after finite 'time' to infinity or can be prolonged to the whole l.
In both cases the curve L approaches the line at infinity. □

From this lemma it follows that the leaves $\widetilde{L}$, which pass near the (punctured) leaf L_∞^*, are dense in the whole $\mathbb{CP}^2$. Indeed, if a point $p \in \mathbb{CP}^2$ belongs to a leaf accumulating at L_∞^*, then it lies in the closure of such $\widetilde{L}$'s. On the other hand, only finitely many leaves accumulate at $L_\infty \setminus L_\infty^*$ (i.e., at the infinite singular points). So, there are leaves near p accumulating at L_∞^* (and then at $\widetilde{L}$'s).
Let L be a phase curve of a vector field from $\mathcal{V}'$. It approaches L_∞ either at L_∞^* or at some of the singular points p_i. In the first case L is dense in a neighborhood of L_∞, and then L is dense in $\mathbb{CP}^2$. The next lemma completes the proof of Theorem 10.12.

5. **Lemma.** *If the closure* $\overline{L} \cap L_\infty^* = \emptyset$, *then* $\overline{L}$ *is an algebraic curve.*

Proof. Let $\Omega = \mathbb{CP}^2 \setminus$ (singular points). We shall show that L is closed in Ω. Because the complement of Ω is of complex codimension 2, that would imply that L is analytic and then algebraic (the *Chow theorem*).
Let $q \in \Omega$ be a point, near which L is not given as a zero of a local analytic function. Then q must be an accumulation point of L. We can assume also that $q \notin L$. Of course, the phase curve $L(q)$ (passing through q) also consists of accumulation points for L.
Let us look at the behavior of $L(q)$ near infinity. If $L(q)$ accumulates at L_∞^*, then L should also accumulate there. If $L(q)$ tends to infinity along a separatrix of some singular point p_j, then L, as a complex spiral, also should accumulate along L_∞^*. □

6. B. Mjuller [**Mju**] proved the following improvement of the Hudai-Verenov theorem.

In the class of holomorphic foliations on $\mathbb{CP}^2$ *of fixed degree the foliations without algebraic leaved are typical.*

We refer the reader to the paper [**LR**] by F. Loray and J. Rebelo for the proof of a generalized version of it; namely, for 1-dimensional foliations in $\mathbb{CP}^n$. □

10.23. Definition. Let L_0 be a leaf of a holomorphic foliation $\mathcal{F}$ in a complex surface and let $\gamma \subset (L_0, p_0)$ be a closed loop. Consider the holonomy map Δ_γ acting on a holomorphic disc transversal to L_0 at the point p_0. We have $\Delta_\gamma = \nu z + \ldots$. It is possible that some iterate of Δ_γ is identity, or not.
In the first case, i.e., $\Delta_\gamma^q = id$, the cycle $\gamma_{L_0} = q \cdot \gamma$ belongs to a family of cycles γ_L at the adjacent leaves L of the foliation. We agree to treat two cycles, which are homologous one to another at one leaf, as the same. We call such a family of cycles the **center**.
In the second case we say that γ is a **limit cycle** of the foliation.

10.24. Theorem of Ilyashenko (Limit cycles of complex foliations). ([**Il2**]) *A generic vector field from* $\mathcal{V}_n$, $n \geq 2$, *has infinitely many homologically independent limit cycles.*

Proof. 1. We put the following conditions on the foliations from $\mathcal{V}_n$, for which we prove the existence of infinitely many limit cycles.

(i) There exist $n+1$ non-degenerate singular points p_i at L_∞ with ratios of eigenvalues $\lambda_i \in \mathbb{C} \setminus \mathbb{R}$.

(ii) $\nu_1 = \mathrm{e}^{2\mathrm{i}\pi\lambda_1}$ and ν_2 generate a dense subgroup in $\mathbb{C}^*$.

(iii) The commutator $[\Delta_1, \Delta_2] = \Delta_1 \circ \Delta_2 \circ \Delta_1^{-1} \circ \Delta_2^{-1} \neq id$.

(iv) Let $Z_1(u)$ be a solution of the variation equation $\mathrm{d}Z_1/\mathrm{d}u = A_1(u)Z_1$ for the vector field near infinity (with the coordinate $z = 1/x, u = y/x$) and let γ_j be simple loops in (L_∞^*, p) surrounding just one point p_j. Then $I_1 \neq I_2$, where

$$I_j = (\nu_j^{-2} - 1)^{-1} \int_{\gamma_j} Z_1^{-2} \mathrm{d}u.$$

The conditions (i) and (ii) are generic. The genericity of the condition (iii), i.e., of the non-commutativity of $\Delta_{1,2}$, follows from their non-commutativity at the level of 2-jets: if $\Delta_j = \nu_j(z + a_j z^2 + \ldots)$, then $[\Delta_1, \Delta_2] = z + [(\nu_2 - 1)a_1 - (\nu_1 - 1)a_2]z^2 + \ldots$. Here the coefficients $a_{1,2}$ are calculated from the equation for the second variation

$$\mathrm{d}z/\mathrm{d}u = A_1(u)z + A_2(u)z^2,$$

with the solution $z = Z_1(u)\left[z_0^{-1} - \int_{u_0}^u Z_1(s)A_2(s)\mathrm{d}s\right]^{-1}$. The expression $Z_1 = \prod(u - u_i)^{\lambda_i}$ is the first variation. After rewriting $A_2(u)$ in the form of a fraction, we get the formula for the second variation (with the Schwarz–Christoffel function)

$$Z_2(u) = Z_1(u) \int_{u_0}^u \prod (s - u_i)^{\lambda_i - m_i} B(s)\mathrm{d}s$$

(m_i – integers, B – polynomial).

The monodromy of the functions of the Darbouxian form, like Z_1, and of the Schwarz–Christoffel form (like the integral in Z_2), will be subject to the analysis in the next chapter. Here we note that the coefficients $a_{1,2}$ in the expansions of $\Delta_{1,2}$ are expressed by means of the increment of $Z_2(u)$ as u varies along the loops $\gamma_{1,2}$. It depends (non-trivially) on the coefficients of $A_2(u)$. The latter are controlled by means of the $(n-1)^{\text{th}}$ homogeneous part of the vector field (in the affine coordinates).

The condition (iv) is a condition on the parameters u_i, λ_i in the formula for Z_1. It is an open and generic condition.

2. Let us pass to the construction of limit cycles. Let ζ be the chart in the disc D linear for Δ_1, $\Delta_1(\zeta) = \nu_1\zeta$, and we assume that $|\nu_1| < 1$.

Repeating the proof of Lemma 3 in the proof of Hudai-Verenov theorem, for any ζ_0 we can approximate $\nu = \Delta_2(\zeta_0)/\zeta_0$ by the ratios ν_1^k/ν_2^m. The maps $F_{k,m,l} = \Delta_1^{-l}\Delta_2^{-m}\Delta_1^k\Delta_1^l$ approach the map $\nu\zeta$. By the condition (iii) (non-commutativity of Δ_1 and Δ_2) the map Δ_2 is nonlinear.

Thus the equation $F_{k,m,l}(\zeta) = \Delta_2(\zeta)$ is a small perturbation (near ζ_0) of the nonlinear equation $\nu\zeta = \Delta_2(\zeta)$.

The solution $\tilde{\zeta}$ is an isolated fixed point of a map F from the monodromy group G. Moreover, one can assume that $F'(\tilde{\zeta})$ is not a root of unity. Now it is clear that $\tilde{\zeta}$ generates a loop δ_N in the leaf $L(\tilde{\zeta})$. The loop δ_N is a limit cycle in the sense of Definition 10.23. Here N will denote the sequence of quadruples (ζ_0, k, m, l), where $\zeta_0 \to 0$ and the other terms are integers tending to infinity.

3. Here we show that the cycles δ_N are homologically independent. Note that they do not intersect one another and do not have self-intersections. If some of them, say $\tilde{\delta}_1 = \delta_{N_1}, \ldots, \tilde{\delta}_r = \delta_{N_r}$, are dependent on a leaf L, then we have $\sum c_j\tilde{\delta}_j = 0$ in $H_1(L, \mathbb{Z})$. Using the specificity of the topology of Riemann surfaces, we can conclude from the above that the coefficients $c_i = 0, \pm 1$. (We cut the leaf L along $\tilde{\delta}_j$'s, we obtain several connected components, whose boundaries consist of $\tilde{\delta}_j$'s with multiplicity ± 1).

Thus, for the independence, it would be enough to show that $|\int_{\tilde{\delta}_r}\omega| > \sum |\int_{\tilde{\delta}_j}\omega|$ for some polynomial holomorphic 1-form ω.

The 1-form is chosen as $\omega = x\mathrm{d}y - y\mathrm{d}x = -\mathrm{d}u/z^2$. We approximate the z by its first variation $z(u) \approx \zeta_0 Z_1(u)$. Thus, the integral of ω along δ_N is approximated by ζ_0^{-2} times the integral of $Z_1^{-2}(u)$ along the projection γ_N of the loop δ_N onto the L_∞^*. The loop γ_N runs several times along the loops $\gamma_{1,2}$, but each time we choose different branch of Z_1; these branches are multiplied by powers of ν_1 and of ν_2. After summing-up independently the contributions above γ_1 and above γ_2, we obtain

$$\int_{\gamma_N} \omega \sim \zeta_0^{-2}(\nu_1^{-2})^{k+l}(I_1 - I_2),$$

where I_j are the quantities defined in the condition (iv). We see that the above integrals tend to infinity exponentially fast.

This completes the proof of Theorem 10.24. □

Remarks. (a) In [**SRO**] A. Shcherbakov, E. Rosales-González and L. Ortis-Bobadilla proved an analogous theorem (about existence of infinitely many limit cycles), but with limited number of conditions. They skip the assumption (ii) in the 1. of the Ilyashenko proof and obtain only finitely many conditions (finite number of algebraic inequalities in $\mathcal{V}_n$).

(b) E. Landis and I. Petrovski [**LP**] tried to solve the 16th Hilbert problem, about real limit cycles of polynomial vector fields in $\mathbb{R}^2$ (see 6.11), by investigating the limit cycles of the corresponding holomorphic foliation in $\mathbb{CP}^2$. As Theorem 10.24 shows, one cannot estimate the number of the latter. The estimates from [**LP**] have turned out not true.

§3 Groups of Analytic Diffeomorphisms

Now we pass to analytic and formal classification of finitely generated subgroups of the group $\text{Diff}(\mathbb{C},0)$, of germs of conformal diffeomorphisms of the complex line. The object $\text{Diff}(\mathbb{C},0)$ is a group, when treated abstractly; the composition is the composition of germs (without fixed domains of definitions). Any subgroup of $\text{Diff}(\mathbb{C},0)$ is treated analogously. When one looks at the dynamics of G, as acting on a neighborhood of 0, then G should be treated as a *pseudo-group* of transformations of $(\mathbb{C},0)$ (see [**Lor3**]).
We shall follow mainly the article [**EISV**] by P. Elizarov, Yu. Ilyashenko, A. Shcherbakov and S. Voronin, the paper [**CeMo**] by D. Cerveau and R. Moussu and the lectures [**Lor3**] of F. Loray.

10.25. Definition. Two subgroups G, G' of Diff $(\mathbb{C},0)$ are called **analytically equivalent** iff they are internally conjugate. It means that there exists an analytic germ h and a homomorphism $G \longmapsto G'$, $f \longmapsto f'$, such that $h \circ f \circ h^{-1} = f'$.
Two subgroups G, G' of Diff $(\mathbb{C},0)$ are called **formally equivalent** iff they are conjugate inside the group $\widehat{\text{Diff}}(\mathbb{C},0)$ of formal power series $\hat{f} \sim \nu z + \sum a_j z^j \in \mathbb{C}[[z]]$, $\nu \neq 0$. It means that there exists a formal power series conjugating the Taylor series of elements from G with the Taylor series of elements from G'.
Recall that a group G is **abelian** iff any two its elements commute. G is **solvable** iff the *central derivative sequence* $G \supset G^{(1)} \supset \ldots G^{(k)} \supset \ldots$, where $G^{(k+1)} = [G^{(k)}, G^{(k)}]$ is the commutator, is finite. We shall agree to call a group solvable iff it is solvable and non-abelian.
We denote $\mathcal{A}_p = \{f = z + az^{p+1} + \ldots,\ a \neq 0\}$ and

$$\mathcal{A} = \{id\} \bigcup_p \mathcal{A}_p = \{f = z + \ldots\},$$

the group of germs tangent to identity. We also denote $\widehat{\mathcal{A}} = \{\widehat{f} \in \mathbb{C}[[z]] : \widehat{f} \sim z + \ldots\}$.

Let the multiplicative group

$$\Lambda_G = \{f'(0) : f \in G\}$$

be the *group of multipliers of G*. It is a subgroup of $\mathbb{C}^*$.

The group G can be abelian, solvable or non-solvable. The classification of abelian and solvable groups is practically finished. The classification of non-solvable groups has only began. We begin with the abelian case.

10.26. Examples of abelian groups. (a) One example is a subgroup of the group $GL(1,\mathbb{C}) = \mathbb{C}^*$ consisting of linear maps $z \longmapsto \lambda z$.

(b) The next example is not that trivial. It is a subgroup of the group

$$G_a(p,\mu) = \{\lambda g_{\boldsymbol{W}}^t : \lambda^p = 1, \quad t \in \mathbb{C}\},$$

where $g_{\boldsymbol{W}}^t$ is the phase flow map generated by the vector field $\boldsymbol{W} = \boldsymbol{W}_{p,\mu} = [z^{p+1}/(1+\mu z^p)]\frac{\partial}{\partial z}$. Here the subscript a stands for 'abelian' and the vector field $\boldsymbol{W}$ is fixed for the whole group. The reader can easily check that $\lambda g_{\boldsymbol{W}}^t \circ \lambda^{-1} = g_{\boldsymbol{W}}^t$ iff $\lambda^p = 1$.

If an abelian group G is formally equivalent to a subgroup of $G_a(p,\mu)$, then we denote

$$T_G = \{t : g_{\boldsymbol{W}}^t \in G\}.$$

It is a subgroup of the additive group $\mathbb{C}$.

If an abelian group G is formally equivalent to a group of the type (a), then we call it **formally linearizable**. If G is formally linearizable and Λ_G is finite, then we say that G is **finite**; (it is finite in fact).

If G is formally equivalent to a group of the type (b) and the additive group T_G is cyclic (isomorphic to $\mathbb{Z}$), then we call it **abelian exceptional**; otherwise G is **abelian typical.** Note that for an abelian exceptional group we have

$$\Lambda_G^p = \{1\}, \quad T_G \simeq \mathbb{Z}.$$

This division is motivated by the following result.

10.27. Theorem (Formal classification of abelian groups). *Let $G \subset \mathrm{Diff}(\mathbb{C},0)$ be a finitely generated abelian group.*

(a) *If $G \cap \mathcal{A}_p \neq \{id\}$, then G is formally conjugated to a subgroup of $G_a(p,\mu)$, $\boldsymbol{W} = \boldsymbol{W}_{p,\mu}$, for some $\mu \in \mathbb{C}$; (we write $\widehat{G} \subset G_a(p,\mu)$).*

(b) *If $G \cap \mathcal{A} = \{id\}$, then G is formally linearizable.*

(c) *If $G \cap \mathcal{A} = \{id\}$ and Λ_G is finite, then G is analytically equivalent to a finite linear group.*

Proof. 1. Assume that $G \cap \mathcal{A} \neq \{id\}$ and contains a germ h. By Theorem 9.26, h is formally equivalent to $g^1_{\boldsymbol{W}}$, $\boldsymbol{W} = \boldsymbol{W}_{p,\mu}$. We fix the formal chart with $\hat{h} = g^1_{\boldsymbol{W}} = z + z^{p+1} + \sum h_j z^{pj+1}$. In this chart other elements of $\widehat{G}$ are series $\hat{f}$ commuting with $\hat{h}$.
Let such an element has the form $\hat{f} = \lambda z + \ldots$. One can see that

$$\hat{f}^{-1} \circ \hat{h} \circ \hat{f} = z + \lambda^p z^{p+1} + \ldots$$

Therefore the abelianess of G implies that $\Lambda^p_G = \{1\}$.
The next lemma gives the point (a) of Theorem 10.27.

2. **Lemma.** *If $\hat{f} = \lambda z + \ldots$, $\lambda^p = 1$, commutes with $\hat{h} = g^1_{\boldsymbol{W}} \in \mathcal{A}_p$, then it equals to $\lambda g^t_{\boldsymbol{W}}$ for some $t \in \mathbb{C}$.*

Proof. Let $\hat{f} = \lambda \cdot (z + \sum a_j z^j)$. We have

$$\begin{aligned}
\hat{f} \circ \hat{h} &= \lambda[\hat{h} + \sum a_j (z^j + j z^{p+j} + \ldots)], \\
\hat{h} \circ \hat{f} &= \lambda(z + \sum a_j z^j) + \lambda^{p+1}(z + \sum a_j z^j)^{p+1} + \sum h_j (\lambda z)^{pj+1} + \ldots \\
&= \lambda[\hat{h} + \sum a_j z^j + \sum (p+1) a_j z^{p+j} + \ldots]
\end{aligned}$$

and

$$\hat{f} \circ \hat{h} - \hat{h} \circ \hat{f} = \sum_j b_j z^{p+j},$$

where $b_j = \lambda(j - p - 1)a_j +$ (terms depending on a_i, $i < j$). It means that we can determine all the coefficients a_j for $j \neq p+1$ in an unique way. The choice of a_{p+1}, which we put in the form λt, is arbitrary.
On the other hand, the series $\lambda g^t_{\boldsymbol{W}}$ commutes with $\hat{h}$. Because it has the same coefficient before z^{p+1} as $\hat{f}$, the two series coincide. □

3. Let $G \cap \mathcal{A} = \{id\}$ and let Λ_G contains a nonresonant element λ_0. Then we can assume that one of the maps from $\widehat{G}$ is linear $\lambda_0 z$. For any other $\hat{f} = z + \sum a_j z^j$ we get $\lambda_0^{-1} \hat{f} \circ \lambda_0 = z + \sum \lambda_0^{j-1} a_j z^j$ and the abelianess implies all $a_j = 0$. This gives the point (b) of the theorem in the case of infinite Λ_G.

4. Let $G \cap \mathcal{A} = \{id\}$ and let all the elements of Λ_G be resonant. Then Λ_G is a cyclic group. Let $f = \lambda z + \ldots$ be such that $\lambda = \mathrm{e}^{2\pi i p/q}$ generates Λ_G.
The formal normal form for f is λz; generally, we have $f \sim \lambda z + \sum b_j z^{qj+1}$, but $id = f^q = z + \sum q b_j z^{qj+1} + \ldots$. Note that here the chart z is analytic.
We assume that $f = \lambda z$. Because other maps from g commute with λz, they are of the form $g = \lambda^k z + \sum c_j z^{qj+1}$. Because $g^q = id$ we get $g = \lambda^k z = f^k$.
Therefore the group is cyclic and analytically equivalent to a finite group of rotations (i.e., the point (c) of the theorem). □

Before describing the analytic classification of abelian groups, we must say a little bit about connection of the Ecalle–Voronin moduli with the centralizer of a germ

tangent to identity (see Section 3 in Chapter 9). Recall that the **centralizer** $Z(f)$ of an element $f \in G$ is the set of g's from G commuting with f.
Let $f \in \mathcal{A}_{p,\mu}$, i.e., f is formally equivalent to $g^1_{\boldsymbol{W}}$, $\boldsymbol{W} = \boldsymbol{W}_{p,\mu}$. We look for $Z(f)$ in $\mathcal{A}$.
If g commutes with f, then it sends whole orbits of the action of f onto such orbits: if $y = f(x)$, then $g(y) = f(g(x))$. Thus g defines the action $\pi_* g$ on the space U^*/f of orbits of f in a punctured neighborhood U^* of 0. The space U^*/f is the collection of $2p$ Riemann spheres $\overline{\mathbb{C}} \times \{j\}$ glued successively (one with the next) by means of local diffeomorphisms ϕ_j (near 0 or ∞). Because the linear part of g is id, each sphere is mapped by $\pi_* g$ onto itself.
The action of $\pi_* g$ on each sphere should be linear $\vartheta_j \longmapsto C_j \vartheta_j$ and should be compatible with the gluing diffeomorphisms, $\phi_j \circ C_j = C_{j+1}\phi_{j+1}$. Computation of the linear parts at 0 or at ∞, leads to $C_j = C_{j+1} = C$. (We have met such situation in the problem of root extracting $g^n = f$ in Theorem 9.32; there we had $C = \mathrm{e}^{2\pi \mathrm{i}/n}$.)
Note that, if $|C| \neq 1$, then the gluing diffeomorphisms ϕ_j would be extended from neighborhoods of 0 (or of ∞) to the whole Riemann spheres $\overline{\mathbb{C}} \times \{j\}$. As such, they should be linear and represent trivial Ecalle–Voronin modulus. In this case f is embeddable.
We have three possibilities:

(a) f is a typical map not admitting any nontrivial root extraction;

(b) f admits the root extraction (of maximal order n), but is not embeddable;

(c) f is embeddable.

In the case (a) the identifying maps ϕ_j do not commute with any C, $|C| \neq 1$ (see the above argumentation). Indeed, by assumption, they do not commute with any root of unity. If they would commute with some other C, $|C| = 1$, then they should commute with all C with $|C| = 1$ (also with the roots of unity). We get $C = 1$.
In the case (b) we have $C = \mathrm{e}^{2\pi \mathrm{i}k/n}$.
In the case (c) any C is admitted, the diffeomorphism f is analytically equivalent to its formal normal form (see Theorem 9.32). Assume then that $f = g^t_{\boldsymbol{W}}$, $\boldsymbol{W} = \boldsymbol{W}_{p,\mu}$. In the charts $t_j = -1/(pz^p) + \mu \ln t \in \widetilde{S}_j$, f acts as the translation $id + 1$ and the spheres $\overline{\mathbb{C}} \times \{j\}$ are obtained as closures of the quotients of this action. The action of $\pi_* g$ means also translation $id + s$. This shows that $g = g^s_{\boldsymbol{W}}$, $s \in \mathbb{C}$.
The above implies the following result.

10.28. Theorem (Centralizer).

(a) *If $f \in \mathcal{A}$ does not admit any root extraction, then its centralizer $Z(f) = \{f^k; k \in \mathbb{Z}\}$.*

(b) *If f admits the smallest root extraction $g = f^{1/n}$, then $Z(f) = \{g^k : k \in \mathbb{Z}\}$.*

(c) *If $f = g^1_{\boldsymbol{W}}$ is embeddable, then $Z(f) = \{g^t_{\boldsymbol{W}} : t \in \mathbb{C}\}$.*

In particular, $Z(f)$ *is an abelian group of rank* 1 *in the cases* (a) *and* (b)*, and of infinite rank in the embeddable case.*

10.29. Theorem (Analytic classification of abelian groups). *Let* G *be an abelian finitely generated subgroup of* $\mathrm{Diff}(\mathbb{C}, 0)$.

(a) *If* G *is abelian typical (i.e.,* $G \cap \mathcal{A}$ *is infinite and not cyclic), then* G *is analytically equivalent to a subgroup of* $G_a(p, \mu)$.

(b) *If* G *is abelian exceptional (i.e.,* T_G *is infinite and cyclic), then* G *is generated by two germs* $h = z + \ldots \in \mathcal{A}_p$ *and* $g = \mathrm{e}^{2\pi \mathrm{i}/q} z + \ldots$*, such that* $p = sq$ *and*

$$g^q = h^k,$$

for some integers k *and* s*. Moreover, the Ecalle–Voronin modulus of the germ* h *defines the modulus of the analytical classification of groups generated by* h, g *with the above relations.*

(c) *If* G *is formally linearizable (i.e.,* $G \cap A = \{id\}$*) and either* Λ_G *contains a multiplier satisfying the Briuno condition or* Λ_G *consists of resonant multipliers, then* G *is analytically linearizable.*

Proof. (a) Let the group $G \cap \mathcal{A}$ be not cyclic. Thus the centralizer of some its element h has rank > 1. This means that h is embeddable (Theorem 10.28). By Theorem 9.32, h is analytically equivalent to its formal normal form $g^1_{\mathbf{W}}$. The (analytic) diffeomorphism, reducing h to $g^1_{\mathbf{W}}$, reduces also the other germs from G to the forms $\lambda g^t_{\mathbf{W}}$.

(b) Let h be a generator of the cyclic group $G \cap \mathcal{A} \subset \mathcal{A}_p$.
If $G = \{h^n\}$ is cyclic, then we put $g = h$, $q = k = 1$.
Assume that G is non-cyclic, i.e., $\Lambda_G \neq \{1\}$. Let λ be the generator of Λ_G and let $g \in G$ has λ as its multiplier.
Because $\lambda^p = 1$, we have $\lambda = \mathrm{e}^{2\pi \mathrm{i}/q}$ where q divides p, $p = sq$.
The germ g^q belongs to $G \cap \mathcal{A}$ and must equal to some iterate of h, $g^q = h^k$.
Because $G/G \cap \mathcal{A} = \Lambda_G$, the germs g, h generate the whole group G.
One can also see that g and the integers p, q, k are determined uniquely by h. This means that the analytic classification of G is the same as analytic classification of h.

(c) If G is formally equivalent to a linear group and one multiplier is Diophantine (i.e., satisfies the Briuno condition (9.42) from Theorem 9.65), then the linearizing transformation for one germ is analytic (theorem of Briuno). This transformation linearizes also the other germs.
The analyticity of normalization in the case when all multipliers are resonant was proved in the proof of Theorem 10.27. □

Now we consider solvable groups, i.e., solvable and non-abelian.

10.30. Examples (Solvable groups). The standard example of solvable group is the group of affine diffeomorphisms of the complex line $\mathrm{Aff}(\mathbb{C})$, $\xi \longmapsto a\xi + b$; it is the semi-direct product of the group of translations, $\mathbb{C}$ and of the linear group $\mathbb{C}^* = GL(1)$. By applying the semi-conjugation $z \longmapsto \xi = z^{-p}$ to $\mathrm{Aff}(\mathbb{C})$, we obtain the group

$$G_s(p) = \{\lambda g^t_{\boldsymbol{W}} : \ \lambda \in \mathbb{C}^*, \ t \in \mathbb{C}\},$$

where $g^t_{\boldsymbol{W}} = z(1 - ptz^p)^{-1/p}$ is the flow map of the vector field $\boldsymbol{W} = \boldsymbol{W}_{p,0} = z^{p+1}\frac{\partial}{\partial z}$. Here the subscript s stands for 'solvable'.
The group $G_s(p)$ can be identified with the semi-direct product $\mathbb{C}^* \ltimes \mathbb{C} = \{(\lambda, t)\}$. Because $g^t_{\boldsymbol{W}} \circ \nu = \nu z(1 - t\nu^p z^p)^{-1/p} = \nu g^u_{\boldsymbol{W}}$, $u = \nu^p t$, the *multiplication table* is

$$(\lambda, t) \cdot (\nu, s) = (\lambda\nu, \nu^p t + s). \tag{10.5}$$

Consider a subgroup G of $G_s(p)$ and its normal subgroup $G \cap \mathcal{A}$. The latter can be identified with a subgroup

$$T_G = \{t : \ g^t_{\boldsymbol{W}} \in G\}$$

of $\mathbb{C}$. G is identified with the semi-direct product of Λ_G (of multipliers) and of T_G, where the action of Λ_G on T_G is given by $(\lambda, t) \longmapsto \lambda^p t$.
In particular, if $T_G \approx \mathbb{Z}$ is cyclic, then $\Lambda^p_G \subset \mathbb{Z}$. Because altogether all $\lambda^{-p} \in \mathbb{Z}$, we get $\Lambda^p_G \subset \{+1, -1\}$. If $\Lambda^p_G = \{1\}$, then G would be abelian, which contradicts our agreement to call by solvable only non-abelian solvable groups.

10.31. Definition. A solvable (non-abelian) group G, which is formally isomorphic to a subgroup of $G_s(p)$ and such that $T_G = \mathbb{Z}$, is called **solvable exceptional.** Otherwise G is called **solvable typical.**
For an exceptional solvable group we have

$$\Lambda^p_G = \{\pm 1\}, \quad T_G \simeq \mathbb{Z}.$$

10.32. Theorem (Formal classification of solvable groups). *A finitely generated solvable subgroup of* $\mathrm{Diff}\,(\mathbb{C}, 0)$ *is formally equivalent to a subgroup of* $G_s(p)$ *for some* p.

Proof. 1. Because G is not abelian, there exists a $h \in G \cap \mathcal{A}_p$ for some p. We reduce h to its formal normal form $\hat{h} = g^1_{\boldsymbol{W}}$, $\boldsymbol{W} = [z^{p+1}/(1 + \mu w)]\frac{\partial}{\partial z}$. Next we fix the chart in which h is in the normal form. We shall also omit the hats.

2. **Lemma.** *If a subgroup* $H \subset \mathrm{Diff}(\mathbb{C}, 0)$ *contains two non-commuting elements from* $\mathcal{A}$, *then it is non-solvable. This means that the central derivative sequence of any solvable subgroup of* $\mathrm{Diff}(\mathbb{C}, 0)$ *consists of two groups.*

Proof. Let $f = z + az^{p+1} + \dots$, $g = z + bz^{q+1} + \dots$, $ab \neq 0$ be the two non-commuting elements from H.
If $p = q$, then the commutator $[f, g] = fgf^{-1}g^{-1} = z + cz^{r+1} + \dots$ with $r > p$. So, we assume that $p < q$.
However, $f \approx g^a_{z^{p+1}} = \exp\left(az^{p+1}\partial/\partial z\right)$, $g \approx g^b_{z^{q+1}} = \exp\left(bz^{q+1}\partial/\partial z\right)$ and the first term of the Campbell–Hausdorff formula says that $[f, g] \approx g^{ab}_{\boldsymbol{V}}$, $\boldsymbol{V} = [z^{p+1}\partial/\partial z, z^{q+1}\partial/\partial z] = (q - p)z^{p+q+1}\frac{\partial}{\partial z}$. Thus $f_1 = [f, g] \in \mathcal{A}_{p+q}$.

Next, we define $g_1 = [f, f_1] \in \mathcal{A}_{2p+q}$, $f_2 = [f_1, g_1]$, $g_2 = [f_1, f_2]$, $f_3 = [f_2, g_2]$, etc. The maps f_j are different from id and belong to the derivative subgroups $H^{(j)}$. The central derivative series is infinite. □

3. If $G \subset \mathcal{A}$, then G would be abelian (by Lemma 2). Therefore $G \cap \mathcal{A}$ is a normal subgroup with abelian quotient. By Lemma 2 from the proof of Theorem 10.27 all elements from the latter subgroup belong to the flow generated by the vector field $\boldsymbol{W}$, $G \cap \mathcal{A} = \{g^t_{\boldsymbol{W}},\ t \in T_G\}$.

4. Take an element $g = \lambda z + \ldots \in G \setminus \mathcal{A}$; by non-abelianess, such g exists. Because $\mathrm{Ad}_g h = ghg^{-1} \in \mathcal{A}$, we have $\mathrm{Ad}_g g^1_{\boldsymbol{W}} = g^s_{\boldsymbol{W}} = g^1_{s\boldsymbol{W}}$ for some $s \in \mathbb{C}^*$. However $\mathrm{Ad}_g g^1_{\boldsymbol{W}} = g^1_{\boldsymbol{V}}$, where $\boldsymbol{V} = \mathrm{Ad}_{g*}\boldsymbol{W}$; this implies $g^1_{\boldsymbol{V}} = g^1_{s\boldsymbol{W}}$.
The exponents $\exp(\boldsymbol{V}) = g^1_{\boldsymbol{V}}$ and $\exp(s\boldsymbol{W})$ of two small vector fields coincide. Because the exponential map (from the Lie algebra of formal vector fields to the Lie group of formal diffeomorphisms) is invertible near zero, we get

$$\mathrm{Ad}_{g*}\boldsymbol{W} = s\boldsymbol{W}, \quad s \in \mathbb{C}^*. \tag{10.6}$$

5. **Lemma.** *If* (10.6) *holds, then* $s = \lambda^{-p}$. *Moreover, if* $\lambda^p \neq 1$, *then* $\boldsymbol{W} = z^{p+1}\frac{\partial}{\partial z}$.

Proof. Let $y = g(z) = \lambda z + \ldots$. We have $\dot{z} = \boldsymbol{W}(z) = z^{p+1}/(1+\mu z^p)$, $\dot{y} = sw(y) = sy^{p+1}/(1+\mu y^p)$. This implies

$$\frac{\mathrm{d}y}{\mathrm{d}z} \cdot \frac{z^{p+1}}{1+\mu z^p} = \frac{sy^{p+1}}{1+\mu y^p}.$$

Comparison of the linear terms, gives $s = \lambda^{-p}$.
We obtain the following identity of formal meromorphic 1-forms

$$(y^{-p-1} + \mu/y)\mathrm{d}y = \lambda^{-p}(z^{-p-1} + \mu/z)\mathrm{d}z \tag{10.7}$$

The residue of such form does not depend on the choice of the coordinates. This implies $\mu = \lambda^{-p}\mu$. The assumption $\lambda^p \neq 1$ gives $\mu = 0$. □

6. Suppose that $\Lambda^p_G = \{1\}$. Then we have $s = \lambda^p = 1$ in Eq. (10.6). So, $\mathrm{Ad}_g g^1_{\boldsymbol{W}} = g^1_{\boldsymbol{W}}$.
This identity would hold for any $g = \lambda z + \ldots$ from G and for any $h = g^1_{\boldsymbol{W}} \in G \cap \mathcal{A}$. Thus $gh = hg$ for any $g \in G$, $h \in G \cap \mathcal{A}$. Because $G/G \cap \mathcal{A} \simeq \Lambda_G$, the later would mean that G is abelian (contradiction).

7. We can assume that some $\lambda^p \neq 1$, and hence $\mu = 0$. The integration of the formula (10.7) gives $-1/(py^p) = \lambda^{-p}[-1/(pz^p) + t]$, or $g(z) = y = \lambda z(1 - ptz^p)^{-1/p} = \lambda g^t_{\boldsymbol{W}}$.
Because this holds for any $g \in G$, Theorem 10.32 is complete. □

10.33. Theorem (Analytic classification of solvable groups). *Let G be a finitely generated solvable non-abelian subgroup of* $\mathrm{Diff}(\mathbb{C}, 0)$.

(a) *If G is typical, then it is analytically equivalent to a subgroup of $G_s(p)$.*

(b) *If G is exceptional, then it is generated by two elements $h = z + z^{p+1} + \dots$, $g = \nu z + \dots$ satisfying the properties*

$$\begin{aligned} gh &= h^{-1}g, \\ \nu &= \mathrm{e}^{\pi \mathrm{i} r/p}, \\ \gcd(r, 2p) &= 1. \end{aligned} \tag{10.8}$$

The (additive) Ecalle–Voronin moduli $\widetilde{\Phi}_j$, $j = 1, \dots, 2p$, of the germ h (see the Definition 5. in the proof of Theorem 9.30) can be chosen to satisfy the relations

$$\widetilde{\Phi}_{j+r}(\tau_j) = -\widetilde{\Phi}_j(-\tau_j).$$

The first germ $\widetilde{\Phi}_1$ constitutes the modulus of analytic classification of exceptional groups (with fixed p and r).

Proof. (a) In this case the group $T_G \approx G \cap \mathcal{A}$ is not cyclic. This group lies also in the centralizer of the distinguished germ h. Thus $Z(h)$ has rank greater than one and, by Theorem 10.28, h is embeddable and analytically equivalent to its formal normal form.
The analytic transformation normalizing h normalizes also all the other germs from G.

(b) Here we have $T_G = \mathbb{Z}$ and $\Lambda_G^p = \{\pm 1\}$ (see Definition 10.31).
Let h be the generator of $G \cap \mathcal{A}$. We choose a germ $g = \nu z + \dots$ such that ν generates Λ_G. Because $\nu^p = -1$, we have $\nu = \mathrm{e}^{\pi \mathrm{i} r/p}$ where r is relatively prime with $2p$. Because $g + G \cap \mathcal{A}$ generates $G/G \cap \mathcal{A}$ and h generates $G \cap \mathcal{A}$, it is clear that h and g generate G.
Next, we have the formal forms $\hat{h} = g^1_{z^{p+1}}$, $\hat{g} = \nu g^a_{z^{p+1}}$. Taking the new formal variable $z_1 = g^{-a/2}_{z^{p+1}} = \exp[(-a/2)z^{p+1}]$, we reduce $\hat{g}$ to νz_1 without changing $\hat{h}$; (since $\exp[bz^{p+1}] \circ \nu \circ \exp[bz^{p+1}] = \nu \circ \exp[(1+\nu^p)bz^{p+1}]$).
In the representation of $G_s(p)$ as $\mathbb{C}^* \times \mathbb{C}$, we have $\hat{h} \sim (1,1)$, $\hat{h}^{-1} \sim (1,-1)$, $\hat{g} \sim (\nu, 0)$. The multiplication table (10.5) from Example 10.30 says that $\hat{g}\hat{h} \sim (\nu, 1)$ and $\hat{h}^{-1}\hat{g} \sim (\nu, -\nu^p) = (\nu, 1)$, i.e., the formal formula (10.8).
From the first equality in Eqs. (10.8) at the formal level, we obtain its validity at the analytic level.
Recall that, in the definition of the Ecalle–Voronin moduli, we divided the punctured neighborhood of $z = 0$ into sectors S_j parameterized by $t_j = -1/(pz^p) + \lambda \ln z \in \widetilde{S}_j$ and we reduced h to the shift $\tilde{h} = id + 1$ by means of diffeomorphisms $\widetilde{H}_j : t_j \longmapsto \tau_j$. The coboundary maps $\widetilde{\Phi}_j = \tau_{j+1} \circ {\tau_j}^{-1}$ constitute the Ecalle–Voronin modulus $\mu_h^+ \in \mathcal{M}^+_{p,\lambda}$ (see Definition 5 in the proof of Theorem 9.30). In our case $\lambda = 0$.
Let us describe the action of $\tilde{g}$ in the variables τ_j. Because the linear part of g is the rotation by the angle $\pi r/p$, the sector $\widetilde{S}_j$ is transformed to the sector $\widetilde{S}_{j+r}$:

$\tau_j \longmapsto \tau_{j+r} = \tilde{g}_j(\tau_j)$. Because $\tilde{h}(\tau_j) = \tau_j + 1$, the first formula in (10.8) reads as

$$\tilde{g}_j(\tau_j + 1) = \tilde{g}_j(\tau_j) - 1.$$

The latter formula allows to extend g_j to the whole $\mathbb{C}$, to affine maps $g_j = -\tau_j + a_j$. From the gluing conditions $\tilde{g}_{j+1}\widetilde{\Phi}_j = \widetilde{\Phi}_{j+r}\tilde{g}_j$ and closeness of $\widetilde{\Phi}_j$ to id one gets equality of all constants a_j. Choosing properly the first chart τ_1, we can assume that all $a_j = 0$. In this way we obtain the relation $\widetilde{\Phi}_{j+r}(-\tau_j) = -\widetilde{\Phi}(\tau_j)$.
It allows to define the entire collection $(\widetilde{\Phi}_1, \ldots, \widetilde{\Phi}_{2p})$ by means of $\widetilde{\Phi}_1$. The latter constitutes the modulus of analytic classification of G. □

10.34. Examples (Solvable monodromy groups appearing in resolution of nilpotent singularities.) 1. The application of the theory of finitely generated subgroups of $\mathrm{Diff}(\mathbb{C}, 0)$ to investigation of germs of planar analytic vector fields with nilpotent linear part was initiated by R. Moussu [**Mou**] in the case of **cusp** $2y\frac{\partial}{\partial x} + 3x^2\frac{\partial}{\partial y} + \ldots$ and by D. Cerveau and R. Moussu [**CeMo**] in the case of **generalized cusp** $2y\frac{\partial}{\partial x} + sx^{s-1}\frac{\partial}{\partial y} + \ldots$. They proved that the problem of orbital analytic classification of germs of vector fields with such singularity is equivalent to analytic classifications of subgroups of $\mathrm{Diff}(\mathbb{C}, 0)$ with two generators and some natural relations; the spaces of moduli are the same. It was shown that the monodromy group of such cusp can be: either finite or abelian exceptional or solvable typical or non-solvable. In the case of odd s, the classification of vector fields with solvable monodromy was completed in the work of F. Loray and R. Meziani [**LM**]; here the abelian exceptional group does not appear. In the papers [**SZ1**] and [**Lor2**] the formal orbital classification of generalized cusps is given and to each formal normal form the corresponding type of monodromy is attached.
In [**Str**] E. Stróżyna considered the case of **generalized saddle-node** $(y - x^r)\frac{\partial}{\partial x} + \ldots$ (see below). There is a theorem about equivalence of orbital analytic classification of germs of vector fields and of analytic classification of their projective monodromies (no equality of the moduli spaces). There the monodromy group is: either finite or abelian exceptional or solvable typical or solvable exceptional or non-solvable. The formal classification is also given.
In [**SZ2**] the formal orbital classification of nilpotent germs $y\frac{\partial}{\partial x} + \ldots$ was completed; namely the so-called generalized saddles $(y + ax^r)\frac{\partial}{\partial x} + by^{2r-1}\frac{\partial}{\partial y} + \ldots$were classified. The papers of Cerveau and Moussu [**CeMo**] and of Loray [**Lor1**] contain also topological classification of generalized cusps with solvable holonomy.

2. We pass to more precise formulations of the results. By Takens theorem [**Tak**], any germ of vector field with nilpotent linear part is formally equivalent to

$$\dot{x} = y + a(x), \quad \dot{y} = b(x),$$

where $a(x) = a_r x^r + \ldots$, $b(x) = b_{s-1}x^{s-1} + \ldots$ with $a_r b_{s-1} \neq 0$ (or some of the $a(x), b(x)$ vanishes identically). The reader can prove this result by refining the proof of Poincaré–Dulac theorem 8.14.

In [**SZ1**] it is proved that the Takens normal form is analytic (see also [**Lor4**]). But the final orbital normal form for cusp, i.e.,

$$\dot{x} = 2(y + x\Phi), \quad \dot{y} = 3(x^2 + y\Phi), \quad \Phi = x + \sum a_j(y^2 - x^3)^j,$$

is divergent in general (see [**C-DS**] and the point 9.65 in the previous chapter). We say that we have: the **generalized cusp case** iff $s < 2r$, the **generalized saddle-node case** iff $2r < s$ and the **generalized saddle case** iff $2r = s$.

3. **The cusp case**. Here the vector field $\boldsymbol{V}$ is close to the Hamiltonian field $\boldsymbol{X}_H$ with the Hamilton function $H = y^2 - x^3$ (after some normalizations). The resolution of singularity of the function H was presented at Fig. 4.32 in Chapter 4. There are three elementary blowing-ups giving three exceptional divisors E_1, E_2, E_3. They are topological spheres (i.e., $\mathbb{CP}^1$) with intersections $p_{1,2} = E_{1,2} \cap E_3$. The cusp $y^2 = x^3$ is represented by the disc Γ intersecting E_3 at p_0. The divisor E_3 has the self-intersection index -1 (in the complex surface M such that $\pi : (M, \bigcup E_j) \longmapsto (\mathbb{C}^2, 0))$, $E_2 \cdot E_2 = -2$ and $E_1 \cdot E_1 = -3$ (because after elementary blowing-up the index of intersection of two curves decreases by 1).
Introducing the quasi-homogeneous filtration in the space of germs of vector fields with weights $\deg(x) = 2$, $\deg(y) = 3$, we see that $\deg(\boldsymbol{X}_H) = 1$ and the germ with cusp singularity is a perturbation of $\boldsymbol{X}_H$ by means of terms of higher quasi-homogeneous degree. This implies that the resolution process for H is good also for the perturbed germ.
The foliation $\mathcal{F}$, defined by the vector field $\boldsymbol{V}$, is transformed to a foliation $\widetilde{\mathcal{F}}$ in M. Using Camacho–Sad Theorem 10.13(a), applied to the divisors E_j and $\widetilde{\mathcal{F}}$ with singular points p_i, we find the ratios of eigenvalues at p_i. Because E_1 contains only p_1, the ratio of the eigenvalue in direction of E_3 to the eigenvalue in direction of E_1 equals to $-3 = E_1 \cdot E_1$; p_1 is $1 : -3$ resonant saddle. Analogously p_2 is $1 : -2$ resonant saddle. Next, because $E_3 \cdot E_3 = -1 = -1/6 - 1/3 - 1/2$, we get that p_0 is $1 : -6$ resonant saddle. The formulas for these ratios can be also seen from the formulas for the resolution of the cusp in Fig. 4.32 in Chapter 4: $y^2 - x^3 = x^2(u^2 - x) = u^3v^2(u - v) = r^6w^3(1 - w)^2(2w - 1)$, where $y = ux$, $x = uv$, $u = rw$, $v = r(1 - w)$.
The saddles $p_{1,2}$ are analytically linearizable. This follows from the fact that the monodromy maps corresponding to loops in $E_{1,2}$ are identities (the loops are contractible). The point p_0 can be non-linearizable.
One associates with the punctured divisor $(E^*, p) = (E_3 \setminus \{p_1, p_2, p_3\}, p)$ and the foliation $\widetilde{\mathcal{F}}$, the *monodromy group* G (see 10.19): $G \subset \mathrm{Diff}\,(D, p) = \mathrm{Diff}(\mathbb{C}, 0)$, where D is a holomorphic disc transversal to E^* at p. It is generated by two maps $f_{1,2}$, corresponding to two simple loops in $\pi_1(E^*, p)$ surrounding the points $p_{1,2}$. The loop around p_0 generates the map $f_0 = f_1 \circ f_2$.
The formulas for ratios of eigenvalues imply that

$$f_1(z) = \mathrm{e}^{-2\pi \mathrm{i}/3} z + \ldots, \quad f_2(z) = -z + \ldots \tag{10.9}$$

and $f_3(z) = \mathrm{e}^{-\pi \mathrm{i}/3} z + \dots$.
Moreover, because the points $p_{1,2}$ are linearizable, the corresponding maps are also linearizable and we have the relations

$$f_1^3 = f_2^2 = id. \tag{10.10}$$

R. Moussu proved that the classification of cusps is the same as the classification of groups generated by $f_{1,2}$ with the restrictions (10.9) and (10.10).

We pass to the analysis of particular examples with the cusp singularity. Firstly we shall show that

The Hamiltonian cusp has finite monodromy group.

In order to see it, one uses special coordinates, dictated by the quasi-homogeneity: $u = x^3/y^2$ (of degree 0) and $z = y/x$ (of degree 1). (Here $u = (1-w)/w, z = r$ in terms of the resolution from Fig. 4.32 in Chapter 4.) We obtain the differential equation for the foliation $\widetilde{\mathcal{F}}$ in the form $\frac{\mathrm{d}z}{\mathrm{d}u} = \frac{(3u-2)z}{6u(1-u)}$, with the solutions of Darbouxian type

$$z(u) = Cu^{-1/3}(u-1)^{-1/6}.$$

We see that the maps $z_0 = z(u_0) \longmapsto f_1(z_0)$ and $z_0 \longmapsto f_0(z_0)$, which are defined as analytic prolongations of the multivalued function $z(u)$ along closed loops with vertex at u_0, are linear maps. □

Consider now the perturbation of $\boldsymbol{X}_H$ by means of the vector field $x^{r-1}\boldsymbol{E}_H$, proportional to the quasi-homogeneous Euler vector field $\boldsymbol{E}_H = 2x\frac{\partial}{\partial x} + 3y\frac{\partial}{\partial y}$. Then, in the variables u, z, we obtain the Bernoulli equation for the phase curves

$$\frac{\mathrm{d}z}{\mathrm{d}u} = \frac{3u-2}{6u(1-u)}z + \frac{x^{r-2}}{6(1-u)}z^{2r-2}.$$

It has the first integral of the form

$$F = \left[zu^{1/3}(1-u)^{1/6}\right]^{3-2r} + \frac{1-2r}{6}\int_0^u \tau^{r/3-1}(\tau-1)^{-(2r+3)/6}\mathrm{d}\tau.$$

Note that this function contains the term of the Darbouxian form and the Schwarz–Christoffel integral; we shall call them the **Darboux–Schwarz–Christoffel integrals**. The properties of F depend on r.
Let $r = 3m$. Then the power in $\tau^{r/3-1}$ is a positive integer and the Schwarz–Christoffel integral is elementary. We obtain $F = (u-1)^{1/2-m}[z^{3-6m}u^{1-2m} + P(u)]$, where $P(u)$ is a polynomial. The integral is of the Darbouxian type. One can deduce from it that the monodromy group is abelian linear.
The explanation of this lies in the fact that the perturbed vector field is orbitally equivalent to the Hamiltonian one. Indeed, in the variables $h = y^2 - x^3$ and y, one obtains the system $\dot{h} = 2h(y^2-h)^{m-1}$, $\dot{y} = 1 + (y^2-h)^{m-1}y$. It is non-singular

vector field with local analytic first integral $h + \ldots = y^2 - x^3 + \ldots$, i.e., the $\mathbf{A}_2$ singularity. After analytic change of coordinates the first integral equals to $\tilde{y}^2 - \tilde{x}^3$. Let $r \neq 0 \pmod 3$. Near $u = 0, z = 0$ we have $F = u^{1-2r/3} \times$ (analytic function). This implies that the monodromy map f_1, expressed in the chart $\zeta = F|_{u=u_0}$ takes the form $\zeta \longmapsto \mu_1 \zeta$.
Near $u = 1$ we have $F = c_1 + (u-1)^{(3-2r)/6} \times$ (analytic function). Note that the Schwarz–Christoffel integral diverges at $u = 1$ and one must use several times the integration by parts formula: $\int^u (\tau - 1)^a \phi(\tau) = (a+1)^{-1}(u-1)^{a+1} - (a+1)^{-1} \int^u (\tau-1)^{a+1}\phi'$. The value c_1 is the 'principal' value of the Schwarz–Christoffel integral from 0 to 1. In order to find this value, one uses analytic continuation of the Euler Beta-function

$$B(a,b) = \int_0^1 \tau^{a-1}(1-\tau)^{b-1} \mathrm{d}\tau = \frac{\Gamma(a)\Gamma(b)}{\Gamma(a+b)}.$$

Because the Euler Gamma-function has no zeroes and has poles at non-positive integers, we get that the constant $c_1 \neq 0$. This means that the map f_0 takes the form $\zeta \longmapsto \mu_0 \zeta + c$, with $\mu_0 = \mathrm{e}^{(3-2r)\pi \mathrm{i}/3}$ and $c = c_1(1-\mu_0) \neq 0$.
This shows that the monodromy group G, in the chart ζ, forms a subgroup of the group $\mathrm{Aff}(\mathbb{C})$, of affine diffeomorphisms of a complex line. It is solvable and non-abelian.
The relation between the chart z and ζ is given by the formula $\zeta = Az^{3-2r} + B$. This means that the maps f_j, expressed in the chart z, have the form $\lambda_j g_{\boldsymbol{W}}^{t_j}$, $\boldsymbol{W} = z^{p+1}\frac{\partial}{\partial z}$, i.e., solvable subgroup of $G_s(p)$ with $p = 2r - 3$. Note that here Λ_G consists of roots of unity of order 6 and, as r is not divisible by 3, we have $\Lambda_G^p \neq \{\pm 1\}$. We can conclude the above in the following statement.

The vector field $\boldsymbol{X}_H + x^{r-1}\boldsymbol{E}_H$, with the cusp singularity, has finite monodromy group in the case $r = 0 \pmod 3$ and solvable typical monodromy group otherwise.

The example of cusp singularity with solvable non-abelian holonomy was used by R. Moussu [**Mou**] in his counter-example to the problem (c) of R. Thom (see 10.20): do the holonomies of the separatrices form a complete system of invariants of the singularity? Indeed, in the solvable cases with different r's ($\neq 0 \pmod 3$), the singular point p_0 is linearizable. Thus, the holonomy map, associated with a loop in the separatrix Γ (through p_0), is trivial in all these cases. On the other hand, the different projective holonomies form the obstacles to the analytic equivalence of these cusps.

4. **The generalized cusp with abelian exceptional monodromy**. Consider the field

$$\boldsymbol{X}_H + xH^n(1+\mu H^n)^{-1}\boldsymbol{E}_H, \quad H = y^2 - x^4. \tag{10.11}$$

(Here $\boldsymbol{E}_H = x\frac{\partial}{\partial x} + 2y\frac{\partial}{\partial y}$ is the quasi-homogeneous Euler field). The resolution of this singularity is the same as the resolution of the $\mathbf{A}_3$ singularity of H and is presented in Fig. 10.1.

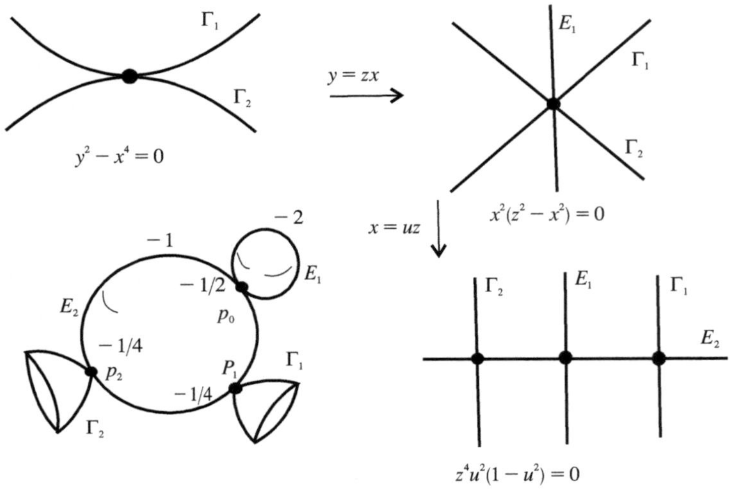

Figure 10.1: Resolution of the generalized cusp singularity

The monodromy group is generated by two maps $f_{1,2} = -\mathrm{i}z + \dots$, $\mathrm{i} = \mathrm{e}^{\pi \mathrm{i}/2} = \sqrt{-1}$ with the relation $f_0^2 = id$, where $f_0 = f_1 \circ f_2$.
In the variables $x, u = x^2/y$ we have $H = x^4(u^2 - 1)$ and the points $p_{1,2}$ correspond to $x = 0, u = \pm 1$ and $p_0 : x = 0, u = \infty$.
It turns out that the vector field has the variables u, H separating. Namely, we have $\mathrm{d}H/\mathrm{d}u = H^{n+1}/[(1 + \mu H^n)(1 - u^2)]$. From this we see that the monodromy maps, expressed in the chart $h = H|_{u=u_0} = \mathrm{const} \cdot x^4$, take the forms $h \longmapsto g_{\boldsymbol{V}}^{s_j}(h)$, where $\boldsymbol{V} = h^{n+1}(1 + \mu h)^{-1} \frac{\partial}{\partial h}$. In the chart x we get $f_j = \lambda_j g_{\boldsymbol{W}}^{t_j}$, which implies the following statement.

The vector field (10.11) *has abelian exceptional monodromy.*

5. **The generalized saddle-node with exceptional solvable monodromy.** Consider the system

$$\dot{x} = y - x^2 + x^{m+1}, \quad \dot{y} = 2x^m y. \tag{10.12}$$

Its two-step resolution is presented at Fig. 10.2. We see that the points $p_{0,1}$ are $1 : -2$ resonant saddles. The ratio of eigenvalues of the point p_2 equals zero.
The monodromy maps take the form $f_{0,1} = -z + \dots$, $f_2 = z + \dots$ with the relation $f_0^2 = id$.
We use the variables $x, u = y/x^2$. We get

$$\dot{x} = x(u - 1) + x^m, \quad \dot{u} = -2u(u - 1).$$

Here $p_0 : x = 0, u = \infty$, $p_1 : x = u = 0$, $p_2 : x = 0, u = 1$. We see that p_2 is saddle-node; this justifies the name of the singularity.

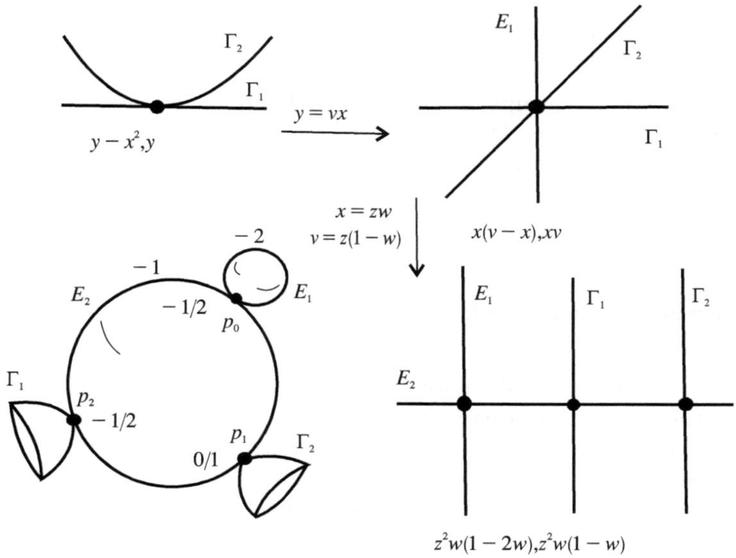

Figure 10.2: Resolution of the generalized saddle–node singularity

The above system has the first integral of the Darboux–Schwarz–Christoffel type

$$F = (xu^{1/2})^{1-m} + \frac{1-m}{2} \int^{u} v^{-(m+1)/2}(v-1)^{-1}\mathrm{d}v.$$

If m is odd, then the first integral is of the form $F = a\ln(u-1) +$ (rational function). It has abelian monodromy and the monodromy group of the vector field is abelian exceptional.

If m is even, then the monodromy of the first integral is solvable. It is generated by the maps $\zeta \longmapsto -\zeta$ and $\zeta \longmapsto \zeta + b$. Any its element is of the form $\pm\zeta + nb$, with integer n. The monodromy group G of the resolution is also solvable. It is a subgroup of $G_s(p)$, $p = m-1$. Because $\Lambda_G = \{\pm 1\}$, we have $\Lambda_G^p = \{\pm 1\}$. On the other hand, the subgroup $G \cap \mathcal{A}$ (of maps with identity linear part) is cyclic $\simeq b\mathbb{Z}$. This means that G is solvable and exceptional, i.e.,

The vector field (10.12) *has solvable exceptional holonomy.*

6. Other examples of vector fields with solvable monodromy constitute the fields

$$\boldsymbol{V}_0 + K\boldsymbol{E}$$

consisting of two homogeneous terms, one of which is proportional to the standard Euler field $\boldsymbol{E} = x\frac{\partial}{\partial x} + y\frac{\partial}{\partial y}$; the field $\boldsymbol{V}_0$ and the function K are homogeneous. The elementary blowing-up reduces this field to a Bernoulli system with a Darboux–Schwarz–Christoffel integral.

10.35. Non-solvable subgroups of $Diff(\mathbb{C},0)$**. Examples.** The standard example of non-solvable subgroup of Diff$(\mathbb{C},0)$ is the group generated by two maps (see Lemma 2 in the proof of Theorem 10.32)

$$f_1 = g^1_{z^{p+1}}, \quad f_2 = g^1_{z^{q+1}}, \quad p < q.$$

One can say even more.

10.36. Theorem of Cohen. ([**Coh**]) *This group is free, i.e., there are no relations between* f_1 *and* f_2.

We do not prove this result. We note only that in the original paper of Cohen it is proved that the group generated by the power z^r and the translation $z+1$ is free. There is no classification of non-solvable subgroups of Diff$(\mathbb{C},0)$. However, they are typical among monodromy groups of algebraic leaves. For example, Yu. Ilyashenko and A. Pyartli [**IlP**] proved that among vector fields of degree n with the line at infinity L_∞ invariant, typical are fields with free holonomy associated with the leaf in L_∞. Also in [**SZ1**] and [**Str**] it was shown that typical nilpotent singularity of the generalized cusp type or the generalized saddle-node type have non-solvable monodromy groups.
Among few results about non-solvable groups we present the A. Shcherbakov's and I. Nakai's generalization of the Hudai-Verenov theorem 10.22 and the Ramis theorem about formal and analytic equivalences of non-solvable groups.

10.37. Density theorem of Shcherbakov and Nakai. ([**Shc2**], [**Nak**]) *If* $G \subset \mathrm{Diff}(\mathbb{C},0)$ *is a non-solvable subgroup, then there is a partition of* $(\mathbb{C},0)$ *into sector-like domains such that G-orbit of any point is dense either in a whole domain of the partition or in a line separating different domains.*

The idea of the proof is the following. Take two non-commuting elements of G: $f_1 = z + z^p + \ldots$, $f_2 = z + z^q + \ldots$, $p < q$. The sequence $f_1^{-m} f_2 f_1^m$ tends to identity and a suitable normalization $c_m(f_1^{-m} f_2 f_1^m - id)$ tends to a definite vector field $\boldsymbol{V}_2$. Replacing f_2 by $f_3 = [f_1, f_2] = z + az^r + \ldots, q < r$, we obtain another vector field $\boldsymbol{V}_3$ which is independent from $\boldsymbol{V}_2$ at generic point. Moving along these vector fields is the same as moving along closures of orbits of G.
Additional information about the dynamics of non-solvable pseudo-groups are given in [**BLL**].

10.38. Theorem of Ramis (Rigidity of non-solvable groups). ([**Ram2**]) *If two non-solvable subgroups* G, G' *of* Diff$(\mathbb{C},0)$ *are formally equivalent, then they are analytically equivalent as well.*

Proof. We sketch the proof following [**EISV**].
Let the germs from G

$$f = z + az^{p+1} + \ldots, \quad g = z + bz^{q+1} + \ldots, \quad p < q, \quad ab \neq 0,$$

be formally conjugated by means of series $\hat{h}$ with analogous germs f', g' from G'. One has to show that the series $\hat{h}$ is the Taylor series of an analytic conjugating map h.

Following the proof of Ecalle–Voronin Theorem, we construct sectorial normalizations of these four germs. However, instead of using the sectorial coverings, we shall use sectorial partitions; i.e., partition into closed sectors which are included in the domains of analyticity of the normalizing maps. It means that we have the functional cochains $H_f = (H_{f,1}, \ldots, H_{f,2p})$, $H_{f'}$, H_g, $H_{g'}$, defined in sectors of partitions Ξ_1, Ξ_2 of $(\mathbb{C}, 0)$ (with angles π/p or π/q), close to identity (as $o(z^{p+1})$ or as $o(z^{q+1})$) and conjugating germs with their formal normal forms. The cochains

$$F = H_{f'}^{-1} H_f, \; \widetilde{F} = H_{g'}^{-1} H_g$$

conjugate f with f' and g with g' respectively. The Taylor series of the maps from F and from $\widetilde{F}$ coincide and are equal to $\hat{h}$.

Take a partition Ξ of $(\mathbb{C}, 0)$, which will serve as a common partition for the cochains F and $\widetilde{F}$, (product of the partitions $\Xi_{1,2}$). Then $G = F - \widetilde{F} = (G_1, \ldots, G_r)$ is a functional cochain decreasing faster than any power of z. Its coboundaries decrease very fast $|\delta G_j| = |G_{j+1} - G_j| < \exp(C/|x|^l)$, where $l = p$ when we are in a line of partition Ξ_1 and $l = q$, when we are in a line of the partition Ξ_2. By the Phragmen–Lindelöf theorem 9.42, the cochain G vanishes identically; therefore, $F = \widetilde{F}$.

Now, because the maps F_j from F and $\widetilde{F}_j$ from $\widetilde{F}$ are analytic in sectors of different magnitude, then we prolong them step-by-step to an univalent function h, representing the conjugating diffeomorphism.

This completes the proof of the Ramis theorem. □

10.39. Remark. Theorems 10.29, 10.33 and 10.38 imply the following property, called the *rigidity*, of some subgroups $G \subset \mathrm{Diff}(\mathbb{C}, 0)$:

Let G, G' be: either finite or abelian typical or solvable typical or non-solvable. If G and G' are formally equivalent, then they are analytically equivalent as well.

This implies the following rigidity property of generalized cusps $\mathrm{d}(y^2 + x^{2k+1}) + \ldots = 0$.

If two such singularities are formally equivalent, then they are also analytically equivalent.

Indeed, by the result of D. Cerveau and R. Moussu, the formal (respectively analytical) equivalence of foliations is equivalent to the formal (respectively analytical) equivalence of their monodromy groups. The latter turn out to be rigid.

R. Meziani [**Mez**] investigated the rigidity property for the cases of generalized saddle and of generalized saddle-node.

10.40. Phragmén–Lindelöf Theorem for functional cochains. (See [**Il3**], [**EISV**]) *If a functional cochain decreases faster than any power of z, then it vanishes identically.*

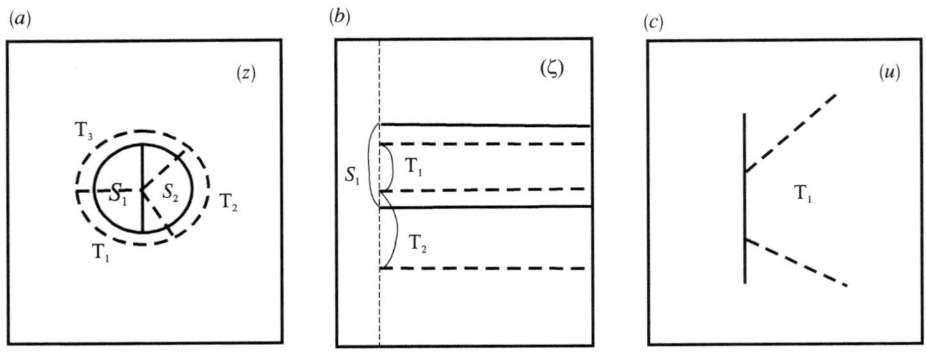

Figure 10.3: Partitions in various coordinate planes used in the proof

Proof. (This theorem constitutes essential part in the Ilyashenko's proof of finiteness of the number of limit cycles for a polynomial vector field, see Theorem 6.12. We use the arguments presented in [**Il3**].)

Recall that the classical *Phragmén–Lindelöf theorem* says that, if a holomorphic function $f(\zeta)$, defined in the sector $|\arg\zeta| < \alpha$ and growing not too fast at infinity, $|f| < \exp(c|z|^{\beta})$, $\beta < \pi/2\alpha$, is bounded by a constant M at the boundary, then the same bound holds inside the sector. (In the proof one replaces f by $f\exp(-\epsilon|z|^{\pi/2\alpha})$, tending to zero at infinity).

This theorem can be applied to domains which are biholomorphically equivalent to sectors, e.g., the half-plane $\mathrm{Re}\zeta > A$ and its image U_A under $\Psi = \zeta + \sqrt{\zeta}$.

Moreover, if f decreases in U_A faster than any exponent, then $f \equiv 0$. Indeed, applying the Phragmén–Lindelöf theorem to $f\mathrm{e}^{Nz}$ we get that $|f| < M\mathrm{e}^{-Nx}$ for any N along the real axis.

Assume that a cochain $G = (G_1, \ldots, G_r)$ decreases faster than any power of z. Take the chart $\zeta = -\ln z$. Then we get a domain like $\mathrm{Re}\zeta > A$, with the partition Ξ into horizontal strips with the boundary $\partial\Xi$ consisting of half-lines $\mathrm{Im}\zeta = \mathrm{const}$. At the lines from $\partial\Xi$ the cochain has jumps δG. The cochain G decreases faster than any exponent of ζ. Its coboundaries $\delta G_j = G_{j+1} - G_j$ decrease as $\exp(-C\mathrm{e}^{p\cdot\mathrm{Re}(\zeta)})$ or as $\exp(-C\mathrm{e}^{q\cdot\mathrm{Re}(\zeta)})$, depending on which line of partition Ξ_1 or Ξ_2 we are. Of course, the functions G_j from G are prolonged to larger strips than the strips of the partition Ξ.

Any cochain $G' = G\cdot\mathrm{e}^{P(\zeta-Q)}$, $P, Q > 0$ has the same properties as G.

Take the functional cochain defined by $H(\zeta) = \frac{1}{2\pi\mathrm{i}}\int_{\partial\Xi}\frac{\partial G'(\xi)}{\xi-\zeta}\mathrm{d}\xi$. (In order to get this expression convergent, one can restrict himself to a region of the type U_A.) The cochain H has the same jumps as G' (the *Plemelj Theorem*). Thus $H - G'$ is an analytic function in U_A. Because it decreases faster than any exponent, it vanishes identically. Thus $G = H\mathrm{e}^{-P(\zeta-Q)}$.

The cochain H can be estimated (using the estimations for δG). This (with varying

$P, Q)$ allows us to show that $|G| < \exp(-Ce^{p \cdot \mathrm{Re}(\zeta)})$, i.e., that

$$|G| < \exp(-C/|z|^p)$$

(see [**Il3**] for details).
Consider now some sector S containing a sector of the partition Ξ_1 (e.g., like S_1 in Fig. 10.3). By means of the chart $u = \mathrm{const} \cdot z^{-p'}$, $p' < p$, it is transformed to the half-plane $\mathrm{Re}(u) > 0$. The cochain G, restricted to the latter sector, has the same properties as the cochain G in the chart ζ, decay properties of the components and of coboundaries. Repeating the above arguments, we obtain that

$$|G| < \exp(-C/|z|^q), \quad z \in S.$$

The further repetition of the above arguments with the sector $T_1 \subset S_1$ and application of the Phragmén–Lindelöf theorem shows that $G \equiv 0$. □

§4 The Ziglin Theory

The Ziglin theory constitutes one of the most beautiful and effective applications of the monodromy theory. It allowed to show non-integrability of certain classical Hamiltonian systems.

10.41. Example (Euler–Poisson system). This is the system describing motion of a rigid body with one fixed point in presence of the constant gravitational field

$$\dot{\boldsymbol{M}} = \boldsymbol{M} \times \boldsymbol{\Omega} + \mu \boldsymbol{\Gamma} \times \boldsymbol{L}, \quad \dot{\boldsymbol{\Gamma}} = \boldsymbol{\Gamma} \times \boldsymbol{\Omega}. \tag{10.13}$$

Here the vector quantities are given in the moving coordinate system, associated with the principal axes of inertia of the body. We denote: $\boldsymbol{M} = (M_1, M_2, M_3)$ – the kinetic momentum with respect to the fixed point; $\boldsymbol{\Omega}$ – the angular velocity, $\boldsymbol{M} = \mathrm{diag}\,(I_1, I_2, I_3)\boldsymbol{\Omega}$, where I_j are the principal inertia moments; $\boldsymbol{\Gamma}$ – the unit vertical vector; $\boldsymbol{L} = (X_0, Y_0, Z_0)$ – the unit vector directed from the fixed point to the center of mass of the body (see Fig. 10.4).
The system (10.13) has three independent polynomial first integrals

$$\boldsymbol{\Gamma}^2, \quad (\boldsymbol{M}, \boldsymbol{\Gamma}), \quad E = (\boldsymbol{M}, \boldsymbol{\Omega})/2 + \mu(\boldsymbol{\Gamma}, \boldsymbol{L});$$

the geometric integral, the area integral and the energy. The *complete integrability* of system (10.13) means the existence of a foliation of the phase space into 2-dimensional tori with periodic or quasi-periodic motion (the *Liouville–Arnold theorem*, see [**Arn1**]). It is equivalent to the existence of an additional first integral.
Such an integral, in the hypersurface $\{\boldsymbol{\Gamma}^2 = 1\}$, exists in three (classical) cases:

- $\mu = 0$; the *Euler case* of free body, with the additional integral $\boldsymbol{M}^2$.

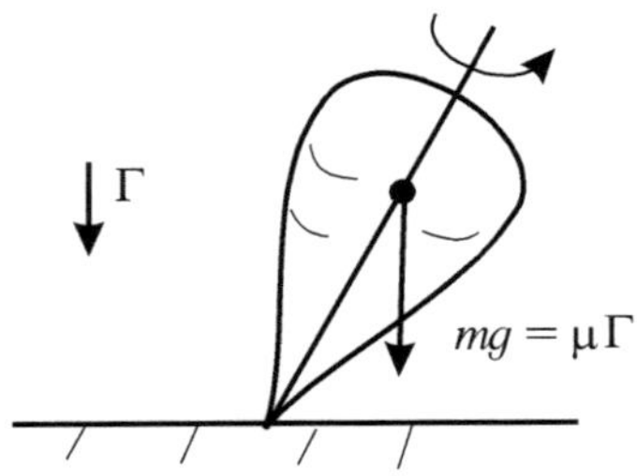

Figure 10.4: The top

- $I_1 = I_2$, $X_0 = Y_0 = 0$; the *Lagrange case*, with the body symmetric with respect to an axis containing the mass center and with the additional integral M_3.
- $I_1 = I_2 = I_3/2$, $Z_0 = 0$; the *Kowalewska case*, with special symmetric body, whose mass center lies in a plane orthogonal to the symmetry axis at the fixed point, and with the integral $|I_1(M_1 + \mathrm{i}M_2) + (\Gamma_1 + \mathrm{i}\Gamma_2)(X_0 + \mathrm{i}Y_0)|^2$.

The problem of complete integrability of system (10.13) remained unsolved for very long time. There were some partial results, concerning particular situations. Only application of the monodromy theory allowed to obtain the complete solution.

10.42. Theorem of Ziglin (Euler–Poisson system). ([**Zig**]) *The above three cases are the only cases, when system* (10.13) *has additional meromorphic first integral.*

We present the general method leading to the proof of this theorem. However, as an application of this method, we present calculations for a Yang–Mills, which is simpler than (10.13). For the proof of the Ziglin's theorem we refer the reader to the Ziglin's work [**Zig**].

10.43. Normal variation equation and the variation monodromy. If $\dot{x} = \boldsymbol{X}(x)$, $x \in \mathbb{R}^m$, is an autonomous system and $x = \phi(t)$ is some its solution, then the system $\dot{y} = [(\partial \boldsymbol{X}/\partial x)(x)]y$, $x = \phi(t)$, is the **variation equation associated with the solution** ϕ. Its solutions measure the divergence of solutions (of the initial system) close to ϕ.

If the curve Γ is the image of ϕ and $N\Gamma = T_\Gamma\mathbb{R}^m/T\Gamma$ is the normal bundle, then the variation equation can be factorized by $T\Gamma$ and defines the **normal variation equation.** Its solutions are sections of the normal bundle.

We will consider the case when the system is Hamiltonian,

$$\dot{x} = \boldsymbol{X}_H$$

in $\mathbb{R}^{2n}$, with polynomial Hamilton function $H(x)$ and with an algebraic curve Γ as the image of the particular solution. We assume $x = (q, p)$ and the standard symplectic structure $\mathrm{d}p \wedge \mathrm{d}q$.

The **complete integrability** of the Hamiltonian system means existence of n functionally independent first integrals in involution, i.e., with vanishing Poisson brackets $\{H_i, H_j\} = \boldsymbol{X}_{H_j}(H_i) = 0$. In particular, if the vector field $\boldsymbol{X}_H$ does not have $n-1$ first integrals functionally independent with H, then this field is not integrable.

Because H and Γ are algebraic, it is natural to consider the system $\dot{x} = \boldsymbol{X}_H$ and the curve Γ in the complex space $\mathbb{C}^{2n}$ and with complex time. Then we get a holomorphic 1-dimensional foliation $\mathcal{F}$ with the algebraic leaf $\Gamma^* = \Gamma\backslash$(singular points).

Using lifts of loops from $\pi_1(\Gamma^*, x_0)$ to the leaves of $\mathcal{F}$, we define the holonomy group of the leaf Γ^*. It is a subgroup of the group of germs of analytic diffeomorphisms of a $2n-1$-dimensional polydisc transversal to Γ^* at x_0. The linear parts of the germs from the holonomy group generate the **variation monodromy group** $\mathrm{Mon}(\Gamma)$. $\mathrm{Mon}(\Gamma)$ is defined by means of the solutions of the normal variation equation.

Because $H(x)$ is a first integral for the Hamiltonian system, the function $\mathrm{d}H \in \Gamma(N^*\Gamma)$ is a first integral of the normal variation system. Note that $\mathrm{d}H$ is linear with respect to the $2n-1$ variables in $N_x\Gamma$ and is algebraic with respect to $x \in \Gamma$. The affine subbundles $V(h) = \{(x, v) \in N\Gamma^* : \langle \mathrm{d}H(x), v\rangle = h\}$, $h \in \mathbb{C}$, are invariant manifolds of the normal variation equation and of the variation monodromy group. The restriction of the action of $\mathrm{Mon}(\Gamma)$ to $V_{x_0}(h)$ is denoted by $\mathrm{Mon}(\Gamma, h)$. The latter consists of affine maps of the form $A_h(v) = A_0 v + hw$, where $A_0 \in Sp(2n-2, \mathbb{C})$ is a symplectic operator, i.e., preserves the natural symplectic structure induced on the fiber $N_{x_0}\Gamma$ by so-called *symplectic reduction.* The linear part $A_0 \in \mathrm{Mon}(\Gamma, 0)$ is uniquely determined by A.

The eigenvalues of A_0 form the system $\lambda_1, \lambda_1^{-1}, \ldots, \lambda_{n-1}, \lambda_{n-1}^{-1}$; because $A_0^\top J A_0 = I$, where $J = \begin{pmatrix} 0 & -I \\ I & 0 \end{pmatrix}$. The operator $A \in \mathrm{Mon}(\Gamma)$ is called **resonant** iff $\lambda_1^{k_1} \ldots \lambda_{n-1}^{k_{n-1}} = 1$ for some integers k_j, $\sum |k_j| \neq 0$.

If A is nonresonant, then it is diagonalizable with eigenvalues $1, \lambda_i, \lambda_i^{-1}$.

10.44. Theorem (Variation monodromy group and integrability). ([**Zig**]) *Assume that* $\mathrm{Mon}(\Gamma)$ *contains a nonresonant operator* A *and the Hamiltonian system has* $n-1$ *meromorphic first integrals functionally independent with* H.

Then any other linear operator $A' \in \mathrm{Mon}(\Gamma)$ *preserves the system of eigenspaces of* A. *Moreover, if any subset of the system of eigenvalues of* A_0' *does not form a regular polygon with center at* 0, *then* A' *commutes with* A.

10.45. Proof of Theorem 10.46. 1. Assume that the field $\boldsymbol{X}_H$ has meromorphic first integrals $H_1, \ldots, H_r$ near Γ, functionally independent with H. We claim the following.

2. **Proposition.** *The variation monodromy group has* r *rational first integrals* $\Phi_j(v)$, $j = 1, \ldots, r$, *which are homogeneous in* v.

3. *Proof of Proposition* 2. Let $H_0 = H$.

If $\Psi(x)$ is a first integral (meromorphic and defined near Γ), then we denote by $\Psi^0(x, v)$ the lowest homogeneous part of $\Psi(x+v)$, $(x \in \Gamma,\ v \in N_x\Gamma)$, with respect to v.
We put $\Phi_0 = H_0^0 = \mathrm{d}H = h$. Next, as H_0 and H_1 are independent, there is a function Ψ_1, which is rational in H_0 and H_1, and such that Φ_0 and Ψ_1^0 are independent. (Take H_1^0; if it is dependent with H_0^0, then $H_1^0 = f(x)h^k$, f rational, $k \in \mathbb{Z}$; we take $H_1 - f(x)H_0^k$, look at its lowest part, etc.) We put $\Phi_1 = \Psi_1^0$.
The further proof goes by induction (see [**Zig**]) and we do not present it.
We obtain first integrals $\Phi_j(x, v)$ of the normal variation system. Their restrictions $\Phi_j(x_0, v)$ to the generic fiber $N_{x_0}\Gamma$ are independent first integrals of the monodromy group $\mathrm{Mon}(\Gamma)$, $\Phi_j \circ A = \Phi_j$. □

4. Let $A \in \mathrm{Mon}(\Gamma)$ be nonresonant, i.e., the symplectic operator A_0 is nonresonant. There is a system of symplectic linear coordinates p_i, q_i in $V_{x_0}(0)$ such that $q_i \circ A_0 = \lambda_i q_i$, $p_i \circ A_0 = \lambda^{-1} p_i$. The functions q_i, p_i can be extended to eigenfunction of A in $N_{x_0}\Gamma$, such that $h, q_1, p_1, \ldots, q_{m-1}, p_{n-1}$ forms a coordinate system in $N_{x_0}\Gamma$ with $h \circ A = h$). Let $z_i = q_i p_i$.
Because λ_j are nonresonant,

Any homogeneous rational first integral of the monodromy is a rational function of $z_1, \ldots, z_{n-1}$ (with coefficients depending on h).

Indeed, this integral is a ratio of two polynomials. Each monomial in these polynomials is multiplied by the same constant after action of the monodromy operator A. Thus ratio of a monomial from the numerator to a monomial from the denominator is A-invariant monomial $q_1^{m_1} p_1^{n_1} \ldots q_{n-1}^{m_{n-1}} p_{n-1}^{n_{n-1}}$. So, $\lambda_1^{m_1-n_1} \ldots \lambda_{n-1}^{m_{n-1}-n_{n-1}} = 1$ and hence $m_j = n_j$.

5. If $\Phi(v)$ is a first integral, then $\mathrm{d}\Phi|_{h=\mathrm{const}} = \mathrm{d}\Phi_h$ belongs to the $(n-1)$-dimensional subspace $L(id)$ of the $(2n-2)$-dimensional space of 1-forms, over the field of rational functions of h, q_j, p_j and generated by $\mathrm{d}z_1, \ldots, \mathrm{d}z_{n-1}$. For $B \in \mathrm{Mon}(\Gamma)$, we denote by $L(B)$ the space generated by the forms $\mathrm{d}(z_1 \circ B), \ldots, \mathrm{d}(z_{n-1} \circ B)$. $\mathrm{d}\Phi_h$ belongs to all $L(B)$, $B \in \mathrm{Mon}(\Gamma)$. This implies the following.

The number of functionally independent first integral of the Hamiltonian system does not exceed the dimension of the space $\bigcap_B L(B)$, $B \in \mathrm{Mon}(\Gamma)$.

In the assumptions of Theorem 10.44 we have $\dim \bigcap_B L(B) = n-1$. This means that any differential $\mathrm{d}(z_i \circ B) \in L(id)$.

6. If A' is another monodromy operator, then it acts linearly in the variables h, q_j, p_j and affinely in the variables q_j, p_j. In particular, $z_j \circ A'_h$ are quadratic functions with linear differentials. Because $\mathrm{d}(z_j \circ A'_h) \in L(id)$, we have $\mathrm{d}(z_j \circ A'_h) = \sum a_k \mathrm{d}z_k$, $a_k = \mathrm{const}$ and hence $z_j \circ A'_h = \sum a_k z_k + a_0(h)$. On the other hand, $z_j \circ A'_h = (q_j \circ A'_h)(p_j \circ A'_h)$ is a product of linear functions, which implies that $a_0 = 0$ and the above sum contains only one summand. Therefore, $z_j \circ A'_h = a_k z_k$ for some k and either $q_j \circ A'_h = \alpha q_k$, $p_j \circ A'_h = \beta p_k$ or $q_j \circ A'_h = \alpha q_k$, $p_j \circ A'_h = \beta q_k$.

This shows that A'_h are linear in the coordinate system q_j, p_j (associated with A) and transform the eigenspaces of A_h to eigenspaces. The operator A'_0 permutes the eigenspaces of A_0.

7. Assume that A'_0 permutes non-trivially some eigenspaces of A_0. Because any permutation is a product of cyclic permutations, A' permutes cyclically some $k \geq 2$ directions. This implies that its characteristic polynomial has divisor of the form $\lambda^k - a$, $a \neq 0$. The zeroes of the latter form vertices of a regular k-gon with center at 0.
Theorem 10.44 is proved. □

10.46. Example (Yang–Mills system). It is the Hamiltonian system with two degrees of freedom with the Hamilton function

$$H = \frac{1}{2}(p_1^2 + p_2^2) + \frac{1}{8}(q_1^2 - q_2^2)^2,$$

or $\dot{q}_{1,2} = p_{1,2}$, $\dot{p}_{1,2} = \mp\frac{1}{2}q_{1,2}(q_1^2 - p_1^2)$. This system is a reduction of a special case of the Euler–Lagrange equations associated with the variational problem $\int F^2$, where F is the curvature of the connection $\nabla = \mathrm{d} - \sum A_\mu \mathrm{d}x^\mu$, $A_\mu \in su(2)$, of a trivial 2-dimensional bundle over the Minkowski space (see [**Zig**]).
If we put $q_2 = p_2 = 0$, then we obtain the invariant algebraic curve $\Gamma: \ p_1^2 = 2H - q_1^4/4$. This curve is an elliptic curve. It is the Riemann surface of the function $p_1 = \sqrt{2H - q_1^4/4}$ with four ramification points $\pm\sqrt[4]{8H}$, $\pm\mathrm{i}\sqrt[4]{8H}$. It is a doubly punctured torus. The closure $\overline{\Gamma} \subset \mathbb{CP}^2$ means adding the point $[0:1:0]$ at infinity. $\overline{\Gamma}$ has the double point singularity at $[0:1:0]$. When we resolve this singularity, then we obtain a smooth torus $\widetilde{\Gamma}$.
The generators of the fundamental group $\pi_1(\widetilde{\Gamma})$ can be chosen in the following way. α is the lift to the Riemann surface of $p_1(\cdot)$ of the loop in the q_1-plane, which surrounds the two points $-\sqrt[4]{8H}$ and $\sqrt[4]{8H}$; β is the lift of the loop, which surrounds $\sqrt[4]{8H}$ and $\mathrm{i}\sqrt[4]{8H}$.
The solution $\phi(t)$, associated with Γ, is expressed by means of elliptic functions. We have the formula

$$t = \int^{q_1} \left[2H - x^4/4\right]^{-1/2} \mathrm{d}x,$$

which defines $\phi(t)$ and the covering of the curve $\widetilde{\Gamma}$ by means of the complex plane, $\widetilde{\Gamma} = \mathbb{C}/\Lambda$.
We replace the doubly punctured torus Γ by a once punctured torus. We use the symmetry $\Pi: (q_1, p_1) \longmapsto (-q_1, -p_1)$. The above Hamiltonian system is well defined on the quotient space $\mathbb{C}^4/\Pi$, and the curve Γ is replaced by $\Gamma' = \Gamma/\Pi$. The symmetry extends to a symmetry of $\widetilde{\Gamma}$ and transposes the punctures. The quotient $\widetilde{\Gamma}' = \widetilde{\Gamma}/\Pi$ is a torus too, with the basic cycles α', β', such that $\Pi_*\alpha = 2\alpha'$, $\Pi_*\beta = \beta'$. The commutator $[\alpha', \beta']$ is a loop around the puncture in $\widetilde{\Gamma}' - \Gamma'$. We have also $\widetilde{\Gamma} = \mathbb{C}/\Lambda'$.

The lattice Λ' is generated by the periods, the integrals $T_1 = \frac{1}{2}\int_\alpha \mathrm{d}q_1/p_1$ and $T_2 = \int_\beta \mathrm{d}q_1/p_1$. Note also that $T_1 > 0$ and the cycle α' can be identified with the segment $[0, T_1]$ (in the covering of $\widetilde{\Gamma}'$ by $\mathbb{C}$). Thus the 'time' along α is real.
Let A and B be the variation monodromy maps associated with α' and β' respectively. Their calculations involve integration of the normal variation system, which takes the form

$$\dot{x_2} = y_2, \quad \dot{y_2} = (1/2)q_1^2(t)x_2, \quad \dot{h} = 0.$$

Here x_2, y_2 are the variations of q_2, p_2 and $h = \Delta H$ is the variation of the Hamilton function. The symplectic operators A_0, B_0 (associated with A, B) are the operators expressed in the variables x_2, y_2.
One can see that the evolution operators $g_s^t : (x_2, y_2)(s) \longmapsto (x_2, y_2)(t),\ s, t \in \mathbb{R}$ (along solutions), are matrices with positive entries. Thus A_0 has positive entries. By the Perron–Frobenius theorem, A_0 has one simple positive eigenvalue, whose eigenvector has positive components. (Proof: A_0 transforms the first quadrant $x_2 > 0, y_2 > 0$ into its strictly proper subset.) Thus A_0 has two different eigenvalues λ, λ^{-1}. This shows that the operator A is non-resonant.
If the Yang–Mills system had additional meromorphic first integral, then, by Theorem 10.44, the operator B_0 (associated with B) should either preserve the eigendirections of A_0, i.e., should commute with A_0, or should exchange these eigendirections. Simple calculations in the basis, where A_0 is diagonal, show that the commutator $A_0B_0A_0^{-1}B_0^{-1} = \mathrm{diag}\,(\lambda^2, \lambda^{-2})$ or $= \mathrm{diag}\,(1,1)$.
On the other hand, $A_0B_0A_0^{-1}B_0^{-1}$ is a monodromy map corresponding to the loop surrounding the puncture in Γ', i.e., the loop around the point at infinity. Putting $z = 1/x$, we obtain the equation $t = -\int_0^{1/q_1}(-1 + 8Hz^4)^{-1/2}\mathrm{d}z$, which gives the solutions $q_1(t) = -2\mathrm{i}/t + O(1/t^2)$ along Γ'.
The normal variation system takes the form $\ddot{x_2} + (2/t^2 + \ldots)x_2 = 0$. It is the equation of the Fuchs class with the Levelt's exponents $\rho_{1,2} = (1 \pm \mathrm{i}\sqrt{7})/2$. The eigenvalues of the monodromy transformation are $-\mathrm{e}^{\pm\pi\sqrt{7}}$. They are not positive, as it was supposed.
We have proven the following result.

10.47. Theorem. *The Yang–Mills system, considered in a domain containing the origin, does not have any meromorphic first integral functionally independent of the Hamilton function.*

10.48. Remarks. 1. Another example, which is treated equally simply as the Yang–Mills case, is the Hénon–Heiles system with the Hamilton function

$$H = \frac{1}{2}(p_1^2 + p_2^2 + q_1^2 + q_2^2) + \frac{1}{3}q_1^3 - q_1q_2^2.$$

We advise the reader to repeat the proof of Theorem 10.47 in this case.

2. Other methods used in the proof of non-integrability of Hamiltonian systems are following (see [**AKN**]).

For perturbations of integrable systems, one finds obstructions to the integrability at the level of formal series expansions; there is no formal first integral. Isolated periodic solutions also say about non-integrability. Sometimes one shows a chaotic character of the system. For example, separatrices of some singular point, instead to form a loop, intersect one another and generate a kind of Smale's horseshoe.

Chapter 11

The Galois Theory

The Differential Galois Theory is a generalization of the Galois Theory. The fields are fields of functions and extensions of fields are defined by means of solutions of linear ordinary differential equations. The differential Galois group is the group of symmetries of the space of solutions.
The monodromy operators introduced in Chapter 9 belong to the differential Galois group. In fact, the monodromy group forms a kind of topological Galois group associated with a linear differential equation. These groups are related one with another (L. Schlesinger).
We present the basic results from the theory of some special extensions of differential fields, the theory of Picard–Vessiot extensions. We apply them to the problem of integration of polynomial vector fields (M. Singer's theorem).
We present the monodromy theory of algebraic functions, with the topological proof of the Abel–Ruffini theorem. We describe the A. Khovanski's generalization of the monodromy group to large class of functions. Finally, we discuss the monodromy properties of the Singer's first integrals.

§1 Picard–Vessiot Extensions

11.1. Introduction. The theory of differential fields is the differential analogue of the theory of number fields. Its origins lie in the papers of E. Picard [**Pic3**] and in the thesis of E. Vessiot [**Ves**].
The number fields are the fields generated by solutions of algebraic equations. Usually the solutions (roots) are indistinguishable, i.e., they are subject to permutations. The latter induce automorphisms of the extension field, called the Galois group; it is the number theoretical Galois group. The solution of algebraic equations in radicals (i.e., by successive root extractions) means solvability of the number theoretical Galois group.
The differential fields, called the Picard–Vessiot extensions, are defined by means of solutions of systems of differential equations. Multivaluedness of these solutions leads to action of certain Lie group, called the differential Galois group. The analogue of the number theoretical radicals form the operations of adjoining of integrals and of exponents of integrals to the initial field of rational functions; i.e., the integration in quadratures. The solvability of a differential system in quadratures means the solvability of its differential Galois group.
In this section we present main elements of the Differential Galois Theory. Unfortunately, the shortest proofs of the main results (existence of Picard–Vessiot

H. Żołądek, *The Monodromy Group*, Monografie Matematyczne 67,
https://doi.org/10.1007/978-3-031-91270-2_11

extensions, algebraicity of the Galois group, the one-to-one correspondence: subfield $\longleftrightarrow$ closed subgroup) are algebraic. Other results (solvability in quadratures and examples) are more natural.

In our exposition we omit many interesting subjects from differential algebra. For example, we do not investigate the inverse Galois problem, i.e., whether there exists a Picard–Vessiot extension with the Galois group equal to a given linear algebraic group.

In the presentation we follow the books of I. Kaplansky [**Kapl**] and A. Magid [**Mag**].

In this version of the book I have added some new material. One of the is the Differential Galois Theory treatment of the Painlevé equations. Another is a whole section devoted to the center-focus problem.

11.2. Definition (Differential fields and their Picard–Vessiot extensions). A **differential field** $K = (K, \partial)$ is a (commutative) field (with the operations of sum and product, with the 0 and the 1), equipped with the **derivation** $\partial : K \longmapsto K$ satisfying the Leibnitz rule $\partial(a+b) = \partial a + \partial b$, $\partial(ab) = (\partial a)b + a(\partial b)$. Sometimes, instead of one derivation, K is equipped with a system $\Delta = \{\partial_1, \ldots, \partial_r\}$ of derivations. We shall use also notations $\partial a = a'$, $\partial^k a = a^{(k)}$. Analogously, one defines the **differential ring** R (over some differential field) and the **differential ideal.**

It is useful to think about a differential field as about some field of holomorphic functions of $x \in \mathbb{C}$ (with singularities) and $\partial = \frac{\mathrm{d}}{\mathrm{d}x}$.

The subfield $C = C_K$ consisting of elements annihilated by derivation(s) is called the **field of constants of** K.

The **homomorphisms** of differential fields (or rings) are the homomorphisms of algebraic fields (rings) commuting with derivation(s).

Let $u \notin K$ be an element from some large differential field $\widetilde{K}$ containing K. Then one defines the *extension of* K *by adjoining* u as $K\langle u\rangle = K(u, u', u'', \ldots)$, $K\langle u\rangle \subset \widetilde{K}$. It can be a finite extension (in the algebraic sense) or a transcendental extension (with finite or infinite transcendental degree).

For extensions of differential fields we use the notations $K \subset M$.

Let

$$D = \partial^n + a_{n-1}\partial^{n-1} + \ldots + a_0, \quad a_i \in K,$$

be a linear 'differential operator' of order n and let $y_1, \ldots, y_n$ be a basis of solutions of the equation $Dy = 0$. It is assumed that y_j lie in some large differential field $\widetilde{K}$. Consider the extension $M = K\langle y_1, \ldots, y_n\rangle$. We say that $K \subset M$ is the **Picard–Vessiot extension of** K **associated with the equation** $Dy = 0$ iff:

(i) the field M does not contain new constants, $C_M = C_K = C$;

(ii) y_i are linearly independent over the field of constants.

The latter means that the **Wronskian**

$$W = W(y_1, \dots, y_n) = \begin{vmatrix} y_1 & y_2 & \dots & y_n \\ y_1' & y_2' & \dots & y_n' \\ \dots & \dots & \dots & \dots \\ y_1^{(n-1)} & y_2^{(n-1)} & \dots & y_n^{(n-1)} \end{vmatrix}$$

is nonzero (in M).
We recall that the Wronskian of any fundamental system of solutions satisfies the linear equation

$$W' + a_{n-1}W = 0.$$

The equivalent definition of Picard–Vessiot extensions uses, instead of the linear 'differential equations of nth order', a system of 'linear differential equations' (or a 'connection') $Dy = (\partial - A)y = 0$, where y is n-dimensional vector and A is $n \times n$ matrix with coefficients in K. Then $M = K(y_{ij})$, where $\mathcal{F} = (y_{ij})$ is a fundamental matrix of the latter system. Of course, we need also $C_K = C_M$.

11.3. Examples. (a) Of course, the number theoretical fields with trivial derivations (i.e., $\partial a = 0$) are examples of differential fields.
The principal natural examples of differential fields constitute the fields: of rational functions of one variable

$$(\mathbb{C}(x), \mathrm{d}/\mathrm{d}x)$$

(or $(\mathbb{C}(x_1, \dots, x_m), \{\partial/\partial x_1, \dots, \partial/\partial x_m\})$) and of germs of meromorphic functions in $(\mathbb{C}, 0)$

$$(\mathcal{M}_0(\mathbb{C}), \mathrm{d}/\mathrm{d}x)$$

(or $(\mathcal{M}_0(\mathbb{C}^m), \{\partial/\partial x_1, \dots, \partial/\partial x_m\})$). (Here the field of germs of meromorphic functions is the field of quotients $Q(\mathcal{O}_0(\mathbb{C}^m))$of the local differential ring $\mathcal{O}_0(\mathbb{C}^m) = \mathbb{C}\{x_1, \dots, x_m\}$; for $m = 1$ one uses also the notation $\mathbb{C}\{x\}[x^{-1}]$.)
The system of independent solutions of a higher order linear differential equation with rational coefficients (or of a first order differential system with rational coefficients) defines the Picard–Vessiot extension of $\mathbb{C}(x)$. Here we have automatically the existence and uniqueness of these extensions.

(b) We present an example showing, why the assumption 'no new constants' in the definition of Picard–Vessiot extensions is necessary.
Let $K = (\mathbb{C}(\mathrm{e}^x), \mathrm{d}/\mathrm{d}x)$ and let u be a *formal* variable whose derivative is equal to u, $u' = u$. We define $M = K\langle u\rangle$. Of course, u is an independent solution of the linear equation $Dy = 0$, $D = \partial - 1$. However, the element e^x/u is a new constant in M, $(\mathrm{e}^x/u)' = 0$.

11.4. Definition (differential Galois group). Let $K \subset M$ be a Picard–Vessiot extension. The group of automorphisms of the differential field M, which are identity on K, is called the **differential Galois group of the extension** $K \subset M$ and is denoted by $\mathrm{Gal}_K M$.

The differential Galois group is a subgroup of the group $GL(V,C) \simeq GL(n,C)$, where V is the space of solutions of the equation (or system) $Dy = 0$. Indeed, let y_i, $i = 1, \ldots, n$, be a basis of solutions and let $\sigma \in \mathrm{Gal}_K M$. Each element σy_j is a solution and is expressed as a linear combination of y_i's, $\sigma y_j = \sum_i y_i d_{ij}$, where the entries $d_{ij} \in M$ are given by the Cramer's formula $d_{ij} = W_1/W_2$ and W_1 and W_2 are the Wronskians of suitable systems of n solutions. Both satisfy the same differential equation $W' + a_{n-1}W = 0$. Thus the derivative of their ratio is equal to zero, which shows that d_{ij} are constants.
Note also that the matrix (d_{ij}) acts on the right on the row vector $(y_1, \ldots, y_n)$. It acts in the same way onto the derivatives $(y_1^{(l)}, \ldots, y_n^{(l)})$.

Agreement. In what follows, we assume that all differential fields are extensions of the field $\mathbb{C}$ consisting of constants. We shall have $C = \mathbb{C}$, which is *algebraically closed and of characteristic zero.* Any eventual exceptions will be underlined.

11.5. Adjoining of an integral. Let $a \in K$ be an element which is not a derivative of another element from K; for example, x^{-1} in $\mathbb{C}(x)$. We want to adjoin to K an element u such that $u' = a$, i.e., we put $M = K\langle u\rangle = K(u)$. We can here treat u as a formal variable with its derivation equal to a. We claim that:

$K(u)$ is a transcendental Picard–Vessiot extension of K with the differential Galois group isomorphic to the additive group C.

Here the notion *transcendental* means that u cannot satisfy any algebraic equation with coefficients in K. Indeed, if $u^n + bu^{n-1} + \ldots = 0$ would be such equation of minimal degree, then its derivation would give the equation of lower degree $(na + b')u^{n-1} + \ldots = 0$. Thus there should be $na + b' = 0$, which contradicts the assumption that a is not derivative.
If M would contain a new constant, represented as rational function of u, $f(u)/g(u)$, then we should have $(f(u))'g(u) - f(u)(g(u))' = 0$, i.e., an algebraic equation for u. Thus M satisfies the first condition in the definition of the Picard–Vessiot extension.
Consider the operator

$$D = \partial^2 - (a'/a)\partial.$$

It is clear that the elements $1, u$ are solutions of the equation $Dy = 0$. Their Wronskian $W(1,u) = a \neq 0$, they are linearly independent. So, $M = K\langle 1, u\rangle$ is Picard–Vessiot.
We see that M, treated as an algebraic field, is the same as $K(u)$, the field of rational functions of u. The formula $u' = a$ provides it with a structure of a differential field.
If $\sigma \in \mathrm{Gal}_K M$, then $\sigma(1) = 1$ and $v = \sigma(u)$ satisfies $v' = \sigma(u') = \sigma(a) = a$. Thus $v - u$ is a constant $c \in C$ and

$$\sigma(u) = u + c.$$

Conversely, any change $u \longmapsto u+c$, $c \in C$, defines an automorphism of the differential ring $K[u]$: $u^k \longmapsto (u+c)^k = u^k + \ldots$, $u' \longmapsto (u+c)'$. This automorphism extends to an automorphism of the field of quotients.

Example. The Galois group of the extension $\mathbb{C}(x) \subset \mathbb{C}(x, \ln(x-x_0))$ is equal to $\mathbb{C}$.

11.6. Adjoining of exponent of an integral. Here $M = K\langle u\rangle = K(u)$, where u satisfies the differential equation $Dy = 0$ with

$$D = \partial - a, \quad a \in K,$$

and with the condition $C_M = C_K$.

The extension $K \subset K(u)$ is a Picard–Vessiot extension with the differential Galois group isomorphic to a subgroup of the multiplicative group $C^* = C \setminus 0$.

The Picard–Vessiot property here is obvious (by definition).
For $\sigma \in \mathrm{Gal}_K M$ and $v = \sigma(u)$ we have $v' = av$. This implies $(v/u)' = 0$ or $v = cu$, $c \in C$, $\sigma : u \longmapsto cu$. Because σ is an automorphism, $c \neq 0$.

Examples. 1. The equation $\frac{\mathrm{d}y}{\mathrm{d}x} = \alpha\frac{y}{x}$ has solutions cx^α. If α is integer, then the above construction gives trivial Picard–Vessiot extension over $\mathbb{C}(x)$. If $\alpha = \frac{p}{q}$ is rational, then the extension is algebraic with the differential Galois group equal to a subgroup of the complex torus $\mathbb{C}^*$, of roots of unity of order q. If α is irrational, then $\mathrm{Gal}_K M = \mathbb{C}^*$.

2. The equation $\frac{\mathrm{d}y}{\mathrm{d}x} = -2xy$ has the solutions ce^{-x^2}. It is Picard–Vessiot extension over $\mathbb{C}(x)$ with the differential Galois group equal $\mathbb{C}^*$.

Now we pass to the presentation of the main general theorems of the Differential Galois Theory. Although their proofs are algebraic, they have the advantage that the results are obtained quickly.

11.7. Theorem (Existence and uniqueness of the Picard–Vessiot extensions). ([**Kol**]) *Let (K, ∂) be a differential field and let D be a linear differential operator with coefficients in K. Then there exists a Picard–Vessiot extension of K associated with $Dy = 0$. Moreover, this extension is unique up to an isomorphism.*

Proof. In fact, in our applications to linear differential equations in complex domain, there is no need to provide the algebraic proof. However, it may occur that the coefficients $a_i \in K$, $a_i = a_i(x)$, have big set of singular points (e.g., dense in $\mathbb{C}$); in that case the classical theorems do not apply. Moreover, the algebraic construction of the Picard–Vessiot extension will be used in proofs of other statements (e.g., of the algebraicity of the Galois group). So, we present this proof. We follow the book of A. Magid [**Mag**].

1. *Existence.* Let us introduce n^2 formal variables

$$u_{1,1}, \ldots, u_{n,1}, \quad u_{1,2}, \ldots, u_{n,2}, \quad u_{1,3}, \ldots, u_{n,n},$$

and denote $u_1 = u_{1,1}, \ldots, u_n = u_{n,1}$. Define the differential ring over K as $K[u] = K[u_{1,1} \ldots, u_{n,n}]$ with the derivation given by $u'_{i,j} = u_{i,j+1}$ and $u'_{i,n}$ defined from the equation $Du_i = 0$. In other words,

$$u' = Au,$$

where $u = (u_{ij})$ and $A \in gl(n, K)$ are matrices.
Let $W(u) = W(u_1, \ldots, u_n)$ be the Wronskian. Define the ring

$$S = K[u][W(u)^{-1}].$$

The ring S has the property that u_i are solutions of the equation $Dy = 0$ and they are independent (in S over the field of constants C_K). The latter is a consequence of the fact that the Wronskian is invertible in S. (In fact, the ring S is the same as the ring $K[X]$ of regular function on the algebraic group $X = GL(n, K)$, treated as an affine algebraic variety over the field K).
One would like to define a differential field by taking the field of quotients of S. However, some new constants would be created in such way. It turns out that one has to divide the ring S by a proper maximal (in the partial order defined by inclusion) ideal I of S and then define the field M as the field of its quotients

$$M = Q(S/I) = Q(K[X]/I). \tag{11.1}$$

Define also $y_i = [u_i] = u_i + I \in M$, i.e., as the cosets of $u_i = u_{i,1}$. They form a basis of solutions of the equation $Dy = 0$ in M. Before proceeding with the further proof, we recall some algebro-geometrical facts and present two examples.

2. **Remark.** Recall that a *divisor of zero* of a ring R is such element $a \neq 0$ that $ab = 0$ for some $b \neq 0$. If R does not contain divisors of zero, then it is called the *integral domain.* The field of quotients $Q(R)$ is defined only when R is an integral domain: then $(a/b) \cdot (c/d) = (ac)/(bd)$ with nonzero denominator. An ideal I is *prime* iff from $ab \in I$ it follows that either $a \in I$ or $b \in I$ (or the quotient ring does not have divisors of zero). Example: in the ring $\mathbb{C}[x_1, x_2]$ the prime ideals correspond to irreducible algebraic varieties; the ideal $(x_1 x_2)$ is not prime. The ideal $I \subset R$ is *proper* iff $0 \neq I \neq R$.
When we have a ring of regular functions on an affine algebraic variety (over an algebraically closed field of characteristic zero), then its maximal prime ideals are in one-to-one correspondence with the points of the variety; (they define the closed points in the spectrum of the ring, an affine K-scheme $\mathrm{spec}_K R$, see [**Bor**], [**HaR**]).
We will also need the notion of *constructible subsets* Y of an affine algebraic variety X (over K) with the Zariski topology. The *Zariski topology* is generated by the open subsets of the form $\{g(x) \neq 0\}$, where $g : X \longmapsto K$ is a regular function (polynomial). The constructible sets are finite unions of *locally closed subsets* Y_i, i.e., such that they are open in their closures $\overline{Y}_i$ (see [**Bor**], Ch. AG, point 1.3). In particular, if $Y \subset K^1$ is constructible, then either it is finite (Zariski closed) or is Zariski open ($Y = K \backslash \{\text{finite points}\}$). If $\alpha : X \longmapsto Z$ is a morphism of

affine algebraic varieties, then the image $\alpha(Y)$ of a constructible subset Y is a constructible subset of Z (theorem of Chevalley, see [**Bor**], Ch. AG, point 10.2). In the case of differential ring the situation is not as clear as in the case of algebraic ring. Firstly, the underlying fields usually are not algebraically closed (e.g., $\mathbb{C}(x)$) and the assumption, that the ideal is closed with respect to the differentiation, constitutes sometimes a serious restriction.

3. **Example.** Consider the extension of $K = \mathbb{C}(x)$ by means of the equation $2xy' = y$, i.e., adjoining of $\sqrt{x}$. We have $X = \mathbb{C}^*$, $K[X] = \mathbb{C}(x)[u, u^{-1}]$ and the ideal I can be chosen as $(u^2 - x)$. Due to the fact that $\mathbb{C}(x)$ is not algebraically closed, the ideal I is maximal. If $\overline{K}$ is the algebraic closure of K (i.e., the field of all algebraic functions of x), then the ideal $\overline{K} \otimes I$ is no longer prime neither maximal in $\overline{K}[X]$. Its zero set consists of two point $\{\sqrt{x}, -\sqrt{x}\} \subset \overline{\mathbb{C}(x)}^*$.
In the case of extension of $K = C(x)$ by means of the equation $xy'' + y' = 0$, i.e., adjoining of $\ln x$, the corresponding ring $K[X]$, $X = GL(2, K)$, does not contain nontrivial prime ideals.

The integrality of $K[X]/I$ and the nonexistence of new constants is a consequence of the following propositions.

4. **Proposition.** *Let S be a finitely generated differential ring, without divisors of zero over a differential field K (i.e., a differential integral domain), and let $I \subset S$ be a maximal differential ideal ($\neq S$). Then the ideal I is prime, i.e., the ring $R = S/I$ has no divisors of zero.*

Proof. The ring R has the property that it does not contain proper ideals.
Assume that $ab = 0$ for $a, b \in R \setminus 0$. Then the identities $(ab)'b = a'b^2 + abb' = a'b^2$ show that $a'b^2 = 0$. Generally $a^{(k)}b^{k+1} = 0$. Take the differential ideal $I_1 = (a, a', \ldots)$ of R. For any $e \in I_1$ we have $eb^m = 0$ for some m. If all $b^m \neq 0$, then $1 \notin I_1$ and I_1 would be a proper ideal. This shows that any zero divisor (e.g., b or a) is nilpotent. In particular, $a^n = 0$ for some minimal n and the formula $na^{n-1}a' = 0$ shows that a' is also zero divisor (and also nilpotent). Repeating this we see that $a^{(j)}$ are all nilpotent zero divisors. Thus a generates ideal I_2 consisting of only nilpotent elements. Because the latter does not contain 1, it should be proper. □

5. **Proposition.** *Let R be a finitely generated differential integral domain over a differential field K (with algebraically closed field of constants C of zero characteristic) without proper differential ideals. Then the field of quotients $M = Q(R)$ does not have new constants, $C_M = C$.*

Proof. (a) First, we notice that the elements from $C_M \setminus C$ cannot be algebraic over K. It follows from the fact that the differentiation in K extends itself uniquely to the algebraic closure $\overline{K}$ of K. If $d \in \overline{K} \setminus K$ satisfies a minimal algebraic equation $p(d) = d^r + a_{r-1}d^{r-1} + \ldots + a_0 = 0$, $p(x) \in K[x]$, then $d' = -p'(d)/\frac{\mathrm{d}p}{\mathrm{d}x}(d)$, where $p'(x) = a'_{r-1}x^{r-1} + \ldots + a'_0$. Thus $d' = 0$ implies $p'(d) = 0$ ($p(x)$ is minimal); so, $p \in C[x]$ and $d \in C$.

(b) Next, we have $C_M \subset R$. Indeed, for any $d = f/g \in C_M$, $f, g \in R$, consider the ideal of denominators of d, $J = \{h \in R : hd \in R\} \subset R$. It is non-zero differential ideal, because $g \in J$ and $h'd = (hd)' \in R$ for $h \in J$.
By assumption, R does not contain proper differential ideals. Thus $J = R$, which means that $d = 1 \cdot d \in R$.

(c) Here we show that, for any $d \in C_M$, there exists an element $c \in C$ such that $d - c$ is not invertible in R. Then the ideal $(d - c) = (d - c)R$ is different from R and therefore it is equal to zero. Thus $d = c \in C$.
We use some methods from algebraic geometry. We replace the field K (of coefficients) by its algebraic closure $\overline{K}$ and the ring R (over K) by $\overline{R} = R \otimes \overline{K}$ (over $\overline{K}$). We want to show that the element $d \otimes 1 - c \otimes 1 \in \overline{R}$ is not invertible for some $c \in C$. Then, of course, also the element $d - c$ will be non-unit in R.
The element $d \otimes 1$ (which we still denote by d) can be treated as a regular function on the space $Y = \mathrm{Spec}_{\overline{K}} \overline{R}$ of maximal ideals of the ring $\overline{R}$, $d : Y \longmapsto K$. Here Y is an affine algebraic variety (equipped with the Zariski topology) and $d(\cdot)$ is a morphism of algebraic varieties. Consider the image $Z = d(Y)$. It is a constructible subset of K (by the theorem of Chevalley, see 2). We have two possibilities: either $\overline{Z} \neq \overline{K}$ or $\overline{Z} = \overline{K}$.
In the first case Z is finite, which means that the function $d(\cdot)$ takes finite number of values. Because Y is irreducible (as R is integral domain), it is a connected space and $d(\cdot) = \mathrm{const}$ (a contradiction with (a))
In the second case Z is Zariski open, equal to $\overline{K} \setminus \{\text{finite set}\}$, and there exists a point $c \in C \cap Z$. The variety $Y_c = d^{-1}(c)$ is a proper subvariety of Y (i.e., $\neq \emptyset, Y$) corresponding to the nonzero ideal $(d - c)\overline{R} = \{f \in \overline{R} : f|_{Y_c} \equiv 0\}$. Thus $d - c$ is a non-unit in $\overline{R}$. □

6. *Uniqueness of the Picard–Vessiot extension.* We have one Picard–Vessiot extension in the form $M_1 = Q(S/I)$. Assume that M_2 is another Picard–Vessiot extension associated with the same equation $Dy = 0$.
We define the ring

$$\widetilde{S} = (S/I) \otimes_K M_2$$

over K; it is the change of the field of coefficients and $\widetilde{S}$ can be treated as a finitely generated algebra over M_2 as well. If $\tilde{I}$ is a maximal ideal in $\widetilde{S}$, then it is prime (Proposition 5) and the field of quotients $M = Q(\widetilde{S}/\tilde{I})$ is an extension of K with the same field of constants (Proposition 6). We have the homomorphisms $\sigma_i : M_i \longmapsto M$, $i = 1, 2$, induced by $s \longmapsto s \otimes 1$, $t \longmapsto 1 \otimes t$.
If V_i, V are the (linear) spaces of solutions of the equation $Dy = 0$ in M_i, M respectively, then $\sigma_i : V_i \longmapsto V$ are embeddings. The equality of dimensions of these spaces over C shows that σ_i realize isomorphisms of the spaces of solutions and hence isomorphisms between the fields M_i and the subfield $K\langle V\rangle \subset M$.
The homomorphisms $\sigma_2^{-1}\sigma_1$ is the isomorphism between M_1 and M_2. □

The algebraic construction of the Picard–Vessiot extension allows to prove easily the following two properties of the differential Galois group.

11.8. Theorem (Normality of Picard–Vessiot extension).

(a) *Let $K \subset M$ be a Picard–Vessiot extension and let $x \in M \setminus K$. Then there exists $\sigma \in \mathrm{Gal}_K M$ such that $\sigma(x) \neq x$.*

(b) *Let $K \subset L \subset M$ be extensions of differential fields, where $K \subset L$ and $K \subset M$ are Picard–Vessiot. Then any $\sigma \in \mathrm{Gal}_K L$ can be extended to an automorphism of M.*

Proof. (a) Take the model $M = Q(S/I)$ with $S = K[u, W(u)^{-1}]$, I – maximal ideal (as in the proof of Theorem 11.7). We have $x = a/b$, $a, b \in S/I$; so, x belongs to the ring $(S/I)[b^{-1}]$. We consider the differential algebra $R = (S/I)[b^{-1}] \otimes_K (S/I)[b^{-1}] \subset M \otimes_K M$. Put

$$z = x \otimes 1 - 1 \otimes x \in R.$$

Because $x \notin K$, $z' \neq 0$ and $z^j \neq 0$, $j = 1, 2, 3, \ldots$; if $z^n = 0$ for a minimal such n, then $0 = nz^{n-1}z' \neq 0$. Take the differential ring $R_z = \{v/z^i : v \in R, i \geq 0\}$, some its maximal prime ideal J and the quotient R_z/J. Note that the element $[z/1]$ is nonzero in R_z/J.

The fields of quotients $Q(R)$ (containing M) and $Q(R_z/J)$ are two models for the Picard–Vessiot extension with the same fields of constants. The two maps $w \longmapsto w \otimes 1$ and $w \longmapsto 1 \otimes w$ define two embeddings σ_1 and σ_2. Moreover, $\sigma_1(M) = \sigma_2(M)$ (because this equality holds for the vector spaces of solutions of the differential equation $Dy = 0$) and $\sigma = \sigma_1^{-1}\sigma_2$ is an automorphism of M. Because $\sigma_1(x) - \sigma_2(x) = [z/1] \neq 0$, we have $\sigma(x) \neq x$.

(b) To show this, one can use the representation $M = Q(S_1/I_1)$, where $S_1 = S \otimes_K L$ and I_1 a maximal prime ideal. The extension is induced by $id \otimes \sigma$. □

11.9. Kolchin's theorem (algebraic structure of Galois group). ([**Kol**]) *If $K \subset M$ is a Picard–Vessiot extension, then its Galois group $\mathrm{Gal}_K M$ is isomorphic to an algebraic subgroup of $GL(n, C)$.*

Proof. We use the representation of the Picard–Vessiot extension in the form (11.1), i.e., $Q(S/I)$, $S = K[u][W(u)^{-1}]$. Because any $\sigma \in \mathrm{Gal}_K M$ preserves the space V of solutions of the equation $Dy = 0$ with the basis $u_1, \ldots, u_n$, we obtain the representation of $\mathrm{Gal}_K M$ in the ring S.

The ring S can be represented as $S = K[X] = K \otimes C[X]$, where $C[X] = C[u_{ij}, W^{-1}]$ is the affine coordinate ring of the affine algebraic variety $X = GL(n, C)$. Because $GL(n, C)$ acts on X by means of right translations, we obtain the induced action of $GL(n, C)$ on the ring $K \otimes C[X]$; i.e., $GL(n, C)$ acts polynomially on this vector space over C (which can be infinite dimensional). The latter action, restricted to the Galois group, coincides with the action of $\mathrm{Gal}_K M$ described before.

If $\sigma \in \mathrm{Gal}_K M$ is represented by a matrix $d_\sigma = (d_{ij}) \in X$, then the formulas $\sigma(y_i^{(l)}) = \sum y_j^{(l)} d_{ji}$ and $u_{i,l} \longmapsto \sum u_{j,l} d_{ji}$, where $y_j^{(l)} = u_{j,l} \pmod{I}$, agree. If $f \in K[X]$ is treated as a function of u, $f = f(u)$, then $\sigma(f) = f \circ d_\sigma = f(ud_\sigma)$.

This means that the Galois group can be defined as

$$\mathrm{Gal}_K M = \{\sigma \in GL(n,C) : \sigma(I) = I\}.$$

Thus $\mathrm{Gal}_K M$ is a stabilizer of the vector subspace I (of the vector space S). It is well known that such groups are algebraic (C. Chevalley).
Using C-bases: $(f_1, f_2, \ldots)$ of I and $(\varphi_1, \varphi_2, \ldots)$ of $\mathrm{Ann}(I) \subset S^\vee$ (functionals vanishing on I) one can write down the equations for $\mathrm{Gal}_K M$ in $GL(n,C)$: $\langle \varphi_i, \sigma(f_j)\rangle = \langle \varphi_i, f_j \circ d_\sigma\rangle = 0$.
(Another approach, by P. Deligne [**Del5**], to the proof of algebraicity of the differential Galois group goes through the Tannakian categories.) □

11.10. Theorem (Subgroups of the Galois group). ([**Kol**], [**Mag**]) *Assume that we have extensions of differential fields $K \subset L \subset M$, where the extension $K \subset M$ is Picard–Vessiot. Then:*

(a) *The extension $L \subset M$ is Picard–Vessiot, too.*

(b) *The extension $K \subset L$ is Picard–Vessiot iff $H = \mathrm{Gal}_L M$ is a normal subgroup of $G = \mathrm{Gal}_K M$; in this case $\mathrm{Gal}_K L = G/H$.*

Proof. (a) The fields of constants are the same and the independent solutions $y_j \in M$ of the equation $Dy = 0$ (with the coefficients in K) constitute independent solutions of the same equations, treated as having coefficients in L. So, $L \subset M$ is Picard–Vessiot with differential Galois group H.
Thus the group H forms a subgroup of G, consisting of those σ which are identity on L. By the previous theorem, H is a Zariski closed subgroup of the algebraic group G.
(b) If, additionally, the extension $K \subset L$ is a Picard–Vessiot extension (associated with an equation $D_1 y = 0$), then elements of G preserve the space of solutions of $D_1 y = 0$. It means that the field L is invariant for G. This implies that H is a normal subgroup of G. Indeed, if $\sigma \in G$ and $\tau \in H$, i.e., $\tau|_L = id$, then $\sigma^{-1}\tau\sigma|_L$ is the composition $L \longmapsto L \overset{id}{\longmapsto} L \longmapsto L$ and is equal to identity. Moreover, we get the homomorphism from G/H to $\mathrm{Gal}_K L$, $\sigma \longmapsto \sigma|_L$. It is clear that this homomorphism is a monomorphism.
On the other hand, by Theorem 11.8(b), any automorphism τ of L (identical on K) can be prolonged to an automorphism of M.
So, we have the homomorphism from $\mathrm{Gal}_K L$ to G/H which, together with the previous, gives the isomorphism between $\mathrm{Gal}_K L$ and G/H.

Proof of the implication: *H is normal subgroup $\Rightarrow$ $K \subset L$ is Picard–Vessiot.* The proof of the Picard–Vessiot property of the extension $K \subset L$ in the case when H is a normal algebraic subgroup of G, is not so elementary. It is omitted in [**Kapl**] and in [**Kol**]; there is only statement that the extension $K \subset L$ is normal, $L^H = K$. Below we present a new proof which uses some ideas from [**Mag**] and [**Zo9**].
Its main idea relies on the following construction. Assume that we have a finitely generated K-algebra $T \subset M$ (without divisors of zero) consisting of elements t such

that the linear space (over C) $\operatorname{span}\{\sigma t : \sigma \in G = \mathrm{Gal}_K M\}$ is finite dimensional. We assume also that T is G-invariant and its field of quotients $Q(T)$ is equal to M. The algebra $K[X]/I$ is a good example.
Take $T^H = \{t \in T : Ht = \{t\}\}$, the set of invariants of the action of the normal subgroup H. The normality of H means that, for any $\tau \in H$, $\sigma \in G$ and $s \in T^H$, we have $\sigma^{-1}\tau\sigma s = s$ or $\tau(\sigma s) = (\sigma s)$. This shows that any $\sigma \in G$ preserves the subdomain T^H, $\sigma(T^H) = T^H$. Thus T^H is a finitely generated G-invariant subalgebra and the restriction of the action of G to T^H coincides with the action of the quotient group G/H on T^H. We claim that:

T^H is generated (over K) by solutions of a linear differential equation with coefficients in K.

Indeed, take a finite dimensional subspace $V_1 \subset T^H$ over C, which generates T^H as a K-algebra and which is G/H-invariant. Let $z_1, \ldots, z_m$ be a basis of V_1, its Wronskian $W(z_1, \ldots, z_m) \in K \setminus 0$. Then any element z from V satisfies the equation $D_1 z = 0$, where

$$D_1 = W(\partial, z_1, \ldots, z_m)/W(z_1, \ldots, z_m).$$

The coefficients of this operator are ratios of determinants, which behave in the same way under the action of the group G. By Theorem 11.8(a), the coefficients of D_1 belong to K. Thus $T^H = K[V_1]$, where V_1 is a space of solutions of a linear differential equation with coefficients in K.

If we knew that $L = M^H = Q(T^H)$, i.e., that the field of invariants of H in M is the field of quotients of the domain of invariants in T, then we would have the proof of the Picard–Vessiot property of $K \subset L$.

Lemma 1. *If we choose $T = K[X]/I = K \otimes C[X]/I$, where $X = GL(n, C)$ and I is a maximal prime ideal, then*

$$M^H = Q(T^H).$$

Proof of Lemma 1. Take any $f \in M^H \setminus 0$. We shall strive to represent it as a ratio of invariants from T. Let $J = \{t \in T : tf \in T\} \subset T$ be the ideal of denominators of f. Since f is H-invariant, J is H-stable ($HJ = J$). Let $s \in J \setminus 0$. The elements τs, $\tau \in H$, generate a finite dimensional space Z (over C). Choose a basis $s_1, \ldots, s_p$ of Z and let $w = W(s_1, \ldots, s_p)$ be the Wronskian. Expansion of this determinant with respect to the first row shows that $w \in J$.
We have the property $\tau w = \det(\tau|_Z) \cdot w$, which means that w is a *semi-invariant* with the *weight* $\chi = \det|_Z$. (The weight is a *character* of the algebraic group H, i.e., an algebraic homomorphism from H to $GL(1, C) = C^*$.)
Let $t = wf$. It belongs to T (because $w \in J$) and is a semi-invariant with the same weight as w. So, we have the representation of f as the ratio of semi-invariants, $f = t/w$. Assume that we can find a nonzero semi-invariant u with the weight

χ^{-1}. Then we would have the desired representation of f as a ratio of invariants $f=(tu)/wu)$.
To show that such u exists, we study in detail the representation of the group H in the space T.
In fact, we consider the group H/H_0, where $H_0 = \bigcap_\omega \ker\omega$ is the intersection of kernels of all characters of H; H_0 is a normal subgroup of H. Because $\omega|_{[H,H]}=\{1\}$, the group H/H_0 is abelian. It is isomorphic to $(H/H_0)_s\times(H/H_0)_u$ (see [**Bor**], Theorem 4.7, Ch.1), where $(H/H_0)_s$ is a semi-simple group (product of finite cyclic groups and an (algebraic) torus $(C^*)^l$) and $(H/H_0)_u$ is unipotent group (isomorphic to the additive group C^q). But there are no nontrivial characters on the unipotent group $(H/H_0)_u$ (there are only transcendental ones, like $a\longmapsto \mathrm{e}^a$). This means that the group $H/H_0=(H/H_0)_s$ is *reductive* and any its representation in a vector space V is diagonalizable. (The reader can prove himself that any algebraic homomorphism from the torus C^*, or from the cyclic group $\mathbb{Z}_p$, to the unipotent group of upper triangular matrices is trivial). The space is split into weight subspaces, $V=\bigoplus_\omega V_\omega$.
The natural spaces, where the group H/H_0 acts, are $C[H/H_0]$, $C[G/H_0]$; moreover H acts on $C[G]$. We claim that

$$C[G]_{1/\chi}\neq 0.$$

Indeed, because the homomorphism $H/H_0\longmapsto G/H_0$ is injective and the homomorphism $G\longmapsto G/H_0$ is surjective, the restriction $C[G/H_0]\longmapsto C[H/H_0]$ is surjective and the homomorphism $C[G/H_0]\longmapsto C[G]$ is an embedding. But, of course, $C[H/H_0]_{1/\chi}\neq 0$ and hence $C[G]_{1/\chi}\neq 0$. □

Recall that we have the K-algebra $T=K[X]/I$, where I is a maximal prime differential ideal. Denote also $\overline{K}$ the algebraic closure of K. We shall use the following property.

Lemma 2. *We have a canonical H-equivariant isomorphism $\overline{T}=\overline{K}\otimes(K[X]/I)\simeq \overline{K}\otimes C[G]$.*

Lemma 2 allows to finish the proof of Theorem 11.10. The group H acts on the second factors in the above tensor products. Thus the non-zero component $C[G]_{1/\chi}$ gives the nonzero component $\overline{K}\otimes C[G]_{1/\chi}$. We have $\overline{T}_{1/\chi}=\overline{K}\otimes T_{1/\chi}=\overline{K}\otimes C[G]_{1/\chi}\neq 0$ and hence $T_{1/\chi}\neq 0$. □

Proof of Lemma 2. We shall prove even more. Consider the set (K-algebra) $\mathcal{T}\subset M$ consisting of such $t\in M$ that the vector space (over C) $\mathrm{span}_C Gt$ has finite dimension. We shall show that:

$\mathcal{T}=T$ and there is an H-equivariant *isomorphism $\overline{K}\otimes\mathcal{T}\simeq\overline{K}\otimes C[G]$.*

Recall that $T=K[X]$. It is clear that $T\subset\mathcal{T}$ and that $\mathcal{T}$ is a K-algebra. Moreover, repeating arguments before Lemma 1, one easily shows that:

$\mathcal{T}$ consists of those $t \in M$, which satisfy a linear differential equation with coefficients in K.

Now we prove the following important property:

$\mathcal{T}$ does not contain proper G-stable ideals.

Indeed, if J were such an ideal with $z_1, \ldots, z_m$ as its generators and $V_1 = \mathrm{span}_C(z_1, \ldots, z_m)$, then the Wronskian $w = W(z_1, \ldots, z_m)$ would be a semi-invariant for G, $\sigma w = (\det \sigma | Z_1) w$.
But for any semi-invariant $v \in M$ the ratio v'/v is an invariant, and equals some $a \in K$ (normality); so v satisfies the linear differential equation $v' - av = 0$ and hence $v \in \mathcal{T}$. Since w^{-1} is a semi-invariant, it also belongs to $\mathcal{T}$ and $1 = w^{-1}w \in J$.

The reason to take $\overline{K}$ (instead of K) is that we want to use the geometrical action of the group $G_{\overline{K}} \subset X_{\overline{K}}$ (change of the coefficients field) on the varieties $A_{\overline{K}} = \mathrm{Spec}_{\overline{K}}\overline{T}$ and $\mathcal{A}_{\overline{K}} = \mathrm{Spec}_{\overline{K}}\overline{\mathcal{T}}$. Here $\overline{T} = T \otimes \overline{K}$, $\overline{\mathcal{T}} = \mathcal{T} \otimes \overline{K}$ and $\mathrm{Spec}_{\overline{K}}\overline{T}$ equals the variety $V(\overline{I}) \subset X_{\overline{K}}$ of zeroes of the ideal $\overline{I} = \overline{K} \otimes I \subset \overline{K}[X]$.
The group $G_{\overline{K}}$ acts on $A_{\overline{K}}$ and on $\mathcal{A}_{\overline{K}}$. On $A_{\overline{K}}$ it acts by right translations (in $X_{\overline{K}} = GL(n, \overline{K})$) and hence is *effective* (with trivial stabilizers of points). It means that $A_{\overline{K}}$ is a disjoint union of $G_{\overline{K}}$-orbits, each isomorphic to $G_{\overline{K}}$ (via an H-equivariant isomorphism). What we aim is to prove that:

$\mathcal{A}_{\overline{K}} \simeq A_{\overline{K}}$ and consists of one orbit of $G_{\overline{K}}$.

It is enough to show that:

(*) *$\overline{\mathcal{T}}$ does not contain proper G-stable ideals.*

Then $\overline{\mathcal{T}}$ does not contain $G_{\overline{K}}$-stable ideals and hence $G_{\overline{K}}$ acts transitively on $\mathcal{A}_{\overline{K}}$. So $\mathcal{A}_{\overline{K}} \simeq G_{\overline{K}}/(\text{stabilizer of a point})$ and there is an injection $\overline{\mathcal{T}} \longmapsto I_{G_{\overline{K}}} = \{f : f|G_{\overline{K}} \equiv 0\}$. Since $\overline{T} \subset \overline{\mathcal{T}}$ and there is a surjection $I_{G_{\overline{K}}} \longmapsto \overline{T}$ (associated with a $G_{\overline{K}}$-orbit $B_{\overline{K}} \subset A_{\overline{K}}$), it should be $\overline{\mathcal{T}} = \overline{T} \simeq I_{G_{\overline{K}}}$.

The proof of (*) uses a representation $x = a_1 \otimes b_1 + \ldots + a_k \otimes b_k$ of an element $x \neq 0$ from a supposed ideal $J \subset \overline{\mathcal{T}} = \mathcal{T} \otimes \overline{K}$; here $a_j \in \mathcal{T}$, $b_j \in \overline{K}$ are independent over K and k is minimal. For fixed b_j's the set $J_1 = \{c_1 : c_1 \otimes b_1 + \ldots + c_k \otimes b_k \in J$ for some $c_2, \ldots, c_k\}$ is an ideal in $\mathcal{T}$; thus, $J_1 = \mathcal{T}$ (see above). Therefore there exists an element $y = 1 \otimes b_1 + \ldots + c_k \otimes b_k \neq 0$ in J. But $\sigma(y) - y = (\sigma(c_2) - c_2) \otimes b_1 + \ldots + (\sigma(c_k) - c_k) \otimes b_k$, $\sigma \in G$, belongs to J and contains fewer terms than x. So $\sigma(y) = y$, and this holds for all $\sigma \in G$. Hence, $y \in \overline{K}$ and is a unit in J. It should be $J = \overline{\mathcal{T}}$. □

Remark. The normality theorem 11.8(a) can be interpreted as absence of fixed elements in $T \setminus K$ for action of the Galois group G (compare the proof). By the way of proving Lemma 2 we have shown even stronger property:

The action of G on T does not fix any proper ideal of T.

Examples. (a) *Extension which is not Picard–Vessiot.* The example is $K = \mathbb{C}(x) \subset M = K\langle \ln(1+\mathrm{e}^x)\rangle$.
Indeed, we have $K \subset K(\mathrm{e}^x) \subset M$, where $K \subset L = K(\mathrm{e}^x)$ is Picard–Vessiot with the Galois group $\mathbb{C}^*$ and $L \subset M$ is Picard–Vessiot with the Galois group $\mathbb{C}$. If $K \subset M$ were Picard–Vessiot, then its Galois group should be a semi-direct product of $\mathbb{C}^* \ltimes \mathbb{C}$. On the other hand, the elements of M, treated as multivalued functions, should have the set of singularities invariant with respect to the action of the Galois group. This means that any automorphism of M should satisfy $\ln(1+\mathrm{e}^x) \longmapsto \ln(1+\mathrm{e}^x) + c$.

(b) *Extension with the Galois group equal to the full linear group.* (We follow [**Kapl**]). The above algebraic ideas allow to construct a Picard–Vessiot extension $K \subset M$ with $\mathrm{Gal}_K M = GL(n, C)$.
Let K_0 be any differential field with the field of constants C.
Let $M = K_0\langle x_1, \ldots, x_n\rangle = K_0(x_1, x_1', \ldots, x_2, x_2', \ldots, x_n, x_n' \ldots)$, where x_i are independent differential variables (without any relations). Then we have $C_M = C$. The action of a matrix $\sigma \in GL(n, C)$ onto the vector $(x_1, \ldots, x_n)$ extends naturally to its action onto the derivatives of this vector, and then onto M. Let $K = M^{GL(n,C)}$ be the subfield of invariants of this action; it is generated by ratios of some Wronskians.
One can see that the differential operator $D = W(\partial, x_1, \ldots, x_n)W^{-1}(x_1, \ldots, x_n)$ has coefficients from K and $y_i = x_i$ form the basis of solutions of the equation $Dy = 0$. Thus $M = K\langle y_1, \ldots, y_n\rangle$, $C_M = C_K$ and $\mathrm{Gal}_K M = GL(n, C)$.

11.11. Theorem (Fundamental theorem of differential Galois theory). *Let $K \subset M$ be a Picard–Vessiot extension with the Galois group G. There is a bijective correspondence between the subfields L, $K \subset L \subset M$, and the Zariski closed subgroups $H \subset G$ given by:*

$$\begin{array}{ccc} L & \longmapsto & \mathrm{Gal}_L M, \\ H & \longmapsto & M^H. \end{array}$$

Proof. 1. One has to show that the compositions

$$\begin{array}{ccccc} L & \longmapsto & H = \mathrm{Gal}_L M & \longmapsto & M^H, \\ H & \longmapsto & L = M^H & \longmapsto & \mathrm{Gal}_L M \end{array}$$

are isomorphisms.

2. The first isomorphism is a consequence of Theorem 11.8(a), applied to the extension $L \subset M$. Indeed, it says that $M^H = \{a \in M : \tau a = a,\ \tau \in \mathrm{Gal}_L M\} = L$.

3. The second isomorphism means that, if $H \subset G$ is a subgroup (maybe not algebraic), then the group $H'' = \mathrm{Gal}_{M^H} M$ forms the Zariski closure of H.
Assume contrary, i.e., that there exists a polynomial f on $GL(n, C)$ such that $f|_H = 0 \neq f|_{H''}$. If $M = K\langle y_1, \ldots, y_n\rangle$ and $u_{1,1}, \ldots, u_{n,1}, u_{1,2}, \ldots u_{n,n}$ are n^2 formal variables, then we define A as the fundamental matrix of the system (y_j),

and B as the analogue of fundamental matrix for the system (u_j). Let also the Galois group act from the right.
Define the polynomial $F(u) = f(A^{-1}B)$, $f \in M[u]$. It has the property that $F(y\sigma) = 0$ for $\sigma \in H$ and $F(yH'') \neq \{0\}$.
Consider the set $\mathcal{E}$ of polynomials with the latter property. Let $E = \sum a_k p_k(u) \in \mathcal{E}$ be a polynomial with minimal number of monomials p_k and with one coefficient equal to 1. Denote by E_τ, $\tau \in H$, the polynomial obtained from E by replacing its coefficients by their τ-images. We have $E_\tau = \sum(a_k\tau)p_k(u) = (\sum a_k p_k(u\tau^{-1}))\tau$ and E_τ belongs to $\mathcal{E}$. Because the polynomial $E - E_\tau$ has fewer monomials than E, we get $(E - E_\tau)(yH'') = \{0\}$. If $E \neq E_\tau$, then, for suitable element $b \in M$, the polynomial $E - b(E - E_\tau)$ belongs to $\mathcal{E}$ and has fewer monomials than E.
Thus $E = E_\tau$ for any τ, which means that the coefficients of E lie in the field M^H. The latter field equals to the field $M^{H''}$; so, the coefficients are invariant with respect to H'' and $E(y\sigma) = E_\sigma(y\sigma) = (E(y\sigma\sigma^{-1}))\sigma = 0$ for any $\sigma \in H''$. This contradiction completes the proof. □

11.11. Definition (Liouvillian and elementary functions). We say that a Picard–Vessiot extension $K \subset M$ is **Liouvillian** iff there exists a sequence of differential fields $K = K_1 \subset K_2 \subset \ldots \subset K_r = M$ such that each K_{i+1} is obtained from K_i by adjoining an integral or exponent of an integral. In the case $K = \mathbb{C}(x)$ the functions from such M are called the **Liouvillian functions**, or the **functions expressed by quadratures**.
We say that a Picard–Vessiot extension $K \subset M$ is **generalized Liouvillian** iff there is an analogous sequence with adjoining of integrals, exponents of integrals and of algebraic elements. The corresponding functions (for $K = \mathbb{C}(x)$) are called the **generalized Liouvillian functions** or the **functions expressed by generalized quadratures.**
An extension $K \subset M$ (not necessarily Picard–Vessiot) is called **elementary** iff $M = K\langle z_1, \ldots, z_s\rangle$, where each z_i is either *logarithm* over $K_i = K(z_1, \ldots, z_{i-1})$, i.e., $z_i' = v_i'/v_i$, $v_i \in K_i$, or is an *exponent* over K_i (i.e., $z_i'/z_i = v_i'$, $v_i \in K_i$), or is algebraic over K_i. In the case $K = \mathbb{C}(x)$ the functions from M are called the **elementary functions**. (We have taken this definition from [**Rit**] and [**Ros2**].)
There is another definition of elementary functions in [**Mag**]. We shall call them **primitively elementary**. They are defined as elements of a Picard–Vessiot extension $M = \mathbb{C}(x)(\ln(x-x_1), \ldots, \ln(x-x_m); z_1, \ldots, z_n)$, where z_j are either algebraic over $K_j = \mathbb{C}(x, z_1, \ldots, z_{j-1})$ or are exponents over K_j.

11.13. Theorem (Solution of linear equations in quadratures).

(a) *A Picard–Vessiot extension is Liouvillian iff its differential Galois group is solvable. In this case all matrices from this group can be simultaneously triangularized.*

(b) *A Picard–Vessiot extension is generalized Liouvillian iff the connected component of identity of its differential Galois group is solvable.*

Proof. 1. Assume that the extension $K \subset M$ is Liouvillian, i.e., we have the sequence $K = K_1 \subset K_2 \subset \ldots \subset K_r = M$ of differential fields such that $K_{j+1} = K_j\langle z_j\rangle$ is obtained by means of adjoining an integral (or of exponent of an integral) z_j to K_j.
If $\sigma \in \mathrm{Gal}_K M$, then σz_1 satisfies the same differential equation and belongs to K_2, $\sigma(K_2) = K_2$. This means that the Galois group $\mathrm{Gal}_{K_2} M = G^{K_2}$ is a normal subgroup of $\mathrm{Gal}_K M$ with the quotient group $\mathrm{Gal}_K K_1$, which is abelian.
Repeating this with respect to the sequence $K_2 \subset \ldots \subset M$, we obtain the solvability of $\mathrm{Gal}_K M$.

2. The property of simultaneous reduction of matrices from a solvable group to triangular form is well known in the theory of Lie groups and Lie algebras (see [**Ser**], [**Kapl**]). It is proved by induction, using the fact that commuting matrices have the same eigenvectors.

3. Consider the case $\mathrm{Gal}_K M \subset GL(1, C) = C^*$. Thus, $M = K\langle u\rangle$. For any $\sigma \in \mathrm{Gal}_K M$ we have $\sigma u = au$, $\sigma u' = au'$ for some $a \in C^*$. We get $\sigma(u'/u) = u'/u$. It means that $u'/u = z \in K$ and $u = \mathrm{const} \cdot \mathrm{e}^{\int z}$.
Assume that all the matrices from $\mathrm{Gal}_K M \subset GL(n, C)$ are triangular:

$$\sigma u_i = a_{i,i} u_i + a_{i,i+1} u_{i+1} + \ldots + a_{i,n} u_n, \quad i = 1, \ldots, n, \tag{11.2}$$

with $a_{i,j}$ constant.
Take the last equation $\sigma u_n = a_{n,n} u_n$. We have $\sigma u_n' = a_{n,n} u_n'$ which implies $u_n'/u_n \in K$. Thus u_n is the exponent of integral of an element from K. Let $K_1 = K\langle u_n\rangle$.
Dividing Eq. (11.2) by the equation $\sigma u_n = a_{n,n} u_n$ and differentiating, we get (with $v_i = (u_i/u_n)'$)

$$\sigma v_i = (a_{i,i}/a_{n,n}) v_i + \ldots (a_{i,n-1}/a_{n,n}) v_{n-1}, \quad i = 1, \ldots, n-1.$$

By induction with respect to n we obtain that $K_1\langle v_1, \ldots, v_{n-1}\rangle$ is a Liouvillian extension of K_1.
Finally, $M = K\langle u_1, \ldots, u_n\rangle$ is obtained from $K_1\langle v_1, \ldots, v_{n-1}\rangle$ by means of adjoining the integrals from v_j.

4. Assume that the extension $K \subset M$ is generalized Liouvillian and the corresponding series of fields begins from $K \subset K\langle z\rangle \subset \ldots$ with z either: (i) algebraic, or (ii) integral or exponent of integral with infinite Galois group.
In the case (i) the extension $K \subset K_1$ is finite, the Galois subgroup $\mathrm{Gal}_{K_1} M$ is of finite index in $\mathrm{Gal}_K M$ and these two groups have the same components of identity.
In the case (ii) the extension $K \subset K_1$ is Picard–Vessiot with the continuous abelian Galois group (equal to C or C^*). It is the quotient group of $\mathrm{Gal}_K M$ by the normal subgroup $\mathrm{Gal}_{K_1} M$. The identity component $(\mathrm{Gal}_{K_1} M)^0$ of $\mathrm{Gal}_{K_1} M$ is also normal subgroup of the identity component $(\mathrm{Gal}_K M)^0$ of $\mathrm{Gal}_K M$ with the same abelian quotient.

Using the induction we can assume that the identity component of $\mathrm{Gal}_{K_1}M$ is solvable. Thus also the identity component of $\mathrm{Gal}_K M$ is solvable.

5. Assume that the identity component G^0 of $G = \mathrm{Gal}_K M$ is solvable. Of course, G^0 is a normal subgroup of G and one associates with it the differential field M^{G^0}. The extension $K \subset M^{G^0}$ is finite Picard–Vessiot with the Galois group equal to G/G^0. So $M^{G^0} \subset M$ is Liouvillian extension.
Completing the resolution sequence (with adjoining of integrals or exponents) for the extension $M^{G^0} \subset M$ by $K \subset M^{G^0}$, we obtain the resolution sequence giving the generalized Liouvillian property of the extension $K \subset M$. □

11.14. Remark. *Let $K \subset M$ be a Picard–Vessiot extension with $\mathrm{Gal}_K M \subset GL(n, C)$ and Wronskian W and let $K_1 = K\langle W^{1/n}\rangle$ be the extension by adjoining exponent of an integral (n^{th} order root of the Wronskian). Then $\mathrm{Gal}_{K_1}M \subset SL(n, C)$ (consists of unimodular matrices).*

It means that, when solving the problem of integrability of equations in quadratures, we can restrict ourselves to the case when the differential Galois group consists of matrices with determinant 1.
Indeed, if M is defined by means of solutions of an equation $y^{(n)} + a_1 y^{(n-1)} + \ldots = 0$ with the Wronskian W, then the change $z = W^{1/n} y$ gives the equation $z^{(n)} + b_2 z^{(n-2)} + \ldots = 0$.

11.15. Theorem (Generalized Liouvillian extensions for $n = 2$). *If $G = \mathrm{Gal}_K M \subset SL(2, C)$ is a Galois group corresponding to a generalized Liouvillian extension, then either:*

(i) *G is finite (the extension is algebraic), or*

(ii) *the identity component G^0 consists of simultaneously diagonalizable matrices and has index 2 in G (M is obtained from a quadratic extension L of K by adjoining exponent of an integral), or*

(iii) *G^0 consists of simultaneously trangularizable matrices (here $K \subset K(u_1) \subset K(u_1, u_2) = M$, where u_1 is exponent of an integral and u_2 is an integral).*

Proof. 1. If G is finite then the theorem holds. So, we can restrict ourselves to the case with infinite G.

2. Assume that G^0 consists of diagonal matrices $\tau = \mathrm{diag}(a, a^{-1})$. Sometimes the a's should satisfy certain algebraic equation, but it is not the case (G^0 infinite). If v is an eigenvector for G^0 (e.g., $\tau v = av$) and $\sigma \in G$, then σv is also eigenvector for G^0 ($\tau\sigma v = \sigma\,(Ad_{\sigma^{-1}}\tau)v = \sigma\tau' v = a'\sigma v$). Thus the quotient group G/G^0 acts on the set of eigenvectors and consists of at most two elements (identity or the transposition).
If $G/G^0 = \mathbb{Z}_2$, then we are in the case (ii), otherwise we have the case (iii).

3. If G^0 consists of triangular matrices, then there exists a unique eigenvector v for G^0. It is preserves by the action of the whole group G.

4. The statements about the differential fields can be easily obtained from the proof of Theorem 11.13. Because the extension $K \subset M$ is finite in the case (i), it is algebraic extension.
In the case (ii) we have $K \subset L \subset M$, where $L = K(z)$, z satisfies a quadratic equation $z^2 + az + b$, $a, b \in K$, and the Galois group of the extension $L \subset M$ consists of diagonal matrices $\sigma = \mathrm{diag}(c, c^{-1})$, $c \in C^*$ (in some basis u_1, u_2). We have $\sigma u_{1,2} = c^{\pm 1} \cdot u_{1,2}$ and $\sigma u'_{1,2} = c^{\pm 1} \cdot u'_{1,2}$. This leads to the equations $u'_{1,2} = a_{1,2} \cdot u_{1,2}$, $a_{1,2} \in K$. But $\sigma(u_1 u_2) = u_1 u_2$, which implies that $u_2 = v u_1^{-1}$, $v \in K$. Therefore $M = L\langle u_1 \rangle$ is obtained by adjoining an exponent of an integral.
If $\sigma = \begin{pmatrix} c & 0 \\ d & 1/c \end{pmatrix}$ is a general automorphism (in a basis y_1, y_2) in the case (iii), then we find that $y_1' = a \cdot y_1$, $a \in K$ and $\sigma(y_2/y_1) = c^{-1}(y_2/y_1) + d$, $\sigma(y_2/y_1)' = c^{-1}(y_2/y_1)'$. So, $(y_2/y_1)' y_1 = v$, $v \in K$ (it is invariant). If we denote $u_1 = y_1$, $u_2 = \int v/y_1$, then we get $K \subset K(u_1) \subset K(u_1, u_2) = M$. □

11.16. Example. *The Airy equation*

$$y'' = xy$$

is not solvable in generalized quadratures.

Proof. Assume the contrary. Because the coefficient before y' vanishes, the Galois group is included into $SL(2, \mathbb{C})$.
The solutions are integer analytic functions $u = \sum a_n x^n$ (not polynomials, because $(n+2)(n+3)a_{n+3} = a_n$). They cannot be also algebraic functions (because the latter have singularities). Thus the corresponding extension $K = \mathbb{C}(x) \subset M$ is not algebraic.
Applying Theorem 11.15 we see that there should exist a solution u which is altogether exponent of integral of an element from K or from L, a quadratic extension of K: $u'/u \in K$ or $u'/u \in L$.
Let $z = -u'/u$. It should be either rational function, $P(x)z - Q(x) = 0$, or it should satisfy a quadratic equation $P(x)z^2 + Q(x)z + R(x) = 0$; here P, Q, R are polynomials.
The function $z(x)$ satisfies the Riccati equation

$$\mathrm{d}z/\mathrm{d}x = z^2 - x. \tag{11.3}$$

Thus Riccati equation (11.3) should have invariant algebraic curve (rational or hyperelliptic). □

Lemma. *Eq.* (11.3) *does not have any finite invariant algebraic curve.*

Proof. Firstly we notice that any solution to (11.3) is a meromorphic function $z = f(x)$. Indeed, in the variable $y = 1/z$ we have $\frac{\mathrm{d}y}{\mathrm{d}x}|_{y=0} = -1 - xy^2|_{y=0} \neq 0$, which shows that the solutions $y(x)$ have simple zeroes and that the function $f(x)$ has simple poles. Thus f is single-valued.

If such a function represents an invariant algebraic curve, then it is rational, $z = P(x)/Q(x)$.
However, when we look for the asymptotic $z \sim Cx^{\alpha}$ of solutions as $x \to \infty$, then we find $\alpha = 1/2$, $C = \pm 1$. This is impossible for rational $z(x)$. □

11.17. Example. *The Bessel equation $x^2y'' + xy' + (x^2 - \nu^2)y = 0$ is integrable in generalized quadratures iff $\nu \in \frac{1}{2} + \mathbb{Z}$ (is a half-integer).*

Proof. First, the change $w = \sqrt{x}y$ transforms the Bessel equation to the equation $w'' + (1 - \mu/x^2)w = 0$, $\mu = \nu^2 - 1/4$, with the Galois group in $SL(2, \mathbb{C})$. Note that for $\nu = \pm 1/2$ we obtain the harmonic oscillator; we assume further that $\nu \neq \pm 1/2$. As in the case of the Airy equation, the problem is reduced to non-existence of finite invariant algebraic curves of the Riccati equation

$$\mathrm{d}z/\mathrm{d}x = z^2 + 1 - \mu/x^2. \tag{11.4}$$

Firstly, we look at the asymptotic $z \sim Cx^{\alpha}$ as $x \to \infty$. We find $\alpha = 0$, $C = \pm \mathrm{i}$. We introduce the coordinate $v = 1/x$ and obtain the system $\dot{v} = v^2$, $\dot{z} = -1 - z^2 + \mu v^2$ with two saddle-nodes $v = 0, z = \pm \mathrm{i}$. The eventual branches of an algebraic curve belong to the center separatrices of these saddle-nodes. As such, they are given by $z = f_{1,2}(x)$ with local single-valued branches $f_{1,2}(x)$, $x \to \infty$.
As in the previous example, we find that, for $x \neq 0$, the solutions to Eq. (11.4) have only simple poles as singularities. So, the functions $f_{1,2}$ are single-valued in $\mathbb{C} \setminus 0$ and hence also in $\mathbb{C}$. Therefore, any algebraic solution to (11.4) is a rational function $z = f_j(x)$.
Eq. (11.4) is invariant with respect to the symmetry $(x, z) \longmapsto (-x, -z)$. This and the above implies that there should exist two algebraic curves $z = f_{1,2}(x)$, $f_2(x) = -f_1(-x)$, $f_1 \not\equiv f_2$.
As $x \to 0$ we have $z \sim \mathrm{const} \cdot x^{-1}$. In the variables $x, u = zx$ we get

$$\frac{\mathrm{d}u}{\mathrm{d}x} = \frac{u^2 + u - \mu + x^2}{x}, \tag{11.5}$$

with two singular points $x = 0, u_{\pm} = (-1 \pm \sqrt{1 + 4\mu})/2$, and the ratios of eigenvalues $\lambda_{\pm} = 2u_{\pm} + 1 = \pm\sqrt{1 + 4\mu} = \pm 2\nu$. The solutions $u = g_{1,2}(x) = xf_{1,2}(x)$ of (11.5) are analytic functions such that $g_2(x) \equiv g_1(-x) \not\equiv g_1(x)$. This means that these solutions represent two separatrices of a singular point which should be a node, that is $(0, u_+)$. Here the ratio $\lambda = \lambda_+$ should be a natural number and we have $u = u_+ + dx^{\lambda} + \ldots$ where the choice of the coefficient d defines the choice of the solution. So, $g_{1,2} = u_+ + d_{1,2}x^{\lambda_+} + \ldots$ By the symmetry, $d_2 = (-1)^{\lambda_+}d_1 \neq d_1$. We find that λ is an odd integer, $\lambda = 2k + 1$, which implies that $\nu = k + \frac{1}{2}$.
It remains to show that for half-integer ν the Bessel equation is solvable in quadratures. We know that it is so for $\nu = \frac{1}{2}$. In the general case we use the formulas $\left(-\frac{\mathrm{d}}{x\mathrm{d}x}\right)^m (x^{-\nu}J_{\nu}(x)) = x^{-\nu-m}J_{\nu+m}(x)$, $\left(\frac{\mathrm{d}}{x\mathrm{d}x}\right)^m (x^{\nu}J_{\nu}(x)) = x^{\nu-m}J_{\nu-m}(x)$, where $J_{\pm\nu}$ are two independent solutions of the Bessel equations (see Example 8.31(b) and [**BE**]). We have $J_{1/2} = \sqrt{2/\pi x}\sin x$, $J_{-1/2} = \sqrt{2/\pi x}\cos x$.

Using this, one can also find recurrent formulas for the rational solutions to the Riccati equation (11.4). □

11.18. Remarks. (a) In [**Kapl**] there is an algebraic proof of non-existence of rational and hyperelliptic solutions of the Airy equation.
Also there is another proof of non-solvability of the Bessel equation for $\nu \notin 1/2+\mathbb{Z}$. It relies on taking into account the Stokes operators (see [**M-R**] and Theorem 11.23 below).

(b) It is not difficult to show that $\mathrm{Gal}_K M = SL(2,\mathbb{C})$ for the Airy equation: the Lie algebra of $\mathrm{Gal}_K M$ is simple and should be equal to $sl(2,\mathbb{C})$. In [**Mag**] one can find association with different algebraic subgroups (the diagonal torus T, the unipotent upper triangular subgroup U, the Borel upper-triangular subgroup $B = UT$) the corresponding subfields of M.

(c) The association of the Riccati equation with the Airy equation is a particular case of the correspondence between second order linear equations $\ddot{x}+a(t)\dot{x}+b(t)x = 0$ and the Riccati equations $\dot{z} = A(t) + B(t)z + C(t)z^2$. The connection is defined by

$$z = c(t)\dot{x}/x.$$

The three function A, B, C are uniquely defined by the three functions a, b, c. A, B, C are rational iff a, b, c are rational.
The 2-parameter family of linear maps $g_s^t \in GL(2,\mathbb{C})$, defining the evolution of the linear equation, are transformed to a 2-parameter family of fractional-linear maps for the Riccati equation. (Here $g_s^t(x_0, \dot{x}_0)$ is the value at the moment t of the solution and its derivative with initial condition $(x_0, \dot{x}_0)$ at the moment s).

11.19. Theorem (Primitive elementary extensions). *Let $K = \mathbb{C}(x) \subset L \subset M$, where $K \subset L$ is Picard–Vessiot and $K \subset M$ is primitively elementary. Then the identity component of the Galois group* $\mathrm{Gal}_K L$ *is abelian.*

Proof. Recall that, by Definition 11.12 of primitive elementary extension, $M = K(\ln(x-x_1), \ldots, \ln(x-x_m); z_1, \ldots, z_n)$, where z_j are either algebraic or exponents over $K(z_1, \ldots, z_{j-1})$. Here $K \subset K' = K(\ln(x-x_1), \ldots, \ln(x-x_m))$ is Picard–Vessiot with the abelian Galois group $\mathbb{C}^m$. We can assume that $K' \subset L$; otherwise we can diminish K' by deleting some logarithms.
Consider the series of Picard–Vessiot extensions $K \subset K' \subset L$. From the proof of Theorem 11.13 it is easy to deduce that $\mathrm{Gal}_{K'} L$ has identity component in the form of an algebraic torus $T = (\mathbb{C}^*)^r$. This torus is a normal subgroup of the identity component G^0 of $\mathrm{Gal}_K L$ with the quotient $\mathbb{C}^m$.
Let $U \subset G^0$ be the unipotent radical of G^0. Because G^0 is solvable, we can treat it as a subgroup of the group of upper-triangular matrices in $GL(n, C)$ (or the Borel subgroup). Its unipotent radical consists of triangular matrices (from G^0) with 1's at the diagonal. The unipotent radical U is a normal subgroup of G^0.
But $T \subset G^0$ is also normal and G^0/T is isomorphic to U. This means that we have the direct product $G^0 = T \times U$ (T and U act on different subspaces of a direct sum decomposition of $\mathbb{C}^n$). Because U is isomorphic to $\mathbb{C}^m$, G^0 is abelian. □

11.20. Example. *The probability function*

$$z(x) = \mathrm{Erf}(x) = \int^x \mathrm{e}^{-t^2}\,\mathrm{d}t$$

is not primitively elementary.

Proof. We follow [**Mag**]. The function $z(x)$ satisfies the equation $y'' + 2xy' = 0$ with the basis of solutions $1, z(x)$ and the Wronskian $W(x) = \mathrm{e}^{-x^2}$ (satisfying $W' = -2xW$). We have the extensions $K = \mathbb{C}(x) \subset L = K(W) \subset M = L(z)$.
Because $\int_{-\infty}^{x} \mathrm{e}^{-t^2}$ and e^{-x^2} decrease faster than any power as $x \to -\infty$, the functions z and W are not algebraic over K.
Also $z \notin L$. Indeed, assume that $z = P(x,W)/Q(x,W)$, where $P = a_n(x)W^n + \dots$ and $Q = W^m + \dots$ are polynomials in W. Then differentiation gives $Q^2W = (P'_x - 2xWP'_W)Q - P(Q'_x - 2xWQ'_W)$, or

$$W^{2m+1} + \dots = [a'_n + 2(m-n)xa_n]W^{m+n} + \dots$$

Consider three cases:

(i) $n < m+1$. Then W^{2m+1} has no counterpart.

(ii) $n = m+1$. Then we get $a'_n - 2xa_n = 1$. This equation has only entire solutions; so $a_n(x)$ should be a polynomial, $a_n(x) = b_r x^r + \dots$. We see that it is impossible.

(iii) $n > m+1$. Then $a'_n = 2(n-m)xa_n$, or $a_n = \mathrm{const} \cdot \mathrm{e}^{2(n-m)x^2}$.

Now it is clear that the Galois group $\mathrm{Gal}_K M$ consists of the maps $(W,z) \longmapsto (\lambda W, \lambda z + a)$. It is isomorphic to the solvable non-abelian group $\mathrm{Aff}(\mathbb{C})$. □

The above statement does not constitute the best result about the function Erf. It is widely known (see [**Ros3**]) that:

Erf(x) *is not elementary function.*

For the proof we use the following Liouville's theorem (see Theorem 11.28 below):

If a function $F(x) \in M$, its derivative $f = F' \in L$ and the extension $L \subset M$ is elementary, then $F = g + \sum c_i \ln h_i$, where $g, h_i \in L$ and c_i are constants. Here we put $L = \mathbb{C}(x,W)$, $W = \mathrm{e}^{-x^2}$.

Suppose that the function Erf is elementary. Then we should have

$$\mathrm{Erf}\,(x) = g(x,W) + \sum c_i \ln h_i(x,W),$$

where g is a rational function and h_i are polynomials. Consider singularities of the components $\ln h_i$, i.e., zeroes of $h_i(x, \mathrm{e}^{-x^2})$. If x_0 is such a zero and we have $h_i = (x-x_0)^{m_i} \cdot \tilde{h}_i(x)$, $\tilde{h}_i(x_0) \neq 0$, then we find a logarithmic singularity Erf $\sim (\sum c_i m_i) \ln(x - x_0)$. Because Erf is entire, we get $\sum c_i m_i = 0$. Thus we can write

$\sum c_i \ln h_i = \ln h(x)$, where $h(x)$ is non-vanishing function of exponential growth of the rank 2 at infinity (because $h_i = O(\exp(\mathrm{const} \cdot |x|^2)))$. This means that $\ln h$ is a polynomial of degree at most 2. We have reduced the proof to the case when Erf is a rational function of x and W. We have shown before that it is not the case. □

There are some connections between differential Galois groups associated with linear meromorphic non-autonomous differential systems $\dot{z} = A(t)z$ on the Riemann sphere and the monodromy group and the Stokes operators associated with the system (see Chapter 8). The first result is classical and relatively simple.

11.21. Theorem of Schlesinger (Galois group). ([**Sch1**) *If the system $\dot{z} = A(t)z$ has only regular singular points, then the algebraic closure of its monodromy group is equal to the Galois group of the corresponding Picard–Vessiot extension of $K = \mathbb{C}(t)$.*

Proof. Let the singular points be $t_1, \ldots, t_m$. They are regular iff the solutions grow at most polynomially as $t \to t_j$. If $\mathcal{F}(t) = (f_{ij}(t))$ is a fundamental matrix of the system, then the Picard–Vessiot extension is $L = K(\{f_{ij}\})$. The monodromy operators describe the variation of $\mathcal{F}$ as the argument varies along loops in $\overline{\mathbb{C}} \setminus \{t_1, \ldots, t_m\}$, $\mathcal{F}(t) \longmapsto \mathcal{F}(t)M_j$, $M_j \in GL(n, \mathbb{C})$.
Because the analytic continuation commutes with differentiation and the rational functions are monodromy invariant, we get that the monodromy operators belong to $\mathrm{Gal}_K L$.
To show that the Zariski closure of the monodromy group Mon is not a proper subgroup of $\mathrm{Gal}_K L$, it is enough to show that any function $f(t) \in L$ invariant with respect to the monodromy maps is rational, i.e., that $L^{\mathrm{Mon}} = K$ (see the fundamental theorem 11.11).
The function $f = f(t)$ is meromorphic outside the singularities t_j. Because it is invariant with respect to the monodromy maps corresponding to small loops around t_j, it is single valued in a punctured disc around t_j. By the regularity assumption, $f(t)$ is meromorphic near t_j. Therefore, f has only poles as singularities in $\overline{\mathbb{C}}$ and must be rational. □

11.22. Differential Galois group in the irregular case. There should exist a generalization of the Schlesinger's theorem to the case of rational differential equations (or systems) with irregular singular points. From the theory developed in Chapter 8 we know some natural operators which should belong to the corresponding differential Galois group, the monodromy and Stokes operators. As we shall see below, they define automorphisms of the corresponding Picard–Vessiot extension. But, generally, the Zariski closure of the group generated by these operators is not the whole differential Galois group. Indeed, the equation $t^2\dot{x} = x$ has $x = C\mathrm{e}^{-1/t}$ as solutions, it has trivial monodromy and no Stokes operators, but $\mathrm{Gal}_{\mathbb{C}(t)}\mathbb{C}(t, \mathrm{e}^{-1/t}) \simeq \mathbb{C}^*$. It turns out that one should take into account so-called *exponential torus* defined below.

We restrict ourselves to the local situation (like in most sources in the literature), i.e., $K = \mathbb{C}\{t\}[t^{-1}] = \mathcal{M}_0(\mathbb{C})$ and the Picard–Vessiot extension $K \subset L$ is defined by means of a linear system

$$t^r \dot{z} = A(t)z, \quad z \in \mathbb{C}^n.$$

Assume also that $r > 1$ (irregularity) and lack of resonances, i.e., $\lambda_i \neq \lambda_j$ for the eigenvalues of the matrix $A(0)$ (see Definition 8.16). Then the formal normal form is diagonal (Theorem 8.20): $t^r \dot{y}_1 = b_1(t)y_1, \ldots, t^r \dot{y}_n = b_n(t)y_n$ with the fundamental matrix

$$W(t) = \operatorname{diag}\left(t^{\mu_1} \exp Q_1(1/t), \ldots, t^{\mu_n} \exp Q_n(1/t)\right),$$

where Q_j are polynomials of degree $r-1$ and without constant terms.
The exponential functions $\exp Q_j(1/t)$ can be algebraically dependent; for example, it may occur that $\prod [\exp Q_j]^{\nu_j} \equiv 1$, $\nu_j \in \mathbb{Z}$. Anyway, they generate some group $\mathcal{G}$ with respect to the multiplication; $\mathcal{G}$ is isomorphic to $\mathbb{Z}^m$, where $m \leq n$ is determined by the number of independent algebraic relations (as above). The **exponential torus** is the group

$$\mathcal{T} = \operatorname{Hom}(\mathcal{G}, \mathbb{C}^*).$$

$\mathcal{T} \simeq (\mathbb{C}^*)^m$ and is identified with a subgroup of the algebraic torus $(\mathbb{C}^*)^n = \{\tau = (\tau_1, \ldots, \tau_n)\}$ defined by relations $\nu\tau = \sum \nu_i \tau_i = 0$ (the same as satisfied by the exponents $\exp Q_i$).
The exponential torus acts on the fundamental matrix of the normalized system as follows:

$$\tau \cdot \mathcal{W}(t) = \operatorname{diag}\left(\tau_1 t^{\mu_1} \exp Q_1(1/t), \ldots, \tau_n t^{\mu_n} \exp Q_n(1/t)\right);$$

it is induced by a constant change of basis in $\mathbb{C}^n$.
The normalizing series, leading to the formal normal form, is divergent in general. But, in sufficiently thin sectors S, we can choose the normalizing changes $H_S(t)$ analytic and with regular growth (see Theorem 8.25). Let us fix one such sector S_0. The matrix $\mathcal{F}_{S_0}(t) = \{f_{ij}(t)\} = H_{s_0}^{-1}(t)\mathcal{W}(t)$ is a fundamental matrix of the initial system $t^r \dot{z} = Az$; its columns $z = \phi_i(t)$ generate the space $\mathcal{L}_{S_0}$ of solutions. The matrix entries $f_{ij}(t)$ generate the differential field L, $L = K(f_{ij})$. The elements of L have the form $F(t, (f_{ij}(t))) = F_1(t, (f_{ij}))/F_2(t, (f_{ij}))$, where $F_{1,2}$ are polynomials in f_{ij}'s with coefficients being holomorphic functions of t. The exponential torus acts on $\mathcal{L}_{S_0}$ as follows

$$\tau \cdot \phi_i(t) = H_{S_0}^{-1}(t)\left(\tau \cdot H_{S_0}(t)\phi_i(t)\right);$$

we see that $\tau \cdot \phi_i(t)$ is a solution.
If S, S' are adjacent sectors and $\mathcal{F}_S = H_S^{-1}\mathcal{W}$, $\mathcal{F}_{S'} = H_{S'}^{-1}\mathcal{W}$, then we have the relation $\mathcal{F}_{S'} = \mathcal{F}_S C_{S'S}$ in $S \cap S'$, where $C_{S'S}$ are the Stokes operators (see Theorem

8.25 and Definition 8.26). By prolongation of $\mathcal{F}_S$ and $\mathcal{F}_{S'}$ from $S \cap S'$ to the sector S_0, the operators $C_{S,S'}$ are regarded as acting on the space $\mathcal{L}_{S_0}$.
Finally, the monodromy operator $M : \mathcal{L}_{S_0} \longmapsto \mathcal{L}_{S_0}$ is defined by analytic prolongation along a loop around $t = 0$.
The action of $\tau, C_{S'S}$ and M onto functions $F(t, \{f_{ij}\})$ is rather obvious: $F(t, \{f_{ij}\}) \longmapsto F(t, \{f'_{ij}\})$, where $f'_{ij}(t)$ are the matrix entries of the transformed fundamental matrix.

11.23. Theorem of Ramis (differential Galois group). *The differential Galois group of the above Picard–Vessiot extension equals the Zariski closure of the group generated by the exponential torus, the Stokes operators and the monodromy operator.*

Proof. There are two statements to demonstrate:

1. the operators indicated in the theorem are automorphisms of the extension $K \subset L$ indeed;
2. any germ from L invariant with respect to these operators is a germ of meromorphic function.

Unlike in the proof of the Schlesinger's theorem, the point 1 is nontrivial. It reduces to showing that the field K is invariant with respect to $\tau \in \mathcal{T}$ and $C_{S,S'}$; moreover, we can assume that the τ's are generic. The point is that the matrix elements $f_{ij}(t)$ may satisfy some algebraic relations of the form

$$G(t, \{f_{ij}(t)\}) \equiv 0,$$

where G is a polynomial in f_{ij}'s with coefficients analytic in t.
If $\{f'_{ij}\}$ is obtained by action of a typical element $\tau \in \mathcal{T}$ (i.e., without additional relations over $\mathbb{Z}$), then we can consider the relation $G \equiv 0$ over the field $\mathcal{M}_{S_0} = \mathcal{O}_{S_0}[t^{-1}]$ (where $\mathcal{O}_{S_0}$ is the space of germs of holomorphic functions in S_0 with regular growth as $t \to 0$). Then we can write $G(t, (f_{ij})) = \widetilde{G}_{S_0}(t, (t^{\mu_i} \exp Q_i(1/t)))$, where $\widetilde{G}_{S_0}$ depends polynomially on $t^{\mu_i} \exp Q_i$ with coefficients from $\mathcal{O}_{S_0}$. By definition of the action of the exponential torus, we have $G(t, (f'_{ij})) = \widetilde{G}_{S_0}(t, (\tau_i t^{\mu_i} \exp Q_i))$. Since there should be $\widetilde{G}_{S_0}(t, (\tau_i t^{\mu_i} \exp Q_i)) \equiv 0$ for any typical τ, $\widetilde{G}_{S_0}$ depends of $t^{\mu_i} \exp Q_i$ only through the 'relational' terms:

$$\widetilde{G}_{S_0} = \sum_k \omega_k^{(S_0)}(t) \prod \left[t^{\mu_i} \exp Q_i\right]^{k_i}, \quad k = (k_1, \dots k_n) \in \mathbb{Z}^n,$$

$\sum k_i \tau_i = \sum k_i Q_i = 0$, where $\omega_k^{(S_0)}(t) \in \mathcal{O}_{S_0}$.
It means that $\widetilde{G}_{S_0} = \sum_k \omega_k^{(S_0)}(t) t^{(\mu,k)}$, $(\mu, k) = \sum \mu_i k_i$. In other sectors S one also writes $G(t, (f_{ij})) = \widetilde{G}_S(t, (t^{\mu_i} \exp Q_i(1/t)))$ and $\widetilde{G}_S = \sum_k \omega_k^{(S)}(t) \prod \left[t^{\mu_i} \exp Q_i\right]^{k_i}$ $= \sum_k \omega_k^{(S)}(t) t^{(\mu,k)}$, where $\omega_k^{(S)}(t) \in \mathcal{O}_S$.
In the sector S_0 we got the identity $\widetilde{G}_{S_0} \equiv 0$. But we cannot claim that the same identity holds in other sectors. This is because the coefficients $\omega_k^{(S)}(t)$ are different from the coefficients $\omega_k^{(S_0)}(t)$. The differences arise from differences between

the normalizing matrix functions $H_s(t)$. More precisely, the differences $\omega_k^{(S)}(t) - \omega_k^{(S')}(t)$ 'exist' only in intersections of adjacent sectors. Moreover, these differences are smaller than any power of $|t|$; since $|H_S(t) - H_{S'}(t)| < \exp(-c\,|t|^{-r})$, also $\left|\omega_k^{(S)}(t) - \omega_k^{(S')}(t)\right| < \exp(d\,|t|^{-r})$. Since $\widetilde{G}_{S_0}(t) \equiv 0$, the other $\widetilde{G}_S(t)$ are exponentially small.

What we obtain is a functional cochain $\left(\widetilde{G}_{S_k}(t)\right)_{k=1,\dots,p}$ associated with a covering of a neighborhood of 0 by sectors S_i. The elements of this cochain decrease faster than any power of $|t|$.

The action of the Stokes operators $\mathcal{F} \longmapsto \mathcal{F}' = \mathcal{F}C_{S,S'}$ also leads to the changes $G(t,(f_{ij})) \longmapsto G(t,(f'_{ij}(t))) =: g_{S'}(t)$. The relation $G \equiv 0$ gives rise to a functional cochain $(g_{S_k}(t))_{k=1,\dots,p}$, which is exponentially small.

But we know how to deal with functional cochains. We met them in Chapter 10 (Section 3) in the proof of another Ramis' theorem about rigidity of no-solvable groups (Theorem 10.38). Namely, we can use the Phragmen–Lindelöf theorem for functional cochains (Theorem 10.39), which states that, if the elements of such a cochain decrease faster than any power, then this cochain vanishes identically.

Therefore $\widetilde{G}_{S_k}(t) \equiv 0$ and $g_{S_k}(t) \equiv 0$, which completes the first part of the proof.

The proof of property 2, i.e., that only the germs $g(t)$ from K are invariant with respect to the exponential torus, the Stokes and monodromy operators, is relatively easy. Using invariance with respect to typical elements $\tau \in \mathcal{T}$ one obtains that $g(t)$ has regular singularity at $t = 0$ (the dependence on $\exp Q_j$'s is trivial). Then invariance with respect to the monodromy implies that g is meromorphic. □

11.24. Remarks. (a) The Ramis's theorem 11.23 is formulated in various sources ([**Ram1**], [**Ram3**], [**M-R**], [**Put**]), but for the proofs the reader is referred to some unpublished preprints of Ramis and M. van der Put.

In [**Put**] a general scheme of the proof is presented. It consists of two steps: a formal one (where it is proved that the corresponding Picard–Vessiot extension over the field $\mathbb{C}[[t]][t^{-1}]$ is determined by the exponential torus and the monodromy) and an analytic one. In the both steps suitable Tannakian categories are defined. In the analytic step the multi-summation property for some Gevrey series (an equivalent version of the Phragmen–Lindelöf theorem for functional cochains) is used.

The proof presented above is inspired by the work [**IlKh**] of Yu. Ilyashenko and A. Khovanski, where it is proved that the Stokes operators belong to the differential Galois group (but nothing is said about the exponential torus).

(b) In [**IlKh**] the assumptions like in Theorem 11.23 are imposed (i.e., $A(0)$ non-resonant). In [**Ram1**], [**Ram3**], and [**Put**] there is no such restriction. According to Theorem 8.22, after a change $t = s^b$ and some Gauge transformation, one gets a situation with separated formal normal form

Moreover, Theorem 11.23 can be extended to the case $K = \mathbb{C}(t) = \mathbb{C}(\mathbb{CP}^1)$ and $K = \mathbb{C}(S)$, i.e., the field of rational functions on a Riemann surface S. In the latter case, instead of a system of linear equations, one considers an equation for horizontal sections of some vector bundle over S with given connection.

(c) There exist theorems characterizing (closed) algebraic subgroups G of $GL(n)$, for which there exists a linear differential system (local near $t = 0$, global in $\mathbb{CP}^1$, or a connection in a bundle over S) with singularities, such that G is the differential Galois group of the corresponding Picard–Vessiot extension. It is the so-called *inverse problem in differential Galois theory.*
In particular, C. Mitschi and M. Singer [**MiSi**] proved that for $K = \mathbb{C}(t)$ any such group is realized. For other situations the answer was given by Ramis in [**Ram3**] and [**Put**].

11.25. Examples. (a) *The Bessel equation for* $\nu \notin \frac{1}{2} + \mathbb{Z}$. By Example 8.31(b) in Chapter 8 the Bessel equation $t^2\ddot{x} + t\dot{x} + (t^2 - \nu^2)x = 0$ has the Stokes multipliers of the form $\begin{pmatrix} 1 & -f \\ 0 & 1 \end{pmatrix}$, $\begin{pmatrix} 1 & 0 \\ f & 1 \end{pmatrix}$, $f = 2\mathrm{i}\cos\pi\nu$ (i.e., at the point $t = \infty$ and in suitable basis). If ν is not half-integer, then they generate a group which contains the two unipotent groups $\begin{pmatrix} 1 & f\mathbb{Z} \\ 0 & 1 \end{pmatrix}$ and $\begin{pmatrix} 1 & 0 \\ f\mathbb{Z} & 1 \end{pmatrix}$. The exponential torus acts via the matrices $\begin{pmatrix} \tau & a \\ 0 & 1/\tau \end{pmatrix}$. These three groups generate the non-solvable connected group $SL(2, \mathbb{C})$. The same is the differential Galois group of the extension $K = \mathbb{C}(t) \subset L = K(J_\nu, J_{-\nu})$.
An analogous analysis (without formulas for the Stokes operators) is presented in [**M-R**].

(b) *The Airy equation* $\ddot{x} = tx$ (Example 11.16). The reader is recommended to calculate the formal normal form at $t = \infty$. It turns out that here one has to apply the change $t = s^{-2}$ (ramification) and one gets $Q_1(s) = s^3$ and $Q_2(s) = -s^3$ (see the proof of Theorem 11.23). The exponential torus acts via diagonal matrices (like in the previous example), the monodromy matrix equals $\begin{pmatrix} 0 & -1 \\ 1 & 0 \end{pmatrix}$ and the Stokes matrices have the form $\begin{pmatrix} 1 & a \\ 0 & 1 \end{pmatrix}$, $\begin{pmatrix} 1 & 0 \\ b & 1 \end{pmatrix}$, $ab \neq 0$ (see [**Put**], [**Hea**]).

(c) In the case of Example 8.31(a), $\dot{z}_1 = -kt^{-k-1}z_1 + \varphi(t)z_2$, $\dot{z}_2 = -at^{-1}z_2$, following [**Zo7**], one can introduce a family of discrete subgroups of $GL(2, \mathbb{C})$, called the *extended monodromy group* and having the property that their algebraic closures are equal to the differential Galois group. Namely, one partially compactifies the punctured neighborhood of $t = 0$, $U^* = U \setminus 0 \simeq \mathbb{S}^1 \times (0, \epsilon)$, by adding some pairs of points $(\theta_i, 0)$, $(\theta_i', 0)$ and later identifying them. The obtained topological space has several holes, and the (simple) loops γ_i around these holes generate the extended monodromy group. With the loop γ_i one associates an operator of the form $\begin{pmatrix} \tau_i & \mu_i \\ 0 & 1 \end{pmatrix}$, where τ_i and μ_i are taken from the exponential torus and the

Stokes matrices respectively. In [**Zo7**] it is proved that the matrices M_i are equal to limits of monodromy transformations of a perturbed systems (with several regular singular points).

Probably that construction can be generalized to other systems with irregular singularity. In [**Gl**] A. Glyutsuk proved that the Stokes matrices can be obtained as limits of the monodromy matrices in a generic deformation, after suitable normalizations of the bases. When the bases are not controlled one should obtain also matrices from the exponential torus.

11.26. Ziglin–Morales-Ruiz–Ramis Theory and Painlevé equations.

(a) **Differential Galois Theory version of the Ziglin Theory.** Recall that the Ziglin theory (§4 in Chapter 10) deals with the normal variation equation associated with a particular phase curve Γ^* of a Hamiltonian system $\dot{x} = \boldsymbol{X}_H(x)$ in $\mathbb{C}^{2n}$. If $\Gamma^* = \{\varphi(t)\}$ for a particular solution $x = \varphi(t)$, then the variation equation along Γ^* is $\dot{\xi} = \boldsymbol{X}_H^{(1)}(\varphi(t))\xi$, $\boldsymbol{X}_H^{(1)} = \partial \boldsymbol{X}_H/\partial x$, and it is naturally projected to the normal bundle $N\Gamma^*$ and becomes the normal variation equation, i.e., $\xi \in N\Gamma^*$. Usually the Hamilton function H is polynomial (but it can be rational as well) and the closure of the phase curve Γ^* is an algebraic curve Γ (usually rational). Ziglin associates with loops from $\pi_1(\Gamma^*, x_0)$ monodromy transformations for solutions of the normal variation equation. Because H is a first integral for the vector field $\boldsymbol{X}_H$, then $\mathrm{d}H$, as a linear section of $N\Gamma$, is a first integral for the normal variation equation, and the monodromy maps A from the monodromy group $\mathrm{Mon}(\Gamma)$ have a special structure.

Recall that the Hamiltonian system $\dot{x} = \boldsymbol{X}_H$ is completely integrable iff it has n functionally independent first integrals in involution (including H). In Theorem 10.44 Ziglin has characterized the monodromy operators A in the cases of complete integrability.

J. Morales-Ruiz and J.-P. Ramis [**M-R, M-RR**] observed that from the Ziglin theorem it follows that the algebraic closure of the monodromy group, i.e., the differential Galois group $G = \mathrm{Gal}_K L$ of the Picard–Vessiot extension $K = \mathbb{C}(x) \subset L$ defined by the normal variation system has abelian component G^0 of the identity. This can be deduced from a detailed analysis of the proof of the Ziglin theorem. The differential Galois group is somehow easier to determine than the monodromy group and this was the reason why the Morales-Ruiz–Ramis approach become popular.

But in application often occurs that the identity component G^0 is abelian, although the Hamiltonian system is not completely integrable. Then one considers higher order normal variation equations. They take he form

$$\dot{\xi}^{(k)} = \boldsymbol{X}_H^{(1)}(\varphi(t))\xi^{(k)} + P(\xi^{(1)}, \ldots, \xi^{(k-1)}),$$

where $\boldsymbol{X}_H^{(1)} = \partial \boldsymbol{X}_H/\partial x$ is as above and P is a polynomial.

The system for $\dot{\xi}^{(1)} = \dot{\xi}$ (from the first variation), $\dot{\xi}^{(2)}, \ldots, \dot{\xi}^{(k)}$ can be arranged as a homogeneous linear system of higher dimension. For example, in the case of second variation one considers additional variables $\eta_{jk} = \xi_j^{(1)}\xi_k^{(1)}$. One has Picard–Vessiot extensions $K = \mathbb{C}(x) \subset L_1 \subset \ldots \subset L_k$ and the differential Galois groups $G_j = \mathrm{Gal}_K L_j$. One has the following generalization of the above statement about these groups (see [**MRRS**]).

Theorem (Morales-Ruiz–Ramis–Simó). *In the completely integrable case the identity components G_j^0 of the groups G_j are abelian.*

(b) **Application to the Painlevé equations.** E. Horozov and Ts. Stoyanova [**HS**, **Stoy**] began systematic application of the Morales-Ruiz–Ramis theory to prove non-integrability of the Painlevé equations in some special cases (see the point 8.58).

Due to works of many mathematicians, such as K. Okamoto [**Oka**], the Painlevé equations can be rewritten in the Hamiltonian form with $1\frac{1}{2}$ degrees of freedom (see also 11.31(b) below). This means that the one-degree-of-freedom Hamiltonian, which we denote $h(x,y,t)$, depends on time and the Hamiltonian system is non-autonomous: $\dot{x} = h'_y(x,y,t)$, $\dot{y} = -h'_x(x,y,t)$. For example, the equations P_1 and P_2 (with x playing the role of y and t playing the role of x) have already Newtonian form with the momentum $y = \dot{x}$ and with the Hamilton functions $h_1 = \frac{1}{2}y^2 - 2x^2 - tx$ and $h_2 = \frac{1}{2}y^2 - \frac{1}{4}x^4 - \frac{1}{2}tx^2 - \alpha x$.

The simple trick relying on treating the time t as a new coordinate, denoted by q, introducing new momentum p and replacing $h(x,y,t)$ with $H(x,y,p,q) = h(x,y,q)+p$ leads to an autonomous Hamiltonian system with 2 degrees of freedom

$$\dot{x} = H'_y,\ \dot{y} = -H'_x,\ \dot{q} = H'_p = 1,\ \dot{p} = -H'_q = -h'_q.$$

Namely, for such system the Morales-Ruiz–Ramis theory is applied. We demonstrate it in one case following the Stoyanova's paper [**Stoy**].

(c) **The fourth Painlevé equation.** This equation P_4 takes the form.

$$\ddot{x} = \frac{1}{2x}\dot{x}^2 + \frac{3}{2}x^3 + 4tx^2 + 2(t^2-\alpha)x + \frac{\beta}{x}, \tag{11.6}$$

We have the Okamoto type Hamiltonian

$$H_4 = xy^2 - x^2y - 2qxy - 2kx - 2ly + p, \tag{11.7}$$

where k and l are some parameters, leading to the system (denoted $\boldsymbol{X}_{H_4}$) :

$$\dot{x} = (2y-x-2q)x - 2l,\ \dot{y} = (-y+2x+2q)y + 2k,\ \dot{q} = 1,\ \dot{p} = 2xy. \tag{11.8}$$

This system is equivalent to the equation P_4 with $\alpha = 1-2k-l$ and $\beta = -2l^2$.

Theorem of Stoyanova. *Assume that $\alpha = m$ and $\beta = -2(1+m+2n)^2$, with $m,n \in \mathbb{Z}$. Then the system (11.8) does not have any meromorphic first integral independent of H_4 and rational in q.*

(d) **The case with $\alpha = 1$ and $\beta = 0$.** Here we have $k = l = 0$ in Eqs. (11.7)–(11.8) (and $n = -1$). Therefore, the curve

$$\Gamma = \{x = y = p = 0\}$$

is invariant; it corresponds to the particular solution $\varphi : t \longmapsto (0, 0, t, 0)$.
The first normal variation equations take the form

$$\dot{\xi}_1 = -2t\xi_1,\ \dot{\eta}_1 = 2t\eta_1, \dot{\rho}_1 = 0;$$

we have $x = 0 + \varepsilon\xi_1 + \varepsilon^2\xi_2 + \ldots$, $y = 0 + \varepsilon\eta_1 + \ldots$, $p = 0 + \varepsilon\rho_1 + \ldots$. Its fundamental matrix equals $\mathcal{F}(t) = \operatorname{diag}(\exp\left(-t^2\right), \exp\left(t^2\right), 1)$. Thus the differential Galois group is abelian and consists of diagonal matrices.
The second variation system, only for ξ_2 and η_2, is following:

$$\dot{\xi}_2 = -2t\xi_2 - \xi_1^2 + 2\xi_1\eta_1,\ \dot{\eta}_2 = 2t\eta_2 + 2\xi_1\eta_1 - \eta_1^2.$$

Denote $u = \xi_2$, $v = \xi_1^2$, $w = \xi_1\eta_1$ and we get the system

$$\dot{u} = -2tu - v + 2w,\ \dot{v} = -4tv,\ \dot{w} = 0; \tag{11.9}$$

an analogous system is related with the equation for η_2. The system (11.9) is regular (no singularities) in the complex plane of t's, but it has an irregular singularity at $t = \infty$.
Let us formally diagonalize this system. Since it is triangular, we simply put $u = u_1 + a(t)v + b(t)w$. We get the conditions

$$\dot{a} = 2ta - 1,\ \acute{b} = -2tb + 2,$$

with the formal solutions $a = \frac{1}{2}t^{-1} + \frac{1}{4}t^{-3} + \ldots$, $b = t^{-1} + \frac{1}{2}t^{-3} + \ldots$ and with the exact solutions

$$a(t) = \mathrm{e}^{t^2} \int_t^\infty \mathrm{e}^{-s^2}\mathrm{d}s,\ b(t) = 2\mathrm{e}^{-t^2} \int_{\mathrm{i}\infty}^t \mathrm{e}^{s^2}\mathrm{d}s.$$

The obtained formal system $\dot{u}_1 = -2tu_1$, $\dot{v} = -4tv$, $\dot{w} = 0$ has fundamental matrix $\mathcal{G}(t) = \operatorname{diag}\left(\mathrm{e}^{-s^2}, \mathrm{e}^{-2s^2}, 1\right)$. It follows that that the identity component of the group generated by the exponential torus and the formal monodromy is abelian and consists of diagonal matrices $\operatorname{diag}(c, c^2, 1)$.
Recall that, by the Ramis theorem 11.23, the differential Galois group is generated by the exponential torus, Stokes operators and the formal monodromy.
But the Stokes matrices are nontrivial, and non-diagonal (compare Ex. 8.31(a) above). Indeed, the general solution is following:

$$u = c_1\mathrm{e}^{-t^2} - c_2\mathrm{e}^{-t^2} \int_\infty^t \mathrm{e}^{-s^2}\mathrm{d}s + 2c_3\mathrm{e}^{-t^2} \int_{\mathrm{i}\infty}^t \mathrm{e}^{s^2}\mathrm{d}s,\ v = c_2\mathrm{e}^{-2t^2},\ w = c_3.$$

So, there are two Stokes matrices, each upper triangular and with one off diagonal entry; that entry equals $\int_{\mathbb{R}} \mathrm{e}^{-s^2} \mathrm{d}s = \sqrt{\pi}$ and $-2\int_{\mathrm{i}\mathbb{R}} \mathrm{e}^{s^2} \mathrm{d}s = -2\mathrm{i}\sqrt{\pi}$ respectively. (In [**Stoy**] the Stokes matrices are computed using the Borel and Laplace transforms.)
This implies that the identity component of the corresponding differential Galois group is non-abelian and hence the system (11.8) is not integrable in the class of meromorphic functions in the Liouville–Arnold sense.

(e) **The other cases.** These cases are $\beta = -2(1+2n+\alpha)^2$, $\alpha, n \in \mathbb{Z}$, from the theorem of Stoyanova. They are reduced to the case $\alpha = 1$, $\beta = 0$ by application of so-called Backlünd transformations. They are discussed in the point 11.31(c) below. □

(f) **Another approach.** This approach comes from the paper [**AHPT**] by P. Acosta-Humánez, M. van der Put and J. Top. In fact, it uses rather directly the Differential Galois Theory.
Firstly, the authors argue that the normal variation system for associated with a particular solution to the Hamiltonian system in $\mathbb{C}^4$ is equivalent to a variation equation for the original Painlevé equation. Thus, if $x = \psi(t)$ is a distinguished solution to an equation $\ddot{x} = R(\dot{x}, x, t)$, then one puts $x = \psi(t) + \varepsilon\xi(t) + \varepsilon^2\eta(t) + \ldots$ to the both sides and expands them in powers of ε.

Example (P_4 with $\alpha = 0$ and $\beta = -2$). Here the particular solution is $x = \psi(t) = -2t$. The first variation equation is

$$\ddot{\xi} - t^{-1}\dot{\xi} - 4t^2\xi = 0,$$

with the basic solutions $\exp(t^2)$ and $\exp(-t^2)$. The corresponding Picard–Vessiot extension is $K = \mathbb{C}(t) \subset L_1 = K(\exp(t^2))$ and the differential Galois group is $\mathrm{Gal}_K L_1 \simeq \mathbb{C}^*$.
So, one considers the second variation equation

$$\ddot{\eta} - t^{-1}\dot{\eta} - 4t^2\eta = \frac{1}{2t}\xi\ddot{\xi} - \frac{1}{4t}\dot{\xi}^2 - 7t\xi^2.$$

This equation is solved for η after putting $\xi = a\exp(t^2) + b\exp(-t^2)$. We obtain three terms in the right-hand side: $a^2(t^{-1} - 6t)\exp(2t^2)$, $ab(t^{-1} - 10t)$ and $b^2(-t^{-1} - 6t)\exp(-2t^2)$. These three terms lead to three solutions $\eta_j(t)$ (using the variation of constants method).
The differential Galois group of the extension $L_1 \subset L_2 = L_1(\eta_1, \eta_2, \eta_3)$ is an unipotent group isomorphic to $\mathbb{C}^3$ and the group $G = \mathrm{Gal}_K L_2 \simeq \mathbb{C}^* \ltimes \mathbb{C}^3$ is non-abelian. Any extension $\bar{\sigma}$ of $\sigma \in \mathrm{Gal}_K L_1$, satisfying $\sigma(\exp(t^2)) = c\exp(t^2)$ with $c^2 \neq 1$, to G satisfies $\bar{\sigma}(\eta_j) = c^2\eta_j + \xi$, where ξ is a solution to the first variation equation. This shows that the action of $\mathbb{C}^*$ on $\mathbb{C}^3$ is nontrivial and the identity component G^0 of G is not commutative.

11.27. First integrals of polynomial planar vector fields. At the end of this section we concentrate on the problem of integration of differential equations

$$\frac{\mathrm{d}y}{\mathrm{d}x} = \frac{Q(x,y)}{P(x,y)}$$

in quadratures. More precisely, we look for the **first integral** $H(x,y)$ of the corresponding vector field $P\frac{\partial}{\partial x} + Q\frac{\partial}{\partial y}$ which satisfies

$$\dot{H} = 0, \quad \nabla H = (\partial_x H, \partial_y H) \neq 0.$$

We ask when H can be chosen elementary, Liouvillian, etc. Of course, the first integral is not chosen uniquely; a function of a first integral is also a first integral. The condition $\dot{H} = 0$ means existence of an **integrating factor** $R(x,y)$ such that $R\left(P\frac{\partial}{\partial x} + Q\frac{\partial}{\partial y}\right) = \boldsymbol{X}_H$ (Hamiltonian vector field).
The classical book of E. Kamke [**Kam**] is full of examples of integrable equations. These examples can be divided into three groups:

(i) with the first integral of the **Darbouxian type** $H = \prod f_i^{a_i}$ (i.e., Darboux function), or of the **generalized Darboux type** $H = \mathrm{e}^g \prod f_i^{a_i}$ ($f_i(x,y)$ – polynomials, $g(x,y)$ – rational function, a_i – constants);

(ii) with the first integral of the **Darboux–Schwarz–Christoffel type**

$$H = V\mathrm{e}^{g(U)} \prod (U - u_i)^{a_i} + \int^U S(u)\mathrm{e}^{g(u)} \prod (u - u_i)^{a_i} \mathrm{d}u$$

($U(x,y)$, $V(x,y), g(u)$ – rational functions, $S(u)$ -polynomial);

(iii) with the first integral expressed by special functions (e.g., hypergeometric) and not represented by quadratures.

11.28. Examples. (a) The equation with separated variables $\frac{\mathrm{d}y}{\mathrm{d}x} = \frac{A(x)}{B(y)}$ has the first integral $\int^x A(s)\mathrm{d}s - \int^y B(s)\mathrm{d}s$, which, after expansion of A and B into elementary fractions and integration, gives $A_0(x) + \sum a_i \ln(x - x_i) + B_0(y) + \sum b_j \ln(y - y_j)$. Its exponent is of the generalized Darbouxian type.
(b) The linear equation $\frac{\mathrm{d}y}{\mathrm{d}x} = A(x)y + B(x)$ has the general solution

$$y = \exp\left(\int^x A(s)\mathrm{d}s\right)\left(C + \int^x \exp\left(-\int A\right) B\right).$$

Performing the integration of A and extracting the constant C, we find the Darboux–Schwarz–Christoffel first integral $y\mathrm{e}^g(x) \prod (x - x_i)^{a_i} - \int^x \mathrm{e}^{g(u)} \prod (u - x_i)^{a_i} \cdot B(u)\mathrm{d}u$.
For example, the equation $\frac{\mathrm{d}y}{\mathrm{d}x} = -xy/(1 - x^2) + 1$ has the first integral $H = y/\sqrt{1 - x^2} - \sin^{-1} x = y/\sqrt{1 - x^2} - \mathrm{i}\ln(x + \mathrm{i}\sqrt{1 - x^2}) = y/\sqrt{1 - x^2} - \int^x \mathrm{d}u/\sqrt{1 - u^2}$ (see also [**PS**]).

(c) Also the Bernoulli equation $\frac{dy}{dx} = A(x)y + B(x)y^n$ has Darboux–Schwarz–Christoffel first integral.

(d) Using the relation between Riccati equations and second order linear equations, presented in Remark 11.18(c), we can present a formula for a first integral of the Riccati equation $\frac{dy}{dx} = A(x) + B(x)y + C(x)y^2$. It equals

$$H = \frac{c(x)u_2'(x) - yu_2(x)}{c(x)u_1'(x) - yu_1(x)},$$

where $u_{1,2}(x)$ are two linearly independent solutions of the associated linear equation $u'' + a(x)u' + b(x)u = 0$ and $y = c(x)\frac{u'}{u}$.

We begin the presentation of the main results in the topics (of first integrals of vector fields) with the classical result of J. Liouville, which generalizes the integral formula from Example 11.27(a).

11.29. Theorem of Liouville (Elementary functions). ([**Lio1**]) *If $K \subset M$ is elementary extension and an element $F \in M$ has derivative in K, then*

$$F = g + \sum c_i \ln h_i,$$

with $g, k_j \in K$ and $c_i \in C$.

The proof of this theorem is not very difficult and uses induction with respect to the length of the chain $K = K_0 \subset K_1 \ldots \subset K_r = M$; where $K_{j+1} = K_j\langle z_j \rangle$ and z_j are either algebraic or are exponents or are logarithms. Note that, if some expression contains an exponent function or an algebraic function (in a rational way), then the derivative of this expression also contains this exponent or this algebraic function. If such expression contains a logarithm and in a nonlinear way, then its derivative also contains this logarithm. For the details of the proof we refer the reader to the book of J. Ritt [**Rit**] and the articles by R. Risch [**Ris**] and by M. Rosenlicht [**Ros2**].

11.30. Theorem of Prelle and Singer (Elementary first integrals). ([**PS**]) *If a polynomial planar vector field has an elementary first integral, then it has a first integral of the form*

$$H = w_0 + \sum a_i \ln w_i,$$

where $w_i(x, y)$ are algebraic functions and a_i are constants.

We shall not prove this result.

11.31. Painlevé equations and elementary functions. The aim of this point is to present some results of the author with G. Filipuk from [**ZF**].
The main result of that paper is following:

(a) **Theorem (Non-integrability of Painlevé equations).** *Any of the equations $P_1 - P_6$, excluding the cases*

(a) $\alpha = \gamma = 0$ *in* P_3,

(b) $\beta = \delta = 0$ *in* P_3,

(c) $\gamma = \delta = 0$ *in* P_5

pdoes not admit a first integral which is an elementary function of x, $\dot{x}$ *and* t.

Moreover, the Hamiltonian vector field $\boldsymbol{X}_H$ *associated with any of the equations* $P_1 - P_6$*, excluding the cases* (a), (b), (c) *above, does not admit any first integral which is an algebraic function of* x, y, q, p *and is independent of* H.

We will prove this theorem only in the P_4 case. But we firstly present some additional results.

(b) **New Hamiltonians.** Of course, P_1 and P_2 are already in the Hamiltonian form, with the Hamilton functions $h_1 = \frac{1}{2}y^2 - 2x^3 - tx$ and $h_2 = \frac{1}{2}y^2 - \frac{1}{2}x^4 - \frac{1}{2}tx^2 - \alpha x$, where $y = \dot{x}$.
The other equations are of the Liénard type:

$$\ddot{x} = A(x,t)\dot{x}^2 + B(x,t)\dot{x} + C(x,t)$$

with rational coefficients (with possible poles at $t = 0$, $t = 1$, $t = \infty$, $x = 0$, $x = 1$, $x = \infty$ and $x = t$).

Proposition. *If the condition*

$$2A'_t = B'_x$$

is fulfilled, then the Liénard equation takes the Hamiltonian form in the variables $(x, y) = (x, \dot{x}/D)$, *where*

$$D(x,t) = \exp\left(\int^{(x,t)} 2A\mathrm{d}x + B\mathrm{d}t\right).$$

The corresponding Hamilton function is given by

$$h(x,y,t) = D(x,t)\frac{y^2}{2} + h_0(x,t), \ h_0 = -\int^x \frac{C}{D}\mathrm{d}x.$$

Moreover, if the $1-$*form* $2A\mathrm{d}x + B\mathrm{d}t$ *has only simple poles with integer residua at them, then the function* $D(x,t)$ *is rational. If, additionally, the* $1-$*form* $\frac{C}{D}\mathrm{d}x$ *has vanishing residua at its poles then the Hamilton function* $h(x,y,t)$ *is rational.*

Proof. With the substitution $y = \dot{x}/D$ we get the following non-autonomous vector field:

$$\boldsymbol{V}(x,y,t) = Dy\frac{\partial}{\partial x} + \left\{(AD - D'_x)\,y^2 + (B - D'_t/D)\,y + C/D\right\}\frac{\partial}{\partial y}.$$

Its divergence equals div$\boldsymbol{V} = (2AD - D'_x)\, y + (B - D'_t/D)$. Therefore, we obtain the following equations for $D(x,t)$:

$$D'_x/D = 2A, \quad D'_t/D = B.$$

This implies the formula for D, which is correct iff the $1-$form $\eta = 2A\mathrm{d}x + B\mathrm{d}t$ is closed. Then locally the function $D(x,t)$ is well defined, i.e., up to a multiplicative constant.

If the poles of η are simple (maybe after a resolution of singularities), then the function $D(x,t)$ near each component $\{f(x,t)=0\}$ of its divisor has the form const $\cdot f^\alpha$, where α is a corresponding residuum. In order that $D(x,y)$ be rational these exponents must be integers.

In the case of the equations $P_3 - P_6$ the conditions from the above proposition hold true. We have the following Hamilton functions:

$$h_{III} = \frac{x^2}{t} \cdot \frac{y^2}{2} - \alpha x + \frac{\beta}{x} - \frac{\gamma}{2} x^2 t + \frac{\delta}{2} \frac{t}{x^2},$$
$$h_{IV} = x \cdot \frac{y^2}{2} - \frac{x^3}{2} - 2tx^2 - 2(t^2 - \alpha)x + \frac{\beta}{x};$$

we skip the formulas for h_V and h_{VI}. □

(c) **Backlünd transformations for P_4.** These are rational transformations

$$(x', y', q', p') \longmapsto (x, y, q, p)$$

of the phase space $\mathbb{C}^4$, which are:

(i) symplectic, i.e., $\mathrm{d}y' \wedge \mathrm{d}x' + \mathrm{d}p' \wedge \mathrm{d}q' = \mathrm{d}y \wedge \mathrm{d}x + \mathrm{d}p \wedge \mathrm{d}q$, and

(ii) transform the Hamilton function, depending on some parameters. to a Hamilton function with other parameters.

We present them in the case of Hamiltonians $H_4 = h_4(x,y,q) + p$ (of Okamoto type) and

$$H_{IV} = h_{IV} + p = H^{(\alpha,\beta)} = \frac{1}{2} x y^2 - \frac{1}{2} x^3 - 2qx^2 - 2q^2 x + 2\alpha x + \frac{\beta}{x} + p.$$

We consider the following extended Okamoto Hamiltonians [**Oka**]:

$$\tilde{H}_{\pm}^{(a,b)} = \tilde{h}_4 + p = \frac{1}{2} x \left[y^2 + 2\left(\pm(x+2q) - \frac{b}{2x} \right) y \right] \mp ax + p$$

from the previous point. Here $\tilde{H}_{-}^{(a,b)}$ equals $4h_4(x, y/4, q) + p$, where h_4 is the Okamoto Hamiltonian from 11.26(c), and $\tilde{H}_{+}^{(a,b)}$ is analogously expressed via a modified Okamoto Hamiltonian denoted by $\bar{H}$. Moreover, we have $\alpha = \mp\left(\frac{1}{2}a - \frac{1}{4}b + 1\right)$, $\beta = -\frac{1}{8}b^2$. The above Hamiltonians are related by means of the following symplectic maps:

$$U_{\pm}:\left(x',y'\pm(x'+2q')-\frac{b}{2x'},q',p'\pm bq'\pm 2x'\right),$$

i.e., $(x,y,q,p)=U_{\pm}(x',y',q',p')$ and $U_{\pm}^{*}H^{(\alpha,\beta)}=\tilde{H}_{\pm}^{(a,b)}$.Moreover,

$$V^{*}\tilde{H}_{-}^{(a,b)}=\tilde{H}_{+}^{(b-a-4,b)},\quad V=V_{b}=U_{-}^{-1}\circ U_{+};\tag{11.10}$$

indeed, we have $b'=b$ and $\alpha=-\frac{1}{2}a+\frac{1}{4}b-1=\frac{1}{2}a'-\frac{1}{4}b+1$. Below we use $\tilde{H}_{-}^{(a,b)}$ as a reference Hamiltonian; thus we will get maps in the (a,b) −plane.
Now we introduce the following symplectic transformations:

$$\begin{aligned}S_1&:\left(x'\frac{x'y'-b}{x'y'-a},y'\frac{x'y'-a}{x'y'-b},q',p'\right),\\S_2&=V_b\circ S_1\circ V_{b-a-4}^{-1},\\S_3&=\left(x'+\frac{a}{y'},y',q',p'+2aq'\right).\end{aligned}$$

Note that $z=xy=x'y'=:z'$ for S_1 and this implies its symplectic property. Next, the representation

$$\tilde{H}_{\pm}^{(a,b)}=\frac{1}{2}y(xy-b)\pm x(xy-a)+(p\pm 2qxy)$$

shows that $S_1^{*}\tilde{H}_{\pm}^{(a,b)}=\tilde{H}_{\pm}^{(b,a)}$. This and Eqs. (11.10) imply

$$S_1^{*}\tilde{H}_{-}^{(a,b)}=\tilde{H}_{-}^{(b,a)},\quad S_2^{*}\tilde{H}_{-}^{(a,b)}=\tilde{H}_{-}^{(-a-8,b-a-4)}.$$

Finally, the formulas

$$\tilde{H}_{-}^{(a,b)}=-x'\cdot xy+\frac{1}{2}y(xy-b)-2q\cdot xy+p$$

and $xy=x'y+a$ (with $x'=x-a/y$) imply

$$S_3^{*}\tilde{H}_{-}^{(a,b)}=\tilde{H}_{-}^{(-a,b-a)}.$$

Therefore, we get the following changes in the parameter plane:

$$s_1:(a,b)\longmapsto(b,a),\ s_2:(a,b)\longmapsto(-a-8,b-a-4),\ s_3:(a,b)\longmapsto(-a,b-a).$$

They are involutions. Next, we have $s_1\circ s_3:(a,b)\longmapsto(b-a,-a)$, i.e., a linear map equivalent to the rotation of order 3. Therefore, the maps s_1 and s_3 generate the Weyl group $\mathbf{A}_2\simeq S(3)$.
We realize this group in a subspace of $\mathbb{R}^3$ with zero sum of coordinates: $\mathbb{X}=\left\{v\in\mathbb{R}^3:v_1+v_2+v_3=0\right\}$. We put $v_1=\frac{1}{12}(a-2b)$, $v_2=\frac{1}{12}(b-2a)$, $v_3=\frac{1}{12}(a+b)$; thus $v\in\mathbb{X}$. Then the above reflections take the following form:

$$s_1:v\longmapsto(v_2,v_1,v_3),\ s_3:v\longmapsto(v_1,v_3,v_2),\ s_2:v\longmapsto(v_1,v_3+1,v_2-1).$$

We see that s_1 and s_3 are orthogonal reflections in the plane $\mathbb{X}$ with respect to the lines $\{v_1 = v_2\}$ and $\{v_2 = v_3\}$, while s_2 is an orthogonal reflection with respect to the line $\{v_2 = v_3 + 1\}$ (parallel to the first line). These maps are the standard generators of the affine Weyl group $W_a(\mathbf{A}_2)$.

(d) **The integrability of P_3 in the cases $\alpha = \gamma = 0$ and $\beta = \delta = 0$.** (The integrability of the P_5 in special case is proved in an analogous way.)
In the case $\beta = \delta = 0$ of the P_3 the first two equations of the Hamiltonian system defined by H_{III} are following:

$$\dot{x} = x^2 y/q, \quad \dot{y} = -xy^2/q + \alpha + \gamma x q,$$

where $q = t$. Let $u = xy$, $v = qx$. Then we get the system

$$\dot{u} = x(\alpha + \gamma v), \quad \dot{v} = x(1 + u)$$

with the first integral

$$F = u + \frac{1}{2}u^2 - \alpha v - \frac{\gamma}{2}v^2 = xy + \frac{1}{2}x^2y^2 - \alpha q x - \frac{\gamma}{2}q^2x^2;$$

note that it does not depend on p.
Analogously, when $\alpha = \gamma = 0$ the corresponding system for $\dot{x}$ and $\dot{y}$ has the first integral

$$G = -xy + \frac{1}{2}x^2y^2 + \beta q/x - \frac{\delta}{2}q^2/x^2.$$

(e) **Proof of the non-integrability of the P_4.**

1. *Introduction of a small parameter.* We apply the following change:

$$x \longmapsto x/\mu, \quad y \longmapsto y/\mu, \quad p \longmapsto p/\mu, \quad q \longmapsto q/\mu.$$

It is not symplectic, but induces the changes $\Omega \longmapsto \Omega/\mu^2$, $H_{IV} \longmapsto H_\varepsilon/\mu^3$, where Ω is the symplectic form and

$$H_\varepsilon = H_0 + \varepsilon p = \left(\frac{1}{2}xy^2 - \frac{1}{2}x^3 - 2qx^2 - 2q^2x + ax + \frac{b}{x}\right) + \varepsilon p$$

with $\varepsilon = \mu^2$, $a = 2\alpha\mu^2$, $b = \beta\mu^4$. We assume that μ (and ε) is small and we treat a and b as other small parameters.
The initial Hamiltonian system is orbitally equivalent (i.e., by a time rescaling) to the Hamiltonian system associated with H_ε, i.e.,

$$\dot{x} = xy, \quad \dot{y} = -\frac{1}{2}y^2 + \frac{3}{2}x^2 + 4qx + 2q^2 - a + \frac{b}{x^2}, \quad \dot{p} = 2x^2 + 4qx, \quad \dot{q} = \varepsilon. \quad (11.11)$$

2. *Unperturbed system.* It is quite obvious that:

For $\varepsilon = 0$ the latter system is completely integrable with the functions $H_0 = H_\varepsilon|_{\varepsilon=0}$ and $H_1 = q$ playing the role of the first integrals in involution.

The common level sets $\{H_0 = h_0,\, q = q_0\}$ are of the form $\Gamma \times \mathbb{C}$ where

$$\Gamma = \Gamma(q_0, h_0) = \left\{(x, y) : xy^2 = x^3 + 4q_0x^2 + cx - 2b/x + 2h_0\right\} \subset \mathbb{C}^2,$$

$c = 4q_0^2 - 2a$ and the line $\mathbb{C} = \{(p, q) : q = q_0\} \subset \mathbb{C}^2$. After the substitution $y = z/x$ we obtain the curve (birationally equivalent with Γ):

$$\Delta = \left\{z^2 = f(x)\right\}$$

with $z = xy$, and $f = x^4 + 4q_0x^3 + cx^2 + 2h_0x - 2b$, i.e., Γ is an elliptic curve (at least for typical values of h_0 and q_0).
Solutions to the equations (11.11) for $\varepsilon = 0$ are the following:

$$x = \mathcal{X}(\tau - \tau_0), \quad y = \mathcal{Y}(\tau - \tau_0), \quad p = \mathcal{P}(\tau - \tau_0), \quad q = q_0.$$

Here $\mathcal{X}(\tau)$, $\mathcal{Y}(\tau) = \dot{\mathcal{X}}/\mathcal{X}$ and $\mathcal{P}(\tau)$ are defined by the following formulas:

$$\tau = \int_{(x_0,y_0)}^{(\mathcal{X},\mathcal{Y})} \frac{\mathrm{d}x}{xy}, \quad \mathcal{P} - p_0 = 2\int_0^\tau \left\{\mathcal{X}^2(s) + 2q_0\mathcal{X}(s)\right\} \mathrm{d}s = 2\int \frac{x + 2q_0}{y}\mathrm{d}x,$$

where the integral $\int_{(x_0,y_0)}^{(\mathcal{X},\mathcal{Y})}$ runs along a path in the complex curve Γ from some initial point (x_0, y_0) to the point $(x, y) = (\mathcal{X}(\tau), \mathcal{Y}(\tau))$. Below we fix the initial conditions for (x, y) by putting $\tau_0 = 0$, $y_0 = 0$ and x_0 as some root of the equation $f(x) = 0$; p_0 is the initial value for p.
We see that $\mathcal{X}(\tau)$ can be expressed via the Weierstrass P-function.

3. *Variation equations.* Take the above special solution of the unperturbed system: $x = \mathcal{X}(\tau)$, $y = \mathcal{Y}(\tau)$, $p = \mathcal{P}(\tau)$, $q = q_0$. We consider a corresponding variation equation along this solution.
Therefore, we substitute

$$x = \mathcal{X}(\tau) + \varepsilon x_1(\tau), \; y = \mathcal{Y}(\tau) + \varepsilon y_1(\tau), \; p = \mathcal{P}(\tau) + \varepsilon p_1(\tau), \; q = q_0 + \varepsilon q_1(\tau),$$

$x_1(0) = y_1(0) = p_1(0) = q_1(0) = 0$, into the system (11.11) and solve it modulo $O(\varepsilon^2)$. It is easy to see that $q_1(\tau) = \tau$ and we skip calculations of other variables. Therefore, we have the following (linear in ε) relations:

$$H_0(x, y, q) + \varepsilon p = h_0 + \varepsilon h_1 + \ldots, q = q_0 + \varepsilon\tau. \tag{11.12}$$

4. *Expansion of an independent first integral.* Suppose that the Hamiltonian vector field has an elementary first integral $F(X, Y, Q, P)$ independent of H. Then system (11.11) has the first integral $F_\varepsilon(x, y, q, p) = F(x/\mu, y/\mu, q/\mu, p/\mu)$, $\mu = \sqrt{\varepsilon}$, independent of H_ε. The following statement is an analogue of Proposition 2 in the proof of the Ziglin's Theorem 10.44; we refer to [**ZF**] for its proof.

Lemma 1. *There exists an elementary first integral $G_\varepsilon(x, y, q, p)$, independent of H_ε and obtained from F_ε by elementary operations involving F_ε, H_ε and ε, such that:*

(i) *G_ε has a uniform expansion*

$$G_\varepsilon = G_0 + G_1(\varepsilon) + \dots,$$

where $G_0 = G_0(x,y,q,p)$, $G_j(\varepsilon) = G_j(\varepsilon;x,y,q,p)$ are elementary functions such that $G_{j+1}/G_j \to 0$ as $\varepsilon \to 0$;

(ii) *the first term in the right-hand side of G_ε is of the form $G_0 = \Psi_0(H_0,q)$ and satisfies*

$$\frac{\partial \Psi_0}{\partial q}(h_0,q_0) \neq 0$$

for typical (h_0,q_0);

(iii) *the condition $F_\varepsilon = \text{const}$ along solutions becomes the condition*

$$G_\varepsilon(x(\tau),y(\tau),q(\tau),p(\tau)) = g_0 + g_1(\varepsilon) + \dots,$$

where g_0 and $g_j(\varepsilon)$ do not depend on τ and are of the same order as G_0 and $G_j(\varepsilon)$ respectively.

Substituting H_0 and q from Eqs. (11.12) into Ψ_0 in the equation $\Psi_0(H_0,q) + G_1(\varepsilon;x(\tau),y(\tau),q(\tau),p(\tau)) + \dots = g_0 + g_1(\varepsilon) + \dots$ (which follows from the Lemma), we obtain the following identity (as a function of τ) :

$$\{G_1(\varepsilon;\mathcal{X},\mathcal{Y},q_0,\mathcal{P}) - g_1(\varepsilon)\} + \dots + \varepsilon\cdot\{A\tau + B\mathcal{P}(\tau) + \Phi(\mathcal{X},\mathcal{Y},\mathcal{P})\} + \dots \equiv 0, \quad (11.13)$$

where

$$A = \frac{\partial \Psi_0}{\partial q}(h_0,q_0) \neq 0, \quad B = -\frac{\partial \Psi_0}{\partial H_0}(h_0,q_0)$$

are constants (see item (ii) in the Lemma). The term $\Phi(\mathcal{X},\mathcal{Y},\mathcal{P})$ is an elementary function corresponding to the term

$$\frac{\partial \Psi_0}{\partial H_0}(h_0,q_0)\cdot \varepsilon h_1 + G_k(\varepsilon;\mathcal{X},\mathcal{Y},\mathcal{P},q_0) - g_k(\varepsilon) = \varepsilon\cdot\{C + \Psi_k(\mathcal{X},\mathcal{Y},\mathcal{P},q_0) - c_k\},$$

when $G_k(\varepsilon) = \varepsilon\cdot\Psi_k(x,y,q,p)$ and $g_k(\varepsilon) = \varepsilon\cdot c_k$. If F (and G) is algebraic then Φ in Eq. (11.13) is also algebraic. If G depends only on x,y,q, then Φ depends only on $\mathcal{X},\mathcal{Y}$.

For the first power (non-obvious) of ε Eq. (11.13) implies the following relation:

$$A\tau + B\mathcal{P}(\tau) \equiv \Phi(\mathcal{X}(\tau),\mathcal{Y}(\tau),\mathcal{P}(\tau)), \quad A \neq 0. \quad (11.14)$$

Here we can say the following about $\Phi(x,y,p)$:

Either it is algebraic (in the 2 degrees of freedom version), or it is an elementary function of only x,y (in the $1\frac{1}{2}$ degrees of freedom version).

5. *Incomplete elliptic integrals.* These are the integrals defining τ and $\mathcal{P}-p_0$ above. In the theory of elementary functions they are regarded as multivalued functions of x. With the substitution $\sqrt{f(x)}$ in place of z, we can rewrite them as follows:

$$I(x)=\int_{x_0}^{x}\frac{\mathrm{d}u}{\sqrt{f(u)}},\quad J(x)=2\int_{x_0}^{x}\frac{u^2+2q_0u}{\sqrt{f(u)}}\mathrm{d}u,$$

i.e., we have $\tau=I(x)$, $\mathcal{P}(\tau)=p_0+J(x)$.
These integrals have *periods* (called also the *complete elliptic integrals*):

$$\begin{array}{ll}\omega_1=\oint_{\gamma_1}\frac{\mathrm{d}x}{z}, & \omega_2=\oint_{\gamma_2}\frac{\mathrm{d}x}{z},\\ \eta_1=2\oint_{\gamma_1}\frac{x^2+2q_0x}{z}\mathrm{d}x, & \eta_2=2\oint_{\gamma_2}\frac{x^2+2q_0x}{z}\mathrm{d}x.\end{array}$$

The loops $\gamma_{1,2}\subset\Delta$ generate the first homology group of the Riemann surface Δ. If x_1, x_2, x_3, x_4 are zeroes of the polynomial $f(x)$, then γ_1 (respectively γ_2) is a lift to the Riemann surface of the function $\sqrt{f(x)}$ of a loop which surrounds the points x_1, x_2 (respectively x_1, x_3) in the $x-$plane.
Hence, the Weierstrass function $\mathcal{X}(\tau)$, inverse to $I(x)$, is doubly periodic,

$$\mathcal{X}(\tau+\omega_1)=\mathcal{X}(\tau),\quad \mathcal{X}(\tau+\omega_2)=\mathcal{X}(\tau).$$

Also $\mathcal{Y}(\tau)=\dot{\mathcal{X}}(\tau)/\mathcal{X}(\tau)$ is doubly periodic with the same periods.
The function $\mathcal{P}(\tau)$ is not periodic, but it satisfies the following relations:

$$\mathcal{P}(\tau+\omega_1)=\mathcal{P}(\tau)+\eta_1,\quad \mathcal{P}(\tau+\omega_2)=\mathcal{P}(\tau)+\eta_2.$$

Lemma 2. *We have*

$$\begin{vmatrix}\omega_1 & \omega_2\\ \eta_1 & \eta_2\end{vmatrix}\neq 0$$

for typical values of the parameters q_0 *and* h_0.

Compare Lemma 5.24 and [**ZF**] for the proof.

Lemma 3. *For* $A\neq 0$ *and any* B *the incomplete elliptic integral*

$$K(x)=AI(x)+BJ(x)$$

is not an elementary function of x.

The proof, given in [**ZF**], uses the Liouville theorem 11.28.

6. *Proof of the non-integrability of* $\mathbf{X}_{H_{IV}}$ *in algebraic functions.* Assume relation (11.14), where Φ is an algebraic function of its arguments. In other words, τ is an algebraic function of the functions $\mathcal{X}(\tau)$, $\mathcal{Y}(\tau)$ and $\mathcal{P}(\tau)$. Let us rewrite the corresponding algebraic equation in the following form:

$$\sum_{m,n}a_{m,n}(\mathcal{X},\mathcal{Y})\tau^m\mathcal{P}^n=0, \tag{11.15}$$

where $a_{m,n}$ are polynomials of $\mathcal{X}$ and $\mathcal{Y}$.
Let us replace the function τ with $\mathcal{R}(\tau) = \tau - \frac{\omega_1}{\eta_1}\mathcal{P}(\tau)$. It has the following properties:

$$\mathcal{R}(\tau+\omega_1) = \mathcal{R}(\tau), \quad \mathcal{R}(\tau+\omega_2) = \mathcal{R}(\tau)+\sigma, \quad \sigma = \omega_2 - \omega_1(\eta_2/\eta_1) \neq 0.$$

Equation (11.15) takes the form

$$\sum_{m,n} b_{m,n}(\mathcal{X},\mathcal{Y})\mathcal{R}^m\mathcal{P}^n \equiv 0. \tag{11.16}$$

Since only the function $\mathcal{P}$ is not invariant with respect to the translation by ω_1, n must be equal to 0 in the above formula. But then also $m=0$, because otherwise the left hand side is not invariant with respect to the translation by ω_2.
On the other hand, the degree with respect to $\mathcal{R}$ of the polynomial in equation (11.16) must be ≥ 1 since equation (11.15) defines τ as an algebraic function of $\mathcal{X}$, $\mathcal{Y}$ and $\mathcal{P}$.
The latter contradiction proves the theorem for the extended Hamiltonian system associated with the fourth Painlevé equation. □

7. *Proof of the non-integrability of P_4 in elementary functions.* Here Eq. (11.14) means that the function $A\tau + B\mathcal{P}(\tau)$, $A \neq 0$, is an elementary function of $\mathcal{X}(\tau)$ and $\mathcal{Y}(\tau)$. Taking into account the algebraic nature of $y = \sqrt{f(x)}$, this implies that the function $K(x)$ defined in Lemma 3 is an elementary function of x. Lemma 3 states that it is not the case. □

11.32. Theorem of Singer (Liouvillian first integrals). ([**Sin**]) *If a polynomial planar vector field has a generalized Liouvillian first integral, then it has an integrating factor of the generalized Darbouxian type.*

Proof. We shall prove only the version of Singer's theorem with Liouvillian first integral. For general case we refer the reader to the original paper of Singer.

1. First, we note that the statement of generalized Darbouxian form of the integrating factor, $R = \mathrm{e}^g \prod f_i^{a_i}$, is equivalent to the representation

$$R = \exp\int U\mathrm{d}x + V\mathrm{d}y,$$

where $U\mathrm{d}x + V\mathrm{d}y$ is a closed rational 1-form. (The author owes this remark to C. Christopher. In the local case it was proved in [**CeMa**].)
Indeed, if R is generalized Darbouxian, then $\frac{\mathrm{d}R}{R}$ is a closed rational 1-form. If $\frac{\mathrm{d}R}{R} = U\mathrm{d}x + V\mathrm{d}y$, then the integral $F(z) = \int_{z_0}^{z} U\mathrm{d}x + V\mathrm{d}y$ is a multivalued holomorphic function on $\mathbb{CP}^2$ with singularities along finite number of irreducible algebraic curves, poles of the 1-form. Restricting the function F to a holomorphic disc transversal to such a curve $f_i = 0$, one finds that the singularity is either of a polar type or a logarithmic singularity. Moreover, the residuum a_i of $U\mathrm{d}x + V\mathrm{d}y$

is defined by means of integral along a loop surrounding $f_i = 0$ and is constant along this curve. The function $F - \sum a_i \ln f_i$ is meromorphic in $\mathbb{CP}^2$ and must be rational.

2. Here we begin the proper proof. Let the vector field $P\frac{\partial}{\partial x} + Q\frac{\partial}{\partial y}$ has a first integral H. The latter is an element of a differential field M and there is a series of extensions $K = (\mathbb{C}(x,y), \{\partial_x, \partial_y\}) = K_0 \subset K_1 \ldots \subset K_r = M$, where $K_i = K_{i-1}(z_i)$ and either $\nabla z_i = (\partial_x z_i, \partial_y z_i) \in K_{i-1}^2$ or $\nabla z_i / z_i \in K_{i-1}^2$.

3. We have $H = \int u\mathrm{d}x + v\mathrm{d}y = \int R(Q\mathrm{d}x - P\mathrm{d}y)$, where R is an integrating factor. The functions u, v satisfy the equations

$$Pu + Qv = 0, \quad \partial_x v = \partial_y u. \tag{11.17}$$

4. **Lemma.** *There is a Liouvillian field L, $K \subset L \subset M$, not containing any nontrivial first integral, and such that Eqs.* (11.17) *have a nonzero solution in L.*

Proof. We can assume that $M = L(z)$, where L does not contain first integral and z is either integral or exponent of an integral. Thus, the first integral equals $H = \frac{S(z)}{W(z)}$, where $S(X) = a_s X^s + \ldots + a_0$, $W(X) = X^w + b_{w-1}X^{w-1} + \ldots$ are polynomials with coefficients in L. The equation $\dot{H} = 0$ means that

$$\dot{S}W - S\dot{W} + (S'W - SW')\dot{z} = 0, \tag{11.18}$$

where the dot means differentiation of the coefficients. By $\frac{\mathrm{d}}{\mathrm{d}t}$ we shall denote the full derivation, i.e., along the vector field $P\partial_x + Q\partial_y$.
Assume that $\nabla z \in L^2$. The coefficient before z^{s+w} in Eq. (11.18) is $\dot{a}_s = \frac{\mathrm{d}}{\mathrm{d}t}a_s$ and a_s must be complex constant (because L does not contain nontrivial first integrals). So, we can assume that $s \neq w$; eventually, we replace H by $H - a_s$. The coefficient before z^{s+w-1} is $\dot{a}_{s-1} + a_s\dot{b}_{w-1} + (s-w)a_s\dot{z} = 0$. Thus $\frac{\mathrm{d}z}{\mathrm{d}t} = \frac{\mathrm{d}a}{\mathrm{d}t}$ for some $a \in L$. The functions $u = \partial_x(z-a)$, $v = \partial_y(z-a)$ belong to L and satisfy Eqs. (11.17).
Assume that $\nabla z / z \in L^2$. We rewrite Eq. (11.18) in the form $\dot{S}W - S\dot{W} + z(S'W - SW')(\frac{\mathrm{d}z}{\mathrm{d}t}/z) = 0$, where $\frac{\mathrm{d}z}{\mathrm{d}t}/z \in L$. Here the coefficient before z^{s+w} is $\dot{a}_s + (s-w)a_s(\dot{z}/z)$. If $s = w$, then $a_s \in \mathbb{C}$ and we replace H by $H - a_s$. If $s \neq w$, then we get $\frac{\mathrm{d}}{\mathrm{d}t}(z^{s-w}a_s) = 0$. One can now see that, if $a = z^{s-w}a_s$, then the functions $u = \partial_x a/a$, $v = \partial_y a/a$ belong to L and satisfy Eq. (11.17). □

5. Lemma 4 implies that the integrating factor $R = \frac{u}{Q} = -\frac{v}{P}$ belongs to L. Let the field L be smallest such field, i.e., with the smallest transcendental degree over K.
There is no problem of writing down the integrating factor in the form $R = \exp\int U\mathrm{d}x + V\mathrm{d}y$. The problem is to which differential field the functions $U = \partial_x R/R$ and $V = \partial_y R/R$ belong. They satisfy the system of equations

$$QV + PU = F, \quad \partial_y U = \partial_x V, \tag{11.19}$$

where $F = \partial_x P + \partial_y Q$ is the divergence of the vector field.
Now we can state about U, V the following. Let $L = L_1(z)$ with z transcendental over a field L_1.

6. **Property.** *L_1 does not contain nontrivial first integrals and nontrivial solutions of system* (11.17).

7. **Lemma.** *System* (11.19) *has a solution in L_1.*

Proof. Let $\nabla z \in L_1^2$. Assume that the integrating factor (from L) takes the form $R = bz^{m_0}\prod \phi_i(z)^{m_i}$, where $b \in L_1$, m_i are integers, and $\phi_i(X)$ are irreducible monic polynomials (with the highest degree coefficient equal to 1). The first equation in (11.19) means that $R^{-1}\frac{\mathrm{d}R}{\mathrm{d}t} \in L_1$. We have

$$\frac{\mathrm{d}R/\mathrm{d}t}{R} = \frac{\mathrm{d}b/\mathrm{d}t}{b} + m_0\frac{\mathrm{d}z/\mathrm{d}t}{z} + \sum m_i \frac{\dot{\phi}_i + \phi_i'\mathrm{d}z/\mathrm{d}t}{\phi_i}. \tag{11.20}$$

Above, $\dot{\phi}_i + \phi_i'\frac{\mathrm{d}z}{\mathrm{d}t}$ have degrees (in z) smaller than $\deg\phi_i$. Because neither z nor $\phi_i(z)$ are first integrals, we have $m_0 = m_i = 0$ and $R \in L_1$. Of course, then also $U, V \in L_1$.
Let $\nabla z/z \in L_1^2$. We again use Eq. (11.20), where we represent the term $(\dot{\phi}_i + \phi_i'\frac{\mathrm{d}z}{\mathrm{d}t})/\phi_i$ as $s(\frac{\mathrm{d}z}{\mathrm{d}t}/z)+[\sum(\frac{\mathrm{d}}{\mathrm{d}t}a_{s-i} - ia_{s-i}\frac{\mathrm{d}z}{\mathrm{d}t}/z)z^{s-i}]/\phi_i$, $s = \deg\phi_i$. Because $a_{s-i}z^{-i}$ are not first integrals, we get $m_1 = \ldots = 0$. Thus, $R = bz^{m_0}$ and $U = \partial_x R/R, V = \partial_y R/R \in L_1$. □

The next lemma completes the proof of Theorem 11.30.

8. **Lemma.** *If the system* (11.19) *has a solution in a Liouvillian extension L_1 (of K) and satisfying the property* 6, *then it has a solution in K.*

Proof. We use the induction with respect to the degree of transcendency over the field K. Let $K_i = K_{i-1}(z)$ contains solutions U, V of Eqs. (11.19) and has property 6.
Let $\nabla z \in K_{i-1}^2$. Differentiating the equation $QV + PU = F$ (from (11.19)) and using $\partial_x V = \partial_y U$, we get

$$\frac{\mathrm{d}U}{\mathrm{d}t} + aU = b, \tag{11.21}$$

where $a, b \in K_{i-1}$. Expanding $U = U(z)$ into elementary fractions, we obtain, from Eq. (11.21), that U must be a polynomial, $U = u_m z^m + \ldots + u_0$. Next, if $m > 1$, then we have $\frac{\mathrm{d}}{\mathrm{d}t}u_m + au_m = 0$ and $mu_m\frac{\mathrm{d}z}{\mathrm{d}t} + u_{m-1} + au_{m-1} = 0$. This means that $\frac{\mathrm{d}}{\mathrm{d}t}(mz + u_{m-1}/u_m) = 0$, which easily implies that $z \in K_{i-1}$; a contradiction.
Thus U (and also V) is linear, $U = u_0 + u_1 z$, $V = v_0 + v_1 z$. Comparison of the terms with z in Eqs. (11.19) gives $Pu_1 + Qv_1 = 0$, $\partial_x u_1 = \partial_y v_1$; i.e., u_1, v_1 satisfy system (11.17), which has only trivial solution (by 6.). So, $U = u_0, V = v_0 \in K_{i-1}$.
Let $\nabla z/z \in K_{i-1}^2$. We expand U, V in decreasing power series of z, $U = u_m z^m + u_{m-1}z^{m-1} + \ldots$, $V = v_m z^m + \ldots$. Then comparison of the terms before z^0 in (11.19), shows that u_0, v_0 satisfy (11.19), too. □

§2 Topological Galois Theory

11.33. Algebraic functions and ramified coverings between surfaces. Assume that an algebraic function $f = f(x)$ is defined by the equation

$$F(x, f) = r_n(x)f^n + r_{n-1}(x)f^{n-1} + \ldots + r_0(x) = 0, \tag{11.22}$$

where $r_j(x)$ are polynomials. We assume that the polynomial $F(x, y)$ is irreducible. The **Riemann surface** M of the algebraic function f can be defined in two ways: by means of the analytic prolongation and as an algebraic curve (defined by Eq. (11.22)).

The first means the following. Let $a \in \mathbb{C}$ be a fixed typical point and let $f_a = f_a(x) = f_{1,a}$ be a fixed local branch of the algebraic function (analytic element of f). It is well defined in the disc D_a of convergence of the Taylor series defining f_a. If δ is a path in the x-plane joining points a and b and avoiding singularities of f, then covering δ by discs (of convergence) D_{a_j} we prolong the germ f_a to a germ f_b (in a disc D_b). The system of these germs (f_b, D_b) defines the Riemann surface of f outside singularities. Its closure (in a suitable topology, including the points at infinity) is just the Riemann surface of f. In particular, when the path δ ends at the point a, then we get some branch $f_{j,a}(x)$, which may be different from the initial branch $f_a = f_{1,a}$. Here we agree to treat a pole of f as not a singularity; it is a germ of a holomorphic map to $(\overline{\mathbb{C}}, \infty)$. We have the natural projection $\pi : M \longmapsto \overline{\mathbb{C}}$. (We have used this approach in Chapter 1, in Definition 1.1).

Let us pass to the algebraic definition of the Riemann surface M. The equation $F(x, y) = 0$ defines an affine algebraic curve $\Gamma_{\mathrm{af}} \in \mathbb{C}^2$. We have the projection $\pi_{\mathrm{af}} : (x, y) \longmapsto x$.

The closure of Γ_{af} in $\mathbb{CP}^2$ defines a projective curve Γ. Assume that Γ does not contain the point $(0 : 1 : 0)$ at infinity. The projection π_{af} is prolonged to the projection π_{pr} of $\mathbb{CP}^2 \setminus (0 : 1 : 0)$ to $\mathbb{CP}^1$.

The curve Γ can be singular. We resolve the singularities of Γ and obtain a smooth curve M in some algebraic surface X. M is the same as the Riemann surface of the algebraic function $f(x)$. Moreover, the projection $\pi_{\mathrm{pr}}|_\Gamma$ is prolonged to a projection $\pi : M \longmapsto \mathbb{CP}^1$.

The critical points of the map π are called the **ramification points**. If $t \in (\mathbb{C}, 0)$ is a local coordinate near ramification point p and $\zeta \in (\mathbb{C}, 0)$ is a local coordinate near the corresponding critical value, then we have $\zeta = \pi(t) = t^\nu(c + \ldots)$, $c \neq 0$, where $\nu = \nu(p)$ is called the **ramification index**. In the affine part the ramification points are included in the set $\{F = F'_y = 0\}$.

The map $\pi : M \longmapsto \mathbb{CP}^1$ is called the **ramified covering**. The degree n of F with respect to y is called the **degree** of the ramified covering π.

The notion of ramified covering is applied to a holomorphic map between smooth Riemann surfaces, $\pi : M \longmapsto N$. Here M can be regarded as a Riemann surface of an algebraic function $f : N^0 \longmapsto \mathbb{C}$, where $N^0 = N \setminus$ (finite subset). $y = f(x)$ is a solution of an algebraic equation, like (11.22) but with $x \in N^0$.

The following formula connects the Euler characteristics $\chi(\cdot)$ of the surfaces.

11.34. Theorem (Riemann–Hurwitz formula). *We have*

$$n\chi(N) = \chi(M) + \sum(\nu(p) - 1).$$

Proof. Take triangulations of the surfaces M and N such that:

- the critical points and the critical values are among vertices of the triangulations;
- the inverse image of a triangle $\Delta' \subset N$ is a sum of triangles $\Delta \subset M$ with homeomorphic projections $\pi|_\Delta$.

The numbers V, E, T and V', E', T' of vertices, edges and triangles of the two triangulations satisfy the relations $E = nE'$, $T = nT'$ and $V = nV' - \sum_p(\nu(p) - 1)$. The latter identity is a consequence of the fact that $\nu(p)$ preimages of a triangle Δ' (with one vertex $\pi(p)$) have only one common vertex p.
Because $\chi(M) = V - E + T$, the Riemann–Hurwitz formula follows. □

Problem. Show that the genus of a generic algebraic curve of degree d in $\mathbb{CP}^2$ is $(d-1)(d-2)/2$.

11.35. Elements of the Algebraic Galois Theory. We recall some classical results from the theory of extensions of algebraic fields (not differential fields in general). We follow the book [**Wae**] of B. van der Waerden. We shall assume that the fields have characteristic zero.

1. **Definition.** An extension $\mathbb{K} \subset \mathbb{L}$ of algebraic fields is called **normal** iff for any element $\beta \in \mathbb{L}$ with its minimal polynomial $g(T) = (T - \beta_1)(T - \beta_2)\ldots(T - \beta_r) \in \mathbb{K}[T]$ (where $\beta_1 = \beta$ and β_j belong to the algebraic closure $\overline{\mathbb{K}}$ of $\mathbb{K}$) all the roots β_j also belong to $\mathbb{L}$.
An extension $\mathbb{K} \subset \mathbb{L}$ is called the **Galois extension** iff it is normal and separated. (The latter means that $g'(\beta) \neq 0$ for the minimal polynomial $g(T)$ of any $\beta \in \mathbb{L}$; in the case of characteristic zero this condition is empty.)
The automorphism group of a Galois extension is called the **Galois group** and is denoted by $\mathrm{Gal}_{\mathbb{K}}\mathbb{L}$. If $\mathbb{K} \subset \mathbb{L}$ is not Galois, then we use the notation $\mathrm{Aut}_{\mathbb{K}}\mathbb{L}$.
For an extension $\mathbb{K} \subset \mathbb{L}$, an element $\alpha \in \mathbb{L}$ is called **primitive** iff $\mathbb{L} = \mathbb{K}(\alpha)$. If a primitive element exists, then the extension is called **simple**.

2. **Theorem.**

(a) *Any extension of the form* $\mathbb{L} = \mathbb{K}(\beta_1, \ldots, \beta_r)$ *is simple.*

(b) *Let* $\mathbb{L} = \mathbb{K}(\alpha)$. *Then any element* σ *of the group* $\mathrm{Aut}_{\mathbb{K}}\mathbb{L}$ *is uniquely defined by the value* $\sigma(\alpha) = \alpha_i$, *where* $\alpha_i \in \mathbb{L}$ *is a root of the minimal polynomial* $g(T)$. *Therefore*

$$|\mathrm{Aut}_{\mathbb{K}}\mathbb{L}| \leq [\mathbb{L} : \mathbb{K}] \stackrel{\mathrm{df}}{=} \dim_{\mathbb{K}} \mathbb{L}$$

and the equality holds only when the extension is Galois.

(c) *If* $\mathbb{K} \subset \mathbb{L} = \mathbb{K}(\alpha)$ *is Galois, then the group* $G = \mathrm{Gal}_{\mathbb{K}}\mathbb{L}$ *acts transitively on the roots of the polynomial* $g(T)$ *and* $\mathbb{L}^G = \mathbb{K}$.

Proof. (a) Assume that $r = 2$ (it is sufficient), i.e., $\mathbb{L} = \mathbb{K}(\beta, \gamma)$. If $f(T) = (T - \beta_1) \dots (T - \beta_k)$ and $g(T) = (T - \gamma_1) \dots (T - \gamma_l)$ are be the the minimal polynomials for $\beta = \beta_1$ and $\gamma = \gamma_1$, then we put the primitive element in the form $\alpha = \beta + a\gamma$, where $a \in \mathbb{K}$ is such that $\beta_i + a\gamma_j \neq \beta + a\gamma$ for all $(i, j) \neq (1, 1)$. Next one shows that $\beta \in \mathbb{K}(\alpha)$. Indeed, the element β is a unique common root of the two polynomials $g(T)$ and $f(\alpha - aT)$, both from $\mathbb{K}(\alpha)[T]$. So, $T - \beta = \gcd(g(T), f(\alpha - aT)) \in \mathbb{K}(\alpha)[T]$.

(b) Since the $\mathbb{K}$-vector space $\mathbb{L} = \mathbb{K}(\alpha) = K[T]/(g(T))$ has the basis $1, \alpha, \dots, \alpha^{n-1}$, any automorphism σ of $\mathbb{L}$ over $\mathbb{K}$ is determined by its value on α. Of course, $\sigma(\alpha) = \alpha_i$ for some i. However, we can take only those σ for which $\alpha_i \in \mathbb{L}$; thus the inequality $|\mathrm{Aut}| \leq n$. The equality holds when all $\alpha_i \in \mathbb{L}$, which means normality.

Finally, (c) follows from (b). □

3. **Theorem (Fundamental theorem of Galois Theory).** *Let* $\mathbb{K} \subset \mathbb{L}$ *be a Galois extension. There exists a one-to-one correspondence between the intermediary fields* $\mathbb{K} \subset \mathbb{M} \subset \mathbb{L}$ *and the subgroups* $H \subset G = \mathrm{Aut}_{\mathbb{K}}\mathbb{L}$ *given by the maps*

$$\mathbb{M} \longmapsto H = \mathrm{Gal}_{\mathbb{M}}\mathbb{L}, \quad H \longmapsto \mathbb{M} = \mathbb{L}^H.$$

Proof. We have to show that the compositions

$$\mathbb{M} \longmapsto H \longmapsto \mathbb{L}^H, \quad H \longmapsto \mathbb{M} \longmapsto \mathrm{Gal}_{\mathbb{M}}\mathbb{L}$$

are identities. Note that above the extensions $\mathbb{M} \subset \mathbb{L} = \mathbb{K}(\alpha) = \mathbb{M}(\alpha)$ are also Galois.

The inclusions $\mathbb{M} \subset \mathbb{L}^H$ and $H \subset \mathrm{Gal}_{\mathbb{M}}\mathbb{L}$ are obvious. The equality $\mathbb{L}^H = \mathbb{M}$ follows from Theorem 2(c). Finally, suppose $H \neq H_1 = \mathrm{Gal}_{\mathbb{L}^H}\mathbb{L}$; then $[G : H_1] = [\mathbb{L} : \mathbb{L}^H]$. But the equality $\mathbb{M} = \mathbb{L}^{\mathrm{Gal}_{\mathbb{M}}\mathbb{L}}$ for $\mathbb{M} = \mathbb{L}^H$ implies that $\mathbb{L}^{H_1} = \mathbb{L}^H$. Thus $[G : H_1] = [G : H]$ and the inclusion $H \subset H_1$ gives $H = H_1$. □

11.36. The monodromy group of an algebraic function. Take $N' = \mathbb{CP}^1 \setminus$ (critical values) and $M' = \pi^{-1}(N')$. The restriction $\pi|_{M'}$, which we again denote by π, is a topological covering. We fix a base point $a \in N'$ and a base point $b \in M'$ (above a), corresponding to the germ f_a (from which the construction of the Riemann surface has started). We have the distinguished fiber $M_a = \pi^{-1}(a)$.

The **monodromy group of the algebraic function** f is the image of the representation of the fundamental group $\pi_1(N', a)$ in the permutation group $S(M_a)$ of the distinguished fiber. A loop $\gamma \in \pi_1(N', a)$ acts on a germ f_a as prolongation of this germ along γ and defines a permutation $\Delta_\gamma \in S(M_a)$. We denote the monodromy group of f by $\mathrm{Mon}(f)$. It is called sometimes the *topological Galois group of the algebraic function* f.

Because we assumed that the polynomial $F(x,y)$ defining f is irreducible, the Riemann surface M is also irreducible. This means that $M \setminus$ (singular points) is connected and hence also M' is connected. Thus any two points in the fiber M_a can be joined by a curve in M'. This means that:

$\mathrm{Mon}(f)$ *acts transitively on the fiber* M_a.

The action of the monodromy can be described, by means of actions of simple loops γ_j around critical values $x_j \in \mathbb{CP}^1$. If $p = (x_j, y_j) \in M$ is a ramification point of the map π of index $\nu = \nu(p)$, then ν branches $f_{i_1}, \ldots, f_{i_\nu}$ are joined together at p. If properly ordered, they undergo the cyclic permutation $f_{j_i} \longmapsto f_{j_{i+1}}$. The decomposition of monodromy maps generated by simple loops into the product of non-intersecting cycles is the same as grouping of the branches above neighborhood of a critical value into singles and clusters glued at the critical points (above this critical value). This situation is presented in Fig. 11.1.

Proposition. *We have* $\mathrm{Mon}(f) = \mathrm{Gal}_{\mathbb{C}(N)}\mathbb{C}(N)(f_1, \ldots, f_n)$, *where the extension* $\mathbb{C}(N) \subset \mathbb{C}(N)(f_1, \ldots, f_n)$ *is Galois (i.e., normal).*

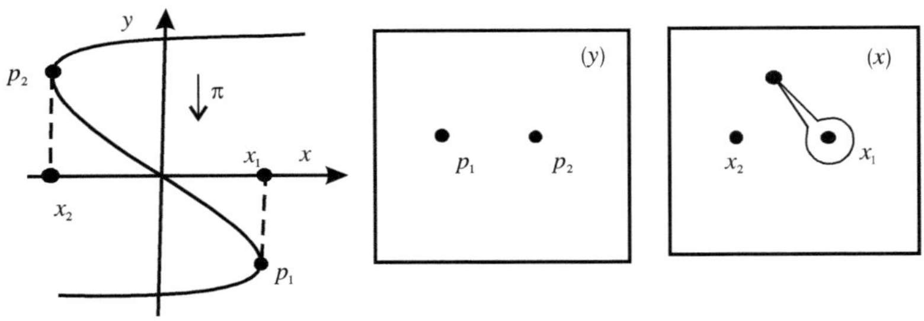

Figure 11.1: Ramified covering

Proof. Here the Galois group is the differential Galois group of the Picard–Vessiot extension L of the field K of rational functions by means of branches $f_{j,a}(x)$. We also treat L is a field consisting of multivalued functions on $\mathbb{CP}^1 \setminus$ (critical values). Any element of the Galois group $\mathrm{Gal}_K L$ is defined by its action on the local branches $f_{j,a}(x)$ (roots of the algebraic equation $F(\cdot, y) = 0$); it acts by permutations of the branches.

It is clear that the monodromy maps define permutations which belong to the Galois group, $\mathrm{Mon}(f) \subset \mathrm{Gal}_K L$. If the Galois group were bigger than the monodromy group, then there should exist a function $g(x) \in L$ which is invariant for the action of $\mathrm{Mon}(f)$ and not invariant with respect to $\mathrm{Gal}_K L$. $g(x)$ should be regular (with power type singularities) and single valued (invariance with respect to the monodromy). Thus the singularities are poles and g should be a rational function.

(*The proof using the Schlesinger theorem.* We can treat L as a Picard–Vessiot extension of $K = \mathbb{C}(x)$ associated with an nth order differential equations with regular singularities and such that the branches f_j form a fundamental system of solutions of this equation. The existence of such an equation was proved by Riemann (see Theorem 8.35 and its Corollary 1). The elements of $\mathrm{Gal}_K L$ are operators in the vector space V (over $\mathbb{C}$) spanned by the branches $f_{j,a}$. Also the monodromy group of f is the same as the monodromy group of the differential equation. The proposition follows from the Schlesinger theorem 11.21: $\mathrm{Gal}_K M$ is the algebraic closure of $\mathrm{Mon}(f)$. Because the both groups are finite, they are equal.) □

11.37. The deck transformations group of a ramified covering. The **deck transformations group** of the algebraic function f (or of the covering $M \longmapsto N$) consists of homeomorphisms $\Phi' : M' \longmapsto M'$ (where $M' = p^{-1}(N\backslash\text{sing.})$ which preserve fibers, i.e., $\Phi' \circ p' = p'$. (It is easy to see that, any such Φ' is analytic and is prolonged to an automorphism Φ of M). This group is denoted by $\mathrm{Deck} = \mathrm{Deck}(f) = \mathrm{Deck}(M \longmapsto N)$.
The covering $M \longmapsto N$ is called the **Galois covering** iff the group Deck acts transitively on fibers.

Proposition. *We have* $\mathrm{Deck}(M \longmapsto N) \simeq \mathrm{Aut}_{\mathbb{C}(N)}\mathbb{C}(M)$, *where* π^* *embeds* $\mathbb{C}(N)$ *into* $\mathbb{C}(M)$ *(but the extension* $\mathbb{C}(N) \subset \mathbb{C}(M)$ *may be not Galois).*

Proof. (We follow the book [**For**].) Any deck transformation Φ defines an automorphism Φ^* of $\mathbb{C}(M)$, which is trivial on functions which are constant on fibers of p; thus $\mathrm{Deck}(f) \subset \mathrm{Aut}_{\mathbb{C}(N)}\mathbb{C}(M)$.
Consider an automorphism $\sigma \in \mathrm{Aut}_{\mathbb{C}(N)}\mathbb{C}(M)$. We apply it to the function $\tilde{f} : M \longmapsto \mathbb{C}$, which defines the algebraic function f by $f = \tilde{f} \circ p^{-1}$. Then $(\sigma\tilde{f}) \circ p^{-1} : N \longmapsto \mathbb{C}$ is a multivalued function on N with algebraic singularities and defines a Riemann surface M_1. The surface M_1 is associated with the same algebraic equation (11.22) as M, i.e., $F(x, y) = 0$. But the Riemann surface M is unique up to a fiber isomorphism. This shows that M_1 is fiber isomorphic to M. That in order defines a deck transformation $\Phi : M \longmapsto M$. □

11.38. Theorem. *Choose a base point* $a \in N'$. *Let* $M_a = \{b_1, \ldots, b_d\}$. *Denote* $\Gamma = \pi_1(N', a)$, Γ_{b_j} *=stabilizer of* b_j *in the monodromy action of* Γ, $\Pi = \pi_*\pi_1(M', b_1) \subset \Gamma$ *and* $N(\Pi) = \{\sigma \in \Gamma : \sigma\Pi\sigma^{-1} = \Pi\}$ *=normalizer of* Π. *Then the following properties hold:*

(a) $\mathrm{Mon} \simeq \Gamma / \bigcap \Gamma_{b_j}$;

(b) $\mathrm{Deck} \simeq N(\Pi)/\Pi$;

(c) *the equality* $\mathrm{Deck} = \mathrm{Mon}$ *holds if and only if* Π *is a normal subgroup of* Γ, *the covering is Galois and* Mon *is a group of order* $d = \deg(M \longmapsto N)$. *It means that the covering is Galois iff the extension* $\mathbb{C}(N) \subset \mathbb{C}(M)$ *is Galois.*

Proof. (We follow [**Zo11**].) (a) Since M is connected, the group Γ acts transitively on M_a. Thus, the formula for Mon expresses the fact that the monodromy group is defined precisely by action of Γ on all elements of M_a.

(b) Let Φ be a deck transformation. It is uniquely defined by its restriction to $V_{b_1} =$ (neighborhood of b_1) $=$ (component of $p^{-1}(U_a)$ containing b_1). Let $V_{b_j} = \Phi(V_{b_1})$. There exists a monodromy map Δ_γ such that $\Delta_\gamma(b_1) = \Phi(b_1) = b_j$. It follows that the deck transformations are defined by those monodromy maps Δ_{γ_0} which can be prolonged from V_{b_1} to fiber diffeomorphisms of M. Moreover, if $\gamma \in \Pi$, then $\Delta_\gamma|V_{b_1} = id|V_{b_1}$ and is prolonged to identical diffeomorphism of M.
The prolongation of $\Delta_\gamma|V_{b_1}$ is realized along paths $\delta \subset M$ which start at b_1. In particular, the prolongation along a closed loop $\delta \in \pi_1(M', b_1)$ should give $\Delta_\gamma|V_{b_1}$ again; this condition is also sufficient. If $\vartheta = \pi(\delta) \in p_*\pi_1(M', b_1) = \Pi$, then the prolongation equals $\Delta_{\vartheta^{-1}\gamma\vartheta}|V_{b_1} = \Delta_{\vartheta^{-1}}|V_{b_j} \circ \Delta_\gamma|V_{b_1} \circ \Delta_\vartheta|V_{b_1}$ and should coincide with $\Delta_\gamma : V_{b_1} \longmapsto V_{b_j}$. Thus, $\Delta_{\vartheta^{-1}}|V_{b_j} = id$, $\Delta_\vartheta|V_{b_1} = id$, and hence $\Delta_{\gamma^{-1}\vartheta\gamma}|V_{b_1} = \Delta_\vartheta|V_{b_1} = id$. This means that $\gamma\vartheta\gamma^{-1} \in \Pi p_*\pi_1(M', b_1)$. Since this holds for any $\vartheta \in \Pi$, we have $\gamma \in N(\Pi)$.

(c) This follows from (a) and (b) and from the points 11.34 and 11.35. □

11.39. Remark. In the literature the notions of the group of deck transformations and of Galois coverings are dominating (see [**For**] for example). They are naturally extended to the case when the Riemann surfaces M and N are algebraic curves defined over number fields of finite characteristics. But there are not so many examples of Galois coverings.
One example of finite Galois ramified covering is given by the equation $y^q - (g(x))^p = 0$, p, q relatively prime, $g(x)$ rational, with the cyclic Galois group $\mathbb{Z}_q$, i.e., the **radical extension**. Another example of Galois covering is given by an algebraic function $y = f(x) = g_1(x)^{p_1/q_1} + \ldots + g_m(x)^{p_m/q_m}$, for different q_i and typical rational functions g_i. Also the universal covering is Galois, but it is infinite.
It is not proved in [**For**] that, for a typical algebraic function (on $\mathbb{C}$), the group of deck automorphisms of the corresponding ramified covering is trivial. Take for example the algebraic function given by $y^3 - 3y - x = 0$ with the critical points $p_{1,2} = (\mp 2, \pm 1)$ and the critical values $x_{1,2} = \mp 2$ (see Fig. 11.1). The Riemann surface M is isomorphic to $\mathbb{CP}^1$ (rational and parametrized by y). Any deck automorphism of the surface $M' = M \setminus$ (ramification points), is prolonged to an automorphism of M. Near a critical point p_j, it either exchanges the two local branches near it (then the third branch is fixed) or it keeps all three branches $f_{1,2,3}(x)$ fixed. Because this automorphism is analytic, it must be identity.
For more informations about relations between the monodromy group and the deck transformations group we refer the reader to [**Zo11**].
V. Arnold in [**Arn8**] says that the Galois theory should be explained via the monodromy group, rather than via the group Deck or via the algebraic extensions. Such approach is presented in the article of A. Khovanski [**Kh4**] and in the book (for high school students) of V. Alexeev [**Al**]. This is done in the next point.

11.40. Non-solvability of a general algebraic equations in quadratures. (We follow mainly [**Al**] and [**Zo10**].)

1. The radical extensions, i.e., defined by $f = (g(x))^{p/q}$ with rational $g(x)$, are extensions by adjoining exponent of integral and have cyclic monodromy group $\mathrm{Mon}(f)$ (equal to the differential Galois group). Thus the extension of Theorem 11.13(a) to the case of finite differential fields extensions gives the following property:

The algebraic equation $F(x,y) = 0$ is solvable in radicals if and only if its monodromy group is solvable.

2. Consider the case of a general algebraic equation. We assume that:

The projective curve $\Gamma \subset \mathbb{CP}^2$ defined by $F = 0$ is smooth and the finite ramification points of the projection $\pi : M \longmapsto \mathbb{CP}^1$ have ramification index 2 with different critical values.

Algebraic functions satisfying this assumption are typical. For example, $F = 5y^5 - 25y^3 + 60y - x$ defines such a function.
The monodromy transformation, corresponding to a simple loop in the x-plane surrounding just one critical value, acts as transposition of two branches joined at the corresponding critical point. Moreover, the smoothness of Γ implies its irreducibility. This implies that:

The monodromy group of this algebraic function is generated by transposition and acts transitively on the fiber.

3. **Lemma.** ([**Kh4**]) *Any subgroup G of the symmetric group $S(n)$, which is transitive and is generated by transpositions, equals to $S(n)$.*

Proof. We say that a subset $A \subset \{1, \dots, n\}$ is *complete* iff each permutation of A extends to some permutations of $\{1, \dots, n\}$ belonging to G. If a transposition $(i, j) \in G$, then the set $\{i, j\}$ is an example of complete subset. Take A_0 a complete subset of maximal cardinality. We claim that $A_0 = \{1, \dots, n\}$.
Suppose that A_0 is a proper subset. There is a transposition $\tau = (ij) \in G$, with $i \in A_0$ and $j \notin A_0$. The group generated by $S(A_0)$ and τ is equal to $S(A_0 \cup \{j\})$ and the set $A_0 \cup \{j\}$ is complete, which contradicts the maximality of A_0. □

4. The above shows that, for typical algebraic functions of degree n, the problem of their representability in radicals is equivalent to the solvability of the group $S(n)$.
The resolvents for the groups $S(2)$, $S(3)$, $S(4)$ are the following:

$$S(2) \supset \{e\},$$
$$S(3) \supset A(3) \supset \{e\},$$
$$S(4) \supset A(4) \supset V \supset \{e\},$$

where $A(n)$ is the subgroup of even permutations (the *alternating group* containing compositions of even number of transpositions), $A(3)$ consists of cyclic permutations, $V = \{e; (12)(34); (13)(24); (14)(23)\}$ is the *Klein's Vierergruppe.*

5. **Theorem.** *The groups $S(n)$, $n \geq 5$, are non-solvable.*

Proof. (We use a proof from [**Bro**].) The non-solvability of $S(n)$, $n \geq 5$, is equivalent to the non-solvability of $A(n)$. (In fact, the groups $A(n)$, $n \geq 5$, are simple, i.e., without proper normal subgroups, but we do not prove it here.) The non-solvability of $A(n)$ follows from the following observation.
If the cycles $\sigma = (123)$ and $\tau = (345)$ (with one common element) belong to a subgroup $H \subset A(n)$, then the elements $[\sigma, \tau] = (\sigma(3)\sigma(4)\sigma(5)) \cdot \tau^{-1} = (145) \cdot (354) = (143)$ and $[\sigma^{-1}, \tau^{-1}] = (253)$ belong to the commutator $[H, H]$. The latter two permutations are also 3-cycles with one common element.
Repeating this argument we see that all derivative groups $A(n)^{(j)}$ contain two 3-cycles with one common element and cannot be trivial. □

We have proved the following result.

6. **Theorem of Abel and Ruffini (Solvability in radicals).** *A general algebraic function with n branches cannot be expressed by means of radicals iff $n < 5$.*

7. **Remark.** In [**Al**] (see also [**Zo10**]) there is a purely topological proof of solvability of the monodromy group of an algebraic function expressed by radicals. It is related with the theory developed in the further points (see below). So, we shortly present the appropriate arguments.
Assume that algebraic functions $f(x)$ and $g(x)$ have solvable monodromy groups $F = \mathrm{Mon}(f)$, $G = \mathrm{Mon}(g)$. We claim that the groups $H = \mathrm{Mon}(f \pm g)$, $\mathrm{Mon}(f \cdot g)$, $\mathrm{Mon}(f/g)$ and $E = \mathrm{Mon}(\sqrt[k]{f})$ are solvable too.
Recall that the Riemann surface of the function f (with n branches $f_1(x), \ldots, f_n(x)$) is constructed from n sheets of the complex plane cut along radii from singular points to infinity. These sheets are next glued along the ridges of cuts in the way the monodromy around singular points dictates. Similarly, the Riemann surface of the function g (with branches $g_1(x), \ldots, g_m(x)$) is realized.
The Riemann surface of the function $f + g$ is constructed in two steps. Firstly we take $n \cdot m$ sheets of $\mathbb{C}$ (labelled by h_{ij}), cut along rays from all singular points (of f and g). We glue these sheets in the obvious way: if after surrounding a singularity a sheet f_{i_1} was glued with f_{i_2} and g_{j_1} was glued with g_{j_2}, then we have to glue $h_{i_1 j_1}$ with $h_{i_2 j_2}$. We obtain a surface M_1. In the second step we identify the sheets h_{ij}, at which the values of the function $f_i(x) + g_j(x)$ coincide. We obtain the Riemann surface $M = M_1/\sim$. For example, the function $y = \sqrt{x} + \sqrt{x}$ has 3 values and satisfies the equation $y^3 - 4xy = 0$. (Analogously, one constructs the Riemann surfaces of the functions $f - g$, fg, f/g).
Also the monodromy group $H = \mathrm{Mon}(f + g)$ is constructed in two steps. The monodromy group associated with the surface M_1 is equal to $F \times G$ and is solvable. Any element from the group H arises from lifting to the surface M of a loop in

$\mathbb{C} \setminus$ (singularities). The same loop admits lifting to the surface M_1 and defines an element from $F \times G$. Therefore we have a surjective homomorphism $F \times G \longmapsto H$. From this the solvability of H easily follows.
Consider now the function $e = \sqrt[k]{f}$. Its Riemann surface is obtained by multiplying each sheet f_i, i.e., by replacing it by k sheets e_{ij}, $j = 0, \ldots, k-1$. The singular points (and the corresponding cuts) correspond to singular points of f and zeroes and poles of f_i. It is also clear how the sheets are glued. Here we have a surjective homomorphism $E = \mathrm{Mon}(e) \longmapsto F$ whose kernel is abelian; this kernel is a subgroup of a product $\mathbb{Z}/k\mathbb{Z} \times \ldots \times \mathbb{Z}/k\mathbb{Z}$. (In [**Zo10**] it was wrongly claimed that this kernel is a subgroup of $\mathbb{Z}/k\mathbb{Z}$.) It is easy to see that E is solvable.

11.41. Khovanski's definition of the monodromy group. We follow the A. Khovanski's paper [**Kh4**]. Here we shall concentrate on functions of one complex variable, but the results can be generalized to functions on $\mathbb{C}^n$ with not too complicated singular set.
A multivalued analytic function is called an S-**function** iff its set of singular points is no more than countable.
Of course, algebraic functions are S-functions. Another example is the elementary function $\ln(1 - x^\alpha)$, where α is irrational, with singularities at 0 and in a dense subset of $\mathbb{S}^1$.
The next result is relatively simple and we present it without proof.

Proposition. *The class of S-functions is closed with respect to differentiation, integration, compositions, solutions of algebraic equations (with S-functions as coefficients), meromorphic operations (i.e., $(f_1, \ldots, f_n) \longmapsto F(f_1, f_1', \ldots, f_1^{(k)}, \ldots, f_n^{(l)})$, F – meromorphic), and solutions of linear differential equations. (Note that the exponentiation belongs to the class of meromorphic operations.)*

If $f = f(x)$ is an S-function, then we define the **Riemann surface of** f as in the case of algebraic functions, i.e., by means of analytic continuations along paths in $\overline{\mathbb{C}} \setminus$ (singular points). Denote it by M and the natural projection onto $\overline{\mathbb{C}}$ by π. Now we cannot claim that $M \longmapsto \overline{\mathbb{C}}$ is a ramified covering. π is well defined on those branches of $f_{j,b}$, which have convergent Taylor expansion, though the point b can be singular for another branch of f.
In what follows we assume that the Riemann surface M is connected. This holds when the construction of M has started from one germ $f_{1,a}$.
For a fixed base point $a \in \overline{\mathbb{C}} \setminus$ (singularities) we have the fixed fiber M_a. If $A \subset \overline{\mathbb{C}}$ is a countable set containing all singular points of f, called the **forbidden set**, then the fundamental group $\pi_1(\overline{\mathbb{C}} \setminus A, a)$ acts on M_a by permutations; we obtain a subgroup of the group $S(M_a)$ of permutations of the fiber M_a. We call it the **monodromy group associated with the forbidden set** A and denote it by $\mathrm{Mon}_A(f)$. (The introduction of the notion of forbidden set is necessary when we want to perform some operations on multi-valued functions.)
In order to eliminate the dependence of the (just defined) monodromy group on the forbidden set, it is useful to consider the closure of the image of $\pi_1(\overline{\mathbb{C}} \setminus A, a)$ in

$S(M_a)$, in the Tikhonov topology in the space $(M_a)^{M_a}$. The latter topology has the following fundamental system of neighborhoods of *id*: $\mathcal{U}_L = \{\sigma : \sigma|_Z = id|_Z\}$, where Z's run through finite subsets of M_a.
The closure of the image of $\pi_1(\overline{\mathbb{C}} \setminus A, a)$ in $S(M_a)$ does not depend on A; it is called the **closed monodromy group of the function** f and is denoted $\overline{\mathrm{Mon}(f)}$.
The above monodromy groups $\mathrm{Mon}_A(f)$ and $\overline{\mathrm{Mon}(f)}$ can be defined also in another way. By connectivity of M, the fundamental group $\Gamma = \pi_1(\overline{\mathbb{C}} \setminus A, a)$ acts transitively on M_a. If $f_j \in M_a$, then we have the *isotropy subgroup* $\Gamma_{f_j} \subset \Gamma$ (or the stabilizer of f_j). The isotropy subgroups are internally conjugated; $\Gamma_{f_k} = \tau \Gamma_{f_j} \tau^{-1}$, where $\tau \in \Gamma$ transforms f_j to f_k. We have

$$\mathrm{Mon}_A(f) = \Gamma / \bigcap_j \Gamma_{f_j} = \Gamma / \bigcap_{\tau \in \Gamma} \tau \Gamma_{f_1} \tau^{-1};$$

it means that the image of Γ in $S(M_a)$ is defined by its action on all elements of the set M_a.
The pair of groups $\left(\pi_1(\overline{\mathbb{C}} \setminus A), \pi_1(\overline{\mathbb{C}} \setminus A)_{f_1}\right)$ is called the **monodromy pair of** f **associated with** A and the pair $\left(\overline{\mathrm{Mon}(f)}, \overline{\mathrm{Mon}(f)}_{f_1}\right)$ is called the **closed monodromy pair of** f.
Often the infinite intersection $\bigcap_{\tau \in \Gamma} \tau \Gamma_0 \tau^{-1}$ can be replaced by finite intersection, over τ's from some finite set P. In this case the image of Γ in $S(M_a)$ forms a discrete subgroup. In such a case one says that f is **almost normal** and we say about the **monodromy pair of** f (it does not depend on the forbidden set).

11.42. Classes of pairs of groups. By a **pair of groups** we mean a pair (Γ, Γ_0) consisting of a group and of its subgroup $\Gamma_0 \subset \Gamma$. The **monodromy group of the pair** (Γ, Γ_0) is the group $\mathrm{Mon}(\Gamma, \Gamma_0) = \Gamma / \bigcap \tau \Gamma_0 \tau^{-1}$. The pair (Γ, Γ_0) is called **almost normal** iff the intersection in the latter expression can be replaced by a finite intersection.

A class $\mathcal{M}$ of pairs of groups is called the **complete class of pairs of groups** iff it satisfies the following conditions:

(i) this class is invariant with respect to images and preimages under homomorphisms (e.g., $(\phi(\Gamma), \phi(\Gamma_0)) \in \mathcal{M}$ for $(\Gamma, \Gamma_0) \in \mathcal{M}$ and $\phi : \Gamma \longmapsto G$) and with respect to topological closures;

(ii) if $(\Gamma, \Gamma_0) \in \mathcal{M}$ and a group Γ_1 satisfies $\Gamma_0 \subset \Gamma_1 \subset \Gamma$, then $(\Gamma, \Gamma_1) \in \mathcal{M}$;

(iii) if $(\Gamma, \Gamma_1) \in \mathcal{M}$, $(\Gamma_1, \Gamma_0) \in \mathcal{M}$, then $(\Gamma, \Gamma_0) \in \mathcal{M}$.

The natural examples of complete classes of pairs of groups are: the class $\mathcal{A}$ containing all abelian groups with their subgroups and the class $\mathcal{F}$ containing all groups with normal subgroups of finite index.
More generally, for a family $\{\mathcal{L}_\alpha\}$ of pairs of groups satisfying the condition (i), the **minimal complete class of pairs** $M\langle\mathcal{L}_\alpha\rangle$ **containing all** $\mathcal{L}_\alpha$ consists of pairs (Γ, Γ_0), for which there is a *resolution tower* of subgroups $\Gamma_0 \subset \Gamma_1 \subset \ldots \subset \Gamma$ such that

each pair (Γ_{i+1}, Γ_i) belongs to one of the $\mathcal{L}_\alpha$'s. If the Γ_i in such tower are normal subgroups in Γ_{i+1}, then the tower is called the **normal tower**.
One can show that, if the monodromy group $(\mathrm{Mon}(\Gamma, \Gamma_0), \{e\})$ of a pair belongs to a complete class, then the pair (Γ, Γ_0) also belongs to the class. The converse is true only for almost normal pairs.
The following minimal complete classes are important:

- $\mathcal{M}\langle\mathcal{A}\rangle$, where $\mathcal{A}$ is the class of all pairs $(G, \{e\})$ with abelian G. Here the resolution towers are normal towers and the monodromy groups of such pairs are solvable.
- $\mathcal{M}\langle\mathbb{C}\rangle$, a subclass of the previous class (i.e., $G \subset \mathbb{C}$).
- $\mathcal{M}\langle\mathcal{F}\rangle$, where $\mathcal{F}$ is the class of pairs consisting of a group Γ and its normal subgroup Γ_0 with finite factor group.
- $\mathcal{M}\langle\mathcal{A}, \mathcal{F}\rangle$, where $\mathcal{F}$ is the class of all finite groups. Here the pairs admit normal towers with quotients either abelian or finite.
- $\mathcal{M}\langle\mathcal{A}, S(n)\rangle$; here the normal towers of pairs have the quotients either abelian or subgroups of $S(n)$.
- $\mathcal{M}\langle\mathbb{C}, \mathcal{F}\rangle$ and $\mathcal{M}\langle\mathbb{C}, S(n)\rangle$.

The reader can observe that the pair $(S(n), \{e\})$ does not belong to the class $\mathcal{M}\langle\mathbb{C}, S(n-1)\rangle$, $n > 4$ (because the group $A(n) \subset S(n)$ is simple).

11.43. Theorem of Khovanski. ([**Kh4**]) *Let $\mathcal{M}$ be some minimal complete class of groups and let $\widehat{\mathcal{M}}$ be the set of S-functions, whose monodromy pairs belong to $\mathcal{M}$. Then the class $\widehat{\mathcal{M}}$ is closed with respect to differentiation, compositions and meromorphic operations.*
If, additionally, $\mathcal{M}$ contains $(\mathbb{C}, \{0\})$, then $\widehat{\mathcal{M}}$ is closed with respect to integrations.
If $\mathcal{M}$ contains $(S(n), \{e\})$, then $\widehat{\mathcal{M}}$ is closed with respect to solving algebraic equations of degree at most n.

Proof. We shall prove only closeness of $\widehat{\mathcal{M}}$ with respect to differentiation. (Other properties are proved in the same way).
If A is the (at most countable) singular set for f, then $\Gamma = \pi_1(\overline{\mathbb{C}} \setminus A, a)$ and Γ_0 consists of those loops γ (with begin and end at a) that the distinguished germ $f_{1,a}$ is invariant with respect to analytic continuation along γ. Because the differentiation operation commutes with analytic prolongation, also the germ $f_{1,a}'$ is invariant with respect to the action of γ. This means that Γ_1, the isotropy subgroup of the germ $f_{1,a}'$ contains Γ_0.
Next one uses the property (ii) from definition of complete class of groups. □

11.44. Corollaries.

(a) *Let a function f be representable by quadratures, compositions and meromorphic operations. Then its closed monodromy pair belongs to the class $\mathcal{M}\langle\mathbb{C}\rangle$ and the closed monodromy group $\overline{\mathrm{Mon}(f)}$ is solvable. In particular, the closed monodromy group of a Liouvillian function is solvable.*

(b) *Let f be a function representable by generalized quadratures, compositions and meromorphic operations. Then its closed monodromy pair belongs to $\mathcal{M}\langle\mathbb{C},F\rangle$. If, additionally, f is almost normal, then its monodromy pair (for any A) belongs to the class $\mathcal{M}\langle\mathbb{C},F\rangle$.*

(c) *Let f be a function representable by n-quadratures (i.e., using solutions of algebraic equations of degree at most n), compositions and meromorphic operations. Then its closed monodromy pair belongs to $\mathcal{M}\langle\mathbb{C},S(n)\rangle$, and, if f is almost normal, then its monodromy pair belongs to $\mathcal{M}\langle\mathbb{C},S(n)\rangle$.*

The advantage of these results over the results from differential Galois theory is obvious. Firstly, the monodromy pairs and closed monodromy pairs constitute more subtle invariants of functions than the differential Galois groups of the corresponding Picard–Vessiot fields. Secondly we admit more operations: meromorphic operations (generalizations of exp) and compositions.
As a particular result we obtain that:

A general algebraic function with $n>4$ branches cannot be representable via rational functions by $(n-1)$-quadratures, compositions and meromorphic operations.

11.45. The monodromy group of the Singer's integrals. (We follow the article [**Zo7**]). Here we are dealing with multi-valued functions on $\mathbb{CP}^n$ of the form

$$H(x)=\int^x R\omega,$$

where $R(x)=\mathrm{e}^g\prod f_i^{a_i}$ is an integrating factor and $\omega=\sum P_i\mathrm{d}x_i$ is a polynomial 1-form. Here $g(x)$ is a rational function and $f_j(x)$, $P_i(x)$ are polynomials and the 1-form $R\omega$ is closed outside singularities. This kind of function has appeared in the theorem of Singer about Liouvillian first integrals 11.30; in the multi-dimensional case it is applied to the Pfaff equation $\omega=0$ where ω is integrable, $\omega\wedge\mathrm{d}\omega=0$.
We see that the integral H has singularities along finite number of algebraic hypersurfaces $S=\sum S_j$, where $S_j=\{f_j=0\}$ or S_j is a polar curve of the rational function $g(x)$. Let M be the Riemann surface of H with the projection $\pi:M\longmapsto\mathbb{CP}^n$. It defines a topological covering outside singularities $\pi:M'\longmapsto N'=\mathbb{CP}^n\setminus S$ (usually of infinite degree). We choose also a base point $a\in N'$ and the distinguished fiber M_a.
The **monodromy group** $\mathrm{Mon}(H)$, in the Khovanski's sense 11.38, is defined as the image in $S(M_a)$ of the fundamental group $\pi_1(N',a)$. Although the fiber M_a is infinite, due to regularity of the singular set S, we should not consider the topological closure of $\mathrm{Mon}(H)$.

The fundamental group of a complement to algebraic set is not easy to determine. It is described in 4.43–4.49. In order to see the action of particular elements of $\pi_1(N')$, one should resolve singularities of the algebraic variety S. In order to avoid technical complications, we assume that the hypersurfaces S_i have only normal intersections as singularities. $\pi_1(N',a)$ is generated by simple loops which lie in a general complex line $L = \mathbb{C} \subset \mathbb{C}^n$ and surround just one point from $L \cap S$. If a loop γ surrounds a point q_j from $L \cap \{f_j = 0\}$, then we can parameterize L by $u = f_j|_L$ and obtain the integral in L of the form

$$H|_L = \int^u \mathrm{e}^{\chi(s)/s^k} s^{a_j} \phi(s) \mathrm{d}s, \tag{11.23}$$

where χ, ϕ are analytic functions.

Assume firstly that $k = 0$, i.e., the exponent e^g is analytic near q_j. Then we have two possibilities:

(i) the exponent a_j is negative integer and the subintegral function has non-negative residuum at $u = 0$,

(ii) $a_j \notin \mathbb{Z}$.

(Otherwise H is meromorphic and non-singular as a map to $\mathbb{CP}^1$).

In the case (i) we have $H = G + c_j \ln u$ with $G(u)$ meromorphic. The corresponding monodromy map takes the form $h_k \longmapsto h_k + 2\pi\mathrm{i}c_j$. Here $h_k = H_{k,a}(a)$ are the 'values' of branches $H_{k,a}(x)$ at $x = a$.

In the case (ii) we have $H = d_j + u^{a_j} G$, G analytic, and the monodromy takes the form $h_k \longmapsto \lambda_j h_k + e_j$, $\lambda_j = \mathrm{e}^{2\pi\mathrm{i}a_j}$, $e_j = d_j(1 - \lambda_j)$. In the case $a_j > -1$ this formula is obvious; in the case $a_j < -1$ we apply the integration by parts formula $(a+1)\int^u s^a \phi(s) = u^{a+1}\phi - \int s^{a+1}\phi'$ (several times). The constant $d_j = H|_{f_j=0}$ is the '*value*' of H at $f_j = 0$.

We see that the monodromy maps are affine diffeomorphisms of the complex line, $\mathrm{Mon}(H) \subset \mathrm{Aff}(\mathbb{C}^1)$. Moreover, these maps are extended to the product $N' \times \mathbb{C}$, $(x,h) \longmapsto (x, \lambda_j h + e_j)$. They preserve the Riemann surface M' (which we can treat as embedded into the product) and realize deck automorphisms of the covering $M' \longmapsto N'$. We have the following result.

11.46. Theorem. ([**Zo7**]) *Let $R = \prod f_j^{a_j}$ and the polynomial 1-form be such that $R\omega$ is closed and let $H = \int^x R\omega$ with the Riemann surface M. Then the monodromy group* Mon(H) *embeds itself as a subgroup into* Aff($\mathbb{C}$). *The covering $M' \longmapsto N'$ is a Galois covering with the group* Deck(M'/N') = Mon(H). *The monodromy group* Mon(H) *is abelian iff the integral H is of the Darbouxian type or of the generalized Darbouxian type.*

Proof. We have to prove only the last point. There are two cases of abelian subgroups of $\mathrm{Aff}(\mathbb{C}^1)$, a subgroup of $\mathbb{C}$ and a subgroup of $\mathbb{C}^*$. The first case occurs when the form $R\omega$ is rational and $H = G + \sum c_j \ln f_j$, G - rational (see the first

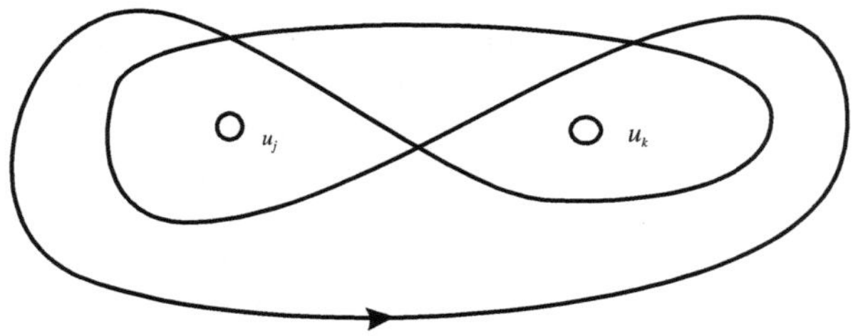

Figure 11.2: Pochhammer cycle

part of the proof of Theorem 11.30). Then e^H is of generalized Darbouxian type. The second case takes place when $H = F \prod f_j^{b_j}$ with rational F.

If the form $R\omega$ has nonzero residuum at f_j and some other exponent a_k is non-integer, then the monodromy group would contain two non-commuting maps (translation and multiplication). Thus we can assume that the only singular hypersurfaces for H are $f_j = 0$ with $a_j \notin \mathbb{Z}$.

We have the Darbouxian case and abelian group $\mathrm{Mon}(H)$, when all the constants d_j, i.e., the 'values' of H at $f_j = 0$ are simultaneously equal (to a constant which can be put $= 0$).

The values d_j are not defined uniquely, but the property: $d_j = d_k$ for all j, k, is correct. The differences $d_j - d_k$ can be calculated as

$$(1-\lambda_j)^{-1}(1-\lambda_k)^{-1} \int_{C(j,k)} (R\omega)|_L,$$

where $C(j,k)$ is the (double) cycle in the complex line L presented at Fig. 11.2. We call this cycle the **Pochhammer cycle**; (it was introduced by L. Pochhammer, see [**WW**]). □

In [**Zo7**] a result analogous to Theorem 11.43 was proved in the case when the exponential function $e^{g(x)}$ is nontrivial. When one replaces the line integral in Eq. (11.23) by a perturbation of an integral with Darbouxian type factors (replacement of e^{χ/s^k} by a product of Darbouxian factors) and takes the limit of suitable monodromy maps, then one obtains the so-called *extended monodromy group*. The extended monodromy group is abelian if and only if H is of generalized Darbouxian type.

11.47. The Darboux–Schwarz–Christoffel integrals. Here we present some examples of polynomial planar vector fields $P\frac{\partial}{\partial x} + Q\frac{\partial}{\partial y}$ (or equations $\frac{\mathrm{d}y}{\mathrm{d}x} = \frac{Q}{P}$, or Pfaff equations $\omega = Q\mathrm{d}x - P\mathrm{d}y = 0$) with Singer first integral. Any student or teacher in mathematics has encountered problems with integration of concrete differential equations.

Examples. 1. We begin with systems with Darbouxian integrals. The following example

$$\frac{\mathrm{d}y}{\mathrm{d}x} = \frac{3 + xy - y^2}{3(x^2 - 4)}$$

from [**Kam**] (Example 1.168) shows how difficult simple equations can be. In [**Kam**] the invariant algebraic curve $f = y^4 - 6y^2 - 4xy - 3 = 0$ was found. The Singer's first integral can be written in the form $H_1 = \int R\omega,\ R = (x^2-4)^{-5/6} f^{-1/2}$. It turns out that there is one more invariant algebraic curve $g = y^4 + 2xy^3 + 6y^2 + 2xy + x^2 - 3 = 0$ and the rational first integral $H_2 = (x^2 - 4) f^3 g^{-3}$. The connection between the two integrals can be obtained from the phase portrait of the vector field. Namely, the point $x = 2, y = -1$ is $1:3$ resonant linearizable node with local first integral $H_3 = (x-2)(y+1+\ldots)^{-3}$. Then we have $H_1 = H_1^{\mathrm{SC}} = \int^{H_3} s^{-5/6}(s-1)^{-1/2}\mathrm{d}s$ (Schwarz–Christoffel) $H_2 = H_2^{\mathrm{D}} = (H_3 - 1)^3 H_3 (14 H_3 - 1)^{-3}$ (Darbouxian).

2. Consider the *Chebyshev integral*

$$f(x) = \int^x s^{\alpha-1}(1-s)^{\beta-1}\mathrm{d}s.$$

One asks when f is of Darbouxian type (modulo a constant). It is so when:

- α and β are both integers;
- either α or β is a positive integer;
- $B(\alpha,\beta) = \int_0^1 s^{\alpha-1}(1-s)^{\beta-1}\mathrm{d}s = 0$, i.e., $\alpha + \beta$ is a negative integer.

Above the Euler Beta-function $B(\alpha,\beta) = \Gamma(\alpha)\Gamma(\beta)/\Gamma(\alpha+\beta)$ is the difference between the 'values' of f at 1 and at 0. In the case $\operatorname{Re}\alpha < 0$ or $\operatorname{Re}\beta < 0$ we use the analytic continuation of the Euler Gamma-functions.
Therefore, $\sin^{-1}(x) = \int_0^x \mathrm{d}s/\sqrt{1-s^2}$ is not Darbouxian, but $\int^x (1-s^2)^{-3/2}\mathrm{d}s$ is.

3. The Darboux–Schwarz–Christoffel integral was introduced in 11.26. Here we define the **multiple Darboux–Schwarz–Christoffel integral** as a function of the form

$$VS(U)^{p/q} + \int^U s(u)^{p/q} T(u)\mathrm{d}u + \sum_{i=1}^{s} \int^{W_i} v^{p/q} \prod_{j=1}^{r_i} (v - v_{ij})^{a_{ij}} \mathrm{d}v, \qquad (11.24)$$

where $U(x,y), V(x,y), W_i(x,y)$ are rational functions of two variables, $S(\cdot), T(\cdot)$ are rational functions of one variable, $v_{ij} \in \mathbb{C}$ and $p/q, a_{ij}$ are rational numbers. Some additional restrictions must be imposed on (11.24), in order to be a first integral of a polynomial Pfaff equation. Note that the poles and zeroes of the function $S \circ U(x,y)$ are branching hypersurfaces of the middle term in (11.24) (and of H). The other integrals also should ramify only at these surfaces. Thus $W_i = S(U)R_i$, where R_i are such that, if $a_{ij} = p_{ij}/q_{ij}$ is not an integer, then the hypersurface $W_i = v_{ij}$ is not a ramification surface, which means that $W_i - v_{ij} =$

$P_{ij}^{q_{ij}}/Q_{ij}$ (P_{ij}, Q_{ij} – polynomials). Also, if $p/q+\sum_j a_{ij}$ is not integer, then the integral does not ramify along $W_i=\infty$.

4. As a more concrete example we take the function, called in [**Zo7**] the *Darboux-hyperelliptic integral,*

$$V\sqrt{S(U)}+\int^U T(u)\sqrt{S(u)}\mathrm{d}u+\sum a_i \ln\frac{R_i-\sqrt{S(U)}}{R_i+\sqrt{S(U)}}.$$

Because

$$\ln\frac{R_i(x)-\sqrt{S(U)}}{R_i(x)+\sqrt{S(U)}}=\int^{S(U)/R_i^2}\frac{\mathrm{d}u}{\sqrt{u}(u-1)},$$

it is included into family (11.24). The reader can easily generalize this example to the case when the square root is replaced with the $(p/q)^{\text{th}}$ power.
In this example the ramification curves are $S(U)=0$, $S(U)=\infty$, the poles of $T(U)$ (with nonzero residuum of the function $T\sqrt{S}$) and the curves $R_i^2=S(U)$ (tangent to $S(U)=0$). The monodromy maps corresponding to loops around the components $S(U)=0$ and $T(U)=\infty$ are the same as in the simple Darboux–Schwarz–Christoffel integral: $h\longmapsto \pm h+c$. The monodromy maps corresponding to the loops around $R_j^2=S(U)$ are of the form $h\longmapsto h\pm 2\pi \mathrm{i} a_j$, where the sign depends on the position of the loop near this hypersurface. (Note, that for fixed $S(U)$, the function R_j can take two different values).
The book of E. Kamke [**Kam**] contains several examples of integrals of the type $V\sqrt{U}+a\ln[(W+\sqrt{U})/(W-\sqrt{U})]$. The above shows that they include themselves into the family of simple Darboux–Schwarz–Christoffel integrals.
It seems that the Darboux-hyperelliptic integrals should play some significant role. The reason for this is that the function $\exp H$ has monodromy group consisting of maps of the form: $h\longmapsto \lambda_i h$, $h\longmapsto \mu_j/h$. One quickly recognizes here a solvable subgroup G of $PSL(2,\mathbb{C})$ such that the commutator $G^{(1)}=[G,G]$ consists of linear maps (corresponding to diagonal matrices) and $G/G^{(1)}=\mathbb{Z}_2$ (compare also Theorem 11.15). Also in the Khovanski's paper [**Kh3**] the case of functions with such monodromy was distinguished.
On the other hand, the Darboux-hyperelliptic integrals form a particular case of the infinite series of analogous integrals with the exponent $1/2$ replaced by p/q.
There is a striking coincidence between the classification of the (solvable) monodromy groups of the above integrals and the classification of solvable subgroups of Diff$(\mathbb{C},0)$ (from Section 3 of Chapter 10). The Darbouxian integrals $\prod f_j^{a_j}$ correspond to linear subgroups of Diff$(\mathbb{C},0)$; here the cases $a_j\in\mathbb{Q}$ (rational integrals) correspond to finite subgroups of Diff$(\mathbb{C},0)$. The logarithms of generalized Darbouxian integrals $g+\sum_1^r a_j\ln f_j$, correspond to formally non-linearizable abelian groups; the latter are typical, if $r>1$, and exceptional, if $r=1$. The Darboux–Schwarz–Christoffel integrals should correspond to solvable groups, but the Darboux-Hyperelliptic integrals $V\sqrt{U}+\sum_1^r a_j\ln[(R_j-\sqrt{U})/(R_j+\sqrt{U})]$, with $r=1$, resemble the exceptional solvable groups.

Also some french mathematicians study the relation between the Singer's theorem and local properties of holomorphic foliations (see [**BT**] and [**Pau**]).

5. This example comes from the work [**Cha**] of J. Chavarriga, where existence of the integrating factor was proved. The equation

$$\frac{\mathrm{d}y}{\mathrm{d}x} = \frac{x(1+x^n+\lambda y^n-2x^2y^{n-2})}{-y(1+x^n+\lambda y^n-2\lambda x^2y^{n-2})}$$

has the multiple Schwarz–Christoffel integral (without Darbouxian summand)

$$\begin{aligned} H &= \int f^a\omega = C_1\int^{S_1}\eta + C_2\int^{S_2}\eta = C_1H_1^{\mathrm{SC}} + C_2H_2^{\mathrm{SC}}, \\ \eta &= u^a(u-1)^b\mathrm{d}u \end{aligned}$$

where $C_1 = (4\lambda)^{2/n}/n$, $C_2 = 4^{2/n}/n, a = -(n+4)/2n$, $b = (2-n)/n$, $f = 1+2(x^n+\lambda y^n)+(x^n-\lambda y^n)^2$, $S_1 = f(1-x^n+\lambda y^n)^{-2}$, $S_2 = f(1+x^n-\lambda y^n)^{-2}$. Here the singular points $1+x^n = y = 0$ and $1+\lambda y^n = x = 0$ are non-elementary (i.e., both eigenvalues vanish) and we must blow-up them. After this done, some exceptional invariant divisors appear. Namely, along these divisors the first integral is ramified.

6. A. Lins-Neto constructed some pencils of foliations defined by forms

$$\omega_\kappa = \omega_0 + \kappa\omega_\infty$$

which have an integrating multiplier R not dependent on the parameter κ. Therefore they have Singerian first integrals

$$H_\kappa^{\mathrm{Sin}} = \int R\omega_\kappa.$$

(i) In [**L-N2**] one finds a remarkable pencil of foliation defined by the form

$$\omega_\kappa = (x^3-1)(x-\kappa y^2)\mathrm{d}y - (y^3-1)(y-\kappa x^2)\mathrm{d}x = \omega_0 + \kappa\omega_\infty$$

with the integrating factor

$$\begin{aligned} R &= \left\{(y-1)(y-\zeta)(y-\bar{\zeta})(x-1)(x-\zeta)(x-\bar{\zeta})(y-x)(y-\zeta x)(y-\bar{\zeta}x)\right\}^{-2/3} \\ &= (F_1\dots F_9)^{-2/3} = \left\{(y^3-1)(x^3-1)(y^3-x^3)\right\}^{-2/3} \end{aligned}$$

where $\zeta = \mathrm{e}^{2\pi\mathrm{i}/3}$. We introduce two rational functions

$$\begin{aligned} S_1 &= F_1F_2F_3/F_4F_5F_6 = \left(y^3-1\right)/\left(x^3-1\right), \\ S_0 &= \left(1-y^{-3}\right)/\left(1-x^{-3}\right) = \left(x^3/y^3\right)S_1. \end{aligned}$$

One can check that the corresponding Singerian integral can be rewritten in form of a multiple Schwarz–Christoffeter integral 2 (without Darbouxian term)

$$H_\kappa^{\mathrm{Sin}} = \frac{\bar\zeta}{3}\int^{S_0}\eta(s) - \frac{\kappa}{3}\int^{S_1}\eta(s) = \frac{\bar\zeta}{3}H_0^{\mathrm{SC}} - \frac{\kappa}{3}H_1^{\mathrm{SC}},$$
$$\eta = \{s(s-1)\}^{-2/3}\,\mathrm{d}s.$$

Indeed, it is easy to check that S_0 (respectively, S_1) is a rational first integral for ω_0 (respectively, ω_∞). Consider the base point $q_3 = [0:0:1]$ of the pencil $\mathcal{P}_0$ defined by the function S_0, $\mathcal{P}_0 = \{S_0(x,y) = c\}_{c\in\mathbb{C}}$. In the blow-up variables $(r,u) = (y, x/y)$ near the base divisor E_3 above q_3 we get

$$\pi^* H_0^{\mathrm{Sin}}|_{E_3} = \int^u (1-\tau^3)^{-2/3}\mathrm{d}\tau = \frac{\bar\zeta}{3}\int^{u^3}\eta(s),$$
$$u^3 = \pi^* S_0|_{E_3}$$

(here and below π is the resolution map). Next, in the blow-up variables $(r,u) = \left(x-1, \frac{y-1}{x-1}\right)$ near E_9 above the base point $q_9 = [1:1:1]$ of the pencil $\mathcal{P}_1$ (defined by S_1) we get

$$\pi^* H_\infty^{\mathrm{Sin}}|_{E_9} = -\frac{1}{3}\int^u \eta(s),$$
$$u = \pi^* S_1|_{E_9}.$$

This demonstrates that the two SC summands in H_κ^{Sin} are obtained from the restriction of $\pi^* H^{\mathrm{Sin}}$ to the base divisors E_3 and E_9 of resolutions of the pencils $\mathcal{P}_0$ and $\mathcal{P}_1$ at q_3 and q_9 respectively.
In [**L-N2**] one can learn about very nice geometry associated with the form.

(ii) For the second pencil from [**L-N3**] we have

$$\omega_0 = 6y(2x-1)\mathrm{d}x + (4x - 9x^2 + y^2)\mathrm{d}y,\ \omega_\infty = 3(y^2 - x^2)\mathrm{d}x + 2y(1-2x)\mathrm{d}y.$$

The form $\omega_\kappa = \omega_0 + \kappa\omega_\infty$ has the integrating factor

$$R(x,y) = F^{-5/6},$$
$$F(x,y) = 4y^2 - 4x^3 - 12xy^2 + 9x^4 + 6x^2y^2 + y^4,$$

and we arrive at the Singerian integral $H_\kappa^{\mathrm{Sin}} = \int R\omega_\kappa$. It is also proved in [**L-N3**] that:

ω_∞ has the first integral $S_1 = F/G_1^3$, $G_1(x,y) = 2x-1$,
ω_1 has the first integral $S_2 = F/G_2^3$, $G_2(x,y) = y - x$,
ω_{-1} has the first integral $S_3 = F/G_3^3$, $G_3(x,y) = -x - y$.

Therefore, one can expect that the Singerian integral $H_\infty^{\mathrm{Sin}} = \int M\omega_\infty$ is a SC integral with the limit $S_1(x,y)$. To check it we analyze the latter integral at the exceptional divisor E_1 (of a resolution π) of the pencil $\mathcal{P}_1 = \{S_1(x,y) = s\}$ at the base point $q_1 = (1/2, 1/2)$.
Firstly, we have $F = z^2 + 4(2x-1)^3$,where $z = y^2 + 3x^2 - 6x + 2 = (y - 1/2) - 3(x - 1/2) + \ldots$; thus $S_1 = z^2/8(x-1/2)^3 + 4$.
Next, $\omega_\infty \approx 3z\mathrm{d}(x-1/2) - 2(x-1/2)\mathrm{d}z$.In the adapted coordinates r, u near E_1 such that $z = r^3 u$ and $x - 1/2 = r^2 u$ (hence $u = (x-1/2)^3/z^2$) we get $\omega_\infty \approx r^5 u\mathrm{d}u$, $F \approx r^6 u^2(32u+1)$, $G_1 = 2r^2 u$, $S_1 \approx 1/8u + 4 =: s(u)$. We find

$$\mathrm{d}\pi^* H_\infty^{\mathrm{Sin}}|_{E_1} = u^{-2/3}(32u+1)^{-5/6}\mathrm{d}u = -\frac{1}{32} s^{-5/6}\left(s-4\right)^{-1/2}\mathrm{d}s.$$

A natural prolongation of this leads to the SC integral

$$-\frac{1}{32} H_1^{\mathrm{SC}} = -\frac{1}{32}\int^{S_1(x,y)} \eta(s),\ \ \eta(s) = s^{-5/6}(s-4)^{-1/2}\mathrm{d}s.$$

Another base point for the pencil $\mathcal{P}_1$ is $q_2 = (1/2, -1/2)$, but ω_∞ is invariant with respect to the reflection $(x,y) \longmapsto (x,-y)$ (time reversibility for the corresponding vector field).
Consider the case $\kappa = 1$. Near the singular point $q_0 = (0,0)$ of the form ω_1 (and the base point of the pencil $\mathcal{P}_2 = \{S_2(x,y) = s\}$) we have $\omega_1 = -6y\mathrm{d}x + (4x + 2y)\mathrm{d}y + \ldots = 6y\mathrm{d}G_2 - 4G_2\mathrm{d}y + \ldots$, $G_2 = y - x$. Next, $F = 4(y - 3G_2^2/2 + \ldots)^2 + 4(G_2 + \ldots)^3 \approx 4(y^2 + G_2^3)$.
Putting $y = r^3 u$ and $G_2 = r^2 u$ (the resolution of q_0 with exceptional divisor E_0) we get $\omega_1 \approx 2r^5 u\mathrm{d}u$, $F \approx 4r^6 u^2(1+u)$ and $S_2 = F/G_2^3 \approx 4\left(1/u + 1\right) =: s(u)$. Therefore,

$$\pi^*\mathrm{d}H_1^{\mathrm{Sin}}|_{E_0} = 2^{-2/3}u^{-2/3}(1-u)^{-5/6}\mathrm{d}u = -s^{-5/6}(s-4)^{-1/2}\mathrm{d}s,$$

i.e.,

$$H_1^{\mathrm{Sin}} = H_2^{\mathrm{SC}} := -\int^{S_2(x,y)} \eta(s),$$

and hence

$$H_\kappa^{\mathrm{Sin}} = \frac{1-\kappa}{32} H_1^{\mathrm{SC}} - H_2^{\mathrm{SC}}.$$

Therefore the Singerian integral is a SC integral of length 2.

(iii) The third pencil from [**L-N3**] is defined by

$$\begin{aligned}\omega_0 &= -2x(y^2-1)\mathrm{d}x + y(x^2-y^2)\mathrm{d}y,\\ \omega_\infty &= 2(x+y)(y^2-1)\mathrm{d}x + (4x - x^3 - x^2 y - 3xy^2 + y^3)\mathrm{d}y,\end{aligned}$$

with the integrating factor

$$R=(F_1F_2F_3F_4)^{-3/4},$$
$$F_{1,2}=x^2+y^2\pm 2x,\quad F_{3,4}=y\pm 1.$$

Moreover:

for $\kappa=0$ one has the rational first integral $S_1(x,y)=F_1F_2/F_3F_4$,

for $\kappa=\infty$ there is the first integral $S_2=F_1F_4/F_2F_3$,

for $\kappa=1/2$ there is the first integral $S_4=F_1F_3/F_2F_4$.

Consider the case $\kappa=0$. We resolve the base point $q_1=(1,1)$ of the pencil $\mathcal{P}_1=\{S_1=s\}$. In local coordinates $\tilde{y}=y-1$ and $z=x-\tilde{y}/3$ we have $F_1\approx 4$, $F_2=2\tilde{y}(1-z+\tilde{y}/3)+z^2\approx 2\tilde{y}+z^2$, $F_3\approx 2$, $F_4=\tilde{y}$, $S_1\approx 2(2\tilde{y}/z^2+1)$ and $\omega_0\approx -4\tilde{y}\mathrm{d}x+2(x-y)\mathrm{d}y=-4\tilde{y}\mathrm{d}z+2z\mathrm{d}\tilde{y}$. The adapted coordinates r,u near a divisor E_1 are such that $z=r$, $\tilde{y}=r^2u$. Then $F_2\approx r^2(2u+1)$, $F_4=r^2u$, $S_1\approx 2\,(2+1/u)=:s(u)$, $\omega_0\approx -2r^3\mathrm{d}u$ and

$$\begin{aligned}\mathrm{d}\pi^*H_0^{\mathrm{Sin}}|_{E_1}&=-2^{-5/4}u^{-3/4}(2u+1)^{-3/4}\mathrm{d}u=\frac{1}{2}s^{-3/4}\,(s-4)^{-1/2}\,\mathrm{d}s\\&=\frac{1}{2}\mathrm{d}\pi^*H_1^{\mathrm{SC}}|_{E_1},\quad H_1^{\mathrm{SC}}=\int^{S_1}s^{-3/4}(s-4)^{-1/2}\mathrm{d}s.\end{aligned}$$

Consider the case $\kappa=\infty$. Here we resolve the point $q_0=(0,0)$ with local coordinates x and $z=y-x$ and with adapted coordinates r,u (near E_0) such that $z=r$ and $x=r^2u$. We have $F_{1,2}\approx \pm 2x+z^2=r^2\,(1\pm 2u)$, $F_{3,4}\approx\pm 1$, $S_2\approx\frac{2u+1}{2u-1}=:\sigma(u)$ and $\omega_\infty\approx 4x\mathrm{d}z-2z\mathrm{d}x=-2r^2\mathrm{d}u$. Therefore

$$\begin{aligned}\mathrm{d}\pi^*H_\infty^{\mathrm{Sin}}|_{E_0}&=-2(4u^2-1)^{-3/4}\mathrm{d}u=(1/\sqrt{2})\sigma^{-3/4}\,(\sigma-1)^{-1/2}\,\mathrm{d}\sigma=\frac{1}{2}\mathrm{d}\pi^*H_2^{\mathrm{SC}},\\ H_2^{\mathrm{SC}}&=\sqrt{2}\int^{S_2}\sigma^{-3/4}(\sigma-1)^{-1/2}\mathrm{d}\sigma=\int^{4S_2}s^{-3/4}(s-4)^{-1/2}\mathrm{d}s.\end{aligned}$$

Hence

$$H_\kappa^{\mathrm{Sin}}=\frac{1}{2}\left(H_1^{\mathrm{SC}}+\kappa H_2^{\mathrm{SC}}\right).$$

The author conjectures that the above examples reveal some general rule. Probably any Singer's integral with non-abelian monodromy group is either a simple Darboux–Schwarz–Christoffel integral or multiple Darboux–Schwarz–Christoffel integral. Moreover, if the exponential factor e^g is nontrivial or if some $a_i\notin\mathbb{Z}$, then we should have the simple Darboux–Schwarz–Christoffel integral. Some progress in this direction was achieved in the paper [**Sca**] by B. Scárdua (the proof in so-called generalized curve case).

7. If a rational **Riccati equation**

$$\mathrm{d}y/\mathrm{d}x=A(x)+B(x)y+C(x)y^2 \tag{11.25}$$

has algebraic invariant curve, then it has a particular solution of the form $y = \phi(x)$, where ϕ is an algebraic function. Introducing the variable $y_1 = y - \phi(x)$, we obtain a Bernoulli equation which is solved in quadratures. This shows that rational Riccati equations with algebraic particular solution are integrable in generalized quadratures; result of Liouville, see [**Lio2**], [**Rit**]. We can say even more, we can guess the form of the first integral.

11.48. Theorem (first integrals for Riccati equation). ([**Zo8**], [**Zo12**]) *Assume that the rational equation* (11.25) *has an invariant algebraic curve not containing any vertical line* $x = \text{const}$. *Then there are four possibilities:*

- *the invariant curve is rational,* $y = W(x)$, *and there is a simple Darboux–Schwarz–Christoffel integral;*
- *the curve is of the form* $[y - W_1(x)][y - W_2(x)] = 0$ *(two rational components) and there is a Darbouxian integral;*
- *the curve is hyperelliptic* $y^2 + W_1(x)y + W_2(x) = 0$ *and there is a Darboux-hyperelliptic integral;*
- *the curve has* > 2 *sheets above the* x*-plane and there is a rational integral.*

Proof. The first three cases are elementary. The fourth case uses the monodromy of solutions $y = \phi(x)$ as x turns around singular points, which are the poles of the right hand side of Eq. (11.25). We fix a base point $x = a$ and consider solutions $y = \phi(x; y_0)$ with the initial condition $\phi(a; y_0) = y_0$. The monodromy group is the group of maps of the line $\{a\} \times \mathbb{C}$. Because the evolution operators for Riccati equation are fractional-linear (see Remark 11.18(c)), the monodromy group Mon of the Riccati equation consists automorphism of the Riemann sphere $\{a\} \times \overline{\mathbb{C}}$, $\mathrm{Mon} \subset PSL(2, \mathbb{C})$.
(The monodromy group of the first integral defined in Example 11.27(d) is the same as the monodromy group Mon of the equation and is also included into $PSL(2, \mathbb{C})$.)
If Φ is an invariant algebraic curve, then it is preserved by the monodromy operations. It means that the finite set $K = \Phi \cap (\{a\} \times \overline{\mathbb{C}})$ is invariant for Mon. By assumption, the cardinality of K is at least 3. This means that the subgroup Mon_K of maps, whose restriction to K are identities, is trivial. Indeed, three points x_0, x_1, x_2 of K can be transformed via a Möbius map to $0, 1, \infty$ (composition of translation $id - x_0$, of $x/(x - x_2)$ and of x/x_1) and any fractional-linear transformation which has $0, 1, \infty$ as fixed points is identity. Now the exact sequence $\{e\} \longmapsto \mathrm{Mon}_K \longmapsto \mathrm{Mon} \longmapsto S(K) \longmapsto \{e\}$ shows that Mon is a finite group.
The finiteness of the monodromy group of Riccati equation means that its solutions behave like algebraic functions. If we knew that these solutions have regular singularities, then we would conclude algebraicity of solutions. However, the example of family $y = Ce^{-1/x^2}$ of univalent functions with non-algebraic singularity shows that we must work a little more.

The corresponding planar vector field takes the form

$$\dot{x} = G(x),\ \dot{y} = D(x) + E(x)y + F(x)y^2$$

where $D, E, F, G = \prod(x - x_j)$. are polynomials. The singularities of solutions of the Riccati equation lie in the points $x = x_j$. The lines $\{x = x_j\} \subset \mathbb{C}^1 \times \mathbb{C}^1$ are invariant lines of the polynomial vector field associated with the Riccati equation. This field defines a holomorphic foliation $\mathcal{F}$ (by complex phase curves) of $\mathbb{CP}^1 \times \mathbb{CP}^1$. The singular points of the foliation $\mathcal{F}$ lie in the projective lines $\{x = x_j\} \times \mathbb{CP}^1$ and $\{x = \infty\} \times \mathbb{CP}^1$. Each such line contains at most 2 singular points. Thus we look at the behavior of phase curves near such singular points.

Consider one such invariant line, which we may assume to be $x = 0$. The curve Φ intersects it at ≥ 3 points, multiplicity counting. Each local branch of Φ represents an algebraic solution $y = \varphi(x)$, where the function φ admits a Puiseux expansion. Firstly, we perform the pull-back $x = u^q$, in order to remove multi-validity of the local branches of Φ. We get $y = cu^l + \ldots$, l – integer, for any such branch and the (projective) line $u = 0$ contains a singular point of the foliation (induced from $\mathcal{F}$) through which at least 2 such branches pass. We choose a pair $y = \psi_1(u)$, $y = \psi_2(u)$ with the greatest order of tangency. Next, we separate the branches $y = \psi_{1,2}(u)$ using a series of changes of the form $y = zu^m$ and/or $y = \psi(u) + z$ (where $y = \psi(u)$ is some branch). Note that the rational Riccati form of the differential equation is preserved during such operations. We arrive to the situation with the branches $z = \chi_{1,2}(u) = c_{1,2} + O(u)$ and $c_1 \neq c_2$. The third branch $z = \chi_3(u)$ of Φ will have the form: either $\chi_3(u) = c_3 + O(u)$, $c_3 \neq c_{1,2}$, or $\chi_3(u) \sim cu^{-j}$, $j > 0$. In the latter case the branch $z = \chi_3(u)$ meets the line $u = 0$ at the point $z = \infty$.

Now we have two possibilities. In the first situation, called the dicritical case, the line $u = 0$ ceases to be invariant and all local solutions are analytic, $z = \chi(u) = c + O(u)$. Therefore all solutions $y = \varphi(x)$ of the initial system are regular near $x = 0$.

In the non-dicritical case the new Riccati system should have at least 3 singular points on $u = 0$: either 3 finite (i.e., $z = c_{1,2,3}$) or 2 finite and one at $z = \infty$. This cannot take place for a Riccati foliation. □

11.49. Remark. Using the classification of solvable finite subgroups of $SL(2, \mathbb{C})$ (see the next chapter) and the above theorem, one can show that the minimal degree (in z) of an invariant algebraic curve for the Riccati equation is 1, 2, 4, 6 or 12. In the first case there is a Darboux–Schwarz–Christoffel integral, in the second case there is a Darboux-hyperelliptic integral and in the latter three cases we have finite monodromy group associated with symmetries of tetrahedron, cube or icosahedron respectively (see the next chapter).

Using this and the relation between the Riccati and second order linear equations J. Kovačic [**Kov**] created an algorithm for solving a second order linear differential equation with rational coefficients (see also [**R-R**]).

§3 Center-Focus Problem

We begin with the following classical result.

11.50. Theorem of Poincaré and Lyapunov. ([**Poi3, Lya**]) *If an elementary singular point for a real analytic vector field is a center, then there exists a local single valued analytic first integral.*

Proof. Since the singular point (which we assume $x = y = 0$ and with standard linear part) is elementary, the eigenvalues are $\pm i\omega$ with the eigenfunctions $x \pm iy$. The existence of a formal first integral is proved using Poincaré–Lyapunov focus numbers, like in the proof of the Bautn's theorem 6.30.
In the polar coordinates r, θ we have the equation

$$dr/d\theta = r^2 f(r, \theta)$$

for the phase curves. On the right-hand side we have a convergent power series in r with coefficients being trigonometric polynomials in θ. The Poincaré return map $P : S = (\mathbb{R}_{\geq 0}, 0) \mapsto S$, $S = (\mathbb{R}_{\geq 0}, 0)$, is defined as the solution after the 'time' $\theta = 2\pi$ of the initial value problem for $\theta = 0$. The center condition means that $P \equiv id$.
But the Poincaré map is analytic, due to the analyticity of the right-hand side of the equation. So, one could put the first integral $G(x, y)$ as $G(r, 0) = r^2$ for $r \geq 0$, and prolong it to a neighborhood of $(0, 0)$ in $(\mathbb{C}^2, 0)$ by the condition that it is constant on the phase curves. But this is not correct, because of possibly wrong values for $(x, 0)$ with $x < 0$.
However, we can consider the monodromy map $\Delta : (\mathbb{R}, 0) \longmapsto (\mathbb{R}, 0)$ such that $\Delta'(0) < 0$ and $P = \Delta \circ \Delta$. Then the function

$$F = \{G + G \circ \Delta\}/2,$$

firstly defined on S, is prolonged to a correct first integral. □

11.51. Analytic solvability or algebraic solvability. The problem of distinguishing of planar vector fields with equilibrium point of the **center** type (i.e., surrounded by periodic solutions) from those of the **focus** type, i.e., the **center-focus problem**, is one of the oldest in the theory of differential equations, (see [**Lya, Poi3, Arn5, AI**]). In the class of smooth or analytic systems the problem is local and was formulated by V. Arnold [**Arn5**] as follows.
The germs of vector fields $\boldsymbol{V} : (\mathbb{R}^2, 0) \to (\mathbb{R}^2, 0)$ with center form a subset of infinite codimension in the space of all germs. So, the problem is divided into steps: find an algorithm deciding whether the nth jet of $\boldsymbol{V}$ has 0 of the focus type (or other type different from center) or is neutral, in which case we should consider $(n+1)^{\text{th}}$ jet of $\mathbf{V}$, etc.
It is known that, when the linear part $j^1\boldsymbol{V}$ is non-zero, then this algorithm is algebraic (see [**AI**]). If $j^1\boldsymbol{V} = 0$ and the codimension of singularity is greater

than 10 then the problem cannot be solved algebraically (see [**Il2**] and the below Example 11.53).
Using the resolution of the singularity, Theorem 9.18, N. Medvedeva [**Med**] proved the following.

11.52. Theorem of Medvedeva. *The center-focus problem is analytically solvable.*

Yu. Ilyashenko [**Il2**] showed that the problem of center cannot be solved algebraically. His proof relies on investigating the monodromy properties of certain function $\phi(\lambda)$ such that the equation $\phi(\lambda)=0$ is the center condition for a family $\boldsymbol{V}_\lambda$ of polynomial vector fields depending algebraically on $\lambda=(\lambda_1,\ldots,\lambda_r)$.
The following explicit example comes from [**LlZ**].

11.53. Example. The family is

$$\begin{aligned}\dot{x} &= -y\left\{(2\mu+1)x^2+y^2\right\}+x^2\left\{\lambda_1 x^2+\lambda_2(x^2+y^2)\right\},\\ \dot{y} &= x\left\{x^2+(1-2\mu)y^2\right\}+xy\left\{\lambda_1 x^2+\lambda_2(x^2+y^2)\right\}.\end{aligned}$$

(It is similar to the family from [**Il2**], but with the function ϕ expressed in terms of some Bessel function.)
In the variables $z=1/x^2, u=y/x$ the equation for phase curves is given by the linear equation

$$\frac{\mathrm{d}z}{\mathrm{d}u}=2\left\{\frac{2\mu u}{(u^2+1)^2}+\frac{u}{u^2+1}\right\}z-2\frac{\lambda_1+\lambda_2(u^2+1)}{(u^2+1)^2}$$

with the solution $z=(u^2+1)\mathrm{e}^{-2\mu/(u^2+1)}[H-F(u)]$, where

$$F(u)=2\int_{-\infty}^{u}(s^2+1)^{-3/2}\mathrm{e}^{2\mu/(s^2+1)}\left\{\lambda_1+\lambda_2(s^2+1)\right\}\mathrm{d}s$$

and H is a constant. In fact, H is the first integral, $H=(x^2+y^2)^{-1}\mathrm{e}^{\mu x^2/(x^2+y^2)}+F(y/x)$. The point $x=y=0$ is center iff $F(y/x)$ is univalent on $\mathbb{R}^2$, i.e., iff $F(\infty)=0$. Therefore we get the condition

$$\phi\left(\mu,\lambda_1,\lambda_2\right):=\lambda_1\Psi'\left(\mu\right)+2\lambda_2\Psi\left(\mu\right)=0,$$

where

$$\Psi(\mu)=\int_0^\infty \mathrm{e}^{2\mu/(u^2+1)}\left(u^2+1\right)^{-2}\mathrm{d}u$$

Introducing the change $u^2+1=\frac{2}{1+t}$ we find

$$\Psi\left(\mu\right)=\frac{1}{4}\int_{-1}^{1}\mathrm{e}^{\mu(1+t)}\sqrt{\frac{1+t}{1-t}}\mathrm{d}t=\frac{1}{4}\frac{\mathrm{d}}{\mathrm{d}\mu}\int_{-1}^{1}\mathrm{e}^{\mu(1+t)}(1-t^2)^{-1/2}\mathrm{d}t.$$

But $\int_{-1}^{1} \mathrm{e}^{-zt} \left(1-t^2\right)^{-1/2} \mathrm{d}t = \pi I_0(z)$ is the modified Bessel function (see Eq. (10) in 7.12 of [**BE**, vol. **2**]). Therefore,

$$\Psi(\mu) = \frac{\pi}{4} \frac{\mathrm{d}}{\mathrm{d}\mu} \left(\mathrm{e}^{\mu} I_0(-\mu)\right).$$

The function $I_0(-\mu)$ has the following properties (see 8.31(b)):

(i) it is an entire function,

(ii) $I_0(-\mu) \sim C_1 \mu^{-1/2} \mathrm{e}^{-\mu}$ as $\mathrm{Re}\mu(\mu) \to -\infty$,

(iii) $I_0(-\mu) \sim C_2 \mu^{-1/2} e^{\mu}$ as $\mathrm{Re}(\mu) \to +\infty$.

We see that Ψ is not of the form polynomial $\times$ exp(polynomial). The same holds for $\phi = \lambda_1 \Psi'(\mu) + 2\lambda_2 \Psi(\mu)$. Therefore the function $\phi(\cdot, \lambda_1, \lambda_2)$ has infinitely many zeroes in $\mathbb{C}$. This means that $\{\phi = 0\}$ is not an algebraic set.

11.54. Local reversibility and local integrability. Note that, if an analytic vector field $\boldsymbol{V}$ in $(\mathbb{R}^2, 0)$ has center at $x = y = 0$, then it has a local single valued real first integral H.
In the elementary case, i.e., with the eigenvalues $\pm i\omega$, the first integral is analytic, $H = x^2 + y^2 + \ldots$ (the Poincaré–Lyapunov theorem [**Poi3**, **Lya**]). In the case of nilpotent singularity $\boldsymbol{V} = y\frac{\partial}{\partial x} + \ldots$ we have $H = y^2 + \ldots$, but it can be non-analytic (see [**AI**] and 11.60 below).
There exist two natural local situations when $\mathbf{V}$ has center at 0.

Firstly, the first integral H can be **generalized Liouvillian**. Thus, following 11.12, there exists a chain $K_0 \subset K_1 \subset \ldots \subset K_n$ of Picard–Vessiot extensions such that:

- $K_0 = \mathbb{C}\{x, y\}$, the field of meromorphic functions near 0 (or the field of quotients of the local ring $\mathcal{O}_0$) with the derivations ∂_x and ∂_y;
- each $K_j = K_{j-1}\langle t_j \rangle$, where either t_j is algebraic over K_{j-1}, or $\partial_x t_j / t_j \in K_{j-1}$ and $\partial_y t_j / t_j \in K_{j-1}$, or $\partial_x t_j \in K_{j-1}$ and $\partial_y t_j \in K_{j-1}$;
- $H \in L_n$.

Another situation, called the **generalized reversibility**, takes place when $\boldsymbol{V}(x, y)$ is a **pull-back** of another germ $\boldsymbol{W}(X, Y)$ of vector field by means of a local map $\Phi : \left(\mathbb{R}^2, 0\right) \longmapsto \left(\mathbb{R}^2, 0\right)$, i.e., the vector fields $\left(\Phi_* \boldsymbol{V}\right)(x, y)$ and $\mathbf{W} \circ \Phi(x, y)$ are parallel, and Φ is a local ramified covering.
In the simplest situation Φ is generically 2 to 1 with a curve $\Gamma = \{\det \Phi_* = 0\}$ of critical points and its image $\Theta = \Phi(\Gamma)$, the curve of critical values; Φ has fold singularity at 0. Moreover, $\boldsymbol{W}$ is non-singular, but the phase curve of $\boldsymbol{W}$ through $X = Y = 0$ is tangent to Θ from outside of $\mathrm{Im}\Phi$. Then the preimages of the pieces of phase curves of W in $\mathrm{Im}\Phi$ become closed phase curves of $\boldsymbol{V}$. Compare Fig. 11.3.

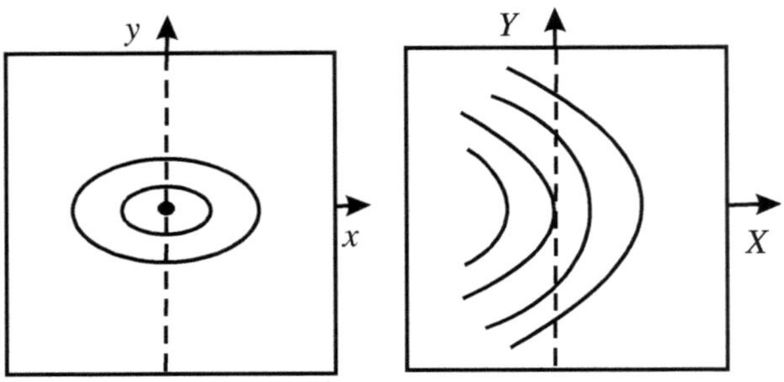

Figure 11.3: Pull-back of a nonsingular vector field using the fold map

For example, in the reversible case Q_3^{R} of center from theorem of Dulac and Kapteyn 6.33, the vector field $\boldsymbol{V} = \left(-y + ax^2 + by^2\right) \frac{\partial}{\partial x} + (x + cxy) \frac{\partial}{\partial y}$ is invariant with respect to the involution $(x, y, t) \longmapsto (-x, y, -t)$, but it is also a pull-back of the vector field $\boldsymbol{W} = 2\left(-Y + aX + bY^2\right) \frac{\partial}{\partial X} + (1 + cY) \frac{\partial}{\partial Y}$ via the map $\Phi : (x, y) \longmapsto (X, Y) = \left(x^2, y\right)$.

It turns out that there is no alternative for center for analytic germs: either generalized Liouvillian integrability of generalized reversibility. Namely, M. Berthier, D. Cerveau and A. Lins-Neto [**BCLN**] proved that there exists a counter-example to this alternative. This counter-example is not constructive.

11.55. Theorem (Berthier–Cerveau–Lins-Neto). *There exists a germ of real analytic foliation in* $\left(\mathbb{R}^2, 0\right)$, *which is not reversible in the generalized sense and does not have a generalized local Liouvillian first integral.*

Proof. 1. *Blowing-up, monodromy map and Poincaré map.* We consider germs of vector fields $\boldsymbol{V}$ (or foliations $\mathcal{F}$), which are real analytic. So, they are prolonged to complex vector fields (or foliations); we do not introduce special notations.
We assume that the resolution of the singularity consists of one blowing-up

$$\pi : (M, E) \longmapsto \left(\mathbb{C}^2, 0\right),$$

where $E \simeq \mathbb{CP}^1$ is the exceptional divisor. The coordinates in a suitable chart in M will be denoted (u, z), where $E = \{z = 0\}$ and u is a coordinate in (the affine part of) E. We have complex conjugation τ in $\mathbb{C}^2$ and corresponding involution $\tilde{\tau}$ in M.
Of course, we have the induced foliation $\pi^*\mathcal{F}$ in M with E invariant for it.
Let $m_j \in E$ and $\tilde{\tau}\left(m_j\right) \in E$, $j = 1, \ldots, p$, be the singular points of the foliation $\pi^*\mathcal{F}$; assume that m_j lie on one side of the real part of E, i.e., $\mathbb{RP}^1 \simeq \mathbb{S}^1$. We have the holonomy transformations $h_j \in \operatorname{Diff}(\mathbb{C}, 0)$ defined by lifting to the leaves of $\pi^*\mathcal{F}$ of loops $\gamma_j \in \pi_1\left(E, u_0\right)$; here $u_0 \in \mathbb{RP}^1$ and $(\mathbb{C}, 0)$ is identified with a disc transversal to E at u_0. The conjugate loops $\tilde{\tau}(\gamma_j)$ define monodromy maps k_j such that $k_j(\bar{z}) = \overline{h_j(z)}$.

The maps h_j and k_j generate the corresponding holonomy group Mon.
We have the monodromy map Δ, associated with the loop along the real divisor $\mathbb{RP}^1$, and the Poincaré return map $P : (\mathbb{R}_{\geq 0}, 0) \longmapsto (\mathbb{R}_{\geq 0}, 0)$; in the center case we have $P = id$. It is clear that $\Delta'(0) < 0$ and

$$P = \Delta \circ \Delta.$$

Moreover, we can assume that $\Delta = h_p \circ \ldots \circ h_1$.

2. *The case with Liouvillian first integral.* Here we have local versions of the Singer theorem 11.32 and of Theorem 11.46. In particular we get the following corollary.

Proposition 1. *If the foliation $\mathcal{F}$ has a local generalized Liouvillian first integral, then the monodromy group Mon is solvable.*

3. *The case with a pull-back foliation.* Suppose that a foliation $\mathcal{F}$ is a pull back of another foliation $\mathcal{G}$ by means of a finite to one local analytic map Φ, $\mathcal{F} = \Phi^*\mathcal{G}$. Here we have the following result, whose proof we skip.

Proposition 2. *Suppose that the foliation $\mathcal{F}$ has three separatrices Λ_j, $j = 1, 2, 3$, which are transversal one to another. Suppose also that $\Phi^{-1}(\Phi(\Lambda_j) = \Lambda_j$, $j = 1, 2, 3$. Then Φ must be a diffeomorphism, i.e., the foliation is irreducible.*

This result is applied to the following situation. Suppose that $\mathcal{F}$ has k separatrices $\Lambda_1, \ldots, \Lambda_k$, $k \geq 3$ which are transversal one to another. Suppose, additionally, that the corresponding holonomy maps h_j around points $m_j = E \cap \Lambda_j$ have linear parts independent over $\mathbb{Z}$, i.e., if $h_j'(0) = e^{\alpha_j}$, then the ratios α_i/α_j are irrational.

Proposition 3. *In this case the foliation $\mathcal{F}$ is irreducible.*

For the (not especially difficult) proof we refer to [**BCLN**].

4. *Choice of the holonomy group.* Let

$$l_1(z) = z\,(1+z)^{-1}, \quad l_2(z) = z\,(1+z^q)^{-1/q},$$

$q > 1$. Using the Cohen theorem 10.36 one can show that these maps generate a free subgroup of the group Diff $(C, 0)$. But we can say more. Consider the maps

$$h_1\,(z) = \lambda l_1(z), \quad h_2(z) = \mu l_2(z),$$

where λ and μ are nonzero complex numbers.
We define the third diffeomorphism by the formula

$$h_1 \circ h_2 \circ h_3(z) = -z.$$

Lemma. *The group $G_{\lambda,\mu}$ generated by h_1 and h_2 is free for a residual set of (λ, μ)'s. Therefore the group generated by the above maps h_j and corresponding maps k_j is not solvable.*

5. *Realization.* The holonomy maps h_j and k_j satisfy the necessary condition for the center $\Delta \circ \Delta = id$, where $\Delta = h_1 \circ h_2 \circ h_3$. The only problem is to construct a foliation whose monodromy group is like above. Here we have the following generalization of Theorem 9.78 of R. Perez-Marco and J.-C. Yoccoz (see also [**P-MY**]).

Proposition 3. *In this case there exists a germ of holomorphic foliation whose monodromy maps are the maps h_j and k_j defined in the previous point.*

In the proof given in [**BCLN**, Theorem 2.3] the authors use known constructions of foliations, with careful control of the reality property. We do not present it here. This completes the proof of the theorem. □

11.56. The center-focus problem. Global aspects. Here we consider the center-focus problem in the case of polynomial vector fields in $\mathbb{R}^2$.

This problem somehow contrasts with the local center problem, because such a vector field depends on finitely many parameters (coefficients in the Taylor expansion), whereas the number of Poincaré–Lyapunov type numbers is infinite.

Therefore one would like to define some mechanism for creating centers in polynomial differential systems. Some time ago I have produced a seminal paper 'The solution of the center-focus problem', where it was claimed that any such center is either 'Darbouxian integrable', or 'rationally reversible' of 'Darboux–Schwarz–Christoffel integrable'. In the first and the third case there exists a first integral of general Darboux type or of Darboux–Schwarz–Christoffel type, like in 11.47. In the second case the vector field with center is a pull-back of a polynomial vector field without center via a rational map with a fold curve, like in 11.54.

My paper was never published. In fact, it was not complete and the above theorem of Berthier–Cerveau–Lins-Neto served as a sort of counterexample. Nevertheless, I think that the idea from my seminal paper can be rather extended, but not abandoned. I present it in the next point.

11.57. Reduction via pull-back. The idea is to use some pull-back operation applied to an integrable 1-form Ω with some simple geometric property to get a 1-form ω associated with a polynomial vector field $\boldsymbol{V}$ with a center which follows from that property of Ω.

Recall that a 1-form Ω in $\mathbb{R}^m$ is **integrable** iff $\mathrm{d}\Omega \wedge \Omega = 0$; equivalently, $\mathrm{d}\Omega$ vanishes on $\ker \Omega$. Moreover, with a polynomial vector field $\boldsymbol{V} = P\frac{\partial}{\partial x} + Q\frac{\partial}{\partial y}$ the 1-form $\omega = Q\mathrm{d}x - P\mathrm{d}y$ is associated, such that the corresponding Pfaff equation $\omega = 0$ defines the phase curves of $\boldsymbol{V}$.

Therefore, we have:

(i) a polynomial integrable 1-form ω in $\mathbb{R}^2$ with corresponding phase curves γ,

(ii) an integrable polynomial 1-form Ω in $\mathbb{R}^2 \times \mathbb{R}^k$ with leaves L of the foliation $\mathcal{F}$ defined by Pfaff equation $\Omega = 0$.

(iii) an algebraic surface $\Sigma \subset \mathbb{R}^2 \times \mathbb{R}^k$ with the natural projection $\pi : \Sigma \longmapsto \mathbb{R}^2$, such that

$$\pi (L \cap \Sigma) = \gamma,$$

i.e., the phase curves of $\boldsymbol{V}$ arise from intersections of leaves of the foliation $\mathcal{F}$ with the surface Σ. Then we say that ω (and $\boldsymbol{V}$) is a **pull-back** of Ω via the projection $\pi : \Sigma \longmapsto \mathbb{R}^2$; if π is 1 to 1 then $\omega(x,y)$ and $\left(\pi^{-1}\right)^* (\Omega|_\Sigma)(x,y)$ are generically parallel as elements of $T^*_{(x,y)}\mathbb{R}^2$.

Let us examine which conditions should be put on π and Ω to get a center for ω. Firstly, we consider the cases of elementary and nilpotent centers; we know that the corresponding local center-focus problem is algebraically solvable. We conjecture that:

In the case of elementary or nilpotent center for the Pfaff equation $\omega = 0$ the form ω is a pull-back of a polynomial integrable 1-form Ω in $\mathbb{R}^2 \times \mathbb{R}^k$ via an algebraic mapping $\pi : \Sigma \longmapsto \mathbb{R}^2$ with a 'fold type singularity', i.e., the closed phase curves of the equation $\omega = 0$ are preimages of intersections with Σ of leaves of the equation $\Omega = 0$.

Let me explain this 'fold type singularity' on some examples.

Examples. 1. *Hamiltonian center.* If $\boldsymbol{V} = \boldsymbol{X}_H$ is a Hamiltonian vector field and $\omega = \mathrm{d}H$, then $\mathbb{R}^2 \times \mathbb{R}^k = \mathbb{R}^3 = \{(x,y,X)\}$, $\Sigma = \{X = H(x,y)\}$ and $\Omega = \mathrm{d}X$. The leaves of the the Pfaff equation $\Omega = 0$ are $L_c = \{X = c\}$, their intersections with Σ are $\{H(x,y) = X = c\}$ and their projections are the level curves of the function H. For a Morse critical point (x_0, y_0) of H (minimum or maximum) with the critical value X_0 the intersections $L_c \cap \Sigma$ near (x_0, y_0, X_0) are closed curves (or empty sets) projected to some ovals of $\{H(x,y) = c\}$.

2. *Rational first integral.* This is the case with a rational first integral $H = f^p/g^q$, where f and g are polynomials and p and q are relatively prime; moreover, the center does not lie on the none of the invariant curves $\{f = 0\}$ and $\{g = 0\}$. On the one hand, it can be considered like in the previous example, i.e., with $\Sigma = \{(x,y,X) : g^q(x,y)X = f^p(x,y)\} \subset \mathbb{R}^2 \times \mathbb{R}$ and $\Omega = \mathrm{d}X$.
On the other hand, we can take $\Sigma = \{(x,y,X,Y) : X = f(x,y), Y = g(x,y)\} \subset \mathbb{R}^2 \times \mathbb{R}^2$. We have $\omega = pg\mathrm{d}f - qf\mathrm{d}g$ (corresponding to $\boldsymbol{V} = pg\boldsymbol{X}_f + qf\boldsymbol{X}_g$) and we take $\Omega = pY\mathrm{d}X - qX\mathrm{d}Y$ as a linear 1-form with the first integral X^p/Y^q. The closed phase curves of $\boldsymbol{V}$ are preimages of segments of intersections of the leaves of $\Omega = 0$ with Σ.

3. *Darbouxian first integral.* Let this first integral be $H = f_1^{a_1} \cdots f_k^{a_k}$,and the center does not lie on the the invariant set $\{f_1 \cdots f_k = 0\}$. Then we take $\Sigma = \{(x,y,X_1,\ldots,X_k) : X_j = f_1(x,y)\} \subset \mathbb{R}^2 \times \mathbb{R}^k$ and $\Omega = \sum a_j d\ln X_j$ is an integrable rational 1-form. Then $\sum a_j \ln X_j$ is a first integral for Ω.

4. *DSC first integral.* Let it be $H = f^\alpha (f-1)^\beta g + \int^f p(u)u^{\alpha-1}(u-1)^{\beta-1}\mathrm{d}u$, where $f(x,y)$, $g(x,y)$ and $p(u)$ are polynomials and again we suppose that the

center does not lie on a 'singular set' of this function. Then we take

$$\Sigma = \{(x, y, X, Y) : X = f(x, y), Y = g(x, y)\} \subset \mathbb{R}^2 \times \mathbb{R}^2 \text{ and}$$
$$\Omega = \{(aX + b)\,Y + p(X)\}\,\mathrm{d}X + X(X - 1)\mathrm{d}Y$$

for suitable constants a and b, such that $X^\alpha(X-1)^\beta Y + \int^X p(u)u^{\alpha-1}\,(u1)^{\beta-1}\,\mathrm{d}u$ is its first integral.

5. *Rational reversibility.* This is the case when, like in the previous section (i.e., when ω is a pull-back of another form Ω by means of a map Φ which is a local ramified covering), but Φ is rational (with rational components). We have $\Sigma = \{(x, y, X, Y) : (X, Y) = \Phi(x, y)\} \subset \mathbb{R}^2 \times \mathbb{R}^2$ and $\Omega = A(X, Y)\mathrm{d}X + B(X, Y)\mathrm{d}Y$ is the 1-form associated with a vector field $\boldsymbol{W}(X, Y)$ in the target.

Finally, I suppose that in the case of center with zero linear part, besides the fold type pull-back we have only the case of a pull-back (via a rational map) of a rational vector field

$$\Omega = \{B\,(X)\,Y + p(X)\}\,\mathrm{d}X + A(X)\mathrm{d}Y$$

with the simplest DSC integral

$$H = D(X)Y + \int^X q\,(u)\,D\,(u)\,\mathrm{d}u.$$

Above $D\,(X)$ is a generalized Darbouxian function of one variable and $q\,(X)$ is a suitable rational function of one variable. Moreover, the center condition reads as

$$\int_{\mathbb{R}} q\,(u)\,D\,(u)\,\mathrm{d}u = 0.$$

This amounts to the property that the domain of integration, i.e., $\mathbb{R}$, is the affine part of a real divisor $\mathbb{RP}^1$ of the resolution of the singularity. I think that there is only one such deciding divisor, it 'supports' the center condition.

Recall that the above Berthier–Cerveau–Lins-Neto example is not constructive and I suppose that it cannot be realized in the class of polynomial vector fields.

11.60. The Liénard system. The standard **Liénard system** is following:

$$\dot{x} = y, \quad \dot{y} = -g(x) + f(x)y; \tag{11.26}$$

here the real functions $f(x)$ and $g(x)$ are analytic near $x = 0$ and $g(0) = 0$, but we will also consider the situation when f and g are polynomials.

The leading part. We are interested in the conditions for center at $(0, 0)$. Of course, the necessary condition is that $f(0) = 0$; we assume it. Moreover, if $b = g'(0) > 0$, then we can have elementary center with the first integral $H_0 = \frac{1}{2}y^2 + \frac{b}{2}x^2 + \dots$

In general, like in 9.65, we have the leading part $\dot{x} = y$, $\dot{y} = -bx^{2r-1} + ax^{r-1}y$ of quasi-homogeneous degree $r - 1$ (where $\deg x = 1$ and $\deg y = r$ and r is half-integer in general).
If (i) $a = 0$, then necessarily r is integer and $b > 0$ and we have the first integral $H_0 = \frac{1}{2}y^2 + \frac{b}{2r}x^{2r}$.
If (ii) $a \neq 0$, then we consider the corresponding linear system $\dot{z} = ry$, $\dot{y} = -bz+ay$, with $z = x^r$. The latter linear system should be a focus (so, $a^2 < 4br$) and $r = 2r_0$ is even; thus the leading system is reversible with respect to the $x-$variable. Moreover, one has a first integral of the form $H_0 = \left(y + \mu x^{2r_0}\right)^{\nu} \left(y + \bar{\mu} x^{2r_0}\right)^{\bar{\nu}}$ for $\mu, \nu \in \mathbb{C} \setminus \mathbb{R}$.
Introduce also the functions

$$G(x) = \int_0^x g(s)\mathrm{d}s, \ F(x) = \int_0^x f(s)\mathrm{d}s.$$

11.61. Theorem (Local center conditions for Liénard system). *The origin is a center for the system* (11.26) *iff there exists an analytic germ* $P(x) = x^2 + \ldots$ *and germs* $\widetilde{G}(u)$ *and* $\widetilde{F}(u)$ *such that*

$$G = \widetilde{G} \circ P, \quad F = \widetilde{F} \circ P.$$

This means that we have the local reversibility by means of the mapping $(x, y) \to \Phi = (P(x), y)$.
In the elementary case, $g'(0) > 0$, *this is equivalent to the property that the system*

$$G(x) = G(z), \quad F(x) = F(z)$$

has a solution $z(x) \in \mathbb{C}\{x\}$ *such that* $z'(0) < 0$.

Proof. Consider first the case of elementary center, following the paper [**Cher**] by L. Cherkas and the book [**ZM**].
We have $G(x) = \frac{b}{2}x^2 + \ldots$. We make the change $x \longmapsto u = \sqrt{b}x + \ldots$ such that $G(x) = \frac{1}{2}u^2$; thus $u = \sqrt{2G(x)}$ and $\frac{\mathrm{d}u}{\mathrm{d}x} = \frac{g(x)}{u}$. Then the system (11.26) becomes orbitally equivalent to the simplified Liénard system

$$\dot{u} = y, \quad \dot{y} = -u + q(u)y, \tag{11.27}$$

where $q(u) = uf(x)/g(x)$. Note that

$$Q(u) = \sum_{k \geq 2} q_k u^k = \int_0^u q(\tau)\mathrm{d}\tau = \int_0^x \frac{\mathrm{d}\tau}{\mathrm{d}s} \frac{\tau}{g(s)} f(s)\mathrm{d}s = F(x).$$

If $Q(u) \equiv 0$, then the system (11.27) is Hamiltonian. More generally, when the function Q is even, i.e., $Q(u) = \widetilde{Q}(u^2)$, then the system (11.27) is reversible with respect to the involution $(u, y, t) \longmapsto (-u, y, -t)$ and we have a center.

Suppose now that the series defining Q is not even, i.e., $q_3 = q_5 = \ldots = q_{2m-1} = 0$, but $q_{2m+1} \neq 0$. Then the polynomial system $\dot{u} = y, \dot{y} = -u + \left(\sum_{j \leq m} 2jq_{2j}u^{2j-1}\right) y$ has a first integral $H(u,y) = \frac{1}{2}y^2 + \frac{1}{2}u^2 + \ldots$, but the whole system (11.27) is its perturbation, with the perturbation $Q_1'(u)y\frac{\partial}{\partial y}$ for $Q_1 = q_{2m+1}u^{2m+1} + \ldots$.
We compute approximately the Poincaré return map by estimating the increment of H. We get

$$\Delta H = \int_0^T H_y' Q_1'(u) y \mathrm{d}t \approx c \oint u^{2m} y \mathrm{d}u = c \int\int u^{2m} \mathrm{d}y \mathrm{d}u \approx \mathrm{const} \cdot h^{m+1} \neq 0,$$

where $c = (2m+1)q_{2m+1}$, T is the return time to the Poincaré section and the second integral runs along the curve $\{H = h\}$.
The map $x \longmapsto z(x)$ from the second statement of the theorem reads as $u \longmapsto -u$. In the general case we apply the change $x \longmapsto u = x + \ldots$ such that $G(x) = \frac{B}{2r}u^{2r}$. We get $\dot{u} = y$, $\dot{y} = -Bu^{2r-1} - q(u)y$ with the leading part $\dot{u} = y$, $\dot{y} = -Bu^{2r-1} - Au^{r-1}y$ having the first integral $H_0(u,y)$ as above. Again we have $F(x) = Q(u) = A_r u^r + \ldots$. If Q is even, then the whole system is reversible. Otherwise, we calculate the increment of the corresponding first integral whose leading part is H_0, as above. Again it turns out that the corresponding contour integral is nonzero. □

Before stating next theorem we recall the following classical result.

11.62. Theorem of Lüroth. *Any transcendental extension L of the algebraic field $\mathbb{C}$, which is a subfield of $\mathbb{C}(x)$, is of the form $\mathbb{C}(K(x))$, where $K(x)$ is a rational function. Moreover, if the subfield L is generated by a finite collection of polynomials, then $K(x)$ is also a polynomial.*

Proof. Since $\mathbb{C}(x) = \mathbb{C}\left(\mathbb{CP}^1\right)$, the field or rational functions on the Riemann sphere, the subfield L is a field of rational functions on another Riemann surface, $L = \mathbb{C}(N)$, and the inclusion $L \subset \mathbb{C}(x)$ is induced by a ramified map $\pi : \mathbb{CP}^1 \longmapsto N$. But the Riemann–Hurwitz formula from Theorem 11.34 implies that $\chi(N) > 0$, i.e., $N \simeq \mathbb{CP}^1$ is also a Riemann sphere. The map π is realized by a rational function $K(x)$.
Next, if $L = \mathbb{C}(P_1, \ldots, P_r)$ where $P_j(x)$ are polynomials, then $\pi^{-1}(\infty) = \infty$. □

The next result in the elementary case $g(0) = 0 < g'(0)$ was firstly obtained by L. Cherkas in [**Cher**].

11.63. Theorem (Center conditions for polynomial Liénard system). *Assume that $f(x)$ and $g(x)$ in the system* (11.26) *are polynomials. Then the origin is a center iff there exist: a polynomial $K(x) = x^2 + \ldots$ and polynomials $\widetilde{G}(z)$ and $\widetilde{F}(z)$, such that*

$$G = \widetilde{G} \circ K, \quad F = \widetilde{F} \circ K.$$

Proof. By Theorem 11.61 the polynomials F and G are local compositions. Consider the field $L = \mathbb{C}(F(x), G(x))$. It is a transcendental over $\mathbb{C}$, but $L \neq \mathbb{C}(x)$. By the Lüroth theorem $L = \mathbb{C}(K)$ and $\mathbb{C}[F,G] = \mathbb{C}[K]$. □

11.64. Infinitesimal problem of center for perturbations of a Liénard system. Consider perturbations of Hamiltonian Liénard system, i.e.,

$$\dot{x} = y + \varepsilon P(x,y), \quad \dot{y} = -G'(x) + \varepsilon Q(x,y),$$

where $P, Q \in \mathbb{C}[x,y]$. Of course, here we deal with the Abelian integral

$$I(h) = \oint \omega, \text{ where } \omega = Q\mathrm{d}x - P\mathrm{d}y$$

and the integration runs over ovals $\Delta(h) \subset \{H = y^2/2 + G(x) = h\}$. C. Christopher and P. Mardesič[**ChMa**] studied the case when $I_\gamma(h) \equiv 0$ and $\{\Delta(h)\}$ is a family of vanishing cycles, say at $(0,0)$; this is the **infinitesimal center problem**.

11.65. Theorem of Christopher and Mardesič. *The function $I(h) \equiv 0$ iff either*

(i) *the form ω is relatively exact, i.e., $\omega = A\mathrm{d}H + \mathrm{d}B$ with $H = \frac{1}{2}y^2 + G(x)$ and $A, B \in \mathbb{C}[x,y]$, or*

(ii) *G is reduced, i.e., $G = \widetilde{G} \circ K$ with $K(x) = x^2 + \ldots \in \mathbb{C}[x]$, and $\omega = \omega_0 + \pi^*\widetilde{\omega}$, where ω_0 is relatively exact, and the cycle $\pi_*\Delta(h)$ is trivial in $\left\{y^2/2 + \widetilde{G}(z) = h\right\}$, where $\pi(x,y) = (z,y) = (K(x), y)$.*

Proof. Assume that $G(x) = \frac{1}{2}x^2 + \ldots$. Following the Petrov's analysis of the elliptic integrals in the proof of Lemma 5.20, in the case of our hyperelliptic integrals, we get the representation

$$\omega = A\mathrm{d}H + \mathrm{d}B + yp(x)\mathrm{d}x.$$

Then we have $I = \oint yp(x)\mathrm{d}x$.
Like in the proof of Theorem 2 we firstly make the change $x \longmapsto u = x + \ldots$ such that $G(x) = \frac{1}{2}u^2$ and we get

$$I(h) = \oint yQ'(u)\mathrm{d}u.$$

Expanding Q in powers of u we find that it contains only even powers, $Q(u) = \widetilde{Q}(u^2)$, and we have local composition. Like in the proof of Theorem 4 we conclude that the primitive P of p is a composition, $P = \widetilde{P} \circ K$ with $K(x) = x^2 + \ldots$. □

Before formulating definitions and results be present (without proof) the following statement from [**Ros1**].

11.66. Theorem of Rosenlicht. *Let $K = \mathbb{C}(x)[z(x)]$ be an algebraic extension of the field $\mathbb{C}(x)$. Let constants $c_j \in \mathbb{C}$, $j = 1, \ldots, n$, be linearly independent over $\mathbb{Q}$ and let $u_1(x), \ldots, u_n(x), v(x) \in K$ satisfy*

$$c_1(\ln u_1)' + \ldots + c_n(\ln u_n)' + v' \equiv \text{const}.$$

Then all $u_j \in \mathbb{C}$ are constant and $v(x) = cx + d$ with $c, d \in \mathbb{C}$.

11.67. The generalized Liénard system. It is the following:

$$\dot{x} = y, \quad \dot{y} = -g(x) + f(x)y + r(x)y^2. \tag{11.28}$$

Above $g = \frac{a(x)}{d(x)}$, $f = \frac{b(x)}{d(x)}$ and $r = \frac{c(x)}{d(x)}$ are rational functions and we assume that $g(0) = 0 < g'(0)$; thus it is a rational system, but orbitally equivalent to a polynomial system, i.e., with $\dot{x} = d(x)y$, etc.

We will use the following transformation due to L. Cherkas [**Cher**].

Lemma of Cherkas. *The change $y = Y \exp R(x)$, with $R(x) = \int_0^x r(s)\mathrm{d}s$, leads to the standard Liénard system $\dot{x} = Y$, $\dot{Y} = -g_1(x) + f_1(x)Y$ with*

$$g_1 = g(x)\exp\left(-2R(x)\right), \quad f_1 = f(x)\exp\left(-R(x)\right).$$

11.68. Theorem of Christopher and Shlomiuk. ([**CS**]) *The system* (11.28) *has center at the origin iff it satisfies one of the following conditions:*

(i) *it is reversible via the symmetry $y \longmapsto -y$;*

(ii) *it is reduced via a pull-back $(x, y) \longmapsto \left(K(x)^{1/r}, L(x)^{1/q}y\right)$ with rational functions $K(x) = cx^{2r} + \ldots$ and $L(x) = d + \ldots$;*

(iii) *there is a first integral of generalized Darbouxian type.*

Proof. If $f \equiv 0$, then the system (11.28) is reversible like in the alternative (i); so, assume that $f_1 \not\equiv 0$.

Then we apply Theorem 11.61 about local center conditions for the Liénard system. Therefore we have a local germ $x \longmapsto z = z(x)$, with $z'(0) < 0$, such that $G_1(z) = G_1(x)$ and $F_1(z) = F_1(x)$, where G_1 and F_1 are the corresponding primitives of g_1 and f_1. We have two possibilities:

either (**A**) $z(x)$ is algebraic, or (**B**) it is transcendental.

Consider the case (**A**). After expansion of the rational function $r(x)$ into partial fraction and some manipulation, one can represent the function R as

$$-R(x) = \sum \alpha_i \ln R_i(x) + R_\infty(x),$$

where $R_i(0) = 1$, $R_\infty(0) = 0$, α_i are independent over $\mathbb{Q}$ and $\alpha_0 = \frac{1}{q} \in \mathbb{Q}$; if none of α_i's is rational, then we take $R_0(x) \equiv 1$.

Next, the Cherkas center condition leads to the conditions $g_1(x)\mathrm{d}x = g_1(z)\mathrm{d}z$ and $f_1(x)\mathrm{d}x = f_1(z)\mathrm{d}z$. Hence

$$\begin{aligned} 1 = \frac{g_1(x)f_1(z)}{g_1(z)f_1(x)} &= \frac{g(x)f(z)}{g(z)f(z)} \exp\left(R(z) - R(x)\right) \\ &= \exp\left\{\frac{1}{q}\ln S_0 + \sum_{i\geq 1} \alpha_i \ln S_i + S_\infty\right\}, \end{aligned}$$

where $S_0 = R_0(x)g^q(x)f^q(z)/R_0(z)g^q(z)f^q(x)$, $S_i = R_i(z) - R_i(x)$ for $0 < i < \infty$, and $S_\infty = R_\infty(z) - R_\infty(x)$. Assuming $z = z(x)$, we differentiate the above and get the identity

$$\sum \alpha_i \left(\ln S_i\right)' + S_\infty{}' = 0 \in \mathbb{C}.$$

By the Rosenlicht theorem it must be: $S_i(x, z(x)) \equiv \text{const} = 1$ and $S_\infty = cx + d$; and hence $S_\infty \equiv 0$.
Now we apply the Lüroth theorem, which implies that:

$$R_0(x)g^q(x)/f^q(x) = \widetilde{R}_0 \circ K \text{ and } R_i = \widetilde{R}_i \circ K \text{ for } i > 0\ ,$$

where $K(x) = cx^2 + \ldots$ and $\widetilde{R}_i(u)$ are rational functions.
After some analysis of the above equations (which we skip) C. Christopher and D. Shlomiuk prove the reversibility, like in the possibility (ii) of the theorem.

In the case (**B**) one differentiates the relation $g_1(x)/f_1(x) = g_1(z)f_1(z)$,using $f_1(x)\mathrm{d}x = f_1(z)\mathrm{d}z$, and arrives at the identity

$$\left\{(g_1/f_1)'/f_1\right\}(x) = \left\{(g_1/f_1)'/f_1\right\}(z)\,.$$

This is an algebraic equation for z. Since $z(x)$ is transcendental, the both sides equal to the same constant, say c.
Further analysis, which we skip, leads to the existence of two invariant curves $y = p(x)$ and $y = ap(x)$. Moreover, the function $C(x,y) = \exp R(x)$ satisfies $\dot{C} = yr(x) \cdot C$, i.e., is a sort of invariant curve. This implies the existence of a generalized Darbouxian first integral $H = \{y - p(x)\}^\alpha \{y - ap(x)\}^\beta \exp R(x)$. □

11.69. Example. For the system

$$\dot{x} = y(1+x), \quad \dot{y} = -x(6+7x)(1+x)\left(1+\lambda x^2 y\right) + y^2/3$$

we have $Y = (1+x)^{-1/3}y = L(x)^{1/3}y$ and $X = x^2(1+x)^{1/3} = K^{1/3}(x)$; here $K(x) = x^6(1+x)$ and $K'(x) = x^5(6+7x)$. The vector field in the target is orbitally equivalent to

$$\dot{X} = Y, \quad \dot{Y} = -3(1+\lambda XY).$$

We can treat this reversibility on a slightly different way. Consider the surface

$$\Sigma = \{(x,y,X,Y) : X^3 = x^6(1+x),\ (1+x)Y^3 = y^3\} \subset \mathbb{R}^2 \times \mathbb{R}^2$$

and the 1-form

$$\Omega = 3(1+\lambda XY)\mathrm{d}X + Y\mathrm{d}Y$$

in $\mathbb{R}^4$. Then the form $\omega = \left\{x(6+7x)(1+x)\left(1+\lambda x^2y\right) + y^2/3\right\}\mathrm{d}x + y(1+x)\mathrm{d}y$ is a pull-back of Ω via the projection $\Sigma \longmapsto \mathbb{R}^2$.

11.70. The center variety. The real planar polynomial vector fields of degree $\leq n$ form a vector space $\mathrm{Vect}(n) = \mathcal{V}_n^{\mathbb{R}} \simeq \mathbb{R}^N$, $N = (n+1)(n+2)$. The vector fields with elementary or nilpotent center form a semi-algebraic subset of this space, called the **center variety** $\mathrm{Cent}(n)$. This variety is an union of irreducible components. It would be highly interesting to define these components. Below I present some of them. The following classical result belongs to Yu. Ilyashenko [**Il1**].

11.71. Theorem of Ilyashenko (Hamiltonian centers). *The Hamiltonian vector fields with center form an irreducible component* $\mathrm{Ham}(n)$ *of the center variety* $\mathrm{Cent}(n)$.

Proof. One has to show that $\mathrm{Ham}(n)$ does not lie in a greater component of $\mathrm{Cent}(n)$. The idea of the proof is to show that the tangent space $T_{\boldsymbol{V}_0}\mathrm{Cent}(n)$ to the space $\mathrm{Cent}(n)$ at a generic point $\boldsymbol{V}_0 = \boldsymbol{X}_{H_0} \in \mathrm{Ham}(n)$ is the same as the tangent space $T_{\boldsymbol{V}_0}\mathrm{Ham}(n)$.
So one takes a perturbation $\boldsymbol{V}_\varepsilon = \boldsymbol{V}_0 + \varepsilon \boldsymbol{V}_1$ of $\boldsymbol{V}_0$ and considers the infinitesimal center problem, i.e., whether the corresponding Abelian integral

$$I(h) = \oint \omega$$

vanishes identically; here ω is the 1-form corresponding to $\boldsymbol{V}_1$ and the integration runs along an oval $\Delta(h) \subset V_h = \{H_0 = h\}$ vanishing at the center. Assume that the Hamilton function H_0 has following properties:

(i) all its critical points (also non-real) are Morse with different critical values,

(ii) the complex algebraic curve $\{H_0 = 0\}$ meets the line at infinity transversally and at $n+1$ different points.

Recall that the first homology group $H_1(V_h, \mathbb{Z})$ is generated by cycles $\Delta_j(h)$ vanishing at some critical points of H_0 and by cycles $\Gamma_k(h)$ around points at infinity in local branches $\Lambda_k = \{y = c_k x + \ldots\} \subset V_h$ around the points at infinity.
The property (i) implies that $\int_{\Delta_j(h)} \omega \equiv 0$ for any cycle $\Delta_j(h)$. Indeed, the monodromy group generated by loops in $\mathbb{C} \setminus \{\text{critical values of } H_0\}$ acts transitively on the subspace of $H_1(V_{h_0}, \mathbb{Z})$ generated by the cycles Δ_j. Therefore, the statement follows from the Picard–Lefshetz formula.
Next, $\int_{\Gamma_k(h)} \omega$ is the residuum at infinity of the restriction of ω to the branch Λ_k. But this restriction is of the form $p_k(x)\mathrm{d}x$, where $p(x)$ grows polynomially as $x \to \infty$. So, this residuum is equal to zero.
Therefore, ω restricted to each complex curve V_h, presents a zero cohomology class. By the de Rham theorem it is exact on V_h, i.e., $\omega|_{V_h} = \mathrm{d}(f|V_h)$, where $f(x,y) = \int_{\eta(x,y)} \omega$, where $\eta(x,y) \subset V_h$ is a path from (x_0, y_0) to (x,y). One has $\omega = \mathrm{d}f \mod \mathrm{d}H$, where we can assume that f is a polynomial of degree $\leq n+1$; one says that ω represents a zero relative de Rham cohomology class. Therefore the equation

$$\omega = \mathrm{d}f + g\mathrm{d}H,$$

for polynomials f and g has a solution (compare the point 1 of 6.36). But, by the comparison of degrees we find that $g \equiv \text{const}$. □

11.72. Remark (Rational first integral). It is worth to mention that H. Movasati proved in [**Mov1**] that polynomial vector fields with rational first integral $H = f^p/g^q$, with polynomials $f(x,y)$ and $g(x,y)$ (of suitable degrees) also form an irreducible component of the venter variety.

11.73. Darbouxian centers. For a given vector of positive integers $d=(d_1,\ldots,d_s)$, such that $d_1+\ldots+d_s=n+1$, consider the subset $\mathrm{Darb}(d)$ consisting of vector fields with the real Darbouxian first integral $H=\prod f_j^{\lambda_j}$, where $f_j(x,y)$ are complex polynomials of degree $\leq d_j$ and $\lambda_j\in\mathbb{C}$.
In [**Mov2**] H. Movasati claimed that the closure set $\overline{\mathrm{Darb}\,(d)}$ of $\mathrm{Darb}(d)$ is an irreducible component of the center variety $\mathrm{Cent}(n)$.
I do not doubt in the validity of his claim, but for me his proof is not complete. Nevertheless, his approach to the problem is interesting and I want to present some of its points.

1. *Tangent cones.* Let us look at the tangent cone to the center variety $\mathrm{Cent}(n)$ at some special points. For example the formula $\omega=\mathrm{d}f+g\mathrm{d}H$ (from the proof of the Ilyashenko' theorem) for the perturbation $\mathrm{d}H+\varepsilon\omega=0$ of the Pfaff equation $\mathrm{d}H=0$, in the case $\deg H=d<n+1$ and $\deg\omega=n$, means that we are at the local intersection of two components of the center variety $\mathrm{Cent}(n):\mathrm{Ham}(n)=\overline{\mathrm{Darb}(n+1)}$ and $\overline{\mathrm{Darb}(d,n+1-d)}$. The perturbation $\mathrm{d}H+\varepsilon\mathrm{d}f=0$ lies in the Hamiltonian component and the perturbation $\mathrm{d}H+\varepsilon g\mathrm{d}H=0$ has Darbouxian first integral $H(1+\varepsilon g)^0$ with the integrating factor $(1+\varepsilon g)^{-1}$.
Also the formula $\omega=A\mathrm{d}H+\mathrm{d}B+yp(x)\mathrm{d}x$, $p(x)=\left(\widetilde{P}\circ K\right)'(x)$, from the proof of the Christopher–Mardesič theorem, means that we are at the intersection of three components of the center variety: Darbouxian, Hamiltonian and pull-back.
H. Movasati [**Mov2**] was interested in the tangent cone to the center variety at a special point. This point corresponds to $\boldsymbol{V}_0=\boldsymbol{X}_{H_0}$ (or $\omega_0=\mathrm{d}H_0$) with

$$H_0=l_0\cdots l_n,$$

such that $l_j(x,y)$ are real affine functions and H_0 has non-degenerate saddle critical points with the critical value 0, and nondegenerate extrema with different critical values.
Especial case of such situation was studied in [**Zo3**].

2. **Example (Hamiltonian triangle).** This is the system $\dot{x}=x(1-x-2y)$, $\dot{y}=-y(1-2x-y)$ with the first integral

$$H_{1,1}=xy(1-x-y).$$

It is a particular case of the integrable Lotka–Volterra system $\dot{x}=\{x\beta-\beta x-(\beta+1)\,y\}$, $\dot{y}=y\,\{-\alpha+(\alpha+1)\,x+\alpha y\}$. The corresponding vector field is denoted by $\boldsymbol{W}_{\alpha,\beta}$ and belongs to the component Q_3^{LV} with the Darbouxian first integral

$$H_{\alpha,\beta}=x^\alpha y^\beta(1-x-y)$$

(see Theorem 6.33). So, we have the component Darb$(1,1,1)$ of the center variety Cent(2). If $\varepsilon_3 = \alpha - 1$ and $\varepsilon_4 = \beta - 1$ are small then we have the Darb$(1,1,1)$ perturbation of the Hamiltonian triangle.
The space of perturbations of the vector field $\boldsymbol{W}_{\alpha,\beta}$ in the generalized Lotka–Volterra class, i.e., which are tangent to the coordinate axes (and which can be non-integrable), is 1–dimensional (see [**Zo3**]). We choose it in the form

$$\varepsilon_0 \boldsymbol{V}_0 = -\varepsilon_0 xy \left(\frac{1}{\alpha}\frac{\partial}{\partial x} + \frac{1}{\beta}\frac{\partial}{\partial y}\right) = -\varepsilon_0 xy \left(\frac{\partial}{\partial x} + \frac{\partial}{\partial y}\right).$$

The other two perturbations are taken in the form

$$\varepsilon_1 \boldsymbol{V}_1 = \varepsilon_1 (\beta x - \alpha y) \left\{(c_1 x + d_1 y)\frac{\partial}{\partial x} + e_1 y \frac{\partial}{\partial y}\right\},$$
$$\varepsilon_2 \boldsymbol{V}_2 = \varepsilon_2 (\beta x - \alpha y) \left\{c_2 x\frac{\partial}{\partial x} + (d_2 x + e_2 y) \frac{\partial}{\partial y}\right\},$$

where $c_1 = -\beta\frac{1-\alpha-\beta}{1+\alpha+\beta}$, $d_1 = -\alpha$, $e_1 = \frac{-2\beta}{1+\alpha+\beta}$, $c_2 = \frac{2\alpha}{1+\alpha+\beta}$, $d_2 = \beta$, $e_2 = \alpha\frac{1-\alpha-\beta}{1+\alpha+\beta}$.
Therefore we have the perturbation

$$\boldsymbol{X}_\varepsilon = \boldsymbol{W}_{\alpha,\beta} + \varepsilon_0 \boldsymbol{V}_0 + \varepsilon_1 \boldsymbol{V}_1 + \varepsilon_2 \boldsymbol{V}_2.$$

We have the pseudo-Abelian integral $h \longmapsto \varepsilon_0 I_0(h) + \varepsilon_1 I_1(h) + \varepsilon_2 I_2(h)$, where $I_j = \oint x^{\alpha-1} y^{\beta-1} \omega_j$ and $\omega_0 = xy\left(\frac{\mathrm{d}y}{\alpha} - \frac{\mathrm{d}x}{\beta}\right)$, $\omega_1 = (\beta x - \alpha y)\left\{e_1 y \mathrm{d}x - (c_1 x + d_1 y)\mathrm{d}y\right\}$, $\omega_2 = (\beta x - \alpha y)\left\{(d_2 x + e_2 y)\mathrm{d}x - c_2 x \mathrm{d}y\right\}$. The Stokes formula implies that

$$I_0 = \int\int x^{\alpha-1} y^{\beta-1} (x+y)\, \mathrm{d}x\mathrm{d}y,$$
$$I_1 = (1-\alpha)\int\int x^{\alpha-2} y^{\beta-1} (\alpha y - \beta x)^2\, \mathrm{d}x\mathrm{d}y,$$
$$I_2 = (1-\beta)\int\int x^{\alpha-1} y^{\beta-2} (\alpha y - \beta x)^2\, \mathrm{d}x\mathrm{d}y,$$

where the double integral is along the compact domain bounded by the oval $\{H = h\}$.
Now we assume that

$$\varepsilon_3 = \alpha - 1, \quad \varepsilon_4 = \beta - 1$$

are small.
Because for $\varepsilon_0 = \varepsilon_1 = \varepsilon_2 = 0$ the displacement $\Delta H_{\alpha,\beta}$ vanishes, we can write $\Delta H_{\alpha,\beta} = \varepsilon_0 A_0 + \varepsilon_1 A_1 + \varepsilon_2 A_2$, where $A_j = A_j(h;\varepsilon)$ are analytic functions.
We have $\varepsilon_0 A_0 = \varepsilon_0 \cdot (I_0 + O(\varepsilon))$; in order to simplify situation we assume now that $\varepsilon_0 = 0$.
Now the vector fields $\boldsymbol{X}_\varepsilon$, with the above restriction, have singular point at $p_{\alpha,\beta}$.
The vector field $\boldsymbol{X}_\varepsilon$ with $\varepsilon_0 = 0$ has center at $p_{\alpha,\beta}$ in the following cases:

(i) $\varepsilon_1=\varepsilon_2=0$, i.e. when $\boldsymbol{X}_\varepsilon\in Q_3^{\mathrm{LV}}$;

(ii) $\varepsilon_1=\varepsilon_4=0$ or $\varepsilon_2=\varepsilon_3=0$ or $\varepsilon_1+\varepsilon_2=\varepsilon_3-\varepsilon_4=0$, i.e. when $X_\varepsilon\in Q_3^{\mathrm{R}}$;

(iii) $\varepsilon_3=\varepsilon_4=0$, i.e. when $X_\varepsilon\in Q_3^{\mathrm{H}}$.

The case (i) was already discussed. Next, we have

$$\begin{aligned}\mathrm{div}M(\boldsymbol{W}_{\alpha,\beta}+\varepsilon_1\boldsymbol{V}_1)&=\varepsilon_1\varepsilon_3\cdot x^{\alpha-1}y^{\beta-2}(\alpha y-\beta x)^2,\\ \mathrm{div}M(\boldsymbol{W}_{\alpha,\beta}+\varepsilon_2\boldsymbol{V}_2)&=\varepsilon_2\varepsilon_4\cdot x^{\alpha-2}y^{\beta-1}(\alpha y-\beta x)^2,\end{aligned}$$

where $M=x^{\alpha-1}y^{\beta-1}=x^{\varepsilon_3}y^{\varepsilon_4}$ is the integrating multiplier for $\boldsymbol{W}_{\alpha,\beta}$. It follows that for $\varepsilon_1=\varepsilon_4=0$ the vector field $\boldsymbol{W}_{\alpha,\beta}+\varepsilon_2\boldsymbol{V}_2$ has integrating factor $M=x^{\alpha-1}$ and a Darbouxian first integral of the type $x^\alpha F$, where $F(x,y)$ is a quadratic polynomial; such form of a first integral is characteristic for reversible quadratic vector fields. If $\varepsilon_2=\varepsilon_3=0$ then the same argument shows that the vector field $\boldsymbol{W}_{\alpha,\beta}+\varepsilon_2\boldsymbol{V}_2$ has first integral of the type $y^{\beta-1}G(x,y)$, $\deg G=2$. In fact, these vector fields lie in the variety Darb$(2,1)$.
If $\varepsilon_1=-\varepsilon_2$ and $\varepsilon_3=\varepsilon_4$ (and $\alpha=\beta$) then the vector field $\boldsymbol{W}_{\alpha,\alpha}+\varepsilon_1\boldsymbol{V}_1-\varepsilon_1\boldsymbol{V}_2$ undergoes a change of sign after the change $x\leftrightarrow y$. So, it is reversible and lies in Darb$(2,1)$.
If $\varepsilon_3=\varepsilon_4=0$ then $M\equiv1$ and the vector field is Hamiltonian (lies in Darb(3)). The fact that these are the only intersections of the family $\{X_\varepsilon\}$ with the center variety follows either from Theorem 6.33 or from the below analysis.
Introduce the following functions of the parameters ε_j :

$$\mu_1=\varepsilon_0,\ \mu_2=\varepsilon_1\varepsilon_3+\varepsilon_2\varepsilon_4,\ \mu_3=(\varepsilon_1\varepsilon_3-\varepsilon_2\varepsilon_4)(\varepsilon_3-\varepsilon_4),\ \mu_4=(\varepsilon_1\varepsilon_3-\varepsilon_2\varepsilon_4)(\varepsilon_1+\varepsilon_2)$$

and the integrals

$$\begin{aligned}J_1(h)&=\int\!\!\int(x+y)\mathrm{d}x\mathrm{d}y,\\ J_2(h)&=\frac{1}{2}\int\!\!\int(y-x)^2\frac{x+y}{xy}\mathrm{d}x\mathrm{d}y,\\ J_3(h)&=\frac{1}{4}\int\!\!\int(x^2-y^2)\frac{(x+y)\ln(x/y)-(x-y)}{xy}\mathrm{d}x\mathrm{d}y,\\ J_4(h)&=\frac{5}{6h}\int\!\!\int(x-y)^4\mathrm{d}x\mathrm{d}y,\end{aligned}$$

where the integration domain is $\{H_{1,1}(x,y)\geq h\}$. In [**Zo3**] it was proved that:

The number of limit cycles of the perturbation X_ε is the same as the number of zeroes of the following third-order Melnikov function

$$J(h)=\mu_1J_1(h)+\mu_2J_2(h)+\mu_3J_3(h)+\mu_4J_4(h);$$

In fact, this number does not exceed 3.

Moreover, $J(h) \equiv 0$ *iff* $\mu_1 = \mu_2 = \mu_3 = \mu_4 = 0.$

3. *Tangent cone to a Darbouxian component.* Movasati ([**Mov2**, Theorem 3.1]) claimed that he tangent cone $TC_{\boldsymbol{V}_0}\mathrm{Cent}(n)$ to the center variety $\mathrm{Cent}(n)$ at $\boldsymbol{V}_0 = \boldsymbol{X}_{H_0}$, with $H_0 = l_0 \cdots l_n$ (as above), is the union of tangent cones $TC_{\boldsymbol{V}_0}\mathrm{Darb}\,(\Pi)$, corresponding to partitions $\Pi = (\pi_1, \ldots, \pi_s)$, $s = s(\Pi)$, of the set $\{0, \ldots, n\}$; here (locally) $\mathrm{Darb}\,(\Pi) \subset \mathrm{Darb}\,(d)$, where $d = (|\pi_1|, \ldots, |\pi_s|)$. It is clear that this implies the thesis of the Movasati's theorem.

Let us explain the statement of this proposition. For a partition $\Pi = (\pi_1, \ldots, \pi_s)$ define $f_{\pi_j}(x,y) = \prod l_k(x,y)$, where the product runs over k's from π_j. For a given partition Π consider the Darbouxian unfolding of the function H_0 :

$$H_\varepsilon = \prod \left(f_{\pi_j} + \varepsilon g_{\pi_j}\right)^{1+\varepsilon\mu(\pi_j)} \approx H_0 \prod \left(1 + \varepsilon\mu(\pi_j)\ln f_{\pi_j} + \varepsilon g_{\pi_j}/f_{\pi_j}\right),$$

where $g_{\pi_j}(x,y)$ are polynomials of degree $\leq \deg f_{\pi_j} = |\pi_j|$ and $\mu\,(\pi_j)$ are pairwise different real numbers. We have

$$\mathrm{d}\ln H_\varepsilon \approx \mathrm{d}\ln H_0 + \varepsilon \sum \left\{\mu\,(\pi_j)\,\mathrm{d}\ln f_{\pi_j} + \mathrm{d}\left(g_{\pi_j}/f_{\pi_j}\right)\right\}.$$

Therefore, the perturbation of $\mathrm{d}H_0$ in the direction of $\mathrm{Darb}\,(\Pi)$ is of the form

$$\mathrm{d}H_0 + \varepsilon H_0 \sum \left\{\mu\,(\pi_j)\,\mathrm{d}\ln f_{\pi_j} + \mathrm{d}\left(g_{\pi_j}/f_{\pi_j}\right)\right\}.$$

4. *Melnikov functions.* Consider now a general unfolding

$$\omega_\varepsilon = \mathrm{d}H_0 + \varepsilon^k\omega_k + \varepsilon^{k+1}\omega_{k+1} + \ldots = 0$$

of the Pfaff equation $\mathrm{d}H_0 = 0$ such it has a center p_ε close to an extremum p of H_0, i.e., the corresponding vector fields lie in the center variety. Movasati claims that there exists a partition $\Pi = (\pi_1 \ldots, \pi_s)$ such that

$$\omega_k = H_0 \left\{ \sum_j \left[\mu_{\pi_j}\mathrm{d}\ln f_{\pi_j} + \mathrm{d}(g_{\pi_j}/f_{\pi_j})\right] \right\}, \tag{11.29}$$

where g_{π_j} are polynomials of degree $\leq \deg f_{\pi_j}$ and the constants μ_{π_j} are pairwise different; from this the statement about the tangent cone follows. (In fact, he presents a double sum $\sum_\Pi \sum_j$ in the braces.) In the proof he somehow claims to explore the vanishing of the Melnikov functions of order $k, k+1, \ldots, 2k$; but Example 2 (where $k = 0$) demonstrates that this cannot be achieved.

Here lies the flaw of the Movasati's argument and I do not know how to fill the gap.

11.74. Theorem of Zare (Pull-back centers). *The set* $\mathrm{Pullback}(p,q)$ *of vector fields* $\boldsymbol{V}(x,y)$ *with elementary center, which are pull-backs of vector fields*

$$\boldsymbol{W} = P(X,Y)\frac{\partial}{\partial X} + Q(X,Y)\frac{\partial}{\partial Y}$$

of degree $\leq p$ via a polynomial map $\Phi : (x,y) \longmapsto (X,Y) = (R(x,y), S(x,y))$, with $\deg R, S \leq q$, *forms an irreducible component of the center variety* $\mathrm{Cent}(n)$, $n = pq + q - 1$.

In the proof in [**Zare**] Y. Zare calculates the tangent cone to $\mathrm{Cent}(n)$ at a special point $\boldsymbol{V}_0$ of $\mathrm{Pullback}(p,q)$.
One has

$$R = R(x) = \prod (x - x_i)\,, S = S(y) = \prod (y - y_i)\,,$$
$$\boldsymbol{W} = \boldsymbol{X}_{\mathcal{H}},\ \mathcal{H} = F(X) + G(Y).$$

The critical values of $F \circ R$ and $G \circ S$ are separated and some additional genericity assumptions are imposed, e.g., to have a center at the critical locus of Φ.
Therefore the corresponding Pfaff form, associated with $\boldsymbol{V}_0$, is Hamiltonian,

$$\omega = \Phi^* \mathrm{d}\mathcal{H} = \mathrm{d}\left(\Phi^* \mathcal{H}\right) = \mathrm{d}H.$$

This means that $\boldsymbol{V}_0$ lies in the intersection of the sets $\mathrm{Ham}(n)$ and $\mathrm{Pullback}(p,q)$.
Zare proves that the tangent cone $TC_{\boldsymbol{V}_0}\mathrm{Cent}$ is the union of the tangent cones to $\mathrm{Ham}(n)$ and $\mathrm{Pullback}(p,q)$.
I do not present details of that proof. □

11.75. Complex centers. Some people consider the cases of center for complex vector fields with elementary singularities which are resonant, i.e., locally

$$\boldsymbol{V} = (-px + \ldots)\partial/\partial x + (qy + \ldots)\partial/\partial y,$$

or we have the Pfaff equation $(qy + \ldots)\mathrm{d}x + (px + \ldots)\mathrm{d}y = 0$. Here the center property means the existence of a local analytic first integral $H = x^q y^p + \ldots$ on the one side, or the existence of a family of vanishing cycles in the complex phase curves of $\boldsymbol{V}$ on the other side.
In [**FSZ**] the center problem was solved completely in the $1:-2$ case for quadratic vector fields. There are 20 cases, components of corresponding center variety.
In [**LlZ**] it was proved that so-called blue sky saddles form an irreducible component of $1:-1$ resonant version of the center variety. These are pull-backs

$$\boldsymbol{V} = \pi^* \boldsymbol{W} = x \cdot (D\pi)^{-1}\, \boldsymbol{W} \circ \pi,$$

via the blowing-up $\pi : (x,y) \longmapsto (X,Y) = (x, xy)$, of vector fields $\boldsymbol{W}(X,Y)$ of degree $\leq m$ without singularity at $X = Y = 0$.
Plausibly there exists a suitable version of the pull-back mechanism for producing $p:-q$ resonant centers of polynomial vector fields.

Chapter 12

Hypergeometric Functions

The Gauss hypergeometric equation is the equation, where the monodromy group demonstrates its full beauty. It connects such domains like regular geometric polyhedra, hyperbolic geometry, elliptic integrals and modular functions. In nineteen century all students and teachers of mathematics knew how the hypergeometric function looks like and how its monodromy is connected with spherical triangles (see the lectures of F. Klein).

Here we present the theory of the hypergeometric equation based on the books of Klein [**Kl1**] and [**Kl2**] and V. Golubev [**Gol**]. We calculate the monodromy and find the hypergeometric equations solved in quadratures.

Next, we shortly present some generalizations of the hypergeometric functions. These generalizations go in two directions. In the Picard–Deligne–Mostow approach the time remains one-dimensional but the number of singularities and dimension of the space of integrals grow. The aim is to obtain new arithmetic subgroups of the group of motions in multi-dimensional complex hyperbolic geometry. In the other approach (I. Gelfand, A. Varchenko and others) the hypergeometric integrals are defined along cycles in complement of complex vector space to a collection of affine hypersurfaces. Here the main effort is concentrated on description of the basis of certain cohomology groups.

In this version of the book I have added a separated section devoted to generalized hypergeometric functions of one variable with results of F. Beukers and G. Heckman and A. Duval and C. Mitschi.

§1 Gauss Hypergeometric Equation

12.1. Second order equations of Fuchs type. The class of these equations was described in Lemma 8.33(b) (in Chapter 8). They are of the form

$$\ddot{x} + p(t)\dot{x} + q(t)x = 0,$$

where $p(t) = P(t)/R(t)$, $q(t) = Q(t)/R^2(t)$, $R(t) = (t - t_1)\dots(t - t_m)$ and P, Q are polynomials of degrees $\leq m - 1$ and $\leq 2(m - 1)$ respectively. We will assume that one of the singular points is $t_{m+1} = \infty$; in this case $\deg P = m - 1$ or $\deg Q = 2(m - 1)$.

For each singular point t_j (or ∞) we have the defining equation for the Levelt exponents $\lambda_{j,1}$, $\lambda_{j,2}$, such that the basis of the space of solutions is of the form $(t - t_j)^{\lambda_{j,1}}(1 + \dots)$ and $(t - t_j)^{\lambda_{j,2}}(1 + \dots)$ (or $t^{\lambda_{\infty,1,2}}(1 + \dots)$ respectively). If

H. Żołądek, *The Monodromy Group*, Monografie Matematyczne 67,
https://doi.org/10.1007/978-3-031-91270-2_12

$p(t) = A_j/(t-t_j)+\dots$, $q(t) = B_j/(t-t_j)^2+\dots$, then the defining equation takes the form

$$\lambda(\lambda-1)+A_j\lambda+B_j=0.$$

If $p \sim A_\infty/t$, $q \sim B_\infty/t^2$, then we get $\lambda(\lambda+1)-A_\infty\lambda+B_\infty=0$. Note that $p=\sum A_j/(t-t_j)$ and $A_\infty=\sum A_j$.
We obtain the following relations for the exponents $\lambda_{j,k}$: $A_j=1-\lambda_{j,1}-\lambda_{j,2}$, $A_\infty = 1+\lambda_{\infty,1}+\lambda_{\infty,2}$. The property $A_\infty=\sum A_j$ leads to the following **Fuchs condition**

$$\sum_{j=1}^{m+1}(1-\lambda_{j,1}-\lambda_{j,2})=2.$$

The above shows that the function $p(t)$ is defined uniquely by the collection of exponents satisfying the Fuchs condition. The relations $\lambda_{j,1}\lambda_{j,2}=B_j$ give only $m+1$ conditions on the $2m$ coefficients of the polynomial $Q(t)$. This means that $Q(t)$ is defined uniquely by the Levelt exponents only when the number of singular points does not exceed three.
The case with one singular point, which we put to ∞, leads to the equation $\ddot{x}=0$. The case with two singular points, at 0 and at ∞, leads to the Euler equation $\ddot{x}+(c/t)\dot{x}+(d/t^2)x=0$ with constant coefficients c, d.

12.2. Riemann P-equation. The equation of Fuchs class with three singular points and with given Levelt exponents can be written explicitly. Assume that the Levelt exponents at 0 are ρ, ρ' at 1 are σ, σ' and at ∞ are τ, τ' and satisfy the Fuchs condition

$$\rho+\rho'+\sigma+\sigma'+\tau+\tau'=1.$$

Then we have $q(t)=[\rho\rho'(t-1)^2+\sigma\sigma' t^2+at(t-1)]/t^2(t-1)^2$ with the condition $\rho\rho'+\sigma\sigma'+a=\tau\tau'$. Finally we get the so called **Riemann equation**

$$\ddot{x}+\left(\frac{1-\rho-\rho'}{t}+\frac{1-\sigma-\sigma'}{t-1}\right)\dot{x}+\left(\frac{\rho\rho'}{t^2}+\frac{\sigma\sigma'}{(t-1)^2}+\frac{\tau\tau'-\rho\rho'-\sigma\sigma'}{t(t-1)}\right)x=0.$$

Riemann suggested the following rule to encode the solutions of the Riemann equation with given singular points a, b, c and with given Levelt exponents (in our case $a, b, c = 0, 1, \infty$)

$$P\left\{\begin{array}{cccc} a & b & c & \\ \rho & \sigma & \tau & t \\ \rho' & \sigma' & \tau' & \end{array}\right\}.$$

The Riemann equation has the following interesting property.

Lemma. *If $x_1(t)$, $x_2(t)$ are two independent solutions of the Riemann equation, then the functions $t^r(t-1)^s x_1(t)$, $t^r(t-1)^s x_2(t)$ are independent solutions of a Riemann equation with changed Levelt exponents. More precisely,*

$$t^r(t-1)^s\cdot P\left\{\begin{array}{cccc} 0 & 1 & \infty & \\ \rho & \sigma & \tau & t \\ \rho' & \sigma' & \tau' & \end{array}\right\}=P\left\{\begin{array}{cccc} 0 & 1 & \infty & \\ \rho+r & \sigma+s & \tau-r-s & t \\ \rho'+r & \sigma'+s & \tau'-r-s & \end{array}\right\}.$$

Proof. The linear space generated by the new functions is invariant with respect to the monodromy transformations. Moreover, these functions are also independent. Thus by Riemann theorem 8.35 they satisfy a second order linear differential equation. From their behavior at singular points we see that this equation must be the Riemann equation. □

12.3. Gauss hypergeometric equation. By applying a transformation from the above lemma, we can reduce two of the Levelt exponents to zeroes. Because of the Fuchs condition, there remain only three independent parameters which we choose in the form

$$P\left\{\begin{matrix} 0 & 1 & \infty & \\ 0 & 0 & \alpha & t \\ 1-\gamma & \gamma-\alpha-\beta & \beta & \end{matrix}\right\}.$$

The corresponding differential equation is the most important linear differential equation in mathematics, is called the **Gauss hypergeometric equation** and takes the form

$$t(t-1)\ddot{x} + [(\alpha+\beta+1)t - \gamma]\dot{x} + \alpha\beta x = 0.$$

This equation has one analytic solution $1 + \sum a_n t^n$, corresponding to the Levelt exponent 0. Simple calculations give the recurrent relations $a_{n+1} = \frac{(\alpha+n)(\beta+n)}{(n+1)(\gamma+n)} a_n$. The obtained series

$$F(\alpha,\beta;\gamma;t) = {}_1F_2\left(\begin{matrix}\alpha,\beta \\ \gamma\end{matrix}\middle| t\right) = \sum_{n=0}^{\infty} \frac{(\alpha)_n(\beta)_n}{(\gamma)_n n!} t^n$$

is called the **hypergeometric series**. Here the **Pochhammer symbol** $(a)_n = \Gamma(a+n)/\Gamma(a) = a(a+1)\dots(a+n-1)$ for $n > 0$ and $(a)_0 = 1$. This series is convergent in the disc with center at $t = 0$ and radius 1 and defines a holomorphic function, the **hypergeometric function**.

The particular cases of hypergeometric function include many known elementary functions. For example, we have

$$\begin{aligned} F(-m,\beta;\beta;t) &= (1+t)^m, \ m \in \mathbb{Z}_+, \\ F(1,\beta;\beta;t) &= (1-t)^{-1}, \\ tF(1,1;2;-t) &= \ln(1+t), \\ \lim_{\beta\to\infty} F(1,\beta;1;t/\beta) &= \mathrm{e}^{-t}. \end{aligned} \tag{12.1}$$

12.4. Three systems of fundamental solutions. We know one solution of the hypergeometric equation. The other solution is of the form $t^{1-\gamma}y(t)$ with analytic y. However, here we can use the lemma from the point 12.2. It says that $y(t)$ belongs to the space of solutions of Riemann equation with shifted Levelt exponents:

$$
\begin{aligned}
t^{\gamma-1}P&\left\{\begin{matrix} 0 & 1 & \infty & \\ 0 & 0 & \alpha & t \\ 1-\gamma & \gamma-\alpha-\beta & \beta & \end{matrix}\right\}\\
&=P\left\{\begin{matrix} 0 & 1 & \infty & \\ \gamma-1 & 0 & \alpha-\gamma+1 & t \\ 0 & \gamma-\alpha-\beta & \beta-\gamma+1 & \end{matrix}\right\}\\
&=P\left\{\begin{matrix} 0 & 1 & \infty & \\ 0 & 0 & \alpha-\gamma+1 & t \\ 1-(2-\gamma) & \gamma-\alpha-\beta & \beta-\gamma+1 & \end{matrix}\right\}.
\end{aligned}
$$

Because $F(\alpha-\gamma+1,\beta-\gamma+1;2-\gamma;t)$ is the unique analytic solution from the latter space, we find that it is equal to $y(t)$.

Thus we have one basis near $t=0$:

$$x_1=F(\alpha,\beta;\gamma;t),\quad x_2=t^{1-\gamma}F(\alpha-\gamma+1,\beta-\gamma+1;2-\gamma;t).$$

After performing the change $t\longmapsto 1-t$, we obtain the hypergeometric equation with exchanged Levelt exponents at the points $t=0$ and $t=1$. The corresponding basis of solutions takes the form

$$
\begin{aligned}
x_3&=F(\alpha,\beta;\alpha+\beta+1-\gamma;1-t),\\
x_4&=(1-t)^{\gamma-\alpha-\beta}F(\gamma-\beta,\gamma-\alpha;1-\alpha-\beta+\gamma;1-t).
\end{aligned}
$$

Finally, near $t=\infty$ we find the basis

$$x_5=t^{-\alpha}F(\alpha,1-\gamma+\alpha;1+\alpha-\beta;1/t),\quad x_6=t^{-\beta}F(\beta,1-\gamma+\beta;1+\beta-\alpha;1/t).$$

In the cases when the Levelt exponents at some point coincide, the basis of solutions is different. For example, if $\gamma=1$, then the other solution takes the form $x_2(t)=x_1(t)\ln t+\psi(t)$ with ψ analytic.

12.5. Euler integral representation. This is the formula

$$F(\alpha,\beta;\gamma;t)=\frac{\Gamma(\gamma)}{\Gamma(\beta)\Gamma(\gamma-\beta)}\int_0^1\frac{s^{\beta-1}(1-s)^{\gamma-\beta-1}}{(1-ts)^{\alpha}}\mathrm{d}s.$$

It easily follows from the expansion of the factor $(1-ts)^{-\alpha}$ into powers and the expression of the Beta-function by means of the Gamma-functions.

The above formula needs some explanations. The integral $\int_0^1 f(s,t)\mathrm{d}s$ is well defined when $\mathrm{Re}\beta>0$ and $\mathrm{Re}(\gamma-\beta)>0$; we must only omit the point $s=1/t$ along a path of integration (from 0 to 1). If $\beta\notin\mathbb{Z}$ and $\gamma-\beta\notin\mathbb{Z}$, then this integral can be replaced by $(1-\mathrm{e}^{2\pi\mathrm{i}\beta})^{-1}(1-\mathrm{e}^{2\pi\mathrm{i}(\gamma-\beta)})^{-1}$ times the integral along the **Pochhammer cycle** $C(0,1)$, presented at Fig. 11.2 in Chapter 11 (with $u_k=0,\ u_j=1$). The cycle $C(0,1)$ surrounds each of the points 0, 1 two times in opposite directions and the subintegral function $f(\cdot,t)$ is single valued on this cycle.

One can even define the function $\int_{C(0,1)} f(s,t)\mathrm{d}s$ as one of the solutions of the hypergeometric equation. Here instead of expansion into power series, we use the identity $\{t(t-1)\partial_t^2+[(\alpha+\beta+1)t-\gamma]\partial_t-\alpha\beta\}f=\alpha\partial_s[s^{\beta}(1-s)^{\gamma-\beta+1}(1-ts)^{-\alpha+1}]$. The solution $\int_{C(0,1)} f$ is well defined even when γ is negative integer (while $F(\alpha,\beta;\gamma;t)=\infty$). Moreover, any of the functions $\int_{C(a,b)} f(s,t)\mathrm{d}s$, where $a,b\in\{0,1,\infty,1/t\}$ and the cycle $C(a,b)$ is defined analogously as $C(0,1)$, defines a solution of the hypergeometric equation.

The function $\int_{C(0,1)} f$ admits continuous prolongation to the case when one (but not two) of the exponents $\beta-1$ or $\gamma-\beta-1$ is negative integer. Then it equals to the factor $2\pi\mathrm{i}(1-\mathrm{e}^{2\pi\mathrm{i}\kappa})$ times the residuum of $f(\cdot,t)$ at the corresponding pole. If both exponents are integers, then the integral is zero.

12.6. Some recurrent formulas. One introduces the new parameters

$$\lambda=1-\gamma,\quad \mu=\beta-\alpha,\quad \nu=\gamma-\alpha-\beta$$

(they are the differences between the Levelt exponents). Analogously, one defines the parameters $\lambda=\rho-\rho'$, $\mu=\sigma-\sigma'$, $\nu=\tau-\tau'$ for the general Riemann equation.

Note that the hypergeometric (respectively Riemann) equation with one triple (λ,μ,ν) has the same monodromy as the analogous equation, for which these parameters differ by a sign. This is consequence of the fact that the pairs of exponents $(\rho,\rho'),\ldots$ are non-ordered pairs.

Other relations allow to represent the hypergeometric series with 'large' parameters by means of linear combinations of the hypergeometric series with 'small' parameters and its derivatives with rational coefficients. In the formulas below, taken from [**BE**], $F=F(\alpha,\beta;\gamma;t)$ and $F(\alpha\pm1),\ldots,F(\gamma\pm1)$ denote the hypergeometric series with the distinguished variable shifted, $F(\alpha\pm1,\beta;\gamma;t)\ldots$. We have

$$\begin{aligned}
F(\alpha+1)&=\frac{1}{\alpha}t^{1-\alpha}\frac{\mathrm{d}}{\mathrm{d}t}[t^{\alpha}F],\\
F(\alpha-1)&=\frac{1}{\gamma-\alpha}t^{1+\alpha-\gamma}(1-t)^{1+\gamma-\alpha-\beta}\frac{\mathrm{d}}{\mathrm{d}t}[t^{\gamma-\alpha}(1-t)^{\alpha+\beta-\gamma}F],\\
F(\gamma+1)&=\frac{\gamma}{(\gamma-\alpha)(\gamma-\beta)}(1-t)^{1+\gamma-\alpha-\beta}\frac{\mathrm{d}}{\mathrm{d}t}[(1-t)^{\alpha+\beta-\gamma}F],\\
F(\gamma-1)&=\frac{1}{1-\gamma}t^{2-\gamma}\frac{\mathrm{d}}{\mathrm{d}t}[t^{\gamma-1}F].
\end{aligned}$$

Because F is symmetric with respect to the parameters α,β, analogous formulas hold for $F(\beta\pm1)$.

Note that the right-hand-sides are of the form $p(t)F+q(t)\dot{F}$ with rational functions p,q. One can also see that the other solution $x_2=t^{1-\gamma}(1+\ldots)$ satisfies the same

relations with shifted parameters. Therefore, using induction with respect to the integer parts of the parameters and the expression of $\ddot{F}$ by means of $F, \dot{F}$, one can express any hypergeometric function in this form with parameters from a fixed cube with the edge of length 1.
(There are also other relations between the hypergeometric series, called the *Gauss relations*, allowing to represent all series, like $F(\alpha+n)$, by means of F and its adjacent functions $F(\alpha\pm 1)$, $F(\beta\pm 1)$, $F(\gamma\pm 1)$. For example, we have $[\gamma-2\alpha-(\beta-\alpha)t]F+\alpha(1-t)F(\alpha+1)+(\alpha-\gamma)F(\alpha-1)=0$. There are 15 of them and can be found in [**BE**] and in [**Kl1**].)
The translation of the parameters α,β,γ by integers means translation of the parameters λ,μ,ν by integers; however, the sum of these three integers must be equal to zero. For example, the change $\alpha\longmapsto\alpha+n$ means the change $\lambda\longmapsto\lambda$, $\mu\longmapsto\mu-n$, $\nu\longmapsto\nu+n$.
If we admit also the changes of signs in λ,μ,ν, then we obtain the following property.

12.7. Proposition. *The monodromy of the hypergeometric equation with the parameters (λ,μ,ν) is the same as the monodromy of the hypergeometric equation with the parameters*

$$(\pm\lambda+l,\pm\mu+m,\pm\nu+n),$$

where l,m,n are integers and $l+m+n$ is even.
If λ,μ,ν are real, then one can restrict the analysis of the hypergeometric equation to the following domain of parameters

$$0\leq\lambda,\mu,\nu<1;\quad 0\leq\lambda+\mu,\lambda+\nu,\mu+\nu\leq 1. \tag{12.2}$$

12.8. Schwarz map. The above integral representations of solutions will be useful in description of the monodromy group of the hypergeometric equation. In order to simplify the explanation, we restrict considerations to the case when the parameters α,β,γ (and λ,μ,ν) are real. (For the general case we refer the reader to [**Gol**], [**IKSY**] and [**BE**]). We also assume that we are in the domain (12.2), defined in Proposition 12.7.
Let $y_1(t)$, $y_2(t)$ be two independent solutions of the hypergeometric equation. Then we have the multivalued **Schwarz map**

$$W:\overline{\mathbb{C}}\setminus\{0,1,\infty\}\longmapsto\mathbb{CP}^1,\quad W(t)=(y_1(t):y_2(t)).$$

In the affine chart in the image, we have

$$w(t)=y_2(t)/y_1(t).$$

The different choices of the basic solutions are related by means of Möbius automorphisms of the target space.
The interpretation of the Schwarz map in terms, which do not involve the choice of basis in the space of solutions, uses the notion of local system. A non-autonomous

linear differential equation on $X = \overline{\mathbb{C}} \setminus$ (singular points) defines a vector bundle $\mathbf{K}$ (of rank d equal to the order of the equation) with constant transition maps. Recall that such a bundle is called a **local system**. (We have met local systems in studying the Gauss–Manin connection, see the point 5.32 in Chapter 5.) A local system is defined uniquely by means of its local horizontal sections. Local horizontal sections of the local system $\mathbf{K}$ are solutions of some differential equation. Thus $\mathbf{K}$ admits an unique flat (i.e., with vanishing curvature) connection, called the Gauss–Manin connection. If $\gamma : [0,1] \longmapsto X$ is a loop, then $\gamma^*\mathbf{K}|_{[0,1)}$ is a trivial bundle $[0,1) \times \mathbb{C}^d$ and can be prolonged to a bundle over the circle $\mathbb{S}^1 = [0,1]/0 \sim 1$ by means of the monodromy operator M_γ. The cocycle $\xi \in H^1(X, GL(d,\mathbb{C}))$, defining the local system, associates to each loop in X the monodromy operator induced by it.
In fact, the bundle $\mathbf{K}$ is a trivial bundle in the topological sense but is not trivial in the geometrical sense (nontrivial flat connection).
Consider the trivial bundle $X \times V \longmapsto X$, $V = \mathbb{C}^d$, where we assume that $X \subset \mathbb{C}$. Above each simply connected domain $U \subset X$ the two bundles, $\mathbf{K}|_U$ and $U \times V$, are isomorphic. These isomorphisms $\mathbf{K}|_U \longmapsto U \times V$ together define a multivalued 'isomorphism' $\Phi : \mathbf{K} \longmapsto X \times V$. Of course, this multivalued isomorphism is uniquely defined by its restriction Φ_a to a distinguished fiber $\mathbf{K}_a$; we assume that Φ_a is fixed. We have a multivalued holomorphic vector function $t \longmapsto \Phi(t) \in V$. If we choose some basis in V, then the coordinates of $\Phi(t)$ form basis of the space of solutions of the differential equation.
Consider the projectivizations of the two bundles, $P\mathbf{K}$ and $X \times P(V)$. The Schwarz map is defined as $W = P \circ \Phi$.
The Schwarz map can be also considered as a kind of period mapping. Indeed, it is expressed by means of integrals of a holomorphic form along varying cycles, like the period mapping from Section 5 in Chapter 7.
In the literature usually the monodromy group of the Schwarz function is called the monodromy group of the Gauss equation. We shall call it the **projective monodromy group of the hypergeometric equation**.

12.9. Theorem of Schwarz (Spherical triangle). ([**Schw**]) *The map W, restricted to the upper half-plane $\mathbb{H} = \{\mathrm{Im} t \geq 0\}$, maps $\mathbb{H}$ biholomorphically onto a spherical triangle Δ with the angles $\pi\lambda, \pi\mu, \pi\nu$, i.e., the intervals $[-\infty, 0]$, $[0,1]$, $[1,\infty]$ are sent to arcs of circles in $\overline{\mathbb{C}}$.*
The prolongation of a point $t \in \mathbb{H}$ to the conjugate point $\bar{t}$ in the lower half-plane along a path going through one of the intervals $(-\infty, 0)$, $(0,1)$, $(1,\infty)$ results in inversion of the point $w = W(t)$ with respect to the circle containing the image of the corresponding interval.
The projective monodromy group of the hypergeometric equation, Mon, *is isomorphic to the subgroup of index 2 of the spherical triangle group, generated by inversions with respect to the sides of the triangle Δ;* Mon *consists of compositions of even number of inversions.*

Recall that the **inversion** in $\overline{\mathbb{C}}$ with respect to a circle $S \subset \mathbb{C}$ with center z_0 and radius r is defined by the equality $|z - z_0| \cdot |z' - z_0| = r^2$ and by the property that

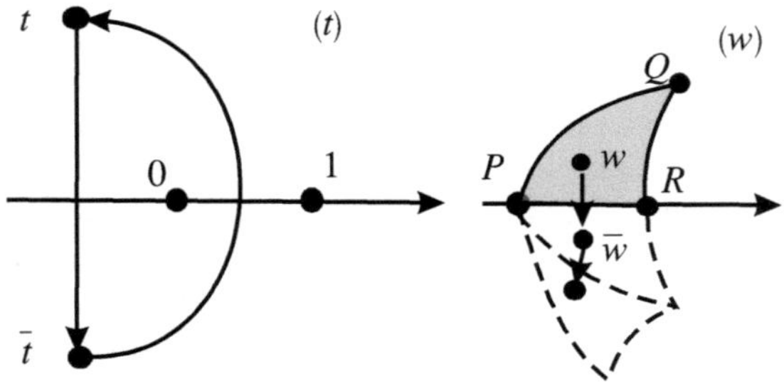

Figure 12.1: Monodromy of the Schwarz map

the point z and its image z' lie on the same radius starting from z_0. The inversion with respect to a circle passing through ∞, i.e., a line in $\mathbb{C}$, is the reflection with respect to this line.

Proof. As the basis of solutions of the hypergeometric equation we choose

$$y_1(t) = \int_0^1 f(s,t)\mathrm{d}s, \quad y_2 = \int_1^\infty f(s,t)\mathrm{d}s.$$

Thus we have $w(t) = y_2(t)/y_1(t)$. Consider its restriction to the upper half-plane $\mathbb{H}$, where we choose the branch, which is real on the interval $[-\infty, 0]$. By the Schwarz reflection principle, the function $w(t)$ can be prolonged through the interval $[-\infty, 0]$ to the lower half-plane with the property $w(\bar{t}) = \overline{w(t)}$. The image of $[-\infty, 0]$ is an interval in the real line in $\mathbb{C}$, i.e., an arc in a circle in $\mathbb{CP}^1 \simeq \mathbb{S}^2$. If we choose another basis of solutions, then the new $\widetilde{w}(t)$ is related with $w(t)$ by means of a fractional-linear map, the image of $[-\infty, 0]$ is still an arc of a circle and $\widetilde{w}(\bar{t}) = \sigma\widetilde{w}(t)$, where σ is an inversion.
By applying this to the other two intervals $[0,1]$ and $[1,\infty]$, we find that the real axis is transformed to boundary of a spherical triangle Δ. Because $w(t)$ is holomorphic with open image, then $w(\mathbb{H}) = \Delta$. Moreover $\frac{\mathrm{d}w}{\mathrm{d}t} = y_1^{-2}(\dot{y}_2 y_2 - \dot{y}_1 y_2) = y_1^{-2} \times (\text{Wronskian}) \neq 0$.
We should also determine the angles of Δ and the action of the monodromy. Assume that the vertices of Δ are $P = w(0)$, $Q = w(1)$ and $R = w(\infty)$. Consider two successive reflections in the t-plane: first $t \longmapsto \bar{t}$ through the interval $(-\infty, 0)$ and the next $\bar{t} \longmapsto t$, but through the interval $(0,1)$. In the w-plane we obtain two inversions: with respect to the arc (P,R) and with respect to the image of the arc (P,Q) (see Fig. 12.1). The result of their composition is the monodromy transformation corresponding to a loop around $t = 0$.
Any two circles in the projective plane $\mathbb{CP}^1$ can be transformed, via a Möbius transformation, to two circles passing through the point at infinity, so that they will form lines in the affine part; it is enough to move one of the intersection points to infinity. Thus we can assume that the arcs (R,P) and (P,Q) form sides

of a straight angle. The composition of reflections with respect to these sides is a rotation by the angle twice greater than the angle of Δ at the vertex p.
Let us find this angle. We know two particular solutions $x_{1,2}(t)$ of the hypergeometric equation. We have $x_2/x_1 = t^{1-\gamma}\phi(t)$ with holomorphic ϕ. The monodromy around $t = 0$ acts as rotation by the angle $2\pi\lambda$, $\lambda = 1 - \gamma$.
Repeating this with other vertices of Δ, we obtain the thesis of Theorem 12.9. □

12.10. Schwarz differential equation. The Schwarz function $w(t)$ is multivalued, but this multivaluedness exhibits regularities. The different branches are related by means of fractional-linear transformations. There is a meromorphic operation (expressed in terms of w and its derivatives) which is invariant with respect to composition of w with Möbius transformations. This operation is called the **Schwarz derivative** and equals

$$\{w, t\} = (\ln w_t)_{tt} - \frac{1}{2}((\ln w_t)_t)^2 = \frac{w'''}{w'} - \frac{3}{2}\left(\frac{w''}{w'}\right)^2.$$

It has the following important properties:

(a) *For the composition* $t \longmapsto \zeta \longmapsto w$ *we have* $\{w, t\} = \{w, \zeta\}\zeta_t^2 + \{\zeta, t\}$.

Indeed, $\ln w_t = (\ln w_\zeta) \circ \zeta + \ln \zeta_t$, $(\ln w_t)_t = (\ln w_\zeta)_\zeta \circ \zeta \cdot \zeta_t + (\ln \zeta_t)_t$ and then $\{w, t\} = (\ln w_\zeta)_{\zeta\zeta}\zeta_t^2 + (\ln w_\zeta)_\zeta \zeta_{tt} + (\ln \zeta_t)_{tt} - \frac{1}{2}[((\ln w_\zeta)_\zeta)^2\zeta_t^2 + 2(\ln w_\zeta)_\zeta \zeta_t (\ln \zeta_t)_t + ((\ln \zeta_t)_t)^2]$. Because $(\ln \zeta_t)_t = \zeta_{tt}/\zeta_t$, the above formula follows.
(Formula (a) implies that a composition of several maps with negative Schwarz derivative has negative Schwarz derivative. This property is recently widely used in one-dimensional real dynamics.)

(b) $\{(at + b)/(ct + d), t\} = 0$.

(The Schwarz derivative constitutes a differential invariant for a class of maps which is wider than the class of Schwarz maps. This class is the following. Namely one considers a conformal map $v : \mathbb{H} \longmapsto \mathbb{CP}^1$ onto a circular polygon $P_1P_2 \ldots P_m$, where $P_j = v(t_j)$ and $t_j \in \mathbb{R}$. Applying the Schwarz principle, we extend v to a multivalued map with ramification points $t_1, \ldots, t_m$.

The above implies the following.

Theorem. *The Schwarz derivative* $\{w, t\}$ *of the mapping* v *(of* $\mathbb{H}$ *onto a spherical polygon) is a single valued function. In particular, we have*

$$\{w, t\} = (1 - \lambda^2)/(2t^2) + (1 - \mu^2)/(2(t - 1)^2) + (\lambda^2 + \mu^2 - \nu^2 - 1)/(2t(t - 1))$$

for the Schwarz map associated with the hypergeometric equation.

The latter identity is a particular case of the formula $\{x_2/x_1, t\} = -p' - p^2/2 + 2q$ for two independent solutions of the second order equation $\ddot{x} + p(t)\dot{x} + q(t)x = 0$. In [**Gol**] there is a formula connecting the Schwarz derivative $\{v, t\}$ with the vertices t_j and the angles of the polygon $P_1P_2 \ldots P_m$. This derivative is determined uniquely for $m \leq 3$; otherwise there remain $2(m-3)$ free parameters. Also the polygons are not determined uniquely (up to the action of $PSL(2,\mathbb{C})$) by $m > 3$ angles.

12.11. Riccati equation associated with hypergeometric equation. As with other second order linear differential equations, the Riccati equation

$$\frac{\mathrm{d}z}{\mathrm{d}t} = -\frac{\alpha\beta}{t(t-1)} - \frac{(\alpha+\beta+1)t-\gamma}{t(t-1)}z - z^2$$

is associated with the hypergeometric equation. Here $z = \dot{x}/x$.
As we know from the proof of Theorem 11.45 (in Chapter 11), the Riccati equation has its own monodromy group; it is a subgroup of the group of automorphisms of the Riemann sphere $\{t_0\} \times \mathbb{CP}^1$, defined by means of prolongation of solutions starting at $\{t_0\} \times \mathbb{CP}^1$ along loops with the beginning and the end at t_0. It is also the monodromy group of the first integral of the Riccati equation.
It turns out that the later group equals to the projective monodromy group of the hypergeometric equation, defined by means of the Schwarz map.
Indeed, let $x_1(t), x_2(t)$ be a basis of solutions of the hypergeometric equation such that

$$x_1(t_0) = 1, \quad \dot{x}_1(t_0) = 0, \quad x_2(t_0) = 0, \quad \dot{x}_2(t_0) = 1,$$

and let $w(t) = x_2/x_1$ be the corresponding Schwarz function. A solution of the Riccati equation with the initial condition $z(t_0) = y$ equals to $z(t) = (\dot{x}_1(t) + y\dot{x}_2(t))/(x_1(t) + yx_2(t))$. If some monodromy operator takes the form $(x_1, x_2) \longmapsto (x_1, x_2)\begin{pmatrix} a & b \\ c & d \end{pmatrix}$, then we get the formulas

$$w \longmapsto \frac{dw+b}{cw+a}, \quad y \longmapsto \frac{dy+c}{by+a}.$$

Therefore:

The Schwarz function is transformed by means of projective maps induced by the monodromy matrices of the linear equation, whereas the monodromy maps for the Riccati equation are induced by the transposed monodromy matrices.

In other words, one can treat the solutions of the Riccati equation as horizontal sections of the local system $P(\mathbf{K}^\vee)$, dual to the local system $P(\mathbf{K})$ associated with the linear second order equation (see the point 12.6).

12.12. Geometry of the tiling generated by Δ. The triangle Δ and its images under successive inversions with respect to its sides generate a system of triangles Δ_j. Usually, i.e., when some of the angles $\pi\lambda$, $\pi\mu$, $\pi\nu$ are not commensurable with π, this system of triangles is very irregular. The triangles with fixed vertex begin to intersect one another in an irregular way.
Geometrically interesting are the cases with rational λ, μ, ν. In this case the coverings are locally finite to one. Notice also that, when the system $\{\Delta_j\}$ realizes a

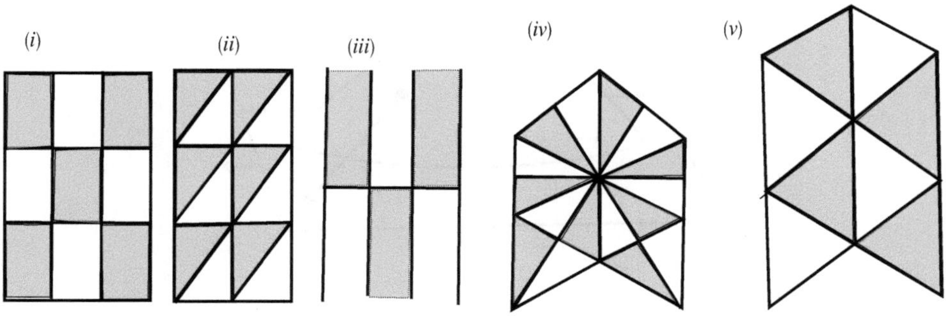

Figure 12.2: Tessellations in the Euclidean case

finite covering, then $w(t)$ takes only finite number of values for fixed t, and, because it has singularities of only algebraic type, it is an algebraic function. Schwarz classified all such situations (see Theorem 12.17 below).

We describe the situations when the system $\{\Delta_j\}$ realizes one-to-one covering of a domain in $\mathbb{CP}^1$, i.e., a regular tiling (or a tessellation). Therefore, we should have $\lambda = 1/n_1$, $\mu = 1/n_2$, $\nu = 1/n_3$ for integer n_j.

The three situations: $\lambda + \mu + \nu = 1$ (**Euclidean**), $\lambda + \mu + \nu > 1$ (**elliptic**) and $\lambda + \mu + \nu < 1$ (**hyperbolic**) are treated separately.

12.13. The euclidean case and the Schwarz–Christoffel function. Assume that two sides of the triangle Δ are straight intervals. If the sum of its angles is π, then the third side is also a straight segment. Therefore, we have a covering by means of euclidean triangles and the monodromy group consists of affine maps $w \longmapsto \widetilde{w} = aw + b$, $\mathrm{Mon} \subset \mathrm{Aff}(\mathbb{C})$.

In this situation we do not need the Schwarz derivative to obtain an invariant of the monodromy group. Because $\widetilde{w}_t = aw_t$, $\widetilde{w}_{tt} = aw_{tt}$, the **nonlinearity**

$$N(w) = w_{tt}/w_t$$

does not depend on the branch of w.

The nonlinearity is an invariant of maps, called the **Schwarz–Christoffel maps**, $v : \mathbb{H} \longmapsto \mathbb{C}$ to euclidean polygon $P_1P_2 \ldots P_m$, $P_j = v(t_j)$ with singularities at $t_j \in \mathbb{R}$. Near such t_j we have $w = \mathrm{const} + (t-t_j)^{\alpha_j}(c + \ldots)$, $w_t = \alpha_j(t-t_j)^{\alpha_j - 1}(c + \ldots)$, $w_{tt} = \alpha_j(\alpha_j - 1)(t - t_j)^{\alpha_j - 2}(c + \ldots)$, where α_j is the angle of the polygon at the vertex P_j. Thus $N(w) = (\alpha_j - 1)/(t - t_j) + \ldots$ near t_j and, generally, $w_{tt}/w_t = \sum(\alpha_j - 1)/(t - t_j)$. After integration we obtain the **Schwarz–Christoffel integral**

$$w(t) = \int_{t_0}^{t} \prod (s - t_j)^{\alpha_j - 1} \mathrm{d}s.$$

It realizes a conformal mapping from the upper half-plane onto an euclidean polygon with angles $\alpha_1, \ldots, \alpha_n$.

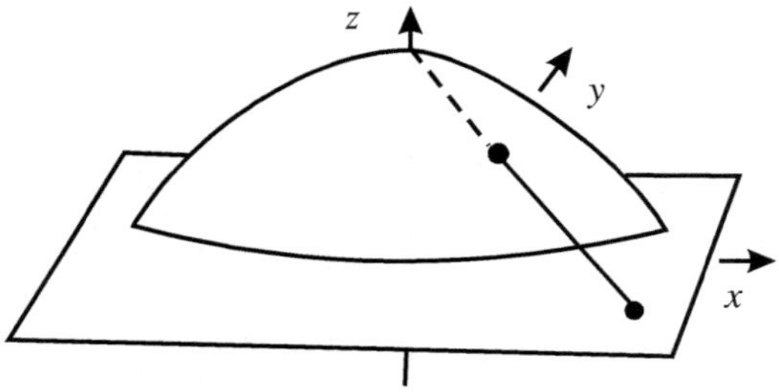

Figure 12.3: The stereographic projection

This map realizes certain covering by means of images of the polygon $P_1 \dots P_m$ under reflections with respect to its sides. In order to get a tiling, we must put $(\alpha_1, \dots, \alpha_m) = (1/n_1, \dots, 1/n_m)$ with integer n_j's. It is easy to check that there are only five possibilities:

$$\begin{array}{rl} \text{(i)} & (1/2, 1/2, 1/2, 1/2), \\ \text{(ii)} & (1/2, 1/4, 1/4), \\ \text{(iii)} & (1/2, 1/2, 1/\infty), \\ \text{(iv)} & (1/2, 1/3, 1/6) \\ \text{(v)} & (1/3, 1/3, 1/3). \end{array}$$

The corresponding tessellations are presented in Fig. 12.2.
The case (i) is realized by means of the incomplete elliptic integral

$$w(t) = \int_{t_0}^{t} \mathrm{d}s / \sqrt{(1-s^2)(1-k^2 s^2)}.$$

The cases (ii), (iii), (iv) and (v) are realized by means of the Schwarz map of the hypergeometric equation. Let us look at the solutions of the Gauss equation in this case. Because $\lambda + \mu + \nu = 1 - 2\alpha$, then $\alpha = 0$. So, we have $F(\alpha, \beta; \gamma; t) \equiv 1$ (see the formula (12.1) in 12.3). As the second solution of the Gauss equation we choose $\int_{1/t}^{\infty} f(s,t) = \int_0^t \tau^{\gamma}(\tau - 1)^{\gamma-\beta-1} \mathrm{d}\tau$. Therefore, the Schwarz map is equal to the Schwarz–Christoffel integral. Here the projective monodromy group of the Gauss equation is isomorphic with its monodromy group.
The monodromy groups of the Schwarz–Christoffel integrals are characterized by the property:

Mon *has one fixed point in* $\mathbb{CP}^1$*; it means that the corresponding matrices from* $GL(2, \mathbb{C})$ *can be simultaneously triangularized.*

12.14. Finite tilings and the regular polyhedra. Here we consider the elliptic case with regular tiling, i.e., $1/n_1 + 1/n_2 + \dots + 1/n_m > m - 2$.

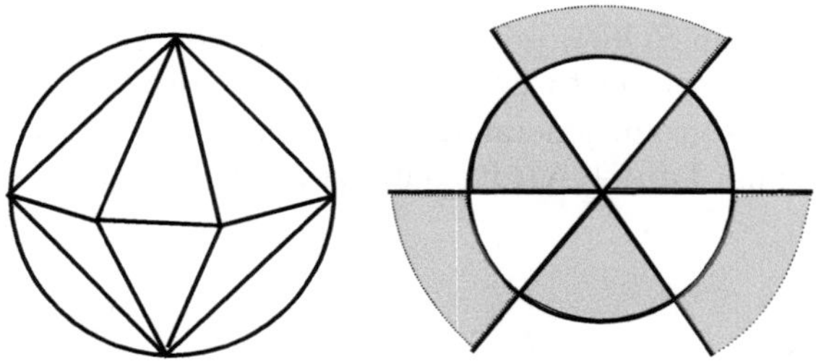

Figure 12.4: Tessellation of the sphere in the case of dihedral group

One can check that there are only four cases:

$$\begin{array}{ll} \text{(i)} & (1/2, 1/2, 1/q), \\ \text{(ii)} & (1/2, 1/3, 1/3), \\ \text{(iii)} & (1/2, 1/3, 1/4), \\ \text{(iv)} & (1/2, 1/3, 1/5). \end{array}$$

The above tilings are connected with some polyhedra inscribed in a sphere in $\mathbb{R}^3$. This connection is obtained in the following way.
Let the sphere be $\mathbb{S} = \{x^2 + y^2 + z^2 = 1\}$ and let $\mathcal{P}$ be a polyhedron inscribed into this sphere. We project the vertices, edges and sides of $\mathcal{P}$ from the center $(0, 0, 0)$ to the sphere $\mathbb{S}$. We obtain some tiling of $\mathbb{S}$. Next, we apply the *stereographic projection* from $\mathbb{S} \setminus (0, 0, 1)$ to $\mathbb{C}$ as in Fig. 12.3: $w = \frac{x+\mathrm{i}y}{1-z}$. The stereographic projection is a conformal map, infinitesimal circles are transformed to infinitesimal circles. (Proof: $\mathrm{d}w = [\mathrm{d}x + \frac{x}{1-z}\mathrm{d}z + \mathrm{i}(\mathrm{d}y + \frac{y}{1-z}\mathrm{d}z)]/(1-z)$ and then $(1-z)^2|\mathrm{d}w|^2 = \mathrm{d}x^2 + \mathrm{d}y^2 + (x^2 + y^2)\mathrm{d}z^2/(1-z)^2 + 2(x\mathrm{d}x + y\mathrm{d}y)\mathrm{d}z/(1-z) = \mathrm{d}x^2 + \mathrm{d}y^2 + \mathrm{d}z^2$, where we have used $x\mathrm{d}x + y\mathrm{d}y + z\mathrm{d}z = 0$.) The image of the tiling of $\mathbb{S}$ under the stereographic projection is the tiling of the complex plane by means of circular polygons. The latter induces the tiling of the complex projective line $\mathbb{CP}^1$.

Case (i) corresponds to a regular polygon $\mathcal{Q}$ with $2q$ vertices inscribed into the circle $x^2 + y^2 = 1, z = 0$. When joined with the two poles $(0, 0, \pm 1) \in \mathbb{S}$, it gives a polyhedron with $2q$ triangular sides. Projected onto $\mathbb{S}$, it gives a partition of $\mathbb{S}$ by means of the equator and q longitudes. In $\mathbb{C}$ we obtain the partition by means of the circle $|w| = 1$ and of the lines $\mathrm{Im}\, w^q = 0$ (see Fig. 12.4).
The function $w \longmapsto z = \frac{1}{2}(w^q + w^{-q})$ sends the curves of the partition to real segments. Thus the Schwarz map is an algebraic function equal to the inverse of the latter map.
The (projective) monodromy group is generated by compositions of even number of inversions with respect to the sides of triangles of the tiling. It is the **dihedral group** (or the Coxeter group $\mathbf{I}_2(q)$ of rank $2q$) generated by two holomorphic involutions, $w \longmapsto 1/w$ and $w \longmapsto \mathrm{e}^{2\pi\mathrm{i}/q}/w$ (see Theorem 4.35). It is also the

group of rotations (from $SO(3)$) preserving the polygon $\mathcal{Q}$; the generators are the axial symmetries with respect to two symmetry axes of $\mathcal{Q}$.
(Note that the triangle group generated by reflections is also Coxeter group, whose Coxeter graph is equal $\mathbf{I}_2(q) + \mathbf{A}_1$; it contains 3 vertices and is not connected.)

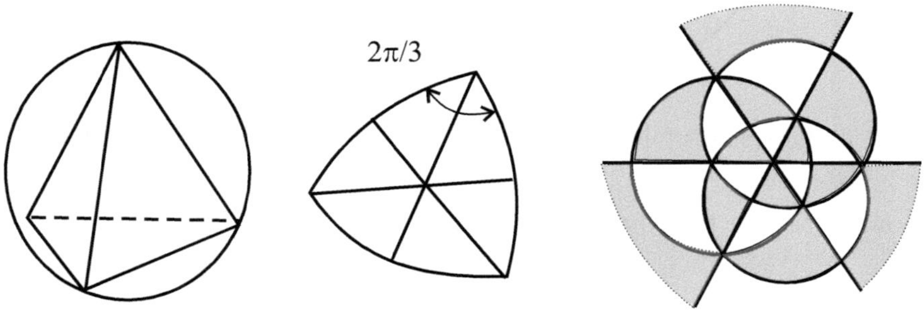

Figure 12.5: Tessellation associated with the regular tetrahedron

Case (ii) is connected with the regular *tetrahedron* inscribed into S. The angles of the triangles of projection of the tetrahedron onto S are equal to $2\pi/3$. Dividing these triangles by means of the great circles containing the sides of the large triangles, we obtain a partition of the sphere, and an induced partition of $\mathbb{C}$ into triangles with the angles $\pi/2$, $\pi/3$, $\pi/3$ (see Fig. 12.5).
The projective monodromy group, i.e., the group of even compositions of inversions, is equal to the group of rotations preserving the tetrahedron. This group consists of rotations by the angles $\pm 2\pi/3$ along the axes passing through vertices and by axial symmetries with respect to the lines passing through centers of opposite edges. The rank of this group is $4 \cdot 2 + 3 \cdot 1 + 1 = 12$ and the group is equal to the alternating group $A(4)$. It can be identified with the group of even permutations of the vertices of tetrahedron.
The group of all isometries of the tetrahedron is the Coxeter group of order 24, equal to $S(4)$, and has the Coxeter graph $\mathbf{A}_3$.
We do not present the algebraic equation defining the algebraic Schwarz function in this case.

Case (iii) is connected with the *cube* inscribed into a sphere. After dividing symmetrically each circular quadrangles into eight triangles, we obtain our tiling (see Fig. 12.6). Here the group of even compositions of inversions is equal to the group of rotations of the cube and consists of: rotations by the angles $\pm 2\pi/3$ around the lines joining antipodal vertices, rotations around the lines joining centers of opposite sides and axial symmetries with respect to the lines joining opposite edges. The rank is equal $4 \cdot 2 + 3 \cdot 3 + 6 \cdot 1 + 1 = 24$. In fact, this group is equal to the group of all permutation of the diagonals of the cube, Mon $= S(4)$. Note that the same

group is the group of rotations of the *octahedron*, which is the dual polyhedron to the cube.
The corresponding triangle group is the Coxeter group of the type $\mathbf{B}_3$. It is generated by the rotations of the cube and by the central symmetry (which acts trivially on the diagonals).

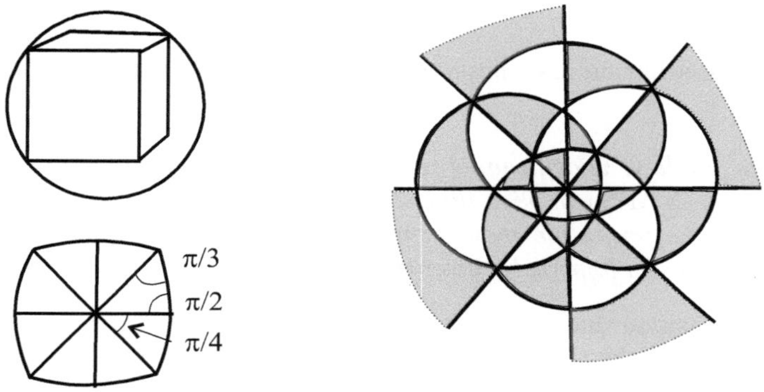

Figure 12.6: Tessellation associated with octahedron (or cube)

Case (iv) is connected with the *dodecahedron* inscribed into the sphere (see Fig. 12.6). The curvilinear pentagons are divided into curvilinear triangles with the angles $\pi/2$, $\pi/3$, $\pi/5$. The (projective) monodromy group is equal to the group of rotations of the icosahedron. It consists of: rotations by $\pm 2\pi/3$ around the 10 axes joining pairs of antipodal vertices, rotations by $2\pi j/5$ around the 6 axes joining centers of opposite pentagons, and of axial symmetries with respect to the 15 axes joining centers of opposite edges. The rank is $10\cdot 2+6\cdot 4+15\cdot 1+1 = 60$. This group is the group of even permutations of the 5 cubes inscribed into the icosahedron, $\mathrm{Mon} = A(5)$. The same is the group of rotations of the *icosahedron*, the dual to icosahedron.
The group of all isometries of the icosahedron is the Coxeter group of the type $\mathbf{H}_3$. It is not equal to the group of all permutations of the inscribed cubes; it is generated by the rotations of the icosahedron and by the central symmetry.

In [**Kl3**] the reader can find the formulas for the algebraic equations, defining the corresponding Schwarz function in all the cases (ii), (iii), and (iv).

12.15. The hyperbolic triangles and Fuchsian groups in hyperbolic geometry. Consider now the case when the triangle Δ, associated with the hypergeometric equation, has the sum of angles smaller than π, i.e., the *hyperbolic case.*

Theorem. *If $\pi/n_1+\pi/n_2+\pi/n_3 < \pi$, then there is a circle $\mathbf{C}$, called the absolute, which is orthogonal to all three circles containing the sides of Δ.*

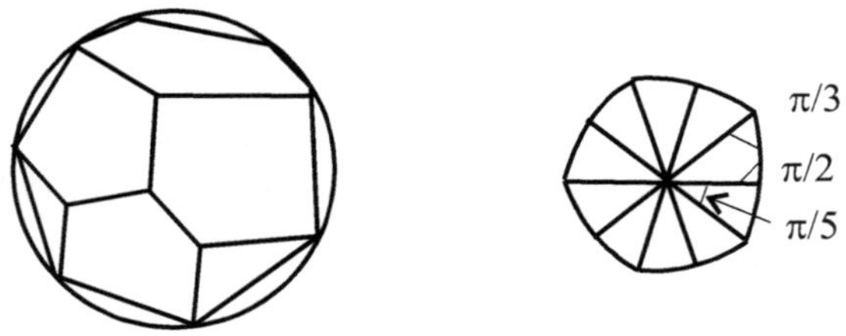

Figure 12.7: Tessellation associated with regular icosahedron (or dodecahedron)

Moreover, the circle **C** *is orthogonal to the circles of all sides of triangles* Δ_j *of the tiling defined by* Δ. *This implies that all the triangles* Δ_j *lie in one disc bounded by* **C** *and that the projective monodromy group forms a discrete subgroup of the group of isometries in hyperbolic geometry. In other words, it is a Fuchsian group.*

Proof. We can assume that two of the sides of $\Delta = P_1P_2P_3$ are straight segments, with vertex at P_1. Take the circle d containing the arc P_2P_3. There are two half-lines starting from P_1 and tangent to the circle d at the points B_2 and B_3. The circle **C** from the thesis of the theorem is the circle with center at P_1 and passing through B_2 and B_3 (see Fig. 12.8).

If a circle e is orthogonal to **C**, then the image of e under reflection with respect to a line passing through P_1 is also orthogonal to **C**. The same holds when the reflection with respect to a line is replaced by inversion with respect to another circle orthogonal to **C** (the inversion preserves angles and **C**). This implies that the images of Δ under inversions lie in one component of $\mathbb{CP}^1 \setminus \mathbf{C}$.

By applying a suitable Möbius transformation, we can assume that $\mathbf{C} = \{\mathrm{Im} w = 0\}$ is the real line in $\mathbb{C}$ and $\Delta \subset \mathbb{H}$.

The upper half-plane is called the **Lobachevski plane**. It is equipped with the *hyperbolic metric* $\mathrm{d}s^2 = |\mathrm{d}w|^2/(\mathrm{Im} w)^2$. The hyperbolic metric is invariant with respect to: (i) translations parallel to the absolute **C**, (ii) dilations $w \longmapsto aw$, $a \in \mathbb{R}$, (iii) symmetries $w \longmapsto -\bar{w}$ and (iv) the inversion $w \longmapsto 1/\bar{w}$ (because $|\mathrm{d}(1/\bar{w})|^2 = |\mathrm{d}w|^2/|w|^2$ and $\mathrm{Im}(1/\bar{w}) = \mathrm{Im} w/|w|^2$). This implies that $\mathrm{d}s^2$ is invariant with respect to all the group $PSL(2,\mathbb{R})$, of fractional-linear transformations preserving **C**. Moreover, from (iii) it follows that the half-line $\mathrm{Re} w = 0$ is geodesic. By means of fractional-linear transformation, the half line $\mathrm{Re} w = 0$ can be transformed to any circle transversal to the absolute C.

The above shows that the monodromy transformations are isometries in the Lobachevski geometry. If the assumption about angles of Δ is satisfied, then $\mathbb{H}$ is covered (without overlapping) by means of quadrangles $R_j = \sigma_j(R)$, the images of an initial quadrangle $R = \Delta \cup \Delta'$ (Δ' the image of Δ under one inversion) under action of the monodromy group Mon. R is the *fundamental domain* for Mon. The above property: existence of a fundamental domain which generates a regular tiling, serves as the definition of discreetness of a subgroup in $PSL(2,\mathbb{R})$.

Theorem is proved. □

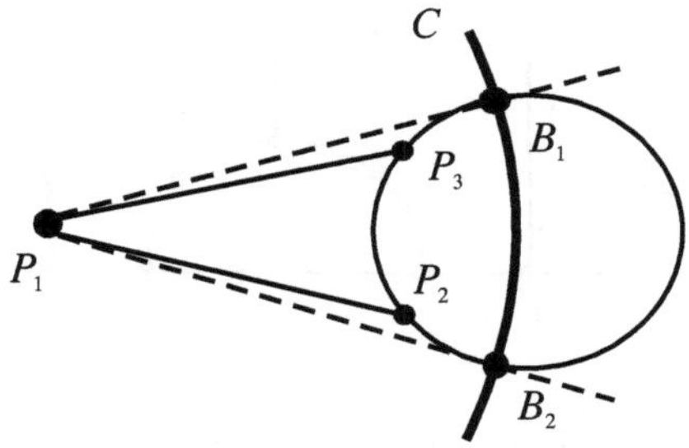

Figure 12.8: Situation in the hyperbolic plane

12.16. The modular function. Special role plays the case when the principal triangle has zero angles, $\lambda = \mu = \nu = 0$. One has $\alpha = \beta = 1/2$, $\gamma = 1$ and the hypergeometric equation becomes the Legendre equation $t(1-t)\ddot{x} + (1-2t)\dot{x} - x/4 = 0$ (see 8.10 in Chapter 8). In 8.10 we observed that the fundamental system of solutions of the Legendre equation is formed by elliptic integrals $\int \frac{\mathrm{d}x}{y}$ along two independent cycles in the elliptic curve $y^2 = (1-x^2)(1-tx^2)$. Thus, the Schwarz function is equal to the ratio of the periods of the holomorphic 1-form $\mathrm{d}x/y$. It can be written as $w(t) = 2\mathrm{i}K'(k)/2K(k)$, $t = k^2$, where $K = \int_0^1 \frac{\mathrm{d}x}{y}$, $\mathrm{i}K' = \int_1^{1/k} \frac{\mathrm{d}x}{y}$.
The (single valued) function $w \longmapsto t$, inverse to the Schwarz function, is called the **modular function**.
The monodromy of the elliptic integrals is defined by means of the Picard–Lefschetz formula. The monodromy matrices corresponding to loops around $t = 0, 1$ are triangular and the corresponding projective monodromy maps are parabolic maps. This explains why the angles of the principal triangle are equal to zero.
The principal triangle for the function $w = \mathrm{i}K'/K$ is presented in Fig. 12.9(a); it has the vertices $w(0) = \infty$, $w(1) = 0$, $w(\infty) = -1$. (Indeed: as $t \to 0$, we have $2K \to \text{const}$, $\mathrm{i}K' \to \int_1^\infty \mathrm{d}x/\sqrt{1-x^2} = \infty$; as $t \to \infty$, $w \sim -\left(\int_\epsilon^1 \frac{\mathrm{d}x}{z}\right) / \left(\int_0^1 \frac{\mathrm{d}x}{z}\right)$, $z = \sqrt{(1-x^2)(\epsilon^2 - x^2)}$, and the latter integrals tend to infinity and differ by a constant; as $t \to 1$ the integral $2\mathrm{i}K'$ runs along vanishing cycle (is finite) and $2K$ tends to infinity.) So, the projective monodromy maps are: the translation $w \longmapsto w + 2$ (composition of reflections with respect to two parallel lines) and $w \longmapsto w/(2w+1)$ (composition of the reflection $w \longmapsto -\bar{w}$ and of the inversion $w - 1/2 \longmapsto (1/2)^2/\overline{(w-1/2)}$.)
One can notice that the corresponding (projective) monodromy group in $SL(2, \mathbb{C})$ is equal to the *congruent group* $\Gamma[2] = \{A \in SL(2, \mathbb{Z}) : A = I \pmod 2\}$. The group $SL(2, \mathbb{Z})$ is very important in the number theory and in the theory of elliptic curves. It is generated by two matrices $\begin{pmatrix} 0 & 1 \\ -1 & 0 \end{pmatrix}$ and $\begin{pmatrix} 1 & 1 \\ 0 & 1 \end{pmatrix}$ with the

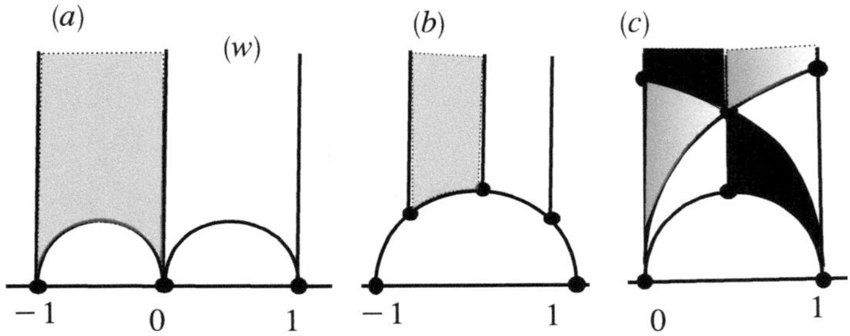

Figure 12.9: Fundamental domains associated with the modular function

fundamental domain D (of its action on $\mathbb{CP}^1$) presented in Fig. 12.9(c). The domain D is the space of moduli of all elliptic curves. Recall that an elliptic curve is a topological torus, but analytically it is equivalent to $\mathbb{C}/\Lambda$, where Λ is a lattice and can be chosen as $\mathbb{Z}+\mathbb{Z}\tau$, where $\tau\in\mathbb{H}$ is the ratio of periods. Two elliptic curves, defined by means of two such τ's, are analytically equivalent iff the τ's are related by means of a fractional-linear map from $PSL(2,\mathbb{Z})$. The corresponding uniformizing parameter $J : D \longmapsto \overline{\mathbb{C}}$ is called the *J-invariant* and is defined as the inverse of the Schwarz map from $\mathbb{H}$ to the triangle with the vertices $\mathrm{i}, \mathrm{e}^{2\pi\mathrm{i}/3}, \infty$ (and the angles $\pi/2, \pi/3, 0$). We see from Fig. 12.9(c) that the fundamental triangle associated with the Legendre equation contains six fundamental triangles associated with the J-invariant. Thus, the index of the congruent subgroup $\Gamma[2]$ in $SL(2,\mathbb{Z})$ is equal to 6.

There is a formula for the J-invariant of an elliptic curve in the Weierstrass form $v^2 = 4z^3 - g_2 z - g_3$: $J = g_2^3/(g_2^3 - 27g_2^2)$. The elliptic curve in the form $y^2 = (1-x^2)(1-tx^2)$ can be transformed to the Weierstrass form (see 8.10(iii)). So, we can calculate the J-invariant by means of the modular function: $27J/4 = (1-t+t^2)^3 t^{-2}(1-t)^{-2}$ (see [**Gol**]).

Finally, we note that E. Picard used the modular function to prove his celebrated **Picard theorem**:

There cannot exist any single valued holomorphic function with isolated essential singular point whose image does not contain three values.

12.17. Theorem (Integrability of the hypergeometric equation). ([**Schw**], [**Kim**])

(a) *The hypergeometric equation is integrable in generalized quadratures iff: either*

  (A) *at least one of the numbers* $\lambda+\mu+\nu$, $-\lambda+\mu+\nu$, $\lambda-\mu+\nu$, $\lambda+\mu-\nu$ *is an odd integer, or*

  (B) *the triple* $(\lambda,\mu,\nu) = (\pm\lambda_0 + l, \pm\mu_0 + m, \pm\nu_0 + n)$, *where* λ_0, μ_0, ν_0 *take values from one of the following* 15 *series, the signs are independent,* *and l, m, n are integers with $l+m+n$ even:*

1.	$(1/2, 1/2, \nu)$	2.	$(1/2, 1/3, 1/3)$	3.	$(2/3, 1/3, 1/3)$
4.	$(1/2, 1/3, 1/4)$	5.	$(2/3, 1/4, 1/4)$	6.	$(1/2, 1/3, 1/5)$
7.	$(2/5, 1/3, 1/3)$	8.	$(2/3, 1/5, 1/5)$	9.	$(1/2, 2/5, 1/5)$
10.	$(3/5, 1/3, 1/5)$	11.	$(2/5, 2/5, 2/5)$	12.	$(2/5, 1/3, 1/5)$
12.	$(4/5, 1/5, 1/5)$	14.	$(1/2, 2/5, 1/3)$	15.	$(3/5, 2/5, 1/3)$

(b) *The hypergeometric equation is integrable in quadratures iff: either the condition* (A) *holds, or the condition* (B)*, restricted to the cases* 1–5*, holds.*

(c) *The hypergeometric equation is integrable in algebraic functions iff all the parameters λ, μ, ν are rational and: either*

- *exactly two of the numbers $\lambda+\mu+\nu$, $-\lambda+\mu+\nu$, $\lambda-\mu+\nu$, $\lambda+\mu-\nu$ are odd integers and none of the singular points $t = 0, 1, \infty$ is logarithmic, or*
- *the condition* (B) *holds.*

Remark. In the original paper of T. Kimura [**Kim**] in the condition (B) above there is no the restriction $l+m+n$ even for the cases 1, 2, 4, 6 (when one of the λ, μ, ν is half-integer); the same is repeated by J. Morales-Ruiz and J.-P. Ramis [**M-RR**]. But $\frac{1}{2} = -\frac{1}{2}+1$ and Proposition 12.7 allows the changes like $\lambda \longmapsto \pm\lambda$. Hence, the cases with odd $l+m+n$ can be reduced to the cases with even $l+m+n$.

Proof of Theorem 12.17. 1. The 15 rational triples $(\lambda_0, \mu_0, \nu_0)$ from the domain (12.2), defined in Proposition 12.7, were found by H. Schwarz [**Schw**]. The corresponding Schwarz triangles Δ are obtained as unions of several elementary triangles (dihedral from case 1, tetrahedral case 2, cubic from case 4 and icosahedral from case 6). For example, the triangle with the angles $2\pi/3, \pi/3, \pi/3$ (case 3) is obtained by gluing two triangles with angles $\pi/2, \pi/3, \pi/3$ (see Fig. 12.5). The triangle 5 (with angles $2\pi/3, \pi/4, \pi/4$) consists of two triangles of the type 4. All the triangles 7–15 are unions of triangles of type 6.
The system of triangles Δ_j (associated with Δ and obtained by successive inversions with respect to the sides) does not form a tiling (because of the overlapping) but is regular in the sense that each point in the w-plane is covered by only finitely many triangles Δ_j. So, the Schwarz function $w(t)$ is algebraic.
On the other hand, if the triple λ, μ, ν satisfies condition (12.2) (i.e., $0 \leq \lambda, \mu, \nu < 1$; $0 \leq \lambda+\mu, \lambda+\nu, \mu+\nu \leq 1$, see Proposition 12.7) and the function $w(t)$ is algebraic, then the numbers λ, μ, ν are rational. Moreover, the circles containing the sides of the (possibly overlapping) system $\{\Delta_j\}$ define a certain regular partition (without overlapping) of the w-plane. The latter partition turns out to be a partition induced by one triangle Δ' and by inversions with respect to its sides. The triangle Δ' must be of one of the types 1, 2, 4, 6.
Detailed analysis of Fig. 12.4–12.7 shows that the list of triples from condition (B) (with $(\frac{1}{2}, \frac{1}{2}, \frac{p}{q})$ in Case 1) contains all the cases with algebraic $w(t)$ and satisfying restriction (12.2).

If the Schwarz function $w(t) = y_2/y_1$ is algebraic, then also the solutions $y_j(t)$ are algebraic. Indeed, we have $\dot{w} = W(y_1, y_2)/y_1^2$, where the Wronskian $W(y_1, y_2) = \text{const} \cdot t^{-\gamma}(t-1)^{\gamma-\alpha-\beta-1}$ is algebraic (because α, β, γ are rational). Because $\dot{w}$ is algebraic, $y_1(t)$ (and then also $y_2(t)$) is algebraic too.
The final remark concerns the (projective) monodromy group of the hypergeometric equation in the cases with algebraic Schwarz function. This group is a subgroup of the group of symmetries of the corresponding regular polyhedron. In particular, in cases 1–5 the monodromy group is solvable. In cases 6–15 the monodromy group contains the non-solvable group $A(5)$. This means that, the Schwarz function $w(t)$ and the solutions $y_{1,2}(t)$ are represented by means of radicals (i.e., quadratures) only in cases 1–5.
The explicit formulas for solutions in the cases 2–5 were found by M. Hukuhara and S. Ohasi in [**HO**] (see also [**Kim**]).

2. In case 1 from the table in Theorem 12.17 there is the solution

$$z(t) = (\sqrt{t} + \sqrt{t-1})^{\nu}$$

of the hypergeometric equation, found by Hukuhara and Ohasi (see [**HO**], [**Kim**]). In this case the other solution, $z_1(t)$ is found from the equation $\dot{z}_1 z - \dot{z} z_1 = W(t)$, where $W(t) = W(z, z_1)$ is the Wronskian with known quadrature form. Thus $z_1(t) = z(t) \int^t z(s)^{-1} W(s) ds$ is also represented by quadratures.

3. The hypergeometric equation is called *reducible* iff it admits a *degenerate solution* of the form

$$y = t^{\epsilon}(1-t)^{\theta} v(t),$$

with rational $v(t)$. (Another definition: y satisfies a first order equation $\dot{y} = r(t)y$, r – rational).
Because the only singularities of $y(t)$ may lie at $0, 1, \infty$, then we find that $v(t)$ must be a polynomial. Moreover, because the asymptotic of the solution $y(t)$ at each singular point is of power type, then y is one of the six functions $x_1(t), \ldots, x_6(t)$ forming the three bases of solutions (see 12.4). Each of the $x_i(t)$'s is of the form: a power function times a hypergeometric series. This means that the corresponding hypergeometric series should be a polynomial.
The series $F(\alpha, \beta; \gamma; t) = \sum \frac{(\alpha)_n (\beta)_n}{(\gamma)_n n!} t^n$ is finite iff $(\alpha)_n = 0$ or/and $(\beta)_n = 0$ for sufficiently large n's. This case occurs when α or/and β is negative integer. Thus $x_1(t) = F(\alpha, \beta; \gamma; t)$ is a degenerate solution, if $\alpha = -j$ or $\beta = -k$ (k, j – positive integers).
Similarly $x_2(t) = t^{1-\gamma} F(\alpha - \gamma + 1, \beta - \gamma + 1; 1 - \alpha - \beta + \gamma; t)$ is degenerate, if $\alpha - \gamma + 1 = -j$ or $\beta - \gamma + 1 = -k$. The analogous statements hold for other $x_j(t)$.
Assume that $\alpha = -j$. In this case the number $\lambda + \mu - \nu = (1-\gamma) + (\gamma + j - \beta) - (-j - \beta) = 2j + 1$ is an odd integer. One can check that, also in other cases of reducible hypergeometric equation, one of the numbers $\lambda + \mu + \nu$, $-\lambda + \mu + \nu$, $\lambda - \mu + \nu$, $\lambda + \mu - \nu$ is an odd integer (i.e., the condition (A) from Theorem 12.17).

If the hypergeometric equation is reducible, then it is solvable in quadratures. Indeed, one solution $y(t)$ is represented by quadratures. The other solution $y_1(t)$ satisfies the linear equation $\dot{y}_1 y - \dot{y} y_1 = W(t)$, where $W(t) = W(y, y_1)$ is the Wronskian.
One can show that the Riccati equation associated with a reducible hypergeometric equation has first integral of the Darboux–Schwarz–Christoffel type.

4. If the hypergeometric equation is reducible, then all its monodromy maps can be simultaneously triangularized; in any basis (y_0, y_1) with degenerate solution y_0. It means that the monodromy group is reducible.
The converse is also true. If the monodromy is reducible, then there is a solution $y_0(t)$ which spans a one-dimensional space, invariant with respect to monodromy. By the Riemann's theorem 8.35, $y_0(t)$ satisfies a first order differential equation, i.e., represents a degenerate solution of the hypergeometric equation.
The monodromy group is diagonalizable iff there are two independent degenerate solutions. This occurs when exactly two of the numbers $\lambda+\mu+\nu$, $-\lambda+\mu+\nu$, $\lambda-\mu+\nu$, $\lambda+\mu-\nu$ are odd integers. If, additionally, the exponents are rational, then the monodromy is diagonalizable and finite and the equation is solvable in algebraic functions.
The Riccati equation associated with a hypergeometric equation with diagonalizable monodromy has Darbouxian type first integral.
Of course, the monodromy group is finite iff the hypergeometric equation is solvable in algebraic functions, (i.e., part 1 of the proof). These algebraic functions are represented by quadratures iff the finite monodromy group is solvable. The latter occurs in the case mentioned above and in cases 1–5 from the table.
In case 1 from the table, i.e., $\lambda = \mu = \frac{1}{2}$, ν arbitrary (see the part (c) of the proof), the monodromy group Mon contains the normal subgroup Mon^0 of index 2, consisting of diagonal matrices. Here also Mon is solvable. The corresponding Riccati equation has Darboux-hyperelliptic first integral (see 11.44 and 12.11).

5. In order to finish the proof of Theorem 12.17, one has to consider the case when the differential Galois group G of the Picard–Vessiot extension, defined by the hypergeometric equation, has solvable identity component, but is neither solvable nor finite.
Let us reduce the situation to the case when $G \subset SL(2, \mathbb{C})$. This is achieved when, instead of the hypergeometric equation, we consider the following Riemann equation

$$P\left\{\begin{array}{cccc} 0 & 1 & \infty & \\ (1-\lambda)/2 & (1-\mu)/2 & -(1+\nu)/2 & x \\ (1+\lambda)/2 & (1+\mu)/2 & -(1-\nu)/2 & \end{array}\right\},$$

associated with it (see 12.2). We have $W(t) \equiv \mathrm{const}$ for the latter Riemann equation.
By Theorem 11.15, the only remaining case is when the identity component G^0 is diagonalizable and $[G : G^0] = 2$.

We shall show that $\lambda = \pm\frac{1}{2} + l$, $\mu = \pm\frac{1}{2} + m$, l, m – integers, in such situation. Let M_0, M_1, M_2 be the monodromy maps corresponding to loops around $t = 0, t = 1, t = \infty$. It is easy to see that $M_j \in G \setminus G^0$ but $M_i M_j \in G^0$. We can also assume that

$$M_0 = \begin{bmatrix} -\mathrm{e}^{-\pi\mathrm{i}\lambda} & a \\ 0 & -\mathrm{e}^{\pi\mathrm{i}\lambda} \end{bmatrix}, \quad M_1 = \begin{bmatrix} -\mathrm{e}^{-\pi\mathrm{i}\mu} & 0 \\ b & -\mathrm{e}^{\pi\mathrm{i}\mu} \end{bmatrix},$$

where $ab \neq 0$. (We chose the basis such that one function is an eigenvector for M_0 and the other is an eigenvector for M_1). We have

$$M_0^2 = \begin{bmatrix} \mathrm{e}^{-2\pi\mathrm{i}\lambda} & -a(\mathrm{e}^{-\pi\mathrm{i}\lambda} + \mathrm{e}^{\pi\mathrm{i}\lambda}) \\ 0 & \mathrm{e}^{2\pi\mathrm{i}\lambda} \end{bmatrix}.$$

It is diagonalizable iff: either (i) $\mathrm{e}^{-2\pi\mathrm{i}\lambda} = \mathrm{e}^{2\pi\mathrm{i}\lambda}$, (then $\mathrm{e}^{-\pi\mathrm{i}\lambda} + \mathrm{e}^{\pi\mathrm{i}\lambda} = 0$ and $\lambda = \pm\frac{1}{2} + l$ is half-integer), or (ii) $\mathrm{e}^{-2\pi\mathrm{i}\lambda} \neq \mathrm{e}^{2\pi\mathrm{i}\lambda}$. Anyway, the vector $v = \binom{1}{0}$ is one of the eigenvectors of M_0^2.
On the other hand, we have

$$M_1 M_0 = \begin{bmatrix} * & * \\ b\mathrm{e}^{-\pi\mathrm{i}\lambda} & * \end{bmatrix}.$$

Because v is also eigenvector for the latter matrix, we obtain $b = 0$ (a contradiction). Therefore, only the first possibility with half-integer λ may occur.
Analogous analysis with M_1^2 and $M_0 M_1$ shows that μ is also half-integer.
Theorem 12.17 is complete. □

12.18. Application to the problem of integrability of Hamiltonian systems. There is a nice application of Theorem 12.17 to Hamiltonian systems with the Hamilton function of the form

$$H = \frac{1}{2}p^2 + V(q),$$

where $(p, q) \in \mathbb{R}^n \times \mathbb{R}^n$ and the potential V is a homogeneous function of integer degree $k \neq 0$ (see [**M-RR**]).
One can find a particular solution of this Hamiltonian system in the form $q = z(t)c$, $p = \dot{z}c$, where $z(t)$ satisfies the equation $\dot{z}^2 = \frac{2}{k}(1 - z^k)$ and the constant vector c is a solution of the algebraic equation $c = \nabla V(c) = V'(c)$.
We get a situation as in the Ziglin's theory (see Section 4 in Chapter 10). We have a solution lying on an algebraic curve Γ. Like in the Ziglin's theory one considers the normal variation equation, which here takes the form

$$\ddot{\eta} = -z(t)^{k-2} V''(c)\eta.$$

After diagonalization of the Hessian V'' (with the eigenvalues λ_i), we obtain a system of independent equations which, after the change $x = z(t)^k$, take the form of the hypergeometric equations

$$x(1 - x)\frac{\mathrm{d}^2\eta_i}{\mathrm{d}x^2} + \left(\frac{k-1}{k} - \frac{3k-2}{2k}x\right)\frac{\mathrm{d}\eta_i}{\mathrm{d}x} + \frac{\lambda_i}{2k}\eta_i = 0.$$

The following result holds (see [**M-RR**] and [**M-R**]).

12.19 Theorem. *If the above Hamiltonian system is completely integrable (i.e., has n functionally independent first integrals in involution), then the differential Galois group associated with the normal variation system has abelian component of identity.*
This takes place when each pair (k, λ_i) *belongs to one of the following list (where* p *is an arbitrary integer and* λ *is an arbitrary complex number):*

$$\begin{array}{ccc}
\left(k, p+p(p-1)\frac{k}{2}\right) & (2,\lambda) & (-2,\lambda) \\
\left(-5, \frac{49-(10/3+10p)^2}{40}\right) & \left(-5, \frac{49-(4+10p)^2}{40}\right) & \left(-4, \frac{9-(4/3+4p)^2}{8}\right) \\
\left(-3, \frac{25-(2+6p)^2}{24}\right) & \left(-3, \frac{25-(3/2+6p)^2}{24}\right) & \left(-3, \frac{25-(6/5+6p)^2}{24}\right) \\
\left(-3, \frac{25-(12/5+6p)^2}{24}\right) & \left(3, \frac{-1+(2+6p)^2}{24}\right) & \left(3, \frac{-1+(3/2+6p)^2}{24}\right) \\
\left(3, \frac{-1+(6/5+6p)^2}{24}\right) & \left(3, \frac{-1+(12/5+6p)^2}{24}\right) & \left(4, \frac{-1+(4/3+4p)^2}{8}\right) \\
\left(5, \frac{-9+(10/3+10p)^2}{40}\right) & \left(5, \frac{-9+(4+20p)^2}{40}\right) & \left(k, \frac{k-1+p(p+1)k^2}{2k}\right)
\end{array}$$

The first property follows from the Ziglin's theory (see Section 4 in Chapter 10) and from Schlesinger Theorem 11.21. The second property follows from Theorem 12.17. We refer the reader to the book of J. Moralez-Ruiz [**M-R**] for the details.

12.20. Algebraic solutions to the Painlevé equation. We describe results of the paper [**DuMa**] by B. Dubrovin and M. Mazzocco about classification of algebraic solutions for a 1-parameter family of the Painlevé equations, defined by

$$\alpha = \frac{1}{2}(2\mu-1)^2, \quad \beta=\gamma=0, \quad \delta=\frac{1}{2},$$

and denoted by Painlevé 6_μ (i.e., in P_6, see 8.54 in Chapter 8).
The equation Painlevé 6 is related with a 2-dimensional linear system

$$\frac{\mathrm{d}Y}{\mathrm{d}z} = A(z)Y, \quad z\in\mathbb{C},\ Y\in\mathbb{C}^2, \tag{12.3}$$

where the matrix $A(z)$ takes the form

$$A = \frac{A_1}{z-u_1} + \frac{A_2}{z-u_2} + \frac{A_3}{z-u_3}$$

and the matrices A_j depend on u_1, u_2, u_3 (and on the parameter μ). The system (12.3) has singular points $z = u_1, u_2, u_3, \infty$. We fix a special fundamental matrix $\mathcal{F}_\infty$, such that $\mathcal{F}_\infty(z) = \mathrm{diag}(z^{-\mu}, z^{\mu})(1 + O(1/z))$ as $z \to \infty$; here we assume that $A_\infty := -A_1 - A_2 - A_3 = \mathrm{diag}(-\mu, \mu)$.Then the monodromy matrix around $z = \infty$ equals

$$M_\infty = \mathrm{diag}(\exp(2\pi\mathrm{i}\mu), \exp(-2\pi\mathrm{i}\mu))$$

and does not depend on the parameters u_j. The other monodromy matrices M_j around u_j also should be constant; i.e., the deformation is isomonodromic. Recall that the corresponding Schlesinger equations lead to the Painlevé equation P_6 (Theorem 8.55).

Dubrovin and Mazzocco assumed additionally that the matrices A_j are nilpotent, i.e.,

$$A_j^2 = 0, \quad j = 1, 2, 3.$$

It turns out that the corresponding Painlevé equation is exactly Painlevé 6_μ. Recall the principal relations between the variables in (12.3) and in the Painlevé 6 (i.e., x, t) :

$$t = \frac{u_2 - u_1}{u_3 - u_1}, \quad x = \frac{q - u_1}{u_3 - u_1},$$

where q is a root of the algebraic equation $a_{12}(q) = 0$ for $A(z) = (a_{ij}(z))_{i,j=1,2}$. Knowing the matrices A_j we will know the solutions of the Painlevé equation. On the other hand, the matrices A_j can be reconstructed from the monodromy matrices M_j (the Riemann–Hilbert problem). In what follows we work only with the matrices M_j.

Since A_j are nilpotent, the M_j are unipotent, i.e., $\mathrm{tr} M_j = 2$ and $\det M_j = 1$. Moreover, $M_3 M_2 M_1 = M_\infty^{-1}$. It turns out that a general triple (M_1, M_2, M_3) satisfying the above conditions is given by the formulas

$$M_1 = \begin{pmatrix} 1 & -x_1 \\ 0 & 1 \end{pmatrix}, M_2 = \begin{pmatrix} 1 & 0 \\ x_1 & 1 \end{pmatrix}, M_3 = \begin{pmatrix} 1 + x_2 x_3/x_1 & -x_2^2/x_1 \\ x_3^2/x_1 & 1 - x_2 x_3/x_1 \end{pmatrix},$$

where

$$x_1^2 + x_2^2 + x_3^2 - x_1 x_2 x_3 = 4 \sin^2 \pi\mu. \tag{12.4}$$

Recall that the braid group $B(3)$ is the fundamental group $\pi_1((\mathbb{C}^3 \setminus \Delta)/S(3))$, where $\Delta = \{(u_1, u_2, u_3) : (u_1 - u_2)(u_2 - u_3)(u_3 - u_1) = 0\}$ and $S(3)$ is the symmetric group. The braid group is generated by two loops: β_1, along which the points u_1 and u_2 exchange their positions, and β_2, leading to the change $u_2 \leftrightarrow u_3$. We have the known relation $\beta_1 \beta_2 \beta_1 = \beta_2 \beta_1 \beta_2$.

The group $B(3)$ acts on the linear system (12.3) via movement of the poles and results in the following action onto the monodromy matrices:

$$\begin{aligned} \beta_1 &: M_1 \longmapsto M_2, \ M_2 \longmapsto M_2 M_1 M_2^{-1}, \ M_3 \longmapsto M_3, \\ \beta_2 &: M_1 \longmapsto M_1, \ M_2 \longmapsto M_3, \ M_3 \longmapsto M_3 M_2 M_3^{-1}. \end{aligned}$$

In the variables (x_1, x_2, x_3) we get very simple action:

$$\begin{aligned} \beta_1 &: (x_1, x_2, x_3) \longmapsto (-x_1, x_3 - x_1 x_2, x_2), \\ \beta_2 &: (x_1, x_2, x_3) \longmapsto (x_3, -x_2, x_1 - x_2 x_3). \end{aligned} \tag{12.5}$$

The braid group plays analogous role in the Painlevé equation as the monodromy group in the Gauss equation. The requirement of algebraicity of a given solution to the Painlevé equation is translated into finiteness of the orbit of the point (x_1, x_2, x_3) with respect to action of the group $B(3)$. (However algebraicity of one solution does not imply algebraicity of other solutions.) Our task is to classify the triples (x_1, x_2, x_3), satisfying Eq. (12.4) and having finite $B(3)$-orbits.

Let us pass to variables r_1, r_2, r_3 from the interval $[0,1]$ and given by the identities $x_j = -2\cos \pi r_j$ (analogy to the change $\alpha, \beta, \gamma \longmapsto \lambda, \mu, \nu$). It turns out that the numbers r_j must be rational. For example, the transformation $\beta_1^2 : (x_1, x_2, x_3) \longmapsto (x_1, x_2 + x_1 x_3 - x_1^2 x_2, x_3 - x_1 x_2)$, for fixed x_1, is equivalent to the rotation by the angle $\pi + 2\pi r_1$ in the plane (x_2, x_3).

The necessary condition for finiteness of the orbit of a triple (x_1, x_2, x_3) with rational r_1, r_2, r_3 is: $\beta(x_1, x_2, x_3) = (-2\cos \pi r_1', -2\cos \pi r_2', -2\cos \pi r_3')$ for any $\beta \in B(3)$ with some rational $r_i' = r_i'(\beta)$. The problem of classification of such triples (r_1, r_2, r_3) leads to the problem of rational solutions of the trigonometric equations $\cos \pi r_k' = \cos \pi r_k + 2\cos \pi r_i \cos \pi r_j$ (for various permutations (ijk) of the indices $1, 2, 3$), or (equivalently) equations

$$\cos 2\pi\phi_1 + \cos 2\pi\phi_2 + \cos 2\pi\phi_3 + \cos 2\pi\phi_4 = 0, \quad \phi_j \in \mathbb{Q}.$$

Analogous equations were investigated by P. Gordan [**Gor**]. Dubrovin and Mazzocco used the Gordan's technique and proved that there exists exactly five finite orbits of the action of $B(3)$. Each such orbit contains a point $(r_1, r_2, r_3) \sim (x_1, x_2, x_3)$ from the below list:

$$\begin{gathered} (1/2, 1/3, 1/3),\ (1/2, 1/3, 1/4), \\ (1/2, 1/3, 1/5),\ (1/2, 1/3, 2/5),\ (1/2, 1/5, 2/5). \end{gathered} \tag{12.6}$$

Different cases correspond the values $-\frac{1}{4}, -\frac{1}{3}, -\frac{2}{5}, -\frac{1}{5}, -\frac{1}{3}$ of the parameter μ. The nonlinear action (12.5) of the group $B(3)$ can be realized by means of a linear action. Take the space $V = \mathbb{R}^3$ with symmetric bilinear form defined via the matrix

$$g = \begin{pmatrix} 2 & x_1 & x_3 \\ x_1 & 2 & x_2 \\ x_3 & x_2 & 2 \end{pmatrix},$$

i.e., $(e_i, e_i) = 2, (e_1, e_2) = x_1$, etc. Define reflections R_i in V, $R_i v = v - (e_i, v)e_i$. The matrices $R_{1,2,3}$ are equal

$$\begin{pmatrix} -1 & -x_1 & -x_3 \\ 0 & 1 & 0 \\ 0 & 0 & 1 \end{pmatrix}, \begin{pmatrix} 1 & 0 & 0 \\ -x_1 & 1 & -x_2 \\ 0 & 0 & 1 \end{pmatrix}, \begin{pmatrix} 1 & 0 & 0 \\ 0 & 1 & 0 \\ -x_3 & -x_2 & 1 \end{pmatrix}.$$

The braid group acts on the reflections

$$\beta_1 : (R_1, R_2, R_3) \longmapsto (R_2, R_2 R_1 R_2, R_3),\ \beta_2 : (R_1, R_2, R_3) \longmapsto (R_1, R_3, R_3 R_2 R_3);$$

this action agrees with Eqs. (12.5).

The group G generated by the reflections $R_{1,2,3}$ is a finite subgroup of the orthogonal group $O(V, g)$. We get symmetry groups of some regular polytopes. The first triple from list (12.6) corresponds the symmetry group of the tetrahedron, the second one – to the octahedron, and the remaining triples correspond to different choices of generators of the symmetry group of the icosahedron. In fact, the last two triples can be associated with the great icosahedron and the great dodecahedron respectively.
In the paper [**DuMo**] the corresponding algebraic solutions to P_6 are presented. For example, in the tetrahedron case this solution is given in the following parametric form:

$$t=\frac{(s-1)^3(1+3s)}{(s+1)^3(1-3s)},\; x=\frac{(s-1)^2(1+3s)(9s^2-5)^2}{(1+s)(25-207s^2+1539s^4+243s^6)}.$$

§2 Generalized Hypergeometric Functions

12.21. Generalized hypergeometric equation. The **generalized hypergeometric function** of one variable is defined by the series

$$_pF_q\left(\alpha_1,\ldots,\alpha_p;\beta_1,\ldots,\beta_q;t\right)=\sum_{k=0}^{\infty}\frac{(\alpha_1)_k\cdots(\alpha_p)_k}{(\beta_1)_k\cdots\left(\beta_q\right)_k k!}t^k. \tag{12.7}$$

Above $p\leq q+1$; otherwise, the series is divergent.
The above notation is standard, see [**BE**, vol. 1], but some authors use different notations. In [**BH**] F. Beukers and G. Heckman consider the functions $_nF_{n-1}$, in [**DMit**] A. Duval and C. Mitschi write q in place of $q+1$. I allow myself to put

$$q+1=n,$$

it is the order of the corresponding differential equation (see Eq. (12.8) below). Usually we will write $F(\alpha_1,\ldots,\alpha_p;\beta_1,\ldots,\beta_{n-1};t)$ or $F(t)$; in the case $p=0$ we write $F(\emptyset;\beta_1,\ldots,\beta_{n-1};t)$ and in the case $p=0$ we write $F(\alpha_1,\ldots,\alpha_p;\emptyset;t)$.
The function (12.7) satisfies the following **generalized hypergeometric equation**:

$$\mathcal{H}u=\{\mathcal{Q}-t\mathcal{P}\}u=0, \tag{12.8}$$

$$\mathcal{Q}=Q\left(\mathcal{D}\right),\quad Q(z)=\prod_{j=1}^{n}\left(z+\beta_j-1\right),$$

$$\mathcal{P}=P(\mathcal{D}),\quad P(z)=\prod_{j=1}^{p}\left(z+\alpha_j\right),$$

where $\mathcal{D}=t\frac{\partial}{\partial t}=t\partial_t$ is the Euler operator and we agree that

$$\beta_n=1.$$

We will denote the operator $\mathcal{H}$ also as $\mathcal{H}(\alpha_1, \ldots, \alpha_p; \beta_1 \ldots, \beta_n) = \mathcal{H}(\alpha; \beta)$.

12.22. Local solutions. The equation (12.8) has singular points at $t = 0$, $t = \infty$ and $t = 1$.
Putting $u = t^\mu + \ldots$, with $t \to 0$, into this equation we get $Q(\mu)\, t^\mu + \ldots = 0$. So, the Levelt exponents are:

$$1 - \beta_1, \ldots, 1 - \beta_{n-1}, 1 - \beta_n = 0.$$

If these numbers are different, then we have n independent solutions

$$u_j(t) = t^{1-\beta_j} \times F_j(t) \tag{12.9}$$

where $F_j(t) = F(\alpha_1 + 1 - \beta_j, \ldots, \alpha_p + 1 - \beta_j; \beta_1 + 1 - \beta_j, .\overset{\vee}{.}., \beta_n + 1 - \beta_j; t)$, are some hypergeometric functions and $.\overset{\vee}{.}.$ means the omission of $\beta_j + 1 - \beta_j$. Note that $u_n(t)$ is the hypergeometric series (12.7). If some of these exponents are equal, then there are solutions with logarithms (see Remark 8.18).
Putting $u = t^\mu + \ldots$, as $\tau = 1/t \to 0$, into Eq. (12.8) we get the equation $-P(\mu)t^{\mu+1} + \ldots = 0$. So, the Levelt exponents are

$$\alpha_1, \ldots, \alpha_p.$$

If these numbers are different, then we have p independent solutions

$$v_j(t) = t^{-\alpha_k} \times G_k(\tau), \tag{12.10}$$

where $G_k(\tau) = G_k(1/t)$ are some hypergeometric series. If $p = n$ and α_k's are different, then these solutions form a fundamental system. If some of them are equal and $p = n$, then there are solutions with logarithms. If $p < n$, then the singular point $t = \infty$ is irregular and we deal with the Stokes phenomenon (see 12.30–12.34).
Consider the situation near $t = 1$. For simplicity assume that $p = n$. Then, putting $s = t - 1$, we can rewrite the operator $\mathcal{H}$ as follows:

$$\begin{aligned}\mathcal{H} &= -s\,[(1+s)\,\partial_s]^n + \left(\sum \beta_j - n - \sum \alpha_j - \alpha s\right)[(1+s)\,\partial_s]^{n-1} + \ldots \\ &= -s\partial_s^n + \left(\sum \beta_j - \sum \alpha_j - n\right)\partial_s^{n-1} + \ldots,\end{aligned}$$

the dots mean terms of higher order (where $\mathrm{ord}\partial_s = -1$). Putting $u(t) = s^\mu + \ldots$, we get

$$-\mu\,(\mu-1)\cdots(\mu-n+2)\left(\mu - \sum \beta_j + \sum \alpha_j - 1\right)t^{\mu-n+1} + \ldots.$$

It follows that the Levelt exponents are

$$0, 1, \ldots, n-2, \gamma = \sum_{j=1}^{n} (\beta_j - \alpha_j) + 1. \tag{12.11}$$

Assuming $\gamma \notin \mathbb{N}$, we have the solutions

$$\begin{aligned} w_j(t) &= s^j \times H_j(s), \ j = 0, \dots, n-2, \\ w_{n-1}(t) &= s^\gamma \times H_{n-1}(s). \end{aligned} \tag{12.12}$$

They are not clearly expressed; the recurrent relations for coefficients in the series defining H_j are of length greater than 2.

12.23. Operators. Now we assume

$$p = n = q + 1.$$

We follow the impressive paper [**BH**] by F. Beukers and G. Heckman.
Let $V(\alpha;\beta)$ be the space of solutions to the equation $\mathcal{H}(\alpha;\beta)u = 0$.
We have the monodromy operators $M_0 = M_0(\alpha;\beta)$, $M_\infty = M_\infty(\alpha;\beta)$ and $M_1 = M_1(\alpha;\beta)$ acting on this space; they are defined by analytic continuation of solutions along loops around the points $t = 0$, $t = \infty$ and $t = 1$ such that $M_\infty M_1 M_0 = Id$. They generate the monodromy group $\mathrm{Mon}(\alpha,\beta)$ of the hypergeometric equation $\mathcal{H}u = 0$; we denote by $\mathrm{Mon}(\alpha;\beta)$ the corresponding representation of the fundamental group $\pi_1\left(\mathbb{CP}^1 \setminus \{0,\infty,1\}\right)$.
We denote $e_j = (0,\dots,0,1,0,\dots,0) \in \mathbb{C}^n$ the standard basic vectors and $(\alpha,\delta) = (\alpha_1,\dots,\alpha_n,\delta)$ for $\delta \in \mathbb{C}$.

Proposition 1. *The operators*

$$\begin{aligned} \mathcal{D} + \alpha_j - 1 &: V(\alpha - e_j;\beta) \longmapsto V(\alpha;\beta), \\ \mathcal{D} + \beta_j &: V(\alpha;\beta + e_j) \longmapsto V(\alpha;\beta) \end{aligned}$$

are interwining operators for the monodromy representation. The operator $\mathcal{D} + \alpha_j - 1$ has nontrivial kernel iff $\alpha_j = \beta_k$ for some k, and the operator $\mathcal{D} + \beta_j$ has nontrivial kernel iff $\beta_j = \alpha_k$ for some k. Moreover, in the case the kernel of some of these operators is nontrivial it has dimension 1.

Proof. We have the easy relations

$$\begin{aligned} (\mathcal{D} + \delta - 1) \circ \mathcal{H}(\alpha;\beta) &= \mathcal{H}(\alpha,\delta;\beta,\delta), \\ \mathcal{H}(\alpha;\beta) \circ (\mathcal{D} + \delta) &= \mathcal{H}(\alpha,\delta;\beta,\delta+1). \end{aligned}$$

They imply

$$\begin{aligned} \mathcal{H}(\alpha;\beta) \circ (\mathcal{D} + \alpha_j - 1) &= (\mathcal{D} + \alpha_j - 1) \circ \mathcal{H}(\alpha - e_j;\beta), \\ \mathcal{H}(\alpha;\beta) \circ \left(\mathcal{D} + \beta_j\right) &= \left(\mathcal{D} + \beta_j - 1\right) \circ \mathcal{H}(\alpha;\beta + e_j). \end{aligned}$$

Next, $\ker(\mathcal{D} + \alpha_j - 1) = \mathbb{C} \cdot t^{1-\alpha_j}$ and $\ker\left(\mathcal{D} + \beta_j\right) = \mathbb{C} \cdot t^{-\beta_j}$. □

Proposition 2. *If $\alpha_k - \beta_k \notin \mathbb{Z}$ for all $j,k = 1,\dots n$, then the representations* $\mathrm{Mon}(\alpha + l;\beta + m)$ *for $l,m \in \mathbb{Z}^n$ and* $\mathrm{Mon}(\alpha;\beta)$ *are equivalent.*

Remark. *If* $\alpha_j - \beta_k \in \mathbb{Z}$ *for some* j, k, *then the monodromy representation* $\mathrm{Mon}(\alpha; \beta)$ *is reducible.*

Proposition 3. *If* $\alpha_k - \beta_k \notin \mathbb{Z}$ *for all* $j, k = 1, \ldots n$, *then the monodromy matrix* $M_1 = M_1(\alpha; \beta)$ *around* $t = 1$ *has the property* $\operatorname{rank}(M_1 - Id) = 1$ *and* $c = \det M_1 = \mathrm{e}^{2\pi \mathrm{i}\gamma}$. *It means that this operator is a* ***complex reflection*** *with special eigenvalue* c.

Proof. We shift the parameters such that the condition $\gamma \notin \mathbb{N}$ is satisfied. In the basis (12.12) the matrix $M_1 = \operatorname{diag}(1, \ldots, 1, c)$. If $M_1 = Id$, then $M_\infty = M_0^{-1}$, but this contradicts the assumption $\alpha_j - \beta_k \notin \mathbb{Z}$; compare Eqs. (12.10)–(12.11). □

12.24. Hypergeometric group. Suppose $a_1, \ldots a_n, b_1, \ldots, b_n \in \mathbb{C}^*$, with $a_j \neq b_k$ for all $j, k = 1, \ldots, n$. The **hypergeometric group** $\mathrm{Hyp}(a, b)$, with numerator parameters $a_1, \ldots, a_n$ and denominator parameters $b_1, \ldots, b_n$, is a subgroup of $GL(n, \mathbb{C})$ generated by elements h_0, h_∞ and h_1 such that

$$h_\infty h_1 h_0 = Id$$

and

$$\det(\lambda - h_\infty) = \prod(\lambda - a_j), \quad \det\left(\lambda - h_0^{-1}\right) = \prod(\lambda - b_j)$$

and h_1 is a complex reflection (like in the Proposition 3).

Proposition 4. *Suppose* $a_j = \exp(2\pi \mathrm{i}\alpha_j), b_k = \exp(2\pi \mathrm{i}\beta_k) \in \mathbb{C}^*$, $j = 1, \ldots n$, *with* $a_j \neq b_k$. *Then the monodromy group* $\mathrm{Mon}(\alpha, \beta)$ *is a hypergeometric group with parameter* a_j, b_k.

Proof. We put $h_j = M_j(\alpha, \beta)$. Then we use the formulas (12.10)–(12.12)). □

Proposition 5. Any hypergeometric group Hyp is irreducible.

Proof. If $V_1 \subset \mathbb{C}^n$ is an Hyp–invariant subspace and $V_2 = \mathbb{C}^n / V_1$, then we get induced groups $H_1 \subset GL(V_1)$ and $H_2 \subset GL(V_2)$. Since h_1 is a reflection, either h_1 restricted to V_1 or h_1 restricted to V_2 is the identity. Hence, if both $V_1 \neq 0$ and $V_2 \neq 0$, then we get a contradiction with the assumption $a_j \neq b_k$. □

Proposition 6. *Let* $\prod(\lambda - a_j) = \lambda^n + A_1\lambda^{n-1} + \ldots + A_n$ *and* $\prod(\lambda - b_j) = \lambda^n + B_1\lambda^{n-1} + \ldots + B_n$ *and let* N_∞ *and* N_0 *be given by the below formulas*

$$\begin{pmatrix} 0 & \ldots & 0 & -A_n \\ 1 & \ldots & 0 & -A_{n-1} \\ \ldots & \ldots & \ldots & \ldots \\ 0 & \ldots & 1 & -A_1 \end{pmatrix}, \quad \begin{pmatrix} 0 & \ldots & 0 & -B_n \\ 1 & \ldots & 0 & -B_{n-1} \\ \ldots & \ldots & \ldots & \ldots \\ 0 & \ldots & 1 & -B_1 \end{pmatrix}.$$

Then the matrices $h_\infty = N_\infty$, $h_0 = N_0^{-1}$ *and* $h_1 = N_\infty^{-1} N_0$ *generate a hypergeometric group with parameters* a_j, b_k. *Moreover, any hypergeometric group with the same parameters is conjugated inside* $GL(n)$ *to this one.*

Proof. The characteristic polynomials of the matrices N_∞ and N_0^{-1} are the same as the characteristic polynomials of h_∞ and h_0^{-1}. Also $h_1 - Id = N_\infty^{-1}(N_\infty - N_0)$ has rank 1.
I refer to [**BH**] for the proof of the second statement. □

12.25. Invariant Hermitian form. From this point we assume that the parameters α_j and β_k are real; we assume

$$0 \le \alpha_1 \le \ldots \le \alpha_n, \quad 0 \le \beta_1 \le \ldots \le \beta_n. \tag{12.13}$$

Therefore $a_j = \bar{a}_j^{-1}$ and $b_k = \bar{b}_k^{-1}$ lie in the unit circle.
We know from the previous section that the Gauss hypergeometric equation, with real parameters $\alpha = \alpha_1$, $\beta = \alpha_2$, $\gamma = \beta_1$ and $\beta_2 = 1$, as a monodromy group which is either contained in $U(2)$ or in $U(1,1) \simeq GL(2,\mathbb{R})$ or in $\mathrm{Aff}(\mathbb{C})$. In fact, the third case takes place when $\alpha = 1$, i.e., $\alpha_1 = \beta_2$; it is the case not considered in [**BH**]. Surprisingly, it turns out that a similar statement holds for the generalized hypergeometric equation as well. The way to demonstrate it is to construct a Hermitian form invariant with respect to the corresponding hypergeometric group.

Proposition 7. *Let* $\mathrm{Hyp}(a,b) \subset GL(n)$ *be a hypergeometric group with parameters* a_j *and* b_k *from the unit circle and such that* $a_j \neq b_k$. *Then there exists a non-degenerate Hermitian form* $F(x,y) = \sum f_{ij} x_i \bar{y}_j$ *on* $\mathbb{C}^n$ *such that*

$$F(hx, hy) = F(x,y), \ \ h \in \mathrm{Hyp}(a,b), \ \ x, y \in \mathbb{C}^n.$$

Sketch of the proof. It suffices to construct a nondegenerate Hermitian form F such that the above property is satisfied by $h = h_\infty$ and $h = h_0^{-1}$. Such a form with matrix $\{F_{ij}\}$ is a solution of the system

$$h_\infty^\top F \bar{h}_\infty, \quad \left(h_0^{-1}\right)^\top F h_0^{-1}, \quad \overline{F} = F^\top.$$

The authors of [**BH**] use the presentation of the matrices h_∞ and h_0^{-1} as N_∞ and N_0 from Proposition 6. It turns out that the entries F_{ij} depend only on $i-j$. By a careful calculation of the number of needed parameters, they show the the above system has a non-trivial solution (I refer to [**BH**] for details).
Finally, the non-degeneracy of F follows from the fact that it is invariant for the group $\mathrm{Hyp}(a,b)$, which acts irreducibly on $\mathbb{C}^n$. □

Proposition 8. *Assume* $a_j = \exp 2\pi \mathrm{i}\alpha_j$ *and* $b_k = \exp 2\pi \mathrm{i}\beta_k$ *with real* α_j *and* β_k *like in Eq.* (12.13). *Let* $m_j = \#\{k : \beta_k < \alpha_j\}$. *Then then the signature* (p,q) *of the Hermitian form* F *from Proposition 7 is given by*

$$|p-q| = \left| \sum_j (-1)^{j+m_j} \right|.$$

I refer to [**BH**] for the proof. As a corollary we have the following.

Proposition 9. *In the assumptions of Proposition 8 the group* $\text{Hyp}(a,b) \subset U(n)$ *iff either*

$$\alpha_1 < \beta_1 < \ldots < \alpha_n < \beta_n \text{ or } \beta_1 < \alpha_1 < \ldots < \beta_n < \alpha_n,$$

i.e., the sets $\{a_1, \ldots, a_n\}$ *and* $\{b_1, \ldots, b_n\}$ *interlace in the unit circle.*

We present without proof another statement.

Proposition 10. *Suppose that the parameters* a_j *and* b_k *are roots of unity, i.e., lie in* $\mathbb{Q}(\exp 2\pi i/h)$ *for some* h. *Then* $\text{Hyp}(a,b)$ *is finite iff for each* $k \in \mathbb{N}$ *with* $\gcd(k,h) = 1$ *the sets* $\{a_1^k, \ldots, a_n^k\}$ *and* $\{b_1^k, \ldots, b_n^k\}$ *interlace in the unit circle.*

12.26. Reflection subgroup. It is the subgroup $H_r(a,b) \subset \text{Hyp}(a,b)$ generated by the reflections

$$h_\infty^k h_1 h_\infty^{-k}, \quad k \in \mathbb{Z}.$$

Proposition 11. *In general the reflection subgroup* H_r *acts* **irreducibly** *on* $\mathbb{C}^n$. *The only exception is the case when there exists a nontrivial root of unity* ζ *such that* $\{\zeta a_1, \ldots, \zeta a_n\} = \{a_1, \ldots, a_n\}$ *and* $\{\zeta b_1, \ldots, \zeta b_n\} = \{b_1, \ldots, b_n\}$.
Moreover, in the reducible case the group $\text{Hyp}(a,b)$ *is* **imprimitive** *in the sense that there exists a direct sum decomposition* $\mathbb{C}^n = V_1 \oplus \ldots \oplus V_d$, *with* $d \geq 2$, *such that* $\text{Hyp}(a,b)$ *permutes the spaces* V_j.

I refer to [**BH**] for a proof.

Proposition 12. *If* $n \geq 3$ *and the reflection subgroup* $H_r(a,b)$ *is irreducible and primitive, then it is a* ***scalar shift*** *of* $\text{Hyp}(a,b)$, *i.e.,* $H_r(a,b) = \text{Hyp}(da, db)$ *for some* $d \in \mathbb{S}^1 \subset \mathbb{C}^*$.

Proof. The authors of [**BH**] use a theorem by A. Cohen, which implies that h_∞ is a scalar times an element of $H_r(a,b)$. □

Proposition 13. *Let* $n \geq 3$ *and let* $\text{Hyp}(a,b)$ *be a primitive hypergeometric group. Then the reflection subgroup* $H_r(a,b)$ *is generally primitive. The opposite situation takes place in five explicit cases (which I do not present).*

I refer to [**BH**] for a proof.

Proposition 14. *If* $H_r = H_r(a,b)$ *is primitive, then the identity component* H_r^0 *of* H_r *in* $GL(n)$ *is either* $\{Id\}$ *or it acts irreducibly on* $\mathbb{C}^n$.

Proof. Assume that H_r^0 act reducibly and $W \subset \mathbb{C}^n$ is an irreducible invariant subspace. Since H_r acts irreducibly on $\mathbb{C}^n$, there exists a reflection $r \in H_r$ with $rW \neq W$. Since H_r^0 is normal in H_r, the intersection $W \cap rW$ is invariant under H_r^0 and we conclude that $rW \cap W = 0$. But r is a reflection, hence $\dim W = 1$. Now, either H_r^0 consists of scalars only, or the decomposition of $\mathbb{C}^n$ into isotypical components of H_r^0 gives a system of imprimitivity for H_r. The latter possibility is excluded by the assumption of the primitivity of H_r. Hence, H_r^0 is contained in the scalars. This fact and the finiteness of H_r/H_r^0 implies that the special eigenvalue of h_1 is a primitive d^{th} root of unity for some $d \geq 2$. Hence the image of the map det : $H_r \longmapsto \mathbb{C}^*$ consists of all $d^{\rm th}$ roots of unity. In particular, this shows that the scalars in H_r consist of $nd^{\rm th}$ roots of unity. Thus we conclude that H_r^0 is finite and it must be $H_r^0 = \{Id\}$. □

Note that $H_r^0 = \{Id\}$ iff H_r is a finite reflection group. The following proposition enables us to understand the Differential Galois Theory aspects of this.

Proposition 15. *Suppose $G \subset SL(n)$ is a connected algebraic group acting irreducibly on $\mathbb{C}^n$ and let $r \in GL(n)$ be a reflection with special eigenvalue c, which normalizes G. Then we have the following possibilities:*

(I) *If $c \neq \pm 1$ then $G = SL(n)$,*

(II) *If $c = +1$ then $G = SL(n)$ or $G = Sp(n)$ (and n is even),*

(III) *If $c = -1$ then $G = SL(n)$ or $G = SO(n)$.*

The proof given in [**BH**] relies on decomposition of the Lie algebra $\mathfrak{g}$ of G into eigenspaces of Ad (r), Each case, (I), (II) and (III), is studied separately. I skip that analysis.

Now it is a place to formulate principal statements of the Beukers–Heckman theory.

12.27. Theorem of Beukers and Heckman (Infinite hypergeometric group). *Let* Hyp(a, b) *be an infinite primitive hypergeometric group, which is not a scalar shift of a finite group, and let* $\overline{\mathrm{Hyp}(a, b)}$ *be its Zariski closure. Then we have two possibilities:*

(i) *There exists d such that $\{da_1, \ldots, da_n\} = \left\{(da_1)^{-1}, \ldots, (da_n)^{-1}\right\}$ and $\{db_1, \ldots, db_n\} = \left\{(db_1)^{-1}, \ldots, (db_n)^{-1}\right\}$. This means that the group* Hyp(da, db) *is self dual with respect to the map $g \longmapsto \left(g^{-1}\right)^*$ and the special eigenvalue of h_1 equals $c = \pm 1$. If $c = +1$, then $\overline{\mathrm{Hyp}(da, db)} = Sp(n)$; if $c = -1$, then $\overline{\mathrm{Hyp}(a, b)} = O(n)$.*

(ii) *In the remaining cases $SL(n) \subset \overline{\mathrm{Hyp}(a, b)}$.*

Proof. From Proposition 13 it follows that $H_r = H_r(a, b)$ is infinite and primitive. By Proposition 14 and the infinity of H_r is follows that $\overline{H_r^0}$ and hence also $\overline{H_r^0} \cap SL(n)$ is irreducible on $\mathbb{C}^n$. Application of Proposition 15, with $G = \overline{H_r^0} \cap SL(n)$ and $r = h_1$, shows that: either $SL(n) \subset \overline{H_r^0}$, or $\overline{H_r^0} \cap SL(n) = Sp(n)$ and $c = 1$, or $\overline{H_r^0} \cap SL(n) = SO(n)$ and $c = -1$.
Suppose we are in case (i). Then the self-duality of Hyp(da, db) means the existence of a non-degenerate bilinear form F in $\mathbb{C}^n$, which is invariant under this group. If $c = +1$, then the form must be antisymmetric and and n is even; if $c = -1$, then the form is symmetric (compare Proposition 15).
I skip the analysis of the second case. □

12.28. Theorem of Beukers and Heckman (Finite hypergeometric group). *Let $n \geq 3$ and let* Hyp(a, b) *be a primitive hypergeometric group, whose parameters* *are roots of unity and generate the cyclotomic field* $\mathbb{Q}(e^{2\pi i/h})$. *Then* Hyp(a,b) *is finite iff, up to a scalar shift, the parameters have the form* $a_1^k, \ldots, a_n^k; b_1^k, \ldots, b_n^k$, *where* $\gcd(k,h) = 1$ *and the exponents of either* $a_1, \ldots, a_n; b_1, \ldots, b_n$ *or* $b_1, \ldots, b_n; a_1, \ldots, a_n$ *are listed in so-called Shephard–Todd table..*

Remark. Above the exponent of $a_j = e^{2\pi i \nu_j}$ equals ν_j.
Next, the Shephard–Todd table classifies finite irreducible reflection groups (which correspond to the reflection subgroups $H_r(a,b)$). They are denoted STk, where $1 \le k \le 77$.
The group $ST1$ is the symmetric group $S(n+1)$ and this is the only finite primitive reflection group in dimension $n \ge 9$. Here $a; b = \zeta_{n+1}, \zeta_{n+1}^2, \ldots, \zeta_{n+1}^n; 1, \zeta_p, \ldots, \zeta_p^{p-1}, \zeta_q, \ldots, \zeta_q^{q-1}$ with $\zeta_r = e^{2\pi i/r}$, $p + q = n + 1$ and $\gcd(p,q) = 1$; thus the exponents are $\frac{1}{n+1}, \ldots, \frac{n}{n+1}; 0, \frac{1}{p}, \ldots, \frac{p-1}{p}, \frac{1}{q}, \ldots, \frac{q-1}{q}$.
The group $ST2$ is the finite imprimitive group $G(m,p,n)$. The group $ST3$ is the cyclic group of order m being a 1-dimensional reflection group. There are 19 2-dimensional finite primitive reflection groups STk with $4 \le k \le 22$ derived from the tetrahedral ($4 \le k \le 7$), the octahedral ($8 \le k \le 15$) and the icosahedral group ($16 \le k \le 22$). In dimension n with $3 \le n \le 8$ there remain 15 exceptional finite primitive reflection groups with $23 \le k \le 37$.
I do not present the whole table.

Proof of Theorem 12.28. Let $H \subset GL(n)$ be a finite primitive hypergeometric group. If its reflection subgroup H_r is imprimitive, then we have the situation like in the second case of Proposition 13 (which I did not presented in detail). Anyway, the authors ensure that it is listed in the Shephard–Todd table, given in [**BH**, Table 8.3] and in [**ShT**].
Suppose H_r is primitive. Then, by Proposition 12, we may as well assume that $H = H_r$. Since H is now a finite reflection group, it is contained in the Shephard–Todd list. In fact, there are 16 such groups, taken from 'A. Cohen's Utrecht University thesis'; see [**BH**, Table 8.1].
To determine the eigenvalues of h_∞ and h_0^{-1} one proceeds as follows. Suppose H equals $ST32$, in dimension 4, with order $2^7 \cdot 3^5 \cdot 5$ and defined over $\mathbb{Q}(\omega)$, $\omega = e^{2\pi i/3}$. So, the characteristic polynomials of h_∞, h_0^{-1} are in $\mathbb{Q}(\omega)[X]$ and have degree 4. There exist finitely many such polynomials and they can be obtained by multiplication of $\mathbb{Q}(\omega)$ −irreducible cyclotomic polynomials. (In [**BH**, Table 8.2] there is a list of irreducible cyclotomic polynomials for various fields.)
So, we have a finite number of possibilities for the eigenvalues of h_∞ and h_0^{-1} and, by using Proposition 10, we can decide which combinations yield a finite group. Using a detailed versions of Propositions 11 and 13 we can weed out the cases, when H is imprimitive, and the remaining cases are cited in the Shephard–Todd table. That table is made such that, if the exponents of $a_1, \ldots, a_n; b_1, \ldots, b_n$ occur, then the exponents of $\zeta a_1^k, \ldots, \zeta a_n^k; \zeta b_1^k, \ldots, \zeta b_n^k$ for $\zeta \in \mathbb{C}^*$ and $\gcd(h,k)$ do not occur in the list. □

Finally, I note that in [**BH**] the cases with imprimitive hypergeometric group are thoroughly discussed. I do not present that analysis.

12.29. Confluent hypergeometric functions. The hypergeometric series (12.7) in the case

$$p < n = q + 1$$

is called the **confluent hypergeometric series** and the corresponding equation (12.8) is called the **confluent hypergeometric equation**; or generalized confluent hypergeometric equations, i.e., when $n > 2$.

Many special functions are more or less directly expressed via confluent hypergeometric functions. The simplest is $F(\emptyset; \emptyset; t) = \mathrm{e}^t$. Next, we have the standard confluent hypergeometric function $F(\alpha; \beta; t)$. Also the Bessel functions (see Example 8.31(b)) can be defined as $J_\nu(z) = \frac{(z/2)^\nu}{\Gamma(\nu+1)} \times F(\emptyset; \nu+1; -z^2/4)$; related with the Bessel functions are Hankel functions, modified Bessel functions and Kelvin functions. Finally, the Airy equation $u'' = zu$ (see Example 11.16) has the solutions $u_1(z) = F(\emptyset; 2/3; z^3/9)$ and $u_2(z) = z \times F(\emptyset; 4/3; z^3/9)$; these solutions are often expressed via suitable Bessel functions or via so-called Airy function (which I do not define).

I am going to present an approach due to A. Duval and C. Mitschi [**DMit, Mit**] to the problem of computing the differential Galois group of these equations.

12.30. Local formal solutions. As in the non-confluent case we agree that

$$\beta_n = 1.$$

The confluent hypergeometric equation has only two singular points: $t = 0$ and $t = \infty$.

Near $t = 0$ the Levelt exponents are the same as in the non-confluent case and, in the case β_j are different, the basic solutions $u_j(t) = t^{1-\beta_j} \times F_j(t)$ are the same as in Eq. (12.9); here the hypergeometric series $F_j(t)$ are convergent in the whole complex plane.

Near $t = \infty$ the situation is more complicated. Firstly, one finds solutions

$$\hat{v}_k(t) = t^{-\alpha_k} \times \widehat{G}_k(t), \quad k = 1, \ldots, p, \tag{12.14}$$

where $\widehat{G}_k(t) = {}_n F_p\left(\tilde{\alpha}_1, \ldots \tilde{\alpha}_n; \tilde{\beta}_1, \ldots, \tilde{\beta}_p; z\right)$, $z = (-1)^r t^{-1}$,

$$r = n - p$$

are suitable hypergeometric series; they are divergent, because the number of $\tilde{\alpha}$'s is greater than 1 plus the number of $\tilde{\beta}$'s.

Other solutions are WKB type series. Indeed, postulating $w(t) = \exp(ct^\kappa) \times t^\lambda \times (1 + \ldots)$ and substituting it into Eq. (12.8), one finds

$$\kappa = \frac{1}{r},$$

$c = c_l = r\mathrm{e}^{2\pi\mathrm{i}l\kappa}$, $l = 1, \ldots, r$, and $\lambda = \kappa(\alpha - \beta + \frac{1}{2}) + \frac{1}{2}$, where $\alpha = \sum_1^p \alpha_j$ and $\beta = \sum_1^n \beta_j$. It follows that there are $n - p$ solutions of the WKB form $\hat{w}_l(t) = \exp(c_l t^\kappa) \times t^\lambda \times \widehat{H}_l(t^{-\kappa})$. In fact, one has

$$\hat{w}_l(t) = \exp\left(r\mathrm{e}^{2\pi\mathrm{i}l\kappa} t^\kappa\right) \times t^\lambda \times \widehat{H}_0\left(\mathrm{e}^{-2\pi\mathrm{i}l\kappa} t^{-\kappa}\right), \quad l = 1, \ldots, r = n - p, \quad (12.15)$$

where $\widehat{H}_0$ is a formal (but divergent) power series in $t^{-\kappa}$.

12.31. Confluence. The name 'confluent' for the confluent hypergeometric function comes from the fact that it is obtained in the process of confluence of two singular points of the non-confluent hypergeometric function, when the parameters and the time are suitably normalized.
Indeed, consider the expression $F(\alpha_1, \ldots, \alpha_n, A, \ldots, A; \beta_1 \ldots, \beta_m; t/A^r)$, where the number of A's is $r = n - p$. It equals

$$\sum_k \frac{(\alpha_1)_k \cdots (\alpha_p)_k}{(\beta_1)_k \cdots (\beta_n)_k k!} \times (A)_k (t^\kappa/A)^k \times \cdots \times (A)_k (t^\kappa/A)^k$$

and tends to $F(\alpha_1, \ldots, \alpha_p; \beta_1, \ldots, \beta_n; t)$ as $t \to \infty$. Here the singular point $t = A^r$ tends to the singular point $t = \infty$.
The corresponding perturbed hypergeometric operator becomes

$$\mathcal{H} = Q(\mathcal{D}) - tP(\mathcal{D}) \times \left(A^{-1}\mathcal{D} + 1\right)^r.$$

12.32. Mellin transform and analytic solutions. The **Mellin transform** of a function $f(t)$ on $\mathbb{R}_+^* = (0, \infty)$ is given by

$$\mathcal{M}f(s) = \int_{\mathbb{R}_+^*} x^s f(t) \mathrm{d} \ln t$$

and the inverse transform is

$$\mathcal{M}^{-1}\phi(t) = \frac{1}{2\pi\mathrm{i}} \int_{c+\mathrm{i}\mathbb{R}} t^{-s}\phi(s)\mathrm{d}s,$$

where $c \in \mathbb{R}$ is such that the integral converges. (Note that, after the substitution $t = \mathrm{e}^x$ and defining $h(x) := f(\mathrm{e}^x)$, we get the Laplace transform $h \longmapsto \int_{\mathbb{R}} h(x)\mathrm{e}^{sx}\mathrm{d}x$, with the known inverse Laplace transform.)
We note the following easy properties of the Mellin transform:

$\mathcal{M}\{tf(t)\}(s) = \mathcal{M}f(s+1),$

$\mathcal{M}\{\mathcal{D}f\}(s) = -s \cdot \mathcal{M}f(s)$, where $\mathcal{D} = t\frac{\partial}{\partial t}$.

They imply that the Mellin transform replaces the differential operator $\mathcal{H}(\alpha; \beta) = Q(\mathcal{D}) - tP(\mathcal{D})$ from Eq. (12.8), acting on $u(t)$, with the following difference operator, acting on $v(s) = \mathcal{M}u(s)$:

$$Q(-s) - (-1)^r P(-s-1)\tau,$$

where $r = n - p$ and

$$\tau : v(s) \longmapsto v(s+1)$$

is the translation operator. The authors of [**DMit**] apply some manipulation and get the difference operator

$$\Delta = \Delta(\alpha, \beta) = Q(s+1)\tau - (-1)^r P(s). \tag{12.16}$$

In [**DMit**] we find the following general solution to the difference equation $\Delta v = 0$:

$$v(s) = \mathrm{e}^{\pi i r s} \frac{\Gamma(s+\alpha_1)\cdots\Gamma(s+\alpha_p)}{\Gamma(s+\beta_1)\cdots\Gamma(s+\beta_n)} \xi(s),$$

where $\xi(s) \in \mathbb{C}\left(\mathrm{e}^{2\pi \mathrm{i} s}\right)$ is 1−periodic.
They use also the classical formula

$$\Gamma(\lambda+s)\Gamma(1-\lambda-s)\sin(\pi(\lambda+s)) = \pi, \tag{12.17}$$

which allows to 'move' some Gamma factors from the numerator to the denominator (and vice versa).
Now, to solve the hypergeometric equation $\mathcal{H}u = 0$ one applies the inverse Mellin transform to the above function $v(s)$ with the choice $\xi(s) = \mathrm{e}^{2\pi \mathrm{i} m s}$, for some integer m, application of Eq. (12.17) and a suitable choice of the integration path $c + \mathrm{i}\mathbb{R}$.
If, for fixed $k \in \{1, \ldots, n\}$, $\beta_j - \alpha_k \notin \mathbb{N}$ for $j = 1, \ldots, m+1$, then we choose a path $\gamma_k \subset \mathbb{C}$ joining $-\mathrm{i}\infty$ with $\mathrm{i}\infty$ and avoiding the points $-\beta_j + \mathbb{N}$, $j = 1, \ldots, m+1$, on the right and the points $-\alpha_k - \mathbb{N}$, on the left. Next, $\gamma_0 \subset \mathbb{C}$ is a path joining $-\mathrm{i}\infty$ with $\mathrm{i}\infty$ and avoiding on the right the points $-\beta_j + N$, $j = 1, \ldots, n+1$.

Proposition 1.

(a) *For $k = 1, \ldots n$, if $\beta_j - \alpha_k \notin \mathbb{N}$ for $j = 1, \ldots, m+1$, then the function*

$$V_k(t) = \frac{1}{2\pi \mathrm{i}} \int_{\gamma_k} \frac{\Gamma(s+\alpha_k) \times \prod \Gamma(1-\beta_j-s)}{\prod\limits_{i\neq k} \Gamma(1-\alpha_i-s)} \mathrm{e}^{\mathrm{i}\pi s} t^s \mathrm{d}s$$

is a solution to the hypergeometric equation $\mathcal{H}u = 0$, holomorphic in the sector $\left\{-2\pi - \frac{r\pi}{2} < \arg z < \frac{r\pi}{2}\right\}$ and with the asymptotic

$$C_k \times t^{-\alpha_k} \times \widehat{G}_k, \quad |t| \to \infty,$$

where C_k is a concrete constant and $\widehat{G}_k$ is the same as in Eq. (12.14).

(b) *The function*

$$V_0(t) = \frac{1}{2\pi \mathrm{i}} \int_{\gamma_0} \frac{\prod \Gamma(1-\beta_j-s)}{\prod \Gamma(1-\alpha_k-s)} t^s \mathrm{d}s$$

is a solution to the hypergeometric equation $\mathcal{H}u = 0$*, holomorphic in the sector* $\left\{-\frac{r\pi}{2} < \arg t < \frac{r\pi}{2}\right\}$*. Moreover, it has the asymptotic*

$$C_0 \times \exp\left(rt^{\kappa}\right) \times t^{\lambda} \times \widehat{H}_0\left(t^{-\kappa}\right), \quad |t| \to \infty$$

where C_0 *is a concrete constant and* $\widehat{H}_0$ *is a formal series in* $t^{-\kappa}$*.*

Proof. The formulas for V_k follow from the above statements about the general solution to the difference equation (12.16) and suitable 'moves' of the Gamma factors using Eq. (12.17). (By the way, the integral formulas for V_j's are special case of Meijer's G-functions, see [**BE**, vol. 1].)

To prove the asymptotic formula for V_k one deforms the integration path γ_k to a path with beginning and end at $s = -\infty$ and surrounding the poles of the function $\Gamma(s+\alpha_k)$. Then one sums over residua at the points $s = -\alpha_k - m$; the leading term comes from the residuum at $s = -\alpha_k$ and equals $\prod \Gamma(1-\beta_j-s) / \prod_{i \neq k} \Gamma(1-\alpha_i - s) \times e^{-i\pi\alpha_k} \times t^{-\alpha_k} = C_k \times t^{-\alpha_k}$.

In the proof of the second asymptotic formula one uses the stationary phase formula from Theorem 5.48. The integral in the formula for V_0, with $s = c + i\tau$ becomes an oscillating integral. Indeed, using the Stirling formula from 5.58, we get the phase

$$\phi(\tau) = \tau\left\{-r\ln\tau + r + \ldots + \ln t\right\},$$

with the critical point at $\tau_c \approx t^{\kappa}$ and the critical value $\phi(\tau_c) \approx rt^{\kappa}$; here $r = 1/\kappa = n - p$. □

12.33. Stokes matrices. The following proposition is given in [**DMit**] without proof. The authors refer to the series of papers by C. Meijer [**Me**], where the functions $V_k(t)$ are expanded in the analytic basis $\{u_j(t)\}$ near $t = 0$.

Proposition 2.

(a) *We have*

$$e^{\pi i\alpha_k} V_k(t) - e^{-\pi i\alpha_k} V_k\left(e^{-2\pi i}t\right) = 2\pi i V_0(t),$$

$-r\frac{\pi}{2} < \arg t < r\frac{\pi}{2}$.

(b) *Let* $m \in \mathbb{Z}$ *be such that* $\varepsilon + \frac{r}{2} \leq 2m \leq 2 - \varepsilon + \frac{3r}{2}$ *and let* $(r-\varepsilon)\pi < \arg t \leq (r+\varepsilon)\pi$*. We have*

$$V_0(t) = \sum_{l=1}^{r-m} A_l V_0(e^{-2\pi i l}t) + \sum_{l=0}^{m-1} B_l V_0\left(e^{-2\pi i(r-l)}t\right) + \sum_{k=1}^{p} D_k V_k\left(e^{-2\pi i(r-m+1)}t\right),$$

where A_l*,* B_l *and* D_k *are concrete constants.*

Next, the authors of [**DMit**] divide a neighborhood of $t = \infty$ into suitable sectors Θ_q, $q = 0, \ldots Q$, where the number $Q + 1$ of sectors depends on the case. They construct a collection $\Sigma_q = \{X_{q,k}(t), Y_{q,j}(t)\}$ of fundamental systems of analytic solutions in the sectors Θ_q. One has

$$X_{q,k}(t) = C_k^{-1} \times \mathrm{e}^{\pi \mathrm{i} \delta_k} \times V_k\left(\mathrm{e}^{\pi \mathrm{i} \sigma_k} t\right),$$

where δ_k and σ_k are suitable constants; i.e., 'shifted' V_k. The solutions $Y_{q,j}(t)$ are either expressed via V_0 (similarly like $X_{q,k}$ via 'shifted' V_k above) or as rather complicated combinations of 'shifted' V_0's and 'shifted' V_k's.

Finally, using Proposition 2, the authors are able to calculate coefficients of the Stokes matrices S_q relating the bases Σ_q and Σ_{q+1} in adjacent sectors. The formulas and their proofs are really very complicated and I failed to simplify them.

The authors applied their formulas for the Stokes operators and the easy expressions for the formal monodromy and the exponential torus to compute the differential Galois group for some confluent equations.

Example. Consider the operator $\mathcal{H}(\alpha;\beta_1,\beta_2,\beta_3) = (\mathcal{D}+\beta_1-1)(\mathcal{D}+\beta_2-1)(\mathcal{D}+\beta_3-1) - t(\mathcal{D}+\alpha)$, i.e., with $n=3$, and $p=1$; usually $\beta_3 = 1$. We put

$$a = 2\pi\mathrm{i}/\prod\Gamma(1+\alpha-\beta_j),\ b = 4\mathrm{i}\pi^2/\prod\Gamma(\beta_j-\alpha),$$
$$c = \sum \mathrm{e}^{-\pi\mathrm{i}\left(\mu+2\beta_j\right)} - \mathrm{e}^{-\pi\mathrm{i}(\mu+2\alpha)}, \mathrm{e}^{-\pi\mathrm{i}(\mu+2\alpha)},\ \mu = \alpha-\beta_1-\beta_2-\beta_3+\frac{3}{2}.$$

It turns out that in the generic case, i.e., $ab \neq 0$, $\alpha \notin \mathbb{Q}$, $\mu \in \mathbb{Q}$, $\mu - \alpha \notin \mathbb{Q}$ and $c \in \mathbb{R}^*$ and such that $\mu + 2\alpha \in \mathbb{Z}$ cannot occur simultaneously, then the differential Galois group equals $GL(3)$.
Also other, non-generic cases are solved in [**DMit**].

In [**Mit**] C. Mitschi continued investigation of the differential Galois groups for some confluent hypergeometric equations. In particular, she proved that in the case $n=7$ and $p=1$, under some restriction for the parameters, the identity component of the differential Galois group equals the exceptional Lie group G_2 (associated with the root system G_2, see Theorem 4.29).

12.34. An approach via confluence. Recall that the hypergeometric equation with $p<n$ is a result of the confluence of a non-confluent hypergeometric equation as the point $t=1$ becomes moving to $t=\infty$, as some parameters vary suitably (see 12.31). One can apply it to determine some coefficients in the Stokes matrices.
Firstly, one determines general shape of the Stokes matrices using the approach of J. Heading from Example 8.31(b). That approach uses equivariance of the Stokes matrices with respect to rotation by the angle $2\pi/r$, $r=n-p$, and (sometimes) to the complex conjugation. There remain some coefficients to be fixed.
But the Stokes operators are suitable limits of the monodromy matrices for suitable deformations of the equations (compare 8.32). These monodromy matrices are somehow easier to handle with.
Sometimes this approach works, but I do not want to go into details.

§3 Picard–Deligne–Mostow Theory

12.35. Picard's generalized hypergeometric functions of many variables. The subject of investigations are the following integrals, introduced firstly by E. Picard [**Pic1**], [**Pic2**], by L. Pochhammer [**Poch**] and by G. Lauricella [**Lau**]:

$$F_{ij}(t_1,\ldots,t_N)=\int_{t_i}^{t_j}(\tau-t_1)^{-\mu_1}\ldots(\tau-t_N)^{-\mu_N}\mathrm{d}\tau=\int\Phi\mathrm{d}\tau. \qquad (12.18)$$

We will also write $F_{ut}(S)=\int_u^t\prod_{v\in S}(\tau-v)^{-\mu_v}\mathrm{d}\tau$, where $u,t\in S=\{t_1,\ldots,t_N\}$. They can be treated as natural generalizations of the Euler integrals, appearing in definitions of hypergeometric functions. Namely, we have

$$\begin{aligned}B(\beta,\gamma-\beta)F(\alpha,\beta;\gamma;t)&=\int_0^1 s^{\beta-1}(1-s)^{\gamma-\beta-1}(1-st)^{-\alpha}\mathrm{d}s\\&=\int_1^\infty \tau^{\alpha-\gamma}(\tau-1)^{\gamma-\beta-1}(\tau-t)^{-\alpha}\mathrm{d}\tau\\&=F_{1\infty}(0,1,t,\infty),\end{aligned}$$

with $\mu_0=\gamma-\alpha$, $\mu_1=\beta-\gamma+1$, $\mu_t=\alpha$, $\mu_\infty=1-\beta$ and $B(\cdot,\cdot)$ as the Euler Beta function.
We will agree to omit the factor $(\tau-t_j)^{-\mu_j}$ in the case when $t_j=\infty$. If all $t_j\neq\infty$ in Eq. (12.18) and the branches of the subintegral 1-form are analytic at ∞, then we have the following restriction onto the exponents:

$$\sum\mu_j=2.$$

Of course, the latter identity extends to the case when one $t_j=\infty$, where $\mu_\infty=2-\sum_{t\in S\setminus\infty}\mu_t$ is the order of the subintegral 1-form at $s=1/\tau=0$.
The functions F_{ij} are multivalued functions, defined on the space $\overline{\mathbb{C}}^N\setminus\Delta$ of ordered subsets $S=\{t_1,\ldots,t_N\}$ of the Riemann sphere; here $\Delta=\bigcup\{t_k=t_l\}$ is the 'generalized diagonal'. The hypersurfaces $t_k=t_l$ are the ramification hypersurfaces of F_{ij}.
We will see that the linear space $\mathcal{V}$, generated by all the functions F_{ij}, is invariant with respect to the monodromy maps corresponding to surroundings of the hypersurfaces $t_k=t_l$ (like in the case of the Gauss hypergeometric functions). One has a representation of $\pi_1(\overline{\mathbb{C}}^N\setminus\Delta)$ in the group $PGL(\mathcal{V})$ with the image Mon (the **projective monodromy group**).
The aim of the Picard's theory [**Pic2**] is to generalize the results of Schwarz (series of discrete subgroups of $PGL(2,\mathbb{C})$ and lattices in $PU(1,1)=PGL(2,\mathbb{R})$) to the group Mon for $N=5$. In particular, he gave series of sufficient conditions which guarantee discreetness and the lattice property of Mon.
G. Mostow and P. Deligne gave in [**DM1**] new proofs of the Picard's results (with a generalization to any N). Mostow and Deligne are also interested in new examples of lattices and of arithmetic subgroups of $PU(1,n)$ (the definitions are given

below). These examples are provided by the monodromy groups of the integrals (12.18) (see also [**Mos2**]).

12.36. Homology theory of local systems. The integrals F_{ij} have two sorts of ramifications. Firstly, the subintegral 1-form is multivalued and, secondly, the path of integration varies as the endpoints t_j change their positions. The proper theory to deal with such kind of objects is the theory of local systems and their (co)homologies.

The subintegral function $\Phi = (\tau - t_1)^{-\mu_1} \dots (\tau - t_N)^{-\mu_N}$, treated as a function of τ, is of Darbouxian type. Its monodromy (as the monodromy of a multivalued function) consists of maps $\Phi \longmapsto \mathrm{e}^{-2\pi \mathrm{i}\mu_t}\Phi$ when the argument varies around $t \in S$. This leads to the idea that Φ may represent a section of certain locally constant bundle.

Following this idea, one introduces two **local system L** and $\mathbf{L}^\vee$ on the space $\overline{\mathbb{C}} \setminus S$, $S = \{t_1, \dots, t_N\}$.

The local system $\mathbf{L}$ is a line bundle (of rank 1) over $\overline{\mathbb{C}} \setminus S$ with constant transition maps, such that a simple loop around t_j generates the transformation $(\tau, v) \longmapsto (\tau, \alpha_j v)$, $\alpha_j = \mathrm{e}^{2\pi \mathrm{i}\mu_j}$, of the fibers. The system $\alpha = (\alpha_j)$ can be treated as the cocycle from $H^1(\overline{\mathbb{C}} \setminus S, \mathcal{O}(GL(1)))$, defining the bundle L.

Any local system admits a natural flat connection, such that the locally constant sections form *horizontal sections.* This connection is called the *Gauss–Manin connection.* The Gauss–Manin connection is given by the formula $\nabla = \mathrm{d} + (\mathrm{d} \ln \Phi) = \mathrm{d} - (\sum \mu_j \mathrm{d} \ln(\tau - t_j))$ and the horizontal multivalued sections s (such that $\nabla s = 0$) are proportional to Φ^{-1} (in the trivialization above the universal covering of $\overline{\mathbb{C}} \setminus S$).

The **dual local system** $\mathbf{L}^\vee$ is the line bundle, defined by means of the analogous transformations $(\tau, v) \longmapsto (\tau, \alpha^{-1} v)$. There is the natural pairing $\mathbf{L} \otimes \mathbf{L}^\vee \longmapsto \mathbb{C}$, where $\mathbb{C}$ denotes the trivial line bundle.

If $\mathbf{K}$ is a local linear system over a manifold X, then one can associate with it the **homology** and **cohomology groups with coefficients in** K, denoted by $H_i(X, \mathbf{K})$, $H^i(X, \mathbf{K})$ and defined as follows. A *simplex of* X *with coefficients in* $\mathbf{K}$ is a pair (δ, s_δ) consisting of a simplex δ in X and a horizontal section s_δ of $\mathbf{K}$ over δ. If $a \in \mathbb{C}$ is a constant, then $a \cdot (\delta, s_\delta) = (\delta, a s_\delta)$. These simplices have naturally defined boundaries. This allows to construct the chain and cochain complexes, whose homology groups are the homology and cohomology groups with coefficients in $\mathbf{K}$. The usual homologies $H_i(X, \mathbb{C})$ are the homologies with coefficients in the trivial local system $\mathbb{C}$.

For any local system $\mathbf{K}$ there is a naturally defined and non-degenerate pairing $H^i(X, \mathbf{K}) \otimes H_i(X, \mathbf{K}^\vee) \longmapsto \mathbb{C}$.

In applications (e.g., to the integrals of type (12.18)) it is natural to consider the **homology groups with closed (locally finite) supports** (with coefficients in $\mathbf{K}$), $H_i^{\mathrm{lf}}(X, \mathbf{K})$ and **cohomology groups with compact supports** $H_{\mathrm{c}}^i(X, \mathbf{K})$. The first ones are defined by means of locally finite (but maybe infinite) chains and the second ones are defined by means of cochains which vanish at too distant simplices. (We treat $H_i(X, \mathbf{K})$ as homology groups with compact support).

For example, the (open) interval $\overline{tu} \subset \overline{\mathbb{C}} \setminus S$, $t, u \in S$, with some horizontal section $s_{\overline{tu}}$ of $\mathbf{L}^\vee$, defines an element $\delta(t,u)$ of $H_1^{\mathrm{lf}}(\overline{\mathbb{C}} \setminus S, \mathbf{L}^\vee)$. On the other hand, the Pochhammer cycle $C(t,u)$ (the double loop around t and u as in Fig. 11.2 in Chapter 11) together with a horizontal section defines element of $H^1(\overline{\mathbb{C}} \setminus S, \mathbf{L}^\vee)$. We have the non-degenerate pairing $H_i^{\mathrm{lf}}(X, \mathbf{K}) \otimes H_c^i(X, K^\vee) \longmapsto \mathbb{C}$.
The intersection of finite and locally finite cycles allows to define the pairing $H_i(X, \mathbf{K}) \otimes H_{m-i}^{\mathrm{lf}}(X, \mathbf{K}^\vee) \longmapsto \mathbb{C}$, $m = \dim X$. This pairing is non-degenerate (the analogue of the Poincaré duality theorem 3.15) and defines the Poincaré isomorphism $H_{m-i}^{\mathrm{lf}}(X, \mathbf{K}^\vee) \simeq H^i(X, \mathbf{K})$.

Let us return to the local system $\mathbf{L}$, defined by means of the function $\Phi = \prod(\tau - t_j)^{-\mu_j}$ on $\overline{\mathbb{C}} \setminus S$, $S = \{t_1, \ldots, t_N\}$.
Integral (12.18) is understood in the following way. We have a 1-simplex $\delta(t_i, t_j) = (\overline{t_i t_j}, s)$ with values in $\mathbf{L}^\vee$. Take the 1-form with values in $\mathbf{L}$

$$\omega = \Phi \cdot e \cdot \mathrm{d}\tau,$$

where e is a horizontal section of $\mathbf{L}$. The form ω defines an element of $H^1(\overline{\mathbb{C}} \setminus S, \mathbf{L})$: its value on a simplex $\delta(t_i, t_j)$ is equal to the integral along this simplex: $F_{ij} = \langle \omega, \delta(t_i, t_j) \rangle = \int_{\overline{t_i t_j}} \Phi \langle e, s \rangle \mathrm{d}\tau = \langle e, s \rangle \int_{t_i}^{t_j} \Phi \mathrm{d}\tau$, where $\langle e, s \rangle = $ const.
Because the section e is multivalued, with monodromies α_i, and the monodromies of Φ are α^{-1}, the form ω defines a single valued section of the bundle $\Omega^1(\mathbf{L})$. (We shall return to this subject later).
Note also that the Pochhammer cycles $C(t,u)$ are related with the locally finite cycles $\delta(t,u)$: $C(t,u) = (1 - \alpha_t^{-1})(1 - \alpha_u^{-1}) \cdot \overline{tu}$. If $\alpha_t, \alpha_u \neq 1$, then this formula allows to regularize the improper integrals.
We say that the system $\mu_1, \ldots, \mu_N$ of exponents in the 1-form ω is **resonant** iff some $\mu_j \in \mathbb{Z}$ (equivalently, iff some $\alpha_j = 1$).

12.37. Theorem. *Assume that the system $\mu_1, \ldots, \mu_N$ is not resonant. Then*

(a) *the natural homomorphism $H_1(\overline{\mathbb{C}} \setminus S, L^\vee) \longmapsto H_1^{\mathrm{lf}}(\overline{\mathbb{C}} \setminus S, L^\vee)$ is isomorphism;*

(b) *the group $H_1^{\mathrm{lf}}(\overline{\mathbb{C}} \setminus S, L^\vee)$ is $N-2$ dimensional with generators defined by means of the cycles $\delta(t_1, t_2)$, $\delta(t_1, t_3), \ldots \delta(t_1, t_{N-1})$ (with some horizontal sections s_j above $\overline{t_1 t_j}$).*

Proof. We begin with calculation of the dimensions of these groups. We have the following equalities between the Euler characteristics:

$$\chi(\overline{\mathbb{C}} \setminus S, \mathbf{L}^\vee) = \chi^{\mathrm{lf}}(\overline{\mathbb{C}} \setminus S, \mathbf{L}^\vee) = 2 - N$$

Indeed, let S_ϵ be the small ϵ-neighborhood of the set S. By the homotopy equivalence, we have $H_*(\overline{\mathbb{C}} \setminus S, \mathbf{L}^\vee) = H_*(\overline{\mathbb{C}} \setminus S_\epsilon, \mathbf{L}^\vee)$ and $H_*^{\mathrm{lf}}(\overline{\mathbb{C}} \setminus S, \mathbf{L}^\vee) = H_*(\overline{\mathbb{C}} \setminus S_\epsilon, \partial(\overline{\mathbb{C}} \setminus S_\epsilon), \mathbf{L}^\vee)$.
The Euler characteristic $\chi(\overline{\mathbb{C}} \setminus S_\epsilon)$ is calculated using any simplicial partition of the compact space $\overline{\mathbb{C}} \setminus S_\epsilon$ and does not depend on $\mathbf{L}$. In the case of constant coefficients

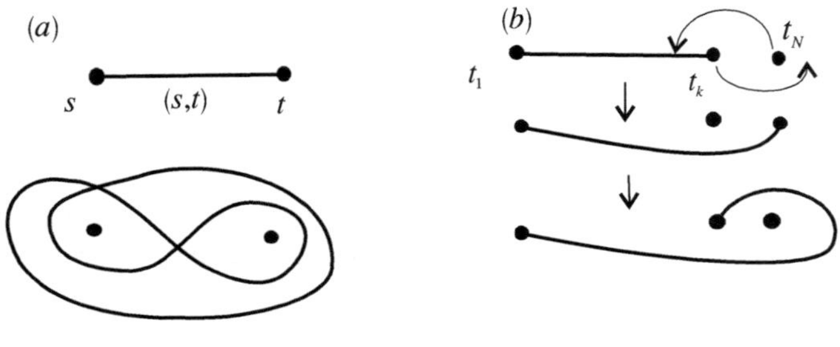

Figure 12.10: Cycles used to define homologies of local systems

$\mathbb{C}$ we have $\chi(\overline{\mathbb{C}} \setminus S_\epsilon) = 1-(N-1) = 2-N$. Similarly, $\chi(\overline{\mathbb{C}} \setminus S_\epsilon, \partial(\overline{\mathbb{C}} \setminus S_\epsilon) = 2-N$.

The space $H_2^{\mathrm{lf}}(\overline{\mathbb{C}} \setminus S, \mathbf{L}^\vee)$ can be identified with the space of global (i.e., single valued) horizontal sections of $\mathbf{L}^\vee$: one associates it to the family of 2-simplices of a locally finite simplicial partition (like in the definition of the orientation). But, if at least one $\alpha_j \neq 1$, then there are no such sections and $H_2^{\mathrm{lf}} = 0$.
Similarly, by the Poincaré duality, $H_0^{\mathrm{lf}}(\overline{\mathbb{C}} \setminus S, \mathbf{L}^\vee) \simeq H_2(\overline{\mathbb{C}} \setminus S, \mathbf{L})^* = 0$ (because $\overline{\mathbb{C}} \setminus S$ is contractible to a 1-dimensional complex). Therefore $\dim H_1^{\mathrm{lf}}(\overline{\mathbb{C}}, \mathbf{L}^\vee) = -\chi = N-2$.
Analogously one shows that $H_1(\overline{\mathbb{C}} \setminus S, \mathbf{L}^\vee) \simeq \mathbb{C}^{N-2}$.

To show that $\delta(t_1,t_2),\ldots,\delta(t_1,t_{N-1})$ are independent cycles in H_1^{lf}, we use the monodromy. Assume that $\delta = \sum a_j \delta(t_1,t_j) = 0$ and let a_k be the first nonzero coefficient.
Consider the homotopy relying on fixing the points t_j, $j \neq k, N$, and rotation of t_k around t_N. From Fig. 12.10 we see that $\delta(t_1,t_k) \longmapsto \delta(t_1,t_k) + \alpha_k^{-1}(\alpha_N^{-1} - 1)\delta(t_N,t_k)$.
By the assumption, $\int_\delta \omega \equiv 0$. But $\langle \omega, \delta(t_k,t_N)\rangle \not\equiv 0$; because, when $t_N = 0$, $t_k = \epsilon \to 0$ we have $\langle \omega, \delta(t_k,t_N)\rangle \sim \mathrm{const} \cdot \epsilon^{1-\mu_k-\mu_N} B(1-\mu_N, 1-\mu_k) \neq 0$. Because $\alpha_k^{-1}(\alpha_N^{-1}-1) \neq 0$, we get a contradiction. This proves the point (b) of Theorem 12.37.
(a) follows from the association to the intervals $\overline{t_1 t_j}$ the Pochhammer cycles $C(t_1,t_j)$. We omit the details. □

12.38. Remark. Another possible choices of the basis of H_1^{lf} are the following. We divide S into two disjoint subsets S_1, S_2 such that $\prod_{t\in S_1} \alpha_t \neq 0$. Let T_1, T_2 be two disjoint tree graphs with the sets S_1, S_2 of vertices. Then the system of 1-cycles $\delta(t_i,t_j)$, such that the edges $\overline{t_i t_j}$ belong to the set of edges of $T_1 \cup T_2$, forms the basis of H_1^{lf} (see [**DM1**]).

12.39. The de Rham cohomologies of local systems. From now on we assume that

$$0 < \mu_j < 1, \quad \sum \mu_j = 2,$$

i.e., $|\alpha_j| = 1$. Here the dual local system $\mathbf{L}^\vee$ is naturally identified with the conjugate local system $\overline{\mathbf{L}}$, (because $\alpha_j^{-1} = \overline{\alpha_j}$). The exponents of $\overline{\mathbf{L}}$ are equal to $1 - \mu_j$ (and lie in the interval $(0, 1)$).

The elements of the first cohomology group of $\overline{\mathbb{C}} \setminus S$ with coefficients in $\mathbf{L}$ (respectively, the first cohomology group with compact support) are represented by means of closed differentiable 1-forms with coefficients in $\mathbf{L}$ (respectively, by closed 1-forms with compact support). It is a version of the de Rham theorem for cohomologies with coefficients in sheaves.

The above form $\omega = \Phi \cdot e \cdot \mathrm{d}\tau$ is closed but has non-compact support. Its value on the non-compact cycle $\delta(t, u)$ can be defined in two ways:

- by means of integration along the corresponding compact Pochhammer cycle $C(t, u)$, or
- by means of replacement of ω by the homologically equivalent form with compact support $\omega_1 = \omega - \mathrm{d}(\sum_j \chi_{t_j} \phi_{t_j})$, where $\phi_{t_j}(\tau)$ are the primitives of $\Phi \cdot e$ near t_j, $\mathrm{d}\phi_{t_j} = \omega$ and χ_j are smooth functions with compact support around t_j and $\chi_{t_j}(\tau) \equiv 1$ near $\tau = t_j$. One can also replace χ_t by the characteristic functions $\chi(|\tau - t| \le \epsilon)$ (treated as a 0-current) and one obtains the so called *Hadamard's regularization of a divergent integral* P.V. $\left(\int_t^u\right) \Phi \mathrm{d}\tau = \int_{t+\epsilon}^{u-\epsilon} \Phi - \phi_u(u - \epsilon) + \phi_t(t + \epsilon)$.

Following [**DM1**] we will define a 'Hodge' type decomposition $H^1(\overline{\mathbb{C}} \setminus S, L) = H^{1,0} \oplus H^{0,1}$ and some Hermitian form (for real exponents μ_j), which is positive in $H^{1,0}$ and negative on the second component.

The space $H^{1,0}$ consists of 1-forms η of the so-called *first kind*, i.e., of the forms $\eta = \Phi \cdot f(\tau) \cdot e \cdot \mathrm{d}\tau$, where f is meromorphic function in $\overline{\mathbb{C}}$ such that at each t_i we have $-\mu_i + \mathrm{ord}_{t_i} f > -1$. This assumption guarantees convergence of the integrals $\langle \eta, \delta(t_i, t_j) \rangle$ and $\int \eta \wedge \bar{\eta}$ along $\overline{\mathbb{C}} - S$.

The expressions $\Phi f \cdot e$ represent sections of the line bundle $\mathcal{O}(\sum \mu_j \cdot t_j)(\mathbf{L})$ on $\overline{\mathbb{C}} = \mathbb{CP}^1$, where $\sum \mu_j \cdot t_j$ is a divisor with real coefficients. Its local sections have orders $\mathrm{ord}_{t_j} f \ge \mu_i$. Each bundle on $\mathbb{CP}^1$ is equal to some $\mathcal{O}(m)$, where m is the degree of the bundle (or the Chern number) and is equal to the sum of orders of zeroes and poles of any its meromorphic section. The section defined by $\Phi \cdot e$ (from ω) is holomorphic and nonzero in the affine part and near infinity it has zero of order $\sum \mu_i = 2$. Therefore $\mathcal{O}(\sum \mu_j t_j)(\mathbf{L}) = \mathcal{O}(2)$.

The forms of first kind are sections of the bundle $\Omega^1(\sum \mu_j t_j)(\mathbf{L}) = \mathcal{O}(-2) \otimes \mathcal{O}(2) = \mathcal{O}(0)$, i.e., the trivial bundle. Thus, $H^{1,0}$ is 1-dimensional. Because the form ω belongs to it and is nonzero (see the previous point), it generates $H^{1,0}$.

The space $H^{0,1}$ consists of anti-holomorphic $\mathbf{L}$-valued 1-forms ρ whose complex conjugates are of the first kind, $\bar{\rho} \in \Omega^1(\sum(1 - \mu_j)t_j)(\overline{\mathbf{L}})$ or $\bar{\rho} = \prod(\tau - t_j)^{\mu_j - 1} \cdot g(\tau) \cdot \bar{e} \cdot \mathrm{d}\tau$. Here g is meromorphic in $\mathbb{CP}^1$ function such that $-(1 - \mu_j) + \mathrm{ord}_{t_j} g > -1$. Repeating the above analysis, we find that $\Omega^1(\sum(1 - \mu_j)t_j)(\overline{\mathbf{L}}) = \mathcal{O}(-2) \otimes \mathcal{O}(\sum(1 - \mu_j)) = \mathcal{O}(N - 4)$. The space of global section of $\mathcal{O}(N - 4)$ is the space of symmetric homogeneous polynomials in two variables of degree $N-4$ and is $(N-3)$-dimensional. So, $H^{0,1} \simeq \mathbb{C}^{N-3}$.

Using the external product of forms, we define the Hermitian form:

$$(\cdot,\cdot) : H^1(\overline{\mathbb{C}} \setminus S, \mathbf{L}) \times H^1(\overline{\mathbb{C}} \setminus S, \mathbf{L}) \longmapsto \mathbb{C}, \quad (\eta, \rho) = \frac{-1}{2\pi \mathrm{i}} \int \eta \wedge \bar{\rho}.$$

It is clear that this form is: (i) positive definite on $H^{1,0}$ ($\mathrm{d}\tau \wedge \mathrm{d}\bar{\tau} = -2\mathrm{i}\mathrm{d}x \wedge \mathrm{d}y$); (ii) negative definite on $H^{0,1}$ and (iii) these two subspaces are orthogonal.
We summarize the above as follows.

12.40. Theorem. *The space $H^1(\overline{\mathbb{C}} \setminus S, L)$ admits a Hermitian form $(\cdot,\cdot)$ of signature $(1, N-3)$ and the orthogonal Hodge expansion $H^{1,0} \oplus H^{0,1}$ onto positive and negative subspaces. Moreover, the space $H^{1,0}$ is generated by the class of the 1-form ω.*

12.41. Generalized Schwarz map. Let $(\overline{\mathbb{C}})^N \setminus \bigcup_{ij}\{t_i = t_j\}$ be the space of ordered subsets S of the Riemann sphere, $|S| = N$. Three elements of S can be fixed, e.g., at $0, 1, \infty$ (by means of the action of $PGL(2, \mathbb{C})$).
Let $Q = \left((\overline{\mathbb{C}})^N \setminus \bigcup_{ij}\{t_i = t_j\}\right)/PGL(2)$ denotes the corresponding space, $Q \simeq (\overline{\mathbb{C}} \setminus \{0, 1, \infty\})^{N-3} \setminus \Delta$, $\Delta = \bigcup_{ij}\{t_i = t_j\}$. The elements of Q are sets $S = \{t_1, \ldots, t_N\}$ such that $t_1 = 0$, $t_2 = 1$, $t_3 = \infty$.
For each $S \in Q$ we have the local system $\mathbf{L} = \mathbf{L}_S$ on $\overline{\mathbb{C}} \setminus S$ and the corresponding $(N-2)$-dimensional space $H^1_S = H^1(\overline{\mathbb{C}} \setminus S, \mathbf{L})$ with $(N-3)$-dimensional projectivization PH^1_S. Let $P : H^1_S \setminus 0 \longmapsto PH^1_S$ denotes this projectivization map.
The system of linear spaces H^1_S, $S \in Q$, forms a vector bundle over Q, denoted by $\mathbf{K} = R^1\pi_*\mathbf{L}$ or by $\mathcal{H}^1$ (i.e., the Leray sheaf or the cohomological bundle). The bundle $\mathbf{K}$ has the flat Gauss–Manin connection and forms a local system over Q, the analogue of the local system $\mathbf{K}$ defined by solutions of the classical hypergeometric equation in 12.8. This holds because, like in the case of the (co)homological Milnor bundle, we have a lattice in $H_1(\overline{\mathbb{C}} \setminus S, \mathbf{L}^\vee)$ generated by the basis $\delta(t_1, t_j)$ of cycles.
The image in PH^1_S of the positive cone $\{\eta \in H^1_S : (\eta, \eta) > 0\}$ is the *ball* $\mathbb{B}_S$ (of real dimension $2N-6$). Indeed, if $x_0, \ldots, x_{N-3}$ are the coordinates in H^1_S such that $(x, x) = |x_0|^2 - |x_1|^2 - \ldots - |x_n|^2$, $n = N-3$, then $x_0 \neq 0$ in the positive cone and, choosing the coordinates $(1 : y_1 : \ldots : y_n)$ in an affine chart of PH^1_S, we get $\mathbb{B}_S = \{\sum |y_i|^2 < 1\}$. The balls $\mathbb{B}_S$ are equipped with complex hyperbolic geometries defined by means of the Hermitian form $(\cdot,\cdot)$, $(\mathbb{B}_S, (\cdot,\cdot) \simeq \mathbb{C}h^n$.
Here $\mathbb{C}h^n$ denotes the complex hyperbolic space. The group of *hyperbolic motions* is generated by the *complex reflections*, which are conjugate to the maps $(x_1, \ldots, x_n) \longmapsto (\mathrm{e}^{2\pi \mathrm{i}\alpha}x_1, x_2, \ldots, x_n)$, $\alpha \in \mathbb{R}$ (i.e., rotations in one complex variable)
As in the case of classical hypergeometric integrals ($N = 4$), we choose a multi-valued trivialization $\Phi : \mathbf{K} \longmapsto Q \times \mathbb{C}^{N-2}$ of the bundle $\mathbf{K}$ by means of a basis of

horizontal sections (see 12.8). In this sense we have also the multivalued trivialization $P \circ \Psi : \bigcup \mathbb{B}_S \longmapsto Q \times \mathbb{B}_{S_0}$ of the bundle of balls $\mathbb{B}_S$.
For each $S \in Q$ we have the 1-form $\omega = \omega_S = \prod_{t \in S}(\tau - t)^{-\mu_t} \cdot e \cdot \mathrm{d}\tau$. The map

$$S \longmapsto W(S) = P \circ \Phi(\omega_S)$$

is a holomorphic map from Q to $\mathbb{B}_{S_0}$. We call it the **generalized Schwarz map.** The map W is also called the *period map*, in analogy with the period map defined by means of integrals along cycles in varying varieties (see [**Var5**]).

12.42. Projective monodromy group. The fundamental group $\pi_1(Q)$ of the space Q is a quotient of the *colored braid group* $\widehat{B}(N-1)$ (of $N-1$ strands in $\mathbb{C}$). This follows from: $\pi_1(\mathbb{C}^{N-1} \setminus \Delta) = \widehat{B}(N-1)$ (see 4.48(d)), $\pi_1(\mathrm{Aff}(\mathbb{C})) = \mathbb{Z}$ and the property $Q = (\mathbb{C}^{N-1} \setminus \Delta)/\mathrm{Aff}(\mathbb{C})$. From 4.48(d) we know that the colored braid group is generated by loops σ_{ij} which surround just one hypersurface $t_i = t_j$ (from the 'diagonal' Δ).
The group $\pi_1(Q, S_0)$ acts on the fixed fiber $P\mathbf{K}_{S_0} = PH^1(\overline{\mathbb{C}} \setminus S_0, \mathbf{L})$. It preserves the Hermitian form $(\cdot,\cdot)$ in the fiber. The image of $\pi_1(Q, S_0)$ in $\mathrm{Aut}(P\mathbf{K}_{S_0})$ is a subgroup of the group $PU(1,n)$. It is called the **projective monodromy group of the hypergeometric integral with exponents** $\mu = (\mu_t, t \in S)$ (or simply the *monodromy group*) and is denoted by $\mathrm{Mon} = \mathrm{Mon}(\mu)$.
It is useful to see how the generators of the group $\pi_1(Q)$ act. Take, for example, the case when the space $\overline{\mathbb{C}} \setminus \Delta$ is deformed in such a way, that one point $t = \epsilon \mathrm{e}^{\mathrm{i}\theta}$ overruns the fixed point $u = 0$ and the other points of S are away from zero and are fixed. It means that we take a loop σ_{ut} in the colored braid group. Choose the basis of $H_1^{\mathrm{lf}}(\overline{\mathbb{C}} \setminus S)$ such that $\delta_0 = \delta(u,t) = (\overline{ut}, s_{\overline{ut}})$ is one of its elements and the other elements δ_j are chosen accordingly to some tree graph T_2 with vertices in $S \setminus \{u,t\}$. As θ runs from 0 to 2π, any point in $\overline{ut}$ undergoes relative rotations around u and around t. Thus the section $s_{\overline{ut}}$ is multiplied by $(\alpha_u \alpha_t)^{-1}$. Other elements δ_j of the basis do not change after rotation.
The generalized Schwarz function W undergoes a similar monodromy transformation. We take the coordinate system in $H^1(\overline{\mathbb{C}} \setminus S)$ defined by the cycles δ_j : $x_j(W(S)) = \int_{\delta_j} \omega$. It is clear that $x_1, x_2, \ldots$ remain unchanged after deformation. However, we have $x_0 = \int_0^t \tau^{-\mu_u}(\tau - t)^{-\mu_t}(C + \ldots) \approx t^{1-\mu_u-\mu_t} \times$ (analytic function). This means that:

The action of the loop σ_{ut} on the ball B_{S_0} means that the coordinate x_0 undergoes multiplication by the number $\mathrm{e}^{-2\pi\mathrm{i}(\mu_u+\mu_t)}$ (of module 1), i.e., σ_{ut} realizes a complex reflection in the complex hyperbolic space $\mathbb{C}h^n$.
This implies that the monodromy group Mon *is a subgroup of the group generated by complex reflections.*

12.43. Integrality condition. This condition states that

$$\mathbf{INT} \quad \begin{cases} 0 < \mu_t < 1, \\ \sum \mu_t = 2, \\ \mu_s + \mu_t < 1, \; s \neq t \Rightarrow \; (1 - \mu_s - \mu_t)^{-1} \in \mathbb{Z} \cup \{\infty\}. \end{cases}$$

From the previous point it follows that condition **INT** is necessary when we want to obtain a complex analogue of the tessellation of the ball $\mathbb{B}_{S_0}$ by means of the generalized Schwarz map. In that case the inverse to the generalized Schwarz map locally prolongs to a map realizing a ramified covering of the configuration space. Assume that **INT** holds. The monodromy group Mon acts on the ball $\mathbb{B}_{S_0}$ and the orbit $\mathrm{Orb}(x_0)$ of a point x_0 forms a discrete subset in the ball. In the case $n = 1$ there is a *fundamental domain*, the fundamental quadrangle, which is isomorphic to the homogeneous space $PU(1,1)/\mathrm{Mon}$. In particular, the *hyperbolic measure* of the fundamental domain is the same as the Haar measure of $PU(1,1)/\mathrm{Mon}$.
In the case $n > 1$ the notion of fundamental domain is elusive, but we have still the isomorphism of the metric spaces $\mathbb{B}_{S_0}/\mathrm{Mon}$ and $PU(1,n)/\mathrm{Mon}$; we do not know if the hyperbolic measure is finite.

12.44. Definition. A subgroup Γ of a Lie group G is called **lattice** iff it is (i) discrete, i.e., a neighborhood of $e \in G$ contains only one element of Γ, and (ii) the Haar measure of G/Γ is finite.

12.45. Theorem of Picard, Deligne and Mostow. ([**DM1**]) *If condition* **INT** *holds, then the monodromy group* $\mathrm{Mon}(\mu)$ *forms a lattice in* $PU(1,n)$.

12.46. Example (Condition INT in the classical case). In the case of classical hypergeometric integrals, i.e., with

$$\mu_0 = \gamma - \alpha, \quad \mu_1 = \beta - \gamma + 1, \quad \mu_t = \alpha, \quad \mu_\infty = 1 - \beta,$$

the space Q is isomorphic to $\overline{\mathbb{C}} \setminus \{0, 1, \infty\}$. The monodromy maps correspond to the deformations, when t runs around $0, 1, \infty$. The parameters $\lambda = 1 - \gamma, \mu = \beta - \alpha, \nu = \gamma - \alpha - \beta$, defining the angles of the principal triangle, are equal to

$$\lambda = 1 - \mu_0 - \mu_t, \quad \mu = 1 - \mu_\infty - \mu_t, \quad \nu = 1 - \mu_1 - \mu_t.$$

In the Schwarz Theorem 12.9 we also assumed that $0 \le \lambda, \mu, \nu < 1$ (see the restriction (12.2) in Proposition 12.7).
The first condition from **INT** implies that $0 < \alpha, \beta < 1$, $\gamma - 1 < \alpha, \beta < \gamma$. The second condition from **INT** means that $1/\lambda, 1/\mu, 1/\nu \in \mathbb{Z} \cup \{\infty\}$. It is compatible with the condition $0 \le \lambda, \mu, \nu < 1$.
Next, we see that $\lambda + \mu + \nu = 1 - 2\alpha < 1$, which shows that we are always in the hyperbolic case. The hyperbolicity is forced by the compatibility of the conditions from Schwarz Theorem 12.9 and of **INT**.

12.47. The scheme of the proof of Theorem 12.45. The discreetness of the monodromy group follows from the geometry of the Schwarz map. The finite Haar measure of $PU(1,n)/\mathrm{Mon}$ is obtained from estimates for the hyperbolic measure in the ball $\mathbb{B}_{S_0}$.
Firstly one shows that the map W is a local diffeomorphism, $\det W_* \neq 0$. (In 12.50 below we give a proof of this in a more general situation.) This means that the universal development $\widehat{W} : \widehat{Q} \longmapsto \mathbb{B}_{S_0}$, corresponding to the universal

covering $\widehat{Q}$ (of Q), is a covering (*étale*) onto the image. One takes the covering $\widetilde{Q}$, equal to the quotient of the universal covering $\widehat{Q}$ by the kernel of the monodromy homomorphism $\pi_1(Q) \longmapsto PU(1,n)$. The corresponding map $\widetilde{W} : \widetilde{Q} \longmapsto B$ is an isomorphism onto the image.
The residual subset $\mathbb{B}_{S_0} \setminus \operatorname{Im} \widetilde{W}$ corresponds to points in a compactification of $Q = (\overline{\mathbb{C}})^N \setminus \Delta)/PGL(2)$.
We complete partially Q by adding to it:

- the diagonals $t_i = t_j$, whenever $\mu_i + \mu_j < 1$ (in the classical case it corresponds to adding vertices to the fundamental quadrangle);
- and (generally) all the multiple diagonals $t_{i_1} = \ldots = t_{i_k}$ with $\mu_{i_1} + \ldots + \mu_{i_k} < 1$.

In this way we obtain the so called **stable compactification** Q_{st}. One defines also the ramified covering $\widetilde{Q}_{\mathrm{st}}$ of Q_{st} as the analogous completion of $\widetilde{Q}$. Under the **INT** assumption, this covering is finite-to-one.
It turns out that, if there are no situations with $\sum_{t \in S_1} \mu_t = 1$, then Q_{st} is a compact manifold, and the covering $\widetilde{Q}_{\mathrm{st}}$ is diffeomorphic to a ball. The map $\widetilde{W}$ is a diffeomorphism between this ball and $\mathbb{B}_{S_0}$. Moreover, there is a compact subset $K \subset \widetilde{Q}_{\mathrm{st}}$, which is projected onto the whole Q_{st}. This gives discreteness of the action of the monodromy in a ball $\mathbb{B}$ (the set $\widetilde{W}(K)$ contains only finite part of any orbit of the group Mon) and the finitude of the hyperbolic measure of $\mathbb{B}/\mathrm{Mon} = \widetilde{W}(K)/\mathrm{Mon}$. In this case the proof is complete.
For example, in the classical case $N = 4$ and strictly positive parameters λ, μ, ν, we obtain $Q_{\mathrm{st}} = \overline{\mathbb{C}}$ and $\widetilde{Q}_{\mathrm{st}}$ forms an infinite ramified covering of $\overline{\mathbb{C}}$, with finite ramification indices.
In the case when some sum $\mu_{i_1} + \ldots + \mu_{i_k} = 1$, one adds to Q_{st} the corresponding diagonals $t_{i_1} = \ldots = t_{i_k}$ and obtains the (compact) space denoted by Q_{sst} (the **semi-stable compactification**). One also has the corresponding ramified covering $\widetilde{Q}_{\mathrm{sst}} \longmapsto Q_{\mathrm{sst}}$. However, here the ramification index above points from $Q_{\mathrm{sst}} \setminus Q_{\mathrm{st}}$ is infinite (because the angle of the corresponding monodromy rotation is zero). The latter points are called the *cusp points*. They are sent by $\widetilde{W}$ to the absolute. Thus $\widetilde{W}$ realizes an isomorphism between $\widetilde{Q}_{\mathrm{sst}}$ and the partial compactification of the ball $\mathbb{B}$, by adding to it the cusp points. Next, it remains to estimate the volume of $\mathbb{B}/\mathrm{Mon}$, which is reduced to estimation of the volume of A/Mon, where $A \subset \mathbb{B}$ is a compact neighborhood of a cusp point. In the case $n = 1$ the local singularity of the fundamental domain is indeed of the cusp type and has finite volume; the analogous property is proved in the general case.
For example, in the case of the Schwarz modular function with $\lambda = \mu = \nu = 0$, Q_{sst} is a topological 2-dimensional sphere, but, from the analytic point of view, the points $0, 1, \infty$ must be treated as singular (cusps).
The space Q_{sst} should be treated not as a manifold in the usual sense, but as an orbifold.

The above idea of adjoining the stable and semi-stable points was taken from the Mumford's construction of analytic quotient spaces of actions of algebraic groups on algebraic varieties (see [**Mum1**]). In our situation we have the action of $PGL(2)$ on $\overline{\mathbb{C}}^N$. □

12.48. The generalizations of the Schwarz's euclidean and elliptic cases. The above theory was adapted to the situation when the monodromy group is a subgroup of the group of hyperbolic isometries of n-dimensional complex ball $\mathbb{B}$. But the group $PU(1,n)$ contains the subgroup $U(n)$ of rotations of the absolute in $\mathbb{B}$ (we prolong the rotations of $\partial\mathbb{B}$ to rotations of $\mathbb{B}$), and the subgroup of euclidean motions of $\mathbb{C}^n$ (when we replace the ball by the half-space $\mathrm{Im}(x_1) > 0$ and consider the motions preserving the coordinate x_1). It turns out that these reductions can be realized by means of hypergeometric monodromy.

(a) *The euclidean case.* One assumes that $S = S_1 \cup \{\infty\}$, $0 \in S_1$ and $\sum_{s\in S_1} \mu_s = 1$, $\mu_s > 0$. Thus $\mu_\infty = 1$ and the form $\omega = \prod(\tau - t)^{-\mu_t} \cdot e \cdot \mathrm{d}\tau$ has first order pole at ∞ (with nonzero residuum).
The (co)homology theory of local systems can be extended to the case when some μ_s are negative integers. In our case one has nontrivial cycle $\delta_\infty = (\gamma, s_\gamma)$ represented by a closed loop γ around ∞ (with unique section above it) and generating a nonzero class in $H_1(\overline{\mathbb{C}}\backslash S, \mathbf{L}^\vee)$. The form ω represents nonzero class in $H^1(\overline{\mathbb{C}}\backslash S, \mathbf{L})$. In particular, we have $\langle \delta_\infty, \omega_S\rangle = 1$ independently of S_1 (we assume $e(\infty) = 1$). The latter property means that the image of the generalized Schwarz map W lies in the hypersurface $E : x_{N-2} = 1$ (for a coordinate system with $x_{N-2} = \langle \delta_\infty, \cdot\rangle$) . The corresponding space Q_{Eucl} is the quotient of the space of S's as above by the action of $\mathbb{C}^* \subset PSL(2,\mathbb{C})$ (of automorphisms which fix 0 and ∞). The map W is homogeneous of degree 0 with respect to this action. Thus, we get the multivalued map $Q_{\mathrm{Eucl}} \longmapsto E$. The affine space E can be identified with the linear space $x_{N-2} = 0$ or with the space $H^1(\overline{\mathbb{C}} \setminus S_1, \mathbf{L}_1)$, where $\mathbf{L}_1$ is the local system with the monodromy defined by μ_t, $t \in S_1$. The calculation of the signature of the corresponding Hermitian form, analogous to the one performed in 12.39, shows that the Hermitian form is negatively defined in $H^1(\overline{\mathbb{C}} \setminus S_1, \mathbf{L}_1)$.
It turns out that, when **INT** condition restricted to S_1 holds, then the monodromy group is a subgroup of the group of isometries of the euclidean space E. It contains a subgroup of translations of finite index.
We should remark here that **INT** conditions restricted to S_1 are not compatible with the conditions from Schwarz theorem 12.9. Indeed, if $\lambda + \mu + \nu = 1$, then $\alpha = \mu_t = 0$. Thus the form ω is holomorphic at t. (Recall that in the basis $F(0,\beta;\gamma;t) = 1$, $F_{0,t} = \int_0^t \tau^{-\gamma}(\tau-1)^{\gamma-\beta-1}\mathrm{d}\tau$, we get the Schwarz map in the form of Schwarz–Christoffel integral.) According to Proposition 12.7, the case with $\alpha = 0$ has the same monodromy as the case with $\alpha = \mu_t = 1$, which fits to the above.

(b) *The elliptic case.* Here we have $S = S_1 \cup \{\infty\}$, $0 \in S_1$, and $\sum_{s\in S_1} \mu_s < 1$. In the calculation of the hermitian form we must replace $\mu_\infty > 1$ by $\mu_\infty = 1$, which

leads to the signature $(0, N-2)$ (see 12.39). The map W is homogeneous of degree $1-\sum_{S_1}\mu_s$ with respect to the $\mathbb{C}^*$-action and defines a map from the corresponding quotient space $Q_{\rm ell}$ to PH^1. The monodromy group forms a subgroup of the compact group $U(N-2)$. By its discreetness, following from **INT** condition in S_1, the group Mon is finite.
Here also the condition from the Schwarz theorem 12.9 $\lambda+\mu+\nu=1-2\mu_t>1$ is not compatible with **INT** ($\mu_t>0$). One should again use Proposition 12.7. (In [**Mos2**] it is said that the euclidean case is equivalent to $\mu_0+\mu_1+\mu_t<1$; it is not exactly so.)

12.49. Other results in the Deligne–Mostow theory. As G. Mostow writes in [**Mos2**], one of the aims to undertake investigations of generalized hypergeometric functions was to look for new examples of *non-arithmetic lattices in Lie groups.*
The simplest example of arithmetic subgroup Γ in a real algebraic Lie group G is $GL(n,\mathbb{Z})\subset GL(n,\mathbb{R})$. A lattice $\Gamma\subset GL(n,\mathbb{R})$ is *arithmetic* iff it is commensurable with $GL(n,\mathbb{Z})$ (see [**Rag**] and [**Ser2**]).
If G is an algebraic group, defined over $\mathbb{Q}$, then a lattice $\Gamma\subset G_{\mathbb{Q}}$ is **arithmetic** iff for any embedding $G\subset GL(n)$ the subgroups Γ and $G_{\mathbb{Q}}\cap GL(n,\mathbb{Z})$ are commensurable. The case of general real Lie group G can be reduced to this one.
Two subgroups $A,B\subset G$ are **commensurable** iff $A\cap B$ has finite index in A as well as in B.
G. Margulis proved that, when G has $\mathbb{R}$-rank $k>1$ (contains a real algebraic torus $(\mathbb{R}^*)^k$, $k>1$), then any its irreducible lattice is arithmetic (see Appendix to the Russian translation of [**Rag**]).
There appeared the problem of finding examples of non-arithmetic lattices, in particular in $PU(1,n)$. G. Mostow and P. Deligne found such examples among the projective monodromy groups associated with hypergeometric integrals satisfying the integrality condition.
In [**DM1**] they proved that, if the **INT** holds, then $\mathrm{Mon}(\mu)$ is arithmetic iff:

For any integer $1<a<d-1$, relatively prime to the least common denominator d of μ_t's, one has $\sum\{a\mu_t\}=1$ or $=N-1$.

(Here $\{\cdot\}$ denoted the fractional part.) For example, the lattice $\mathrm{Mon}(\frac{3}{4},\frac{2}{4},\frac{1}{4},\frac{1}{4},\frac{1}{4})\subset PU(1,2)$ is arithmetic, but $\mathrm{Mon}(\frac{6}{12},\frac{5}{12},\frac{5}{12},\frac{4}{12},\frac{4}{12})$ is non-arithmetic.
In [**DM1**] there is a list of hypergeometric monodromy groups for low N's and d's.

Mostow in [**Mos1**] weakened the **INT** condition. Namely, if μ_t take the same value for t from a subset $S_1\subset S$, ($\mu_t=\mu_s$, $s,t\in S_1$), then we can define the new configuration space $Q'=Q/\Pi$, where $\Pi=S(S_1)$ denotes the group of permutations of S_1, with the fundamental group $\pi_1(Q')$ identified with a subgroup of $\pi_1(Q)$. Mostow proved that:

If $1<(1-\mu_t-\mu_s)^{-1}\in\frac{1}{2}\mathbb{Z}$ for $s,t\in S_1$ and $1<(1-\mu_s-\mu_t)^{-1}\in\mathbb{Z}$ otherwise, then the image $\mathrm{Mon}_{S(S_1)}(\mu)$ of $\pi_1(Q')$ is a lattice in $PU(1,N)$.

In the book [**DM2**] the idea of replacing Q by the quotient $Q' = Q/\Pi$ by a group Π of symmetries of S, corresponding to partition of S into subsets with equal exponents μ_t, has found further development. Series of commensurabilities among lattices (obtained in this way) were established. For example, $\mathrm{Mon}_{\Pi_1}(a, a, b, b, 2 - a - b)$, $\Pi_1 = \langle (1,2),(3,4) \rangle$ turns out to be conjugate in $PGL(3)$ to $\mathrm{Mon}_{\Pi_2}(1 - b, 1 - a, a + b - \frac{1}{2}, a + b - \frac{1}{2}, 1 - a - b)$, $\Pi_2 = \langle (3,4) \rangle$ (see [**DM2**, Theorem 10.6]). Also the lattices $\mathrm{Mon}_{\Pi}(\frac{1}{2} - \frac{1}{k}, \frac{1}{2} - \frac{1}{k}, \frac{1}{2} - \frac{1}{k}, \frac{1}{6} + \frac{1}{k}, 2(\frac{1}{6} + \frac{1}{k}))$, $\Pi = S(1,2,3)$, are commensurable (modulo conjugation in $PGL(3)$) to $\mathrm{Mon}_{\Pi}(\frac{1}{6}, \frac{1}{6}, \frac{1}{6}, \frac{5}{6} - \frac{1}{l}, \frac{4}{6} + \frac{1}{l})$, $\frac{1}{k} + \frac{1}{l} = \frac{1}{6}$.
Some connections of the geometry of hypergeometric monodromy lattices with other topics in algebraic geometry are given, e.g., with the theory of line arrangements and the ball quotients.

§4 Multiple Hypergeometric Integrals

12.50. Hypergeometric integrals associated with a configuration of hyperplanes. Let $T_j = \{f_j = 0\}$, $j = 1, \ldots, N$, be a configuration of *hyperplanes* in general position in $\mathbb{CP}^n$ and let $\mu_0, \ldots, \mu_N$ be a collection of complex numbers (*exponents*) with $\sum \mu_i = n + 1$. We may (and we will) assume that T_N is the hyperplane at infinity and $\mathbb{C}^n = \mathbb{CP}^n \setminus T_N$.
The *multiple hypergeometric integral* is the integral of the form

$$\int_\delta \prod f_j^{-\mu_j} \cdot \mathrm{d}^n x \tag{12.19}$$

along a cycle δ, which will be specified later.
As in the single hypergeometric integral, we introduce a local system $\mathbf{L}$ above $\mathbb{CP}^n - S$, $S = \bigcup T_j$, with monodromies around T_j defined by means of the multiplicators $\alpha_j = \mathrm{e}^{2\pi \mathrm{i} \mu_j}$. The dual system $\mathbf{L}^\vee$ is defined by means of α_j^{-1}.
The local system $\mathbf{L}$ is a topologically trivial line bundle, equipped with the Gauss–Manin connection, which is flat (zero curvature) and is defined as $\nabla = \mathrm{d} + \sum \mu_j \mathrm{d}(\ln f_j)$.
The homology and cohomology theories with coefficients in local systems, described in 12.36, are valid in our situation. In particular, the cycle δ from Eq. (12.19) is an elements from the group $H_n^{\mathrm{lf}} = H_n^{\mathrm{lf}}(\mathbb{CP}^n - S, \mathbf{L}^\vee)$ (locally finite cycle) or (better) from $H_n = H_n(\mathbb{CP}^n \setminus S, \mathbf{L}^\vee)$. δ is a combination (with zero boundary) of n-dimensional cells (σ, s_σ) with horizontal sections above them. The integral (12.19) is understood as $\langle \delta, \omega \rangle$, where

$$\omega = \prod f_j^{-\mu_j} \cdot e \cdot \mathrm{d}^n x,$$

with a horizontal (with respect to ∇) section e of the bundle $\mathbf{L}$; it represents an element of $H^n = H^n(\mathbb{CP}^n \setminus S, \mathbf{L})$ $(= (H_n)^*)$.

Assume that the configuration S is *real*, i.e., that the affine functions f_i have real coefficients.
Then $\mathbb{R}^n \setminus S$ is an union of n-dimensional convex domains. Each such domain, together with a section of $\mathbf{L}^\vee$ above it, defines a locally finite cycle from H_n^{lf}.

12.51. Theorem of Varchenko (Basis). ([**Var5**) *Assume that the system* $\mu=(\mu_1,\dots,\mu_N)$ *is nonresonant, i.e., none of* μ_j*'s is integer.*
Then the natural homomorphism $H_n \longmapsto H_n^{\mathrm{lf}}$ *is an isomorphism and the target group is generated by the cycles* δ_j *represented by bounded components of* $\mathbb{R}^n \setminus S$*; in particular* $\dim H_n^{\mathrm{lf}} = \binom{N-1}{n}$*, which equals* $(-1)^n\chi(\mathbb{C}^n \setminus S)$.

It is a generalization of Theorem 12.37. In particular, there is an analogue of the Pochhammer loop construction (i.e., the cycle $C(u,t)$) in the multidimensional case). There are generators $C_j \in H_n$ related with $\delta_j \in H_n^{\mathrm{lf}}$ by the formula

$$C_j = \left(\prod{}'(1-\alpha_i^{-1})\right)\cdot\delta_j,$$

where the product $\prod'$ is taken over the sides T_i in the boundary of δ_j (see [**AVGL**], part II, and [**Ph1**]).

In [**Var5**] there is the following construction of the basis in H^n. Let us change the notations a little bit. Instead of the system of exponents μ_j, we choose a system of exponents ν_j, $j=1,\dots,N$, with $\sum\nu_j=0$ and as before we put $\alpha_j=\mathrm{e}^{2\pi\mathrm{i}\nu_j}$. Moreover, for simplicity, we assume that $\mathrm{Re}\nu_j>0$, $j=1,\dots,N-1$.
For $I=\{i_1,\dots,i_n\}\subset\{1,\dots,N-1\}$ one defines

$$\omega(I)=\prod_{i\notin I} f_i^{-\nu_i}\cdot e\cdot \mathrm{d}f_{i_1}^{-\nu_{i_1}}\wedge\ldots\wedge\mathrm{d}f_{i_n}^{-\nu_{i_n}}.$$

It turns out that:

The system $\omega(I)$, $I\subset\{2,\dots,N-1\}$, *forms a basis in* H^n.

There are other results, e.g., [**AT**], [**Ao**], [**Kan**], devoted to description of bases of (co)homologies of local systems. We do not present them here.

12.52. Generalized Schwarz map (or the period map). Consider the space of configurations of N hyperplanes in general position. It is $Q=\left((\mathbb{CP}^{n*})^N\setminus\Sigma\right)/PGL(n+1)$, where Σ is the discriminant locus, containing singular configurations (with non-normal intersections), and $\mathbb{CP}^{n*}$ is the dual projective space. The space $\bigcup_{S\in Q}(\mathbb{CP}^n\setminus S)\times\{S\}$ is a bundle over Q (with projection π) and above it we have the line local system $\mathbf{L}=\bigcup_S\mathbf{L}_S$ ($\mathbf{L}_S$ denotes the previous $\mathbf{L}$ over $\mathbb{CP}^n\setminus S$). The system of cohomology groups $H^n(\mathbb{CP}^n\setminus S,\mathbf{L}_S)=H_S^n$ defines a local system $\mathbf{K}=R^n_{\pi*}\mathbf{L}$. (We shall also denote it by $\mathcal{H}^n$ together with the bundle $\mathcal{H}_n^{\mathrm{lf}}=\bigcup H_n^{\mathrm{lf}}(\mathbb{CP}^n\setminus S,\mathbf{L}_S^\vee)$.) $\mathbf{K}$ is a topologically trivial bundle with flat

Gauss–Manin connection. The basis of its (multivalued) horizontal sections defines a multivalued trivialization $\mathbf{K} \longmapsto Q \times H^n_{S_0}$, defined by means of mappings $\Psi(S) : H^n_S \longmapsto H^N_{S_0} = \mathbb{C}^M$. The distinguished n-form $\omega = \omega(1,\ldots,n)$ defines a section of $\mathbf{K}$. The **generalized Schwarz map** (or the **period map**) $W : Q \longmapsto \mathbb{CP}^{M-1}$is defined as usual

$$W = P \circ \Psi \circ \omega,$$

where P denotes the projectivization.

12.53. Torelli theorem for multidimensional hypergeometric integrals. ([**Var5**]) *The map W is a local immersion.*

Proof. Let us assume that $f_1 = x_1, \ldots, f_n = x_n$, $f_{n+1} = 1 - x_1 - \ldots - x_n$, $f_i = 1 + \sum_j a_{ij}x_j$, $i \geq n+2$. Thus Q is parameterized by a_{ij} and has dimension $n(N-n-2)$.
Define the *'Hodge' filtration* $0 \subset F^n \subset \ldots \subset F^0 = H^n_S$:

$$F^k = F^k_S = \operatorname{sran}\{\omega(i_1,\ldots,i_k) : i_1,\ldots,i_k \in \{1,\ldots,n\}\}.$$

Therefore $F^n = \mathbb{C}\omega$ and one can show that $\dim F^k - \dim F^{k+1} = \binom{n}{k}\binom{N-n-2}{n-k}$.
The spaces F^k_S define holomorphic subbundles $\mathcal{F}^k \subset \mathcal{H}$.
The Hodge filtration is not invariant with respect to the Gauss–Manin connection. However we have the following analogue of the Griffiths transversality theorem 7.52(b)

$$\nabla\mathcal{F}^k = \{\nabla_v s : v \in T_S Q,\ s \text{ is a germ of section of } \mathcal{F}^k\} = \mathcal{F}^{k-1}.$$

Consider, for example, the case $k=n$ with the section $\omega(1,\ldots,n)=e\prod_{i=n+1}^{N-1} f_i^{-\nu_i} \cdot \mathrm{d}x_1^{-\nu_1}\ldots\mathrm{d}x_n^{-\nu_n}$ of $\mathcal{F}^n$. Then simple calculations give

$$\frac{\mathrm{d}\omega}{\mathrm{d}a_{ij}} = \frac{-\nu_i}{a_{ij}}(-1)^{j-1}\omega(1,\ldots,\check{j},\ldots,n,i).$$

In this way we obtain all sections of $\mathcal{F}^{n-1}$. The other identities are shown in analogous way.
Now, because $\dim F^{n-1} = n(N-n-2)+1 = \dim Q + 1$, we obtain that the derivative $W_*(S)$ is a linear embedding.
Finally we note that this Hodge filtration for $n=1$ coincides with the Deligne–Mostow filtration $F^1 = F^n \subset F^0$ defined by means of the decomposition $H^1 = H^{1,0} \oplus H^{0,1}$ in 12.39. In that case one obtains a local isomorphism of W (see 12.47). □

Thus, we have the analogue of the first point of the Picard–Deligne–Mostow theorem. This is also an analogue of the Torelli theorem, which says that an algebraic curve can be reconstructed from its periods (see Theorem 7.48).
Before passing to the description of the monodromy group, we describe the analogues of the hypergeometric equations satisfied by the integrals $\langle\delta,\omega\rangle$.

12.54. Generalization of the differential hypergeometric equation. Here we follow the paper of I. Gelfand, M. Karpanov and A. Zelevinsky [**GKZ**].
Consider the integrals of the form

$$F = \int_\delta x^{-\mu} f^{-\nu} \mathrm{d}^n x,$$

where we use the multi-index notations: $\mu = (\mu_1, \ldots, \mu_n)$, $\nu = (\nu_{n+1}, \ldots, \nu_{N-1})$, $f_i = a_{i0} + \sum a_{ij} x_j$, $i = n+1, \ldots, N-1$. F is a function of $(a_{ij}) \in (\mathbb{C}^{n+1})^{N-n-1}$.
We shall derive the differential equations for F.
Let $\Gamma \subset \mathbb{Z}^{N-1} = \mathbb{Z}^n \times \mathbb{Z}^{N-n-1}$ be defined as

$$\bigcup_{j=n+1}^{N-1} \{0 \times e_j \,,\, e_1 \times e_j \,, \ldots ,\, e_n \times e_j\}$$

(e_j are the basic vectors). We have the following properties:

(i) Γ generates $\mathbb{Z}^{N-1}$,

(ii) $\Gamma \subset \phi^{-1}(1)$ for some homomorphism $\phi : \mathbb{Z}^{N-1} \longmapsto \mathbb{Z}$ (e.g., $\phi = x_{n+1} + \ldots + x_{N-1}$).

We can treat the system of parameters a_{ij} as an element of $\mathbb{C}^\Gamma$; we denote it as $a = (a_\gamma, \gamma \in \Gamma)$.
Denote by $L = L(\Gamma) \subset \mathbb{Z}^\Gamma$ the lattice of relations among elements from Γ. L consists of the systems $m = (m_\gamma)$ $m_\gamma \in \mathbb{Z}$, such that

$$\sum_{\gamma \in \Gamma} m_\gamma \cdot \gamma = 0. \tag{12.20}$$

We underline also the relations

$$\sum_{\gamma \in \Gamma} m_\gamma = 0, \tag{12.21}$$

following from application of the homomorphism ϕ from (ii) to (12.20).
For any $m \in L$ we define the differential operator

$$\Delta_m = \prod_{m_\gamma > 0} \left(\frac{\partial}{\partial a_\gamma}\right)^{m_\gamma} - \prod_{m_\gamma < 0} \left(\frac{\partial}{\partial a_\gamma}\right)^{-m_\gamma}.$$

Define also the Euler operators

$$\boldsymbol{E}_i = \sum_\gamma \gamma_i a_\gamma \frac{\partial}{\partial a_\gamma},$$

where γ_i is the ith coordinate of γ in $\mathbb{Z}^{N-1}$.

Theorem of Gelfand, Karpanov and Zelevinsky.

(a) *The hypergeometric integrals satisfy the system of equations*

$$\Delta_m F = 0 \text{ for } m \in L, \quad E_i F = \mathrm{d}_i F, \tag{12.22}$$

where $(d_1, \ldots, d_{N-1}) = (\mu_1 - 1, \ldots, \mu_n - 1, -\nu_{n+1}, \ldots, -\nu_{N-1})$.

(b) *The number of independent solutions of the hypergeometric system* (12.22) *is equal to the* $(N-1)$*-dimensional volume of the convex hull of* Γ *in* $\phi^{-1}(1)$ *(where the volume of elementary simplex is* 1*).*

(c) *The natural mapping* $\delta \longmapsto \langle \delta, \omega \rangle$ *from the local system* $\mathcal{H}_n = (H_n(\mathbb{CP}^n \setminus S, \mathbf{L}^\vee)$ *over* Q *to the system* Hyp *of solutions of the system* (12.22) *is an isomorphism.*

(d) *The monodromy representation of* $\pi_1(Q)$ *in* Hyp *is irreducible.*

We will not give the proof of this theorem. It involves advanced techniques from sheaf theory ($\mathcal{D}$-modules, perverse and constructible sheaves, etc.).
We note only that hypergeometric equations (12.22) are immediate. Indeed, because $\Gamma = \bigcup \Gamma_j \times \{e_j\}$, each $m \in L(\Gamma)$ equals to $(m(1), \ldots, m(N-n-1))$, where each $m(j) \in L(\Gamma_j)$; i.e., it satisfies the relations analogous to (12.20) and (12.21): $\sum_{\delta \in \Gamma_j} m(j)_\delta \delta = 0$, $\sum_{\Gamma_j} m(j)_\delta = 0$. Denote

$$e(j) = \sum_{m(j)_\delta > 0} m(j)_\delta, \quad \eta(j) = \sum_{m(j)_\delta > 0} m(j)_\delta \cdot \delta.$$

We have

$$\begin{aligned} &\prod_{m_\gamma > 0} \left(\frac{\partial}{\partial a_\gamma} \right)^{m_\gamma} x^{-\mu} f^{-\nu} \\ &= x^{-\mu} f^{-\nu} \prod_j (-\nu_j)(-\nu_j - 1) \ldots (-\nu_j - e(j) + 1) f_j^{-e(j)} x^{\eta(j)} \\ &= \prod_{m_\gamma < 0} \left(\frac{\partial}{\partial a_\gamma} \right)^{-m_\gamma} x^{-\mu} f^{-\nu}. \end{aligned}$$

The equations $\boldsymbol{E}_j F = d_j F$ follow from the homogeneity of the integrals with respect to dilations $x_i \longmapsto \lambda_i x_i$.

In [**GKZ**] the authors consider more general situation, when f_j are general Laurent polynomials; the hypergeometric system and the main theorem remain the same. Finally, note that F. Beukers in [**Beuk**] has described the monodromy group of the above hypergeometric system.

12.55. Local systems associated with quantum deformations of Kac–Moody algebras and Verma moduli. We follow the A. Varchenko's book [**Var6**].

The configurations S of hyperplanes in $\mathbb{C}^n$ are the following:

$$x_i = x_j, \quad i, j = 1, \dots, n; \qquad x_i = z_k, \quad i = 1, \dots, n, \quad k = 1, \dots, m.$$

They are not in general position. The corresponding space of parameters is $Q = \mathbb{C}^m \setminus \Delta$.
The exponents (weights) of the local system $\mathbf{L}$ in $\mathbb{C}^n \setminus S$ are chosen in the form

$$\mu_{ij} = -\frac{(\alpha_i, \alpha_j)}{\kappa}, \qquad \nu_{ik} = \frac{(\alpha_i, \Lambda_k)}{\kappa},$$

i.e., associated with the n-form $\prod(x_i - x_j)^{-\mu_{ij}} \prod(x_i - z_k)^{-\nu_{ik}} \mathrm{d}^n x$.
Here $\alpha_1, \dots, \alpha_n$ are independent elements of $\mathfrak{h}^*$, dual space to a finite dimensional vector space $\mathfrak{h}$, equipped with a non-degenerate symmetric bilinear form $(\cdot, \cdot)$; then $\mathfrak{h}^*$ has also the induced bilinear form $(\cdot, \cdot)$. Λ_k are elements of $\mathfrak{h}$ and κ is a complex number. Define the vectors $h_i \in \mathfrak{h}$ by the formula $(h_i, \cdot) = \alpha_i$.
When one puts $a_{ij} = 2(\alpha_i, \alpha_j)/(\alpha_j, \alpha_j)$ and assumes that:

$$\text{(i) } a_{ij} \in \mathbb{Z}; \qquad \text{(ii) } a_{ii} = 2,\ a_{ij} \leq 0 \text{ for } i \neq j; \qquad \text{(iii) } a_{ij} = 0 \Rightarrow a_{ji} = 0,$$

then the matrix $A = (a_{ij})$ is a *generalized Cartan matrix*.
One can construct from the generalized Cartan matrix the *Kac–Moody algebra* $\mathfrak{g}$, generated by $\mathfrak{h}, e_i, f_i$ with the relations

$$[h, e_i] = \alpha_i(h) e_i, \quad [h, f_i] = -\alpha_i(h) f_i, \quad [e_i, f_j] = \delta_{ij} h_i,$$

$$(\text{ad } e_i)^{1-a_{ij}} e_j = (\text{ad } f_i)^{1-a_{ij}} f_j = 0.$$

When $A > 0$ (positively definite), then one obtains a finite dimensional Lie algebra $\mathfrak{g}$ with $\mathfrak{h}$ as its Cartan subalgebra, with α_i as simple roots and with finite Weyl group generated by the reflections $s_i : \beta \longmapsto \beta - [2(\beta, \alpha_i)/(\alpha_i, \alpha_i)]\alpha_i$ (see Section 3 in Chapter 4).
In the general case the obtained algebra is infinite dimensional with infinite Weyl group. The usual construction of the Kac–Moody algebra starts from a generalized Cartan matrix which is symmetrizable (i.e., after multiplication by a diagonal matrix becomes symmetric). Moreover, some supplementary extension should be performed in the case $\det A = 0$ (see [**Kac**]).
Any group G (e.g., Lie group) can be investigated from the point of view of its group algebra $\mathbb{C}(G)$, i.e., algebra of functions on G with values in $\mathbb{C}$. It is commutative algebra equipped with the *comultiplication* $\Delta : \mathbb{C}(G) \longmapsto \mathbb{C}(G) \otimes \mathbb{C}(G)$ (dual to the multiplication), with the *counit* $\epsilon : \mathbb{C}(G) \longmapsto \mathbb{C}$ (dual to the embedding of the trivial group) and with the *antipode* $a : \mathbb{C}(G) \longmapsto \mathbb{C}(G)$ (dual to the inversion). Thus $\mathbb{C}(G)$ becomes the commutative *Hopf algebra.* The analogue of $\mathbb{C}(G)$ for the Lie algebra $\mathfrak{g}$ is its *universal algebra* $U\mathfrak{g}$, consisting of tensor powers of $\mathfrak{g}$, modulo the ideal generated by $u \otimes v - v \otimes u - [u, v]$.
The *quantum group* is a non-commutative Hopf algebra.

With the Kac–Moody algebra, the following *Drinfeld–Jimbo quantum group* $U_q^{\mathrm{DJ}}\mathfrak{g}$ (see [**Dr**]) is associated. It is a universal algebra generated by $\mathfrak{h}, 1, e_i, f_i$ with the relations

$$[h, e_i] = \alpha_i(h)e_i, \quad [h, f_i] = -\alpha_i(h)f_i, \quad [e_i, f_j] = \delta_{ij}(h_i)_q,$$

$$\sum_{k=0}^{1-a_{ij}} \binom{1-a_{ij}}{k}_q (-1)^k e_i e_j e_i^{1-a_{ij}-k} = \sum_{k=0}^{1-a_{ij}} \binom{1-a_{ij}}{k}_q (-1)^k f_i f_j f_i^{1-a_{ij}-k} = 0, \tag{12.23}$$

where $(a)_q = q^{a/2} - q^{-a/2}$, $\binom{a}{b}_q = [(a)_q(a-1)_q \dots (a-b+1)_q)]/[(1)_q \dots (b)_q]$. Here $q = \mathrm{e}^{2\pi\mathrm{i}/\kappa}$, $q^a = \mathrm{e}^{2\pi\mathrm{i}a/\kappa}$, where κ is the parameter appearing in the definition of the exponents μ_{ij} of the local system.

In [**Var6**] one defines the algebra $U_q\mathfrak{g}$, as above, but without the *Serre relations* (12.23). It is a Hopf algebra with the comultiplication, counit and antipode: $\Delta h = h \otimes 1 + 1 \otimes h$, $\Delta e_i = e_i \otimes q^{h_i/4} + q^{-h_i/4} \otimes e_i$, $\Delta f_i = f_i \otimes q^{h_i/4} + q^{-h_i/4} \otimes f_i$; $\epsilon h = \epsilon e_i = \epsilon f_i = 0$; $ah = -h$, $ae_i = -q^{(\alpha_i,\alpha_i)}e_i$, $af_i = -q^{-(\alpha_i,\alpha_i)}f_i$.

For $\Lambda \in \mathfrak{h}^*$ one defines the *Verma module* $M(\Lambda)$ over $U_q\mathfrak{g}$ (with the highest weight Λ), as the module generated by one vector v subject to the relations

$$hv = \Lambda(h)v; \quad fv = 0,$$

for any $h \in \mathfrak{h}$ and any f from the subalgebra generated by 1 and f_j's. The Verma moduli are infinite dimensional (with the generators $v, e_iv, e_je_jv, \dots$) and are uniquely defined from the relations defining $U_q\mathfrak{g}$.

One can conclude the above as follows:

The weights μ_{ij}, ν_{ik} are associated with the quantum deformation $U_q\mathfrak{g}$ of the Kac–Moody algebra $\mathfrak{g}$ and with a collection of the Verma moduli $M(\Lambda_1), \dots, M(\Lambda_m)$ over $U_q\mathfrak{g}$.

12.56. Switched local system on the space of configurations and representation of its monodromy via R-matrices. The homology groups $H_n(\mathbb{C}^n - S, \mathbf{L}_S^\vee)$ of the local system $\mathbf{L} = \bigcup_S \mathbf{L}_S$ on the space of configurations Q organize themselves to a local system $\mathcal{H}_n$ on Q. We switch this system by the line local system $\mathbf{M}$ on Q with the weights $\eta_{kl} = -(\Lambda_k, \Lambda_l)/\kappa$. We investigate the bundle $\mathcal{H}_n \otimes \mathbf{M}$, which corresponds to the situation when one integrates the form $\prod(z_k - z_l)^{-\eta_{kl}} \prod(x_i - x_j)^{-\mu_{ij}} \prod(x_i - z_k)^{-\nu_{ik}} \mathrm{d}^n x$.

The fundamental group of $Q = \mathbb{C}^m \setminus \Delta$ is the colored braid group of m strands, $\pi_1(Q, S_0) = \widehat{B}(m)$. It acts on the distinguished fiber $\widetilde{\mathcal{H}}_{S_0} = H_n(\mathbb{C}^n \setminus S_0, \mathbf{L}_{S_0}^\vee) \otimes \mathbf{M}_{S_0}$.

The homology groups $(\mathcal{H}_n \otimes \mathbf{M})_{S_0}$ are calculated in [**Var6**] using certain (rather complicated) chain complex. The action of the monodromy group is described in terms of the action of the colored braid group $\widehat{B}(m)$ on this complex.

(In fact, there are many more complexes in [**Var6**]. Moreover, it is not said that one deals with action of the colored braid group $\widehat{B}(m)$; it is said only about action of

the whole braid group $B(m)$. But the group $\widehat{B}(m)$ is more suitable here, compare 12.41.)
It turns out that the chain complex leading to $H_n(\mathbb{C}^n\setminus S)$ is quasi-isomorphic with so-called *Hochschild complex*; recall that quasi-isomorphism induces isomorphism of the homologies. The usual Hochschild complex of an A-module M consists of the chain moduli $A^{\otimes j}\otimes M$ and the differentials $a_j\otimes\ldots\otimes a_1\otimes m\longmapsto -a_j\otimes\ldots\otimes a_2\otimes a_1 m+\sum(-1)^i a_j\otimes\ldots\otimes a_i a_{i-1}\otimes\ldots\otimes m$ and its homology groups are called the Hochschild homology groups.
The (algebraic) complex considered in [**Var6**] has the j-chains of the form $b_1\otimes\ldots\otimes b_{m+j}$, where j of the elements b_i's belong to the quantum group $A=U_q\mathfrak{g}$ and the other are elements of the Verma modules $M(\Lambda_k)$. We omit the definition of the differential.
The action of the loop σ_p (where the points z_p and z_{p+1} turn around themselves) on the Hochschild complex is described by means of the so called *R-matrix*, which acts on the parts $M(\Lambda_p)$ and $M(\Lambda_{p+1})$ of the chain moduli.
(In the simplest case $U_q sl(2)$, generated by h,e,f, and two moduli V_1,V_2, we get the operator from $V_1\otimes V_2$ to $V_2\otimes V_1$, defined as $P\circ R$, where P transposes the factors and $R\in U_q\mathfrak{g}\otimes U_q\mathfrak{g}$ is given by

$$R=q^{h\otimes h/4}\sum_{k=0}^{\infty}q^{-k(k+1)/4}\frac{(1)_q^k}{(1)_q\ldots(k)_q}q^{kh/4}e^k\otimes q^{-kh/4}f^k.)$$

In the general situation one has $R=q^{\Omega_0/2}\sum_{\boldsymbol{k}}q^{h_{\boldsymbol{k}}\otimes 1-1\otimes h_{\boldsymbol{k}}-D_{\boldsymbol{k}}/4}\Omega_{\boldsymbol{k}}$, where the sum runs over $\boldsymbol{k}=(k_1,\ldots,k_n)$, Ω_0 represents the bilinear form $(\,,\,)$, $h_{\boldsymbol{k}}=\sum k_i h_i$, $D_{\boldsymbol{k}}$ is a complex number and $\Omega_{\boldsymbol{k}}$ are certain operators in $M(\Lambda_p)\otimes M(\Lambda_{p+1})$. The fact, that the action of PR is compatible with the relation in the braid group, is a consequence of the *Yang–Baxter equations*

$$R_pR_{p+1}R_p=R_{p+1}R_pR_{p+1},$$

where R_s acts on the s and $s+1$ factors in $M(\Lambda_1)\otimes\ldots\otimes M(\Lambda_m)$.
The rough summarization of all this is following:

The action of the fundamental group $\pi_1(Q,S_0)$ on the fiber of the local system $H_n\otimes\mathbf{M}$ is determined by the R-representation of the colored braid group $\widehat{B}(m)$ in the Hochschild homology groups of the Verma moduli $M(\Lambda_k)$ over the quantum groups $U_q\mathfrak{g}$.

For example, in the classical hypergeometric case ($n=1,m=3$) associated with the form $\omega=\prod(z_k-z_l)^{\Lambda_k\Lambda_l/\kappa}\prod(\tau-z_k)^{-\Lambda_k/\kappa}d\tau$, we have $\mathfrak{g}=sl(2)$, $H_1(\mathbb{C}-\{z_1,z_2,z_3\},\mathbf{L}^\vee)\approx\{x\in M(\Lambda_1)\otimes M(\Lambda_2)\otimes M(\Lambda_3):ex=0,hx=(\Lambda_1+\Lambda_2+\Lambda_3-2)x\}$, and the action of the colored braid group $\widehat{B}(3)$ on the latter module is given by means of the operators R_i.

12.57. Khizhnik–Zamolodchikov equations. The classical Riemann equation from 12.2 (whose particular case is the hypergeometric equation) is equivalent to the following system of first order equations. Put

$$\omega_j = \omega/(\tau - z_j), \quad I_j = \int_\gamma \omega_j,$$

where γ is a cycle in $H_1(\mathbb{C} \setminus S, \mathbf{L}^\vee)$, (e.g., the Pochhammer cycle $C(z_1, z_2)$). We have $\Lambda_1 I_1 + \Lambda_2 I_2 + \Lambda_3 I_3 = 0$.

One can check that the vector function $I = (I_1, I_2, I_3)$ satisfies the following **Khizhnik–Zamolodchikov system:**

$$\frac{\partial I}{\partial z_i} = \sum_{j \neq i} \frac{\Omega_{ij}}{z_i - z_j} I, \quad i = 1, 2, 3,$$

where Ω_{ij} are constant matrices.

It turns out [**Var6**] that, also in the general case, one can choose a basis of n-forms in $\mathcal{H}^n \otimes \mathbf{M}^\vee$ such that their integrals $I = (I_1, \ldots, I_M)$ along cycles from $\mathcal{H}_n$ are functions which satisfy another Khizhnik–Zamolodchikov system, of the same form as the above one.

The Khizhnik–Zamolodchikov equations were firstly discovered in physics; they are obeyed by correlation functions in the Wess–Zumino–Witten model of conformal quantum field theory (see [**KhZ**]). They began to play an important role in the knot theory (see [**Kon**]).

12.58. The twisted Picard–Lefschetz formula. Let $F = F(x, \lambda) = F_\lambda(x) : (\mathbb{C}^n \times \mathbb{C}^k, 0) \longmapsto (\mathbb{C}, 0)$ be a mini-versal deformation of a singularity $f = F(\cdot, 0)$ with the multiplicity k. The classical Picard–Lefschetz formula says that the monodromy group (i.e., the image of the action of $\pi_1(\mathbb{C}^k \setminus \Sigma)$ on $H_{n-1}(V_\lambda)$, $V_\lambda = F_\lambda^{-1}(0) \cap \mathbb{B}$, $\mathbb{B}$ – ball), is generated by the reflections $h \longmapsto h + (-1)^{n(n+1)/2} \langle h, \Delta_k \rangle \Delta_k$. Here $\langle \cdot, \cdot \rangle$ is the intersection form and Δ_k form the distinguished basis of vanishing cycles (see Theorem 4.18).

This formula is not invariant with respect to the operation of suspension (or stabilization) of the singularity $F \longmapsto F'(x, y, \lambda) = F(x, \lambda) + y^2$ in $\mathbb{C}^{n+1} \times \mathbb{C}^k$. If Δ_k is the corresponding basis in the suspension, then we have $\langle \Delta_i, \Delta_j \rangle = \text{sign}(j - i)(-1)^n \langle \Delta_i, \Delta_j \rangle$: anti-symmetric form is replaced by symmetric and vice versa.

In [**Givl**] A. Givental proposed the following way to connect the two kinds of monodromy action. Namely, one should consider the homology group of the fiber $E_\lambda = \mathbb{B} \setminus V_\lambda$ with coefficients in the linear local system $\mathbf{L}(q)$, defined by the multiplier q arising after surrounding the hypersurface $F_\lambda(x) = 0$.

Theorem of Givental.

(a) *We have* $H_n(E_\lambda, \mathbf{L}(-1)) \simeq H_n(V'_\lambda, \mathbb{C})$, $V'_\lambda = (F'_\lambda)^{-1}(0)$.

(b) *Let V be the upper-triangular matrix with the numbers $(-1)^{n(n-1)/2}$ on the diagonal and $\langle \Delta_i, \Delta_j \rangle$, $i \leq j$, above. There exist distinguished bases $e_1, \ldots, e_k$*

in $H_{n-1}(E_\lambda, \mathbf{L}(q))$ *and* $\hat{e}_1, \dots, \hat{e}_k$ *in* $H^{\mathrm{lf}}_{n-1}(E_\lambda, \mathbf{L}(q^{-1}))$ *such that the intersection matrix* $\langle e_j, \hat{e}_i \rangle$ *equals to* $qV - (-1)^n V^\top$ *and the monodromy group is generated by the reflections* $M_k : h \longmapsto h + (-1)^{n(n+1)/2} \langle h, \hat{e}_k \rangle e_k$.

The point (a) shows that there is a relation between $H_{n-1}(V_\lambda, \mathbb{C}) \simeq H_n(E_\lambda, \mathbf{L}(1))$ and $H_n(V'_\lambda, \mathbb{C}) \simeq H_n(E_\lambda, \mathbf{L}(-1))$.
The point (b) implies that $(M_k - (-1)^n q)(M_k - 1) = 0$: because $M_k e_k = (-1)^n q e_k$ (calculate) and M_k is identity in the hyperplane orthogonal to $\hat{e}_k$.
In [**GivS**] A. Givental and V. Schechtman applied this result to the case of $\mathbf{A}_k$ singularity. Here the fundamental group of the complement of the bifurcational diagram Σ is equal to the braid group $\pi_1(\mathbb{C}^k \backslash \Sigma) = B(k)$. One obtains the following result.

Theorem of Givental and Schechtman. *The representation of the braid group* $B(k)$ *in* $H_n(E_\lambda, \mathbf{L}(q))$ *is isomorphic to the representation of the braid group in the Hecke algebra* $H_k(\alpha)$, $\alpha = (-1)^{n+1} q$.

The **Hecke algebra** $H_k(\alpha)$ is generated by elements $\tau_1, \dots, \tau_{k-1}$ with the relations

$$\begin{aligned} \tau_i \tau_j &= \tau_j \tau_i, \quad |i - j| > 1, \\ \tau_i \tau_{i+1} \tau_i &= \tau_{i+1} \tau_i \tau_{i+1}, \\ (\tau_i + \alpha)(\tau_i - 1) &= 0. \end{aligned}$$

In the case $\alpha = 1$ we get the relations for the permutation group $S(k)$.

It is worth to say that the first twisted Picard–Lefschetz formula was obtained by F. Pham in [**Ph1**] (see also [**AVLG**], Part II). One integrates monomial forms along the hypersurface $x_1^{m_1} + \dots x_n^{m_n} = t$ in $\mathbb{R}^n$, associated with the Pham singularity. The change $y_j = x_j^{m_j}$ leads to integration of multivalued forms. Thus we have a local system in $\mathbb{C}^{n-1} = \{\sum y_j = t\}$, deprived of the hyperplanes $y_i = 0$.
We do not present this formula.

12.59. Generalized hypergeometric integrals as functions of exponents. It is natural to investigate the hypergeometric integrals $I(\mu) = \int_\delta \prod_1^M f_j^{-\mu_j}$ as functions of the exponents μ_j.
It turns out that $I(\mu)$ is a meromorphic function of μ with poles in hypersurfaces in $\mathbb{C}^M$ corresponding to the so-called resonant exponents. In the case, when the surfaces $f_j = 0$ are affine and in generic position, the resonances correspond to $\mu_j = 1, 2, 3, \dots$. We obtain a series of one-sided sequences of parallel hyperplanes in $\mathbb{C}^M$ defined by means of equations with integer coefficients.
In the general case an analogous statement is true. $I(\mu)$ has poles in a finite system of series of parallel hyperplanes defined by integer equations. The proof uses the resolution of singularities of the configuration of hypersurfaces (see [**BG**] and [**AVGL**]).
This gives some connection with the results from Chapter 5 about asymptotic of integrals along vanishing cycles and of oscillating integrals.

Bibliography

[AC1] N. A'Campo, *Le nombre de Lefschetz d'une monodromie*, Indag. Mat. Proc. Kon. Ned. Acad. Wet. Ser. A **35** (1973), 113–118.

[AC2] N. A'Campo, *La fonction zêta d'une monodromie*, Comment. Math. Helvetici **50** (1975), 233–248.

[AHPT] P. Acosta-Humánez, M. van der Put and J. Top, *Variations for some Painlevé equations*, Sym. Integr. Geom. Meth Appl. (SIGMA) **15** (2019), 088.

[Ahl] L. Ahlfors, "Lectures on Quasi-Conformal Mappings", Van Nostrand–Reinhold, Princeton, 1966.

[AT] A. Alexandrov and S. Tanabé, *Gauss–Manin connections, logarithmic forms and hypergeometric functions*, in "Geometry from the Pacific Rim", Berrick/Loo/Wang (eds.), Walter de Gruyter & Co., Berlin–New York, 1997, pp. 1–21.

[Al] V. Alexeev, "The Abel's Theorem in Problems and Solutions", Nauka, Moscow, 1976 (Russian).

[AB] D. Anosov and A. Bolibruch, "The Riemann–Hilbert Problem", Aspects of Mathematics. Vieweg, Wiesbaden, 1994.

[Ao] K. Aomoto, *On the structure of integrals of power products of linear functions*, Sci. Papers College Gen. Ed. Univ. Tokyo **27** (1977), 49–61.

[Arn1] V. Arnold, "Mathematical Methods of Classical Mechanics", Springer–Verlag, New York, 1989, [Russian: Nauka, Moscow, 1974].

[Arn2] V. Arnold, *Normal forms of functions in neighborhoods of non-degenerate critical points*, Russ. Math. Surveys **29** (1974), 10–50, [Russian: **29** (1974), 11–49].

[Arn3] V. Arnold, *Wave front evolution and equivariant Morse Lemma*, Commun. Pure Appl. Math. **29** (1976), 557–582.

[Arn4] V. Arnold, *Index of a singular point of vector field, Petrovski–Oleinik inequalities and mixed Hodge structures*, Funct. Anal. Appl. **12** (1978), No 1, 1–14 [Russian].

[Arn] V. Arnold, "Geometrical Methods in the Theory of Differential Equations", Springer–Verlag, Berlin–Heidelberg–New York, 1983, [Russian: Nauka, Moscow, 1978].

[Arn] V. Arnold, *Singularities of ray systems*, "Proc. of the Int. Congr. of Math. 1982, Warsaw", v. **1**, PWN, Warszawa, 1983, pp. 27–49.

H. Żołądek, *The Monodromy Group*, Monografie Matematyczne 67,
https://doi.org/10.1007/978-3-031-91270-2

[Arn7] V. Arnold, *The theory of catastrophes*, in: "Bifurcation Theory and Catastrophe Theory", V. Arnold, S. Afraimovich, Yu. Ilyashenko and L. Shilnikov, Springer-Verlag, Berlin, 1999, [Russian: Fundamental Directions **5**, VINITI, Moscow, 1986, pp. 219–282].

[Arn8] V. Arnold, *On teaching of mathematics*, Russ. Math. Surv. **53** (1998), No 1, 229–236; [Russian: **53** (1998), No 1, 229–234].

[Arn9] V. Arnold, *Singularities of ratios and behavior of polynomials at infinity*, Proc. Steklov Inst. Math. **221** (1998), No 2, 40–59; [Russian: **221** (1998), 48–68].

[AI] V. Arnold and Yu. Ilyashenko, *Ordinary differential equations*, in "Ordinary Differential Equations and Smooth Dynamical Systems", Springer–Verlag, Berlin–Heidelberg–New York, 1997, pp. 1–148, [Russian: Fundamental directions **1**, VINITI, Moscow, 1985, pp. 1–146].

[AKN] V. Arnold, V. Kozlov and A. Neishtadt, *Mathematical aspects of classical and celestial mechanics*, Springer–Verlag, Berlin 1997, [Russian: Fundamental Directions **3**, VINITI, Moscow, 1985, pp. 5–303].

[AVG] V. Arnold, A. Varchenko and S. Gusein-Zade "Singularities of Differentiable Mappings", Monographs in Mathematics, v. **82**, **83**, Birkhäuser, Boston, 1985, 1988, [Russian: v. 1, 2, Nauka, Moscow, 1982, 1984].

[AVGL] V. Arnold, V. Vassiliev, V. Goryunov and O. Lyashko, *Singularities. I. Local and global theory*, Encycl. Math. Sci. **6**, Springer–Verlag, Berlin, 1993, [Russian: Fundamental Directions **6** VINITI, Moscow, 1986, pp. 1–256]; *II. Classification and applications*, Encycl. Math. Sci. **39**, Springer–Verlag, Berlin, 1993, [Russian: Fundamental Directions **39**, VINITI, Moscow, 1989, pp. 5–255].

[Art] E. Artin, *Theory of braids*, Annals of Math. **48** (1947), 101–126.

[BV] D. Babbitt and V. Varadarajan, *Local moduli for meromorphic differential equations*, Asterisque **169–170** (1989).

[BaiB] W. Baily and A. Borel, *Compactifications of arithmetic quotients of bounded symmetric domains*, Annals of Math. **84** (1966), 442–528.

[BE] G. Bateman and A. Erdelyi, "Higher Transcendental Functions", v. **1, 2, 3**, McGraw–Hill Book C., New York, 1953.

[BauB] P. Baum and R. Bott, *On zeroes of meromorphic vector fields*, in "Essays on Topology and Related Topics. (Mémoires dediés à Georges de Rham)", Springer–Verlag, Berlin–Heidelberg–New York, 1970, pp. 29–47.

[Baut] N. Bautin, *On the number of limit cycles which appear with the variation of the coefficients from an equilibrium point of focus or center type*, Amer. Math. Soc. Transl. Ser. I **5** (1962), 396–413, [Russian: Mat. Sbornik **30** (1952), 181–196].

[BLL] M. Belliart, I. Liousse et F. Loray, *Sur l'éxistence de points fixes attractifs pour les sous-groupes de* Aut($\mathbb{C}$, 0), C. R. Acad. Sci. Paris Sér. I **324** (1997), 443–446.

[Ben] I. Bendixson, *Sur les courbes définies par des équations différentielles*, Acta Math. **25** (1901), 1–88.

[BKh] F. Berezovskaya and A. Khibnik, *On the problem of bifurcations of self-oscillations close to 1:4 resonance*, Selecta Math. **13** (1994) , No 2, 107–128.

[BG] J. Bernstein and S. Gelfand, *Meromorphy of the function* P^λ, Funct. Anal. Appl. **3(1)** (1969), 84–85 [Russian].

[BCLN] M. Berthier, D. Cerveau and A. Lins-Neto, *Sur les feuilletages analitiques réels et le probléme du centre*, J. Diff. Equat. **131** (1996), 244–266.

[BT] M. Berthier et F. Touzet, *Sur l'intégration des équations différentielles holomorphes réduites en dimension deux,* Bol. Soc. Brasil Mat. (N. S.) **30** (1999), No 3, 247–286.

[Ber] D. Bertrand, *Travaux récents sur les points singuliers des équations différentielles linéaires*, in "Sem. Bourbaki 1978/79", Lect. Notes in Math., v. **770**, Springer–Verlag, Berlin–Heidelberg–New York, 1980, pp. 228–243.

[Beuk] F. Beukers, *Monodromy of A–hypergeometric functions*, J. reine angew. Math. **718** (2016), 183–206.

[BH] F. Beukers and G. Heckman, *Monodromy for the hypergeometric function* ${}_nF_{n-1}$, Invent. Math. **95** (1993), 45–65.

[BNY] G. Binyamini, D. Novikov and S. Yakovenko, *On the number of zeroes of Abelian integrals. A constructive solution of the infinitesimal Hilbert sixteenth problem*, Invent. Math. **181** (2010), 227–289.

[Bir1] G. Birkhoff, *The generalized Riemann problem for linear differential equations and the allied problem for linear difference and q–difference equations*, Proc. Amer. Acad. Arts Sci. **49** (1913), 531–568.

[Bir2] G. Birkhoff, *Déformations analytiques et fonctions auto-équivalentes*, Annales **9** (1939), 51–122.

[BZ1] M. Bobieński and H. Żołądek, *Limit cycles for multidimensional vector fields. The elliptic case*, J. Dynam. Control Syst. **9** (2003), No 2, 265–310.

[BZ2] M. Bobieński and H. Żołądek, *Limit cycles for three dimensional polynomial vector fields*, Nonlinearity **18** (2003), 175–209.

[Bog] R. Bogdanov, *The versal deformations of singular point on vector field on the plane in the zero eigenvalues case*, Proc. of Petrovski Sem. **2** (1976), 37–65 [Russian].

[Bor] A. Borel, "Linear Algebraic Groups", W. A. Benjamin, New York–Amsterdam, 1969.

[BN] A. Borel and E. Narasimhan, *Uniqueness conditions for certain holomorphic mapping*, Invent. Math **2** (1967), 247–255.

[Bou] N. Bourbaki, "Groupes et Algèbres de Lie", v. **XXXIV**, Herman, Paris VI, 1968.

[Bra] B. Braaksma, *Multisummability and Stokes multipliers of linear meromorphic differential equations*, J. Diff. Equat. **92** (1991), 54–75.

[BS] J. Briançon et H. Skoda, *Sur la clôture d'un ideal de germes de fonctions holomorphes en un point de* $\mathbb{C}^n$, C. R. Acad. Sci. Paris Sér. A **278** (1974), 949-951.

[Brie] E. Brieskorn, *Die Monodromie der isolierten Singularitäten von Hyperflähen*, Manuscr. Math. **2** (1970), 103–161.

[BM] E. Briestone and P. Milman, *Canonical desingularization in characteristic zero by blowing-up the maximal strata of a local invariant*, Invent. Math. **128** (1997), 207–302.

[BB] C. Briot et J. Bouquet, *Propriétés des fonctions définies par des équations différentielles*, J. Éc. Polytech. **36** (1856), 133–198.

[Briu] A. Briuno, *Analytic form of differential equations*, in: "Trans. of Moscow Math. Soc. for the Year 1971", Amer. Math. Soc., Providence; 1973, [Russian: **25** (1971), 119–262]; *II*, in: "Trans. of Moscow Math. Soc. for the Year 1972", Amer. Math. Soc., Providence; 1974, [Russian: **26** (1972), 199–239].

[Bro] J. Browkin, "The Theory of Fields", PWN, Warszawa, 1978 [Polish].

[CS1] C. Camacho and P. Sad, *Topological classification and bifurcations of flows with resonances in* $\mathbb{C}^2$, Invent. Math. **67** (1982), 447–472.

[CS2] C. Camacho and P. Sad, *Invariant varieties through singularities of holomorphic vector fields*, Annals of Math. **115** (1982), 579–595.

[C-DS] M. Canalis-Durand and R. Shäfke, *On the normal form of a system of differential equations with nilpotent linear part*, C. R. Acad. Sci. Paris, Ser. I **336** (2003), 129–134.

[COGP] P. Candelas, X. de la Ossa, P. Green and L. Parkes, *A pair of Calabi–Yau manifolds as an exactly soluble superconformal theory*, Nucl. Phys. B **359** (1991), 21–74; also in: "Essays on Mirror Symmetry" S.-T. Yau (ed.), International Press, Hong Kong, 1992, 31–95; also in: AMS/IP Studies in Adv. Math. **9** (1998), 31–95.

[Can] F. Cano, *Reduction of singularities of foliations and applications*, in "Singularities Symposium - Łojasiewicz 70", Banach Center Publ., v. **44**, PWN, Warszawa, 1998, pp. 51–70.

[Cha] J. Chavarriga, *Particular solutions of homogeneous polynomial vector fields*, Applic. Math. **23** (1995), 339–350.

[CeMa] D. Cerveau et J. Mattei, *Formes intégrables holomorphes singuliéres*, Asterisque **97** (1982).

[CeMo] D. Cerveau et R. Moussu, *Groupes d'automorphismes de* $(\mathbb{C}, 0)$ *et équations différentielles* $ydy + \ldots = 0$, Bull. Soc. Math. France **116** (1988), 459–488.

[Cher] L. Cherkas, *On the conditions for a center for certain equations of the form* $yy' = P(x) + Q(x)y + R(x)y^2$, Differ. Equations **8** (1972), 1104–1107 [Russian: Differ. Uravnenya **8** (1972), 1435–1439].

[ChMa] C. Christopher and P. Mardesič, *The monodromy property and tangential center problem*, Funct. Anal. Appl. **44** (2010), 22–35 [Russian]

[CS] C. Christopher and D. Schlomiuk, *On general algebraic mechanism for producing centers in polynomial differential systems*, J. Fixed Point. Th. Appl. **3** (2008),331–351.

[Cle] C. Clemens, *Picard–Lefschetz theorem for families of nonsingular algebraic varieties acquiring ordinary singularities*, Trans. Amer. Math. Soc. **136** (1969), 93–108.

[CL] E. Coddingtom and N. Levinson, "Theory of Ordinary Differential Equations", McGraw–Hill Book C., New York, 1955.

[Coh] S. Cohen, *The group of translations and rational powers is free*, Quart. J. Math. Oxford Ser. 2 **46** (1995), 21–93.

[Cox] H. Coxeter, "Introduction to Geometry", J. Wiley & Sons, New York, 1961.

[CoKa] D. Cox and S. Katz, "Mirror Symmetry and Algebraic Geometry", Math. Surveys Monographs **68**, Amer. Math. Soc., Providence, 1998.

[CH] T. Crespo and Z. Hajto, "Algebraic Groups and Differential Galois Theory", Graduate Studies in Math., v. **112**, Amer. Math. Soc., Providence, 2011.

[Cre] H. Cremer, *Über die Häufigkeit der Nichtzentren*, Math. Ann. **115** (1938), 573–580.

[Dar] G. Darboux, *Mémoire sur les équations différentielles algébriques du premier ordre et du premier degré*, Bull. Sci. Math. (Mélanges) (1878), 60–96; *II*, 123–144; *III*, 151–200.

[Dav] J. Davenport, "On the Integration of Algebraic Functions", Springer–Verlag, Berlin–Heidelberg–New York, 1981.

[DW] R. Dedekind and H. Weber, *Theorie der algebraischen Funktionen einer Veränderlichen*, J. reine angev. Math. **92** (1882), 181–290.

[Dek] W. Dekkers, *The matrix of a connection having regular singularities on a vector bundle of rank 2 on* $P^1(\mathbb{C})$, in "Équations Différentielles et Systèmes de Pfaff dans le Champ Complexe", R. Gérard et J.-P. Ramis (eds.), Lect. Notes in Math., v. **712**, Springer–Verlag, Berlin-Heidelberg-New York, 1979, pp. 33–43.

[Del1] P. Deligne, *Théorème de Lefschetz et critéres de dégénérescence de suites spectrales*, Publ. Math. IHES **35** (1968), 107-126.

[Del1] P. Deligne, "Équations Différentielles a Points Singuliers Reguliers", Lect. Notes in Math., v. **163**, Springer–Verlag, Berlin–Heidelberg–New York, 1970.

[Del3] P. Deligne, *Théorie de Hodge I*, "Actes Congr. Int. Math 1970, Nice," v. 1, Gauthier–Villars, Paris, 1970, pp. 425–430; *II*, Publ. Math. IHES **40** (1971), 5–58; *III*, Publ. Math. IHES **44** (1975), 5-77.

[Del4] P. Deligne, *Le groupe foundamental du complément d'une courbe plane n'ayant que des points doubles ordinaires est abélien*, in "Sem. Bourbaki 1979/80", Lect. Notes in Math., v. **842**, Springer–Verlag, Berlin–Heidelberg–New York, 1981, pp. 1–10.

[Del5] P. Deligne, *Catégories tannakienes*, "The Grothendieck Festschrift", P. Cartier at all (eds.), v. **2**, Birkhäuser, Boston, 1990, pp. 111–195.

[DK] P. Deligne and N. Katz, "Séminaire de Géométrie Algébrique du Bois–Marie. (SGA7(2)). (Groupes de Monodromie en Géométrie Algébrique)", Lect. Notes in Math., v. **340**, Springer–Verlag, Berlin–Heidelberg–New York, 1973.

[DM1] P. Deligne and G. Mostow, *Monodromy of hypergeometric functions and non-lattice integral monodromy*, Publ. Math. IHES **63** (1986), 5–89.

[DM2] P. Deligne and G. Mostow, "Commensurabilities Among Lattices in $PU(1,n)$", Annals of Math. Studies, v. **132**, Princeton Un-ty Press, Princeton, 1993.

[DeMu] P. Deligne and D. Mumford, *The irreducibility of the space of curves of given genus*, Publ. Math IHES **36** (1969), 75–100.

[dlVP] C. de la Valleé Poussin, *Sur l'équation différentielle linéaire du second ordre. détermination d'une intégrale par deux valeurs assignées. extension aux équations d'ordre n*, J. Math. Pures Appl. **8** (1929), 125–144.

[DH] A. Douady and J. Hubbard, "Étude Dynamique des Polynômes Complexes I, II," Publ Math. d'Orsay **84-2/85-4**, Univ. Paris Sud, Orsay, 1984/1985.

[Dr] V. Drinfeld, *Quantum groups*, in "Proc. of the Int. Congr. of Math. 1986, Berkeley", v. **1**, Amer. Math. Soc., Providence, 1987, pp. 798–820.

[DNF] B. Dubrovin, S. Novikov and A. Fomenko "Modern Geometry – Methods and Applications. Part I. The Geometry of Surfaces, Transformation Groups and Fields", Springer–Verlag, New York, 1992, [Russian: Nauka, Moscow, 1986].

[Dub] B. Dubrovin, *Geometry of 2D topological field theories*, in: "Integrable Systems and Quantum Groups", M. Francaviglia and S. Greco (eds.), Lect. Notes in Math., v. **1620**, Springer–Verlag, Berlin–Heidelberg–New York, 1996, pp. 120–348.

[DM] B. Dubrovin and M. Mazzocco, *Monodromy of certain Painlevé–VI transcendents and reflection groups*, Invent. Math. **141** (2000), 55–147.

[Dul1] H. Dulac, *Sur les cycles limites*, Bull. Soc. Math. France **51** (1923), 45–188.

[Dul2] H. Dulac, *Determination et intégration d'une classe d'équations différentielles ayant pour point singulier un centre*, Bull. des Sci. Math., Sér. 2 **32** (1908), 230–252.

[Dum] F. Dumortier, *Singularities of vector fields in the plane*, J. Diff. Equat. **23** (1977), 53–106.

[DMit] A. Duval and C. Mitschi, *Matrices de Stokes et groupe de Galois des équations hypergéometriques confluentes généralizées.* Pacific J. Math. **138** (1989), No 1, 25–56.

[Ec1] J. Ecalle, *Théorie itérative. Introduction à la théorie des invariants holomorphes*, J. Math. Pures et Appl. **54** (1975), 183–258.

[Ec2] J. Ecalle, "Sur les Fonctions Résurgentes ", v. I, II, III, Publ. Math. d'Orsay, Orsay, 1981.

[Ec3] J. Ecalle, "Introduction aux Fonctions Analysables et Preuve Constructive de la Conjecture de Dulac", Actualités Mathématiques, Herman, Paris, 1992.

[Ehr] C. Ehresmann, *Sur la topologie de certains espaces homogenes*, Ann. Math. (2) **35** (1934), 396–443.

[Eli] P. Elizarov, *The orbital topological classification of analytic differential equations in neighborhood of a degenerate elementary singular point in the two-dimensional complex plane*, J. Sov. Math. **50** (1990), No 2, 1447–1468, [Russian: Proc. of Petrovski Sem. **13** (1988), 137–165].

[Eli2] P. Elizarov, *Tangents to moduli maps*, in "Nonlinear Stokes Phenomena", Yu. Ilyashenko (ed.), Adv. in Sov. Math., v. **14**, Amer. Math. Soc., Providence, 1993, pp. 107–137.

[EISV] P. Elizarov, Yu. Ilyashenko, A. Shcherbakov and S. Voronin, *Finitely generated groups of germs of one-dimensional conformal mappings and invariants of complex singular points of analytic foliations of the complex plane*, in "Nonlinear Stokes Phenomena", Yu. Ilyashenko (ed.), Adv. in Sov. Math., v. **14**, Amer. Math. Soc., Providence, 1993, pp. 57–105.

[Ess] A. van den Essen, *Reduction of singularities in differential equations* $Ady = Bdx$, in "Équations Différentielles et Systèmes de Pfaff dans le Champ Complexe", R. Gérard et J.-P. Ramis (eds.), Lect. Notes in Math., v. **712**, Springer–Verlag, Berlin–Heidelberg–New York, 1979, pp. 44-59.

[Eul1] L. Euler, *De seriebus divergentibus*, "Leonhardi Euleri Opera Omnia. I", v. **14**, Teubner, Leipzig–Berlin, 1925, pp. 601–602.

[Eul2] L. Euler, *Specimen transformationinis singularis sewrierum*, Nova Acta Petropolitana **XII** (1801), 58–78.

[Fab] E. Fabry, "Sur les Intégrales des Équations Différentielles à Coefficients Rationnels", Thése, Paris, 1885.

[FN] H. Flaschka and A. Newell, *Monodromy and spectral preserving deformations*, Commun. Math. Phys. **76** (1980), 65–116.

[For] O. Forster, "Riemannsche Flächen", Springer–Verlag, Berlin–Heidelberg–New York, 1977.

[Fr] J.-P. Françoise, *Successive derivatives of a first return map, application to study of quadratic vector fields*, Ergodic Th. Dyn. Syst. **16** (1996), 87–96.

[FSZ] A. Fronville, A. Sadovski and H. Żołądek, *Solution of the 1:-2 resonant center problem in the quadratic case*, Fund. Math. **157** (1998), 191–207.

[FY] J.-P. Françoise and Y. Yomdin, *Bernstein inequalities and applications to analytic geometry and differential equations* J. Funct. Anal. **146** (1997), 185–205.

[Fuch] R. Fuchs, *Über lineare homogene Differentialgleichungen zweiter Ordnung mit drei Endlichen gelegenen wesentlich singulären Stellen*, Math. Ann. **63** (1907), 301–321.

[Gab] A. Gabrielov, *On projections of semi-analytic sets*, Funct. Anal. Appl. **2** (1968), 282–291, [Russian: **2(4)** (1968), 18–30].

[Gar1] R. Garnier, *Sur les équations différentielles du troisième ordre dont l'intégrale est uniforme et sur une classe d'équations nouvelles d'ordre supérieur dont l'intégrale génerale a ses points critiques fixes*, Ann. Éc. Norm. Sup. **29** (1912), 1–126.

[Gar2] R. Garnier, *Sur une classe de systems différentielles abéliens deduits de la théorie des équations linéaires*, Rend. Circ. Mat. Palermo **43** (1919), 155–191.

[Gau] K. Gauss, "Werke", T. 3, König. Gessel. der Wiss., Göttingen, 1876.

[Gav1] L. Gavrilov, *The infinitesimal 16th Hilbert problem in the quadratic case*, Invent. Math. **143** (2001), No 3, 449–497.

[Gav2] L. Gavrilov, *Higher order Poincaré–Pontryagin functions and iterated path integrals*, Ann. Fac. Sci. Toulouse Math. (6) **14** (2005), 663–682.

[GKZ] I. Gelfand, M. Karpanov and A. Zelevinsky , *Generalized Euler integrals and A–hypergeometric functions*, Adv. in Math. **84** (1990), 255–271.

[Gel] O. Gelfond, *Sur le septième problème de Hilbert*, Izv. Akad. Nauk USSR OMEN **7** (1934), No 4, 623–634.

[Gev] M. Gevrey, *Sur la nature analytique des solutions des équations aux dérivées partielles*, Ann. Sci. Éc. Norm. Sup. Sér. 3 **25** (1918), 129–190.

[Giv1] A. Givental, *Twisted Picard–Lefschetz formulas*, Funct. Anal. Appl. **22** (1988), No 1, 10–18, [Russian: **22** (1988), No 1, 13–22].

[Giv2] A. Givental, *Equivariant Gromov–Witten invariants*, Intern. Math. Research Not. **13** (1996), 613-663.

[GivS] A. Givental and V. Schechtman, *Monodromy groups and Hecke algebras*, Russ. Math. Surv. **42** (1987), No 4, 138–139 [Russian].

[GJ] J. Glimm and A. Jaffe, "Quantum Physics. A Functional Integral Point of View", Springer–Verlag, Berlin–Heidelberg–New York, 1981.

[Gl1] A. Glyutsuk, *Stokes operators via limit monodromy of generic perturbation*, J. Dynam. Contr. Syst. **5** (1999), No 5, 101–135.

[Gl2] A. Glyutsuk, *Confuence of singular points and the nonlinear Stokes phenomena*, Trans. Moscow Math. Soc. **62** (2001), 49–95 [Russian: Trudy Mosk. Mat. Obshch. **62** (2001), 54–101].

[GlIl] A. Glyutsuk and Yu. Ilyashenko, *Restricted version of the infinitesimal Hilbert 16th problem*, Moscow Math. J. 7 (2007) No 2, 281–325, 351.

[God] P. Godement, "Topologie Algébrique et Théorie des Faisceaux", Hermann, Paris, 1958.

[Gol] V. Golubev, "Lectures on Analytic Theory of Differential Equations", Gos. Izdat., Moscow, 1950 [Russian], [also: "Vorlesungen über Differentialgleichungen im Komplexen", VEB Deutsch. Verlag Wissen., Berlin, 1958].

[Gor] P. Gordan, *Über endliche Gruppen linearer Transformationen einer Veränderlichen*, Math. Ann. **12** (1877), 23–46.

[GrRe] H. Grauert and R. Remmert, "Theorie der Steinsche Räume", Springer–Verlag, Berlin–Heidelberg–New York, 1977.

[Gri1] P. Griffiths, *Periods of integrals on algebraic manifolds. I*, Amer. J. Math. **90** (1968), 568–626; *II*, 805–865.

[Gri2] P. Griffiths, *Periods of integrals on algebraic manifolds: Summary of main results and discussion of open problems*, Bull. Amer. Math. Soc. **76** (1970), 228–296.

[Gri3] P. Griffiths, *On the periods of certain rational integrals. I*, Ann. Math. **90** (1969), 460–495; *II*, 498–541.

[Gri4] P. Griffiths (Ed.), "Topics in Transcendental Algebraic Geometry", Annals of Math. Studies, v. **106**, Princeton Un-ty Press, Princeton, 1984.

[GH] P. Griffiths and J. Harris, "Principles of Algebraic Geometry", v. 1, 2, J. Wiley & Sons, New York, 1978.

[GS] P. Griffiths and W. Schmid, *Locally homogeneous complex manifolds*, Acta Math. **123** (1969), 253–302.

[GL] V. Gromak and N. Lukashevich, "Analytic Properties of Solutions of the Painlevé Equations", Univ. Izd., Minsk, 1990 [Russian].

[Gro1] A. Grothendieck, *Sur la classification des fibrés holomorphes sur la sphère de Riemann*, Amer. J. Math. **79**, (1956), 121–138.

[Gro2] A. Grothendieck, "Séminaire de Géométrie Algébrique du Bois–Maire (SGA 7(1)). (Groupes de Monodromie en Géométrie Algébrique)", Lect. Notes in Math., v. **288**, Springer–Verlag, Berlin–Heidelberg–New York, 1972.

[GuRo] R. Gunning and H. Rossi, "Analytic Functions of Several Complex Variables", Prentice–Hall Inc., Englewood Cliffs, 1965.

[HaP] P. Hartman, "Ordinary Differential Equations", J. Wiley & Sons, New York, 1964.

[HaR] R. Hartshorne, "Algebraic Geometry", Springer–Verlag, Berlin–Heidelberg–New York, 1977.

[Hau] H. Hauser, *The Hironaka theorem on resolution of singularities (Or: A proof we always wanted to understand)*, Bull. Amer. Math. Soc. **40** (2003), No 3, 323–403.

[Hea] J. Heading, "An Introduction to Phase–Integral Methods", J. Wiley & Sons, New York, 1962.

[Hil] D. Hilbert, *Mathematische probleme*, in "II Int. Congr. Math., Paris 1900", Nachr. Ges. Wiss. Gottingen Math.–Phys. Kl., 1900, pp. 253–297.

[Hir] H. Hironaka, *Resolution of singularities of an algebraic variety over a field of characteristic zero. I, II*, Annals of Math. **79** (1964), 109–326.

[Hor] E. Horozov, *Versal deformations of equivariant vector fields for the cases of symmetry of order 2 and 3*, Proc. of Petrovski Sem. **5** (1979), 163–192 [Russian].

[HS] E. Horozov, and Ts. Stoyanova, *Non-integrability of some Painlevé VI equations and dilogarithms*, Regul. Chaotic Dyn. **12** (2007), 620–627.

[Hua] K. Huang, "Statistical Mechanics", J. Wiley & Sons, New York, 1963.

[HV] M. Hudai-Verenov, *A property of the solutions of a differential equation*, Math. Sbornik **56** (1962), 301–308 [Russian].

[Huk] M. Hukuhara, *Sur les points singuliers des équations différentielles linéaires. II*, J. Fac. Sci. Hokkaido Univ. **5** (1937), 123–166.

[HKM] M. Hukuhara, T. Kimura and T. Matuda, "Équations Différentielles Ordinaires du Premier Ordre dans le Champ Complexe", Math. Soc. of Japan, Tokyo, 1961.

[HO] M. Hukuhara and S. Ohasi, *On Riemann P–functions which are expressible in terms of elementary functions.*, Sûgaku **2** (1949-50), 227–230; *II*, **8** (1956), 27–29 [Japanese].

[Hum] J. Humphreys, "Reflection Groups and Coxeter Groups", Cambridge Studies in Adv. Math., v. **29**, Cambridge Un-ty Press, Cambridge, 1990.

[Il1] Yu. Ilyashenko, *The origin of limit cycles under perturbation of equation* $\frac{\mathrm{d}w}{\mathrm{d}z} = -R_z/R_w$, *where* $R(z, w)$ *is a polynomial*, Math. Sboornik **7** (1969), No. 3 [Russian].

[Il2] Yu. Ilyashenko, *Algebraic non-solvability and almost algebraic solvability of the center–focus problem*, Funct. Anal. Appl. **6** (1972) No 3, 30–37 [Russian.

[Il3] Yu. Ilyashenko, *On zeroes of special Abelian integrals in real domain*, Funct. Anal. Appl. **11(4)** (1978), 78–79 [Russian].

[Il4] Yu. Ilyashenko, *The topology of phase portraits of analytic differential equations in the complex projective plane*, Selecta Math. Sov. **5** (1986), 141–199, [Russian: Proc. of Petrovski Sem. **4** (1978), 83-136].

[Il5] Yu. Ilyashenko, *Finitness theorems for limit cycles*, Russ. Math. Surv. **45** (1990), No 2, 129–203, [Russian: **45** (1990), No 2, 143–200].

[Il6] Yu. Ilyashenko, "Finitness Theorems for Limit Cycles", Amer. Math. Soc., Providence, 1992.

[Il7] Yu. Ilyashenko, *Nonlinear Stokes phenomena*, in "Nonlinear Stokes phenomena", Yu. Ilyashenko (ed.), Adv. in Sov. Math., v. **14**, Amer. Math. Soc., Providence, 1993, pp. 1–53.

[Il8] Yu. Ilyashenko, *Nonlinear Riemann–Hilbert problem*, Proc. Steklov Inst. Math. **213** (1997), 10–34, [Russian: **213** (1996), 6–29].

[Il9] Yu. Ilyashenko, *The centennial history of the 16th Hilbert problem*, Bull. Amer. Math. Soc. **39** (2002), No 3, 301–354.

[IlKh] Yu. Ilyashenko and A. Khovanski, *Galois groups, Stokes operators and the Ramis theorem*, Funct. Anal. Appl. **24** (1990), No 4, 286–296 (1991), [Russian: **24** (1990), No 4, 31–42].

[IlP] Yu. Ilyashenko and A. Pyartli, *The monodromy group at infinity of a generic polynomial vector field on the complex projective plane*, Russian J. of Math. Phys. **2** (1994), 275–315, [Russian: Proc. Steklov Math. Inst. **213** (1997), 56–73].

[IlY1] Yu. Ilyashenko and S. Yakovenko, *Double exponential estimate for the number of zeroes of complete abelian integrals*, Invent. Math. **121** (1995), 613–650.

[IlY2] Yu. Ilyashenko and S. Yakovenko (eds.), "Concerning XVI–th Hilbert Problem", Adv. in Math. Sci., v. **27**, Amer. Math. Soc., Providence, 1995.

[IlY3] Yu. Ilyashenko and S. Yakovenko,"Lectures on Analytic Differential Equations", Graduate Studies in Math., v. **86**, Amer. Math. Soc., Providernce, 2008.

[Inc] E. Ince, "Ordinary Differential Equations", Dover Publ., New York, 1944.

[IKSY] K. Iwasaki, T. Kimura, S. Shimomura and M. Yoshida, "From Gauss to Painlevé. A Modern Theory of Special Functions", Aspects of Mathematics, Vieweg, Wiesbaden, 1991.

[Jea] P. Jeanquartier, *Development asymptotique de la distribution de Dirac*, C. R. Acad. Sci. Paris Sér. A **271** (1970), 1159–1161.

[JMU] M. Jimbo, T. Miwa and K. Ueno, *Monodromy preserving deformations of linear ordinary differential equations with rational coefficients. I. General theory and τ–function*, Physica 2D **2** (1981), 306–352 .

[Jou] J. Jouanolou, "Équations de Pfaff Algébriques", Lect. Notes in Math., v. **708**, Springer–Verlag, Berlin–Heidelberg–New York, 1979.

[Kac] V. Kac, "Infinite Dimensional Lie Algebras", Cambridge University Press, Cambridge, 1985.

[Kam] E. Kamke, "Differentialgleichungen Lösungsmetoden und Lösungen", Akademische Verlag, Leipzig, 1959.

[Kan] J. Kaneko, *The Gauss–Manin connection of the integral of the deformed difference product*, Duke Math. J. **92** (1998), 355–379.

[Kapl] I. Kaplansky, "An Introduction to Differential Algebra", Hermann, Paris, 1957.

[Kapt] W. Kapteyn, *Niew onderzoec omtrent de middel punten der integralen van differentiaalvergelijkingen van de eerste orde en den eersten graad I*, Konikl. Nederland Akad. **DL.XX** (1912), 1354–1365; *II* **DL.XXI** (1912), 27–33.

[Kat1] N. Katz, *Nilpotent connections and the monodromy theorem: application of a result of Turrittin*, Publ. Math. IHES **39** (1970), 207–238.

[Kat2] N. Katz, *The regularity theorem in algebraic geometry*, in "Actes Congr. Int. Math 1970, Nice", v. 1, Gauthier–Villars, Paris, 1970, pp. 437–443.

[Kat3] N. Katz, "Rigid Local Systems", Annals of Math. Studies, v. **139**, Princeton Un-ty Press, Princeton, 1996.

[KhZa] V. Khizhnik and A. Zamolodchikov, *Current algebras and Wess–Zumino models in two dimensions*, Particle Phys. **B247** (1984), 83–103.

[Kh1] A. Khovanski, *On representation of functions by quadratures*, Usp. Mat. Nauk **26** (1971), No 4, 251–252 [Russian].

[Kh2] A. Khovanski, *Real analytic manifolds with finitude properties and complex abelian integrals*, Funct. Anal. Appl. **18** (1984), No 2, 119–128 [Russian].

[Kh3] A. Khovanski, "Fewnomials", Trans. Math. Monographs **88**, Amer. Math. Soc., Providence, 1991.

[Kh4] A. Khovanski, *Topological obstructions to representability of functions by quadratures*, J. Dynam. Contr. Syst. **1** (1995), 91–123.

[Kim] T. Kimura, *On Riemann's equations which are solvable by quadratures*, Funkcial. Ekvac. **12** (1969), 269–281.

[Kir] A. Kirillov, "Elements of the Theory of Representations", Springer–Verlag, New York, 1976, [Russian: Nauka, Moscow, 1978].

[Kl1] F. Klein, "Über die Hypergeometrische Function", Teubner, Leipzig, 1906.

[Kl2] F. Klein, "Vorlesungen über die Entwicklung der Mathematik im 19. Jahrhundert", Teil 1, Verlag von Julius Springer, Berlin, 1926.

[Kl3] F. Klein, "Vorlesungen über das Ikosaedr und die Auflösung der Gleichungen von fünften Grade", Teubner, Leipzig, 1884.

[Kob] S. Kobayashi, "Transformation Groups in Differential Geometry", Springer–Verlag, Berlin–Heidelberg–New York, 1972.

[KN] S. Kobayashi and K. Nomizu, "Foundations of Differential Geometry", v. 1, 2, Interscience Publishers, New York, 1963, 1969.

[KNS] K. Kodaira, L. Nirenberg and D. Spencer, *On the existence of deformations of complex analytic structures*, Annals of Math. **68** (1958), 450–459.

[Kol] E. Kolchin, "Differential Algebra and Algebraic Groups", Academic Press, New York, 1973.

[KoZe] R. Kooij and A. Zegeling, *Uniqueness of limit cycle in polynomial systems with algebraic invariants*, Bull. Austral. Math. Soc. **49** (1994), 7–20.

[Kon] M. Kontsevich, *Vassiliev's knot invariants*, "I. M. Gelfand Seminar", S. Gelfand and S. Gindikin (eds.), v. 2, Adv. in Sov. Math., Amer. Math. Soc., Providence, 1991, pp. 137–150.

[Kos] V. Kostov, *Fuchsian systems on* $\mathbb{CP}^1$ *and the Riemann–Hilbert problem*, C. R. Acad. Sci. Paris, Sér. I **315** (1992), 143–148.

[Kov] J. Kovačic, *An algorithm for solving second order linear homogeneous differential equations*, J. Symb. Comput. **2** (1986), 3–43.

[Kra] B. Krauskopf, *Bifurcation sequences at 1:4 resonance: an inventory*, Nonlinearity **7** (1994), No 3, 1073–1091.

[Kuc] M. Kuczma, "Functional Equations in a Single Variable", Monogr. Matem., v. **46**, PWN, Warszawa, 1968.

[Kul] V. Kulikov, "Mixed Hodge Structures and Singularities", Cambridge Unty Press, Cambridge, 1998.

[KK] V. Kulikov and P. Kurchanov, *Complex algebraic varieties: periods of integrals, Hodge structures*, Encycl. Math. Sci. **36**, Springer–Verlag, Berlin, 1998, [Russian: Fundamental Directions **36**, VINITI, Moscow, 1989, pp. 3–231].

[Kur] M. Kuranishi, *On the locally complete families of complex analytic structures*, Annals of Math. **75** (1962), 536–577.

[LR2] C. Lambert and C. Rousseau, *Complete system of analytic invariants for unfolded differential linear systems with an irregular singularity of Poincaré rank 1*, Moscow Math. J. **12** (2012), 77–138.

[LR2] C. Lambert and C. Rousseau, *Moduli space of unfolded differential linear systems with an irregular singularity of Poincaré rank 1*, Moscow Math. J. **13** (2013), 529–550.

[LP] E. Landis and I. Petrovski, *On the number of limit cycles of the equation* $\mathrm{d}y/\mathrm{d}x = P(x,y)/Q(x,y)$, *where* P *and* Q *are polynomials*, Amer. Math. Soc. Transl (2) **14** (1960), 181–199, [Russian: Math. Sbornik **43** (1957), 149–168.

[Lan] A. Landman, *On the Picard–Lefschetz transformation for algebraic manifold acquiring general singularities*, Trans. Amer. Math. Soc. **181** (1973), 89–126.

[L-D] I. Lappo-Danilevskij, "Mémoires sur la Théorie des Systemes des Équations Différentielles Linéaires", v. 1, 2, 3, Chelsea Publ., New York, 1953.

[Lau] G. Lauricella, *Sulle funzioni hipergeometriche a plu variabili*, Rend. di Palermo **VII** (1893), 111–158.

[Lea] L. Leau, *Étude sur les équations fonctionelles à une ou plusieurs variables*, Ann. Fac. Sci. Touluse **11** (1897), 1–110.

[Lef] S. Lefschetz, "L'analysis Situs et la Géométrie Algébrique", Gauthier–Villars, Paris, 1924.

[LZ] P. Leszczyński and H. Żołądek, *Limit cycles appearing after perturbation of certain multidimensional vector fields*, J. Dyn. Diff. Equat. **13** (2001), No 4, 689–709.

[Lev] A. Levelt, *Jordan decomposition for a class of singular differential operators*, Arkiv. Math. **13** (1975), 1–27.

[L-N1] A. Lins-Neto, *Algebraic solutions of polynomial differential equations and foliations in dimension two*, "Holomorphic Dynamics, Mexico 1986", Lect. Notes in Math. v. **1345** Springer–Verlag, Berlin–Heidelberg–New York, 1988, pp. 193–232.

[L-N2] A. Lins-Neto, *Construction of singular holomorphic foliations in dimension two*, J. Diff. Geom. **26** (1987), 1–31.

[L-N3] A. Lins-Neto, *Some examples for the Poincaré and Painlevé problems*, Ann. Sci. Éc. Norm. Sup. 4e sér. **35** (2002), 231–266.

[L-N4] A. Lins-Neto, *Curvature of pencils of foliations*, in: "Analyse Complexe, Systémes Dynamiques, Sommabilité des Series Divergentes et Théories Galoisiennes I", Asterisque **296** (2004), 167–190.

[L-NS] A. Lins-Neto and M. Soares, *Algebraic solutions of one-dimensional foliations*, J. Diff. Geom. **43** (1996), 652–673.

[Lio2] J. Liouville, *Premier mémoire sur la détermination des intégrales dont la valeur est algébrique*, J. Éc. Polytech. **14** (1833), 124–148; *Second mémoire sur la détermination des intégrales dont la valeur est algébrique*, J. Éc. Polytech. **14** (1833), 149–193.

[Lio2] J. Liouville, *Mémoire sur l'integration d'une classe d'équation différentielle du second ordre en quantities finies explicites*, J. Math. Pures Appl. **IV** (1839), 423–456.

[LlZ] J. Llibre and H. Żołądek, *The Poincaré center problem*, J. Dynam. Contr. Syst. **14** (1995), No 4, 505–535.

[Loj1] S. Łojasiewicz, "Ensembles Semi-Analytiques", IHES, Bures-sur-Yvette, 1965.

[Loj2] S. Łojasiewicz, "Introduction to Complex Analytic Geometry", Birkhäuser, Basel, 1991; [Polish: PWN, Warszawa, 1988].

[Lor1] F. Loray, *Rigidité topologique pour des singularités de feuilletages holomorphes*, Publ. Inst. Rech. Math. Rennes **37** (1995), 1–33.

[Lor2] F. Loray, *Réduction formelle des singularités cuspidales de champs de vecteurs analytiques*, J. Diff. Equat. **158** (1999), 152–173.

[Lor3] F. Loray, "5 Leçons sur la Structure Transverse d'une Singularité de Feuilletage Holomorphe en Dimension 2 Complexe", Monogr. Red. TMR Eur. Sing. Ec. Diff. Fol., v. **1**, 1999, pp. 1–92.

[Lor4] F. Loray, *A preparation theorem for codimension one foliations*, Ann. Math. (2) **163** (2006) No 2, 709–722.

[LM] F. Loray and R. Meziani, *Classification des certaines feuilletages associes à un cusp*, Bol. Soc. Bras. Mat **25** (1994), 93–106.

[LR] F. Loray and J. Rebelo, *Minimal rigid foliations by curves on* $\mathbb{CP}^n$, J. Eur. Math. Soc. **5** (2003), No 2, 147–2001.

[LZ1] P. Lubowiecki and H. Żołądek, *The Hess–Appelrot system. I. Invariant torus and its normal hyperbolicity*, J. Geometric Mechanics **4** (2012), 443–467.

[LZ2] P. Lubowiecki and H. Żołądek, *The Hess–Appelrot system. II. Perurbations and limit cycles*, J. Diff. Equat. **232** (2012), 1701–1722.

[Lya] A. Lyapunov, "Probleme General de la Stabilité du Mouvement", Annals of Math. Studies, v. **17**, Princeton Un-ty Press, Princeton, 1947, [Russian: Gostekhizdat, Moscow, 1950].

[MMNS] A. Maciejewski, J. Moulin-Ollagnier, A. Nowicki and J.-M. Strelcyn, *Around the Jouanolou non-integrability theorem*, Indag. Math. (New Series) **11** (2000), No 2, 239–254.

[Mag] A. Magid, "Lectures on Differential Galois Theory", Amer. Math. Soc., Providence, 1994.

[Mah] G. Mahoux, "Introduction to the Theory of Isomonodromic Deformations of Linear Ordinary Differential Equations with Rational Coefficients", CRM Ser. Math. Phys., Springer–Verlag, New York, 1999, pp. 35–76.

[Mal1] B. Malgrange, "Ideals of Differentiable Functions", Oxford Un-ty Press, London, 1967.

[Mal2] B. Malgrange, *Intégrales asymptotiques et monodromie*, Ann. Sci. Éc. Norm. Sup. Sér. 4, **7** (1974), 405–430.

[Mal3] B. Malgrange, *Remarques sur les équations différentielles à points singulier irrégulier*, in "Équations Différentielles et Systèmes de Pfaff dans le Champ Complexe", R. Gérard et J.-P. Ramis (eds.), Lect. Notes in Math., v. **712**, Springer–Verlag, Berlin–Heidelberg–New York, 1979, pp. 77–86.

[Mal4] B. Malgrange, *Sur les déformations isomonodromiques. I. Singularités regulierès*, in "Séminaire Éc. Norm. Sup.", Birkhäuser, Boston, 1982.

[Mal5] B. Malgrange, *Introduction aux travaux de J. Ecalle*, L'Enseign. Mathém. **31** (1985), 261–282.

[Man1] Yu. Manin, *Rational points of algebraic curves over function fields*, Amer. Math. Soc. Trans. (2) **50** (1966), 189–234, [Russian: Izv. Akad. Nauk USSR, Ser. Mat. **27** (1963), 1395–1440].

[Man2] Yu. Manin, *Moduli Fuchsiani*, Ann. Sc. Norm. Sup. Pisa **19** (1965), 113–126.

[Man3] Yu. Manin, *Sixth Painlevé equation, universal elliptic curves and mirror of* $\mathbb{P}^2$, in: "Geometry of Differential Equations", Amer. Math. Soc. Transl.. (2) **186** (1998), 131–151.

[MRR] P. Mardesič, R. Roussarie and C. Rousseau, *Modulus of analytic classification of unfoldings of generic parabolic diffeomorphisms*, Moscow Math. J. **4** (2004), 455–502, 535.

[MR1] J. Martinet and J.-P. Ramis, *Problèmes de modules pour des équations différentielles nonlinéaires du premier ordre*, Publ. IHES **55** (1982), 63–164.

[MR2] J. Martinet and J.-P. Ramis, *Classification analytique des équations différentielles non-linéaires resonantes du premier ordre*, Ann. Sci. Éc. Norm. Sup. **16** (1983), 571–621.

[MM] J. Mattei and R. Moussu, *Holonomie et intégrales premiéres*, Ann. Sci. Éc. Norm. Sup. **13** (1980), 469–523.

[MS] J. Mattei and E. Salem, *Complete system of topological and analytical invariants for a generic foliation of* $(\mathbb{C}^2, 0)$, Math. Res. Letters **4** (1997), No 1, 131–141.

[Med] N. Medvedeva, *On the analytic solvability of the problem of distinguishing between center and focus*, Proc. Steklov Inst. Math. **254** (2006), 7–93 [Russian].

[Mei] C. Meijer, *On the G–functions (I–VIII)*, Indag Math. **8** (1956), 124–134; 213–225; 312–324; 391–400; 468–475; 595–602; 661–670; 713–723.

[Mel] V. Melnikov, *On the stability of a center for time-periodic perturbations*, Trans. Moscow Math. Soc. **12** (1963), 1–57, [Russian: **12** (1963), 3–52].

[Mez] R. Meziani, *Classification analytique d'équations différentielles* $ydy + \ldots = 0$ *et espace de modules*, Bol. Soc. Bras. Mat. **21** (1996), 23–53.

[Mil1] J. Milnor, "The Morse Theory", Annals of Math. Studies, v. **51**, Princeton Un-ty Press, Princeton, 1965.

[Mil2] J. Milnor, "Singular Points of Complex Hypersurfaces", Annals of Math. Studies, v. **61**, Princeton Un-ty Press, Princeton, 1968.

[Mit] C. Mitschi, *Differential Galois groups of confluent generalized hypergeometric equations: an approach using Stokes multipliers*, Pacific J. Math. **176** (1998), No 2, 365–405.

[MiSi] C. Mitschi and M. Singer, *The inverse problem in the differential Galois theory*, in: "The Stokes Phenomenon and Hilbert's 16th Problem", B. Braaksma, G. Immink and M. van der Put (eds.), World Sci. Publ. Co., Inc., River Edge, N. J., 1996, pp. 185–196.

[Miw] T. Miwa, *Painlevé property of monodromy preserving deformation equation and the analyticity of τ–function*, Publ. Res. Inst. Math. Sci. Kyoto Univ. **17** (1981), 703–721.

[Mju] B. Mjuller, *The density of the solutions of certain equations in* $\mathbb{CP}^2$, Math. Sbornik **98** (1975), No 3, 363–377, [Russian].

[M-R] J. Morales-Ruiz, "Differential Galois Theory and Non-Integrability of Hamiltonian Systems", Progress in Math., v. **179**, Birkhäuser, Boston, 1999.

[M-RR] J. Morales-Ruiz and J.-P. Ramis, *A note on the non-integrability of some Hamiltonian systems with a homogeneous potential*, Banach Center Publ., v. **58**, Polish Ac. Sci. (2002), pp. 165–175.

[MRRS] J. Morales-Ruiz, J.-P. Ramis and C. , *Integrability of Hamiltonian systems and differential Galois group of higher variation equations*, Sci. l'Éc. Norm. Sup. **40** (2006), 845–884.

[Mos1] G. Mostow, *General Picard lattices arising from half-integral conditions*, Publ. Math. IHES **63** (1986), 91–106.

[Mos2] G. Mostow, *Braids, hypergeometric functions and lattices*, Bull. Amer. Math. Soc. **16** (1987), 225–246.

[Mov1] H. Movasati, *On deformation of foliations with a center in the projective space*, An. Acad. Bras. Cienc. **73** (2001), No 2, 191–196.

[Mo2] H. Movasati, *Center conditions: rigidity of logarithmic differential equations*, J. Diff. Equat. **197** (2004), 197–217.

[Mou] R. Moussu, *Holonomie évanescente des équations différentielles dégénérées transverses*, in "Singularities and Dynamical Systems", S. N. Pnevmaticos (ed.), North–Holland, Amsterdam, 1985, pp. 161–173.

[Mum1] D. Mumford, "Geometric Invariant Theory", Springer–Verlag, Berlin-Heidelberg-New York, 1965.

[Mum2] D. Mumford, *Semi-stable reduction*, in "Toroidal Embeddings I.", G. Kempf, F. Knudsen, D. Mumford and B. Saint-Donat (eds.), Lect. Notes in Math., v. **339**, Springer–Verlag, Berlin–Heidelberg–New York, 1973, pp. 53–108.

[Nak] I. Nakai, *Separatrices for non solvable dynamics on* $\mathbb{C}, 0$, Ann. Inst. Fourier **44** (1994), 569–599.

[Nei] A. Neishtadt, *Bifurcations of phase portrait of certain system of equations arising in the problem of loss of stability of self-oscillations near resonance 1:4*, J. Appl. Math. Mech. **42** (1978), 830–840 [Russian].

[NN] A. Newlander and L. Nirenberg, *Complex analytic coordinates in almost complex manifolds*, Annals of Math. **65** (1954), 391–404.

[NY] D. Novikov and S. Yakovenko, *Simple exponential estimate for the number of zeroes of complete abelian integrals*, C. R. Acad. Sci. Paris Sér. I **320** (1995), 853–858.

[Oka] K. Okamoto, *Studies on the Painlevé equations. I*, Ann. Mat. Pura et Applic. **145** (1987), 337–381; *II*, Jap. J. Math. **13** (1987), 47–76; *III*, Math. Ann. **275** (1986), 221–256; *IV*, Funkcial. Ekvac. **30** (1987), 305–332.

[OSS] G. Okonek, H. Schneider and H. Spindler, "Vector Bundles on Complex Projective Spaces", Birkhäuser, Boston, 1980.

[Oor] F. Oort, *Alterations can remove singularities*, Bull. Amer. Math. Soc. **35** (1998), No 4, 31–331.

[Pai] P. Painlevé, "Œuvres", v. I, II, III, SNRS, Paris, 1976.

[Pau] E. Paul, *Feuilletages holomorphes singuliers à holonomie résoluble*, J. reine angew. Math. **514** (1999), 9–70.

[P-M] R. Perez-Marco, *Solution complète au problème de Siegel de linéarization d'une application holomorphe au voisinage d'un point fixe. (D'aprés J.-C. Yoccoz)*, in "Sém. Bourbaki, Fév. 1992", Asterisque, v. **206** (1992), 273–310.

[P-MY] R. Pérez-Marco and J.-C. Yoccoz, *Germes de feuilletages holomorphes à holonomie prescrite*, Asterisque **222** (1994), 345–371.

[Pet1] G. Petrov, *On the number of zeroes of complete elliptic integrals*, Funct. Anal. Appl. **18** (1984), No 2, 73–74 [Russian].

[Pet2] G. Petrov, *The Chebyshev property of elliptic integrals*, Funct. Anal. Appl. **22** (1988), No 1, 72–73, [Russian: **22** (1988), No 1, 83–84].

[Pet3] G. Petrov, *Non-oscillation of elliptic integrals*, Funct. Anal. Appl. **24** (1990), No 4, 105–210 (1991), [Russian: **24** (1990), No 4, 45–50].

[PO] I. Petrovski and O. Oleinik, *On topology of real algebraic surfaces*, Amer. Math. Soc. Transl. (1952), No 70, [Russian: Izv. Akad. Nauk USSR. Ser. Mat. **13** (1949), 389–402].

[Ph1] F. Pham, *Formules de Picard–Lefschetz généralisées et ramification des intégrales*, Bull. Soc. Math. France **93** (1965), 333–367.

[Ph2] F. Pham, "Introduction a l'Étude Topologique des Singularités de Landau", Gauthier–Villars, Paris, 1967.

[Ph3] F. Pham, "Singularités des Systémes Différentielles de Gauss–Manin", Progress in Math., v. **2**, Birkhäuser, Boston, 1979.

[Pic1] E. Picard, *Sur une extension aux fonctions de deux variables du problème de Riemann relatif aux fonctions hypergéométriques*, Ann. Éc. Norm. Sup. **10** (1881), 305–322.

[Pic2] E. Picard, *Sur les fonctions hyperfuchsiennes provenant des séries hypergéométriques de deux variables*, Ann. Éc. Norm. Sup. Sér. III **2** (1885), 357–384; *II*, Bull. Soc. Math. France **15** (1887), 148–152.

[Pic3] E. Picard, "Analogies entre la Théorie des Équations Différentielles Linéaires et la Théorie des Équations Algébriques", Gauthier–Villars, Paris, 1936.

[PiSi] E. Picard and G. Simart, "Théorie des Fonctions Algébriques de Deux Variables Independantes", Gauthier–Villars, Paris, 1897.

[Ple] J. Plemelj, "Problems in the Sense of Riemann and Klein", Interscience, New York, 1964.

[Poch] L. Pochhammer, *Über hyperheometrische Funktion n^{ter} Ordnung*, J. reine angew. Math. **71** (1870), 316–362.

[Poi1] H. Poincaré, *Analysis situs*, J. Éc. Polytech. Norm. **1** (1895), 1–121.

[Poi2] H. Poincaré, *Sur les groupes des équations linéaires*, Acta Math. **5** (1884), 201–312.

[Poi3] H. Poincaré, *Mémoire sur les courbes définies par une équation différentielle*, in "Œuvres de Henri Poincaré", v. 1, Gauthiers–Villars, Paris, 1951.

[PoSz] G. Polya and G. Szegö, "Aufgaben und Lehrsätze aus der Analysis", Springer–Verlag, Berlin, 1964.

[Pon] L. Pontryagin, *On dynamical systems close to Hamiltonian systems*, in: "Selected Works", v.1, Gordon&Breach, New York, 1986, [Russian: Zh. Exper. Teoret. Fiziki **4** (1934), 234–238].

[PSi] M. Prelle and M. Singer, *Elementary first integrals of differential equations*, Trans. Amer. Math. Soc. **279** (1983), 215–229.

[Put] M. van der Put, *Recent work on differential Galois theory,* Sem. Bourbaki 1997–98, Asterisque **252** (1998), 341–367.

[Pya] A. Pyartli, *Generation of complex invariant manifolds near the singular point of a vector field depending on a parameter*, Funct. Anal. Appl. **6** (1972), No 4, 95-96 [Russian].

[Rag] M. Raghunathan, "Discrete Subgroups of Lie Groups", Springer–Verlag, Berlin–Heidelberg–New York, 1972.

[Ram1] J.-P. Ramis, *Phénomene de Stokes et filtration Gevrey sur le groupe de Picard–Vessiot*, C. R. Acad. Sc. Paris Ser. I **301** (1985), 165–167.

[Ram2] J.-P. Ramis, *Confluence et resurgence*, J. Fac. Sci. Univ. Tokyo. Sec. IA Math. **36** (1989), 706-716.

[Ram3] J.-P. Ramis, *About the inverse problem in differential Galois theory: the differential Abhyankar conjecture,* in "The Stokes Phenomenon and Hilbert's 16th Problem", B. Braaksma, G. Immink and M. van der Put (eds.), World Sci. Publ. Co., Inc., River Edge, N. J., 1996, pp. 261–278.

[RS] M. Reed and B. Simon, "Methods of Modern Mathematical Physics. v. 1. Functional Analysis", Acad. Press, New York, 1972.

[Rie] B. Riemann, "Œuvres Mathématiques", Gauthier–Villars, Paris, 1898.

[Ris] R. Risch, *The solution of the problem of integration in finite terms*, Bull. Amer. Math. Soc. **76** (1970), 605–608.

[Rit] J. Ritt, "Integration in Finite Terms. Liouville's Theory of Elementary Methods", Columbia Un-ty Press, New York, 1948.

[Roh] H. Röhrl, *Das Riemann–Hilbertsche Problem der Theorie der linearen Differentialgleichungen*, Math. Ann. **133** (1957), 1–25.

[Ros1] M. Rosenlicht, *On the explicit solvability of certain transcendental equations*, Publ. Math. **36** (1969), 15–22.

[Ros2] M. Rosenlicht, *Integration in finite terms*, Amer. Math. Monthly **79** (1972), 963–972.

[Ros3] M. Rosenlicht, *On Liouville's theory of elementary functions*, Pacific J. Math. **65** (1976), 485–492.

[Rou] R. Roussarie, "Bifurcations of Planar Vector Fields and Hilbert's Sixteenth Problem", Progress in Math., v. **164**, Birkhäuser, Basel, 1998.

[RT] C. Rousseau and L. Teyssier, *Analytic moduli for unfoldings of saddle–node vector fields*, Moscow Math. J. **8** (2008) No 3, 547–614

[RZ] C. Rousseau and H. Żołądek, *Zeroes of complete elliptic integrals in real domain*, J. Diff. Equat. **94** (1991), 41–54.

[Sai] M. Saito, *Hodge filtrations on Gauss–Manin systems. I*, J. Fac. Sci. Univ. Tokyo Sec. IA Mat. **30** (1984), 489–496.

[SMJ1] M. Sato, T. Miwa and M. Jimbo, *Aspects of holonomic quantum fields – isomonodromic deformations and Ising model,* "Complex Analysis, Microlocal Calculus and Relativistic Quantum Theory", D. Iagolnitzer (ed.), Lect. Notes in Phys., v. **126**, Springer–Verlag, Berlin–Heidelberg–New York, 1979, pp. 429–449.

[SMJ2] M. Sato, T. Miwa and M. Jimbo, *Holonomic quantum fields. I*, Publ. Res. Inst. Mat. Sci. Kyoto Univ. **14** (1978), 223–267; *II,* **15** (1979), 201–278; *III,* **15** (1979), 577–629; *IV,* **15** (1979), 871–972; *V,* **16** (1980), 531–584.

[Sau] L. Sauvage, *Sur les solutions régulièrs d'un system*, Ann. Sci. Éc. Norm. Sup. **3** (1886), 391–401.

[Sca] B. Scárdua, *Integration of complex differential equations*, J. Dynam. Contr. Syst. **5(1)** (1999), 1–50.

[SS] J. Scherk and J. Steenbrink, *On the mixed Hodge structure on the cohomology of the Milnor fibre*, Math. Ann. **271** (1985), 641–665.

[Schm] W. Schmid, *Variation of Hodge structures: the singularities of the period mappings*, Invent. Math. **22** (1973), 211–319.

[Sch1] L. Schlesinger, "Handbuch der Theorie der Lineardifferentialgleichungen", v. 2, Teubner, Leipzig, 1897.

[Sch2] L. Schlesinger, "Einführung in die Theorie der gewöhnlichen Differentialgleichungen auf funktionentheoretischer Grundlage", Gruyter, Berlin–Leipzig, 1922.

[Schn] T. Schneider, *Transcendenzuntersuchungen periodischer Funktionen. I*, J. reine angew. Math. **172** (1934), 65–69; *II*, 70–74.

[Schw] H. Schwarz, *Über diejenige Fälle in welchen die Gaussische hypergeometrsche Reihe eine algebraische Function ihres viertes Elementes darstellt*, J. reine angew. Math. **75** (1873), 292–335.

[ST] M. Sebastiani eand R. Thom, *Un résultat sur la monodromie*, Invent. Math. **13** (1971), 90–96.

[Sei] A. Seidenberg, *Reduction of singularities of differential equation* $Ady = Bdx$, Amer. J. of Math. **90** (1968), 248–269.

[Ser1] J.-P. Serre, "Algèbres de Lie Semi-Simples Complexes", W. A. Benjamin, New York, 1966.

[Ser2] J.-P. Serre, "Abelian l–adic Representations and Elliptic Curves", W. A. Benjamin, New York, 1968.

[Ses] C. Seshadri, *On a theorem of Weitzenböck in invariant theory*, J. Math. Kyoto Univ. **1–3** (1962) 403–409.

[Sha] I. Shafarevich, "Basic Algebraic Geometry", Springer–Verlag, Berlin-Heidelberg-New York, 1974, [Russian: Nauka, Moscow, 1988].

[Shc1] A. Shcherbakov, *Topological classification of germs of conformal mappings with an identical linear part*, Vestnik of Moscow Un-ty, Ser. I Mat. Mech. **37** (1982), No 3, 52–57 [Russian].

[Shc2] A. Shcherbakov, *Topological and analytic conjugation of non-commutative groups of conformal mappings*, Trans. of the Petrovski Sem. **10** (1984), 170–192 [Russian].

[SRO] A. Shcherbakov, E. Rosales-González and L. Ortis-Bobadilla, *Countable set of limit cycles for the equation* $\mathrm{d}w/\mathrm{d}z = P_n(z, w)/Q_n(z, w)$, J. Dynam. Contr. Syst. **4** (1998), 539–581.

[ShT] G. Shephard and J. Todd, *Finite unitary reflection groups*, Canad. J. Math **6** (1954), 274–304.

[Shi] M. Shishikura, *Bifurcation of parabolic fixed points*, in: "The Mandelbrot Set, Theme and Variations", London Math. Soc. Lecture Note Ser., v. **274**, Cambridge Un-ty Press, Cambridge, 2000, pp. 325–363.

[Sibi] K. Sibirski, "Introduction to Algebraic Theory of Invariants of Differential Equations", Manchester Un-ty Press, New York, 1988 [Russian: Shtiintza, Kishinev, 1976].

[Sib1] Y. Sibuya, *Simplification of a system of linear ordinary differential equations about a singular point*, Funkcial. Ekvac. **4** (1962), 29–56.

[Sib2] Y. Sibuya, *Stokes phenomena*, Bull. Amer. Math. Soc. **83** (1977), 1075–1077.

[Sie] C. Siegel, *Iterations of analytic functions*, Annals of Math. **43** (1942), 607–612.

[Sin] M. Singer, *Liouvillian first integrals of differential equations*, Trans. Amer. Math. Soc. **333** (1992), 673–688.

[So] M. Soares, *On algebraic sets invariant by one-dimensional foliations on* $\mathbb{CP}(3)$, Ann. Inst. Fourier **43** (1993), 143–162.

[Spa] E. Spanier, "Algebraic Topology", McGraw Hill Book C., New York, 1966.

[Ste1] J. Steenbrink, *Limits of Hodge structures*, Invent. Math. **31** (1975), 229–257.

[Ste2] J. Steenbrink, *Mixed Hodge structures on the vanishing cohomology*, in "Real and Complex Singularities", P. Holm (ed.), Sijthoff & Noordhoff Int. Publ., Alphen van den Rijn, 1977, pp. 523–563.

[Ste3] J. Steenbrink, *Intersection form for quasi-homogeneous singularities*, Compos. Math. **34** (1977), 211–223.

[Sto1] G. Stokes, *On the critical values of the sums of periodic series*, Trans. Cambr. Philos. Soc. **8** (1847), 533–610.

[Sto2] G. Stokes, Trans. Cambr. Philos. Soc. **11** (1871), 412–425.

[Sto3] G. Stokes, *On the discontinuity of arbitrary constants that appear as multipliers of semi-convergent series*, Acta Math. **26** (1902), 393–397.

[Stoy] Ts. Stoyanova, *Non-integrability of the fourth Painlevé equation in the Liouville–Arnold sense*, Nonlinearity **27** (2014), 1029–1044.

[Str1] E. Stróżyna, *The analytic and formal normal forms for the nilpotent singularity. The generalized saddle–node case*, Bull. Sci. Math. **126** (2002), 555–579.

[Str2] E. Stróżyna, *Normal forms for germs of vector fields with quadratic leading part. The polynomial first integral case*, J. Diff. Equat. **259** (2015), 6718–6748; *Normal forms for germs of vector fields with quadratic leading part. The remaining cases*, Studia Math. 239 (2017), 6718–48.

[SZ1] E. Stróżyna and H. Żołądek, *The analytic and formal normal forms for the nilpotent singularity*, J. Diff. Equat. **179** (2002), 479–537 .

[SZ2] E. Stróżyna and H. Żołądek, *Orbital formal normal forms for general Bogdanov–Takens singularity*, J. Diff. Equat. **193** (2003), 239–259.

[SZ3] E. Stróżyna and H. Żołądek, *The complete normal form for the Bogdanov–Takens singularity*, Moscow Math. J. **15** (2015), No 1, 141–178.

[SZ4] E. Stróżyna and H. Żołądek, *Analytic properties of the complete formal normal form for the Bogdanov–Takens singularity*, Nonlinearity **34** (2021), 3046–3082.

[SZ5] E. Stróżyna and H. Żołądek, *Invariants of group actions, dimension/ degree duality and normal forms of vector fields*, Bull. Sci Math.

[Suz] M. Suzuki. *Sur les relations d'équivalence ouvertes dans les espaces analytiques*, Ann. Sci. École Norm. Sup., Sér. 4, **7** (1974), 531–542.

[Tak] F. Takens, *Singularities of vector fields*, Publ. Math. IHES **43** (1974), 47–100.

[Tey] L. Teyssier, *Classification analytique des champs de vecteurs nœud-cols*, C. R. Acad. Sci. Paris, Ser. I **336** (2003), No 8, 619–624.

[Tho] R. Thom, "Structural Stability and Morphogenesis", Benjamin–Addison Wesley, New York, 1975.

[Tor] R. Torelli, *Sulle varietá di Jacobi*, Rend. Accad. Lincei **22** (1914), 98–103.

[Tur] H. Turrittin, *Convergent solutions of ordinary differential equations in the neighborhood of an irregular singular point*, Acta Math. **93** (1955), 27–66.

[Vara] V. Varadarajan, *Linear meromorphic differential equations: A modern point of view*, Bull. Amer. Math. Soc. **33** (1996), 1–42.

[Var1] A. Varchenko, *Asymptotic Hodge structure in vanishing cohomology*, Math. USSR Izv. **18** (1982), 469–512, [Russian: Izvestia Akad. Nauk USSR. Ser. Mat. **45** (1981), 540-591].

[Var2] A. Varchenko, *On the monodromy operator in vanishing cohomology and the operator of multiplication by f in the local ring*, Sov. Math. Dokl. **24** (1981), 248–252, [Russian: Dokl. Akad. Nauk USSR **260** (1981), No 2, 272-276].

[Var3] A. Varchenko, *Estimate of the number of zeroes of Abelian integrals depending on parameters and limit cycles*, Funct. Anal. Appl. **18** (1984), No 2, 98–108 [Russian].

[Var4] A. Varchenko, *Euler Beta–function, van der Monde determinant, Legendre equation and critical values of linear functions on configuration of hypersurfaces. I*, Izv. Akad. Nauk USSR. Ser. Mat. **35** (1990), No 3, 543–571, [Russian: **53** (1989), 1206–1253]; *II,* **36** (1991), No 1, 155–167, [Russian: **54** (1990), 146-158].

[Var5] A. Varchenko, *Hodge filtration of hypergeometric integrals associated with an affine configuration of general position and a local Torelli theory*, in "I. M. Gelfand seminar", (S. Gelfand and S. Gindikin, (eds.), v. **2**, Adv. in Sov. Math., Amer. Math. Soc., Providence, 1991, pp. 167–178.

[Var6] A. Varchenko, "Multidimensional Hypergeometric Functions and Representation Theory of Lie Groups and Quantum Groups", Adv. Ser. Math. Phys., World Scientific, River Edge, N. J., 1995.

[VC] A. Varchenko and S. Chmutov, *Finite irreducible groups generated by reflections are the monodromy groups of suitable singularities*, Funct. Anal. Appl. **18** (1984), No 3, 1–14 [Russian].

[VGZ] V. Vassiliev, I. Gelfand and A. Zelevinsky, *General hypergeometric functions on complex grassmannians*, Funct. Anal. Appl. **21** (1987), No 1, 23–38 [Russian].

[Ves] E. Vessiot, "Sur l'Intégration des Équations Différentielles Linéaires", Thése, Gauthier–Villars, Paris, 1892.

[Voi] C. Voisin, "Symmétrie Miroir", Panoramas et Synthèses, v. **2**, Soc. Math. de France, Paris, 1996.

[Vor] S. Voronin, *Analytic classification of germs of conformal maps* $(\mathbb{C},0)\mapsto(\mathbb{C},0)$ *with identical linear part*, Funct. Anal. and Appl. **15** (1981), No 1, 1–17 [Russian].

[Wae] B. van der Waerden, "Algebra I", Springer–Verlag, Berlin, 1971.

[Was] W. Wasow, "Asymptotic Expansions for Ordinary Differential Equations", J. Wiley & Sons, New York, 1965.

[Wel] R. Wells, "Differential Analysis on Complex Manifolds", Prentice–Hall, New Jersey, 1973.

[WW] E. Whittaker and G. Watson, "A Course of Modern Analysis", Cambridge Un-ty Press, Cambridge, 1935.

[Wit] E. Witten, *Mirror manifolds and topological field theory*, in "Essays on Mirror Manifolds", S.-T. Yau, (ed.), Intern. Ser. in Math. Phys., Internat. Press, New York, 1992, pp. 120–158.

[Wl] J. Włodarczyk, *Simple Hironaka resolution in characteristic zero*, J. Amer. Math. Soc. **18** (2005), No 4, 779–822.

[Yak] S. Yakovenko, *Abelian integrals as rational envelopes*, Nonlinearity **7** (1994), 1237-1250.

[Yoc] J.-C. Yoccoz, *Théorème de Siegel, nombres de Briuno et polynômes quadratiques*, Asterisque **231** (1995), 3–88.

[Zare] Y. Zare, *Center conditions; pull-back of differential equations*, Trans. Amer. Math. Soc. **372** (2019), No 5, 3167–3189.

[Zar] O. Zariski, "Algebraic Surfaces", Springer–Verlag, Berlin–Heidelberg–New York, 1971.

[Zig] S. Ziglin, *Branching of solutions and non-existence of first integrals in Hamiltonian mechanics. I,* Funct. Anal. Appl. **16** (1982), No 3, 30–41; *II,* **17** (1983), No 1, 8–23 [Russian].

[Zo1] H. Żołądek, *On versality of certain family of symmetric vector fields on the plane*, Math. Sbornik **48** (1984), 463–492, [Russian: Mat. Sbornik **120** (1983), 473–499].

[Zo2] H. Żołądek, *Bifurcations of certain family of planar vector fields tangent to axes*, J. Diff. Equat. **67** (1987), 1–55.

[Zo3] H. Żołądek, *Quadratic systems with center and their perturbations*, J. Diff. Equat. **109** (1994), 223–273.

[Zo4] H. Żołądek, *On certain generalization of the Bautin's theorem*, Nonlinearity **7** (1994), 273–279.

[Zo5] H. Żołądek, *Algebraic solutions of algebraic Pfaff equations*, Studia Math. **114** (1995), 117–126.

[Zo6] H. Żołądek, *New examples of holomorphic foliations without algebraic leaves*, Studia Math. **131** (1998), 137–142.

[Zo7] H. Żołądek, *The extended monodromy group and Liouvillian first integrals*, J. of Dynam. Contr. Syst. **4** (1998), 1–28.

[Zo8] H. Żołądek, *Algebraic invariant curves for the Liénard equation*, Trans. Amer. Math. Soc. **350** (1998), 1681–1701.

[Zo9] H. Żołądek, *Two remarks about Picard–Vessiot extensions and elementary first integrals,* Colloquium Math. **84/85** (2000), 173–183.

[Zo10] H. Żołądek, *The topological proof of Abel–Ruffini theorem*, Topol. Methods Nonlinear Anal. **16** (2000), 253–265.

[Zo11] H. Żołądek, *Note on the deck transformations group and the monodromy group*, Topol. Methods Nonlinear Anal. **19** (2002), 237–256.

[Zo12] H. Żołądek, *Polynomial Riccati equations with algebraic solutions,* in: "Differential Galois Theory", T. Crespo and Z. Hajto (eds.), Banach Center Publications, v. **58**, Warsaw, 2002, pp. 219–231.

[Zo13] H. Żołądek, *Multi-dimensional Jouanolou system*, J. reine angew. Math. **556** (2003), 47–78.

[Zo14] H. Żołądek, *Remarks about moduli*, preprint, Un-ty of Warsaw, 2018.

[ZF] H. Żołądek and G. Filipuk, *Painlevé equations, elliptic integrals and elementary functions*, J. Diff. Equat. **258** (2015), 1303–1353.

[ZM] H. Żołądek and R. Montero, "Qualitative Theory of ODEs. An Introduction to Dynamical Systems Theory", World Sci. Publ. Europe Ltd., Singapore, 2023.

Index

H. Żołądek, *The Monodromy Group*, Monografie Matematyczne 67,
https://doi.org/10.1007/978-3-031-91270-2

MIX
Papier aus verantwortungsvollen Quellen
Paper from responsible sources
FSC® C105338

If you have any concerns about our products,
you can contact us on
ProductSafety@springernature.com

In case Publisher is established outside the EU,
the EU authorized representative is:
Springer Nature Customer Service Center GmbH
Europaplatz 3, 69115 Heidelberg, Germany

Printed by Libri Plureos GmbH
in Hamburg, Germany